玉米研究文选
（续 2012—2017 年）

——北京市农林科学院玉米研究中心
成立二十周年文章选编

◎ 赵久然 主编

中国农业科学技术出版社

图书在版编目（CIP）数据

玉米研究文选：续 2012—2017 年／赵久然主编．—北京：中国农业科学技术出版社，2017.9
ISBN 978-7-5116-3208-1

Ⅰ．①玉… Ⅱ．①赵… Ⅲ．①玉米-文集 Ⅳ．①S513-53

中国版本图书馆 CIP 数据核字（2017）第 189425 号

责任编辑	崔改泵　李　华
责任校对	贾海霞

出 版 者	中国农业科学技术出版社
	北京市中关村南大街 12 号　邮编：100081
电　　话	（010）82109708（编辑室）　（010）82109702（发行部）
	（010）82109709（读者服务部）
传　　真	（010）82106631
网　　址	http://www.castp.cn
经 销 者	各地新华书店
印 刷 者	北京富泰印刷有限责任公司
开　　本	889mm×1 194mm　1/16
印　　张	53.25
字　　数	1 360 千字
版　　次	2017 年 9 月第 1 版　2017 年 9 月第 1 次印刷
定　　价	300.00 元

版权所有·翻印必究

《玉米研究文选》（续 2012—2017 年）
——北京市农林科学院玉米研究中心成立二十周年文章选编
编委会

主　　编：赵久然

副 主 编：王荣焕

编　　委：（按姓氏笔画为序）

　　　　　　王元东　王凤格　王荣焕　王晓光　卢柏山

　　　　　　田红丽　任　雯　刘　亚　刘春阁　刘新香

　　　　　　苏爱国　宋　伟　陈怡凝　赵久然　徐田军

　　　　　　徐　丽

前　言

北京市农林科学院玉米研究中心成立于 1997 年，是经北京市政府编办批准成立的专门从事玉米研究及开发的科研机构。坚持以科研为基础、以需求为导向、以创造价值为核心，立足北京、服务全国，集玉米种质创新、新品种选育、种子检测鉴定、良种良法配套及示范推广、生产技术指导和科技咨询等五位一体，坚持产学研相结合和育繁推一条主线。

目前有 14 个部门，100 多名员工，资产 1 亿多元。其中，高级职称 23 人，博士 18 人，在站博士后 2 人。在北京建有玉米育种与示范基地 800 多亩，海南南繁基地 200 亩，鉴定示范点遍布全国各地。入选全国农业杰出科研人才及其创新团队、科技北京百名领军人才及其创新团队、北京市农林科学院院级优秀创新团队等。拥有北京学者、国务院特殊津贴专家、全国粮食生产先进工作者、种业十大杰出人物、鲜食玉米育种领军人物等高层次人才。是农业部玉米专家指导组组长单位、国家玉米产业技术体系岗位科学家单位、国家早熟耐旱宜机收玉米育种创新基地、国家玉米品种区域试验站、国家农业科学实验站、农业部首批农业转基因试验基地、农业部玉米原原种基地（北京）、玉米 DNA 指纹及分子育种北京市重点实验室等。2007 年被农业部评为"十五"全国农业百强科研所，综合实力在全国 1 077 个院所中列第 54 位；2016 年被农业部评为"全国农业先进集体"。

通过多年育种实践，集成创新了以"高大严选系、IPT 配合力测配、DH 单倍体诱导及加倍、MAS 分子标记辅助、多年多点多生态区鉴定"等技术为主要内容的"五位一体"工程化玉米品种育种技术，并获得国家发明专利。并运用该技术创制选育出京 724、京 2416 等 30 多个优良玉米自交系；创新选育并通过审定玉米新品种 100 多个，其中国审品种 30 个，有 5 个被农业部推荐为主导品种，居全国之首。京科 968 年种植面积已达 3 000 万亩，成为我国生产面积位列第三的玉米主导大品种；京科糯 2000 十多年来一直是我国面积最大的鲜食玉米主导品种，种植面积约 500 万亩，约占全国糯玉米总面积的一半，并已走出国门，成为越南、韩国等一带一路国家的主导品种；京科糯 928、农科玉 368 是一种甜加糯-高叶酸营养强化型鲜食玉米，具有突出创新性；京农科 728 是我国首批通过国家审定的机收籽粒品种，实现夏玉米籽粒直收，并成为京津冀协同发展首个共同推广的农业技术，已累计推广 1 000 多万亩；京科青贮 301、516 等系列青贮玉米品种被多家企业选定为专用青贮种植品种。

玉米品种标准 DNA 指纹库构建关键技术研究与应用处于国际前沿水平。牵头制定玉米 DNA 指纹鉴定标准，构建了已有 28 000 多个品种全球数量最大的玉米标准 DNA 指纹库，研发出功能强大、查询方便的数据库管理系统并实现数据联网。制定了 5 项 DNA 指纹检测标准，引领和推进了玉米及其他多种作物分子检测的标准化进程。为国家玉米品种审定、品种权保护、DUS 测试、种子执法及农民权益维护等鉴定样品总计 50 000 多份（次），为多方面提供了强有力的科技支撑。

建议并致力于实施和技术指导玉米"一增四改"、"雨养旱作"、高产创建等节本增效、高产高效集成技术。玉米"一增四改"技术被农业部连续多年列为主导技术，为我国粮食持续增产发挥了突出作用。玉米"雨养旱作"技术已基本实现京郊玉米生产全覆盖，并辐射到周边地区。

主持完成国家及北京市重大科研项目课题 100 多项。获国家及省部级等科技奖励 40 多项、国家发明专利及植物新品种权 90 多项，发表学术论文 200 多篇，其中 SCI 论文 20 多篇，编著专业书籍 50 余部。

在玉米研究中心成立二十周年之际，为总结和回顾中心近年来所取得的成绩，我们收集和整理了玉米研究中心 2012—2017 年近 5 年来的主要科研论文并结集成册，以不断总结经验，进一步提高科技创新和科技服务能力，促进北京市、京津冀及全国玉米科研和生产发展。

玉米研究中心自成立以来，各项事业的发展得到了农业部、科技部、北京市科委、北京市农委、北京市农业局、北京市科协、国家自然科学基金委和北京市农林科学院等上级领导部门和各界同行的关心及大力支持和帮助，值此中心成立二十周年之际深表谢意！在未来发展中，恳请继续给予支持和帮助，为我国玉米事业再做新贡献！

因编写时间仓促，书中不妥之处在所难免，敬请广大读者批评指正。

编　者

2017 年 7 月

目 录

Functional Divergence and Origin of the *DAG-like* Gene Family in Plants
.... Luo Meijie　Cai Manjun　Zhang Jianhua et al （1）
Mapping of a Major QTL for Salt Tolerance of Mature Field-grown Maize Plants
　　Based on SNP Markers .... Luo Meijie　Zhao Yanxin　Zhang Ruyang et al （17）
Identification of Genes Potentially Associated with the Fertility Instability of S-Type
　　Cytoplasmic Male Sterility in Maize via Bulked Segregant RNA-Seq
　　.. Su Aiguo　Song Wei　Xing Jinfeng et al （31）
RNA-seq Analysis Reveals MAPKKK Family Members Related to Drought Tolerance in Maize
　　.. Liu Ya　Zhou Miaoyi　Gao Zhaoxu et al （45）
Genome-Wide Epigenetic Regulation of Gene Transcription in Maize Seeds
　　.. Lu Xiaoduo　Wang Weixuan　Ren Wen et al （68）
Development of MaizeSNP3072, a High-throughput Compatible SNP Array, for DNA
　　Fingerprinting Identification of Chinese Maize Varieties
　　.. Tian Hongli　Wang Fengge　Zhao Jiuran et al （86）
Paenibacillus zeae sp. nov., Isolated from Maize (*Zea mays* L.) Seeds
　　.. Liu Yang　Zhai Lei　Wang Ronghuan et al （98）
Paenibacillus Chinensis sp. nov., Isolated from Maize (*Zea mays* L.) Seeds
　　.. Liu Yang　Zhao Ran　Wang Ronghuan et al （106）
High-throughput Sequencing-based Analysis of the Composition and Diversity of Endophytic
　　Bacterial Community in Seeds of "Beijing" Hybrid Maize Planted in China
　　.. Liu Yang　Wang Ronghuan　Li Yinhu et al （114）
Identification and Antagonistic Activity of Endophytic Bacterial Strain *Paenibacillus* sp. 5 L8
　　Isolated from the Seeds of Maize (*Zea mays* L., Jingke 968)
　　.. Liu Yang　Wang Ronghuan　Cao Yanhua et al （123）
Molecular Mapping of Quantitative Trait Loci for Grain Moisture at Harvest in Maize
　　.. Song Wei　Shi Zi　Xing Jinfeng et al （133）
Effect of Saline Stress on the Physiology and Growth of Maize Hybrids and their Related Inbred
　　Lines .... Luo Meijie　Zhao Yanxin　Song Wei et al （141）
Exploring Differentially Expressed Genes Associated with Fertility Instability of S-type Cytoplasmic
　　Male-sterility in Maize by RNA-Seq .... Su Aiguo　Song Wei　Shi Zi et al （156）

Improvement of *Agrobacterium*-mediated Transformation Efficiency of Maize (*Zea mays* L.)
Genotype Hi-II by Optimizing Infection and Regeneration Conditions
　　　　　　　　　　　　　　　　　　　　　　　　　　 Xu You　Ren Wen　Liu Ya et al（171）
中国玉米生产发展历程、存在问题及对策 ························· 赵久然　王荣焕（183）
我国玉米产业现状及生物育种发展趋势 ··················· 赵久然　王荣焕　刘新香（190）
玉米及其制品质量安全风险及控制 ······························ 赵久然　刘月娥（198）
我国玉米品种标准 DNA 指纹库构建研究及应用进展 ······ 赵久然　王凤格　易红梅等（203）
中国玉米品种审定现状分析 ······························ 杨扬　王凤格　赵久然等（210）
农作物品种 DNA 指纹库构建研究进展 ··················· 王凤格　赵久然　田红丽等（223）
DNA 指纹技术在玉米区域试验品种真实性及一致性检测中的应用
　　　　　　　　　　　　　　　　　　　　　　　　　　 王凤格　易红梅　赵久然等（232）
SSR 和 SNP 两种标记技术在玉米品种真实性鉴定中的比较分析
　　　　　　　　　　　　　　　　　　　　　　　　　　 李雪　田红丽　王凤格等（239）
中国玉米审定品种标准 SSR 指纹库的构建 ··············· 王凤格　杨扬　易红梅等（246）
中国 328 个玉米品种（组合）SSR 标记遗传多样性分析
　　　　　　　　　　　　　　　　　　　　　　　　　　 王凤格　田红丽　赵久然等（260）
玉米纯度分子鉴定标准研制相关问题的探讨 ··············· 王凤格　易红梅　赵久然（270）
玉米真实性 SSR 鉴定标准的研制及应用 ·················· 王凤格　易红梅　赵久然（277）
玉米 SSR-DNA 指纹库构建方案在高粱中的通用性 ········ 周青利　王蕊　张春宵等（285）
玉米 InDel 分子标记 20 重 PCR 检测体系的建立 ········· 冯博　许理文　王凤格等（295）
玉米品种京科 968 纯度鉴定引物的确定 ················· 冯博　许理文　王凤格等（307）
非优异 SSR 原因分析及解决方案 ······················· 刘文彬　许理文　冯博等（316）
基于两种荧光毛细管电泳平台筛选评估玉米新型 SSR 引物
　　　　　　　　　　　　　　　　　　　　　　　　　　 刘文彬　许理文　王凤格等（327）
玉米 SSR 标记 DNA 指纹数据库中 N+1 峰的消除 ········· 薛宁宁　许理文　易红梅等（335）
一种适于 SNP 芯片分型的玉米种皮组织 DNA 提取方法
　　　　　　　　　　　　　　　　　　　　　　　　　　 刘可心　王璐　蔚荣海等（346）
利用核心 SNP 位点鉴别玉米自交系的研究 ··············· 宋伟　王凤格　田红丽等（352）
利用 SNP 标记对 51 份玉米自交系进行类群划分 ········· 吴金凤　宋伟　王蕊等（361）
分子标记辅助选择玉米自交系京 24 两种抗病主效基因的聚合
　　　　　　　　　　　　　　　　　　　　　　　　　　 余辉　宋伟　赵久然等（369）
分子标记辅助选择育成的玉米自交系京 24 单抗丝黑穗病和茎腐病改良材料性状分析
　　　　　　　　　　　　　　　　　　　　　　　　　　 余辉　宋伟　赵久然等（377）
京 724 玉米自交系 S 型细胞质雄性不育系分子标记辅助选育研究
　　　　　　　　　　　　　　　　　　　　　　　　　　 宋伟　苏爱国　邢锦丰等（384）
SSR 和 SNP 标记在玉米分子标记辅助背景选择中的应用比较
　　　　　　　　　　　　　　　　　　　　　　　　　　 宋伟　赵久然　王凤格等（389）
基于 SNP 标记的玉米容重 QTL 分析 ··················· 许理文　段民孝　田红丽（397）
玉米生物诱导孤雌生殖 DH 群体的 SNP 偏分离分析 ······· 许理文　段民孝　宋伟等（403）

同一来源玉米 DH 系的性状分离分析……………………………邢锦丰　张如养　赵久然等（411）
秋水仙素浸芽法处理玉米单倍体的加倍效果研究………………段民孝　刘新香　赵久然等（417）
单倍体技术在玉米种质改良和育种中的应用方向
　　……………………………………………………………………张如养　段民孝　赵久然等（423）
不同基因型玉米材料的单倍体诱导和加倍研究…………………张如养　段民孝　赵久然等（428）
6 个玉米单倍体诱导系诱导率的差异性研究……………………张如养　段民孝　赵久然等（434）
8 个玉米单倍体诱导系诱导率的配合力研究……………………张如养　段民孝　赵久然等（441）
玉米单倍体与其加倍后代性状的相关性研究……………………张如养　段民孝　赵久然等（447）
玉米单倍体诱导系诱导性状的杂种优势分析……………………张如养　段民孝　赵久然等（452）
高产、高淀粉玉米品种 NK718 的选育与栽培技术 ……………王卫红　王元东　张春原等（459）
适于全程机械化生产的玉米新品种选育探讨……………………王元东　张华生　段民孝等（463）
利用外来新种质 X1132x 选育优良玉米自交系的研究 …………王元东　张华生　段民孝等（467）
高产早熟耐密抗倒伏宜机收玉米新品种"京农科 728"的选育与配套技术研究
　　…………………………………………………………………………段民孝　赵久然　李云伏等（473）
玉米新品种 NK971 的选育及栽培制种技术要点 ………………王元东　张华生　段民孝等（479）
玉米新品种 MC220 的选育、特征特性及栽培技术要点 ………赵久然　王元东　王荣焕等（482）
机收玉米新品种 MC812 的选育 …………………………………张华生　段民孝　陈传永等（485）
玉米新品种 MC703 选育与配套技术 ……………………………张华生　范会民　张雪原等（488）
玉米新品种 MC1002 的选育 ……………………………………张雪原　赵久然　王元东等（491）
玉米新品种 MC4592 选育及栽培制种技术要点 ………………王元东　张华生　段民孝等（494）
玉米新品种京科 193 的选育及配套技术 ………………………邢锦丰　段民孝　王元东等（497）
青贮玉米品种利用现状与发展……………………………………杨国航　吴金锁　张春原等（500）
青贮玉米新品种京科 932 选育及配套技术 ……………………邢锦丰　段民孝　王元东等（505）
杂交玉米新品种京科 968 种子生产技术 …………………………………………冯培煜（508）
京科 968 等系列玉米品种"易制种"性状选育与高产高效制种关键技术研究
　　…………………………………………………………………………王元东　赵久然　冯培煜等（510）
雄性不育制种京科 968 的遗传背景及其表型分析 ……………宋　伟　赵久然　邢锦丰等（516）
玉米品种京农科 728 北京密云地区制种技术 …………………王荣焕　王元东　赵久然等（522）
京科青贮 516 玉米高产制种技术 …………………………………………冯培煜　宋瑞连（526）
玉米制种单粒播种技术及相关配套措施…………………………………冯培煜　宋瑞连（528）
春玉米区玉米制种预防高温危害的方法与措施………………冯培煜　宋瑞连　王晓光（530）
以玉米籽粒粒形进行分离精选种子的意义及其新方法…………………冯培煜　宋瑞连（533）
玉米杂交制种测产方法及应用……………………………………………冯培煜　宋瑞连（536）
适贮条件下不同基因型玉米种子活力及生理特性研究…………张海娇　成广雷　赵久然等（540）
临界胁迫贮藏条件下不同基因型玉米种子活力及生理变化
　　…………………………………………………………………………成广雷　张海娇　赵久然等（546）
北京优质杂交玉米种子内生细菌种类多样性……………………李南南　刘　洋　赵　燃等（559）
常压室温等离子体对玉米种子及花粉萌发的影响………………骆美洁　赵衍鑫　宋　伟等（570）
转基因玉米种子萌发期抗旱性鉴定………………………………冷益丰　张　彪　赵久然等（576）

标题	作者	页码
玉米萌发幼苗期的抗旱性鉴定评价	徐田军　吕天放　赵久然等	(585)
转 TsVP 基因玉米抗旱性鉴定研究	张志方　刘　亚　任　雯等	(593)
远红外成像技术在植物干旱响应机制研究中的应用	刘　亚	(600)
2002—2009 年东北早熟春玉米生育期及产量变化	王玉莹　张正斌　杨引福等	(607)
不同种植方式对玉米农艺性状和产量的影响	赵　杨　钱春荣　王俊河等	(615)
不同种植模式对玉米产量与农艺性状影响分析	赵　杨　钱春荣　王俊河等	(618)
应对及预防京郊玉米倒伏的技术措施	王荣焕　赵久然　徐田军等	(624)
国审玉米品种京单 68 的主要特点及高产栽培技术	刘春阁　路明远　刘海伍等	(627)
不同播期条件下"京单 68"和"郑单 958"的籽粒灌浆特性研究	徐田军　王荣焕　赵久然等	(630)
密度和播期对京单 68 冠层结构和产量的影响	徐田军　刘秀芝　赵久然等	(638)
玉米新品种京科 968 配套高产栽培技术	赵久然	(646)
京科 968 的灌浆特征与产量性能分析	陈传永　赵久然　王荣焕等	(650)
京单 38 不同播期籽粒灌浆特性研究	王荣焕　赵久然　陈传永等	(656)
播期对玉米干物质积累转运和籽粒灌浆特性的影响	徐田军　吕天放　陈传永等	(663)
玉米籽粒灌浆特性对播期的响应	徐田军　吕天放　赵久然等	(671)
不同熟期玉米品种的籽粒灌浆特性及其与温度关系研究	钱春荣　王荣焕　赵久然等	(681)
玉米果穗不同位势籽粒灌浆特性分析	钱春荣　王荣焕　赵久然等	(693)
不同生育时期遮光对玉米籽粒灌浆特性及产量的影响	陈传永　王荣焕　赵久然等	(702)
遮光对玉米干物质积累及产量性能的影响	陈传永　王荣焕　赵久然等	(711)
玉米新品种京农科 728 全程机械化生产技术	王元东　张华生　段民孝等	(719)
不同播期对玉米品种京农科 728 产量及机收籽粒相关性状的影响	王元东　王荣焕　张华生等	(722)
播期和密度对玉米籽粒机收主要性状的影响	王荣焕　徐田军　赵久然等	(726)
品种和氮素供应对玉米根系特征及氮素吸收利用的影响	程　乙　王洪章　刘　鹏	(732)
我国糯玉米育种及产业发展动态	赵久然　卢柏山　史亚兴等	(744)
京科甜系列水果型优质玉米品种选育及应用	卢柏山　史亚兴　徐　丽等	(751)
鲜食玉米的发展与前景——探索我国甜玉米的北方市场	史亚兴　张保民	(756)
不同温度胁迫对双隐性甜糯玉米出苗的影响	卢柏山　史亚兴　樊艳丽等	(764)
不同播种深度对双隐性甜糯玉米出苗的影响	史亚兴　卢柏山　樊艳丽等	(768)
基于 SNP 标记技术的糯玉米种质遗传多样性分析	史亚兴　卢柏山　宋　伟等	(773)
利用 SNP 标记划分甜玉米自交系的杂种优势类群	卢柏山　史亚兴　宋　伟等	(781)
不同收获期糯玉米杂交种的种子萌发和幼苗生长	卢柏山　史亚兴　徐　丽等	(788)
甜玉米杂交种种子成熟度对种子萌发和幼苗生长的影响	卢柏山　史亚兴　徐　丽等	(795)
优质高产超甜玉米品种京科甜 158 的选育及栽培技术	史亚兴　卢柏山　樊艳丽等	(802)
果蔬型甜玉米新品种京科甜 179 的选育	史亚兴　卢柏山　徐　丽等	(805)
甜糯玉米新品种"京科糯 928"	卢柏山　史亚兴　赵久然等	(810)

优质甜糯玉米新品种京科糯 2010 的选育及栽培技术要点
                    史亚兴 卢柏山 赵久然等（812）
新型甜加糯鲜食玉米品种农科玉 368 的选育 …………………… 卢柏山 史亚兴 徐 丽等（814）
中国种子企业研发模式与发展策略研究 ……………………………… 杨海涛 赵久然等（817）
种子企业技术购买研发模式的决策研究 ……………………… 杨海涛 赵久然 鲁利平等（823）
我国玉米品种权保护现状、问题与建议 ……………………… 杨海涛 赵久然 杨凤玲等（830）
基于案例分析的玉米品种权维权问题、启示与建议 ………… 杨海涛 赵久然 陈 红等（836）

Functional Divergence and Origin of the *DAG-like* Gene Family in Plants

Luo Meijie[1]　Cai Manjun[2]　Zhang Jianhua[2]　Li Yurong[2]
Zhang Ruyang[1]　Song Wei[1]　Zhang Ke[2]　Xiao Hailin[2,3]　Yue Bing[2]
Zheng Yonglian[2]　Zhao Yanxin[1,2]　Zhao Jiuran[1]　Qiu Fazhan[2]

(1. *Beijing Key Laboratory of Maize DNA Fingerprinting and Molecular Breeding*, *Maize Research Center*, *Beijing Academy of Agriculture and Forestry Sciences*, *Beijing*, 100097, *China*; 2. *National Key Laboratory of Crop Genetic Improvement*, *Huazhong Agricultural University*, *Wuhan*, 430070, *China*; 3. *Present address*: *Life Science and Technology Center*, *China National Seed Group Co.*, *Ltd.*, *Wuhan*, 430075, *China*)

Abstract: The nuclear-encoded *DAG-like* (*DAL*) gene family plays critical roles in organelle C-to-U RNA editing in *Arabidopsis thaliana*. However, the origin, diversification and functional divergence of *DAL* genes remain unclear. Here, we analyzed the genomes of diverse plant species and found that: *DAL* genes are specific to spermatophytes, all *DAL* genes share a conserved gene structure and protein similarity with the inhibitor I9 domain of subtilisin genes found in ferns and mosses, suggesting that *DAL* genes likely arose from I9-containing proproteases via exon shuffling. Based on phylogenetic inference, *DAL* genes can be divided into five subfamilies, each composed of putatively orthologous and paralogous genes from different species, suggesting that all *DAL* genes originated from a common ancestor in early seed plants. Significant type I functional divergence was observed in 6 of 10 pairwise comparisons, indicating that shifting functional constraints have contributed to the evolution of *DAL* genes. This inference is supported by the finding that functionally divergent amino acids between subfamilies are predominantly located in the DAL domain, a critical part of the RNA editosome. Overall, these findings shed light on the origin of *DAL* genes in spermatophytes and outline functionally important residues involved in the complexity of the RNA editosome.

Key words: Maize; Plant; *DAL* gene; RNA editing; Inhibitor I9

Introduction

C-to-U RNA editing (deamination of cytidine to uridine) is an essential step of RNA maturation in chloroplasts and mitochondria of land plants from bryophytes to angiosperms. U-to-C RNA editing is also observed in ferns and mosses. More than 400 editing sites in mitochondria and 30~

Luo Meijie and Cai Manjun contributed equally to this work. Correspondence and requests for materials should be addressed to Y. Z. (E-mail: rentlang2003@163.com) or J. Z. (E-mail: maizezhao@126.com) or F. Q. (E-mail: qiufazhan@mail.hzau.edu.cn)

40 editing events in chloroplasts are typically found in flowering plants. RNA editing occurring in plant organelle mRNAs can restore functionally conserved amino acids at the post-transcriptional level, create functional proteins and play important roles in efficient splicing of introns and processing of precursor tRNA molecules. In plants, some mutants with impaired RNA editing at specific nucleotide sites cause deleterious phenotypes and even lethality. The site specificity of the cytidines to be edited in a plant organelle is determined by a crucial *cis*-acting regulatory sequence and the RNA editosome that will bind to it. The RNA editosome is composed of nuclear-encoded *trans*-acting factors that recognize the *cis*-element and perform RNA editing. Recent extensive genetics studies have revealed that these *trans*-factors enlisted in the RNA editosome include DYW-type pentatricopeptide repeat (PPR) proteins, RNA-Editing Factor Interacting Protein (RIP) family or Multiple Organelle RNA Editing Factor (MORF) family proteins, RNA-recognition motif (RRM) -containing proteins, protoporphyrinogen IX oxidase 1 (PPO1) and organelle zinc-finger 1 (OZ1). PPR proteins are characterized by tandem 35-amino acid PPR motifs. The DYW-PPR proteins each recognize one or a few editing sites that have similar *cis*-elements and thereby bind directly to the *cis*-acting sequences. The DYW domain of DYW-PPR proteins has a sequence similar to the active sites of known cytidine deaminases and editing enzymes, and may be responsible for deamination of cytidine to uridine.

The *RIP/MORF* gene family, which controls multiple organelle RNA editing sites, was identified in *Arabidopsis thaliana* and designated the *RIP* gene family and the *MORF* gene family by two research groups. Here, we adopt the name, the *DAG-like* (*DAL*) gene family, based on the first identified member (*DAG*) of the gene family in *Antirrhinum majus*. *Arabidopsis* DAL proteins are all targeted to mitochondria or chloroplasts and required for RNA editing at all sites in both organelles. The mutation of *DAL* genes in plants results in abnormal development of plants, even lethality. Yeast two-hybrid analysis confirmed that DAL/RIP/MORF proteins can interact selectively with diverse PPR proteins by the binding of the DAL domain to PPR motifs, and moreover, DAL proteins can connect to form hetero-and homodimers. A variation of the *DAL* gene (*ORRM*1) was identified and functionally analyzed; it harbors a pair of truncated RIP domains (RIP-RIP) at its N terminus and an RRM domain at its C terminus. ORRM1 is an essential plastid editing factor that can interact selectively with PPR proteins via its RIP-RIP domain, and the ORRM1 protein can also bind to sequences near at least some of its RNA targets *in vitro*. Furthermore, the RRM domain can rescue the editing defect in *orrm*1 protoplasts independent of RIP domains, and three other RRM-containing proteins were identified because of their roles in organelle RNA editing, suggesting that the RRM domain participates in the RNA editosome. Together, DAL proteins may be connectors between the site-specific PPRs and the as-yet-unknown deaminase or other components in the RNA editosome, such as RRM-containing proteins, PPO1 and OZ1.

Compared with the RNA editosomes responsible for C-to-U or A-to-I (deamination of adenosine to inosine) RNA editing in mammals, the plant organelle RNA editosomes have more diverse components. In addition to the interpretation that more RNA-edited sites in plant organelles require more *trans*-acting editing factors, the diverse composition of the organelle RNA editosome in plants probably overcomes the deficiency in RNA editing caused by the mutation of PPR protein or changes in the *cis*-acting sequences of edited sites, especially for those edited sites in plant mitochondrial genomes which evolve much more quickly. Thus, the origin, classification and evolution analysis of *trans*-acting factors is important for understanding the evolution and molecular mechanism of the RNA editosome in plants. In the RNA editosome, *DYW-PPR* genes undergo purifying selection at sites targeted for RNA editing because they are important for recognizing *cis*-element sequences. However, the functional evolution and origin of the *DAL* gene family is unknown.

In this study, we identified the DAL proteins in various plant lineages, including green algae, moss, ferns, gymnosperms and flowering plants, to investigate functional divergence and origin of the *DAL* gene family in plants. The result indicated *DAL* genes are specific to spermatophytes other than to lower plants. Plant *DAL* genes shared a strong conserved gene structure and appear to have evolved from the I9-containing proprotease via exon shuffling. Functional divergence analysis revealed that there was significant functional divergence between different DAL clades which may be associated with differences in the roles different *DAL* genes play in RNA editing and RNA metabolism. The evolutionary and functional divergence analysis of the *DAL* genes in plants presented here provides useful information for further probing the molecular mechanism by which DAL proteins contribute to the RNA editosome.

Results

Identification and sequence analysis of *DAL* genes in maize

To identify putative *DAL* genes in the maize genome, we searched the maize genome annotation data with known plant DAL proteins as a query. In total, we obtained 7 putative *DAL* genes in maize named *ZmDAL1—ZmDAL7* based on their order on the chromosomes (Figure 1a and Supplementary Table S1). *ZmDAL*s were distributed on 5 of 10 maize chromosomes, and chromosomes 9 and 10 both had two *ZmDAL* genes (Supplementary Figure S1). The gene model of *ZmDAL1* was reannotated correctly by analyzing the similarity between *ZmDAL* genes and their orthologs (Supplementary Figure S2). The veracity of each gene model of *ZmDAL* genes was assessed using reverse transcription polymerase chain reaction (RT-PCR) assays with the gene-specific primers listed in Supplementary Table S2, as 4 of 7 *ZmDAL* genes had more than one transcript for each *ZmDAL* gene in the MaizeGDB database (http://www.maizegdb.org/). The RT-PCR results indicated that seven *ZmDAL* genes were expressed in maize seedlings and only a single transcript was found for each *ZmDAL* gene (Supplementary Figure S3). All identified maize *DAL* genes encoded proteins ranging from 215 (ZmDAL1) to 412 amino acids (aa) (ZmDAL2), and their isoelectric points (Ip) were similar (>8.0).

No known motif was found in the maize DAL proteins by screening the PFAM and INTERPRO databases, except the MORF box (called the DAL domain in this study) which had been identified previously. Novel putative motifs were explored using the MEME server with different motif lengths. By selecting a motif length between 10 and 50 aa, we identified 4 conserved motifs, and all 4 motifs were located in the DAL domain (Figure 1b, c, d), suggesting that the DAL domain is a conserved sequence among *Arabidopsis* and maize DAL proteins. To obtain an intact motif containing the DAL domain, we enlarged the MEME motif length and identified one motif containing 114 aa (Supplementary Figure S4). Like their homologs in *Arabidopsis*, maize DAL proteins were predicted using TargetP and Predotar to enter mitochondria or chloroplasts. Of them ZmDAL1 and ZmDAL6 were also detected in the plastid nucleoid proteome by searching the maize organelle proteomics database (PPDB, http://ppdb.tc.cornell.edu/) (Figure 1a).

The gene structures of the *ZmDAL* genes were constructed by aligning the extracted genomic sequences to predicted cDNA sequences of maize *DAL* genes. This showed that *ZmDAL* genes have a conserved gene structure (Figure 2); each of the *ZmDAL* genes has 3 introns with the intron phases 2, 1 and 1 separating DAL domain-encoding exons 1, 2, 3 and 4 (Figure 2b). Motifs 1 and 2 are encoded by exon 1; motif 3 is encoded by exons 2 and 3; and motif 4 is located in exon 4 (Figure 2c and d). Furthermore, the length of exons 2 (98 basepairs, bp) and 3 (66bp) is conserved among all five *ZmDAL* genes, even though the size of the introns between the exons varies between different *ZmDAL* genes.

Identification and phylogenetic analysis of plant *DAL* genes

To mine more DAL domain-encoding genes in plants, we used the HMMER 3.0 package to build a hidden Markov model (HMM) file (dal.hmm, Supplementary Data File S1) with 17 DAL domain sequences of those DAL proteins from *A. majus*, maize and *Arabidopsis* (Supplemental Data File S2). We then used the dal.hmm algorithm to query the genomes of a variety of plants representing the major evolutionary lineages, including *Chlamydomonas reinhardtii*, *Physcomitrella patens*, *Selaginella moellendorffii*, *Picea abies*, *Brachypodium distachyon*, *Oryza sativa* Japonica, *Zea mays*, *Sorghum bicolor*, *Aquilegia coerulea*, *Vitis vinifera*, *A. thaliana*, *Arabidopsis lyrata* and *Populus trichocarpa*. The result showed that putative *DAL* genes were only identified in seed plants but not in lower plants (*C. reinhardtii*, *P. patens* and *S. moellendorffii*) (Figure 3). The numbers of *DAL* genes of higher plants used here are comparable, ranging from 6 (in *A. coerulea*) to 11 (in *A. lyrata*). In total, 79 *DAL* genes were identified in 10 plant genomes (Supplementary Table S3). In addition, we identified *ORRM*1-*like* genes in this study that were also specific to seed plants, and these genes encoded two tandem truncated DAL domains at the N terminus and one RNA recognition motif (RRM) at the C terminus, except the MA_10436715g0010 protein found in *P. abies*, which had no C-terminal RRM domain (Supplementary Figure S5).

To investigate the phylogenetic relationship among plant *DAL* genes, an unrooted neighbor-joining (NJ) tree containing all 79 DAL proteins was generated based on the conserved DAL domain alignment (Figure 4 and Supplementary Figure S5). On the basis of the phylogeny, the *DAL* gene family in plants was subdivided into five groups, named group I to group V (Figure 4). In the NJ tree shown in Figure 4, *DAL* genes of each group were all from diverse plant species. In groups I, II and V, species-specific gene duplication events occurred after the lineages

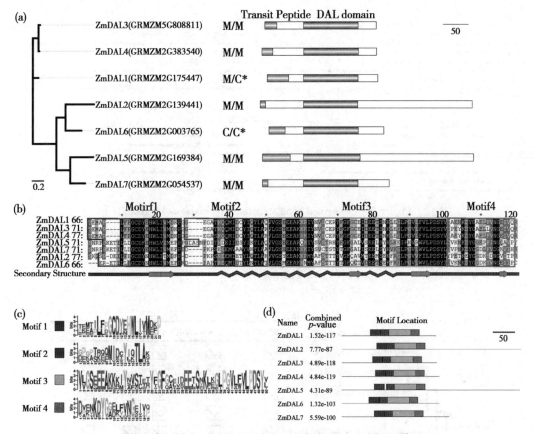

Figure 1　Maize DAL proteins and their conserved motifs

(a) Multiple sequence alignment of maize DAL proteins was carried out using ClustalW2.0, and the NJ tree was built using MEGA v5.0. The chloroplast and mitochondrial transit peptides of maize DAL proteins were predicted using Predotar (http://urgi.versailles.inra.fr/predotar/predotar.html) and (/) TargetP (http://www.cbs.dtu.dk/services/TargetP/). M, mitochondria; C, chloroplast. The DAL proteins marked by an asterisk were observed in the plastid proteome. (b) Alignment of conserved DAL domains in maize DAL proteins was conducted using ClustalW2.0 and displayed with GeneDoc. The secondary structure of the DAL domain was inferred using MINNOU (http://minnou.cchmc.org/). (c, d) Putative motifs were explored using the MEME server with the parameters of between 10 and 50 aa in length and sharing of each motif among all ZmDAL proteins

diverged, resulting in the inclusion of more than one *DAL* gene per species (Figure 4). Since the *DAL* genes were found to be specific to spermatophytes, we can infer that the ancestral *DAL* gene appeared after the divergence of seed plants and ferns.

The exon-intron organization analysis of 79 plant *DAL* genes indicated that plant *DAL* genes all share a conserved gene structure, with the 2-1-1 intron phase pattern separating DAL domain-encoding exons, as observed in maize *DAL* genes, except *Al_477997* and *MA_489006g0010*, which have intron phase patterns 0-1-1 and 0-2-2, respectively (Supplementary Figure S6).

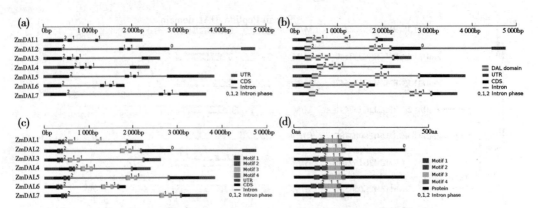

Figure 2 Conserved gene structures of maize *DAL* genes

The gene structures of *ZmDAL* genes were built using GSDraw (http://wheat.pw.usda.gov/piece/GSDraw.php) through both alignment of DNA obtained from MaizeGDB (http://www.maizegdb.org/) and coding sequences (CDS) of *ZmDAL* genes

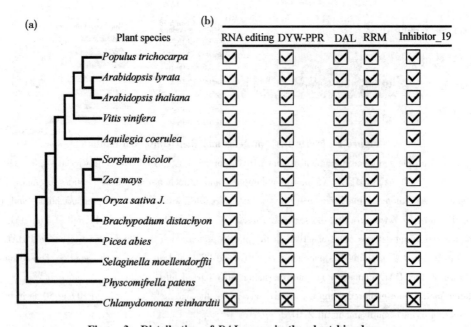

Figure 3 Distribution of *DAL* genes in the plant kingdom

(a) A schematic diagram of plant evolution tree was constructed according to the plant tree shown in PGDD (http://chibba.agtec.uga.edu/duplication/). (b) RNA editing sites (organelle C-to-U RNA editing), DYW-type *PPR* genes, *DAG-like* genes, *ORRM* genes, and Inhibitor I9 genes were identified in the plants listed on the left. The checkmark in the box denotes that the above genes can be found in the corresponding genomes, and the cross in the box indicates none of above genes are found in these genomes

The source of plant *DAL* genes

Putative genes or gene fragments homologous to *DAL* genes were identified in lower plants to

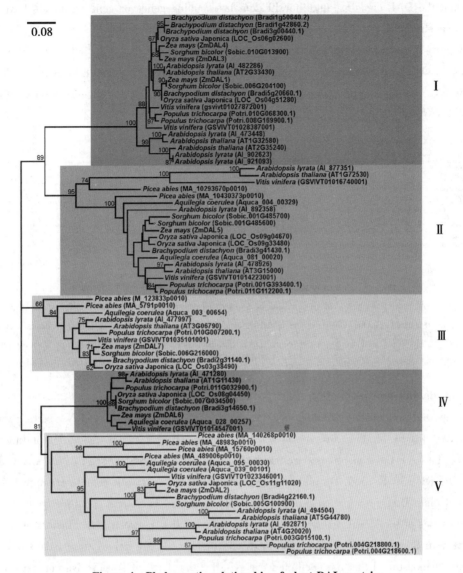

Figure 4 Phylogenetic relationship of plant DAL proteins

The neighbor-joining (NJ) phylogenetic tree of 79 plant *DAL* genes was constructed based on the multiple protein sequence alignment of the conserved regions (Supplemental Data File S3) according to the Poisson model. Bootstrap values >60% are indicated at each node, and the scale bar denotes the substitute rate per site. The species names are displayed before plant *DAL* genes

identify the origin of *DAL* genes in higher plants by lowering the HMMER search threshold (E-value of full sequence < 0.01). Peptidase S8 propeptide / proteinase inhibitor I9 domain of subtilisins were identified as putative homologs of DAL proteins in *P. patens* and *S. moellendorffii* but not in *C. reinhardtii* (Supplementary Table S4). The proteinase inhibitor I9 domain is the propeptide of the serine peptidase family S8A (subtilisin family) and is responsible for the modulation of folding and activity of these proenzymes. In addition to the protein similarity of the inhibitor I9 domain and the DAL domain, inhibitor I9 domain-encoding genes or gene fragments have conserved gene structure with *DAL* genes, including the 2-1-1 intron phase pattern and the 98bp exon (Figure 5 and S7), which suggests that *DAL* genes probably originated from inhibitor I9 domain-encoding DNA sequences. The combination of inhibitor I9 domain-encoded exons and other exons, such as RRM-encoding exons, could be responsible for the appearance of *DAL* genes and *ORRM1-like* genes in higher plants.

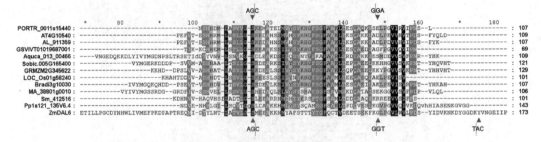

Figure 5 Alignment of inhibitor I9 domains and the DAL domain

A multiple protein sequence alignment was performed using hmmalign in the HMMER 3.0 package and was displayed using GeneDoc (http://www.nrbsc.org/gfx/genedoc/ebinet.htm). The DAL domain of the ZmDAL6 protein was used to represent plant DAL domains. The codons harboring intron splicing sites denoted by red dashed lines were from *Pp1s121_135V6.4* (the upper ones) and *ZmDAL6* (the basal ones).

Functional divergence evaluation between plant *DAL* subfamilies

As reported previously, *Arabidopsis DAL* genes play inequable roles in RNA editing for different RNA sites by binding diverse DYW-PPR proteins. To investigate the different functional constraints between these members, we conducted a maximum likelihood test of functional divergence using DIVERGE v3.0. The unrooted NJ tree was generated with complete amino acid sequences of plant DAL proteins excluding those of *P. abies* and *V. vinifera* (Supplementary Figure S8). Two types of functional divergence (type I and type II) between gene clusters of the *DAL* gene family in flowering plants were examined. The theta (θ) ML values were calculated, and the results demonstrated that the coefficients of type I (θ_I) for 6 of 10 pairwise comparisons between DAL subfamilies were significantly greater than zero (Bonferroni corrected $P<0.05$), and only one pairwise subfamily comparisons showed significant divergence with the coefficients of type II (θ_{II}) test (Bonferroni corrected $P<0.05$) (Table 1). Functional divergence-related sites were identified based on the posterior probabilities with a cut-off value of 0.85, and most were located in the DAL domain (Supplementary Figure S9). These observations indicate that there were significant site-specific shifted selective constraints on most members of the *DAL* gene family. Furthermore, we also observed that the values of θ_I were much larger than the estimates of θ_{II} in each pairwise comparison (Table 1), indicating that type I functional divergence predominantly contributed to the diversified evolution of plant *DAL* genes. In addition, we checked the functional divergence of intragroup members of the *DAL* genes, such as Sub. Ia vs. Sub. Ib (Supplementary Figure S8), but there was no significant functional divergence in any intragroup comparison (Supplementary Table S5), suggesting that intragroup *DAL* genes might play similar conserved roles in different plant lineages.

Table 1 Functional divergence between the subfamilies of *DAL* genes in the NJ tree based on complete protein alignment

Comparison	θ	θ_{SE}	z-Score[a]	P-value[b]
Type I (Gu99)	θ_I			
Sub. I vs. Sub. II	0.431 0	0.156 8	2.748 7	0.006 0
Sub. I vs. Sub. III	0.673 6	0.213 9	3.149 1	0.001 6*
Sub. I vs. Sub. IV	0.438 4	0.149 4	2.934 4	0.003 3
Sub. I vs. Sub. V	0.744 0	0.235 5	3.159 2	0.001 6*
Sub. II vs. Sub. III	0.622 4	0.152 5	4.081 3	< 0.000 1*
Sub. II vs. Sub. IV	0.517 8	0.138 0	3.752 1	0.000 2*
Sub. II vs. Sub. V	0.648 0	0.146 7	4.417 2	< 0.000 1*
Sub. III vs. Sub. IV	0.798 4	0.177 3	4.503 1	< 0.000 1*
Sub. III vs. Sub. V	0.372 8	0.189 4	1.968 3	0.049 0
Sub. IV vs. Sub. V	0.183 2	0.127 1	1.441 4	0.149 5
Type II	θ_{II}			
Sub. I vs. Sub. II	0.157 1	0.150 0	1.047 3	0.295 0
Sub. I vs. Sub. III	0.236 7	0.106 3	2.226 7	0.026 0
Sub. I vs. Sub. IV	0.322 8	0.101 0	3.196 0	0.001 4*
Sub. I vs. Sub. V	0.433 5	0.146 5	2.959 0	0.003 1
Sub. II vs. Sub. III	0.160 0	0.153 7	1.041 0	0.297 9
Sub. II vs. Sub. IV	0.234 9	0.148 4	1.582 9	0.113 4
Sub. II vs. Sub. V	0.386 2	0.184 1	2.097 8	0.035 9
Sub. III vs. Sub. IV	0.108 5	0.118 9	0.912 5	0.361 5
Sub. III vs. Sub. V	0.095 4	0.193 1	0.494 0	0.621 3
Sub. IV vs. Sub. V	-0.148 4	0.193 7	-0.766 1	0.443 6

[a] z-Score is the ratio of ThetaML (θ) to SE Theta (θ_{SE})

[b] P-value is evaluated based on the normal distribution test of the z-score

* Bonferroni corrected $P < 0.05$

GC content of *DAL* genes in monocots and dicots

The GC content, an important genomic feature, plays a critical role in determining the physical properties of DNA molecules and genome regulation by providing substrates for DNA methylation. The base composition analysis of plant *DAL* genes revealed that the *DAL* genes of monocots (grass) have a higher GC content than those of dicots (Figure 6a). To investigate DNA methylation of the GC-rich *DAL* genes, CpG islands of *ZmDAL6* and *At1g*11430 as representatives were predicted, and there were two CpG island regions identified in *ZmDAL6* but not in *At1g*11430 (Figure 6b). We analyzed the DNA methylation of the first CpG island of *ZmDAL6*, which was located at the exon 1-harboring region, using bisulfite sequencing and observed that 7 cytosines within a 19bp region (nucleotides 1 222~1 240 of the *ZmDAL6* DNA sequence in Figure 6c) were methylated, including three CHH sites, two CHG sites and one CG site (Figure 6c). In the 607bp CpG island, however, most cytosines were not substantially modified by DNA methylation. Moreover, the codon base composition of the *DAL* genes was analyzed, and the result showed that the GC content of each base of one codon in monocot *DAL* genes was higher than that of dicot *DAL* genes (Student t-test, $P<0.01$). For each group, only the codon second base (GC2) of group Ⅲ *DAL* genes showed no significant difference in GC content between dicots and monocots. The largest difference of GC content was found in the codon third base (GC3) of groups I and IV between dicots and monocots (Figure 6d), and in these two groups, more than 75% of the monocot *DAL* GC3 bases were guanine or cytosine nucleotides. Given that no significant functional divergence was observed between intragroup *DAL* genes, we inferred that the higher GC content of *DAL* genes in monocots could be caused by GC-biased gene conversion because the base composition of *DAL* genes was consistent with the high GC content of monocot genomes.

Expression analysis of *ZmDAL* genes by RT-PCR

Given the important roles of *DAL* genes in plant organelle RNA editing, the preferential expression patterns of *ZmDAL* genes were analyzed. To analyze the expression pattern of *ZmDAL* genes, a NimbleGen maize microarray data (ZM37) was performed for 60 tissues representing 11 major organ systems and various developmental stages of the B73 maize inbred line using *ZmDAL* probes. The median log signal values for all 7 *ZmDAL* genes were extracted. Four *ZmDAL* genes (*ZmDAL2*, *ZmDAL3*, *ZmDAL4*, and *ZmDAL5*) showed a constitutive expression pattern in 60 different tissues, with a CV value < 5% (Supplementary Figure S10). *ZmDAL1* had a much higher expression level in the leaves than in other tissues, while *ZmDAL6* showed a lower expression level in the roots and first internode compared with that in other tissues or organs. The predominant expression levels of *ZmDAL7* were observed in the developing seed, embryo and endosperm. Of these *DAL* genes, *ZmDAL1* and *ZmDAL6* were predicted to localize to chloroplasts and were preferentially expressed in the leaves, despite different expression patterns of the two genes (Supplementary Figure S10). The expression pattern was similar for paralogous genes (*ZmDAL3* and *ZmDAL4*), indicating they were formed by segmental duplication and retained their function (Supplementary Figure S1).

To confirm the organ-specific expression of *ZmDAL* genes shown by microarray data, RT-PCR was performed with total RNA isolated from the roots, leaves, ears, and immature tassels. The RT-PCR analysis revealed that the results for *ZmDAL2*, *ZmDAL3*, *ZmDAL4*, and *ZmDAL5* match with the DNA chip data but that the other *ZmDAL* genes do not (Figure 7). Howev-

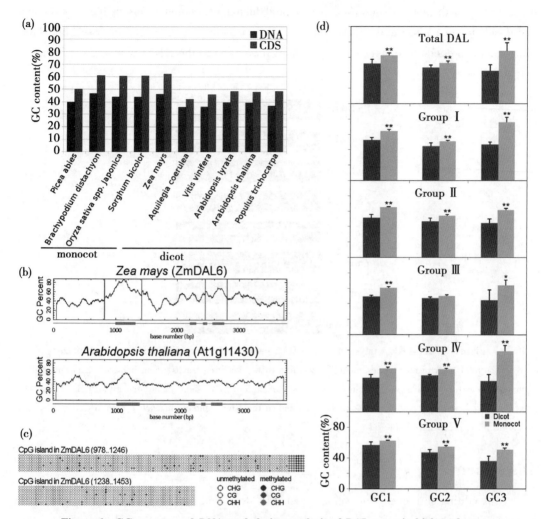

Figure 6 GC content and *DNA* methylation analysis of *DAL* genes in higher plants

(a, b) GC content and CpG islands of plant *DAL* genes were identified and displayed using CpGplot of EMBOSS (http://www.ebi.ac.uk/Tools/emboss/) with a window size of 100bp and the following set options: Observed/Expected ratio>0.60, Percent C+ Percent G >50.00 and Length >200. The DNA sequences of the *DAL* genes used here contain 1 000bp before the start codon and 1 000bp after the stop codon. Two CpG islands within the 607bp (nucleotides 817~1 423) and 359bp (nucleotides 817~1 423) regions (817~1 423) were observed and delimited by two close red lines, respectively. (c) The DNA methylation of the 607bp CpG island was analyzed using bisulfite sequencing with the primers listed in Supplementary Table S2, and the data were displayed using Kismeth software (http://katahdin.mssm.edu/kismeth/revpage.pl). (d) Codon base composition of *DAL* genes in flowering plants. The significant differences of codon base composition between dicot and monocot *DAL* genes were statistically analyzed according to Student's t-test. $*P < 0.05$; $**P < 0.01$. GC1, the GC contents at the first base of one codon; GC2, the GC contents at the second base; GC3, the GC contents at the third base

er, the expression levels of *ZmDAL2*, *ZmDAL3*, *ZmDAL4*, and *ZmDAL5* were abundant in the ears and tassels, where more biological energy from mitochondria is required. *ZmDAL1* was expressed little in the four tissues. ZmDAL7 showed higher expression levels in the leaves, ears, and tassels in comparison to the roots. ZmDAL6 was predominately expressed in tassels but showed little or no expression in roots, leaves and ears, according to the RT-PCR analysis (Figure 7).

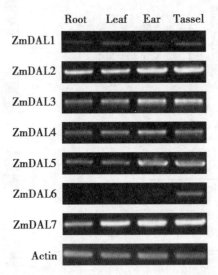

Figure 7 RT-PCR analysis of maize *DAL* genes in the four different tissues of maize

The total RNA of four tissues including seedling roots, seedling leaves, 5cm immature ears and non-emerged immature tassels was isolated and used to perform the semi-quantitative RT-PCR of *ZmDAL* genes. *Actin1* was used for internal controls to normalize the RNA contents in each sample. Primers used are shown in Supplementary Table S2

Discussion

RNA editing of a single nucleotide, such as C-to-U and A-to-I substitution, requires *trans*-factors to recognize the nucleotide to be edited and remove the amino group. In mammals, A-to-I RNA editing is catalyzed by a family of enzymes called adenosine deaminases that act on RNA (ADARs), while C-to-U editing of apolipoprotein B mRNAs is performed by the zinc-dependent RNA-editing enzyme apolipoprotein B editing catalytic subunit 1 (APOBEC-1), which interacts with APOBEC-1 complementation factor (ACF) for site-specific editing. In plant organelles, C-to-U RNA editing is mediated by site-specific DYW-PPR proteins and by several other nuclear-encoded factors, including DAL proteins, PPO1 and OZ1. However, the precise roles of each component of the plant RNA editosome in the complicated editing machinery are not yet well known. In this study, we performed a comprehensive analysis of *DAL* genes in plants and uncovered their seed plant-specific distribution, origin and the potential role of functional divergence.

Using a custom-built HMM file derived from multiple DAL domain alignments, we screened a plant annotation database for putative *DAL* genes and found that *DAL* genes were specific to seed plants. The absence of *DAL* genes in lower plants, in which there are thousands of organelle RNA-edited sites, indicates that the RNA editosome differs between higher and lower plants. Given that the presence of *DYW-PPR* genes in plants is associated with C-to-U RNA editing events and that DAL proteins interact with PPR proteins, *PPR* genes may not be necessary for the emergence of *DAL* genes in plants (Figure 3), although DAL proteins can interact with *P. patens* PPR proteins. Furthermore, the homologs of OZ1, ORRM and PPO1 proteins, which have been proven to interact with DAL proteins, were found in *P. patens* and *S. moellendorffii* (Figure 3), and in particular, putative PPO1 proteins in *P. patens* (*Pp1s28_304V6. 1*) and *S. moellendorffii* (*Sm_82264*) all have the 22aa regions that are essential for their interaction with DAL proteins, suggesting that these additional RNA editing factors are also not required for the evolution/function of *DAL* genes, although there is no evidence for the roles of these proteins in RNA editing in lower plants. However, we cannot exclude the possibility that unidentified alternative genes aside from those of the *DAL* gene family may play similar roles in RNA editing in lower plants. In addition to the regulation of RNA editing via bridging multiple subunits in the RNA editosome, DAL proteins probably have other functions required in higher plants.

When we searched DAL proteins with the dal. hmm information as a query, propeptide inhibitor I9 domains of plant subtilisins were found to show high similarity to part of the DAL domain. In addition, the inhibitor I9 domain-encoding genes and *DAL* genes have a conserved gene structure, including the intron phases 2, 1, and 1 and the 98bp exon (Supplementary Figure S7), and *inhibitor I9* genes were present prior to the divergence of seed plants from ferns (Figure 3). Therefore, it is inferred that the *DAL* genes originated from *inhibitor I9* genes by combining I9 domain-encoding exons with another unidentified sequence via exon shuffling. Additional evidence for this hypothesis stems from the fact that the *ORRM1-like* genes encode two tandem DAL domains followed by an RRM domain at the C terminus, which could have arisen from the combination of two I9 domain-encoding exons and an extra RRM domain-encoding sequence. In addition to the protein sequence similarity and conserved gene structure between the *DAL* genes and I9 domain-encoding genes, they also have similar molecular function, as the DAL domain mediates the protein-protein interaction of DAL proteins with other RNA editing factors, while the I9 domain inhibits proenzymes by hiding substrate-binding domains.

Plant *DAL* genes were assigned to five distinct subfamilies based on their phylogenetic relationships (Figure 4). In *Arabidopsis*, *DAL/RIP* genes play nonequivalent roles in RNA editing. The major factor is *RIP1* (*At3g15000*), which belongs to group II and controls 75% of RNA-edited sites in mitochondria and 20% of RNA-edited sites in chloroplasts. *RIP2* (*At2g33430*), *RIP3* (*At3g06790*), *RIP9* (*At1g11430*) and *RIP8* (*At4g20020*) have moderate effects on RNA editing and are in groups I, III, IV and V, respectively (Figure 4). As minor genes, *RIP4* (*At5g44780*), *RIP5* (*At1g32580*), *RIP6* (*At2g35240*) and *RIP7* (*At1g72530*) each shares clades with the above major or moderate factors, suggesting that these *DAL* genes have recently duplicated and have evolved to only control few RNA editing sites. We analyzed the functional divergence between DAL groups to identify the sites distinguishing the different DAL members. Five groups resulted in 10 pairwise comparisons; of these comparisons, 6 showed significant type I functional divergence, and one showed significant type II functional divergence (Table 1). In addition, we analyzed the functional divergence between intragroup members. These intragroup *DAL* genes were divided into two subgroups: a dicot subgroup (a) and a monocot subgroup (b) (Supplementary Figure S8). No significant functional divergence was observed between intragroup members of *DAL* genes, although the GC contents of *DAL* genes in monocots and dicots were different. The sites involved in functional divergence between DAL groups were predominantly localized to the DAL domains, and relaxed selection on these sites would serve to increase the complexity for determination of RNA editing because DAL proteins act on RNA editing in the form of heterodimers in addition to homodimers. Also, these sites probably account for the interaction of each DAL with different PPR proteins in the RNA editosome. Therefore, it is understandable that different dimers formed with homogenous or heterogeneous DAL proteins, which confer RNA editing to corresponding sites, have increased the diversity of RNA editing regulation in higher plants. However, further studies on the biochemical character of DAL proteins and the crystal structure of the RNA editosome are required to parse the roles of DAL proteins with the functionally diverged sites. In addition, the putative effects of DAL proteins on other RNA processing events in addition to RNA editing should be further investigated.

Materials and Methods

Identification and sequence analysis of putative *DAG-like* genes in plants

Known *MORF/RIP* genes At4g20020, At2g33430, At3g06790, At5g44780, At1g32580, At2g35240, At1g72530, At3g15000, and At1g11430 from *A. thaliana* and *DAG* (NCBI Protein ID: Q38732) from *A. majus* were used to query the maize filtered gene set (ZmB73_5b_FGS_translations.fasta downloaded from www.maizesequence.org) using a local BLASTP program with an E-value < 1e-10 and a bit score > 100. ZmDAL protein sequences were analyzed using ExPASy tools available at http://us.expasy.org/tools/. Multiple sequence alignments of ZmDAL proteins and the above known DAL proteins were performed using ClustalW (http://www.ebi.ac.uk/Tools/msa/clustalw2/). To mine the conserved domain, the alignment results (Supplementary Data File S1) were used to build a protein HMM file, dubbed dal.hmm (Supplementary Data File S2) by the hmmbuild program in HMMER 3.0 package.

To investigate the evolution of *DAL* genes in the plant kingdom, the dal.hmm information was used as a query to search the genome annotation data of the following representative species from Phytozome v8.0 (http://www.phytozome.net/), except those of *P. abies*, which were from ConGenIE (http://congenie.org/), in HMMER 3.0 package: *C. reinhardtii*, *P. patens*, *S. moellendorffii*, *P. abies*, *B. distachyon*, *O. sativa* Japonica, *S. bicolor*, *A. formosa*, *V. vinifera*, *A. lyrata* and *P. trichocarpa*. Protein hits with an E-value < 1e-10 and sequence score of "best 1 domain" > 100 were collected. The homologs of the *OZ1*, *PPO1* and *ORRM* genes in the above plants were identified using a local BLASTP program with the protein sequences of known *Arabidopsis* OZ1, PPO1, ORRM2, ORRM3 and ORRM4 proteins as queries.

The PFAM (http://pfam.sanger.ac.uk/) and INTERPRO (http://www.ebi.ac.uk/

interpro/) databases were screened to detect known motifs in ZmDAL proteins and the DAL proteins of other plants. The MEME program (http://meme.nbcr.net/meme/cgi-bin/meme.cgi) was used to investigate the putative conserved motifs among these ZmDAL proteins with the following parameters: length between 10 and 50 aa, maximum number of motifs to find = 5, and one per sequence. To obtain the intact conserved DAL domain, different limits for length of each motif were taken that were between 100 and 120 aa.

Gene structures of plant DAL genes

The DNA and transcript sequences of *ZmDAL* genes obtained from the maize sequence annotation database MaizeGDB (http://www.maizegdb.org/) were used to analyze the gene structures of *ZmDAL* genes. Several *ZmDAL* genes had more than one gene model annotated in MaizeGDB. To confirm the putative alternative splicing transcripts, transcript-specific primers (Supplementary Table S2) were designed to amplify corresponding DNA isolated from B73 seedlings and cDNA derived from B73 seedling RNA. Gene structures producing validated transcripts were drawn and displayed using the online GSDraw program of the PIECE server (http://wheat.pw.usda.gov/piece/GSDraw.php). Conserved DAL domains were also displayed using the GSDraw program. The gene structures of *DAL* genes from other plant species were obtained from the Phytozome v8.0 annotation database (http://www.phytozome.net/) and displayed using the GSDraw program.

Subcellular localization prediction of ZmDAL proteins

Two *in silico* programs, Predotar and TargetP, were used to predict the putative organelle localization of ZmDAL proteins. The maize organelle proteomics database (PPDB, http://ppdb.tc.cornell.edu/) was screened to detect the accumulation of ZmDAL proteins.

Phylogenetic dendrogram

The multiple sequence alignment analysis of conserved DAL domains collected from 79 DAL proteins identified in maize and in other higher plants was carried out using MUSCLE v3.8.31, and the resulting alignment was used to build the NJ distance phylogenetic tree using MEGA v5.0 by applying the Poisson substitution model, 1 000 bootstrap samples, and pairwise deletion for gaps/missing data. The tree was displayed using FigTree v1.4.0 (http://tree.bio.ed.ac.uk/software/figtree/).

Functional divergence analysis

To investigate the functional alteration of duplicated *DAL* genes in plants, the GU99 method within DIVERGR v3.0 was used to calculate the coefficients of type I and type II functional divergence (θ_I and θ_{II}, respectively) between two groups after gene duplication and to predict functionally divergent amino acids based on their different evolutionary rates. Within two duplicated groups of a gene family, type I functional divergence helps identify the relaxation of functional constraint in one group relative to that of another, while type II identifies shifting patterns of functional constraint.

GC content and DNA methylation analysis with bisulfite sequencing

The entire DNA sequences of plant *DAL* genes together with 1kb of upstream and downstream flanking sequences were used for calculation of GC content and prediction of CpG islands in the EMBOSS CpGplot online server (http://www.ebi.ac.uk/Tools/seqstats/emboss_cpgplot/). To confirm the DNA methylation of *ZmDAL* DNA sequences, leaf DNA of B73 seedlings was isolated and treated with bisulfate using the EpiTect® bisulfite kit (QIAGEN, USA) according to the manufacturer's instructions. The primers for detection of DNA methylation were designed using MethPrimer (http://www.urogene.org/methprimer/) and modified using Primer3web (http://primer3.ut.ee/). PCR products were cloned into a pGEM-T vector (Promega, USA) and subsequently sequenced using an ABI3730 DNA sequencer (Shanghai Sunny Bio., China). DNA methylation states via bisulfite sequencing were analyzed and displayed using Kismeth software (http://katahdin.mssm.edu/kismeth/revpage.pl).

Expression analysis of *ZmDAL* genes in different tissues

To investigate the spatiotemporal expression patterns of *ZmDAL* genes, the log2-transformed and RMA-normalized data for *ZmDAL* genes were downloaded from PLEXdb (http://www.plexdb.org/), and cluster analysis of these expression data were performed using Cluster v3.0 with the hierarchic method. A heat map was produced using Java TreeView v1.1.5. The coefficient of variation (CV) was calculated according to the following equation to estimate the expression variation of each *ZmDAL* gene in different tissues: CV = sd/mean, where sd indicates the standard deviation of a gene in different tissues and the mean represents the average expression level.

Semi-quantitative reverse transcription PCR (semiq-RT-PCR)

Total RNA was isolated from different tissues of the B73 inbred lines, including seedling roots, leaves, 5cm ears and immature tassels, using the Trizol® reagent (Invitrogen, USA) according to the manufacturer's protocol. First-strand cDNA was produced from 1μg of total RNA (25μl reaction volume) using M-MLV reverse transcriptase (Invitrogen, USA) at 37℃ for 1h. All gene-specific primers were designed as shown in Supplementary Data Table S2. Specific primers for the maize *Actin*1 gene (GenBank ID: NM_001155179) were used as an internal control. Reactions were performed with Taq Polymerase (Dalian Takara Biotechnology, China) on a Bio-Rad Thermal Cycler (Bio-Rad, USA) using the following procedure: 5min at 94℃ to start; 32 cycles of 30s at 94℃, 30s at 58℃ and 1min at 72℃; and a final extension step of 72℃ for 10min to complete the reaction, and the *Actin1* transcript was amplified with 28 PCR cycles. Each PCR pattern was performed in triplicate, mixtures without a template were employed as negative controls, and the maize *Actin1* amplicon served as an internal control for each gene investigated.

References (omitted)

Scientific Reports, 2017, 7 (1): 5 688

Mapping of a Major QTL for Salt Tolerance of Mature Field-grown Maize Plants Based on SNP Markers

Luo Meijie[1]* Zhao Yanxin[1]* Zhang Ruyang[1] Xing Jinfeng[1]
Duan Minxiao[1] Li Jingna[1] Wang Naishun[1] Wang Wenguang[1]
Zhang Shasha[1] Chen Zhihui[2] Zhang Huasheng[1]
Shi Zi[1]** Song Wei[1]** Zhao Jiuran[1]**

(1. Beijing Key Laboratory of Maize DNA Fingerprinting and Molecular Breeding, Maize Research Center, Beijing Academy of Agricultural and Forestry Sciences (BAAFS), Beijing, China;
2. Institute of Crops Research, Hunan Academy of Agricultural Sciences, Changsha, China)

Abstract: Background: Salt stress significantly restricts plant growth and production. Maize is an important food and economic crop but is also a salt sensitive crop. Identification of the genetic architecture controlling salt tolerance facilitates breeders to select salt tolerant lines. However, the critical quantitative trait loci (QTLs) responsible for the salt tolerance of field-grown maize plants are still unknown.

Results: To map the main genetic factors contributing to salt tolerance of mature maize, a double haploid population (240 individuals) and 1 317 single nucleotide polymorphism (SNP) markers were employed to produce a genetic linkage map covering 1 462.05cM. Plant height of mature maize cultivated in the saline field (SPH) and plant height-based salt tolerance index (ratio of plant height between saline and control fields, PHI) were used to evaluate salt tolerance of mature maize plants. A major QTL for SPH was detected on Chromosome 1 with the LOD score of 22.4, which explained 31.2% of the phenotypic variation. In addition, the major QTL conditioning PHI was also mapped at the same position on Chromosome 1, and two candidate genes involved in ion homeostasis were identified within the confidence interval of this QTL.

Conclusions: The detection of the major QTL in adult maize plantestablishesthe basis for the map-based cloning of genes associated with salt tolerance andprovides a potential target for marker assisted selection for the breeding of maize varieties with salt tolerance.

Key words: Salt tolerance; QTL mapping; SNP; Plant height; Maize

Background

Elevated salt content in the soil leads to the suppression of plant growth and the reduction of

---

* These authors contributed equally to this work
** Correspondence: Shi Zi, shizi_baafs@126.com; Song Wei, songwei1007@126.com;
Zhao Jiuran, maizezhao@126.com

productivity. About 6% of the land on earth and 20% of the total cultivated land worldwide are affected by high salt. In many areas, salinity problemis further aggravated by unsustainable agricultural practices. Salt stress causes ion and hyperosmotic imbalance in plants, whichcauses secondary oxidative damage. These changes occur at the molecular, cellular, and whole-plant levels, resulting in the plant growth arrest and death. Therefore, understanding the genetic architecture of salt tolerance in plant is of great significance for the selection, utilization, and breeding of salt tolerant varieties.

A major salt tolerance strategy in plant is to re-establish cellular ion homeostasis. A high concentration of sodium ions (Na^+) inhibits many key enzymes, so Na^+ influx to the cell cytoplasm and organelles is sophisticatedlyregulated. Therefore, tremendous efforts have been made to identify the transporters modulating the influx and efflux of Na^+ in plant cells. Vacuolar Na^+/H^+ antiporters manage the compartment Na^+ in the vacuole to prevent Na^+ toxicity in cytosol, which has been shown as a strategy in many naturally salt tolerant plants (halophytes). In addition to the control of Na^+ influx, Na^+/H^+ antiporters on the plasma membrane are also important in exporting Na^+ to maintain low Na^+ concentration in the cytoplasm. In *Arabidopsis*, a salt overly sensitive (SOS) signal transduction pathway has been identified tomediate ion homeostasis and Na^+ tolerance. *SOS1* encodes a plasma membrane Na^+/H^+ antiporter, and *sos1* mutation renders *Arabidopsis* the extreme sensitivity to Na^+ stress. Besides the SOS-dependent pathway, other mechanisms have also been reported to play roles in salt tolerance of plants, including the accumulation of osmolytes to establish osmotic homeostasis and the increase of antioxidants to mediate oxidative protection.

Quantitative trait locus (QTL) mapping has been applied to detect the genetic basis of salt tolerance in many plants, which provides valuable information for further map-based cloning of salt tolerance genes and marker-assisted selection (MAS) in crop breeding. Using restriction fragment length polymorphism (RFLP) markers and an $F_{2:3}$ population derived from the cross between salt tolerant and salt sensitive rice varieties, two major QTLs explaining 48.5% and 40.1% of the total phenotypic variance (PVE) were detected in rice. The QTL detection led to the cloning of the gene responsible for salt tolerance, *SKC1*, which encoded a high affinity K^+/Na^+ transporter. Taking advantage of the amplified fragment length polymorphism (AFLP), RFLP and simple sequence repeat (SSR) markers, a QTL, *Nax1*, was identified on the Chromosome 2AL of durum wheat using an F_2 and an $F_{2:3}$ population, which accounted for 38% of the phenotypic variation, and the SSR marker closely linked to the QTL was proven to be useful for the MAS in the breeding program. Later, the *Nax2* locus was discovered and a Na^+-selective transporter gene *TmHKT*1;5-A was subsequently characterized in durum wheat. In soybean, utilizing random amplified polymorphic DNA (RAPD), insertion-deletion (InDel), and SSR markers with an $F_{2:3}$ population and recombinant inbred lines (RILs), a salt tolerance QTL was mapped within a 209kb region on Chromosome 3 and its flanking markers were used for MAS of soybean breeding. In maize, nine conditional QTLs for salt tolerance at the seedling stage were identified on Chromosomes 1, 3, and 5 using single nucleotide polymorphism (SNP) markers and $F_{2:5}$ RILs, three of which explained more than 20% of phenotypic variation. However, studies of QTLs for salt tolerance in maize are still very limited, and QTLs for salt tolerance has not been reported in mature field-grown maize.

In this study, we identified a major QTL for salt tolerance in mature maize grown in a saline field using a permanent double haploid (DH) population and high-density SNP markers, and two candidate genes harbored in this QTL might be involved in the SOS pathway. Our resultsnot only shed light on the mechanism of salt tolerance in field-grown maize, but will also facilitate the breeding of maize varieties with salt tolerance.

Methods

Plant materials and treatment

The parental maize inbred lines, PH6WC and PH4CV, were obtainedfrom DuPont Pioneer (Johnston IA, USA). A DH population consisting 240 lines derived from the F_1 hybrid of PH6WC × PH4CV was developed by pollinating with the parthenogenetic-inducing line Jingkeyou006 to obtain the haploid plants, and then followed by artificial doubling with colchicine. Jingkeyou006 was obtained from the Maize Research Center of Beijing Academy of Agricultural and Forestry Sciences.

The 240 DH lines and their parents PH6WC and PH4CV were used in field experiments. For the salt stress treatment, plant materials were planted in the saline field at Tongzhou (TZ, N39°41′49.70″, E116°40′50.75″ in Google Earth™), Beijing, China, in the spring of 2014, 2015, and 2016. Plants cultivated in the normal field at Changping (CP, N40°10′50.38″, E116°27′15.40″), Beijing, in the spring of 2014, 2015 and 2016 served as control. All field experiments were performed in the accordance with local legislation. At each location, the experiments were conducted in a randomized complete block designwith two replicates in each year. For each block, 20 plants of each DH line were planted for a whole row, with the row length of 5m and the spacing between rows of 60cm.

Soil sampling and analyses

According to the five-point sampling method, soil samples were collected from five representative locations (upper left, upper right, lower left, lower right and the middle) of the fields in both TZ and CP for composition analysis. Total salt content was determined using the residue-drying method. Soil pH was measured using a pH meter (Hach, Loveland, CO, USA), and Na^+ content was determined by flame emission spectroscopy (PerkinElmer, Norwalk, CT, USA).

Phenotype analysis

At harvest stage, in the summer of 2014, 2015, and 2016, plant height of the DH lines and their parents in the saline field (SPH) and control field (NPH) were recorded. Plant height was measured from the top of the main inflorescence down to the ground. For 240 DH lines, five randomly selected plants in each row were measured as one replicate, and two replicates were performed each year. After collecting plant height data in TZ and CP, salt tolerance index (PHI) based on plant height for each line was calculated using the following formula, PHI = H_{SPH}/H_{NPH}, where H_{SPH} represents the average height of five mature maize plants of each DH line grown in TZ, H_{NPH} represents the average height of 10 mature maize plants (5 of each replicate) of the same line grown in CP.

Linkage map construction and QTL identification

Total genomic DNA was extracted from leaves using the CTAB method and then the 240 DH lines were genotyped using the MaizeSNP3072 chip. A comparative linkage map was constructed using the Kosambi function in JoinMap4 software with a minimum LOD score of 2.0. Composite interval mapping was carried out using Windows QTL Cartographer software V2.5 which was developed by the Department of Statistics, NCSU with a walk speed of 1.0cM and the LOD threshold of 3.0.

Candidate gene analysis

Salt tolerance-related genes previously characterized in other plant species, such as *AtSOS1*, *AtSOS2*, *AtSOS3*, *OsSKC1*, and *TmHKT1;5*, were employed to query the maize genome database (MaizeGDB, http://www.maizegdb.org/, B73 RefGen_v2) using the tool of local BLASTP with the E-value cutoff of 1e-4. The maize homologs fell in the confidence interval of the identified major QTL were considered as candidate genes associated with salt tolerance.

qRT-PCR analysis

Seeds of PH6WC and PH4CV were surface sterilized with 1% NaClO for 10min, following by rinsing with sterile water for three times. The resulting sterile seeds were sown in maize seedling identifying instrument (Chinese patent, patent number: ZL200920177285.0) according to its manufacturer's instructions, and then were placed in a greenhouse under 12h light/12h dark at 25℃ with the light density of 150~180μmol m^{-2} s^{-1} and the relative humidity of 70%. Seeds were grown in sterile water for three days and then in the Hoagland's nutrient solution (Phyto Technology Laboratories Co., Ltd., USA) which was replaced with fresh Hoagland's solution for every 2 days. After 7 days, non-germinated seeds and seedlings exhibited abnormal growth were discarded and the remaining seedlings were treated with 100mM NaCl at day 12. Shoot and root samples were collected in three biological replicates each with five seedlings, and were immediately frozen in liquid nitrogen for RNA extraction.

Total RNA of shoot and root was extracted using the TRIZOL reagent (Invitrogen, Carlsbad, CA, USA). After RNA was treated with RNase-free DNase1 (Fermentas, Thermo scientific, USA), reverse transcription was conducted using the PrimeScript™ II 1st strand cDNA Synthesis Kit (Takara Bio Inc, Shiga, Japan) according to the user's manual. qRT-PCR was carried out using the QuantStudio™ 6 Flex (ABI Life Technologies, USA). Gene-specific forward and reverse primers were designed (Additional file 1) using the PrimerQuest tool on IDTDNA (http://www.idtdna.com/Primerquest/Home/Index). The reaction was performed in the 20μl PCR system using the SYBR Premix Ex TaqII (Takara Bio Inc, Shiga, Japan) according to the user's manual. *ZmActin1* served as the internal reference and mRNA relative expression levels were calculated using the $2^{-\Delta\Delta Ct}$ method.

Statistical analysis

Using the Graphpad Prism software (http://www.graphpad.com/), the average of total salt content, Na⁺ content, and pH in fields of TZ and CP were compared by t-tests. Two-way ANOVA with Bonferroni test was carried out forplant height of two parental lines grown in two soil conditions as well as the gene expression analysis. The frequency distribution, linear regression, and correlation analysis of the average of SPH, NPH and PHI across three environmentswere determined with the non-linear Gaussian regression, linear regression and correlation function in the column analysis. The combined ANOVA of SPH, NPH and PHI for the DH population and the heritability of three traits were obtained using the ANOVA tool in IciMapping 4.1 program.

Results

Soil composition

Soil in agricultural land is usually subject to the compound effect of salt and alkali. To understand the main stress factor in soil in this study, we analyzed the soil composition in TZ and CP fields at the depths of 0~20cm and 20~40cm (Figure 1). The total salt content and Na⁺ content of soil from both the 0~20cm and 20~40cm layers in TZ were significantly higher than those in CP, but no substantial difference in the pH was observed between two locations, indicating that salt, rather than pH, was the major stress factor in TZ soil.

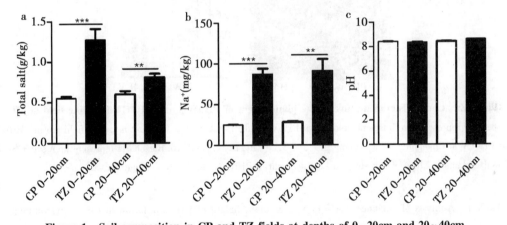

Figure 1 Soil composition in CP and TZ fields at depths of 0~20cm and 20~40cm

The bar charts represent the mean ± SE of five soil samples fromeach field. a. Total salt content; b. Na⁺ content. c. pH value. *, **, and *** indicate significant difference at $P<0.05$, $P<0.01$, and $P<0.001$, respectively. CP indicates Changping. TZ indicates Tongzhou

Phenotypic variation and correlations

To evaluate salt tolerance of mature maize in the field, plant height in the saline field (SPH) and salt tolerance index (PHI) were used as salt tolerance indicators. At the mature stage, although the plant height of both PH6WC and PH4CV were significantly reduced by salt stress (Figure 2a and b), the PHI of PH6WC was significantly higher than that of PH4CV, indicating that PH6WC is less sensitive to salt stress compared to PH4CV (Figure 2c). The variance effects of genotype (G), environment (E), and their interaction were extremely significant for all three traits (Table 1), and variance of the replicates within seasons for SPH and PHI was also significant. Broad sense heritability (H^2) of SPH, NPH and PHI were 74.7%, 86.4% and 74.9% respectively, indicating that genotypes play an important role in the determination of these phenotypes. Phenotypic frequencies of all three traits showed a normal or near-normal distribution (Figure 3). The correlation coefficient of the three-year average of SPH and NPH was low ($r = 0.397$), but it was as high as 0.686 between SPH and PHI, suggesting that the phenotypic results of SPH and PHI were highly positively correlated. As expected, NPH was negatively correlated with PHI with a low correlation coefficient of −0.229 (Figure 3).

Figure 2 Comparison of mature maize plant height of PH6WC and PH4CV in saline field at TZ

a. The representative image of field grown PH6WC and PH4CV in TZ; b. Plant height of field grown PH6WC and PH4CV at Changping (CP) (control) and TZ (salt stress) in 2016. Bar charts represent the mean ± SE of 15 maize plants. c Effect of salt stress on plant height of PH6WC and PH4CV. Scale bar = 25cm

Table 1 Analysis of variance (ANOVA) for plant height of DH population in three environments

Trait	Source of variation	F	H^2
SPH	Environment (E)	31.8862***	0.7471
	Genotype (G)	14.1894***	
	Replication	3.0553*	
	G × E	4.4658***	
NPH	Environment (E)	4.7790**	0.8639
	Genotype (G)	39.2024***	
	Replication	0.7548	
	G × E	6.0168***	

(continued)

Trait	Source of variation	F	H^2
PHI	Environment (E)	12.3846***	0.7492
	Genotype (G)	16.2502***	
	Replication	3.6290*	
	G × E	5.1059***	

Note: H^2 indicates broad-sense heritability. *, ** and *** represent significant levels at $P<0.05$, $P<0.01$ and $P<0.001$, respectively

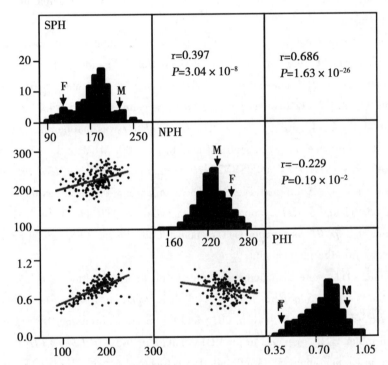

Figure 3 Frequency distribution and correlation analysis of NPH, SPH and PHI traits across three seasons

Frequency distribution and correlationanalysis were conducted using Graphpad Prism software (http://www.graphpad.com/). r indicates correlation coefficient. F represents PH4CV andM represents PH6WC

Genetic linkage map construction based on SNP markers

A genetic linkage map was developed using the 240 DH lines and MaizeSNP3072 chip. Of the 3 072 SNPs, 1 337 were polymorphic between the parents. For the 1 337 SNPs, twenty SNPs were removed because of inconsistencies between their physical and genetic positions. With the 1 317 SNPs, the linkage map constructed covered 10 chromosomes and spanned 1 462.05cM of the maize genome with an average distance of 1.11cM between marker loci.

Identification of salt tolerance related QTLs in maize

Based on the average plant height across three seasons at TZ, we identified one major QTL on Chromosome 1 with a LOD score of 22.40 and the PVE of 31.24%, which was designated as qSPH1 (Table 2; Figure 4). The LOD peak of qSPH1 was mapped at 88.51cM on Chromosome 1 and its confidence interval covered the region of 77.61~98.18cM between the SNP markers *PZE101094436* and *PZE101150513*. In fact, qSPH1 was detected across all seasons (Additional file 2), indicating that it is highly stable and not sensitive to environmental factors. Five QTLs controlling NPH were detected on chromosomes 4, 5, 8 and 9, respectively, with the PVE ranged from 5.0% to 11.9%. Except for qNPH8 (LOD, 3.88; PVE, 6.36) which was identified across all three environments, qNPH4, qNPH5, qNPH9-1 and qNPH9-2 were detected only in single or two environments (Table 2; Figure 4). There was no QTL controlling NPH mapped on Chromosome 1, revealing that qSPH1 is the locus associated with salt tolerance, but not the plant height of maize.

In addition, to exclude the impact of inherited factor on plant height under salt stress, we compared the plant height between the saline and control fields for each maize line to obtain PHI. Using the average of PHI across three seasons, a major QTL, named qPHI1, was identified on Chromosome 1 with the LOD score of 16.2, which accounted for 25.94% of the total phenotypic variation (Table 2; Figure 4). The LOD peak of qPHI1 was mapped at 90.21cM on Chromosome 1 and its confidence interval spanned from 80.11~100.71cM of the genetic map between the SNP markers *PZE101109084* and *SYN25920*. Similar to qSPH1, qPHI1 was also identified across all three seasons (Additional file 2). The similar confidence intervals, positions, and LOD scores of qSPH1 and qPHI1, along with the fact that no QTL detected for NPH on Chromosome 1 indicated that qSPH1 is a major QTL responsible for the salt tolerance of mature maize but not the loci controlling the plant height.

Several minor QTLs were also detected under saline field condition, such as qSPH5-1 and qSPH5-2 on Chromosome 5. qSPH5-2 (LOD, 4.26; PVE, 6.17%) was only identified in 2016, and qSPH5-1 was detected both in 2015 and the mean of three-year SPH data (Table 2). Four additional minor QTLs responsible for PHI, named qPHI3, qPHI4, qPHI9, and qPHI10, were identified on Chromosomes 3, 4, 9 and 10, respectively, with the LOD score ranged from 3.29~4.63, which contributed 4.47%~13.64% of phenotypic variation. qPHI3 was only detected in 2014, while other three minor QTLs were identified only based on the three-year average of PHI (Table 2). Their inconsistency detection in different planting seasons indicated that these QTLs were susceptible to environmental conditions.

Table 2 QTLs controlling plant height of mature maize plants grown in saline soil in three planting seasons.

Traits[a]	Years	QTL[b]	Chr.	Position (cM)	Marker interval (Coordinates)	LOD	PVE[c] (%)	Add[d]
NPH	2014	qNPH4	4	49.71	PZE104023902-SYN4889 (27048623-39360273)	4.39	8.33	8.40
		qNPH8	8	56.01	PZE108041337-PZE108090114 (66870236-146632672)	3.67	6.90	8.12

(continued)

Traits[a]	Years	QTL[b]	Chr.	Position (cM)	Marker interval (Coordinates)	LOD	PVE[c] (%)	Add[d]
	2015	qNPH4	4	46.01	PZE104023902-PZE104081530 (27048623-155939590)	6.28	10.11	7.78
		qNPH5	5	73.51	PZE105045981-PZE105128581 (34001093-185707663)	3.19	4.98	5.55
		qNPH8	8	61.61	PZE108028588-PZE108103365 (26352213-158556937)	5.03	7.07	6.94
		qNPH9-2	9	71.31	PZE109064469-SYN27201 (107573682-139631117)	4.56	7.24	6.61
	2016	qNPH8	8	57.31	PZE108041337-PZE108097446 (66870236-152623958)	4.29	7.75	8.11
		qNPH9-1	9	45.81	PZE109011840-PZE109040519 (12582417-62366526)	5.79	10.74	8.93
	mean	qNPH4	4	49.71	PZE104023902-PZE104081530 (27048623-155939590)	6.97	11.90	8.13
		qNPH8	8	57.31	PZE108028588-PZE108090114 (26352213-146632672)	3.88	6.36	6.40
		qNPH9-2	9	71.31	PZE109064469-SYN27201 (107573682-139631117)	4.80	7.97	6.66
SPH	2014	qSPH1	1	95.21	SYN309-SYN25920 (91198154-194798124)	4.47	13.28	13.79
	2015	qSPH1	1	88.51	PZE101094436-PZE101150513 (92353978-194525458)	11.46	16.99	16.86
		qSPH 5-1	5	77.31	SYN16675-PZE105128581 (165361318-185707663)	3.50	5.01	8.82
	2016	qSPH1	1	90.21	PZE101109084-SYN25920 (116462939-194798124)	19.47	35.03	24.69
		qSPH5-2	5	71.01	PZE105049283-PZE105117757 (40588124-174670738)	4.26	6.17	19.86
	mean	qSPH1	1	88.51	PZE101094436-PZE101150513 (92353978-194525458)	22.40	31.24	19.14
		qSPH5-1	5	77.31	SYN1390-PZE105128581 (163779167-185707663)	3.32	3.96	6.55
PHI	2014	qPHI1	1	94.81	PZE101109084-SYN25920 (116462939-194798124)	8.94	28.10	0.10
		qPHI3	3	23.51	SYN28626-PZE103019163 (2813337-11407445)	4.63	13.64	0.67
	2015	qPHI1	1	90.21	SYN5444-SYN25920 (96632104-194798124)	10.59	19.51	0.08
	2016	qPHI1	1	90.21	PZE101109084-SYN25920 (116462939-194798124)	14.27	27.51	0.09

(continued)

Traits[a]	Years	QTL[b]	Chr.	Position (cM)	Marker interval (Coordinates)	LOD	PVE[c] (%)	Add[d]
	mean	qPHI1	1	90.21	PZE101109084-SYN25920 (116462939-194798124)	16.20	25.94	0.07
		qPHI4	4	56.91	SYN4889-SYN4250 (39360273-170439029)	4.14	5.63	0.03
		qPHI9	9	130.34	SYN24345-SYN5732 (153690286-155054745)	3.39	5.59	0.03
		qPHI10	10	108.16	PZE110100655-SYN19213 (144384594-147029157)	3.29	4.47	0.03

Note: [a] NPH, plant height under normal condition (Changping, Beijing). SPH, plant height under salt stress (Tongzhou, Beijing). PHI, plant height-based salt tolerance index. [b] QTLs are designated as "q" followed by trait name: "NPH" (plant height under normal field), "SPH" (plant height under salt stress) and "PHI" (plant height-based salt tolerance index), and then the chromosome number. [c] Phenotypic variance explained by QTL. [d] Additive effect of tolerance allele

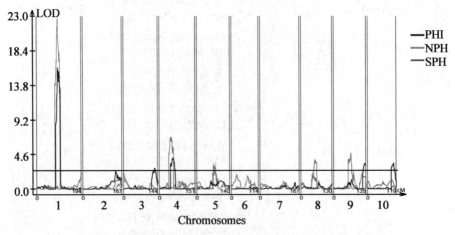

Figure 4 QTL locations of salt tolerance in field grown maize

qSPH represents QTL based on average plant height under salt stress across threeseasons; qNPH represents QTL based on average plant height under normal condition across three seasons; qPHI represents QTL based on theaverage plant height salt tolerance index across three seasons

Candidate genes in the major QTL

AtSOS1, AtSOS2, AtSOS3, OsSKC1, and TmHKT1; 5 have been shown to play major roles in the salt tolerance of plants. Using the protein sequences of these genes as queries, we screened the annotated maize genome database, and discoveredtwo candidate genes within the confidence interval of the detected major QTL: GRMZM2G007555 and GRMZM2G098494 (Additional files 3 and 4). No homologs of AtSOS2, OsSKC1, and TmHKT1; 5 were identified within the QTL region. The product of maize gene GRMZM2G007555 showed 61.38% identity to AtSOS3 which was predicted to encode a calcineurin B-like (CBL) protein, and thus it might be involved in regulating calcium signaling under salt stress. GRMZM2G098494 was a homolog of AtSOS1 and was annotated as a Na^+/H^+ antiporter.

To further validate the potential candidate genes, their relative expression levels in the shoot and root of salt treated seedlings were determined by qRT-PCR (Figure 5). In shoot, the expression of both candidate genes in the salt tolerant parent PH6WC was extremely significantly higher than that of the salt susceptible parent PH4CV at all three tested time points, 0h, 1h and 2h after salt treatment, though the expression showed fluctuation in both lines (Figure 5a and b). In root, although the basal expression of *GRMZM2G098494* in PH4CV was comparable to that in PH6WC, its induced expression was exceptionally higher in PH6WC compared to that of PH4CV (Figure 5c). While the expression of *GRMZM2G07555* in PH6WC held at a steadily high level, it was dramatically decreased in PH4CV as the duration increase of salt treatment (Figure 5d). Therefore, the detection of these two candidate genes in the major QTL along with their expression pattern upon salt treatment in both parents suggests that ion homeostasis regulation may play important roles in the salt tolerance of field-grown maize.

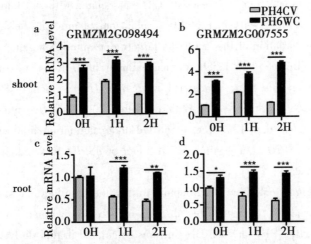

Figure 5 The relative expression level of potential candidate genes in maize seedlings of two inbred lines

Total RNA samples were collected at 0h, 1h and 2h after salt treatment from shoot and root tissues, respectively. qRT-PCR was performed using maize Actin as the internal reference. Bar charts represent the mean ± SE of three biological replicates, each containing tissues from five individual seedlings. Asterisks show statistically significant difference in two lines by ANOVA. *, ** and *** indicate $P<0.05$, $P<0.01$ and $P<0.001$, respectively. Expression of *GRMZM2G098494* in shoot (a), *GRMZM2G007555* in shoot (b), *GRMZM2G098494* in root (c) and *GRMZM2G007555* in root (d)

Discussion

Utilizing salt tolerant maize germplasms is an important practice to battle the challenge of field saline stress, so understanding the genetic basis controlling salt tolerance in the mature field-grown maize can guide the breeding of maize varieties with salt tolerance. Therefore, QTL mapping for salt tolerance of mature maize in field conditions is of great significance, as it is closely related to crop production. In the present study, we identified a major QTL associated with salt tolerance in field-grown maize using a DH population (Table2; Figure 4).

Salt tolerance in plant is a complicated phenomenon and many parameters have been used as indicators to evaluate its effect, including agronomic traits, physiological indicatorsand phenotypic classification. As an important index for plant salt tolerance, yield served as one of the phenotypic factors in the QTL analysis of salt tolerance in tomato and bread wheat. However, grain yield in maize typically exhibits low heritabilityand is affected by many external conditions. On the contrary, plant height is an important agronomic trait with relatively high heritability, and it has been shown to positively correlate to grain yield, such as in the V12 growth stage of maize, coffee and wheat. Therefore, along with its easy determination at the early growing stage, plant height may serve as a desired proxy of the grain yield. Actually, it has been employed in the identification of QTLs conditioning salt tolerance in other crop species, including a barley DH population and rice RILs. Thus, we believe that plant height of maize growing in the saline soil is the appropriate trait to investigate QTLs associated with salt tolerance.

In this study, the broad sense heritability of plant height in saline fields was 74.7%, indicating that phenotypic variation of this trait was largely determined by genotype. It is comparable with the high heritability of plant height in maize under heat stress and the tall fescue under drought stress. Based on this trait, a major QTL of *qSPH1* was identified on Chromosome 1, which explained 31.2% of phenotypic variation (Table 2). Only five minor QTLs controlling plant height with PVE less than 12% were identified under normal environment and none of them was co-localized with *qSPH1*, suggesting that the loci of *qSPH1* was linked to salt tolerance of mature field-grown maize.

To eliminate the inherent effect of agronomic traits, the ratio between stress and normal conditions has been used to evaluate the degree of stress tolerance in maize, rice, and cowpea. Accordingly, the PHI for each DH line was calculated in this study and was used to map QTLs for salt tolerance. A major QTL, *qPHI1*, explaining 25.94% of PVE was identified at the same location as *qSPH1*, further verifying that this main QTL is responsible for salt tolerance in maize.

Although seedlings and mature plants are different at developmental stages, and their physiological characters and stress resistance mechanisms usually vary, but the genetic factors controlling a certain trait can still be the same in different stages. Using an $F_{2:5}$ RILs population, Cui et al, identified a major QTL, *QFgr1* or *QFstr1*, flanked by the marker *PZE101140869* and *PZE101138116*, which was responsible for the maize germination rate in saline fields or seedling salt tolerance ranking. Comparing the QTL locations, we found that the position of *qSPH1* and *qPHI1* was comparable to *QFgr1* or*QFstr1*. Thus, it seems that this major QTL confers salt tolerance in two different populations, and it controls salt tolerance of both maize seedling stage and mature stage. In this regard, this QTL location may serve as a good target site for MAS in maize breeding to increase the salt tolerance of salt susceptible and elite inbred lines during both seedling and adult stages.

Tremendous efforts have been made to identify genes controlling salt tolerance in plants. In *Arabidopsis*, the SOS pathway has been demonstrated to mediate salt tolerance which comprises three members: SOS1, SOS2, and SOS3. After binding Ca^{2+}, SOS3 interacts with SOS2 to form a kinase complex, which further activates SOS1, a plasma membrane Na^+/H^+ antiporter that exports Na^+ from the cell. In addition, *OsSKC1* and *TmHKT1*; 5 were responsible for salt tolerance in rice and wheat, respectively, and they were both cloned as Na^+ transporters. To investigate whether genes in the confidence interval of the major QTL in the present study share any similarity to these genes cloned in other species, the blastp was performed and two proteins-GRMZM2G007555 and GRMZM2G098494 were shown to be homologous to AtSOS3 and AtSOS1 with the E-value of 1.740e-62 and 0, respectively. Such low E-values showed substantial protein identity, suggesting that two candidate genes may share similar function to *AtSOS3* and *AtSOS1* in the resistance to salt stress. The induced level of both genes was not substantially elevated at 1h and 2h after salt treatment (Figure 5), which probably due to the sample collecting time points and the possible discrepancy in the salt response time between *Arabidopsis* and maize seedlings, but the expression of both *GRMZM2G098494* and *GRMZM2G007555* in the salt tolerant PH6WC was significantly higher than that of PH4CV in both shoot and root tissues of maize seedlings in basal and induced conditions, except for *GRMZM2G098494* at 0h in root. In addition, the expression pattern of both genes in PH4WC was declined as the increase in the treatment duration, whereas PH6WC exhibited a stable or slightly increase level, suggesting that these two genes may be associated with the difference in salt tolerance of both parents. Further investigation with more time points is required to fully understand their expression in the course of salt treatment. Therefore, we speculate that the regulation of ion homeostasis by the SOS pathway may be the genetic control for salt tolerance of mature field-grown maize. However, further research is needed to verify the function of these genes in salt tolerance, which will potentially facilitate the breeding of salt tolerant maize varieties.

Conclusions

In the present study, we have constructed a SNP-based maize genetic linkage map and mapped a major QTL at the 88~95cM region on Chromosome 1 based on plant height of mature maize grown in a saline field, which accounted for 31.24% of PVE. This main effect QTL for salt tolerance was further validated by QTL mapping based on plant height salt tolerance index, and two candidate genes with predicted functions in the SOS pathway have been identified in the locus, indicating that ion homeostasis regulation may play an important role in the salt tolerance of mature maize. Therefore, this locus will be valuable for marker assisted selection in maize breeding.

Additional files

Additional file 1: **Table S1**. The primers used in the qRT-PCR.

(DOCX 16kB)

Additional file 2: Figure S1. Chromosomal locations and logarithm of odds (LOD) scores of the major QTL conditioning SPH and PHI on Chromosome 1 using data from individual years and overall means. Mean represents the average of three growing seasons.

(DOCX 1 512kb)

Additional file 3: Figure S2. Protein sequences alignments by BLAST against B73 filtered gene set translations 5b.60 for RefGen_v2 (maizesequence.org) database. (a) Alignment of AtSOS1 and GRMZM2G098494. (b) Alignment of AtSOS3 and GRMZM2G007555.

(DOCX 10 769kb)

Additional file 4: Table S2. The identity of two candidate genes identified in the major QTL region to *Arabidopsis SOS* genes and their position and annotation.

(DOCX 22kb)

Abbreviations

AFLP: Amplified fragment length polymorphism; CBL: Calcineurin B-like; CP: Changping; DH: Double haploid; E: Environment; G: Genotype; H^2: Broad sense heritability; InDel: Insertion-deletion; LOD: Logarithm of odds; MAS: Marker-assisted selection; Na^+: Sodium ions; NPH: Plant height in control field; PHI: Salt tolerance index; PVE: Phenotypic variance; QTL: Quantitative trait loci; RAPD: Random amplified polymorphic DNA; RFLP: Restriction fragment length polymorphism; RILs: Recombinant inbred lines; SNP: Single nucleotide polymorphism; SOS: Salt overly sensitive; SPH: Plant height in the saline field; SSR: Simple sequence repeat; TZ: Tongzhou

References (omitted)

Bmc Plant Biology, 2017, 17 (1): 140

Identification of Genes Potentially Associated with the Fertility Instability of S-Type Cytoplasmic Male Sterility in Maize via Bulked Segregant RNA-Seq

Su Aiguo[*] Song Wei[**] Xing Jinfeng Zhao Yanxin Zhang Ruyang
Li Chunhui Duan Minxiao Luo Meijie Shi Zi[**] Zhao Jiuran[**]

(Maize Research Center, Beijing Academy of Agriculture and Forestry Sciences, Beijing Key Laboratory of Maize DNA Fingerprinting and Molecular Breeding, Beijing 100097, China)

Abstract: S-type cytoplasmic male sterility (CMS-S) is the largest group among the three major types of CMS in maize. CMS-S exhibits fertility instability as a partial fertility restoration in a specific nuclear genetic background, which impedes its commercial application in hybrid breeding programs. The fertility instability phenomenon of CMS-S is controlled by several minor quantitative trait locus (QTLs), but not the major nuclear fertility restorer (*Rf3*). However, the gene mapping of these minor QTLs and the molecular mechanism of the genetic modifications are still unclear. Using completely sterile and partially rescued plants of fertility instable line (FIL) -B, we performed bulk segregant RNA-Seq and identified six potential associated genes in minor effect QTLs contributing to fertility instability. Analyses demonstrate that these potential associated genes may be involved in biological processes, such as floral organ differentiation and development regulation, energy metabolism and carbohydrates biosynthesis, which results in a partial anther exsertion and pollen fertility restoration in the partially rescued plants. The single nucleotide polymorphisms (SNPs) identified in two potential associated genes were validated to be related to the fertility restoration phenotype by KASP marker assays. This novel knowledge contributes to the understanding of the molecular mechanism of the partial fertility restoration of CMS-S in maize and thus helps to guide the breeding programs.

Introduction

Cytoplasmic male sterility (CMS) is a common phenomenon in higher plants and is characterized by maternal inheritance, pollen sterility and normal pistil development. Plant CMS is mediated by nuclear-mitochondrial interactions in which the sterility results from the expression of mitochondrial genes and can be restored by nuclear restoring fertility (*Rf*) genes. The CMS

---

  [*] These authors contributed equally to this work
  [**] These authors also contributed equally to this work
  [**] shizi_baafs@126.com (ZS); maizezhao@126.com (JZ)

system is widely applied in crop hybrid breeding to avoid extra efforts for artificial emasculation. Therefore, CMS has important theoretical and commercial value in hybrid seed production. Maize (*Zea mays* L.) has three major types of CMS, designated as T (Texas), C (Charrua) and S (USDA). The S-type of CMS (CMS-S) is the largest group among the three types and has wide cytoplasmic sources. Unfortunately, S-type is also the most fertility instable type and exhibits incomplete male sterility in sterile lines under specific genetic backgrounds. This fertility instability requires careful supervision in the field and therefore negatively affects the commercial application of CMS-S in maize hybrid breeding.

In maize CMS-S, a 1.6kb transcript of the mitochondrial genome R region, containing the chimeric gene sequence *orf355-orf77*, contributes to pollen sterility during pollen development, while the nuclear fertility restorer *Rf3* affects its transcription, resulting in fertility restoration in the sterile lines. In addition, the fertility restoration of CMS-S is mainly controlled by the major restorer *Rf3*, which is mapped to chromosome 2 within bin2.09.

The fertility instability of CMS-S is characterized by the appearance of partially rescued plants in the offspring of sterile lines, in which the pollen becomes partially viable. Fertility instability deviates under specific nuclear genetic backgrounds. The ratios of instability of CMS-S are observed at a relatively high level with Wf9 and M825 nuclear genomes. Previous research indicates that multiple genes might be involved in the fertility instability of CMS-S. Recently, Feng et al identified 30 significant loci for pollen fertility, anther exsertion and pollen shedding in CMS-S maize using a mapping panel of 513 plants in three environments, suggesting that in addition to the major restorer *Rf3*, multiple minor genetic loci are involved in fertility instability. In addition, Tie et al detected 6 quantitative trait locus (QTLs) with significant effects on the male fertility of CMS-S, and Kohls et al also identified 7 QTLs for the partial fertility restoration of CMS-C. Moreover, the expression of another restoring gene, *Rf9*, which is a less effective fertility restorer compared to *Rf3*, is influenced by both nuclear background and temperature. Nevertheless, the molecular mechanism of the fertility instability of CMS-S is still unknown because this phenomenon is considerably complicated and can be easily influenced by many environmental factors.

Bulk segregant RNA-Seq (BSR) is a modification of bulked segregant analysis (BSA) and utilizes the RNA-seq reads to map genes in a rapid and efficient manner. BSR-Seq provides not only the position of the target genes but also the differential expression pattern of potential associated genes between the bulks. It can also facilitate in de novo SNP discovery for marker development in breeding programs. Thus, BSR-Seq has been successfully utilized in maize genetic mapping for the accumulation of epicuticular waxes, oil biosynthesis and other traits.

The maize inbred line Jing724 is the maternal parent of Jingke968, which is one of the leading varieties in China. Fertility instable line (FIL) -B was obtained using marker-assisted selection by backcrossing the CMS-S sterile line MD32, which possesses the desired agronomic traits and complete pollen sterility, to Jing724. In the BC3 population, FIL-B showed partial fertility restoration, exhibiting anther exsertion with no anther dehiscence or pollen shedding. Thus, understanding the genetic architecture of the fertility instability in CMS-S is crucial for the preferably application of the three-line breeding in maize.

The objectives of this study were to (1) determine the genomic regions containing minor genetic loci associated with partial fertility restoration; (2) predict the potential associated genes involved in fertility instability using the available genomic resources; and (3) examine their differential expression patterns in anthers and validate the SNPs identified from RNA-Seq.

Materials and Methods

Plant materials

Jing724 is an inbred line that was developed and released by the Maize Research Center, Beijing Academy of Agricultural and Forestry Science. By crossing to the CMS-S sterile line MD32, the sterile F_1 was obtained, which was consecutively backcrossed to Jing724. A few families of the BC_3 population exhibited partial fertility instability with anther exsertion. Molecular markers were applied to select the Jing724 nuclear background in each backcross generation (S4 Table), which made the sterile and partially rescued families as near isogenic lines. The BC_4 population was acquired from a backcross of sterile male individuals from the fertility segregating family in the BC_3 to inbred Jing724, which was designated as FIL-B. The plant materials of the inbred line Jing724 and FIL-B were planted in the Hainan Maize Propagation Base in Yacheng, Hainan (HN-YC, 18.3°N, 109.5°E) for sample collection.

Identification of fertility

Anther exsertion and anther dehiscence were observed for individual plants in the field from the start of the tasseling stage to the end of the pollinating stage. Three spikelets, respectively positioned at the top, in the middle and at the bottom of the tassel, were collected from each of the 30 plants of both completely sterile plants and plants with partial anther exsertion. The spikelets were then stored in Carony's solution [75% (v/v) ethanol in acetic acid] in 1.5ml eppendorf tubes. The pollen was stained and scored as previously described by Zhu et al and was examined using a light microscope (Olympus IX73, Japan) at 40 × magnification. Three representative fields of each sample were used to calculate the proportion of viable pollen in the sterile plants and the partially rescued plants.

Pollen microspore development

Spikelets of FIL-B were sampled every three days for the whole course of tassel development. The samples were fixed in Carony's solution and stored in 70% ethanol. The anther samples were dehydrated, embedded and sliced as previously described by Zhu et al. The slices were stained with safranin-green and sealed in neutral gum after toasting at 60℃. The images were obtained using an Olympus IX73 microscope (Olympus, Tokyo) at 100 × magnification a week after drying.

Analysis of genes with differential expression

The anthers of the completely sterile and partially rescued plants were collected from 30 biological replicates and stored in a 10× volume RNA wait solution (Solarbio, Beijing). The RNA was extracted by Trizol, and the integrity and concentration were determined by Agilent2010

(Agilent, Santa Clara, CA) to ensure that the RIN value was greater than 6.8. Thirty RNA samples from the partially rescued plants were equally mixed and designated as bulk F, while 30 samples from the sterile plants were mixed and designated bulk S. The transcriptomes of both bulks were sequenced at Biomarker Technologies Co., LTD (Beijing, China) using Illumina HiSeq2500 (Illumina, San Diego, CA).

The clean data were aligned to a maize reference genome (B73 RefGen_ v3 ftp://ftp.ensemblgenomes.org/pub/plants/release-24/fasta/) to create the mapped data. The differentially expressed genes (Fold change ≥2 and FDR<0.01) were identified by EBSeq. Multiple databases were utilized to annotate the differentially expressed genes, including NR, Swiss-Prot, GO, COG, and KEGG.

QTLs mapping by BSR-Seq

The sequencing reads of both bulks were aligned to the reference genome by STAR (2.3.0e, https://github.com/alexdo bin/STAR/releases), and the SNPs were identified by GATK (3.1-1, https://www.broadinstitute.org/gatk/index.php). Because the frequencies cannot be accurately determined at low read counts, a minimum cutoff of 3 reads was used.

The Euclidean Distance (ED) algorithm was used to obtain the genetic distance to the associated QTLs. The calculation was performed as previously described. Basically, the ED for each SNP was calculated using the following equation:

$$ED = \sqrt{(A_F - A_S)^2 + (C_F - C_S)^2 + (G_F - G_S)^2 + (T_F - T_S)^2}$$

The higher the ED value is, the closer the distance is to the targeted genes. The letters represent their corresponding bases, and A_F and A_S represent the frequencies of A in the bulk F and bulk S, respectively. The ED values were raised to a power of 5 (ED^5) to decrease the noise generated by small variations in the estimations. The data are fitted using a Loess curve with the fitted values as the median of the up and downstream 50 SNPs.

Expression analysis of potential associated genes

RNA was extracted from completely sterile and partially rescued plants as described above. QPCR was conducted using 3 biological replicates each with 3 technical replicates in a 20μl system for each potential associated genes by employing the StepOnePLUS system (SinoGene, Beijing). The 20μl QPCR mixture contained 1μl of diluted-cDNA, 7.5μl of 2×SG Green PCR Master Mix, and 0.25μl of each of the 10μM primers. The amplification was performed as follows: 95℃ for 10min followed by 40 cycles of 95℃ for 20s; 60℃ for 30s and 72℃ for 30s. The maize Actin gene (GQ339773.1) was served as the internal control. Primers used in the QPCR are included in the S1 Table. The quantification was carried out using the relative ΔΔCt method on Prizm 5 software (http://www.graphpad.com/quickcalcs/).

KASP assay design

Genomic DNA was extracted from 50 individual sterile plants and 50 partially rescued plants as previously described. Kompetitive Alleles Specific PCR (KASP) assays were developed for 4 SNPs identified in two potential associated genes from RNA-Seq. KASP primers sequences were listed in S3 Table. KASP assays were carried out with 4μl reaction system including 2μl genomic

DNA at 20ng/μl, 0.106μM of each primer and 2μl of KASP master mix (Kbiosciences, Herts England). PCR conditions for the assays were set up as previously described. PCR fluorescent endpoint readings were evaluated using the BMG Pherastar (LGC, Middlesex, UK), and the visualization of the clusters with the SNP allele callings was obtained by Kluster Caller software (LGC, Middlesex, UK).

Results

Identification of fertility instable plants

Anther exsertion was first observed in FIL-B population 3 days after tasseling, and 3 additional days later, approximately 1/3 of the plants showed anther exsertion (Figure 1A), which categorized the FIL-B as grade II fertility according to the classification in Feng et al. Unlike Jing724, anthers of FIL-B showed no anther dehiscence or pollen shedding (Figure 1B). The anther size of the partially rescued FIL-B was smaller than the Jing724 but was comparable to the sterile anthers. The anthers of the partially rescued plants exhibited an intermediate color and shape when compared to the Jing724 and sterile individuals. Anther exsertion served as the phenotypic classification of fertility in the BC_4 population, which characterized 14.9% of the 462 plants as partially rescued. These data suggest that the fertility instability of FIL-B is regulated by multiple loci.

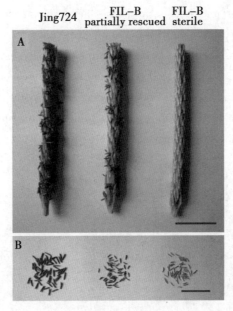

Figure 1 Partially restored anther exsertion in the FIL-B population

A. Representative image of the tassels of Jing724 and FIL-B individuals with partial anther exsertion and complete sterility at six days after tasseling. The scale bars represent 3cm. B. Representative image of the anthers harvested from A. The scale bars represent 3cm

Although a small portion of pollen from the partially rescued plants showed normal development with black staining (Figure 2), the majority exhibited an irregular shape without staining, which was similar to the pollen from the sterile plants. No pollen was stained in the sterile plants, but 5.56% of the pollen from the partially rescued plants was stained to show normal starch accumulation, indicating that pollen of FIL-B is fertility instable with possible fertility restoration.

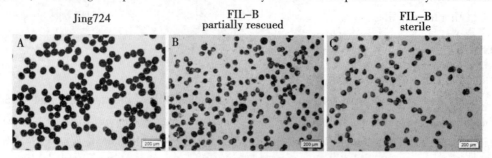

Figure 2 Pollen staining with I2-KI revealed the normal development
of a small portion of pollen from the partially rescued FIL-B

Representative images of the stained pollen of Jing724 (A) and the partially rescued (B) and sterile (C) individuals of FIL-B. Pollen was collected when the anthers exserted in the partially rescued plants. Round pollen with black staining was recorded as normal. The scale bars represent 200μm

Pollens and microspores in the partially rescued plants show partial restoration

To test pollen development during the microspore stage in the anthers, we observed the anther dissection under a light microscope. However, the development differed in the sterile and partially rescued anthers during telophase of the uninucleate microspores (Figure 3). Although the development and structure of tapetum were the same in both genotypes, all of the microspores collapsed in the sterile anthers, while a small proportion of the microspores had normal development in the partially rescued plants. The individual normal microspores from the FIL-B partially rescued plants might lead to normal mature pollen, which is consistent with the pollen staining data and further confirms the fertility instability of FIL-B.

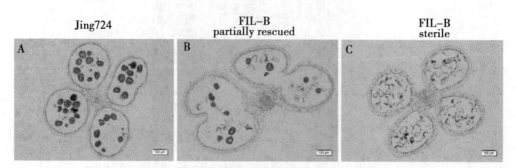

Figure 3 Paraffin slides show normal microspore
development in the partially rescued FIL-B

Microscopic images of anther transverse sections of Jing724 (A) and the partially rescued (B) and sterile (C) individuals of FIL-B. The black round microspores were considered as normal. The scale bars represent 100μm

Transcriptome analysis identifies 3 672 differentially expressed genes

To map the loci controlling the fertility instability in FIL-B, we performed BSR-Seq to identify the differentially expressed genes in the two bulks with distinct fertility phenotypes. Eighty-nine and 61 million clean reads were obtained for bulk F and bulk S, respectively, with an average read length of 252bp, and the percentage of bases (Q30) over 87.09%. The clean reads were aligned to the maize reference genome B73, revealing an even distribution on the 10 chromosomes of B73, and the unique mapped reads ratio for bulk F and S were 72.79% and 69.17%, respectively. By the RNAseq, the total of 31 905 and 31 895 gene models were expressed in partially rescued and sterile bulk, respectively. We identified a total of 3 672 genes with differential expression (Fold change $\geqslant 2$ and FDR<0.01) between the two bulks, among which 3 540 genes were annotated in the NR, Swiss-Prot, GO, COG and KEGG databases, and 3 548 genes were up regulated and 124 down regulated in the bulk F.

TopGO analysis was performed using the Gene Ontology (GO) annotations of the genes with differential expression between the two bulks. This analysis compares the relative proportion of genes represented in the whole maize genome for specific GO terms to the proportion of same terms represented in the differentially expressed genes. This approach leads to the calculation of a KS score, which indicates the statistical significance of the enrichment for a specific GO class. We performed a TopGO analysis for all three ontologies (cellular component, molecular function and biological process), and the ten most significantly enriched terms presented with the smallest KS scores are included in S2 Table. Notably, the majority of the enriched GO categories were associated with pollen tube development (e.g., "pollen tube tip" (GO: 0090404) and "pollen tube growth" (GO: 0009860)), cell wall modification (e.g., "pectinesterase activity" (GO: 0030599) and plant-type cell wall modification (GO: 0009827)) and carbohydrate metabolism (e.g., "galacturan 1, 4-alpha-galacturonidase activity" (GO: 0047911) and "polygalacturonase activity" (GO: 0004650)), highlighting the different expression patterns between the sterile plants and partially rescued plants. Previously, it was shown that genes involved in energy metabolism are related to pollen fertility, indicating that the enrichment of those genes may contribute to the fertility restoration of the FIL-B in the partially rescued individuals.

To further demonstrate the over-representation of carbohydrate metabolism in the partially rescued plants, a pathway enrichment analysis was carried out for the genes with differential expression between the two bulks using the KEGG pathway database (http://www.genome.jp/kegg/). Eight pathways were significantly enriched (Table 1), and a majority of these were carbohydrate metabolism pathways with the enrichment fold ranging from 2.1 to 3.4, including "Amino sugar and nucleotide sugar metabolism" (ko00520), "Starch and sucrose metabolism" (ko00500), "Glycolysis/Gluconeogenesis" (ko00010) and "Citrate cycle" (ko00020). These results further confirmed that the overexpression of carbohydrate metabolism pathways and signal transduction pathways in the partially rescued individuals were associated to the fertile instability of FIL-B.

Table 1 Enrichment of the KEGG pathways in the differentially expressed genes

Pathway	KO	Enrichment Fold	Q-value
Galactose metabolism	ko00052	3.4	1.2E-06
Amino sugar and nucleotide sugar metabolism	ko00520	2.4	6.2E-06
Glycerophospholipid metabolism	ko00564	2.6	1.5E-05
Starch and sucrose metabolism	ko00500	2.2	1.9E-05
Glycolysis / Gluconeogenesis	ko00010	2.1	3.9E-05
Pyruvate metabolism	ko00620	2.5	4.0E-05
Citrate cycle (TCA cycle)	ko00020	2.4	7.8E-05
Phosphatidylinositol signaling system	ko04070	3.2	2.2E-04

KO: KEGG pathway ID

Enrichment Fold: the ratio of the genes annotated for a specific pathway in the differentially expressed genes to the genes annotated for the same pathway in the whole genome

Q-value: P value after adjustment for multiple hypothesis testing

Identification of six potential associated genes for fertility instability

The sequencing reads were aligned to the B73 reference genome to obtain the SNPs in both of the bulks. A total of 171 594 SNPs and 170 820 SNPs were identified in bulk F and bulk S, respectively. To map the QTLs controlling fertility restoration, the Euclidean Distance (ED) algorithm was applied to measure allele segregation and to identify the linked genomic loci based on the SNPs between the two bulks. A statistically significant peak of ED was observed on chromosome 2 (Figure 4A and 4B), indicating its association to fertility instability in FIL-B. Seven genomic regions were associated with fertility instability by using the 99% percentile of the $ED5$ (0.176 78) as the significant cutoff (Figure 4C and 4D), among which two regions consisted of only a couple of bases. The other 5 regions were all located on the short arm of chromosome 2 (Table 2). To identify the potential associated genes within these 5 genetic loci, the annotated genes were further screened based on their expression pattern within the two bulks. A total of 12 genes (Table 3) exhibited differential expression, 8 genes were located in region 2 and 4 genes were located in region 5, with no genes identified in the other 3 regions (Table 2). Of these genes, *GRMZM2G315401* and *GRMZM2G434669* encode the serine/arginine-rich protein 45, *GRMZM2G430362* encodes an ATP-dependent RNA helicase SUV3, *GRMZM2G474783* and *GRMZM2G010338* encode the leucine-rich repeat extensin-like protein 5 and a putative SPRY-domain family protein, respectively, and *GRMZM2G127173* encodes an uncoupling protein regulating ATP biosynthesis. The 6 above mentioned genes were selected as potential associated genes for the following expression study because of their possible function as regulatory genes. Moreover, *GRMZM2G012328* is predicted as pectinesterase, *GRMZM2G145758* possibly encodes histone H3, and *AC191050.3_FG003* encodes a potential precursor of the globulin-1 S allele, all of which may involve in cell wall metabolism and modification. The other three genes were annotated as hypothetical proteins and thus need further investigation.

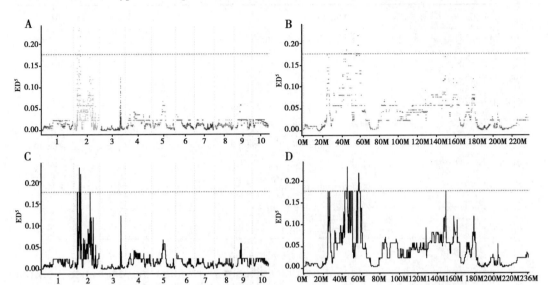

Figure 4 Identification of genomic regions contributing to the fertility instability in FIL-B by BSR-Seq using the Euclidean distance (ED) algorithm

A. The ED scores raised to the fifth power across the genome. Each dot represents each SNP identified from the RNA-Seq, and the different colors designate the different chromosomes as indicated on the X-axis. For all of the panels, the gray vertical dotted lines delineate the chromosome edges, and the width of the chromosome represents the relative numbers of SNPs identified. The pink horizontal dotted lines represent the significant threshold of the 99% percentile of the ED^5. B. The ED^5 scores of a close-up of chromosome 2. C. The Loess fit curve calculated from A. D. The Loess fit curve of a close-up of chromosome 2 with the physical position indicated on X-axis. Each peak represents a possible associated genomic region

Table 2 The physical position and number of genes in the five identified genomic regions on chromosome 2

Region	Physical Position	Size	Gene	Differentially expressed gene
1	26,652,602-26,654,212	1.61Kb	0	0
2	42,383,737-45,424,781	3.04MB	85	8
3	46,980,442-47,717,638	0.74MB	13	0
4	48,752,208-50,549,409	1.79MB	29	0
5	55,113,525-58,997,475	3.88MB	63	4

Table 3 Twelve genes with differential expression from the 5 genomic regions identified on chromosome 2

Gene ID	Annotation	FDR	Fold Change	Regulated
GRMZM2G010338	Putative SPRY-domain family protein	8.97E-09	10.3	up
GRMZM2G012328	Pectinesterase	1.35E-12	19.5	up

(continued)

Gene ID	Annotation	FDR	Fold Change	Regulated
GRMZM2G127173	Mitochondrial uncoupling protein 3	0.00053	4.6	up
GRMZM2G145758	Histone H3	9.40E-05	5.6	up
GRMZM2G315401	Serine/arginine-rich protein 45	7.96E-10	11.9	up
GRMZM2G347489	Hypothetical protein	2.85E-11	16.7	up
GRMZM2G430362	ATP-dependent RNA helicase SUV3	1.02E-06	7.4	up
GRMZM2G474783	Leucine-rich repeat extensin-like protein 5	5.52E-11	14.4	up
AC191050.3_FG003	Globulin-1 S allele	0.00073	5.8	up
GRMZM2G406026	Hypothetical protein	3.07E-12	18.0	up
GRMZM2G434669	Serine/arginine-rich protein 45	0.001	4.3	up
GRMZM5G858609	Hypothetical protein	2.63E-07	9.2	up

Expression analysis of the potential associated genes validates the difference between the two bulks

QPCR was performed for the six genes using individual plants to validate the RNA-Seq data and to obtain more quantitative transcript level measurements. The relative expression of all of the six potential associated genes showed a 4-to 8-fold up regulation in the partially rescued plants compared to the sterile plants (Figure 5). The trends in regulation exhibited a consistent pattern between methods and thus, confirming the qualitative values of the RNA-Seq.

Validation of SNPs identified from RNA-Seq

To further assess the effect of the potential associated genes, SNPs identified in six potential associated genes were evaluated. Two SNPs in each of the *GRMZM2G315401* and *GRMZM2G430362* appeared to be conclusive, whose alleles and positions on B73 were included in Table 4. To further determine the correlation between these SNPs and the fertility phenotype, KASP marker assays were performed with 50 individuals of sterile plant and 50 partially rescued plants from the BC_4 population. As expected, all the sterile individuals possessed the homozygous loci, and all the fertile plants exhibited heterozygous loci for all four SNPs with no exceptions (Table 4 and Figure 6). The PCR-based SNP assays validated the SNPs identified from RNA-Seq, further confirming the potential function of two associated genes in fertility instability.

Table 4 SNPs in genes potentially associated with fertility instability identified from RNA-Seq

KASP Assay	Gene	Alt	Allele of recurrent parent Jing724	SNP positions
FIL-1	GRMZM2G315401	C/G	C	42994131
FIL-2	GRMZM2G315401	G/C	G	42995108
FIL-3	GRMZM2G430362	C/G	C	42743133
FIL-4	GRMZM2G430362	A/G	A	42743749

Alt: allele in sterile plant/ allele in fertile plant

SNP positions: physical positions of SNP on B73 genome

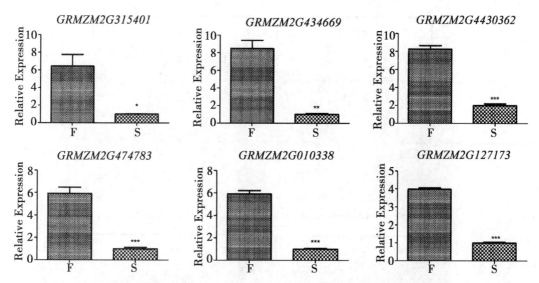

Figure 5   Comparison of the transcript levels of the six potential associated genes from the partially rescued (F) and sterile (S) plants of FIL-B, as detected by QPCR

Each bar represents the mean±SE of the biological replicates. The values are calculated using Actin as an internalcontrol. The asterisks show the statistically significant difference compared to the partially rescued plants, as determined by the analysis of variance: * ($P<0.05$), ** ($P<0.01$), *** ($P<0.001$)

Discussion

CMS is widely exploited in maize hybrid breeding to avoid excessive labor for artificial emasculation and possible hybrid contaminations. Therefore, tremendous efforts are made to understand the molecular mechanism of CMS-S in maize. The maize cytoplasmic gene *orf77* controls male sterility in CMS-S and contains three *atp9* chimeric sequences, which have affected the function of ATP9. Due to the importance of ATP9 in the energy metabolism of the mitochondria, the chimeric sequences are considered to lead to pollen sterility. The partial fertility restoration in FIL-B suggests that multiple minor effect loci may contribute to the process, which eliminates the energy loss due to sterile gene expression. CMS-S in maize is gametophytic male sterility, whose fertility is conditioned by the pollen genotype instead of its maternal genetic background. Therefore, we select anther tissues as study materials, where the formation and development of pollen occur. In this study, we revealed over 3 500 differentially expressed genes between the two bulks of FIL-B, each bulk consisted of 30 individual plants based on the anther exsertion phenotypes. The differentially expressed genes were categorized as part of the starch and sucrose metabolism pathway, the gluconeogenesis pathway, the TCA pathway, etc. (Table 1). In addition, an overrepresentation of the genes involved in cell wall modification, pollen tube growth and regulatory enzymatic activity was also observed (S2 Table). Using BSR-Seq, we located the minor effect QTLs to 5 genomic regions on the short arm of chromosome 2 (Figure 4). The five genomic regions identified in this study are not consistent with the previous study of fertil-

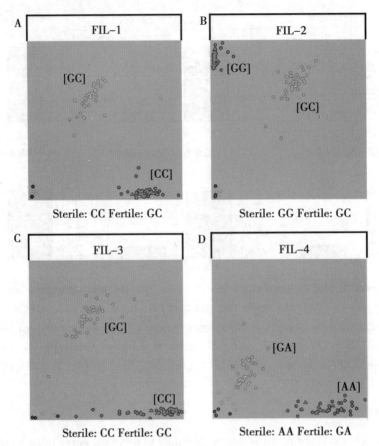

Figure 6 Graphs of four KASP marker assays

KASP assays were developed for SNPs identified in tSwo potential associated genes from RNA-Seq. Fertility instable line (FIL) -1 and FIL-2 were SNPs resided in *GRMZM2G315401*. FIL-3 and FIL-4 were SNPs detected in *GRMZM2G430362*. The triangles represent the recurrent parent of Jing724. Black data points are negative control and pink ones are ambiguous calling

ity instability of CMS-S. The discrepancy is probably due to the different methods and different mapping population used, which further suggests that fertility instability is conditioned by many minor effect loci.

Among the 12 differentially expressed genes within these 5 genomic regions, 6 were identified as potential associated genes with regulatory annotations. The expression level of these 6 genes were confirmed with QPCR (Figure 5), and the four SNPs of two potential associated genes identified from RNA-Seq were validated with KASP marker assays (Figure 6). Two of the potential associated genes, *GRMZM2G315401* and *GRMZM2G434669*, were annotated as Serine/arginine-rich protein 45 (SR45), which plays a crucial role in constitutive and alternative mRNA splicing and metabolism. SR45 is involved in floral organ morphogenesis and carbohydrate metabolism in Arabidopsis. *GRMZM2G474783* was annotated as a Leucine-rich repeat (LRR) extensin-like protein 5, and the LRR motif is believed to bind to a specific ligand during pollen tube development in tomato. *GRMZM2G010338* was predicted to be a putative SPRY-domain family protein, which is considered to be involved in protein-protein interactions. By combining the corresponding receptors, the SPRY structure domain participates in the cytokine signaling pathway through protein ubiquination and degradation in primates. These genes may regulate pollen starch accumulation, floral organ development, carbohydrate metabolism and other related biological process, resulting in fertility restoration with anther exsertion in the partially rescued plants. In addition, *GRMZM2G430362* encodes an ATP-dependent RNA helicase SUV3, a key control element in nuclear-mitochondrial interactions, whose function has been implicated in a variety of mitochondrial posttranscriptional and translational processes in yeast and some plants. The higher expression of this gene in the partially rescued plants indicates its possible modification of the cytoplasmic sterility factor, leading to the restoration of pollen fertility.

Interestingly, *GRMZM2G127173* encodes a mitochondrial uncoupling protein 3 (UCP3), which is a mitochondrial transporter located in the inner membrane of the mitochondria, and this protein was also up regulated in the partially rescued plants. Pollen maturation requires a great amount of energy, and it has been proposed that the cytoplasmic sterility gene compromises the electron transport chain of the mitochondria, which is then unable to provide sufficient ATP for normal pollen development in beets and rice. Previous research has demonstrated that the UCPs reduce the proton potential across the mitochondrial inner membrane and eliminate ATP biosynthesis in plants and mammals. It is possible that the over expression of UCP3 in the partially rescued plants serves as a negative feedback mechanism. Thus, unlike the direct inhibition of mitochondrial *orf355-orf77* expression by the major effect loci of *Rf3*, the extent of anther exsertion and the percentage of fertility restoration in FIL-B are resulted from a fine-tuned coordination of a set of minor effect loci and their interactions.

Unlike the major restorer *Rf3*, the contributions of minor effect loci are susceptible to environmental factors. Temperature, humidity and total transpiration change the fertility stability of CMS-S. A significantly different ratio of fertile restoration was observed in FIL-B at two locations with a temperature difference of 3~5℃ (data not shown), and the complicated interaction with the environment made the illustration of the mechanism a tedious topic. Despite of the fertility restoration in FIL-B, some of the families selected from the backcross population between MD32 and Jing724 were stably sterile, which were already successfully applied in the commercial seed production of Jingke968.

Based on the study of the fertility instability of FIL-B, we conclude that (1) multiple minor effect loci are involved in the partial fertility restoration, especially genes functioning in the regulation of floral organ differentiation and development, energy metabolism and carbohydrates biosynthesis; (2) the regulatory mechanism of pollen sterility is considerably complicated, and the fertility instability phenotype in a specific nuclear background may result from the interaction of positive and negative effect minor genes; and (3) minor effect fertility restoration genes are easily affected by environmental factors.

The identification of genes potentially associated with fertility instability in this study sheds light on the theoretical knowledge of its mechanism, which could facilitate the commercial promotion of CMS-S in maize breeding. Future investigation is needed to elucidate their functions and possible interactions at the molecular level.

Supporting Information

S1 Dataset. Raw RNA-Seq data: http://pan.baidu.com/s/1jIGp5Qm.
(DOCX)
S1 Table. Primers used for QPCR.
(XLSX)
S2 Table. Top 10 most significantly enrichedGO classes for each ontology group by TopGO analysis.
(XLSX)
S3 Table. Primer sequences used for KASP assays.
(XLSX)
S4 Table. SSR fingerprint information of the Jing724 backcross progeny.
(XLSX)

Author Contributions

Conceptualization: AS ZSWS JZ.
Data curation: AS ZSWS.
Formal analysis: AS ZSWS CL.
Funding acquisition: AS JZ.
Investigation: ASWS JX YZ RZ ML.
Methodology: AS ZSWS.
Project administration: AS.
Resources: JX MD JZ.
Supervision: JZ.
Writing-original draft: AS ZS.
Writing-review & editing: AS ZSWS JZ.

References (omitted)

Plos One, 2016, 11 (9): e0163489

RNA-seq Analysis Reveals MAPKKK Family Members Related to Drought Tolerance in Maize

Liu Ya[1]*   Zhou Miaoyi[1]*   Gao Zhaoxu[2]*   Ren Wen[1]
Yang Fengling[1]   He Hang[2]**   Zhao Jiuran[1]**

(1. Maize Research Center, Beijing Academy of Agricultural and Forestry Science, Beijing 100097, China; 2. School of Life Sciences, Peking University, Beijing 100871, China)

Abstract: The mitogen-activated protein kinase (MAPK) cascade is an evolutionarily conserved signal transduction pathway that is involved in plant development and stress responses. As the first component of this phosphorelay cascade, mitogen-activated protein kinase kinase kinases (MAPKKKs) act as adaptors linking upstream signaling steps to the core MAPK cascade to promote the appropriate cellular responses; however, the functions of MAPKKKs in maize are unclear. Here, we identified 71 MAPKKK genes, of which 14 were novel, based on a computational analysis of the maize (Zea mays L.) genome. Using an RNA-seq analysis in the leaf, stem and root of maize under well-watered and drought-stress conditions, we identified 5 866 differentially expressed genes (DEGs), including 8 MAPKKK genes responsive to drought stress. Many of the DEGs were enriched in processes such as drought stress, abiotic stimulus, oxidation-reduction, and metabolic processes. The other way round, DEGs involved in processes such as oxidation, photosynthesis, and starch, proline, ethylene, and salicylic acid metabolism were clearly co-expressed with the MAPKKK genes. Furthermore, a quantitative real-time PCR (qRT-PCR) analysis was performed to assess the relative expression levels of MAPKKKs. Correlation analysis revealed that there was a significant correlation between expression levels of two MAPKKKs and relative biomass responsive to drought in 8 inbred lines. Our results indicate that MAPKKKs may have important regulatory functions in drought tolerance in maize.

Introduction

The world population is increasing at an alarming rate, while abiotic stresses play crucial roles in crop failures and reductions in field crop productivity. Drought is the most important environmental stress affecting agricultural production. Maize (Zea mays L.), which is frequently exposed to drought stress conditions, is one of the most important cereal crops in the world, together with rice and wheat. In recent years, some studies have found that several genes encoding pro-

---

* These authors contributed equally to this work
** hehang@pku.edu.cn (HH); maizezhao@126.com (JZ)

tein kinases can activate responses to abiotic stresses, such as drought. Drought stress is one of the major limiting factors for maize production. Some studies have revealed that drought stress can affect maize, especially during the reproductive stage. Their working hypothesis involves signaling events associated with increased ABA levels, decreased glucose levels, the disruption of ABA/ sugar signaling, and the activation of programmed cell death/senescence through the repression of a phospholipase C-mediated signaling pathway. A total of 524 non-synonymous single nucleotide polymorphisms (nsSNPs) that were associated with 271 candidate genes involved in plant hormone regulation, carbohydrate and sugar metabolism, signaling molecules regulation, redox reaction and acclimation of photosynthesis to environment were detected by common variants (CV) and cluster analyses with the availability of maize B73 reference genome and whole-genome resequencing of 15 maize inbred lines. These selected genes will not only facilitate our understanding of the genetic basis of the drought stress response but will also accelerate genetic improvement through marker-assisted selection in maize.

The mechanisms of drought tolerance in plants are complex. Effects of drought on plant hormone signal transduction, protein modification and photosynthesis have been reported. It has been demonstrated that the accumulation of proline can increase tolerance to water stress in plants. Transgenic plants with increased proline content produce higher biomass under drought. An analysis of *edr1* mutation determined that the *EDR1* gene functions in negative regulation after salicylic acid treatment. The stress-induced increase in reactive oxygen species (ROS) in plant cells results from an imbalance between generation and degradation. Drought stress can successively induce stomatal closure, moderate increases in ROS and decreases in photosynthesis. As described above, the response of plants to drought stress involves numerous genes and pathways related to diverse mechanisms.

Some molecular and biochemical studies have revealed that drought stress can elicit defense responses through MAP kinase pathways. For example, a functional analysis of Arabidopsis demonstrated that the expression of *AtMPK3*, which shows high sequence similarity to *WIPK*, was increased when plants underwent drought. The expression of *OsMAPK5*, *OsMSRMK2*, *OsMAPKK44* and *OsMKK1* was inducible by drought treatment in rice. An analysis of transgenic rice plants strongly revealed that *OsMAPK5* could regulate drought tolerance. Moreover, in maize, the transcript level of *ZmMPK3* was increased by drought stress. An investigation in alfalfa indicated that MKK4 kinase is transiently induced by drought treatment. In addition, it has been reported that the tomato MAPK gene *SIMPK4*, which shows homology with Arabidopsis *AtMPK4*, can improve tolerance to drought stress. In this study, we investigated the relationship between MAPKKK genes and drought stress in maize. Several previous studies have indicated that MAPKKK genes can be activated by drought treatment. In Arabidopsis, the mRNA level of the MAPKK kinase AtMEKK1, which can activate its downstream factors ATMKK2 and MEK1, can be increased by drought. A MAPKKK gene that responds to drought has also been identified in rice. Furthermore, NPK1 isolated from tobacco has been reported to positively regulate drought tolerance. *NPK1* transgenic maize and rice both have higher yields under drought conditions compared with their non-transgenic counterparts.

The mitogen-activated protein kinase (MAPK) cascade, which is highly conserved, is a major signal transduction pathway in all eukaryotes, including yeasts, animals and plants. The MAPK signaling pathway, which plays a pivotal role in plant cellular responses, is involved in cell division, differentiation, apoptosis, and responses to a diversity of environmental stimuli, including cold, heat, drought and pathogen attacks. MAPK, which consist of MAPKs, MAP kinase kinases and MAP kinase kinase kinases, are serine/threonine-specific protein kinases. To transduce external stimuli into cellular responses, MAPK cascades transfer these signals via phosphorylation. By scanning the Arabidopsis genome, 80 MAPKKKs, 10 MAPKKs and 20 MAPKs have been identified. Previous analyses of the rice genome have identified 75 MAPKKKs, 8 MAPKKs and 17 MAPKs. Currently, 9 MAPKKs and 19 MAPKs have been characterized in maize. In plants, MAPKKKs have been divided into three groups, the MEKK-like family, the Raf-like family and the ZIK-like family. Among these groups, MEKK-like family is most similar to animal MEKKs and yeast MAPKKKs. Most of the Raf family proteins have a C-terminal kinase domain and extend N-terminal regulatory domain. Whereas most of the ZIK family members have N-terminal kinase domain and members of MEKK family has less conserved protein structure with kinase domain located either at N-or C-terminal or central part of the protein.

In addition, it is well documented that MAPKKKs are related to many stress-response pathways, such as plant hormone signal transduction and oxidative signaling. The Raf-like MAPKKK gene in rice, *DSM1*, functions as an early signaling component to regulate the response to drought stress through ROS (reactive oxygen species) scavenging. The results were identified by examining *dsm1* mutants. *DSM1*-RNA interference lines were hypersensitive to drought stress and were more sensitive to oxidative stress. *DSM1* may also function in the drought stress signaling response through an ABA-independent pathway that involves oxidative stress signaling. Moreover, *ANP1* has been shown to be involved in auxin signaling transduction and the oxidative stress signaling pathway. In turn, oxidative stress can activate MAPK cascades to mediate the induction of oxidative stress-responsive genes. The over expression of *ZmMKK1* in Arabidopsis can enhance drought tolerance and enhance the expression of ROS-scavenging enzyme-related genes. Analyses in maize have revealed the involvement of *ZmMPK3* and *ZmMPK5* in ROS signaling pathways. Data from recent studies provide evidence that *ZmMKK1*, *ZmMKK3* and *ZmMKK4* can enhance drought tolerance through ROS scavenging. Additionally, antioxidant enzyme activities can be enhanced in ZmMKK4-overexpressing plants compared with wild-type plants. In brief, MAPK cascades can respond to drought stress through different pathways. However, the involvement of MAPKKKs in the drought response in maize is not yet clear.

In maize, 74 MAPKKKs had been reported by searching against the maize genome database and NCBI. In this work, we will search against the maize B73 genome in the Maize GDB to identify MAPKKK genes in maize using the BLAT program with a rigorous limiting condition. To further explore whether the identified maize MAPKKKs are related to drought stress, a variety of maize grown widely in South West China, ZD619, and eight maize inbred lines were used in the

RNA-seq, qRT-PCR and biomass analyses respectively. Additionally, important metabolic processes and regulatory process pathways related to the identified MAPKKK genes were analyzed to investigate the MAPKKKs involved in the drought stress response in maize.

Materials and Methods

Plant materials and stress treatments

In this experiment, the RNA-seq used ZD619, a variety of growing widely in SouthWest China, which showed strong drought resistance in the field. Eight inbred lines consisting of J24, J853, X178, Q319, B73, E28, C8605-2 and 200B as well as ZD619 were used to further analyze on their expression pattern by qRT-PCR and phynotype discrimination responsive to drought stress. The maize seeds were pre-germinated in the plant incubator. The uniformly germinated seeds were chosen and sown in plastic pots (23cm×16cm) which filled with a 1∶1∶1 mix of soil∶vermiculite∶nutrient soil in greenhouse under 16h of light (25℃) and 8h of dark (20℃). Four uniformly strong plants finally retained at the 2~3 leaf stage.

The experiment included control (80% of relative soil water content, RSWC) and drought stress (35% of RSWC). Each treatment was comprised of two replications. The RSWC for the control plants exceeded to 80%. The drought treatment plants did not receive water from three-week-old seedling until the RSWC decreased to 35%, the drought treatment takes about one week, then kept stable for 2~3 days.

Plant biomass determination

Each plant except forqRT-PCR analysis was rapidly collected and weighed for the fresh biomass of two replicates in the greenhouse. Then the entire plant was put into a paper packet and was dried at approximately 75℃ in a drying cabinet. The dry biomass was determined until 48 hours.

RNA isolation and qRT-PCR

The samples were collected and were immediately frozen in liquid N_2 for further use. Each total RNA sample of three tissues, leaf (the top three leaves), stem and root, was extracted from four uniformly plants using an RNeasy Plant Mini Kit followed by RNase-Free DNase Set (Qiagen, Hilden, Germany). The extracted RNA was assessed by a NanoDrop 2000 Spectrophotometer, 1% agarose gel electrophoresis. The poly (A) mRNA was isolated from purified total RNA using oligo (dT) magnetic beads. The mRNA was fragmented into short pieces by adding the fragmentation buffer. The first-strand cDNA was synthesized by reverse transcriptase and random primers using mRNA fragments as templates. Second-strand cDNA synthesis was produced by RNase H and DNA polymerase I. Double-stranded cDNA fragments went through end repair, 3′ dA tailing and adapter ligation. The required fragments were then purified and enriched by PCR to create the final cDNA library. The sequences were generated by Illumina HiSeq 2000 sequencer.

In qRT-PCR three biological replicates were conducted and each biological replicate was technically repeated three times. The qRT-PCR was carried out using Maxima SYBR Green/ROX qPCR Master Mix (2×) (Thermo Scientific) performed with a 7300 Real-Time PCR System

(Applied Biosystems, Foster City, CA) according to the supplier's protocols. Each PCR reaction mixture contained 12.5μl of 2×real-time PCR mix, 1μl of gene-specific primers, 1μl reverse transcribed cDNA product and water. The thermal cycle applied was as follows: 95℃ for 10min followed by 40 cycles of 95℃ for 15s, 60℃ for 30s, 72℃ for 30s, 82℃ for 5s. At the end of the PCR cycles, melting curve analysis was performed using a single cycle consisting of 95℃ for 15s and 60℃ for 1min followed by a slow temperature increase to 95℃ at the rate of 0.3℃/s. Relative gene expression was calculated according to the delta-delta Ct method of the system.

Definition of the maize MAPKKKs and phylogenetic analysis

Maize protein sequences were downloaded from MaizeGDB database (http://www.maizegdb.org/). The maize MAPKKKs was identified through BLAT searches against Arabidopsis, rice MAPKKK query sequences. 25% identity was taken as the threshold, and the query sequences which aligned 5 BLAST alignments passed the threshold in MAPKKK gene family in Arabidopsis and rice were collected as MAPKKKs in maize.

Multiple sequence alignments were conducted on the amino acid sequences of MAPKKK proteins in maize, Arabidopsis and rice genomes using ClustalW with default settings. Phylogenetic tree was constructed by MEGA5.0 software based on alignments using Maximum Likelihood method with 1 000 bootstraps to investigate the evolutionary relationship among MAPKKK proteins in three species. Analyze the conserved domain of the full protein sequences of MAPKKK genes by BioEdit.

Subcellular localization and chromosomal locations

The predicted subcellular localization of maize MAPKKKs was retrieved using the CELLO v2.5 server (http://cello.life.nctu.edu.tw/). Information about the physical locations of maize MAPKKK genes on chromosomes was obtained from MaizeGDB database. The chromosomal locations of the MAPKKK genes in maize were mapped on chromosomes using PERL script and Adobe Illustrator Artwork software.

RNA-seq analysis and identification of DEGs

The RNA-seq reads for each tissue were mapped to maize reference genome B73 using the Tophat. Then the differentially expressed genes between drought processed sample and well-water processed sample in three tissues were extracted after processing Cufflinks and Cuffdiff (http://cufflinks.cbcb.umd.edu/manual.html).

We identified differentially expressed genes (DEGs) by fold change greater than two and P-value ≤ 0.05.

Our raw data and the processed RNA-seq data have been deposited in the National Center for Biotechnology Information Gene Expression Omnibus (GSE71377).

GO Analysis and Functional classification

We used an online tool Venn (http://bioinfogp.cnb.csic.es/tools/venny/index.html) to compare the DEGs in three tissues. The cluster was done by cluster3.0 using the Complete linkage method. The heatmap was drew by treeview and we chose the log2 (fold change) absolute value one as scale. Functional classification was performed by using the tool agriGO (http://bioinfo.cau.edu.cn/agriGO/). The functional clusters enrichment analysis was calculated by comparing with the whole maize genome V5a by Singular Enrichment Analysis (SEA) method, and the highest classification stringency was chosen for clustering by FDR≤0.05.

We used an interactive ontology tool PageMan to generate overview graphs for profiling experimentswith maize database and a user-driven tool MAPMAN to display genomics data sets onto diagrams of metabolic pathways and other biological processes. The Wilcoxon test likes a t-test was used to test the hypothesis that objects within one functional class behave differently from the rest of the objects. The significance of the change is reflected in the intensity and the direction by the color. The number used to do the heatmap was calculated by P-value. All P-values above 0.05 are set to a z-score of 0 to avoid misinterpretation. A highly saturated color indicates a high absolute value, whereas smaller values are indicated by lower color saturation. The two different colors can be selected to distinguish between categories where the average of the signals for all the genes in a category increases (red) or decreases (blue).

Co-expression analysis

We carried out pathway enrichment analysis by KAAS (http://www.genome.jp/tools/kaas/) for our DEGs in three tissues with Arabidopsis database. 14 common regulated pathways and 5 tissue specific regulated pathways were used for the heat map show. An R package WGCNA was used for our sequenced maize transcripts. We kept the transcripts by at least FPKM>0.1 in six samples. Firstly, we obtained the DEGs in each selected pathways and the co-expressed MAPKKKs and the co-expression score between each two genes. Secondly, the cutoff of 0.2 was used to select co-expression level in all the three tissues and a standard Z-score for each term is given by the following formula: $Z = (Sm - \mu) \times \sqrt{m}/\delta$, where Sm is the mean co-expression value for genes selected with the cutoff, μ is the mean co-expression value of the entire gene list from WGCNA, m is the number of MAPKKK genes co-expressed with the DEGs, and δ is the SD of the co-expression value in the entire gene list. A cutoff value of 0.2 was used to select highly co-expression events.

Genes specific primers

First-strand cDNA was used as a template in a qRT-PCR synthesized from 5μg total RNA using the SuperScript III First-Strand Synthesis System for RT-PCR (Invitrogen Corp.). The endogenous reference zSSIIb from *zea mays* starch synthase II was used as an internal control to normalize the data. The eight genes specific primers which designed based on the cDNA sequences were used to detect the gene expression level. The qRT-PCR primers were check the specification by NCBI primer-BLAST and validated by PCR (S4 File).

Results

Physical location and phylogenetic analysis of MAPKKKs in maize

To identify MAPKKK genes in maize, the Arabidopsis and rice MAPKKK protein sequences were employed to perform a BLAST search against the protein sequences of maize B73. In maize,

71 MAPKKKs were identified, including 57 reported MAPKKKs and 14 novel ones (Table A in S1 File), which were basically consistent with previous research. These reported MAPKKKs were named according to the foregoing study, while since there was no standard nomenclature followed for MAPKKKs neither in Arabidopsis nor in rice, the 14 putative MAPKKKs were designated according to their group (Table A in S1 File). We obtained the chromosomal locations of these 71 MAPKKKs, and the predicted subcellular localization was determined from the CELLO v2.5 server (Table A in S1 File). The physical location data showed that 70 MAPKKKs were distributed on all 10 maize chromosomes (Figure 1D), except the *GRMZM2G011070* (*ZmMAPKKK29*) gene, which was situated on an undetermined chromosome (chromosome unknown). Among the 14 novel MAPKKKs, *GRMZM2G459824* (*ZmMAPKKK75*) was located on chromosome 1; *GRMZM2G404078* (*ZmMAPKKK76*) and *GRMZM2G335826* (*ZmMAPKKK77*) were located on chromosome 3; and *GRMZM2G158860* (*ZmMAPKKK78*), *GRMZM2G032619* (*ZmMAPKKK81*) *GRMZM2G002531* (*ZmMAPKKK85*), *GRMZM2G028709* (*ZmMAPKKK86*) and *GRMZM5G852329* (*ZmMAPKKK87*) were located on chromosome 4. *GRMZM2G127632* (*ZmMAPKKK88*) was located on chromosome 5. Chromosome 6 contained *AC204050.4* (*ZmMAPKKK79*). Chromosome 7 contained *GRMZM2G034779* (*ZmMAPKKK82*) and *GRMZM2G023444* (*ZmMAPKKK83*). *GRMZM2G072395* (*ZmMAPKKK84*) was located on chromosome 8. *GRMZM2G378852* (*ZmMAPKKK80*) was located on chromosome 9.

The phylogenetic analysis was performed using the protein sequences in Arabidopsis, rice and maize. We aligned the full protein sequences of all the MAPKKKs in Arabidopsis, rice and maize using ClustalW, and we built a phylogenetic tree using MEGA5.0 (Figure 1). We identified 80 MAPKKKs in Arabidopsis and 75 MAPKKKs in rice, which could be subdivided into three major subtypes, Raf, MEKK and ZIK. Based on the subtypes subdivided in a recent analysis, the MAPKKKs in maize could also be divided into three major groups, including Raf, MEKK and ZIK. The MEKK group included 26 MAPKKKs in maize, 22 MAPKKKs in rice, and 21 MAPKKKs in Arabidopsis. The ZIK group included 8 MAPKKKs in maize, 10 MAPKKKs in rice and 11 MAPKKKs in Arabidopsis. The Raf group included 37 maize MAPKKKs, 43 rice MAPKKKs and 48 Arabidopsis MAPKKKs. As the results indicated, MAPKKKs are highly conserved among plants. Furthermore, the degree of conservation in the three subfamilies was also very high, and the numbers were similar in species that may have similar functions. Similar to the conservation of the MAPKKK motif in Arabidopsis and rice, the ZIK family in maize shared the conserved domain GTPEFMAPE (L/V) (Y/F) [Figure A (a) in S5 File], and the MEKK family in maize had the conserved domain G (T/S) Px (W/F) MAPEV [Figure A (b) in S5 File]. Furthermore, the conserved domain GTxx (W/Y) MAPE was detected in the Raf family in maize [Figure A (c) in S5 File].

RNA-seq analysis of well-watered and drought-stressed maize transcriptomes in three tissues

The maize variety ZD619 was selected to analyze differentially expressed genes in response to drought stress (Figure 2A). RNA samples from leaf, root and stem tissues of ZD619 under well-

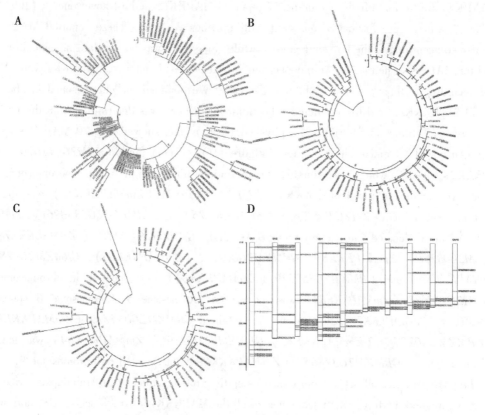

Figure 1 Phylogenetic tree and genomic locations of maize, rice and Arabidopsis MAPKKK family genes

The different colors represent three species. A. Raf subfamily; B. MEKK subfamily; C. ZIK subfamily; D. Physical locations of MAPKKK genes on maize chromosomes

watered and drought stress conditions were isolated for next-generation sequencing using the Illumina HiSeq 2000 platform. The raw data included 40 million reads from each sample. The quality of the sequencing data is very well. The Q20 which represents for the percentage of the base whose quality is greater than 20 is approximately 98%. And the Q30 can also be around 93%. (Table B in S1 File) After trimming the adaptor and low-quality reads, approximately 85% of the processed data could be aligned to the B73 reference using TopHat (Table c in S3 File). Then, using Cuffdiff, 5 866 differentially expressed genes (DEGs) were identified in leaf, stem and root. Among them, 2 319 DEGs were identified in leaf, including 1 206 that were up-regulated and 1 113 that were down-regulated. 2 371 DEGs were identified in stem, including 1 057 that were up-regulated and 1 314 that were down-regulated. 2 181 DEGs were also identified in root, including 1 544 that were up-regulated and 637 that were down-regulated (Table 1, S3 File, Figure c in S5 File).

Through comparing the DEGs in leaf, stem and root, we found that only 74 genes were commonly regulated in all three tissues in response to drought, which suggested that these genes may play key roles during drought stress and that most of these genes were tissue-specific

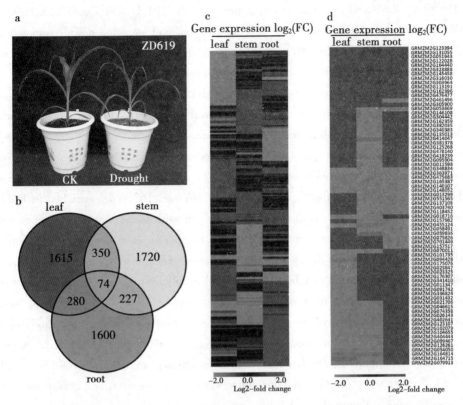

Figure 2　Differentially expressed genes in leaf, stem, and root based on RNA-seq data

a. The growth phenotypes of ZD619 in the eight leaf stage under drought and well-water conditions. The well water-treated plant was shown on the left and drought-treated plant was shown on the right in the picture; b. Venn diagram of DEGs in leaf, stem, and root; c. Heat map of 5,866 DEGs in the three tissues. The genes in at least one of the subsets were analyzed. The bar represents the log2 of the drought/control ratio; d. Heat map of 74 DEGs in the three tissues. The bar represents the log2 of the drought/control ratio

(Figure 2B). The heat map of the DEGs also demonstrated significant differences among the three tissues (Figure 2C). Out of those genes, there are 74 genes commonly regulated in the three tissues, approximately half of them were oppositely regulated in different tissues, which also indicated a large difference in the manner of regulation in different tissues (Figure 2D). Among the 74 highly significant co-regulated genes, *GRMZM2G476477* (*ZmMAPKKK20*) is a member of the MAPKKKs, which showed significant up-regulation in all three tissues. In addition, it has been reported that *GRMZM053669*, *GRMZM011598* and *GRMZM059836* were up-regulated under dehydration conditions. *GRMZM2G051943* is a gene related to plant defense. *GRMZM2G162359* is a gene related to *Aspergillus flavus* pathogenesis. *GRMZM2G137108* and *GRMZM2G455124* are required for leaf development. *GRMZM2G419239* and *GRMZM2G341410* are involved in cell wall-related biogenesis.

In maize, 9 MAPKKs and 19 MAPKs have been identified. Among the 5 866 DEGs, *GRMZM2G344388* (ZmMKK10-2) is a member of MAPKK family, *GRMZM2G053987* (ZmMPK3-1) and *GRMZM2G062761* (ZmMPK15) are two members of MAPK family. However, ZmMKK10-2 had no interaction with ZmMPK3-1 or ZmMPK15. ZmMKK10-2, ZmMPK3-1 and ZmMPK15 were up-regulated in leaf and stem but down-regulated in root under drought conditions. Significantly, we have found highlight other members of the cascade, MAPKK and MAPK differentially expressed in their transcriptome levels.

Then, a GO analysis was conducted with respective DEG terms in the three tissues using the agriGO tool (see methods). Four GO classifications were commonly enriched after comparison among the three tissues, including response to stimulus, oxidation-reduction, polysaccharide metabolic process and metabolic process. The numbers of genes enriched in the three tissues for each commonly enriched GO classification were similar, and only a fraction of these overlapped among the three tissues. For example, there were 190 167 and 167 genes enriched in leaf, stem and root, respectively, in the "response to stimulus" classification, but only 8 genes were the same, which accounted for approximately 5% of the number in each gene set. Correspondingly, there were 5.8%, 6.3%, and 4.4% commonly enriched genes in leaf, stem and root for oxidation-reduction, polysaccharide metabolic process and metabolic process, respectively. Seven GO classifications were enriched in two tissues. For example, response to chemical stimulus, response to stress and carbohydrate metabolic process were enriched in leaf and root; and cellulose biosynthetic process, cellulose metabolic process, multicellular organismal process and response to abiotic stimulus (which contains drought stress) were enriched in leaf and stem (Figure 3). Furthermore, 74 GO classifications were tissue specific, especially metabolism-related classifications (Figure B in S5 File). Our results suggested that different tissues of plants may have similar stress signal reception and conduction pathways, yet the tissue-specific genes had a greater impact on plant growth and development.

Table 1 Differentially expressed genes (drought vs. control)

Tissue	Up-regulated	Down-regulated
leaf	1 206	1 113
stem	1 057	1 314
root	1 544	637

MAPKKK genes revealed apparently different expression profiles in different tissues

Among all the MAPKKK genes, four genes (*ZmMAPKKK18*, *ZmMAPKKK19*, *ZmMAPKKK20* and *ZmMAPKKK56*) in leaf, three genes (*ZmMAPKKK19*, *ZmMAPKKK20* and *ZmMAPKKK21*) in root, and seven genes (*ZmMAPKKK18*, *ZmMAPKKK19*, *ZmMAPKKK20*, *ZmMAPKKK21*, *ZmMAPKKK22*, *ZmMAPKKK26* and *ZmMAPKKK73*) in stem revealed significantly differential expression in response to drought stress. A comprehensive analysis of these three tissues revealed that eight MAPKKK genes may be involved in the response to drought stress. We conducted a cluster analysis of the three MAPKKK sub-families' expression profiles, including the ZIK family genes (Figure 4a), the MEKK family genes (Figure 4b), and the Raf family genes (Figure 4c). The data suggested that most of the differentially expressed MAPKKKs be-

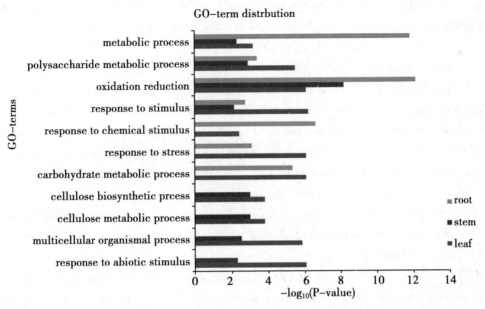

Figure 3 Functional categorization of differentially expressed genes (DEGs) in three tissues

The top four GO terms are functional categorizations (biological process) of the DEGs that were enriched in all three tissues, and the lower seven GO terms are functional categorizations (biological process) of the DEGs enriched in two tissues. The enrichment figure was constructed from significantly enriched GO terms only. The X-axis is the -log10 (P-value), which represents the level of enrichment, and the cutoff of the P-value is 0.05. Blue represents leaf tissue, red represents stem tissue, and green represents root tissue

longed to the MEKK family. We detected an important gene, *GRMZM2G165099* (*ZmMAPKKK19*), which was predicted to function similar to NPK1 (*Nicotiana* protein kinase), in the MEKK sub-family. The expression of Nicotiana protein kinase (NPK1) has been shown to enhance drought tolerance in transgenic maize. In our results, *ZmMAPKKK19* showed an up-regulation pattern in all three tissues, especially in root, which is in accordance with the previous research.

According to the subcellular localization results (Table A in S1 File) and the expression patterns shown in Figure 4B, the highly up-regulated MAPKKKs in leaf, *ZmMAPKKK17*, *ZmMAPKKK18*, *ZmMAPKKK19*, *ZmMAPKKK20* and *ZmMAPKKK79*, were all located in the chloroplast, which is similar to the genetic mode of their expression in leaf. The results indicated that MAPKKKs may play an important role in drought-related stress. In addition, the pattern of the functional MAPKKKs displays clearly high expression under drought.

Some of the differentially expressed genes in leaf, root, and stem of ZD619 that were identified using RNA-seq were confirmed by qRT-PCR to validate whether these genes were connected with the drought response (Figure 4d). In leaf, the expression levels of *ZmMAPKKK 18* and *ZmMAPKKK56* were markedly up-regulated, with greater than 4-fold increases, whereas the changes in *ZmMAPKKK19* and *ZmMAPKKK20* were not significant and were less than two-fold. The expression of genes in stem showed a slight increase in *ZmMAPKKK20*. The expression levels of *ZmMAPKKK18*, *ZmMAPKKK19*, *ZmMAPKKK21* and *ZmMAPKKK22* were up-regulated in stem.

DEGs were enriched in different pathways in response to drought stress

As the above results showed, different tissues showed different responses to drought stress, and the DEGs in the three tissues represented distinct functional categories. Several MAPKKKs were including among the DEGs affected by drought. We investigated how drought regulated the differentially expressed genes, particularly the MAPKKKs, using PageMan to categorize the drought-mediated gene expression data for leaf, stem and root tissues into known metabolic pathways and different regulatory processes (Figure 5).

There were many DEGs enriched inhormone signaling, photosynthesis, CHO metabolism, secondary metabolic processes and regulation of transcription (Figure 5). Genes enriched in hormone signaling included more up-regulated genes in leaf and more down-regulated genes in root. Genes enriched in photosynthesis included many down-regulated genes in leaf. Genes enriched in CHO metabolism included many up-regulated genes in root. Genes enriched in secondary metabolic processes included more down-regulated genes in the three tissues. Genes enriched in regulation of transcription included more up-regulated genes in leaf and stem. To identify important enriched pathways in our DEGs, we analyzed an overview of enrichment in the metabolic processes TCA, starch synthesis and decomposition; proline metabolic pathways (Figure D in S5 File); and hormone signaling pathways (Figure E in S5 File). Genes that were significantly involved in metabolism were more down-regulated in leaf but more up-regulated in root. Genes that were significantly involved in regulation were more down-regulated in stem but up-regulated in leaf and root. These results suggested that the differential regulation of genes in different tissues may lead to particular patterns of regulation.

Furthermore, the overview of regulation revealed three regulated MAPKKKs (*ZmMAPKKK19*, *ZmMAPKKK20* and *ZmMAPKKK56*) in leaf, four regulated MAPKKKs (*ZmMAPKKK20*, *ZmMAPKKK21*, *ZmMAPKKK22* and *ZmMAPKKK26*) in stem, and four regulated MAPKKKs (*ZmMAPKKK17*, *ZmMAPKKK19*, *ZmMAPKKK20* and *ZmMAPKKK21*) in root. *ZmMAPKKK19* and *ZmMAPKKK20* were common in at least two tissues. They exhibited distinctly differential expression after drought stress (Figure 4B). According to the phylogenetic tree (Figure 1B), *ZmMAPKKK17*, *ZmMAPKKK19*, *ZmMAPKKK20*, *ZmMAPKKK21* and *ZmMAPKKK22* belong to the MEKK subfamily. These results indicated that MAPKKKs may have important regulatory functions in drought tolerance in maize.

Co-expression analysis of DEGs and MAPKKKs

To examine patterns of correlation between the differentially expressed genes and MAPKKKs, we investigated their co-expression using WGCNA. All 26 901 filtered transcripts were clustered using WGCNA to obtain an overview of their expression relationships. Networks and models for our data sets were constructed to determine the co-expression value between each pair of genes from our filtered data sets.

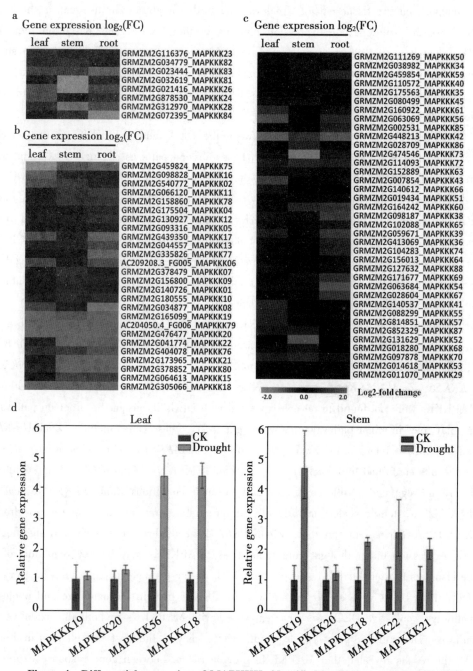

Figure 4　Differential expression of MAPKKKs identified by RNA-seq and qRT-PCR

a, b and c Expression patterns of 71 MAPKKK genes in the three tissues. a. ZIK family genes; b. MEKK family genes; c. Raf family genes; d. Relative expression levels of the MAPKKK genes in various tissues based on qRT-PCR analysis. The expression in leaf is shown on the left. The expression in stem is shown on the right

We focused on the MAPKKKs that were especially co-expressed with DEGs that were enriched in important pathways. After filtering with at least 10 genes, the most significant pathways could be identified, and most of these pathways coexisted in all three tissues. Then, we chose fourteen common and five tissue-specific pathways for further analysis. The heatmap in the left of Figure 6 shows the different co-expression patterns in the three tissues based on their Z-scores. The plant hormone signaling pathway categories oxidation stress, photosynthesis, starch and sucrose metabolism, arginine and proline metabolism, TCA cycle, auxin, cytokinin, gibberellin, ABA, ethylene, BR, jasmonic and salicylic acid presented high Z-scores in the co-expression analysis.

To better understand the relationships between the DEGs and MAPKKKs in the selected pathways, a heat map was constructed to study the expression patterns of single DEGs that were co-expressed with MAPKKKs corresponding to the nineteen enriched pathways (right in Figure 6). The data revealed that the DEGs were highly related to the MAPKKKs (Table D in S1 File). The DEGs that were co-expressed with MAPKKKs in these pathways also showed a broad differential expression pattern. It was found that most of the DEGs were up-regulated in root. In the high Z-score pathways, the photosynthesis pathway contained most of the down-regulated pathway-enriched DEGs, whereas the proline, ethylene and salicylic acids pathways contained many up-regulated pathway-enriched DEGs, especially in root.

Furthermore, thirteen co-expressed MAPKKKs were co-expressed with DEGs in the above nineteen enriched pathways; these MAPKKKs were *ZmMAPKKK11*, *ZmMAPKKK12*, *ZmMAPKKK13*, *ZmMAPKKK22*, *ZmMAPKKK34*, *ZmMAPKKK38*, *ZmMAPKKK39*, *ZmMAPKKK51*, *ZmMAPKKK56*, *ZmMAPKKK65*, *ZmMAPKKK72*, *ZmMAPKKK74* and *ZmMAPKKK83*. Although some MAPKKKs were not found to be co-expressed with DEGs in any of the nineteen pathways, they may play roles in minor pathways under drought stress; these genes included *ZmMAPKKK19* and *ZmMAPKKK21*, among other. *ZmMAPKKK26* and *ZmMAPKKK73* play roles in several pathways. Our results confirmed that *ZmMAPKKK19*, *ZmMAPKKK21*, *ZmMAPKKK22* and *ZmMAPKKK56* were up-regulated in different tissues under drought conditions. *ZmMAPKKK56* in leaf and *ZmMAPKKK22* in stem revealed significantly differential expression in response to drought stress. *ZmMAPKKK21* in stem and root, *ZmMAPKKK19* in all three tissues showed significantly differential expression under drought conditions. These MAPKKKs may be key regulators of the tolerance of plants to drought. The co-expressed DEGs in these nineteen pathways were very peculiar. Only three DEGs were co-expressed with MAPKKKs both in the oxidative and photosynthesis pathways, only five DEGs were co-expressed with MAPKKKs in both the oxidative and TCA pathways. The oxidative pathway, especially ROS, may be the most core element in drought stress. All the results revealed that MAPKKKs provide a link between gene response expression and drought tolerance. The drought stress-responsive MAPKKK genes may be crucial to maize under drought conditions.

Differential expression levels of the eight MAPKKK genes in response to drought stress in maize

To understand the differences of eight MAPKKK genes in response to drought stress among various maize varieties, a hybrid ZD619 and eight inbred lines of maize were used to further qRT-PCR analysis and biomass determination (Table 2). Plant biomass and kernel yield are usu-

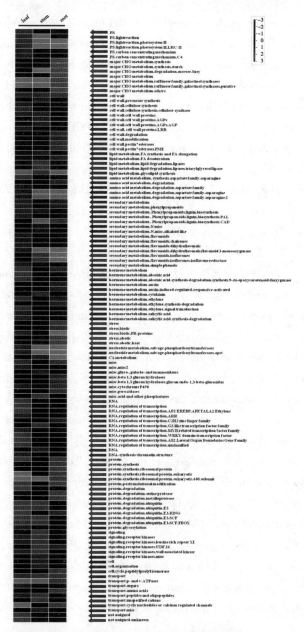

Figure 5 Pathway enrichment of differentially expressed genes involved in different regulatory processes under drought stress

A q-value cutoff of 0.05 was used to select enriched gene sets in all three tissues. The heat map represents the Z-scores obtained from a parametric analysis of gene set enrichment q-values for term enrichment. Red represents enriched genes in the treatment group that were over-represented compared with the control set. Blue represents the enriched genes in the treatment group that were under-represented compared with the control set. The absolute values represent the enrichment level. The bar represents the Zscore region from -3 to 3

ally used as integrative indicators to judge the drought-tolerant potential of plants. In this study, seedling biomass was regarded as a criterion to identify drought tolerance capability. The maize seedling biomass was measured as fresh weight and dry weight and subsequently calculated as drought tolerance index (DTI) to compare the drought response performance of various inbred lines. Results indicated that drought stress exhibited enough influence to elicit differential phenotypes. Each inbred line showed different drought resistance capability, among them X178 was the most tolerant to drought stress and E28 was the most sensitive to drought stress.

Eight MAPKKK genes for expression analysis by qRT-PCR were shown in Figure 7. The expression levels of the eight genes differed among the nine varieties. The expression levels of ZmMAPKKK18 and ZmMAPKKK56 were up-regulated in response to drought stress in all varieties. The expression levels of ZmMAPKKK19, ZmMAPKKK20 and ZmMAPKKK73 were up-regulated under drought conditions in seven inbred lines except E28. In addition, ZmMAPKKK21 was up-regulated in seven inbred lines except B73. The expression levels in various inbred lines had significantly different. The inbred line X178, in expression level increased by 58-fold in ZmMAPKKK18, 20-fold in ZmMAPKKK19, 11-fold in ZmMAPKKK56, 8-fold in ZmMAPKKK20, 7-fold in ZmMAPKKK22, and 4-fold in ZmMAPKKK21. In the inbred lines E28 and B73, the expression levels were not significantly changed, and some genes were even down-regulated. In E28, the expression levels of ZmMAPKKK19, ZmMAPKKK20 and ZmMAPKKK73 were markedly down-regulated. The results indicated that different expression levels of differential expression MAPKKK genes had certain relation with the variety's drought resistance ability.

We then performed a correlation analysis of the different expression levels of eight MAPKKK genes in eight inbred lines. Pearson's correlation coefficients were calculated between each gene's expression pattern and the plant biomass DTI values in the eight inbred maize lines (Table 2). The results suggested that the expression levels of the genes GRMZM2G305066 (ZmMAPKKK18) and GRMZM2G063069 (ZmMAPKKK56) were significantly correlated with changes in plant biomass in response to drought ($P < 0.05$). The expression levels of GRMZM2G165099 (ZmMAPKKK19) and GRMZM2G476477 (ZmMAPKKK20) were slightly correlated with plant biomass ($P<0.1$). This analysis suggested that the eight MAPKKK genes, especially ZmMAPKKK18 and ZmMAPKKK56, were obviously related to biological characteristics of maize under drought stress. Among the eight genes, according to this analysis, ZmMAPKKK18 and ZmMAPKKK56 may have crucial functions in the response of maize to drought stress.

Discussion

MAPKKK genes are related to drought stress in maize

Plants have developed complicated signaling pathways to adapt to environmental stress. In some species, it has been verified that MAPKKK can be rapidly activated during suffering to drought stress. In Arabidopsis, the mRNA levels of AtMEKK1, which is structurally related to MAPKKK, and AtMPK3, which is structurally related to MAPK, has increased in response to dehydration stress. In rice, a Raf-like MAPKKK gene, DSM1, can respond to drought stress through ROS scavenging. The NPK1 (MAPKKK) isolated from tobacco is conserved among different organisms. Transgenic rice and maize with NPK1 showed significantly improved tolerance to

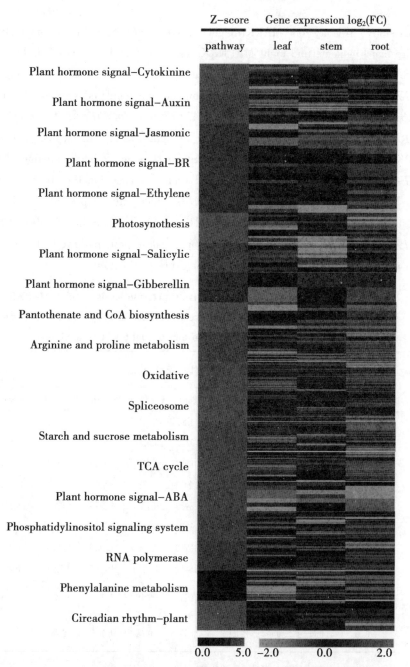

Figure 6　Co-expression analysis of tissue-specific DEGs and MAPKKK genes

　　The weight value obtained from the WGCNA package was used as a parameter for the parametric analysis of gene co-expression levels. A cutoff of 0.2 was used to select highly co-expressed genes in all three tissues. The left side of the heat map represents the Z-scores obtained from a parametric analysis of gene co-expression. The lower left bar represents the degrees of the Z-score scale. The right side of the heat map represents the expression patterns of the DEGs co-expressed with MAPKKKs that were enriched in the corresponding pathways on the left. The lower right bar represents the log2 of the drought/control ratio drought.

In this study, seventy-one MAPKKK genes have been obtained by a computational analysis of the entire maize genome and eight MAPKKK genes that exhibited significantly differential expression in response to drought stress are identified by an RNA-seq analysis. Among these genes, *ZmMAPKKK26* and *ZmMAPKKK73*, which show down-regulated expression in stem under drought stress, are located in the nucleus. *ZmMAPKKK18*, *ZmMAPKKK19*, *ZmMAPKKK20* and *ZmMAPKKK22*, which are up-regulated under drought stress, are located in the chloroplast. Four MAPKKKs are predicted with a serine/threonine-protein kinase function. Notably, *ZmMAPKKK19* is predicted to have NPK1-related protein kinase function. In addition, *ZmMAPKKK19* is highly homologous with *ZmMAPKKK18*. Indeed, the expression levels of these two genes are up-regulated after drought treatment. The results indicate that *ZmMAPKKK18* and *ZmMAPKKK19* may be involved in the drought response. These proteins have high sequence similarity, which may means the correlation with similar expression patterns.

Table 2 The correlation between relative expression levels of the 8 MAPKKK genes based on qRT-PCR analysis and relative fresh and dry biomass in 8 maize inbred lines under well-watered and drought-stress conditions

	Relative expression in drought based on qRT-PCR								the ratios between drought and control	
	MAPKKK18	MAPKKK19	MAPKKK20	MAPKKK21	MAPKKK22	MAPKKK26	MAPKKK56	MAPKKK73	freshbiomass	Drybiomass
X178	58.485 2	20.299 0	14.757 1	4.208 6	7.378 5	1.388 3	11.712 7	1.777 7	0.739	0.812
J24	7.981 5	4.993 3	1.417 5	1.533 3	1.327 2	4.428 0	1.681 8	7.310 7	0.616	0.698
Q319	3.498 3	4.387 3	5.585 4	1.892 1	1.840 4	1.477 7	4.552 5	3.819 4	0.606	0.695
J853	23.533 9	11.631 8	5.388 9	25.281 3	32.748 0	1.178 3	7.638 7	1.477 7	0.566	0.634
C8605	4.510 6	5.476 8	5.133 7	1.093 0	0.840 9	0.728 7	4.277 2	1.089 2	0.503	0.576
B73	1.887 7	3.638 5	2.417 2	0.468 7	0.320 9	0.906 5	1.705 3	1.280 5	0.491	0.569
200B	8.835 6	13.454 3	7.498 8	1.602 1	1.057 0	0.404 3	1.488 0	1.494 8	0.483	0.554
E28	2.821 9	0.091 4	0.203 8	2.620 8	0.640 2	0.814 1	1.666 3	0.518 8	0.421	0.503
Pearson Correlation for fresh biomass	0.78*	0.66	0.71	0.13	0.23	0.45	0.77*	0.45		
Pearson Correlation for dry biomass	0.75*	0.61	0.67	0.09	0.19	0.48	0.74*	0.48		

* indicate significant correlation at $P<0.05$ respectively

Furthermore, qRT-PCR is performed with eight MAPKKK genes to compare their expression levels in different varieties between drought conditions and well-watered conditions. The results show that the expression levels of these eight MAPKKKs in the nine maize varieties are different. In all the varieties, the relative expression levels of *ZmMAPKKK18* are markedly up-regulated when maize seedlings are under drought stress condition. Additionally, the relative expression levels of *ZmMAPKKK19*, which is predicted to have *NPK1*-related protein ki-

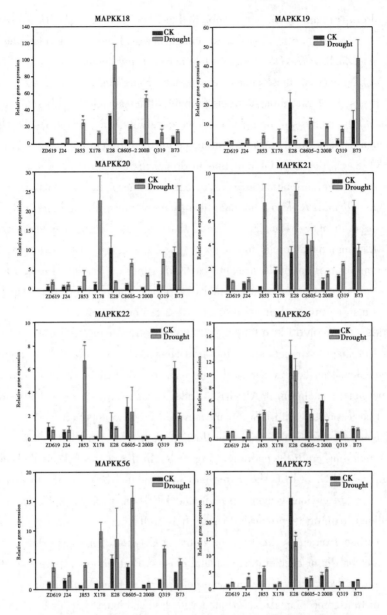

Figure 7 Relative gene expression of the 8 MAPKKK genes in
various inbred lines based on qRT-PCR analysis

To determine whether the relative expression levels of the drought stress-responsive MAPKKK genes differed among varieties induced by drought, ZD619 and the 6 inbred maize lines J24, J853, X178, E28, C8605-2, 200B, Q319 and B73 were used. Lines X178, J24 are drought-resistant lines; 200B and E28 have poor drought tolerance. The eight MAPKKK genes are GRMZM2G305066 (ZmMAPKKK18), GRMZM2G165099 (ZmMAPKKK19), GRMZM2G476477 (ZmMAPKKK20), GRMZM2G173965 (ZmMAPKKK21), GRMZM2G041774 (ZmMAPKKK22), GRMZM2G021416 (ZmMAPKKK26), GRMZM2G063069 (ZmMAPKKK56), GRMZM2G474546 (ZmMAPKKK73). * indicates significant differences in comparison with the control at $P<0.05$ respectively. Error bars indicate standard deviation for three replicates nase function, are up-regulated in seven inbred lines. The genes *ZmMAPKKK18* and *ZmMAP-KKK19* may play a crucial role in response to drought stress. However, to elucidate the roles of *ZmMAPKKK18* and *ZmMAPKKK19* in drought tolerance, many details need to be clarified by further research. A comparison of these genes' expression among the different maize lines show that the abilities of genes to tolerate drought stress probably differ among the different lines.

In our study, the analysis of drought tolerance capability which identified by DTI values indicates that drought stress exhibit enormous influence for the different phenotypes. Among the eight inbred lines, X178 was the most tolerant line to drought stress and E28 was the most sensitive line to drought stress. The correlation analysis reveals that the eight MAPKKK genes, especially *ZmMAPKKK18* and *ZmMAPKKK56*, are significantly correlated with the biological characteristics under drought stress in maize. The data suggests that, among the eight genes, *ZmMAPKKK18* and *ZmMAPKKK56* may have crucial functions in the response of drought stress in maize.

In maize, the expression levels of eight MAPKKK genes are regulated under drought stress, and the results are confirmed in different maize varieties. These observations suggests that the MAPKKK genes may be involved in the drought stress response in maize.

MAPKKK genes are involved in multiple pathways under drought stress

Plants tend to adapt to drought stress by serial changes. The mechanisms of drought tolerance in plants are complex. The data in our study suggests that drought stress can lead to expression changes in some genes, including MAPKKKs. Under drought stress, several MAPKKKs, especially those that co-expressed with DEGs, are enriched in some important pathways, such as the photosynthesis pathway, hormone signal transduction pathway, oxidation pathway, and protein modifications (including proline) pathway (Figure 8). In the photosynthesis pathway, there are more down-regulated DEGs in leaf. There are many up-regulated DEGs in the three tissues, especially in root. There are obviously more up-regulated DEGs in oxidation and protein modification (including proline) pathways, primarily in root. It is well known that drought stress encumbers photosynthetic carbon fixation by limiting the entry of CO_2 into leaves via down-regulating the photosynthetic metabolism or accelerating stomatal closure. Furthermore, proline synthesis moderately increases under drought stress, which can help plants to resist drought stress. The accumulation of proline in transgenic plants is associated with higher biomass under drought stress. In our study, the enriched DEGs are down-regulated in the photosynthesis pathway in leaf and are up-regulated in proline modification, which are in accordance with previous studies. Some MAPKKKs are found to be co-expressed with DEGs in these pathways, which suggests that MAPKKKs are involved in the photosynthesis pathway and proline modification pathway under drought stress condition. Among them, the down-regulated *ZmMAPKKK26* and *ZmMAPKKK73* are co-expressed with DEGs in the photosynthesis pathway, while the up-regulated *ZmMAPKKK19*, *ZmMAPKKK21*, *ZmMAPKKK22* and *ZmMAPKKK56* are co-expressed with DEGs in proline modification pathway.

Moreover, the accumulation of ethylene and salicylic acid may be caused by water stress. Several earlier investigations have found that some ethylene response factor genes and en-

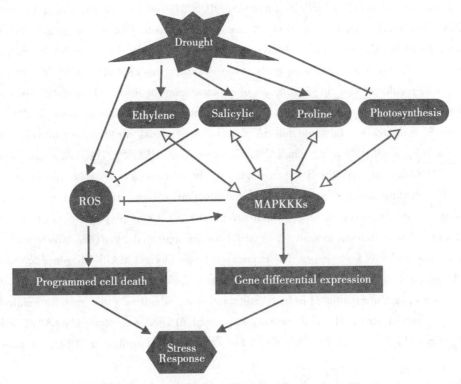

Figure 8 A model of drought stress effects on maize

The colored shapes represent groups of glyphs with similar processes or events. The arrows ending in a solid triangle indicate positive effects. The arrows ending in a transverse line indicate a clear negative influence. The arrows ending in a hollow triangle indicate co-expression. The curved arrows indicate positive feedback

dogenous salicylic acid levels can induce water stress tolerance in plants. Seed which is pretreated with SA promotes plant vigor index under drought stress condition. A study in *Medicago* and Arabidopsis plants showed that a MAPK pathway was involved in ethylene signaling. Drought inhibits maize growing by causing oxidative damage to biological membranes and disturbing the water status in tissues. Reactive oxygen species (ROS) are central signaling molecules in response to drought. To reduce the toxicity of ROS, plant cells have developed an antioxidative system. The increasing ROS to toxic levels can activate PCD to remove damaged cells. Ethylene and salicylic acid response factor proteins may be connected to the ROS pathway through defending cells against ROS attacks. Salicylic acid can enhance the antioxidant status and the tolerance to drought stress. Ethylene plays an important role in lowering ROS generation and protecting plants under drought stress. Ethylene can modulate the programmed cell death induced by O_3 exposure. Many studies have reported that MAPK cascades play crucial roles in ROS signaling pathway under water stress. MAPK, which is induced by oxidative stress, can promote ROS scavenging. Arabidopsis *MKK4* increases plant drought tolerance and decreases the production of ROS. Tolerance to drought stress, which is enhanced by *NPK1*, may activate the oxidative signaling pathway. In tobacco, *GbMPK3* increases ROS scavenging under drought stress. The MAPKKKs that are activated under drought stress may promote ROS scavenging to control the concentrations of ROS. In our study, DEGs are enriched in the ethylene and SA hormone signal transduction pathways. The expression levels of most of the genes related to salicylic acid and ethylene are up-regulated. The up-regulation of proteins may be involved in redox homeostasis, ROS scavenging and its damaging effects on proteins and DNA. In addition, some MAPKKKs are co-expressed with DEGs in ethylene and SA plant hormone signal transduction pathways, such as *ZmMAPKKK19*, *ZmMAPKKK21*, *ZmMAPKKK22* and *ZmMAPKKK56*. The data supports the theory that MAPKKKs may promote ROS scavenging by the ethylene and SA pathways in response to drought. The ethylene and SA hormone signal transduction pathway-related genes, especially in root, show an up-regulation pattern, which reveals that roots may play a key role in response to drought stress. Under drought condition, MAPKKKs are activated by ROS, which leads to the differential expression of some genes. The expression levels of eight MAPKKK genes are increased under drought treatment. The genes *ZmMAPKKK18* and *ZmMAPKKK19*, which are highly homologous to NPK1, are up-regulated in leaf, stem and root, which may play roles in regulating the response to drought stress. The differentially expressed MAPKKKs, *ZmMAPKKK19*, *ZmMAPKKK21*, *ZmMAPKKK22* and *ZmMAPKKK56* are co-expressed with some DEGs in most pathways.

In conclusion, seventy-one MAPKKK genes in maize are identified in our study. Among them, 8 differentially expressed MAPKKK genes responsive to drought stress are obtained. The results suggest that MAPKKKs may react to drought stress through various pathways. To our knowledge, this is the first study of the expression of maize MAPKKK genes in relation to drought stress. The results of our study should provide further information about the characteristics of MAPKKKs in maize and their roles in the response to drought stress. Further biological functional analyses of MAPKKKs and their downstream MAPKK and MAPK targets will help elucidate the mechanisms underlying the response of MAPK cascades to drought in maize.

Supporting Information

S1 File. Supplementary tables. S1 File contains five Supplementary tables: Table A. 71 predictive MAPKKKs in maize. Table B. The quality of the RNA-seq data for Q20 and Q30. Table C. RNA-seq data and mapping rates. Table D. The co-expressed MAPKKKs with DEGs in the enriched pathway. Table E. The common co-expressed MAPKKKs with our DEGs in the enriched pathway.

(DOC)

S2 File. Differently Expressed Genes Responded to drought in leaf, stem and root.

(XLSX)

S3 File. Genes with different regulated pattern in three organs.

(XLSX)

S4 File. Primers for qPCR.

(XLSX)

S5 File. Contains five supplementary figures: Figure A. Conserved domains of MAPKKKs in three sub-families. (a). The conserved motif GTPEFMAPE (L/V) (Y/F) in the ZIK family. (b). The conserved motif G (T/S) Px (W/F) MAPEV in theMEKK family. (c). The conserved motif GTxx (W/Y) MAPE in the Raf family. Figure B. Functional categorization (biological process) of DEGs that are enriched in only one tissue. Blue represents leaf. Red represents stem. Green represents root. The enrichment figure was constructed of significantly enriched GO terms only. The X-axis is the -log10 (P-value), which represents the level of enrichment, and the p-value cutoff was 0.05. Figure C. Venn analysis for the DEGs with up-regulated and downregulated in different organs. Each circle represents an organ with different regulated pattern. Figure D. Differentially expressed transcripts involved in differentmetabolic processes under drought stress. (a, b and c) represent drought-mediated expression changes in differentmetabolic processes in the leaf, stem and root meristem, respectively. The images were obtained using MapMan and show the different functional categories that fulfilled the criteria for differential expression (a P-value less than 0.05 and greater than 2-fold change). Figure E. Differentially expressed transcripts involved in different regulatory processes under drought stress. (a, b and c) represent drought-mediated expression changes in different regulatory processes in the leaf, stem and root meristem, respectively. The images were obtained using MapMan and show the different functional categories that fulfilled the criteria for differential expression (a P-value less than 0.05 and greater than 2-fold change).

(ZIP)

Acknowledgments

We thank Dr Yimiao Tang for his helpful advice on this work. This work was supported by the National Major Program of China (2014ZX0800303B) and the Program of Beijing New Scientific Star (Z111105054511066).

Author Contributions

Conceived and designed the experiments: YL HH JZ. Performed the experiments: MZ FY WR. Analyzed the data: YL MZ ZG HH JZ. Contributed reagents/materials/analysis tools: YL HH JZ. Wrote the paper: YL MZ ZG HH JZ.

References (omitted)

Plos One, 2015, 10 (11): e0143128

Genome-Wide Epigenetic Regulation of Gene Transcription in Maize Seeds

Lu Xiaoduo[1,2]*   Wang Weixuan[2,3]*   Ren Wen[4]*   Chai Zhenguang[1]
Guo Wenzhu[2,3]   Chen Rumei[2,3]   Wang Lei[2,3]   Zhao Jun[2,3]
Lang Zhihong[2,3]   Fan Yunliu[2,3]   Zhao Jiuran[4]**   Zhang Chunyi[2,3]**

(1. School of Life Sciences, Qilu Normal University, Jinan, 250200, China;
2. Department of Crop Genomics & Genetic Improvement, Biotechnology Research Institute,
Chinese Academy of Agricultural Sciences, Beijing 100081, China; 3. National Key Facility
for Crop Gene Resources and Genetic Improvement (NFCRI), Beijing 100081, China;
4. Maize Research Center, Beijing Academy of Agriculture and Forestry
Sciences, Beijing 100097, China)

Abstract

Background

Epigenetic regulation is well recognized for its importance in gene expression in organisms. DNA methylation, an important epigenetic mark, has received enormous attention in recent years as it's a key player in many biological processes. It remains unclear how DNA methylation contributes to gene transcription regulation in maize seeds. Here, we take advantage of recent technologies to examine the genome-wide association of DNA methylation with transcription of four types of DNA sequences, including protein-coding genes, pseudogenes, transposable elements, and repeats in maize embryo and endosperm, respectively.

Results

The methylation in CG, CHG and CHH contexts plays different roles in the control of gene expression. Methylation around the transcription start sites and transcription stop regions of protein-coding genes is negatively correlated, but in gene bodies positively correlated, to gene expression level. The upstream regions of protein-coding genes are enriched with 24nt siRNAs and contain high levels of CHH methylation, which is correlated to gene expression level. The analysis of sequence content with in CG, CHG, or CHH contexts reveals that only CHH methylation is affected by its local sequences, which is different from Arabidopsis.

Conclusions

In summary, we conclude that methylation-regulated transcription varies with the types of

---

* These authors contributed equally to this work
** maizezhao@126.com (Jiuran Zhao); zhangchunyi@caas.cn (CZ)

DNA sequences, sequence contexts or parts of a specific gene in maize seeds and differs from that in other plant species. Our study helps people better understand from a genome wide viewpoint that how transcriptional expression is controlled by DNA methylation, one of the important factors influencing transcription, and how the methylation is associated with small RNAs.

Introduction

Cytosine methylation, an epigenetic marker, is important for transposable element (TE) silencing, gene expression and gene imprinting in vertebrates, flowering plants, and some fungi. Global demethylation of genomic DNA strongly reactivates TE transcription in mammals and plants. Decreased DNA methylation in *Arabidopsis thaliana* leads to retrotransposon mobilization and TE activation and results in the increase of TE copy number. In mammals, DNA methylation patterns are established and maintained by DNA methyltransferase 3 (DNMT3) and methyltransferase DNMT1, respectively. In plants, DOMAINS REARRANGED METHYLTRANSFERASE2 (DRM2), the plant homologue of DNMT3, catalyzes *de novo* methylation; MET1, the plant homologue of DNMT1, maintains CG methylation. CHG methylation is maintained by CHROMOMETHYLASE 3 (CMT3), a plant-specific DNA methyltransferase. *de novo* methylation mechanism by DRM2 is responsible for the maintenance of CHH methylation.

Endogenous small interfering RNAs (siRNAs) are the best characterized small RNAs that defend eukaryotic cells against TE mobilization in plants. siRNAs regulate TE activity primarily through RNA-directed DNA methylation (RdDM). Two plant-specific RNA polymerases, Pol IV and Pol V, are involved in RdDM. Pol IV initiates 24nucleotide (nt) siRNA biogenesis by transcribing long single-stranded RNAs (ssRNAs). RNA-dependent RNA polymerase 2 (RDR2) utilizes the ssRNAs as templates to generate double-stranded RNAs (dsRNAs) which are processed into 24nt siRNAs by DICER-like 3 (DCL3). 24nt siRNAs are loaded into AGO4 which interacts with NUCLEAR RNA POLYMERASE E1 (NRPE1), a Pol V subunit. Pol V functions to produce intergenic noncoding (IGN) transcripts which are essential for DNA methylation and silencing of surrounding loci, but not to produce 24nt siRNAs. A complex comprising the AGO4-siRNAs and a number of other proteins (including DRM2) triggers local DNA methylation.

Maize seeds are not only one of the most important crop materials which provide resource for food, feed, biofuel and raw material for processing, but also an important model organism for fundamental research of genetics and genomics. Maize seed development initiates from double fertilization that two of the pollen sperms fuse with egg cell and central cell to produce embryo and endosperm, respectively. Epigenetic regulation of gene expression is crucial for seed development. Recently, we reported that the epigenetic machinery is probably operating in the early developing maize seed, indicating an important role of epigenetic control in governing maize seed development. To advance our understanding of epigenetic networking in maize seed, highly integrated epigenome maps for 9-DAP (days after pollination) embryo and endosperm of maize B73 are constructed via deep sequencing of the cytosine methylome (methylC-seq), transcriptome (mRNA-seq), and small RNA transcriptome (sRNA-seq). The dataset will aid to understand the epigenetic mechanisms underlying gene expression in the early developing maize seeds.

Results

Bisulfite sequencing of the maize seed genome

To decipher DNA methylation landscapes during early seed development, we isolated genomic DNA from 9-DAP embryo and endosperm of maize inbred line B73, and performed MethylC-seq to identify cytosines that are methylated. MethylC sequencing yielded 433 715 164 and 456 749 505 reads for the embryo and endosperm, respectively (Table A in S1 File). Among those, 165 million reads (38.11%, embryo) and 191 million reads (41.93%, endosperm) were aligned to unique locations of the B73 reference genome. The cytosines (2 936 910 521 from the embryo and 3 523 921 294 from the endosperm aligned to unique positions and covered 33.65% and 35.64% of the total genomic cytosines with average read depths of 9-and 10-fold coverage of each DNA strand, respectively (Table B in S1 File). Like other flowering plants, cytosine methylation occurs in CG, CHG (H is A, C or T) and CHH sequence contexts in both embryo and endosperm of maize. The bulk cytosine methylation frequency was 80.26% for CG, 63.81% for CHG, and 2.51% for CHH in embryo, and 78.40% for CG, 57.60% for CHG and 1.82% for CHH in endosperm (Table A in S1 File), indicating the maize endosperm genome was hypo-methylated compared to the embryo genome (Figure 1; Table c in S1 File). 87% of the CG contexts were methylated, out of which more than 70% were heavily methylated (80%~100%). Similar to CG, over 80% of CHG was methylated in both the embryo and endosperm, the majority of which were heavily methylated (80%~100%), while CHH was markedly less methylated compared to CG (Figure A and Table B in S1 File).

Methylation profiles of 9-DAP maize embryo and endosperm

Overall, the maize endosperm genome was hypomethylated compared to the embryo genome (Figure 1; Table C in S1 File), which is in agreement with previous reports. Higher CG methylation in the embryo compared to endosperm was found mainly in the transcribed regions of protein-coding genes and TEs as well as in repeat regions (Figure 1A, 1D and 1G). However, CHG methylation was slightly higher in the endosperm than the embryo in the middle part of the transcribed region of protein-coding genes (Figure 1B), and significantly higher in the embryo than the endosperm in upstream to downstream repeat regions and TEs (Figure 1E and 1H). CHH methylation was consistently higher in the embryo than the endosperm (Figure 1C, 1F, 1I and 1L). There is no significant difference at CG context between embryo and endosperm, while the methylation level at CHG and CHH context is lower in endosperm than embryo. This pattern is similar to rice, another monocotyledon plant, and different from Arabidopsis, a dicotyledon plant. 87% methylated CGs were observed, among which more than 70% were highly methylated (80%~100%). Unlike CGs, CHHs were either demethylated or hypomethylated both in embryo and endosperm (Figure A in S1 File).

To further identify sequences that are differentially methylated in the embryo compared with the endosperm, differential methylation region (DMR) analysis was conducted. We calculated fractional methylation in each context within 50 base pair (bp) windows and subtracted endosperm methylation from embryo methylation. The results showed that DNA methylation differences between the embryo and endosperm varied at genomic loci subsets (Figure B and Table C in S1 File). 421 137 and 415 490 discreet DMRs corresponding to 24 341 600 and 24 041 950bp in CG methylation were identified in sense and antisense strand, respectively. 285 017 (67.68%) and 281 796 (67.82%) of those DMRs were highly methylated in embryo in sense and antisense strand, respectively (Table D in S1 File). In CHG context, 738 334 (47 402 500bp) and 736 262 (47 335 050bp) loci were more methylated in sense and antisense strand, respectively. About 78% (580 449 loci in sense strand and 578 949 loci in antisense strand) of these DMRs were more methylated in embryo. We also found 577 714 (31 658 700bp) and 577 009 (31 664 700) loci with change in CHH methylation in sense and antisense strand, respectively. 63.1% (364 486 loci in sense strand and 364 239 loci in antisense strand) of the loci were more methylated in embryo (Table D in S1 File). Notably, around 22% and 37% of identified loci were hypermethylated at CHG and CHH, respectively, in endosperm. Surprisingly, about one third of identified loci were hypermethylated at CG in endosperm, which is much higher than that in Arabidopsis.

Higher CG methylation inthe embryo compared to endosperm was found mainly in the transcribed regions of protein-coding genes and TEs, as well as in repeat regions (Figure 1A, 1D and 1G). However, CHG methylation was slightly higher in the endosperm than the embryo in the middle part of the transcribed region of protein-coding genes (Figure 1B), and significantly higher in the embryo than the endosperm in upstream to downstream repeat regions and TEs (Figure 1E and 1H). CHH methylation was consistently higher in the embryo than the endosperm (Figure 1C, 1F, 1I and 1L).

CG, CHG and CHH methylation were lowest from 600bp downstream of the transcription start site (TSS) to 700bp within the transcript, and the same pattern was also present at the 3' end of genes (Figure 1A, 1B and 1C), which differs from rice, *Arabidopsis*, and human. CG and CHG methylation patterns were somewhat similar between repeats and TEs (Figure 1D, 1E, 1G and 1H), while CHH methylation differed significantly (Figure 1F and 1I). Interestingly, the CG and CHG methylation patterns in the transcribed regions of pseudogenes were similar to those of protein-coding genes, but the methylation level of pseudogenes was significantly higher than that of protein-coding genes (40%~80% in pseudogenes *vs.* 20%~60% in protein-coding genes for CG; 20%~60% in pseudogenes *vs.* 10%~30% in protein-coding genes for CHG; Figure 1A, 1B, 1J and 1K), suggesting a correlation between enhanced methylation and pseudogene inactivation.

We observed that CHH methylation pattern differed from CG or CHG. Both CG and CHG were increasingly methylated from the 5'end inwards and decreasingly methylated towards the 3' end in protein-coding genes and pseudogenes (Figure 1A, 1B, 1J and 1K); CG and CHG were evenly methylated in repeat regions (Figure 1D and 1E), but less evenly methylated in

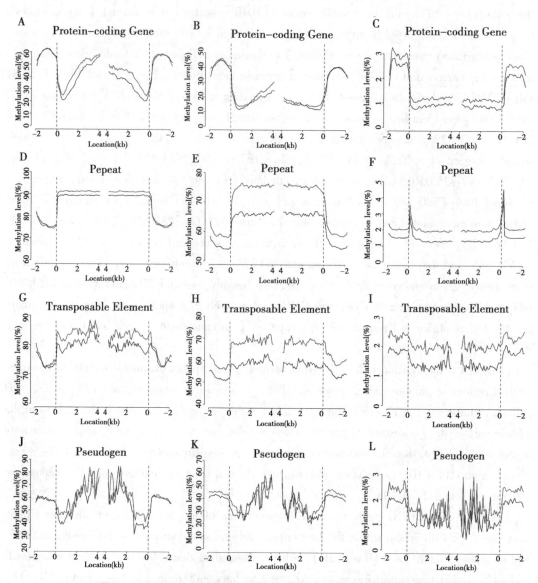

Figure 1　DNA methylation profiles in the embryo and endosperm

(A to L) Maize B73 RefGen_ V2-annotated protein-coding genes [(A), (B), and (C)], repeat regions [(D), (E), and (F)], transposable elements [(G), (H), and (I)] and pseudogenes [(J), (K), and (L)] are aligned at the 5′ end (left) or the 3′end (right), and average methylation levels for each 50nt interval are plotted from 2 kb away from the gene (negative numbers) to 4 kb into the gene (positive numbers). Embryo methylation is represented by the red trace and endosperm by the blue trace. The dashed line at zero represents the point of alignment. CG methylation is shown in (A), (D), (G), (J), CHG in (B), (E), (H), (K), and CHH in (C), (F), (I) and (L)

transcribed regions of TEs (Figure 1G and 1H). In contrast, CHG methylation was almost absent in transcribed regions in Arabidopsis and rice. Unlike CG or CHG, CHH was methylated at the lowest frequencies in the transcribed regions of protein-coding genes and TEs as well as in repeat regions compared to other regions of the genes (Figure 1C, 1F and 1I), peaking at the two ends of repeats (Figure 1F).

Local sequence effects on DNA methylation

To explore the local sequence effects on DNA methylation, the upstream two nucleotides and downstream four nucleotides surrounding cytosines were assessed in terms of their effects on cytosine methylation (Figure 2; S1 Table). Strong effects were found in a CHH context. A cytosine immediately followed by another cytosine was less likely to be methylated than a cytosine neighboring a thymidine or adenine; in contrast, a cytosine immediately followed by an adenine was more likely to be methylated (Figure 2C). This was clearly demonstrated by the observation that CAH sites were methylated at a level twofold higher than CCH sites in both the embryo and the endosperm (Figure 2C). As opposed to the slightly repressive effect of cytosines at positions +1, +2 or +3, adenosines at the 3' end of the CHH context were associated with an increase in cytosine methylation frequency. This effect was strongest at the +2 positions where a CHA was methylated twofold more frequently than CHC or CHT (Figure 2). The sequence effect in the CHH context on DNA methylation was also observed in the endosperm (Figure C in S1 File; S1 Table), and was conserved between maize and *Arabidopsis*. However, only minor effects were observed for CHG or CG context, which is different from Arabidopsis.

The association of small RNAs with DNA methylation

Previously it was demonstrated that a subset of small RNAs (sRNAs) pool targets DNA methylation through RdDM, an essential process for the establishment of DNA methylation and its maintenance in asymmetric contexts. To characterize the relationship between sRNAs and genome methylation in maize seed, we first performed deep sequencing of sRNAs from the embryo and endosperm, respectively, and then investigated the correlation between sRNA production and DNA methylation. We found that 24nt sRNAs were significantly more abundant in the upstream and downstream regions of genes in the embryo than in the endosperm (Figure 3); in contrast, 21-, 22-or 23-nt sRNAs were produced at higher levels in the endosperm than in the embryo (Figure D-F in S1 File). A significant positive correlation between CHH methylation and 24nt sRNA accumulation was found mainly in the upstream region of protein-coding genes and pseudogenes (Figure 3E and 3H) and in the two ends of repeats (Figure 3F), but we did not observe any correlation between CG/CHG methylation and 24nt sRNA production (Figure 3A-3D). Similar relationships were also observed for 21nt, 22nt, or 23nt sRNAs (Figure D-I in S1 File), suggesting that the functions of those sRNAs may differ from those of 24nt sRNAs.

siRNA-regulated gene expression in maize seed

siRNAs regulate gene expression through directing DNA methylation or degrading mRNAs. In maize outer layer of mature ear prior to fertilization, the 24nt siRNAs accumulated at gene ends. In our dataset, all of the sRNAs ranging from 21nt to 24nt in length accumulated predominantly at the ends of protein-coding genes and in the upstream or downstream regions of TEs and pseudogenes both in embryo and endosperm (Figure 4; Figure D-J in S1 File).

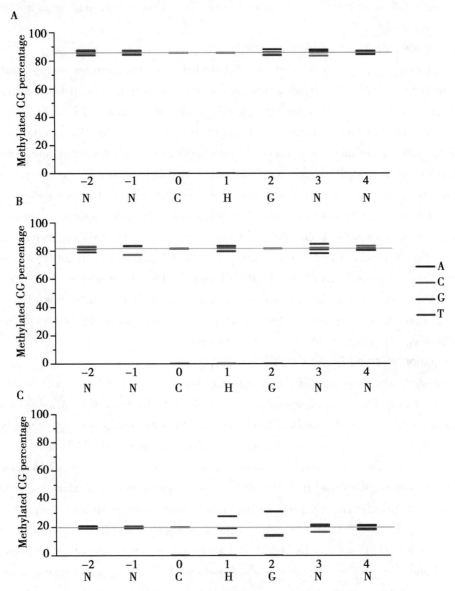

Figure 2 Local sequence effects on DNA methylation in the embryo

Sequence contexts that are preferentially methylated in the embryo for 7-mer sequences, in which the methylated cytosine is in the third position. (A), CG context in the embryo; (B), CHG context in the embryo; (C), CHH context in the embryo. The y axis indicates the methylation level and the x axis indicates the base composition and position

We asked whether sRNAs production is associated with gene expression. The protein-coding genes and pseudogenes and TEs were grouped into five levels by expression (see "Material and Method"), and a genome-wide association of sRNA accumulation with gene expression was performed in both the embryo and endosperm (Figure 4; Figure J in S1 File). In the transcripts of protein-coding genes, significant accumulation of 21~24nt sRNAs was detected mainly in

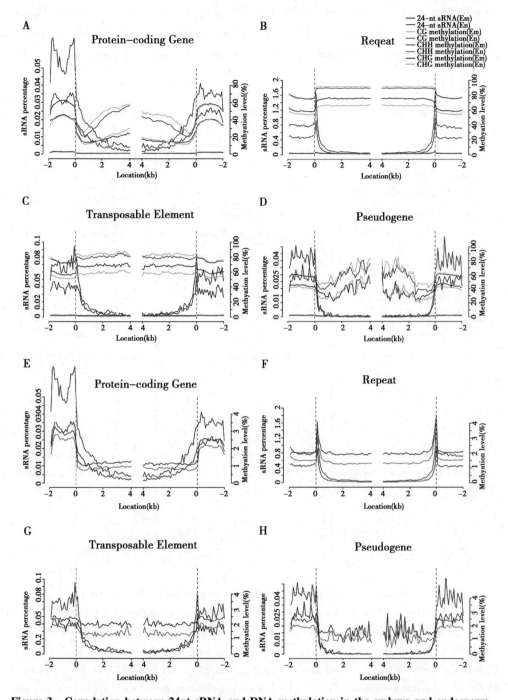

Figure 3　Correlation between 24nt sRNA and DNA methylation in the embryo and endosperm

（A to D）Correlations between 24nt sRNA and DNA methylation in protein-coding genes（A）, repeats（B）, transposable elements（C）and pseudogenes（D）.（E to H）Correlations between 24nt sRNA and CHH methylation in protein-coding genes（A）, repeats（B）, transposable elements（C）and pseudogenes（D）. The dashed line at zero represents the point of alignment genes with high levels of expression (RPKM > 100; Figure 4A, 4D, 4G and 4J; Figure J in S1 File). However, in TEs and pseudogenes, high accumulation of sRNAs was detected mainly in genes with low expression (Figure 4B, 4C, 4E, 4F, 4H and 4K; Figure J in S1 File).

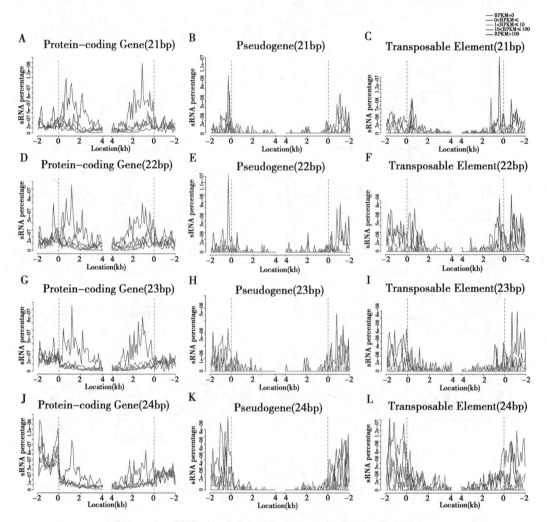

Figure 4 sRNA associated with gene expression in the embryo

(A to L) show 21~24nt sRNAs that regulate protein-coding gene expression [(A), (D), (G) and (J)], pseudogene gene expression [(B), (E), (H), and (K)] and transposable element activity [(C), (F), (I), and (L)]; 21nt sRNAs are shown in (A), (B), and (C), 22nt sRNA in (D), (E), and (F), 23nt sRNA in (G), (H), and (I), and 24nt sRNA in (J), (K), and (L). Gene expression was classified into five levels according to the number of reads per kilobase per million reads (RPKM, see Materials and Methods), and the correlation between sRNA accumulation and gene expression was investigated. The dashed line at zero represents the point of alignment. Note that there are only 13 pseudogenes and 20 TEs whose RPKM value is higher than 100 in the embryo, meaning that the sample size was too small to be statistically significant

The association of DNA methylation with gene expression

Cytosine methylation plays important roles in regulating gene expression and TE silencing in plants and animals. To understand the relationship between cytosine methylation and gene expression in maize seeds, we evaluated correlations of mRNA-seq data with methylC-seq data (see "Materials and Methods"). The effects of methylation on gene expression were sequence context- or gene-dependent. We observed that CG, CHG, or CHH methylation in the upstream or downstream of genes was not correlated with protein-coding gene expression (Figure 5A, D, G); however, CG methylation in transcribed regions seemed to be positively correlated with the expression level, whereas CHG methylation negatively correlated, suggesting an opposite role in gene expression regulation between CG and CHG methylation. Interestingly, protein-coding gene expression varied inversely with CG, CHG, or CHH methylation around the TSS or TTS (Figure 5A, D, G). For example, it's evident that genes with highest abundance of transcripts (RPKM > 100) at TSS or TTS had lowest CHG methylation level; in contrast, genes with lowest abundance of transcripts (RPKM = 0) had highest CHG methylation level (Figure 5D). Another interesting observation was the presence of two CHH islands, which exhibited high density of CHH methylation, within 2kb upstream of protein-coding genes, and CHH methylation in the TSS-proximal CHH island was positively correlated with transcription (Figure 5G). In addition, the correlation between methylation at TSS and TTS regions with transcription was also observed in pseudogenes, albeit it was not as high as that in protein-coding genes (Figure 5B, E).

TEs were opposite to protein-coding genes regarding the effects of CG methylation on gene expression, as demonstrated by the observation that TEs with low expression showed high levels of CG methylation evenly across entire regions from upstream to downstream (RPKM < 1; Figure 5C; Figure 6C). Similar effects of CHG or CHH were also observed for the TEs with low expression level (RPKM < 1; Figure 5F and 5I; Figure 6F and 6I). These observations indicated that expression of protein-coding genes and TEs may be differentially regulated by DNA methylation. In addition, high level of DNA methylation within pseudogenes at CHG or CHH context led to low expression (Figure 5E and 5H; Figure 6E and 6H).

Discussion

In this study, we used next-generation sequencing technology to identify single-base DNA methylome, transcriptome and smRNAome in maize seed at early developing stage. High throughput analysis of these data deciphered a complex landscape of gene expression profiling regulated by cytosine methylation and sRNAs.

DNA methylation, an epigenetic modification, has been found in diverse eukaryotic organisms and plays a key role in embryogenesis, genomic imprinting, and tumorigenesis in mammals, and in transposon silencing and gene regulation in plants. The single-base cytosine methylation maps of some organisms, including human, Arabidopsis, rice and silkworm have been reported. During the preparation of the manuscript, single-base DNA methylation

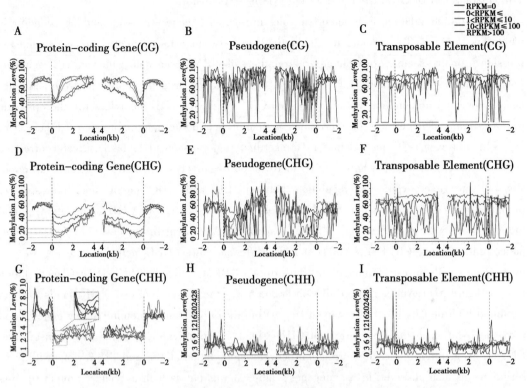

Figure 5 DNA methylation associated with gene expression in the embryo

(A to I) The relationships between DNA methylation and protein-coding gene expression [(A), (D), and (G)], pseudogene gene expression [(B), (D), and (H)] and transposable element activity [(C), (F), and (I)]; CG methylation is shown in (A), (B), and (C), CHG in (D), (E), and (F), and CHH in (G), (H), and (I). The dashed line at zero represents the point of alignment. As shown in Figure 4, the number of pseudogenes and TEs with RPKM > 100 are only 13 and 20 in the embryo, respectively

sequencing of the outer layer of mature maize ears prior to fertilization was reported. We sequenced the 9-DAP maize embryo and endosperm DNA methylome using the bisulfite-based whole-genome sequencing. Like rice and Arabidopsis, the maize endosperm is hypomethylated compared to the embryo, and the CG methylation pattern is highly similar amongst the three plant species. However, some drastic differences in CHG and CHH methylation between the different species were observed. For example, both rice and Arabidopsis gene bodies contained almost exclusively CG methylation, whereas maize contained not only CG but also CHG methylation, implying that maize genome may have evolved a more complex regulatory mechanism underlying protein-coding gene expression than rice and Arabidopsis. The single-base resolution of bisulfite-Seq technology allows determination of the precise boundaries between methylated and unmethylated regions. For example, we observed that the boundary between repeats and flanking DNA showed an apparent peak of CHH methylation, which was not detected in other DNA sequences including protein-coding genes, pseudogenes and TEs (Figure 1F). This apparent peaking methylation was correlated with sRNA accumulation (Figure 3F; Figure G-I in S1 File), suggesting that the CHH methylation in the boundary regions is probably regulated by sRNAs through the RdDM pathway.

DNA methylation repressed gene expression by blocking transcription factors binding to the promoters. High methylation levels of promoters are correlated with low or no transcription. However, this was not found in the maize seeds: the transcript abundance in protein-coding genes was not correlated with the DNA methylation of the promoters. Early days of DNA methylation research on human revealed that transcribed genes are featured with gene body methylation. Thereafter, extensive studies have illustrated positive correlations between transcription and gene body methylation in plant and animal genomes. We found that CG and CHG as well as CHH methylation within gene body all influenced transcription: high level of CG methylation or low level of CHG or CHH methylation was corresponding to active transcription (Figure 5A and 5D; Figure 6A and 6D), indicating that CG methylation of gene body may stimulate transcription elongation, whereas CHG/CHH methylation of gene body may block transcription elongation. Rice methylation patterns closely resemble those of Arabidopsis in many salient features: modestly expressed genes are most likely to be methylated. In contrast, inactive genes exhibited high levels of CHG/CHH methylation in maize (Figure 5D and 5G). Previously it was also reported that in cancer cells there existed inverse relationship between methylation of non-CG islands and expression genome wide. Taking into account all the studies performed in variety of eukaryotes, it can be concluded that gene body methylation other than promoter methylation is an ancient property of the genomes, and transcription elongation seems to be under opposite control by CG and CHG/CHH methylation, respectively, in maize seeds.

It is demonstrated that CG methylation around the TSS and TTS negatively affects gene expression in rice. In this study, we found that not only CG methylation but also CHG/CHH methylation of the TSS-or TTS-proximal regions were inversely correlated with gene expression (Figure 5; Figure 6). This suggests that lack of methylation in both TSS and TTS is important for gene expression, and it's likely that the epigenetic mechanisms underlying gene expression are more complex in maize than in rice. Previous studies reported that methylated CG islands at TSSs cannot initiate transcription after the DNA has been assembled into nucleosomes which are the substrates for *de novo* methylation. It can also be concluded that in maize both transcription initiation and transcription termination seem to be sensitive to DNA methylation silencing. In addition, functioning of CG methylation varies with the position where CG methylation occurs: within gene body CG methylation may play a stimulating role in the regulation of gene expression, and methylation at TSS or TTS CG may negatively influence gene expression. Very recently it was reported that major classes of transposons close to cellular genes exhibited a peak of CHH methylation in maize, which was named CHH islands. Likewise, two peaks of CHH methylation were found in the 9-DAP maize seeds in this study (Figure 5G). The difference in the number of CHH islands may reflect the dynamics of CHH methylation in different tissues or different development stage. We ob-

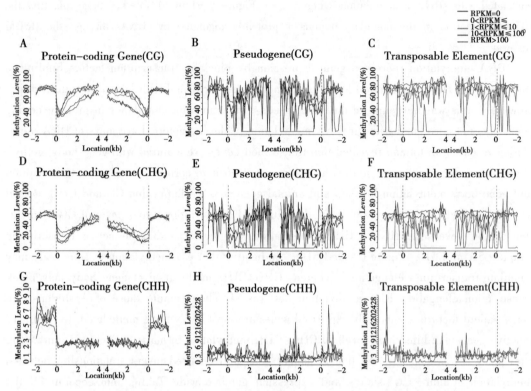

Figure 6 DNA methylation associated with gene expression in the endosperm

(A to I) show the relationships between DNA methylation and proteincoding gene expression [(A), (D), and (G)], pseudogene gene expression [(B), (D), and (H)] and transposable element activity [(C), (F), and (I)]; CG methylation is shown in (A), (B), and (C), CHG in (D), (E), and (F), and CHH in (G), (H), and (I). The dashed line at zero represents the point of alignment. As shown in Figure J in S1 File, the number of pseudogenes and TEs with RPKM>100 are only 14 and 15 in the embryo, respectively

served a positive correlation between methylation in CHH islands and gene transcription, indicating that genes with high transcription intend to confer high CHH methylation to the intergenic regions close to genes.

It is estimated that the maize genome contain more than 10 000 pseudogenes. Three classes of pseudogenes have been identified: nonprocessed pseudogenes derived from gene duplication, processed pseudogenes originated from retrotransposition, and unitary pseudogenes arising from spontaneous mutations in protein-coding genes. Since pseudogenes are generally noncoding, they are considered "junk DNA". Even though recent studies indicate that the pseudogenes have diverse functions to regulate their parental gene expression or unrelated genes, the regulatory mechanism of pseudogenes remains unclear. Nevertheless, the methylation level of pseudogenes was higher than genes and repeats in Arabidopsis. Our data also demonstrated that the level of methylation in pseudogenes was higher than that in protein-coding genes in maize seeds (Figure 1), suggesting a possible link between the enhanced DNA methylation and loss of protein coding. Moreover, the observation that the methylation profiles of protein-coding genes and pseudogenes are similar in shape is suggestive of a common epigenetic mechanism governing the transcription of protein-coding genes and pseudogenes (Figure 1). How the DNA methylation and sRNAs interact to regulate pseudogene expression needs to be further elucidated.

siRNAs cause RNA-directed DNA methylation. Previous studies demonstrated that only a fraction of the siRNA clusters (i.e. endogenous loci corresponding to high local concentrations of siRNAs) are heavily methylated in Arabidopsis, suggesting that a large amount of DNA methylation is maintained without persistent targeting by siRNAs. In this study we found that there were no correlations between CG/CHG methylation and sRNA accumulation, but in some positions (i.e. upstream regions of protein-coding genes and pseudogenes) higher accumulation of 24nt sRNAs corresponded to denser methylation (Figure 3). This is consistent with the previous reports. We also observed that although higher abundance of sRNAs, was present in the upstream region of TEs as compared to TEs themselves, but CHH methylation occurred evenly from upstream to downstream. This may be due to the fact that TEs were not grouped for the correlation analysis by their proximity to cellular genes. Small RNAs and DNA methylation interacted to induce the silencing of TEs. In maize embryo and endosperm, high level of DNA methylation corresponded to the low TE expression (Figure 5; Figure J in S1 File), and high level of sRNA accumulation in the upstream or downstream of TEs corresponded to l (Figure 4; Figure I in S1 File). However, the DNA methylation level is not consistent with the sRNA accumulation (Figure 3; Figure C-H in S1 File). These results indicate that sRNA and DNA methylation may repress TE expression through different mechanisms in maize.

In summary, we suggest that maize has evolved more complex epigenetic machinery than rice and Arabidopsis, and different DNA context methylation has different role in gene expression regulation. Moreover, the mode of methylation-regulated gene expression varies with gene type, sequence context or position of a given gene.

Materials and Methods

Plant material

The maize inbred line B73 was grown in the field during the summer of 2009 in Langfang, Hebei province, China. Ears were bagged before silk emergence. Each set of inbred kernels were generated on the same day by self-pollination. At 9 days after pollination (DAP), the endosperm and embryo were isolated using tweezers and collected in 300mM sorbitol solution with 5mM MES (pH 5.7) from the ovules, and were then transferred into tubes, snap-frozen in liquid nitrogen and stored at -80℃ for further use.

MethylC-Seq library generation

Genomic DNA (10μg) was extracted from the embryo and endosperm using the DNeasy Mini Kit (Qiagen). The DNA was fragmented by sonication to 280~350nt with a Bioruptor (Diagenode). The DNA was end-repaired using a mixture of T4 DNA polymerase, Klenow DNA polymerase and T4 PNK (Enzymatics), and a 3' overhang A was added using the Klenow exo-enzyme (Enzymatics). The resultant fragments were ligated with the Illumina methylation adapters by DNA T4 ligase (Enzymatics) according to the Illumina protocol. Adapter-linked DNA fragments were bisulfated using the EZ DNA Methylation Kit (Zymo), as per the manufacturer's protocol. The treated DNA was amplified by PCR for 11 cycles. The DNA fragments were purified, quantified and then sequenced for 100 cycles using the Illumina protocol.

RNA-Seq library generation

Total RNA (10μg) from each sample was extracted using RNeasy Mini Kit (Qiagen), according to the manufacturer's protocol. mRNA was isolated from total RNA using 7μl of oligo dT on Sera-magnetic beads and 50μl of binding buffer. mRNA was fragmented by metal hydrolysis in RNA fragment buffer (Ambion) for 2min at 70℃. The reaction was stopped by adding 2μl of fragmentation stop solution (Ambion). The fragmented RNA was converted to double-stranded cDNA. After polishing the ends of the cDNA, an adenine base was added at the 3' ends, after which Illumina multiplex adaptors were ligated. The ligated DNA was separated on 2% agarose gel and 300nt targeted DNA was extracted. DNA was purified from the gel using the Qiagen Gel extraction kit. The purified DNA was amplified by 15 cycles of PCR, and the PCR DNA was then purified on the Qiagen PCR purification kit to obtain the final seq library for sequencing. The DNA concentration of the seq library was determined on Qubit (Invitrogen).

sRNA library generation

Total RNA (10μg) from each sample was extracted using the RNeasy Mini Kit (Qiagen) according to the manufacturer's protocol. Novex 15% TBE-Urea gel (Invitrogen) was used to isolate small RNA fragments (30nt in length) from total RNA. The purified small RNAs were ligated to a 5' adaptor (Illumina) and the ligation products were purified in Novex 15% TBE-Urea gels. Next, a 3' adaptor (Illumina) was ligated to the 5' ligation products and further purified in a Novex 10% TBE-Urea gel (Invitrogen). Reverse transcriptase PCR was used to reverse transcribe these ligation products. Then, a 6% TBE-Urea gel (Invitrogen) was used to purify the amplification products. The DNA fragments were purified, quantified and then sequenced for 36 cycles using the protocol provided by Illumina.

High-throughput sequencing

MethylC-Seq, RNA-Seq and sRNA-seq libraries were sequenced using the Illumina HiSeq 2000, as per the manufacturer's protocol. The paired-end protocol was used for RNA-Seq sequencing, while the single ends sequencing dataset was used for MethylC-Seq sequencing. Read lengths of RNA-seq and MethylC-Seq were up to 100nt. Image analysis and base calling were performed with the standard Illumina pipeline.

MethylC-Seq analysis

The raw data in FastQ format produced by the Illumina pipeline were first pre-processed, including: a) Filtering of low quality reads and b) trimming reads to before the first occurrence of a low-quality base (quality score < 20). Remaining short sequences were mapped to the maize reference genome (RefGen ZmB73 Release 5b) using Bismark version 0.4.1, allowing up to four mismatches per read. Only uniquely aligning reads were retained for the next procedure. Three types of methylation calls (CG, CHG, CHH), which were covered by at least 10 reads excluding any duplication, were extracted. For each sequence context, bulk fractional methylation were calculated using the formula #C/(#C+#T). Fractional methylation within a 50nt sliding window was also calculated to identify the differential methylation region (DMR) between the maize endosperm and embryo. The upstream two nucleotides and downstream four nucleotides surrounding cytosines were analyzed to determine whether they have local sequence effects on DNA methylation of the CG, CHG, and CHH contexts. The annotations of genes, repeat regions, transposable elements and pseudogene regions were retrieved from the B73 filter gene set (release 5b).

RNA-Seq analysis

RNA-seq datasets were aligned to the maize reference genome using tophat. The resulting alignment files were subjected to Cufflinks to generate a transcriptome assembly and make the annotation. Reads per kilobase of transcript per million reads (RPKM) were calculated. Five ranges of RPKM values representing different expression levels were collected and associated with DNA methylation and sRNA accumulation.

sRNA-Seq analysis

Read sequences produced by the Illumina analysis pipeline were mapped to the maize reference sequence using bwa. Up to two mismatches were allowed in the alignment. Information from the B73 filter gene set release 5b was used to make the annotation. sRNAs were then separated according to length (21~24nt) to identify the accumulation at different regions. sRNAs of specific lengths were normalized (divided by the total number of sRNAs), and the sRNA percentage (2kb distal from to 4kb into the gene) for each 100nt interval was calculated.

Sequence Data

The data for this article have beendeposited at the National Center for Biotechnology Information under accession number SRP056646.

Supporting Information

S1 File. Figure A, Distribution of the percentage methylation in the CG, CHG and CHH contexts. The y axis indicates the fraction of the total methylcytosines that display each percentage of methylation (x axis), defined as the fraction of reads at a reference cytosine containing cytosines following bisulfite conversion. Fractions were calculated within bins of 20%, as indicated on the x axis. Figure B, DMR distributions of repeats, transposable elements, pseudogenes, and protein-coding genes in the embryo and endosperm. (A to C) DMR distributions in repeats of type I transposons (LTR and LINE) and type II transposons (TIR). (D to F) represent the DMR distributions in TEs, pseudogenes and proteincoding genes, respectively. Figure C, Local sequence effects on DNA methylation in the endosperm. Sequence contexts that are preferentially methylated in the endosperm for 7-mer sequences, in which the methylated cytosine is in the third position. (A), CG context; (B), CHG context; (C), CHH context. The y axis indicates the methylation level, and the x axis indicates the base composition and position. Figure D, Correlation between 21nt sRNA and DNA methylation. (A to D) indicate the correlations between 21nt sRNA and DNA methylation in protein-coding genes (A), repeats (B), TEs (C) and pseudogenes (D). The dashed line at zero represents the point of alignment. Figure E, Correlation between 22nt sRNA and DNA methylation. (A to D) indicate the correlations between 22nt sRNA and DNA methylation in protein-coding genes (A), repeats (B), TEs (C) and pseudogenes (D). The dashed line at zero represents the point of alignment. Figure F, Correlation between 23nt sRNA and DNA methylation. (A to D) indicate the correlations between 23nt sRNA and DNA methylation in protein-coding genes (A), repeats (B), TEs (C) and pseudogenes (D). The dashed line at zero represents the point of alignment. Figure G, Correlation between 21nt sRNA and CHH methylation. (A to D) indicate the correlations between 21nt sRNA and CHH methylation in protein-coding genes (A), repeats (B), TEs (C) and pseudogenes (D). The dashed line at zero represents the point of alignment. Figure H, Correlation between 22nt sRNA and CHH methylation. (A to D) indicate the correlations between 22nt sRNA and CHH methylation in protein-coding genes (A), repeats (B), TEs (C) and pseudogenes (D). The dashed line at zero represents the point of alignment. Figure I, Correlation between 23nt sRNA and CHH methylation. (A to D) indicate the correlations between 23nt sRNA and CHH methylation in protein-coding genes (A), repeats (B), TEs (C) and pseudogenes (D). The dashed line at zero represents the point of alignment. Figure J, sRNA associated with gene expression in the endosperm. (A to L) show that 21~24nt sRNAs regulate protein-coding gene expression [(A), (D), (G) and (J)], pseudogene gene expression [(B), (E), (H), and (K)] and TE activity [(C), (F), (I), and (L)]; 21nt sRNAs are shown in (A), (B), and (C), 22nt sRNAs in (D), (E), and (F), 23nt sRNAs in (G), (H), and (I), and 24nt sRNAs in (J), (K), and (L). The dashed line at zero represents the point of alignment. Note that there are only 14 pseudogenes and 15 TEs whose RPKM value is above 100 in the embryo, meaning that the sample size was too small to be statistically significant. Figure K, DNA methylation patterns of Fie1 and floury-1. (A) DNA methylation pattern of Fie1. (B) DNA methylation pattern of floury-1. Table A, Statistics of DNA methylation in embryo and endosperm. Table B, Methylation fraction distribution in embryo and endosperm. Table C, DMR distribution in Embryo and Endosperm. Table D, Statistics of DMR between embryo and endosperm. Table E, Gene expression in embryo and endosperm.

(PDF)

S1 Table. Sequence preferences for methylation.

(XLS)

Acknowledgements

The authors would like to thank Berry Genomics, in particular Feng Tian and Jun Wang, for their help with data processing and figure drawing.

Author Contributions

Conceived and designed the experiments: CZ Jiruan Zhao YF. Performed the experiments: XL WRZC WG. Analyzed the data: WW. Contributed reagents/materials/analysis tools: RCLW ZL. Wrote the paper: CZ XL Jun Zhao.

References (omitted)

Plos One, 2015, 10 (10): e0139582

Development of MaizeSNP3072, a High-throughput Compatible SNP Array, for DNA Fingerprinting Identification of Chinese Maize Varieties

Tian Hongli Wang Fengge Zhao Jiuran Yi Hongmei
Wang Lu Wang Rui Yang Yang Song Wei

*(Maize Research Center, Beijing Academy of Agriculture & Forestry Sciences,
Beijing Key Laboratory of Maize DNA Fingerprinting and Molecular Breeding,
Shuguang Garden Middle Road No. 9, Beijing 100097, China)*

Abstract: Single nucleotide polymorphisms (SNPs) are abundant and evenly distributed throughout the maize (*Zea mays* L.) genome. SNPs have several advantages over simple sequence repeats (SSRs), such as ease of data comparison and integration, high-throughput processing of loci, and identification of associated phenotypes. SNPs are thus ideal for DNA fingerprinting, genetic diversity analysis, and marker-assisted breeding. Here, we developed a high-throughput and compatible SNP array, maizeSNP3072, containing 3 072 SNPs developed from the maizeSNP50 array. To improve genotyping efficiency, a high-quality cluster file, maizeSNP3072_GT.egt, was constructed. All 3 072 SNP loci were localized within different genes, where they were distributed in exons (43%), promoters (21%), 3′ untranslated regions (UTRs; 22%), 5′ UTRs (9%), and introns (5%). The average genotyping failure rate using these SNPs was only 6%, or 3% using the cluster file to call genotypes. The genotype consistency of repeat sample analysis on Illumina GoldenGate vs. Infinium platforms exceeded 96.4%. The minor allele frequency (MAF) of the SNPs averaged 0.37 based on data from 309 inbred lines. The 3 072 SNPs were highly effective for distinguishing among 276 examined hybrids. Comparative analysis using Chinese varieties revealed that the 3072SNP array showed a better marker success rate and higher average MAF values, evaluation scores, and variety-distinguishing efficiency than the maizeSNP50K array. The maizeSNP3072 array thus can be successfully used in DNA fingerprinting identification of Chinese maize varieties and shows potential as a useful tool for germplasm resource evaluation and molecular marker-assisted breeding.

Key words: Maize variety; Variety identification; Single nucleotide polymorphism (SNP); DNA fingerprinting; Molecular breeding

Introduction

Maize (*Zea mays* L.) is an important crop widely grown throughout the world for food, feed, and fuel production. In China, maize is the top-ranked crop in terms of cultivated area and total yield and plays a key role in the country's agricultural economic structure. As a hybrid crop,

---

Tian Hongli and Wang Fengge have contributed equally to this work, e-mail: maizezhao@126.com maize is additionally a model plant species for genetic studies because of its high recombination rate and rich genetic diversity. Since the 1980s, the number of maize varieties in China has steadily increased. As of 2013, the total number of varieties approved by national and provincial governments is 6 291, including 503 approved by the national government. Because thousands of maize samples are inspected annually in China (Yang et al, 2014), the large number of maize varieties complicates variety management. At the same time, convergence of breeding resources and patterns has resulted in a gradual narrowing of the maize germplasm genetic base, causing a negative effect on breeding and seed production. The identification of new and existing varieties has consequently become challenging. Traditional field identification and protein electrophoresis cannot meet the need for rapid and accurate identification because of their limited ability to distinguish varieties and the long turnaround time required. DNA fingerprinting technology has become an important approach for distinguishing maize varieties, as this technique is rapid, accurate, and independent of the environment.

Over the past two decades, several different DNA marker technologies, including those based on restriction fragment length polymorphisms, inter-simple sequence repeats (ISSRs), amplified fragment length polymorphisms, simple sequence repeats (SSRs), and single nucleotide polymorphisms (SNPs), have been widely applied in research areas such as DNA fingerprinting of varieties, genetic diversity analysis, association studies, and molecular marker-assisted breeding (Nandakumar et al, 2004, Coombs et al, 2004, Wang et al, 2011a; Barcaccia et al, 2003, Garcia et al, 2004, Clerc et al, 2005, Lu et al, 2009, Semagn et al, 2012; Khampila et al, 2008, Weng et al, 2011, Lu et al, 2011; Chen et al, 2011, Chai et al, 2012, Thomson et al, 2012). DNA fingerprinting refers to the identification of the different compositions, orders, and lengths of DNA sequences among varieties, which, like human fingerprints, are specific. Compared with other markers, SSR and SNP markers have the advantages of codominant inheritance and known chromosomal location, and can be used in high-throughput analyses. Consequently, the International Union for the Protection of New Varieties of Plants (UPOV, 2010, 2011) recommends SSRs and SNPs as preferred markers for DNA fingerprinting and database construction.

SSR markers have been used for variety identification for more than 10 years because of their high discriminatory power and associated relatively well-developed experimental techniques, which are easily performed without expensive instrumentation. With the development of new technologies and increased identification requirements, however, SSR markers have been shown to possess some disadvantages. For example, the throughput of locus processing cannot be easily increased, and data comparison and integration between different detection platforms is difficult. Compared with SSRs, SNPs offer several advantages. They display a higher and more even distribution density in the genome; in maize, for example, SNP loci occur every 44~75bp (Gore et al, 2009), whereas SSR loci are present approximately every 8kbp (Wang et al, 1994). In addition, SNPs are bi-allelic, making them easy to read, compare, and integrate between different data sources, and they are facilitate high-throughput processing. Finally, SNP loci are more likely to be distributed within a gene region associated with a phenotype. With the recent publication of whole genome sequences of crops such as maize and rice (Huang et al, 2009; Schnable et al, 2009; Lai et al, 2010; Jiao et al, 2012), numerous SNP loci have been developed (Gore et al, 2009; Chia et al, 2012; Chen et al, 2011) and various SNP genotyping platforms have been introduced. SNPs have become the most promising markers for DNA fingerprinting and database construction. A variety of SNP genotyping platforms are available, such as the high-throughput GoldenGate (Fan et al, 2003) and Infinium platforms (Steemers and Gunderson 2007), TaqMan by Life Technologies (Livak et al, 1995), and the KASPar platform (KBiosciences' Competitive Allele-Specific PCR system). When a relatively small number (i. e., dozens) of SNP loci are to be assayed, a relatively flexible platform such as TaqMan, Sequenom, or KASPar is recommended; when more than 100 loci are available, the high-throughput GoldenGate or higher-throughput Infinium platform should be chosen. For DNA fingerprinting and database construction, platforms based on chip technology, such as GoldenGate and Infinium, are appropriate choices.

Numerous studies have focused on the development, evaluation, and application of SNP loci in maize. Many SNP markers have been developed by whole genome or transcriptome sequencing (Jones et al, 2009; Gore et al, 2009; Lai et al, 2010; Mammadov et al, 2010; Jiao et al, 2012; Chia et al, 2012). In addition, Infinium and GoldenGate platforms have been used to develop a variety of SNP arrays, such as the high-density SNP array maizeSNP50 (Ganal et al, 2011) that has been successfully applied to genome-wide association and quantitative trait locus (QTL) mapping studies (Weng et al, 2011; Wang et al, 2012). Moreover, various sized SNP arrays have been used to assess genetic diversity in maize (Yan et al, 2009; Lu et al, 2009; Yan et al, 2010; Hao et al, 2011; Semagn et al, 2012).

Although many SNP markers are available in maize, only a small percentage of polymorphic loci can be combined in an SNP array. Consequently, selection and evaluation of SNP sets are important steps in maize DNA fingerprinting. Fingerprint databases that have been constructed using only fixed locus sets are also valuable. The criteria used to select loci for variety identification are quite different from those used in research areas such as genetic diversity and association analyses. Previously reported SNP sets cannot be directly applied to maize variety identification. In this study, we consequently selected and evaluated SNP loci for maize DNA fingerprinting analysis from the maizeSNP50 array, which contains 56 110 SNPs, using Chinese maize varieties with broad genetic backgrounds. We also examined the stability of this SNP array through sample amplification, and examined its discriminatory power, compatibility, and applicability to maize DNA fingerprinting identification.

Materials and methods

Plant materials and DNA extraction

A total of 96 samples were selected to evaluate the SNP markers, including 40 hybrids and 56 inbred individuals (Electronic Supplementary Material Table S1). The 40 hybrid samples included varieties with large planting areas, control varieties from different maize regional trial groups, and some specialized hybrids. The 56 inbred samples included the corresponding parents of the above hybrids and other elite inbred lines (22 triplets with their parents and the F_1 generation), four groups of similar lines, and a group of doubled haploid lines. Total genomic DNA was extracted from 50 pooled leaf samples using the CTAB procedure according to Wang et al, (2011b). DNA quality and concentration were measured with a NanoDrop 2000 UV spectrophotometer (Thermo Scientific, MA, USA), and working solutions were prepared at a concentration of 100ng/μl.

MaizeSNP50 SNP genotyping

The 96 DNA samples were analyzed using the maizeSNP50 BeadChip containing 56 110 SNP loci (Ganal et al, 2011). Raw data were obtained by scanning the chip with hybridized signals using an iScan instrument (Illumina). The genotype data from each sample were analyzed with GenomeStudio software (v2010; Illumina) using the maize SNP50_ B. egt cluster file.

Selection of SNP loci

We first mapped 56 110 SNP loci to the maize B73 sequence and identified the physical location of each locus. SNP selection was performed using a three-step process. First, we selected candidate loci on the basis of GenomeStudio GenTrain scores. These scores can range from 0.00 to 1.00, reflecting the accuracy of the data: the higher the score, the more reliable the data. The GenTrain score of each SNP is calculated according to the following SNP genotype cluster characteristics: angle, dispersion, overlap, and intensity. Genotypes with lower scores are located further from the cluster center and have a lower reliability. From the 56 110 SNPs, a total of 35 894 (64%) SNPs with GenTrain scores between 0.70 and 1.00 were designated as candidate loci. Second, 20 212 SNPs were selected on the basis of reproducibility, missing data rate, signal strength, and their utility for defining the three genotypes. If a sample data point fell outside of a shaded call region on the GenomeStudio SNP graph, it was treated as missing data. SNPs with a missing data rate of more than 5% of sites were removed. As the three genotypes of an ideal SNP should have obvious boundaries on the graph and be easy to differentiate, any SNP having a shifted cluster or a non-obvious boundary was deleted. Third, the candidate loci were further screened based on copy number, minor allele frequency (MAF), and even distribution. SNPs with copy numbers greater than or equal to 2 or MAF values under 0.2 were deleted. The remaining loci were screened according to the physical map. The best SNP in each genic region was chosen on the basis on coding region priority and good experimental quality principles.

The maizeSNP3072 array

The 4 050 candidate SNPs were submitted to Illumina (http://www.illumina.com) to assess their designability score values based on the Golden Gate assay. The key factors influencing these scores were the accuracy and conservation of the DNA sequences located 100bp upstream and downstream of the SNP site. A score higher than 0.6 indicated that the SNP had a relatively higher probability of success, and SNPs with scores below 0.4 were deleted. A total of 3 072 SNPs were obtained. The probe pool was developed according to the flanking sequences of the

3 072 SNPs, and the maizeSNP3072 array chip was ordered based on GoldenGate technology.

Evaluation of the maizeSNP3072 array

The 96 samples were genotyped using the maizeSNP3072 array to verify the repeatability of the 3 072 SNPs. To assess the stability and discriminatory power of the 3 072 SNPs, 309 inbred lines, including 217 elite lines commonly used in China and 92 US samples, were genotyped using this array. The 217 Chinese inbred lines had a wide genetic background that included six heterotic groups: Tang – si – ping – tou (STPT), P, Improved Reid, Lancaster, Waxy, Landrace. In addition, 276 hybrid samples representing varieties approved by the Chinese Ministry of Agriculture were used to assess the effects of the 3 072 SNPs on maize DNA fingerprinting and database construction.

Data analysis

Clusterdifferentiation of the three possible genotypes (AA, BB, and AB) was performed for the 3 072 SNPs based on genotype data from 22 triplets, 309 inbred lines, and 276 hybrids. Repeatability, miss rate, polymorphism, and variety – distinguishing efficiency of the 3 072 SNPs were analyzed using the data from 96 selected samples, 309 inbred lines, and 276 hybrids. To assess the compatibility of the 3 072 SNPs, we analyzed the genotype consistency of repetitive samples between Infinium and GoldenGate platforms. Polymorphism of each locus was analyzed based on the inbred line data. The percentage of different loci was analyzed according to pairwise comparisons among 309 inbreds and 276 hybrids.

Results

Selection process for the 3 072 SNPs

GenTrain scores were calculated for 56 110 SNPs across the 96 selected samples. There were 16.2% loci with scores less than 0.6, 19.8% with scores between 0.6 and 0.7, and 64% with scores higher than 0.7. The different patterns of AA, AB, and BB genotype calls obtained using GenomeStudio software are shown in Figure 1. Five different types of loci were removed during the screening process: loci with weak signal values (Figure 1A), loci with more than five failed data points (Figure 1B), loci for which more than three parent/F_1 triplets showed pedigree inconsistency (Figure 1C), loci for which more than five inbred samples showed the AB genotype (Figure 1D), and loci for which one or all three genotypes exhibited an obvious shift towards one side of the diagram (Figure 1E). To be considered an ideal SNP, the three genotypes for that locus should fall into clearly defined clusters (Figure 1F).

Establishment of an accurate genotype calling procedure through clusterdifferentiation

The raw data obtained from theiScan system were imported to GenomeStudio software for genotype analysis. SNPs were automatically called for AA, AB, and BB genotypes as shown in Figure 2A and 2C. If a rare AB genotype was identified or some data points were shifted to one side, the automatic SNP calling frequently produced errors. To resolve this problem, we used parent/F_1 triplets and inbred and hybrid samples to construct a high-quality standard cluster file, maize SNP3072_ GT. egt (Figure 2B, D) which was used to define regions corresponding to the three genotypes. Characteristics of all sample data points for each SNP were visualized using SNP graphs in GenomeStudio. In these graphs, the x-axis represented normalized theta (angle deviation from pure A signal), with 0 corresponding to pure A signal, 1.0 to pure B signal, and 0.5 to the AB cluster theta mean (Figs. 1 and 2). The cluster file was defined using these graphs according to the following criteria: (1) clear boundaries could be drawn between different genotypes; (2) the missing data rate was minimized; (3) center points of AA, AB, and BB clusters were positioned at 0, 0.5, and 1.0, respectively; and (4) center points could be shifted based on the actual evaluation results of the triplets and inbred and hybrid samples. The missing data rate for the 309 inbred lines, 6%, was only 3% when automatic analysis was performed using the combined cluster file.

Characteristics of the maizeSNP3072 array

Designability scores of candidate SNP loci, which were provided by Illumina, ranged from 0 to 1.0. A score higher than 0.6 indicates that an SNP has a relatively high probability of success when used in a GoldenGate assay, whereas a score below 0.4 indicates that the SNP is predicted to have a poor success rate. The designability score distribution of the 3 072 SNPs was 1.5% (0.40~0.60), 27.5% (0.61~0.80), and 71% (0.81~1.00) (Figure 3A). Because of the wide genetic background of the 96 samples used to evaluate loci, the polymorphism rate of the 3 072 SNPs exhibited little change when loci were assessed instead using 309 inbred lines. All MAF values were greater than 0.20, with an average of 0.37. The percentage of MAF values between 0.20 and 0.25 was relatively low (8%), while percentages for other intervals were between 15 and 21% (Figure 3B, C).

Two 92-sample parallel experiments were performed to assess the compatibility of the 3 072 SNPs on Infinium and GoldenGate platforms. The consistency percentage of the 3 072 loci was 96.4% based on 92-sample genotype data calling on the two platforms. The conversion rate of the 3 072 SNPs between the two platforms was thus higher than 95%. The parent-hybrid heritability of 22 combinations was analyzed using all SNPs that were scorable in each hybrid and its two parents (Electronic Supplementary Material Table S2). The parents had diverse genetic backgrounds that included Improved Reid, P, STPT, Luda Red Cob, SSS, NSSS, and landrace groups. Pedigree consistency values higher than 95% were uncovered for 17 combinations, with a value of 100% obtained for two combinations having B73 and/or Mo17 as parents. Consistency values below 95% were observed in five combinations; this higher inconsistency was due to different seed sources for a hybrid and parents or the low purity rates of inbred lines such as Shen137 and Zong31.

Analysis of several similar inbred groups uncovered no differences among T877-series maize samples (Electronic Supplementary Material Table S1; ID nos. N91-N93). The genetic backgrounds of T877 lines were anticipated to be highly similar, as the series was produced by radiation-induced mutagenesis. Unlike T877 lines, Qi319, P25, and F349 lines were obtained through backcross breeding. Although some different loci were consequently uncovered in Qi319, P25, and F349 series (Table S1; ID nos. N84-N90), all genetic similarity values were greater

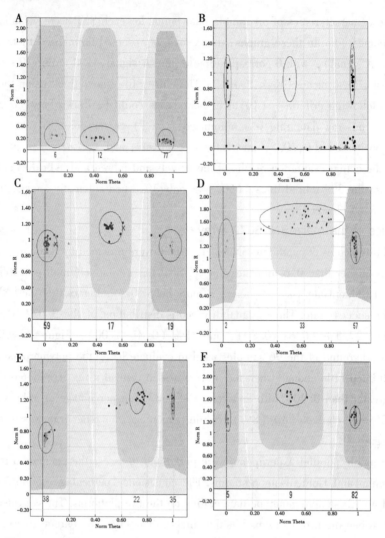

Figure 1　Different patterns of clustering of AA, AB, and BB genotypes based on GenomeStudio analysis

　　Each point is an actual call. The three shaded areas correspond to calculated limits, with darker colors indicating higher levels of confidence. Ellipses are used to adjust the position of the allele-calling areas. Three different genotypes are called: homozygous for allele A (red), heterozygous AB (purple), and homozygous for allele B (blue). Allele calls that fall in the lighter-colored areas in between or below these areas are set to "failed". Filled green circles represent inbred-line data points. If any parent/F_1 errors are found in the data, the F_1 hybrid appears as an "X" and the parent as an "O". A Only weak sample signals detected; B Three clusters observed, but numerous failed samples (at the bottom) not called; C Pedigree inconsistency exhibited by more than three parent/F_1 combinations; D Heterozygous genotypes shown by more than five inbred lines; E One or all three clusters shifted towards one side of the graph; F Perfect genotyping locus in which the three genotypes fall into clearly defined clusters across all 96 samples. The x-axis represents normalized theta (angle deviation from pure A signal, with 0 indicating pure A signal and 1.0 representing pure B signal), and the y-axis corresponds to the distance of the point from the origin

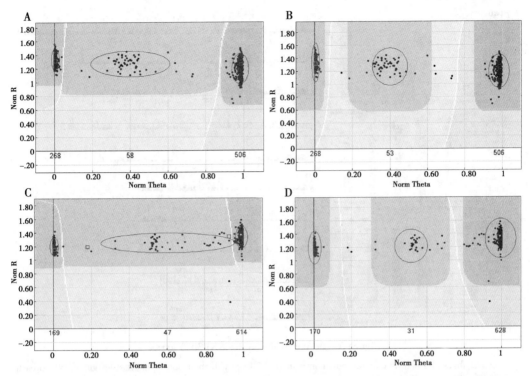

Figure 2 MaizeSNP3072 cluster file constructed to improve the genotyping efficiency of the 3 072 loci

Samples with reproducibility errors appear as squares. A and C Automatic SNP calling using GenomeStudio software; B and D Corrected SNP calling using a maizeSNP3072 cluster file

than 97%. To verify the ability of the 3 072 SNPs to discriminate among maize varieties, we performed 37 950 and 47 586 pairwise comparisons among 276 hybrids and among 309 inbred lines. Differential locus percentages of between 5 to 70% were observed in 99.9% of these pairwise comparisons for both inbred and hybrid lines. The most frequent differential locus percentages uncovered were 50% among inbred lines (in 4 888 comparisons) and 60% among hybrids (in 4 216 comparisons) (Electronic Supplementary Material Figure S1).

Distribution characteristics of the 3 072 SNP loci

The physical distribution of the 3 072 loci on the 10 maize chromosomes was determined using their mapped positions on the B73 genome sequence. Each chromosome was divided into 1 000kbp-sized windows, and the number of SNPs per window was counted (Figure 3D). Almost all of the SNPs were found to be distributed evenly throughout the genome. SNPs were relatively sparse around centromeres and relatively abundant near telomeric regions (Figure 3D). All 3 072 SNPs were in genic regions, where they were distributed in exons (43%), promoters (21%), 3′ untranslated regions (UTRs; 22%), 5′UTRs (9%), and intron regions (5%).

Comparative analysis of maizeSNP3072 and maizeSNP50K arrays

The performance of maizeSNP3072 and maizeSNP50K arrays was compared using the 96 samples evaluated in this study (Table 1, Figure 4). On the Chinese materials, the 3072SNP array showed a better marker success rate and higher average MAF values, evaluation scores, and

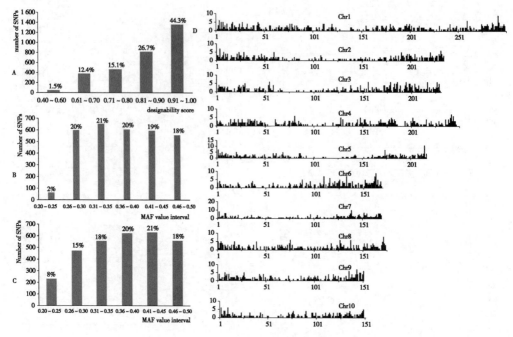

Figure 3　Design of the maize SNP3072 array

A Numbers of single nucleotide polymorphisms (SNPs) with their corresponding designability scores for 3 072 SNPs evaluated by Illumina. B MAF values of the 3 072 SNPs based on data from 96 samples. C MAF values of the 3 072 SNPs based on data from 309 inbred lines. D Distribution of the 3 072 SNPs on 10 chromosomes. The window size is 1 000kbp, the x-axis represents the order of the widows, and the y-axis corresponds to the number of SNP loci

variety-distinguishing efficiency than the maizeSNP50K array. Differences in the distinguishing efficiency of the two arrays are shown in Figure 4 for Chinese inbred and hybrid samples. Differential locus rates among inbreds ranged from 15 to 72% (average of 49%) and 12 to 52% (average of 35%) for 3 072 and 56 110 SNPs, respectively; among hybrids, the corresponding rates were 16 to 66% (average of 58%) and 12 to 57% (average of 45%).

Table 1　Comparative analysis of maizeSNP3072 and maizeSNP50K chips based on data from 3 072 and 56 110 single nucleotide polymorphisms in 96 evaluated maize samples

Comparative item	maizeSNP3072	maizeSNP50K
marker success rate	94%	67%
average MAF value	0.37	0.17
GenTrain score (0.6)[a]	100%	84%
GenTrain score (0.7)[a]	89%	64%
variety distinguishing efficiency[b]	inbred (49%), hybrid (58%)	inbred (35%), hybrid (45%)

[a] Percentage of loci with GenTrain scores greater than 0.6 or 0.7; [b] average differential locus rate

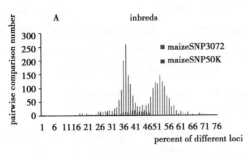 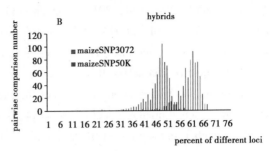

Figure 4 Distribution of different locus percentages obtained by pairwise comparative analysis of evaluated samples based on 3 072 and 56 110 SNPs

A. Comparison using genotyping data from 53 inbred lines; B. Comparison using genotyping data from 38 hybrids

Discussion

SNP array development and characterization for maize DNA fingerprinting

Fixed SNP sets are preferred for maize DNA fingerprinting and database construction. SNP marker development is well underway in maize, with numerous SNPs listed in databases such as Panzea and MaizeGDB. Not all SNPs are suitable for DNA fingerprinting, however, and some loci do not meet array chip design requirements. The selection of fixed SNP locus sets with high discriminatory ability, good stability, and even distribution is thus the most important step in SNP marker-based fingerprinting research. To construct an SNP array for maize DNA fingerprinting, a set of evaluation materials with a broad genetic basis, reasonable SNP selection principles, and an accurate cluster genotyping file are required. Polymorphism bias will be present if the genetic background of the selected materials is concentrated. In addition, maize DNA fingerprinting must be able to differentiate among hybrids quickly and accurately. Consequently, representative hybrids must be selected to validate the variety-discriminatory efficiency and heterozygous calling accuracy of candidate SNPs. Common assessment indices for selecting a set of SNPs include repeatability, discriminatory power, uniformity of distribution, and conservatism of flanking sequences. Automatic and accurate genotype calling also are quite important. To ensure that three genotype clusters can be easily distinguished, the selected SNP should be a single-copy locus, and both inbred and hybrid lines should be used to evaluate cluster independence and stability. Automatic SNP calling using GenomeStudio software is sometimes prone to mistakes, especially when a rare AB genotype cluster is present. To improve the accuracy and efficiency of genotype calling, a standard genotyping cluster file based on the characteristics of each SNP should therefore be established.

Comparison of published SNP arrays and the maizeSNP3072 array

Published SNP genotyping arrays include the high-density maizeSNP50 array (Ganal et al, 2011), a 768-SNP array reported by the Pioneer Co. for commercial maize resource identification (Jones et al, 2009), a 1 536-SNP array used for germplasm resource assessment

(Lu et al, 2009; Yan et al, 2010; Semagn et al, 2012), and an SNP array used in an association study (Yan et al, 2009). SNP selection criteria for DNA fingerprinting are different from those used for association and QTL analyses (Yan et al, 2009), with the latter focused on loci related to objective characters. In this study, we selected 3 072 SNPs using inbred and hybrid lines as evaluation materials on the basis of their stability, genotype-calling accuracy, discriminatory power, copy number, and evenness of distribution.

The number and sources of SNPs initially employed in this study were largely different from those used to produce the 768- and 1 536-SNP sets described above. In particular, 3 072 SNPs were selected from 56 110 SNP loci in our study, whereas approximately 2 000 initial loci were used to construct the 768- and 1 536-SNP sets. The 768- and 1 536-SNP sets were selected primarily from PHM loci (developed by Pioneer) and PZA loci in the Panzea database, of which only a small proportion were candidate loci in the current study. The 3 072 SNP loci thus have few similarities with the 768 and 1 536 loci. In conclusion, compared with the previously reported SNP arrays, the maizeSNP3072 set is more suitable for Chinese maize variety DNA fingerprinting and database construction.

Selection of SNPs compatible with Infinium and GoldenGate platforms

Although many types of SNP genotyping platforms exist, such as GoldenGate, Infinium, TaqMan, and KASPar, not all SNPs are transferable across different platforms. A success rate of approximately 89% has been obtained on GoldenGate and Infinium platforms (Mammadov et al, 2012). Although both platforms are based on bead-chip technology, they differ somewhat in regard to probe design, reagents, and experimental processes. Some SNPs may therefore not be transferable across the two platforms. The SNPs identified in this study, which were evaluated using Infinium and GoldenGate platforms, had a conversion rate between the two platforms of over 95%.

Applications of the maizeSNP3072 array

The maizeSNP3072 array can be used for Chinese maize DNA fingerprinting, germplasm resource evaluation, and molecular marker-assisted breeding. SSRs are currently the primary markers used to identify maize varieties (Wang et al, 2011a). With new developments in molecular technology and increasing identification requirements, SSRs have been found to suffer from various drawbacks such as data sharing and integration problems. These shortcomings can be overcome by using SNP markers. With bead-chip technology, locus throughput ranges from 1 to million SNPs. Because only two alleles exist per SNP, data integration between different laboratories or platforms is easy. The 3 072 SNP loci reported in this study were selected on the basis of DNA fingerprinting criteria. The obtained SNP set has high variety-distinguishing efficiency, good reproducibility, and a uniform distribution throughout the genome. The maizeSNP3072 array can be directly applied to maize variety identification, database construction, or genuineness verification using core SNPs selected from the 3 072 SNPs. In addition, our assessment of the maizeSNP3072 array demonstrated that the 3 072 locus set displayed a superior genotyping performance in inbred lines and, when used in conjunction with a cluster file, could automatically differentiate among three genotypes. The 3 072 SNPs showed high levels of polymorphism: 77% of loci had MAF values greater than 0.30, and 39% had values greater than 0.40. The maizeSNP3072 array can therefore be used to assess the genetic diversity of maize germplasm resources. Furthermore, the 3 072 SNPs can serve as powerful markers for molecular breeding studies, including QTL mapping, background scanning of breeding materials and homozygosity identification.

Acknowledgments

We thank Professor Shoucai Wang for providing someco-isogenic strain samples. This research was supported by grants from the Beijing Science and Technology Project (D131100000213002), "The Twelfth Five-Year" Rural Area National Science and Technology Project from the Ministry of Science and Technology of China (2011BAD35B09) and Beijing New-Star plan of Science and Technology (Z121105002512038).

References (omitted)

Molecular Breeding, 2015, 35 (6): 136

Paenibacillus zeae sp. nov., Isolated from Maize (*Zea mays* L.) Seeds

Liu Yang[1,2]　Zhai Lei[2]　Wang Ronghuan[1]　Zhao Ran[2]　Zhang Xin[2]
Chen Chuanyong[1]　Cao Yu[2]　Cao Yanhua[2]　Xu Tianjun[1]
Ge Yuanyuan[2]　Zhao Jiuran[1]　Cheng Chi[2]

(1. Maize Research Center, Beijing Academy of Agriculture and Forestry Sciences, Beijing 100097, China; 2. China Center of Industrial Culture Collection (CICC), China National Research Institute of Food and Fermentation Industries, Beijing 100015, China)

Abstract: Four Gram-stain-positive bacterial strains, designated $6R2^T$, 6R18, 3T2 and 3T10, isolated from seeds of hybrid maize (*Zea mays* L., Jingke 968) were investigated using a polyphasic taxonomic approach. Cells were aerobic, motile, spore-forming and rod-shaped. Phylogenetic analysis based on 16S rRNA gene sequences indicated that the isolates may represent a novel species of the genus *Paenibacillus*, the four closest neighbours being *Paenibacillus lautus* NRRL NRS-666^T (97.1% similarity), *Paenibacillus glucanolyticus* DSM 5162^T (97.0%), *Paenibacillus lactis* MB 1871^T (97.0%) and *Paenibacillus chibensis* JCM 9905^T (96.8%). The DNA G+C content of strain $6R2^T$ was 51.8 mol%. Its polar lipid profile consisted of diphosphatidylglycerol, phosphatidylglycerol and phosphatidylethanolamine. The predominant respiratory quinone was menaquinone 7 (MK-7) and the major fatty acids were anteiso-$C_{15:0}$ and iso-$C_{14:0}$. Strains $6R2^T$, 6R18, 3T2 and 3T10 were clearly distinguished from the above type strains using phylogenetic analysis, DNA-DNA hybridization, and a range of physiological and biochemical characteristics. It is evident from the genotypic and phenotypic data that strains $6R2^T$, 6R18, 3T2 and 3T10 represent a novel species of the genus *Paenibacillus*, for which the name Paenibacillus zeae sp. nov. is proposed. The type strain is $6R2^T$ ($5KCTC\ 33674^T5CICC\ 23860^T$).

The genus *Paenibacillus* was identified as comprising group 3 bacilli, based on 16S rRNA gene sequence analysis (Ash et al, 1993). At the time of writing, the genus contains 165 species and four subspecies listed on the LPSN database (Prokaryotic names with standing in nomenclature: www.bacterio.net). Members of the genus *Paenibacillus* are aerobic or facultatively anaerobic, endospore-forming, rod-shaped bacteria that are widely distributed in nature (Priest, 1977; Claus & Berkeley, 1986; Slepecky & Hemphill, 1992; Weon-Taek et al, 1999; Chung et al, 2000; Daane et al, 2002; Scheldeman et al, 2004; Liu et al, 2010; Xiang et al, 2014).

During the characterization of endophytic bacteria in hybrid maize (*Zea mays* L., Jingke 968), four bacterial strains, $6R2^T$, 6R18, 3T2 and 3T10, were isolated from maize seeds (about 100g) collected from the scientific research base of Beijing Academy of Agriculture and Forestry Sciences (BAAFS) in Zhangye city, Gansu province (39°04′57.94″N 100°12′49.40″E, north-western China) in September 2014. The characterization and description of a novel species is reported here.

Strains 6R2^T and 6R18 were originally isolated on R2A agar plates (Luqiao), while strains 3T2 and 3T10 were originally isolated on trypticase soy agar (TSA) plates (Luqiao), which had been seeded with a suspension of hybrid maize seeds and incubated at 30℃ for 48h. The seed suspension was prepared according to the following procedure as described by Qiu et al (2007). The new isolates were maintained on nutrient agar (NA) plates after incubation at 30℃ for 48h. Motility was determined by the handing-drop method (Skerman, 1967). Gram-staining was performed as described by Gerhardt et al (1994). The harvested cells were washed twice in 0.1M PBS buffer (pH 7.2) and fixed using 2.5% glutaraldehyde at 4℃. Cells fixed overnight were washed three times in 0.1M PBS buffer (pH 7.2) and dehydrated using a graded ethanol series (50, 70, 85 and 95% ethanol once and 100% ethanol three times). The sample was then subjected to critical point drying with CO_2 (Bal-Tec CPD030) and metalspraying apparatus (Bal-Tec SCD005). Cell morphology and size were determined using a scanning electron microscope (Quanta 200; JEOL). Cell morphology was also examined by transmission electron microscopy according to Liu et al (2010).

To determine optimal conditions for strains 6R2^T, 6R18, 3T2 and 3T10, cultures were grown in nutrient broth (NB) at 4, 10, 15, 20, 25, 30, 37, 42, 50 and 55℃ and pH 4.0~13.0 (at 1 pH unit intervals). NaCl tolerance was determined using NB medium supplemented with 0~6.0% (w/v) NaCl at 0.5% intervals for 48h at 30℃. Growth on NA medium under anaerobic conditions was checked in an anaerobic chamber filled with a mixture of gases (N_2/H_2/CO_2 at 90∶5∶5) for 1 week (Xiang et al, 2014). Growth was also tested on R2A, Luria-Bertani (LB) and MacConkey agars (all from Luqiao). Oxidase and catalase activity were determined according to Liu et al (2010). Methyl red reactions were tested according to Dong & Cai (2001), and the lecithinase test on egg yolk agar was performed on NA with 10% egg yolk emulsion. Assimilation and acid production from carbohydrates and enzyme activities were tested by using the API 20E, API 20NE and API 50CH kits (all bioMérieux) according to the manufacturer's instructions, and these tests (three replicates for each) were performed at 30℃ for 48h.

Preparation of genomic DNA was carried out according to the method of Marmur (1961). The G+C content of the DNA was determined using the HPLC method (Mesbah et al, 1989). The 16S rRNA gene was PCR-amplified using the universal primers 27F (59-AGAGTTTGATC-CTG GCTCAG-39) and 1492R (59-GGTTACCTTGTTACG ACTT-39) (Lane, 1991). The amplified products were purified and cloned into vector Top10 (Tiangen) for sequence determination carried out using an automated DNA sequencer (model ABI 3730; Applied Biosystems). The sequencing primers were SP6 (59-ATTTAGGTGACA CTATAGAATAC-39) and T7 (59-TAATACGACTCACTA TAGGG-39). The 16S rRNA gene sequences of the four studied bacterial strains and those of other *Paenibacillus* species retrieved from GenBank were aligned using the program CLUSTAL X v1.8 (Thompson et al, 1997). A distance matrix method (with distance options according to Kimura's two-parameter model; Kimura, 1983), including clustering using the neighbour-joining (Saitou & Nei, 1987), maximum-likelihood (Felsenstein, 1981) and maximum-parsimony methods (Eck & Dayhoff, 1966; Fitch, 1971) from the MEGA 5 program (Tamura et al, 2011), was used to infer the phylogenetic evolutionary trees (Figure 1 and Figure S1, available in the online Supplementary Material). The percentage of replicate trees in which the associated taxa clustered together in the bootstrap test (1 000 replicates) is shown next to the branches (Felsenstein, 1985). Cells of the four novel bacterial strains and the type strains of *Paenibacillus* species grown on TSA to the same physiological age were used in the following comparisons. Fatty acid methyl esters were extracted, separated and identified according to the instructions of the Microbial Identification System (using the RTSBA6 database).

Purified cell-wall preparations were obtained according to the method of Schleifer & Kandler (1972) and the amino acids in peptidoglycan hydrolysates were analysed as described by Schleifer (1985). Freeze-dried cells (100mg) were treated with chloroform/methanol (2:1, v/v) overnight to extract isoprenoid quinones. Preparative TLC was performed using Kieselgel 60 F_{254} plates (20cm×20cm, 0.5mm thick; Merck) with petroleum ether/diethyl ether (9:1, v/v) as the solvent. The resulting bands were scraped from the plate under short-wavelength UV light, and then redissolved in acetone. Finally the quinone profile was analysed using reversed-phase HPLC (LC20AD system; Shimadzu) with a Zorbax Eclipse XDB-C18 column (250mm× 4.6mm, 5mm; Agilent) and a UV detector at 270nm. Polar lipids were extracted, separated using two dimensional TLC and identified according to published procedures (Komagata & Suzuki, 1987).

Comparative 16S rRNA gene sequence (over 1 500bp) analysis revealed that levels of similarity among strains $6R2^T$, 6R18, 3T2 and 3T10 were more than 99%. The 16S rRNA gene sequence of strain $6R2^T$ was a continuous stretch of 1 518bp. Sequence similarity calculations indicated that strain $6R2^T$ shared similarities with the type strains of recognized species of the genus *Paenibacillus* ranging from 96.8% (*Paenibacillus chibensis*) to 97.1% (*Paenibacillus lautus*). Despite the high 16S rRNA gene sequence similarity of strain $6R2^T$ to *P. lautus* NRRL NRS-666T (97.1%), *Paenibacillus glucanolyticus* DSM 5162^T (97.0%), *Paenibacillus lactis* MB 1871^T (97.0%) and *P. chibensis* JCM 9905T (96.8%), the phylogenetic trees reconstructed using the neighbour-joining, maximum-likelihood and maximum-parsimony methods suggested that strains $6R2^T$, 6R18, 3T2 and 3T10 are mem-bers of the genus *Paenibacillus* but are not close phylogenetic relatives of recognized species of the genus and represent a distinct clade (Figs 1 and S1).

The status of strains $6R2^T$, 6R18, 3T2 and 3T10 was confirmed by genomic DNA-DNA reassociation studies. Strain $6R2^T$ showed 74.0, 81.0 and 90.6% DNA-DNA relatedness to strains 6R18, 3T2 and 3T10, respectively; strain 3T2 showed 86.9 and 93.2% relatedness to strains 6R18 and 3T10, respectively; and strains 6R18 and 3T10 showed 90.6% mutual DNA-DNA relatedness. DNA-DNA hybridization experiments were further performed with strain $6R2^T$ and

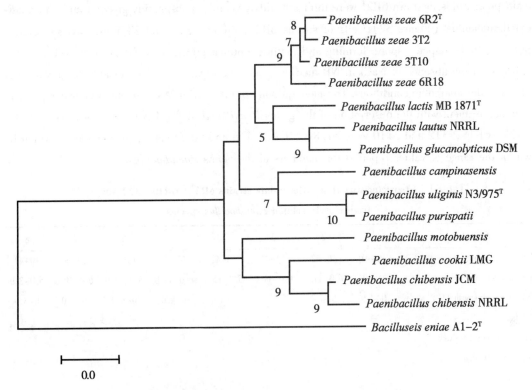

Figure 1 Phylogenetic tree showing the relationships of the four studied strains to species of the genus *Paenibacillus*

The tree was based on an alignment of almost complete 16S rRNA gene sequences and was reconstructed using the neighbou-joining method (Saitou & Nei, 1987). The phylogenetic tree was produced following a bootstrap sampling of 1 000 datasets. Numbers at nodes indicate percentages of bootstrap support based on a neighbour-joining analysis of 1 000 resampled data-sets. Bar, 0.01 substitutions per nucleotide position

P. lautus NRRL NRS-666T, *P. glucanolyticus* DSM 5162^T, *P. lactis* MB 1871T and *P. chibensis* JCM 9905T using the method described by Ezaki et al (1989). The results indicated low levels of relatedness between strain 6R2^T and its closest phylogenetic neighbours: *P. lautus* NRRL NRS-666T (50.9%), *P. glucanolyticus* DSM 5162^T (36.9%), *P. lactis* MB 1871^T (33.6%) and *P. chibensis* JCM 9905T (50.1%). These values are clearly below the 70% cut-off value recommended for the assignment of strains to the same genomic species (Wayne et al, 1987).

Morphological, cultural, physiological and biochemical characteristics of strains 6R2^T, 6R18, 3T2 and 3T10 are given in the species descriptions and in Table 1, which showed the distinctive features of the four strains as well as those of closely related species of the genus *Paenibacillus*. The fatty acid profiles of the four strains are shown in Table S1. The major fatty acids of strain 6R2^T were anteiso-$C_{15:0}$ (40.8%) and iso-$C_{14:0}$ (10.0%), similar to those of the type strains compared in this study, but there were differences in the proportions of some fatty acids. The diagnostic cell-wall diamino acid of strain 6R2^T was *meso*-diaminopimelic acid. The main polar lipids in strain 6R2^T were diphosphatidylglycerol, phosphatidylglycerol and phosphatidylethanolamine (Figure S2). Cells of strains 6R2^T, 6R18, 3T2 and 3T10 were Gram-stain-positive, rod-shaped, spore-forming and motile; optimal growth was observed at 30℃, at pH 7.0 and in the absence of NaCl in NB medium, and the organisms were not able to grow on NA plates under anaerobic conditions. The menaquinone profile, with MK-7 as the major component, is in agree-ment with the description of the genus *Paenibacillus*. The DNA G+C contents of strains 6R2^T, 6R18, 3T2 and 3T10 were 51.8, 50.9, 51.6 and 50.2 mol%, respectively, which lie within the range of values reported for members of the genus *Paenibacillus*.

Table 1 Characteristics that differentiate strains 6R2^T, 6R18, 3T2 and 3T10 from closely related *Paenibacillus* species

Characteristic	1	2	3	4	5	6	7	8
Temperature range for growth (℃)	10~42	10~42	10~42	10~42	10~42	10~42	20~50	10~42
pH range for growth	6.0~10.0	6.0~11.0	6.0~11.0	5.0~11.0	6.0~11.0	6.0~11.0	6.0~10.0	5.0~8.0
NaCl range for growth (%)	0~4.0	0~4.0	0~4.0	0~4.0	0~4.0	0~4.0	0~4.0	0~3.0
Oxidase	-	-	-	+	-	-	-	-
Lecithinase test (egg yolk)	-	-	-	-	-	+	-	-
Methyl red reaction	-	-	-	-	-	-	-	+
Hydrolysis of:								
Casein	-	-	-	-	+	+	-	-
Starch	-	-	-	-	+	-	+	+
DNA	+	+	+	+	-	-	+	-
Utilization of (API 50CH):								
Glycerol	-	-	-	-	+	-	-	-
Erythritol	-	-	-	-	-	-	-	-
D-Arabinose	-	-	-	-	+	-	W	-
L-Arabinose	-	-	-	-	+	+	+	+
D-Ribose	W	+	-	W	+	+	+	+
D-Xylose	-	-	-	-	+	+	+	+
L-Xylose	-	-	-	-	+	-	-	-
D-Adonitol	-	-	-	-	+	-	-	-
Methylb-D-xylopyranoside	-	-	-	-	+	-	W	+
D-Galactose	-	+	W	W	+	+	+	+
D-Glucose	-	+	W	+	+	+	+	+
D-Fructose	W	+	W	+	+	+	+	+
D-Mannose	W	+	W	+	+	+	+	+
L-Sorbose	-	-	-	-	+	-	-	-
L-Rhamnose	-	-	-	-	+	-	-	-
Dulcitol	-	-	-	-	+	-	-	-
Inositol	-	-	-	-	+	-	-	-

(continued)

Characteristic	1	2	3	4	5	6	7	8
D-Sorbitol	-	-	-	-	+	-	-	-
Methyla-D-mannopyranoside	-	-	-	-	+	+	-	+
Methyla-D-glucopyranoside	-	-	-	-	+	+	-	+
Arbutin	W	+	+	+	+	+	+	+
Salicin	+	+	W	+	+	+	+	+
Raffinose	+	+	W	W	+	+	+	+
Xylitol	W	-	-	-	-	-	-	+
Gentiobiose	+	+	-	+	+	+	+	+
Turanose	+	-	W	+	+	+	-	+
L-Fucose	-	-	-	-	+	-	-	-
Activity of (API 20E and API 20NE):								
Urease	-	+	-	-	-	-	-	-
Voges-Proskauer reaction	-	-	-	-	-	+	-	+
Nitrate reduction	-	-	-	-	-	-	+	+
Assimilation of (API 20NE):								
Glucose	-	-	-	-	-	+	-	+
Arabinose	-	-	-	-	-	+	-	+
Mannose	-	-	-	-	-	+	-	+
Mannitol	+	+	+	+	-	+	-	+
N-Acetylglucosamine	+	+	+	+	-	+	-	+
Maltose	-	+	-	-	-	+	-	+
Gluconate	-	-	-	-	-	+	-	+
Malic acid	-	-	-	+	-	+	-	+
DNA G+C content (mol%)	51.8	50.9	51.6	50.2	51.0	48.0	51.6	52.8

Strains: 1, 6R2^T; 2, 6R18; 3, 3T2; 4, 3T10; 5, *P. lautus* NRRL NRS-666T; 6, *P. glucanolyticus* DSM 5162^T; 7, *P. lactis* MB 1871^T; 8, *P. chibensis* JCM 9905T. All data were generated in this study, except for the G+C content of *P. lautus* NRRL NRS-666T (Heyndrickx et al, 1996), *P. glucanolyticus* DSM 5162T (Alexander and Priest, 1989), *P. lactis* MB 1871T (Scheldeman et al, 2004) and *P. chibensis* JCM 9905T (Shida et al, 1997). All strains have the same optimum temperature (30℃), pH (7.0) and NaCl (0%, w/v). All are positive for the following: motility; growth on LB and R2A agar medium; activity of catalase, ONPG hydrolysis and galactosidase; and assimilation of aesculin and acid production from D-mannitol, N-acetylglucosamine, amygdalin, aesculin, cellobiose, maltose, lactose, melibiose, sucrose, trehalose, starch and glycogen. All are negative for the following: growth under anaerobic conditions and on MacConkey agar; hydrolysis of tyrosine, Tween 20 and Tween 80; acid production from inulin, mele-zitose, D-lyxose, D-tagatose, D-fucose, D-arabitol, L-arabitol, potassium gluconate, potassium 2-ketogluconate, glucose, mannitol, inositol, sorbitol, rhamnose, sucrose, melibiose, amygdalin, arabinose and glucose; assimilation of capric acid, adipic acid, citrate and phenylacetic acid; and activity of arginine dihydrolase, lysine decarboxylase, ornithine decarboxylase, citrate utilization, H2S production, tryptophan deaminase, indole production, indole production and gelatinase. +, Positive; 2, negative; W, weakly positive

Based on the combination of morphological, physiological, chemotaxonomic and phylogenetic data discussed here, it appears that strains 6R2^T, 6R18, 3T2 and 3T10 belong to the genus *Paenibacillus*. However, they could be distinguished from closely related species in some phenotypic and chemotaxonomic features, such as a negative reaction for the utilization of L-arabinose and D-xylose, and the presence of iso-$C_{14:0}$. Additionally, phylogenetic distinctiveness distinguishes the four strains from recognized *Paenibacillus* species. On the basis of the data described above, the four strains represent a novel species of the genus *Paenibacillus*, for which the name *Paenibacillus zeae* sp. nov. is proposed.

Description of Paenibacillus zeae sp. nov.

Paenibacillus zeae [ze9ae. L. gen. n. *zeae* of spelt, of *Zea mays*, referring to its isolation from corn (*Zea mays* L.)].

Cells are Gram-stain-positive, aerobic, motile, spore-forming rods (2.0~3.0μm | × 0.3~0.5μm) (Figs S3 and S4). Colonies on NA plates are off-white, circular and raised with entire edges. Growth occurs at 10~42℃ (optimum 30℃) and at pH 6.0~10.0 (optimum pH 7.0). Grows with 0~4.0% (w/v) NaCl (optimum 0%) in NB, and unable to grow on NA plates under anaerobic conditions. Cells are positive for activity of catalase, ONPG hydrolysis, DNA hydrolysis and galactosidase; assimilation of aesculin, mannitol and *N*-acetylglucosamine; and acid production from D-mannitol, *N*-acetylglucosamine, amygdalin, aesculin, cellobiose, maltose, lactose, melibiose, sucrose, treha-lose, starch, glycogen, salicin, raffinose, gentiobiose and turanose. Negative for growth under anaerobic conditions and on MacConkey agar; lecithinase and methyl red reac-tion; hydrolysis of tyrosine, casein, starch, Tween 20 and Tween 80; activity of arginine dihydrolase, lysine decarboxylase and ornithine decarboxylase, citrate utiliz-ation, Voges-Proskauer reaction, H2S production, nitrate reduction, tryptophan deaminase, indole production and gelatinase; acid production from glycerol, erythritol, D-arabinose, L-arabinose, D-xylose, L-xylose, D-adonitol, methyl b-D-xylopyranoside, D-galactose, D-glucose, L-sorbose, L-rhamnose, dulcitol, inositol, D-sorbitol, methyl a-D-mannopyranoside, methyl a-D-glucopyranoside, L-fucose, inulin, melezitose, D-lyxose, D-tagatose, D-fucose, D-arabitol, L-arabitol, potassium gluconate, potassium 2-ketogluconate, glucose, mannitol, inositol, sorbitol, rhamnose, sucrose, melibiose, amygdalin, arabinose and glucose; and assimilation of glucose, arabinose, mannose, maltose, gluconate, malic acid, capric acid, adipic acid, citrate and phenylacetic acid. The diagnostic diamino acid in the cell-wall peptidoglycan is meso-diaminopimelic acid and the predominant respiratory quinone is MK-7. The major fatty acids are anteiso-$C_{15:0}$ and iso-$C_{14:0}$. The major polar lipids are diphosphatidylglycerol, phos-phatidylglycerol and phosphatidylethanolamine.

The type strain, 6R2T (5KCTC 33674T5CICC 23860T), was isolated from seeds of hybrid maize (*Zea mays* L., Jingke 968) sampled from the scientific research base of BAAFS in Gansu province, China. The DNA G+C content of the type strain is 51.8 mol%. 6R18, 3T2 and 3T10 are additional strains of the species, isolated from the same source.

Acknowledgements

This work was supported by the Beijing Nova Program (No. Z141105001814095), the Beijing Nova Interdisciplinary Coopera-tional Program (No. Z1511000003150150), the National Natural Science Foundation of China (No. 31300008), the Chinese Postdoc-toral Science Foundation (No. 2015M570969), the Fund of National Infrastructure of Microbial Resources (No. NIMR2015-4), and the Scientific and Technological Development Project of China National Research Institute of Food and Fermentation Industries (No. 2012kJFZ-BS-01).

References (omitted)

Antonie Van Leeuwenhoek, 2016, 109 (2): 207-213

Paenibacillus Chinensis sp. nov. , Isolated from Maize (*Zea mays* L.) Seeds

Liu Yang　Zhao Ran　Wang Ronghuan　Yao Su　Zhai Lei
Zhang Xin　Chen Chuanyong　Cao Yanhua　Xu Tianjun
Ge Yuanyuan　Zhao Jiuran　Cheng Chi

(China Center of Industrial Culture Collection (CICC), China National Research Institute of Food and Fermentation Industries, Beijing 100015, China)

Abstract: Four Gram-stain positive bacterial strains, designated as 4R1^T, 4R9, 4L13 and 4L18, isolated from seeds of hybrid maize (*Zea mays* L., Jingke 968), were investigated using a polyphasic taxonomic approach. The cells were found to be facultatively aerobic, motile, spore-forming and rod-shaped. Phy-logenetic analysis based on 16S rRNA gene sequences showed that the isolates should be recognised as a species of the genus *Paenibacillus*, with two close neighbours being *Paenibacillus nicotianae* YIM h-19^T (98.41% similarity) and *Paenibacillus hordei* RH-N24^T (98.37%). The DNA G+C content of strain 4R1T was determined to be 51.6 mol%. Its polar lipid profile was found to consist of diphosphatidylglycerol, phosphatidylglycerol, phosphatidylethanolamine and an unidentified lipid. The predominant respiratory quinone was identified as MK-7 and the major fatty acids were found to be anteiso-$C_{15:0}$, anteiso-$C_{12:0}$, anteiso-$C_{13:0}$ and anteiso-$C_{11:0}$. Strains 4R1^T, 4R9, 4L13 and 4L18 were clearly distinguished from the reference type strains using phylogenetic analysis, DNA-DNA hybridization and a range of physiological and biochemical characteristics. It is evident from the genotypic and phenotypic data that strains 4R1T, 4R9, 4L13 and 4L18 represent a novel species of the genus *Paenibacillus*, for which the name *Paenibacillus chinensis* sp. nov. is proposed. The type strain is 4R1^T (=KCTC 33672T = CICC 23864T).

Key words: *Paenibacillus chinensis* sp. Nov; Maize; 6S rRNA gene; Polyphasic taxonomy

Introduction

The genus *Paenibacillus* was established from members of group 3 *Bacillus* basing on 16S rRNA gene sequence analysis (Ash et al, 1993). At the time of writing, there are more than 150 validly named species and 4 subspecies in the genus *Paenibacillus* listed at LPSN (Prokaryotic names with standing in nomenclature: www.bacterio.net) and the type species is *P. polymyxa* (Ash et al, 1993). Most of these *Paenibacillus* strains are aerobic or facultatively anaerobic, endospore-forming, rod-shaped bacteria and motile by means of peritrichous flagella (Kim et al, 2013). Many *Paenibacillus* species have been isolated from a wide range of environ-

---

Liu Yang and Zhao Ran have contributed equally to this work

ments including water springs (Tang et al, 2011), naked barley (Kim et al, 2013) and selenium mineral soil (Yao et al, 2014). Some *Paenibacillus* strains are very important from industrial and economic points of view (Sle-pecky and Hemphill 1992; Priest 1977; Chung et al, 2000; Weon-Taek et al, 1999).

During the characterisation of endophytic bacteria in hybrid maize (*Zea mays* L., Jingke 968), 4 bacterial strains $4R1^T$, 4R9, 4L13 and 4L18 were isolated from maize seeds (about 100g) collected from the scientific research base of Beijing Academy of Agriculture and Forestry Sciences (BAAFS) in Zhangye city, Gansu province (39°04′57.94″N, 100°12′49.40″E, north western China) in September 2014. The 4 strains were found to be the dominant members of the of endophytic bacterial population in the maize seeds during this study. According to polyphasic taxonomic data presented here, the new isolates should be affiliated to the genus *Paenibacillus* as a new species, for which the name *Paenibacillus chinensis* sp. nov. is proposed.

Materials and methods

Strain and culture conditions

Strains $4R1^T$ and 4R9 were originally isolated on plates of R2A agar (Luqiao, China), whilst strains 4L13 and 4L18 were originally isolated on plates of LB agar (Luqiao, China), which had been seeded with a suspension of hybrid maize seeds and incubated at 30℃ for 48h. The seed suspension was prepared according to the procedure as described by Qiu et al (2007). The new isolates were maintained on nutrient agar (NA) plates after incubation at 30℃ for 48h.

Morphological, physiological and biochemical characteristics

Gram-staining was performed as described by Ger hardt et al (1994). Cell morphology was examined by light microscopy and the presence of flagella was examined by transmission electron microscopy (TEM). The harvested cells were washed twice with 0.1M PBS buffers (pH 7.2) and fixed using 2.5% glutaraldehyde at 4℃. Cells fixed overnight were washed 3 times with 0.1M PBS buffers (pH 7.2) and dehydrated using an ethanol gradient with increasing concentration (50, 70, 85, 95% ethanol once and 100% ethanol 3 times). Then, the sample was prepared by critical point drying with CO_2 (BAL-TEC CPD030) and metal-spraying apparatus (BAL-TEC SCD005). The morphology and size of the cells were observed using a scanning electron microscope (Quanta 200; JEOL). Spore production was carried out by the procedure described by Doetsch (1981).

To determine optimal growth conditions for strains $4R1^T$, 4R9, 4L13 and 4L18, cultures were grown in nutrient broth (NB) at 4, 10, 15, 20, 25, 30, 37, 42 and 55℃ and pH 3.0~10.0 (at one unit intervals). NaCl tolerance was determined using NB medium (pH 7.0) supplemented with 0~6.0% (w/v) NaCl at 0.5% intervals for 48 h at 30℃. Growth on NA medium under anaerobic conditions was checked in an anaerobic chamber filled with a gas mixture (N_2 : H_2 : CO_2 at 90 : 5 : 5) for 1 week (Xiang et al, 2014). Oxidase and catalase activities were determined according to Liu et al (2010). Hydrolysis of casein, starch and tween 80 were assessed using the methods of Smibert and Kreg (1994). Assimilation and acid production from carbohydrates and enzyme activities were tested by using the API 20E, API 20NE and API 50CH kits (bioMérieux, France) according to the manufacturer's instructions; these tests (each 3 repetitions) were performed at 30℃ for 48h.

Chemotaxonomy

Purified cell wall preparations were obtained by the method of Schleifer and Kandler (1972) and the amino acids in peptidoglycan hydrolysates were analysed as described by Schleifer (1985). Freeze dried cells (100mg) were treated with chloroform/methanol (2 : 1, v/v) overnight to extract isoprenoid quinones. Prepar-ative thin layer chromatograpy (TLC) was performed using Kieselgel 60 F254 plates (20cm×20cm, 0.5mm thick; Merck) with petroleum ether/diethyl ether (9 : 1, v/v) as the solvent. The resulting bands were scraped from the plate under short-wavelength UV light and solubilised in acetone. Finally the quinone profile was analysed using reversed-phase HPLC (LC20AD system; Shimadzu) with a ZORBAX Eclipse XDB-C18 column (250 9 4.6mm, 5-Micron, Agilent) and a UV detector at 270nm. Polar lipids were extracted, separated using two-dimensional TLC and identified according to published procedures (Komagata and Suzuki, 1987). Cells of the 4 novel bacterial strains and the type strains of *Paenibacillus* species grown on tryptose soy agar to the same physiological age were used in the comparison. Fatty acid methyl esters were extracted, separated and identified according to the instructions of the Microbial Identification System (MIDI; Microbial ID).

Molecular analysis

Preparation of genomic DNA was carried out according to the method of Marmur (1961). The G+C content of the DNA was determined using the HPLC method (Mesbah et al, 1989). The 16S rRNA gene was PCR-amplified using the universal primers 27F (50-AGAGTTTGATC-CTGGCTCAG-30) and 1492R (50-GGTTACCTTGTTACGACTT-30) (Lane, 1991). The amplified products were purified and cloned into vector Top10 (Tiangen, China) for sequence determination carried out using an automated DNA sequencer (model ABI 3730; Applied Biosystems). The sequenc-ing primers were SP6 (50-ATTTAGGTGACACTA-TAGAATAC-30) and T7 (50-TAATACGACTCAC TATAGGG-30). The 16S rRNA gene sequences of the 4 novel bacterial strains and those of other *Paenibacillus* species retrieved from GenBank were aligned using the program CLUSTAL X v1.8 (Thompson et al, 1997). A distance matrix method (with distance options according to Kimura's two-parameter model) (Kimura, 1983), including clustering using neighbour-joining (Saitou and Nei, 1987), maximum likelihood (Felsenstein, 1981) and maximum-parsimony methods (Eck and Dayhoff 1966; Fitch, 1972) from the MEGA 5 (Tamura et al, 2011) was used to infer the phylogenetic evolutionary trees. The percentage of replicate trees in which the associated taxa clustered together in the bootstrap test (1 000 replicates) is shown next to the branches (Felsenstein, 1985).

The status of the 4 strains 4R1[T], 4R9, 4L13 and 4L18 was confirmed by genomic DNA-

DNA reassociation studies, and DNA-DNA hybridization exper-iments were further performed with 4R1^T with *P. nicotianae* YIM h-19^T and *P. hordei* JCM 17570^T using the method described by Ezaki et al (1989).

Result and discussion

Phenotypic and chemotaxonomic characteristics

Cells of 4R1^T, 4R9, 4L13 and 4L18 were observed to be Gram-positive, rod-shaped (3.0~4.0μm) × (0.6~0.8μm); Supplementary Figure S1 and S2 and spore-forming bacteria, motile and facultatively aerobic. Optimal growth of the strains was observed at 30℃, at pH 7.0 and in 0% (w/v) NaCl in NB. Morphological, cultural, physiological and biochemical characteristics of strains 4R1^T, 4R9, 4L13 and 4L18 are given in the species description and in Table 1, which shows the distinctive features of the 4 strains, as well as the closely related strains *P. nicotianae* YIM h-19^T and *P. hordei* JCM 17570^T. The fatty acid profiles of the 4 strains are shown in supplementary Table S1. The major fatty acids of strain 4R1^T were identified as anteiso-$C_{15:0}$ (30.5%), anteiso-$C_{12:0}$ (12.6%), anteiso-$C_{13:0}$ (8.7%) and ante-iso-$C_{11:0}$ (5.0%). Overall, the profiles of the 4 strains were similar to those of the reference type strains *P. nicotianae* YIM h-19^T and *P. hordei* JCM 17570^T but the proportions of some fatty acids were different. The diagnostic cell wall diamino acid of strain 4R1^T wasidentified as meso-diaminopimelic, along with the amino acids aspartic acid, glycine, arginine and alanine. The main polar lipids in strain 4R1^T were found to be diphosphatidylglycerol, phosphatidylglycerol, phosphatidylethanolamine and an unidentified lipid (Sup plementary Figure S3). The menaquinone profile, with MK-7 as the major component, is in agreement with the description of the genus *Paenibacillus* (Liu et al, 2010). The DNA G+C content of strains 4R1^T, 4R9, 4L13 and 4L18 were determined to be 51.6, 52.2, 53.1 and 53.6 mol% respectively values which lie within the range of values reported for members of the genus *Paenibacillus* (Liu et al, 2010).

Table 1 Characteristics that differentiate strains 4R1^T, 4R9, 4L13 and 4L18 from other recognised *Paenibacillus* species

Characteristic	1	2	3	4	5	6
Growth in NaCl (%)	0~1	0~1	0~1	0~1	0~3	0~3
Utilisation of substrate in API 50CH						
Glycerol	-	-	-	?	?	?
D-Mannose	?	?	-	?	?	?
L-Rhamnose	-	-	-	?	?	-
Methyl a-D-glucopyranoside	-	?	-	-	?	?
Melezitose	-	?	-	-	-	-
D-Arabitol	-	-	-	?	-	-

(continued)

Characteristic	1	2	3	4	5	6
Utilisation or activity in API 20E and API 20NE						
Mannitol	–	–	–	w	?	w
Arabinose	w	w	–	w	?	–
Nitrate reduction	–	–	–	–	–	1
Assimilation of substrate in API 20NE						
Mannose	–	1	1	1	1	1
Gluconate	1	1	–	1	?	1
Malic acid	–	–	1	–	–	–
Citrate	–	–	1	–	–	–
DNA G+C content (mol%)	51.6	52.2	53.1	53.6	54.5	53.0

Strains 1, 4R1T; 2, 4R9; 3, 4L13; 4, 4L18.5 P. nicotianae YIM h – 19T; 6. P. hordei JCM 17570T. All data were generated in this study. All the 4 novel strains have the same growth temperature 4~37℃, optimum temperature 30℃, growth pH 5~9, pH value 7.0 and NaCl 0%, w/v. All the 4 strains tested are positive for the following: growth in anaerobic conditions; activity of catalase; hydrolysis of starch; acid production from L-arabinose, D-ribose, D-xylose, D-galactose, D-glucose, D-fructose, D-mannitol, N-acetyl-glucosamine, amygdalin, arbutin, aesculin, salicin, D-cellobiose, D-maltose, D-lactose, melibiose, sucrose, D - trehalose, inulin, raffinose, starch, glycogen, gentiobiose, D - turanose; ONPG hydrolysis and galactosidase; Voges-Proskauer reaction; assimilation of esculin, glucose, arabinose, mannitol, Nacetyl glucosamine, maltose. All the 5 strains tested are negative for the following: activity of oxidase; hydrolysis of casein and tween 80; acid production from erythritol, D-arabinose, L-xylose, D-adonitol, methyl b-D-xylopyranoside, L-sorbose, dulcitol, inositol, D-sorbitol, methyl a-D-mannopyranoside, D-lyxose, D-tagatose, D-fucose, L-fucose, L-arabitol, potassium gluconate, potassium 2-ketogluconate, glucose, inositol, sorbitol, rhamnose, sucrose, melibiose, amygdalin,; assimilation of capric acid, adipic acid and phenylacetic acid; activity of arginine dihydrolase, lysine decarboxylase, ornithine decarboxylase, citrate utilisation, H2S production, tryptophan deaminase, indole production, urease and gelatinase. +, Positive; –, negative; w, weakly positive

Molecular analysis

The 16S rRNA gene sequence obtained for strain 4R1^T is a continuous stretch of 1516 bp. The GenBank/ EMBL/DDBJ accession numbers for the 16S rRNA gene sequences of strains 4R1^T, 4R9, 4L13 and 4L18 are KP965572, KP965575, KP965573 and KP965574 respectively. A comparative 16S rRNA gene sequence (over 1 500bp) similarity analysis revealed that the similarities among strains 4R1^T, 4R9, 4L13 and 4L18 were more than 99%, and the strain 4R1^T is closely related to P. nicotianae YIM h–19^T and P. hordei JCM 17570^T, with similarities of 98.41 and 98.37%, respectively. The phylogenetic trees constructed using the neighbour-joining method (Figure 1), the maximum likelihood and the maximum-parsimony methods (Supplementary Figure S4) suggested that 4R1^T, 4R9, 4L13 and 4L18 form a branch with P. nicotianae

YIM h-19^T and *P. hordei* JCM 17570^T, which shows that they are members of the genus *Paenibacillus* but represent a distinct species.

The status of the 4 strains 4R1^T, 4R9, 4L13 and 4L18 was confirmed by genomic DNA-DNA reasso-ciation studies. Strain 4R1^T showed 76.6%±0.55%, 75.0%±0.69% and 88.5%±0.62% DNA-DNA related-ness with strains 4R9, 4L13 and 4L18, respectively; and strain 4R9 showed 76.1%±0.61% and 78.3%±0.57% with strains 4L13 and 4L18; and the two strains 4L13 and 4L18 showed 76.1%±0.74% mutual DNA-DNA relatedness. DNA-DNA hybridization experiments were further performed between strain 4R1^T and *P. nicotianae* YIM h-19^T and *P. hordei* JCM 17570^T using the method described by Ezaki et al (1989). Strain 4R1^T showed 56.2%±0.87% and 43.0%±0.71% DNA-DNA related-ness with *P. nicotianae* YIM h-19^T and *P. hordei* JCM 17570^T, respectively. These values are clearly below the 70% cut-off point recommended for the assignment of strains to the same genomic prokaryotic species (Wayne et al, 1987).

In view of the combination of morphological, physiological, chemotaxonomic and phyloge-netic data discussed here, it is evident that strains 4R1^T, 4R9, 4L13 and 4L18 belong to the genus *Paenibacillus*. Differences in some phenotypic characteristics and their phylogenetic distinc-tiveness distinguish the 4 strains from previously described *Paenibacillus* species. On the basis of the data described above, the 4 strains represent a novel species of the genus *Paenibacillus*, and strain 4R1^T is designated the type strain, for which the name *Paenibacillus chinensis* sp. nov. is proposed.

Description of *Paenibacillus chinensis* sp. nov

Paenibacillus chinensis (chin. en'sis. N. L. masc. adj. *chinensis*, pertaining to China, where the type strain was isolated and studied).

Cells are Gram-positive, facultatively aerobic, motile with peritrichous flagella, spore-forming and rod-shaped [(3.0~4.0) μm×(0.6~0.8) μm]. Colonies on NA plate are white-cream with light pink undersides, circular and raised with irregular edges. Growth occurs between 4~37℃ (optimum 30℃) and at pH5.0~9.0 (optimum pH 7.0). Grows in 0~1.0% (w/v) NaCl (optimum 0%) in NB. Cells are positive for growth in anaerobic conditions on NA plates, activity of catalase, ONPG hydrolysis, Voges-Proskauer reac-tion, hydrolysis starch and galactosidase; assimilation of esculin, glucose, arabinose, mannitol, N-acetyl-glu-cosamine, maltose and gluconate; acid production from L-arabinose, D-ribose, D-xylose, D-galactose, D-glucose, D-fructose, D-mannitol, N-acetyl-glu-cosamine, amygdalin, arbu-tin, aesculin, salicin, D-cellobiose, D-maltose, D-lactose, melibiose, sucrose, D-trehalose, inulin, raffinose, starch, glycogen, gentio-biose, D-turanose and D-mannose. Cells are negative for activity of oxidase; hydrolysis of casein and tween 80; activities of nitrate reduction, arginine dihydro-lase, lysine decarboxylase, ornithine decarboxylase, mannitol, citrate utilisation, H$_2$S production, trypto-phan deaminase, indole production, urease and gelati-nase; acid production from erythritol, D-arabinose, L-xylose, D-adonitol, methyl b-D-xylopyranoside, L-sorbose, dulcitol, inositol, D-sorbitol, methyl a-D-man-nopyranoside, D-lyxose, D-tagatose, D-fucose, L-fucose, L-arabitol, potassium

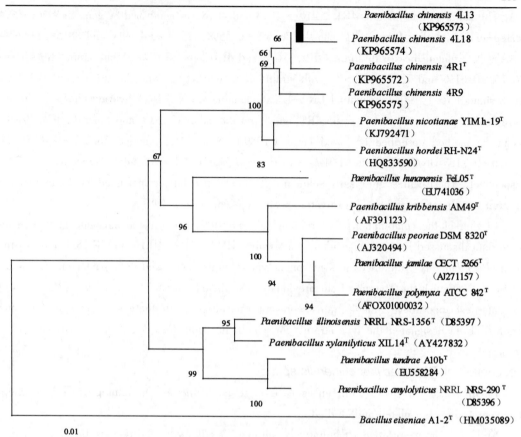

Figure 1 Phylogenetic tree showing the relationships of the novel strains to closely related species within the genus*Paenibacillus*

The trees were based on an alignment of almost-complete 16S rRNA gene sequences and constructed using the neighbour-joining method (Saitou and Nei, 1987). The phylogenetic tree was produced following the bootstrap sampling of 1 000 datasets. The numbers at the nodes indicate the percentages of bootstrap support based on neighbour-joining analysis of 1 000 resampled datasets. The scale bar indicated 0.01 substitutions per nucleotide position

gluconate, potassium 2-ketogluconate, glucose, inositol, sorbitol, rhamnose, sucrose, melibiose, amygdalin, glycerol, L-rhamnose, methyl a-D-Glucopyranoside, melezitose and D-ara-bitol; assimilation of capric acid, adipic acid, pheny-lacetic acid, mannose, malic acid and citrate. The diagnostic cell wall diamino acid is meso-di-aminopimelic, with the amino acids aspartic acid, glycine, arginine and alanine present. The predomi-nant respiratory quinone is MK-7. The major fatty acids are anteiso-$C_{15:0}$, anteiso-$C_{12:0}$, anteiso-$C_{13:0}$ and anteiso-$C_{11:0}$ (>5%). The major polar lipids are diphosphatidylglycerol, phosphatidylglycerol, phos-phatidylethanolamine and an unidentified lipid. The DNA G+C content of the type strain is 51.6 mol%.

The type strain 4R1^T (=KCTC 33672^T = CICC 23864^T), was isolated from seeds of hybrid maize (*Zea mays* L., Jingke 968) sampled from the scientific research base of Beijing Academy of Agriculture and Forestry Sciences (BAAFS) in Gansu province, China. The GenBank/EMBL/DDBJ accession number for the 16S rRNA gene sequence of strain 4R1^T is KP965572. Strains 4R9 (= KCTC 33673 = CICC 23865), 4R13 (= KCTC 33668 = CICC 23862) and 4R18 (=KCTC 33669 = CICC 23863) are additional strains of the species, isolated from the same source.

Acknowledgments

This work was supported by the Beijing nova program (No. Z141105001814095), the Beijing nova interdisciplinary cooperational program (No. Z1511000003150150), the National Natural Science Foundation of China (No. 31300008), the Chinese Postdoctoral Science Foundation (No. 2015M570969), the fund of National Infrastructure of Microbial Resources (No. NIMR2015-4), and the Scientific and Technological Development Project of China, National Research Institute of Food and Fermentation Industries (No. 2012KJFZ-BS-01).

References (omitted)

Antonie Van Leeuwenhoek, 2016, 109 (2): 207-213

High-throughput Sequencing-based Analysis of the Composition and Diversity of Endophytic Bacterial Community in Seeds of "Beijing" Hybrid Maize Planted in China

Liu Yang[1,2]　Wang Ronghuan[1]　Li Yinhu[3]　Cao Yanhua[2]
Chen Chuanyong[1]　Qiu Chuangzhao[3]　Bai Feirong[2]
Xu Tianjun[1]　Zhang Xin[2]　Dai Wenkui[3]　Zhao Jiuran[1]　Cheng Chi[2]

(1. *Maize Research Center*, *Beijing Academy of Agriculture and Forestry Sciences*, *Beijing* 100097, *China*; 2. *China Center of Industrial Culture Collection*, *China National Research Institute of Food and Fermentation Industries*, *Beijing* 100015, *China*; 3. *Department of Microbial Research*, *BGI Shenzhen*, *Shenzhen* 518083, *China*)

Abstract: Maize (*Zea mays* L.) is the largest food crops in China with the plangting area and total yield of 37.076 million hectares and 215.67 million tons respectively in 2014. The technology of cross breeding was the primary method to cultivate new maize varieties and promote the yield level. In recent years, more and more agriculturalists discovered the existence of endophyte in maize and their close relationship with soil environmental adaption which affect the production of maize. In this study, the seeds of six different maize varieties which were self-developed and cultivated from capital city of China "Beijing" and extensively planted in China were collected, this is the first time to acquire all of the "Beijing" hybrid maize to investigate their endopytes. We clarified eight species exists in all the varieties and the relative abundance of top three species including *Pantoea agglomerans*, *Enterobacter cloacae* and *Aeribacillus pallidus* taken about 60% of the whole endophyte. Besides these, we also discovered the correlations between the endophytic bacteria which might affect the growth of maize. On the other hand, the distributions of *E. cloacae* and *A. pallidus* between maize varieties with different male parent were apparently different. So we deduced the endophyte affect the environmental adaptation of different maize varieties and the results showed the light on the future maize variety cultivation from the angle of endophyte.

Key words: Beijing; Maize seed; Endophyte; Diversity; *Pantoea agglomerans*

Introduction

Maize (*Zea mays* L.) is a native of Mexico and Peru along the Andean areas in South America. Since Columbo discovered America in 1492, it had been brought to every corner of the

---

Liu Yang, Wang Ronghuan and Li Yinhu have contributed equally to this work world and became one of the important crops. Compatible with the different soil environment, various maize varieties were cultivated. In order to obtain higher production, cross breeding was adopted in common. In recent years, the endopyte has received significant attention from agriculturists and the studies on their relationship with maize production have been launched.

In previous researches, many plants were proved to host an abundant microbial community in rhizosphere, phyllosphere and endosphere (Lebeis et al, 2012; Turner et al, 2013; Bulgarelli et al, 2013; Berg et al, 2014). These microbes colonizing plant surfaces and interior issues were proved vital for plant health and productivity (Bonfante, 2010; Berendsen et al, 2012; Ferrara et al, 2012; Berg et al, 2014), but some of them could lead to disease development of plants (James and Olivares 1998; Monteiro et al, 2012; Van Overbeek et al, 2014) as prior reports also indicated that plant microbiome could impose substantial effect on human health through consumption of raw plants (Blaser et al, 2013; Van Overbeek et al, 2014). Therefore, understanding microbial composition and interaction networks between them and plants will advance the development of sustainable agriculture (Berg, 2009; Lugtenberg and Kamilova, 2009).

Several scientific teams documented the feasibility of promoting environment-friendly agriculture through manipulation of plant microbiome (Bloemberg and Lugtenberg, 2001; Adesemoye et al, 2009; Philippot et al, 2009; Singh et al, 2010; Bakker et al, 2012). Bloemberg *et al.* revealed that plant microbiome could reduce incidence of plant disease (Bloemberg and Lugtenberg, 2001), and research conducted by Bakker et al showed contribution of plant microbiome to agricultural production (Bakker et al, 2012). Plant microbiome also holds the potential to keep plant productivity with decreased chemical inputs (Adesemoye et al, 2009) and function as a key player in biogeochemical cycles (Philippot et al, 2009; Singh et al, 2010). Although endophytic microbes were ever considered contaminants in some of prior reports (Ryan et al, 2008; Reinhold-Hurek and Hurek, 2011; Mitter et al, 2013), they should be the most stable microbial partners of plants. Various researches identified endophytic bacteria in plants and implicated their significance in promoting plant growth and control phytopathogens (James, 2000; James et al, 2002; Reinhold-Hurek and Hurek, 2011; Sessitsch et al, 2012; Suarez-Moreno et al, 2012). But it is challenging to isolate and inoculate these inner bacteria, making it difficult to get a whole-picture of interaction network among various bacteria and between them and hosts. Therefore culture-independent strategy is increasingly used to uncover endophytic bacterial community like those in rice and sugarcane (Sessitsch et al, 2012; Fischer et al, 2012).

In this study, we performed 16S rRNA gene analysis on 6 different maize varieties which were self-developed and cultivated from capital city of China "Beijing" and extensively planted in China since 2009. The components and abundances of endophytic bacteria which corresponding to different varieties have been clarified and compared. From the research, we want to understand the effect of endophye on the maize production and the relationship between endophytes distribution and maize varieties.

Materials and Methods

Maize seed sampling

Seeds were collected from four hybrid combinations of maize (*Zea mays* L.) (Jinghua8, Jingnongke728, Jingdan68, NK718, Jingke968, Jingke665) supplied by Professor Jiuran Zhao at Beijing Academy of Agriculture and Forestry Sciences (BAAFS), which were self-developed and cultivated from capital city of China "Beijing" and extensively planted in China since 2009 (Supplementary Figure1). The genetic relationships among the samples are shown in Table1. All the samples were collected from the scientific research base of BAAFS in Zhangye, Gansu (39°04′57.94″ N, 100°12′49.40″ E, northwestern China) in 2013 and stored at 4℃ in aseptic bags, and the seeds were stored for 2 weeks before the experiments were carried out. Maize seed samples were surface-sterilized according to the Lundberg's et al and Edwards's et al method (Lundberg et al, 2012; Edwards et al, 2015).

Table 1 General information on maize varietie

Variety	Male parent	Female parent	Market Release Year	Raw starch content (%)
Jinghua8	Jing2416	JingX005	2009	75.29
Jingnongke728	Jing2416	JingMC01	2012	73.33
Jingdan68	Jing2416	CH8	2010	73.65
NK718	Jing2416	Jing464	2011	75.32
Jingke968	Jing92	Jing724	2011	75.42
Jingke665	Jing92	Jing725	2013	74.54

DNA extraction, amplification, and sequencing

About 5.0g of surface-sterilized samples were frozen with liquid nitrogen and quickly ground into a fine powder with a precooled sterile mortar. Then, the CTAB procedure was used to extract bacterial DNA (Liu et al, 2012), which was used as template to amplify V6 region of the 16S rDNA by primers 967F (5′-CAACGCGAAGAACCTTACC-3′) and 1046R (5′-CGACAGCCATGCANCACCT-3′). The purified PCR products were mixed in equal concentration, and sequenced by Miseq (Illumina, USA.) following the manipulation instructions at BGI Shenzhen (China).

Acquisition of OTUs

The reads with low quality bases, adapter sequences, N bases or amplification primer sequences were filtered and high quality paired-reads were connected into tags by overlapping. The minimum overlapping length was 30bp without mismatch or N base. Non-redundant tags were produced by Mothur (Schloss et al, 2009) (version 1.27.0), and the unique tags were the typical tags representing all the similar tags. Unique tags were listed based on abundance and pre-clustered by single-linkage pre-clustering (SLP) following 98% similarity. Then the unique tags were annotated and clustered into operational taxonomy units (OTUs) following 97% identity (Edgar, 2013).

Taxonomy assignment and abundance analysis

Unique tags were classified by alignment to Silva RefSSU database using BLAST (version 2.2.23, and the key parameters were "-p blastn-m 8-F F-a 2-e 1e-5-b 50"), and the best alignments were selected. If more than 66% of the unique tags in a OTU were aligned to the same species, the OTU was assigned to the species and then the analysis went into the next taxonomic rank. The abundance of OTUs in different classification levels was calculated according to the alignment results.

PCA analysis

Principal component analysis (PCA) was used to summarize factors mainly responsible for this difference, similarity is high if two samples are closely located. Based on the OTU abundance information, the relative abundance of each OTU in each sample will be calculated, and the PCA of OTU was done with the relative abundance value. The software used in this step was package "ade4" of software R (v3.0.3).

Analysis on sample complexity and similarity

Sample complexity was measured by indexes including chao1, ACE, Shannon and Simpson. Values of rarefaction was calculated by Mothur (Schloss et al, 2009) (version 1.27.0) and the rarefraction curve was drawn by R (version 2.11.1) in which the extracted tags were used as X-axis and the OTUs number was used as Y-axis. Heatmap was a graphical representation of data where the individual values contained in a matrix were represented as colors: species clustering was established based on their abundance. Longitudinal clustering indicated the similarity of all species among different samples, and the horizontal clustering indicated the similarity of certain species among different samples. Heatmaps were generated using the package "gplots" of software R (v3.0.3) and the distance algorithm is "euclidean", the clustering method is "complete".

Comparison on bacterial distribution

Using the OTUs distribution in different maize varieties, the Venn diagram was drawn by VennDiagram of software R (v3.0.3), then the common and specific OTU ID were summarized. On the other hand, the bacterial distribution was calculated for maize with different male parent on the taxonomy level of genus and species. Then the box plot diagram was drawn by package of "ggplot2" in software R (v3.0.3).

Results

Information on sample collection

In the study, the maize seeds were collected by cross breeding and most of them have been planted extensively in the north of China since 2009. 6 groups of maize seeds with 3 repeats were obtained and the samples can also be classified into 2 groups by their male parent. The detail information on the maize varieties was listed in Table1.

Overall of sequencing data and microbial community in maize seeds

The microbial community in the maize seeds were sequenced by MiSeq and nearly 20 000 high quality 16S rRNA gene sequence were produced for each sample. In total, 155 operational taxonomic units (OTUs, ≥97% similarity) were obtained from 18 samples and the number of OTUs ranged from 31 to 103. The sequencing data covered almost all the species in the community of the samples under the indication of the rarefaction curve (Figure 1). On the other hand, the distributions of endophytic microbial composition in different maize seeds were detected on different taxonomy level (Figure 1) and other bacterial diversity indices including chao1, ace, shannon and simpson were applied for the estimating of endophytic community complexity (Supplementary1). Under the indication of the results, the diversity of endopytic community in Jinghua8 was much less than other varieties and P. agglomerans was the primary species in all the varieties. As Gond et. al. reported, *P. agglomerans* could enhance the salt tolerance capability which was significant for the cultivation of maize (Gond et al, 2015).

Similarity in endophytic bacteria of different maize varieties

With the bacterial components and their corresponding abundances, the similarity in endophytic bacteria was detected between the varieties. The results of principal component analysis (PCA, Figure 2) exhibited that Jinghua8 contained different endophytic bacterial composition from other varieties and the situation was also found in Jingke665. But it was hard to distinguish the other 4 maize varieties through their bacterial components with the PCA results. Then the samples were clustered by the similarity of their species which weighted by species abundances on different taxonomy level. Based on the results, it was found that the samples from Jingke968 and Jingke665 which have the same male parent were clustered together, and other samples were located on the other branch except samples from Jinghua8 (Figure 2). So we deduced that the components of endophytic bacteria might be affected by their varieties.

Comparison between maize varieties with different male parent

In order to detect the effect of different male parent on the composition of endophytic bacteria, the samples were divided into 2 large groups including group1 which contained samples from Jinghua8, Jingnongke728, Jingdan68 and NK718, and group2 which contained samples from Jingke968 and Jingke665. With the bacterial distribution results, we found that the relative abundances of the most genera between these two groups were same except for *Enterobacter* which was higher in group1 and *Pantoea* which was slightly lower. Under the indication of species composition, the key species which affect the species abundance were *Enterobacter* cloacae and *P. agglomerans* separately (Figure 3). With the previous reports, we knew that E. cloacae could mineralize the endosulfan in the soils and it was helpful for plant growth (Abraham et al, 2015).

Special species contained in maize varieties

To detect the special endophytic bacteria which might affect the production or growth of maize varieties with different male parent, core-pan analysis which based on OTUs was carried out. The results exhibited that four varieties in group1 shared 45 OTUs and two varieties in group2 shared 89 OTUs. Then we found that all the OTUs shared by group1 were also existed in group2 but

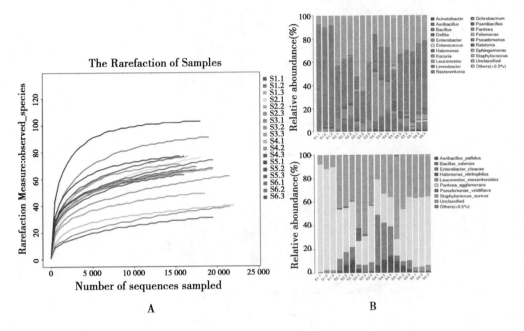

Figure 1 Rarefaction curve and bacterial distribution in different maize seeds

A. The rarefaction curve of observed species which evaluated by 16s sequences; B. The composition distribution and relative abundance in maize varieties on genus-level and species-level. The upper picture and lower pictures stand for genus-level and species-level separately. S1.1, S1.2 and S1.3 from Jinghua8; S2.1, S2.2, and S2.3 from Jingnongke728; S3.1, S3.2, and S3.3 from Jingdan68; S4.1, S4.2, and S4.3 from NK718; S5.1, S5.2, and S5.3 from Jingke968; S6.1, S6.2, and S6.3 from Jingke665

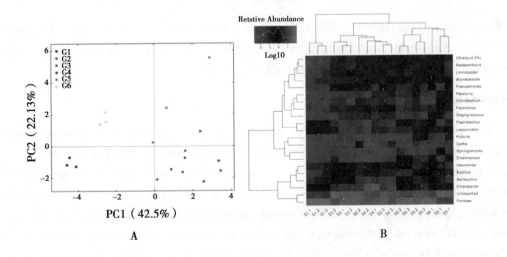

Figure 2 Similarity in bacterial composition of different maize varieties

A. PCA of endophytic bacteria based on OTUs; B. Heatmap of sample clustering on genus level. The species of which abundance is less than 0.5% in all samples were classified into "others"

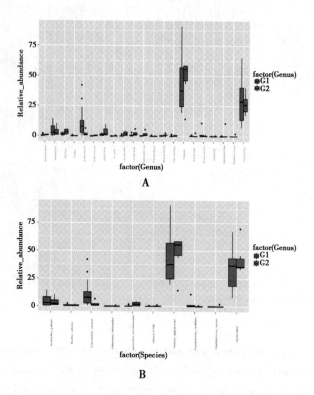

Figure 3 Comparison on relative abundances of endophyte in different maize varieties

A. The relative abundances of genus in two groups; B. The relative abundances of species in two groups. In the figure, G1 stand for group1 including Jinghua8, Jingnongke728, Jingdan68 and NK718; G2 stands for group2 including Jingke968 and Jingke665

group2 contained 5 special OTUs which did not exist in any of the samples in group1 (Figure 4). The OTUs corresponded to Actinobacteria, Bacteroidetes, Firmicutes and Proteobacteria which often existed in soil and were close related with human health.

Discussion

Endophytic bacteria refer to those microbes that are able to colonize in the tissues of healthy plants and subsequently establish a harmonious relationship with plants. They are able to affect maize growth directly or indirectly through biological control, plant growth-promoting effects, endophytic nitrogen-fixing activity and so on (Liu et al, 2012). In order to boost maize yield, the technique of cross breeding has been adopted many years and the plant endophyte has gained significant attentions since the extensive application of second generation sequencing which could help us to trace the rare microbes in plant tissues. Nitrogen-fixation microbe was one of the famous endophyte which known for the growth of leguminous plants, and more attention have been focused on the endophyte in other crops. Gond et al has found the existence of endophyte in maize and the role of *P. agglomerans* in the metabolism of salt and water in maize has been clarified. But the domain endophytes in maize varieties and variation of endophyte in different maize varieties

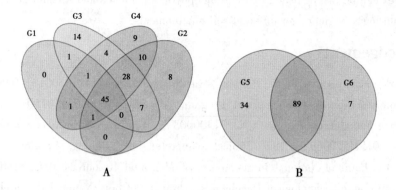

Figure 4 Core-pan analysis on OTUs

A. Core-pan analysis on group1; B. Core-pan analysis on group2. In the figure, G1 stand for Jinghua8; G2 stand for Jingnongke728; G3 stand for Jingdan68; G4 stand for NK718; G5 stand for Jingke968; G6 stand for Jingke665

have seldom accounts.

In the research, 6 maize varieties self-developed and cultivated from capital city of China "Beijing" and extensively planted in China since 2009 and the endophyte in these maize seeds were investigated. With the analyzing results of 16S rRNA gene, 45 OTUs which could be annotated to 8 species including *A.pallidus*, *B.safensis*, *E.cloacae*, *H.nitritophilus*, *L.mesenteroides*, *P.agglomerans*, *P.viridiflava* and *S.aureus* were shared by these maize varieties. *P.agglomerans* was the primary endophyte whose average relative abundance reached 46.55% in all varieties. *E. cloacae* which charges for the mineralization of endosulfan account for 8.52% of all the endophyte and the average relative abundance of *A.pallidus* was 4.38%. Besides these, *L.mesenteroides* was helpful for nitrogen fixation (Martínez-Rodríguez et al, 2015), *B.safensis* could secret indoleacetic acid, *H.nitritophilus* could balance the cell osmotic pressure, *P.viridiflava* could antagonize the maydis pathogen. And the existence of *A.pallidus* and *S.aureus* in the maize was closely related with human health (Radha et al, 2014).

Based on the relative abundances of endophytes in different maize verities, the correlations within endophyte microorganisms were detected. The Person-correlation coefficients were-0.67 between *P.agglomerans* and *H.nitritophilus*. On the other hand, *A.pallidus* was positive correlated with *B.safensis* (R=0.94) and *H.nitritophilus* (R=0.78) in maize seeds. Although the function of *A.pallidus* was not quite clear, its positive relation with other endophyte was helpful the growth of maize. So we deduced that the high relative abundance of *P.agglomerans*, *E.cloacae* and *A.pallidus* could promote the production of maize and *H.nitritophilus* which negatively correlated with *P. agglomerans* should also be needed to balance the cell osmotic pressure. On the other hand, the comparison on the relative abundances of endophyte exhibited that *E.cloacae* was higher and *P.agglomerans* was lower in the maize with Jing2416 as male parent. This is the first time to acquire all of the "Beijing" hybrid maize to investigate their endopytes and the results demonstrated that maize varieties can be distinguished by their endophytic abundance and it could be help to culture the compatible maize verieties for different soil environments.

Acknowledgements

This work was supported by the Beijing Nova Program (No. Z141105001814095), the Beijing Nova Interdisciplinary Cooperational Program (No. Z1511000003150150), the National Natural Science Foundation of China (No. 31300008), the Chinese Postdoctoral Science Foundation (No. 2015M570969), the Project supported by Beijing Postdoctoral Research Foundation, the Fund of National Infrastructure of Microbial Resources (No. NIMR2016-4), and the Scientific and Technological Development Project of China National Research Institute of Food and Fermentation Industries (No. 2012KJFZ-BS-01). We also thank Dr. Zhengqiu Cai at Brigham and Women's Hospital (USA) for assistance with the English.

References (omitted)

Identification and Antagonistic Activity of Endophytic Bacterial Strain *Paenibacillus* sp. 5 L8 Isolated from the Seeds of Maize (*Zea mays* L., Jingke 968)

Liu Yang[1,2] Wang Ronghuan[1] Cao Yanhua[2] Chen Chuanyong[1]
Bai Feirong[2] Xu Tianjun[1] Zhao Ran[2] Zhang Xin[2]
Zhao Jiuran[1] Cheng Chi[2]

(1. *Maize Research Center, Beijing Academy of Agriculture and Forestry Sciences, Beijing* 100097, *China*; 2. *China Center of Industrial Culture Collection, China National Research Institute of Food and Fermentation Industries, Beijing* 100015, *China*)

Abstract: Endophytes are related to the health and growth of plants. In this study, the endophytic bacterial strain *Paenibacillus* sp. 5 L8 was isolated from the seeds of maize (*Zea mays* L., Jingke 968). Strain 5 L8 was identified as *Paenibacillus polymyxa* by polyphasic taxonomy based on morphological and physiological methods, and 16S rRNA gene phylogenetic analysis. Via antagonistic testing against the common plant pathogens *Fusarium graminearum*, *Bipolaris maydis*, *Bipolaris sorokiniana*, *Cochliobolus heterostrophus*, *Aspergillus aculeatus*, *Phomopsis chimonanthi* and *Verticillium dahliae*, strain 5 L8 showed comprehensive and good antagonistic activity. The draft genome of *Paenibacillus* sp. 5 L8 was sequenced by Illumina HiSeq 2000, and three coding sequences (CDSs) for the glucanase gene were annotated and correlated to antagonistic activity. Using specific primers to amplify the biocontrol gene in the glucanase family, the $\beta-1, 3-1, 4-$glucanase gene (*gluB*) was found. This study supplies a basis for further study of industrial research and application.

Key words: Maize "Jingke 968"; Endophytes; *Paenibacillus* sp; 5L8; Identification; Antagonistic activity

Introduction

As a reproductive organ of spermatophytes, a seed is an embryonic plant that carries parental genes, and it is taken as a specific and important development stage of individual plants (Guan, 2009). Seeds are not only preservers and messengers of plant genetic information, but are also an adaptive strategy for plants to guarantee species reproduction under environmental stress (Song et al, 2008). Meanwhile, seeds are the basis of agricultural production, playing an important role in food security. As a special and irreplaceable means of agricultural production, the degree of seed development and application is an important indication of a country's development of its agricultural industry (Hu, 2006).

Endophytic bacteria are microbes capable of colonizing the tissues of healthy plants and subsequently establishing a harmonious relationship with their hosts (Kloepper and Beauchamp, 1992). They are able to affect plant growth directly or indirectly through biological control, plant growth-promoting effects, endophytic nitrogen-fixing activity and so on (Glick et al, 1999). Plant seeds are carriers of various beneficial bacteria and pathogens. As one of the important sources of endophytic and rhizosphere microorganisms, plant seeds affect plant traits such as growth and development, health conditions, quality and production output, and the generation of functional components (Liu et al, 2013).

Maize is the most widely grown grain crop throughout the world, and it is also the major crop in China and in Beijing. Meanwhile, maize has become a fastest-growing main crop in terms of its crop area and production output in China in recent years. Development of maize production is of great importance for China's food security, livestock husbandry development as well as improving the dietary pattern and living standard of its citizens (Zhao and Wang, 2013). In recent years, maize disease caused by the colonization and dispersal of pathogenic fungus such as *Fusarium* sp., *Bipolaris* sp. and *Verticillium* sp. have become the main obstacle for affecting the production of maize and restricting the development of maize industry. Using the microbial culture method, (Liu et al, 2015, unpublished) conducted a study on the structure and diversity of endophytic bacterial communities in seeds of the new maize variety, Jingke 968, planted in the northern region of East China. It was found that *Paenibacillus* sp. was the first dominant bacterium in an endophytic bacterial community of Jingke 968 seeds. Moreover, no bacterial colony was found around bacterial strain 5 L8 on LB agar for 72h at 30℃. Accordingly, an initial conclusion can be made that the bacterial strain 5 L8 features biological antagonism activity (Liu et al, 2015, unpublished). This paper conducted polyphasic taxonomy identification on the bacterial strain from the aspects of morphological characteristics, cultural characteristics, physiological and biochemical characteristics and 16S rRNA gene sequencing. By employing techniques such as the antagonism and bio-control tests, whole-genome sequencing and bio-control gene cloning, this paper presents further study on the antagonistic activity and mechanism of bacterial strain 5 L8. This study has laid a foundation for developing and utilizing biological bio-control bacteria agents, enhancing the capacity of crop resistance to fungal diseases, and improving crop production output.

Materials and methods

Strain 5 L8

Strain 5 L8 was isolated from seeds of hybrid maize "Jingke 968" (*Zea mays* L. Jingke 968) collected from the scientific research base of BAAFS in Zhangye, Gansu (39°04′57.94″ N, 100°12′49.40″ E, northwestern China) in September 2014, and the strain has been deposited in the China Center of Industrial Culture Collection (CICC) under the number CICC 23866.

Pathogenic strains for antagonistic testing

All the pathogenic strains selected in this study, *Fusarium graminearum* CICC 2697, *Bipolaris maydis* CICC 2530, *Bipolarisso ro kiniana* CICC 2531, *Cochliobolus heterostrophus* CICC

2699, *Aspergillus aculeatus* CICC 41132, *Phomopsis chimonanthi* CICC 2611 and *Verticillium dahliae* CICC 2534, were supplied by the CICC.

Morphological characteristics observation of strain 5 L8

Colonial properties of strain 5 L8 were observed on LB agar (L^{-1}: 10g casein peptone, 5g yeast extract, and 10g NaCl; pH 7.0). Cell morphology was examined by light microscopy, scanning electron microscopy (SEM) and transmission electron microscopy (TEM) according to Liu et al (2010, 2015).

Biochemical characteristics identification of strain 5 L8

Carbon source utilization and enzyme activities of strain 5 L8 were tested by using analytical profile index (API) kits (API 20E and API 50CH, all bioMérieux) according to the manufacturer's instructions. API 20E and API 50CH profiles were investigated, and these tests (three repetitions each) were per formed under 30℃ for 48h.

16S rRNA gene phylogenetic analysis of strain 5 L8

A biomass loop was scraped off the agar plate, suspended in 20ml ddH$_2$O and lysed by boiling for 10min and freezing for 5min. Following centrifugation, the supernatant was used as the template for polymerase chain reaction (PCR) analysis. The 16S rRNA gene was amplified using the universal primers 27F (5'-AGAGTTTGATCCTGGCTCAG-3') and 1492R (5'-GGTTAC-CTTGTTACGACTT-3'; Lane, 1991). The amplified products were purified and cloned into vector Top10 (Tiangen) for sequence determination. Automated sequencing was performed using an Applied Biosystems "BigDye Primer" cycle sequencing ready reaction kit and an ABI 3730 DNA sequencer. The sequencing primers were SP6 (5'-ATTTAGGTGACACTATAGAATAC-3') and T7 (5'-TAATACGACTCACTATAGGG-3'). The 16S rRNA gene sequences of the strain 5 L8 and those of other Paenibacillus species retrieved from GenBank were aligned using the program CLUSTAL_X v1.8 (Thompson et al, 1997). Tree-making algorithms, namely the neighbour-joining method (Saitou and Nei, 1987) from molecular evolutionary genetics analysis software (MEGA5, Tamura et al, 2011), were used to infer the phylogenetic evolutionary trees. The percentage of replicate trees in which the associated taxa clustered together in the bootstrap test (1 000 replicates) is shown next to the branches (Felsenstein, 1985).

Antagonistic test of strain 5 L8 with pathogens

After two events of transfer-incubation and excitation of pathogens selected in this study, 10mL of sterile water was added into the pathogen's tube; the spore and hypha were carefully scraped off with an inoculating needle, and poured into a triangular flask with glass beads. The pathogenic blocks were then shaken on an oscillator to break them apart, and then stewed for 5min. The supernatant was then poured into a 300ml potato dextrose agar (PDA) culture medium which had been cooled down to 45℃; after shaking, it was poured onto a plate. After cooling solidification, small circular sterile filter papers were placed (two pieces/vessel), cultivating for one night at the temperature of 3.0℃. The bacterial strain 5 L8 to be tested was put into lysogeny broth (LB) medium, cultivated at a 150rpm shaking speed and 30℃ for one night. The density of the strain suspension was 108~109 cell/ml and it was dropped onto the filter paper with pathogenic culture medium (3μl/piece), cultivated 24~36h at 30℃, and the inhibition zone was observed.

Whole genome sequencing of strain 5 L8

The draft genome of strain 5 L8 was determined by Illumina HiSeq 2000 according to the method of Liu et al (2014). The genome sequence was annotated by the "Rapid Annotations Using Subsystems Technology" (RAST) server (Aziz et al, 2008). tRNAs and rRNAs were predicted by using tRNAscan-SE v. 1. 23 (Lowe and Eddy, 1997) and RNAmmer 1. 2 (Lagesen et al, 2007), respectively.

Biocontrol gene amplification of strain 5 L8

According to the results of whole genome sequencing of strain 5 L8, specific primers W1 (5'-TTGCGAATGTGTTCTGGGAACC-3') and W2 (5'-TACTAATTGCTCGTATATTTTACCCA-3'; Yao et al 2004) were selected and used to clone the biocontrol gene of β-1, 3-1, 4-glucanase (gluB). A 50μl volume of PCR mixture containing 50ng of DNA extract, 1×Taq reaction buffer, 20 pmol of each primer, 200μmol each dNTP, and 1.5 units of Taq enzyme (Fermentas, Thermo Fisher Scientific, Vilnius, Lithuania). The cycling procedure consisted of an initial denaturation cycle at 94℃ for 5min, followed by 30 cycles of denaturation at 94℃ for 30s, annealing at 56℃ for 50s, and elongation at 72℃ for 1min, with a final extension at 72℃ for 10min. The PCR products were then electrophoretically separated and the target band was excised and purified by the "Wizard SV Gel and PCR Clean-up System" (Promega, Madison, WI) as described by the manufacturer. Automated sequencing was performed using an ABI "BigDye Primer" cycle sequencing ready reaction kit and ABI 3730 DNA sequencer, and the sequencing results were blasted in the GenBank database.

Results

Morphological observation

Cells were Gram-positive, aerobic or facultatively anaerobic, motile by means of peritrichous flagella, endospore-forming, rod-shaped bacterium [0.5μm×(1.0~2.0) μm] observed by light microscope, SEM and TEM. After 24~48h growing on LB, colonies were 3~7mm in diameter, circular, raised, with entire edges and offwhite to yellow (Figure 1).

Biochemical characteristics identification

The results of physiological characterization of strain 5 L8 were shown in Table1. Strain 5 L8 tested by API 20E and API 50CH kits was positive for gelatin and ortho-nitrophenyl-β-galactoside (ONPG) hydrolysis, and utilization of D-glucose, D-mannitol, sucrose, melibiose, amygdalin, L-arabinose, D-ribose and so on.

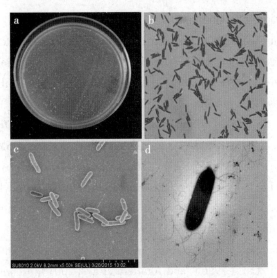

Figure 1 Morphological observation of strain 5 L8

a. Colonies; b. Cells under light microscope; c. Cells under SEM; d. Cells under TEM

Table 1 Biochemical characteristics of strain 5 L8

Characteristic	Result	Characteristic	Result	Characteristic	Result	Characteristic	Result
D-Glucose	+	D-Mannitol	+	Inositol	-	D-Sorbitol	-
L-Rhamnose	-	Sucrose	+	Melibiose	+	Amygdalin	+
L-Arabinose	+	Erythritol	-	D-Arabinose	-	D-Ribose	+
D-Xylose	+	L-Xylose	-	D-Adonitol	-	Methylβ-D-xylopyranoside	-
D-Galactose	+	D-Fructose	+	D-Mannose	+	L-Sorbose	-
Dulcitol	-	Methylα-D-mannopyranoside	-	Methylα-D-glucopyranoside	+	N-Acetyl-glucosamine	-
Arbutin	+	Aesculin	+	Salicin	+	D-Cellobiose	+
D-Maltose	+	D-Lactose	+	D-Trehalose	+	Inulin	+
D-Melezitose	-	D-Raffinose	+	Starch	+	Glycogen	+
Xylitol	-	Gentiobiose	+	D-Turanose	+	D-Lyxose	-
D-Tagatose	-	D-Fucose	-	L-Fucose	-	D-Arabitol	-
L-Arabitol	-	Potassium gluconate	-	Potassium 2-keto-gluconate	-	Potassium 5-keto-gluconate	-
Hydrolyses gelatin	+	ONPG hydrolysis	+	Arginine dihydrolase	-	Lysine decarboxylase	-
Ornithine decarboxylase	-	Citrate utilization	-	H_2S production	-	Urease	-
Tryptophan deaminase	-	Indole production	-				
Voges-Proskauer reaction	+						

Note: +, Positive; -, negative; ONPG ortho-nitrophenyl-β-galactoside

16S rRNA gene phylogenetic analysis

The 16S rRNA gene sequence of strain 5 L8 was a continuous stretch of 1 466bp, and the GenBank/EMBL/DDBJ accession number for the 16S rRNA gene sequence of strain 5 L8 was KP843070. Sequence-similarity calculations (over 1 450bp) indicated that strain 5 L8 was most closely related to *Paenibacillus jamilae* CECT 5266T (98.49% similarity), and shared sequence similarities of 97.06%~98.36% with other members of genus *Paenibacillus*, including *Paenibacillus polymyxa* ATCC 842T (98.36%), *Paenibacillus peoriae* DSM 8320T (97.87%), *Paenibacillus brasilensis* PB172T (97.82%), *Paenibacillus kribbensis* AM49T (97.67%) and *Paenibacillus terrae* AM141T (97.06%). Based on the results of the phylogenetic analysis, high sequence similarity was found between strain 5 L8 and its closest related type strains, so the strain belongs to the genus *Paenibacillus* (Figure 2).

Based on phenotypic and genotypic data, as well as phylogenetic analysis, and according to the report of Heyndrickx et al (1996) and Aguilera et al (2001), it was suggested that the endophytic bacterial strain 5 L8 was clas-sified as a member of the genus *Paenibacillus*, and it belongs to *Paenibacillus polymyxa*.

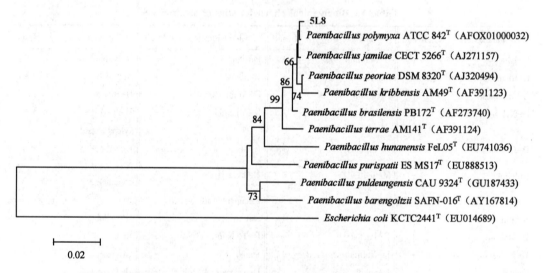

Figure 2 Phylogenetic tree showing the relationships of strain 5 L8 to closer species within the genus Paenibacillus and Escherichia coli as an outgroup

The tree was based on an alignment of almost-complete 16S rRNA gene sequences and constructed using the neighbour-joining meth-od (Saitou and Nei 1987). The phylogenetic tree was produced following the bootstrap sampling of 1000 datasets. The numbers at the nodes indicate the percentages of bootstrap support based on neighbour-joining analysis of 1000 resampled datasets. The scale bar indicated 0.02 substitutions per nucleotide position

Antagonistic test

Through the antagonistic test with common plant pathogens *Fusarium graminearum* CICC 2697, *Bipolaris maydis* CICC 2530, *Bipolaris sorokiniana* CICC 2531, *Cochliobolus* heterostrophus CICC 2699, *Aspergillus aculeatus* CICC 41132, *Phomopsis chimonanthi* CICC 2611 and *Verticillium dahliae* CICC 2534, the results showed that the strain 5 L8 had comprehensive and fine antagonistic activity against path-ogens chosen for this research (Figure 3).

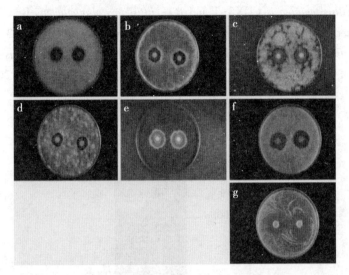

Figure 3 Antagonistic test of strain 5 L8 with pathogens

a. Against Fusarium graminearum CICC 2697; b. Against Bipolaris maydis CICC 2530; c. Against Bipolaris, sorokiniana CICC 2531; d. Against Cochliobolus heterostrophus CICC 2699; e. Against Aspergillus aculeatus CICC 41132; f. Phomopsis chimonanthi CICC 2611; g. Verticillium dahlia CICC 2534

Whole genome sequencing

The draft genome of strain 5 L8 was determined by Illumina HiSeq 2000 (Liu et al 2014). The genome sequence contained 5 100 526 reads for shotgun sequencing and 6 634 428 reads for pair-end sequencing. The reads were assembled using the SOAPdenovo 2.04 system (Li et al, 2008, 2010) into 22 contigs (>200bp) with a length of 6 098 914bp and a G + C content of 45.5%. The genome sequence of strain 5 L8 contained 5 370 protein coding sequences (CDSs). 125 tRNAs and 40 rRNAs were identified. Three CDSs for glucanase, 4 CDSs for cellulase and 2 CDSs for chitinase were annotated and correlated to antagonistic activity (Thrane et al, 1997; Dubois-Dauphin et al, 2011; Duzhak et al, 2012), which was in accordance with experimental results in this study.

Biocontrol gene amplification

Strain 5 L8 genomic DNA and recovery of β-1, 3-1, 4-glucanase gene (*gluB*) after purification indicated the length of the *gluB* band was about 550bp (Figure 4). The *gluB* sequence of 5 L8 was submitted to the GenBank database with the accession number of KR336541. The phylogenetic tree constructed by the *gluB* sequence of 5 L8 and the similar sequences in the GenBank are illustrated in Figure 5. Sequence similarities between strain 5 L8 *gluB* and relatives were as follows: *Paenibacillus polymyxa* ATCC 842^T *gluB* (99.4% similarity), *Paenibacillus barcinonensis* BP-23 *gluB* (75%), *Bacillustequilensis* CGX 5-1 (67.4%), *Bacillus amyloliquefaciens*

BS5582 *gluB* (62.7%) and *Paenibacillus mucilaginosus* K02 *gluB* (62.4%). The sequences form two distinct clusters. One cluster includes the *gluB* sequences of *Paenibacillus polymyxa* ATCC 842^T, *Paenibacillus barcinonensis* BP-23 and *Paenibacillus mucilaginosus* K02. The *gluB* sequences of *Bacillus tequilensis* CGX5-1 and *Ba-cillus amyloliquefaciens* BS5582 were assigned to the other cluster. The biocontrol gene acquired in this study was the β-1, 3-1, 4-glucanase gene of *Paenibacillus polymyxa* 5 L8. β-1, 3-1, 4-glucanase of *Paenibacillus polymyxa* was an important factor of the antagonistic effect against plant pathogens (Yao et al, 2004; Song et al, 2011).

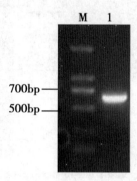

Figure 4 Electrophoresis of strain 5 L8β-1, 3-1, 4-glucanase gene. M. D2000 Marker Ladder; 1. 5 L8 β-1, 3-1, 4-glucanase gene

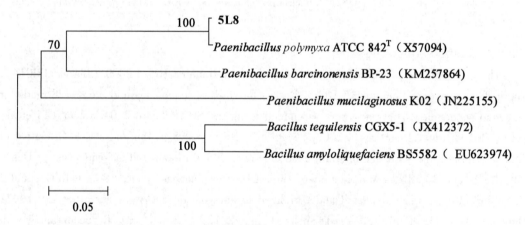

Figure 5 Phylogenetic tree showing relationships of strain 5 L8 and related species based on their β-1, 3-1, 4-glucanase gene sequences

The consensus phylogenetic tree was produced following bootstrap sampling of 1000 datasets. Numbers at nodes indicate percentages of bootstrap support based on a neighbour-joining analysis of 1 000 resampled datasets. Bar 0. 005 substitutions per nucleotide position

Discussion

Endophytic bacteria establish an harmonious relationship with plants. Some of them featur a biological control effect, which can promote plant growth. Up to now, endophytic bacteria with a good control effect could be isolated from many plants such as rice, cotton, banana, potato, pepper and so on (Hao, 2010; Zou et al, 2011). However, as far as we know, in maize, there was no relevant report about endophytic bacteria with a biological control effect. Disease control mechanisms of endophytic bacteria mainly include secreting antifungal substances, competing for an ecological niche and nutrients, inducing systemic resistance and being used as an exogenous gene carrier (Huang, 1986; Downing and Thomson, 2000; Ramamoorthy et al, 2001; Liu et al, 2012). Our previous re-search showed that, with *Enterobacter cloacae* inoculation, rice seedlings with a true leaf could induce rice's resistance to *Xanthomonas oxyzae*, showing that initial inoculation can induce plants to generate resistance to pathogen infection (Yang et al, 1999).

At present, a large area in the east part of North China is used to vigorously promote the new-generation maize variety Jingke 968. Jingke 968 passed the examination and approval of the Chinese government in 2011. It has been listed as the leading maize variety of China by the Ministry of Agriculture since 2012. It was listed as a new generation of maize variety in Beijing in 2013. Meanwhile, Jingke 968 has been listed as the main high-yield maize type to be promoted. Jingke 968 features a high yield potential and high resistance to various maize diseases. Breeding and promotion of the variety played an important role in improving maize yield, increasing farmer income and accelerating the development of Chinese maize industry. Study on the beneficial microorganisms of Jingke 968 was conducted to analyse the internal mechanism of the variety with excellent resistance to diseases from the perspective of the seed's microbiological ecology.

Taking the endophytic bacterial strain 5 L8 separated from the seeds of Jingke 968 as the seed material, and using methods of morphological observation, physiological and bio-chemical characteristic identification, and phylogenetic analysis to research the taxonomic status of its strains, eventually identifies it as *Paenibacillus polymyxa*. Via biological control and antagonistic testing, whole-genome sequencing, and biocontrol gene cloning, it was found that the strain has a wide-spectrum of good antagonistic activity to pathogenic bacteria. In addition, this study also revealed its essence as an antagonistic active molecule at the genome level. In the 1990s, as members of our research team, Wang et al successfully separated *Paenibacillus polymyxa* WY110 from rice roots. *Paenibacillus polymyxa* WY110 featured high invitro specificity for antagonistic activity to three types of major pathogenic bacteria of rice blast, rice sheath blight, and rice bacterial blight, with a wide antagonistic spectrum. In greenhouse experiments, it also showed a good plant disease prevention effect, usually above 50%, with good research and application value (Wang et al, 2007); In addition, Yang et al also further separated and purified the antifungal protein P2 with antagonistic activity to rice blast. Based on physicochemical indexes and amino acid sequencing, it was identified as $\beta-1,3-1,4$-glucanase. The gene of the antifungal protein was cloned, and prokaryotic expression was achieved (Yao et al, 2004). Following the *Paenibacillus polymyxa* WY110, the research object of this study was another *Paenibacillus polymyxa* separated from gramineous plants with $\beta-1,3-1,4$-glucanase antagonistic activity. In the meantime, bacterial strain 5 L8 was also the first maize endophytic bacteria with biocontrol function, and was the first strain in the world for which whole-genome sequencing was completed.

This study provides new insight into the mutual interactive relation between microorganisms and their host plants in the plant microecosystem, and provides microbial ecological evidence for further disclosing the broad spectrum of disease resistance of the new corn variety, Jingke 968. Meanwhile, it laid a foundation for exerting and utilizing the biocontrol and biological activity of this bacterial strain. Properly developed and applied microorganism antibiological inoculants are beneficial to seeds, and improve a crop's ability to resist fungal diseases, thereby increasing production output.

Acknowledgements

This work was supported by the Beijing Nova Program (No. Z141105001814095), the Beijing Nova Interdisciplinary Cooperational Program (No. Z1511000003150150), the National Natural Science Foundation of China (No. 31300008), the Chinese Postdoctoral Science Foundation (No. 2015 M570969), the Fund of National Infrastructure of Microbial Resources (No. NIMR2015-4), and the Scientific and Technological Development Project of the China National Research Institute of Food and Fermentation Industries (No. 2012KJFZ-BS-01). We also thank Dr. Zhengqiu Cai at Brigham and Women's Hospital (USA) for assistance with the English.

References (omitted)

Annals of Microbiology, 2016, 66 (2): 653-660

Molecular Mapping of Quantitative Trait Loci for Grain Moisture at Harvest in Maize

Song Wei[*]　Shi Zi[*]　Xing Jinfeng　Duan Minxiao　Su Aiguo
Li Chunhui　Zhang Ruyang　Zhao Yanxin
Luo Meijie　Wang Jidong　Zhao Jiuran[**]

(*The Maize Research Center, Beijing Academy of Agriculture & Forestry Sciences,
Beijing Key Laboratory of Maize DNA Fingerprinting and Molecular Breeding,
Shuguang Garden Middle Road No. 9, Beijing* 100097, *China*)

Abstract: In maize, high grain moisture (GM) at harvest causes problems in harvesting, threshing, artificial drying, storage, transportation, and processing. Understanding the genetic basis of GM will be useful for breeding low-GM varieties. A quantitative genetics approach was used to identify quantitative trait loci (QTLs) related to GM at harvest in field-grown maize. The GM of a double haploid (DH) population consisting of 240 lines derived from Xianyu335 was evaluated in three planting seasons and a high density genetic linkage map covering 1 546.4cM was constructed. The broad-sense heritability of GM at harvest was 71.0%. Using composite interval mapping, six QTLs for GM at harvest were identified on five chromosomes (Chr). Two QTLs located on Chr1, *qgm1-1* and *qgm1-2*, explained 5.0% and 10.8% of the phenotypic variation in GM at harvest, respectively. The QTLs *qgm2*, *qgm3*, *qgm4*, and *qgm5* accounted for 3.3%, 8.3%, 5.4%, and 11.0% of the mean phenotypic variation, respectively. Because of their consistent detection over multiple planting seasons, the detected QTLs appear to be robust and reliable for the breeding of low-GM varieties.

Key words: Grain moisture; QTL mapping; SNP; Maize

Introduction

Grain moisture (GM) at harvest is one of the main properties that affects maize production, especially in the northern maize growing area (Wang et al, 2012). In seed production, the GM level at harvest affects seed quality, production costs, and mechanized harvesting processes (Sala et al, 2006). High GM at harvest is one of the major problems in maize production, resulting in difficulties in harvesting, threshing, artificial drying, storage, transportation, and processing (Wang et al, 2015; Wang et al, 2012). Therefore, breeding and cultivation of low-GM maize varieties has become a major focus for breeders and producers.

Because GM at harvest is a complicated quantitative trait, traditional breeding of low-GM

---

[*] These authors contributed equally to this work
[**] Corresponding authors　E-mail: Jiuran Zhao (maizezhao@126.com)

maize varieties is labor-and time-consuming and expensive. However, molecular marker assisted selection (MAS) may provide an accurate and efficient way to facilitate the breeding of low-GM maize varieties. Consequently, there have been significant efforts to understand the genetic architecture of GM in maize to aid in the development of molecular markers. It has been demonstrated that the inheritance of maize GM at harvest is mainly determined by an additive effect with high heritability (Dudley et al, 1971; Hallauer and Miranda, 1981; Sentz, 1971). A significant difference in maize GM has been detected in a reciprocal hybrid, indicating that maternal effects may also play a role in GM (Zhang et al, 2005). In addition, QTLs controlling GM have been reported in several recent studies. With an $F_{2:4}$ population of B73x Mo17 and restriction fragment length polymorphism (RFLP) markers, six QTLs explaining 5%~22% of phenotypic variation were identified (Beavis et al, 1994). Later, using RFLP and simple sequence repeat (SSR) markers, Ragot et al, identified nine QTLs on chromosomes (Chr) 3, 5, and 9 (Ragot and Sisco 1995), Melchinger et al, detected more than 10 QTLs in two mapping populations (Melchinger et al, 1998), and Austin et al, identified 14 QTLs using six testcross populations under eight different environmental conditions (Austin et al, 2000). More recently, more than 40 QTLs controlling GM at harvest were mapped using RFLP markers with four $F_{3:6}$ and testcross populations, but a large discrepancy was observed among the different populations (Mihaljevic et al, 2005). With an $F_{2:3}$ population of 181 individuals and SSR markers, six QTLs were discovered on six chromosomes, accounting for 10.4%~19.7% of phenotypic variation in GM (Sala et al, 2006).

Despite the large number of studies on mapping GM QTLs in maize, the detected QTLs are not identical among those studies, probably because of differences in the mapping populations, marker types, environmental conditions, and moisture measurement methods. Therefore, the accurate identification of QTLs will be useful to facilitate MAS in early generations, and to greatly accelerate the forward-and back-crossing breeding of maize lines with desirable GM traits.

The aim of this study was to map QTLs related to maize GM at harvest using a permanent mapping population. To obtain reliable data for mapping QTLs, a double haploid population derived from Xianyu335, a hybrid resulting from the cross of "PH4CV" and "PH6WC", was tested in three planting seasons in the North Plain of China.

Materials and methods

Plant materials

The haploid plant was obtained using the parthenogenetic inducing line Jingkeyou006 (Zhang et al, 2013a) as the male parent, which was developed by the Maize Research Center of Beijing Academy of Agriculture and Forestry Sciences, to induce the single-cross hybrid Xianyu335 (PH6WC×PH4CV, F_1). The double haploid (DH) population comprised 240 individuals and was established after artificial doubling with colchicine. The parents PH6WC and PH4CV were both developed by DuPont Pioneer, and showed different GM levels at harvest.

Crop management and field data collection

Field experiments included the DH population and both parents. These experiments were conducted at Tongzhou, Beijing over three planting seasons: spring 2014, spring 2015, and summer 2015. The life cycles of 240 DH lines and two parents were included in Supplement Table 1. For each environment, two replicates for each line were arranged in a randomized complete block design and planted with 5m row length, 60cm spacing between rows, and 27.5cm spacing between plants within a row. To synchronize ear development for each DH line, the female inflorescences were bagged before tasseling, and pollination was carried out at same time for each line. The GM level measured at 45 days after pollination was designated as GM at harvest. For each DH line, three circles of grains from each of five randomly collected ears were sampled, and GM was measured by the whole grain drying method. The fresh weight of the kernels was immediately measured after sampling, and then the kernels were air-dried. The air-dried kernels were then placed in paper bags and oven-dried at 60℃ for 1 week until no more condensation was observed on the oven door, and the dry weight was measured. The grain moisture was calculated as follows: GM = (fresh weight-dry weight) / fresh weight × 100%.

Linkage map construction and QTL detection

The MaizeSNP3072 (Tian et al, 2015) chip was used to genotype the DH population, and the genetic map was constructed as previously described (Xu et al, 2015). In total, 1 337 SNPs were selected because of their polymorphism between the parents, and the comparative linkage map was created using Kosambi's regression function in JoinMap 4.1 software (Van Ooijen, 2006) with an LOD score of 3.0 as the cutoff to establish linkage. Composite interval mapping (CIM) was carried out using Windows QTL Cartographer 2.5 software (Basten et al, 2002) for the DH population with the standard marker model and a walk speed of 2cM. The LOD cutoff was determined using 1 000 permutations and a 0.05 level of significance.

Statistical analysis

Combined analysis of variance was conducted with the ANOVA tool in QTL IciMapping (http://www.isbreeding.net/).

Results

Phenotypic analysis

The GM at harvest was significantly higher for the female parent PH6WC than for the male parent PH4CV in all three planting seasons (Figure 1). Analyses of the GM data showed that the phenotypic frequencies showed a normal or near-normal distribution in all three planting seasons (spring 2014, spring 2015, and summer 2015). There were statistically significant variance effects of genotype and environment on GM at harvest in the DH lines planted in three seasons (Table 1). The range for GM at harvest was 16.3% ~ 52.5%. The broad sense heritability of this trait was 71.0%, indicating that it is largely determined by genotype.

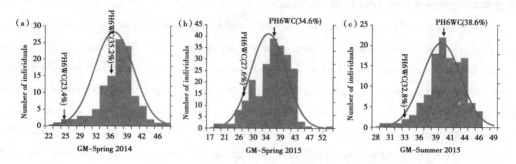

Figure 1 Representative phenotypic frequencies of grain moisture (GM) for the double haploid maize population tested in three seasons

A spring 2014, B spring 2015, and C summer 2015. Black arrows indicate groups with GM values similar to those of the parents (female: PH6WC and male: PH4CV). Red lines represent theoretical normal distribution curve

Table 1 Variation statistics and heritability (H^2) for the DH population

Source	Df	GM
Genotype	239	10 127.9**
Season	2	498 104.0**
Linex Season	478	3 651.1**
Reps (season)	3	6 861.4**
Error	717	1 398.7
LSD (0.01)	103.1	
Heritability%		71.0
Range%		16.3~52.5

** significant level at $P=0.01$

QTL mapping

Of the 3 072 SNPs on the MaizeSNP3072 chip, 1 337 SNPs were informative because they were polymorphic between the parents. A "Similarity of loci" analysis was performed for all SNPs in JoinMap 4.1. In this analysis, only one SNP was retained at one location if multiple SNPs showed >0.99 similarity, resulting in the random removal of 325 SNPs. In addition, 48 SNPs were manually removed from the genome either because of inconsistencies between their physical and genetic positions or because they could not be assigned to any of the linkage groups. Therefore, the linkage map was 1 546.4cM in total, and consisted of 964 SNPs.

Using mean data across all three planting seasons, we identified six QTLs responsible for GM at harvest by CIM with winQTLcart2.5. The six QTLs were mapped to Chr1, Chr2, Chr3, Chr4, and Chr5 (Figure 2 and Table 2), and were designated as *qgm1-1*, *qgm1-2*, *qgm2*, *qgm3*, *qgm4*, and *qgm5*, respectively. The phenotypic variance explained (% PVE) by the six mapped QTLs ranged from 3.3% to 11.0% with LOD scores of 3.3~9.7. Of the six QTLs, *qgm1-1* and *qgm5* were detected across all three environments, with PVEs of 5.0% and 11.0%

and LOD scores of 4.6 and 9.7, respectively. The QTL *qgm1-2* (PVE, 10.8%; LOD 9.6) was identified in two planting seasons, spring 2014 and spring 2015. Similarly, *qgm3* was detected in two seasons, spring 2015 and summer 2015 (PVE, 8.3%; LOD, 7.5). Two other QTLs, *qgm2* and *qgm4* (PVE of 3.3% and 5.4%, respectively), were detected only in spring 2015, aside from the overall mean data (Table 2). One QTL on Chr3 and one on Chr8 were only detected in a single season and not from the overall mean data across three seasons (Figure 2); therefore, these QTLs were not included in Table 2.

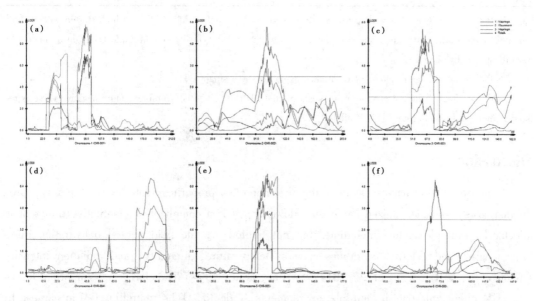

Figure 2 Chromosomal location and logarithm of odds (LOD) scores of QTLs related to grain moisture (GM) in a double haploid population tested during three seasons in 2014—2015

QTL likelihood plots were generated by WinQTLCart2.5 using composite interval mapping (CIM) analysis with 1 000 permutations and significance level of 0.05. Genetic distance of each chromosome (Chr) is indicated on x-axis. 15springA represents mean of two replicates from spring 2015, 15summerA represents mean of two replicates from summer 2015, 14springA represents mean of two replicates from spring 2014, and TotalA represents the overall mean of all three seasons. A. Chr1; B. Chr2; C. Chr3; D. Chr4; E. Chr5; F. Chr8

Table 2 Quantitative trait loci (QTLs) for GM at harvest identified by composite interval mapping (CIM) in the DH population grown in three seasons of 2014—2015

QTL[a]	Chr	Position (cM)	Spring 2014			Spring 2015			Summer 2015			Mean		
			LOD	%PVE[b]	Add[c]	LOD	%PVE	Add	LOD	%PVE	Add	LOD	%PVE	Add
qgm1-1	1	37.53	2.6	3.7	-0.9	5	5.9	-1.4	7.1	10.6	-1.2	4.6	5.0	-0.9
qgm1-2	1	86.04	7.1	10.6	-1.5	6.9	8.0	-1.5				9.6	10.8	-1.3
qgm2	2	96.99				4.8	5.1	-1.1				3.3	3.3	-0.6

(continued)

QTL[a]	Chr	Position (cM)	Spring 2014			Spring 2015			Summer 2015			Mean		
			LOD	%PVE[b]	Add[c]	LOD	%PVE	Add	LOD	%PVE	Add	LOD	%PVE	Add
qgm3	3	58.49				6.1	7.2	1.4	4.7	6.3	1.0	7.5	8.3	1.1
qgm4	4	90.52				7.0	8.3	-1.4				5.0	5.4	-0.8
qgm5	5	74.15	3.6	5.1	-0.9	7.3	8.7	-1.5	10.1	14.4	-1.4	9.7	11.0	-1.2

[a] QTLs were designated according to the previous rule (McCouch et al, 1997), which started with a "q" followed by the trait name "gm" and the Chr number on which the QTL located. Multiple QTLs on a single Chr were illustrated by -1 and -2

[b] %PVE = percentage of phenotypic variance explained by a single QTL

[c] additive effect, where a positive value suggests that PH6WC allele is favorable, and a negative value is related to increased effects from PH4CV allele

Discussion

In maize, GM at harvest is one of the most important properties related to seed quality, production costs, and mechanized harvesting. However, it is a complicated quantitative trait that is affected by multiple factors including, but not limited to, the field natural dehydration rate, bract length, and external conditions such as temperature, humidity, and radiation intensity (Wang et al, 2012; Feng et al, 2014; Sweeney et al, 1994). Therefore, it is important to acquire GM phenotypic data in multiple environments to identify QTLs controlling GM at harvest. In this study, we collected GM data in three planting seasons in 2014—2015 and identified some QTLs that were consistent among seasons (Figure 2).

It is important to identify molecular markers tightly linked to the QTLs related to GM at harvest for MAS to breed low-GM maize varieties. Currently, SSR and RFLP markers are mainly used in QTL mapping; Ragot and Sisco, 1995; Melchinger et al, 1998; Austin et al, 2000; Mihaljevic et al, 2005; Sala et al, 2006; Sala et al, 2012). However, the marker intervals in genetic maps constructed based on those markers are large, typically 10 ~ 20cM (Ding et al, 2011; Sala et al, 2006). In addition, the genotyping procedures for both SSR and RFLP markers are labor - and time - consuming and cost ineffective. In contrast, SNP markers have been used increasingly in genetic analyses, because of their high density on the whole genome and easy operation in a high throughput setting (Song et al, 2012; Tenaillon et al, 2001; Pham et al, 2014; Tian et al, 2015; Shi et al, 2015). The average density of SNP markers on the maize genome is one every 48 ~ 104bp (Tenaillon et al, 2001; Ching et al, 2002). Therefore, we took advantage of the MaizeSNP3072 chip (Tian et al, 2015) to genotype the 240 DH lines, and constructed a high density genetic linkage map covering 1 546.4cM using data of 964 polymorphic SNPs. This map was comparable to those previously established using SSR markers (Ding et al, 2011; Zhang et al, 2013b; Sala et al, 2006).

However, the average marker interval of the SNP-based linkage map is 1.6cM, which is significantly smaller than that in SSR-based maps, allowing for a more accurate QTL detection. Using this map and GM data obtained with the whole grain drying method, six QTLs were identified on Chr1, 2, 3, 4, and 5 (Figure 2 and Table 2), explaining 3.3%~11.0% of the phenotypic variation in GM at harvest (Table 2).

There seems to be some discrepancy between our mapping results and previously reported results, but it is difficult to compare the consistency between previously mapped QTLs and those detected here, because most of the previous studies were performed in testcross populations and were based on linkage maps constructed using RFLPs (Beavis et al, 1994; Mihaljevic et al, 2005; Melchinger et al, 1998; Ragot and Sisco, 1995). However, as demonstrated in Beavis et al (1994) and Mihaljevic et al (2005), there is some consistency between QTLs mapped in lines and testcross populations. In addition, the size and the distribution of the genetic linkage maps are similar. Therefore, to some extent, our QTLs are comparable to those published previously. The QTLs responsible for GM were mapped to Chr1, 2, 3, 4, 5, 8 and 10 in five previous studies (Sala et al, 2006; Beavis et al, 1994; Ragot and Sisco, 1995; Melchinger et al, 1998; Mihaljevic et al, 2005), and some of those QTLs were also detected in the present study (Table 2). The consistency of QTLs detected on the same chromosome from such distinct populations and environmental conditions indicates that these are important loci for GM at harvest. More research is required to further narrow down the QTL interval to understand which loci, and which genes, control GM at harvest.

The broad sense heritability of GM at harvest was determined to be 71.0% (Table 1), consistent with the moderately high heritability reported in previous studies (Wang et al, 2015; Dudley et al, 1971; Hallauer and Miranda, 1981). These findings suggest that genotype plays an important role in this trait. However, the PVE values were lower than 11% for all of the detected QTLs, indicating that multiple environment-sensitive QTLs are responsible for GM at harvest. Out of the six detected QTLs, four ($qgm1-1$, $qgm5$, $qgm1-2$, and $qgm3$) were detected in at least two planting seasons. The QTLs with non-significant environmental interactions for GM at harvest were stable under different environments, and are therefore the logical targets for MAS. The consistency of some of the QTLs identified here not only provides basic theoretical knowledge about the genetic architecture of GM in maize, but also identifies potential candidate regions for marker development. Further research is required to develop robust markers to accelerate MAS for the breeding of low-GM maize varieties that are suitable for machine harvesting.

Acknowledgements

We appreciate the project funding supported by the National Natural Science Foundationof China (31440067), the National Plan for Science and Technology Support (2014BAD01B09), the Special Subject for Scientific and Technological Innovation from Beijing Agriculture and Forestry Sciences (KJCX20140202). We would like to thank Gang Liu and Bin Tang for their constructive suggestions during the course of data analysis.

Conflict of Interest

The authors declare that they have no conflict of interest.

References (omitted)

Plant Breeding, 2016, 136

Effect of Saline Stress on the Physiology and Growth of Maize Hybrids and their Related Inbred Lines

Luo Meijie*   Zhao Yanxin*   Song Wei
Zhang Ruyang   Su Aiguo   Li Chunhui
Wang Xiangpeng   Xing Jinfeng   Shi Zi**   Zhao Jiuran**

(*Beijing Key Laboratory of Maize DNA Fingerprinting and Molecular Breeding, Maize Research Center, Beijing Academy of Agricultural and Forest Sciences, Shuguang Garden Middle Road No. 9, Haidian District, Beijing 100097, China*)

Abstract: Salinity is one major abiotic stress that restrict plant growth and crop productivity. In maize (*Zea mays* L.), salt stress causes significant yield loss each year. However, indices of maize response to salt stress are not completely explored and a desired method for maize salt tolerance evaluation is still not established. A Chinese leading maize variety Jingke968 showed various resistance to environmental factors, including salt stress. To compare its salt tolerance to other superior maize varieties, we examined the physiological and growth responses of three important maize hybrids and their related inbred lines under the control and salt stress conditions. By comparing the physiological parameters under control and salt treatment, we demonstrated that different salt tolerance mechanisms may be involved in different genotypes, such as the elevation of superoxide dismutase activity and/or proline content. With Principal Component Analysis of all the growth indicators in both germination and seedling stages, along with the germination rate, superoxide dismutase activity, proline content, malondialdehyde content, relative electrolyte leakage, we were able to show that salt resistance levels of hybrids and their related inbred lines were Jingke968 > Zhengdan958 > X1132 and X1132M > Jing724 > Chang7-2 > Zheng58 > X1132F, respectively, which was consistent with the saline field observation. Our results not only contribute to a better understanding of salt stress response in three important hybrids and their related inbred lines, but also this evaluation system might be applied for an accurate assessment of salt resistance in other germplasms and breeding materials.

Key words: Maize; Jingke968; Salt stress evaluation; Principal component analysis

Introduction

Salinity is one major abiotic stress that seriously limits plant growth and productivity throughout the world (Chinnusamy et al, 2005; Deinlein et al, 2014; Parida et al, 2009; Shabala et al, 2014). Till now, over 800 million hectares of land have been affected by salinity, accounting

---

* These authors contributed equally to this work
** These authors also contributed equally to this work
** Corresponding author: E-mail: shizi_baafs@126.com; maizezhao@126.com

for more than 6% of the world's total land area (Munns and Tester, 2008). Moreover, the amount of salt-affected land continues to increase due to the persistent non-sustainable farming practices. Hence, developing and breeding salt tolerant crops are essential for sustaining crop production.

Plants respond to salt stress througha series regulation of physiological and morphological adaptations (Bhaskar Gupta, 2014; Deinlein et al, 2014; Shabala et al, 2014). As a result of osmotic and ionic effects, plants accumulate various antioxidant enzymes and solute induced by salt stress to improve salt resistance. Proline (Pro) is an effective osmotic adjustment agent, which accumulates in plant to maintain water uptake and regulate physiological metabolism, and to protect cells from damage (Gangopadhyay et al, 2000; Gao et al, 2014). Salt stress results in overproduction of reactive oxygen species (ROS), leading to oxidative damage to plant cells (Yu et al, 2015), which triggers antioxidant defense system and results in the a rise of superoxide dismutase (SOD, EC 1.15.1.1) enzyme activity, and these antioxidant enzymes can effectively remove ROS and alleviate oxidative damage (Bhatia et al, 2010). In addition, many physiological factors are used to assess the severity of salt stress, including malondialdehyde (MDA) and relative electrolyte leakage (REL). MDA is the product of membrane lipid peroxidation, which will seriously damage the biological membrane (Miao et al, 2010; Yazici et al, 2007). REL is an important indicator to measure the permeability of cell membrane, and the greater the value, the more the leakage of the electrolyte is together with the more the cell membrane is affected (Cui et al, 2015). Salinity also impairs seed germination, retards plant development and reduces grain yield (Gao et al, 2014). However, these responses vary in different genotypes.

Maize is an important food, feed and economic crop in most part of the world. Unfortunately, maize is a salt-sensitive crop with the salt tolerance limit of 1.02 ‰. Maize grain yield would reduce by 12% with every 0.6‰ increase in salt concentration (Zheng et al, 2010). Maize shows the most sensitivity to salt at germination and seedling stages. Salt tolerance evaluation of important maize hybrids in agricultural production and their related inbred line from their response to salt stress is a critical step for maize breeding. Growth and physiological parameters in seedling stage are frequently used for salt tolerance evaluation in Maize (Giaveno et al, 2007).

In the present study, we selected three major leading maize varieties Jingke968 (Wang et al, 2016), Zhengdan958 (Du et al, 2006) and X1132, all of which possess many desired traits: such as high quality, high productivity, and wide resistance, as well as their related inbred lines. However, information regarding the effects of salt stress on those hybrids and their related inbred lines have not been specifically studied. Here, we analyzed the effect of salinity stress on the physiology and growth of the three important maize hybrids and their related inbred lines at different growth stages, and their salt tolerance levels were comprehensively evaluated by the principal component analysis. These results provided a better understanding and a more integrated picture of phenotypic response to salt stress and shed light on a useful salt tolerance evaluation approach in maize.

Materials and methods

Plant materials and treatments

Local maize (*Zea mays* L.) cultivar Jingke968 and its female parent inbred line Jing724, local maize cultivar Zhengdan958 and its parental inbred lines Zheng58 (Female parent) and Chang7-2 (Male parent), foreign maize cultivar X1132 and its parental inbred lines X1132M and X1132F, and inbred line B73 were collected from germplasm bank of Maize Research Center, Beijing Academy of Agriculture and Forestry Sciences, China. These three cultivars used in this study have been largely grown in China because of its broad stress resistance. Seeds were sown and hydroponically cultured in maize seedling identifying instrument (Chinese patent, patent number: ZL200920177285.0) in Hoagland's nutrient solution (Phyto Technology Laboratories Co., Ltd., USA) under long day condition [12 light/12 dark, 150~180μmol m^{-2} s^{-1}, (26 ± 1)℃, 70% humility]. For salt treatment in germination stage, seeds were sown in Hoagland medium with no (control) and 100mM NaCl (Coolaber Co., Ltd., Beijing, China) (Stress), and cultured for seven days. For salt treatment in seedling stage, three-leaf maize seedlings were cultured in Hoagland medium containing no (control) and 100mM NaCl (Stress), and shoot and root samples were collected for analysis after seven days.

Physiological assays

Activities of SOD were measured bynitro blue tetrazolium (NBT) photo reduction method (Beauchamp and Fridovich, 1971). Free Pro content of seedling shoot was extracted and quantified following the ninhydrin-based colorimetric assays (Hu et al, 1992). MDA content was determined by the thiobarbituric acid (TBA) assay (Schmedes and Hølmer, 1989). REL assay was conducted according to the method described previously (Liu et al, 2012). All the physiological assays were performed with six biological replicates for each treatment.

Growth parameter measurements

In order to determinegermination rate (GR), thirty maize seeds were cultured in control and in saline condition for seven days with two biological replicates for each treatment. Seed was recorded as germinated when both radicle and plumule had emerged, and the plumule length reached more than 1/2 of seed length. Seed germination number was counted seven days after culture and GR was calculated using the following formula: GR = (seed germination number / total seed number) × 100%. After seeds were cultured in control and in salt treatment for seven days after culturing, shoot fresh weight (SFW), shoot length (SL), root fresh weight (RFW) and root length (RL) was measured with six biological replicates each with five seedlings for both control and salt treatment. After dried at 80℃ for three days, the shoot dry weight (SDW) and root dry weight (RDW) of seedlings in germination stage were measured. For growth parameters analysis in seedling stage, three-leaf seedlings were cultured in control and in saline condition for seven days. Then SL, RFW and RL were determined using four biological replicates each with five seedlings for each treatment. SDW and RDW of seedlings in seedling stage were determined after dried at 80℃ for three days.

Data analysis

Salinity tolerance index (STI) of GR, SOD, Pro, MDA, REL, SFW, SDW, SL, RFW, RDW and RL were calculated using the following formulas (Yang et al, 2011):

Formula 1: GR_{STI}, SFW_{STI}, SDW_{STI}, SL_{STI}, RFW_{STI}, RDW_{STI} and RL_{STI} = value measured in salt stress / value measured in control

Formula 2: SOD_{STI} = (SOD activity in salt stress − SOD activity in control) / SOD activity in control

Formula 3: Pro_{STI} = (Pro content in salt stress − Pro content in control) / Pro content in control

Formula 4: MDA_{STI} = 1 − (MDA content in salt stress − MDA content in control) / MDA content in control

Formula 5: REL_{STI} = 1 − (REL in salt stress − REL in control) / REL in control

For comprehensive evaluation of salt tolerance level, Principal Component Analysis (PCA) was performed based on the STI of all physiological and growth traits using SPSS software.

Differences in the average of physiological and growth parameters were compared by two-way ANOVA and Bonferroni post-tests using GraphPad Prism software. Values are shown as means ± SEM. * means $P<0.05$, ** means $P<0.01$, *** means $P<0.001$.

The principal components (PC) of the PCA analysis were calculated using the following equation:

$$PC_i = \sum_{j=1}^{m} (\text{loading matrix})ij * STIij$$

where i = 1, 2, 3… and j = GR, SOD, Pro, MDA, REL, $SFW_{(seedling\ stage)}$, $SDW_{(seedling\ stage)}$, $SL_{(seedling\ stage)}$, $RFW_{(seedling\ stage)}$, $RDW_{(seedling\ stage)}$, $RL_{(seedling\ stage)}$, $SFW_{(germination\ stage)}$, $SDW_{(germination\ stage)}$, $SL_{(germination\ stage)}$, $RFW_{(germination\ stage)}$, $RDW_{(germination\ stage)}$, $RL_{(germination\ stage)}$.

To determine the comprehensive STI (CSTI) for each genotype using the principal components, the following formula was employed:

$$CSTI = \sum_{k=1}^{n} (\text{contribution rate})k * PC_i, \text{ where } k = PC1, PC2, PC3$$

Results

GR and physiological responses of maize hybrids to saline stress

To determine the effect of salt stress on the physiology of maize hybrids, we investigated GR, SOD activity, Pro content, MDA content and REL in control and salt treatment (Figure 1A−E). Generally, GR and REL did not show significant difference among three maize hybrids (Figure 1A, E). In X1132, salt stress led to a significant increase in Pro concentration ($P<0.001$) and no significant difference was observed in SOD activity and MDA content (Figure 1B−D). In contrast, for Jingke968, salt stress significantly increased SOD activity ($P<0.001$) and significantly decreased Pro ($P<0.05$) and MDA content ($P<0.001$) (Figure 1B−D). Similarly, for Zhengdan958, SOD activity was substantially increased ($P<0.05$), and MDA

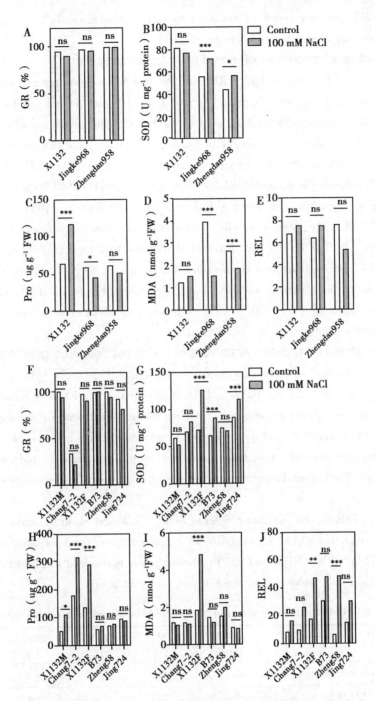

Figure 1 GR and physiological analysis of maize hybrids and their related inbred lines under control and saline conditions

A. GR; B. SOD activity; C. Pro content; D. MDA content; E. REL. For germination rate, the bar charts represent Mean± SE of two biological replicates each with 30 seeds. For other parameters, the data represent Mean± SE of six biological replicates. * represents the level of significance of 0.05, ** 0.01, and *** 0.001, respectively content was significantly decreased ($P<0.001$) by salt treatment, while Pro content didn't significantly change compared to control (Figure 1B–D).

GR and physiological responses of inbred lines to saline stress

The changes of GR, SOD activity, REL and Pro and MDA concentrations were also evaluated in control and salt stressed six maize inbred lines (Figure 1F–J). Maize inbred lines X1132M, Jing724 and Zheng58 are female parent of X1132, Jingke968 and Zhengdan958, respectively, and maize inbred lines X1132F and Chang7-2 are the male parents. In all of these inbred lines, GR did not significantly change compared to the control (Figure 1F). In X1132M and Chang7-2, salt stress led to a substantial increase in Pro content ($P<0.05$ and $P<0.001$, respectively) and no significant difference was observed in SOD activity, MDA content and REL (Figure 1F–J). In X1132F, except for GR, salt stress resulted in a significant elevation in all of these parameters ($P<0.001$ for SOD, Pro and MDA; $P<0.01$ for REL) (Figure 1F–J). In B73 and Jing724, SOD activity was significantly increased with the salt treatment ($P<0.001$) but the Pro content, MDA content and REL stayed the same (Figure 1G–J). In Zheng58, REL was significantly increased ($P<0.001$) while SOD activity, MDA content and Pro content did not vary compared to control (Figure 1G–J).

Effect of saline stress on growth parameters of maize hybrids in the germination stage

To study the effects of salt stress on growth parameters of three maize hybrids in germination stage, SFW, SDW, SL, RFW, RDW and RL of three maize hybrids were measured, and all of them exhibited a significant decrease upon saline solution treatment ($P<0.001$) (Figure 2A–F). In order to evaluate the inhibition degree of growth in maize hybrids, we compared the growth parameters between salt treatments and their control counterparts and calculated their percentage values. The higher the percentage value was, the represents less degree of growth inhibition was.

As shown in Table 1, the inhibition order of SFW, SDW and SL in the three hybrids during the germination stage was X1132 < Zhengdan958 < Jingke968, while it was Jingke968 < Zhengdan958 < X1132 for RFW, RDW and RL. In addition, results revealed that growth parameters of shoot were more severely affected than those of root upon salt stress.

Table 1 Effect of salt stress on the growth of maize hybrids and their related inbred lines

Maize lines		SFW* (%)	SDW* (%)	SL* (%)	RFW* (%)	RDW* (%)	RL* (%)
Germination stage							
Hybrids	X1132	40.9	48.8	44.2	66.0	74.0	70.9
	Jingke968	27.6	42.1	31.6	71.9	83.1	73.1
	Zhengdan958	34.0	44.7	41.6	72.1	76.3	70.3

(continued)

Maize lines		SFW* (%)	SDW* (%)	SL* (%)	RFW* (%)	RDW* (%)	RL* (%)
Germination stage							
Inbred lines	X1132M	29.5	31.7	31.8	63.2	70	95.8
	Chang7-2	31.1	35.1	31.9	58.4	89.0	67.1
	X1132F	9.6	14.3	9.0	37.8	68.8	29.3
	Zheng58	26.3	30.8	31.3	53.6	55.1	65.3
	Jing724	30.8	37.1	34.7	59.3	65.9	70.5
Seedling stage							
Hybrids	X1132	59.4	75.8	74.0	82.5	111.1	88.1
	Jingke968	72.0	83.6	79.6	94.4	106.4	59.7
	Zhengdan958	48.1	75.1	76.5	92.4	111.1	64.6
Inbred lines	X1132M	91.8	104.8	91.5	112.1	130.8	83.9
	Chang7-2	46.3	64.2	67.5	85.0	117.1	75.7
	X1132F	67.4	91.7	73.0	117.7	119.3	65.3
	Zheng58	65.4	79.7	80.1	87.5	124.3	85.1
	Jing724	63.5	89.7	77.2	90.8	117	77.7

* displayed value = (observed value in salt treatment / observed value in control) ×100%

Effect of saline stress on growth parameters of maize hybrids in seedling stage

Effects of salt stress on growth parameters of three maize hybrids in seedling stage were also investigated. In general, salt stress significantly decreased SFW and SL but no significant change in the RFW and RDW of three maize hybrids was observed (Figure 2G, I, J, K). In X1132, salt stress did not affect SDW and RL (Figure 2H, L), but they were substantially reduced in Zhengdan958 (Figure 2H, L). In addition, RL was significantly decreased upon salt stress in Jingke968, but SDW wasn't affected (Figure 2H, L).

Growth parameters between salt applications and their control counterparts were compared and their percentage values were calculated to illustrate the growth inhibition degree. As shown in Table 1, the inhibition order of SFW, SDW and SL among the three hybrids in seedling stage was Jingke968 < X1132 < Zhengdan958, however, the inhibition degree of RFW, RDW and RL among three hybrids was not consistent. For RDW and RL, the inhibition order was X1132 < Zhengdan958 < Jingke968, while it was Jingke968 < Zhengdan958 < X1132 for RFW. Similar to the germination stage, the shoot was more sensitive to salt stress than root during the seedling stage, but the susceptibility was reduced compared with the germination stage.

Effect of saline stress on growth parameters of maize inbred lines in germination and seedling stage

To demonstrate the salt sensitivity of the parent inbred lines of three hybrids, the effect of salt

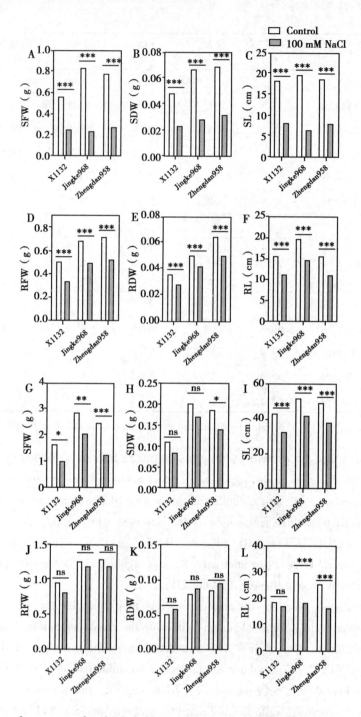

Figure 2 Growth response of maize hybrids to salt stress in the germination and seedling stages

A. SFW; B. SDW; C. SL; D. RFW; E. RDW; F. RL. The bar charts represent mean±SE of six biological replicates each with five seedlings for all parameters for the germination stage or four biological replicates for the seedling stage. * represents the level of significance of 0.05, ** 0.01, and *** 0.001, respectively stress on growth parameters of five inbred lines in germination stage and in seedling stage were also studied. In germination stage, except for RDW of Chang7-2 and RL of X1132M, salt stress led to a significant decrease in all growth parameters (S_ figure and Figure 3A-F). As shown in Table 1, the inhibition order of SFW, SDW and SL among the five inbred lines was Jing724 < Chang7-2 < X1132M < Zheng58 < X1132F, it was X1132M < Jing724 < Chang7-2 < Zheng58 < X1132F for RFW and RL, and Chang7-2 < X1132M < X1132F < Jing724 < Zheng58 for RDW. In addition, as in the hybrid, root was much more resistant than the shoot in the inbred lines.

In seedling stage, RFW was not substantially altered in all the tested inbred lines by salt stress (Figure 3J). In X1132M, only RL was significantly reduced by salt stress (Figure 3L), but all growth parameters, except for RFW, were significantly affected by salt stress in Zheng58 (Figure 3G-L). In Chang7-2, SFW, SDW, SL and RL were significantly reduced by salt stress (Figure 3G-I, L). Additionally, in X1132F and Jing724, SFW, SL and RL were significantly decreased by salt stress (Figure 3G, I, L). As shown in Table 1, for SFW, SDW and SL, the inhibition order of five inbred lines was X1132M < X1132F / Zheng58 / Jing724 < Chang7-2, and it was X1132M / X1132F < Zheng58 / Jing724 / Chang7-2 for RFW, RDW and RL. Furthermore, similar to the hybrid, shoot was more susceptible to salt stress than root in both germination and seedling stages, and the germination stage was more sensitive to the stress compared to its older stage.

Comprehensive evaluation of salt tolerance in maize hybrids and inbred lines

The salt tolerance levels of three hybrids and their related inbred lines varied according to different evaluation indices. To comprehensively evaluate salt tolerance levels of maize hybrids and their related inbred lines, PCA was performed based on STI of SFW, SDW, SL, RFW, RDW and RL in both germination stage and in seedling stage, along with the STI of GR, SOD, Pro, MDA with REL. For maize hybrids, two PCs were obtained with the eigen value of 12.326 and 4.674, respectively, explained 100% of observed variations (Table 2). According to the loading values of different traits (Table 2), the value of each PC was obtained, and the CSTI was further determined according to the contribution rate of each PC (Table 2 and Table 3).

Based on the formula, CSTI of X1132, Jingke968 and Zhengdan958 was 1.283, 3.767 and 3.201 (Table 3), respectively, indicating that the order for salt tolerance levels of the three hybrids was Jingke968 > Zhengdan958 > X1132.

For inbred lines, four PCs were obtained by PCA and their eigen values were 8.031, 5.408, 2.526 and 1.036, respectively, explained 100% of observed variations. According to loading values and contribution rate (Table 2), four PCs and CSTIs were calculated, which revealed the CSTI of 3.900, 2.943, 1.575, 2.721 and 3.359 for X1132M, Chang7-2, X1132F, Zheng58 and Jing724, respectively (Table 3), indicating that the order for salt tolerance levels of was X1132M > Jing724 > Chang7-2 > Zheng58 > X1132F using our evaluation system.

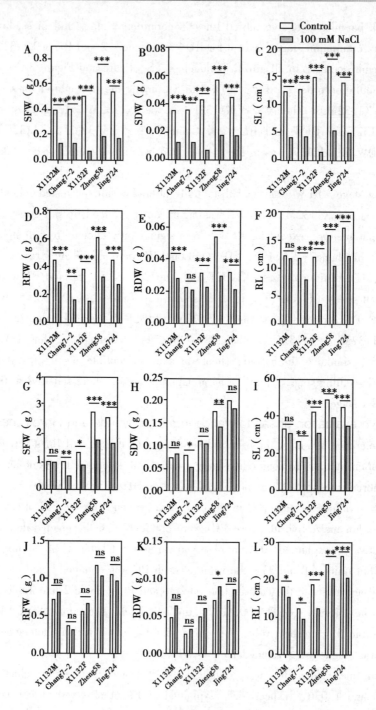

Figure 3　Growth response of maize inbred lines to salt stress in the germination and the seedling stage

A. SFW; B. SDW; C. SL; D. RFW; E. RDW; F. RL. Data rep-resent mean ± SE of six biological replicates each with five seedlings for all parameters for the germination stage or four biological replicates for the seedling stage. * represents the level of significance of 0.05, ** 0.01, and *** 0.001, respectively

Table 2 Loading matrix of each trait and contribution rate of principal components

Maize lines	Eigen value	Contribution rate (%)	Cumulative contribution rate (%)	Loading matrix														
				Seedling stage											Germination stage					
				GR	SOD	Pro	MDA	REL	SFW	SDW	SL	RFW	RDW	RL	SFW	SDW	SL	RFW	RDW	RL
Hybrids																				
1	12.326	72.505	72.505	0.351	0.873	−0.904	0.992	−0.085	0.543	0.871	0.998	0.934	−0.866	−0.935	−1.000	−0.992	−0.947	0.851	0.962	0.746
2	4.674	27.495	100.00	0.936	0.488	−0.428	0.127	0.996	−0.840	−0.491	−0.062	0.358	0.500	−0.354	−0.022	−0.128	0.321	0.525	−0.275	−0.666
Inbred lines																				
1	8.031	47.240	47.240	−0.146	−0.912	−0.474	0.964	−0.101	0.123	−0.072	0.431	−0.601	0.341	0.900	0.964	0.934	0.982	0.961	−0.012	0.913
2	5.408	31.810	79.050	0.867	−0.238	0.257	−0.175	0.024	0.977	0.916	0.901	0.705	0.849	0.294	−0.222	−0.302	−0.158	0.008	−0.584	0.292
3	2.526	14.859	93.908	−0.422	0.060	0.697	0.167	0.873	0.171	0.182	0.039	0.376	0.077	−0.255	0.131	0.051	−0.027	0.268	0.786	0.285
4	1.036	6.092	100	0.219	0.329	−0.473	0.113	0.476	0.043	0.35	0.014	0.022	−0.396	−0.196	0.072	0.185	0.103	0.07	−0.201	−0.03

Table 3 Salt tolerance levels of maize hybrids and inbred lines

Maize lines	STI																PC1	PC2	PC3	PC4	CSTI	Salt tolerance level
	Seedling stage										Germination stage											
	GR	SDO	Pro	MDA	REL	SFW	SDW	SL	RFW	RDW	RL	SFW	SDW	SL	RPW	RDW	RL					
Hybrids																						
X1132	0.95	−0.06	0.82	0.80	0.89	0.59	0.75	0.74	0.83	1.11	0.88	0.41	0.49	0.44	0.66	0.74	0.71	1.441	0.866			1.283	3
Jingke968	0.97	0.27	−0.26	1.62	0.85	0.72	0.84	0.80	0.94	1.06	0.60	0.28	0.42	0.32	0.72	0.83	0.73	4.628	1.496			3.767	1
Zhengdan958	1.00	0.26	−0.17	1.30	1.30	0.48	0.75	0.77	0.92	1.11	0.65	0.34	0.45	0.42	0.72	0.76	0.70	3.580	2.203			3.201	2
Inbred lines																						
X1132M	0.93	−0.14	1.00	1.13	0.23	0.92	1.05	0.92	1.12	1.31	0.84	0.30	0.32	0.32	0.63	0.70	0.96	3.903	5.401	2.417	−0.328	3.901	1
Chang7-2	0.63	0.16	0.74	1.15	−0.55	0.46	0.64	0.68	0.85	1.17	0.76	0.31	0.35	0.32	0.58	0.89	0.67	3.547	3.409	1.508	−0.673	2.943	3
X1132F	0.92	0.72	1.06	−0.52	−0.55	0.67	0.92	0.73	1.18	1.19	0.65	0.10	0.14	0.09	0.38	0.69	0.29	−0.184	4.771	1.250	−0.674	1.575	5
Zheng58	0.93	−0.06	0.03	0.75	−4.51	0.65	0.80	0.80	0.88	1.24	0.85	0.26	0.31	0.31	0.54	0.55	0.65	4.070	4.277	−2.888	−2.200	2.721	4
Jing724	0.87	0.26	−0.14	1.21	0.11	0.64	0.9	0.77	0.91	1.17	0.78	0.31	0.37	0.35	0.59	0.66	0.71	3.942	4.038	1.306	0.303	3.359	2

Discussion

Salt stress is a complicated phenomenon, involving a series of growth and physiological responses (Gao et al, 2014; Wu, 2007; Ye et al, 2005). Tremendous efforts have been made to establish the system for screening and evaluation of salt tolerance based on multiple traits and indices. Pires comprehensive analyzed salt stress response of 56 rice phenotype seedlings with 15 phenotypic traits (Pires et al, 2015). Based on growth and physiological parameters, the salinity levels of Avicennia marina seedlings were evaluated (Yan et al, 2007). Zhang and Yang used five physiological and biochemical characters, including GR, SOD activity, Pro content, MDA content and REL, to evaluate seedling tolerance levels of 26 maize hybrids and 67 maize inbred lines (Yang et al, 2011; Zhang et al, 2010). However, the evaluation of the salt tolerance only using the data from the seedling stage had its limitation, because the effect of salt may be considerable in the germination stage for some genotypes but only marginal during seedling stages. Therefore, in the present study, we investigated multiple growth parameters in both stages. To take different parameters into the assessment of a certain trait, PCA is widely used for plant tolerance comprehensive evaluation (Gandour et al, 2013). In this study, salt tolerance indices including SFW, SDW, SL, RFW, RDW and RL in both germination and in seedling stages, along with GR, SOD, Pro, MDA and REL were comprehensively evaluated by PCA for the salt tolerance evaluation of three important maize hybrids and their related inbred lines, and we found that the order of the salt tolerance levels for hybrids and inbred lines was Jingke968 > Zhengdan958 > X1132 and X1132M > Jing724 > Chang7 - 2 > Zheng58 > X1132F, respectively, which was consistent with the field observation under salt stress, indicating that our method is useful for evaluating maize salt tolerance.

Salt stress mainly affects plant via ionic stress, osmotic stress and oxidative stress (Zhang and Shi, 2013). The accumulation of Pro is important for osmotic stress adaptation (Gangopadhyay et al, 2000). A substantial increase of Pro content was found in stressed Salicornia europaes by Aghaleh (Aghaleh et al, 2011), and they also demonstrated a significant correlation between salt tolerance and the increase in Pro content (Aghaleh et al, 2011). SOD can alleviate oxidative damage (Sun and Tao, 2011). It was reported that salt tolerant cultivars of alfalfa had a higher SOD activity (Wang and Han, 2009). In our study, physiological responses varied among different genotypes. Explicitly, SOD activity was significantly increased by salt treatment in Jingke968 and Zhengdan958, while only Pro content was substantially elevated in the hybrid of X1132, suggesting that different salt tolerance mechanisms play roles in different maize varieties (Zhu, 2001). Similarly, in the inbred lines, SOD activity was increased in B73 and Jing724, whereas Pro content was dramatically improved in X1132M and Chang7-2. In addition, both SOD activity and Pro content were significantly increased in the line of X1132F, indicating that probably multiple mechanisms were involved in its salt tolerance (Pires et al, 2015). MDA content and REL represent the degree of cell membrane damage, and they are negatively correlated with plant salt tolerance (Cui et al, 2015; Miao et al, 2010; Yazici et al, 2007). In our study, no increase in MDA and REL was observed upon salt stress in all three maize hybrids indicating that probably no cell membrane damage in these hybrids, which supports the concept that these three hybrids are resistant to salt stress (Du et al, 2006; Wang et al, 2016).

Maize is a salt sensitive crop and their plant growth and development are greatly affected by salinity. In the present study, different growth and physiological parameters upon salt treatment at different developmental stages did not share the same variation pattern among genotypes. The SFW, SDW, SL, RFW, RDW and RL were significantly reduced in all three hybrids and nearly all their related inbred lines in the germination stage, suggesting that maize is sensitive to salt stress at 100mM in the germination stage. However, except for RDW of Zheng58, the RFW and RDW of all maize genotypes did not respond to salt stress in seedling stage, which suggests that maize is more susceptible to salt in the germination stage compared to the seedling stage, similar to the results found in salt treated wheat (Al-Naggar, 2015). Moreover, similar to the result in avocado (Bernstein et al, 2001), we observed that the shoot growth was more sensitive to salinity than root growth in all tested genotypes during both germination and seedling stages. Interestingly, we found that the RDW of Zheng58 inbred lines was significantly increased by salt treatment, and this phenomenon was also reported in sorghum (Chaugool et al, 2013).

Conclusion

Salt tolerance levels of three important hybrids and their related inbred lines were comprehensively evaluated based on theirgrowth and physiological parameters, and the results were consistent with the observation of these lines growing in the saline soil, and thus this evaluation system might be applied for an accurate assessment of salt resistance in other germplasms and breeding materials. In addition, we showed that maize was more susceptible to salt stress at germination stage than at seedling stage, and that the shoot growth was more sensitive to salinity than root growth during both germination and seedling stages. Our data revealed different maize genotypes might utilize different salt tolerance mechanisms in maize seedling, but all of them involved the response of Pro and SOD activity.

Author contributions

LMJ and YXZ collected samples and performed the laboratory work. JRZ, LMJ, YXZ, WS, RYZ, ZS, AGS, CHL, XPW developed the manuscript. LMJ, YXZ and ZS prepared the manuscript.

Acknowledgment

This work was supported by grants from BAAFS Innovation Team of Corn Germplasm Innovation and Breeding of New Varieties (JNKYT201603) and 2016 Postdoctoral Talent Training Fund of BAAFS.

Conflict of Interest

The authors declare that they have no conflict of interest.

References (omitted)

Exploring Differentially Expressed Genes Associated with Fertility Instability of S-type Cytoplasmic Male-sterility in Maize by RNA-Seq

Su Aiguo*   Song Wei*   Shi Zi   Zhao Yanxin
Xing Jinfeng   Zhang Ruyang   Li Chunhui   Luo Meijie
Wang Jidong   Zhao Jiuran

(*Maize Research Center, Beijing Academy of Agriculture and Forestry Sciences/Beijing Key Laboratory of Maize DNA Fingerprinting and Molecular Breeding, Beijing* 100097, *China*)

Abstract: The germplasm resources for the S-type male sterility is rich in maize and it is resistant to *Bipolaris Maydis* race T and C1, but the commercial application of S-type cytoplasmic male sterility (CMS-S) in maize hybrid industry is greatly compromised because of its common fertility instability. Currently, the existence of multiple minor effect loci in different nuclear genetic background was considered as the molecular mechanism for this phenomenon. In the present study, we evaluated the fertility segregation of the different populations with the fertility instable material FIL-H in two environments of Beijing and Hainan. Our results indicated that the fertility instability of FIL-H was regulated by multiple genes, and the expression of these genes was sensitive to environmental factors. Using RNA sequencing (RNA-Seq) technology, the transcriptome of the sterile plants and partially fertile plants resulted from the backcross of FIL-H × Jing724 in Hainan was analyzed and 2108 genes with different expression were identified, including 1951 up-regulated and 157 down-regulated genes. The cluster analysis indicated that these differentially expressed genes (DEGs) might play roles in many biological processes, such as the energy production and conversion, carbohydrate metabolism and signal transduction. In addition, the pathway of the starch and sucrose metabolism was emphatically investigated to reveal the DEGs during the process of starch biosynthesis between sterile and partially fertile plants, which were related to the key catalytic enzymes, such as ADP-G pyrophosphorylase, starch synthase and starch branching enzyme. The up-regulation of these genes in the partially fertile plant may promote the starch accumulation in its pollen. Our data provide the important theoretical basis for the further exploration of the molecular mechanism for the fertility instability in CMS-S maize.

Key words: CMS-S; fertility instability; RNA-Seq; DEGs

Introduction

Cytoplasmic male sterility (CMS) is the phenomenon in higher plant characterized by pollen

---

Su Aiguo, E-mail: sx_201.su@163.com;
Correspondence Zhao Jiuran, Tel: +86-10-51503936, E-mail: maizezhao@126.com
* These authors contributed equally to this study

sterility, normal ear development and function, which has been reported in more than 150 plant species in the natural environment (Laser and Lersten, 1972). CMS has been utilized in hybrid production in many important crops, so it has an important commercial value (Bohra et al, 2016; Kubo and Newton, 2008). Taking the advantage of heterosis, maize is now an important food and feed crop worldwide. The discovery and creation of maize CMS material with different origins, complete sterility and stable fertility to establish the efficient technology system of hybrid seed production not only allows the decrease of production cost and the improvement of production efficiency, but also facilitates the utilization of maize heterosis in an effective manner. Currently, maize has three major types of CMS, designated as T (Texas), C (Charrua) and S (USDA) (Laughnan and Gabaylaughnan, 1983). T and C type CMS and their hybrids are susceptible to *Bipolaris Maydis* race T& C (Charles and Levings, 1993; Liu et al, 1991), while S-type cytoplasmic male sterility (CMS-S) is resistant. In addition, CMS-S is the largest group among three types with abundant germplasm resources (Vančetović et al, 2010), and thus it has a significant production and application prospects. However, fertility instability is widely observed in CMS-S, which greatly constrains its application in the commercial hybrid seed production (Weider et al, 2009).

The fertility of CMS-S in maize is determined by the interaction between nucleus and cytoplasm (Zhang et al, 2005). It has been demonstrated that the 1.6kb transcript encoded by the *orf355-orf77* in mitochondria genome highly expresses in the pollen of sterile plants during the microspore biogenesis, which may interfere the normal function of ATP synthase complex, leading to abortive pollen (Gabay-Laughnan et al, 2009; Wen and Chase, 1999; Zabala et al, 1997). *Rf3*, the main fertility restorer gene, can inhibit the expression of cytoplasmic sterile genes and restore the fertility (Kamps and Chase, 1997; Xu et al, 2009).

The fertility instability in maize CMS-S presents as the shift of some pollens from sterile to fertile in S (*rfrf*) plants. The molecular genetics mechanism is very complicated for this phenomenon, affected by different genetic background and environmental factors (Bennetzen et al, 2009; Buckmann et al, 2014; Matera et al, 2011; Weider et al, 2009). Gabay-Laughnan et al. identified a minor effect fertility restorer, *Rf9*, which is non-allelic to *Rf3*, and its expression is greatly influenced by the nuclear genetic background and temperature (Gabay-Laughnan et al, 2009). However, it is not sufficient to explain the common phenomenon of fertility instability in the CMS-S lines. Recent studies indicate that multiple minor effect loci may be involved in the fertility restoration ability (Feng et al, 2015; Zheng and Liu, 1992). Tie et al has mapped two minor effect quantitative trait locus for the fertility restoration of CMS-S on Chromosome 6 and 9 (Tie et al, 2006). Feng et al identified 30 significant loci related to pollen fertility, anther exsertion and pollen shedding by genome-wide association study. The expression of these genes in CMS-S may compensate the mitochondrial deficiency caused by the cytoplasmic sterile genes, resulting in the incomplete pollen abortion (Zheng and Liu, 1992). However, it is difficult to fine map and clone these genes using traditional methods because of their complex genetic basis, instable phenotype and susceptibility to environmental factors. In addition, the mechanism of the interaction among minor effect restorers, cytoplasmic sterile genes and main fertility restorers has not been reported.

The maize CMS-S line MD32 andthe line Jing724 were used as the donor for cytoplasmic sterile gene and recurrent parent, respectively, to obtain the thorough pollen abortion line S-Jing724 with desired traits by multiple generation backcrossing and marker assisted background selection (Song et al, 2016). Currently, S-Jing724 has been successfully applied in the hybrid seed production with CMS. However, very few families during the backcross process exhibited fertility instability, showing anther exsertion with neither anther dehiscence nor pollen shedding, and a small amount of normal pollen was also observed.

In the present study, we selected one complete sterile BC_4 plant from the fertility instable family and designated it as FIL-H. The materials from the backcross of FIL-H × Jing724 were planted in Beijing and Hainan two different locations to identify their fertility instability. Meanwhile, RNA sequencing (RNA-seq) was used to analyze genome-wide transcriptome changes of the anther tissues collected from both sterile and partially fertile plants in Hainan, where the fertility instability was easily induced. The results will facilitate us to understand the molecular mechanism of the fertility instability of CMS-S.

Materials and methods

Experimental material

The fertility instable CMS-S line FIL-H and the inbred line Jing724 were released by the Maize Research Center of Beijing Academy of Agricultural and Forestry Sciences. Plant materials were planted and identified in the International Seed Industry Park in Tongzhou, Beijing (BJ-TZ, 39°N, 116°E) and the Maize Propagation Base in Yacheng, Hainan (HN-YC, 18.3°N, 109.5°E).

Thepopulation of FIL-H×Jing724 was planted in both BJ-TZ and HN-YC to identify the fertility stability. In HN-YC, the plant with partially restored fertility and the sterile plants were used as maternal parent to backcross to Jing724, and the fertility instability was observed in different locations.

Identification of fertilityinstability

The fertility identification was performed using the population of field observation of the tassel and the microscopic examination of the pollen in the lab. From the beginning of heading stage to the end of pollen shedding stage, the status of anther exsertion, anther dehiscence and pollen dispersal for individual plants were evaluated. In the present study, the plants with completely no anther exsertion and the pollen exhibited no staining from I_2-KI were considered as sterile plants. Plants with visible anther exsertion and partially normal pollen staining were considered as partially fertile plants.

The pollen microscopic examination was carried out by I_2-KI staining method. After fertility segregation, three spikelets respectively positioned at the top, in the middle and at the bottom of the tassel were collected from both partially fertile plant and sterile plant, and stored in the

Cayrony's fix solution [75% (v/v) ethanol in acetic acid] in 1.5ml eppendorf tube. The anther was removed from the spikelet and put onto slide, and the pollens were squeezed out by tweezers. The pollens were then stained with 1% I_2-KI solution containing 0.5% (w/v) iodine and 1% (w/v) potassium iodide, and assessed by the light microscope (Olympus IX73, Japan) at 40 × magnification. Three representative fields for each sample were selected for the observation and data calculation.

Sample collection and RNA handling

For the backcross of FIL-H × Jing724 planted in HN-YC, sterile plants and partially fertile plants served as control and experimental group, respectively, each with three biological replicates. The tassel was detached from plant, soaked in water and brought back to the lab. Under RNase-free condition, the anther was quickly removed from tassel and stored in 10× volume of RNAwait solution (Solarbio, Beijing). The anther RNA was extracted using Trizol (Invitrogen, Beijing) and its purity, concentration and integrity were examined by electrophoresis and Agilent2100 (Agilent, Santa Clara, CA).

Library preparation for transcriptome sequencing

A total amount of 3μg RNA per sample was used as input material for the sterile plants and partially fertile plants. Sequencing libraries were generated using RNA Library Prep Kit (NEB, USA) following manufacturer's recommendations and index codes were added to attribute sequences to each sample. Briefly, mRNA was purified from total RNA using poly-T oligo-attached magnetic beads. Fragmentation was carried out using divalent cations under elevated temperature in next first strand synthesis reaction buffer (5×). First strand cDNA was synthesized using random hexamer primer and M-MuLV reverse transcriptase. Second strand cDNA synthesis was subsequently performed using DNA polymerase I and RNase H. Remaining overhangs were converted into blunt ends via exonuclease/polymerase activities. After adenylation of 3′ ends of DNA fragments, next adaptor with hairpin loop structure were ligated to prepare for hybridization. In order to select cDNA fragments of preferentially 150~200bp in length, the library fragments were purified with AMPure XP system (Beckman Coulter, Beverly, USA). Then 3μl USER enzyme (NEB, USA) was used with size-selected, adaptor-ligated cDNA at 37℃ for 15min followed by 5 min at 95℃ before PCR. Then PCR was performed with phusion high-fidelity DNA polymerase, Universal PCR primers and index primer. At last, PCR products were purified (AMPure XP system) and library quality was assessed on the Agilent Bioanalyzer 2100 system. The transcriptomes were sequenced using Illumina HiSeq 2500 (Illumina, San Diego, CA) at Biomarker Technologies Co., LTD (Beijing, China).

Quality control and sequence alignment

Raw data werefirstly processed through in-house perl scripts. In this step, clean data (clean reads) were obtained by removing reads containing adapter, reads containing ploy-N and low quality reads from raw data. At the same time, Q30, GC-content and sequence duplication level of the clean data were calculated. Clean sequence data were aligned with the maize reference genome B73 RefGen_ v3 downloaded from ftp: //ftp.ensemblgenomes.org/pub/plants/release-

24/fasta/ to obtain the mapped data. Analyses of differentially expressed genes were based on mapped data.

Analysis of differentially expressed genes

The Pearson's Correlation Coefficient r (Schulze et al, 2012) was calculated and served as the evaluation indictor for the correlation between biological replicates. The differentially expressed genes were identified using DESeq (Anders and Huber, 2010) with the criteria of the fold change ≥2 and the false discovery rate (FDR) <0.01. To ensure the true representation of fragment numbers for the gene expression level, the normalization of the reads numbers and gene length was performed using Fragments Per Kilobase of transcript per Million fragments mapped (FPKM) (Trapnell et al, 2010) in Cuffdiff software. The annotation and enrichment analysis of the differentially expressed genes (DEGs) were performed with multiple databases, NR (Deng et al, 2006), Swiss-Prot (Rolf et al, 2004), GO (Ashburner, 2000), COG (Tatusov et al, 2000) and KEGG (Kanehisa, 2004).

The expression analysis of the key starch biosynthesis genes

Three sterile plants and three partially fertile plants were selected from the backcross of FIL-H × Jing724, and the anther RNA was extracted as described above. The expression level of three differently expressed genes (*GRMZM2G144002*, *GRMZM2G163437*, *GRMZM2G032628*) related to starch biosynthesis was evaluated by quantitative PCR (QPCR) with *Actin* (NM 001155179) as the internal reference. The primers for QPCR were included in Supplemental Table S1.

QPCR was performed onStepOnePLUS (SinoGene, Beijing) with three technical replicates for each sample. The 20μl QPCR reaction system contained 1μl of diluted-cDNA, 7.5μl of 2×SG Green PCR Master Mix, and 0.25μl of each of the 10μM primers, and the amplification condition was as follows: 95℃ for 10min, followed by 40 cycles of 95℃ for 20s, 60℃ for 30s and 72℃ for 30s. The quantification of the expression level was carried out using the relative $\Delta\Delta Ct$ method on Prizm 5 software (http://www.graphpad.com/quickcalcs/).

Results

Genetic analysis of the fertility instability

The fertility status was evaluated for three different populations in both Beijing (BJ) and Hainan (HN), as included in Table 1. For the population of FIL-H×Jing724, the proportion of partially fertile plants in HN was significantly higher than that of BJ, which stayed the same for the other two populations, partially fertile × Jing724 and sterile × Jing724. The proportion of fertility restoration of partially fertile × Jing724 was greater comparing to sterile × Jing724 in HN, but the proportions of two populations were comparable in BJ, indicating that the fertility stability is susceptible to environmental factors.

Table 1 The evaluation of the fertility stability for different populations

Generation	Planting time	Location	Population	Number of total plants	Number of partially fertile plants	Proportion of partially fertile plants (%)
BC_5	2014.05	BJ	FIL-H×Jing724	84	2	2.4
BC_5	2014.11	HN	FIL-H×Jing724	70	14	20
BC_6	2015.05	BJ	Partially fertile × Jing724	158	1	0.6
BC_6	2015.05	BJ	Sterile × Jing724	162	2	1.2
BC_6	2015.11	HN	Partially fertile × Jing724	62	29	46.8
BC_6	2015.11	HN	Sterile × Jing724	61	15	24.6

The identification of the plant with fertility instability

The backcross of FIL-H× Jing724 was planted in HN, and no anther exertion was observed in all individuals within two days the tassel emergence. Two to three days later, the anther exsertion reached its maximum in partially fertile plants with about 1/3 of the total anthers, which is categorized as grade II anther exsertion according to the classification by Feng (Feng et al, 2015). In the partially fertile plants, the anther was exserted without dehiscence, while the spikelets were clung to the main tassel branch with no anther exposed in the sterile plants (Figure 1A). Comparing the upper anther of the spikelet at the same developmental stage, the anther of the partially fertile plant exhibited a darker color and a better plumpness than that of the sterile plant, but their sizes were comparable (Figure 1B).

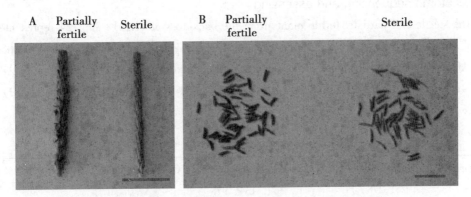

Figure 1 The identification of fertility instability in FIL-H

A. representative tassel image of partially fertile and completely sterile individuals of FIL-H at 3 days after the emergence of tassel. The scale bar represents 3cm. B. representative image of anthers collected from A. The scale bar represents 3cm

A little portion of the pollens from partially fertile plant showed normal round shape with blackI_2-KI staining (Figure 2), which was considered as normal pollen, but the majority was exhibited irregular shape with only partial I_2-KI staining, which was the characteristics of abortive pollen. About 6.5% pollens from the partially fertile plants were stained, while no pollen was stained in sterile plants, which suggests the fertility instability of CMS-S in FIL-H.

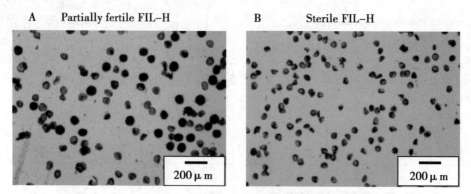

Figure 2　Representative pollen staining images of partially fertile (A) and completely sterile FIL-H (B)

Pollens were harvested at pollen dispersal stage and stained with iodine-potassium iodide dyeing method (I2-KI). Pollens exhibited round shape with black staining were considered as viable. The scale bars represent 200μm

Theanther exsertion corresponded to the pollen staining in the partially fertile plants, so the anther exsertion could be used as the proxy of normal pollen development on the tassel. In addition, because of the low proportion of normal pollen from the shrunken and non-dehiscent anther, it was hard to get seed from its selfing.

Transcriptome sequencing and assessing

The sterile and partially fertile plants from the backcross of FIL-H× Jing724 represented the control and experimental group to perform the transcriptome sequencing of their anther tissues. After removing the sequence of the adapter primers and sequences with low quantity, the total of 43.8Gb clean data was obtained for two groups with the percentage of bases (Q30) over 86.2% (Table 2).

Table 2　The evaluation of the sequencing data

Samples	Read Number	Base Number	GC Content (%)	≥Q30 (%)
S-1	25,278,797	6,369,134,978	57.0	86.5
S-2	26,667,392	6,718,765,357	57.3	86.5
S-3	32,414,750	8,166,378,442	57.2	86.5
F-1	29,722,660	7,488,365,948	56.5	86.3
F-2	30,895,939	7,783,637,957	57.1	86.5
F-3	29,176,100	7,350,252,711	56.9	86.2

Note: S-1, S-2 and S-3 represent sterile plants. F-1, F-2 and F-3 represent partially fertile plants. Read Number: the total number of pair-end reads in the clean data. Base number: the total number of bases in the clean data. GC content: the GC content of clean data. ≥Q30: the percentage of bases with quality greater than 30 in the clean data

The clean reads of each sample were aligned to the reference genome B73_ v3. The average of uniq mapped reads ratio for the control and experimental groups was 60.1% and 61.8%, respectively, both of which were evenly distributed on the 10 chromosomes of B73_ v3. The pearson's correlation coefficient r^2 for both groups was greater than 0.83, indicating a high correlation between replicates, so the sequencing data were reliable for the analysis of differentially expressed genes.

Hierarchical cluster analysis and sequence annotation

Based on the selection criteria of fold change ≥2 and FDR<0.01, 2108 genes with differential expression were identified. Compared to control, 1 951 genes were up-regulated and 157 genes were down-regulated in the experimental group. Based on the similarity of gene expression, hierarchical clustering was performed for the DEGs between two samples (Figure 3). Visual inspection of these expression groups indicated the possible existence of diverse and complex regulation patterns.

Taking advantages of the databases of NR, Swiss-Prot, GO, COG and KEGG, the DEGs were annotated. 2 052 out of 2 108 genes (97.3%) had significant matches in the NR database, followed by 1 687 genes (80%) in GO, 1 563 genes (74.1%) in Swiss-Prot, 655 genes (31.1%) in COG and 168 genes (8%) in KEGG. Overall, 2 052 genes (97.3%) were annotated in at least one database.

Gene ontology analysis of DEGs

The gene ontology (GO) analysis was performed for annotated 1 687 DEGs (Figure 4). At least one DEG was seen in 47 out of the total 54 GO terms (Supplement Table S2). In three GO categories, cellular component, molecular function and biological process, significant enrichment of DEGs was observed in the terms of photosystem I, pectinesterase activity, plant-type cell wall modification, response to abiotic stimulus, pollen tube growth via topGO (Alexa and Rahnenfuhrer, 2010) analysis. The top 10 most significantly enrich terms in each of the three GO categories were included in Supplement Table S3. GO analysis indicated that the identified genes with different expression between two samples might mediate the certain biological processes, such as the production and conversion of energy, carbohydrate metabolism and signal transduction.

168 DEGs were annotated with KEGG database, which is commonly used to explore the gene enrichment in metabolism pathway and signal transduction pathway (Kanehisa, 2004). The statistical analysis showed that 8 pathways were significantly enriched (Table 3 and Supplement Table S4).

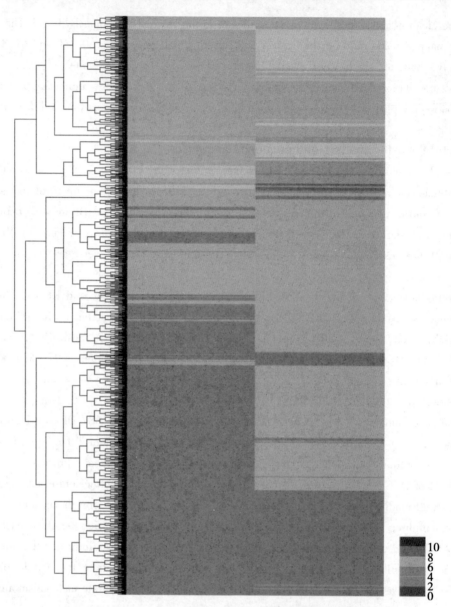

Figure 3   Hierarchical clustering analysis of gene expression in completely sterile and partially fertile plants based on log ratio fragments per kilobase of transcript per million fragments mapped (RPKM) data

The columns represent different samples and rows represent different genes. Various colors represent the gene expression level (log2FPKM+1). The expression data were obtained as the average of three biological replicates for each sample

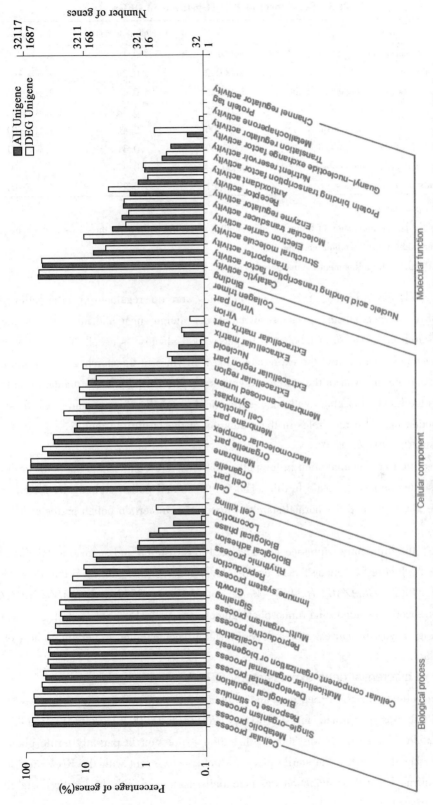

Figure 4 Gene ontology (GO) functional analysis of differentially expressed genes (DEGs) in sterile and partially fertile plants
Three major GO categories and their secondary terms were as indicated at X-axis. The right Y-axis represents the proportion of annotated genes in a specified GO term to all the GO term in the whole category, and the right Y-axis represents the number of annotated genes

Table 3 Enrichment of KEGG pathway in DEGs

Pathway	KO	Enrichment Fold	Q_ value
Photosynthesis-antenna proteins	ko00196	0.1	1.20E-09
Galactose metabolism	ko00052	0.29	2.60E-02
Amino sugar and nucleotide sugar metabolism	ko00520	0.4	3.90E-02
Photosynthesis	ko00195	0.38	4.80E-02
Starch and sucrose metabolism	ko00500	0.44	1.10E-01
Terpenoid backbone biosynthesis	ko00900	0.31	1.60E-01
Arginine and proline metabolism	ko00330	0.37	3.00E-01
Fatty acid biosynthesis	ko00061	0.29	3.30E-01

KO: KEGG pathway ID

Enrichment Fold: the proportion of the genes annotatedfor a specific pathway in the DEGs to the genes annotated for the same pathway in the whole genome

Q-value: P value after adjustment for multiple hypothesis testing

In the partially fertile plants, 11, 17 and 17 genes were up-regulated in three pathways related to carbohydrate metabolism, galactose metabolism, amino sugar and nucleotide sugar metabolism and starch and sucrose metabolism, respectively. Galactose is usually found in plants as the form of vegetable gelatin, and the expression of key enzymes in the starch and sucrose metabolism pathway may be involved in the starch biosynthesis in the partially fertile plants. In addition, the DEGs with the function of photosynthesis, amino acid metabolism, terpenoid biosynthesis and signal transduction may also play roles in the regulation of the fertility instability.

Starch and sucrose metabolism

The normal nutrient accumulation in the pollen not only represents as the hallmark for its maturity, but also determines the male fertility. The 6.5% normal pollen staining by I_2-KI in the partially fertile plant revealed the normal starch accumulation in certain pollen grains of FIL-H × Jing724.

Therefore, the differently expressed genes related to the starch and sucrose metabolism were further analyzed (Figure 5). Among the 17 up-regulated genes in this pathway, the up regulation of *GRMZM2G144002*, *GRMZM2G163437*, *Maize_ new Gene_ 3044* and *GRMZM2G032628*, which possibly encode the adenosine diphosphate glucose pyrophosphorylase, starch synthase and starch branching enzyme in maize, may induce the starch accumulation in certain anthers of partially fertile plant.

Confirmation of differential gene expression by QPCR

To further validate the RNA-Seq data and to obtain more quantitative transcript level measurements, QPCR was performed for 3 DEGs, *GRMZM2G144002*, *GRMZM2G163437* and *GRMZM2G032628* (Figure 6). It was showed that their expression in partially fertile plants was substantially higher than that of the sterile plant, which is consistent with the RNA-Seq result, and further confirmed that their induction was potentially associated with the fertility instability in the partially fertile plant.

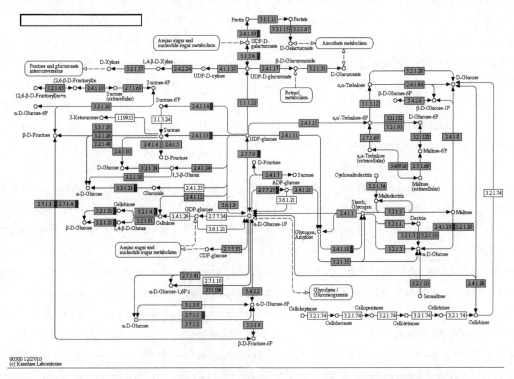

Figure 5 Starch and sucrose metabolism pathways (KO00500) from KEGG and the position of 17 annotated DEGs. Red boxes indicate all 17 DEGs were up regulated in the partially fertile plants

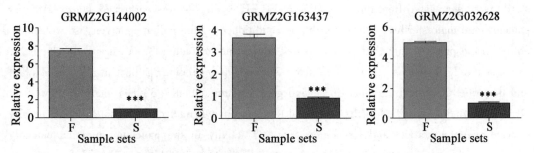

Figure 6 QPCR validation of the transcript level for three DEGs identified in the partially fertile (F) and sterile (S) plants

The gene loci were as indicated on top of the histogram. Each bar represents the mean±SE of three biological replicates. The values are determined using *Actin* as an internal reference. The asterisks show the statistically significant difference of $P<0.001$ determined by ANOVA

Discussion

In the CMS resulted from theinteraction between nucleus and cytoplasm, the mutation of the cytoplasmic gene induces the interruption of the mitochondrial metabolic pathway, which leads to the pollen abortion (Hanson and Bentolila, 2004). In the sterile plant, the defect of mitochondrial function caused by the cytoplasmic sterile gene may be slightly compensated by the minor effect fertility-related genes, resulting in the formation of normal fertile pollen (Tie et al, 2006; Zheng and Liu, 1992). In the breeding process, the accumulation of these minor effect genes in certain sterile lines may occur due to the difficulty of their identification and limited screening by molecular markers. Moreover, the effect of these minor effect loci may be amplified under certain environmental circumstances.

In the present study, because FIL-H was selected from the CMS-S fertile instable family, all three backcrossing populations, FIL-H × Jing724, partially fertile plant × Jing724 and sterile plant × Jing724, exhibited various levels of fertility instability in Beijing and Hainan. At the same location, the segregation ratio for three backcrosses suggests the phenomenon may be the composite regulation outcome of multiple minor effect genes. In addition, a significant difference in the fertility instability of FIL-H was observed in Hainan, probably due to the environmental factors, especially in certain temperature and humidity conditions. Previous studies demonstrate that climate factors, particularly the air temperature, evapotranspiration and water vapor, are closely related to the fertility of CMS-S maize 10 days before anthesis as well as during anthesis (Weider et al, 2009). The fertility instability is easily induced by relatively low temperature and high humidity (Duvick, 1965). Around the heading stage of FIL-H, the day average temperature in Hainan was 5℃ lower than that in Beijing, but the humidity was substantially higher than that in Beijing because it is adjacent to the South China Sea. Our data is consistent with previous results and confirms that the fertility stability of FIL-H is susceptible to the environment.

With the aid of RNA-Seq technology, the transcriptome analysis of partially fertile plant and sterile plant segregated from the population of FIL-H × Jing724 was performed. Maize CMS-S is gametophytic male sterility, whose fertility is conditioned by the pollen genotype, so we selected anther in the present study (Li et al, 2004), the tissue for pollen formation and development.

For the 2 108 DEGs, 1 951 (92.6%) showed up regulation in the partially fertile plant, and they were intensively grouped in several gene clusters (Figure 3). GO analysis showed that DEGs were mainly enriched in the functional terms of guanyl-nucleotide exchange factor activity, enzyme regulator activity and translation regulator activity in the molecular function category, which may play an important role in the regulating the fertility instability of FIL-H. For example, 7 DEGs belonging to the GO functional term of guanyl-nucleotide exchange factor activity, encode Rho guanine nucleotide exchange factor, and all were up regulated in the partially fertile plant. Studies demonstrate that small GTPase encoded by Rho significantly participates in signal transduction, growth and development of tissues and organs and transmembrane transport (Anders and Juergens, 2008; Kuhlmann et al, 2015). In addition, substantial enrichment of the DEGs was also observed in the functional term of membrane part in cellular component, and locomotion and multi-organism process in the biological process (Supplement Table S2). These functional GO terms revealed the transciptomic discrepancy between sterile plants and partially fertile plants.

In the present study, we emphasized on the function of DEGs in the starch and sucrose metabolism pathway (ko00500). The starch was synthesized through a series of enzymatically catalyzed reaction in the amyloplast, including the adenosine diphosphate glucose pyrophosphorylase (AGPase), starch synthase and starch-branching enzyme and starch debranching enzyme (Baguma et al, 2003). AGPase is the first key enzyme in starch biosynthesis, and is also the rate limiting enzyme, responsible for converting the 1-phosphate glucose and ATP generated from photosynthesis to ADP-glucose (Ballicora et al, 2003; Thorbjornsen et al, 1996). It has been demonstrated that the starch production is significantly elevated in potato and corn by overexpressing the AGPase gene *glgC* from *E. coli* (Stark et al, 1992; Wang et al, 2007). *GRMZM2G144002*, *Maize_ new Gene_ 3044* and *GRMZM2G163437* encode the AGPase, the expression of which may facilitate the synthesis of ADP - glc, the immediate precursor of starch. *GRMZM2G538064* encodes one subunit of 1, 4-alpha-glucan-branching enzyme 2. The branching enzyme catalyzes the transfer of gluan chain to the C6 hydroxyl group of the glucose residue on the other glucan chain to form the α-1, 4-glucosidic bond, which plays a key role in the structure of the branched starch and starch granules (Satoh et al, 2003). In addition, the active expression of genes encoding glucanotransferase, sucrose synthase and endoglucanase may mediate the metabolism of starch and other carbohydrates in the pollen. QPCR was performed for 3 DEGs to validate the confidence of the RNA-Seq, and it showed a consistent result. Further investigation is needed to illustrate the function of these genes in maize CMS-S.

Currently, the major bottleneck of utilizing CMS-S in maize hybrid seed production is the fertility instability of the sterile line. With our breeding and researches, this issue might be improved in the following aspects: 1) Enhance selection pressure. The fertile instability in CMS-S is probably conditioned by multiple genes in certain nuclear genetic background, so the stable sterile CMS-S line can be obtained with properly screening. Moreover, because of the environmental sensitivity of the minor effect genetic loci, the test results from multiple locations and years should be appropriately considered in the process of identification and backcross. In fact, we have obtained the stable sterile line S-Jing724 after the genetic screening in multiple locations and years, and it has been applied in the hybrid seed production with CMS for the leading maize variety Jingke968 in China. 2) Develop related molecular markers to assist the field selection of sterile plant by taking the advantages of the study on the minor effect loci responsible for the fertility instability. Overall, despite of the instable fertility in CMS-S maize, its utilization in three-line breeding still has broad application prospects in the commercial maize hybrid seed production.

Conclusion

The phenomenon of instable fertility is common in CMS-S maize, which greatly limits its application in the commercial hybrid seed production. In the present study, we used the fertility instable material FIL-H, which has been undergone several rounds of backcross, to preliminarily explore its genetic mechanism. The field phenotype indicated that the fertility instability was conditioned by multiple nuclear genes, and that it was also sensitive to environmental factors. The gene expression profile analysis of the anther tissues from both partially fertile and sterile plants showed genes with differential expression were significantly enriched in the biological processes of energy production and conversion, carbohydrate metabolism and signal transduction. The expression levels of the key catalytic enzymes in the starch and sucrose metabolism pathway, AGPase, starch synthase and starch branching enzyme, were elevated in the partially fertile plants, which may induce the starch accumulation in the pollen of the partially fertile plants, and thus exhibiting normal I_2-KI staining. Our data provide an important basis for the further investigation on the molecular mechanism of fertility instability in CMS-S maize.

List of abbreviations

CMS, Cytoplasmic male sterility; CMS-S, S-type cytoplasmic male sterility; RNA-Seq, RNA sequencing; DEGs, Differentially expressed genes; FPKM, Million fragments mapped; FDR, false discovery rate; QPCR, quantitative polymerase chain reaction; AGPase, Adenosine diphosphate glucose pyrophosphorylase.

Supporting information

Supplement Table S1 Primers used in QPCR.

Supplement TableS2 GO functional classification analysis for all annotated unigenes in the whole genome and annotated DEGs.

Supplement Table S3 Top 10 most significantly enriched GO terms for each GO category by TopGO analysis.

Supplement Table S4 Details of the enriched KEGG pathway.

Acknowledgements

This work is supported by the Beijing Postdoctoral Research Foundation (2014ZZ-68), the Postdoctoral Scientific Fund of Beijing Academy of Agricultural and Forestry Science (2014013), the National Technology Supporting Project (2014BAD01B09) and the Sci-technology Innovation Project of Beijing Academy of Agricultural and Forestry Science (KJCX20140202), the innovative team construction project of BAAFS (JNKYT201603).

References (omitted)

Journal of Integrative Agriculture, 2017, 16 (0): 60 345-60 347

Improvement of *Agrobacterium*-mediated Transformation Efficiency of Maize (*Zea mays* L.) Genotype Hi-II by Optimizing Infection and Regeneration Conditions

Xu You[1,2], Ren Wen[1], Liu Ya[1]*, Zhao Jiuran[1]*

(1. *Maize Research Center*, *Beijing Academy of Agricultural and Forestry Science*, *Beijing* 100097, *China*; 2. *Taizhou Vocational College of Science and Technology*, *Taizhou* 318000, *China*)

Abstract: Improving Agrobacterium tumefaciens mediated transformation of maize would be beneficial for maize transgenic research. Although, maize genotype Hi-II is used in various maize transformation research efforts, improvements in the transformation frequency of this genotype are still needed. In the present study, the immature maize embryos were given heat-shock pre-treatment prior to Agrobacterium-mediated transformation. About 9% increase in the transformation frequency was achieved when the embryos were subjected to 42℃ for 3min or 38℃ for 9min.

It was also found that the use of 100mg/L Casein Hydrolysate (CH) accelerated somatic embryogenesis in the initial period of regeneration. The combined use of 0.5mg/L Naphthaleneacetic acid (NAA) and 2.0mg/L Multi-effect triazole (MET) at different regenerate stage could effectively enhance root development and seedlings vigour to improve the survival rate of the transformed plantlets. Through integrating several methods can effectively improve the Hi-II genetic transformation efficiency.

Key words: Maize (*Zea mays* L.); Hi-II; *Agrobacterium tumefaciens*; Transformation

Introduction

Transgenic technology plays a key role in the increase of maize production. Transformation procedures produce genetically modified hybrid corn for commercial purposes and provide a basic tool for evaluating foreign gene expression in maize. Functional genomics studies in maize require the functional identification of a large number of candidate genes, making the use of high-throughput transformation systems essential. *Agrobacterium*-mediated transformation in maize was first reported in 1988. Optimization of the types of plant materials used for infection with *Agrobacterium tumefaciens*, the choice of vectors and strains of *Agrobacterium tumefaciens* employed and the tissue culture techniques have all contributed towards the increase in transformation efficiency.

* Author for Correspondence
* srlyyd@gmail.com; maizezhao@126.com

The use of the super binary vector during the *Agrobacterium* infection stage has enabled very high-frequency transformation of maize in the industrial sector. The use of L-cysteine has significantly enhanced Hi-II maize transformation employing standard binary vectors in public laboratories.

Among the several target maize genotypes, maize genotype Hi-II is the predominant genotype that has been used for genetic transformation of this crop. Hi-II can be used to generate type II embryogenic callus at a high frequency and is very suitable for high-throughput *Agrobacterium*-mediated transformation at an adequate level of efficiency. Although, Hi-II has become a preferred genotype for use in various maize transformation efforts, its transformation frequency still has a potential to further improve to accelerate advances in maize genetic improvement and gene functional verification research.

Various methods for improving *Agrobacterium*-mediated genetic transformation efficiency have previously been reported. The most common method is the application of physical force to the plant material during the *Agrobacterium* infection stage such as centrifugation, sonication and vacuum infiltration, all of which can effectively increase transformation efficiency. Vacuum infiltration along with sonication-assisted *Agrobacterium*-mediated transformation results in higher transformation efficiency as compared to the transformation alone in wheat, citrus and cowpea Centrifugation-assisted transformation has been employed in banana transformation and led to a significant enhancement in transformation frequency.

Theinduction frequency of calli from immature maize embryos was higher in response to heat-shock treatments, the transformation frequencies in rice and maize can be increased by treating immature embryos with both heat-shock and centrifugation prior to *Agrobacterium* transformation. Immature embryo-derived callus is more efficient for plant regeneration than calli from other explant tissues. Additional red light can be conductive to plant regeneration. 6-benzylaminopurine (BA) can effectively promote the emergence of shoot, the embryogenic callus readily formed plantlets. CH has a significant effect on the formation of maize embryoids which may be caused by containing L-Glu and L-Asp of two kinds of amino acids.

The objective of the present study was to optimize the parameters affecting *Agrobacterium*-mediated gene transformation in maize genotype Hi-II. The heat-shock pre-treatment of immature embryos was studied at varying temperatures and time durations affecting the transformation frequency of Hi-II. We also tested the effects of CH, NAA and MET on somatic embryo re-differentiation and rooting of seedlings. Our experiments will help both the production of transgenic maize and gene functional verification.

Material and Methods

Agrobacterium strains and plasmids

Agrobacterium strain LBA4404 and standard binary vector pCAMBIA3301-*EPSPS* were used as described previously. The vector contained a *GUS* gene, with an intron in the coding region, under the control of CaMV35S promoter. This *GUS* gene is expressed in plants but not in *Agrobac-*

terium tumefaciens. The vector also contained an *EPSPS* gene under the control of the maize ubiquitin promoter (Figure 1). The construct was mobilized into *Agrobacterium* and its integrity was confirmed by restriction digest analysis.

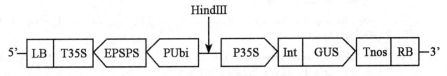

Figure 1 T-DNA region of the binary vector pCAMBIA3301-*EPSPS*

Plant materials

Hi-II genotype as maize transformation recipient material was grown in a greenhouse and the ears were harvested between 9~12 days after pollination. The immature embryos used in the experiments were 1.0~1.5mm in length. Maize ears were first sterilized with 20% sodium hypochlorite (5% active chlorine) plus 0.1% Tween-20 for 20min. Immature embryos 1.0~1.5mm in length were aseptically excised and employed in the experiments.

Medium formulations: The media used in this study were as described by Frame et al[5] with a minor modification. The concentration of 2, 4-D in the infection medium was increased from 1.5mg/L to 2.0mg/L. Different concentrations of plant growth regulators, 100mg/L CH were supplemented into the regeneration medium I, 0.5mg/L of NAA and 2.0mg/L of MET were added into the regeneration medium III respectively to test the effect of Hi-II calli regeneration. All media were sterilized before use (Table 1).

Table 1 Composition of media and solutions

Medium	Composition
YEP (solid)	Yeast extract 5g/L, NaCl 5g/L, peptone 10g/L, agar 15g/L, pH 6.8
YEP (liquid)	Yeast extract 5g/L, NaCl 5g/L, peptone 10g/L, pH 6.8
Infection	1/2 N6 salts, N6 vitamins, 2, 4-D 2.0mg/L, L-pro 0.7g/L, sucrose 68.4g/L, D-glucose 36g/L, MES 0.5g/L, myo-inositol 0.1g/L, AS (acetosyringone) 100μM, pH 5.2
Cocultivation	1/2 N6 salts and N6 vitamins, 2, 4-D 2.0mg/L, L-pro 0.7g/L, sucrose 30g/L, MES 0.5g/L, myo-inositol 0.1g/L, DTT (dithiothreitol) 0.154g/L, L-cys 0.3g/L, silver nitrate 0.85mg/L, AS 100μM, agar 8g/L, pH 5.8
Resting	N6 salts and N6 vitamins, 2, 4-D 1.5mg/L, L-pro 0.7g/L, Sucrose 30g/L, MES 0.5g/L, Myo-inositol 0.1g/L, Silver nitrate 0.85mg/L, Phytagel 3g/L, pH 5.8
SelectionI	N6 salts and N6 vitamins, 2, 4-D 1.5mg/L, L-pro 0.7g/L, Sucrose 30g/L, MES 0.5g/L, Myo-inositol 0.1g/L, Silver nitrate 0.85mg/L, glyphosate 100mg/L, Phytagel 3g/L, pH 5.8
SelectionII	N6 salts and N6 vitamins, 2, 4-D 1.5mg/L, L-pro 0.7g/L, Sucrose 30g/L, MES 0.5g/L, Myo-inositol 0.1g/L, Silver nitrate 0.85mg/L, glyphosate 200mg/L, Phytagel 3g/L, pH 5.8
Regeneration I	MS salts and N6 vitamins, Myo-inositol 0.1g/L, Sucrose 60g/L, CH (Casein Hydrolysate) 100mg/L, phytagel 3g/L, pH 5.8

(continued)

Medium	Composition
Regeneration II	MS salts and N6 vitamins, Myo-inositol 0.1g/L, Sucrose 30g/L, 6-BA 0.5mg/L, phytagel 3g/L, pH 5.8
Regeneration III-NAA	MS salts and N6 vitamins, Myo-inositol 0.1g/L, Sucrose 30g/L, NAA 0.5mg/L, phytagel 3g/L, pH 5.8
Regeneration III-MET	MS salts and N6 vitamins, Myo-inositol 0.1g/L, Sucrose 30g/L, MET (multi-effect triazole) 2mg/L, phytagel 3g/L, pH 5.8
GUS buffer	Sodium phosphate buffer (pH 7.0) 50mM, $K_3[Fe(CN)_6]$ 100mM, Triton X-1 000.001% (v/v), $K_4[Fe(CN)_6]$ 100mM, Na_2 EDTA 10mM, methyl alcohol 20%, X-gluc 0.5mg/ml

Pre-treatment with heat-shock

For heat-shockpre-treatment, embryos were placed in an "Infection+AS (acetosyringone)" medium in micro-centrifuge tubes and incubated in a H_2O_3 Dry Bath Incubator at the designated temperature (34, 38, 40, 42 or 46℃) for the specific time period (0, 3, 9 or 15) minutes, followed by immediate cooling on ice for 1 min. Immature embryos subjected to *Agrobacterium*-mediated transformation at 25℃ served as the control.

Five micro-centrifuge tubes were used for each experiment, with each tube containing approximately 50~100 embryos that had been washed twice with infection+AS (100μM) medium. One of five tubes was placed on a co-cultivation medium without heat-shock pre-treatment and infection and served as an experimental negative control. Another one tube was subjected to infection, without heat-shock pre-treatment to serve as a transformation frequency control. The other three tubes were subjected to heat-shock and infection as described above.

Agrobacterium culture initiation

Agrobacterium tumefaciens LBA4404 harboring the expression vector pCAMBIA3301-*EPSPS*, which was stored as a glycerol stock at-80℃, was streaked onto a YEP solid medium plate containing 50μg/ml rifampicin and 50μg/ml kanamycin. The plate was incubated at 28℃ for 3 days until single colonies developed. A single colony was inoculated in 5ml YEP liquid medium (in a 15ml tube) containing 50μg/ml rifampicin and 50μg/mL kanamycin and was cultured at 28℃ at 200rpm for 20~24hours. Thereafter, a 500μl of bacterial culture was pipetted into 50ml YEP liquid medium (in a 150ml Erlenmeyer flask) and cultured at 28℃ at 200rpm for 16~20 hours.

Inoculation and co-cultivation

The bacterial culture was suspended in infection+AS (100μM) medium at a cell density OD_{550} = 0.3. 1.0ml of *Agrobacterium* suspension was added to each 1.5 ml tube containing 50~100 embryos. The inoculation duration time was for 10min. The embryos were then transferred into co-cultivation medium plates with the scutella facing upwards. The plates were incubated at 20℃ for 3 days in the dark.

Selection and regeneration

The infected immature embryos were inoculated into the resting medium at 28℃ in the dark for 7 days after co-cultivation. Immature embryos were then transferred into the selection medium I to conduct a preliminary selection at 28℃ in the dark for 2~3 weeks until callus formation. The selection medium II was used to select the transgenic callus to eliminate non-transgenic callus under greater selective pressures. The plates were incubated at 28℃ in the dark for 2~3 weeks.

The screened transgenic calli were inoculated into regeneration medium I at 26℃ in the dark for 10 days to develop somatic embryos, then were transferred into regeneration medium II at 26℃ in the light to induce shoots. After the shoots elongated more than 4cm, the seedlings were put into regeneration medium III to produce roots. After about 3 weeks the transgenic plantlets were transplanted into the soil to grow.

Analysis of transformants

The expression of *GUS* gene in the immature embryos was assayed immediately after co-cultivation by performing histochemical staining with 5-bromo-4-chloro-3-indolyl glucuronide (X-gluc)[22], sterile buffer was used in the dye solution to avoid interference from contaminating bacteria. The embryos were examined for changes in color and the GUS-stained embryos were counted. The explants with *GUS* positive expression frequency was defined as the ratio of numbers of GUS positive explants to total numbers of evaluated explants.

For PCR and Southern blot analysis of the transformants, genomic DNA was isolated from regenerated maize leaves by the DNA extraction Kit. PCR assays were performed with specific primers for the *EPSPS* gene: F: 5'-ggagagcgacaggattcaggtgatg-3'; R: 5'-aggaagttcgggaag ctggtgg-3'. Thermal cycling for 15s at 94℃, 30s at 63℃, 30s at 72℃ was performed for 35 cycles. For the Southern blot analysis, a 639bp probe for *EPSPS* was amplified using the primers 5'-ACAC-CGAGAGGATGCTGAG-3' and 5'-CGCAGTCGTGGATTCTGATC-3'. Genomic DNA digested with *Hin*dIII from the transgenic plants (60μg) and the plasmid controls (5ng) was separated by electrophoresis on 0.8% agarose gels and transferred to a nylon membrane. The membrane was hybridized with the DIG labeled probes. Expression of the *EPSPS* gene was also assayed by using herbicide "Roundup" leaf painting in the regenerated plants.

Results

Heat-shock at different temperatures

Heat-shock pre-treatment at 42℃ for 3 min improved *GUS* expression in the transgenic immature Hi-II embryos (Figure 2). However, heat-shock at 34℃ or 38℃ for 3 minutes had almost no effect on *GUS* expression compared with the control. Heat-shock pre-treatment at 42℃ for 3 minutes enhanced *GUS* expression in immature Hi-II embryos by an average of 10.9% compared with the control group ($P<0.05$). However, heat-shock pre-treatment at 46℃ for 3 minutes reduced the level of *GUS* expression causing an average drop in expression by 19.6% compared with that of the control group ($P<0.05$). The results of these preliminary experiments suggest that heat-shock pretreatment at 42℃ for 3 minutes had the most positive effect on *GUS* expression of all of the heat-shock temperatures examined. Therefore, pre-treatment with heat-shock at 42℃ for 3 minutes can significantly improve the transformation frequencies of Hi-II immature embryos.

Heat-shock at 42℃ or 38℃ for different time periods

The immature embryos were subjected to heat-shock pretreatment at 42℃ or 38℃ for different periods of time prior to infection. Four time durations of heat-shock were tested including 0 min (CK), 3min, 9min and 15min (Table 2).

Table 2　Effects on pretreatment of immature Hi-II maize embryos with heat-shock at 42℃/38℃ for various periods of time prior to *Agrobacterium*-mediated transformation

Group (42℃)	Number of independent transgenic plants* (a)	Number of inoculated immature embryos (b)	Frequency of GUS expression (a/b)%	Group (38℃)	Number of independent transgenic plants* (a)	Number of inoculated immature embryos (b)	Frequency of GUS expression (a/b)%
42-0-1	35	49	71.4	38-0-1	41	55	74.5
42-0-2	35	49	71.4	38-0-2	40	54	74.1
42-0-3	43	59	72.9	38-0-3	64	89	71.9
CK (25℃) total	113	157	71.9	CK (25℃) total	145	198	73.2
42-3-1	46	58	79.3	38-3-1	33	43	76.7
42-3-2	38	46	82.6	38-3-2	38	51	74.5
42-3-3	51	62	82.3	38-3-3	61	84	72.6
42℃ 3min total	134	166	81.4	38℃ 3min total	132	178	74.2
42-9-1	34	50	68	38-9-1	37	45	82.2
42-9-2	29	39	74.4	38-9-2	45	53	84.9
42-9-3	51	73	69.9	38-9-3	75	93	80.6
42℃ 9min total	114	162	70.4	38℃ 9min total	157	192	82.6
42-15-1	26	50	52	38-15-1	33	44	75.0
42-15-2	28	47	59.6	38-15-2	30	41	73.2
42-15-3	35	64	54.7	38-15-3	59	90	65.6
42℃ 15min total	89	161	55.3	38℃ 15min total	122	175	69.7

*Epsps-resistant and GUS-positive plants

The *GUS* expression was significantly higher in embryos treated at 42℃ for 3min as compared to the control ($P < 0.01$) which concurred with the results from the preliminary experiments. Furthermore, it shows the reproducibility of the technique for practical applications. The *GUS* expression frequency increased 9.5% compared with the control (Figure 2c). The difference between the 42℃ 9min group and the control group was not significant. However, the 42℃ 9min group showed lower *GUS* expression than the 42℃ 3min group. Pretreatment with 42℃ heat-shock for 15min reduced the level of *GUS* expression as compared to that of the control group ($P<0.01$). These results clearly showed that the levels of *GUS* expression varied across different pre-treatments.

The *GUS* expressions in two groups, heat-shock at 38℃ for 3min and heat-shock at 38℃ for 15min, were very similar to those of the control group, with no significant differences in *GUS* ex-

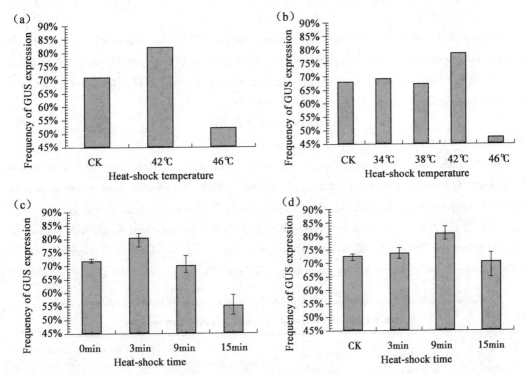

Figure 2 Frequency of *GUS* expression in immature Hi-II maize embryos pretreated with heat-shock at various temperatures for different time

a, b. heat-shock at different temperate for 3min; c. heat-shock at 42℃ for different periods of time; d. heat-shock at 38℃ for different periods of time

pression observed. However, heat-shock pre-treatment at 38℃ for 9min significantly increased the *GUS expression of Hi-II immature embryos* ($P<0.01$) which averaged 9.4% higher than that of the control group (Figure 2d). The effect of heat-shock at 38℃ for 9min was similar to the effect of heat-shock at 42℃ for 3min.

Therefore, both heat-shock at 42℃ for 3min and 38℃ for 9min can significantly improve the transformation of immature Hi-II embryos and that the error range gradually increased with increase in pretreatment time.

Effective accumulated temperature

The effective accumulated temperature is defined as the sum of the effective temperatures required to improve plant transformation by *Agrobacterium tumefaciens*. This definition is similar to of the definition of this term used in ecology. The effective accumulated temperature can be used to explain the effects of heat-shock on plant transformation. The effective temperature is defined as the difference between the heat-shock temperature and the minimum threshold temperature. Only when the heat-shock temperature is greater than the minimum threshold temperature, heat-shock will play a role in the transformation. The effective accumulated temperatures for Hi-II maize (K_{Hi-II}) and the minimum threshold temperature (C_{Hi-II}) were calculated as follows:

$$K_{Hi-II} = N(T - C_{Hi-II})$$

where N is the duration of heat-shock pretreatment and T is the heat-shock temperature. By inserting our experimental data into this formula, the values K_{Hi-II} (K_{Hi-II} = 18min · ℃) and C_{Hi-II} (C_{Hi-II} = 36℃) were obtained. The appropriate combination of heat-shock, temperature and pretreatment time for Hi-II maize can be predicted as follows:

$$N = 18min \cdot ℃ / (T - 36℃)$$

Induction of transgenic embryonic calli

The frequencies of embryonic callus induction of immature Hi-II maize embryos with heat-shock were 100% without transformation and selection which showed that there were no negative effects on embryonic callus induction with heat-shock. Two effective heat-shock groups, 38℃ 9min and 42℃ 3min, were used for embryonic callus induction after heat-shock with transformation and selection. The results showed that pre-treatment of heat-shock at 38℃ for 9min can improve embryonic callus induction by an average of about 9% compared to the control (Table 3).

Table 3 Effects on transgenic callus induction of immature Hi-II maize embryos after heat-shock pre-treatment with transformation and selection

Group	Heat-shock temperature (℃)	Heat-shock duration (min)	Number of immature embryos inducting transgenic callus* (a)	Number of inoculated immature embryos (b)	Frequency of transgenic callus induction (a/b)%
CK-0-1	CK (25℃)	0min	4	44	9.1
CK-0-2			5	56	8.9
CK (25℃) total			9	100	9.0
38-9-1	38℃	9min	6	36	16.7
38-9-2			8	44	18.2
38℃ 9min total			14	80	17.5
42-3-1	42℃	3min	5	41	12.2
42-3-2			6	52	11.5
42℃ 3min total			11	93	11.8

* Epsps-resistant and GUS-positive plants

Regeneration of transgenic seedlings

Somatic embryos in the regeneration medium I with 100mg/L CH had a larger size and faster reproductive speed than those in the medium without CH. Furthermore, a large number of somatic embryos showed rooting in the regeneration medium I with CH after 14 days (Figure 3).

The concentrations of NAA, 0.5mg/L, significantly enhanced the emergence of adventitious roots in the regeneration medium III-NAA. However, the root system growth was slow which made the seedlings not suitable for transplanting to the fields after 3~4 weeks. The regeneration medium III-MET with 2.0mg/L MET promoted lateral root emergence and enhanced root system sturdily. New lateral roots were thinner at the beginning but became thicker gradually. Roots were also observed elongated within 24 hours and continued to extend doubled over a period of 3 days. The

Figure 3 Induction of Somatic embryos in two kinds of media for 7 or 14 days
On the left is the normal Regeneration medium I, on the right is the Regeneration medium I with CH

seedlings with medium III-MET showed more advantageous characteristics such as stronger stems, darker color and larger leaves. However, new adventitious roots did not appear even after 14 days in the regeneration medium III-MET.

Based the above results, we combined use of NAA and MET, translated the seedlings induced in the regeneration medium III-NAA into the regeneration medium III-MET to cultivate for a period of time. When the seedlings exhibited the strong roots, larger leaves and thicker shoots, they would be transplanted after seedling adaptation. The transplantation survival rate of these plantlets was about 98% using the above method. Though these transgenic plants were generally becoming lower, they almost had the normal characteristics of flowering and fruiting. Contrastingly, only about 60% of seedlings were able to survive after transplanting using the not modified regeneration medium.

Analysis of transformed plantlets: In this study, several hundred transformed calli were obtained, each derived from a separate embryo. We randomly chose 100 transformed calli to regenerate and averagely 3 T_0 plants were produced from each embryo. PCR analysis results showed that over 90% of these T_0 plants were detected *EPSPS* gene (Figure 4). To further confirm the presence of *EPSPS* gene in the obtained transgenic maize plants, some of those were analyzed by Southern blot (Figure 5). There is one *Hin*dIII site existing in plasmid vector, so digestion with *Hin*dIII will linearize the plasmid about an 11kb fragment. Only one band was detected on transgenic maize line 3, 5, 6 and 7, indicating that only one copy of *EPSPS* was integrated into the maize genome of these transgenic lines. Two bands were detected on transgenic line 1 and 2, indicating two copies of *EPSPS* for those transgenic lines.

All of regenerated plants were assayed for EPSPS viability by 41% glyphosate (Roundup) leaf painting in the leaf tissues. These assays showed that 80% of transplants expressed *EPSPS*

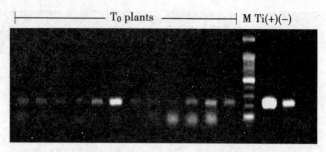

Figure 4 PCR results of *EPSPS* gene for some T_0 plants.

M. Transgen 100bp DNA ladder; Ti. Ti plamid containing *EPSPS*; (+).Positive samples; (-).Negative samples; Other lanes: T_0 regenerated plants. Fragment length of PCR was 231bp

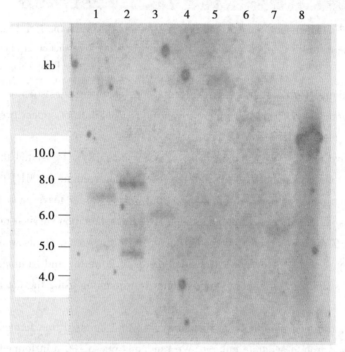

Figure 5 Southern blot analysis of transgenic maize lines on the genomic DNA digested with *Hind*III

Lane 1-7. transgenic maize; Lane 8. plasmid pCAMBIA3301-*EPSPS*

and were resistant to 10% glyphosate, only 10% of transplants expressed *EPSPS* and were resistant to 41% glyphosate.

LB, RB. T-DNA left and right borders respectively; PUbi. maize ubiquitin promoter; T35S. CaMV35S terminator; *EPSPS*. glyphosate resistance gene for selecting transformants; *Int-GUS*. intron-containing β-glucuronidase (*GUS*) gene; Tnos. nopaline synthase gene terminator.

Six random positive transgenic T_0 plants were self-crossed, successive propagation and T_1 plants were employed for progeny analyzing, the PCR amplification and EPSPS assay results showed that four out of six events followed the expected 3∶1 segregation ratio for transformed foreign gene and other two events were deviated from this ratio (data not shown) which could be attributable to the small number of progeny plants examined or *EPSPS* gene silencing.

Discussion

The efficiency of plant transformation depends on many factors. For *Agrobacterium*-mediated genetic trans formation, good T-DNA delivery and plant tissue survival following *Agrobacterium* inoculation are the two critical components for achieving high transformation efficiency. In the present study, we focused on the *Agrobacterium* infection and regenerate step in maize genotype Hi-II. Specifically, we examined the frequency of transformation of immature embryos of Hi-II subjected to heat-shock pretreatment.

The results showed that thepre-treatment of immature embryos with heat-shock at 38℃ for 9min or 42℃ for 3min could effectively improve the efficiency of *Agrobacterium*-mediated transformation in Hi-II. From the results it can be drawn that the lower was the heat-shock temperature, the longer was the pretreatment time required for improved transformation. Our study also showed that *GUS* expression variation had a larger fluctuation with increase in pre-treatment time at a particular temperature. The stability of transformation decreased with increase in the time of heat-shock pre-treatment. Short heat-shock pretreatment could be used to stably improve the genetic transformation efficiency.

We used the concept ofeffective accumulated temperature described in the results to attempt to explain this phenomenon. By calculating the effective accumulated temperature using the results of our experiment, we determined that heat-shock pre-treatment of immature embryos at 40℃ for approximately 4.5min might effectively improve the genetic transformation efficiency of Hi-II. Our study suggested that not all materials are suitable for a specific pre-treatment condition. Explants of different genotypes also have different minimum threshold temperatures.

The effective accumulated temperatures also differ between genotypes. In this study the two heat-shock groups had a similar *GUS* transient expression efficiency and no negative effects on callus induction, it can be speculated that a suitable heat-shock group can promote the integration and expression of foreign genes or the induction of transgenic callus. Therefore, we found out that an effective combination of heat-shock and pre-treatment time can promote the expression efficiency of foreign genes using effective accumulated temperature theory by selecting the optimum combination in practice.

The present study suggests thatpre-treatment with heat-shock of maize materials is a very effective option in the improvement of maize transformation efficiency. It would be inferred that treatment with a particular stress, such as heat-shock, centrifugation, ultrasound, vacuum or others may increase the resistance of explants material to adversity and then improve the survival of the explants.

Regeneration is a very important step in the long tissue culture process of maize genetic transformation. By optimizing experimental conditions to improve the regeneration efficiency of recipient material was also very critical to maize transformation efficiency. From our former study and others research, shoots were induced for regeneration first, then were roots, this procedure was very beneficial to normal development of maize seedlings. In this study we complied with this procedure and modified some components in regeneration media (Table 1).

We found that 100mg/L CH can significantly accelerate somatic embryos into the period of redifferentiation. It had a higher somatic embryo induction rate in the regeneration medium added CH than that not-added. NAA concentrations of 0.5mg/L enhanced the development of adventitious roots, but did not promote root grow stronger which restricted the production efficiency. 2.0mg/ml MET promoted lateral root emergence and root system sturdily but did not enhance the development of adventitious roots. In this study we combined use of NAA and MET, there was a great effect on embryogenic calli regeneration and the survival rate of the transplanted tube plantlets. The transplanted seedlings exhibited generally strong root system and thick stems.

Conclusion

The integrated and improved maize genotype Hi-II transformation system presented here is simple and reproducible. We have already produced hundreds of transgenic maize plants using the new method. Maize is a major feed and food crop worldwide, genetic transformation technology can be used to impart a maize variety of more superior traits. Therefore, a high efficiency transformation system is very essential to obtain amount of transgenic plants. The use of our system should facilitate the study on the maize functional verification of candidate genes as well as the maize genetic improvement in important agronomic traits.

Acknowledgement

This work was supported by the National Transgenic Major Program of China (2014ZX0800303B) and the Program of Beijing New Scientific Star (Z111105054511066).

References (omitted)

Research Journal of Biotechnology, 2016, 11 (3): 1-8

中国玉米生产发展历程、存在问题及对策

赵久然　王荣焕

(北京市农林科学院玉米研究中心，北京　100097)

摘　要：玉米已成为当前中国种植面积最大且总产量最高的第一大作物，发展玉米生产对保障国家粮食安全具有重要战略意义。全面阐述了改革开放以来中国玉米生产的发展历程及各发展阶段的生产特点，并重点从自然资源分布、农田基础设施、玉米品种与种子质量、栽培技术、生产经营方式、收储流通等方面深入分析了当前中国玉米生产存在的主要问题。在此基础上，从增加科研投入、加强基础设施建设、强化政策扶持、加强技术服务、加大高产创建力度和改善产后储运流通等方面提出了促进中国玉米进一步增产的政策建议，以及加强玉米新品种选育、抗旱节水、保护性耕作、提高播种质量和群体整齐度、机械化栽培和病虫草害防治等技术措施。

关键词：玉米；发展历程；问题；政策建议；技术措施

Development Process, Problem and Countermeasure of Maize Production in China

Zhao Jiuran　Wang Ronghuan

(Maize Research Center, Beijing Academy of Agriculture and Forestry Sciences, Beijing 100097, China)

Abstract: Maize has become the 1st major crop with the largest planting area and total yield in China, and it is strategically significant to promote the maize production for ensuring the national food security. In this paper, the development process of maize production in China since the reform and opening up and the typical features for each phases were introduced systematically, and the main problems and factors limiting the maize production such as the distribution of natural resource, agronomic infrastructure, maize variety and seed quality, cultivation technology, the production and management mode, the storage and circulation conditions were also analyzed. At last, some policy suggestions including increasing and strengthening the scientific research input, basic infrastructure, supporting policy, technical services, high-yield construction, storage and circulation conditions, and technical measures such as maize new variety breeding, water-saving agricultural technology, conservation tillage, improving the sowing quality and population uniformity, mechanized cultivation, diseases & insects and weeds protection etc. for the further increasing of maize yield level in the future

---

基金项目：农业部2013年农业技术试验示范专项、国家玉米产业技术体系项目(nycytx-02)资助

作者简介：赵久然，研究员，博士，主要从事玉米遗传育种及高产栽培研究。E-mail：maizezhao@126.com

were purposed, respectively.

Key words: Maize; Development process; Problem; Policy suggestion; Technical measure

玉米已成为当前中国种植面积最大、总产量最高的作物。据国家统计局统计，2012 年，中国玉米种植面积 3 494.90万 hm^2，玉米总产量 2.08 亿 t，平均单产 5 955.00kg/hm^2，再创历史新高，实现了自 2004 年以来的"九连增"，且玉米总产量首次超过水稻（较水稻产量高 383 万 t）成为中国种植面积和总产量都是第一的粮食作物。

玉米是近年来中国主要粮食作物中面积和产量增长最快的作物，已成为粮食增产的主力军。2004—2012 年，中国玉米种植面积增加了 1 088.10万 hm^2，增幅为 45.2%，占粮食面积增量的 91.6%；玉米产量增加了 923 亿 kg，增幅为 79.7%，玉米增产对全国粮食增产的贡献率高达 58.1%。玉米在中国农业生产中占有重要地位，大力发展玉米生产对保障国家粮食安全具有重要战略意义。本文回顾了改革开放 30 多年来中国玉米生产的发展历程，深入分析了当前玉米生产中存在的主要问题，并在此基础上提出今后进一步促进玉米可持续增产的对策，可为推动中国玉米生产快速发展提供重要参考。

1 改革开放以来中国玉米生产发展历程

自 1978 年改革开放以来，在党中央和国务院一系列支农惠农政策的大力支持、各级农业主管部门的有力推动、优良品种与先进技术及装备的推广应用以及强劲玉米市场需求和高价位拉动下，广大农民种植玉米的积极性和生产水平不断提高，中国的玉米种植面积、单产和总产均大幅提高，玉米生产得到了迅速发展。这期间，中国玉米生产发展大体可分为稳步增长阶段、快速增长阶段、调整下降阶段和恢复增长共 4 个阶段（图1）。

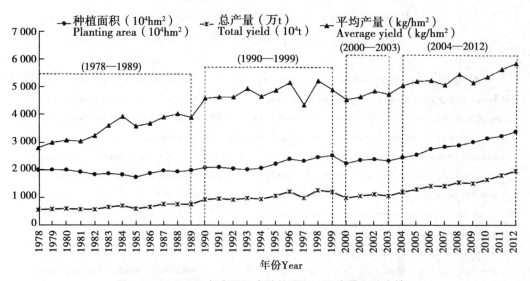

图 1 1978—2012 年中国玉米种植面积、总产量和单产情况

Figure 1 Maize planting area, total yield and average yield in China during 1978—2012

数据来源：中国人民共和国国家统计局

Data source: National Bureau of Statistics of China

1.1 稳步增长阶段（1978—1989 年）

十一届三中全会后，家庭联产承包责任制的实施大大调动了农民的生产积极性，且中单 2 号、丹玉 6 号、掖单 2 号、烟单 14 号和四单 8 号等一批优良高产杂交种及配套技术，以及科学施肥技术、病虫害防治技术的研究与推广为推动本时期玉米单产水平的提高发挥了重要作用。1978 年，全国玉米种植面积、单产和总产分别为 1 996.10 万 hm^2、2 805.00kg/hm^2、0.56 亿 t，至 1989 年已发展至 2 035.30 万 hm^2、3 945.00 kg/hm^2、0.80 亿 t。该阶段玉米总产增幅较大（增幅为 62.82%），总产增加主要由于单产提高较快（单产增幅为 57.01%）。

1.2 快速增长阶段（1990—1999 年）

1990 年后，畜牧养殖业和玉米加工业的发展导致玉米需求量逐步增大，玉米种子、农药、化肥等生产资料投入加大，高产抗病品种的选育与推广、地膜覆盖、育苗移栽及病虫草害防治技术，尤其是 20 世纪 90 年代中后期农业部紧凑型玉米活动的实施推广促进了玉米生产的快速发展。1990 年，全国玉米种植面积、单产和总产分别为 2 140.10 万 hm^2、4 620.00kg/hm^2、0.99 亿 t，1999 年玉米面积发展至 2 590.40 万 hm^2，1998 年玉米单产和总产水平均达到了历史最高水平，分别为 5 269.00kg/hm^2 和 1.33 亿 t。该阶段玉米总产量快速增长，增幅为 59.29%。总产增幅较大是由于单产和面积均增加较快，种植面积和单产增幅分别为 27.27% 和 25.35%。

1.3 调整下降阶段（2000—2003 年）

因前一阶段玉米生产相对过剩，玉米价格降低，农民种植玉米积极性下降。2000 年，政府调整了种植结构，玉米种植面积、单产和总产均大幅下降，玉米种植面积、单产和总产分别为 2 305.60 万 hm^2、4 598.00kg/hm^2、1.06 亿 t。玉米总产降幅为 9.59%，总产降低主要是由于面积大幅减少（面积降幅为 7.09%），单产略有降低（降幅为 2.67%）。

1.4 恢复增长阶段（2004—2012 年）

自 2004 年起，国家相继出台了一系列支农惠农政策，并且因玉米市场需求强劲，玉米价格持续高位，充分调动了农民种植玉米的积极性，玉米生产开始恢复性增长。2004 年，全国玉米种植面积、单产和总产分别为 2 544.60 万 hm^2、5 119.95 kg/hm^2 和 1.30 亿 t，至 2012 年已发展至 3 494.9 万 hm^2、5 955.00kg/hm^2 和 2.08 亿 t。这期间，玉米总产大幅增加（增幅为 79.72%），总产大幅增加主要是由于玉米种植面积增加较快（面积增幅为 45.21%），同时玉米单产也开始恢复性增长，2011 年达到 5 735.85kg/hm^2，2012 年达到 5 955.00kg/hm^2，单产增幅为 23.73%。

2 当前玉米生产存在的主要问题

自改革开放以来的 30 多年间，以中单 2 号、丹玉 13 号、掖单 13 号、农大 10 和郑单 958 为典型代表的玉米主栽品种引领了中国玉米品种的 5 次大规模更新换代。高产优良品种及高产高效栽培技术的研究与推广有效地促进了玉米生产的快速发展。但在新的历史形势下，中国玉米生产仍面临着严峻的挑战，存在一些亟需研究和解决的问题。

2.1 自然资源分布不均，灾害频发

2.1.1 水资源匮乏

中国玉米主要分布在干旱、半干旱区域，雨养种植面积约 65%，干旱是影响玉米高

产稳产的最主要自然气候因素。另外,水资源紧张且时空分布不均,旱涝灾害频繁发生,极端性天气增加,玉米生产遭遇旱灾、涝灾、冷害、冻害、高温等灾害的几率增多,由此带来的农业病虫害影响也将加大。

2.1.2 光热资源不足

不同产区的光热资源条件存在较大差异,并从不同方面影响玉米的生长发育和产量形成,如东北地区的苗期低温和后期秋霜冻害、黄淮海地区长时间的低温寡照以及东北、西北及西南等地高纬度或高海拔冷凉地区积温不足、热量资源紧张等。

2.1.3 耕地资源恶化

随着工业化和城镇化进程的加快,人增地减矛盾将更加突出。据农业部统计,高产田仅占耕地面积的30%,中产田占38%,低产田占32%。目前玉米生产大多是在中低产田条件下进行的,并且施肥技术和化肥利用率总体较低,农田地力水平不断下降。另外,因连年大规模使用小四轮拖拉机等小型农机具导致土壤耕层变浅,犁底层不断加厚,土壤蓄墒保水保肥能力和抗逆性大幅降低。此外,部分地区还存在土壤沙化和石漠化的问题,限制了土壤高产潜力的发挥。

2.2 农田基础设施不健全,抵抗旱涝灾害能力较差

目前,农业基础设施薄弱的局面没有根本改变,一半耕地处于水资源紧缺的干旱、半干旱地区,农田有效灌溉面积仅50%,约1/3耕地位于易受洪水威胁地区,大部分玉米产区农田水利设施落后,田间排灌设施陈旧老化、沟渠道路不配套,抗御自然灾害能力较差。如2009年东北辽宁等省大旱、2010年西南大旱、2010年吉林涝害以及2012年台风导致京津冀及东北部分地区洪涝灾害等,在一定程度上因农田水利设施落后而未能确保及时做到旱浇涝排,对玉米生产造成不同程度的影响。

2.3 优良品种相对缺乏,种子质量有待进一步提高

自2000年以来,中国相继选育出郑单958、浚单20等一批高产、耐密型玉米品种,并已在生产中大面积推广,这对促进全国玉米生产整体水平提高发挥了重要作用。据全国农技推广服务中心统计,2011年全国玉米种植面积仍以郑单958为最高(452.13万hm^2),其次是先玉335和浚单20(分别为238.13万hm^2和220.07万hm^2),种植面积6.66万hm^2以上的品种共75个。虽然玉米品种数量较多,但种质基础相对狭窄,少数骨干自交系利用频率过高,育种技术相对落后且低水平重复严重,育成品种同质化严重,高产、优质、多抗、广适的优良品种相对较少,特别是大部分现有品种在机械化生产、高产抗逆等方面不能适应和满足当前玉米生产发展需求。在种子质量方面,虽然种子的纯度和净度基本达标,但其发芽率和活力还有待进一步提高。目前玉米种子的发芽率指标仅为≥85%,难以实现单粒播种保全苗。

2.4 栽培技术措施到位率较低,未能充分发挥品种的增产潜力

大部分地区玉米栽培技术落后或不到位:一是密度普遍偏低,未达到合理密度,个别区域还存在不间苗、盲目增加种植密度现象;二是因未科学合理搭配肥料种类、比例、数量、时间及采用地表撒肥等不合理的施肥方法,化肥利用率总体水平较低;三是黄淮海夏玉米区还存在一定面积的麦田套种,限制了密度进一步提高,玉米粗缩病等发生较重,群体整齐度和产量降低,并且夏玉米普遍收获偏早,品种和资源的产量潜力未得到充分发挥。此外,随着工业化和城镇化的快速发展,农村青壮年劳动力大多外出务工,留乡务农

人员以中老年人为主，素质普遍偏低，对先进科学技术的理解和接受能力较差，先进科学技术的推广应用也因此受到较大限制，在一定程度上限制了玉米生产的发展。

2.5 生产经营方式落后，玉米机械化水平仍较低

玉米生产规模小，机械化程度低（其中机耕和机播水平相对较高而机收水平很低），机械收获仍是薄弱环节，成为制约中国玉米生产机械化的最主要因素。一方面，农机农艺措施不协调、不配套，各地玉米种植模式复杂、种植行距多样化以及套作模式等在很大程度上制约了机械收获的发展。另一方面，机械化水平较低也限制了玉米先进实用增产技术的推广和应用，这也是导致当前玉米增产缓慢的主要原因之一。近年在国家政策支持及各级政府共同努力下，中国玉米机械收获快速发展，机收率不断提高。据农业部初步统计，2012年全国玉米机收率超过40%，已连续4年增幅超过6%。在新的历史形势下，玉米全程机械化作业水平还有待进一步提高。

2.6 收储条件不足、流通环节不畅

在玉米收获后的储存环节，目前仍以农户储存为主，但大部分农户的玉米晾晒场所和储存设施简陋，特别是在东北地区因籽粒含水量较高，如储存不当则易造成玉米霉变或虫害，进而造成玉米大量损失。并且，在玉米的收购、储运环节，仓储物流设施陈旧老化也造成一定的玉米损失。此外，玉米流通环节不畅往往导致流通环节和成本增加，因此还需发展现代物流体系等。

3 促进玉米可持续增产的对策

为促进玉米生产快速发展并保持持续增产，结合当前全国玉米生产现状，从政策层面和技术层面提出如下建议。

3.1 政策建议

3.1.1 继续加大对玉米科研支持力度，增强科技创新能力

建立以政府为主导的多元化、多渠道的科研投入体系，增加科研投入，特别是加强农业基础性、前沿性科学研究，力争在玉米高产优质品种选育、高产高效栽培技术模式和资源高效利用等方面取得新突破，提高科技兴粮合力。

3.1.2 加强基础设施建设，提高玉米综合生产能力

继续加强基础设施建设，新建和修复包括农田水利在内的农业农村基础设施，加强田间末级灌排沟渠、机井、泵站等小型农田水利设施建设，努力改善农业生产条件，增强玉米生产抵御干旱和洪涝灾害的能力。

3.1.3 加强宏观调控，进一步出台和强化支农惠农政策

以多种形式加大对玉米生产的补贴力度，通过政府的政策导向充分调动和保护农民的生产积极性；继续加大对玉米深加工企业的支持力度，积极采取拉动玉米需求的各项有效措施，充分发挥政策的引导和激励作用，推动玉米整体产业健康稳步发展。

3.1.4 建立健全社会化技术服务体系，打通科技推广"最后一公里"

通过形式多样的技术服务使无形的技术推广变成直观的现场，积极引导农民选择优良品种、应用先进适用技术，促使先进适用技术向现实生产力快速转化，提高农民科学种粮技能和水平。此外，积极引导和鼓励玉米企业、农民专业合作经济组织等开展技术创新和推广活动，积极为农民提供科技服务。

3.1.5 加大高产创建力度，推动整建制均衡增产

继续开展集高产优质良种、高产高效栽培技术和优质高效投入品为一体的玉米高产创建活动，强化部门联合和协作公关，充分调动农业、科研、教学、推广、企业、农民等各方面的积极性，挖掘不同规模的高产典型，并以高产创建为抓手，大面积示范和推广玉米优良品种及高产高效种植技术模式，促进大面积均衡增产。

3.1.6 加强玉米产后储运和流通环节建设，提高市场宏观调控能力

进一步改善玉米收获后的保管和晾晒条件，提高籽粒的商品质量和竞争力；加强现代仓储物流设施与体系建设，减少产后损失；加强市场宏观调控能力，建立以市场机制为基础的现代玉米产业链，促进玉米生产长期稳步发展。

3.2 技术措施

3.2.1 加快选育高产耐密优良品种，确保良种良法配套

进一步加强高产稳产耐密型玉米新品种的选育和良种良法配套栽培模式的研究与示范推广，通过良种良法配套充分挖掘品种及区域生产潜力；推广早熟耐密型高产稳产品种，防止品种越区种植；在黄淮海夏玉米区大力推广适时晚收技术，在东北地区大力推广促早熟、防早霜等技术措施，进一步提高籽粒的成熟度和商品品质。

3.2.2 推广旱作节水灌溉技术，提高水分利用率

优化集成多种抗旱节水农艺措施，大力发展节水灌溉，鼓励和引导农民因地制宜使用喷灌、滴灌、微灌和"坐水种"等高效节水灌溉技术。同时，积极发挥旱作农业的技术优势，推广地膜覆盖、双垄沟播覆膜、膜下滴灌、深松深耕、保护性耕作等技术，进一步提高灌溉水和自然降水的利用率。

3.2.3 实施保护性耕作并科学施肥，提高地力及肥料利用率

进一步加强深松改土、秸秆还田等保护性耕作措施，并重视增施有机肥，培肥地力，提高土壤生产潜力、实现可持续增产。同时进一步实施测土配方科学施肥并推广化肥机械深施，提高化肥利用率、节本增效。

3.2.4 提高种子质量和播种质量，确保合理密度和群体整齐度

进一步提高种子质量特别是发芽率和种子活力。同时通过机械和农艺等措施提高播种质量，保障合理密度，并提高群体整齐度和果穗均匀度；提高和完善现行玉米种子质量标准，建议将一级种子发芽率标准提高到≥95%、二级和三级种子发芽率标准分别提高到≥90%和≥85%。同时加强对品种真实性的检测和监督管理，建立健全种子质量保障体系，强化种子质量检测和监督管理力度。

3.2.5 研究推广机械化轻简栽培技术，促进生产方式转变

进一步研究与示范推广以机械化为核心的区域化玉米轻简栽培技术体系，并通过与种粮大户、家庭农场及各种农民合作组织的发展紧密结合努力推进玉米生产规模化和机械化，逐步实现玉米生产全程机械化，促进玉米生产向现代生产方式的转变。

3.2.6 积极防控自然灾害和生物灾害，减少灾害损失

大力推广种子包衣、生物防治、化学除草等技术，加强病虫草鼠害的防治和灾害性自然天气的预测预报和预警工作，积极探索社会化专业化服务机制，提高防灾减灾能力，努力减少灾害性损失。

4 展望

近些年来,在党中央和国务院政策的利好形势下,农民种植玉米的积极性较高,且玉米科技支撑能力进一步增强,玉米生产取得了长足的进步和发展,对满足玉米市场需求和保障国家粮食安全发挥了重要作用。农业部韩长赋部长指出,随着工业化、城镇化快速发展和人民生活水平不断提高,中国已进入玉米消费快速增长阶段;并且,从未来发展看,玉米将是中国需求增长最快、也是增产潜力最大的粮食品种,抓好玉米生产就抓住了粮食持续稳定发展的关键。在当前形势下,通过国家及各级政府的政策支持和带动,进一步稳定玉米种植面积、努力调动各方面的积极性、有效改善玉米生产条件,并充分挖掘和发挥先进科学技术的增产潜力,推进玉米生产机械化、规模化和信息化,加快玉米传统生产方式向现代生产方式转变,将是促进中国玉米生产再上新台阶和保持玉米可持续增产的重要举措。

参考文献(略)

本文原载:中国农业科技导报,2013,15(3):1-6

我国玉米产业现状及生物育种发展趋势

赵久然　王荣焕　刘新香

(北京市农林科学院玉米研究中心，北京　100097)

摘　要：玉米是全世界也是我国种植范围最广、用途最多、总产量最高的作物。发展玉米生产对保障我国粮食安全和满足市场需要发挥着至关重要的作用。分析了我国玉米产业的现状及面临的新形势，提出了通过生物育种实现节本增效提高玉米产业国际竞争力的技术途径，并对玉米主要育种技术及优势进行了阐述。

玉米是一种既普遍又神奇、既古老又年轻的作物。玉米被人类驯化种植已经有约1万年的历史，但从美洲传播到世界各地广泛种植也仅有500多年。目前，玉米已成为全世界种植范围最广、用途最多的作物，也是总产量最高的作物，全球玉米年种植面积20亿亩(1亩≈667m^2，全书同)以上、总产量已达10亿t。同时，玉米也是全世界杂种优势应用最早、最普及的作物，是目前全球种业市值最大的作物，也是国际种业巨头竞争的主要业务范围和生物技术育种的主要竞争领域。

我国是全球第二大玉米生产国和消费国，近年来玉米总产量占全球玉米总产量的20%以上。新中国成立以来，特别是改革开放以来，我国玉米生产取得了长足的进步和发展，但玉米产业发展面临着生产经营方式转变、生态环境束缚、资源制约、库存压力大、缺乏市场竞争力等严峻形势，还需依靠科技创新特别是先进的生物育种技术提高我国玉米产业的国际竞争力。

1　我国玉米产业现状及面临的新形势

1.1　我国玉米产业现状

新中国成立以来，我国玉米生产得到了快速发展，全国玉米平均亩产水平由1949年的64.10kg提升至目前的约400kg(最高水平为2013年的401.80kg)。改革开放以来，我国玉米生产发展迅猛，经历了稳步增长阶段(1978—1989年)、快速增长阶段(1990—1998年)、调整下降阶段(1999—2003年)和恢复增长阶段(2004—2015年)4个阶段。在调整下降阶段末期的2003年，我国玉米种植面积3.61亿亩，平均亩产320.87kg，总产量1.16亿t，处于玉米生产过程中的一个低谷期。

2004年以来，随着国家和各级政府一系列支农惠农政策的发布实施，农业部提出"关于加快玉米生产发展工作方案"并大力推广玉米"一增四改"关键技术，组织开展高产创建和增产模式攻关等活动，以及市场强劲需求和高价位拉动等多种因素的综合作用下，我国玉米生产开始恢复性增长，且发展势头强劲。2007年，我国玉米种植面积4.42亿亩，首次超过水稻成为我国种植面积最大的作物；2011年，玉米种植面积首次突破5亿亩，达到5.01亿亩；2012年，玉米总产量突破2亿t，首次超过水稻，成为我国种植面积和总产量双第一的名副其实的第一大粮食作物；2013年，玉米种植面积进一步增加，达到5.42亿亩，总产量达到2.18亿t，平均亩产首次突破400kg，达到401.80kg；2015年，我国玉米生产又迎来一个丰收年，种植面积和总产量均再创历史新高，分别达到5.72亿亩、2.25亿t，平均亩产达到392.79kg，接近历史最高值（表1）。2015年，我国玉米总产量占全国粮食总产量的36.14%，2004—2015年我国玉米增产对粮食增产的贡献率近60%，居粮食作物之首，玉米已成为我国2004年后粮食生产"十二连增"的主力军。

表1 2004—2015年我国玉米种植面积、总产量和单产

年份 项目	2004	2005	2006	2007	2008	2009	2010	2011	2012	2013	2014	2015
种植面积 （亿亩）	3.82	3.95	4.27	4.42	4.48	4.67	4.87	5.01	5.25	5.42	5.56	5.72
总产量 （亿t）	1.30	1.39	1.52	1.52	1.66	1.64	1.77	1.92	2.06	2.18	2.16	2.25
单产 （kg/亩）	341.33	352.47	355.07	344.50	370.40	350.40	364.10	382.39	391.31	401.80	387.80	392.79

我国玉米生产的迅猛发展与玉米良种及配套高产高效栽培技术的研究与推广密不可分。从我国审定玉米品种的数量来看，1972—2013年省级以上审定的玉米品种共7 468个，生产中推广应用的有5 590个，除去不同省份重复审定的品种，生产中推广应用的省级以上审定品种4 882个；从玉米品种的审定年份分布来看，绝大多数品种是2002—2013年审定，1972—2001年审定的玉米品种仅占现存总品次的8%；从玉米品种的选育主体来看，以科研院所和种业公司为主，分别占48.3%和46.6%，而政府机构和个人所占比例较小，分别为2.2%和2.9%。以中单2号、丹玉13、掖单13、农大108、郑单958等为代表的主栽玉米品种引领了近30多年来我国玉米生产品种的5次大规模更新换代，再加上地膜覆盖、育苗移栽、保护性耕作、化学除草、病虫害防治、机械精量播种、"一增四改"等技术的推广应用，通过良种良法配套，充分挖掘和发挥了品种和配套技术的增产潜力。

虽然我国玉米生产自2004年开始恢复性增长，玉米播种面积和总产量持续增加，但平均单产水平一直徘徊在亩产400kg左右，与美国相比，虽同处北半球、自然条件也有许多相似之处，但美国的玉米科研和生产水平最高，产量、消费量和出口量最大，差距较大。美国玉米最高平均单产为2014年的715.33kg/亩，是我国玉米最高平均单产（2013年的401.80kg/亩）的1.78倍；美国玉米最高单产纪录是2015年创造的2 187.53kg/亩，我国玉米最高产纪录是2013年在新疆创造的1 511.74kg/亩，美国玉米最高产纪录是我国的1.45倍。与美国相比，我国的玉米生产在许多环节都存在着差距，尤其在通过生物育种技术手段实施的玉米种质创新、品种选育等方面差距更加明显。

1.2 我国玉米产业面临的新形势

近年来，随着农村劳动力进一步转移、土地流转进程加快以及种粮大户、家庭农场、专业合作社等新型农业生产主体和经营方式出现，玉米生产规模化、机械化和轻简化成为

现代农业发展的必然趋势。目前，我国玉米生产过程中耕地、整地和播种环节已基本实现机械化作业，但机械收获率较低，特别是机械直接收获玉米籽粒更是刚刚起步，占比很低。这已不能满足玉米生产对省时省工的迫切要求，收获环节成为限制玉米生产全程机械化的主要瓶颈。影响玉米机械收获的主要原因除农机不配套外，主要还是缺乏籽粒脱水快、植株抗倒能力强、适宜机械收获作业特别是机械直收籽粒的玉米品种。

近年我国粮食生产实现了连年增产，但是以耕地质量不断下降、环境污染日趋加重、地下水超采严重、生态环境承载压力不断加大等为牺牲和代价的，农业生产发展方式主要采取的是追求产量增长和依赖资源消耗的粗放型经营，农业发展面临生态环境束缚和资源制约两道"紧箍咒"，拼资源、拼投入品、拼生态环境的传统发展方式难以为继，必须走资源节约、生态友好的农业可持续发展之路，向数量质量效益并重、注重提高竞争力、注重可持续的集约发展方式转变。

此外，随着我国玉米种植生产中劳动力成本增加、化肥施用量偏大且利用率偏低及规模化经营中的土地流转成本偏高等因素，玉米生产总体成本增加，国际市场竞争力偏弱，我国玉米价格较国际玉米价格高近50%。并且，随着我国玉米种植面积逐年扩大，玉米连年增产，单产和总产水平不断提高，再加上消费乏力及玉米、玉米替代品的大量进口等多种因素叠加，导致当前我国玉米库存偏高，农民卖粮难。2015年度，根据《农产品进口关税配额管理暂行办法》我国允许进口玉米配额720万t，但饲用大麦、高粱、木薯以及DDGS（玉米干酒糟及其可溶物）则未受配额限制。据海关数据显示，2015年我国玉米以及DDGS和高粱等玉米代替物的进口总量创4 235万t纪录高位（2014年为2 770万t）。其中，进口价格相对低廉的DDGS进口总量为创纪录的682万t，较2014年增加了26%；饲用大麦进口总量为创纪录的1 070万t，较2014年增加了98%；饲用高粱进口总量为创纪录的1 069万t，较2014年增加了85%；木薯进口总量为937万t，较2014年增加了8.4%。

因此，我国玉米产业的当务之急是调结构、转方式、去库存、降成本，提高国际竞争力。

2 玉米生物育种技术途径为节本增效、提高竞争力提供科技支撑

良种在玉米增产中的贡献率约占40%，发挥着至关重要的作用。因此，未来的玉米育种应该针对玉米生产和市场中的突出问题，从遗传改良途径为玉米高产高效、节本增效、提质提效、资源高效、节水节肥节药和环境友好的可持续生产目标提供科技支撑，增加科技贡献率，改变完全以追求高产为目标的导向。玉米生产成本主要包括人工费用、化肥投入和土地流转费用等，共占总投入的85%左右，而其他如种子、农药等所占的比例很小。从节本增效角度来看，首先要朝全程机械化方向发展，降低人工播种、定苗、除草、收获等成本，特别是解决机械直接收获籽粒的技术瓶颈。要选育和推广种植耐旱、耐低氮、耐瘠薄、抗多种病虫害等资源高效利用、环境友好的优良品种和配套种植技术。

2.1 总体育种目标仍应是"高产、优质、多抗、广适、易制种"

育种过程是一个不断改良优化、不断超越已有、聚合优良性状的过程。一个品种要成为能够在生产上大面积推广、让广大农民认可接受、有市场竞争力，需要经受多种环境和

生产条件的检验，因此必须具备"高产、优质、多抗、广适、易制种"等多方面综合优点。任何片面强调或夸张某一个特点而不及其余，都是经不起生产实践考验的。

2.2 "早熟、耐密、适宜机收籽粒"是未来玉米育种的主导目标之一

降低劳动力成本、提高劳动生产效率、实施全程机械化特别是机械直收籽粒成为必然趋势。因此，在具备"高产、优质、多抗、广适、易制种"等综合优点基础上，特别要突出熟期适宜、耐密抗倒、脱水快、适宜机收籽粒等特点。机械直收籽粒在欧美等国早已实现，我国黑龙江农垦和新疆兵团农场也已经大面积实施，近两年在东华北春玉米区和黄淮海夏玉米区也开始试验示范。但机收籽粒需要适宜的品种、收获机械、烘干仓储甚至包括种植模式等多方面条件和配套，切勿盲目跟风，一哄而上。适宜籽粒机收的品种主要有3个方面的要求：第一是要抗倒伏，不但前期抗倒，生理成熟后也要抗倒，总倒伏倒折率低于5%；第二是后期脱水速度快，在适宜收获期内含水率应降到25%以下，黄淮海夏玉米可适当放宽标准到28%以下；第三是仍要有较高的产量水平，比主栽对照增产或产量水平相当。同时，在全程机械化生产中还需要种子的发芽率高、活力高等，符合单粒精量播种和一播全苗的生产需求。此外，当前和今后使用苗后除草剂越来越普遍，伴随的是生产上除草剂药害普遍发生，给玉米生产带来很大影响，因此选育耐除草剂品种也已成为当务之急。

2.3 "一控两减、资源高效利用"的需求目标是大趋势并呈常态化

为实现生态环保、环境友好的可持续发展以及节本增效目标需求，玉米育种更应着重选育：耐干旱和水分生产利用效率高的品种；耐低氮、耐瘠薄，肥料利用效率高的品种；抗多种病虫害、农药使用量尽量少的品种。国审玉米品种京科968，近几年之所以种植推广面积快速上升，成为仅次于郑单958和先玉335之后的第三大品种，除了具有多方面的综合优点外，更兼具耐干旱、耐瘠薄、氮肥高效以及抗大斑病、穗粒腐病、丝黑穗病等多种病害的特点，还是非转基因的抗虫品种，抗玉米螟、黏虫、蚜虫等多种虫害。

2.4 加强鲜食玉米和青贮玉米等专用玉米品种选育

玉米按照收获物和用途可分为三类：籽粒用玉米（即通常所说的大田玉米）、鲜食玉米（收获和食用玉米的鲜嫩果穗）、青贮玉米（收获玉米鲜绿全株，经切碎发酵作奶牛等草食牲畜饲料）。我国鲜食玉米发展很快，种植面积近2 000万亩，已成为全球最大的鲜食玉米种植生产国。我国在鲜食糯玉米育种方面处于国际领先地位，所有生产上种植推广的鲜食糯玉米品种均由我国自主选育，并已走出国门推广种植到韩国、东南亚及非洲等国家，产品更是远销全球。同时，还创制加甜糯这一新的鲜食玉米品种并已大规模产业化。

鲜食玉米需要在具备"高产稳产、多抗广适"等综合优点基础上，更加突出品质。特别是食用品质和口感、风味等，此外增强其功能营养品质如维生素A、维生素E和叶酸等可增加其价值潜力。同时，应尽量延长其适采期和货架期，并在籽粒颜色、熟期、穗型、粒型等方面满足市场多样化需求。如我国创制出的加甜糯玉米品种，即在一个玉米果穗上，同时具有甜玉米籽粒和糯玉米籽粒，一般是1∶3的比例，消费者一口吃下去可以同时品尝到两种口味的鲜食玉米，这正在成为一种发展趋势。

青贮玉米是奶牛养殖业的饲料基础和支柱，目前我国已种植约2 000万亩，根据我国奶牛保有数量，至少需要3 000万亩的种植面积。青贮玉米育种除了"高产稳产、多抗广适"外，更要求生物产量高、饲用品质好、制种产量高等。饲用品质的主要指标有中性

洗涤纤维和酸性洗涤纤维（代表不易消化的纤维素和木质素等）含量，含量越低则消化利用率越高，淀粉和粗蛋白含量高则说明营养价值高。

2.5 专用品质及特殊功能品质品种育种

玉米的主要用途定位是"高淀粉能量型作物"，玉米的加工要综合利用，但主要是利用其淀粉。因此选育高淀粉品种既能增加其品质产值，又与产量不相冲突，而选育高油或高赖氨酸等类型的品种往往导致产量降低。

2.6 注意推广品种的遗传多样性和生态稳定性

避免一个模式一刀切，针对不同区域应因地制宜，良种良法配套，发挥耕作与栽培的资源配置作用。

3 玉米主要育种技术及优势

3.1 常规育种技术

玉米常规育种就是依据孟德尔遗传规律和杂种优势技术的理论，通过控制授粉等进行杂交育种的技术手段，可将不同育种材料的优良性状聚合在一起。这是一个基因重组过程，玉米有10对染色体，其基因重组变化的可能性非常多，再加上成对的同源染色体之间还可以进行大量的基因交换，使得常规杂交育种所能产生的遗传变异就更多了。我国生产上应用的品种大部分仍是采用常规育种方法选育的。如目前我国玉米生产中的两个主导品种郑单958和浚单20就是通过常规育种手段选育而成。郑单958由河南省农业科学院粮食作物研究所培育，年种植面积已突破6 000万亩；浚单20是由河南省鹤壁市农业科学院选育的夏播玉米品种，年种植面积最高在4 000万亩以上。这两个品种已经连续十多年成为我国玉米生产中两个主导品种。因此，常规育种技术仍有很大潜力。但重要的是，还要不断地有优异种质资源的引入和利用，使优良性状的基因不断地聚合和积累，不断地创制新的育种材料。

3.2 DH单倍体育种技术

DH单倍体育种技术即利用孤雌生殖诱导系做父本进行杂交，诱导相应的母本产生大大高于自然频率的单倍体，同时诱导系本身还具备鉴别孤雌生殖单倍体的遗传标记系统。目前在孤雌生殖诱导系中普遍采用的是 $A1A2C1C2BPIR-nj$ 显性遗传标记系统。该标记系统由两个不同的遗传标记性状组合而成，具有较高的鉴别效率和可靠性。首先根据籽粒 Navajo 标记性状（由 $A1A2C1C2BPIR-nj$ 基因控制）分拣出可能的单倍体籽粒，再经过胚根色素或苗期叶鞘色素（由 $A1A2BPI$ 基因控制）的有无确定是否为单倍体植株。

由于DH单倍体育种技术在缩短育种周期、提高获得纯合基因型效率、排除显性干扰、提高选择准确性和利于亲本保纯等方面的优势而具有很大育种潜力，可直接用于实际育种工作。目前，全世界有250多个作物物种应用了单倍体育种技术，已成为美国孟山都公司、德国KWS公司及俄罗斯等国的一些玉米种子公司玉米育种的重要方法，以此加快育种速度，缩短新杂交种推出时间，占有和扩大市场。在我国，中国农业大学、北京市农林科学院、吉林省农业科学院、四川农业大学、北京奥瑞金种业股份有限公司等单位和企业也都先后从事了这方面研究，并选育出了一系列高诱导率的诱导系，如农大高诱1号、吉高诱3号、京科诱044、京科诱045等。

3.3 诱变育种技术

诱变育种技术是指利用物理、化学等因素诱导，使基因组发生单个或多个位点的突变进而发生变异，然后从变异群体中选择符合当前育种目标要求的单株或个体，进而培育成新的品种或种质的育种方法，是常规育种技术的重要补充。物理诱变应用较多的是辐射诱变，即用α射线、β射线、γ射线、X射线、中子和其他粒子、紫外辐射、微波辐射及空间诱变等物理因素诱发变异；化学诱变剂主要有甲基磺酸乙酯（EMS）、乙烯亚胺（EI）、亚硝基乙基脲烷（NEU）、亚硝基甲基脲烷（NMU）等。

20世纪70年代以来，我国利用γ射线等诱变技术育成了原武02、原辐17、丹340、丹360、鲁原92等自交系，组配了掖单13、鲁单5号、鲁单50等鲁单系列杂交种，推动了玉米生产发展。但诱变育种技术存在的主要问题是有益突变频率较低，常常产生畸形植株，变异的方向和性质尚难控制。因此，提高诱变效率、迅速鉴定和筛选突变体以及探索定向诱变的途径也是今后研究的重点。

3.4 分子标记辅助育种技术

分子标记辅助选择（Molecular Marker-assisted selection，MAS）是分子育种的重要组成部分。其利用与目标性状基因紧密连锁的分子标记进行间接选择，是对目标性状在分子水平的选择，不受环境影响，不受等位基因显隐性关系干扰，选择结果可靠，一般可在育种早代进行选择，从而大大缩短育种周期。目前在玉米育种工作中，主要利用RFLP、AFLP、CAPS、SSR、SNP和InDel标记等技术进行品种纯度和真实性检测、分析遗传多样性、定位功能基因、划分杂种优势群以及转基因检测等。如高伟等利用cDNA-SRAP标记对玉米氮素营养进行诊断，孙世孟等利用RAPD标记研究玉米亲本自交系间相似系数并划分类群，郝转芳等基于SNP标记的关联分析对玉米耐旱基因进行研究等。因此，分子标记辅助育种技术在保持玉米遗传多样性的基础上，通过分子生物学等手段对产量、品质、抗性等进行分子水平研究，对玉米常规育种具有重要辅助作用。

3.5 转基因育种技术

转基因育种技术是根据育种目标，从供体生物中分离目的基因，经DNA重组与遗传转化或直接运载进入受体作物，经筛选获得稳定表达的遗传工程体，并经田间试验与大田选择育成转基因新品种或种质资源。其过程涉及目的基因的分离与改造、载体的构建及其与目的基因的连接等DNA重组技术；通过农杆菌介导、基因枪轰击等方法使重组体进入受体细胞或组织以及转化体的筛选、鉴定等遗传转化技术和相配套的组织培养技术；获得携带目的基因的转基因植株；遗传工程体在有控条件下的安全性评价以及大田育种研究直至育成品种。转基因育种技术体系的建立大大拓宽了可利用的基因资源，实现了动物、植物、微生物中分离克隆的基因在三者间相互转移利用；可对植物目标性状进行定向变异和定向选择，同时随着对基因认识的不断深入和转基因技术手段的完善，对多个基因进行定向操作也将成为可能，为培育高产、优质、高抗、适应各种不良环境条件的品种提供了崭新的育种途径，可大大提高选择效率，加快育种进程；可将植物作为生物反应器生产药物等生物制品。

转基因技术自发现至今仅30年就得到了快速发展，已有一批抗虫、抗除草剂的转基因玉米品种获得批准而在美国等多个国家实现了商业化。孟山都及陶氏益农公司共同研发的玉米新品种Smart-Stax具有广谱抗玉米根部虫害及全面控制广泛的杂草优势。孟山都公司开发的抗旱转基因玉米（来自枯草芽孢杆菌的 *CspB* 基因）在西部大平原进行的抗旱玉米田间试验中已达到 6%~10%的增产目标。未来，转基因育种的主要发展方向集中在转基因技术本身的安全性、功能基因组研究、载体构建、遗传转化体系的优化和改善及多基因叠加的复合性状转基因玉米的选育等方面。

3.6 分子设计育种技术

分子设计育种就是高通量地鉴定和聚合优良等位基因，设计出新品种的育种技术。随着玉米全基因组测序的完成，玉米分子设计育种的发展获得了新的机遇。中国农业大学国家玉米改良中心带领国内玉米优势单位对 6 个中国玉米骨干自交系进行全基因组重测序，得到 125 万个高质量的单核苷酸多态性（SNP）位点，发现了 101 个低序列多态性区段，为玉米分子设计育种奠定了坚实的基础。美国先锋公司和孟山都公司已拥有上万个 SNP 标记，每天处理的分子标记数据高达 20 万~30 万个。为开发更多标记，先锋公司测定了 600 个优良玉米自交系的 10 000 个基因组，分子设计育种的投入给这两家种子企业带来了巨额的回报。与此同时，控制玉米复杂农艺性状的主效 QTL 位点的发掘也成为分子设计育种的重点，在玉米数据库中收录的玉米 QTL 已超过 2 000 个，分子设计育种技术将快速、高效地聚合优良等位基因，加快玉米优良种质的创制，大大提高育种效率。

3.7 基因编辑技术

基因组编辑是近年出现的能够精确改造生物基因组 DNA 的技术，有望成为"非转基因"生物育种的一种关键技术。我国在植物基因组编辑技术中处于国际前沿水平，拥有水稻、小麦、玉米三大粮食作物的 TALEN 和 CRISPR/Cas 技术体系及相关多项应用专利。2012 年后，CRISRP/Cas9 系统引发了基因组编辑的研究热潮，迄今已在水稻、小麦、玉米等多种农作物中成功应用。美国农业部已对磷高效玉米等多个基因组编辑农产品下达了转基因法规监控的豁免权；阿根廷、瑞典等国已明确宣布基因组编辑的作物不在转基因立法管辖范围之内。目前，基因组编辑技术及应用在全球每年有 100 多个专利申请，基因组编辑技术的储备及应用对于分子设计育种的高效实施有重要意义。

3.8 "高大严"等五位一体工程化育种技术

五位一体工程化育种技术集成了自交系"高大严"选系技术、IPT（以单株配合力测试为核心的选育技术）、DH、MAS 和多生态区鉴定五种杂交种培育方法为一体，是一种高效实用的育种方法。随着育种研究不断深入，DH 育种技术、MAS 分子标记辅助育种等高新技术或方法不断涌现。但大多数育种者依然采用常规系统选育方法或结合其中某一项新技术，依靠个人或少数几个人的实践经验，在内部试验基地或鉴定点进行自交系选育、杂交种组配及鉴定工作，存在盲目性和偶然性且效率低，影响育种工作的效率、稳定性和可持续发展。为改变育种工作现状，笔者等依据遗传育种理论并结合多年育种实践，提出了一套五位一体品种选育方法，实现了品种培育方法的又一次突破。

目前，利用五位一体工程化育种技术已选育出了京科 968、京科 665、京农科 728 等一大批耐密、高产、优质、稳产、抗性好等综合农艺性状优良的玉米杂交种，并在生产上得到大面积推广应用。如京科 968 自 2011 年通过国家审定以来累计推广种植面积近 5 000 万亩，特别是在最近几年种业市场种子数量过剩较为疲软的严峻形势下，该品种逆势增长发展成为年推广面积 1 600 万亩以上的品种，成为我国玉米生产主推品种。京农科 728 于 2012—2015 年累计推广面积 200 万亩以上，可机械直收籽粒，实现了人们期盼已久的

"像收小麦一样收玉米"。除早熟、宜机收优势外，该品种还具有高产、优质、耐旱、耐密、抗倒等特点，非常适合在华北平原旱作耕作区种植，预计几年后种植面积将达到1 000万亩。

3.9 SPT新型不育系技术

SPT（Seed Production Technology）新型不育系技术即雄性核不育制种技术，区别于传统的三系配套技术，通过构建一个转基因载体（载体上含有一个绿色荧光标记基因和一个控制核育性基因）对玉米进行遗传转化，在转基因后代果穗上可分离出两类种子，一类种子为核不育系，不含任何转基因成分，直接用于玉米杂交种生产；另一类种子为保持系，含有转基因成分（绿色荧光标记基因和核育性基因），可用于不育系和保持系的继续繁殖。该技术所用的不育系为核不育基因控制，不育系败育彻底，不受遗传背景限制，能大大降低制种成本，提高杂交种质量。通过该技术所产生的不育系由于不含有任何转基因成分，美国农业部于2011年解除了对其转基因管制审批。

3.10 表型快速鉴定评价技术

玉米新品种选育过程中主要对基因型和表型进行选择，随着测序技术的不断发展，基因型分析通量和表型鉴定通量也不断提高。表型鉴定又分为可控环境条件下的表型鉴定技术、半可控环境条件下的表型检测技术和不可控环境条件下的表型鉴定技术。其中，可控环境条件下的表型鉴定技术主要依赖于机器视觉技术的发展，能够近距离对玉米植株不同部位的动态生长进行监测，进而获得植株生长信息和表型参数，目前，这项技术是国内外研究的重点。如德国Lemna Tec公司建立的高通量植物表型检测平台，中国农业科学院生物技术研究所于2014年建成的我国第一个全自动高通量3D成像植物表型组学研究平台，还有美国杜邦先锋公司、中国农业大学以及北京市农林科学院分别研制的玉米果穗考种系统等。半可控环境条件下的表型检测技术可通过接种辅助检测相关表型，如接种病原菌、人工接虫等，目前是对部分农艺性状鉴定的有效技术，如美国先锋公司研制的大型移动人工风洞。不可控环境条件下的表型检测技术是指在完全开放的环境条件下，没有任何人工的参与进而快速获得多个环境中植物表型的方法，尽管准确性差，但在多个环境条件下能获得农艺性状表型参数的综合表现值，具有重要参考价值，这也是当前我国玉米育种过程中主要采取的方法，常用的多环境测试是国家、省级和各大种子公司的品种试验示范和区域试验。

综上所述，在政策拉动、科技支撑和市场带动等多重因素的作用下，我国玉米生产在过去的几十年中取得了快速发展和进步，但当前及今后一段时间还面临着生产经营方式转变、资源环境约束、国际竞争力缺乏等严峻形势和挑战。今后还需依靠育种技术，特别是先进的生物育种技术，实现新形势下的玉米生产调结构、转方式、节本增效，提高我国玉米产业的国际竞争力。

参考文献（略）

本文原载：生物产业技术，2016（3）：45-52

玉米及其制品质量安全风险及控制

赵久然　刘月娥

（北京市农林科学院玉米研究中心，北京　100097）

摘　要：玉米是全世界也是我国种植范围最广、总产量最大、用途最多的粮食作物。玉米按收获物和用途可分为籽粒用玉米、鲜食玉米、青贮玉米等3类，其中籽粒用玉米是我国玉米的主要产品。指出我国籽粒用玉米及其制品在质量安全方面存在的主要问题是真菌毒素浸染，同时也在一定程度上存在着农药残留和重金属污染等问题，并从玉米育种和作物栽培等方面给出相应的控制对策：①选育并推广抗病虫害的玉米新品种，积极采取措施预防和防治病虫害。②选育并推广脱水快、成熟度好、利于籽粒直收的玉米品种，规范玉米籽粒晾晒及储存过程。③完善相关标准体系和检测体系。④综合防治病虫草害，合理规范使用杀虫剂、杀菌剂、除草剂，生物防治为主，化学防治为辅，使用低毒农药。⑤加强标准农田建设，减少面源污染，降低重金属在玉米中吸收和积累。

关键词：玉米；玉米制品；质量安全；多抗玉米品种；京科968

玉米是全球第一大粮食作物，总产量和单产量居粮食作物之首。近年全球每年玉米种植面积20亿亩（$1.3×10^6 km^2$）以上、总产量已达10亿t。在保障世界粮食安全方面，玉米的作用意义重大。我国是全球第二大玉米生产国和消费国，总产量占全球玉米总产量的20%以上。2015年我国玉米的总产量和种植面积分别为2.25亿t和5.72亿亩（$3.8×10^5 km^2$），远高于水稻和小麦，居我国粮食作物之首；玉米按收获物和用途可分为籽粒用玉米、鲜食玉米、青贮玉米等3大类，其中籽粒用玉米是我国玉米的主要产品。近年来，我国籽粒用玉米总产量的10%作为口粮，68%作为饲料，20%以上作为工业加工原料。玉米产量已成为影响我国粮食供求和畜牧业发展的重要因素。同时，玉米及其制品的质量与安全事关我国农业的可持续发展和国民健康。

我国玉米籽粒的质量因受到土壤状况、病虫害、环境条件（光、温、水等）以及收获和仓储条件的影响，质量安全面临诸多问题，其中最主要的问题是真菌毒素浸染，同时也在一定程度上存在着农药残留和重金属污染等问题。本文对上述几个方面存在的问题进行综述，并从玉米育种和栽培等角度针对玉米及其制品中存在的质量安全问题进行对策分析，对提高我国玉米及其制品质量、保证我国玉米的可持续发展和人畜健康具有重要意义。

1　真菌毒素的污染现状及控制对策

1.1　真菌毒素的污染现状

真菌毒素是农产品的首要天然污染物，具有极强的致病性，不仅造成原粮减产和品质

作者简介：赵久然，男，研究员，博士，主要从事玉米育种和栽培方面的研究

降低，对食用被真菌毒素污染食品和饲料的人、畜也会造成很大危害，由其引起的食源性疾病和食品贸易争端一直是全球关注的热点，是影响玉米及其制品质量安全的最主要因素。据世界粮农组织估计，世界范围内有25%的农作物不同程度受真菌毒素的污染。对我国玉米及其制品而言，由于受环境因素、收获及仓储条件的影响，玉米及其制品受到真菌毒素的影响较为严重。

国内针对玉米黄曲霉毒素、伏马毒素、呕吐毒素和玉米赤霉烯酮等真菌毒素对玉米及其制品的污染现状做了较多研究。陈必芳等从全国104个饲料加工厂和饲养场采集饲料样品627份，对我国饲料霉变现状进行调查。调查结果表明，配合饲料及饲料原料污染率均分别高达100%和99%。高秀芬等对吉林、河南、湖北、四川、广东、广西的279份玉米样品的黄曲霉素进行检测，检出率为75.63%；戴玉瑞在河南省的鹤壁、濮阳、信阳、南阳、郑州以及驻马店采集25份玉米样品，发现伏马菌毒素的阳性检出率高达100%，其次是呕吐毒素84%，玉米赤霉烯酮80%，黄曲霉素B1 71%。伏马菌毒素、呕吐毒素、黄曲霉毒素B1和玉米赤霉烯酮均有不同程度的超标，超标率分别为96%、36%、25%和20%。敖志刚等对我国14个省市玉米样品的霉菌污染状况进行检测，发现黄曲霉毒素、玉米赤霉烯酮及呕吐毒素的检出率分别为82.0%、81.8%和100%，超标率分别为2.6%、36.4%和72.7%。对我国饲料原料和配合饲料中玉米赤霉烯酮进行检测，检出率高达100%，超标率为23.21%～40%。伏马毒素的检出率分布在19.2%～85.7%，其中山东和河南地区检出率较高，宁夏地区检出率较低；甄阳光等对我国7个地区11个省份的饲料原料及配合饲料样品的呕吐毒素含量进行检测，发现北方地区（西北、东北、华北）呕吐毒素污染情况较南方地区（西南、华南）严重；饲料原料中玉米的呕吐毒素含量和超标率较其他原料高，检出率为97.9%～98.8%，超标率为34.5%～57.6%。北京地区玉米呕吐毒素的检出率和超标率分别为92.9%和57.1%。

1.2 控制对策

1.2.1 选育并推广抗病虫害的玉米新品种，积极采取措施预防和防治病虫害

串珠镰孢是一种世界性的植物病原菌，在环境条件适宜时可侵染玉米植株，引发根腐病、茎腐病和穗粒腐病，玉米镰孢菌穗粒腐病病原的串珠镰孢是通过花丝进入穗内，籽粒表皮破裂或其他伤口都可以增加病菌的侵染。所以，受到虫害的原粮玉米容易受到霉菌的侵害，加重真菌毒素的污染，含有较高水平的真菌毒素。并且，玉米穗腐病的病菌导致伏马毒素、呕吐毒素和玉米赤霉烯酮的大量发生，严重影响我国玉米及其制品的安全，所以玉米的生产过程中病虫害的防治至关重要。

（1）选育并推广抗病虫害的玉米新品种。为了保证玉米的高产稳产性、对不良逆境因子的抵抗性及真菌毒素抗性，应该加强对综合性优良玉米品种的选育，选育并推广抗病虫害的玉米新品种是保证玉米及其制品质量安全的重要途径。但是我国在这方面的玉米新品种更新换代步伐相对滞后，我们可以通过常规育种、DH单倍体育种、诱变育种、分子标记辅助育种和转基因育种等技术选育高产、稳产、广适、多抗的玉米新品种，特别是抗多种病虫的品种。于艳秋等研究发现，Bt玉米不但能够抗玉米螟，同时在农田研究中也发现Bt玉米对串珠镰孢霉菌感染也具有一定抗性，并且含有较低水平的伏马毒素。而未来转基因育种的主要发展方向除关注抗病虫害玉米品种的选育方面，更应集中在转基因技术本身的安全性、功能基因组研究、载体构建、遗传转化体系的优化和改善及多基因叠加

的复合性状转基因玉米的选育等。

（2）积极采取措施预防和防治病虫害。除选育并推广抗病虫害的玉米新品种外，还应积极采取措施预防和防治病虫害。在玉米播种时，玉米种子尽量用种衣剂包衣，种衣剂具有防虫抗病、增产增收的作用，尤其对苗期害虫，玉米丝黑穗病等，具有良好的防治效果。通过种子包衣减少玉米的病虫害，从而减少真菌污染，提高玉米质量。在玉米整个生长发育期间应密切关注病虫害的动态变化，大力开展科学防灾减灾工作，积极推广抗病虫品种，本着"预防为主，综合防治；生物防治为主，化学防治为辅"的原则，依靠生产条件的改善和科技的进步减少玉米病虫害损失以及对玉米籽粒的感染，并及时准确做好预测预报工作。

1.2.2 选育并推广脱水快、成熟度好、利于籽粒直收的玉米品种，规范玉米籽粒晾晒及储存过程

玉米收获和晾晒过程的温度和湿度是影响真菌毒素的主要因素，若收获季晾晒过程遭遇阴雨天气，空气湿度过大，会加重真菌毒素的污染，所以一定要严格控制玉米收获、晾晒和储存时玉米籽粒的水分含量，从而减少玉米籽粒受真菌毒素污染的机会。

（1）选育并推广脱水快、成熟度好、利于籽粒直收的玉米品种。玉米收获及晾晒过程的气候条件对玉米籽粒的质量影响巨大，以往我国玉米的收获主要是以人工收获和机械收穗为主，收获持续时间及晾晒时间都较长，大大加重了玉米籽粒感染真菌毒素的机会。现在农业部正在大力推广机械直收籽粒技术，但是，当前我国玉米主产区玉米品种在收获时普遍含水量较高、籽粒破损率较大，极易感染真菌毒素，缺乏丰产稳产性好、耐密抗倒性好、成熟度好、适宜机械直收籽粒的优良玉米品种。因此，我们应加强选育并推广脱水快、成熟度好、利于籽粒直收的玉米品种，并结合不同的玉米生态区，合理品种配置，优化玉米生产品种布局，最终缩短玉米收获及晾晒时间，减小气候因素对玉米籽粒质量的影响，最终提高玉米籽粒质量。例如，玉米品种京科968，具有成熟时籽粒含水量低、脱水快的特点。与相同生育期品种郑单958相比，京科968生理成熟时的含水量和脱水速率都显著低于郑单958，适于机械直收籽粒。

（2）玉米籽粒晾晒及储存过程的规范。在玉米晾晒的过程中，应注意通风、清洁，减少霉变籽粒的产生、真菌毒素的污染以及杂质的掺入。有条件的可以使用烘干设备，但是烘干温度不能过高，容易产生玉米爆腰和烘糊粒，在操作时，籽粒被加热的温度不易超过50℃，介质温度不超过140℃。在玉米储存时要严格控制籽粒含水量，一般要求玉米籽粒的水分含量≤14%。

1.2.3 完善相关标准体系和检测体系

（1）相关标准体系的完善。在我国，玉米质量的标准主要有3个，分别是最基础的玉米标准、饲用玉米标准和工业用标准，广泛的应用于商品玉米的收购、储存、运输、加工以及销售。为了我国玉米的质量安全，2009年我国发布了GB1353-2009《玉米》，2009年9月1日起开始实施。《玉米》新标准历经几次修订后仍以容重进行定等级，等级由原来的3级改为5级，并增加了等外级，完善了等级设定指标。同时我国还建立了饲用玉米、高油玉米等相应的国家标准，以及爆裂玉米、糯玉米、优质蛋白玉米、高淀粉玉米、高油玉米、甜玉米等相应的行业标准，完善了我国的玉米标准体系。

真菌毒素是农产品中一种重要的天然污染物，具有极强的致病性，对人、畜造成极大的伤害。因此，加强从农田到餐桌各环节粮食中真菌毒素的监测，同时结合我国居民的膳食消费量进行风险评估，制定粮食制品中真菌毒素的限量标准将是未来的研究重点。鉴于真菌毒素对人体健康的危害，我国应加强相关标准体系的建设，对相应的有害物质进行严格的限定并建立相应的法律法规，保证我国玉米及其制品的质量安全。

（2）相关检测体系的完善。在我国，为了玉米及其制品的安全，相关检测体系的完善在我国粮食产业中同样有着举足轻重的作用。例如，在我国粮食收购过程中，粮食的品质分级非常重要，而粮食物理特性指标检测是粮食品质评价的重要组成部分，检测指标一般包括品种分类、不完善粒、霉变粒、整精米率、出糙率以及容重等。粮食霉变不仅降低粮食的营养和商品价值，更重要的是影响粮食及其制品的可食性和安全性，并且有些霉菌还能产生具有毒性的二级代谢产物，严重影响粮食的品质安全，因此，将其作为粮食品质检测中的重要指标。

我国粮食质量检验技术的基础研究和自主创新能力还很薄弱，还处在跟踪学习发达国家先进技术阶段。我国在粮食品质检测中大多采用简单低效的设备和主观的感官方法，现代检测方法在粮食品质的检验中应用较少，提高我国粮食质量的检测技术水平及新技术在检测中的应用至关重要。目前，粮食检测技术正朝着快速、高效、高准确性的方向快速发展，未来粮食检测的发展趋势是尽量以更多的仪器检测代替感官评价，使感官指标仪器化、标准化、智能化。随着粮食检测与识别技术的完善与成熟，我国粮食质量分级标准将会更加规范，从而满足人们对粮食质量分级提出的越来越高的要求。

2 我国玉米及其制品中农药残留现状及控制对策

2.1 农药使用现状

众所周知，农药在保证农业生产和粮食安全中发挥了重要作用，但随着农药的大量和不合理应用，农药所造成的环境污染以及残留农药对人体健康造成的危害不断加剧，这一问题已越来越受到社会的高度关注和重视。但与林果和蔬菜相比，玉米在生长过程中所用的农药较少。在玉米生产上，玉米播种时经常使用种衣剂拌种，但是用量较少；利用杀虫剂、杀菌剂和除草剂预防玉米大斑病、茎腐病、玉米螟、黏虫、二点委夜蛾等病虫害以及各种杂草，但这些农药只有在发生严重年份使用，发生较轻的年份使用不普遍。近年来由于除草剂使用比较普遍，在生产上药害频繁发生，玉米对除草剂的敏感性研究也日益受到人们的重视。但以上杀虫剂、杀菌剂和除草剂在玉米生长的过程中发生降解，玉米籽粒中残留量较低，没有发现玉米籽粒中农药残留超标的报道，因此，在玉米上使用农药从而影响玉米质量安全的问题不突出。

2.2 控制对策

2.2.1 玉米新品种的选育

玉米新品种尤其是抗病虫、农药使用量尽量少和抗除草剂品种的选育是解决玉米药害和农药残留的重要途径。但是我国在这方面的玉米新品种相对较少，尤其是抗除草剂的玉米品种，所以我们可以通过先进的育种技术加强这方面的研究工作。转基因技术已在抗虫、抗除草剂的转基因玉米品种取得了很大进展，并在美国等多个国家实现了商业化。孟山都及陶氏益农公司共同研发的玉米新品种 Smart-Stax 具有广谱抗玉米根部虫害及全面控制广泛的杂草优势。

2.2.2 综合防治病虫草害，合理规范使用杀虫剂、杀菌剂、除草剂

在玉米整个生长发育期间应密切关注病虫草害的动态变化，大力开展科学防灾减灾工作，积极推广抗病虫品种，综合防治病虫草害。本着"预防为主，综合防治；生物防治为主，化学防治为辅"的原则，依靠生产条件的改善和科技的进步减少玉米病虫草害损失，并及时准确做好预测预报工作。同时，合理规范使用杀虫剂、杀菌剂、除草剂等，化学防治病虫草害应采用低毒高效农药，同时还应加强生物防治，努力提高防治效率。近年来，生物防治玉米螟技术取得较大发展，目前玉米生产中应用较多的玉米螟生物防治技术主要有赤眼蜂防治、白僵菌封垛防治与苏云金芽孢杆菌（Bt）乳剂防治。

3 重金属污染现状及控制对策

3.1 重金属污染现状

随着工业的发展和农药、化肥的广泛应用，含重金属的物质污染物通过各种途径进入环境，造成土壤，尤其是农田土壤重金属污染日益严重，我国耕地的重金属污染率高达16.67%，主要分布在辽宁、河北、江苏、广东、山西、湖南、河南、贵州、陕西、云南等地，特别是辽宁和山西的耕地土壤重金属污染尤其严重。玉米具有富集重金属的能力，甚至可作为重金属污染土壤修复的植物，因此很有可能存在植物内重金属超标的现象。张勇等对辽宁省部分县玉米籽粒中的重金属进行检测，发现大部分玉米籽粒中能检出铅、镉和砷等重金属，部分地区铅和镉的超标率分别高达40%和80%。俞华齐对徐州市主要谷物的铅和镉的污染状况进行了检测，研究发现玉米籽粒中铅的超标率高达100%。但我国重金属残留的问题只是局部地块问题，并不是我国玉米质量安全的普遍问题，但关于我国玉米重金属超标的问题我们还应严格防范。

3.2 加强标准农田建设，减少面源污染，降低重金属在玉米中吸收和积累

在我国，随着重工业的发展，农药和化肥的大量使用，导致土壤受重金属污染的情况日益严重，并且随着无机肥和农药的过量施用，玉米及其制品中重金属超标问题严重，所以为了我国玉米及其制品的食品安全问题，应根据我国相应的标准加强标准农田建设，减少面源污染，降低重金属在玉米中吸收和积累。在我国，关于农田的主要标准是国标 GB 15618—1995《土壤环境质量标准》规定了土壤中污染物的最高允许浓度指标值及相应的检测方法，同时我国发布 NY/T 395—2012《农田土壤环境质量检测技术规范》和 NY/T 1119—2012《耕地质量检测技术规程》，对我国的农田土壤环境质量进行检测；国标 GB 5084—2005《农田灌溉水质标准》，规定了农田灌溉水质要求、检测和分析方法，适用于全国以地表水、地下水和处理后的养殖业废水及以农产品为原料加工的工业废水作为水源的农田灌溉水。

参考文献（略）

本文原载：食品科学技术学报，2016，34（4）：12-17

我国玉米品种标准 DNA 指纹库构建研究及应用进展

赵久然　王凤格　易红梅　田红丽　杨扬

(北京市农林科学院玉米研究中心/玉米 DNA 指纹及分子育种北京市重点实验室，北京　100097)

摘　要：介绍了我国玉米品种标准 DNA 指纹库构建的进展及应用情况。我国玉米品种标准 DNA 指纹库规模已达到 22 089 份，经与国内外作物品种 DNA 指纹库比较，是全球品种数量和建库规模最大的 DNA 指纹库，并且在区域试验、品种权保护及种子质量管理等方面都得到了广泛应用。至今已开展玉米品种真实性鉴定等达 40 000 多份（次）。同时，对 DNA 指纹库构建未来发展进行预测，建议继续发挥 SSR 标记的作用、加快推进 SNP 标记的研发、发掘 InDel 标记的应用潜力并持续关注测序技术的发展。

关键词：玉米；品种；DNA 指纹库；SSR；SNP

随着我国玉米种植面积的扩大和产业的快速发展，玉米品种数量呈逐年上升趋势，截至 2013 年我国已通过国家或省级审定的品种 6 291 个，国家及各省区域试验参试品种数量每年约 2 000 个；截至 2014 年底，申请品种权保护 4 050 个，已授权 1 718 个品种。生产上年种植面积在 10 万亩（0.67hm^2）以上玉米品种有 879 个（2013 年统计数据）。由于少数骨干亲本的集中应用，以及在原品种基础上仅改良少数甚至单个性状的回交育种方式广泛使用，育成品种间的遗传差异越来越小，如何有效管理和甄别玉米品种成为品种管理的难题；目前，玉米种子假冒伪劣情况禁而不绝，种子市场秩序比较混乱，严重侵害了农民和合法经营企业的权益，如何控制品种多、乱、杂，净化玉米种子市场环境成为种子质量管理的难题。

作物品种的真实性、纯度、特异性及一致性鉴定，传统的方法是田间种植，依据形态差异进行鉴别区分，在玉米 DUS 测试指南中列出了 40 个必测性状，这些都属于能够直接观测的性状。但田间种植鉴定有很大的局限性：一是时间较长，需要至少一个生长季，多的可达数年；二是受环境影响，同一品种在不同生态区种植，其表型有很大差异，甚至连育种者本人都不能辨别；三是即使在同一地点种植，因土壤地力和水肥措施的不均匀，也可导致生长不一致的问题；四是形态鉴定只能鉴别直观目测的性状，对内在的生理和病理差异未能考虑。在生化标记基础上发展起来的同工酶及贮藏蛋白电泳鉴定方法，简单快速，成本低，至今仍在玉米纯度及真实性鉴定上广泛应用，但由于多态性有限，随着玉米品种数量逐年增多，特别是大量近似品种的存在，难以进行有效甄别。

基金项目："十二五"国家支撑计划项目（2011BAD35B09）
作者简介：赵久然，研究员，主要从事玉米遗传育种研究。王凤格为并列第一作者，副研究员，主要从事玉米品种分子检测研究

DNA 是生物遗传信息的直接载体，分子标记是以个体间遗传物质内核苷酸序列变异为基础的标记，是 DNA 水平遗传多态性的直接反映。最早发展起来的分子标记是 RFLP（第一代分子标记），之后有 RAPD、AFLP、ISSR、SSR 等基于 PCR 的标记（第二代分子标记），以及近年来发展起来的基于单碱基突变的 SNP 标记（第三代分子标记）。分子标记的位点数量多，多态性高，信息量大，理论上可形成无限的区分能力，基于分子标记所标示的 DNA 特定位点信息组合，能够像人的指纹一样，将不同品种区别开，形象的称之为 DNA 指纹；将大量的不同品种的 DNA 指纹信息集合在一起，并形成具有比对、查询等功能的数据库，称之为 DNA 指纹库；而采用国家或行业标准规定的标准方法，为政府品种管理部门提供的品种标准样品建立的 DNA 指纹库，可相应称之为品种标准 DNA 指纹库。

1 我国玉米品种标准 DNA 指纹库构建进展

1.1 建库关键技术研发

我国玉米品种 DNA 指纹库构建技术研发工作可以追溯到 1993 年，起步于同工酶技术的研发；1995 年开始 RAPD 标记在玉米品种真实性及纯度鉴定上的研发和应用；2001 年开始从 RAPD 升级为 SSR 标记；2007 年开始 SNP 标记的研发。针对玉米品种标准 DNA 指纹库构建的关键技术，经过 20 多年的研发，实现了技术专利化、专利标准化和标准实用化。

在 SSR 技术研发方面，提出和确定核心引物组合法 DNA 指纹库构建方案，解决了品种区分和数据整合共享的难题；针对我国玉米品种建库和鉴定的需要，进行系统的筛选确定和重新设计了玉米 DNA 指纹库构建的 SSR 核心引物组合；建立了高效的 10 重荧光毛细管电泳检测体系，将建库效率至少提高 10 倍以上；开发适合玉米品种 SSR 指纹分析器 SSR analyser，统一了 SSR 指纹分析工具；研制出国内首个和唯一的适合玉米品种真实性检测的 DNA 指纹复合扩增试剂盒专利产品。

在 SNP 技术研发方面，通过跟踪国际研究进展，探索 48 重的 SNPlex 检测系统；及时掌握基因芯片技术的研究进展，从 5 万多个 SNP 位点全基因组芯片中筛选出玉米品种鉴定和建库 SNP 专用系列芯片 MaizeSNP3072、MaizeSNP384 等，并获得国家发明专利授权；与北大未名凯拓、中国农业大学合作定制多功能位点高通量的 200K 芯片，同时探索样品高通量的 KASP 和 array tape 检测平台，实现 SNP 技术在位点高通量和样品高通量上的提升。

在数据库管理系统研发方面，确立了玉米 DNA 指纹数据库标准化规范，实现多个实验室 DNA 指纹数据的整合和联合共建；将数据库管理系统拓展为兼容多种标记（SSR、SNP）、多物种（玉米、水稻、小麦等）的植物品种 DNA 指纹管理系统；建立数据库比对网站平台，实现数据共享。

在技术标准化方面，2007 年制定颁布了首个玉米 DNA 指纹鉴定行业标准；2009 年开始制定《植物品种鉴定 DNA 指纹方法总则》，并于 2014 年颁布实施；同时参与 UPOV-BMT 分子测试指南国际标准的修订，实现国家标准与国际标准对接。玉米和植物鉴定总则标准的颁布，引领和推动了其他植物 DNA 指纹鉴定标准化的进程，在总则的框架下，陆续出台了水稻、小麦、大豆等 14 个作物的 DNA 指纹鉴定标准。

1.2 指纹库构建

汇总我国玉米区域试验、审定及品种权保护品种建库工作,截至 2014 年年底,我国已完成 22 089 份玉米品种标准 DNA 指纹库,并将以每年 3 000 份左右的速度扩容。

在区域试验品种方面,自 2002 年起,在农业部区域试验管理部门推动下,建立国家区域试验所有玉米品种 DNA 指纹库,2005 年起将区域试验建库范围逐步扩大到国内 20 多个省份,至 2014 年年底,全国区域试验玉米品种 DNA 指纹库库容已达到 15 398 份,并以每年 2 000 份左右的速度递增。

在品种权保护品种方面,自 2004 年起,在农业部品种保护部门推动下,建立我国品种权保护玉米品种 DNA 指纹库,至 2014 年年底,已完成 1 861 份授权品种 DNA 指纹库的构建,2015 年将建成库容 3 768 份的所有申请品种权保护玉米品种 DNA 指纹库,并以每年 400 份左右的速度递增。

在审定品种方面,自 2010 年起,随着农业部征集玉米审定品种标准样品工作的开展,启动玉米审定品种标准 DNA 指纹库构建,至 2014 年年底,玉米审定品种 DNA 指纹库库容已达到 4 632 份,并以每年 500 多份的速度递增。

随着 SNP 技术的发展成熟,上述样品除了利用 40 个 SSR 核心引物构建 SSR 指纹库外,正在利用 MaizeSNP384 芯片构建 SNP 指纹库,2015 年上半年将完成 4 632 份审定品种的 SNP 指纹库构建。

1.3 与国内外作物品种 DNA 指纹库的比较

国内外已构建品种 DNA 指纹库的作物中,粮食作物主要有玉米、水稻、小麦、马铃薯,经济作物主要有棉花、大豆、番茄、大白菜、黄瓜、葡萄、西瓜、月季、康乃馨、烟草、茶等,目前国内外其他作物的品种建库规模一般不超过 5 000 份,而我国玉米品种建库规模已达到 22 089 份,是全球建库规模最大的指纹库。国外品种标准 DNA 指纹库构建工作主要是在 UPOV 框架下开展植物品种权保护品种的建库,其中建库规模较大的有:德国建立 480 份欧洲小麦品种 SSR 指纹库,法国建立 1 537 份玉米品种 SSR 和 SNP 指纹库,荷兰建立 734 份月季、521 份番茄品种 SSR 指纹库,英国建立 892 份马铃薯品种 SSR 指纹库,意大利建立 1 005 份葡萄品种 SSR 指纹库等。我国品种标准 DNA 指纹库构建主要是在农业部品种管理部门的统筹下开展品种权保护品种及区域试验审定品种的建库,其中玉米、水稻率先开展建库工作,随着植物品种鉴定 DNA 指纹方法总则和各作物 DNA 指纹标准的颁布,小麦、棉花、油菜、大豆、高粱、西瓜、大白菜、黄瓜、辣椒等作物的建库工作正在陆续开展,除了水稻、小麦品种建库规模较大,分别达到 3 829 份和 3 000 份外(内部交流),其他作物的建库规模一般在 1 000 份以内。

即使不局限于品种标准 DNA 指纹库构建,将国内外在种质资源 DNA 指纹库构建方面的工作考虑在内,我国玉米品种 DNA 指纹库的规模也是最大的。在种质资源方面建库规模较大的如 CIMMYT 构建了 800 多份玉米、5 000 多份小麦种质资源指纹库,美国构建 2 815 份玉米、2 256 份棉花种质资源指纹库,中国构建了 820 份玉米、2 625 份大麦、3 318 份黄瓜、788 份西瓜等种质资源指纹库,这些种质资源建库的数量也远远低于玉米品种指纹库。

2 我国玉米品种标准 DNA 指纹库的应用

2.1 应用概况

玉米品种 DNA 指纹库的构建为政府、企业等多方面需求提供了强有力的科技支撑，至今已开展玉米品种真实性鉴定等 40 000 多份（次），包括区域试验审定样品 15 000 多份，品种权 DUS 测试样品 6 000 多份，维权打假样品 1 800 多份，种子质量监督抽查样品 7 800 多份，以及企业、科研单位、农民的科研育种、种子生产、生产经营、维权打假等方面的样品 17 000 多份。同时，北京市农林科学院玉米研究中心也获得多方面的认可或认证，成为农业部首批品种真实性检测资质单位，高级法院认定的品种分子检测司法鉴定单位，国家玉米区试 DNA 指纹检测技术牵头单位，辽宁、吉林、浙江等 20 余省玉米区试 DNA 检测的委托鉴定单位，农业部玉米审定和品种权标准样品 DNA 指纹构建指定单位，农业部种子执法年行动指定唯一的玉米品种真实性鉴定单位，农业部种子质量监督抽查指定真实性鉴定单位，跨国种业先锋、孟山都、先正达、利马格兰、KWS 在中国检测单位，中种、奥瑞金、登海、德农、屯玉等 50 余家龙头企业依托检测单位，UPOV 组织 BMT 技术工作组成员单位。受农业部委托，多次举办国际国内技术培训班，并为企业和科研单位随时上门进行培训学习，累计培训 3 000 多人次，推动了玉米 DNA 指纹库及其关键技术在检测实践中的广泛应用。

2.2 在区域试验中的应用

在国家玉米品种区域试验工作整体安排下，我国 2002 年开始组织实施了"国家玉米区域试验品种一致性和真实性检测"工作，从源头上在分子水平给每个参加国家级试验的玉米品种一个特有的身份标志，十多年来，检测品种数量不断增加，检测作物类型、组别逐步扩大，检测力度逐步加强，检测标准化和规范化程度逐步推进，检测单位数量和规模不断扩大，检测内容逐步丰富，检测技术不断提升，真实性和一致性问题逐步得到有效控制。已经对参加国家区域试验的 5 108 个玉米品种进行了有效的检测，并逐渐辐射到各省级区域试验管理，对辽宁、北京、吉林、内蒙古自治区（以下称内蒙古）、宁夏回族自治区（以下称宁夏）、山东、河北、山西、天津、河南、云南、陕西、四川、重庆、江苏、安徽、广西壮族自治区（以下称广西）、新疆维吾尔自治区（以下称新疆）、湖北、湖南、福建、贵州、甘肃等 23 个玉米生产省区域试验样品进行了检测，截至 2014 年年底，已检测国家及各省区域试验样品 15 596 份，剔除或停试了一批与已知品种仿冒雷同、区组间或年度间随意更换组合、一致性差或较差的参试品种。同时还制定了《国家区试玉米品种一致性及真实性 DNA 指纹检测技术》标准规范及《国家玉米品种试验 DNA 指纹鉴定管理办法》，建立了全国玉米区域试验 DNA 指纹联合数据库，实现了国家级和省级玉米区域试验的联合监控。

分析 10 余年的区域试验 DNA 监测数据，具有以下规律：①一致性检测方面，国家普通玉米区域试验在 2005 年之前一致性不合格样品较多，而经过 3 年全面监控，一致性得到显著改善，不合格情况几乎绝迹，一致性达到 1 级的比例显著提高；甜糯玉米从 2006 年开始一致性检测，当年一致性不合格样品达到 24 个，2007 年下降到 9 个，之后显著下降到 0~2 个，表明甜糯品种一致性与其品种类型无关，可以采用与普通玉米相同的标准；青贮玉米自 2006 年开始一致性检测以来，一致性不合格样品始终未完全杜绝，分析表明与青贮玉米经常存在三交种等其他杂交类型有关，对这类品种可适当放宽一致性检测标准。②同名更换检测方面，2002—2005年重点检测国家区域试验内部不同年份、区组的同名更换，经过几年检测，这类问题显著减少，由原来明显的差异较大的样品更换为主变为近似姊妹系更换为主；随着国家预试验及与国家区域试验库联合，开始检测出一批预试验到区域试验发生同名更换的样品；随着省区域试验逐步和国家区域试验联合检测，国家区域试验与各省区域试验的同名更换成为主体。③仿冒雷同检测方面，2006年以前，由于采用PAGE电泳系统，很难形成整合数据库，检测出的仿冒雷同品种只有1~2个，且均是根据田间表现怀疑有问题的进一步在实验室进行成对比较复核验证；2006年以后，随着ABI3730xl DNA分析仪的使用及大规模品种DNA指纹数据库的构建，仿冒雷同品种的筛查均是通过与已知品种DNA指纹数据库比较筛查出来的，且随着库容不断增大，检出的仿冒雷同数量也在增加。

2.3 在品种权保护中的应用

在辅助品种权测试方面，2004年起开始构建申请品种权保护样品标准DNA指纹数据库并探索在DUS测试中的应用，通过利用玉米DNA指纹数据库管理系统的品种查询比对功能，筛查出与每个新申请品种最近似的品种名单，作为推荐需参考形态性状数据进行综合判定的名单或需进一步进行田间特异性测试名单。2004—2010年完成了1 237份2009年之前申请品种权保护样品DNA检测；2011年完成了350份2009—2011年申请品种权保护样品DNA检测；2013年完成了274份2012—2013年申请品种权保护样品；2014年开始对所有申请品种权保护的玉米品种及其近似品种共6 000份进行全面DNA检测和分析。使用玉米品种DNA指纹库辅助筛查近似品种，提高了特异性测试的准确性和效率，加快了品种权保护从申请到授权的进度。

在辅助政府维权打假方面，2009年开展甘肃制种基地品种权执法，2010—2013开展"种子执法年"打击侵犯品种权和销售假冒伪劣种子专项行动，累计检测玉米样品1 851份，检测出一批真实性有问题的品种并进行处理，实现了打击侵权从制种基地到种子市场的多环节全方位监控。

2.4 在种子质量管理中的应用

2010年起，农业部将种子真实性纳入到国家种子质量管理中，当年的春季种子市场监督抽查中首次包括了真实性检测项目，检测真实性样品1 330份；2011年在扩大春季种子市场监督抽查规模的基础上，增加了冬季种子企业督查行动，检测真实性样品887份；2012年在继续扩大春季和冬季检测的基础上，增加了夏季制种基地巡查，检测真实性样品984份；2013年继续扩大夏季基地巡查的规模，并增加了典型县市的种子市场暗访摸底行动，检测真实性样品2 052份；2014年将抽查单位从农业部向各省扩展，全年检测真实性样品3 996份，仅农业部和各省春季抽查样品就达到2 549份。在2010—2014年的5年期间，已为种子质量管理累计检测样品7 802份，至此，我国玉米种子真实性管理实现了从制种基地源头到企业加工中间环节到种子销售市场终端的全程监控。

此外，多年来为企业、科研单位、农民的科研育种、种子生产、生产经营、维权打假等需求提供真实性检测服务，作为重要的维权证据或线索，已累计检测样品17 000多份，对违规造假行为产生极大的威慑作用，净化了种子市场。

3 未来发展趋势

3.1 继续发挥SSR标记在品种分子检测方面的作用

与其他技术相比，SSR标记的优势主要体现在以下几个方面：位于非功能区，一般不受选择压的影响；就单个位点而言，展示了丰富的多态性；技术相对成熟，研究基础较强；检测快速准确、方便灵活，单个样品检测成本低。与SNP标记等相比，SSR标记主要有两个方面局限性：一是不同平台、实验室间数据整合较难，需要设立参照样品；二是检测通量中等，毛细管电泳最多可达10重，但当引物位点数量翻倍时，检测工作量和成本都相应翻倍。

尽管如此，SSR标记仍具有SNP标记不可比拟的优势：SNP标记虽然在纯系和单株个体样品的DNA鉴定上具有明显优势，但在杂交种以及一致性较差品种的混合样品DNA鉴定上，比SSR标记更容易出现数据缺失或分型不准确的情况；SNP标记尽管有位点高通量和样品高通量的检测平台，可以检测几十万个位点或上万个样品，但对检测位点数在20~100个、检测量在几十到几百个样品的品种真实性鉴定需求，SSR标记仍具有灵活、简便、成本低的优势。SSR标记技术作为目前最为成熟的DNA检测技术，并没有过时，应继续发挥SSR标记的作用。

3.2 加快推进SNP标记应用于品种检测的研发

SNP标记技术平台按照检测通量可分为三类：①传统的低通量SNP分型方法，如传统Sanger测序方法、限制性内切酶酶切法、单链构象多态性技术等。这些分型方法速度慢、自动化程度低、试验条件需反复调试，因此只适合检测样本量和位点数目均较少的项目。②中通量的SNP分型方法，如SNPlex、SNaPshot、TaqMan、DHPLC、Invader assay、SNP-MassARRAY、质谱法等。AB公司在中通量SNP分型技术上做了大量开发工作并推出基于ABI遗传分析仪的SNPlex和SNaPshot分型系统及基于定量PCR仪的TaqMan等位基因检测法。SNaPshot适合于几百个样本、十到几十个位点的检测；TaqMan不能同时检测两个及以上探针，因此适合于大量样本、少数几个高频位点的检测；SNPlex系统可以同时检测48个或96个位点，但由于检测成本较高、流程较复杂等原因已经停产。这些中通量分型方法的主要问题是单个数据的价格都较高，性价比较低。③高通量的SNP分型方法。主要有两类，一是位点高通量平台，如Illumina公司、Affymetrix公司推出的芯片平台，一次检测位点可达到几十万至上百万；二是样本高通量平台，如LGC公司推出KASP，Douglas公司推出的Array Tape平台等。

SNP标记是未来最有发展潜力的建库技术之一，与SSR标记相比具有以下优势：在基因组中分布密度更高更均匀，容易筛选到一套均匀分布核心SNP位点组合；通常为二等位变异位点，易实现数据间整合比较；借助于新型SNP分型平台，更易实现高通量检测；有可能实现与功能基因甚至表型相关。对于建库及真实性鉴定而言，理想的SNP检测方法应具有高度自动化、高通量、高准确性、较高的性价比等特征，未来应加快推进SNP技术的标准化研发进程。

3.3 发掘InDel标记的应用潜力

InDel是真核生物基因组中一类特殊的二等位遗传标记，表现为基因组中插入或缺失了一小段DNA，在玉米基因组上平均126bp就有一个INDEL，频率仅次于SNP。InDel作为鉴定标记兼具 SSR 和 SNP 的优点：与 SSR 相似，InDel 本质上属于长度多态性，可以在 PAGE 电泳及毛细管电泳技术平台上进行分析，基因分型清楚，PCR 扩增简单，容易实现多重 PCR 和多重电泳分析；与 SNP 相似，InDel 也属于二等位基因多态，等位基因固定并且已知，基因分型清楚，数据统计准确，容易多平台数据共享。

InDel 标记最大的优势就在于兼容 PAGE 电泳、毛细管电泳和芯片、KASP 等分析检测平台，容易实现多平台分析数据的整合共享。InDel 标记技术对检测平台没有依赖，可以充分利用现有的 SSR 和 SNP 的检测设备，非常有利于大范围推广应用。我们已经开始启动适合多平台的核心 InDel 标记的开发，InDel 有望成为除 SSR 和 SNP 标记之外的另一个有应用潜力的建库标记。

3.4 跟踪关注测序技术的发展

全基因组测序在 SSR、SNP、InDel 等分子标记开发以及功能基因挖掘等研究中应用比较广泛。但全基因组测序成本仍然比较昂贵、试验流程比较复杂，基于全基因组序列建立的数据库，绝大部分数据属于单态位点，所建数据库区分效率非常低，进行整库比较时耗时较长，分析成本较高，因此该技术难以直接应用于品种建库及真实性检测中。

基于测序技术检测 SNP 位点的方法具有一定发展潜力，比如基于第二代测序技术进行 SNP 基因分型的 restriction-site associated DNA sequencing（RAD-seq）技术，不受作物有无参考基因组的限制，简化基因组的复杂性，通过一次测序可获得数以万计的多态性标记。随着测序技术的高速发展及测序成本的不断降低，有可能开发出适合品种建库和真实性鉴定的检测方法，今后对测序技术应保持持续关注。

参考文献（略）

本文原载：作物杂志，2015（2）：1-6

中国玉米品种审定现状分析

杨 扬* 王凤格* 赵久然* 刘亚维

(北京市农林科学院玉米研究中心/玉米DNA指纹及分子育种北京市重点实验室，北京 100097)

摘 要：【目的】品种审定是玉米品种获准进入市场并进行推广的凭证，是相关部门规范、治理种子市场的依据，也是农民选择品种和维权的参考。根据玉米品种管理等业务的需求对玉米品种审定信息进行全面分析，深入挖掘品种数据的内在联系和隐含信息，可以为种子管理部门、科研单位、种子企业和农民等相关单位和个人提供更加系统、清晰的参考和依据。【方法】通过资料搜集、整理和校验，梳理了 42 年来中国国家级审定和省级审定的玉米品种信息，讨论了市场上玉米品种随年份和地域变化的趋势、审定品种中亲本自交系重复使用率以及品种审定与退出、保护之间的关系等问题。【结果】1972—2013 年，国家和各省、直辖市、自治区审定的玉米品种总数为 6 291 个，其中，国家审定的品种数为 503 个。2013 年现存的审定品种数为 4 882 个，其中，国审品种数为 332 个。1 726 个单位有审定品种记录，科研院所和种业公司分别占 48.3%和 46.6%，平均每个单位拥有的审定品种数分别为 7 个和 3 个。审定品种共使用了 8 754 个亲本自交系，其中 19.4%的自交系存在重复使用，使用频率最高的 3 个自交系依次为 Mo17、昌 7-2、丹 340。在过去 42 年审定的玉米品种中，共有 1 876 个品次退出，已经停止推广的品种平均使用年限为 13.3 年；29.2%的审定品种申请了植物新品种权，15.1%获得了授权。【结论】在近 10 年间，各级审定的玉米品种数量年均增幅趋于平稳，品种保有量仍在逐年增加，审定数量上呈现北多南少、增加速度上呈现东高西低的趋势；审定品种数量目前以科研院所居前列，较大型种业公司的审定品种数量上升明显；审定品种中新命名的亲本自交系数量迅速增加，使统计上的亲本重复使用率逐年下降。列入停止推广的审定品种数量年际间趋于稳定，审定品种与品种权保护品种的重叠程度有所下降。

关键词：玉米；品种审定；选育单位；亲本组合；品种退出；新品种权保护

Analysis of the Current Situation of Accredited Maize Varieties in China

Yang Yang Wang Fengge Zhao Jiuran Liu Yawei

(Maize Research Center, Beijing Academy of Agricultural and Forestry Sciences/Beijing Key Laboratory of Maize DNA Fingerprinting and Molecular Breeding, Beijing 100097)

基金项目：国家"十二五"农村领域国家科技计划（2011BAD35B09）、北京市科技计划（D131100000213002）

* 作者简介：杨扬，E-mail：caurwx@163.com。王凤格，E-mail：gege0106@163.com。杨扬和王凤格为同等贡献作者。通讯作者赵久然，E-mail：maizezhao@126.com

Abstract:【Objective】Variety accreditation acts as the permiting for commercialization of maize varieties, the basis for the regulation and administration of seeds market, and the reference for variety selection and right protection in China. Therefore, comprehensive analysis on accredited maize varieties according to demands of businesses such as maize variety management, and in-depth exploration of the internal link and implicit message of these variety data can provide more systematic and distinct reference to seeds management departments, scientific research institutions, seeds enterprises and farmers.【Method】The state and provincial authorized maize variety data of China in 42 years were obtained through material search, collection and data verification, and the trend of change of maize varieties on the market in different years and regions, the repeated rate of parental inbred lines of all the accredited maize varieties and the relationship between the accreditation, protection and withdrawl from promotion of maize varieties were discussed.【Result】It is shown that during the period of 1972-2013, the total number of state and provincial accredited maize varieties was 6291, among which 503 were of state level, while the present state and provincial accredited variety number was 4 882, among which 332 were of state level. A total of 1 726 breeding organizations have records of accredited maize varieties, and scientific research institutions and seeds enterprises accounted for 48.3% and 46.6%, respectively, with an average of 7 and 3 varieties for each organization. A total of 8 754 different parental inbred lines have been used in hybridization match of all accredited maize varieties, 19.4% of which were used repeatedly. The three most frequently used inbred lines were Mo17, Chang7-2 and Dan340. A record count of 1 876 of all accredited maize varieties in 42 years have been withdrawn from promotion and the average employment period of such withdrawn varieties was 13.3a. Applications for new variety property have been filed for 29.2% of all accredited maize varieties in history and approved for only 15.1% of them.【Conclusion】During the recent 10 years, the number of accredited maize varieties at all levels increased steadily year by year, so was the maize variety inventory. In terms of trend, the northern area has more accredited varieties than the southern area and the number of accredited varieties in the eastern area increased faster than that in the western area. Of all breeding organizations, scientific research institutions acted as the main entity in terms of the quantity of accredited maize varieties and the number of accredited maize varieties of large-scale seeds enterprises increased remarkably. The number of newly-named parental inbred lines has increased rapidly, so the repeating rate of them has decreased year by year. The yearly number of accredited varieties withdrawn from promotion tends to be steady and fewer varieties are overlapped in both accreditation and property applications.

Key words: Maize; Variety accreditation; Breeding organization; Parental combination; Varieties out of promotion; New variety property

引言

【研究意义】品种是决定农作物产量和品质的内在因素,品种创新能力是一个国家农业核心竞争力的主要标志之一。中国地域辽阔,生态环境和农业生产条件复杂多样,因地制宜地推广优良品种是确保粮食增产的基本保障措施,也是国家对品种审定制度立法的初衷。品种审定是指国家或省级农业行政部门的农作物品种审定委员会根据选育单位和个人的申请,对其提交的新育成或引进的品种进行相应级别的区域试验和生产试验,按审定标准审查参试品种的试验结果,决定其能否通过审定并确定适宜种植区域的行政管理措施。

对主要农作物的强制审定制度符合中国当前的国情。品种审定信息主要是指品种审定委员会发布的品种审定公告信息，内容包括审定品种名录、引种和认定品种名录、停止推广品种目录等。对品种审定信息进行分析和整理，可以为科研单位进行品种选育、品种试验和品种检测提供数据支持，为种子企业进行品种研发和市场布局提供信息服务，为农业种植户进行品种选购、种子维权和种植管理提供信息指导，同时也是种子管理部门进行品种管理、市场监管和制定产业发展规划的信息基础。国家统计局2012年度统计数据显示，中国玉米的播种面积为 3.5×10^7 hm^2，占粮食作物的31.5%，比水稻高出4.4%；玉米的总产量为 2.0×10^8 t，占粮食作物的34.9%，比水稻高出0.2%。玉米的两大生产指数均居粮食作物第一位，而其播种面积实际上从2007年至今一直稳居粮食作物之首，可见玉米已经成为中国第一大粮食作物。全国农业技术推广服务中心2013年度统计资料显示，全国各省的玉米种子商品化率接近100%。2012年玉米种子市场价值为253.87亿元，占六大农作物的41.5%，比水稻高出10.6%，使得玉米成为中国主要农作物种子市场中占有率最高的粮食作物，也是中国主要种子企业生产经营的主要作物品种。因此，玉米品种审定信息具有一定的种业市场代表性和产业研究价值。研究玉米品种审定信息的年份演变特征和地域分布规律，是深入了解当前中国玉米种业市场情况最直接和有效的方法。【前人研究进展】根据《种子法》的规定，中国实行国家和省级两级审定，从1972年以后就有关于玉米品种审定的详细资料可查，已经有一些研究者对中国玉米品种审定的实践进行不同角度和不同范围的分析和评价，但多年来，系统地汇总、分析和利用这些资料的研究不多。胡宁霞对比"十五"期间江苏省品种审定和品种保护的作物构成和同期比重，认为品种保护工作滞后的原因是惯性思维、法律意识薄弱、受益不显著等。李健英等分析了山西省2000—2007年玉米品种审定概况、品种特点和种植推广情况，认为品种数量不少但推广面积大的品种却不多。冯勇等分析了内蒙古审（认）定玉米品种的地区分布、单位类型、育种方法和应用状况，总结出该区育种条件提升对策应该是加强高产、优质、多抗玉米新品种选育和综合技术集成研究。李登海统计了"十五"期间国审玉米品种的区域分布、育种单位类型和品种类型，从发展种业公司、以高产为育种导向等角度探讨了玉米种业未来。王凤格等整理了全国玉米主要审定品种5年推广面积数据，分析了不同省份、品种的推广情况和趋势。连灵燕等利用VB和SQL Server技术完成了国家玉米审定品种数据库的结构建设、功能设计和系统实现。杨国航等介绍了中国农作物审定品种退出机制的实施现状和必要性。李晓飞、杨忠萍从性质、目的、受理对象、法律依据、具备条件、流程、期限等方面比较了品种保护、审定、认定的异同。刘海燕、张雪清等从申请数量、品种权人、区域分布、自交系利用等角度分析了中国玉米品种保护的现状及存在问题。【本研究切入点】目前，对玉米品种基本信息的统计分析研究已经涉及到品种保护、品种推广等多领域，研究内容主要是针对单个省份品种审定情况进行统计，研究重点集中在分析品种审定信息与品种保护等环节的关系，而针对全国范围内多年玉米品种审定信息进行全面梳理统计并深入分析其时空演变规律的研究未见报道。【拟解决的关键问题】目前，品种审定是玉米品种合法进入销售市场的主要途径之一，停止推广公告是玉米退出市场销售的唯一法律依据。本研究以上述两类信息为基础数据，研究了玉米审定品种的年份和地域分布规律，统计了玉米审定品种的选育单位和亲本来源情况，并分析了玉米品种审定与停止推广、新品种权保护之间的联系，旨在通过对玉米品种审定信息进行全面而深入的分析，为促进种子市场监管、规范依法经营行为、加快优良品种推广、品种标准样品征集、审定品种标准 DNA 指纹库构建等提供信息支持。

1　材料与方法

1.1　数据来源

收集整理的玉米审定品种信息分为三个部分：①通过审定的品种名录，数据主要来源于全国农业技术推广服务中心编撰的书籍资料和权威网站发布的国家、各省历年主要农作物审定品种目录公告。②停止推广的品种名录，数据出自于各级政府网站发布的退出种子市场品种名录公告。③植物新品种权申请、授权的品种名录，数据源自农业部植物新品种保护办公室网站发布的申请、授权公告。以上三类信息均为 2013 年 12 月 31 日前公布或公开发表的数据，其具体内容所覆盖的时间跨度为 1972—2013 年。品种审定信息按照级别划分为国家级品种审定和省级品种审定，后者涵盖了全国 28 个省、直辖市和自治区的审定情况。品种审定信息的内容主要包括品种名称、审定编号、审定年份、审定省份、选育单位、杂交组合等；品种类型包括普通玉米和特用玉米（甜、糯、爆裂、青贮等），以杂交种为主。

数据来源说明：本文从网络上搜集的数据主要来源于以下网站：①中华人民共和国农业部种子管理局：http：//www.zzj.moa.gov.cn/。②农业部植物新品种保护办公室：http：//www.cnpvp.cn/。③中国种业信息网：http：//www.seedchina.com.cn/。④玉米新品种信息网：http：//www.newcorn.com.cn/。⑤农作物品种查询系统：http：//202.127.45.197/。⑥种业商务网：http：//www.chinaseed114.com/seed/sspz/。⑦第 1 种业：http：//www.a-seed.cn/index.asp。⑧中华人民共和国国家统计局：http：//data.stats.gov.cn/index。⑨各省种业信息网、农业厅局网站。

1.2　数据预处理

首先对品种审定信息进行规范化和标准化的预处理，具体内容如下：①参考 2012 年 3 月 14 日颁布的《农业植物品种命名规定》，按照一定的品种命名规则，对品种名称不统一的原始数据进行订正。②补充或修改其他缺失或有误的关键信息内容。③修正各表数据关联存在的逻辑错误，保证不同品种名录之间的逻辑关系符合以下条件：审定名录包含了退出市场的所有品种；品种权申请名录包含了授权名录的所有品种。由于云南省本身已有省级审定的品种名录，其历年的品种审定还包括特定区域的审定（如"滇特（昆明、玉溪）审玉米 2012001 号"），在下文针对国家级和省级品种审定的统计中不再将该特定区域的数据列入。本研究以预处理后形成的 3 张品种信息表为基础，分析了当前中国玉米种业市场品种的总体推广情况、不同年份和不同省份审定品种的差异。

1.3　统计指标

品次数：在指定级别的品种审定目录中，统计实际列明的记录数，不区分一个品种多省审定的情况。

品种数：在指定级别的品种审定目录中，统计品种名称不重复的记录数，即剔除因一个品种多省审定而重复出现的记录。

育种效率：在指定级别的品种审定目录中，某一类育种单位审定品种数的平均值，计算公式如下：

$$E = \frac{1}{n}\sum_{i=1}^{n} V_i \qquad (1)$$

式中，n 为某类育种单位的总数量，V_i 为第 i 个育种单位获得审定的品种数，E 为该类育种单位的育种效率。用该指标代表某类育种单位实际育成并审定品种能力的平均水平。

亲本重复使用率：对指定范围内不同杂交种的亲本自交系的利用情况进行统计，判定重复使用相同自交系作为亲本的杂交种占所有杂交种的最大比例，计算公式如下：

$$R_p = \frac{P_t - P_a}{0.5 P_t} \qquad (2)$$

式中，P_t 为理论亲本数目，表示参与统计的杂交种全由不同亲本自交系杂交育成时的亲本数，P_a 为实际亲本数目，表示参与统计的杂交种存在相同亲本自交系杂交育成时的不重名亲本数量，R_p 为亲本重复使用率，代表了假设存在重复使用亲本的杂交种都仅含有一个重复自交系的情况下，含有重复亲本的杂交种所占的最大比例。

退出率：在指定级别的品种审定目录中，停止推广的品种占全部品种历史记录的百分比，其计算公式如下：

$$R_d = \frac{V_s}{V_h} = 1 - \frac{V_e}{V_h} = 1 - R_e \qquad (3)$$

式中，V_s 为指定审定级别的停止推广品种的品次数，V_h 为指定审定级别的所有历史品种的品次数，V_e 为指定审定级别的现存品种的品次数，R_d 为指定审定级别的品种退出率，R_e 为指定审定级别的品种现存率。

实际使用年限：审定品种在市场上合法存在的限定时间长度，其计算公式如下：

$$T_u = Y_o - Y_i \qquad (4)$$

式中，Y_o 为审定品种的退出年份，Y_i 为审定品种的审定年份，T_u 代表了该品种的实际使用年限。

2 结果

2.1 玉米审定品种的时空分布

截至 2013 年，1972 年以来省级以上审定的总品次数为 7 468 个。除去停止推广的品种，现存省级以上审定的总品次数为 5 590 个。有 774 个品种存在一品多省审定的情况，除去重复审定的品种，省级以上审定的总品种数为 6 291 个，现存品种中有 489 个出现一品多省审定的情况，除去重复审定的品种，现存省级以上审定的总品种数为 4 882 个。国家和省级审定品种的历年变化情况如图 1、图 2 所示，省级审定品种的空间差异情况如图 3 所示。

经统计，国家级玉米品种审定已有 24 年的数据，从 1990 年到 2013 年共审定了 513 个品次，实际品种数为 503 个。省级玉米品种审定从 1972 年起有系统纪录，至 2013 年共审定了 6 955 个品次，实际品种数为 6 065 个。2004—2013 年，通过国家审定的玉米品种数量逐年下降，主要是因为国家从 2007 年开始实施从严准入、"精品制" 战略，使其年均减少 4 个品次，后 5 年平均通过数量为 20 个品次，通过数量最多的年份是 2006 年，达到 68 个品次。1999—2006 年的 8 年间，通过省级审定的玉米品种数量快速上升，年均增加 64 个品次；而 2007—2013 年通过数量趋于平稳，年均通过数量为 498 个品次，通过数量最多的年份是 2006 年，达到 564 个品次。历史上通过审定次数最多的前 3 个品种是掖单 13（15 次）、丹玉 13（13 次）和掖单 4 号（14 次）。

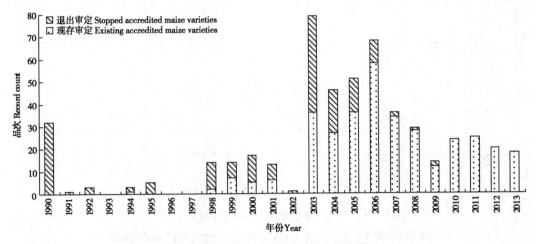

图 1　国家审定玉米总品次的年际变化

Figure 1　Inner-annual variations of total record count of national accredited maize varieties

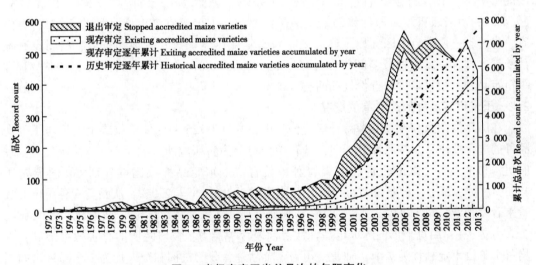

图 2　省级审定玉米总品次的年际变化

Figure 2　Inner-annual variations of total record count of provincial accredited maize varieties

对当前市场上合法存在的玉米审定品种统计得出，国审品种为 339 个品次，实际品种数为 332 个，通过审定时间最早的为 1994 年，2003 年以前（含）通过审定的品种仅存 56 个品次，其余的为近 10 年通过审定的品种，占现存国审品种的 83.4%；现存的省级审定品种为 5 253 个品次，实际品种数为 4 754 个，通过审定时间最早的为 1975 年，2000 年以前（含）通过审定的品种仅存 311 个品次，其余的为近 13 年通过审定的品种，占现存省审品种的 94.1%。现存品种中通过审定次数最多的是郑单 958（9 次）、燕禾金 2000（9 次）和先玉 335（8 次）（图 3）。

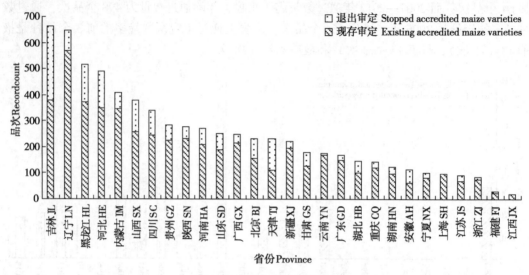

图 3 各省审定的玉米总品次
Figure 3 Total record count of each province accredited maize varieties

计算各省历年累计审定品种的情况得出,在历年的审定玉米品种中,累计通过数量最多的省份是吉林,占全部省级审定品种的 9.5%,其次为辽宁省（9.3%）、黑龙江省（7.4%）;有省级审定品种且数量最少的是江西省,占全部省级审定品种的 0.3%;海南省、青海省、西藏自治区（以下称西藏）没有审定玉米品种,台湾、香港、澳门没有统计数据。现存审定品种最多的省份是辽宁省,占全部现存省级审定品种的 10.9%,有省级审定品种且数量最少的省份是江西省（0.3%）。

2.2 玉米审定品种的选育单位情况

在历年的国家和省级玉米品种审定公告中,共有 1 726 个育种单位或个人参与了审定品种的选育或申请。按照性质不同,选育单位可分为科研院所（含大学）、种业公司、政府机构和个人四大类（图 4）,四类育种单位育成的审定品种数比重分别为:科研院所 48.3%,排名第一,种业公司 46.6%,政府机构 2.2%,个人 2.9%。从四类育种单位的数量来看,种业公司 1 000 家,占 57.9%;科研院所 472 家,占 27.3%;政府机构占 7.0%,个人占 7.7%;用每个单位的审定品种数目代表育种能力:科研院所的育种效率最高,平均每个单位审定品种为 7 个,种业公司为 3 个品种,政府机构和个人均为 1 个品种。如图 5 所示,根据育种单位所处的行政区域进行划分,审定品种数最多的前 3 个省份或直辖市分别为辽宁省、吉林省和北京,其审定品种数占比依次为 11.8%、8.6% 和 7.9%;对各省育种单位的数量进行统计,辽宁省、四川省、吉林省和河北省位列前四,其单位数占比分别为 7.9%、7.2%、7.1% 和 7.1%;对各省育种单位的实际育种水平进行比较可以看出,育种效率最高的省份或直辖市分别为北京、辽宁省和天津,每个育种单位平均育成 6 个品种。此外,外国种业公司有 17 家,审定品种共有 64 个品次。

2.3 玉米审定品种的亲本来源统计

玉米审定品种的亲本自交系数目的历年变化如图 6 所示。两级审定的玉米品种目录共登记了 13 780 个亲本自交系,实际涉及的不重名亲本数量为 8 754 个,其中有 19.4% 的自

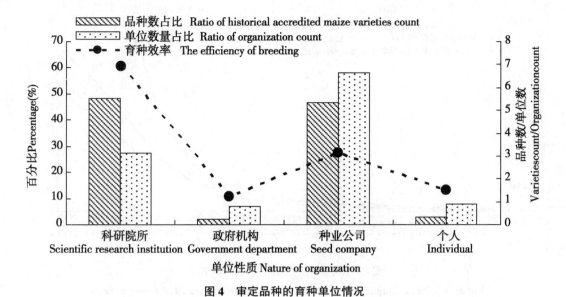

图 4 审定品种的育种单位情况
Figure 4 Breeding organization situation of accredited maize varieties

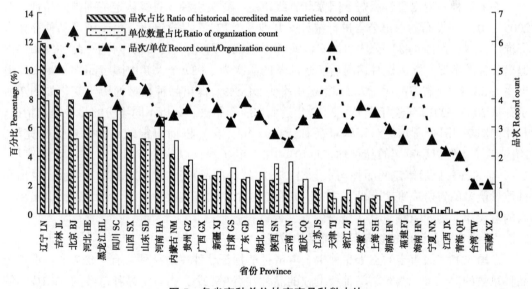

图 5 各省育种单位的审定品种数占比
Figure 5 Ratio of accredited maize varieties record count of each province breeding organization

交系被不同审定杂交种重复使用，使用频率最高的前十个亲本自交系为 Mo17、昌 7-2、丹 340、自 330、吉 853、黄早 4、掖 478、丹 598、K12 和 8112，其使用次数依次为 100、80、77、75、67、61、58、43、39 和 36 次。1972—1999 年亲本自交系数量的增幅较为平稳，年平均增加 41 个；2000—2006 年亲本自交系数量以每年 100 个左右的速度迅速增加，2006 年达到历史最高，为 804 个；2006—2013 年亲本自交系的年平均增加量逐渐下降到 524 个。

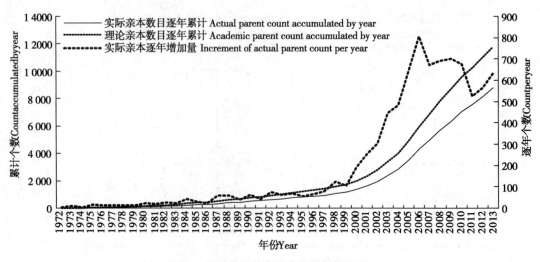

图 6 审定品种的亲本数目的年际变化

Figure 6 Inner-annual variations of parent count of accredited maize varieties

2.4 玉米审定品种的退出（停止推广）情况

玉米品种的退出公告涵盖了国家和省级审定、省级引种、省级认定范围内的品种。对 2005—2013 年发布的退出公告进行统计，共涉及 2 175 个品次退出。由于国审品种存在多个推广区域，而停止推广公告可能只涉及部分区域，因此，会出现针对同一品种发布多次退出公告的情况。除去这种情况后实际涉及的品次为 2 037 个，其中国审品次 174 个，退出率为 33.9%；省审品次 1 702 个，退出率为 24.5%。退出品种数量的历年变化如图 7 所示，从 2005—2012 年逐年上升，2012 年达到 583 个品次。分析不同年份的品种退出情况，具体如图 7、图 8 所示，对于国审品种，1997 年（含）以前审定的品种基本退出完毕，2010 年（含）以后审定的品种均未退出。1983 年（含）以前的省审品种基本清退完毕，2011 年（含）以后审定的品种均未退出。不同省份的品种退出情况如图 3 所示，退出品种数量最多的省份或直辖市是吉林省，共有 298 个品次，其次是天津、河北省，退出率最高的是天津，为 53.0%，其次是安徽省、吉林省。

2.5 玉米审定品种的植物新品种权保护情况

在国家和省级审定的玉米品种中，申请过植物新品种权的品种共有 1 837 个，占审定品种总数的 29.2%；获得新品种权授权的品种共有 946 个，占审定品种总数的 15.1%。统计审定品种的审定年份和新品种权保护的申请年份发现：有 336 个品种的申请年份早于审定年份，740 个品种的申请与审定同年，761 个品种的申请年份晚于审定年份。统计审定年份与授权年份的先后：有 16 个品种的授权年份早于审定年份、33 个品种的授权与审定同年、897 个品种的授权年份晚于审定年份。

图 8 显示了不同审定年份的审定品种中新品种权申请与授权的逐年变化。申请新品种权保护的审定品种中审定年份最早的是 1990 年，到 2003 年，审定品种的新品种保护权申请比例从 3% 一直增长到 48%；审定品种获得新品种权保护授权的比例从 3% 一直增长到 43%；从 2003 年起，申请和授权比例逐年下降，2013 年申请比例下降到 22%，授权比例下降到 1%。

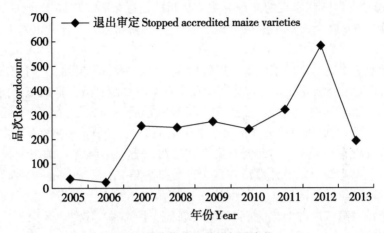

图7 停止推广品种的年际变化

Figure 7　Inner-annual variations of record count of stopped accredited maize varieties

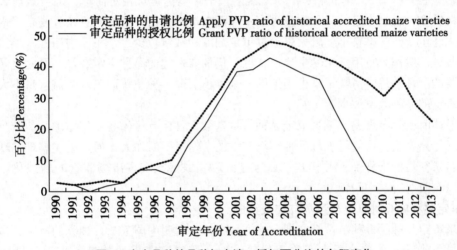

图8　审定品种的品种权申请、授权百分比的年际变化

Figure 8　Inner-annual variations of accredited maize varieties related to plant variety property（PVP）

3　讨论

3.1　审定品种的年份与地域分布

从年份上来看，1999 年前，历年两级审定的玉米品种累计总品次呈直线缓慢上升趋势（图 8 虚线所示），1980—1999 年平均每年增加 56 个品次；2002—2013 年呈直线快速上升趋势，平均每年增加 487 个品次。假设未来十年内历史累计总品次的发展趋势保持不变，用 OLS 法构建年份与历史累计总品次的一元线性回归模型，预测的累计总品次在 2018 年将突破 10 000 个。图 8 中的实线展示了近 42 年来现存玉米品种两级审定的累计总品次数，该曲线的变化趋势与前者基本一致，但仅说明了现存品种数量随审定年份变化的分布情况。前 30 年（1972—2001 年）审定的品种仅占现存总品次的 8%，说明现存两级

审定品种以 2002—2013 年审定的品种为主。从目前情况来看，市场上合法存在的玉米审定品种数量接近 5 000 个，每年两级审定玉米品种的新增数量大于停止推广品种的减少数量。

参照传统意义上的玉米种植区划，以省份为不可拆分的基本单元，将中国划分为以下 6 个区域：东北地区（黑龙江省、吉林省、辽宁省、内蒙古自治区）的玉米播种面积约占全国的 38.6%（2012 年统计数据），历年来省级审定品种的数量占全国省级审定品种的 1/3，省级审定品种的更新换代速度位居全部区域的第二；同样是中国玉米主产区的华北及黄淮海地区（北京、天津、河北省、山东省、河南省、山西省）的玉米播种面积约占全国的 31.9%，其省级审定品种数量在六个区域排名第二，占全国省级审定品种的 1/4，品种更新换代速度位居全国第一；西南地区（四川省、重庆、贵州省、云南省）的省级审定品种数量排名第三，占 1/7；西北地区（新疆、甘肃省、宁夏、陕西省）的省级审定品种数量排名第四，占 1/9；中南地区（湖北省、湖南省、广东省、广西）的省级审定品种数量排名第五，华东地区（安徽省、江苏省、上海、浙江省、福建省、江西省）排在最后，两地区共占 1/6。华北和黄淮海地区的种植环境优越且相对简单，适合该地区的育种新材料较多，通过审定的品种数量最多；而西南地区种植环境苛刻且条件复杂，培育适合多样化环境条件的材料难度较大，通过审定的品种数量较少；华东、中南地区虽然不是玉米主栽区，但作为特用玉米的主要消费区，仍拥有一定的品种审定数量；西北地区作为玉米主要的制种基地，消耗了不少当地的玉米种植资源，种植普通玉米的需求量不大，所以通过审定的品种数量最少。

由上述时空分布可见，玉米审定品种数量基数大且保有量仍在逐年增加，地域分布不均且情况复杂多变，这使得玉米品种市场监管的工作量不断加大，通过上述整理的现存审定品种名录征集标准样品，并构建玉米审定品种指纹库，实现市场监管的精细化、信息化，可以有效解决这一问题。

3.2 审定品种的育种单位与亲本自交系组成

玉米审定品种的育种单位组成情况能够从大体上说明中国玉米育种界的概况：虽然从单位数量上来看，种业公司占有绝对优势，但从获得审定品种数量上来看，科研院所（含大学）依然是育种的主力，政府机构已基本退出品种选育领域，而个人选育的审定品种仅零星存在。目前科研院所的平均育种效率大约是种业公司的 2.5 倍，但不可否认极少数种业公司的育种水平也较高。从育种单位注册登记省份与其省级审定品种所在省份的关系来看，两者相同的比例为 72.6%，说明各省审定品种基本以当地的育种单位培育的品种为主。

审定品种登记的品种来源一项记录了亲本材料的选育过程，可反映审定杂交种对亲本自交系的使用情况。实际亲本数量从 2000 年开始迅速上升，2000—2013 年的 14 年中实际亲本数量占过去 40 年（1972—2013 年）实际亲本总数量的 89.5%。但对比 2000 年前后的亲本重复使用率，1972—1999 年为 94.0%，2000—2013 年下降为 71.0%，可见，不同审定品种的亲本自交系的交叉使用程度在下降。假如不同名的亲本材料为有真实差异的自交系，可能说明中国的自交系资源日渐丰富，但亲本自交系使用背后的乱象是不争的事实。另外，审定信息的品种来源中自交系名称命名的不规范，也使得了解其系谱来源，确定亲本信息的真实性难度很大。因此，规范亲本自交系的名称命名并建立管理体系是有意义的。

3.3 审定品种的退出与保护情况

农作物审定品种退出机制的实施使玉米审定品种可分为三类：审定品种、退出品种和现存品种。品种退出机制从2005年执行至今，两级玉米审定品种的整体退出率为25.1%，1999年以前（含）审定的玉米品种中有79.8%被退出，低于主要农作物的平均退出率90%。退出品种的审定时间已经推进到2010年，说明审定品种的退出范围不局限于审定年份。从退出品种的实际使用年限的分布来看，使用年限最长为36年，最短为1年，整体平均值为13.3年，实际使用年限为10年的品种数量最多，7~11年为主，长于农作物一般品种的最佳使用年限5~6年，说明总体上玉米审定品种退出机制的成效是显著的，但执行力度有待进一步的加强。

1999年4月中国新品种权保护工作启动，1990—1998年通过审定的品种大部分不符合《中华人民共和国植物新品种保护条例》的新颖性条件，只有极少数申请并获得新品种权保护。审定品种和新品种权保护申请品种的交集在1999—2003年迅速上升，2003—2013年则缓慢下降至1999年的水平。审定品种与新品种权授权品种的交集在1998—2008年期间呈现正态分布，交集最大的年份是2003年，以后呈逐年下降，2009年以后审定的品种还只有极少数获得新品种权保护授权（可能与审查期延长有关）。审定品种与新品种权保护品种的交集逐年下降可能有2个原因：首先，审定品种申请保护亲本的增加，直接保护杂交种的减少，因为保护亲本得到的权益更大；其次，保护杂交种虽然在维权方面更直接，但往往由于品种在区试时就被育种单位频繁交易，达不到品种权的新颖性要求，导致申请获得通过的难度加大。随着品种审定和品种保护的重复品种数量的下降，齐头并进地构建审定品种DNA指纹数据库与新品种权保护的品种DNA指纹数据库的必要性日趋显现。在品种审定与品种保护重叠的这部分品种中，得到授权的品种数占申请品种数的51.5%，高于新品种权保护授权率的总体水平。从品种审定与品种保护的先后顺序来看，在申请新品种权保护的审定品种中，只有1/5的品种申请早于审定，2/5的品种申请与审定同步，2/5的品种申请晚于审定（平均晚1.8年）；在得到新品种权保护授权的审定品种中，94.8%的品种授权晚于审定，平均晚3.9年。由此可见，在历史上的品种管理工作中，品种保护工作的执行时间总体落后于品种审定工作。为了防止"一品多名"现象扰乱种子市场，2013年新修订的《主要农作物品种审定办法》规定，品种审定将新增DUS测试并与区域试验同步进行，这将促使未来的品种保护工作和品种审定工作趋于同步。

4 结论

在过去的42年（1972—2013年）中，全国农作物品种推广工作共审定了6 291个玉米品种，除去停止推广的品种，仍有4 882个玉米审定品种合法流通于种子市场。玉米审定品种的时间演变规律为：近十年间，各级审定玉米品种数量的年均增幅趋于平稳，全国玉米审定品种保有量仍在逐年增加。玉米审定品种的空间分布规律为：玉米品种的审定数量北多南少、增加速度东高西低，与玉米种植面积的空间分布相吻合。从玉米审定品种的数量看，目前科研院所仍是育种单位的主体，但个别规模较大的种业公司实力也较强。从审定品种的来源看，近年来亲本自交系按照名称统计的数量增加迅速，但重复使用率在逐年下降。

品种审定的退出机制执行 9 年以来，共清退了 25.1% 的历史审定品种记录，其特征有两点：一是停止推广对象以 1999 年（含）以前的审定品种为主；二是列入停止推广的审定品种数量近 7 年趋于稳定。

在新品种权保护工作开展初期的 1999—2003 年，品种审定与品种保护的重叠程度在迅速上升，2004 年以后（含）二者的重叠程度逐渐下降。从品种审定与品种保护的先后顺序来看，大多数育种单位都采用先审定再保护方式，不利于在审定品种推广初期进行维权。随着新的品种审定办法开始实施，这两项工作将逐渐趋于同步。

参考文献（略）

本文原载：中国农业科学，2014，47（22）：4 360-4 370

农作物品种 DNA 指纹库构建研究进展

王凤格 赵久然* 田红丽 杨 扬 易红梅

(北京市农林科学院玉米研究中心/玉米DNA指纹及分子
育种北京市重点实验室，北京 100097)

摘 要：构建品种标准DNA指纹库应满足三个基本要求，即标准样品、核心引物和标准
程序。本文综述了近20年来国内外农作物品种DNA指纹库构建进展情况，结果显示国外品
种标准DNA指纹库构建主要是在UPOV框架下开展植物品种权保护品种的建库，而中国品种
标准DNA指纹库构建主要是在农业部品种管理部门的统筹下开展品种权保护品种及区试审定
品种的建库。所用建库技术主要为SSR技术，少数作物如玉米、水稻、大豆等已开始采用
SNP技术，玉米、水稻、小麦等作物品种建库数量均已达到3000份以上，其中玉米建库数量
已经超过2万份。进一步分析发现，不同作物分子标记开发情况、作物的物种特性、不同国
家对不同作物建库需求的迫切程度、所建DNA指纹库的级别是影响不同作物DNA指纹库构
建进展的四个主要因素。最后本文对农作物品种未来DNA指纹库构建工作的开展进行展望。

关键词：农作物；品种；DNA指纹库；SSR；SNP

The Progress of the Crop Varieties DNA Fingerprint Database Construction

Wang Fengge Zhao Jiuran* Tian Hongli Yang Yang Yi Hongmei

(*Maize Research Center, Beijing Academy of Agriculture & Forestry Sciences, Beijing
Key Laboratory of Maize DNA Fingerprinting and Molecular Breeding, Beijing* 100097)

Abstract: The three essential requirements of constructing standard crop varieties DNA fingerprinting databases (the Construction) include standard samples, core primers, and standard procedure. This article reviews the progress of the Construction in China and abroad over the past 20 years. The findings indicate that, in foreign countries, the Construction is dominated by database construction for the varieties protected by the Plant Variety Rights under the UPOV framework; while in China, the Construction mainly comprises of database construction both for the varieties protected by the Plant Variety Rights and for those examined by regional trial under the arrangement and supervision by related authorities at Ministry of Agriculture. The SSR technology is applied in most of the Construction, and the SNP technology has been adopted by a few crops such as maize, rice, and soybean. Databases constructed for maize, rice, and soybean have exceeded 3 000 in number, and maize

---

基金项目：本研究由"十二五"国家支撑计划项目（2011BAD35B09）和北京市科技计划项目
（D131100000213002）共同资助

* 通讯作者：jiuran@126.com

databases amount to over 20 000. Further analysis suggests that there are four principal factors influencing the progress of the Construction for different crops: the development of molecular markers for different crops, the characteristics of different species, the urgency degree of database construction in different countries, and the level of established DNA fingerprinting databases. Last but not least, this article presents an outlook on the future crop DNA fingerprinting database construction.

Key words: Crops; Variety; DNA fingerprint database; SSR; SNP

构建农作物品种标准 DNA 指纹库,确定每个已知品种的分子身份证,能够解决品种鉴定中"什么是真"的问题。通过提供已知品种标准 DNA 指纹,一方面可以避免对标准样品的反复取用和反复检测,另一方面可以解决没有数据库时不易解决的问题,即如何确定待测样品是什么或与什么品种最相似等。品种标准 DNA 指纹库的主要用途在于真实性鉴定,主要包括有:在品种区域试验中,检测参加区试品种不同年份或不同区组的送检样品是否发生组合更换,以及在数据库中筛查与已知品种是否雷同;在种子质量监督抽查、维权打假及侵权案司法鉴定中,鉴定样品是否为假种子或套牌侵权种子,并在数据库中筛查出套牌侵权对象;在品种权保护中,辅助筛查申请品种的最近似品种,代替原来由申请者自行提供近似品种。

DNA 指纹库构建技术发展迅速,从 RFLP(Restriction Fragment Length Polymorphism,限制性片段长度多态性)、RAPD(Random Amplification Polymorphic DNA,随机扩增多态性)、AFLP(Amplified Fragment Length Polymorphism,扩增片断长度多态性)、ISSR(inter-simple sequence repeat,简单重复序列间多态性)到 SSR(Simple Sequence Repeat,简单重复序列多态性)、SNP(single nucleotide polymorphism,单核苷酸多态性)等,新技术层出不穷。RAPD、AFLP 技术曾被应用到建库上,但因技术的稳定性、复杂性及数据整合等问题未广泛应用。目前最成熟的建库技术还是 SSR 技术,在主要农作物如玉米、水稻、小麦、棉花、大豆、油菜等都作为重点建库技术进行研究应用;新型建库技术-SNP 技术的研究正在进行中,目前超高通量的 SNP 检测平台已经出现,玉米、水稻等作物的 SNP 标准研制及数据库构建工作也正在进行中。本文概括了构建农作物品种标准 DNA 指纹库的基本要求,汇总近 20 年来国内外主要农作物品种 DNA 指纹库构建进展情况,分析影响不同农作物 DNA 指纹库构建进展的主要因素,并介绍中国玉米品种 DNA 指纹库构建历程和经验,最后对主要农作物未来建库工作开展进行了展望。

1 构建作物品种标准 DNA 指纹库的基本要求

构建作物品种标准 DNA 指纹库需具备三个基本条件,一是需要有来源可靠的建库标准样品;二是需要有一套固定的核心引物作为该作物的建库引物;三是需要有统一的标准化建库程序。

1.1 标准样品

由于品种标准 DNA 指纹库构建的主要目的是为品种管理服务,因此,建库品种主要包括区试审定品种和品种权保护品种。其中审定品种采用省级以上农业行政主管部门指定机构收集保存的经认定代表审定品种特征特性的样品;品种权保护品种采用申请单位提交的经品种保护部门 DUS 测试合格的样品;区域试验品种采用国家和省级区试管理部门保留的经测试不存在组合更换和仿冒雷同等问题的样品。

从全球范围看，标准样品的主要来源是品种权保护样品，特别是加入国际植物新品种保护联盟（简称UPOV）组织的成员国，都进行了品种权保护品种的长期入库保存。中国的标准样品来源则包括了品种权保护样品和区试审定样品，其中植物品种权保护工作自1999年开始启动来，至2014年年底已有11 800多份申请品种权保护品种进行入库保藏和测试（http://www.cnpvp.cn），其中仅玉米和水稻就占60%以上（分别为3 800多份和3 500多份）。中国在区试中保留标准样品的工作自2002年在国家级区域试验中开始，在各个省级区域试验中更晚。为解决早期审定品种没有标准样品的问题，中国农业部2010年启动玉米、水稻、小麦、棉花、大豆、油菜等主要农作物审定品种标准样品征集工作（农技种［2010］43号），要求凡已审定通过的品种，品种选育单位需提供规定数量的标准样品进行封存，同时启动了"国家主要农作物区试标准样品库建设"项目，对主要农作物区试样品进行长期保存留样。

1.2 核心引物

核心引物是指多态性、稳定性、重复性等综合特性好，可作为DNA指纹鉴定优先选用的一套引物，核心引物组合法是由赵久然等（2003）、王凤格（2003）等于2003年率先提出的替代特征谱带法和引物组合法的建库方案。不同实验室通过使用固定的一套核心引物组合，其指纹图谱可以相互比较和整合。在核心引物组合法的建库思想指导下，中国主要农作物的DNA指纹库构建工作进入了全新的阶段，玉米率先筛选出40个SSR核心引物应用于建库（Wang et al, 2007），随后水稻、小麦、棉花、大豆、油菜、大白菜等十几个作物都陆续开展了核心引物筛选工作，核心引物的确定为这些作物的大规模建库工作开展奠定了技术基础。

国际上已开展的针对建库引物联合筛选评估工作，主要有欧盟针对玉米、小麦、番茄等作物SSR引物筛选评估，CIMMYT针对玉米、小麦种质资源鉴定SSR引物的筛选评估等（George et al, 2004），均涉及到多个国家不同实验室检测平台，为开展国际间联合数据库构建奠定基础。

1.3 标准程序

BMT分子测试指南是UPOV组织为了在分子检测过程中获取高质量的分子数据，满足不同实验室利用不同技术方法构建植物品种DNA指纹数据库的需要而制订的技术方案。该指南对分子标记的选择、分析材料的来源类型及样品量、分析方法的标准化、数据库构建的标准化进行了原则上的规范，对各国开展作物品种DNA指纹库构建具有重要指导意义。

中国自2005年开始参加BMT历届会议，并参与BMT分子测试指南修订。借鉴国际标准的经验，2007年首先在玉米上形成了农业行业标准，基于玉米的成功经验，并综合考虑不同作物的特殊性，2010年启动了《植物品种鉴定DNA指纹方法总则》的编制，起到了规范我国各作物品种DNA指纹标准研制的作用（王凤格等，2014）。基于DNA指纹标准制定总体布局安排，在第一批作物（玉米、水稻）的SSR指纹鉴定标准2007年颁布后，第二批14个作物（小麦、大豆、油菜、高粱、大麦、棉花、番茄、大白菜、西瓜、黄瓜、辣椒、甘蓝、百合、苹果）的SSR指纹鉴定标准于2013—2014陆续颁布实施。此外，玉米、水稻、大豆的SNP指纹鉴定标准也已列入研制计划。这些技术标准的颁布实施，极大地推动了各作物品种 DNA 指纹库构建工作开展。

2 国内外作物品种建库概况

国内外已构建品种 DNA 指纹库的作物中，粮食作物主要有玉米、水稻、小麦、马铃薯，经济作物中纤维类主要是棉花、麻类，油料类主要有大豆、油菜、橄榄、芝麻，蔬菜类主要有番茄、大白菜、黄瓜、辣椒等，水果类主要有葡萄、西瓜、甜瓜、柑橘、桃、梨等，花卉类主要有月季、康乃馨、百合、莲等，特用类主要有烟草、茶、可可等。所用建库技术主要为 SSR 技术，少数作物如玉米、水稻、大豆等已开始采用 SNP 技术，此外 IN-DEL、ISSR、SCAR、AFLP、RAPD 等标记也在少量使用，三大作物玉米、水稻、小麦的建库品种数量均已超过 3 000 份（表 1）。

国外品种标准 DNA 指纹库构建工作主要是在 UPOV 框架下开展植物品种权保护品种的建库，如德国建立 480 份欧洲小麦品种 SSR 指纹（R O Der et al, 2002），法国建立 1 537 份玉米品种 SSR 指纹（Van Inghelandt et al, 2010），荷兰建立 734 份月季、521 份番茄品种 SSR 指纹（Bredemeijer et al, 2002），英国建立 892 份马铃薯品种 SSR 指纹（Reid et al, 2011），意大利建立 1 005 份葡萄品种 SSR 指纹等（Cipriani et al, 2010）。中国品种标准 DNA 指纹库构建主要是在农业部品种管理部门的统筹下开展品种权保护品种及区试审定品种的建库，其中玉米、水稻率先开展建库工作，至今已完成 22 089 份玉米和 3 829 份水稻区试、审定及品种权保护标准样品 DNA 指纹库构建。随着植物品种鉴定 DNA 指纹方法总则和各作物 DNA 指纹标准的颁布，小麦、棉花、油菜、大豆、高粱、番茄、西瓜、大白菜、黄瓜、辣椒等作物的建库工作正在陆续开展（匡猛等，2012；赖运平，2014；郑永胜等，2014）。

如果不局限于品种标准 DNA 指纹库构建，国内外在种质资源 DNA 指纹库构建方面也有大量的研究，如 CIMMYT 构建了玉米、小麦种质资源 DNA 指纹库（Prasanna, 2012），美国构建了玉米、棉花等种质资源 DNA 指纹库（Romay et al, 2013；Hinze et al, 2015），中国构建了玉米、水稻、小麦、大豆、大麦、高粱、花生、烟草、茶等种质资源 DNA 指纹库等（张赤红等，2008；黄建安等，2009；李丽等，2009；姜慧芳等，2011；吕婧，2011；徐军，2011；刘志斋等，2012）。此外，企业为育种需要也进行了自育品种 DNA 指纹库构建，如美国杜邦-先锋公司、中国北京金色农华公司等均建立自育品种 DNA 指纹库。

表 1 作物 DNA 指纹库构建情况汇总
Table 1 Summary of the DNA fingerprint database construction in crops

国家或单位 State or units	作物种类 Crop species	品种数量 number of varieties	标记及数量 Marker and number	样品来源 Sample source
法国 France	玉米 Maize	1 537	SSR, 359; SNP, 8 244	自交系 Inbred lines
美国 U.S.A	玉米 Maize	2 815	SNP, 681 257	种质资源 Germplasm resources

(续表)

国家或单位 State or units	作物种类 Crop species	品种数量 number of varieties	标记及数量 Marker and number	样品来源 Sample source
北京市农林科学院 Beijing Academy of Agriculture and Forestry	玉米 Maize	22 089	SSR，40；SNP，384*	区试审定及品种权保护 Varieties in regional trials、Authorized varieties、varieties of PBR
中国农业科学院 Chinese Academy of Agricultural Sciences	玉米 Maize	820	SSR，40	种质资源 Germplasm resources
中国水稻研究所 China Rice Research Institute	水稻 Rice	3 829	SSR，48	审定和品种权保护 Authorized varieties、varieties of PBR
德国 Germany	小麦 Wheat	502	SSR，20	品种及种质资源 Cultivars、Germplasm resources
中国农业科学院、CIMMYT China Academy of Agricultural Sciences、CIMMYT	小麦 Wheat	5 000	SSR，78	种质资源 Germplasm resources
山东农业科学院 Shandong Academy of Agricultural Sciences	小麦 Wheat	1 625	SSR，42	品种权保护 varieties of PBR
北京市农林科学院 Beijing Academy of Agriculture and Forestry	小麦 Wheat	3 000	SSR，21	区试品种 Varieties in regional trials
英国 United Kingdom	马铃薯 Potato	892	SSR，9	登记品种 Registered variety
中国农业科学院 Chinese Academy of Agricultural Sciences	马铃薯 Potato	217	SSR，11	育成品种 Cultivars
中国农业科学院 Chinese Academy of Agricultural Sciences	大麦 Barley	2 625	SSR，35	种质资源 Germplasm resources
山东农业科学院 Shandong Academy of Agricultural Sciences	高粱 Sorghum	253	SSR，32	品种及种质资源 Cultivars、Germplasm resources
美国 U.S.A	棉花 Cotton	2 256	SSR，105	种质资源 Germplasm resources
中国农业科学院 Chinese Academy of Agricultural Sciences	棉花 Cotton	138	SSR，36	育成品种 Cultivars

(续表)

国家或单位 State or units	作物种类 Crop species	品种数量 number of varieties	标记及数量 Marker and number	样品来源 Sample source
中国农业科学院 Chinese Academy of Agricultural Sciences	大豆 Soybean	248	SSR, 100	种质资源 Germplasm resources
中国农业科学院 Chinese Academy of Agricultural Sciences	油菜 Rapeseed	163	SSR, 40	区试品种 Varieties in regional trials
中国农业科学院 Chinese Academy of Agricultural Sciences	花生 Peanut	212	SSR, 25	种质资源 Germplasm resources
南京农业大学 Nanjing Agricultural University	芝麻 sesame	192	SRAP, 31; SSR, 25	种质资源 Germplasm resources
荷兰 Holland	番茄 Tomato	521	20	品种权保护 varieties of PBR
北京市农林科学院 Beijing Academy of Agriculture and Forestry	大白菜 Chinese cabbage	686	SSR, 13	种质资源 Germplasm resources
中国农业科学院 Chinese Academy of Agricultural Sciences	黄瓜 Cucumber	3 318	SSR, 23	种质资源 Germplasm resources
中国农业科学院 Chinese Academy of Agricultural Sciences	辣椒 Chili	374	SSR, 28	种质资源 Germplasm resources
中国农业科学院 Chinese Academy of Agricultural Sciences	小豆 Adzuki bean	375	SSR, 13	种质资源 Germplasm resources
中国农业科学院 Chinese Academy of Agricultural Sciences	菜豆 Kidney beans	377	SSR, 36	种质资源 Germplasm resources
意大利 Italy	葡萄 Grapes	1 005	SSR, 34	登记品种 Registered variety
吉林农业大学 Jilin Agricultural University	葡萄 Grapes	360	SSR, 18	种质资源 Germplasm resources
中国农业科学院 Chinese Academy of Agricultural Sciences	桃 Peach	237	SSR, 16	种质资源 Germplasm resources
东北农业大学 Northeast Agricultural University	甜瓜 Melon	105	SSR, 18	育成品种 Cultivars

(续表)

国家或单位 State or units	作物种类 Crop species	品种数量 number of varieties	标记及数量 Marker and number	样品来源 Sample source
北京市农林科学院 Beijing Academy of agriculture and Forestry	西瓜 Watermelon	788	SSR，23	育成品种 Cultivars
西北农林科技大学 Northwest Agriculture and Forestry University	枣 Jujube	237	SSR，19	种质资源 Germplasm resources
中国农业科学院 Chinese Academy of Agricultural Sciences	柑橘 Citrus	70	SSR，12；ISSR，2	育成品种 Cultivars
荷兰 Holland	月季 Rose	734	SSR，11	品种权保护 varieties of PBR
荷兰 Holland	康乃馨 Carnation	82	SSR，11	品种权保护 varieties of PBR
中国农业科学院 Chinese Academy of Agricultural Sciences	百合 Lily	96	SSR，20	品种及种质资源 Cultivars、Germplasm resources
浙江大学 Zhejiang University	杨梅 Red bayberry	123	SSR，14	育成品种 Cultivars
福建农林大学 Fujian Agriculture And Forestry University	甘蔗 Sugar cane	181	SSR，15	种质资源 Germplasm resources
中国农业科学院 Chinese Academy of Agricultural Sciences	烟草 tobacco	71	SSR，36	审定品种 Authorized varieties
中国农业科学院 Chinese Academy of Agricultural Sciences	茶 tea	59	SSR，64	品种及种质资源 Cultivars、Germplasm resources

注：*SNP 正在应用于 4 000 份玉米审定品种及 1 000 份品种权保护自交系的建库

Note：*SNP is being applied to the database construction of the 4 000 authorized maize varieties and 1 000 maize inbred lines of PBR

3 影响不同作物 DNA 指纹库构建进展的因素

3.1 不同作物的分子标记开发情况

BMT 测试指南中推荐 SSR 和 SNP 作为作物品种 DNA 指纹库构建的标记方法。然而，不同作物在 SSR、SNP 标记开发和核心引物的筛选评估上的研发基础相差很大。玉米、水稻、小麦、大豆等作物全基因组测序和分子标记的开发较早，为其核心引物筛选及建库奠定了坚实基础（Sharopova et al, 2002; Song et al, 2005）；而一些较小的作物因基因组测序信息有限，标记开发难度较大，提供的可用 SSR 标记较少，早期只能使用 AFLP、ISSR、RAPD 等物种通用型标记建库（Laurentin and Karlovsky，2007）。随着测序技术发展和测序成本的大幅下降，越来越多的作物开展了全基因组测序及重测序，标记开发更加简单容易，目前主要农作物基本上都有可用的 SSR 标记，并逐步将通用型标记替代，部分作物还开发了大量 SNP 标记（Unterseer et al，2014）。

3.2 不同作物的物种特性

影响不同作物建库难度的内在特性主要有两点。第一是物种的倍性，玉米、水稻、大豆等二倍体作物 SSR 和 SNP 指纹比较简单，容易统计；而棉花、油菜等异源四倍体作物和小麦等异源六倍体作物的指纹会出现多种类型，统计比较复杂且不易标准化，而如果要求其筛选具有二倍体特征的标记，则会大大降低可用标记的数量，标记开发难度较大（Li et al，2013）。第二是繁殖类型，不同繁殖类型的作物品种预期样品一致性高低不同，而样品一致性越低，建库难度和复杂性越大（Bredemeijer et al，2003），其中无性繁殖、严格自花授粉、人工自交、单交种品种因样品预期一致性较高，建库时可采用混合样品，而常异花授粉、天然异花授粉、除单交种外的其他人工杂交品种因样品预期一致性较低，建库时需采用至少 20 个以上个体样品。这也是玉米、水稻容易开展大规模建库（二倍体、人工自交、自花授粉及单交种品种，样品预期一致性高），而棉花等作物开展大规模建库难度较大（异源四倍体、常异花授粉品种，样品预期一致性较低）的内在原因（匡猛，2012）。

3.3 不同国家对不同作物建库需求的迫切程度

不同作物在不同国家的分布情况及价值意义不同，因此不同国家建库需求也各有侧重：德国、法国、荷兰等欧洲国家是在 UPOV 框架下开展建库，其中法国侧重玉米、小麦等大田作物，荷兰侧重番茄、月季、康乃馨等蔬菜花卉作物；美国的商业化作物品种由企业自行建库，种质资源由国家统一建库；CIMMYT 主要开展玉米、小麦种质资源管理及评价。中国情况比较复杂，一方面审定和保护品种数量庞大，对玉米、水稻、小麦等粮食作物和棉花、油菜、大豆等经济作物的品种管理需求非常迫切；另一方面，作为种质资源大国，中国作物类型多样且数量巨大，种质资源管理的任务也很艰巨，除了六大主要农作物（玉米、水稻、小麦、大豆、棉花、油菜）外，大麦、高粱、谷子、花生、芝麻、大白菜、西瓜、黄瓜、辣椒、茶、甘蔗、枣等作物种质资源都很丰富，因此都需要开展种质资源建库。

3.4 所建 DNA 指纹库的级别

建成的作物品种标准 DNA 指纹库是仅仅作为辅助筛查的工具，还是作为品种的标准指纹在品种鉴定中代替标准样品使用，这决定了指纹库构建的级别和使用价值。其中第一个目的对指纹库的质量要求不高，但只能起到缩小问题品种范围的作用；第二个目的对建库数据准确性和完整性的要求很高，实现难度较大，如果能实现，则可以不用反复调取标准样品及反复试验，使指纹库的价值得到充分体现。早期构建的作物品种 DNA 指纹库由于建库引物筛选标准不高，建库程序标准化程度较低，且大部分是基于 PAGE 凝胶电泳检测平台建立的 SSR 指纹库，造成不同实验室数据可比性较差，数据库级别仅仅是作为辅助筛查的工具。随着标准化程序的建立，特别是荧光毛细管电泳检测平台的广泛使用，新构建的玉米、水稻等 SSR 指纹库数据库在数据准确性上有了大幅度提高，有可能实现代替标准样品使用的目标。

4 未来发展建议

随着 DNA 指纹技术的发展，重要农作物的建库研究开始转向 SNP 标记，但 SSR 标记在品种鉴定应用层面仍然发挥重要作用，今后应继续发挥 SSR 技术的作用，同时积极推进 SNP 技术，并探索其他适合建库的标记技术，具体建议如下：①总结两大重要农作物玉米、水稻 SSR 指纹库构建的成熟经验，迅速向其他农作物推广，使各作物的建库效率更加科学高效，充分发挥 SSR 指纹库对现阶段品种管理的技术支撑作用。②以玉米、水稻为突破点积极推进高通量 SNP 技术在建库中的应用，注意借鉴 SSR 建库的经验，如核心位点组合法的建库思路、核心位点选择的指标等。③不断探索其他有价值的建库标记，如 INDEL 标记具有二态性和长度多态的特点，在建库上兼具 SSR 和 SNP 的优势，值得关注（Bhattramakki et al, 2002；孙宽等, 2013）。④持续关注测序技术的进展及在品种 DNA 指纹库构建中应用的可能性，如基于二代测序技术进行基因分型的 RAD-seq、Ion AmpliSeq 等技术，一次测序可以实现几百到上万位点的检测，且能够提供更详细的分型信息，有一定的应用潜力（王洋坤等, 2014；Ellison et al, 2015）。

参考文献（略）

本文原载：分子植物育种，2015，13（9）：2 118-2 126

DNA 指纹技术在玉米区域试验品种真实性及一致性检测中的应用

王凤格[1]　易红梅[1]　赵久然[1]*　孙世贤[2]　杨国航[1]　任　洁[1]　王　璐[1]

(1. 北京市农林科学院玉米研究中心/玉米 DNA 指纹及分子育种北京市重点实验室，北京　100097；2. 全国农业技术推广服务中心，北京　100125)

摘　要：DNA 指纹技术在玉米区域试验品种管理中具有重要应用价值。我国自 2002 年起开始开展玉米区域试验品种真实性和一致性检测工作，截至 2014 年年底，已累计检测国家及各省玉米区域试验品种 15 596 份。十多年来玉米区域试验品种 DNA 指纹检测规模不断扩大，检测技术不断提升，真实性和一致性问题得到有效控制。本文回顾了我国玉米区域试验品种 DNA 指纹检测的发展历程，总结了玉米区域试验品种 DNA 指纹检测的发展规律，分析了区域试验、审定、品种权保护及种子质量管理之间的协调和衔接情况，最后对玉米区域试验品种 DNA 指纹检测中面临的新品种的判定标准、区域试验品种的命名及标准样品的留取等问题进行了深入探讨。

关键词：玉米；DNA 指纹；品种区试；品种管理

The Application of DNA Fingerprint Technology in Maize Varieties' Authenticity and Consistency Identification in Maize Regional Test

Wang Fengge[1]　Yi Hongmei[1]　Zhao Jiuran[1]*
Sun Shixian[2]　Yang Guohang[1]　Ren Jie[1]　Wang Lu[1]

(1. *Maize Research Center, Beijing Academy of Agriculture & Forestry Sciences, Beijing Key Laboratory of Maize DNA Fingerprinting and Molecular Breeding, Shuguang Garden Middle Road No. 9, Beijing 100097; 2. The National Service Center of Agricultural Technique Popularization, Agricultural Ministry, Beijing 100125*)

Abstract: DNA fingerprint technology has important application value in variety management of regional test. Maize varieties' authenticity and consistency identification with DNA fingerprint technology in national regional test were organized and implemented since 2002. By the end of 2014, 15 596 maize varieties in the national and provincial regional test have been detected, maize varieties detected in regional test continues to expand the scale, the detection technology continues to improve, the problem in varieties' authenticity and consistency gets effective control. This paper reviewed the 10

基金项目：本研究由"十二五"国家支撑计划项目 (2015BAD02B02) 资助
* 通讯作者：maizezhao@126.com

years' development process and summed up the regular pattern of DNA identification in maize regional trial, then analyzed the coordination and convergence among the regional test, validation, variety right protection and seed quality management, finally discussed in detail the problems including the criteria of new varieties, the varieties nomenclature in regional test and the extraction of standard samples.

Key words：Maize；DNA fingerprint；Regional test；Variety management

DNA 指纹技术在玉米区域试验品种管理中具有重要应用价值。自1972年以来，中国通过国家或省级审定的玉米品种总品次达到7 468个（杨扬等，2014），主要依据国家和各省种子管理部门组织的多年多点区域试验的筛选结果进行审定。每年国家和省级区域试验玉米品种数量约2 000个，玉米品种获得审定一般需要两年区域试验和一年生产试验，从参试品种的送检样品提供到试验结果汇总只有几个月的时间。为了保证区域试验全过程的真实性和品种筛选结果的有效性，需要采用快捷高效的 DNA 指纹技术进行试验品种真实性和一致性检测，其结果作为试验品种淘汰、继续试验或推荐审定的依据之一。

中国自2002年起陆续启动了玉米（孙世贤等，2009）、水稻（程本义等，2007）、小麦（王立新等，2010）、棉花（孙宁等，2011）、油菜（许鲲等，2014）、大豆（关荣霞等，2012）等主要农作物区域试验品种真实性和一致性 DNA 指纹检测。在所有作物中，玉米区试品种 DNA 指纹检测工作开展的最早，10多年来，玉米区试检测规模不断扩大，技术不断提升，品种真实性和一致性问题得到有效控制，截至2014年年底，已累计检测国家及各省区试玉米组合15 398份，并构建了玉米区试品种 DNA 指纹数据库（赵久然等，2015），为玉米品种区域试验管理提供了有力的技术支撑。

1 DNA 指纹检测在玉米区域试验中的应用历程

2002年起农业部率先在国家玉米区试京津唐、黄淮海、西南和武陵山区四个组中开展 DNA 指纹检测。2003年起逐步将检测范围扩大到国家玉米区试所有普通玉米组，并由区试扩展到预试。2005年在玉米区试检测发展历程中是最重要的一年，一是将区试检测范围从国家区试逐步扩大到省区试；二是制定并试行 DNA 指纹检测标准及管理办法；三是所有玉米区试样品开始入库长期储存；四是 SSR 检测技术从 PAGE 电泳向荧光毛细管电泳升级。2006年起将国家区试检测范围进一步扩大到鲜食、青贮、爆裂等特用玉米组，实现了对国家玉米区试品种检测的全覆盖。2007年起，检测范围由国家区试逐步扩大各省区试，至2014年基本实现了玉米区试品种 DNA 指纹检测的全覆盖。

2 DNA 指纹检测在玉米区域试验中的发展规律

2.1 检测规模不断扩大

玉米区试检测品种数量呈明显增加趋势：自2002年起，每年检测数量从不足100份上升到约2 000份，截至2014年已累计检测全国玉米区试品种15 398份。2005年以来，国家玉米区试品种数量基本稳定在400~500份，检测数量增加主要依赖于检测省份数量的扩大，截至2014年开展检测的省份已从辽宁一个省上升到20多个省。

玉米区试检测品种类型及范围呈逐年扩大趋势：检测品种类型从普通玉米推进到鲜

食、青贮、爆裂等特用玉米类型；检测范围从区试推进到预试和生试，并从国家级向省级扩大。随着 2014 年开始的企业绿色通道试验的开展，将进一步实现国家级、省级、企业级试验的联合检测。

2.2 检测技术不断提升

2002 年启动玉米区试品种检测工作时，检测单位采用的是 SSR 技术，但当时 SSR 技术体系尚不完善，为加快 SSR 技术的成熟化和标准化，提出了核心引物组合法的 DNA 指纹库构建方案，设计开发了适于玉米 DNA 指纹检测的 40 对 SSR 核心引物（Wang et al, 2011），建立了十重荧光毛细管电泳复合检测体系（Wang et al, 2007），研制出适合玉米品种真实性检测的复合扩增试剂盒，开发了适合植物品种 SSR 指纹分析的专用软件，研制了兼容多标记及多物种的植物品种 DNA 指纹库管理系统，建立了玉米品种标准 DNA 指纹数据比对平台（http：//www.maizedna.org）。此外，检测单位还不断探索 SNP 标记在玉米真实性检测中的应用，研制了适于玉米品种真实性鉴定和建库的 SNP 系列芯片 MaizeSNP3072、MaizeSNP384 等。

基于上述研发工作，2005 年制定了《国家玉米品种试验 DNA 指纹检测管理办法》并在玉米区试中试行，2007 年颁布并于 2014 年修订了《玉米品种鉴定技术规程 SSR 标记法》（NY/T 1432—2014），2014 年颁布了《植物品种鉴定 DNA 指纹方法总则》（NY/T 2594—2014），2015 年启动玉米 SNP 检测相关标准的研制。

2.3 真实性和一致性问题得到有效控制

从同名品种组合更换情况看，由于同名更换行为主要是选育单位的主观故意行为，因此 DNA 指纹监测的效果非常明显：①经过 2002—2005 年对国家区试不同年份不同区组的同名品种的监测，同名更换品种数量显著下降，并由早期以差异较大样品的更换为主变为相似度较高的姊妹系更换为主。②随着 2005 年国家预试 DNA 检测工作启动及国家区试库的联网，检测出一批预试与区试间同名更换的品种（李俊芳等，2006），并呈逐年下降趋势。③随着 2006 年以来各省区试 DNA 检测工作启动及全国区试库的联网，国家区试与各省区试之间的同名更换成为主体，并呈现逐年下降趋势。

从品种仿冒雷同情况看，随着玉米区试、审定及品种权保护品种 DNA 指纹库的构建和联合监测机制的逐步形成，对仿冒雷同品种的检出能力逐步增强，仿冒雷同得到逐步控制：① 2010 年以前，仿冒雷同主要来源于同一单位将同一组合采用不同名称在不同省参试，或选育单位将同一组合转让给多家单位后用不同名称参试，随着全国区试 DNA 指纹库联网，这种情况已逐步减少。②模仿优秀已知品种一直是仿冒雷同的重要来源，特别是郑单 958、先玉 335 等优良品种成为主要模仿对象，通过 DNA 指纹检测，这种情况已得到较好控制。③参试单位利用差异较小的姊妹系配制成不同的组合参试，这一类型正在成为仿冒雷同的重要来源，今后需要在鼓励育种单位持续改良和自主创新之间找到平衡点。

从一致性情况看，经过多年持续监测，对国家区试玉米品种总体一致性情况得到显著改善（张雪原等，2009），但不同品种类型略有不同：①普通玉米品种在 2005 年之前一致性不合格样品较多，经过几年全面监测，一致性得到显著改善，不合格情况几乎绝迹。②甜糯玉米品种自 2006 年启动一致性检测，当年一致性不合格品种达到 24 个，2007 年下降到 9 个，2007 年以后显著下降到 0～2 个/年。③青贮玉米品种自 2006 年开始一致性检测以来，不合格品种始终未完全杜绝，经进一步分析，与青贮玉米品种经常存在三交种等其他杂交类型有关。

3 区试、审定、品种权保护及种子质量管理的协调和衔接

3.1 国家级、省级及企业级区域试验的协调

当真实性有问题的品种发生在国家级、省级及企业级区试不同单位之间时，将带来国家级和省级种子管理部门之间或企业与政府管理部门之间的工作协调问题。从以往玉米区试DNA指纹检测结果看，国家区试内部和省区试内部的真实性问题品种检出量逐年下降，甚至基本绝迹，但涉及国家和省之间以及不同省之间的问题品种检出量呈逐年增加趋势。跨不同单位的问题品种的处理难度远高于单位内部的问题品种的处理难度，需要国家和各省品种管理部门之间共同协商形成统一的管理办法，防止出现同一品种在不同单位处理结果不同的情况。

绿色通道企业自行组织试验的制度实施后，由于涉及企业较多，进一步增加了不同单位之间协调和联合管理的复杂性。当两个不同名品种判定为相同或极近似后，如何确定两个品种在国家及不同省、不同企业区试的最早参试时间，进而确定哪份品种属于原始品种，这已成为目前面临的比较困难的事。此外，国家级区试的级别是否高于省级，省级区试的级别是否高于企业级？如果三者是相同级别，则只需要依据参试时间早晚确定哪份为原始品种，而如果国家级区试级别高于省级，或省级高于企业级，则情况将变得更加复杂。

3.2 区试与审定的衔接

区试与审定的衔接主要涉及到样品、命名及信息三个方面的衔接。

首先是样品的衔接。根据农业部下达的相关文件及通知，今后将把区试留样作为审定品种的唯一标准样品，避免征集标准样品可能面临的样品不真实的问题，但同时带来了区试样品大量保存的问题。由于参试样品需要至少保存2~3年，要保证样品能够妥善保藏且不损失芽率，应建立区试样品低温周转库并在样品入库前进行干燥处理。由于每年每省区试样品数量仅玉米就有平均100~200份，每份保存3kg，平均保存3年的话，需要低温周转库库容能常年存放900~1800kg的标准样品，这对全国区试样品库建设带来较大的挑战。

其次是命名的衔接。目前区试品种名称用的是区试代号，没有纳入命名管理范围，允许区试品种通过审定后重新提供正式审定命名。这种规定为区试品种的管理带来极大的困难：首先，由于许多省的正式审定公告中没有提供审定名称与区试代号的对应关系表，区试品种的保藏样品及DNA指纹无法与审定品种的建立关联，造成区试中已构建的DNA指纹库无法提供给审定品种使用，还需要对审定品种重新建库。其次，由于对区试代号没有明确规定，造成同一品种在多个省同时参加区试时使用不同的区试代号，或已经审定的品种再次参加区试时使用了与审定名称不同的区试代号，从而对区试品种真实性判定带来干扰。随着全国区试DNA指纹库的联网，每年均检测出大量这样的区试品种并花费大量精力与参试单位核实，最后只能统一名称而不作为问题品种处理，真正的问题品种却没有充足的处理时间。鉴于此，提出如下建议：一是在审定公告中应公布审定名称和区试代号的对应关系；二是由农业部牵头制定区试品种命名规范，避免区试品种命名的随意性，比如

规定同一品种在多省参试时区试代号应相同，参试品种如果已审定或申请品种权保护则必须用审定或保护时的名称，不得再编其他区试代号等。

最后是信息的衔接。实现区试品种的样品入库信息、DNA 指纹信息及产量试验信息能够传递到审定品种的信息库中，关键是品种管理上真正实现一体化。解决信息衔接的最佳方案是品种自参加区试之日起就正式命名，审定时没有合理理由不予改名，唯有如此，才能最大程度的降低管理成本。

3.3 区试与品种权保护的衔接

玉米区试审定与品种权保护的衔接主要涉及杂交种，具体包括三方面内容：一是试验衔接，即在区试的同时做 DUS 测试，使审定和授权基本同步；二是样品衔接，即审定和品种权保护的标准样品均为区试中的同一份留样；三是技术衔接，即 DNA 检测和 DUS 测试两种技术方法结合使用，使检测结果更具说服力。在实践中可以采取如下操作方式：区试第一年，构建参试品种 DNA 指纹库并筛查出怀疑真实性有问题的品种名单，同时将该结果通知参试单位，对无异议的，按真实性有问题进行处理，对有异议的，依据产量试验结果推荐是否进入续试；第二年，对有异议的且进入续试的品种安排田间 DUS 测试，进行并排种植比较，根据 DUS 测试结果判定是否具有形态性状差异，综合 DNA 和 DUS 测试结果对品种真实性和特异性进行最终判定，并作为是否推荐审定和授权的依据之一。

随着品种管理的精细化，玉米审定和品种权保护的衔接也将会涉及到自交系。建议今后应推进玉米自交系登记制度，将区试品种的亲本自交系作为已知自交系，采取强制登记制度。目前申请品种权保护的玉米自交系已达到 1 000 多份，如果将区试品种的亲本自交系与品种权保护自交系进行比较，就可以排查出冒用他人材料、自交系系谱说明不实等问题，保证区试工作的规范性和严肃性。

3.4 区试与种子质量管理的衔接

区试品种通过审定进入推广销售阶段后，就纳入种子质量管理的范围，种子质量管理的一项重要内容就是种子真实性。品种随着不断的种子繁殖和生产，有可能出现正常范围内的种性变化，因此有必要建立长期跟踪品种审定后种性变化的工作机制，并作为种子质量管理时确定真假种子阈值界限的参考依据，从而将种子企业故意制售假种子的行为和品种正常范围内的种性变化进行有效的区分，使种子质量管理更加切合生产实际。

4 需探讨的几个问题

4.1 新品种的判定标准问题

对区试品种的真实性鉴定，包括两方面内容：一是检测同一品种在不同区试年份或不同区试组别中遗传是否一致，即是否发生样品更换；二是检测参试品种与已有品种（已审定、已申请品种权保护或已参试在先的品种）及同时参试的品种间 DNA 指纹有无明显差异，即品种是否具有特异性。玉米区试中采用行标规定的 40 个核心引物进行真实性检测，成对比较时差异两个或以上位点判定为不同，差异 1 个位点判定为近似，无差异位点判定为相同或极近似，对差异 1 和 0 的非同名品种视同无明显差异。

在检测实践中，对这部分无明显差异的非同名品种的处理方式经历了三个阶段：早期阶段将其直接判定为问题品种，并终止试验；第二阶段允许参试单位提出异议，并对提出异议的品种进一步进行田间 DUS 测试，综合 DNA 检测和 DUS 测试结果进行最终判定；第三阶段则除了接受 DUS 测试结果外，还允许参试单位提供足够的其他差异位点进行综合判定。尽管处理方式越来越灵活，但对品种规范化管理造成一定干扰，没有彻底解决品种真实性判定的问题。出现这一情况的原因在于，现行标准是适应我国加入 UPOV 的 78 年文本的背景而制定的，标准中将品种之间的关系分为不同和无明显差异两大类，而无明显差异的类型（即判定为近似，相同或极近似的）实际上包含着同一品种和派生品种两种情况，同一品种当然可以作为仿冒雷同进行处理，而派生品种由于形态上存在真实差异，特别是当差异出现在穗轴颜色等质量性状及抗病性等重要经济性状上时，如果也作为仿冒雷同进行处理就会面临育种单位的质疑。

要想真正解决这一问题，需要引入派生品种的概念（陈红等，2009），从而可以将品种之间的关系分为不同、派生和相同三大类，判定为不同的，可正常继续试验；判定为相同的，可终止试验；判定为派生的，参试单位应提供与原品种权人达成的使用协议才可以继续试验。正在修订的新种子法中，已将派生品种的概念纳入其中，但仍有一些操作层面的细节需要确定（李菊丹和尹锋林，2013），比如，派生品种的鉴定方法和鉴定标准如何确定？派生品种是否需要审定？如需审定，是否可以简化审定程序？派生品种的命名是否要求显示出与原始品种的渊源？

4.2 区试品种的命名问题

关于区试品种的命名，有两类不同的意见：一类意见认为，由于区试过程中参试品种面临大量淘汰，只有极少数品种能够通过审定，因此只有品种审定名称才是正式名称，区试品种只有区试代号，允许审定后重新命名；而另一类意见认为，品种在首次参加国家或省级区域试验时就应该确定正式的名称，更改名称必须有充分合理的理由并提交正式的改名申请。这两个意见各有其合理性：由于未通过审定的区试品种对育种单位是没有意义的，大部分育种单位只有在品种审定后才为其正式命名，因此第一个意见比较符合当前育种单位的实际；而第二个意见更符合区试管理部门的实际，因为只有在区试时将品种名称固定下来，才容易实现区试、审定、品种权保护三库合一的统一化管理。

4.3 标准样品的留取问题

从区试整体情况看，尽管每年参加区试的玉米品种有上千个，但从预试进入区试、第一年区试进入第二年区试、第二年区试进入审定（部分省份第二年区试和生产试验分成不同年度进行）的过程中会淘汰大量品种。全国每年审定的玉米品种只有几百个，只有这些审定品种才有必要长期储存大量标准样品用于今后种子质量管理，其他大量区试品种只需短期保存少量检样用于区试过程的监测。

《农业部办公厅关于加强主要农作物品种审定标准样品工作的通知》（农办种[2014]30号）中提出了从区域试验第一年的试验种子中留取标准样品的方案，尽管减少了持续监测同名品种更换的成本，但也带来了其他问题：品种管理部门面临区试阶段大量样品保存带来的发芽率下降风险和样品周转库建设问题，参试单位面临尽管大部分参试品种第一年就会被淘汰，但仍需一次繁殖两年用种的浪费问题。DNA 指纹技术自 2002 年起应用于区试品种真实性及一致性鉴定以来，对区试品种同名更换的监测效果极为显著，参试单位很少提出异议，因此通过 DNA 指纹技术对预试、区试第一年、区试第二年及生产试验的种子进行持续检测，对同名更换的品种及时排除，在生产试验阶段再留样作为区试品种通过审定后的标准样品，这一方案是完全可行的，其整体区试管理成本也是较低的。

参考文献（略）

本文原载：分子植物育种，2016，14（2）：456-461

SSR 和 SNP 两种标记技术在玉米品种真实性鉴定中的比较分析

李 雪[1,2]*　田红丽[2]*　王凤格[2]　赵久然[2]**
李云伏[2]　王 蕊[2]　扬 扬[2]　易红梅[2]

(1. 北京农学院，北京　102206；2. 北京市农林科学院玉米研究中心，北京　100097)

摘　要：SSR 和 SNP 被推荐为玉米品种 DNA 指纹鉴定优选标记技术。本研究选用大面积推广的 11 套玉米杂交种及其亲本作为试验材料，基于 SNP 芯片分型平台和 SSR 荧光毛细管电泳平台分析比较两种标记在玉米品种真实性鉴定中的应用。基于上述两种检测平台获得 3 072 个 SNP 位点和 40 对 SSR 引物的基因型数据，分析比较各个参数。结果显示：（1）两种标记数据获得率均较高，3 072 个 SNP 位点平均数据获得率为 98.4%，40 对 SSR 引物平均数据获得率为 99.47%。（2）两种标记均具备较高品种区分能力，3 072 个 SNP 位点平均 MAF（minor allele frequency）值为 0.357，MAF 值大于 0.30 占 71%，平均 DP（discrimination power）值为 0.515，DP 大于 0.5 的占 56%；40 对 SSR 引物平均 PIC（polymorphism information content）值为 0.605，PIC 大于 0.51 有 32 对，平均 DP 值为 0.745，DP 值大于 0.71 有 29 对。（3）两种标记位点鉴定样品杂合率、亲子关系方面结果是一致的，样品杂合率均介于 0.4~0.6，平均杂合率分别为 0.525 和 0.514；SNP 数据显示符合亲子关系位点百分比介于 95.2%~99.9%，SSR 数据显示符合亲子关系位点介于 36~40 个。综上所述，SSR 和 SNP 两种标记在数据完整性、区分品种能力、位点稳定性等方面差别较小。

关键词：玉米；SSR；SNP；真实性鉴定

Comparison of SSR and SNP Markers in Maize Varieties Genuineness Identification

Li Xue[1,2]*　Tian Hongli[2]*　Wang Fengge[2]　Zhao Jiuran[2]**
Li Yunfu[2]　Wang Rui[2]　Yang Yang[2]　Yi Hongmei[2]

(1. Beijing University of Agriculture, Beijing 102206; 2. Maize Research Center, Beijing Academy of Agriculture and Forestry Sciences, Beijing 100097)

Abstract: SSR and SNP are recommended to varieties of maize DNA fingerprint selection markers. This article selects the widespread promotion of 11 sets of maize hybrids and their parents as re-

---

基金项目：本研究由北京市农林科学院院青年基金项目（QNJJ201207）和北京市优秀人才培养项目（2012D002020000004）共同资助
* 第一共同作者
** 通讯作者：maizezhao@126.com

search material, based on SNP chips parting platform and SSR fluorescence capillary electrophoresis analysis to compare two tags in the application of corn varieties authenticity identification. Based on the above two kinds of test platform for 3 072 SNPS loci and 40 genotype data of SSR primers, all parameters and comparative analysis. The result shows that: (1) Two tag data acquisition rate is higher, 3 072 SNPS loci data obtained at a rate of 98.4%, about 40 SSR primers data obtained at a rate of 99.47% on average. (2) Both markers are varieties of high distinguish ability, the average 3 072 SNPS loci MAF (minor allele frequency) value of 0.357, MAF value is greater than 0.30 (71%), average DP (discrimination power) value of 0.515, DP is greater than 0.5 (56%); Average PIC about 40 SSR primers (polymorphism information content) value of 0.605, has 32 of PIC is greater than 0.51, the average DP value of 0.745, 29 of the DP value is greater than 0.71. (3) Hybrid two kinds of marker loci to identify sample rate, the result is consistent with the parent-child relationship, sample hybrid rate between 0.4 to 0.6, the average hybrid rate was 0.525 and 0.525, respectively; SNP data showed that conform to the parent-child relationship loci percentage between 95.2% to 99.9%, SSR data showed that conform to the parent-child relationship loci between 36~40. Above all SSR and SNP marker in data integrity, the ability to distinguish between varieties, site stability etc., and the difference is smaller.

Key words: Maize; SSR; SNP; Genuineness identification

玉米是中国重要农作物之一,是具有粮、经、果、饲、能等多元用途的作物。就种植面积而言目前已经是第一大作物,品种数量逐年上升,随着种业的发展,品种真实性鉴定需求也急剧增加。传统田间种植鉴定由于存在受种植周期长、易受环境影响等因素而限制其应用。DNA 指纹鉴定技术具有不受环境影响、快速、准确的优点,其中 SSR(simple sequence repeats)和 SNP(single nucleotide polymorphisms)由于均为共显性、反映的是具体片段或序列信息的标记而被 UPOV-BMT 分子测试指南推荐作为品种真实性鉴定和数据库构建的优选标记(UPOV, 2010, www.upov.int/edocs/mdocs/upov/en/bmt/10/bmt_guidelines_proj_6.pdf)。与 SSR 相比,SNP 被视为是在品种真实性鉴定和数据库构建方面最具发展潜力的标记技术,主要具有四个方面的优势:一是在基因组中密度更高分布更均匀;二是更易实现数据整合比较;三是更高通量;四是可能与功能基因相关。

SSR 指纹技术主要采用变性聚丙烯酰胺凝胶和荧光毛细管两种电泳检测平台,由于荧光毛细管电泳检测平台具有自动化程度高、检测通量高、省时省力、数据读取精确等优势,因而被认为是主流检测平台。SNP 基因分型检测平台较多,如 SNPlex、SNaPshot、TaqMan、ArrayTape、光纤微珠芯片平台等。综合比较,各个检测平台中 Illumina 公司推出的 SNP 芯片检测技术在玉米品种真实性鉴定方面具备一定位点通量优势。

本研究选取在中国推广面积较大的 11 套玉米杂交种及其父母本作为研究材料,基于荧光毛细管电泳平台的 SSR 基因分型方法和光纤微珠芯片 GoldenGate SNP 基因分型方法,评估 SSR 和 SNP 两种标记技术在玉米品种真实性鉴定中应用。

1 结果与分析

33 份亲子组合样品 3 072 个 SNP 位点平均数据获得率(Call rate 值)为 98.4%,每个样品数据获得率介于 95%~100%;40 对 SSR 引物平均数据获得率为 99.47%,每个样品的数据获得率介于 97.5%~100%。3 072 个 SNP 位点和 40 对 SSR 引物的数据获得率均较高，数据具有较高的完整性。

基于 11 个杂交种 3 072 个 SNP 位点数据分析每个位点 MAF（minor allele frequency）值和 DP（discrimination power）值分布情况，结果显示：平均 MAF 值为 0.357，MAF 值小于等于 0.2 的位点占 7%，0.21~0.30 的占 22%，0.31~0.40 的占 33%，0.41~0.5 的占 38%；平均 DP 值为 0.515，DP 值小于等于 0.3 的占 4%，0.31~0.40 的占 8%，0.41~0.5 的占 32%，大于 0.5 的占 56%。基于 11 个杂交种 40 对 SSR 核心引物的数据分析每个位点 PIC（polymorphism information content）值和 DP 值分布情况，结果显示：平均 PIC 值为 0.605，PIC 值小于等于 0.2 有 1 对引物，0.21~0.50 有 7 对，0.51~0.70 有 22 对，大于 0.70 的有 10 对；平均 DP 值为 0.745，DP 值介于 0.30~0.50 有 2 对引物，介于 0.51~0.70 有 9 对，介于 0.71~0.80 有 12 对，大于 0.80 的有 17 对。综合比较各参数显示，两种标记位点均具备较高的品种区分能力。表 1 所示的是 11 个杂交种基于 3 072 个 SNP 位点数据和基于 40 对 SSR 核心引物数据分析其杂合率情况，SNP 和 SSR 分析所得出的 11 个杂交种的杂合率均介于 0.400~0.600，平均杂合率分别为 0.525 和 0.514，表明两种标记位点分析的样品杂合率是一致的。表 2 所示的是利用 SNP 和 SSR 两种技术分析 11 套亲子材料，在每个位点上符合亲子遗传规律的情况。基于 SNP 数据显示，符合亲子关系位点百分比介于 95.2%~99.9%；基于 SSR 数据显示，符合亲子关系位点介于 36~40 个。

表 1 基于 SNP 和 SSR 数据分析 11 份杂交种的杂合率
Table 1 The heterozygosity of 11 maize hybrids based on SNPs and SSRs data

编号 No.	样品名称 Sample name	杂合率基于 SNP 数据 SNP hybrid rate	杂合率基于 SSR 数据 SSR hybrid rate
1	先玉 335 Xianyu335	0.487	0.475
2	郑单 958 Zhengdan958	0.533	0.500
3	农大 108 Nongda108	0.600	0.575
4	鲁单 981 Ludan981	0.585	0.525
5	浚单 20 Xundan20	0.545	0.425
6	蠡玉 6 Liyu 6 hao	0.516	0.474
7	沈单 16 Shendan16	0.523	0.600
8	东单 60 Dongdan60	0.499	0.525

(续表)

编号 No.	样品名称 Sample name	杂合率基于 SNP 数据 SNP hybrid rate	杂合率基于 SSR 数据 SSR hybrid rate
9	京玉 7 Jingyu 7 hao	0.498	0.500
10	掖单 13 Yedan13	0.530	0.550
11	京科糯 120 Jingkenuo120	0.463	0.500

表 2　基于 SNP 和 SSR 数据分析 11 套杂交种与其亲本之间系谱一致性情况
Table 2　Pedigree consistency analysis of 11 Parent/F₁ combinations based on SNPs and SSRs data

编号 No.	杂交种 Hybrids	母本 Female parents	父本 Male parents	符合亲子关系位点 百分比（SNP）（%） Percent of pedigree consistency loci（SNP）（%）	符合亲子关系 位点数目（SSR） Pedigree consistency loci（SSR）
1	先玉 335 Xianyu335	PH6WC	PH4CV	99.9	37
2	郑单 958 Zhengdan958	郑 58 Zheng58	昌 7-2 Chang7-2	96.8	40
3	农大 108 Nongda108	X178	黄 C HuangC	95.5	37
4	鲁单 981 Ludan981	齐 319 Qi319	Lx9801	95.9	38
5	浚单 20 Xundan20	浚 9058 Xun9058	浚 92-8 Xun92-8	97.8	39
6	蠡玉 6 Liyu 6 hao	连 87 Lian87	543	97.2	39
7	沈单 16 Shendan16	K12	沈 137 Shen137	93.5	40
8	东单 60 Dongdan60	A801	丹 598 Dan598	99.0	38
9	京玉 7 Jingyu 7 hao	京 24 Jing24	京 501 Jing501	96.8	39
10	掖单 13 Yedan13	478	丹 340 Dan340	95.2	40
11	京科糯 120 Jingkenuo120	京糯 6 号 Jingnuo 6 hao	白糯 6 号 Bainuo 6 hao	98.9	36

2 讨论

经过多年的研究与应用，SSR标记技术在玉米品种真实性鉴定中的应用已趋向成熟，如40对核心引物、试验流程、判定标准等均已经标准化，并且在全国各检测机构被广泛应用（王凤格和赵久然，2011，中国农业科学技术出版社，pp.140~142；王凤格等，2013，中国农业科学技术出版社，pp.7~10；王凤格等，2014，中华人民共和国农业行业标准 NY/T 1432—2014）。目前面对鉴定样品量的急剧增加，各种鉴定需求的出现，对检测技术在数据共享、位点通量等方面提出了更高的要求。与SSR相比，SNP具备两个方面的优势：一是容易实现数据间的整合共享，SNP是二等位基因，代表基因组中最小的遗传变异单元，数据统计简单、准确，不存在数据兼容性问题；二是更易实现位点高通量检测，如本文选用的光纤微珠芯片技术平台。

根据本研究结果显示，SSR和SNP两种标记在数据完整性、准确性、鉴定品种间差异性等方面差别较小。但是两种标记技术在数据整合共享、位点通量、DNA样品制备、试验流程和数据分析上具有一定差异。数据整合共享和位点通量上正好体现SNP标记技术的优势。DNA制备方面：由于40对SSR引物需要逐个进行扩增，SNP只需进行一次扩增，因此SSR标记DNA试验用量明显高于SNP的用量。在试验流程方面：对于96个样品而言，40对SSR引物通常一天即可完成扩增、电泳试验，若引物数目增加，时间相应增长；SNP芯片检测平台试验周期为3d，不需考虑位点数目增加的问题。在数据分析方面：SSR标记需逐个位点、逐个样品即逐个数据点独立分析；而SNP标记能实现每个位点所有样品数据叠加在一起分析，易于同时比较不同样品间数据点特征，数据分析效率明显高于SSR标记。

随着DNA指纹技术的发展，SSR指纹技术已经在农作物品种鉴定领域成熟并广泛应用，SNP指纹技术的应用研发也逐渐成熟，并表现出了一定优势。任何DNA指纹技术都不是万能，两种标记技术各自具有优缺点，并且优势互补，应继续发挥SSR技术检测灵活性、成本低、易推广的优势，同时积极推进SNP检测技术，根据各自优势，在不同检测项目发挥作用。

3 材料与方法

3.1 供试材料

本研究采用在中国推广种植面积较广的11套玉米杂交种及其父母本为研究对象，材料信息见表3。

表3 11套杂交种及亲本材料
Table 3 Eleven sets of hybrids and their parents used in the study

编号 No.	杂交种 Hybrids	母本 Female parents	父本 Male parents
1	先玉335 Xianyu335	PH6WC	PH4CV
2	郑单958 Zhengdan958	郑58 Zheng58	昌7-2 Chang7-2

(续表)

编号 No.	杂交种 Hybrids	母本 Female parents	父本 Male parents
3	农大108 Nongda108	X178	黄C HuangC
4	鲁单981 Ludan981	齐319 Qi319	Lx9801
5	浚单20 Xundan20	浚9058 Xun9058	浚92-8 Xun92-8
6	蠡玉6号 Liyu 6 hao	连87 Lian87	543
7	沈单16 Shendan16	K12	沈137 Shen137
8	东单60 Dongdan60	A801	丹598 Dan598
9	京玉7号 Jingyu 7 hao	京24 Jing24	京501 Jing501
10	掖单13 Yedan13	478	丹340 Dan340
11	京科糯120 Jingkenuo120	京糯6号 Jingnuo 6 hao	白糯6号 Bainuo 6 hao

3.2 基因组 DNA 提取

利用常规 CTAB 提取法提取基因组 DNA，并去除 RNA，具体试验操作步骤按照王凤格等实施（王凤格和赵久然，2011，中国农业科学技术出版社，pp.140~142）。DNA 质量和浓度用 NanoDrop 2000（Thermo Scientific）紫外分光光度计进行测定，根据测量值调节工作液浓度。

3.3 SSR 标记试验

PCR 扩增体系：模板 DNA $3\mu l$，$0.25\mu mol/L$ 引物，$1×PCR$ Buffer，$0.15\mu mol/L$ dNTP，$2.5\mu mol/L$ $MgCl_2$，1 单位 *Taq* 酶，反应总体积为 $20\mu l$。每对引物中的一条 5′端用荧光染料标记，选用 PET、NED、VIC、FAM 四种荧光染料（Applied biosystems，USA 公司合成）。PCR 反应程序：94℃预变性 5min，1 个循环；94℃变性 40s，60℃退火 35s，72℃延伸 45s，共 35 个循环；72℃延伸 10min；4℃保存。选用的 40 对引物是由北京市农林科学院玉米研究中心报道的第Ⅰ组的 20 对基本核心引物和第Ⅱ组的 20 对扩展核心引物，引物名称、序列等具体信息参考 Wang 等（2011）已发表的文献。

PCR 产物在荧光毛细管电泳系统 AB3730XL DNA 分析仪（Applied Biosystems，USA）上进行采用 10 重 PCR 产物电泳检测的方法，即按照扩增片段大小和荧光颜色差异将 10 对引物的扩增产物混合在一起电泳（Wang et al，2011）。在 96 孔电泳板的单个孔中分别加入 $1.5\mu l$ 10 重 PCR 产物的混合物，$8.5\mu l$ 甲酰胺，$0.1\mu l$ 内标（GeneScanTM-500 LIZ，Applied Biosystems，USA）。上述混合样品在 PCR 仪上运行 95℃变性 5min，将变性后的电泳产物取出，1 000r/min 离心 1min 后，于 AB 3730XL DNA 分析仪上进行电泳。预电泳时间为 2min，15kV，电泳时间为 20min，15kV。利用 Date Collection v1.0 软件收集原始数据，利用 GeneMarker 软件分析原始数据，统计每个样品在 40 对 SSR 标记位点上的基因型。

3.4 SNP 标记试验

利用北京市农林科学院玉米研究中研发的 maizeSNP3072 芯片，包含 3 072 个 SNP 位点（王凤格等，2012），对 33 份亲子组合样品，按照 Illumina 公司提供的试验操作指南（http：//supportres.illumina.com/documents/myillumina/2bbe5885－34be－4e2e－b655－4848b7d3e3d7/goldengate_ genotyping_ assay_ guide_ 15004065_ b.pdf）进行基因分型试验。试验流程分为 3d，第一天试验操作主要包括基因组 DNA 激活，探针与目的 DNA 链杂交、杂交产物延伸、连接，PCR 扩增；第二天主要包括 PCR 产物纯化、产物与芯片杂交；第三天主要包括芯片清洗、风干，利用 iScan SNP 芯片扫描仪（Illumina，USA）分析杂交信号。将原始数据文件导入 GenomeStudio software ver.2010 软件，统计分析每个样品在 3 072 个 SNP 位点上的基因型。

3.5 数据分析

基于 11 套杂交种及其亲本材料的 3 072 个 SNP 位点和 40 个 SSR 位点数据，分析数据的 Call rate 值（数据获得率）、样品杂合率、符合亲子关系位点百分比，以及 SNP 和 SSR 位点的多态性和区分效率，3 072 个 SNP 位点主要分析其 MAF 和 DP 值，40 个 SSR 位点分析其 PIC 和 DP 值。

参考文献（略）

本文原载：分子植物育种，2014，12（5）：1 000-1 004

中国玉米审定品种标准 SSR 指纹库的构建

王凤格[1]　杨扬[1]　易红梅[1]　赵久然[1]　任洁[1]　王璐[1]
葛建镕[1]　江彬[2]　张宪晨[2]　田红丽[1]　侯振华[1]

(1. 北京市农林科学院玉米研究中心/玉米 DNA 指纹及分子育种北京市重点实验室，北京　100097；2. 北京华生恒业科技有限公司，北京　100083)

摘　要：【目的】对数量庞大的已知玉米品种构建可共享的作物品种标准 DNA 指纹库。【方法】基于荧光毛细管电泳检测平台和植物品种 DNA 指纹库管理系统，利用筛选的 40 对 SSR 核心引物对 3 998 份中国玉米审定品种标准样品进行建库，通过多实验室、多检测平台进行建库数据的质量评估。【结果】绘制了 40 个玉米建库引物的等位基因频率分布图作为每个引物的特征图谱，在建库试验中发挥了相当于参照样品的作用。形成了一套十重荧光毛细管电泳组合，并在 SSR 指纹分析器中建立了一套系统默认 PANEL 作为不同实验室建库时的标准 PANEL。统计玉米审定品种指纹库构建的试验情况，每份样品具有 2~5 套原始试验数据及对应的指纹图谱，其中 61% 的样品做了 2 组独立试验，33% 的样品做了 3 组独立试验，最终建成的标准指纹库累计缺失和差异位点仅占 0.2%，数据完整性达到 99.8%。在不同实验室、不同电泳检测平台的评估结果表明，同一荧光电泳检测平台上获得 SSR 指纹数据一致性高，而不同的电泳检测平台获得的数据存在一定偏差，为实现不同实验室的 SSR 指纹数据共享，需要统一荧光引物、分析软件及电泳检测平台。对所有审定品种指纹数据进行整体两两比较，表明中国玉米审定品种之间差异比较大，品种间差异位点百分比集中在 80%~95%（占 78.28%），品种间差异位点百分比在 50% 以上的已达到 99.21%，而低于 20% 的只有 0.09%；对玉米审定品种杂合率水平进行分析，平均品种杂合率达到 64%，主要集中在 50%~80%（占 89%）。通过玉米品种标准指纹比对服务平台（网址：http://www.maizedna.org/）实现了指纹库的共享。【结论】形成了构建作物品种 SSR 指纹库的标准化程序，构建了 3 998 份玉米审定品种的 SSR 指纹库，通过多实验室联合比较试验，保证建库数据的准确性和数据库的可共享性；建立了玉米品种标准指纹比对服务平台网站，实现玉米审定品种指纹库在全国种子检验系统的共享。

关键词：玉米；品种鉴定；DNA 指纹库；SSR

基金项目：国家"十三五"科技支撑计划（2015BAD02B02）、北京市农林科学院院科技创新能力建设专项（KJCX20161501）

通讯作者：赵久然，E-mail：maizezhao@126.com

Construction of an SSR-Based Standard Fingerprint Database for Corn Variety Authorized in China

Wang Fengge[1]　Yang Yang[1]　Yi Hongmei[1]　Zhao Jiuran[1]
Ren Jie[1]　Wang Lu[1]　Ge Jianrong[1]　Jiang Bin[2]　Zhang Xianchen[2]
Tian Hongli[1]　Hou Zhenhua[1]

(1. *Maize Research Center, Beijing Academy of Agriculture & Forestry Sciences, Beijing Key Laboratory of Maize DNA Fingerprinting and Molecular Breeding, Beijing* 100097; 2. *Beijing TodaySoft Incorporation, Beijing* 100083)

Abstract:【Objective】It is of great importance to construct a shareable high-quality crop variety standard DNA fingerprint database for effectively managing the huge number of known varieties.【Method】Based on fluorescence capillary electrophoresis detection platforms and the plant variety DNA fingerprint database management system, a database containing 3 998 maize authorized accessions was built with 40 SSR primer pairs. Multi-laboratories and multi-detecting platforms were used to conduct the quality evaluation of the database.【Result】Allele frequency distribution graphs of the 40 corn primers were plotted as characteristic spectrums of each primer, which played the role of the similarity of reference samples in the database construction. A decuplet fluorescent capillary electrophoresis combination was formed and a set of system default PANEL was established in the SSR analyzer. Statistics were conducted on the experimental conditions of the database construction. Of the total samples, 61% of them were subjected to two group independent trials and 33% of them were subjected to three group independent trials. Each sample had 2~5 sets of original experimental data and the corresponding fingerprint maps. The accumulated loss and variable sites of the final built standard fingerprint database accounted for only 0.2%, the data integrity reached 99.8%. The assessment results in different laboratories and different electrophoresis platforms showed that the SSR fingerprint data obtained high agreement on the same electrophoresis fluorescence detection platform, but showed a certain bias in different electrophoretic detection platforms. In order to realize the sharing of SSR fingerprint data in different laboratories, a unified fluorescent primer, analysis software and electrophoresis detection platform were needed. Overall pairwise comparisons were conducted on all the fingerprint data, the results showed that there existed a relative big overall difference among the certification varieties of corn in China. Percentage of different sites among the authorized varieties was mainly concentrated between 80% and 95% (accounted for 78.28%), the percentage of different sites with more than 50% reached 99.21%, and less than 20% of only 0.09%. The average hybrid rate of the authorized varieties reached 64% and mainly concentrated between 50% and 80% (accounted for 89%). By using the corn variety standard fingerprint matching service platform (URL: http://www.maizedna.org/), a shared fingerprint database is realized.【Conclusion】A standard procedure in constructing crop variety SSR fingerprint database was formed in this study and the SSR fingerprint database was constructed with a scale of nearly 4 000 corn authorized varieties. Through joint multi-laboratory comparison tests, the accuracy of database building and the sharing property of database were ensured. A service platform website for corn variety standard fingerprint matching was established in this study, thus achieving sharing of corn authorized variety fingerprint database in national seed identification system, and providing an important reference for other crop species in building high-quality SSR fingerprint database.

Key words：Maize；Variety identification；DNA fingerprint Database；SSR

引言

【研究意义】主要农作物品种审定登记制度实施以来，审定品种的数量呈逐年增多趋势，截至2013年底，玉米审定品种数量就达到了6 291个，对品种高效管理带来了挑战；随着玉米品种资源遗传基础日趋狭窄及少数骨干亲本被集中应用，出现了一批在原品种基础上仅做细微改良的派生品种，对品种鉴定技术提出了更高要求；品种审定后进入生产推广环节，出现标签名称与实际包装的种子不符等品种真实性问题，增加了种子市场监管的难度。构建高质量的玉米审定品种标准DNA指纹库对解决上述问题和需求具有重要意义，将为政府的品种管理、企业的品种维权、农民的利益维护提供强有力的技术支撑。【前人研究进展】许多作物都开展了品种DNA指纹库构建，其中开展较早或规模较大的主要有玉米、水稻、小麦、油菜、马铃薯、番茄、月季、葡萄等，所用建库技术主要为SSR技术，SNP、INDEL、AFLP、ISSR等标记也有少量使用。早期由于SSR引物开发难度大，引物筛选评价不够充分，试验程序标准化程度较低，且大部分是基于PAGE凝胶电泳检测平台建立起来的，不同实验室数据可比性较差；近年来随着引物开发数量的增多，对引物筛选和评估的强度提高，建库程序的逐步标准化，以及荧光毛细管电泳检测平台的广泛使用，所建指纹库的质量得到一定提高。【本研究切入点】总结以往作物品种指纹库构建的工作，在以下几个方面尚待进一步完善：一是未能彻底解决不同实验室数据整合和共享的问题，不同实验室数据采集不一致情况未能提出根本的解决方案；二是所建指纹库多数为选取几十到几百个代表性品种的规模较小的研究型数据库，建库样品为自行收集的材料或种质资源，并非来自官方登记备案的标准样品；三是所建数据库停留在研究阶段，未经过大量检验实践的充分验证。【拟解决的关键问题】为从根本上解决玉米品种SSR指纹数据整合共享的问题，通过对试验程序、引物特征参数的设置、引物分组的设置、检测平台、分析软件、数据库管理系统进一步规范，确立大规模构建高质量可共享指纹库的切实可行方案；构建规模达到近4 000份的玉米审定品种标准样品SSR指纹库，通过多实验室联合比较试验，保证建库数据的准确性和数据库的可共享性，并通过建立指纹比对平台网站，实现玉米审定品种指纹库在全国种子检验系统的应用。

1 材料与方法

1.1 试验材料

建库样品共3 998份，由国家种质库提供，系国家和省级农作物品种审定委员会向品种选育单位收集或在区域试验种子中留取的玉米审定品种标准样品，涉及3 806份玉米审定品种，包括3 361个普通玉米品种（3 540份样品）和445个甜糯玉米品种（460份样品），其中168个品种有2~4个不同来源的同名样品，这些品种基本涵盖了我国曾经推广或正在推广使用的玉米杂交种。

1.2 基因组 DNA 提取

采取改良的 CTAB 法提取 DNA：每份供试样品均随机抽取 50 粒种子形成混合样品，充分磨碎后移入 2.0ml 离心管；加入 700μl CTAB 提取液，65℃水浴 60min；加入等体积的三氯甲烷/异戊醇（24∶1）并充分混合，静置 10min 后 12 000r/min 离心 15min；上清液加入等体积的预冷异丙醇沉淀 DNA，12 000r/min 离心 10min，弃上清液；加入 70%乙醇清洗 2 次，晾干后加入 100μl 超纯水，充分溶解后备用。除了供试样品外，另外选用一份玉米 DH 系作为参照样品进行大量 DNA 提取。同一批次提取的 DNA 样品，随机抽取约 5%进行 DNA 质量和浓度测定，根据测量值估算稀释成工作液的倍数后对该批 DNA 统一进行稀释。采用 NanoDrop 2000（Thermo Scientific）紫外分光光度计进行 DNA 质量和浓度测定。

1.3 SSR 标记分析

40 对建库引物序列信息、采用的 PCR 反应体系和反应程序以及荧光毛细管电泳检测程序见行业标准《玉米品种鉴定技术规程 SSR 标记法》（NY/T 1432—2014）和已发表的论文。每对引物其中一条的 5′端用一种荧光染料进行标记，共选用了 PET、NED、VIC、FAM 4 种荧光染料（Applied Biosystems，USA 公司合成）。PCR 反应在 Veriti 384 well Thermal cycler（Applied Biosystem）上进行，荧光毛细管电泳在 ABI 3730XL DNA analyzer（Applied Biosystem）上进行，分子量内标采用 LIZ 500。

采用仪器自带的 Date Collection Ver. 1.0 软件收集原始数据并形成 FSA 文件。采用北京市农林科学院玉米研究中心与北京华生恒业公司合作开发的 SSR 指纹分析器（软件登记号：2015SR161217）对 FSA 文件进行分析，分析时设置的引物特征参数见表 1。

表 1 建库引物指纹分析特征参数设置
Table 1 Fingerprint analysis parameter setting of primers for database construction

引物编号 Primer number	引物名称 Primer name	染色体位置 Chromosome position	标记荧光染料 Labeling fluorescent dye	重复单元碱基数 repeating units	引物扩增范围 Primers range	拖尾峰设置* Tailing peak setting	邻峰设置 Adjacent peak setting	N+1峰设置 N+1 peak setting	PANEL 分组 PANEL grouping
MP01	bnlg439w1	1.03	NED	2	320~380	是，339	70/50	是，最高峰	MQ1
MP02	umc1335y5	1.06	PET	2	234~256	是，240	50/30	是，最高峰	MQ3
MP03	umc2007y4	2.04	FAM	2	238~296	是，262	70/50	是，最高峰	MQ1
MP04	bnlg1940k7	2.08	PET	2	335~386	是，367	70/50	是，最高峰	MQ2
MP05	umc2105k3	3.00	PET	2	286~352	是，305	70/50	否	MQ1
MP06	phi053k2	3.05	NED	4	333~366	否	30/20	是，右峰	MQ2
MP07	phi072k4	4.01	VIC	4	410~433	否	50/30	是，右峰	MQ2
MP08	bnlg2291k4	4.06	VIC	2	364~418	是，382	70/50	是，最高峰	MQ1
MP09	umc1705w1	5.03	VIC	2	267~331	是，281	70/50	是，最高峰	MQ1
MP10	bnlg2305k4	5.07	NED	2	244~312	是，274	70/50	是，最高峰	MQ2
MP11	bnlg161k8	6.00	VIC	2	145~219	是，183	70/50	是，最高峰	MQ1
MP12	bnlg1702k1	6.05	VIC	2	265~323	是，291	70/50	是，最高峰	MQ2
MP13	umc1545y2	7.00	NED	4	191~246	否	50/30	是，最高峰	MQ1
MP14	umc1125y3	7.04	VIC	4	146~175	否	30/20	是，最高峰	MQ4

(续表)

引物编号 Primer number	引物名称 Primer name	染色体位置 Chromosome position	标记荧光染料 Labeling fluorescent dye	重复单元碱基数 repeating units	引物扩增范围 Primers range	拖尾峰设置* Tailing peak setting	邻峰设置 Adjacent peak setting	N+1峰设置 N+1 peak setting	PANEL分组 PANEL grouping
MP15	bnlg240k1	8.06	PET	2	221~245	否	50/30	是,最高峰	MQ4
MP16	phi080k15	8.08	PET	5	202~231	否	30/20	是,左峰	MQ1
MP17	phi065k9	9.03	NED	2	393~413	否	30/20	是,最高峰	MQ1
MP18	umc1492y13	9.04	PET	3	273~301	否	50/30	是,最高峰	MQ4
MP19	umc1432y6	10.02	PET	4	217~262	是, 252	50/30	是,右峰	MQ2
MP20	umc1506k12	10.05	FAM	4	167~194	否	50/30	是,最高峰	MQ1
MP21	umc1147y4	1.07	NED	2	154~170	否	50/30	是,最高峰	MQ3
MP22	bnlg1671y17	1.10	FAM	2	174~253	是, 186	70/50	是,右峰	MQ3
MP23	phi96100y1	2.00	FAM	4	245~277	否	30/20	是,最高峰	MQ2
MP24	umc1536k9	2.07	NED	2	212~241	否	50/30	是,最高峰	MQ3
MP25	bnlg1520k1	2.09	FAM	2	157~210	是, 179	70/50	是,最高峰	MQ2
MP26	umc1489y3	3.07	NED	3	232~270	否	30/20	否	MQ4
MP27	bnlg490y4	4.04	NED	2	266~330	否	30/20	是,最高峰	MQ3
MP28	umc1999y3	4.09	FAM	3	176~200	否	30/20	是,最高峰	MQ4
MP29	umc2115k3	5.02	VIC	5	271~300	否	30/20	是,最高峰	MQ4
MP30	umc1429y8	5.03	VIC	3	120~154	否	30/20	是,最高峰	MQ3
MP31	bnlg249k2	6.01	VIC	2	259~313	是, 265	70/50	是,最高峰	MQ3
MP32	phi299852y2	6.07	VIC	2	211~255	否	50/30	是,右峰	MQ4
MP33	umc2160k3	7.01	VIC	2	199~252	否	50/30	是,最高峰	MQ2
MP34	umc1936k4	7.03	PET	2	154~185	是, 180	30/20	是,最高峰	MQ2
MP35	bnlg2235y5	8.02	VIC	3	169~193	否	30/20	是,最高峰	MQ3
MP36	phi233376y1	8.09	PET	3	204~222	否	30/20	是,右峰	MQ3
MP37	umc2084w2	9.01	NED	4	177~214	否	30/20	是,最高峰	MQ4
MP38	umc1231k4	9.05	FAM	2	229~292	否	30/20	是,最高峰	MQ4
MP39	phi041y6	10.00	PET	4	295~333	否	50/30	是,左峰	MQ3
MP40	umc2163w3	10.04	NED	2	279~360	是, 322	70/50	是,右峰	MQ4

* 拖尾峰设置时先设置是否存在拖尾峰,如果是,进一步确定拖尾峰的起点,用等位基因值表示

1.4 建库及数据统计分析

建库过程中采用了植物品种DNA指纹库管理系统(以下简称指纹库管理系统,登记号:2015SR085905),安排至少两组独立平行试验,选取60%以上组数的试验数据一致且试验质量较高的数据和指纹图谱进入标准指纹库,对多组试验中数据不一致而指纹图谱一致的位点,通过人工选择一组试验质量较高的数据和指纹进入标准指纹库,对数据缺失严重的样品,启动新一组独立试验,直到形成位点缺失率控制在5%以下的标准指纹库。

采用Power-Marker ver. 3.25进行SSR引物位点的等位基因信息统计,采用Microsoft Excel 2010进行引物等位基因频率分布图及差异位点分布图的绘制。

1.5 建库数据的质量评估

为评估所构建SSR指纹库的质量,检验指纹库共享的可能性,开展如下几项评估试

验：①从建成的标准指纹库中随机抽取 200 份样品的 DNA（约占总数 5%，名单略），匿名编号后提供给两个实验室进行检测并与标准指纹库的数据进行对比，采用与构建标准指纹库时相同的建库程序和电泳检测平台，只是仪器型号略有不同，评估不同实验室在同一荧光电泳检测平台上数据采集的一致性和利用指纹库进行成对比较的结果一致性。②从标准指纹库中挑选 8 份代表性样品的 DNA（名单略），由北京玉米种子检测中心实验室统一进行 PCR 扩增，扩增产物提供给具有贝克曼 GenomeLabGeXP 遗传分析仪的三家实验室进行检测并与标准指纹库的指纹进行对比，评估在不同荧光电泳检测平台上数据采集的一致性和利用指纹库进行成对比较的结果一致性。

1.6 指纹比对平台网站的建立

为了实现指纹库在全国种子检验系统中的共享使用，建立玉米品种标准指纹比对服务平台，平台上可以查看玉米审定品种标准指纹库，并提供了全库比较、疑似比较、同名比较、范围内互比等便捷的比对功能。

2 结果

2.1 引物等位基因频率分布及多态性评估

对建库引物的总体多态性情况进行了全面的评估，从等位基因个数、基因型个数、PIC 值、个体识别率、遗传多样性等多个参数上看，40 个建库引物总体多态性较高（表2），这套核心引物是通过代表性玉米自交系和杂交种品种筛选评估后入选的高多态性的引物，表明玉米审定品种大规模建库对引物的评估结果与以往的筛选评估具有较高的一致性。

表2 40 个建库引物多态性评估
Table 2 Polymorphism Evaluation of 40 Primers for database construction

引物编号 Primer No.	引物名称 Primer name	等位基因数 Allele No.	基因型数 Genotype No.	PIC 值 PIC	杂合率 Heterozygosity	个体识别率 DP	遗传多样性 Gene diversity	主要等位基因及其频率 Major allele and its frequency
MP01	bnlg439w1	26	160	0.83	0.81	0.95	0.85	350, 0.31
MP02	umc1335y5	9	30	0.58	0.24	0.73	0.61	241, 0.59
MP03	umc2007y4	29	202	0.82	0.84	0.94	0.84	256, 0.33
MP04	bnlg1940k7	35	285	0.85	0.68	0.95	0.86	348, 0.26
MP05	umc2105k3	30	220	0.84	0.72	0.96	0.85	291, 0.32
MP06	phi053k2	12	45	0.69	0.66	0.88	0.73	336, 0.41
MP07	phi072k4	20	118	0.69	0.54	0.84	0.71	411, 0.50
MP08	bnlg2291k4	25	145	0.75	0.79	0.92	0.78	364, 0.36
MP09	umc1705w1	32	166	0.82	0.76	0.95	0.84	273, 0.26
MP10	bnlg2305k4	27	148	0.83	0.75	0.95	0.84	252, 0.25
MP11	bnlg161k8	33	253	0.90	0.92	0.98	0.90	165, 0.18
MP12	bnlg1702k1	29	199	0.84	0.75	0.96	0.85	265, 0.30
MP13	umc1545y2	22	133	0.83	0.79	0.95	0.85	208, 0.30

(续表)

引物编号 Primer No.	引物名称 Primer name	等位基因数 Allele No.	基因型数 Genotype No.	PIC 值 PIC	杂合率 Heterozygosity	个体识别率 DP	遗传多样性 Gene diversity	主要等位基因及其频率 Major allele and its frequency
MP14	umc1125y3	7	19	0.69	0.80	0.86	0.73	173, 0.40
MP15	bnlg240k1	10	44	0.77	0.74	0.91	0.79	237, 0.33
MP16	phi080k15	11	36	0.63	0.51	0.84	0.66	217, 0.52
MP17	phi065k9	4	9	0.56	0.67	0.78	0.63	413, 0.46
MP18	umc1492y13	9	16	0.27	0.20	0.43	0.28	278, 0.84
MP19	umc1432y6	24	83	0.48	0.49	0.72	0.50	222, 0.69
MP20	umc1506k12	13	50	0.73	0.74	0.91	0.76	185, 0.39
MP21	umc1147y4	4	10	0.30	0.31	0.52	0.32	154, 0.81
MP22	bnlg1671y17	36	220	0.89	0.69	0.97	0.90	184, 0.19
MP23	phi96100y1	9	40	0.73	0.62	0.90	0.77	253, 0.36
MP24	umc1536k9	12	50	0.71	0.71	0.89	0.75	222, 0.39
MP25	bnlg1520k1	28	129	0.69	0.80	0.86	0.73	173, 0.36
MP26	umc1489y3	8	25	0.67	0.73	0.86	0.72	233, 0.35
MP27	bnlg490y4	16	59	0.72	0.40	0.88	0.76	271, 0.37
MP28	umc1999y3	7	20	0.52	0.62	0.75	0.59	176, 0.55
MP29	umc2115k3	17	86	0.83	0.57	0.95	0.85	276, 0.21
MP30	umc1429y8	9	17	0.52	0.62	0.76	0.59	126, 0.55
MP31	bnlg249k2	31	174	0.81	0.81	0.95	0.83	263, 0.35
MP32	phi299852y2	14	46	0.74	0.52	0.91	0.78	234, 0.29
MP33	umc2160k3	21	84	0.76	0.83	0.92	0.79	207, 0.30
MP34	umc1936k4	9	32	0.56	0.58	0.79	0.61	170, 0.56
MP35	bnlg2235y5	10	40	0.71	0.76	0.90	0.74	183, 0.43
MP36	phi233376y1	8	29	0.69	0.64	0.89	0.73	215, 0.35
MP37	umc2084w2	13	55	0.78	0.71	0.93	0.81	185, 0.29
MP38	umc1231k4	8	16	0.40	0.35	0.68	0.51	261, 0.54
MP39	phi041y6	22	101	0.74	0.76	0.91	0.77	309, 0.36
MP40	umc2163w3	36	162	0.81	0.52	0.93	0.83	310, 0.27
平均 Mean	—	18	94	0.70	0.65	0.86	0.73	—

通过统计审定品种在 40 个引物位点上的等位基因频率分布情况，绘制了 40 个玉米建库引物的等位基因频率分布图（图 1），由于不同引物在等位基因数、等位基因频率上具有不同的分布特征，引物的等位基因频率分布图可以作为每个引物的特征图谱。

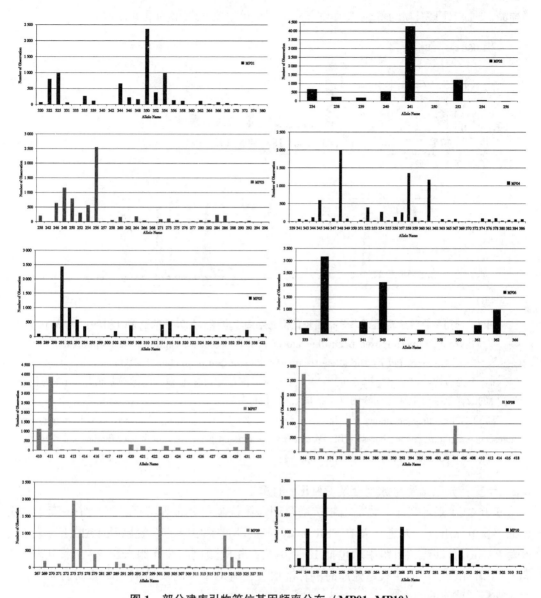

图 1 部分建库引物等位基因频率分布（MP01-MP10）

Figure 1 The allele frequency distribution of part primers for maize fingerprint database construction (MP01-MP10)

一般情况下，荧光毛细管电泳叠加峰图与相应引物的等位基因频率分布图的图型特征应基本相似，因此在具体建库试验中，引物等位基因频率分布图可以发挥与参照样品相似的作用。以引物 MP35 为例（图 2），该引物的等位基因频率分布图（A）的基本特征可以概括为 175、183 频率高，175、180、188、189、193 频率中等，178、181、186、192 频率低，除此之外的其他等位基因应没有或极其稀有；未校正前的原始荧光毛细管电泳叠加峰图（B1）则 178、186 频率高，而 175、183 频率中或低，并出现了 2 个新等位基因。这显然与引物 MP35 的特征不符；对峰图进行整体向右移动 3bp 的校正后的荧光毛细管电泳叠加峰图（B2）则完全符合引物 MP35 的等位基因频率分布图特征，表明即使没有参照样品，通过对照引物的等位基因频率分布图也可以实现数据的准确读取。

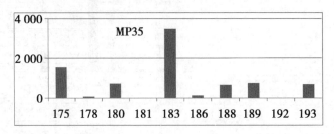

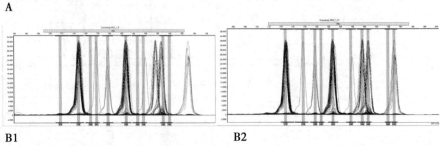

图 2 利用引物等位基因频率分布图校正荧光毛细管电泳叠加峰图的位置（引物：MP35）

Figure 2　Superimposed peak diagram of with fluorescence capillary electrophoresis corrected by the allele frequency distribution of primer（MP35）

注：A 为引物等位基因频率分布图；B1 为未校正位置的荧光毛细管电泳叠加峰图；B2 为校正位置后的荧光毛细管电泳叠加峰图（整体向右移动 3bp）

Note：A. the allele frequency distribution of primers；B1. Uncorrectedsuperimposed peak diagram of with fluorescence capillary electrophoresis；B2. corrected superimposed peak diagram of with fluorescence capillary electrophoresis（overall moved to the right 3bp）

2.2　十重荧光毛细管电泳组合及 PANEL 的建立

基于 40 个玉米建库引物的引物扩增范围、等位基因分布等特征，形成了一套十重荧光毛细管电泳组合，并进一步在 SSR analyzer 中设置了相应的系统默认 PANEL（表 3），该套 PANEL 分四组，每组中均包括 10 个引物，相同颜色的荧光组合了 2~3 个引物。这套 PANEL 相当于不同实验室建库时的标准语言，有助于不同实验室建库数据的共享，并大大提高了建库的效率。

表 3　引物 Panel 组合信息

Table 3　The Panel information of 40 primers for maize fingerprint database construction

Panel 编号 No. of Panel	荧光类型 Fluorescent dye type	引物编号（等位变异范围，bp） No. of primer（range of allel，bp）		
		1	2	3
MQ1	FAM	P20（166~196）	P03（238~298）	
	VIC	P11（144~220）	P09（266~335）	P08（364~420）
	NED	P13（190~248）	P01（319~382）	P17（391~415）
	PET	P16（200~233）	P05（287~354）	

（续表）

Panel 编号 No. of Panel	荧光类型 Fluorescent dye type	引物编号（等位变异范围，bp） No. of primer（range of allel，bp）		
		1	2	3
MQ2	FAM	P25（157~211）	P23（244~278）	
	VIC	P33（198~254）	P12（263~327）	P07（409~434）
	NED	P10（243~314）	P06（332~367）	
	PET	P34（153~186）	P19（216~264）	P04（334~388）
MQ3	FAM	P22（173~255）		
	VIC	P30（119~155）	P35（168~194）	P31（260~314）
	NED	P21（152~172）	P24（212~242）	P27（265~332）
	PET	P36（202~223）	P02（232~257）	P39（294~333）
MQ4	FAM	P28（175~201）	P38（227~293）	
	VIC	P14（144~174）	P32（209~256）	P29（270~302）
	NED	P37（176~216）	P26（230~271）	P40（278~361）
	PET	P15（220~246）	P18（272~302）	

2.3 指纹库构建情况分析

统计审定品种指纹库构建的试验情况，61%的样品做了2组独立试验，33%的样品做了3组独立试验，6%的样品做了4或5组独立试验，每份样品一般具有2~5套原始试验数据及对应的指纹图谱。样品指纹负责人对多组试验数据进行审核后，缺失的位点占0.15%；不同组试验存在数据差异的位点占6.08%。分析数据差异的原因，表明大部分是由于样品在个别位点上一致性较差形成高低峰或多峰，介于阈值附近的峰在不同组试验中峰高存在细微差异，造成不同组自动采集的数据不同，对这些差异位点进一步人工比较指纹图谱，6.03%的位点指纹图谱一致并经人工审核后入标准指纹库。最终建成的审定品种标准指纹库累计缺失和差异位点仅占0.2%，数据完整性达到99.8%，标准指纹库的数据和图谱均来自至少两组指纹图谱一致的独立试验数据，数据准确性高。

不同实验室在相同电泳检测平台上的试验结果显示，37个建库引物表现出高度的试验稳定性和重复性，不同实验室采集数据的差异低于5%，采集的指纹图谱一致，指定样品的成对比较结果也是一致的；3个建库引物（M02、MP18、MP38）在不同实验室采集数据的差异略高于5%，今后可以考虑筛选出更好的建库引物进行替换。不同实验室在不同电泳平台上试验的结果显示，同一样品DNA在3台GenomeLabGeXP遗传分析仪上获得的原始数据的差异在0.5bp以内；而同一样品DNA在GenomeLabGeXP遗传分析仪与ABI 3730XL遗传分析仪上获得的原始数据存在1~5bp的差异。评估结果表明，同一荧光电泳检测平台上获得SSR指纹数据一致性高，而不同的电泳检测平台获得的数据存在一定偏差，因此，要想实现不同实验室的数据共享，获得高度一致的SSR指纹图谱和高度一致的样品成对比较结果，需要统一荧光引物、统一分析软件及统一电泳检测平台。

2.4 审定品种指纹库整体情况

对审定品种指纹数据进行两两比较,共涉及 7 332 535 对比较,结果表明,中国玉米审定品种之间差异比较大,品种间差异位点百分比集中在 80%~95%(占 78.28%),品种间差异位点百分比在 50% 以上的已达到 99.21%,而低于 20% 的只有 0.09%(表 4)。由于这些审定品种代表了中国生产上推广的几乎全部品种,这一信息对今后的实际品种鉴定具有一定参考价值:由于中国玉米审定品种之间差异较大,因此,实际品种鉴定中当 2 个样品之间差异位点百分比低于 50% 时,就需要考察 2 个样品是否存在亲缘关系或具有共同亲本。

对玉米审定品种杂合率水平进行分析(表略),平均品种杂合率达到 64%,主要集中在 50%~80%(占 89%),表明对大部分玉米审定品种而言,40 核心引物中均能筛选到 20~30 个双亲互补型引物,这一结果对利用指纹库筛选玉米审定品种的纯度鉴定引物具有重要价值。

表 4 玉米审定品种指纹数据两两比较情况
Table 4 Pairwise comparison of maize varieties with fingerprint data

差异位点百分比 Percentage of different loci (%)	涉及成对比较的对数 The number of pairwise comparisons	成对比较的百分比 Percentage of pairwise comparisons (%)	成对比较的累计百分比 Cumulative percentage of pairwise comparisons (%)
95~100	471 610	6.43	6.43
90~95	1 805 540	24.62	31.06
80~90	3 934 675	53.66	84.72
70~80	888 814	12.12	96.84
60~70		1.81	98.65
50~60		0.56	99.21
40~50		0.33	99.54
30~40		0.23	99.77
20~30		0.14	99.91
10~20		0.07	99.98
5~10		0.02	99.99
0~5		0.01	100.00

322 份样品(约占 8%)存在 2~3 个样品指纹相同的情况,对这 322 份样品已经提交审定标准样品管理机构,正在进一步核实是否来自于同一品种,如果来自同一品种,应在标准样品库中将指纹相同的样品保留一份,其他进行合并或清理,如果来自不同品种,应进一步提供其他差异。标准样品库的清理工作完成后,就能够保证一个品种名称只有一个惟一的指纹图谱且相同的指纹图谱只对应一个品种名称,从而大大提高审定品种标准指纹

库在真实性检测实践中的应用价值。

2.5 指纹库的应用效果

玉米品种标准指纹比对服务平台（网址：http://www.maizedna.org/）自 2015 年正式上线（网站界面见图 3），平台上提供了已建成的全部玉米审定品种 SSR 指纹。该平台已向农业部授权的省级种子检测机构开放，甘肃、北京等省种子管理站已经开始使用，在市场监督抽查、区试、品种权执法、企业维权等真实性检测方面已发挥了重要作用：在没有指纹比对平台之前，开展玉米真实性检测时，需要将待测样品和审定品种的标准样品一起进行成对比较试验，这意味着检测机构每次都要向种质库申请相应的标准样品进行试验，不仅会造成标准样品频繁出库，而且使室内检验的工作量翻倍；有了指纹比对平台后，则只需对待测样品进行试验，然后将待测样品指纹与平台的标准指纹进行比对就可以直接出具检测结果。

图 3　指纹比对服务平台网站界面
Figure 3　A service platform website for corn variety standard fingerprint matching

3　讨论

3.1　指纹库数据的采集

指纹库构建时采用什么样的数据记录方式不仅影响数据库构建的难易程度，还影响数据库使用的方便程度。汇总以往指纹库构建时采用的数据记录方式，主要分为以下 2 种。

第一种是 0/1 编号，即按照电泳谱带的有无记录，有带记录为 1，无带记录为 0，这种方式在早期构建的指纹库中被使用过，现在的建库中已经较少使用。

第二种是等位基因代码，根据等位基因代码的编号方式进一步分为 3 类：第 1 类是按等位基因扩增产物大小的顺序用阿拉伯数字依次编号，部分研究者进一步按基因型进行了编码，这在小规模建库时具有简单直观的优点，但当建库规模扩大时面临新增等位基因无法编号的问题以及多实验室数据不易整合的问题。第 2 类是主要依据重复序列的重复次数对等位基因命名，这种方式具有命名严谨规范的优点，在人类指纹库构建中采用了这种命名方式，选择的建库引物主要是 4~5 碱基重复类型的，个别作物如葡萄等通过重新开发和筛选合适的引物也开始使用这种方式，但这种命名方式对标记的选择比较严苛，2 碱基重复类型的引物由于拖尾峰等峰型的干扰和扩增区间出现插入缺失的干扰不易准确确定碱基重复次数，不适合采用这种命名方式。由于大部分作物在筛选建库引物时主要考虑了引物多态性和试验重复稳定性，而 2 碱基重复类型的 SSR 引物的开发数量和多态性普遍高于 3 碱基以上的引物，造成建库引物中 2 碱基重复类型的 SSR 引物比例较高，因此大部分作物目前不适合采用重复次数的命名方式；第 3 类是依据等位基因扩增产物大小对等位基因命名，这种方式不仅有利于不同实验室数据的整合，而且对标记的要求有所降低，在大规模建库中对扩增区间出现的插入缺失具有较高的容忍度，是适合目前大多数作物品种 SSR 指纹库构建的方式，本研究对玉米审定品种建库即采用了这种数据采集方式。

3.2 指纹库的级别

根据指纹库的实际使用价值不同，可将指纹库分为 3 个级别：初级指纹库仅起辅助筛查的作用；中级指纹库能够利用指纹库直接出具部分检测结果，仍有部分样品需进一步增加试验才能出具结果；高级指纹库能够利用指纹库直接出具几乎全部检测结果。

除了人类和少数动植物物种外，目前基于 SSR 标记建立的大部分作物指纹库仍只能起到辅助筛查的作用，而能否调取原始指纹图谱，是影响作物 SSR 指纹库利用价值的关键因素：与人类以个体为基础的指纹库不同，作物品种指纹库构建时存在品种内部不同个体的一致性问题，如果只采集数据，则只能反映该品种的主要等位基因信息，如果同时采集指纹图谱则不仅能反映所有等位基因及其大致比例，也能反映试验质量和数据的可靠程度，因此只有整合了数据和指纹图谱的作物品种 SSR 指纹库才有可能发挥直接出具检测结果的价值。本研究构建的玉米审定品种 SSR 指纹库通过同时采集数据和指纹图谱入库，实现了能够直接出具检测结果的目标，对其他作物品种 SSR 指纹库构建重要借鉴价值。

以往的研究者还在水稻、高粱、油菜、百合、大豆等许多作物上开展了分子身份证的研制，通过将指纹数据转换为条形码或二维码的形式制作分子身份证，部分分子身份证还加入了商品信息等其他辅助信息，但尚未见综合利用分子数据和指纹图谱信息的报道。下一代的分子身份证可采取与指纹库网站建立链接的方式，建立每个品种在网站上的二维码，通过扫描二维码，即可调取该品种的分子数据信息及指纹图谱信息。

3.3 可共享 SSR 指纹库的构建

SSR 标记具有多等位基因的特点，这一特点对指纹库构建带来正反两方面的影响。从正面影响看，一是标记多态性高，区分品种能力强，一个 SSR 标记大约相当于 10 个 SNP 标记的区分能力，用较少标记就能对大量品种进行有效区分；二是基因型类型较多，可以采用混合样品建库，当进行样品成对比较时，只要在比较的位点上不存在共同等位基因，即使样品一致性差，也可以明确判定为差异位点，而 SNP 标记基因型最多只有 3 个，在样品一致性差的情况下无法采用混合样品建库及成对比较；三是杂交种的杂合率比较高，可以很容易筛选到双亲互补型的引物用于杂交种纯度鉴定。从负面影响看，一是 SSR 标记等位基因数一般在 2 个以上，不属于二态的标记类型，基因分型主要采用基于长度多态的电泳检测平台，检测通量一般在 10 个位点以内，很难实现位点高通量的检测；二是等位基因分布比较密集带来不同实验室、不同检测平台数据整合的难度。对于 SSR 检测通量低问题，今后通过扩增产物测序的方式有可能实现一次检测几千个 SSR 位点。实现几百到几千个位点的复合检测。对于指纹库数据整合的问题，目前采取在建库方案、取样方案、检测平台、指纹分析工具和引物设置等方面进行统一规范的方式实现 SSR 数据整合，今后可通过开发重复次数严格按照重复单元的倍数增减的 3~5 碱基重复类型的玉米 SSR 引物，并在指纹数据采集时依据重复序列的重复次数对等位基因命名，从而大大降低共享数据库构建的难度。

在建库程序标准化的基础上，为进一步消除不同实验室或同一实验室不同时期试验的系统误差，还需要通过参照样品进行数据的校正。荧光毛细管电泳检测平台上只需提供 1 份参照样品就可发挥校正系统误差的作用，在玉米指纹库构建过程中，为试验设计方便，将每块电泳板的最后一个孔位作为放置参照样品扩增产物的位置，SSR 指纹分析器在指纹分析过程中，依据该位置的已知标准指纹对 PANEL 进行左右整体位移，直到参照样品的读取指纹与其已知标准指纹一致。参照样品在作物品种共享指纹库构建中的价值已经得到了认可，然而选用什么样的参照样品，以及如何使用参照样品还有待商榷：部分研究者推荐的参照样品是从国家库中选取的一套代表建库引物的主要等位基因的标准样品。从今后的检测实践看，参照样品应同时具有样品高度一致和容易大量提供的特点，由指定实验室统一繁殖并统一提供比从国家库取用更加方便，一致性高的自交系（或 DH 系）比杂交种更适合做参照样品。

4 结论

通过对试验程序、引物特征参数的设置、引物分组的设置、检测平台、分析软件、数据库管理系统进一步规范，形成了构建作物品种 SSR 指纹库的标准化程序。构建了近 4 000 份的高质量玉米审定品种 SSR 指纹库，建立了玉米品种标准指纹比对服务平台网站，实现玉米审定品种指纹库在全国种子检验系统的共享。

参考文献（略）

本文原载：中国农业科学，2017，50（1）：1-14

中国 328 个玉米品种（组合）SSR 标记遗传多样性分析

王凤格　田红丽　赵久然　王　璐　易红梅　宋　伟　高玉倩　杨国航

（北京市农林科学院玉米研究中心，北京　100097）

摘　要：【目的】从育成年份、种植区域角度分析中国 328 个玉米品种遗传多样性情况，在分子水平上分析各适宜种植区域品种的遗传分化特点，为玉米品种区域试验、审定管理以及育种策略提供一定的理论依据。【方法】利用均匀分布于玉米基因组的 40 对核心 SSR 引物，采用 10 重荧光毛细管电泳检测技术对 328 个代表性育成品种进行基因分型；通过 Power-Marker ver. 3.25 软件评估 40 对 SSR 引物多态性，对参试品种按年份、区试组分析遗传多样性情况；利用多变量统计分析软件 MVSP ver. 3.13 对 328 个品种按区试组进行主坐标分析。【结果】基于 328 个玉米品种数据，检测到 40 对 SSR 引物等位基因变异范围为 3~16 个，平均 8.10 个，多态信息指数（PIC）值变化范围为 0.18~0.85，平均为 0.63。参试品种不同年份间遗传多样性变化不大，PIC 值在 0.60 左右摆动；8 个区试组品种的总等位基因和总基因型变异范围为 170~262 和 200~511，京津唐和西南分别表现出了最低值和最高值，京津唐 PIC 值最低为 0.51，其余组均接近于 0.60，平均杂合度均在 0.60 左右。8 个区试组主坐标分析显示鲜食、极早熟品种遗传分布偏离普通玉米区域，具有明显的种质特异性；青贮玉米由于早期对照品种为普通玉米导致遗传分布向普通玉米延伸，仅有少数品种分布偏离普通玉米；京津唐、西南、东北早熟三组品种遗传分布相对集中；极早熟、京津唐、西南三组之间品种遗传分布几乎无重叠区域；东华北和黄淮海两区参试品种遗传分布具有明显的分化但也有部分重叠，原因与对照品种曾经相同、育种同质化有关。【结论】利用 SSR 标记分析 328 份玉米品种的遗传多样性及遗传分化特点，表明近年来育成品种遗传多样性指数在不同年份间变化不大，除京津唐区之外在其他区试组间差异性也不大。参试品种主坐标分析显示各区试组划分、对照品种设置起到了育种导向的作用。

关键词：玉米；SSR 标记；区域试验；遗传多样性

Genetic Diversity Analysis of 328 Maize Varieties (hybridized combinations) Using SSR Markers

Wang Fengge　Tian Hongli　Zhao Jiuran　Wang Lu
Yi Hongmei　Song Wei　Gao Yuqian　Yang Guohang

(*Maize Research Center*, *Beijing Academy of Agriculture & Forestry Sciences*, *Beijing* 100097)

基金项目：国家国际科技合作项目（2010DFB33740）和国家"十二五"农村领域国家科技计划课题（2011BAD35B09）

作者简介：王凤格，E-mail：gege0106@163.com；田红丽，E-mail：tianhongli9963@163.com；王凤格和田红丽同为同等贡献作者。通讯作者赵久然，E-mail：maizezhao@126.com

Abstract: 【Objective】The objective of this study is to analyze the genetic diversity of 328 maize varieties from breeding year, planting region, and to research the genetic differentiation characteristics of all the varieties from every appropriate planting region. This study will provide a theoretical basis for the maize regional trial, validation management and breeding strategic adjustment. 【Method】Forty core SSR primers covered the entire maize genome, were used to scan 328 samples by high-throughput 10-plex capillary electrophoresis platform. The polymorphism of 40 SSR loci and genetic diversity of 328 samples were evaluated via the software Power-Marker ver. 3.25, and the Principal Coordinate Analysis (PCoA) of these samples from every regional trial groups was executed using statistical analysis software MVSP ver. 3.13. 【Result】The detected alleles of 40 SSR primers ranged from 3 to 16, with an average of 8.10, the polymorphism information index (PIC) ranged from 0.18 to 0.85, with an average of 0.63. There were no significant genetic diversity changes among varieties from different years, all the PIC value were about 0.60; the total allele and gene-type number of 8 regional trial groups ranged from 170 to 262 and from 200 to 511. The varieties from Jing-Jin-Tang and Xi-Nan displayed the minimum and maximum values. The varieties from Jing-Jin-Tang group displayed the lowest PIC value of 0.51, and the remaining groups were near 0.60, and the average heterozygosity was about 0.60. Principal coordinate analysis (PCoA) of varieties from eight regional trial groups showed that the genetic distribution of varieties from Xian-Shi (maize for table use) and Ji-Zao-Shu (very early maturity maize) groups deviated from the common maize region, displaying an obvious germplasm specificity. As the common maize varieties were used as control samples at early stage, the genetic distribution of Qing-Chu (silage corn) extended to common maize. The genetic distribution of varieties from Jing-Jin-Tang, Xi-Nan and Dong-Bei-Zao groups was relatively concentrated. The genetic distribution of varieties from Ji-Zao-Shu, Jing-Jin-Tang and Xi-Nan groups is almost no overlap. There is an obrious gentic variation of varieties from Dong-Hua-Bei and Huang-Huai-Hai groups, but also there are some overlaps, and the reasons might be the same control variety and the same breeding resource. 【Conclusion】Genetic diversity and genetic differentiation of 328 maize varieties was analyzed using 40 SSR markers. The results showed that there were no significant genetic diversity changes among varieties in different years, and there were also no significant changes among different groups except Jing-Jin-Tang. Principal coordinate analysis showed that the setting of regional trial groups and control varieties has played a guiding role in breeding.

Key words: Maize; SSR marker; Regional trial; Genetic diversity

引言

【研究意义】玉米是具有粮、经、果、饲、能多元用途的中国重要农作物之一，就种植面积而言已经是第一大作物，其总产的增加对全国粮食增产贡献率位居各大粮食作物之首。除了面积、产量快速增加之外，随着育种进程的加快，玉米品种数量也急剧增加。统计1999—2011年数据，通过国家和各省市审定的玉米品种数目大约5 600个，其中，国家审定417个，而且近几年每年参加国家和各省市的区域试验样品数千个。虽然玉米已经日益凸现出重要性，但是就杂交种而言其遗传多样性研究报道极少，玉米杂交种多样性研究主要体现在已经大面积推广应用的杂交种和参加区域试验的品种。玉米区域试验是玉米育种和新品种推广的重要环节，参试品种尤其是进入生产试验品种代表了当前玉米育种的动向和水平，因此，对参试品种的遗传多样性进行分析，可以客观、全面了解当前玉米品种现状，对于品种管理、品种选育以及种质资源收集和保护具有重要意义。【前人研究进展】随着分子生物技术的发展，分子标记种类和检测手段日趋完善，各种分子标记已经广泛应用玉米遗传多样性研究中。众多分子标记中，简单重复序列（simple sequence repeats, SSR）标记因具有简便、快捷、重复性高、多态性高、共显性标记等优点而被广泛应用。中国水稻、小麦、大豆等重要农作物均已利用 SSR 标记对核心种质或育成品种进行了分析评估，为遗传研究和育种工作提供重要的依据和参考。对玉米而言，全基因组测序研究基础较好，开发的大量标记位点可共享如 maizeGDB 和 panzea 网站公布大量 SSR 位点。这些研究为玉米品种 DNA 指纹数据库构建、遗传多样性分析等研究奠定了良好的基础。迄今，国内外关于玉米遗传多样性的报道，主要集中在自交系种质资源评估和杂种优势群划分或地方品种遗传多样性分析。【本研究切入点】相对于玉米自交系而言，育成杂交种的报道极少，在中国仅有早期基于形态或系谱对杂交种的分析，2005 年国家预试品种分析或省内推广品种的分析；国外玉米杂交种的研究主要是欧洲近 50 年来育成品种的遗传多样性分析。育成玉米杂交种代表了当前育种水平和动态，品种管理者和育种家十分关注杂交种的具体情况，其审定品种的 SSR 标准指纹均已经构建，但是玉米杂交种的遗传多样性没有进行系统的分析。因此除应继续加强 DNA 标准指纹数据库构建之外，可按国家、各省逐步开展玉米杂交种的遗传多样研究，以便及时对中国玉米杂交种的遗传多样性进行动态监测。【拟解决的关键问题】本研究利用 40 对 SSR 引物对 2004—2009 年来参加国家玉米区试生产试验品种及大面积推广品种共计 328 个品种，从育成年份、适宜种植区域多个角度进行遗传多样性分析，试图在分子水平上摸清中国育成品种的遗传多样性情况，各种植区域品种的遗传分化特点，主要目的为玉米品种区域试验、审定管理提供理论依据，同时对育种策略的调整具有一定的参考价值。

1 材料与方法

1.1 试验材料

供试材料包括 2004—2009 年参加国家玉米区试生产试验的品种 289 份（表1）；各区组对照品种 19 份；为了全面反映育成品种现状，并且数据分析时具有参照样品，同时补充了在全国推广面积前 20 位的品种（按 2000—2009 年累计推广面积统计），共计 328 份品种。

1.2 基因组 DNA 提取

供试品种总 DNA 提取采取 CTAB 大量提取法，具体步骤按照 CIMMYT（International Maize and Wheat Improvement Center）试验操作流程。DNA 质量和浓度用 NanoDrop 2000（Thermo Scientific）紫外分光光度计进行测定，根据测量值调节工作液浓度。

1.3 SSR 标记分析

选用的 40 对引物是本实验室已报道的第 I 组的 20 对基本核心引物和第 II 组的 20 对扩展核心引物，引物名称、序列等具体信息参考已发表的文献，这 40 对引物为国家玉米品种区域试验 DNA 指纹鉴定指定引物。每对引物其中一条的 5′端用一种荧光染料标记，共选用了 PET、NED、VIC、FAM 四种荧光染料（Applied Biosystems，USA 公司合成）。PCR 反应体系为 20μl，包括 4μl DNA、0.25μmol·L^{-1} 引物、0.15μmol·L^{-1} dNTP、2.5mmol·L^{-1} MgCl$_2$、1 单位 Taq 酶（Genacea, 美国）、1×PCR Buffer。PCR 反应在 BIO-RAD 公司的 PTC-100 型 PCR 仪器上进行，程序为 94℃ 5min，94℃ 40s，60℃ 35s，72℃

45s，35个循环；72℃ 10min，4℃保存待测。PCR产物在毛细管荧光电泳系统AB 3730XL DNA分析仪（Applied Biosystems，USA）上检测。采用10重PCR产物电泳检测的方法，即按照扩增片段大小和荧光染料的种类可将10对引物的扩增产物混合在一起电泳。在96孔电泳板的单个孔中分别加入1.5μl 10重PCR产物的混合物，8.5μl甲酰胺，0.1μl内标（GeneScan™-500 LIZ, Applied Biosystems，USA）。上述混合样品在PCR仪上运行95℃变性5min，将变性后的电泳产物取出，1 000r/min离心1min后，于AB 3730XL DNA分析仪上进行电泳。预电泳时间为2min，15kV，电泳时间为20min，15kV。Date Collection Ver. 1.0软件收集原始数据。

1.4 数据统计分析

用GeneMapper Ver. 3.7（Applied Biosystems，USA）进行原始数据统计和校正，得到328样品×40对SSR引物的原始数据表。根据40对核心引物等位基因信息表对上述原始数据进行等位基因确定，并形成标准DNA指纹数据，该等位基因信息表由北京市农林科学院玉米研究中心通过在毛细管荧光电泳系统分析大量自交系和杂交种数据总结整理得到。选用Power-Marker ver. 3.25软件，基于328份玉米品种数据分析每对引物的等位基因数目和PIC值（polymorphism information content）。同时对参试品种按年份、按区试组别分别对等位基因变异丰富度（即等位基因数目变化情况）、PIC值、杂合度（Heterozygosity）等情况进行分析。选用群体遗传分析软件Popgene ver. 1.31分析各区试组之间的Nei遗传距离。参考Reif等针对于杂交种分析方法，在本研究中选用多变量统计分析软件MVSP ver. 3.13（Kovach computing services），将8个区组分别作为独立群体进行主坐标分析（principal coordinate analysis，PCoA），直观反映8个区试组之间的关系，为了进一步确定东华北与黄淮海两区，西南与东北早两区之间参试样品关系，将其独立进行主坐标分析。

2 结果

2.1 40对SSR引物多态性分析

基于328份品种数据分析40对SSR引物的等位基因和PIC指数变化（图1A），40对引物检测到的等位基因变异范围为3~16个，平均8.10个，多态信息指数PIC值变化范围为0.18~0.85，平均0.63，反映PIC指数变化与等位基因变异丰富度密切相关，随着等位基因数目的增加PIC值也在增加，但并非线性关系，有的位点（如N19）等位基因数目较多，但PIC值却较低。对N19（等位基因数=6，PIC=0.36）和N22（等位基因=16，PIC=0.85）2个位点的所有等位基因在所有品种中的分布频率进行统计分析（图1B和图1C），位点N19的6个等位基因中，等位基因222分布频率非常高达78%，因此，6个等位基因在杂交种的分布不均匀。与N19的情况相反，位点N22的等位基因分布频率相对均匀，有多个高频率的等位基因。

2.2 中国选育玉米品种的遗传多样性和遗传结构分析

不同年份选育玉米品种的遗传多样性比较：289份区域试验品种依据年份划分为6组，进行遗传多样性分析评价（表1）。2005—2007年、2009年品种的总等位基因数、总基因型数、平均等位基因数、平均基因型数均高于2004年和2008年相应的值，但可能受参试品种数目变化影响，各年度间PIC值以及平均杂合度变化不大，PIC值和杂合度均在0.60左右摆动，说明近些年来参加国家玉米品种生产试验样品的遗传多样性没有较大变化。

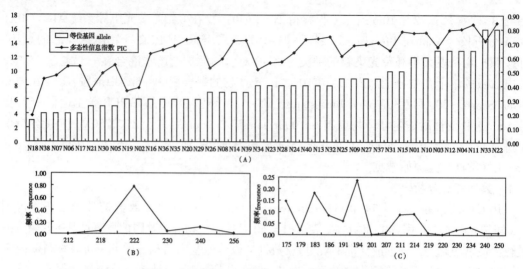

图 1　本文选用 40 对核心 SSR 引物多态性分析

Figure 1　Polymorphism of the 40 core SSR primers used in this study

注：A. 基于 328 份玉米品种数据，40 对 SSR 引物等位基因和 PIC 指数变化情况，横坐标引物位置按照等位基因数目从小到大顺序排列；B. 引物 umc1432y6（N19）检测到的等位基因变异及其频率；C. 引物 bnlg1671y17（N22）检测到的等位基因及其频率

Note: A. Change of alleles and PIC index of the 40 SSR primers based on 328 samples data, the order of primers displayed on the abscissa is according to the allele number; B. Alleles variation and frequencies detected in primers umc1432y6（N19）; C. Alleles variation and frequencies detected in primers bnlg1671y17（N22）

表 1　参加生产试验玉米品种不同年份之间的遗传多样性

Table 1　Genetic diversity comparison of maize varieties tested in national production trial from different years

年度 Year	样品数目 Sample number	总等位基因数 Total allele number	总基因型数 Total gene type number	平均等位基因数 Average allele number	平均基因型数 Average gene type number	PIC 值 PIC index	平均杂合度 Average heterozygosity
2004	21	201	323	5.03	8.08	0.59	0.58
2005	78	254	526	6.35	13.15	0.60	0.60
2006	75	259	547	6.48	13.68	0.61	0.62
2007	42	249	466	6.23	11.65	0.61	0.59
2008	23	212	351	5.30	8.78	0.61	0.57
2009	50	255	501	6.38	12.53	0.62	0.63

不同区试组别玉米品种的遗传多样性比较：国家玉米品种区域试验主管部门依据生态区划、农业区划、品种类型、结合生产实际、播期类型等划分各个试验区组。328 份杂交种按照用途和区域试验组别划分为 10 组（根据 2009 年国家玉米品种区域试验方案划分），其中，西北春玉米组由于每年进入生产试验的样品量较少，东南玉米组是 2007 年才增加的一个区试组别，所以这两个组中样品量较少没有列入分析（表 2）。参与比较分析

的 8 个区试组品种的总等位基因和总基因型变异范围为 170~262 和 200~511，京津唐和西南组分别表现出了最低值和最高值，就平均等位基因和平均基因型数而言，京津唐和极早熟组相对较低，PIC 值京津唐组最低为 0.51，其余均接近于 0.60，平均杂合度均在 0.60 左右。说明除京津唐之外，各区组之间遗传多样性差异不大。

表 2 8 个玉米区试组之间的品种的遗传多样性比较
Table 2 Genetic diversity comparison among different maize regional trial groups

区试组别（简称） Regional trial groups (shortened form)	样品数目 Sample number	总等位基因数 Total allele number	总基因型数 Total gene type number	平均等位基因数 Average allele number	平均基因型数 Average gene type number	PIC 值 PIC index	平均杂合度 Average heterozygosity
京津唐夏播早熟玉米 JingJinTang（JJT）	14	170	200	4.25	5.00	0.51	0.59
东北早熟春玉米 DongBeiZao（DB）	39	217	386	5.43	9.65	0.58	0.62
东华北春玉米 DongHuaBei（DHB）	85	244	504	6.10	12.60	0.59	0.61
黄淮海夏播玉米 HuangHuaiHai（HHH）	45	224	395	5.60	9.88	0.58	0.60
西南玉米 XiNan（XN）	59	262	511	6.55	12.78	0.60	0.60
极早熟玉米 JiZaoShu（JZS）	11	173	229	4.33	5.73	0.56	0.58
鲜食甜糯玉米 XianShi（XS）	41	252	473	6.30	11.83	0.63	0.56
青贮玉米 QingChu（QC）	32	227	371	5.68	9.28	0.58	0.61

杂交种各区试组之间的遗传距离：以 8 个区试组别为单位分析各组别之间的 Nei 遗传距离（表3）。就各组之间的遗传距离而言，京津唐与其他组之间遗传距离最远，如京津唐与极早熟、鲜食、东华北、西南之间的遗传距离相对于其他组别之间最远，西南与东华北之间遗传距离最近，各组之间遗传距离相对较近的均集中在西南、东北华与青贮、黄淮海、东北早熟之间。

表 3 8 个区试组别之间的 Nei（1972）遗传距离
Table 3 Nei（1972）genetic distance among eight regional trial groups

区试组别 Shortened form	东北早 DB	东华北 DHB	黄淮海 HHH	京津唐 JJT	极早熟 JZS	青贮 QC	西南 XN	鲜食 XS
东北早 DB	0.000							
东华北 DHB	0.053	0.000						
黄淮海 HHH	0.081	0.059	0.000					

(续表)

区试组别 Shortened form	东北早 DB	东华北 DHB	黄淮海 HHH	京津唐 JJT	极早熟 JZS	青贮 QC	西南 XN	鲜食 XS
京津唐 JJT	0.164	0.209	0.124	0.000				
极早熟 JZS	0.146	0.182	0.194	0.277	0.000			
青贮 QC	0.114	0.064	0.086	0.183	0.220	0.000		
西南 XN	0.073	0.040	0.076	0.207	0.187	0.051	0.000	
鲜食 XS	0.184	0.164	0.156	0.257	0.178	0.178	0.159	0.000

杂交种各区试组的 PCoA 分析：对 8 个区试组进行主坐标分析，直观反映它们之间的关系（图 2）。鲜食与普通玉米能明显区分开，虽然在分布上有一定的重叠区域，但基本上各自占据相应的位置。青贮绝大部分品种与其他普通玉米分布重叠没有区分开，少数品种与普通玉米能区分开，但青贮玉米的区试对照品种（雅玉青贮 26 和雅玉青贮 8 号）分布偏离普通玉米区域。除了鲜食玉米之外，极早熟玉米与其他普通玉米在分布上也有一定的隔离（图 2A）。极早熟、京津唐、西南、东北早熟这四个组的品种遗传变异相对集中，极早熟、京津唐、西南三组之间样品遗传分布几乎无重叠区域；西南与东北早两区在地理分布上无任何交叉但遗传分布重叠（图 2B 和图 2C）。东华北、黄淮海的遗传变异较大（图 2D）。根据 8 个区试组别的具体分布情况将主坐标分析图划分成 4 个区域Ⅰ、Ⅱ、Ⅲ、Ⅳ，西南、东北早熟组分布在Ⅰ区，青贮玉米分布在在Ⅰ和Ⅱ区，但主要集中在Ⅰ区，鲜食和极早熟组主要分布在Ⅱ区，京津唐组分布在Ⅳ组，东华北和黄淮海在Ⅰ和Ⅳ区都有分布但东华北分布主要集中在Ⅰ区。根据 8 个区组分析结果显示，西南和东北早两区样品遗传分布几乎完全重叠，东华北和黄淮海两区样品遗传分布有重叠，但是两区具有一定的分化。将东华北与黄淮海两区，西南与东北早两区之间参试样品独立进行主坐标分析显示，东华北和黄淮海两区样品遗传分布有部分重叠但明显有分化，西南与东北早两区样品遗传分布几乎无重叠，并且对照品种分布于各自区域内（图 3）。

3 讨论

3.1 玉米品种遗传多样性分析

评价玉米品种遗传多样性可以从两个方面着手，一是参加区域试验样品的遗传多样性，二是生产上已经大面积推广应用的品种。本研究利用 40 对核心 SSR 引物系统分析 2004—2009 年参加生产试验和大面积推广玉米品种的遗传多样性状况。328 份样品在某些引物中等位基因的分布规律表现出了一定的种质利用趋同趋势，即在部分引物（如 N02、N19）表现出在一个或两个等位基因的分布频率非常高，而在其他的等位基因频率非常低的特点，这可能是由于农艺性状的选择导致了等位基因分布不平衡，也可能与某些资源尤其是优良骨干自交系的集中利用有关。近年来育成品种遗传多样性在不同年份间，不同区试组之间波动都不大。8 个区试组比较显示京津唐、极早熟区等位基因、基因型数目上相对较低，西南区显示出了最高值，就 PIC 值而言除京津唐之外其余区组差别不大。这可能与京津唐区域地理跨度最小，气候、土壤、耕作栽培特点等差异小有关，相反西南区域地理跨度较大，并且气候复杂，耕作栽培特点多样化，故品种多样化程度较高。

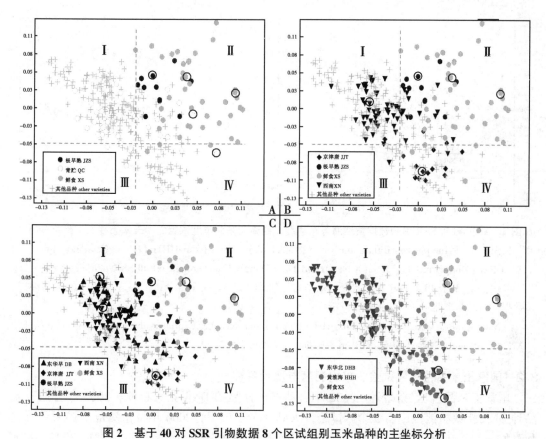

图 2 基于 40 对 SSR 引物数据 8 个区试组别玉米品种的主坐标分析
Figure 2 Principal coordinate analysis (PCoA) of varieties from eight regional trial groups based on 40 SSR primers data

注：A. 鲜食、极早熟、青贮、其他普通品种之间的相互关系；B. 极早熟、京津唐、西南各区品种之间的相互关系；C. 极早熟、京津唐、西南、东北早各区品种之间的相互关系；D. 东华北、黄淮海区域品种之间的相互关系。被黑色空心圆圈标示的为各区对照样品

Note：A. This figure reflected the relationship among XianShi, JiZaoShu, QingChu and other common maize varieties, B. This figure reflected the relationship among JiZaoShu, JingJinTang, XiNan groups, C. This figure reflected the relationship among JiZaoShu, JingJinTang, XiNan and DongBeiZao groups, D. This figure reflected the relationship between DongHuaBei and HuangHuaiHai groups. control varieties of each groups marked by a black circle

小麦、水稻育成品种的遗传多样性均有揭示品种遗传基础狭窄化的状况。本研究结果显示玉米育成品种遗传基础狭窄问题不明显，这应该与所研究材料有一定的关系，因小麦、水稻分析比较的材料时间跨度比较大，早期推广应用的地方原始品种比较多一些，从纵向上来比较遗传多样性可能在下降。另一方面也与作物本身的特征有一定的关系，小麦、水稻属于自交授粉并且大部分品种属于常规育种，而玉米属于杂交授粉而且目前全部是单交种，因此玉米类群本身遗传变异相对大，根据 Buckler 等的报道玉米种内的遗传变异非常高以至于两份玉米品种之间的平均差异度比人类和灵长类之间差异还大。

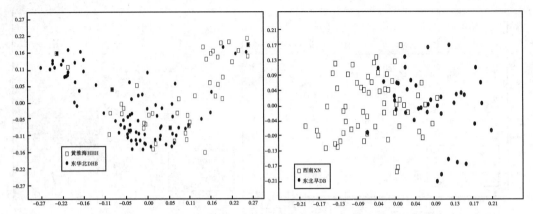

图 3 基于 40 对 SSR 引物数据黄淮海与东华北、西南与东北早区试组之间玉米品种的主坐标分析

Figure 3 Principal coordinate analysis (PCoA) of varieties from HHH (HuangHuaiHai) and DHB (DongHuaBei), XN (XiNan) and DB (DongBeiZao) based on 40 SSR primers data

注：A. 基于 40 对 SSR 引物数据，东华北与黄淮海 2 个区主坐标分析；B. 基于 40 对 SSR 引物数据，西南与东北早 2 个区主坐标分析

Note: A. PCoA analysis of DHB and HHH groups based on 40 SSR primers data; B. PCoA analysis of XN and DB groups based on 40 SSR primers data

3.2 国家玉米品种区域试验、审定制度与现代育种

国家玉米品种区域试验于 2002 年启动参试样品的 DNA 指纹检测，2004 年普通玉米品种 DNA 指纹检测范围由 5 个区试组扩大到 8 个区试组，2006 年将检测范围扩大到鲜食和青贮，并且在 2005 年要求所有参试样品均入长期库保存。本文选取了参加 2004—2009 年生产试验玉米样品，基本上代表了最新育成的品种情况，在一定程度上反映了中国当前一个时期玉米育种的水平和特点。对这些品种的遗传多样性和种植区域遗传分化进行分析评估，一是服务于中国玉米品种区试、审定管理工作，二是服务于品种选育工作，为两者提供数据支持。

国家玉米品种区域试验各区组育成品种与区组划分情况：玉米品种按用途分为普通玉米、鲜食、青贮和爆裂（2013 年新增加爆裂玉米）四大类，普通玉米适宜种植区域划分在 2009 年稳定为京津唐、东北早熟、东华北、黄淮海、西南、西北、东南、极早熟 8 个区域。划分玉米各区组是为了针对不同品种类型和生态特点筛选区域适应性的品种，从而使玉米品种的选育推广具有针对性。本研究结果显示育成的鲜食玉米具有类型特异性，可与普通玉米明显区分开。青贮玉米育种的遗传基础较宽泛，与普通玉米重叠分布，原因可能是与早期对照样品为粮饲兼用型玉米农大 108 有关，2007 年之后对照样品修改为青贮专用型玉米，因此，预期 2009 年之后的选育的品种会向青贮专用资源转变。极早熟样品具有区域特殊性，遗传分布相对集中且与其他普通玉米之间的遗传差异较大。东华北和黄淮海两个区不仅在地理分布上有重叠，主坐标分析结果显示其育成品种遗传分布也有部分重叠，但两区的参试样品具有的遗传分化。表明两区的育种资源具有部分相似性，如两区组中黄改系的应用；除此之外同一品种同时参加两区审定相对较多，如富友 9 号、安玉 13、农华 101 等，造成东华北和黄淮海两区品种存在种质利用趋同情况；但是根据 2 个区样品主坐标分析结果显示两区遗传分化是比较明显的（图 3A），这应该与两大主产区所涉及地理区域和生态区域差异性、适宜品种类型差异性、主要育种资源、育种模式差异性相关。

国家玉米品种区域试验各区组育成品种与对照品种关系：对照品种代表了各区最适宜推广种植的品种类型，对各区玉米品种选育发挥导向作用。本研究显示鲜食、京津唐、极早熟、东北早、西南区组的对照品种基本上能代表各自区域特异性。青贮区组对照品种2007年以来改为青贮专用型品种雅玉青贮8号和雅玉青贮26，以引导青贮专用玉米选育，主坐标分析结果显示雅玉青贮8号和雅玉青贮26的分布偏离普通玉米的整体分布，用它们作为青贮对照能够起到育种导向从粮饲兼用型向青贮专用型转变的作用。根据主坐标分析显示，东华北和黄淮海两区具有比较明显的分化，也有部分重叠。这与两区对照品种设置历史是相符的，两区对照品种早期（2004年以前）均为农大108，之后黄淮海区修改为郑单958，东华北除郑单958之外还设置其他对照品种。同时也说明2个区虽然有一定的重叠，但所涉及地理区域和生态区域差异比较大，育种资源、育种模式具有一定差异性，导致参试品种呈现一定的分化、倾向性。因此应该选择各自分化区域的代表类型作为其对照品种，从而在育种导向上明确这两个区的区域特殊性。

4 结论

通过利用SSR标记分析328份玉米品种的遗传多样性及遗传分化特点，发现近年来育成品种遗传多样性指数在不同年份间变化不大，除京津唐区之外在其他区试组间差异性也不大。参试品种主坐标分析显示各区试组划分、对照品种设置起到了育种导向的作用。

参考文献（略）

本文原载：中国农业科学，2014，47（5）：856-864

玉米纯度分子鉴定标准研制相关问题的探讨

王凤格 易红梅 赵久然

(北京市农林科学院玉米研究中心/玉米DNA指纹及分子育种北京市重点实验室,北京 100097)

摘 要:通过分析纯度鉴定和一致性鉴定的关系,探讨与纯度鉴定标准制定相关的8个问题,包括是否覆盖所有品种类型、是否鉴定所有杂株类型、是否提供杂交种的制种亲本、是否用田间小区种植鉴定结果进行校正、是否指定检测平台、是否在纯度鉴定前先鉴定真实性、是否所有样品都能鉴定、是否适用于转基因品种纯度鉴定,确定一套切实可行的纯度分子鉴定技术规范和标准。以玉米杂交种纯度分子鉴定为例,确立了鉴定引物选择和鉴定程序,并分析了弱苗、病苗、非典型株、种子发芽率对田间小区种植和室内分子鉴定结果吻合性的影响。

关键词:玉米;纯度鉴定;分子鉴定;SSR

Discussion about the Purity of Maize Standard Molecular Identification Development

Wang Fengge　Yi Hongmei　Zhao Jiuran

(*Maize Research Center, Beijing Academy of Agriculture & Forestry Sciences, Beijing Key Laboratory of Maize DNA Fingerprinting and Molecular Breeding, Shuguang Garden Middle Road No.9, Beijing 100097, China*)

Abstract: To determine a set of feasible technical standard of the purity molecular identification, the relationship between the purity and uniformity identification was analyzed, and discussed eight problems associated with purity identification standards including whether to cover all variety types, if identifying all miscellaneous plant types, whether asking for parent inbred lines, whether correction with field plot planting appraisal result, whether specified test platform, whether authenticity identification before purity identification, whether all the samples could be identified, whether applying to purity identification of genetically modified varieties. The appraisal primer selection and evaluation procedure of maize hybrid purity molecular identification is established. Four factors impacting the field plot planting appraisal results and molecular identification were analyzed, including weak seedlings, disease seedlings, atypical plants, seed germination rates.

基金项目:国家科技支撑计划项目(2011BAD35B09)

作者简介:王凤格(1975—),女,河北沧州人,副研究员,博士,主要从事玉米分子鉴定工作 Tel: 010-51503558;E-mail: gege0106@163.com

赵久然为本文通讯作者。E-mail: jiuran@263.net

Key words：Maize；Purity identification；Molecular identification；SSR

随着现代生物技术的不断发展，玉米种子纯度鉴定技术由传统田间小区种植鉴定逐步发展到集形态鉴定、蛋白电泳鉴定和分子标记鉴定为一体的综合鉴定技术体系，并出台了一系列纯度鉴定标准和规范。然而，以往提出的纯度分子鉴定方案，在检测实践中仍存在不足：一是为解决个别杂交种的纯度鉴定，需要进行大量的引物筛选工作，才能找到适合的鉴定引物，前期筛选工作量大，且仅限于少数杂交种的纯度鉴定；二是对同一品种、不同样品采用同一引物进行鉴定，造成某些鉴定结果与实际情况不符。玉米品种纯度分子鉴定面临的情况比较复杂，如何确定一套纯度鉴定的快速、准确的分子检测技术规范和标准，有很多问题需要深入探讨。

1 纯度鉴定与一致性鉴定的关系

由于纯度鉴定和一致性鉴定均是鉴定样品内部不同个体间是否整齐一致，且国内种子市场中仍存在个别品种在审定时一致性尚未完全达标就进入杂交种制种生产阶段，造成一致性问题和纯度问题并存，因此二者经常被混为一谈。但二者还是有明显区别的：从概念上来讲，一致性是指申请品种的植物新品种经过繁殖，除可以预见的变异外，其相关的特征或者特性一致；纯度是指品种在特征特性方面典型一致的程度，用本品种的种子数占供检本作物样品种子数的百分率表示。从检测目的上来讲，一致性鉴定主要是为了检测一个待测品种遗传稳定性，是成为一个新品种的必要条件；纯度鉴定是为了检测一个品种经过生产繁育后其特征特性方面是否保持一致，是反映种子质量好坏的重要指标之一。从检测对象上来讲，一致性鉴定的对象一般是未登记、审定或获得保护的试验用种；种子纯度鉴定的对象一般是大田生产用种。从鉴定方式上来讲，一致性鉴定由于主要鉴别特定品种本身（自交系或杂交种）特征特性的一致性，一般不考虑制种过程造成的混杂，因此对杂交种中明显的自交株或自交系中明显的杂交种是剔除不统计在内的；纯度鉴定由于自交株对实际产量影响最大，杂株对实际产量影响也会有不同程度的影响，因此主要监控杂交种生产过程中因去雄不彻底引起的自交株，以及因隔离不够外源花粉串粉造成的杂株等。从所处阶段上讲，目前一致性鉴定一般在品种区试阶段和申请品种权保护阶段进行，作为判定是否成为一个品种的前提条件之一；纯度鉴定一般在种子进入销售市场阶段进行，作为判定种子质量是否合格的指标之一。从鉴定结果来看，一致性合格的品种其生产的种子纯度未必全部都达标，两者不存在直接的相关性。

生产上普遍使用的玉米单交种是由两个自交系杂交组配而成的，品种一致性高低取决于亲本的情况，如果两个亲本自交系均是高度一致的，则组配的杂交种理论上也是高度一致的，如果只有一个亲本自交系一致性高而另一个亲本自交系一致性较差，或两个亲本自交系一致性均较差，则组配的杂交种一致性就会降低。

在纯度鉴定中，如果杂交种一致性差，筛选纯度鉴定引物的难度较大，否则对杂株判定时不易区分是由于制种过程中混杂造成的异型株，还是由于亲本遗传不稳定造成的非典型株。而如果杂交种一致性高，则只需要筛选到少数双亲互补型引物，就能对自交苗和异型株进行准确的判定，大大降低工作量和成本并提高结果的准确性。

2 制定玉米纯度分子鉴定标准时需考虑的几个问题

2.1 是否覆盖所有品种类型

玉米品种类型主要包括杂交种、自交系和地方品种等，其中杂交种按照杂交方式不同又分为单交种、三交种、双交种等。在这些品种类型中，单交种纯度鉴定相对简单，只需要鉴定出自交苗和明显的杂株，目前玉米审定品种类型主要为单交种，占总品种数的95%以上，解决了单交种的纯度鉴定问题，就解决了生产中的大部分问题，因此纯度标准可以优先解决单交种的纯度鉴定。影响自交系纯度的因素不同于单交种，广义的自交系纯度包括两个指标：位点纯合度（纯合稳定的位点所占比例）和杂株比例（杂株个体所占比例），这两个指标应分别设定判定标准，因此自交系的纯度鉴定标准应与单交种的分开。而地方品种及杂交种中的三交种、双交种等由于群体内部DNA水平上的一致性不高，纯度鉴定的难度较大。

2.2 是否要求鉴定出所有杂株类型

玉米单交种纯度鉴定中涉及的杂株类型主要包括以下几类：一是自交苗，包括母本自交系、父本自交系和其他自交系，一般是制种过程中去雄不彻底等原因造成；二是其他类型杂株，包括父本再利用造成的回交苗、制种过程中隔离不好导致外源花粉串粉造成的异型株及机械混杂带来的异品种等。从鉴定的难度和对生产的影响来看，自交苗对产量的影响最大，并且鉴定最容易，而其他类型杂株鉴定难度较大，且对产量的影响低于自交苗。因此，单交种的纯度鉴定标准应将重点放在自交苗的鉴定上，对其他类型杂株，可以作为选择鉴定项。

2.3 是否要求提供杂交种的制种亲本

是否提供制种亲本对鉴定结果没有明显影响，只是会造成提供的信息量减少，以及鉴定引物筛选的工作量增加：如果提供制种亲本，则可以预先了解亲本的遗传稳定性情况以决定是否继续进行纯度鉴定，鉴定时可以准确区分母本自交系和父本自交系。如果不提供制种亲本，则只能通过送检样品的小样本推测其亲本遗传稳定性情况，鉴定时能够鉴定出亲本自交系，但不能进一步区分是母本还是父本。实际检测中，如果是企业自检，一般能够得到制种的双亲，如果是政府监督抽查，很难得到制种双亲，因此纯度鉴定标准制定时不应要求必须提供制种亲本。

2.4 是否需要用田间小区种植鉴定结果进行校正

在行业标准《玉米种子纯度盐溶蛋白电泳鉴定方法》中，对盐溶蛋白电泳获得的纯度检测值进一步用回归方程 $Y = 52.9 + 0.461X$ 进行校正后作为最终检测结果，以国标GB4404.1判定被检样品是否合格。然而，电泳检测值是根据电泳谱带差异进行鉴定，田间小区种植鉴定是用形态性状差异进行鉴定，二者是不完全相同的，通过电泳值与田间种植鉴定值进行相关分析推出的回归方程进行纯度值校正也很难达到高度相符。在制定利用SSR标记法进行纯度鉴定的标准时，同样面临与田间小区种植鉴定结果的关系问题。一个更加可行的方案是将其作为独立检测系统，独立出具检测结果，并配套规定SSR标记法下的纯度质量标准。

2.5 是否指定检测平台

品种真实性鉴定是不同样品间的比较，为了便于数据库整合共享，玉米品种真实性鉴定标准中推荐采用变性PAGE电泳和荧光毛细管电泳检测平台。而纯度鉴定是样品内部不同个体的比较，没有数据库整合共享的要求，因此不需要指定检测平台，包括琼脂糖电泳、非变性PAGE电泳等各种检测平台都可以适用。可以综合考虑通量、成本、时间等因素选择合适的检测平台。

2.6 是否在纯度鉴定前先鉴定真实性

纯度鉴定和真实性鉴定是两个独立的鉴定项目，一般而言，真实性鉴定涉及的是样品之间的关系，纯度鉴定涉及的是样品内部不同个体之间的关系。如果纯度鉴定的前提是样品必须具有真实性，或者真实性鉴定的前提是样品必须纯度达标，则大大增加了鉴定的复杂度，无法简单快捷的开展纯度鉴定。实践中大量存在样品匿名提供或样品的标签名称与样品不符的情况，如果这些样品因真实性无法确定而不能继续进行纯度鉴定的话，将不利于实践中纯度鉴定的开展。

2.7 是否所有的样品都能鉴定

在纯度鉴定前，需要通过制种亲本或样品的小样本了解样品遗传稳定情况，如果样品只是在个别位点上存在遗传不稳定，则可以将这些位点剔除，选取遗传稳定的位点进行纯度鉴定；如果样品在多数位点上均存在遗传不稳定，严重干扰了对自交苗和异型株的准确判定，可终止试验，不再继续进行纯度鉴定。因此，并不是所有样品都能进行纯度鉴定。

2.8 是否适用于转基因品种的纯度鉴定

虽然玉米的转基因品种尚未批准使用，但对转基因品种的纯度鉴定研究需要提前储备。与非转基因品种的纯度鉴定不完全相同，转基因品种的纯度鉴定除了上述提到的鉴定内容外，还应包括特定转基因性状的纯度鉴定，即鉴定样品中表达所转的特定性状的个体占比。例如，对于抗虫转基因玉米品种，鉴定样品中具有抗虫基因的个体占多大比例，这需要针对所转基因设计特异标记进行专项鉴定，目前阶段的纯度鉴定标准尚不适用于此类鉴定需求，但一旦转基因品种批准使用，就需要及时提供这类品种的纯度鉴定技术标准。

3 玉米杂交种纯度的分子鉴定

3.1 鉴定引物的选择

3.1.1 纯度鉴定与真实性鉴定在引物选择上的异同

玉米杂交种纯度鉴定与真实性鉴定在核心引物选择上的基本要求相同：一是杂合率高，从而能够比较容易筛选到双亲互补型的引物；二是多态性高，从而具有较高的区分异型株的能力。但侧重点有所不同：从核心引物选择指标上看，纯度鉴定侧重于引物杂合率的指标，其次是多态性指标，而真实性鉴定侧重于引物多态性的指标，并要求所用的多个引物间没有连锁或连锁低，对杂合率指标没有特殊要求；从所用引物数量上看，纯度鉴定每份样品一般检测至少100个个体，增加检测的引物数量会造成工作量的成倍增加，因此侧重于选择适于特定品种纯度鉴定的特异引物，一般只需要1~2个引物即可，而由于玉米品种数量很大，要找到所有品种的特异引物是不可能的，因此真实性鉴定只能用多个引物的组合进行检测。

3.1.2 引物选择的3个层次

第一个层次是适于玉米品种纯度鉴定的通用核心引物的确定，其筛选过程是利用一套具有广泛代表性的玉米品种从大量引物名单中筛选确定出一套纯度鉴定通用核心引物，既不针对特定品种，也不针对特定样品，这类引物的要求一是杂合率高，从而能够比较容易筛选到双亲互补型的引物，二是多态性高，从而具有较高的区分异型株的能力。第二个层次是适于特定品种纯度鉴定的候选引物的确定，其筛选过程是利用适于玉米品种纯度鉴定的通用核心引物构建该品种的 DNA 指纹，并从中筛选出双亲互补型引物作为该品种纯度鉴定的候选引物，这些候选引物是针对该品种的，并不针对该品种的具体样品，所以数量要多于实际样品的鉴定需要。第三个层次是适于具体样品纯度鉴定的特异引物的确定。其筛选过程是选用适于该品种纯度鉴定的候选引物对待测样品的小样本进行分析，评估其纯度问题是由于自交苗、回交苗、其他类型杂株造成的还是由于遗传不稳定造成的，并确定适于该样品纯度鉴定的引物用于下一步实际鉴定，这些特定引物是针对该样品的最适引物，并不一定适合同一品种名称的其他样品。

3.1.3 引物选择的两个方案

玉米杂交种纯度鉴定的引物选择可采取固定一套鉴定引物组合的方案，或采取针对特定品种筛选特异引物的方案。这两种方案分别适用于不同的鉴定情况。

如果是企业内部质量控制，由于种子企业生产经营的品种数量较少，甚至多年仅生产 1~2 个品种，且制种亲本一次繁种后足够多年使用。因此，如果能够针对所生产的品种筛选特异引物，并在对亲本遗传稳定性了解的情况下，进一步筛选出适用于这批亲本生产出来的杂交种种子的纯度鉴定特异引物。这种方案虽然存在引物筛选过程，但对特定品种只需进行一次引物筛选就可应用到该品种大批量的种子纯度鉴定中，并保证了用最少的引物实现对杂交种种子纯度鉴定的准确可靠，因此企业一般会选择这种鉴定方案。

如果是政府监督抽查或检测机构承担委托检验，由于面对的是全国生产经营的品种，而每年市场上推广销售的品种数量上千种，且即使对同一品种，不同企业所用亲本一致性情况也千差万别。因此，如果针对每个品种筛选特异引物，并进而针对该品种的每个送检样品筛选鉴定引物的话，将面临引物筛选的时间和工作量超过用所选引物进行实际样品鉴定时间，而这种引物筛选推荐的引物只会应用到少数几份送验样品的鉴定中。此外，如果面对的是大批量送验样品，且样品的品种名称各不相同，即使为每个样品筛选到了合适的鉴定引物，由于为不同样品推荐的鉴定引物名单不同，也很难设计高通量的纯度鉴定。因此，如果能够根据已知品种 DNA 指纹库的信息筛选出一套普适性的引物组合应用于外控样品的纯度鉴定，就省掉了纯度鉴定特异引物的筛选环节，如果针对这套引物组合能实现多重 PCR 或多重电泳，则更便于组织大批量样品的纯度鉴定。

3.1.4 DNA 指纹库在纯度鉴定引物选择中的价值

以已知品种 DNA 指纹库中筛选出的引物名单为基础，在实际检测中进一步筛选适合具体样品纯度鉴定的引物，大大减少纯度鉴定的引物筛选工作量。在利用 SSR 标记进行玉米杂交种纯度鉴定时，第一步就是筛选双亲互补的纯度鉴定引物，而通过建成的已知品种标准 DNA 指纹库，首先可以初筛出每个杂交种的具有杂合带型的引物，一般在 40 对核心引物中，每个杂交种至少在 20 个引物位点上具有杂合带型；其次由于片段大小已经提供，可以进一步筛选出兼容多个检测平台的引物，对于琼脂糖凝胶电泳、非变性 PAGE 电泳，由于分辨率比变性 PAGE 电泳平台略低，因此可以选择杂合的且两条谱带片段大小差异较大的引物。

3.2 鉴定程序的确定

玉米杂交种纯度鉴定有两个目的：一是鉴定出自交苗，这是最主要的目的，仅需要1对双亲互补型引物就可以；二是鉴定出杂株，这需要用具有较强区分能力的特异引物或引物组合。

拿到待测样品后，如果该单交种为已知品种，首先从已知品种DNA指纹库中挑选在该品种上具有杂合带型的所有引物位点作为纯度鉴定候选引物，选用若干候选引物对待测样品的小样本进行初检，判断其纯度问题是由于自交苗、回交苗、其他类型杂株造成的还是由于遗传不稳定造成的，同时确定下适合待测样品纯度鉴定的引物名单。如果该单交种没有在已知品种DNA指纹库中，则利用推荐的纯度鉴定核心引物进行筛选和评估，确定该纯度问题是以自交苗、回交苗、其他类型杂株造成的还是由于遗传不稳定造成的，并挑选出合适待测样品纯度鉴定的引物名单。鉴定引物根据具体情况选用：自交苗造成的，选用1对引物；回交苗造成的，选用能够综合判定出回交苗的2~4对引物；其他品种造成的，选用能够鉴定出异品种的引物1~2对；遗传不稳定造成的，剔除遗传不稳定的位点后选用一致性表现良好的1~2对引物。

具体样品鉴定。从待测样品中随机抽取100个个体（一般为种子），用纯度鉴定引物进行鉴定。根据引物对每个个体的检测结果，统计并计算待测样品的纯度。如果提供了父、母本，统计时可将自交苗进一步区分为母本苗和父本苗；如果没有提供父母本的话，统计时中只需记录自交苗和异型株。

4 影响田间小区种植和室内分子鉴定结果吻合性的因素

4.1 弱苗

田间小区种植纯度鉴定时，受地力不均匀的影响，容易出现弱苗。实际现场鉴定中，如果是弱苗，往往会不予判定，即无法判定是正常株还是自交株，但在计算纯度时，总株数并不将这些未做判定的株排除，实际上就相当于默认是正常株了，这会导致田间鉴定的纯度值偏高；相反，如果田间鉴定经验不足，将正常杂交种的弱苗误判为自交苗，则会导致田间鉴定的纯度值偏低。在田间纯度鉴定中，往往会出现将正常株的弱苗判定为自交苗的情况。因此，田间小区种植鉴定对地力均匀度和管理的精细程度要求比较高，当田间的地力严重不均时是不适合纯度鉴定的。

4.2 病苗

植株在田间生长过程中，容易受病害影响而感病，出现病苗，感病类型不同，对纯度鉴定的影响也不同。如果是粗缩病，因植株矮小，则造成或者无法判定，或者误判为自交苗；如果是丝黑穗病，因穗部性状无法鉴定，而植株最大量的表型性状都集中在穗部，因此往往无法判定；如果是锈病、大小斑病等叶部病害，则视发病严重程度，对判定有不同程度的影响。因此，田间小区种植鉴定的地块必须做好防病虫工作，否则就会造成无法准确鉴定的情况。

4.3 非典型株

如果纯度问题主要由于亲本未纯合稳定一致或外源花粉串粉，则会导致很难找到典型的正常杂交种和自交苗。田间容易将弱苗判定为自交苗，其实是非典型株或正常株；室内虽然能够通过多引物组合，辨别出非典型株，但会大大增加纯度鉴定的工作量。

4.4 种子发芽率

室内 DNA 指纹鉴定是通过对种子的基因组 DNA 进行鉴定,无论种子是否能够正常发芽,都能提取 DNA 并得到鉴定;田间小区种植鉴定是通过对生长的植株表型进行鉴定,当种子发芽率低时,能够正常发芽的种子可以鉴定,发芽势弱的种子容易长成弱苗造成鉴定的难度,不能发芽的死种子未能得到鉴定。当拿到一份样品后,如果样品发芽率高,则室内 DNA 指纹鉴定和田间小区种植鉴定面对的鉴定群体基本相同,鉴定结果吻合度较高;如果样品发芽率低,由于样品中的杂交种子的整体发芽率往往比亲本种子的整体发芽率高,造成室内 DNA 指纹鉴定比田间小区种植鉴定的纯度偏低。在检测实践中,如果送检样品送样后未及时鉴定,放置较长时间导致发芽率明显降低,则 DNA 指纹鉴定结果与样品放置时间长短无关,而田间小区种植鉴定结果可能会随着样品放置时间延长而变化。

目前,种子检验中常用的纯度检测方法有田间小区种植、蛋白电泳、分子标记鉴定等,田间小区种植鉴定技术比较简单、直观,但周期长,检测量很难提高,海南异地鉴定成本较高;蛋白电泳鉴定技术简单快速、成本较低,但多态性有限,不能适合所有品种;SSR 标记鉴定技术程序比蛋白电泳复杂,成本比蛋白电泳高,通量中等,但由于多态性高,理论上能解决所有品种的纯度鉴定问题,是目前较理想的纯度鉴定分子标记;SNP 和 INDEL 等新型标记具有统计简单和高通量的优势,将成为今后研究的重点。应根据实际需要灵活选择不同的鉴定技术。

参考文献(略)

本文原载:玉米科学,2015,23:(4):48-53

玉米真实性 SSR 鉴定标准的研制及应用

王凤格 易红梅 赵久然 田红丽 任 洁 葛建镕 王 璐

(北京市农林科学院玉米研究中心，北京 100097)

摘 要：全面总结真实性鉴定的类型及与其他鉴定项目的区别，系统分析我国玉米真实性 SSR 鉴定标准的研制过程中的关键问题，比较不同作物品种真实性鉴定标准研制的异同，为其他作物类型及其他分子标记类型的真实性鉴定标准研制提供借鉴。玉米真实性鉴定标准在我国已得到快速推广和应用，为种子市场的繁荣和稳定提供了有力的技术支撑。

关键词：玉米；真实性鉴定；分子鉴定；SSR

Development and Application of Maize Authenticity Identification Standard by SSR Technology

Wang Fengge Yi Hongmei Zhao Jiuran
Tian Hongli Ren Jie Ge Jianrong Wang Lu

(*Maize Research Center, Beijing Academy of Agriculture & Forestry Sciences, Beijing 100097, China*)

Abstract: The types of authenticity identification and differences with other identification project were fully summarized, and systematically analyzed the key issues in the development process for standard of maize authenticity SSR identification in China as well as compared the similarities and differences of developing authenticity identification standard with other corps, aiming to offer a good reference to development of authenticity identification standard for other types of corps and other types of molecular markers. Maize authenticity identification standard has been widely promoted and applied in China, providing strong technical support to prosperity and stability of seed market.

Key words: Maize; Authenticity identification; Molecular identification; SSR

《种子法》颁布实施以来，在繁荣种子市场的同时，品种未审先推、制售假劣种子及套牌侵权等违法行为日益增多，造成了种子市场的混乱，严重影响了种业健康持续发展，这些行为的发生，归根结底还是品种真实性鉴定问题。田间种植鉴定由于费时费力和受环境影响

基金项目：国家科技支撑计划项目（2015BAD02B02）
作者简介：王凤格（1975— ），女，河北沧州人，副研究员，博士，主要从事玉米分子鉴定工作。
　　　　　Tel：010-51503558；E-mail：gege0106@163.com
　　　赵久然为本文通讯作者。E-mail：jiuran@263.net

较大等缺点，在实际检测中很少被采用；蛋白质电泳方法由于区分能力有限，面对品种数量越来越多和品种间差异越来越小的现状，已不能满足真实性鉴定需要；DNA 指纹鉴定技术具有不受环境影响及快速准确的优点，成为目前玉米品种真实性鉴定的重要手段，其中 SSR 检测技术较为成熟，具备普遍推行的条件。制定玉米品种真实性 SSR 分子标记检测技术规范，从源头上保证种子质量，保障农业生产安全，具有非常重要的现实意义。本文概述了真实性鉴定的类型及与其他鉴定项目的区别，系统总结并深入探讨了我国玉米真实性 SSR 鉴定标准的研制过程中的关键问题，比较了不同作物品种真实性鉴定标准研制的异同，以期对其他作物类型及其他标记类型的真实性鉴定标准研制提供有益借鉴。

1 真实性鉴定的类型

目前真实性鉴定需求主要来自种子质量管理、品种管理、法院司法鉴定、企业及个人维权等方面，从检测目的看，可分为真实性验证和真实性身份鉴定两大类。

真实性验证是指待测样品与其对应品种名称的标准样品比较，检测该样品的品种名称标注是否名副其实。这种检测需求最为普遍，具体分以下几种情况：①与来自农业部品种权保护的标准样品比较，主要目的是鉴定是否发生侵权，送样单位一般为司法机构或品种权人。②与已审定品种标准样品比较，主要目的是进行市场打假，规范市场秩序，送样单位一般为各省种子管理站、工商管理局等行政执法部门。③与模仿对象进行比较，育种单位采用模仿育种的方式进行育种，对目前主推品种郑单 958、先玉 335 等的亲本进行细微改良，鉴定改良后的品种与原品种在 DNA 水平上是否发生了明显变化，是否可以作为不同品种使用。④姊妹系或近等基因系之间比较，育种单位采用连续回交、诱变、突变、转基因等方式选育的自交系与其原始自交系之间，或采用二环系方式选育的连续自交 6 代后的不同姊妹系之间，遗传相似度一般比较高，需要确定是否可以作为新自交系使用。⑤跟踪品种种性是否发生变化，主要目的是监控品种在多年的繁育过程中，或者育种者持续提纯复壮可能会带来的种性上的改变，特别是一些推广年限较长的品种，有可能与审定时相比已有明显改变，应作为不同品种重新进行区试审定。⑥同一品种不同来源样品之间比较，在区试中主要监控同一品种在不同参试年份或不同组别是否发生和更换，在品种权保护中主要监控由不同申请单位提供的同名的近似品种样品是否相同，在育种中主要监控收集的不同来源同名自交系是否相同。

真实性身份鉴定是指待测样品通过与已知品种标准指纹数据比对平台筛查比较，确定该样品的真实品种名称。这种检测需求的实现需要构建比较完备的已知品种标准指纹数据库，具体分以下几种情况：①在种子市场打假中，鉴定市场抽检的品种仿冒了什么已知品种。市场销售的种子会存在以甲品种的名称销售乙品种的种子的情况，郑单 958 等品种由于丰产性好，农民比较认可，但这些品种已经获得品种权，未经品种权人许可不能销售，在这种情况下，就会出现以其他品种名义实际销售该品种的种子的情况。②在国家和各省区试中，鉴定区试品种是否与已知品种雷同，并作为品种能否推荐审定的必要条件，区试品种应是新品种，与所有已知品种不同，否则不能通过审定。③在品种权保护中，辅助筛查申请品种的最近似品种，代替原来由申请者自行提供的近似品种，由于申请者自己掌握的已知品种数量有限，申请者提供的近似品种并不一定是申请品种的最近似品种，利用完备的已知品种库，可以将待测品种与库内所有品种进行比较，并筛选出最近似的品种。

2 真实性鉴定与其他鉴定项目的区别

2.1 与特异性鉴定的比较

真实性鉴定和特异性鉴定均是检测不同品种（样品）之间的关系，但二者还是有明显区别的：从概念和应用范畴上讲，真实性是指一批种子所属品种、种或属与文件描述或与备案的标准样品是否相符，是种子检测系统中使用的名词；特异性是指申请品种权的植物新品种应当明显区别于在递交申请以前已知的植物品种，是品种权保护系统中使用的名词；从检测目的上讲，真实性鉴定是检测种子是否与标示的品种名称相符，是否具有真实性是判定种子质量是否合格的指标之一；特异性鉴定是检测申请品种是否与所有已知品种都不同，是否具有特异性是判定是否成为新品种的必要条件；从所处阶段及检测对象上讲，真实性鉴定一般在种子进入销售市场阶段进行，检测对象是进入生产和销售环节的商品种子，特异性鉴定一般在区域试验阶段和申请品种权保护阶段进行，检测对象是尚未通过审定或获得授权的试验用种；从检测方案上讲，真实性鉴定主要是待测样品与标准样品比较，如果不符则判定为真实性有问题，特异性鉴定是待测品种与所有已知品种比较，如果均不同则判定具有特异性。

两者在检测技术方法选择上是具有共性的，在检测实践中两种检测需求也存在一定交叉。在区域试验阶段，一方面需要保证待测组合与所有已知品种不同，即具有特异性；一方面也需要保证待测组合在不同年份、不同区组或不同省份参试时不会发生更换组合，即具有真实性。由于国内种子市场中仍存在个别品种在审定时不具有特异性就进入生产销售阶段，往往造成真实性和特异性的问题并存。

2.2 与纯度鉴定的比较

真实性和纯度均是种子检测系统中使用的名词，检测对象是进入生产和销售环节的商品种子，是判定种子质量是否合格的指标。检测实践中许多人往往将真实性鉴定和纯度鉴定混为一谈，在制定真实性鉴定标准时，有人提出在真实性鉴定前必须先进行纯度鉴定，剔除不代表本品种的个体后，用标准个体进行真实性鉴定；在制定纯度鉴定标准时，有人提出在纯度鉴定前必须先进行真实性鉴定，只有真实性没有问题的样品才能进行纯度鉴定。实际上，真实性鉴定和纯度鉴定是两个独立的鉴定项目，二者具有明显的区别，真实性鉴定涉及不同样品之间的关系，检测结果用不同、近似、相同等表示；纯度鉴定涉及同一样品内部不同个体之间的关系，检测结果用本品种的种子数占供检样品种子数的百分率表示。

2.3 与转基因鉴定的比较

真实性鉴定和转基因鉴定是目前种子检测机构正在开展的两类分子鉴定项目，两类项目采用的技术手段相似，在实验室硬件条件和人员培训上的要求基本相同，因此许多实验室都在同时开展这两项鉴定。二者的差异从鉴定标记上看，真实性鉴定是对样品整体遗传背景的检测，选用的 SSR 标记具有物种特异性，玉米真实性鉴定的核心引物并不适用于其他物种；转基因鉴定是对目的基因的鉴定，检测样品中是否含有外源基因，选用的鉴定标记只与所鉴定基因有关，具有物种通用性。从检测方式看，真实性鉴定涉及样品之间的关系，预先构建可共享的已知品种标准 DNA 指纹库对开展真实性鉴定具有重要价值；转基因鉴定仅涉及待测样品本身，不需要预先构建 DNA 指纹库。从试验污染控制上看，真实性鉴定的试验污染主要出现在样品准备阶段，系同批次样品的相互交叉污染，环境中的外源核酸一般不会对检测产生影响；转基因鉴定的试验污染主要出现在PCR扩增及电泳阶段，环境中的外源核酸容易导致检测结果出现假阳性，因此转基因鉴定的试验污染控制更加严格。

3 真实性鉴定标准研制中的关键指标

3.1 样品取样方式和取样量的确定

选择最佳的取样方式和最优的样本数量在真实性鉴定中具有重要意义，不仅关系到检测结果是否准确性，还关系到工作量和检测成本等实际问题。样品取样方式可分为混合取样和个体取样两种，选取何种取样方式取决于样品一致性情况。根据种子繁殖方式不同，可分为无性繁殖、严格自花授粉、人工自交、人工杂交、常异花授粉、天然异花授粉等类型，无性繁殖、严格自花授粉、人工自交或单交的样品预期一致性较高；常异花授粉、天然异花授粉、除单交外的其他人工杂交类型的样品预期一致性较低。样品一致性高低还与制种生产过程有关，因此，预期一致性较高的样品也会出现一致性较低的情况。对一致性高的样品采用混合取样，可以减少工作量和试验耗费，提高工作效率。

玉米品种根据种子繁殖方式不同，可分为自交系、单交种、多交种、开放授粉品种等，对于自交系或单交种，实际检测中应分析至少20个个体的混合样品，或分析至少5个个体的单个样品。对于三交种、双交种及开放授粉品种等，应分析至少20个个体的单个样品，实际检测中一般推荐采用30个个体的单个样品。对于玉米品种而言，从其种胚、幼苗、叶片、苞叶中提取的DNA代表了该品种的基因组序列特征，但从杂交种种子上的果皮提取的DNA代表了母本自交系的基因组序列特征，从杂交制种田母本植株所结的果穗籽粒提取的DNA代表了下一代杂交种的基因组序列特征。

3.2 核心引物筛选确定和使用

3.2.1 核心引物筛选确定

适于玉米品种真实性鉴定的引物组合的筛选原则：品种间多态性高；能够有效区分等位变异，数据容易统计；扩增重复性好，非特异扩增片段少；引物位点在染色体上均匀分布，避免紧密连锁；突变率低；避免选择零等位变异；根据不同物种染色体数目的多少确定核心引物的数量；引物扩增产物大小范围合适，具有多重电泳的潜力。某些单个引物评价较好，但是却没有入选到核心引物组合中，原因在于一套核心引物组合需兼顾引物间无紧密连锁、具备组合多重扩增或多重电泳的潜力等原则。

核心引物筛选程序，根据国内外研究文献、遗传信息数据库等列出的引物位点及其多态性评价结果等信息，进行引物多态性的筛选，为保证筛选的引物适用于我国品种的鉴定，筛选材料选取时应包括国内有代表性品种；对初筛入选的引物再增加样品量进行筛选，综合引物组合区分效率、均匀分布、扩增片段大小等原则确定候选引物组合。最后对候选引物进行多实验室联合评估，检验引物在不同的检测设备环境中的重复性和稳定性。

若选用的引物数目太多，工作量会成倍增加，不符合鉴定工作的实际；若选用的引物数目太少，又不能有效区分鉴定所有品种，易引起误判。经过上述筛选程序，玉米筛选确定40对SSR引物作为真实性鉴定的核心引物，并据此构建了玉米已知品种的SSR指纹数据库。

3.2.2 核心引物的使用

品种真实性验证强调的是对结果的否定,对引物数量要求不高,可以先用第一组 20 对引物进行检测,检测到可以判定不符结果的差异位点数时,可终止检测,否则继续利用第二组 20 对引物进行检测。品种真实性身份鉴定强调的是对结果的肯定,通过与玉米品种 SSR 指纹数据比对平台进行比较,可以筛查至具体品种,检测时可直接采用全部 40 对 SSR 引物进行检测,经比较后仍与几个已知品种存在没有位点差异而无法得出结论的,必要时可采用其他能够区分的 SSR 标记进行检测。

3.3 检测程序的选择

3.3.1 DNA 提取环节

DNA 提取方法有很多种,常见的有 CTAB 法、SDS 法、吸附柱法、磁珠法、碱煮法等提取方法,以及以这些方法为基础的提取纯化试剂盒及各种改良方法。无论采用何种 DNA 提取方法,只要 DNA 总量和质量符合 PCR 扩增的要求,对真实性检测结果一般不会造成影响,因此在玉米 SSR 检测标准中未对其进行强制规定。实际检测中可综合考虑提取质量、总量、效率、成本等因素选择合适的 DNA 提取方法。如果 DNA 用量较大,质量要求较高,需要长期保存,一般选用 CTAB 法,并用幼嫩叶片或幼苗提取 DNA;如果 DNA 用量较少,质量要求不高,不需要长期保存,但要求提取方法成本低效率高,一般选用 SDS 法或碱煮法;试剂盒提取法具有快捷高效的优点,但成本仍然偏高,适用于 DNA 用量较少,但 DNA 质量要求较高的情况。

3.3.2 PCR 扩增环节

反应体系和反应程序可根据扩增仪型号、酶、引物等不同进行优化调整。引物既可以进行单引物扩增,也可以通过优化体系进行多引物组合扩增;反应程序可以采用统一的程序,也可以根据不同引物或不同类型扩增仪的特殊要求推荐不同的反应程序。为了便于安排试验及保证试验的规范性,玉米 40 对核心引物的扩增采用了相同的反应体系及反应程序。

3.3.3 扩增产物检测环节

检测平台的选择是开放的,尽管目前主流的 SSR 检测平台都是电泳检测平台,但随着技术的发展,已出现了基于测序的 SSR 检测平台,只要适合检测目的都可以选择使用。考虑到不同检测平台的分辨率、灵敏度等差异较大,对检测结论会造成一定影响,因此在出具检验报告时须注明所使用的检测平台。玉米 SSR 鉴定标准中推荐了变性 PAGE 凝胶电泳和荧光毛细管电泳两种检测平台,其中变性 PAGE 凝胶电泳平台的检测通量和自动化程度较低,数据整合难度较大,适于少量样品的成对比较,一般只用于真实性验证;荧光毛细管电泳检测平台的通量和自动化程度较高,数据整合较容易,适合大量样品检测及数据库比较,能够满足真实性身份鉴定的需求。

3.4 指纹数据的采集记录

SSR 指纹数据的采集记录方式对数据库构建及数据统计分析具有重要影响,数据采集记录方式的发展经历了 3 个阶段:①传统的 0/1 编号方式,这种方式是将每个标记位点上的不同等位变异(不同谱带)均设为不同的字段,统计每个样品是否存在该等位变异,赋以 0(无)或 1(有)的值。该方式是 SSR 指纹数据统计最早使用的一种方法,但存在几个缺陷:一是不同引物位点上扩增产物的基因型统计的位数不同,不便于数据的规范化;

二是随着样品的增加，会不断出现新的等位基因，原有编号方式不利于新等位基因的添加；三是不同实验室由于分析的材料不同，等位基因的有无和相对位置不同，数据库不易整合。②二位代码描述方式，根据 SSR 标记的重复序列及侧翼序列，可以确定该 SSR 位点上具有最小片段值的等位基因，将该等位基因命名为 01，其他等位基因按照 SSR 位点的重复基元依次累加，分别命名为 02、03……，二倍体植物物种的纯合位点的基因型数据记录为 X/X，杂合位点的基因型数据记录为 X/Y，其中 X、Y 分别为该位点上两个不同等位基因的命名。该方式记录的数据格式标准化，降低了数据库整合的难度，在玉米、水稻、小麦等作物中都曾推荐使用，但需要每次将所有 SSR 位点上出现的等位基因作为分子量标准同时电泳，才能对每个等位变异进行统计，否则容易误判等位基因的相对位置。③等位变异大小统计法，直接记录待测品种在每个引物位点上的电泳谱带的片段大小，每个样品在每个引物位点上的谱带号用两条电泳谱带的片段大小描述，按从小到大的顺序记录，中间用"/"隔开。由于统计的数据为等位变异的片段大小，数据更加直观，更加有利于不同来源数据的整合和数据的标准化，目前玉米、水稻等作物在指纹采集中已开始采用该方法。

3.5 参照样品的筛选和使用

参照品种是对应于 SSR 位点上不同等位变异的一组品种，使用参照品种的目的是辅助确定待测样品的等位变异，以及校正仪器设备的系统误差。如果采用荧光毛细管电泳检测，只需从核心参照品种名单中选择 1～2 份参照样品使用，如果采用变性 PAGE 凝胶电泳检测，则需根据所检测的核心引物而定，一般需选取该引物所有等位变异对应的参照样品组合。

参照样品筛选时，为了提高参照样品使用时的便利性，一方面需要考虑参照样品的代表性和易获取性，一方面需要考虑如何用最少的参照样品代表真实性鉴定引物位点的全部等位变异。玉米真实性鉴定参照样品的筛选以 2 895 个农业部审定品种和 1 861 个品种权保护品种作为目标来源库，核心参照品种的选取方法参考了核心种质构建中的逐步聚类法，采用分层取样方式进行逐步聚类，兼顾品种的代表性和常用性，用农业部公布的玉米品种推广面积作为判断品种常用性的依据，得到了包含基因频率 0.05 以上的所有等位变异的 20 个品种作为核心参照品种，为了补充某些缺少的稀有等位基因，又增添了 20 个扩展参照品种。最终确定的 40 个玉米参照品种覆盖了几乎全部的等位变异，分为核心和扩展两组，核心参照品种共 20 个，包含了基因频率在 0.05 以上的所有等位变异；扩展参照品种共 20 个，主要补充基因频率在 0.05 以下的稀有等位变异。

3.6 结果判定

3.6.1 位点差异的判定

位点差异情况分为下列 4 种：存在差异、完全相同、数据缺失、无法判定。采用混合样品分析时，样品一致性差是造成个别位点无法判定的主要原因，这些无法判定的位点只有当影响到样品比较结论的出具时，才需要利用多个个体统计基因型频率的方式进一步明确判定，否则只需记录为无法判定即可。

3.6.2 判定指标的选择

从国际上看，无论是人类的个体鉴定还是大部分植物的品种鉴定，主要采用差异位点数作为判定样品之间关系的指标，很少采用品种选育或种质资源分析中常用的遗传相似度指标。分析其原因,主要基于以下几点考虑。

①品种真实性鉴定与品种遗传关系分析的目的并不相同,差异性状/位点是判定样品异同的衡量指标,而遗传相似度则作为材料遗传关系远近的衡量指标。在品种权 DUS 形态测试中,也是采用评价个别性状是否存在差异进行特异性判定。而随着分子育种技术的发展,在遗传背景高度相似的情况下,个别性状/位点的差异有可能会作为区分不同品种的指标。②结果表示的严谨性,鉴定引物数目较少时,如 40 对 SSR 引物,采用遗传相似度进行判定,每个引物结果占比就达到 2.5%,使得结果的判定及描述不够严谨,而采用差异位点数目就不存在上述问题。③采用差异位点数的判定方式,还有利于试验的灵活安排,减少不必要的试验,在实践中可操作性更强。在玉米品种真实性鉴定标准中推荐先做 20 对核心引物,如果判定待测样品与标准样品不符,就可以停止试验,不需要继续做完剩余 20 对引物,从而减少不必要的检测工作量。而如果采取遗传相似度的判定方式,只有将标准推荐的 40 个引物全部做全,才能保证不同实验室得出的遗传相似度数值完全一致。

4 不同作物真实性鉴定标准的比较

不同作物真实性鉴定标准研制具有很强的共性,从国际上看,国际植物新品种保护联盟(UPOV)为不同植物物种建立了一套统一的分子检测试验方法,并形成了 BMT 分子测试指南,该指南在分子标记的选择、分析材料的来源类型及样品量、分析方法的标准化、数据库构建的标准化方面对不同植物分子鉴定进行了原则上的规范。从国内情况看,在所有作物中,玉米率先开展了真实性鉴定标准的研制并得到广泛的推广应用,基于玉米真实性鉴定的成功经验,并参考 BMT 分子测试指南,制定了《植物品种鉴定 DNA 指纹方法总则》,对各个作物分子鉴定标准研制进行了规范,在此基础上,我国已经颁布的 14 个作物分子鉴定标准,并不断进行修订和完善。

尽管如此,不同作物真实性鉴定标准研制仍具有较强的物种特异性,从样品准备上看,由于不同作物主要品种的繁殖类型和繁殖方式不同,样品预期一致性情况也不一样,因此适合的取样方式不同,玉米、水稻品种以单交种和纯系为主,小麦、大豆品种以纯系为主,马铃薯、甘薯品种以无性繁殖为主,这类样品预期一致性高,主要采用混合样品的取样方式,许多牧草品种仍以综合种为主,样品预期一致性较低,主要采用多个个体样品的取样方式。从核心引物选择上看,由于 SSR 引物具有种属特异性,不同作物所选用的核心引物不具有通用性,每种作物均需研制适合本物种的核心引物,由于不同作物的倍性不同、染色体数目不同、基因组大小不同、可用引物数量和多态性不同,不同作物所选用的核心引物数量和入选标准也不同。

5 实际应用

大规模开展真实性分子鉴定需具备 4 个条件:①具备已知品种的标准样品;②制定真实性分子鉴定标准方法;③构建已知品种标准指纹数据库;④具备有真实性检测能力和资质的种子检测机构。经过 10 余年的努力,我国玉米真实性分子鉴定的上述 4 个条件均已具备,自 2010 年起开展了玉米审定品种标准样品收集工作并基本征集齐全;2007 年颁布实施玉米品种 DNA 指纹鉴定标准,并于 2014 年进一步修订完善;已完成 4 013 份审定品种和 1 861 份品种权保护品种的标准 DNA 指纹库构建；已认定了 16 家具备种子真实性检测能力和资质的部级种子检测机构。

  玉米真实性鉴定标准在我国已得到快速推广和应用，为种子市场的繁荣和稳定提供了有力的技术支撑。从政府需求看，自 2010 年起玉米真实性纳入到农业部种子质量管理中，在春季种子市场监督抽查中首次包括了真实性检测项目；在扩大春季种子市场监督抽查规模的基础上，2011 年起增加了冬季种子企业督查行动；2012 年起增加了夏季制种基地巡查；2013 年起增加了典型县市的种子市场摸底行动。至此，我国玉米种子真实性管理实现了从制种基地源头到企业加工中间环节到种子销售市场终端的全程监控体系。从社会需求看，随着玉米真实性检测标准的颁布实施，来自企业维权、工商打假、司法鉴定、科研育种、农民维权等方面的检测需求呈逐年上升趋势。

参考文献（略）

本文原载：玉米科学，2016（4）：61-66

玉米 SSR-DNA 指纹库构建方案在高粱中的通用性

周青利[1,2]* 王蕊[1]* 张春宵[3] 周海涛[4] 易红梅[1] 王凤华[4]
李晓辉[3] 田红丽[1] 葛建镕[1] 席章营[2]** 王凤格[1]**

(1. 北京市农林科学院玉米研究中心/玉米 DNA 指纹及分子育种北京市重点实验室，北京 100097；2. 河南农业大学农学院，河南 450002；3. 吉林省农业科学院作物资源研究所，长春 136100；4. 吉林省农业科学院/农业部植物新品种测试分中心，公主岭 136100)

摘 要：近年来，随着主要农作物品种 SSR 分子鉴定技术规程的相继颁布，玉米、水稻已经建立了标准样品 SSR-DNA 指纹数据库，其他作物正在启动，整合一套稳定性好、通用性高、操作便捷的建库方案对多种作物 DNA 指纹库的构建具有重要意义。本研究以 6 份高粱和 3 份玉米标准样品为例，以高粱和玉米品种真实性检测标准为基础，研究跨作物应用玉米 DNA 指纹库构建方案的可行性。研究表明将玉米 DNA 指纹库构建方案应用于高粱中是可行的，这为后续探究其他作物 DNA 指纹库构建通用性方案研究奠定基础。

关键词：高粱；玉米；DNA 指纹库；通用性

A Study on Universal Application of Maize SSR-DNA Fingerprint Database in Sorghum

Zhou Qingli[1,2]* Wang Rui[1]* Zhang Chunxiao[3] Zhou Haitao[4]
Yi Hongmei[1] Wang Fenghua[4] Li Xiaohui[3] Tian Hongli[1]
Ge Jianrong[1] Xi Zhangying[2]** Wang Fengge[1]**

(1. Beijing Key Laboratory of Maize DNA Fingerprinting and Molecular Breeding, Maize Research Center, Beijing Academy of Agriculture & Forestry Sciences, Beijing 100097; 2. Agricultural College of Henan Agricultural University, Henan 450002; 3. Crop Resources Institute, Jilin Academy of Agricultural Science, Changchun 136100; 4. Gongzhuling Station for Testing of New Varieties of Plant, MOA, Jilin Academy of Agricultural Sciences, Gongzhuling 136100)

Abstract: In recent years, along with the main crop varieties of SSR molecular identification procedures successively promulgated, maize (*Zea mays* L.), rice (*Oryza sativa*) has established a

---

基金项目：本研究由科技部国家科技支撑计划（2015BAD02B02）和北京市农林科学院院科技创新能力建设专项（KJCX20161501）共同资助

* 同等贡献作者

** 通讯作者：XIZHANGYING@163.com；gege0106@163.com

standard sample SSR-DNA fingerprint database, and other crops are starting. Some crops have been launched the DNA fingerprint database. It is important to integrate a general solution for constructing crops DNA fingerprint database. Based on sorghum [*Sorghum bicolor* (L.) Moench] and maize standards, this paper studies the universality of constructing maize DNA fingerprint database cross species with 6 sorghum and 3 maize varieties. Result shows the application of the solution of maize DNA fingerprint database to sorghum is feasible, lays the foundation for the universal solution of constructing other crops DNA fingerprint databases.

Key words: Sorghum; Maize; DNA fingerprint database; Universality

随着分子标记技术的发展，使用 SSR 标记的品种鉴定技术已相对成熟。近年来玉米（NY/T 1432—2014）、水稻（NY/T 1433—2014）等多种农作物品种鉴定技术规程相继颁布，涵盖作物不仅包括中国几种主要农作物，也包括瓜果、蔬菜以及重要经济作物，相关技术规程的颁布为各作物构建 DNA 指纹建库奠定了基础。然而目前中国大部分农作物依据前期研究分别研制品种鉴定标准，研制过程中较少考虑程序在多平台的稳定性、通用性、便捷性等因素，如各作物品种鉴定技术规程的 DNA 提取方式多样化、规程间 PCR 体系程序均不一致、部分规程内部引物间退火温度也不尽相同、多种电泳方式检测等问题，以致检测机构在检测多个作物任务时需要同时执行多套标准、多种体系和程序，同时额外增加的试验流程和仪器设备，对于检验人员增加了试验繁琐程度，因此有必要建立农作物通用性 DNA 指纹库的构建方案。

目前关于高粱品种建库已有一些报道，如王晶等（2012）基于 119 份高粱材料确立了适于高粱 DNA 指纹图谱构建的核心引物，张春宵等（2012）使用这些核心引物对 78 份高粱种质遗传关系和遗传结构进行了解析，王瑞等（2015）利用 36 对 SSR 标记构建了 20 个高粱主推品种的 SSR 指纹图，Cuevas and Prom（2013）使用 20 对 SSR 引物对 137 份北美高粱种质资源进行评估等，但关于高粱兼容其他作物的建库方案尚未有相关研究和报道。目前，玉米根据其鉴定规程利用荧光毛细管电泳检测平台提高了品种检测效率，并且在玉米 DNA 指纹库的构建中已形成一套成熟的建库方案，建立了中国玉米审定品种标准 DNA 指纹库（王凤格等，2006；赵久然等，2015；王凤格等，2016），这为高粱建库方案确立提供了良好的借鉴。

本研究将玉米 DNA 指纹库构建方案应用于高粱品种建库流程中，拟解决跨作物通用性建库方案中的几个关键性问题，包括取样方式、DNA 制备、PCR 反应体系和程序、数据分析及读取方式等。本研究利用玉米建库方案对 6 份高粱标准样品和 3 份玉米标准样品建库，研究玉米建库方案在高粱中的可行性，为建立兼容其他作物 DNA 指纹库构建方案奠定基础。

1 结果和分析

1.1 样品处理及 DNA 提取

对利用 CTAB 法提取的幼苗组织总基因组 DNA 进行 1%琼脂糖凝胶电泳检测，试验结果显示利用该方法提取的 DNA 纯度高、浓度高、基本无杂质、无 RNA 残留（图1），说明高粱和玉米在 DNA 提取上使用此方案符合构建通用性 DNA 指纹库的要求。

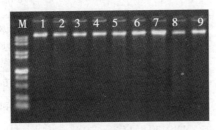

图 1 高粱和玉米基因组 1% 琼脂糖检测
Figure 1 Sorghum and Maize genome DNA 1% agarose detection

注：M. 分子量标准 DL2000 marker；1~9. 对应表 2 序号
Note：M. DNA marker DL2000 marker；1~9. corresponds to table 2 No.

1.2 高粱和玉米行业标准的初步验证

基于荧光毛细管电泳检测平台，本试验在验证高粱行业标准（NY/T 2467—2013）中 40 对引物的基础上，形成多重荧光电泳引物组合。考虑到高粱在日常检测中需求量较玉米、水稻少，而国外荧光引物合成成本较高，周期较长，国内荧光基团 TAMRA 容易对其他荧光产生影响等因素，本次建库引物荧光标记为 FAM、HEX、ROX 三色。根据引物扩增情况、染色体分布、片段大小和标记荧光颜色等，对调整后的引物形成 6 组荧光电泳引物组合（表 1）。

表 1 高粱引物信息
Table 1 Sorghum primer information

引物序号 Primer No.	引物组 Panel	引物名称 Primer Name	荧光染料 Fluorescent dye	染色体位置 Chromosome location	等位变异范围（bp） Allelic variation Range（bp）
1	Q1	SBKAFGK1	FAM	SBI05	116~135
2	Q1	TXP168	FAM	SBI07	175~180
3	Q1	TXP426	FAM	SBI03	238~257
4	Q1	SB868	HEX	SBI01	123~150
5	Q1	TXP015	HEX	SBI05	197~219
6	Q1	TXP021	ROX	SBI04	169~188
7	Q1	SB5014	ROX	SBI09	223~238
8	Q1	TXP41	ROX	SBI04	263~293
9	Q2	TXP343	FAM	SBI04	110~154
10	Q2	TXP481	FAM	SBI07	203~223
11	Q2	TXP289	FAM	SBI09	269~299
12	Q2	TXP010	HEX	SBI09	132~147
13	Q2	TXP23	HEX	SBI05	174~199

（续表）

引物序号 Primer No.	引物组 Panel	引物名称 Primer Name	荧光染料 Fluorescent dye	染色体位置 Chromosome location	等位变异范围（bp） Allelic variation Range（bp）
14	Q2	SB3811	HEX	SBI06	240~246
15	Q2	SB934	ROX	SBI02	163~182
16	Q2	TXP130	ROX	SBI10	244~251
17	Q3	TXP65	FAM	SBI05	119~133
18	Q3	TXP230	FAM	SBI09	169~201
19	Q3	TXP159	HEX	SBI07	152~181
20	Q3	SB4273	HEX	SBI08	255~268
21	Q3	TXP436	ROX	SBI03	136~150
22	Q3	CUP07	ROX	SBI10	189~270
23	Q4	TXP430	FAM	SBI02	153~165
24	Q4	TXP424	FAM	SBI03	225~239
25	Q4	TXP17	HEX	SBI06	160~186
26	Q4	SB5360	HEX	SBI10	234~250
27	Q4	TXP321	ROX	SBI08	186~229
28	Q4	TXP99	ROX	SBI07	272~287
29	Q5	TXP123	FAM	SBI05	254~272
30	Q5	TXP7	FAM	SBI02	212~233
31	Q5	SB5407	HEX	SBI10	121~150
32	Q5	SB3683	HEX	SBI06	229~244
33	Q5	TXP482	ROX	SBI01	221~231
34	Q5	SB1-10	ROX	SBI04	281~310
35	Q6	GPSB089	FAM	SBI01	164~172
36	Q6	TXP80	FAM	SBI02	280~305
37	Q6	TXP494	HEX	SBI03	195~222
38	Q6	SB3727	HEX	SBI06	278~294
39	Q6	GPSB067	ROX	SBI08	160~179
40	Q6	SB2507	ROX	SBI04	241~280

试验表明玉米样品使用玉米行业标准体系和程序能够有效扩增其引物，如图2符合亲子关系的杂交种京科968和父母本京724、京92在多次重复试验中样品DNA指纹都表现出明显的双亲互补带型，并且分析稳定一致。高粱样品在利用高粱行业标准扩增其引物

时，在不同试验批次验证中，SBKAFGK1、TXP010、TXP430、TXP321、TXP7 等位点会出现一定的数据缺失和扩增不一致现象，不一致现象表现为出现非特异条带的出现。如图3 所示，SBKAFGK1 位点的三组重复试验中，前两组重复中杂交种及其双亲样品的扩增均无131bp 片段，最后一组凤杂4 号样品131bp 片段为明显的非特异性条带。非特异性条带的出现在一定程度上影响了自动读取指纹数据的准确性，进一步影响建库指纹的稳定性和

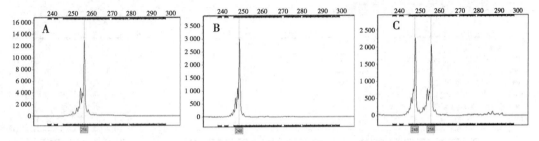

图 2　标准体系和程序扩增的玉米 DNA 指纹图谱

Figure 2　Maize DNA fingerprint profiling in standard PCR system and program

注：玉米引物 umc2007y4；A. 京科 724；B. 京 92；C. 京科 968

Note：Maize Primer umc2007y4；A. Jing724；B. Jing92；C. JingKe968

准确性。

1.3　高粱和玉米通用 PCR 反应体系和程序的验证

采用玉米行业标准中的 PCR 反应体系和程序扩增 6 份高粱标准样品，并同之前高粱采用高粱行业标准扩增效果比较。研究显示之前高粱在采用高粱行业标准时个别引物在 55℃退火温度下存在一定的非特异性扩增，而当高粱采用玉米行业标准将退火温度提高到 60℃时，有效去除了非特异性扩增（图4）；从峰值来看，60℃比 55℃扩增的峰值整体相对偏低，但在 DNA 质量较好的情况下，不影响建库指纹质量。使用玉米行业标准对高粱标准样品进行扩增与高粱预期特异性指纹扩增一致，并且相比较高粱行业标准中的体系和程序能够获得更稳定、准确的建库指纹数据，利于快速、准确、自动化读取指纹数据。

1.4　数据读取及参数设置

玉米采用 SSR 指纹分析器（V1.2.4）及建库分析参数能够实现指纹数据的自动化读取（图2）。而高粱样品荧光电泳数据在使用 SSR 指纹分析器（V1.2.4）时如果没有特定的建库分析参数时，往往会出现连续峰被识别为多峰、连续峰峰值低不被识别、N+1 峰被识别为多峰、数据读取困难等现象（图5A1，5B1，5C1，5D1，5E1，5F1），最终导致建库样品指纹错误，数据分析自动化程度低，同时因读值为具体片段大小而不利于不同批次和不同实验室数据整合的问题，图5 在无参数分析时目标片段往往读取为带有小数的形式，如 164.8bp，164.6bp 等。

使用 SSR 指纹分析器（V1.2.4）及玉米建库分析参数对高粱荧光电泳原始数据进行分析，研究显示，使用玉米建库分析参数能够准确识别高粱原始荧光电泳 DNA 指纹中大部分异常峰的目标片段，并且能够自动化读取样品指纹（图5A2，5B2，5C2，5D2，5E2，5F2）。同时根据对位点区间设置 BIN（目标片段左右各延伸 0.5bp 范围），读值以整数作为该等位基因大小，利于不同批次的数据整合，很大程度上方便了不同实验室对试验数据的整合和比对。

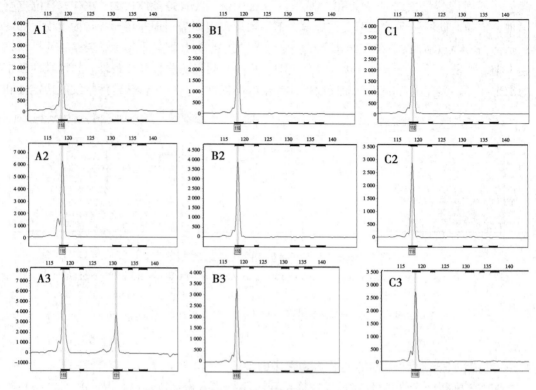

图3　标准体系和程序扩增的高粱DNA指纹图谱
Figure 3　Sorghum DNA fingerprint profiling in standard PCR system and program
注：高粱引物SBKAFGK1；A1，A2，A3.凤杂4号；B1，B2，B3.3148A；C1，C2，C3.南133
Note：Sorghum Primer：SBKAFGK1；A1，A2，A3.FengZa4；
B1，B2，B3.3148A；C1，C2，C3.Nan133

2　讨论

试验中对样品的处理都采用无光照培养幼苗组织的方法，其原因有两个：一是无光照对于植物来说会减少其叶片组织内糖分的积累，利于DNA的提取；二是幼苗组织可通过一系列的呼吸转化，将种子中的蛋白质、淀粉等物质转化为可溶物质（孙耀中等，1999）。总DNA的提取采用CTAB法，利用幼叶组织配合CTAB法能够有效去除植物组织中的多糖和多酚，避免DNA降解（王卓伟等，2001；陆单等，2010；罗志勇等，2001）。同时本试验也证明了高粱和玉米采用幼苗组织配合CTAB法获得的DNA质量和浓度均能够满足后续构建DNA指纹库的需求。

使用高粱PCR反应体系和程序发现个别位点的重复试验数据出现不一致的情况，分析可能的原因一是早期标准研制过程中引物设计退火温度条件设置不一致，造成个别引物在低退火温度下易出现非特异性条带，二是后期对引物评估多是基于变性PAGE凝胶电泳，所用引物未经过多实验室、多体系、多平台、多人员验证，导致个别引物重复性和稳定性差并且在荧光毛细管电泳平台检测容易产生异常峰（王凤格等，2007），影响品种指纹数据的自动化读取和整合。建立农作物品种鉴定通用PCR程序，一是对于退火温度相

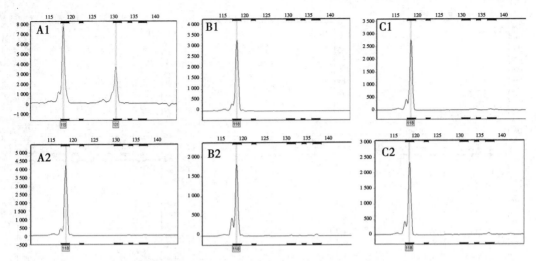

图4 高粱凤杂4号及其父母本在不同退火温度的DNA指纹图谱
Figure 4 The DNA fingerprint profiling of FengZa4 and parents in different annealing temperature

注：A1, A2. 凤杂4号；B1, B2. 3148A；C1, C2. 南133；高粱引物．SBKAFGK1；A1, B1, C1. 退火温度55℃；A2, B2, C2. 退火温度60℃

Note：A1, A2. FengZa4；B1, B2. 3148A；C1, C2. Nan133；Primer. SBKAFGK1；A1, B1, C1. Annealing temperature 55℃；A2, B2, C2. Annealing temperature 60℃

近的引物，寻找通用退火温度，验证相关引物在通用退火温度下扩增的准确性和稳定性，但对于退火温度相差太大的引物，整合难度较大；二是根据当前引物退火温度采用降落PCR程序，此方法对引物退火温度适用范围较广，但该程序时间较长，对仪器有特殊要求，实际应用时存在较大的局限性（王凤格等，2008）；三是基于全基因组测序技术，完善引物设计条件，针对不同作物统一标准，把退火温度控制在固定范围内，那么多作物通用PCR反应程序得以建立。例如根据目前建库经验较为成熟的玉米引物设计参数，设计其他作物引物参数，从而统一PCR反应程序和体系条件。

使用玉米建库分析参数对高粱电泳数据分析得到了良好的效果，说明这套分析参数对于其他二倍体作物的指纹分析具有参考意义，可经过进一步验证后直接或微调使用。而像棉花、小麦等多倍体作物SSR指纹图谱更为复杂，统计难度大且不易标准化，如何解决多倍性数据分析问题，使之兼容二倍体数据分析系统，成为多倍体作物建库的一个限制因素（王凤格等，2015）。目前一些多倍体作物部分引物虽然已经实现了二倍体化，但在后续研究中还需要完善其他引物设计并加以验证。

3 材料和方法

3.1 试验材料

试验从166份高粱标准建库样品中选取具有代表性的材料6份，由吉林农科院提供材料，含2组杂交种及其父母本，包括杂交种、不育系、恢复系，以验证扩增条带是否真实可靠，其中材料南133设置为样品重复；玉米材料3份，本单位收集整理，含1组杂交种及其父母本，包括杂交种和自交系（表2）。

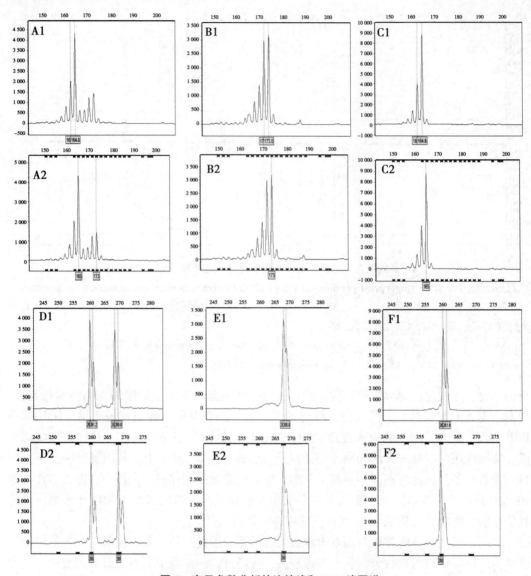

图 5 有无参数分析的连续峰和 N+1 峰图谱
Figure 5 The profiling of continuous peak and N+1 peak with or without analysis parameter

注：A1, A2. 凤杂 4 号；B1, B2. 3148A；C1, C2. 南 133；高粱引物 . SB4273；D1, D2. 四杂 25；E1, E2. TAM428A；F1, F2. 南 133；高粱引物 . TXP159；A1, B1, C1, D1, E1, F1. 无参数分析；A2，B2，C2，D2，E2，F2. 玉米参数分析

Note：A1, A2. FengZa4；B1, B2. 3148A；C1, C2. Nan133；Primer. SB4273；D1, D2. SiZa25；E1, E2. TAM428A；F1, F2. Nan133；Primer. TXP159；A1, B1, C1, D1, E1, F1. No parameter analysis；A2, B2, C2, D2, E2, F2. Parameter analysis of corn

表 2 样品信息
Table 2 The sample information

序号 No.	样品名称 Sample name	作物名称 Crop name	样品类型 Sample type
1	凤杂 4 号 Fengza4hao	高粱 Sorghum	杂交种 Hybrids
2	3148A	高粱 Sorghum	不育系 Sterile line
3	南 133 Nan133	高粱 Sorghum	恢复系 Restorer line
4	四杂 25 Siza25	高粱 Sorghum	杂交种 Hybrids
5	TAM428A	高粱 Sorghum	不育系 Sterile line
6	南 133 Nan133	高粱 Sorghum	恢复系 Restorer line
7	京科 968 Jingke968	玉米 Maize	杂交种 Hybrids
8	京 724 Jing724	玉米 Maize	自交系 Inbred line
9	京 92 Jing92	玉米 Maize	自交系 Inbred line

3.2 样品的准备及 DNA 提取

高粱和玉米每个品种各取 50 粒种子，采用沙培的方式无光照培养 3d（高粱 26℃，玉米 30℃），取 200~300mg 幼苗备用。基因组 DNA 提取采用 CTAB 法，具体试验流程参见《玉米品种鉴定技术规程 SSR 标记法》（NY/T 1432—2014）。对提取的基因组 DNA 采用 1%琼脂糖凝胶电泳进行 DNA 浓度和质量测定。

3.3 SSR 引物的选择和 PCR 反应体系及程序

本试验以高粱农业行业标准（NY/T 2467—2013）和玉米农业行业标准（NY/T 1432—2014）中规定的 40 对荧光标记 SSR 核心引物和 PCR 反应程序及体系为基础（表 3）。其中调整高粱 40 对核心引物中的 Gap57 为 SB2507（F：5′-3′CTTTCTCTCTCCAC-CTTTTCACGC；R：5′-3′GGTGTGGATTCTCTGAGGTTTGCT）其他引物和行业标准完全相同，荧光在引物 5′标记为 FAM、HEX、ROX 三色（表 1），引物由北京梓熙生物科技有限公司合成。玉米行业标准 PCR 反应程序：94℃预变性 5min，1 个循环；94℃变性 40s，60℃退火 35s，72℃延伸 45s，共 35 个循环；72℃延伸 10min，4℃保存；高粱行业标准 PCR 反应程序：94℃预变性 4min，1 个循环；94℃变性 45s，55℃退火 45s，72℃延伸 60s，共 36 个循环；72℃延 10min，4℃保存。

表3 玉米和高粱PCR反应体系对比

Table 3 Comparison of PCR reaction system between Maize and Sorghum

作物名称 Crop name	玉米 Maize	高粱 Sorghum
DNA	2.5ng/μl	20~40ng
PCR buffer	1×PCR buffer	1×PCR buffer
Mg^{2+}	2.5mmol/L	2.5mmol/L
dNTP	0.10mmol/L 每种（Each）	0.10mmol/L 每种（Each）
正反向引物 Forward and reverse primers	0.25μmol/L 每种（Each）	0.24μmol/L 每种（Each）
Taq DNA 聚合酶 Taq DNA polymerase	0.04U/μl	0.4U
总体积 Total volume	20μl	10μl

3.4 电泳及数据分析

分别取荧光标记扩增产物按引物分组等体积混合（表1），吸取1μl混合产物液加入DNA分析仪专用的上样板中。板中各孔加入0.1μl分子量内标（LIZ 500，Applied Biosystem）和8.9μl去离子甲酰胺。将样品放置PCR仪上95℃变性5min，取出并立即置于碎冰上，冷却10min以上。离心后置于ABI 3730XL DNA analyzer（Applied Biosystem）上进行荧光毛细管电泳，利用Date Collection V1.0软件收集原始数据。

选用SSR指纹分析器（软件登记号：2015SR161217，北京市农林科学院玉米研究中心开发）分析原始数据，并形成不同品种DNA指纹图谱（指纹图谱横坐标单位：bp；纵坐标单位：RFU，相对荧光值）。SSR指纹分析器分析内容包括：pull-up峰消除、二倍体过滤、识别连续峰、N+1峰、邻峰等（王蕊等，2012）。

参考文献（略）

本文原载：http://kns.cnki.net/kcms/detail/46.1068.S.20170406.1210.020.html

玉米 InDel 分子标记 20 重 PCR 检测体系的建立

冯 博[1,2]　许理文[1]　王凤格[1]*　薛宁宁[1]　刘文彬[1,2]　易红梅[1]
田红丽[1]　吕远大[3]　赵 涵[3]　金石桥[4]　张力科[4]　蔚荣海[2]　赵久然[1]

（1. 北京市农林科学院玉米研究中心／玉米 DNA 指纹及分子育种北京市重点实验室，北京　100097；2. 吉林农业大学农学院，长春　130118；3. 江苏农业科学院农业生物技术研究所，南京　210014；4. 全国农业技术推广服务中心，北京　100125）

摘　要：为构建玉米多重 PCR 检测体系，提高分子标记检测效率，利用 10 份代表性玉米材料对 238 对 InDel 引物进行单重 PCR 评估，共得到 192 对扩增效率高、稳定性好的引物。根据软件评估结果、扩增质量、产物范围、染色体均匀分布原则从 192 对引物中优选出 30 对综合表现较好的引物形成扩增产物范围在 80～200bp 和 200～400bp 的两组核心引物组合，每套组合中有 10 对引物分布在不同染色体上。在核心引物组合的基础上综合考虑染色体分布、碱基片段范围、引物荧光颜色，逐一添加引物，最终形成两组玉米 20 重 PCR 体系，一组 40 重荧光标记毛细管电泳。

关键词：玉米；InDel；多重 PCR

Establishment of 20 PCR Detection System with InDel Molecular Markers in Maize

Feng Bo[1,2]　Xu Liwen[1]　Wang Fengge[1]*　Xue Ningning[1]
Liu Wenbin[1,2]　Yi Hongmei[1]　Tian Hongli[1]　Lv Yuanda[3]　Zhao Han[3]
Jin Shiqiao[4]　Zhang Like[4]　Yu Ronghai[2]　Zhao Jiuran[1]

（1. *Maize Research Center*, *Beijing Academy of Agriculture & Forestry Sciences／Beijing Key Laboratory of Maize DNA Fingerprinting and Molecular Breeding*, *Beijing* 100097; 2. *College of Agriculture*, *Jilin Agricultural University*, *Changchun* 130118; 3. *Provincial Key Laboratory of Agrobiology*, *Jiangsu Academy of Agricultural Sciences*, *Nanjing* 210014; 4. *National Agricultural Technical Extension and Service Center*, *Beijing* 100125）

Abstract: In order to improve the detection efficiency with molecular marker, the multiple PCR detection system was constructed. In this study, 10 major materials were used to evaluate the single

---

基金项目：本研究由科技部国家科技支撑计划（2015BAD02B02）和北京市农林科学院院科技创新能力建设专项（KJCX20161501）资助
联系方式：冯博，E-mail：fengbo02220108@163.com；许理文，E-mail：xulw0408@126.com
　　　　　同等贡献
* 通讯作者：王凤格，E-mail：gege0106@163.com

pair PCR with 238 pairs of InDel primers. According to the software quality evaluation results, product range, and the principle of chromosome uniform distribution, 30 pairs of primers were selected from 192 primers with better performance to form two groups of core primer combinations with amplified products in the range of 80~200bp and 200~400bp, There were10 pairs of primers distributing in different chromosomes for each primer combination. Based on core primer combination and comprehensive consideration on chromosome distribution, base fragment, and primers fluorescent color, we established two groups of corn test 20 PCR system and a group of 40 fluorescent capillary electrophoresis.

Key words: Maize; InDel; Multiplex PCR

多重PCR（Multiplex PCR）是在常规PCR的基础上改进的，在同一个PCR反应体系中加入多对引物，对多个DNA模板或同一模板的不同区域同时扩增多个目的片段的技术，这个概念由Chambercian等率先提出。多重PCR技术可以同时扩增多个目的基因，具有节省珍贵的试验样品、节省时间提高效率、降低成本的优点，所以自从多重PCR技术报道以来，受到众多研究者的重视，并且迅速发展至多个研究领域，多重PCR技术已经成为生命科学各个领域一种重要的检测手段。

目前，在动物和人类上得到了较广泛深入的研究及应用。以前的多重体系较多基于SSR引物，人类STR（short tandem repeat）中已经成功的组建了26重PCR体系，玉米中成功组建了10重PCR体系。随着InDel新型标记的开发，人们首先在动物及人类上构建多重PCR复合扩增体系，已在人类上成功组建3组16重PCR，48重荧光毛细管电泳。在玉米等植物上研究较少，主要原因是在玉米上开发的InDel标记较少，本实验室基于玉米核心种质自交系的重测序数据全面的挖掘玉米基因组上的InDel位点，并评估其多态性，分析位点两侧序列的保守性，获得一批InDel引物（文章未发表）。本试验准备借鉴多重PCR在动物及人类上的研究成果探索以InDel标记构建玉米多重PCR体系的方案。该体系的建立具有现实意义，既可为大规模、高通量的玉米指纹库构建创造有利条件，又可提高玉米遗传研究和标记辅助育种的效率。

1 材料与方法

1.1 材料

共10份（表1），涵盖了普通玉米、甜玉米、糯玉米，具广泛代表性。其中包括2份玉米自交系和8份常用玉米杂交种。

表1 供试材料
Table1 Test materials in this study

样品编号 Sample code	样品名称 Sample name	样品类型 Sample type	种质类型 Germplasm types
1	先玉335 Xianyu335	杂交种 Hybrid	普通玉米 Common corn
2	郑单958 Zhengdan958	杂交种 Hybrid	普通玉米 Common corn
3	京科968 Jingke968	杂交种 Hybrid	普通玉米 Common corn
4	农大108 Nongda108	杂交种 Hybrid	普通玉米 Common corn

(续表)

样品编号 Sample code	样品名称 Sample name	样品类型 Sample type	种质类型 Germplasm types
5	正大619 Zhengda619	杂交种 Hybrid	普通玉米 Common corn
6	鲁丹981 Ludan981	杂交种 Hybrid	普通玉米 Common corn
7	京科糯2000 Jingkenuo2000	杂交种 Hybrid	糯玉米 Waxy corn
8	京科甜183 Jingketian183	杂交种 Hybrid	甜玉米 Sweet corn
9	郑58 Zheng58	自交系 Inbred line	普通玉米 Common corn
10	昌7-2 Chang7-2	自交系 Inbred line	普通玉米 Common corn

1.2 试验方法

1.2.1 DNA 提取

采用王凤格等的改良 CTAB 法。

1.2.2 PCR 扩增

单重 PCR 体系 200μmol L^{-1}含 DNA 模板 2μl，2×Taq Plus Master Mix 10μl，0.25 μmol L^{-1}引物 2μl，总体积 20μl。PCR 程序为 95℃预变性 5min；94℃变性 40s，60℃退火 35s，72℃延伸 45s，35 个循环；72℃延伸 10min；4℃保存。在多重 PCR 体系中将单重 PCR 体系引物浓度调整为 0.5μmol L^{-1}，引物用量为 0.2μl，在体系优化过程中调整部分引物浓度，其余与单重 PCR 相同。

1.2.3 荧光标记毛细管电泳

对于单重 PCR，在 96 孔板的各孔中分别加入 9μl 去离子甲酰胺、0.2μl GS3730-500 分子量内标和 2μl 稀释 10 倍的 PCR 产物。95℃变性 5min，于 4℃保存 10min，2 000 转/min 离心 1min，于 ABI3730XL DNA 分析仪上进行荧光毛细管电泳。对于多重 PCR，采用产物原液，其余与单重 PCR 电泳一致。

1.2.4 数据收集与分析

采用 Date Collection 软件收集原始数据，用本单位自主研发的 SSR Analyser（V1.2.4）指纹分析器统计与分析数据。

1.2.5 InDel 引物开发设计与评估

引物位点开发阶段的筛选条件为 InDel 长度在 3~10bp、两侧序列保守性高、染色体上均匀分布。引物设计标准是利用 Primer3 设计引物，为有效构建多重 PCR，所有引物设计时采用统一设置参数，主要参数包括 Tm 值（60±2）℃、扩增产物片段范围（80~400bp）、GC 含量（40%~60%）。引物扩增质量筛选原则是荧光标记毛细管电泳图谱中峰值高于 1 000、无非特异峰、多态性高、稳定性好。

1.2.6 InDel 引物多重组合

多重 PCR 核心引物组的筛选：在 192 对候选引物中，将扩增产物片段分为 80~200bp 和 200~400bp 两组，每组引物按染色体均匀分布、片段范围平均差异 20bp 左右选择一对引物，挑选出 10 对引物，形成核心引物组合。扩增产物片段范围在 80~200bp 的引物按 4 色（ABI 合成）或 2 色荧光（国内合成）分别挑选出一套核心引物组合。对核心引物组合及候选引物进行软件评估，并对所评估的引物按与核心引物组合相互作用的大小进行排序。试验评估在核心引物组合的基础上进行逐一添加引物，优先选择引物之间干扰弱的引物。

2 结果与分析

2.1 引物筛选与确定

根据引物扩增质量筛选原则，对于单重 PCR 扩增，从 238 对荧光引物中筛选出 192 对，其在染色体上的分布情况（表2），大多数引物在染色体上均匀分布，平均相差 20bp 有一对引物。第 1 染色体上引物分布最多，共 30 对；第 4 和第 7 染色体上分布最少，共 13 对。扩增产物片段范围在 80~300bp 的引物分布均匀，350bp 以上的分布较少。表 2 左侧数列为扩增产物片段范围，引物的荧光颜色可根据需要及组合方式自行标记。可由表 2 信息根据染色体分布和扩增产物片段大小自由组建多重引物组合。

表 2 192 对候选引物名单
Table 2 List of 192 candidate primers

扩增产物片段范围 Amplified fragment range	第1染色体 Chr.1	第2染色体 Chr.2	第3染色体 Chr.3	第4染色体 Chr.4	第5染色体 Chr.5	第6染色体 Chr.6	第7染色体 Chr.7	第8染色体 Chr.8	第9染色体 Chr.9	第10染色体 Chr.10
80~100bp	IDP197 (80, 86)	IDP214 (87, 90)	IDP221 (80, 84)				IDP214 (87, 90)		IDP339 (97, 103)	
	IDP196 (87, 90)	IDP268 (98, 101)	IDP220 (87, 90)						IDP406 (94, 98)	
	IDP201 (87, 90)		IDP277 (99, 106)							
100~120bp	IDP246 (102, 105)	IDP124 (101, 106)	IDP052 (102, 107)		IDP136 (102, 105)	IDP139 (105, 110)	IDP147 (107, 112)	IDP152 (102, 107)	IDP154 (104, 107)	IDP122 (105, 110)
	IDP113 (107, 110)	IDP128 (102, 105)			IDP137 (102, 106)		IDP411 (113, 121)	IDP150 (115, 123)		IDP029 (115, 118)
	IDP117 (119, 125)	IDP037 (108, 111)			IDP067 (103, 112)					
		IDP040 (109, 115)								
120~140bp	IDP250 (123, 127)	IDP036 (122, 130)	IDP280 (125, 132)	IDP416 (135, 141)		IDP086 (124, 129)	IDP145 (125, 129)	IDP333 (122, 126)		IDP264 (125, 132)
	IDP350 (123, 130)	IDP123 (127, 130)					IDP362 (129, 133)	IDP153 (123, 127)		IDP369 (125, 132)
		IDP125 (132, 136)								
		IDP032 (138, 148)								
		IDP182 (138, 147)								

（续表）

扩增产物片段范围 Amplified fragment range	第1染色体 Chr. 1	第2染色体 Chr. 2	第3染色体 Chr. 3	第4染色体 Chr. 4	第5染色体 Chr. 5	第6染色体 Chr. 6	第7染色体 Chr. 7	第8染色体 Chr. 8	第9染色体 Chr. 9	第10染色体 Chr. 10
140~160bp	IDP024 (153, 156) IDP015 (158, 164)		IDP054 (142, 150) IDP279 (147, 151) IDP110 (148, 151)	IDP063 (153, 157)	IDP409 (153, 157)	IDP313 (145, 148)		IDP104 (141, 147) IDP103 (143, 146)	IDP156 (143, 146) IDP343 (149, 154) IDP157 (152, 158)	IDP028 (140, 143) IDP206 (155, 160) IDP259 (155, 160) IDP181 (159, 164)
160~18bp	IDP204 (160, 163)	IDP273 (167, 175)	IDP281 (171, 174) IDP051 (173, 181) IDP163 (174, 180)	IDP062 (164, 168)	IDP302 (162, 165) IDP079 (168, 175)	IDP161 (169, 175)	IDP320 (172, 181)	IDP402 (163, 166) IDP097 (167, 170) IDP098 (171, 176)	IDP108 (171, 174)	
180~200bp	IDP114 (185, 188)	IDP030 (183, 188)	IDP130 (194, 197) IDP053 (198, 203)		IDP167 (190, 194)	IDP417 (181, 185) IDP191 (187, 191)	IDP403 (190, 198)	IDP151 (196, 199)		IDP348 (189, 193) IDP261 (198, 201)
200~220bp	IDP251 (203, 211) IDP115 (206, 209)	IDP374 (211, 231)	IDP278 (219, 224) IDP412 (222, 225)	IDP290 (202, 211) IDP060 (209, 212)	IDP064 (207, 210) IDP165 (213, 216)	IDP424 (205, 209)	IDP146 (207, 212)	IDP099 (219, 227)	IDP105 (203, 212) IDP168 (215, 218) IDP106 (219, 223)	
220~240bp	IDP022 (229, 235) IDP101 (222, 225) IDP177 (231, 237) IDP203 (231, 240)	IDP031 (237, 245)		IDP295 (238, 241)	IDP076 (227, 236)		IDP324 (224, 232) IDP388 (234, 237)	IDP335 (224, 228)	IDP158 (222, 227) IDP340 (223, 226) IDP391 (235, 242)	IDP121 (226, 233)
240~260bp	IDP017 (244, 248) IDP120 (250, 254)	IDP271 (257, 264)	IDP283 (244, 247)	IDP293 (242, 246) IDP132 (248, 251) IDP057 (252, 255) IDP133 (255, 258)	IDP073 (258, 262) IDP298 (258, 265)	IDP311 (243, 246) IDP142 (250, 253) IDP083 (256, 264)	IDP148 (259, 268)		IDP109 (247, 255) IDP344 (255, 258)	IDP026 (242, 245)

（续表）

扩增产物片段范围 Amplified fragment range	第1染色体 Chr. 1	第2染色体 Chr. 2	第3染色体 Chr. 3	第4染色体 Chr. 4	第5染色体 Chr. 5	第6染色体 Chr. 6	第7染色体 Chr. 7	第8染色体 Chr. 8	第9染色体 Chr. 9	第10染色体 Chr. 10
260~280bp	IDP020 (264, 274) IDP018 (266, 273) IDP372 (268, 276) IDP248 (279, 283)	IDP045 (277, 280)	IDP284 (266, 277) IDP164 (274, 280) IDP048 (275, 281)		IDP070 (262, 269) IDP018 (266, 273)	IDP084 (261, 266) IDP316 (263, 266)		IDP004 (260, 263)		IDP027 (260, 270) IDP367 (171, 175)
280~300bp		IDP034 (282, 285) IDP267 (286, 291) IDP216 (287, 292)	IDP186 (285, 291)	IDP294 (290, 293)		IDP089 (291, 297) IDP085 (292, 298)	IDP325 (280, 287)	IDP100 (292, 298)		IDP207 (297, 303)
300~320bp	IDP254 (311, 314)	IDP185 (315, 321)		IDP227 (302, 306) IDP226 (307, 311)	IDP075 (303, 307) IDP306 (312, 315)	IDP236 (301, 304) IDP234 (304, 309) IDP235 (308, 311)		IDP135 (302, 305)	IDP193 (316, 320)	IDP180 (301, 311) IDP266 (306, 310) IDP256 (308, 311)
320~340bp	IDP021 (324, 328)		IDP187 (324, 331)		IDP300 (332, 335) IDP101 (336, 341)			IDP415 (328, 311)		IDP025 (323, 311)
340~360bp	IDP255 (342, 345) IDP200 (353, 363) IDP178 (354, 361)				IDP074 (347, 354)			IDP127 (350, 353)		
360~380bp	IDP014 (368, 375)	IDP210 (362, 368) IDP275 (370, 378)					IDP091 (378, 385)	IDP096 (368, 372)		
380~400bp		IDP046 (383, 387) IDP047 (386, 390)	IDP049 (392, 398)							

2.2 10重核心引物组合

10重核心引物组合经软件评估引物间无干扰。10重 PCR 电泳结果表明 10 对引物可以同时成功扩增（图1）。10 份材料验证多重组合稳定性较高，又在实验室选取 200 份材料对多重组合进行验证，均取得了较好的试验结果。10 重 PCR 的扩增片段大小与单重 PCR 扩增出的片段大小一致、没有非特异峰、同色荧光引物扩增片段无交叉；虽然 10 对引物的扩增效率差异较大，但随后扩增体系优化的过程中各引物浓度有所调整。10 重核心引物组合名单详见表3。

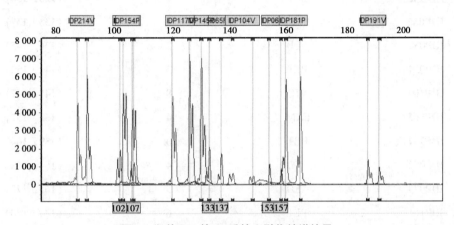

图 1　郑单 958 的 10 重核心引物扩增结果
Figure 1　Amplification results of 10 core primers of Zhengdan958

2.3 20重引物组合的确定

2.3.1 80~200bp 四色荧光引物组合

在 10 重核心引物组合的基础上添加至 20 重引物组合，20 对引物均成功扩增。10 份材料验证多重组合稳定性较高，又在实验室选取 200 份材料对多重组合进行验证，均取得了较好的试验结果。在染色体上分布均匀、多重 PCR 的扩增片段大小与单重 PCR 扩增出的片段大小一致、引物扩增片段无交叉、扩增效率较高、扩增效果较稳定。由于引物扩增片段大小以及引物自身扩增效率的原因采用相同引物浓度无法实现等效扩增，根据引物扩增效率对扩增体系进行优化，即对各引物扩增浓度进行调整，调整浓度后各引物扩增效率与调整前有明显改善（图2）。

表 3　20 对引物基本信息
Table 3　Basic information of 20 primers

引物名称 Primer name	荧光颜色 Fluorescent color	染色体 Chromosome	等位基因 Allelic gene
IDP238	FAM	7	(84, 88)
IDP201	NED	1	(87, 90)
IDP214	**VIC**	**2**	**(88, 91)**
IDP052	**FAM**	**3**	**(102, 107)**

（续表）

引物名称 Primer name	荧光颜色 Fluorescent color	染色体 Chromosome	等位基因 Allelic gene
IDP147	NED	7	(103, 108)
IDP154	**PET**	**9**	**(103, 106)**
IDP117	**VIC**	**1**	**(120, 126)**
IDP145	**PET**	**7**	**(126, 130)**
IDP065	**FAM**	**5**	**(133, 137)**
IDP032	NED	2	(136, 145)
IDP054	PET	3	(141, 150)
IDP104	**VIC**	**8**	**(141, 147)**
IDP063	**FAM**	**4**	**(153, 157)**
IDP181	**PET**	**10**	**(159, 164)**
IDP062	VIC	4	(164, 168)
IDP079	NED	5	(165, 172)
IDP161	PET	6	(174, 180)
IDP348	NED	10	(186, 190)
IDP191	**VIC**	**6**	**(187, 191)**
IDP151	FAM	8	(196, 199)

注：加粗部分为 10 个核心引物。FAM 为蓝色荧光，NED 为黑色荧光，VIC 为绿色荧光，PET 为红色荧光（国外合成）

Note：The bold represent 10 core primers. FAM is blue-fluorescence, NED is black-fluorescence, VIC is green-fluorescence, and PET is red-fluorescence（Foreign synthesis）

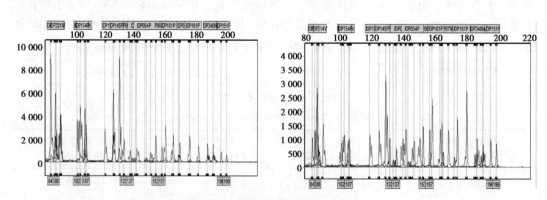

图 2　郑单 958 的 80~200bp 20 重引物 PCR 效果

Figure 2　Effect chart of 80~200bp 20 primers multiplex-PCR in Zhengdan958

2.3.2　两色荧光组合

借鉴组建 80~200bp 20 重引物组合的经验，利用 FAM 和 ROX 两色荧光组建了扩增片段范围在80~200bp、200~400bp的20重PCR组合，所有引物均成功扩增（图3）。

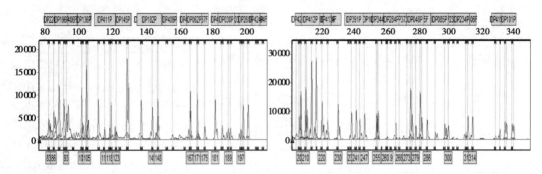

图3 郑单958的80~200bp及200~400bp 20重引物PCR效果

Figure 3 Effect chart of 80~200bp and 200~400bp 20 primers multiplex-PCR in Zhengdan958

2.4 40重电泳

将采用FAM和ROX两色荧光标记的80~200bp、200~400bp的两个20重PCR组合电泳，获得了二色荧光标记40重电泳组合（图4）。10份材料验证多重组合稳定性较高。又在实验室选取200份材料对多重组合进行验证，均取得了较好的试验结果。ABI3730XLDNA分析仪为5色荧光毛细管电泳仪，1色为内标，4色可以标记引物，因此理论上可以做到80重荧光标记毛细管电泳组合，将极大的提高InDel基因型分型的效率。40对引物基本信息见表4。

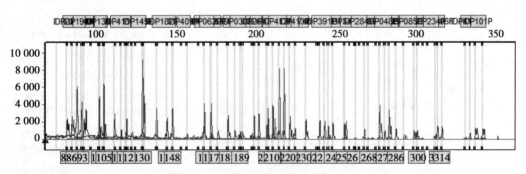

图4 40重荧光标记毛细管电泳效果

Figure 4 Effect of 40 capillary tube electrophoresis in ZhengDan958

表4 40对引物基本信息

Table 4 Basic information of 40 primers

引物名称 Primer name	荧光颜色 Fluorescent color	染色体 Chromosome	等位基因 Allelic gene	引物序列 Primer sequence（5'-3'）
IDP220	FAM	3	87, 90	CCGCTGAAGTCGTTGTAGGAGAG; AACCGCCTGTCTGGGAATCTTTC
IDP406	FAM	9	95, 99	TGCCCACCCTCTGTTGACG; AGGACAGAAGCCATCGGATACG

(续表)

引物名称 Primer name	荧光颜色 Fluorescent color	染色体 Chromosome	等位基因 Allelic gene	引物序列 Primer sequence (5'-3')
IDP128	FAM	2	105, 108	GCACCAGCACAGCACTATACAAG; TGCCGCACCCTAGTTAGTTTCTG
IDP029	FAM	10	115, 118	GCAAGCCACTCAGATGTAGGAAAC; GTTGGGAATGGCACTACGAAAGC
IDP350	FAM	1	123, 130	GACAGTTGACAGTGGCTGATGC; ACATGGTCCAGGTAGCTCCTTG
IDP416	FAM	4	135, 141	TTCTGCAAGCTCTGGAACAACTG; TTCATCCTCAGGGTCGCTATCG
IDP110	FAM	3	145, 148	TTTCCAGTGTCTTGCCAGTTTGC; ATCAACGTCCTCCGATGCTTCTC
IDP402	FAM	8	163, 166	CATGAACCGCCTCGAATCCTAC; GGTCTAGTGTGTATGCCTCCAAG
IDP367	FAM	10	171, 175	TGCCATCCCATTTGATGTCCAAG; TTGCCATGCGTGTTCGTAGTG
IDP417	FAM	6	181, 185	GCTGCGGTGCTCATCTTCTG; AGCTCGTGTGAAACAACTTCCTC
IDP403	FAM	7	190, 198	GTTGTGGAGCAGCACGAAGATC; ACGGTGTTCGCCAGGAAGG
IDP064	FAM	5	207, 210	AAGGCATCAGGCAGGTGCTAAG; CGCAGGTGACCGTGGTGATC
IDP374	FAM	2	221, 231	GCAATTAGTCCACGGTGAAGGG; AACGGCTGGGATTCGGTGTC
IDP295	FAM	4	238, 241	ACATCAAGCACCCACAGAAACAG; GCGTGGATCGGTGGATGAATC
IDP109	FAM	9	247, 255	GCCTTGGTATGTGAAATGCTGTCC; CGCCTTCCTCTAGTCCTGACATTC
IDP004	FAM	8	260, 263	AGACCATCCAAGTTCACGCATCTC; AGCATCACCTCCCTGGAAACAAAG
IDP372	FAM	1	268, 276	ATGTTTGCTCCGTCTTCTCTTCG; AGATAAGCAGGTTCACACAAGGC
IDP325	FAM	7	281, 288	CATCCACACGCTGCTGATTCG; CGCTGGTCATGTACACTATCGC
IDP236	FAM	6	301, 304	TGGCTCAACTTCTGGTGCTGAC; ATGCCTTGTATCGCCGTGGATC
IDP306	FAM	5	312, 315	AAGACGGCAAGGGCAGAGAG; ACTGTGGCAAGGGAAGAGACTC
IDP196	ROX	1	84, 87	CTCCGACGCCGTGTTCCTAAG; GGGTGCCACTGTCTGCTTAATTC

（续表）

引物名称 Primer name	荧光颜色 Fluorescent color	染色体 Chromosome	等位基因 Allelic gene	引物序列 Primer sequence（5'-3'）
IDP136	ROX	5	102, 105	TACGAACCAGGCAGAACCAAGC; TGTGGGAAAGGCAAGGTGTGG
IDP411	ROX	7	113, 121	GCCACAGCACACCAACCATG; CTGTTGCGTGCCATTTGAGTTTG
IDP145	ROX	7	126, 130	TAGGTAGCGTTGCGACAAAGGAG; AAGGCTCAGACTCGTCGTTCTTC
IDP182	ROX	2	138, 147	GGCTGATCCTCGACACTCCAAC; GCTGGCAGGTGAGTGGTGAC
IDP409	ROX	5	153, 157	GGCAGGCACAGATGAGGAAAG; GCACCATTTAGGGACATTGAACG
IDP062	ROX	4	164, 168	GCCGACCACATTTAAGATGCTCTG; TGCTCGGTATTGGCGTCGTAAC
IDP030	ROX	2	183, 188	GTTTCTCTTGCCGTTGACCTGATC; ACCAACCGCCATTCCTACTGC
IDP261	ROX	10	198, 201	GCTGCTGGAACACCCAACTTG; GCTGATGTCATCCTCGTCTTGTG
IDP424	ROX	6	205, 209	CAGGTCCTCCGTTCCGTTCC; GAGCTTCAAGCCTGCTGTGTC
IDP412	ROX	4	212, 215	GCCTTGACCTTACTACGAGAACC; GATGTGCTTGCTGTCCCTGAG
IDP413	ROX	1	222, 225	CGGAGCCTGTGCCACCTC; GACCAGGAACAAGCCATCATCAG
IDP391	ROX	9	235, 242	TGTGCTGGTGCTGCTACAGG; CCATTCCGGCTGCTCATCAATC
IDP344	ROX	9	255, 258	GGAACCTGAGTGAAGAAGCCATC; ACAGGGAACCACAGTGCTACG
IDP284	ROX	3	263, 268	CGAATGCGGCAAGGTTGAGAAC; GTCAGTAAGCCACACAGCAATCC
IDP048	ROX	3	275, 281	CTTCCTCCTCCACGGCAAGC; AGAAGCGGTTGACGATGGTGAG
IDP085	ROX	6	292, 298	GGTGTCGTCATGCTCCTCTTCC; GAGTTGCGATTTGGACAGGAACC
IDP234	ROX	6	304, 309	TGGTTGACATCGGTCGCTTCG; AGATTCCTGATTCCCACCTGACG
IDP415	ROX	8	328, 331	TCCCTCACACACCTACATCAGC; ATGCCAGATTTGTCGCCACATC
IDP101	ROX	8	336, 341	CCGAACTCCACCGCCATATCC; GCGAAACCAAGCCAACAATACCC

注：FAM 为蓝色荧光，ROX 为红色荧光（国内合成）
Note：FAM is blue-fluorescence, ROX is red-fluorescence（Domestic synthesis）

3 讨论

InDel 遍布于真核生物整个基因组的频率仅次于 SNP，位居第二，在玉米基因组也是如此，平均 126bp，大约是 SNP（60.8）的 2 倍。InDel 在玉米基因组上分布不均，在非编码区平均 85bp 一个，在编码区平均每 2 500bp 一个。InDel 引物在组建多重 PCR 时自身二态性标记的特点有明显优势，每个引物只有 2 个等位基因、不会出现新的等位基因、目标产物片段大小确定，所以在组建多重 PCR 时能够避免引物片段大小有交叉、能够保证引物充分扩增易于实现多重 PCR 以及多重电泳。所用 InDel 引物设计时均采用相同的退火温度，多个引物可采用同一程序同时扩增。InDel 引物通用性强、稳定性高，应用目前 SSR 引物的扩增程序即可成功扩增，不涉及到扩增程序上的优化，为组建多重 PCR 提供了更大的可能性。InDel 引物没有连续多峰和三峰的现象，便于数据的统计与分析；兼容多个检测平台，荧光毛细管电泳平台、芯片平台、KASP 平台，易于数据的整合。

利用 InDel 标记将候选引物按扩增产物片段范围、染色体分布进行分类，构建 10 对核心引物组合后逐一添加引物的方案使本试验最终成功的组建了两组 20 重 PCR，一组 40 重荧光标记毛细管电泳。本试验成功组建 20 重 PCR 的试验方法、试验程序以及试验流程，为今后多重 PCR 体系的构建提供了丰富的经验。后续试验进一步研究的空间还很大，可以继续组建 20 重以上的多重复合体系，理论上可以实现 80 重荧光标记毛细管电泳，在标记上具有较高的通量。理论上的 80 重荧光毛细管电泳，在一个孔中即可实现对 80 个位点的基因型分型分析，一块 96 孔板就可以实现 7 680 个位点的基因型分型分析、一块 384 板则可实现 30 720 个位点的基因型分型分析。20 个引物同时扩增提高了试验的准确性；减少了繁琐的试验操作、节省了试验样本、试验耗材、试验时间，工作效率提高了 20 倍。本研究获得的 20 重 PCR 组合以及 40 重荧光标记毛细管电泳将在较大范围内具有应用价值，今后可以应用到更多有关玉米品种真实性和纯度鉴定中，同时也可以为分子植物育种提供技术支撑。

4 结论

利用 10 份玉米材料结合 InDel 标记，成功组建了两组 20 重 PCR 体系、一组 40 重荧光毛细管电泳以及一套组建 PCR 复合体系的试验方案，为今后的利用 InDel 以及其他标记组建多重 PCR 复合体系，分子辅助育种的背景选择以及将多重 PCR 复合体系应用到玉米品种鉴定中提供了重要的思路和研究方法。

参考文献（略）

本文原载：作物学报，43（8）：1 139-1 148

玉米品种京科968纯度鉴定引物的确定

冯博[1,2]* 许理文[2]** 王凤格[2]** 蔚荣海[1]** 易红梅[2] 刘文彬[1,2]
葛建镕[2] 田红丽[2] 侯振华[2] 王璐[2] 任洁[2] 王元东[2] 赵久然[2]**

(1. 吉林农业大学农学院,长春 130118; 2. 北京市农林科学院玉米研究中心/玉米DNA指纹及分子育种北京市重点实验室,北京 100097)

摘 要:为了确定一套京科968玉米品种纯度鉴定的引物组合,本研究从现有的玉米40对SSR引物中根据多态性、等效扩增、染色体分布挑选出15对候选引物。综合考虑多重电泳,对扩增产物片段大小相差10bp以下的10对引物分别跑非荧光和荧光标记毛细管电泳;对扩增产物片段大小相差10bp以上的引物跑琼脂糖凝胶电泳。最终根据三种电泳方式确定了相对应的纯度鉴定引物组合。此套纯度鉴定组合适于京科968玉米品种的纯度鉴定,并为其他玉米大品种的纯度鉴定提供了思路、尤其是对主要经营一些玉米大品种的企业尤为适用。

关键词:京科968;纯度鉴定;分子标记

Determination of Primers for Purity Identification of Maize Variety Jingke 968

Feng Bo[1,2]* Xu Liwen[2] Wang Fengge[2]** Yu Ronghai[1]**
Yi Hongmei[2] Liu Wenbin[1,2] Ge Jianrong[2]
Tian Hongli[2] Hou Zhenhua[2] Wang Lu[2] Ren Jie[2]
Wang Yuandong[2] Zhao Jiuran[2]**

(1. College of Agriculture, Jilin Agricultural University, Changchun 130118;
2. Maize Research Center, Beijing Academy of Agriculture & Forestry Sciences/Beijing Key Laboratory of Maize DNA Fingerprinting and Molecular Breeding, Beijing 100097)

Abstract: In order to make sure a set of primers combination of maize Jingke 968's variety and purity identification, 15 pairs of primers were selected from the existing maize 40 SSR primers according to the polymorphism, the equivalent amplification and the chromosome distribution. Considering multiple electrophoresis synthetically, 10 pairs of primers were used to run the non-fluorescent and fluorescently labeled capillary electrophoresis which the size of the amplified fragment was below 10bp, PCR products were used to run the agarose gel electrophoresis which the size of the amplified fragment

---

基金项目:本研究由科技部国家科技支撑计划(2015BAD02B02)和北京市农林科学院院科技创新能力建设专项(KJCX20161501)共同资助

* 同等贡献作者

** 共同通讯作者:gege0106@163.com;1010081487@qq.com;jiuran@263.net was more than 10bp. Finally, according to the three electrophoresis methods to identify the corresponding primers. This collection of purity identification suited for the Jingke 968's purity identification. It also provides the ideas for the purity identification of other corn varieties and especially suits for some enterprises with managing some corn varieties.

Key words: Jingke 968; Purity identification; Molecular marker

随着农业的发展越来越多的人关注玉米种子的质量，因为玉米种子的质量关乎农业生产，关乎农民收益（莫伟健等，2007）。纯度在一定程度上反应玉米品种的质量（刘景云，2011；王凤格等，2015a）。纯度是指品种在特征特性方面典型一致的程度，用本品种的种子数占供检本作物样品种子数的百分率表示（梅眉和路璐，2005）。因种子纯度不达标造成的减产损失能够抵消一个耗费大量物力、财力、人力和时间而选育出的优良主推品种的增产潜力而带来的经济效益，因此强化玉米种子质量检测，加强玉米种子的纯度检验工作迫在眉睫（罗黎明等，2011）。随着玉米品种鉴定技术的发展，玉米品种纯度鉴定早已由传统的田间种植鉴定技术发展到分子标记技术（高文伟等，2004）。其中SSR标记是应用较多的技术标记（李素玲等，2007；杨旭等，2016）。玉米杂交种纯度鉴定的引物要求一是杂合率高，从而能够比较容易筛选到双亲互补型的引物；二是多态性高，从而具有较高的区分异型株的能力。

对于一些大型种子企业内部质量控制，由于种子企业生产经营的品种数量较少，但每个品种的制种量大。为其主推品种量身定制一套纯度鉴定方案至关重要，不仅省时省力而且减少了不必要的试验耗材。同时也为今后玉米大品种的纯度鉴定提供了思路。例如京科968这种玉米大品种（赵久然，2013）的纯度问题。因其用种量大，如果每次做纯度鉴定都筛选引物的话，那么工作量会很大。本试验借助三种电泳检测平台筛选适用于玉米品种京科968的引物组合。今后对京科968的品种鉴定只需要根据不同的电泳平台选择不同的引物即可，无需每次试验时都进行繁琐的引物筛选工作。本试验的方案及流程为今后的玉米特定品种纯度鉴定引物的筛选提供了技术路线与思路。

1 结果与分析

本试验根据三种电泳检测平台成功的确立了玉米品种京科968的纯度鉴定引物组合，三种平台各有一套适用于京科968玉米品种纯度鉴定的引物名单。根据待测品种京科968的40对核心引物DNA指纹图谱和数据信息（王凤格等，2015），剔除掉单峰（纯合带型）的引物位点及表现为高低峰（两条谱带高度差异较大）、多峰（两条以上的谱带）等异常峰型的引物位点，挑选出具有双峰（杂合带型）的15对引物作为纯度鉴定候选引物（表1）。

表1 15对候选引物名单
Table 1 List of 15 candidate primers

引物编号 Primer numbers	引物序列 5′→3′ Primer sequence 5′→3′	染色体 Chromosome	京科968的片段大小（bp） The size of fragment in Jingke 968 (bp)
P03	TTACACAACGCAACACGAGGC	2	248/256

（续表）

引物编号 Primer numbers	引物序列 5′→3′ Primer sequence 5′→3′	染色体 Chromosome	京科 968 的片段大小（bp） The size of fragment in Jingke 968（bp）
P06	GCTATAGGCCGTAGCTTGGTAGACAC CCCTGCCTCTCAGATTCAGAGATTG	3	336/343
P08	TAGGCTGGCTGGAAGTTTGTTGC GCACACCCGTAGTAGCTGAGACTTG	4	364/380
P11	CATAACCTTGCCTCCCAAACCC TCTCAGCTCCTGCTTATTGCTTTCG	6	165/172
P12	GATGGATGGAGCATGAGCTTGC GATCCGCATTGTCAAATGACCAC	6	265/277
P13	AGGACACGCCATCGTCATCA AATGCCGTTATCATGCGATGC	7	191/201
P16	GCTTGCTGCTTCTTGAATTGCGT TGAACCACCCGATGCAACTTG	8	217/222
P17	TTGATGGGCACGATCTCGTAGTC CGCCTTCAAGAATATCCTTGTGCC	9	408/413
P22	GGACCCAGACCAGGTTCCACC CCCGACACCTGAGTTGACCTG	1	184/193
P24	CTGGAGGGTGAAACAAGAGCAATG TGATAGGTAGTTAGCATATCCCTGGTATCG	2	222/238
P25	GAGCATAGAAAAAGTTGAGGTTAATATGGAGC CACTCTCCCTCTAAAATATCAGACAACACC	2	173/179
P30	GCTTCTGCTGCTGTTTTGTTCTTG CTTCTCCTCGGCATCATCCAAAC	5	134/144
P32	GGTGGCCCTGTTAATCCTCATCTG AGCAAGCAGTAGGTGGAGGAAGG	6	228/234

(续表)

引物编号 Primer numbers	引物序列 5′→3′ Primer sequence 5′→3′	染色体 Chromosome	京科 968 的片段大小（bp） The size of fragment in Jingke 968（bp）
P34	AGCTGTTGTGGCTCTTTGCCTGT GCTTGAGGCGGTTGAGGTATGAG	7	156/170
P37	TGCACAGAATAAACATAGGTAGGTCAGGTC ACTGATCGCGACGAGTTAATTCAAAC TACCGAAGAACAACGTCATTTCAGC	9	185/199

对于设备条件较好的实验室优先选择使用荧光标记毛细管电泳，荧光标记毛细管电泳适于大样本、复杂样本以及引物的筛选。根据京科968所出的条带位置，综合考虑染色体分布、片段大小、多重电泳等因素从适于荧光标记毛细管电泳检测平台的10对候选引物中确定了8对引物。P03、P06、P13、P16、P17、P25、P30、P32本试验中8对引物组合在一起区分杂株的能力增强、节省试验耗材、提高工作效率。孔H3、H4、C6分别为京科968的父母本及杂交种；孔G7为杂株，不同样品间区分效果很明显（图1）。

如果使用琼脂糖凝胶电泳或非变性凝胶电泳等分辨率较低的电泳检测平台进行纯度检测，则在上述候选引物中进一步挑选出两个谱带片段大小相差较大的引物。利用入选的引物对待测杂交种小样本进行初检（杂交种取20粒），判断其纯度问题是由于自交苗、回交苗、其他类型杂株还是遗传不稳定造成的。并进一步确定该样品的纯度鉴定引物对其大样本进行鉴定（王凤格等，2015a）。

对于储备引物较少，样品少的实验室优先选择使用非变性凝胶毛细管电泳平台。根据所用仪器的分辨率挑选出9对引物。预试验中，该平台的碱基分辨率为4bp。P03、P06、P08、P11、P13、P16、P22、P30、P32这9对引物单重电泳（图2）均取得了较好的试验研究。

试验中将3对引物按扩增产物片段大小组合在一起电泳（图3~图5）。除个别孔位扩增失败外其他孔位能清晰看到目标条带，与单重电泳相比多重电泳时出现小杂峰较多。3对引物组合在一起电泳节省了电泳时间、试验成本。对于自交系以及杂株区分能力较强。

对于一般的实验室可以由研发单位提供适合琼脂糖引物的名单，利用琼脂糖电泳进行纯度检测。本试验从5对候选引物中筛选到4对引物，试验发现扩增产物片段大的引物需要较长时间的电泳才能区分开，所以将P08剔除，P12、P24、P34、P37用于京科968及其父母本的纯度鉴定。本套引物只适用于混杂父母本引起的纯度问题，对于含有其他类型杂株的纯度鉴定则不适用，因为这4对引物并不是在所有其他品种的玉米杂交种中均双亲互补。P24、P34、P37区分父母本及杂交种能力较强，P12弱一些（图6）。

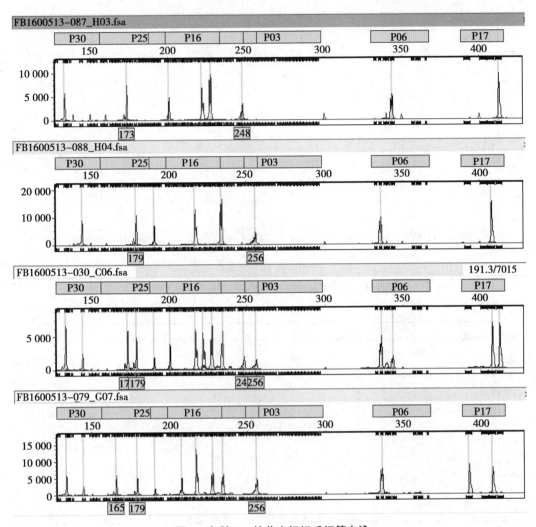

图1 京科968的荧光标记毛细管电泳

Figure 1　Fluorescence labeling capillary electrophoresis of Jingke 968

2　讨论

三种电泳方法优缺点的比较。本试验主要针对玉米品种京科968纯度鉴定筛选引物。本试验采用三种电泳方式，三种电泳方式各有优缺点，可根据试验需求自行选择。

（1）荧光标记全自动毛细光电泳，用于扩增样品的高通量电泳。SSR引物双亲互补的均可使用，最后确定P03、P06、P13、P16、P17、P25、P30、P32这8对引物，因为他们可以组合电泳，即8重电泳。引物位置读取精准可以精确到1bp，另外本实验室拥有自主研发的SSR Analyser分析软件以及数据管理纯度分析系统，不用繁琐的数据统计只需简单的分析即可得到纯度鉴定分析；大部分的纯度均需要多个引物组合对自交苗、回交苗、其他类型杂株、遗传不稳定的判断，8重电泳提高了通量；电泳速度快，35min即可完成一块96孔板的电泳检测；但要求荧光引物，试验所用耗材偏贵。

（2）非荧光全自动毛细管电泳，用于扩增样品的高通量电泳。可用的引物较多，对

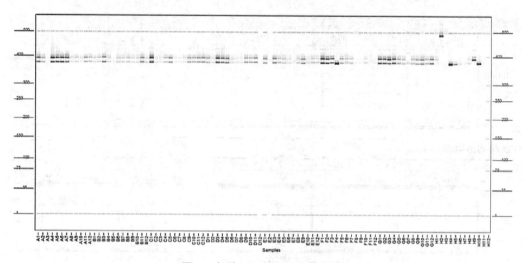

图 2　京科 968 在 P08 上的电泳
Figure 2　Electrophoresis of Jingke 968 on P08

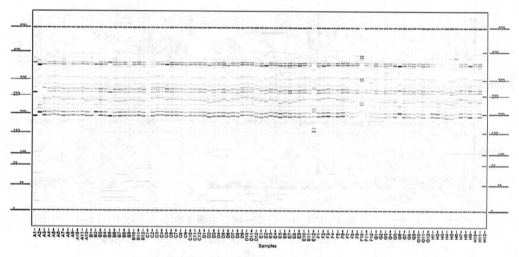

图 3　京科 968 的三重电泳组合（P03，P06，P22）
Figure 3　Three Electrophoresis combination of Jingke 968（P03，P06，P22）

大部分 SSR 引物适用；引物不需要荧光标记所以试验成本相对于荧光标记毛细管电泳要少一些；试验中扩增产物不需要变性，减少了试验程序，避免接触有毒试剂；三重电泳节省了试验成本。但扩增带型存在片段大小读取不准确，孔内误差 2bp、数据不易整合、平均 80min 一块 96 孔电泳板，试验耗时较长。

（3）琼脂糖凝胶电泳。分辨率低，要求引物扩增区间大至少相差 10bp，扩增产物片段读取的是大概值，没有具体的数值；不适合组合多重电泳，通量小；但设备价格便宜，操作简单，大多数实验室均具备。适合鉴定由于自交苗引物的纯度问题，即只需要 1~2 对引物即可完成纯度鉴定（王凤格等，2015a）；对于其他杂株的鉴定则难度较大，因为存在杂株两条带相差片段较小或者是单带，这样的话就很难区分无法判断。

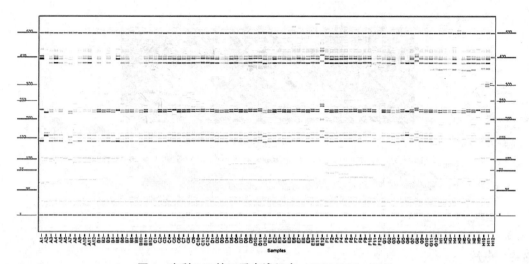

图 4 京科 968 的三重电泳组合 (P08, P16, P30)
Figure 4 Three Electrophoresis combination of Jingke 968 (P08, P16, P30)

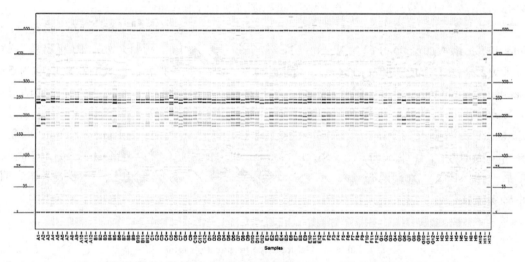

图 5 京科 968 的三重组合电泳 (P11, P13, P32)
Figure 5 Three Electrophoresis combination of Jingke 968 (P11, P13, P32)

3 材料与方法

3.1 候选引物

本研究以京科 968 及其父母本,采用王凤格和赵久然(2011)的改良 CTAB 法。玉米行业标准中的 40 对引物,依据《国家审定品种 SSR 指纹图谱》挑选出 15 对引物 P03、P06、P08、P11、P12、P13、P16、P17、P22、P24、P25、P30、P32、P34、P37。

3.2 PCR 扩增

单重 PCR 反应体系 200μmol/L,DNA 模板 2μl,2× Taq Plus Master Mix 10μl,0.25μmol/L 引物 2μl,总体积 20μl。PCR 程序:95℃预变性 5min;94℃变性 40s,60℃退

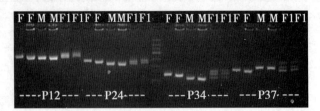

图 6　京科 968 及其父母本琼脂糖电泳
Figure 6　Agarose gel electrophoresis of Jingke 968 and its parents

注：F 为父本，M 为母本，F1 为京科 968 杂交种
Note：F as the male parent, M as female parent and F1 as Jingke 968 hybrids

火 35s，72℃延伸 45s，35 个循环；72℃延伸 10min；4℃保存。

3.3　电泳

荧光标记毛细管电泳在 96 孔板的各孔中分别加入 9μl 去离子甲酰胺、0.2μl GS3730-500 分子量内标和 PCR 产物的混合物 2μl，95℃变性 5min，于 4℃保存 10min，2 000r/min 离心 1min，于 ABI3730XL DNA 分析仪上进行荧光标记毛细管电泳。

非荧光标记毛细管电泳。a. 胶和染料混合物制备（表 2）；b. Inlet Buffer 准备：1× Inlet Buffer 溶液加至 96 孔深孔板，每孔加 1ml，每板 Inlet Buffer 可以做 6～10 板试验，故每天更换一次 1×Inlet Buffer 溶液；c. Capillary Conditioning 溶液准备：试剂盒提供的为 5× Capillary Conditioning 溶液，试验前加超纯水稀释至 1× 浓度待用，溶液不足时添加即可；d. Marker 准备：Marker 板每孔加 30μl 的 Marker，之后每孔加一滴矿物油。注：添加 Marker 时要涡旋混匀，加好之后务必进行离心，确保无气泡；e. Ladder/Sample 准备：Sample 板前 95 个孔加 22μl 的 1X TE dilution buffer 溶液和 2μl 样品（22μl TE+2μl 样品），最后一孔（H12）加 24μl 的 Ladder。注：样品和稀释溶液务必要混匀，Ladder 添加之前也必须涡旋震荡，混匀之后方可使用，最后务必进行离心，确保样品板每孔无气泡。

表 2　胶和染料配制
Table 2　Gel and dye preparation table

96 孔板程序类型 96-well plates to banalyzed (FC = Full Conditioning; GP = Gel Prime Only)	ZAG dsDNA 凝胶近似体积 Approximate Volume of ZAG dsDNA Gel	嵌入染料量 Volume of Intercalating dye
1（1FC Only）	25ml	1.25μl
2（1FC+1GP）	30ml	1.50μl
5（1FC+4GP）	45ml	2.25μl
8（1FC+7GP）	60ml	3.00μl
10（1FC+9GP）	75ml	3.75μl

琼脂糖凝胶电泳 PCR 扩增产物在质量百分比为 6% 的琼脂糖凝胶中，80V 稳压电泳分离，最后在凝胶成像系统中观察拍照。

参考文献（略）

　　本文源载：http：//kns.cnki.net/kcms/detail/46.1068.S.20170406.1155.018.html

非优异 SSR 原因分析及解决方案

刘文彬[1,2]*　许理文[2]*　冯博[1,2]　易红梅[2]　任洁[2]　田红丽[2]
赵涵[3]　吕远大[3]　蔚荣海[1]**　赵久然[2]　王凤格[2]**

(1. 吉林农业大学农学院，长春　130118；2. 北京市农林科学院玉米研究中心/
玉米 DNA 指纹及分子育种北京市重点实验室，北京　100097；
3. 江苏省农业科学院农业生物技术研究所，南京　210014)

摘　要：SSR 标记在真实性检测、纯度鉴定、以及指纹库构建中起着至关重要的作用。在其筛选过程中会遇到许多具有 N+1 峰等异常峰型、非特异性峰、多态性低、不适于软件自动分析等非优异 SSR 引物。为了解并掌握非优异 SSR 引物表现以及形成原因，本研究以 22 份初筛材料、27 套三联体材料及 95 份自交系对新型 SSR 引物进行评估筛选，从中选取有代表性的 11 对 SSR 引物，现在已对 10 种非优异 SSR 引物表现进行描述，分析形成原因，通过测序对一对引物进行重新设计，并获得较好效果。

关键词：优异 SSR；非特异峰；N+1 峰；测序；重新设计

The Reason Analysis and Solution about Imperfect SSR

Liu Wenbin[1,2]*　Xu Liwen[2]*　Feng Bo[1,2]　Yi Hongmei[2]　Ren Jie[2]
Tian Hongli[2]　Zhao Han[3]　Lv Yuanda[3]　Yu Ronghai[1]**
Zhao Jiuran[2]　Wang Fengge[2]**

(1. College of Agriculture, Jilin Agricultural University, Changchun 130118;
2. Beijing Key Laboratory of Maize DNA Fingerprinting and Molecular Breeding, Maize Research Center, Beijing Academy of Agriculture & Forestry Sciences, Beijing 100097;
3. Institute of Agricultural Biotechnology, Jiangsu Academy of Agricultural Sciences, Nanjing 210014)

Abstract: SSR markers play an important role in authenticity detection, purity identification and fingerprint library building. In the process of SSR primers screening will meet imperfect SSR primers, such as N+1 peak and other abnormal peak, nonspecific peak, low polymorphism and not suitable for automatic analysis software. In order to understand and master the performance and the reasons about

基金项目：本研究由科技部国家科技支撑计划（2015BAD02B02）和北京市农林科学院院科技创新能力建设专项（KJCX20161501）共同资助

* 同等贡献作者
** 通讯作者，1010081487@qq.com；gege0106@163.com

imperfect SSR primers, in this study, using 22 materials about first screening, 27 sets of mother-father-child trios and 95 inbred lines for evaluate new SSR primers screening and choose 11 pairs representative primer, now in 10 kinds of imperfect SSR primers was described expression and analyzes the formation reasons. Redesigned a pair of primer by sequencing and obtain satisfactory result.

Key words：Perfect SSR；Nonspecial Peak；N+1 Peak；Sequencing；Redesign

SSR 分子标记已被广泛用于玉米、大豆、水稻、棉花等作物基因定位及资源分析等方面（楚渠和孟刚，2016），优异的 SSR 引物在玉米真实性、纯度、种质资源鉴定中以及 DNA 指纹库构建等研究中起着至关重要的作用（王凤格等，2006）。于新艳等（2007）对不符合构建指纹库的引物进行了重新设计，在 Tm 值、引发效率、引物二聚体等方面进行改善，已使引物达到了预期要求；潘菲等（2008）对多重 PCR 荧光毛细管电泳中微卫星不稳定性的常见问题进行了分析；王凤格等（2007）对荧光毛细管电泳中检测到的三峰、连续多峰等异常峰型进行了分析；刘珊珊等（2013）对非特异性扩增进行了研究；目前标准中正在使用的 40 对玉米 SSR 引物多为 2 碱基重复，不利于数据整合，本单位正在基于重测序数据开发出的 SSR 位点而设计的新型 SSR 引物进行评估筛选，力求筛选出适于软件自动分析以及数据整合的优异 SSR，用于今后的真实性鉴定及指纹库构建等方面，本研究对此过程中引物表现的情况进行现象描述及原因分析，对部分表现不好的引物 PCR 产物进行测序，从中挑选 2 对进行举例说明，并对引物 SR26829 进行重新设计且获得较好效果。

1 结果与分析

1.1 不可以改善的非优异 SSR 引物表现

部分非优异 SSR 引物只能剔除，是引物保守性差造成的，如不完全以基序为重复单位；引物特异性差，如多态性低；有 InDel 位点插入和缺失，如在三联体材料上不符合孟德尔遗传规律，在简单重复序列内部的插入和缺失是不能够改善的。

一般用多态性信息指数（PIC 值）来评估多态性，PIC 值的变化与等位基因变异频率是密切相关的，一般情况下，某个位点的变异数目较多，其 PIC 值也会高，呈正相关关系，但如果该位点各个等位基因变异频率不均匀，则 PIC 值并不高。图 1-1 SR19853 有四个等位基因，但是分布不均匀，图 1-2 SR36461 只有两个等位基因，因此两对引物的多态性表现都较低，也可能是由于样品亲缘关系较近，代表性不强。

符合孟德尔遗传规律的引物即双亲互补型引物，在该标记位点上父母本为不同的基因型，对应的杂交种表现为父母本基因型的组合。图 1-3 中 SR3111 杂交种 BGG156 在 196bp 位置有条带，其双亲分别在 190bp 和 196bp 位置，图 1-4 中在杂交种 BGG1408 上 186bp、198bp 位置上有条带且有严重不对称扩增现象，其双亲分别在 186bp 和 190bp 位置，皆无法证明其双亲关系。有两种原因，一是样品混杂导致的，二是有插入或缺失位点，是否能改善要根据 InDel 位点所在位置决定。

1.2 可以改善的非优异 SSR 引物表现

部分非优异 SSR 引物是可以改善的，主要原因分为以下几个方面：引物设计质量问题，表现为 PCR 失败、不对称扩增及扩增效率低；引物位点不纯合导致双带，引物保守

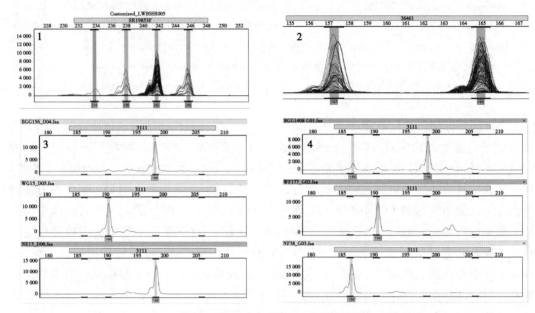

图 1　不可以改善的 SSR 引物图谱
Figure 1　The figureprinting about can't improve primers
注：1-1 和 1-2. 多态性低；1-3 和 1-4. 不符合孟德尔遗传规律
Note：1-1 and 1-2. Polymorphism is low；1-3 and 1-4. Do not conform Mendeian genetic law

性低导致有非特异性扩增；扩增滑移导致 N+1 峰；在引物设计区域有 InDel（冯芳君等，2005）（Insertion/Deletion 插入或缺失多态性）或 SNP（single nucleotide polymorphisms，单核苷酸多态性，主要是指由于单个核算的变异而引起基因组水平上的 DNA 序列多态性，形式包括单碱基的缺失，插入，转换及颠换等）位点的插入等可以通过重新设计引物或其他方法改善去解决（唐立群等，2012）。

1.2.1　PCR 失败

在电泳平台上检测不到目标片段有以下原因：①引物设计问题，上下游引物不能正常稳定地结合；②引物合成存在问题；③试验操作误差导致 PCR 失败。在 ABI3730 电泳平台上扩增目标位点无荧光信号还有以下原因（潘菲等，2008）：①引物上标记的荧光素衰竭；②电泳迁移率变化或 LIZ 峰值太低导致数据分析不能通过图 2-1 为 QS274 PCR 失败的表现，45bp 为引物二聚体，QS274 的产物扩增范围在 220~235bp，未出现目的条带。

1.2.2　不对称扩增

由于杂合子的杂合位点上，杂合子不对称扩增使两个特异峰的峰值存在差异，形成高低峰，高低峰峰值相差较大可以影响品种鉴定。不对称扩增可能是由 DNA 质量以及引物设计质量引起的（王凤格等，2007），也可能是非特异扩增，若峰高相差 20%~30%，应加大模板量并 PCR 确定其特异峰型（易红梅，2006）。图 2-7 和图 2-8 为不对称扩增。

1.2.3　扩增效率低

扩增效率是表征 PCR 反应扩增能力的一个重要参数，Chatterjee 认为模板复性可能是扩增效率降低的主要因素，并构建 PCR 反应在动力学上的分析方程，易健明（2014）亦证明模板复性是主要因素，而引物及底物的耗竭，以及酶失活及饱和不是主要因素。峰高是分子鉴定中重要的判定依据，峰值过低无法判断是否是真实峰，若引物自身的引发效率低导致扩增效率低，进而导致峰值低，容易造成误判，且扩增效率结果更容易受其他因素影响。相同 PCR 体系下其他引物峰值均在 8 000 以上，图 2-5 中 SR40017 峰高在 2 000~4 000。

引物位点不纯合会导致自交系出现双带的现象，图 2-2 为引物 QS288 在自交系京 724 上有 183bp 和 187bp 两条带。

1.2.4 非特异扩增

PCR 反应过程中 PCR 产物检测到的电泳条带脱离正常范围内的条带为非特异扩增，非特异扩增会对特异性扩增进行干扰，并且导致假阳性。非特异性扩增是由气雾胶污染、试验操作污染，或者是引物保守性低，除了在靶目标片段上结合，也可以在别的位置结合并进行 PCR，如引物之间互为模板（刘珊珊等，2013）Taq DNA 聚合酶在 PCR 反应中仍具有聚合酶活性，易产生非特异扩增及引物二聚体，周晓薇等（2011）通过化学修饰法改造 Taq DNA 聚合酶，使其具有更高特异性及灵敏性。图 2-3 为 QS304 的 panel，产物扩增范围为 165~171bp，119bp 位置产物为非特异性扩增产物，图 2-4 为 QS304 在双金 11 上检测到 119bp 和 168bp 两条带。

1.2.5 N+1 峰

N+1 峰指 PCR 反应过程中普通 Taq 酶在产物 3′端自动加入一个碱基引起，表现为在同一位置上出现了相差 1bp 的两个峰，在 PAGE 电泳表现为相距很近的两条带（王凤格等，2007）。QS288 在京 724 上表现为左高型 N+1 峰（图 2-2），SR40562 表现为右高的 N+1 峰（图 2-6），还有等高型的 N+1 峰。相同的引物在不同的试验条件下有时会出现 N+1 峰，SR26829 在不同批次试验下，三联体材料上表现为单峰（图 2-9），自交系材料上表现为 N+1 峰（图 2-10）。

1.2.6 Indel 插入或缺失

引物 SR26829 在三联体材料上 155~171bp 范围内表现较好（图 2-9），完全四碱基重复且峰型好，峰值较高，分布较均匀，符合优异 SSR，在 120~142bp 范围内，不以基序为重复单位，且峰高差异较大。SR26829 在 95 份自交系材料上的表现（图 2-10），不仅有不符合基序的等位基因，同时峰型表现为 N+1 峰。QS068 在 95 份材料上的表现（图 2-11），有五个样品在在 221bp、224bp、227bp 上有带，亦是未以基序为重复单位。

1.3 引物改善方案

通过重测序数据获得侧翼序列，若是在简单重复序列两端有 indel 位点，则可以进行重新设计，若在简单重复序列上有 indel 位点，只能剔除，下面以引物 QS068 和 SR26829 为例，着重介绍 SR26829。

通过 SSR 指纹分析器软件对引物进行分析，找出目标片段不符合基序的样品及对应孔位（表1），利用该样品进行 PCR 扩增（SSR60），并送至测序公司进行测序，产物体积 20μl，上游引物原液浓度 15μmol/L，体积 20μl，并选取部分产物进行 2% 琼脂糖凝胶电泳鉴定（图 3），鉴定结果 PCR 产物合格，测序样品如表一所示。图 4 为测序样品 WG81 在 120bp 并符合亲子关系，是真实条带，测序结果表明 QS068 有 2bp 的插入（图 5），SR26829 在简单重复序列之前有大的 indel 片段插入或缺失，测序及分析结果见图 6、

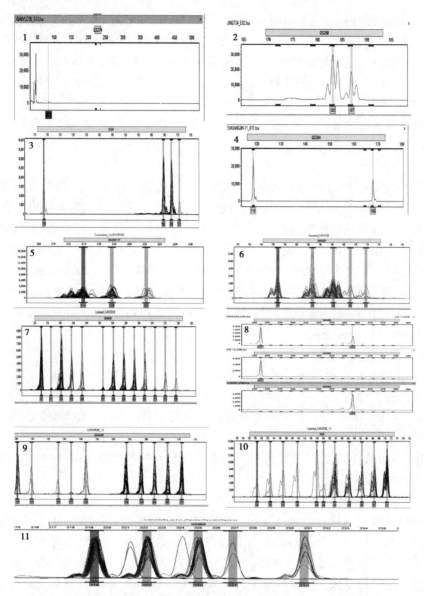

图 2 可以改善的引物表现图谱
Figure 2 The figureprinting about can improve primers

注：2-1. QS274 PCR 失败；2-2. QS288+京 724 自交系出现双带；2-3. QS304 在 24 个样品上的 panel；2-4. QS304+双金 11，有非特异性扩增；2-5. SR40017 扩增效率低；2-6. N+1 峰；2-7 和 2-8. SR36996 不对称扩增；2-9. SR26829 在三联体材料上的表现；2-10. SR26829 在 95 份自交系上的表现；2-11. QS068 不以基序为重复单位

Note：2-1. QS274 PCR failure；2-2. QS288+Jing724 selfing line appear dual-band；2-3. The panel about QS304 on 24 samples；2-4. QS304+Shuang Jin 11 have nonspecific amplification；2-5. SR40017 amplification efficiency is low；2-6. N+1peak；2-7 and 2-8. SR36996 asymmrtry amplification；2-9. The show about SR26829 on mother-father-child trios；2-10. The show about SR26829 on 95selfing lines；2-11. QS068 do not repeat unit for motif

表2。通过重新设计将序列向左平移,避开 indel 位点,获得新的引物,命名为26829WF,26829WF 引物在 95 份自交系上的电泳图谱 panel 如图 7 所示,整体效果较好,个别等位基因不符合完全重复序列,但解决了改善前引物具有的问题。

表 1　测序样品信息（SR26829）
Table1　The information of sequencing samples（SR26829）

DNA 板 DNA plate	孔位 Well	样品编号 Sample number	样品名称 Sample name	片段大小 Fragment
三联体 Mother-father-child trios	D12/E2	WG79/WG81	DH40/DH65232	120
三联体 Mother-father-child trios	B12	WF549	豫自 87-1	125
95 份自交系 95 inbred lines	B5	CX123	parviglumis	129
95 份自交系 95 inbred lines	A5/H5	WG073/WF246	CML162/巴 816	133
三联体 Mother-father-child trios	E11/E12	WG99/WG42	F06/19F	142
95 份自交系 95 inbred lines	B4	CX122	huehuetenangensis	148
95 份自交系 95 inbred lines	B6/B7	CX124/CX125	nucaraguensis/luxurians	154
95 份自交系 95 inbred lines	F10/F11	WF111/WF113	M276/B567	163
95 份自交系 95 inbred lines	H8/H9	WF276/WF278/	东 D201/辽 3162/ B73	171

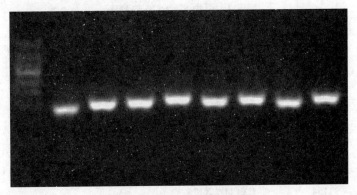

图 3　测序样品电泳检测结果

Figure 3　The results of electrophoresis detection about sequencing samples

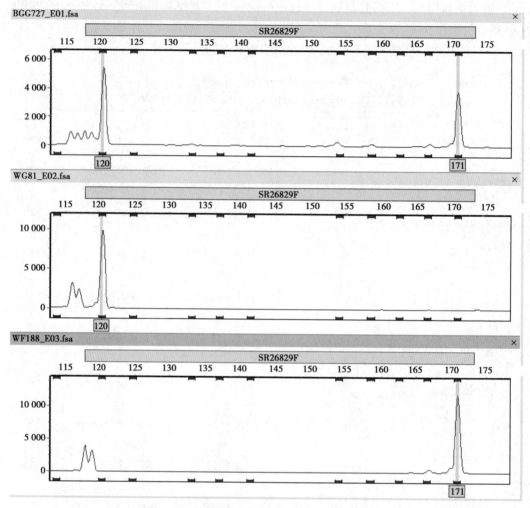

图 4 三联体 BGG727，WG81，WF100 的图谱

Figure 4　The figureprinting about mother-father-child trios materials BGG727, WG81 and WF100

表 2　引物 SR26829 的测序
Table 2　The sequencing results about primer SR 26829

顺序号 SN	片段 Fragment	样品编号 Sample number	基序 motif	插入/缺失（37bp） InDel（37bp）
1	120	WG79/WG81	4	D
2	125	WF549	5	D
3	129	CX123	6	D
4	133	WG073/WF246	7	D
5	142	WG99/WG42	9	D
6	148	CX122	2	I
7	154	CX124/CX125	3	I
8	163	WF111/WF113	5	I
9	171	WF276/WF278/B73	7	I

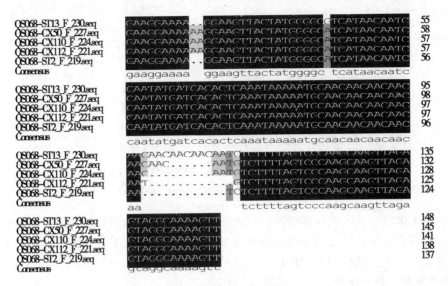

图 5 引物 QS068 测序

Figure 5　The sequencing results about primer QS068

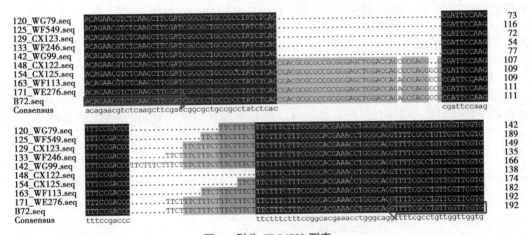

图 6 引物 SR26829 测序

Figure 6　The sequencing results about primer SR26829

2 讨论

2.1 引物入选标准

引物的筛选条件多为多态性高、杂合率高、重复性好、条带清晰，未将基序大小纳入引物筛选条件之中。在现有标准中的 40 对玉米 SSR 引物中，有部分 2 碱基重复的引物在建库过程及真实性检测中，不同样品之间的条带可能相差 2bp，或者杂交种在该位点上相差 2bp，不同的电泳平台上分辨率不同，如聚丙烯酰胺凝胶电泳无法准确判断出是哪一条带，和荧光毛细管电泳获得的数据无法准确结合。而本单位在引物筛选过程中严格按照基序≥3 碱基，且完全按照基序进行重复的原则，避免了 2 碱基重复的引物，有望实现不同平台之间的数据整合，实现数据共享。

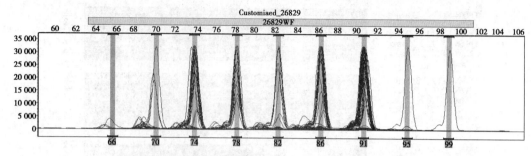

图 7　重新设计的引物 26829WF 在 95 份自交系上的图谱
Figure 7　The figureprinting about redesign primer 26829WF on 95 selfing lines

在荧光毛细管电泳上连续多峰及三峰的读取相对较难，不同检验员之间可能会读取不同的条带，在引物筛选阶段解决这些特异峰型是必不可少的，引物严格筛选会为今后的分子检测及建库等工作减少很大工作量。

2.2　引物重新设计与优化

引物筛选工作量大且繁琐，能筛选出一对综合表现良好，多态性高、重复性好的引物较不易，大部分引物表现存在着缺陷，影响引物的实用性。通过一些技术方法将非优异 SSR 引物优化为优异 SSR 引物，可以有效增加优异引物的获得率。

3　材料与方法

3.1　试验材料

包含玉米杂交种及自交系在内 22 份材料（初筛）见表 3；27 套包含父母本及杂交种在内的三联体材料（复筛），样品名单略；95 份自交系（放大样品），样品名单略。

表 3　22 份材料名单
Table 3　22 materials list

样品编号 Sample No.	样品名称 Sample name	样品编号 Sample No.	样品名称 Sample name
M1	京科 968 Jingke968	M12	黄早四 Huangzaosi
M2	京 724 Jing724	M13	黄 C HuangC
M3	京 92 Jing 92	M14	丹 598 Dan598
M4	郑单 958 Zhengdan 958	M15	齐 319 Qi319
M5	郑 58 Zheng 58	M16	丹 340 Dan340
M6	昌 7-2 Chang7-2	M17	SW1611

（续表）

样品编号 Sample No.	样品名称 Sample name	样品编号 Sample No.	样品名称 Sample name
M7	京科糯 2000 Jingkenuo 2000	M18	PHG42
M8	京糯 6 Jingnuo 6	M19	Mo17
M9	白糯 6 Bainuo6	M20	B73
M10	先玉 335 Xianyu335	M21	双–11 Shuang–11
M11	掖 478 Ye478	M22	衡 522 Heng522

3.2 引物

11 对新型 SSR 引物，来源于北京市农林科学院与江苏省农业科学院共同研发，基于重测序数据挖掘 SSR 位点，在引物评估筛选过程中挑选的 11 对表现非优异的引物，引物基本信息见表 4。

表 4 11 对引物基本信息
Table 4 11 primers basic information

引物名称 Primer name	染色体 Chromosome	上游引物序列 5t-3eq Forward primer sequence 5t-3eq	下游引物 5t-3eq Reverse primer sequence 5t-3eq	基序 Motif	
SR3111	chr1	TGTCAGAGCCTTGTTCTCAGATG	GACCAGATGTGAAATTGCCAGAC	（ATCT)	6
SR19853	chr4	ACTCTCCAGAAGTCCTGATTTGC	CGCCAATGTCTCGACTCTCC	（AATA）	5
SR26829	chr5	ACAGAACGTCTCAAGCTTCGATC	CACCAACCAACAGGCGAAAAC	（TTCT)	7
SR36461	chr8	GCTCTCGTGTTGGTGTGGAG	GCATCTCTGTCGTCTCGTACTC	（AAAG）	4
SR36996	chr8	CATCTAACCGCACCGCTTAC	CCTTCATTCTCCAAGTACTAGCC	（AGAT）	4
SR40017	chr9	ACCAGAATTAGGGACCAATGCG	TGGTGCGTTGGGTTTGGAAG	（GCCG）	3
SR40562	chr9	ACGGATGAAGCAGGCAGAATTG	GCACGGTGTCCCTTGGTTC	（CGGC）	3
QS068	Chr3	AGAAGCGACATTGCCACATCTG	GCCTTGAACAGGAGTGTCTGAC	（CAA）	6
QS274	chr5	TCTTCGGACTTGGAGCCAGTG	GCAAGGCATGAATTGTCACCAC	（GAA）	9
QS288	chr7	AGGGTCGTCGGGACATAGTAAG	CGAAGCATACTCAGGAGCCATG	（GTAT）	6
QS304	chr5	CAAACGCCTGCCCACCATG	TCGATGTCGCTGCCTTCCG	（AGG）	6

3.3 方法

改良 CTAB 法单株提取 DNA。

PCR 程序：SSR60：95℃预变性 5min，1 个循环；94℃变性 40s；60℃退火 35s；72℃

延伸45s，共35个循环；72℃延伸10min；4℃保存。

电泳及数据分析：利用ABI3730荧光毛细管平台进行电泳检测，ABI3730获取的数据使用本单位开发的SSR指纹分析器软件（SSR Analyser V1.2.4）进行数据分析；利用琼脂糖凝胶电泳平台对测序样品进行检测，并送至擎科生物进行测序。

参考文献（略）

本文原载：http://kns.cnki.net/kcms/detail/46.1068.S.20170406.1328.022.html

基于两种荧光毛细管电泳平台筛选评估玉米新型 SSR 引物

刘文彬[1,2]　许理文[1]　王凤格[1]　赵久然[1]
冯　博[1,2]　赵　涵[3]　吕远大[3]　蔚荣海[2]

(1. 北京市农林科学院玉米研究中心/玉米 DNA 指纹及分子育种北京市重点实验室，北京　100097；2. 吉林农业大学农学院，长春　130118；3. 江苏省农业科学院农业生物技术研究所，南京　210014)

摘　要：以 93 份代表自交系及 27 套包括杂交种及双亲在内的三联体材料为试材，分别用 Qiagen 和 ABI3730 两种 DNA 分析仪，从 111 对新型玉米 SSR 引物中筛选出多态性高、稳定性好、适于自动化分型的引物 9 对，确定 1 套适合 SSR 引物评估筛选的方案，可以为今后引物设计至评估入选提供高效的筛选方案提供参考。

关键词：玉米；DNA 分析仪；新型 SSR 引物；引物评估筛选；入选率

Evaluating and Screening New Maize SSR Primer Based on Two Kinds of Fluorescent Capillary Eelectrophoresis Platform

Liu Wenbin[1,2]　Xu Liwen[1]　Wang Fengge[1]　Zhao Jiuran[1]
Feng Bo[1,2]　Zhao Han[3]　Lv Yuanda[3]　Yu Ronghai[2]

(1. Maize Research Center, Beijing Academy of Agriculture & Forestry Sciences, Beijing Key Laboratory of Maize DNA Fingerprinting and Molecular Breeding, Shuguang Garden Middle Road No.9, Beijing 100097; 2. College of Agriculture, Jilin Agricultural University, Changchun, 130118; 3. Institute of Agricultural Biotechnology, Jiangsu Academy of Agricultural Sciences, Nanjing 210014)

Abstract: Using the 93 inbred lines and 27 sets of mother-father-child trios containing hybrid and parents as the experimental material, separately using two kinds of DNA analyzer to screen

---

基金项目：国家科技支撑计划（2015BAD02B02）、北京市农林科学院院科技创新能力建设专项（KJCX20161501）

作者简介：刘文彬（1992— ），女，吉林伊通人，硕士，从事作物遗传育种及品种鉴定标记开发工作。Tel: 18513704211；E-mail: 18513704211@163.com
许理文为共同第一作者。Tel: 18600168592；E-mail: xulw0408@126.com
赵久然和王凤格为本文通讯作者

primers about Qiagen and ABI3730, screening 9 pairs of primers with high polymorphism, good stability and suitable for automatic classification from 111 pairs of primers, determined a set of suitable for SSR primers screening evaluating scheme. in order to lay a foundation for provide high efficiency screening scheme about primers design, evaluate and choose in future.

Key words: Maize; DNA analyser; New SSR primer; Primers screening and evaluating; Selected ratio

优异的SSR（simple sequence repeats，简单重复序列，亦称微卫星）引物在玉米真实性与纯度鉴定、DNA指纹库构建、种质资源鉴定、遗传多样性分析、杂种优势群划分等研究中起着至关重要的作用。在玉米重测序背景下，SSR引物位点得到大量挖掘，由早期数以千计至数以百万计，但是引物优劣需要进行评估。已有的SSR引物筛选评估方案主要是利用聚丙烯酰胺凝胶电泳，少量是利用琼脂糖凝胶电泳，但是评估筛选的引物数量较少，评估时间较长，精准度较低，不适于大批量引物的评估。利用荧光毛细管电泳平台进行引物评估的较少，一般只进行了少量的评估，没有进行大规模的引物评估。本研究在以荧光毛细管平台做大量建库的背景下，通过Qiagen和ABI3730两种荧光毛细管电泳平台，基于重测序开发引物进行评估，为今后的引物筛选提供方案支持。

1 材料与方法

1.1 供试材料

选用黄早四、Mo17、丹340、B73、掖478、京糯6共6个有代表性的自交系材料，用于在Qiagen平台上进行引物初筛；27套包括杂交种及双亲在内的三联体材料用于ABI3730平台上进行复筛（表1）；87份自交系材料用于扩大样品试验。

表1 27套三联体材料及其对应关系
Table 1 Twenty-seven sets of triplets material and corresponding relationship

编号 No.	样品名称 Sample Name			编号 No.	样品名称 Sample Name		
	杂交种 Hybrid	母本 Female parent	父本 Male parent		杂交种 Hybrid	母本 Female parent	父本 Male parent
1	郑单958	郑58	昌7-2	9	沈单10	沈137	Q1261
2	农大108	X178	黄C	10	浚单20	浚9058	浚92-8
3	鲁单981	齐319	Lx9801	11	东单60	A801	丹598
4	四单19	444	Mo17	12	丹玉39	C8605-2	丹598
5	中单2号	Mo17	自330	13	蠡玉6号	543	连87
6	吉单209	8902	吉853	14	龙单13	K10	龙抗11
7	豫玉22	综3	豫自87-1	15	登海11	DH65232	DH40
8	沈单16	K12	沈137	16	登海9号	DH65232	8723

(续表)

编号 No.	样品名称 Sample Name			编号 No.	样品名称 Sample Name		
	杂交种 Hybrid	母本 Female parent	父本 Male parent		杂交种 Hybrid	母本 Female parent	父本 Male parent
17	农大3138	综31	P138	23	京科968	京724	京92
18	会单4号	SW1611	掖107	24	登海605	DH351	DH382
19	正大619	F06	F19	25	农大84	F349	P25
20	京单28	郑58	京24	26	农华101	NH60	S121
21	苏玉糯1号	通系5号	衡白522	27	米哥	0762SF-S-2	831h-01MCh
22	渝糯7号	S147	S181				

1.2 试验方法与数据分析

采用改良CTAB法单株提取DNA。

PCR反应程序为，95℃预变性5min，94℃变性40s，60℃退火35s，72℃延伸45s，共进行35个循环；72℃延伸10min，4℃保存。

数据分析：Qiagen获取的数据使用QIAxcel ScreenGel进行数据分析；ABI3730获取的数据使用本单位开发的SSR指纹分析器软件（SSR AnalyserV1.2.4）进行数据分析。

1.3 引物评估过程

北京市农林科学院与江苏省农业科学院共同合作，基于重测序数据挖掘SSR位点，开发新型玉米SSR引物共111对。

（1）利用Qiagen进行引物初步筛选，统计片段大小、峰高、有无非特异性峰、杂峰情况、无主带现象和等位基因数。根据Qiagen结果，初步筛选出综合表现较好，等位基因数≥3、无非特异峰及双峰等现象的引物。

（2）利用ABI3730及27套三联体材料对初选引物进行复筛，统计等位基因数、双亲互补情况（双亲不互补的<4），杂合率≥0.5，是否完全以基序为重复单位，筛选出综合表现良好、多态性较高、分布较均匀、适于软件自动化分型的引物。

（3）对候选引物用87份自交系进行放大样品试验，统计等位基因数目，是否完全以基序为重复单位，计算PIC值。进一步统计其多态性，筛选出PIC值≥0.6的引物。$PIC = 1 - \sum Pi^2$，式中，Pi表示等位基因i的频率。

2 结果与分析

2.1 Qiagen电泳检测平台初筛

111对引物经过初筛，共66对引物等位基因数≥3且扩增质量较好，无特殊条带，图1为入选引物图谱。在未通过初选引物中，40对引物等位基因数≤2；具双带引物有4对，包括3对等位基因数为3的引物（M062、M075、M093），1对等位基因数为2的引物（M060）；5对引物（M027、M047、M053、M074、M075）扩增质量差，有非特异性峰；2对引物（M049、M064）无主带；引物M078PCR扩增失败。

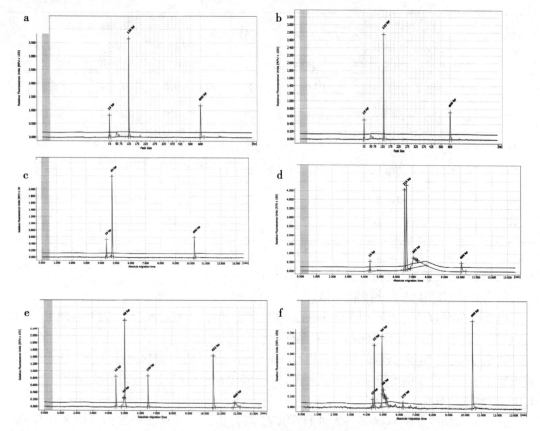

图 1 部分初选及淘汰引物指纹图谱

Picture 1 The part of finger-printing for original selection and eliminatation

注：a. M099+黄早四；b. M099+Mo17；c. M078+B73，PCR 失败，6 个样品均无条带；d. M034+B73，双峰；e. M018+黄早四，有非特异峰，主带为 156bp，峰值低于引物峰及非特异峰；f. M049+478，无主带。

Note: a. M099+ HUANGZAO-SI; b. M099+Mo17; c. M078+B73, PCR failure, six samples without peak; d. M034+ B73, double peak; e. M018+HUANGZAO-SI, nonspecific peak, main peak at 156bp and lower than primerpeak and nonspecific peak; f. M049+478, without main peak

2.2 基于 ABI3730 电泳检测平台的复筛

考虑到组成多重电泳的需要，在 66 对初选引物中合成 49 对荧光引物（有 5 对初筛等位基因数为 2 的引物）进入复筛阶段。

以 27 套三联体材料为试材进行复筛。19 对为 N+1 峰，N+1 峰未严重至影响软件分析。36 对引物未入选，主要是因为有 22 对引物不以基序为重复单位（图 2），其中，有 7 对不对称扩增（包含 3 对不以基序为重复单位），1 对不符合孟德尔遗传定律中自由组合定律，9 对杂合率低。在 Qiagen 上等位基因数为 2 的 5 对引物中，有 3 对等位基因数仍为 2（M012、M033、M041），1 对不以基序为重复单位（M044），只剩下 1 对引物（M043）进入下一步筛选过程。最终，共 13 对引物 M003、M025、M032、M042、M043、M045、M050、M058、M065、M088、M096、M108 和 M109 满足筛选条件（表2）。

表2 49对引物在27套三联体材料的评估结果
Table 2 Forty nine pairs of primers in 27 sets of mother-father-child trios material evaluation results

引物编号 Primer number	基序 Motif	片段大小（bp） Fragment size	等位基因数（个） Allele num	不以基序为重复单位 Not in the motif as the repeat unit	不对称扩增 Asymmetric amplification	不符合孟德尔遗传定律 Do not Conform Mendelian gentic law	杂合率低 Low hybrid rate	N+1峰 N+1 peak
M001	(CGGC) 3	157~172	5					√
M002	(GCCG) 3	212~222	3	√				
M003	(CCAT) 5	186~196	4					
M008	(ATAC) 6	181~201	6	√				
M010	(AGAT) 4	227~279	13		√			
M012	(AAAG) 4	157~165	2				√	
M015	(CATT) 8	226~252	6	√	√			
M016	(GAAA) 5	151~163	4				√	
M019	(ATGT) 7	221~241	6	√				
M021	(AGGG) 3	168~184	3	√				
M025	(TTCT) 8	185~200	7					
M026	(TGGA) 3	193~205	4	√	√			
M028	(TTCT) 7	121~171	10	√				
M029	(AATA) 8	213~225	4				√	
M030	(TAGA) 5	151~194	7	√				
M032	(TCTA) 9	145~179	11					
M033	(GTAC) 5	155~158	2				√	√
M035	(TTAT) 4	192~200	3	√				
M036	(TTCG) 3	155~167	4				√	
M037	(CAGC) 4	224~232	3	√				√
M040	(AATA) 5	234~246	4				√	√
M041	(GAAT) 4	180~192	2				√	
M042	(ATAG) 6	204~216	6					
M043	(AAGG) 5	197~207	4					√
M044	(CGCA) 6	234~246	3	√				√
M045	(TCTA) 3	185~195	3					
M050	(ATCA) 5	196~210	4					√

（续表）

引物编号 Primer number	基序 Motif	片段大小（bp） Fragment size	等位基因数（个） Allele num	不以基序为重复单位 Not in the motif as the repeat unit	不对称扩增 Asymmetric amplification	不符合孟德尔遗传定律 Do not Conform Mendelian gentic law	杂合率低 Low hybrid rate	N+1峰 N+1 peak
M052	(AAGA)5	137~157	5				√	
M056	(CTCC)4	195~203	3		√			
M058	(ACGG)5	151~165	4					
M065	(ATTT)5	211~225	5					
M069		155~163	3				√	
M077		116~129	4	√				√
M079		110~118	8	√				
M083		184~205	5		√			√
M084		113~125	4	√	√			
M085		87~109	3		√			√
M086		94~120	4	√				
M088		110~128	6					√
M090		93~132	6	√				√
M092		130~142	4	√				
M096		104~121	5					
M097		103~123	6	√				√
M098		91~107	3					√
M099		120~128	3	√				
M105		107~119	4	√				√
M107	(ATCT)6	186~206	4			√		
M108	(GGCA)5	158~179	4					
M109	(CATG)3	139~156	4					√

2.3 扩大材料后的多态性评估

对复筛入选的13对引物，采用87份自交系进一步验证，统计等位基因数并计算PIC值（Polymorphism information content，多态性信息量）。等位基因数与三联体材料整体一致，结果具可重复性，放大样品后其PIC值更加客观，5对引物PIC值>0.74，4对引物PIC值介于0.5~0.6（M043、M045、M050、M108），M043即Qiagen平台上等位基因数为2的引物因PIC值低淘汰。

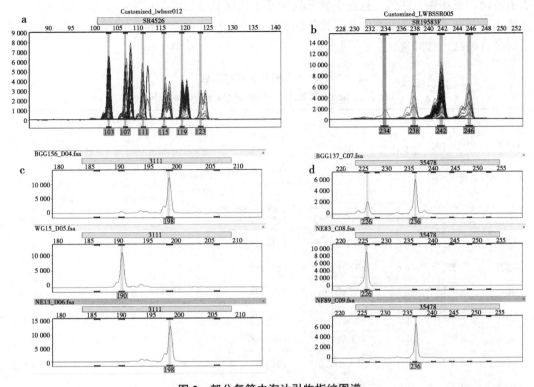

图 2 部分复筛中淘汰引物指纹图谱

Picture 2 A part of finger-print about secondary selection and eliminatation

注：a. M097 不以基序为重复单位；b. M040 杂合率低；c. M107 不符合孟德尔遗传定律；d. M015不对称扩增

Note：a. M097 is not in the motif as the repeat unit；b. M040 hybrid rate is low；c. M107 donot conform Mendelian genetic law；d. M015Asymmetric amplification

本研究对 5 对初筛等位基因数为 2 的引物进行复筛及放大样品试验，最终由于多态性低被淘汰，剔除等位基因数为 2 以下的可以筛除掉大部分多态性低的引物，效果直观。

本研究引物入选率较低，主要淘汰原因是多态性低，初筛中等位基因数为 2 的占 39.63%，复筛中不以基序为重复单位的占 44.90%，不对称扩增及杂合率低分别占比 14.29%和 18.37%，不符合孟德尔遗传定律占 2.00%。

2.4 SSR 引物高效筛选方案的确定

经过 Qiagen 初筛与 ABI3730 复筛及放大样品试验，基于上述筛选评估最终入选 9 对引物（表3）。

基于利用 ABI3730 荧光毛细管平台进行大规模建库的前提下，提出高效筛选方案，引物开发阶段引物信息未知时，先用 Qiagen 平台，对 6 个代表性自交系进行初步评估筛选，候选引物扩增质量好，剔除无主带、扩增失败、具非特异性峰、杂峰、双带等特殊带型的引物，进行初步过滤；对候选引物使用三联体材料 ABI3730 平台进行精准评估，剔除不对称扩增严重、杂合率低、不符合孟德尔遗传定律的引物；对候选引物进行放大样品试验，剔除不以基序为重复单位、有特异峰型以及 PIC 值低的引物（设定 PIC 值为 0.6）。

当引物设计成功率≥80%，且在引物开发阶段生物信息学已经提供明确的引物信息，包括扩增片段范围、PIC值、等位基因数等信息，可直接组成多重电泳，利用荧光毛细管电泳平台进行精准评估筛选。

表3 入选引物的基本信息
Table 3 The basic information of the selected primers

引物编号 Primer number	引物名称 Primer name	染色体 Chr.	基序 Motif	片段大小（bp）Fragment size	初筛等位基因数（个）Allele No of Preliminary Screening	复筛等位基因数（个）Allele No of secondary screening	放大样品等位基因数（个）Allele No of amplified sample	多态信息含量 PIC
M032	25 132	chr5	(TCTA) 9	145~179	3	11	11	0.803
M088	4 553	chr5	(AGTC) 3	110~128	4	6	6	0.782
M025	29 530	chr6	(TTCT) 8	185~200	4	7	8	0.766
M042	18 731	chr3	(ATAG) 6	204~216	3	6	4	0.764
M109	616	chr1	(CATG) 3	139~156	4	4	4	0.747
M003	38 923	chr9	(CCAT) 5	186~196	4	4	5	0.675
M096	4 527	chr1	(AAAG) 6	104~121	3	5	6	0.628
M065	8 452	chr10	(ATTT) 5	211~225	3	5	4	0.624
M058	11 232	chr2	(ACGG) 5	151~165	3	4	4	0.609

3 讨论

与已有的引物筛选方案相比，本方案更加精确高效。尽管试验是以SSR标记为研究对象，但在其他标记类型上同等有效，具有一定的通用性。SSR引物基序多为2~6bp，多态性较高，以前方案由于电泳平台精度不够导致获得的信息量不足，ABI3730荧光毛细管电泳分辨率达到1bp。根据峰值能获得引物相对扩增效率，且通量较高，为5色荧光，可以同时利用其中4~5种荧光进行多重电泳，电泳成本显著降低，且耗时较短，每块板只需30~40min便可分析96个样品，数据整合容易。聚丙烯酰胺凝胶电泳操作较为烦琐，耗时长，通量低，不适于大量引物的评估筛选。本研究以利用SSR标记与ABI3730荧光毛细管电泳平台作大规模建库为前提，适于大量玉米和其他作物SSR引物的评估筛选。本单位已根据重测序数据挖掘了1 000个SSR引物位点，需要利用此方案进行大规模的SSR引物评估筛选，同时可以为以后SSR引物评估筛选及不同标记的引物评估筛选提供借鉴。

参考文献（略）

本文原载：玉米科学，2017，25（2）：24-30

玉米 SSR 标记 DNA 指纹数据库中 N+1 峰的消除

薛宁宁[1,2]*　许理文[1]*　易红梅[1]　葛建镕[1]　王凤格[1]**　陆大雷[2]**

(1. 北京市农林科学院玉米研究中心/玉米 DNA 指纹及分子育种北京市重点实验室，北京　100097；2. 扬州大学，江苏省作物遗传生理重点实验室，扬州　225009)

摘　要：为解决玉米 SSR 标记 DNA 指纹数据库构建时，N+1 峰带来的指纹数据标准化采集问题，本研究对 40 个玉米品种鉴定核心引物进行评估，其中 13 对引物的扩增产物产生 N+1 峰。对于这些引物，通过增加 PCR 程序最后一步的延伸时间和下游引物 5′端加一个 G（鸟嘌呤）两种方法进行 N+1 峰的消除。结果表明：与标准中提供的程序和引物序列相比，将 PCR 程序最后一步的延伸时间增加至 90min，消除了 9 对引物扩增产物的 N+1 峰，下游引物 5′端加一个 G 能消除 12 对引物扩增产物的 N+1 峰。最终提出两种消除 N+1 峰的方案，均能有效改善玉米标准 DNA 指纹数据库的图谱质量。

关键词：N+1 峰；SSR 标记；DNA 指纹数据库

Elimination of N+1 Peak in the Construction of Maize DNA Fingerprint Database of SSR Marker

Xue Ningning[1,2]*　Xu Liwen[1]*　Yi Hongmei[1]
Ge Jianrong[1]　Wang Fengge[1]**　Lu Dalei[2]**

(1. *Maize Research Center*, *Beijing Academy of Agricultural & Forestry Sciences*, *Beijing Key Laboratory of Maize DNA Fingerprinting and Molecular Breeding*, *Beijing* 100097; 2. *Yangzhou University*, *Key Laboratory of Crop Genetics and Physiology of Jiangsu Province*, *Yangzhou* 225009)

Abstract: In order to solve the impacts of N+1 peak on data collection standardization in construction of maize DNA fingerprint database, 40 SSR Core Primers in maize DNA fingerprint database were assessed in study, finding that 13 primers produce N+1 peak. Two methods were chose to eliminate N+1 peak for the 13 primers; First, increasing the extension time of the last step in PCR procedure; Second, adding a G (guanine base) in the reverse primer 5′ end; The results show that 9 primers' N+1 peak were eliminated by increasing the extension time to 90min, and 12 primers' N+1 peak were eliminated by adding a G in the reverse primer. Finally, we put forward two optimized pro-

---

基金项目：本研究由国家科技支撑计划课题（2015BAD02B02）和北京市农林科学院创新能力建设专项（KJCX20161501）共同资助

* 同等贡献作者

** 通讯作者，gege0106@163.com；dllu@yzu.edu.cn

grams of N+1 peak elimination, which were brought out to improve the data quality in maize DNA fingerprint database.

Key words：N+1 peak；SSR marker；DNA fingerprint database

SSR 标记是玉米品种鉴定和 DNA 指纹数据库构建的常用标记，结合荧光毛细管电泳技术构建的玉米 SSR-DNA 指纹数据库对于品种鉴定产生了巨大的推动作用（赵久然等，2015）。该数据库不仅提供 SSR-DNA 指纹数据，并且提供指纹图谱，使数据更形象更直观，为使用者提供了更准确科学的信息（王凤格等，2010）。

在玉米 SSR 标记 DNA 指纹数据库构建过程中，对荧光毛细管电泳数据分析时，部分引物会出现异常峰型，限制了数据分析的自动化，并且降低了数据统计的准确性，影响了指纹图谱的质量。N+1 峰是其中一种常见类型，它往往造成 PCR 产物片段大小读取不准确，存在至少 1bp 的误差，降低了数据分析效率，影响数据的整合与共享。而另外一种二碱基重复 SSR 标记所特有的峰型，即 N+1 峰与连续多峰叠合峰型，对片段读取的准确性影响更大，经常造成等位基因的多读、误读。

N+1 峰的产生是 PCR 反应过程中普通 *Taq* 酶在非模板扩增片段的 3′端自动加一个腺苷酸引起的，表现为在同一等位基因位置上出现了相差 1bp 的两个峰（Clark，1988；王凤格等，2007）。对于 N+1 峰的消除，在先前的报道中已经提出了多种方法，包括增加 PCR 程序最后一步的延伸时间（Smith et al，1995），下游引物 5′端添加一个 G（Hu，1993）或"PIG-tail"（Brownstein et al，1996），减少 DNA 模板量或引物浓度，增加 *Taq* 酶浓度以及更换高保真 DNA 聚合酶等（Guichoux et al，2011）。通过前期试验发现，减少 DNA 模板量或引物浓度，增加 *Taq* 酶浓度等方法试验操作较复杂，而更换高保真 DNA 聚合酶，试验成本较高，因此，本研究的重点是通过增加 PCR 程序最后一步的延伸时间和下游引物 5′端添加一个 G 这两种方法来进行消除 N+1 峰。通过探索两种方法是否能对玉米 SSR 标记 DNA 指纹数据库构建中的 N+1 峰进行消除，从而改善 DNA 指纹图谱质量，建立更精准的玉米 SSR 标记 DNA 指纹数据库。

1 结果与分析

1.1 40 对引物的指纹图谱统计与分析

经过统计分析可知，18 份材料采用标准方法得到的 40 对标准引物的指纹图谱中，13 对引物的指纹图谱出现 N+1 峰（表1），将其分为左高型、接近等高型和右高型等 3 种类型，P14、P16、P20、P37、P39 的 N+1 峰为左高型，P06、P19、P21、P32 的 N+1 峰为接近等高型，P36 的 N+1 峰为右高型，其余 3 个为一种二碱基重复 SSR 标记所特有的峰型，即 N+1 峰与连续多峰叠合峰型（P12、P22 和 P40）。图 1 是 4 个具有 N+1 峰的引物的指纹图谱（图1）。

1.2 两种方法对引物的 N+1 峰的消除结果

对产生 N+1 峰的 13 对引物，采用两种方法消除其 N+1 峰。

方法一是通过调整 PCR 程序，增加最后一步延伸时间至 90min。比较该方法和标准方法中产生 N+1 峰的引物的指纹图谱，其中，P16、P20、P21 的 N+1 峰均变为右高型，P40 的 N+1 峰没有产生明显变化，其余引物的 N+1 峰被消除。结果表明：该方法消除了

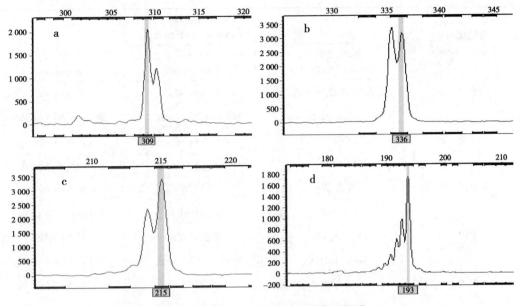

图 1 4 个具有 N+1 峰的引物的指纹图谱

Figure 1 Fingerprints of four primers to produce N+1 peak

注：a，b，b，d 分别是 P39、P06、P36、P22 采用标准方法获得的指纹图谱，代表了左高型 N+1 峰、接近等高型 N+1 峰、右高型 N+1 峰以及 N+1 峰与连续多峰叠合峰型

Note：a，b，c and d are respectively fingerprints used standard method of P39，P06，P36 and P22，represented as N+1 peak of left tall，equal height，right tall and accompany stuttering

9 对引物的 N+1 峰，对 4 对引物的 N+1 峰没有消除，而这 4 对引物（P16、P20、P21、P40）的基因分型错误率明显降低，不需要增加更长延伸时间进行消除。

方法二是对于产生 N+1 峰的引物，在其下游引物 5′末端加一个 G，使用标准反应体系和反应程序进行扩增。其中，P40 的 N+1 峰没有产生明显变化，其余引物的 N+1 峰被消除。结果表明：12 对引物的 N+1 峰被消除，P40 的基因分型错误率明显降低。

比较方法一与方法二对 13 对引物的 N+1 峰消除结果（表 1、图 2），可知：两种方法对其中 9 对引物的 N+1 峰消除取得相同效果（P06、P12、P23、P36），对 P40 的效果不明显，而对于剩余的 3 对引物，方法二优于方法一。

表 1 13 对引物在三种方法处理下的 N+1 峰表现

Table 1 Feature of N+1 peak of thirteen primers used three methods

引物编号 Primer name	N+1 峰表现 Feature of N+1 peak		
	标准方法 Standard method	方法一 Method one	方法二 Method two
P06	接近等高型 Equal height	无 N+1 峰 Without N+1 peak	无 N+1 峰 Without N+1 peak

(续表)

引物编号 Primer name	N+1 峰表现 Feature of N+1 peak		
	标准方法 Standard method	方法一 Method one	方法二 Method two
P12	右高型 & 连续多峰 Right tall & Stuttering	无 N+1 峰 Without N+1 peak	无 N+1 峰 Without N+1 peak
P14	左高型 Left tall	无 N+1 峰 Without N+1 peak	无 N+1 峰 Without N+1 peak
P16	左高型 Left tall	右高型 Right tall	无 N+1 峰 Without N+1 peak
P19	接近等高型 Equal height	无 N+1 峰 Without N+1 peak	无 N+1 峰 Without N+1 peak
P20	左高型 Left tall	右高型 Right tall	无 N+1 峰 Without N+1 peak
P21	接近等高型 Equal height	右高型 Right tall	无 N+1 峰 Without N+1 peak
P22	右高型 & 连续多峰 Right tall & Stuttering	无 N+1 峰 Without N+1 peak	无 N+1 峰 Without N+1 peak
P32	接近等高型 Equal height	无 N+1 峰 Without N+1 peak	无 N+1 峰 Without N+1 peak
P36	右高型 Right tall	无 N+1 峰 Without N+1 peak	无 N+1 峰 Without N+1 peak
P37	左高型 Left tall	无 N+1 峰 Without N+1 peak	无 N+1 峰 Without N+1 peak
P39	左高型 Left tall	无 N+1 峰 Without N+1 peak	无 N+1 峰 Without N+1 peak
P40	右高型 & 连续多峰 Right tall & Stuttering	右高型 & 连续多峰 Right tall&Stuttering	右高型 & 连续多峰 Right tall&Stuttering

1.3 消除 N+1 峰的两种优化方案

依据以上结果，对于玉米 SSR 标记 DNA 指纹数据库中 N+1 峰的消除，我们提出两种优化方案（表 2）。优化方案一中，所有引物都采用优化的 PCR 程序，其他没有产生 N+1 峰的引物采用该方案，峰型没有产生异常变化。优化方案二中，13 对产生 N+1 峰的引物进行下游引物 5′端加一个 G，其余引物不修改，反应程序采用标准程序。

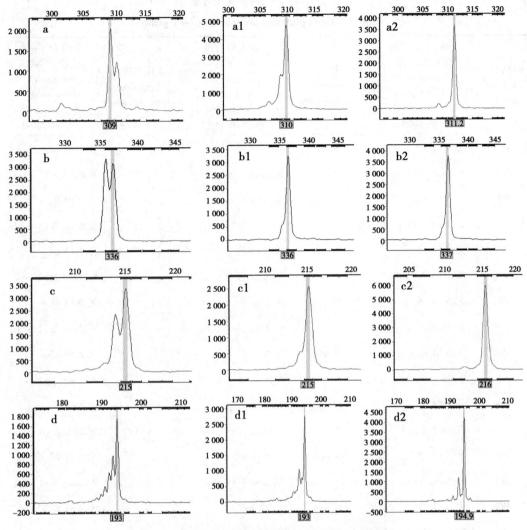

图 2 产生 N+1 峰的引物在三种方法下的指纹图谱

Figure 2　Fingerprints of primers to produce N+1 peak used three methods

注：A，B，C，D 分别是 P39，P06，P36，P22 采用标准方法的指纹图谱；A1，B1，C1，D1 是 4 对引物采用方法一的指纹图谱；A2，B2，C2，D2 是 4 对引物采用方法二的指纹图谱

Note：A，B，C and D are respectively represented as fingerprints used standard method of P39, P06, P36 and P22; A1, B1, C1 and D1are respectively represented as fingerprints used method one of four primers; A2, B2, C2 and D2 are respectively represented as fingerprints used method two of four primers

利用 96 份玉米杂交种对两种优化方案进行验证，结果表明：两种优化方案均能消除 N+1 峰。

表2　N+1峰的两种优化方案
Table 2　Two optimized programs of N+1 peak

引物编号 Primer number	N+1峰表现（标准方案） Feature of N+1 peak (Standard program)	优化方案一 Optimized program one	N+1峰表现（优化方案一） Feature of N+1 peak (Optimized program one)	优化方案二 Optimized program two	N+1峰表现（优化方案二） Feature of N+1 peak (Optimized program two)
P01	无N+1峰 Without N+1 peak	90min	无N+1峰 Without N+1 peak		无N+1峰 Without N+1 peak
P02	无N+1峰 Without N+1 peak	90min	无N+1峰 Without N+1 peak		无N+1峰 Without N+1 peak
P03	无N+1峰 Without N+1 peak	90min	无N+1峰 Without N+1 peak		无N+1峰 Without N+1 peak
P04	无N+1峰 Without N+1 peak	90min	无N+1峰 Without N+1 peak		无N+1峰 Without N+1 peak
P05	无N+1峰 Without N+1 peak	90min	无N+1峰 Without N+1 peak		无N+1峰 Without N+1 peak
P06	等高型 Equal height	90min	无N+1峰 Without N+1 peak	加G Add G	无N+1峰 Without N+1 peak
P07	无N+1峰 Without N+1 peak	90min	无N+1峰 Without N+1 peak		无N+1峰 Without N+1 peak
P08	无N+1峰 Without N+1 peak	90min	无N+1峰 Without N+1 peak		无N+1峰 Without N+1 peak
P09	无N+1峰 Without N+1 peak	90min	无N+1峰 Without N+1 peak		无N+1峰 Without N+1 peak
P10	无N+1峰 Without N+1 peak	90min	无N+1峰 Without N+1 peak		无N+1峰 Without N+1 peak
P11	无N+1峰 Without N+1 peak	90min	无N+1峰 Without N+1 peak		无N+1峰 Without N+1 peak
P12	右高型&连续多峰 Right tall & Stuttering	90min	无N+1峰 Without N+1 peak	加G Add G	无N+1峰 Without N+1 peak
P13	无N+1峰 Without N+1 peak	90min	无N+1峰 Without N+1 peak		无N+1峰 Without N+1 peak

(续表)

引物编号 Primer number	N+1 峰表现（标准方案）Feature of N+1 peak (Standard program)	优化方案一 Optimized program one	N+1 峰表现（优化方案一）Feature of N+1 peak (Optimized program one)	优化方案二 Optimized program two	N+1 峰表现（优化方案二）Feature of N+1 peak (Optimized program two)
P14	左高型 Left tall	90min	无 N+1 峰 Without N+1 peak	加 G Add G	无 N+1 峰 Without N+1 peak
P15	无 N+1 峰 Without N+1 peak	90min	无 N+1 峰 Without N+1 peak		无 N+1 峰 Without N+1 peak
P16	左高型 Left tall	90min	右高型 Right tall	加 G Add G	无 N+1 峰 Without N+1 peak
P17	无 N+1 峰 Without N+1 peak	90min	无 N+1 峰 Without N+1 peak		无 N+1 峰 Without N+1 peak
P18	无 N+1 峰 Without N+1 peak	90min	无 N+1 峰 Without N+1 peak		无 N+1 峰 Without N+1 peak
P19	等高型 Equal height	90min	无 N+1 峰 Without N+1 peak	加 G Add G	无 N+1 峰 Without N+1 peak
P20	左高型 Left tall	90min	右高型 Right tall	加 G Add G	无 N+1 峰 Without N+1 peak
P21	等高型 Equal height	90min	右高型 Right tall	加 G Add G	无 N+1 峰 Without N+1 peak
P22	右高型 & 连续多峰 Right tall & Stuttering	90min	无 N+1 峰 Without N+1 peak	加 G Add G	无 N+1 峰 Without N+1 peak
P23	无 N+1 峰 Without N+1 peak	90min	无 N+1 峰 Without N+1 peak		无 N+1 峰 Without N+1 peak
P24	无 N+1 峰 Without N+1 peak	90min	无 N+1 峰 Without N+1 peak		无 N+1 峰 Without N+1 peak
P25	无 N+1 峰 Without N+1 peak	90min	无 N+1 峰 Without N+1 peak		无 N+1 峰 Without N+1 peak
P26	无 N+1 峰 Without N+1 peak	90min	无 N+1 峰 Without N+1 peak		无 N+1 峰 Without N+1 peak

(续表)

引物编号 Primer number	N+1 峰表现 (标准方案) Feature of N+1 peak (Standard program)	优化方案一 Optimized program one	N+1 峰表现 (优化方案一) Feature of N+1 peak (Optimized program one)	优化方案二 Optimized program two	N+1 峰表现 (优化方案二) Feature of N+1 peak (Optimized program two)
P27	无 N+1 峰 Without N+1 peak	90min	无 N+1 峰 Without N+1 peak		无 N+1 峰 Without N+1 peak
P28	无 N+1 峰 Without N+1 peak	90min	无 N+1 峰 Without N+1 peak		无 N+1 峰 Without N+1 peak
P29	无 N+1 峰 Without N+1 peak	90min	无 N+1 峰 Without N+1 peak		无 N+1 峰 Without N+1 peak
P30	无 N+1 峰 Without N+1 peak	90min	无 N+1 峰 Without N+1 peak		无 N+1 峰 Without N+1 peak
P31	无 N+1 峰 Without N+1 peak	90min	无 N+1 峰 Without N+1 peak		无 N+1 峰 Without N+1 peak
P32	等高型 Equal height	90min	无 N+1 峰 Without N+1 peak	加 G Add G	无 N+1 峰 Without N+1 peak
P33	无 N+1 峰 Without N+1 peak	90min	无 N+1 峰 Without N+1 peak		无 N+1 峰 Without N+1 peak
P34	无 N+1 峰 Without N+1 peak	90min	无 N+1 峰 Without N+1 peak		无 N+1 峰 Without N+1 peak
P35	无 N+1 峰 Without N+1 peak	90min	无 N+1 峰 Without N+1 peak		无 N+1 峰 Without N+1 peak
P36	右高型 Right tall	90min	无 N+1 峰 Without N+1 peak	加 G Add G	无 N+1 峰 Without N+1 peak
P37	左高型 Left tall	90min	无 N+1 峰 Without N+1 peak	加 G Add G	无 N+1 峰 Without N+1 peak
P38	无 N+1 峰 Without N+1 peak	90min	无 N+1 峰 Without N+1 peak		无 N+1 峰 Without N+1 peak
P39	左高型 Left tall	90min	无 N+1 峰 Without N+1 peak	加 G Add G	无 N+1 峰 Without N+1 peak
P40	右高型 & 连续多峰 Right tall & Stuttering	90min	右高型 & 连续多峰 Right tall & Stuttering	加 G Add G	右高型 & 连续多峰 Right tall & Stuttering

2 讨论

两种优化方案对玉米 SSR 标记 DNA 指纹数据库构建中引物产生的 N+1 峰都能消除。优化方案一不需要重新设计引物，只需对标准 PCR 程序进行调整，方便简洁，但是该方案延长了试验时间，降低了试验效率。优化方案二对部分引物进行重新设计，相比方案一，该方案对 N+1 峰的消除效果更彻底，不需要延长 PCR 扩增时间，并且有利于多重 PCR 体系的构建。综合比较两种方案，在进行大规模玉米 SSR 标记 DNA 指纹数据库构建时，推荐采用优化方案二。

两种消除 N+1 峰的方法都是促进引物扩增产物充分加 A，提高加 A 产物比例。方法一通过增加延伸时间，促使 TaqDNA 聚合酶有充足的时间在更多扩增产物的 3′末端加 A，方法二利用 TaqDNA 聚合酶特性，即当非模板链 3′末端碱基为 C 时，该酶极易额外添加一个 A（Hu，1993），通过改变下游引物 5′末端碱基，发挥该酶的特性促进扩增产物完全加 A。

本研究的两种方案是针对玉米 SSR 标记单重 PCR，其可能适用于其他作物或其他标记的 DNA 指纹数据库构建中 N+1 峰的消除，以及在多重 PCR 反应中 N+1 峰的消除，我们将继续探索。

3 材料与方法

3.1 材料

玉米（zea mays L.）材料为 9 份骨干自交系和 9 份常用杂交种（表3）。

表3 供试 18 份玉米材料
Table 3 Eighteen maize samples used in study

编号 No.	样品名称 Sample name	样品类型 Sample type	编号 No.	样品名称 Sample name	样品类型 Sample type
1	京 724 Jing724	自交系 Inbred line	7	黄早四 Huangzaosi	自交系 Inbred line
2	京 92 Jing 92	自交系 Inbred line	8	Mo17	自交系 Inbred line
3	京科 968 Jingke 968	杂交种 Hybrid	9	丹 340 Dan 340	自交系 Inbred line
4	京糯 6 Jingnuo 6	自交系 Inbred line	10	B73	自交系 Inbred line
5	白糯 6 Bainuo 6	自交系 Inbred line	11	478	自交系 Inbred line
6	京科糯 2000 Jingkenuo 2000	杂交种 Hybrid	12	郑单 958 Zhengdan 958	杂交种 Hybrid

(续表)

编号 No.	样品名称 Sample name	样品类型 Sample type	编号 No.	样品名称 Sample name	样品类型 Sample type
13	农大108 Nongda 108	杂交种 Hybrid	16	京科甜183 Jingketian 183	杂交种 Hybrid
14	京单28 Jingdan 28	杂交种 Hybrid	17	正大619 Zhengda 619	杂交种 Hybrid
15	浚单20 Xundan 20	杂交种 Hybrid	18	鲁单981 Ludan 981	杂交种 Hybrid

3.2 DNA 提取

采用 CTAB 法提取全基因组 DNA，去除 RNA 后用 TE 溶解，然后利用 NanoDrop2000 紫外分光光度计进行 DNA 质量和浓度测定。

3.3 引物

采用农业部公布的 NY/T 1432—2014《玉米品种鉴定技术规程 SSR 标记法》中的 40 对引物（表4）。对产生 N+1 峰的引物，在其下游引物 5′末端加一个 G。

表4 供试 40 对 SSR 引物信息
Table 4 Information of forty SSR primers used in study

引物编号 Primer number	引物名称 Primer name	标记荧光 Fluorescence labeled	引物编号 Primer number	引物名称 Primer name	标记荧光 Fluorescence labeled
P01	bnlg439w1	NED	P17	phi065k9	NED
P02	umc1335y5	PET	P18	umc1492y13	PET
P03	umc2007y4	FAM	P19	umc1432y6	PET
P04	bnlg1940k7	PET	P20	umc1506k12	FAM
P05	umc2105k3	PET	P21	umc1147y4	NED
P06	phi053k2	NED	P22	bnlg1671y17	FAM
P07	phi072k4	VIC	P23	phi96100y1	FAM
P08	bnlg2291k4	VIC	P24	umc1536k9	NED
P09	umc1705w1	VIC	P25	bnlg1520k1	FAM
P10	bnlg2305k4	NED	P26	umc1489y3	NED
P11	bnlg161k8	VIC	P27	bnlg490y4	NED
P12	bnlg1702k1	VIC	P28	umc1999y3	FAM
P13	umc1545y2	NED	P29	umc2115k3	VIC
P14	umc1125y3	VIC	P30	umc1429y7	VIC
P15	bnlg240k1	PET	P31	bnlg249k2	VIC
P16	phi080k15	PET	P32	phi299852y2	VIC

(续表)

引物编号 Primer number	引物名称 Primer name	标记荧光 Fluorescence labeled	引物编号 Primer number	引物名称 Primer name	标记荧光 Fluorescence labeled
P33	umc2160k3	VIC	P37	umc2084w2	NED
P34	umc1936k4	PET	P38	umc1231k4	FAM
P35	bnlg2235y5	VIC	P39	phi041y6	PET
P36	phi233376y1	PET	P40	umc2163w3	NED

3.4 PCR 扩增

反应液体积为 20μl，其中含有 2μl DNA 模板，10×Buffer，2.5mmol/L MgCl$_2$，0.15mmol/L dNTPs，1U *Taq* 酶，0.25μmol/L 引物。

标准反应程序列于表中（表5）（易红梅等，2006）。改良反应程序：将标准反应程序中的 72℃ 延伸 10min 调整为延伸 90min。

表5 PCR 扩增标准反应程序
Table 5 Standard reaction procedure of PCR amplification

循环次数 Cycle index	程序 Program	温度（℃） Temperature（℃）	时间 Time
1	预变性 Pre denaturation	94	5min
35	变性 Denaturation	94	40s
	退火 Annealing	60	35s
	延伸 Extension	72	45s
1	延伸 Extension	72	10min
—	保存 Save	4	—

3.5 电泳检测

本试验采用 ABI3730XL DNA 分析仪进行 PCR 产物的检测，电泳分析流程及数据收集参照（易红梅等，2006），数据分析利用本单位研发的 SSR Analyser（V1.2.4）。

参考文献（略）

本文原载：分子植物育种，2017（3）：965-973

一种适于 SNP 芯片分型的玉米种皮组织 DNA 提取方法

刘可心[1]　王璐[2]　蔚荣海[1]　王凤格[2]　田红丽[2]*

(1. 吉林农业大学，长春　130118；2. 北京市农林科学院玉米研究中心/玉米 DNA 指纹及分子育种北京市重点实验室，北京　100097)

摘　要：以京科 968、郑单 958 两套三联体样品为材料，分别采用 CTAB 和改良 CTAB 法提取叶片和杂交种种皮的 DNA，利用 maizeSNP384 芯片（包括 384 个 SNP 位点）进行 DNA 指纹分析。结果显示：玉米叶片和种皮提取 DNA 获得基因型数据在成功率、数据质量上一致，均可获得较好的基因型数据；通过将杂交种种皮 DNA 与母本叶片 DNA 基因型数据的比较分析，三联体样品亲子关系分析，显示种皮 DNA 基因型能够代表杂交母本 DNA 指纹。因此玉米种皮提取 DNA 结合 SNP 芯片检测技术能够在品种权保护和品种鉴定中发挥一定作用。

关键词：玉米；种皮；DNA 提取；SNP 分析

A DNA Extraction Method from Maize Seed Capsule Tissue Suitable for SNP Chip Genotyping

Liu Kexin[1]　Wang Lu[2]　Yu Ronghai[1]　Wang Fengge[2]　Tian Hongli[2]*

(1. *Jilin Agricultural University*, *Changchun* 130118; 2. *Beijing Academy of Agriculture and Forestry Sciences*, *Maize Research Center*, *Key Laboratory of maize DNA fingerprinting and molecular breeding of Beijing*, *Beijing* 100097)

Abstract: Using Jingke 968 and Zhengdan 958 two sets triplets as research materials, the DNA was extracted using CTAB and modified CTAB methods from leaves and hybrid seed capsules, and DNA fingerprint was constructed using maizeSNP384 chip (including 384 SNP loci). The results showed that good genotype data could be obtained from both leaves and seed capsules DNA, the data success rates and quality were consistent. Comparative analysis of the genotype data from hybrid seed capsules and maternal leaves DNA, and parent-child relationship analysis showed that seed capsules DNA genotype could represent the maternal DNA fingerprint of hybrid. Therefore, the detection technology of maize seed capsules DNA extraction combined with SNP chip could play a certain role in the protection of intellectual property and variety identification.

Key words: Maize; Seed capsule; DNA Extraction; SNP analysis

基金项目：本研究由北京市农林科学院创新能力建设专项（KJCX20161501）和国家科技支撑计划课题（2015BAD02B02）共同资助

* 通讯作者：tianhongli9963@163.com

玉米（*Zea mays* L.）是我国第一大作物，也是种业市场份额最大的作物，在中国农业经济结构中具有关键作用。随着种业市场的发展，玉米品种权保护与管理日趋重要。基于 DNA 指纹方法的玉米品种侵权/确权鉴定时需要送检方提供杂交种子的同时提供双亲种子，以便于亲本追溯。但是送检方经常无法提供亲本种子，或者出现使用了亲本自交系却更名换姓从而无法提供其真实亲本的情况。玉米籽粒种皮是全部继承母本遗传信息的 F1 代组织，通过分离 F_1 代种皮并从中提取 DNA、构建指纹图谱，能够获得母本的指纹；结合 F_1 杂交种和母本指纹可分析推算父本指纹。早期有适于 SSR 指纹分析的，玉米种皮 DNA 提取方法报道（赵久然等，2004）。

随着测序技术发展，玉米海量 SNP 标记位点被开发，同时各种检测平台的陆续推广，玉米 SNP 品种鉴定技术也在快速发展。玉米品种鉴定、派生品种鉴定 SNP 位点组合已经报道（Tian et al，2015），各相关研究单位、种子企业、检测中心也在积极研发农作物品种 SNP 指纹鉴定技术。SNP 芯片为位点高通量检测平台，一个样本孔可以同时分析几百至几万位点；所需总 DNA 量少，一般所需浓度 20~50ng/μl，体积为 5μl。因此与 SSR 标记检测平台相比，种皮提取 DNA 更适合在 SNP 芯片平台上进行基因分型分析。

基于上述背景，本研究拟从种皮组织 DNA 提取出发，针对于存在的种皮难剥离、杂质多、DNA 含量少三大问题，以玉米种子为材料，从种皮剥离、提取过程两个层面进行研究，并与叶片提取 DNA 进行比较分析，探索一种适于 SNP 芯片分型的玉米种皮 DNA 提取方法。

1 结果与分析

1.1 提取 DNA 浓度和质量

叶片组织提取 DNA 浓度和质量：浓度范围介于 1 274~2 048ng/μl，OD_{260}/OD_{280} 比值范围为 1.90~2.00，OD_{260}/OD_{230} 比值均大于 2.00。种皮组织提取 DNA 浓度范围为 20~35ng/μl，满足 SNP 芯片基因分型对 DNA 总量的要求，OD_{260}/OD_{280} 比值为 2.33、2.05，OD_{260}/OD_{230} 比值为 1.59、0.98（表1）。

表 1 种皮或叶片组织提取 DNA 的浓度和质量
Table 1 The concentration and quality of extracted DNA from seed capsule and leaf tissue

样品（提取组织） Sample（tissue）	DNA 浓度（ng/μl） Concentration of DNA（ng/μl）	OD_{260}/OD_{280}	OD_{260}/OD_{230}
京科 968（叶片） Jingke 968（leaf）	1 274	1.97	2.16
京 724（叶片） Jing 724（leaf）	1 530	2.00	2.39
京 92（叶片） Jing 92（leaf）	2 048	1.97	2.34
郑单 958（叶片） Zhengdan 958（leaf）	1 575	2.00	2.23
郑 58（叶片） Zheng 58（leaf）	1 369	2.01	2.36

（续表）

样品（提取组织） Sample（tissue）	DNA 浓度（ng/μl） Concentration of DNA（ng/μl）	OD_{260}/OD_{280}	OD_{260}/OD_{230}
昌 7-2（叶片） Chang 7-2（leaf）	1 753	1.99	2.34
京科 968（种皮） Jingke 968（seed capsule）	34	2.33	1.59
郑单 958（种皮） Zhengdan 958（seed capsule）	23	2.05	0.98

1.2 两次重复试验数据的稳定性和准确性分析

两次重复试验数据缺失率均为 0.50%，位点 MG009 和 MG193 由于基因型分型效果不好作为数据缺失；GenomeStudio（illumina，USA）软件评估总体信号值介于 0.34~2.13，数据质量分值（score 值）介于 0.32~0.97；两次数据之间重复率为 100%。两套三联体样品两次重复试验在位点 MG064 上的三种基因型分布情况具体如下（图 1）：每个彩色的点代表每个 DNA 在 MG064 位点上的数据点，红蓝紫三个圈和阴影区域分别代表两种纯合、一种杂合基因型即 AA、BB、AB，每种基因型的数据点纵坐标值越高、分布愈集中表示数据质量越高。绿色和红色数据点为郑单 958、父母本、种皮 DNA 的两次重复，黄色和蓝色是京科 968、父母本、种皮 DNA 的两次重复。并且可以看出 16 个数据点质量较高，三种基因型区分明确，两次重复数据一致。

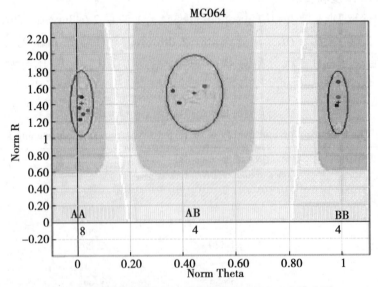

图 1 两套三联体样品两次重复在位点 MG064 上的分型

Figure 1 Two sets of triplets materials genotyping map on the MG064 locus with 2 repeats

1.3 玉米叶片和种皮 DNA 在 SNP 芯片平台上的基因分型比较

对玉米叶片和种皮提取 DNA 在 maizeSNP384 芯片中试验的结果进行比较分析（表

2)。玉米叶片和种皮提取 DNA 数据缺失率相同;信号值前者范围为 0.34~2.02,平均值为 1.37,后者范围为 0.45~2.13,平均值为 1.33;评估分值前者范围为 0.37~0.97,平均值为 0.76,后者范围为 0.36~0.95,平均值为 0.76。

种皮提取 DNA 分别与其母本叶片提取 DNA 基因型数据比较显示,京科 968 种皮与京 724 叶片提取 DNA 基因型相同;郑单 958 种皮与郑 58 叶片提取 DNA 之间有 7 个位点数据不同,这可能是由于郑单 958 在制种时用的母本与本试验中母本来源不同导致的。但是将种皮提取 DNA 的基因型数据(代表了母本基因型)与京科 968、郑单 958 进行亲子关系分析,排除亲子关系位点数目为 0,即种皮提取 DNA 基因型完全代表了在制种时选用的母本指纹信息。

表 2 玉米叶片和种皮组织提取 DNA 在 SNP 芯片分型平台中试验结果比较
Table 2 Comparison of the experimental results of DNA on the SNP chip of maize leaf and seed capsule extraction

提取 DNA 组织 Extract DNA tissue	数据缺失率 (%) Data missing rate (%)	信号值 (平均值) Signal value (average)	Score 值分布情况 Score Distribution		
			<0.6	0.6~0.8	0.8~1.0
叶片提取的 DNA Extract DNA from leaf	0.50%	0.34~2.02(1.37)	14	216	154
种皮提取的 DNA Extract DNA from capsule	0.50%	0.45~2.13(1.33)	18	222	144

2 讨论

2.1 玉米籽粒种皮提取 DNA 在 SNP 芯片平台进行基因分型的可行性

玉米基因组 DNA 提取方法比较多,与报道的快速提取法、SDS 提取方法相比(赵久然等,2004;郭景伦等,2005),本研究选取了提取步骤多,所需时间长的 CTAB 方法原因有两点。一是由于种皮杂质多,选取了提取质量稳定、去除多糖等杂质效果较好的 CTAB 方法。二是考虑在玉米甚至其他主要农作物 DNA 指纹检测中,提取种皮 DNA 的数量达不到规模化的需求,相反需要精细化的提取以便获得满足试验要求的 DNA,利用 CTAB 法提取种皮 DNA 的通量满足检测需求。

玉米籽粒种皮组织提取 DNA 存在种皮难剥离、杂质多、DNA 含量少三大问题。本研究采用 NaOH 浸泡结合 CTAB 法提取玉米种皮 DNA,试验结果表明 30 粒种子剥取种皮提取 DNA 的总量满足玉米 SNP 芯片基因分型的需求。关于种皮难剥离:从玉米籽粒上成功剥离种皮是 DNA 提取的关键,利用 NaOH 浸泡种子使种皮中的果胶降解,种皮与胚和胚乳之间浸入溶液,易于剥离,并且污染较少。杂质含量高:由于种皮中淀粉、糖类杂质较多,为了保证 DNA 提取的稳定性,本研究采用了提取质量稳定、去除多糖等杂质效果较好的 CTAB 提取法。DNA 含量少:采取 30 粒种子混合提取 DNA 的方式,保证了提取 DNA 的总量;同时与 SSR 标记单位点扩增方法不同,SNP 芯片为多重扩增,所需 DNA 总量远远低于 SSR 标记(李雪等,2014)。综上所述,种皮提取 DNA 的浓度和质量已经满足SNP芯片试验需求，并且经验证能够获得较好的基因分型数据。

2.2 种皮提取DNA结合SNP芯片技术在玉米等主要农作物分子检测中的辅助应用

本研究种皮与叶片组织提取DNA在SNP芯片试验结果比较显示两者数据质量、准确性没有差异，种皮组织DNA的基因型能够代表杂交种母本DNA指纹。虽然种皮组织提取DNA浓度和质量低于叶片组织提取DNA，但是完全适用于SNP芯片平台对DNA浓度和质量的要求。获得母本DNA指纹信息主要是在品种侵权/确权鉴定以及种子纯度鉴定中应用。2015年4月农业部发布了《农作物品种DNA身份鉴定体系构建实施方案》，方案中提出解决品种身份鉴定、育种材料确权、种质资源评价三个方面的研究内容。其中育种材料确权主要鉴定待测品种亲本是否涉及侵权及亲本追溯，需送检方提供待测杂交种以及母本种子，但是在实际检测中往往是拿不到亲本的。因此需要从F_1代杂交种出发，获取亲本DNA指纹信息，同时结合F1指纹推出父本指纹。

利用SNP芯片技术平台，相比于SSR标记，一次可以获得高通量位点的基因型数据，增加了亲本追溯的准确性。种子纯度鉴定时需要区分母本自交系苗，在只提供杂交种没有双亲，甚至是匿名杂交种情况下，增加了纯度鉴定的难度。需要首先获得母本DNA指纹信息才能准确清晰的分析出送检杂交种的纯度以及自交苗、异型株的比例。综上所述，种皮组织提取DNA的方法能够在玉米等主要农作物品种侵权/确权鉴定以及种子纯度鉴定中发挥作用。

3 材料和方法

3.1 试验材料

试验材料包含两套双亲/F_1三联体样品，分别为京科968、京724、京92，郑单958、郑58、昌7-2。

3.2 试验材料一致性验证

为了消除种子质量对试验结果的影响，6份玉米样品利用SSR标记进行一致性验证。保证选用一致性较高的样品用于本研究中，控制异型株的比率低于5%。每份样品随机分析20个单株，利用《玉米品种鉴定技术规程SSR标记法》行业标准推荐的40对SSR引物中的前10对引物进行一致性分析（王凤格等，2007；王凤格等，2014）。DNA提取和SSR扩增、电泳、数据分析均按照标准中发布的流程进行操作。

3.3 玉米叶片组织DNA提取

6份试验样品，每份样品随机选取50粒种子于培养箱内发苗5d，然后随机选取30个单株叶片，混合后磨成粉末。采用常规CTAB法提取总DNA，总DNA提取方法按照玉米品种鉴定行业标准操作（王凤格等，2014）。利用紫外分光光度计NanoDrop 2000（Thermo Scientific，USA）测定提取的DNA的质量和浓度，并调节其浓度制备形成工作液。

3.4 玉米杂交种种皮组织DNA提取

3.4.1 种皮剥离

用纯水将种子表面洗净，取30粒种子（经过SSR标记检测，一致性达95%的样品）浸泡1%NaOH溶液中30min，能看到种皮与胚之间有NaOH溶液存在。用尖头镊子和切刀将种皮剥离下并用纯水冲洗3遍（将残留的NaOH洗去，避免对后续试验产生干扰），将种皮晾干备用。

3.4.2 改良 CTAB 法提取种皮 DNA

取干燥后的种皮置于研钵中，加入液氮，快速充分的将种皮磨碎，移入 2.0ml 离心管中；加入 65℃ 预热的 CTAB 提取液 1ml，充分混匀，65℃ 水浴 1h；加入 850μl 三氯甲烷/异戊醇（24∶1），抽提 15min；12 000rpm/min 离心 10min；吸取上清液移至另一个 2.0ml 离心管中，再加入等体积预冷的异丙醇，颠倒混匀，12 000rpm/min 离心 10min；将上清液倒掉，加入 70% 的乙醇 1.5ml，震荡后弃上清；沉淀物经室温干燥后最后加入 50μl 1×TE 缓冲液（pH=8.0）充分溶解；利用紫外分光光度计 NanoDrop2000（Thermo Scientific，USA）测定 DNA 浓度和质量。

3.5 玉米 SNP 芯片基因分型

种皮或叶片组织提取的 8 份 DNA，利用 maizeSNP384 芯片进行 SNP 基因分型试验，该芯片为北京市农林科学院玉米研究中心研制，包含 384 个 SNP 位点（王凤格等，2014，中国专利，201410756086.0）。为了验证种皮提取 DNA 在 SNP 基因分型试验中的稳定性，将 8 份 DNA 分别点在不同芯片上做了平行的试验重复。具体试验按照 illumina 公司提供的试验指南操作（GoldenGate Genotyping Assay Manual，Part#15004070 Rev. B）。

试验流程包括三方面主要步骤。步骤一：基因组 DNA 与生物素结合，384 种探针与 DNA 链杂交，杂交产物延伸、连接形成 384 种特异片段，将特异片段作为模板进行 PCR 扩增；步骤二：PCR 产物进行纯化，并将纯合后产物点到芯片上进行杂交（杂交条件 60℃ 30min，45℃ 16~18h）；步骤三：芯片的清洗、真空风干和扫描，利用 iScan SNP 芯片分析仪（Illumina，USA）扫描芯片上每个样品的杂交信号，收集原始数据（包括 idat 和图片格式文件）。将原始数据文件（idat 格式）导入 GenomeStudio ver.2010（illumina，USA）软件，分析获得每个样品在 384 个 SNP 位点上的基因型数据，以及评估数据质量的各种参数。

3.6 数据分析

采用本实验室开发 SNP 基因型数据分析软件 SNP Analyzer 对 384 个位点的原始数据进行分析评估。主要分析角度有：数据获得率，两次试验重复之间的一致性，种皮和叶片两种组织提取 DNA 在 SNP 芯片平台上试验结果比较，郑单 958、先玉 335 两套三联体样品亲子关系分析等。

参考文献（略）

本文原载：分子植物育种，2017（1）：195-199

利用核心 SNP 位点鉴别玉米自交系的研究

宋 伟 王凤格 田红丽 易红梅 王 璐 赵久然

(北京市农林科学院玉米研究中心，北京 100097)

摘 要：根据多态性水平、染色体位置等信息从公共数据库上筛选了 48 个玉米核心 SNP 位点。利用 48 重-SNPLex 分型系统对 105 份玉米自交系进行 SNP 基因分型分析，探索 SNP 标记在玉米品种鉴定中的应用前景。结果表明，48 个 SNP 位点中有 42 个位点峰型正常；PIC 值在 0.019~0.375，平均为 0.242；任何两份自交系间的遗传距离均在 0.015 以上，即 42 个 SNP 位点的基因分型数据信息足以将 105 份自交系材料区分开。

关键词：玉米；自交系；SNP；品种鉴定

Identification of Maize Inbred Lines Using Core SNP Loci

Song Wei　Wang Fengge　Tian Hongli
Yi Hongmei　Wang Lu　Zhao Jiuran

(*Maize Research Center, Beijing Academy of Agricultural and Forestry Sciences, Beijing* 100097)

Abstract: Based on the level of polymorphism and chromosome location, 48 maize core SNP loci were screened from public databases. To explore the application prospect of SNP marker in maize variety identification, SNP genotyping analysis of 105 maize inbred lines was finished on 48 X-SNPLex™ genotyping system. The results showed that 42 of 48 SNP loci have normal peak. The polymorphism information content (PIC) of passed loci varied from 0.019 to 0.375 with an average of 0.242. Analysis showed that genetic distance between any two inbred lines was more than 0.015. In other words, while more SNP loci may be needed for variety discrimination in populations with high consanguinity level, such as essential derivation varieties, the 42 SNP loci panel presents enough power for identity control of the 105 major inbred lines.

Key words: Maize; Inbred line; SNP; Variety identification

SNP 标记作为第三代分子标记，具有其他标记技术无法比拟的优势：①分布密度高。

基金项目：973 计划项目（2009CB118405）；北京市优秀人才培养资助项目（2011D002020000003）
作者简介：宋伟（1979— ），女，山东人，副研究员，博士，从事玉米分子育种工作。Tel：010-51503558，E-mail：songwei1007@126.com
赵久然为本文通讯作者

SNP 是目前为止分布最为广泛、存在数量最多的一种多态性类型，其标记密度比 SSR 标记更高。②与功能基因的关联度高，更容易开发到与性状相关的 SNP 功能标记。③遗传稳定性强。SNP 是基于单核苷酸的突变，其突变率低，一般仅为 10^{-9}，具有极高的遗传稳定性。④检测通量高，易实现自动化分析。SNP 标记一般只有两种等位基因型，数据统计简单，且不依赖检测平台，容易实现不同来源数据的整合和标准化。本研究根据多态性水平、染色体位置等信息从公共数据库上筛选了 48 个玉米 SNP 位点，利用 48 重-SNPLex 分型系统，对 105 份玉米自交系材料进行 SNP 基因分型分析，探索 SNP 标记在玉米品种鉴定中的应用前景。

1 材料与方法

1.1 试验材料

选用 105 份玉米自交系（92 份国内骨干系、13 份国外系）为试验材料，详见表 1。

表 1 105 份自交系名称及系谱来源

Table1 Name and pedigree of 105 inbred lines

序号 ID	名称 Name	系谱来源 Pedigree
1	478	U8112×沈 5003
2	1141	选自美国杂交种 78599
3	8001	488×3189
4	B73	BSSSC$_5$
5	C8605-2	7922×5003
6	H21	黄早 4×H84
7	K12	黄早 4×潍春
8	P138	选自美国杂交种 78599
9	丹 340	白骨旅 9×有稃玉米
10	黄早 4	选自塘四平头
11	Lx9801	掖 502×H21
12	沈 137	选自美国杂交种 6JK111
13	沈 5003	选自美国杂交种 3147
14	铁 7922	选自美国杂交种 3382
15	邢 K36	选自冀库 6 号
16	昌 7-2	（黄早 4×潍 95）×S901
17	掖 107	选自 Dekalb 杂交种 XL80
18	掖 502	丹 340×黄早 4

(续表)

序号 ID	名称 Name	系谱来源 Pedigree
19	1145	选自美国杂交种 78599
20	铁 9010	丹抗 1×丹 340
21	吉 853	黄早 4×自 330
22	X178	选自美国杂交种 78599
23	J0045	478×P78599
24	京 501	选自 SR 群 CO
25	京 5237	黄早 4×丹 340
26	D9046	铁 7922×沈 5003
27	京 404	(黄早 4×墨白 02)×黄早 4
28	自 330	Oh43×可利 67
29	郑 58	选自掖 478
30	京 89	掖 478×78599
31	SW1611	引自泰国
32	U8112	3382×3147
33	MC0303	(9042×京 89)×9046
34	京 24	早熟 302×黄野四
35	502196	黄早 4×丹 340
36	综 31	选自自 330
37	444	A619×黄早 4
38	81162	(矮金 525×掖 107)×106
39	吉 842	吉 63×Mo17
40	龙抗 11	Mo17×自 330
41	武 314	(黄早 4×武 302D)×黄爆裂
42	中 106	也门矮玉米×综合种
43	冀 53	冀群 2Co-2
44	原辐黄	选自黄早 4
45	3189	U8112×沈 5003
46	A801	丹 9042×(丹 9046×墨黄 9)
47	吉 1037	(Mo17×Suwan1)×Mo17
48	D 黄 212	D729×黄早 4
49	长 3	选自英粒子
50	大八趟	中国地方品种

(续表)

序号 ID	名称 Name	系谱来源 Pedigree
51	丹黄 25	选自美国杂交种 78599
52	墩白	中国地方品种
53	多 29	选自美国杂交种 78599
54	浚 9058	6JK 选系×8085
55	京 594	黄早 4×P78599
56	5872	郑 58×Mo17
57	英粒子	引自欧洲
58	早 673	不详
59	哲 446	不详
60	哲 773-2	吉 63×黄早 4
61	MC30	1145×1141
62	835	V8112×718
63	丹 598-1	选自丹 598
64	齐 319	选自美国杂交种 78599
65	黄野四	（野鸡红×黄早 4）×墩子黄
66	4F1	选自 Mo17
67	冲 72	3147×B37Ht
68	Q126	黄早 4×潍春
69	辐 80	旅 9×有稃玉米
70	K10	5003×长 3
71	获唐黄	获白×唐 203
72	JN15	J0045×齐 319
73	E28	（A619Ht1×旅 9 宽）×旅 9 宽
74	Mo17	187-2×103
75	白糯 6	选自紫糯 3 号
76	SH-251	选自超甜 1 号
77	紫糯 3	选自紫糯 3 号
78	京糯 6	选自中糯 1 号
79	9901	糯玉米自交系
80	紫玉-3	糯玉米自交系
81	335	糯玉米自交系
82	紫糯 5B	糯玉米自交系

(续表)

序号 ID	名称 Name	系谱来源 Pedigree
83	P12	选自美国杂交种 78599
84	434	466×桦 94
85	鲁原 92	原齐 122×1137
86	齐 318	选自美国杂交种 78599
87	苏 80-1	金黄 55×原武 02
88	吉 846	吉 63×Mo17
89	D375	02428×南农籼 2 号
90	合 344	白头霜×Mo17
91	CA335	选自 Pool 33
92	承 18	顶上玉米×（公 70× 60-22）
93	PHZ51	PH814 × PH848
94	NS701	A632 × B73Ht
95	PHW65	PH861 × PH595
96	PHG83	PH814 × PH207
97	PHT77	PH814 × PH995
98	PHT55	A33GB4 × A34CB4
99	NK790	NK235 × B73
100	PHK76	PHAD18 × PHB02
101	PHG84	PH848 × PH595
102	LH156	Va85 × Pa91
103	LH132	（H93 × B73）× B73
104	LH123HT	Pioneer Hyb 3535
105	NK764	NK235 × B73

1.2 试验方法

1.2.1 基因组 DNA 提取

玉米基因组 DNA 提取参照 Rogers 和 Bendich 的 CTAB 抽提法。

1.2.2 SNP 位点筛选

根据 www.panzea.org 等公共数据库上提供的 SNP 标记位点信息，依据以下原则对 SNP 位点资料进行汇总，筛选候选 SNP 位点。①位点多态性高，PIC 值≥0.45，检测样品数≥1 000；②染色体位置已知且均匀分布；③目标 SNP 位点上、下游有 100bp 可读，其中临近 SNP 位点的上、下游 25bp 保守性高即没有多态性，且 GC 含量在 40%~60%。利用美国应用生物系统公司（AB）网站提供的试验设计辅助工具——自动化试验设计流程对候选位点进行组合优化，筛选获得符合试验设计标准的48重-SNP位点组合。

1.2.3 SNP位点分型与检测

按照AB公司提供的SNPlex基因分型试验操作指南进行。首先对提取的基因组DNA进行片段化处理，产生符合下一步连接反应要求的DNA片段；生成的基因组DNA片段、连接子及探针在连接酶的作用下通过磷酸二酯键连接成链；再经核酸外切酶消化处理得到纯化的目的DNA片段连接产物；将连接产物作为PCR反应的模板在有生物素标记的通用引物作用下进行复合扩增；扩增产物与固定有链酶亲和素的杂交板进行杂交；洗脱掉没有生物素标记的DNA链；与连接产物中等位基因特异性探针上ZipCode序列互补的ZipChute探针和杂交板上固定的扩增产物杂交；将多余探针除去，洗脱杂交成功的ZipChute探针；最后在AB3730xl DNA分析仪上进行毛细管电泳。

1.2.4 数据统计分析

利用GeneMapper软件对Data Collection软件收集的原始数据进行统计。采用PowerMarker3.25软件计算48个核心SNP位点的多态性信息含量（PIC）和待检自交系间的遗传距离，对各个SNP位点进行评价。

2 结果与分析

2.1 48个SNP位点的筛选确定

依据位点筛选原则，从www.panzea.org等公共数据库上共筛选出符合标准的候选位点271个。根据SNPlex™ Genotyping System 48-plex Assay Design and Ordering Guide的序列提交原则，将候选SNP位点的上下游序列转换成FASTA格式提交，经格式检测的初步筛查和设计流程的优化组合，最终获得了符合试验设计要求的48个核心SNP位点组合。位点相关信息见表2。其中未通过设计程序的SNP位点主要是由以下3种情况造成的：一是由于SNP位点多重间相互作用对试验存在干扰；二是由于SNP位点自身序列会形成二级结构或其他相互作用；三是由于在目标SNP位点的上下游15bp内存在其他多态性位点或目标SNP两侧16~40bp存在2个以上的"N"。

表2 48个SNP位点信息
Table 2 Information of 48 selected SNP loci

编号 Number	名称 SNP Name	等位基因（A1/A2） Allele（A1/A2）	染色体位置 BIN	PIC值 PIC value
M001	PZA03071.15	G/T	10.05	0.153
M002	PZA00381.3	C/G	1.08	0.119
M003	PZA00505.5	C/T	7.04	0.359
M004	PZA00678.1	A/G	10	0.335
M005	PZA00042.5	C/T	6.06	0.309
M006	PZA00565.3	A/G	3.04	0.000
M007	PZA00703.1	G/T	1.11	—
M008	PZA00112.4	A/G	5.02	0.337

（续表）

编号 Number	名称 SNP Name	等位基因（A1/A2） Allele（A1/A2）	染色体位置 BIN	PIC 值 PIC value
M009	PZA00656.14	A/G	10.03	—
M010	PZD00011.3	C/T	6.05	0.375
M011	PZA00497.1	C/T	2.03	0.233
M012	PZA00393.1	C/T	1.01	0.159
M013	PZA00240.4	C/T	1.03	0.351
M014	PZA00606.3	G/T	6	—
M015	PZA02955.3	A/C	8.02	0.360
M016	PZA00686.4	C/G	8.06	0.037
M017	PZA02872.1	A/G	7.01	0.355
M018	PZA00070.5	C/G	9.03	0.330
M019	PZA00547.6	C/T	5.04	0.320
M020	PZA03083.7	C/T	2.06	0.019
M021	PZA00474.14	C/T	7.04	0.157
M022	PZA00142.4	A/G	8.05	0.022
M023	PZA03090.31	G/T	6.01	0.215
M024	PZD00083.4	A/G	1.09	0.080
M025	PZA00653.5	A/T	9.03	0.349
M026	PZA00166.1	A/G	3.06	0.345
M027	PZA00448.5	C/G	5.07	0.252
M028	PZA02824.5	A/G	3.09	0.142
M029	PZA00100.2	A/G	3.02	0.374
M030	PZA00715.3	C/G	5.06	—
M031	PZA00548.3	A/G	2.07	—
M032	PZA00226.6	C/T	2.04	0.000
M033	PZA02964.7	A/G	2.08	0.207
M034	PZA03012.7	C/T	8.04	0.375
M035	PZA00453.2	A/G	4.06	0.363
M036	PZA00551.4	C/G	10.07	—
M037	PZA00379.2	C/T	8.03	0.117
M038	PZD00067.3	A/G	4.05	0.364
M039	PZA00182.4	C/T	6.04	0.370

(续表)

编号 Number	名称 SNP Name	等位基因（A1/A2） Allele（A1/A2）	染色体位置 BIN	PIC 值 PIC value
M041	PZA00436.7	C/T	4.02	0.298
M042	PZA00193.1	A/C	4.08	0.368
M043	PZA03013.8	C/T	4	0.215
M044	PZA00603.1	C/T	5.05	0.334
M046	PZA00213.7	A/T	9.05	0.252
M048	PZA00533.4	G/T	7.03	0.019

2.2 48重SNP位点的总体评价

由表2可以看出，48个入选SNP位点在染色体上分布均匀，除第5、第9、第10染色体上分别有6个、3个和4个SNP位点外，其他7条染色体上均有5个SNP位点。通过对105份国内外骨干系进行SNP分型检测发现，48个SNP位点的通过率为88%，其中M007、M009、M014、M030、M031、M036等6个位点由于检测信号较弱，无法进行正常的数据统计，其余42个位点峰型正常，分型结果稳定可靠，且重复性好（图1）。

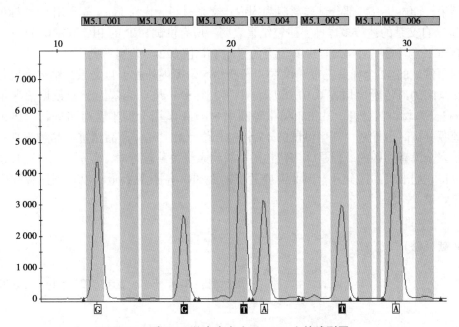

图1 正常SNP位点在自交系X178上的峰型图

Figure 1 The genotyping result of some passed SNP loci on inbred line X178

注：横坐标为各SNP位点的等位基因分型结果；纵坐标为峰值高低。

Note: The abscissa was SNP genotyping data, the ordinate was peak level.

2.3 48重SNP位点的PIC值和品种鉴定能力分析

根据42个峰型正常SNP位点在105份自交系上的SNPlex基因分型数据，采用Power-

Marker3.25软件计算各个SNP位点的PIC值和待检自交系间的遗传距离。结果表明，42个SNP位点的PIC值在0.019~0.375，平均为0.242。自交系间的遗传距离在0.015 0~0.586 7，说明利用供试的42个SNP位点对105份自交系进行鉴定，在任何两份自交系间均可以检测到差异。

3　结论与讨论

　　与SSR标记相比，SNP标记具有更高的检测通量，数据统计易于实现自动化，不同检测平台之间可以实现数据共享等优点，在水稻、玉米、大麦、葡萄藤等植物遗传图谱构建及遗传多样性分析等方面逐渐显现出广阔的应用前景。与此同时，SNP标记的各种检测方法也应运而生。本研究利用美国应用生物系统公司48重-SNPlex检测平台，一次反应可以同时分析48个SNP位点，适用于大批量样品的多个位点检测。在入选的48个SNP位点中，通过多色荧光毛细管电泳检测，42个位点峰型正常且重复性好，位点通过率为88%。这与Massimo Pindo等的研究结果一致，他们利用SNPlex基因分型系统对葡萄藤进行SNP分析检测时发现，17组48重SNP位点组合的位点通过率均在85%以上。

　　根据105份国内外自交系的SNP分型数据计算得出，42个SNP位点的PIC值在0.019~0.375，与我们最初筛选候选SNP位点的原则PIC值≥0.45存在一定差距。究其原因主要可能是所用试验材料不同，公共数据库上的供参考信息主要是基于国外自交系的分析结果，而本研究大部分试验材料为国内自交系材料。E. S. Jones等利用SNP标记分析58份玉米自交系的遗传多样性，研究结果表明，所有供试SNP的PIC值在0.02~0.5，平均为0.26，与本研究中42个SNP位点的平均PIC值0.242差别不大。

　　利用供试的42个峰型正常SNP位点分析105份国内外骨干自交系，结果表明任何两份自交系间的遗传距离均在0.015以上，即42个SNP位点的基因分型数据信息足以将105份自交系材料区分开。M. Ballester等利用SNPlex基因分型平台对猪进行个体识别和遗传溯源分析时也推断，利用供试的46个SNP位点对大部分商品猪进行品种识别和亲子鉴定是完全没有问题的。但是，当待检的材料间亲缘关系较近时，仅凭42个SNP位点可能很难将它们准确的区分开，因此需要适当加大SNP位点的数量，以达到比较理想的品种鉴定效果。

参考文献（略）

本文原载：玉米科学，2013，21（4）：28-32

利用 SNP 标记对 51 份玉米自交系进行类群划分

吴金凤[1]　宋　伟[2]　王　蕊[2]　田红丽[2]
李　雪[3]　王凤格[2]　赵久然[2]　蔚荣海[1]

(1. 吉林农业大学农学院，长春，130118；2. 北京市农林科学院玉米研究中心/玉米 DNA 指纹及分子育种北京市重点实验室，北京，100097；
3. 北京农学院，北京，102206)

摘　要：利用 1 041 个 SNP 位点对 51 份玉米自交系进行基因型分析，最小等位基因频率平均值为 0.359、多态性信息含量 (PIC) 的变化范围为 0.186～0.375，平均值为 0.345。Roger's 遗传距离的变化范围是 0.009 2～0.704 1。根据 Roger's 遗传距离信息，利用 NJ 聚类法将 51 份玉米自交系划分为 7 个杂种优势群，包括改良瑞德群、瑞德群、兰卡斯特群、旅大红骨群、塘四平头群 (或称黄改群)、P 群和糯质群，划分结果和系谱来源基本一致。7 个杂种优势群的群体间遗传距离变化范围为 0.285 0～0.432 1，瑞德群和改良瑞德群之间的遗传距离最近；旅大红骨群和 P 群之间的遗传距离最远。

关键词：玉米；自交系；SNP；类群划分

Heterotic Grouping of 51 Maize Inbred Lines by SNP Markers

Wu Jinfeng[1]　Song Wei[2]　Wang Rui[2]　Tian Hongli[2]
Li Xue[3]　Wang Fengge[2]　Zhao Jiuran[2]　Yun Ronghai[1]

(1. Faculty of Agronomy, Jilin Agricultural University, Changchun, 130118;
2. Maize Research Center, Beijing Academy of Agriculture and Forestry Sciences, Beijing Key Laboratory of maize DNA Fingerprinting and Molecular Breeding, Beijing, 100097;
3. Beijing University of Agriculture, Beijing, 102206)

Abstract: The genotype of 51 maize inbred lines was analyzed by 1 041 SNP marker locus, the average MAF was 0.359, the PIC ranging from 0.186～0.375, average PIC was 0.345, the Roger's

---

基金项目：北京市科技计划课题（D131100000213002）、国家科技支撑项目（2011BAD35B09）
作者简介：吴金凤（1988— ），女，吉林人，在读硕士，作物遗传育种专业。Tel：010-51503558，
　　　　　E-mail：ff18643994357@163.com
　　　　宋伟（1979— ），女，山东人，副研究员，博士，从事玉米分子育种工作。Tel：010-
　　　　　51503558，E-mail：songwei1007@126.com
　　　　吴金凤和宋伟为本文共同第一作者
　　　　王凤格为本文通讯作者

genetic distance ranging from 0.009 2~0.704 1. According to information of Roger's genetic distance, NJ clustering method is used to divide the 51 maize inbred lines into seven heterotic groups. It includes Improved Reid, Reid, Lancaster, Luda Red Cob, Tang-Si-Ping-Tou (or Improved Huangzao4), Waxy groups. The clustering results were nearly consistent with the pedigree information. The changing rang of genetic distance among seven heterotic groups was from 0.285 0 to 0.432 1, in which the genetic distance between Reid group and Improved Reid group was the most nearest, and the genetic distance between Luda Red Cob group and P group was the farthest.

Key words: Maize; Inbred line; SNP; Heterotic group

杂种优势是玉米新品种选育的重要理论依据，合理划分杂种优势群并且构建相应的杂种优势模式，有利于拓展种质资源，降低杂交组合组配的盲目性，提高育种成功率，对玉米育种具有重要的指导意义。20世纪末，育种家们通过系谱分析、地理来源、配合力分析以及同工酶标记等方法对玉米种质资源进行分类。随着分子生物学的发展，DNA分子标记技术为杂种优势群的划分提供了新的方法。

分子标记是以个体间遗传物质内核苷酸序列变异为基础的遗传标记，是DNA水平遗传多态性的直接反映。目前，主要应用的分子标记有RFLP、RAPD、AFLP、SSR、SNP等。随着基因芯片技术和测序技术的发展，SNP分子标记技术迅速成为继RFLP和SSR之后最有前途的第三代分子标记。它具有分布密度高，遗传稳定性好，易于实现自动化分析等优点，被广泛用于遗传图谱构建、性状关联分析、分子标记辅助育种等研究中。

近年来，研究者利用RFLP、RAPD、SSR、SNP等不同类型的分子标记开展了各类玉米种质资源的遗传多样性和类群划分研究。由于研究的标记种类、标记数量和种质材料来源等存在差异，类群划分的结果也不尽相同。袁力行等利用RFLP、SSR、AFLP和RAPD 4种分子标记研究对15份玉米自交系进行遗传多样性比较研究，结果表明SSR和RflP两种分子标记方法适合进行玉米种质遗传多样性的研究。Reif等分别用SSR对不同的群体进行遗传多样性分析，证明了用分子标记分析玉米群体遗传多样性及划分杂种优势群的可行性。王凤格等研究表明，SSR分子标记不仅能追踪自交系的系谱，还能检测出因选择、遗传漂移及基因突变造成的遗传差异，因此比系谱法更全面的分析自交系间的遗传关系。Jones等研究表明，与SSR相比，SNP可以提高标记的数量和质量；Van等研究表明，假设预算相同，利用SNP分子标记的方法计算的修正Roger's距离和基因多样性更精确。本研究利用1 041个SNP位点，对51份已知系谱来源的玉米常用自交系进行类群划分，验证了SNP标记进行类群划分的有效性，同时为种质改良及杂交种组配提供DNA水平上的有益信息。

1 材料与方法

1.1 试验材料

本研究选取了51份常用玉米自交系为试验材料，名称及系谱来源见表1。

表 1 51 份自交系名称及系谱来源
Table 1 Name and pedigree source of 51 inbred lines

自交系 Lines	系谱来源 Source	自交系 Lines	系谱来源 Source
丹 598	{[(OH43Ht3×丹 340)×丹黄 02]×丹黄 11}×78599	9901	糯玉米自交系
MC30	1145×1141	9902	糯玉米自交系
JN15	J0045×齐 319	白糯 6	糯玉米自交系
沈 137	选自美国杂交种 6JK111	京糯 6	糯玉米自交系
1145	选自美国杂交种 78599	香糯 8	糯玉米自交系
多 29	选自美国杂交种 78599	紫糯 3	糯玉米自交系
HOF2	选自美国杂交种 78599	紫糯 5B	糯玉米自交系
P138	选自美国杂交种 78599	紫玉-3	糯玉米自交系
齐 318	选自美国杂交种 78599	U8112	3382×3147
齐 319	选自美国杂交种 78599	C8605-2	铁 7922×沈 5003
X178	选自美国杂交种 78599	4112	A619×8112
掖 478	U8112×沈 5003	B73	BSSSC5
沈 5003	选自美国杂交种 3147	PH6WC	PH01N×PH09B
郑 58	选自掖 478	A801	丹 9042×(丹 9046×墨黄 9)
浚 9058	[6JK108×8085(泰)]×掖 478	丹 9046	铁 7922×沈 5003
沈 5005	选自 8147	铁 7922	选自美国杂交种 3382
吉 1037	(Mo17×素湾 1)×Mo17	昌 7-2	(黄早四×潍 95)×S901
Mo17	187-2×103	浚 92-8	昌 7-2×5237
PH4CV	PH7V0×PHBE2	京 24	早熟 302×黄野四
合 344	白头霜×Mo17	黄野四	(野鸡红×黄早四)×墩子黄
F349	沈 5003×丹 340	H21	黄早四×H84
龙抗 11	Mo17×自 330	吉 853	黄早四×自 330
自 330	Oh43×Keli67	原辅黄	选自黄早四
丹 340	白骨旅 9×有稃玉米	黄早四	选自塘四平头
综 31	选自自 330	Lx9801	掖 502×H21
335	糯玉米自交系		

1.2 试验方法

1.2.1 DNA 提取

室内幼苗培养后取样，CTAB 法提取基因组 DNA，去除 RNA。琼脂糖电泳和紫外分光光度计分别检测所提取 DNA 的质量。琼脂糖电泳显示 DNA 条带单一，完整无降解。紫外分光光度计检测 A260/280 介于 1.8~2.0（DNA 样品没有蛋白、RNA 污染），A260/230 介于 1.8~2.0（DNA 样品盐离子浓度低），稀释 DNA 浓度至 100ng/μl。

1.2.2 SNP 位点

利用北京市农林科学院玉米研究中心开发的 MaizeSNP3072 芯片对供试自交系进行 SNP 基因分型。该芯片是基于玉米品种鉴定和分子育种而研发的芯片，包含 3 072 个均匀分布于全基因组的 SNP 位点，SNP 的详细信息可从 http：//www.illumina.com.cn 和 http：//www.panzea.org 网站查询。本研究中选用的 1 041 个 SNP 位点是该实验室根据分型效果、数据缺失率等从上述 3 072 个位点中优选出来的。

1.2.3 GoldenGate 技术芯片试验操作流程

待提取的基因组 DNA 与激活的生物素充分结合后，通过离心沉淀使 DNA 纯化；将探针与纯化的目的 DNA 链杂交，漂洗杂交产物除去非特异结合试剂或过量试剂后进行延伸连接反应；延伸后产物经漂洗变性后作为模板进行 PCR 扩增，PCR 产物与磁珠结合后过滤纯化；将纯化的 PCR 产物与芯片杂交，杂交结束后进行芯片清洗、真空抽干；风干后的芯片立刻在 iScan 仪上进行扫描，最后利用 GenomeStudio 软件进行数据分析。

详细试验流程参照 Illumina 公司提供的试验操作指南 GoldenGate© Genetyping Assay，Manual Part#15004070 Rev. B。

1.2.4 数据统计

利用 PowerMarker V3.25 软件对 1 041 个 SNP 位点的数据进行分析，计算其最小等位基因频率（MAF）、多态性信息含量（PIC）；应用 Roger's 1972 算法计算 51 份玉米自交系两两之间的遗传距离，利用 Mega4.0 软件构建 NJ 聚类图，进行杂种优势群划分；根据聚类结果，应用 Roger's 1972 算法计算不同杂种优势群之间的遗传距离。

2 结果与分析

2.1 SNP 位点的总体评价

利用 PowerMarker V3.25 软件分析 51 份自交系的 SNP 基因分型数据，对 1 041 个 SNP 位点进行评价。结果表明，最小等位基因频率（MAF）的变化范围为 0.118~0.500，平均值为 0.359，其中 MAF 值≤0.2 的 SNP 位点有 41 个，占 3.9%，MAF 值≥0.45 的 SNP 位点有 195 个，占 18.7%（图 1）；多态性信息含量（PIC）的变化范围为 0.186~0.375，平均值为 0.345，其中 PIC 值≥0.25 的 SNP 位点有 1 037 个，占 99.7%（图 2）。

2.2 聚类分析结果

通过 PowerMarker V3.25 软件应用 Roger's 1972 算法计算 51 份玉米自交系两两之间的遗传距离，构建 NJ 聚类图，进行聚类分析（图 3）。自交系间遗传距离的变化范围为 0.009 2~0.704 1，平均值为 0.446 3，遗传距离最小的两个自交系浚 92-8 和昌 7-2，遗传距离最大的是 B73 和 Mo17。

类群划分结果表明，51 份玉米自交系被划为 7 个杂种优势群：第 Ⅰ 类群为改良瑞德群，包括掖 478、浚 9058、沈 5003、沈 5005 和郑 58；第 Ⅱ 类群为瑞德群，包括 U8112、B73、PH6WC、4112、丹 9046、A801、C8605-2 和铁 7922 共 8 个自交系；第 Ⅲ 类群为兰卡斯特群，包括 Mo17、吉 1037、合 344 和 PH4CV；第 Ⅳ 类群为旅大红骨群，由自 330、龙抗 11、F349、丹 340 和综 31 组成；第 Ⅴ 类群为塘四平头群（或称黄改群），包括原辅

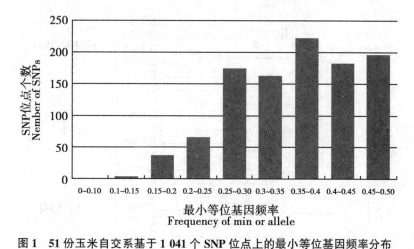

图1 51份玉米自交系基于1 041个SNP位点上的最小等位基因频率分布

Figure 1 Frequency distribution of minor allele among 51 maize inbred lines based on 1 041 SNPs

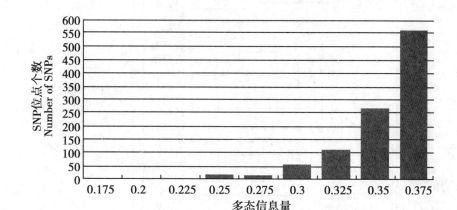

图2 51份自交系基于1 041个SNP位点上的多态性信息含量

Figure 2 Polymorphic information content among 51 maize inbred lines based on 1 041 SNPs

黄、黄早四、黄野四、H21、吉853、浚92-8、昌7-2、Lx9801、京24和丹598共10个由塘四平头衍生而来的自交系；第Ⅵ类群为P群，包括沈137、P138、X178、齐319、JN15、齐318、多29、HOF2、1145和MC30共10个自交系，其中大部分由美国杂交种P78599选育而来；第Ⅶ类群为糯质群，由京糯6、紫糯3、香糯8、9902、9901、京糯6、紫糯5B、335和紫玉-3共9个糯玉米自交系组成。

根据聚类结果，计算7个杂种优势群的群体间遗传距离（表2）。7个杂种优势群之间的遗传距离变化范围为0.285 0~0.432 1，其中瑞德群和改良瑞德群间的遗传距离最近，旅大红骨群和P群的遗传距离最远。

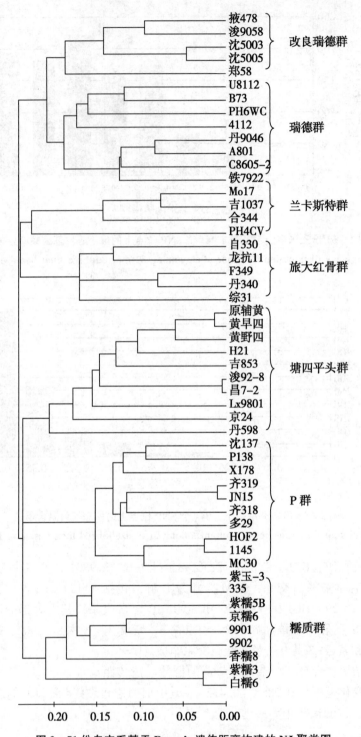

图 3 51 份自交系基于 Roger's 遗传距离构建的 NJ 聚类图

Figure 3 Neighbor-joining (NJ) trees for 51 inbred lines based on Roger's genetic distance

表2 杂种优势群的群体间遗传距离
Table 2 Genetic distance between different heterotic groups

类群 Group	糯质群 Waxy	P群 P	塘四平头群 Tang-Si-Ping-Tou	旅大红骨群 Luda Red Cob	兰卡斯特群 Lancaster	改良瑞德群 Improved Reid	瑞德群 Reid
糯质群 Waxy	0.000 0	0.352 5	0.315 0	0.347 8	0.369 7	0.350 2	0.317 4
P群 P		0.000 0	0.423 3	0.432 1	0.399 6	0.369 5	0.359 7
塘四平头群 Tang-Si-Ping-Tou			0.000 0	0.390 8	0.424 4	0.424 1	0.388 1
旅大红骨群 Luda Red Cob				0.000 0	0.358 7	0.389 6	0.383 6
兰卡斯特群 Lancaster					0.000 0	0.404 6	0.417 4
改良瑞德群 Improved Reid						0.000 0	0.285 0
瑞德群 Reid							0.000 0

3 结论与讨论

本研究利用1 041个SNP位点对51份自交系进行类群划分，SNP位点多态性信息含量在0.186~0.375，平均值为0.345。Lu等利用SNP标记对770份国内外玉米自交系进行遗传多样性分析，结果表明，1 034个SNP位点的多态性信息含量在0.003~0.375，平均值为0.259。本研究选用的1 041个SNP位点具有较高的多态性和品种区分能力，供试51份自交系被划分为7个杂种优势群，划群结果和系谱来源具有较好的一致性，利用SNP标记进行种质资源的杂种优势群划分能够获得比较理想的划群结果，特别是对于系谱不清的新系，通过类群划分指导杂交种组配，可以避免大量盲目的测配工作，提高新系利用效率。

Wang等利用60对SSR标记研究231份国内外自交系的遗传多样性时发现，糯质自交系与普通玉米自交系遗传关系较远，可单独成群。张金渝等、田孟良等的聚类分析结果表明，在分子水平上糯玉米与普通玉米存在较大遗传差异。本研究中9份糯玉米自交系成为一个独立的杂种优势群，与前人的研究结果一致。

本研究中群体间遗传距离结果显示，普通玉米自交系的6个杂种优势群中瑞德群和改良瑞德群间的遗传距离最近，这与其系谱关系也是相吻合的；塘四平头群、旅大红骨群2个国内种质群与具有国外种质血缘的改良Reid群、P群、Reid群和兰卡斯特群间的遗传距离较远。由聚类分析结果可以看出，我国育种家自主培育的两大玉米主推品种郑单958（郑58×昌7-2）、浚单20（浚9058×浚92-8）的杂优模式均为改良Reid群×塘四平头群，是国外群×国内群的成功范例。另外，鲁单 981（齐 319/Lx9801）、农大 3138（P138/综 31）和吉单 180（吉 853/Mo17）的双亲通过聚类分析后发现，三者分别属 P 群×塘四平头群、P 群×旅大红骨群、塘四平头群×兰卡斯特群的杂种优势模式，均是利用国内群×国外群杂优模式组配的优良杂交种。可见，国内群×国外群目前仍然是我国玉米育种中利用的主要杂种优势模式。

参考文献（略）

本文原载：玉米科学，2014，22（5）：29-34

分子标记辅助选择玉米自交系京24 两种抗病主效基因的聚合

余 辉[1,2]*　宋 伟[1]*　赵久然[1]**　王凤格[1]　吴金凤[3]

（1. 北京市农林科学院玉米研究中心，北京，100097；2. 首都师范大学生命科学学院，北京，100089；3. 吉林农业大学农学院，长春，130118）

摘 要：本研究以分子标记辅助选择（MAS）手段和回交转育方法相结合培育出的单抗丝黑穗病和单抗茎腐病京24抗病改良系为基础材料，经杂交、自交一代，结合MAS和抗性接种鉴定方法，以实现抗丝黑穗病和抗茎腐病主效QTL在京24基因组上的聚合。利用毛细管荧光电泳检测手段，在供受体间筛选到2对丝黑穗病前景选择引物（STS148、MZAI）和2对茎腐病前景选择引物（449B、483），可用于抗病基因聚合中对主效抗性QTL的分子标记辅助选择。对单抗丝黑穗病改良材料和单抗茎腐病改良材料杂交得到的F_1代群体各单株进行田间抗性接种鉴定，F_1代群体丝黑穗病和茎腐病发病株率分别比受体京24降低44.8%和37.2%。利用MAS，从205个F_2代分离植株中检测到9个纯合双抗株。结果验证了利用MAS辅助选择抗病主效基因聚合的可行性，获得的双抗改良材料作为抗病育种的新材料为抗病品种选育和抗病基因进一步聚合提供了材料基础。

关键词：分子标记辅助选择；玉米；抗病主效基因；丝黑穗病；茎腐病；聚合

Two Major Resistance Genes Pyramiding on Maize Inbred Line Jing24 with Marker Assisted Selection

Yu Hui[1,2]*　Song Wei[1]*　Zhao Jiuran[1]**　Wang Fengge[1]　Wu Jinfeng[3]

（1. *Maize Research Center*, *Beijing Academy of Agriculture and Forestry Sciences*, *Beijing* 100097；2. *College of Life Science*, *Capital Normal University*, *Beijing* 100089；3. *College of Agriculture*, *Jilin Agricultural University*, *Changchun* 130118）

Abstract: By the way of marker assisted selection (MAS) in backcross breeding, improved materials of inbred line Jing24, single-resistance to head smut or stalk rot has been obtained. In this research, the two kinds of improved single-resistance materials were crossed in order to achieve major genes pyramiding resistant to head smut and stalk rot on the genetic background of Jing24. Through fluorescence capillary electrophoresis, the primers for foreground selection in MAS were identified, of which STS148 and MZAI were for head smut major resistant QTL qHSR1, 449B and 483 for stalk rot

---

基金项目：本研究由北京市科技计划课题（D131100000213002）、北京市科技新星项目（2009A28）和国家科技部973计划课题（2009CB118405）共同资助

* 第一共同作者

** 通讯作者：maizezhao@126.com

major resistant QTL qRfg1. F_1 population, crossed between improved materials severally resisting head smut and stalk rot, was resistance evaluated under artificial inoculation. Compare to Jing24, the incidence rates of head smut and stalk rot of F_1 are reduced by 44.8% and 37.2%, respectively. Nine plants were screened to be homozygous on both major resistance QTLs of head smut and stalk rot from 205 F_2 segregating plants using MAS. It was sufficiently verified that resistance genes pyramiding by MAS was feasible and effective. . The double-resistance improved materials of Jing24 would be used in breeding and further resistance gene pyramiding.

Key words: Marker assisted selection; Maize; Major resistance gene; Head smut; Stalk rot; Pyramiding

玉米丝黑穗病是中国北方春玉米产区的主要病害，近年来在范围和程度上均有逐年加重的趋势（王燕，2009；邢跃先，2012）。玉米茎腐病是一种毁灭性的土传病害，在世界主要玉米种植区均有发生（Yang et al，2010）。该病不但会造成大幅减产（王波等，2013），还会因茎秆受损倒伏引起收获困难（Ledencan et al，2003）。目前最经济有效的抗病防治手段是培育抗病品种（李志明，2010）。研究表明（李莉，2012），利用分子标记辅助选择（Marker assisted selection，MAS）结合常规育种方法，可快速有效地定向改良骨干系的抗病性，为抗病新品种选育提供优良的亲本材料。

开展 MAS 育种工作的前提是对抗病 QTL（Quantitative trait loci，QTL）的定位和标记开发。针对丝黑穗病和茎腐病这两种玉米主要病害，中国农业大学徐明良教授实验室开展了系统的精细定位和基因克隆工作。该实验室（Chen et al，2008）利用丝黑穗病高抗亲本吉 1037 和高感亲本黄早四构建作图群体，在第 2 号和第 5 号染色体上检测到两个丝黑穗病抗性 QTL，分别解释表型变异的 36% 和 9%；利用茎腐病高抗亲本 1145 和高感亲本 Y331 构建作图群体，在第 1 号和第 10 号染色体上检测到两个茎腐病抗性 QTL，分别解释表型变异的 8.9% 和 36.3%。周洪昌等（2011）利用徐明良教授实验室提供的 13 对与主效抗性 QTL（qHSR1，Bin2.09）紧密连锁的候选引物，在供体吉 1037 和受体京 24 间筛选得到前景选择引物，用于丝黑穗病 MAS 育种。在抗茎腐病方面，Yang 等（2010）。李少博（2012）利用徐明良教授实验室提供的 10 对与主效抗性 QTL（qRfg1，Bin10.03）紧密连锁的候选引物，筛选出在供体 1145 和受体京 24 间具有明显多态性的前景选择引物 248 和 337，并应用到目标性状的 MAS 中。随着精细定位和基因克隆工作的深入开展，徐明良教授实验室针对上述两个抗性主效 QTL 开发了连锁更加紧密的新标记，为分子标记辅助前景选择提供了更加准确的候选标记信息。

在玉米实际生产过程中，往往多种病害同时或相继发生，因此兼抗多种病害的优良抗病品种才能具有广泛的适应性和长久的生命力。玉米自交系京 24 是我单位选育的优良骨干亲本，具有耐旱广适、耐密抗倒、高配合力等优点，但高感丝黑穗病和茎腐病。本研究以我单位利用 MAS 和回交转育方法相结合培育出的单抗丝黑穗病和单抗茎腐病京 24 抗病改良系为基础材料，经过杂交自交，结合 MAS 及抗性接种鉴定方法，实现抗丝黑穗病和抗茎腐病主效 QTL 在京 24 基因组上的聚合。

1 结果与分析

1.1 前景选择引物在抗感亲本间的筛选

利用中国农业大学徐明良教授实验室提供的 6 对丝黑穗病前景选择候选引物对供体吉

1037和受体京24进行PCR扩增和毛细管荧光电泳检测，筛选得到扩增稳定、与目的基因遗传距离较近、分别位于抗病基因上下游、在供受体间具有明显差异的两对引物STS148和MZAI。研究结果表明，前景选择引物STS148在吉1037与京24上的扩增片段大小分别为118bp和129bp，MZAI在吉1037与京24上的扩增片段大小分别为407bp和324bp（图1、图2）。两个引物扩增片段的峰值都在4 000以上，说明二者在吉1037与京24两亲本间具有明显的多态性，且扩增效果稳定便于数据统计。

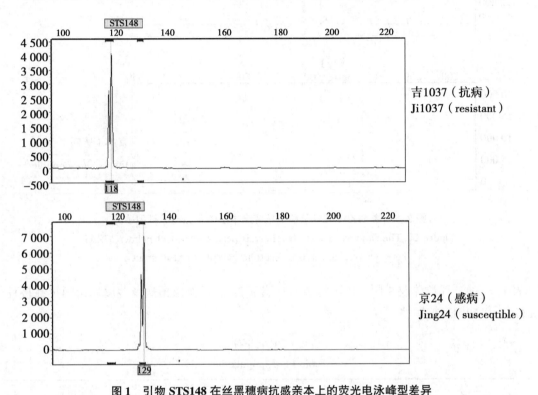

图1 引物STS148在丝黑穗病抗感亲本上的荧光电泳峰型差异
Figure 1 The fluorescence electrophoresis peak contrast of primer STS148 between resistant and susceptible parents to head smut

注：横坐标表示片段大小，纵坐标表示峰型高低；分子内标为GS3730-500；以下同
Note: The abscissa represents fragment size, the ordinate represents peak level; the molecular internal standard is GS3730-500; the same below

从中国农业大学徐明良教授实验室提供的7对茎腐病前景选择候选引物中筛选出在供体1145和受体京24之间扩增有明显差异的、与目的基因遗传距离较近、位于抗病基因上下游、峰型单一且尖锐的两对引物449B和483。研究结果表明：前景选择引物449B在1145与京24上的扩增产物峰型尖锐，片段大小分别为270bp和231bp，引物483在1145与京24上的扩增片段大小分别为446bp和436bp（图3、图4）。两个引物扩增片段的峰值都在2 000以上，即二者在京24与1145两亲本间具有明显的多态性，且扩增效果稳定便于数据统计。

1.2 F_1代群体抗性接种鉴定

单抗丝黑穗病京24改良材料和单抗茎腐病京24改良材料杂交得到9个F_1代群体

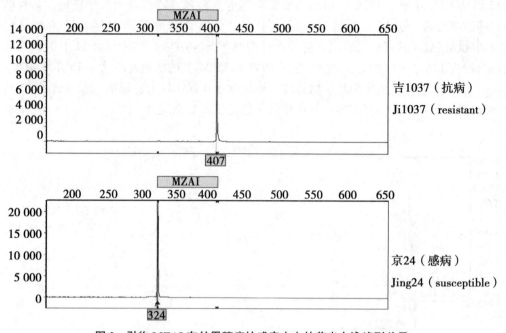

图 2 引物 MZAI 在丝黑穗病抗感亲本上的荧光电泳峰型差异
Figure 2 The fluorescence electrophoresis peak contrast of primer MZAI between resistant and susceptible parents to head smut

(表1)。对 F_1 代群体及相同条件下种植的对照京24进行丝黑穗病及茎腐病田间抗性接种鉴定，单株调查抗病性。

表 1 F_1 代群体
Table 1 F_1 population

编号 Serial No.	杂交组合 Hybrid combination	编号 Serial No.	杂交组合 Hybrid combination
①	A75×A124	⑥	A79×A124
②	A75×A130	⑦	A79×A130
③	A77×A124	⑧	A80×A124
④	A77×A130	⑨	A80×A130
⑤	A78×A124		

丝黑穗病田间抗性接种鉴定的 F_1 代群体共218株，其中感病14株，抗病204株，发病株率为6.4%；原受体京24发病株率为51.2%。茎腐病田间抗性接种鉴定的 F_1 代群体共274株，其中感病64株，抗病210株，发病株率23.4%；原受体京24发病株率为60.6%。

F_1 代群体与受体京24的抗性接种鉴定结果显示，F_1 代群体的丝黑穗病发病株率比受体京24降低44.8%；茎腐病发病株率比受体京24降低37.2%。根据丝黑穗病及茎腐病病

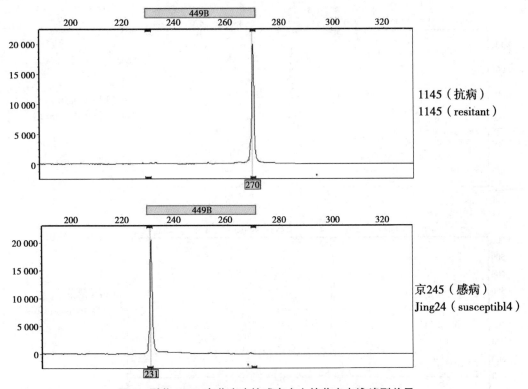

图3 引物449B在茎腐病抗感亲本上的荧光电泳峰型差异
Figure 3 The fluorescence electrophoresis peak contrast of primer 449B between resistant and susceptible parents to stalk rot

情分级标准，F_1代群体对这两种病害均表现抗病，受体京24均表现高感，即F_1代群体的抗病性与受体京24相比得到了明显提高。

1.3 F_2代分离群体室内分子检测结果分析

对205个F_2代分离株利用丝黑穗病和茎腐病前景选择入选引物进行单株分子检测。鉴定结果如下：丝黑穗病纯合抗病50株，杂合抗病108株，纯合感病47株，$\chi^2_{1:2:1}$ = 0.68；茎腐病纯合抗病43株，杂合抗病109株，纯合感病53株，$\chi^2_{1:2:1}$ = 1.80；纯合双抗9株，纯合双感12株，其他184株，$\chi^2_{1:14:1}$ = 2.93。上述结果符合孟德尔遗传定律。根据室内检测结果，选择纯合双抗的9个单株进行自交扩繁，即为兼抗丝黑穗病和茎腐病的京24抗病改良材料。

2 讨论

常规育种技术在玉米性状改良方面虽然取得了显著成就，但由于其选择效率较低、周期较长，已不能满足当前玉米生产对优良品种的需求（李建生，2007）。随着分子标记技术的发展，MAS为常规育种提供了一条快速、准确的辅助手段。本研究紧密跟踪丝黑穗病和茎腐病抗病主效QTL精细定位的最新进展，筛选适用于特定回交转育群体的、与抗病基因连锁更加紧密的前景选择标记，实现了两种抗病主效QTL的有效聚合。利用MAS

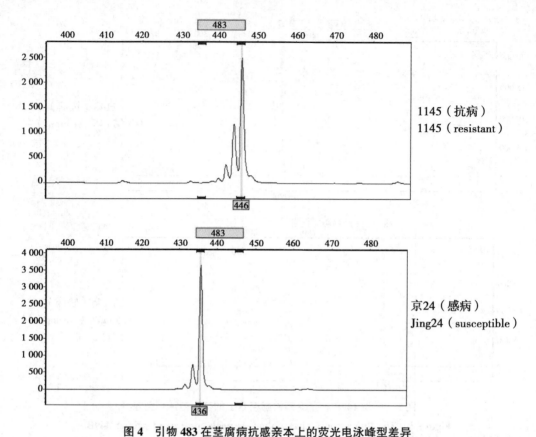

图 4 引物 483 在茎腐病抗感亲本上的荧光电泳峰型差异
Figure 4 The fluorescence electrophoresis peak contrast of primer 483 between resistant and susceptible parents to stalk rot

和传统育种技术相结合，本实验室通过回交三代自交两代获得了两种单抗改良材料；在此基础上将两种单抗改良材料通过杂交一代、自交纯合一代，完成了两种抗病主效 QTL（qHSR1 和 qRfg1）的聚合，探索了一套抗丝黑穗病和抗茎腐病主效 QTL 聚合的实操方法，建立了一种高效实用的分子标记辅助抗病基因聚合的育种模式。

丝黑穗病抗性和茎腐病抗性均属于数量性状，由多个抗性 QTL 控制。本实验室已对京 24 单抗改良材料进行了抗性接种鉴定，两种单抗改良材料的发病株率均比京 24 有明显降低；对于 F_1 代群体，丝黑穗病发病株率比京 24 降低 44.8%，茎腐病发病株率比京 24 降低 37.2%。无论是单抗改良材料还是 F_1 代群体，抗病性均得到了明显提高。因此，针对玉米第 2 号染色体丝黑穗病主效抗性 QTL（qHSR1）和第 10 号染色体茎腐病主效抗性 QTL（qRfg1）开展分子标记辅助选择抗病改良工作是切实可行的。

F_2 代分离群体室内分子检测结果显示，无论是纯合单抗材料还是纯合双抗材料，检测到的实际株数均略少于理论株数。我们推测除群体数量稍小外，自然淘汰可能是产生这种结果的主要原因。纯合抗病植株虽然抗病能力相对于其他植株有明显提高，但是由于其纯合度高于杂合抗病株，芽率、生活力等方面的不足增大了被自然淘汰的概率。对于纯合的 F_2 代植株，如果自然淘汰发生在分子检测取样之前，这类植株将不能被统计在内，就会导致检测的实际结果与理论结果偏离。

3 材料与方法

3.1 供试材料

受体自交系京24、抗丝黑穗病供体自交系吉1037、抗茎腐病供体自交系1145，沙培发苗，于3~5叶期提取DNA。选用的5份单抗丝黑穗病京24改良材料（均携带纯合主效抗性QTL qHSR1）分别为：A75、A77、A78、A79和A80，2份单抗茎腐病京24改良材料（均携带纯合主效抗性QTL qRfg1）分别为：A124、A130。两种京24改良材料杂交得到F_1代群体（表1），于4月下旬和5月中旬两个时期重复种植在北京市农林科学院玉米研究中心小汤山育种基地，分别进行丝黑穗病和茎腐病田间接种抗性鉴定。选择抗性好、综合农艺性状优良的F_1代单株自交，得到F_2代分离群体。F_2代群体于同年冬季种植在北京市农林科学院玉米研究中心三亚育种基地，大喇叭口期单株叶片田间取样提取DNA。

3.2 前景选择候选引物

丝黑穗病前景选择候选引物共计6对，与Bin2.09上的主效抗性QTL（qHSR1）紧密连锁：MZAI、STS661、E148、N6AC6、9AC4和STS148；茎腐病前景选择候选引物共计7对，与Bin10.03上的主效抗性QTL（qRfg1）紧密连锁：248、337、449B、373、TE1A、PINS3和483。上述候选引物序列信息由中国农业大学徐明良教授实验室提供。

3.3 DNA提取和引物扩增

提取DNA采用CTAB法（Rogers and Bendich, 1985）。

引物扩增反应体系总体积20μl，包括：DNA溶液4μl、ddH_2O 12.35μl、10mmol/L dNTP 1.2μl、10×Buffer（2.5mmol/L $MgCl_2$）2μl、Primer（F+R）0.25μl、5U/μl Taq酶0.2μl。扩增反应程序：94℃预变性5min；94℃变性40s，60℃退火35s，72℃延伸45s，总共35个循环；72℃延伸10min，PCR结束；4℃保存。PCR扩增在Veriti 384 Well Thermal Cycler仪上完成。

3.4 扩增产物检测

对PCR产物进行毛细管荧光多重电泳，即按照扩增片段大小差异和引物荧光颜色差异将多对引物的PCR产物混在一起进行电泳。电泳板上的每孔分别加入4μl PCR产物的混合物、0.08μl GS3730-500分子量内标（Applied Biosystems, USA）和7.92μl去离子甲酰胺。95℃变性5min，在PTC-100 PCR仪上进行，4℃冷却10min，2 500rpm低速离心1min，于AB 3730XL DNA分析仪上进行荧光电泳。预电泳：15kV，2min；电泳：15kV，20min。最后用Date Collection v1.0软件收集原始数据。

利用GeneMarker-BMSTC v1.0软件对原始数据进行分析。选择分子内标GS500LIZ_3730，选择分析类型Fragment（Plant），得出分子标记位点多态性数据，excel存档保存。

3.5 丝黑穗病和茎腐病田间抗性接种鉴定及抗性评价标准

对4月下旬种植的F_1代群体和受体京24进行丝黑穗病人工接种抗性鉴定，抗病对照Mo17、感病对照黄早四。接种方法：于上一季节典型病株上采集丝黑穗病病穗，铺展开并通风干燥保存，在播种前一周将捻碎，散出冬孢子团，过筛，收集纯病原菌菌粉。将菌粉与过筛的田土充分混合拌匀，配成0.1%的带菌土用于接种。播种后将菌土100g覆盖于种子上，后覆盖田土。

对5月下旬种植的F_1代群体和受体京24进行茎腐病人工接种抗性鉴定，抗病对照齐

319、感病对照液478。接种方法：高温灭活玉米籽粒当做培养基，播种前将病原菌菌剂在培养基上培养繁殖。播种后将带有病原菌的玉米籽粒培养基盖在种子上面，再覆土。接种后适量灌溉1次，此后保持土壤水分充足。茎腐病病菌由河北省农林科学院植物保护研究所提供。

丝黑穗病和茎腐病的调查方法：发病期对 F_1 代群体和京24逐株调查记载抗感情况。F_1 代群体及京24的发病株率（%）计算公式：（感病株数/总株数）×100%。玉米丝黑穗病和茎腐病田间抗性划分标准见表2。

表2 玉米丝黑穗病和茎腐病田间抗性划分标准
Table 2　The dividing standards of resistance of maize head smut and stalk rot in field

丝黑穗病 Head smut		茎腐病 Stalk rot	
发病株率（%）Incidence rate（%）	抗性 Resistance	发病株率（%）Incidence rate（%）	抗性 Resistance
0～5.0	高抗 Highly resistant	0～5.0	高抗 Highly resistant
5.1～20.0	抗 Resistant	5.1～10.0	抗 Resistant
20.1～50.0	感 Susceptible	10.1～30.0	中抗 Moderately resistant
50.1～100.0	高感 Highly susceptible	30.1～40.0	感 Susceptible
		40.1～100.0	高感 Highly susceptible

参考文献（略）

本文原载：分子植物育种，2014，12（2）：240-245

分子标记辅助选择育成的玉米自交系京 24 单抗丝黑穗病和茎腐病改良材料性状分析

余 辉[1,2]* 宋 伟[1]* 赵久然[1]** 王凤格[1] 吴金凤[3]

(1. 北京市农林科学院玉米研究中心,北京 100097;2. 首都师范大学生命科学学院,北京 100089;3. 吉林农业大学农学院,长春 130118)

摘 要:玉米自交系京 24 是北京市农林科学院玉米研究中心选育的具有黄改系血缘的骨干亲本,具有高配合力、耐密抗倒和耐旱广适等优点,但对丝黑穗病和茎腐病表现为高感。本中心分别选用自交系吉 1037 和 1145 作为丝黑穗病和茎腐病的抗源,通过回交转育方法,结合分子标记辅助选择技术,选育出了一批单抗丝黑穗病和单抗茎腐病的京 24 抗病改良材料。本研究通过对其中的 5 份单抗丝黑穗病改良材料、2 份单抗茎腐病改良材料以及受体京 24 进行遗传背景检测、抗性接种鉴定和表型性状调查,检验单抗改良材料的改良状况,发掘以京 24 为遗传背景的抗病新种质。40 对 SSR 核心引物的 DNA 指纹数据分析结果表明,7 份单抗改良材料与京 24 的遗传相似度均在 95%～100%,改良材料的平均纯合度为 94.3%。田间抗性接种鉴定结果表明,单抗丝黑穗病改良材料的发病株率比京 24 降低了 37.3%～51.2%,单抗茎腐病改良材料的发病株率比京 24 降低了 40.0% 以上,改良材料的抗性得到了明显提高。表型性状调查数据显示,7 份改良材料的穗部性状和株高、穗位高等田间表型性状均与京 24 无显著性差异。可见,改良材料在保持京 24 自身优良性状和具备较高纯合度的前提下,对丝黑穗病和茎腐病的抗性得到了明显提高,证实分子标记辅助抗病性状改良是切实可行的。京 24 单抗改良材料为品种选育和抗病基因聚合提供了材料基础。

关键词:玉米;自交系;单抗;抗病改良;分子标记辅助选择

Characters Analysis on Resistance Improved Materials of Jing24 Single-resistance to Head Smut and Stalk Rot Bred with Molecular Marker-assisted Selection

Yu Hui[1,2]* Song Wei[1]* Zhao Jiuran[1]** Wang Fengge[1] Wu Jinfeng[3]

(1. *Maize Research Center, Beijing Academy of Agriculture and Forestry Sciences, Beijing* 100097; 2. *College of Life Science, Capital Normal University, Beijing* 100089; 3. *College of Agriculture, Jilin Agricultural University, Changchun* 130118)

Abstract: Maize inbred line Jing24 is one of foundation parents with Huangzaosi improved lines

---

基金项目:本研究由北京市科技计划课题(D131100000213002)、北京市科技新星项目(2009A28)和国家科技部 973 计划课题(2009CB118405)共同资助

* 同等贡献作者

** 通讯作者:maizezhao@126.com

germplasm, which was bred by Maize Research Center of Beijing Academy of Agriculture and Forestry Sciences. It has advantage of high combining ability, condensing tolerance, lodging resistance, drought tolerance and wide adaptability, but shows high susceptibility to head smut and stalk rot. Using inbred lines Ji1037 and 1145 as head smut and stalk rot resistance sources respectively, the center has obtained a number of improved materials of Jing24 single-resistance to head smut and stalk rot, by the way of backcross breeding combined with molecular marker-assisted selection. In this research, the improving condition of the improved single-resistance materials was tested by means of genetic background detection, resistance evaluation with artificial inoculation and phenotypic characters investigation among 5 improved materials single-resistance to head smut, 2 improved materials single-resistance to stalk rot and the receptor Jing24. DNA fingerprint data of 40 pairs of SSR core primers indicated that genetic similarity between the 7 improved single-resistance materials and Jing24 was from 95.0% to 100%, with the average homozygosity of improved materials being 94.3%. Results of field resistance evaluation with artificial inoculation show that incidence rates of the improved materials severally resisting to head smut and stalk rot are reduced by 37.3% ~ 51.2% and over 40.0%. It was obviously that the resistance of the improved materials was enhanced. Investigation data of phenotypic traits revealed that ear characters and field traits, such as plant height, ear height et al., have no significant difference between the 7 improved materials and Jing24. It was proven that resistance improving of Jing24 by molecular marker-assisted selection was feasible. And resistance improved materials of Jing24 would be the basis of variety breeding and resistance gene pyramiding.

Key words: Maize; Inbred line; Single-resistance; Resistance improving; Molecular marker-assisted selection

玉米丝黑穗病是中国春玉米生产区的重要病害，主要分布于东北、华北、华中和西北地区（曹士亮，2009）。该病对玉米生产构成极大威胁，一般发病率在7%~35%，严重的地方达到62%（邱红波等，2008），一旦发生将会引起大幅度减产并造成严重经济损失。防治玉米丝黑穗病的方法有以下几种：加强田间管理，实行轮作和药剂防治等，但这些措施均不能从根本上解决玉米丝黑穗病的为害，选育和利用抗病品种是抑制玉米丝黑穗病发生的最经济有效的方法（李志明，2010）。

玉米茎腐病是一种毁灭性的土传病害，在世界主要玉米种植区均有发生（Yang et al, 2010）。从1980年开始，该病在中国各玉米产区相继发生，一般年份发病率为10%~20%，个别年份和地区可达50%以上；一般减产25%~30%，严重时甚至绝收（王波等，2013）。茎腐病不但会造成大幅减产，还会严重降低玉米品质和因茎秆破损倒伏而带来收获问题（Ledencan et al, 2003）。由于玉米茎腐病病原种类多，发病原因复杂，故应采取选育抗病品种的农业手段为主，其他防治手段为辅的综合防治策略（温瑞等，2000）。

采用常规育种的方式选育抗病品种，存在目标性状选择效率低、选育周期长的问题，已不能满足当前玉米生产对优良品种的需求（李建生，2007）。分子标记作为常规育种的辅助手段，弥补了常规抗病育种中的诸多不足。在丝黑穗病（周洪昌，2011）和茎腐病（李少博，2012）改良育种方面，北京市农林科学院玉米研究中心以骨干亲本京24为受体材料，分别选用自交系吉1037和1145为抗源，通过三代回交和两代自交，结合分子标记辅助前景选择跟踪目标性状抗性主效QTL（quantitative trait loci），同时借助分子标记辅助背景选择加速遗传背景回复，实现了京24在这两种病害上快速精确的定向改良，分别获得了以京 24 为遗传背景的单抗丝黑穗病改良材料和单抗茎腐病改良材料。

本研究选用 5 份单抗丝黑穗病和 2 份单抗茎腐病京 24 改良材料为试验材料,比较分析单抗改良材料与受体京 24 在遗传背景、抗病性和表型性状三方面上的差异,以检验单抗改良材料的实际改良情况,发掘以京 24 为遗传背景的抗病新种质,为品种选育和抗病基因聚合提供材料基础。

1 结果与分析

1.1 单抗改良材料与京 24 的遗传背景比较分析

利用 40 对 SSR 核心引物对单抗改良材料和京 24 进行遗传背景检测,分析改良材料与京 24 之间的遗传相似度。由表 1 可知,7 份单抗改良材料中与京 24 在 40 对核心引物上差异位点数≥2 的材料有 2 份,差异位点数 = 1 的材料有 3 份,差异位点数 = 0 的材料有 2 份。所有单抗改良材料与京 24 的遗传相似度均在 95.0%以上,单抗改良材料在 40 对核心引物上的 DNA 指纹与受体相差不大。根据 40 个核心位点中纯合位点占总检测位点的比例,计算并分析各改良材料的纯合度。从表 1 可以看出,7 份材料的纯合度在 87.5% ~ 97.5%之间,平均为 94.3%,即经过 3 代回交和 2 代自交,结合分子标记辅助背景选择,7 份改良材料均具有较高的纯合度。

表 1 单抗改良材料纯合度以及与京 24 在 40 对核心引物上的遗传背景差异
Table 1 The homozygosity of improved single-resistance materials and genetic background contrast to Jing24 in 40 pairs of core primers

材料 Materials	差异位点数 The number of variation loci	遗传相似度(%) Genetic similarity (%)	纯合度(%) Homozygosity (%)
A75	1	98.8	95.0
A77	4	95.0	87.5
A78	0	100.0	97.5
A79	2	97.5	92.5
A80	0	100.0	97.5
A124	1	98.8	95.0
A130	1	98.8	95.0

1.2 单抗改良材料与京 24 的抗病性状比较分析

对单抗改良材料及京 24 进行丝黑穗病和茎腐病田间抗性接种鉴定,调查抗感情况并计算发病株率。5 份单抗丝黑穗病改良材料和京 24 的丝黑穗病发病株率分别为:7.2%、13.9%、10.0%、13.3%、0%、51.2%,改良材料的发病株率比京 24 降低了 37.3% ~ 51.2%;2 份单抗茎腐病改良材料和京 24 的茎腐病发病株率分别为:19.5%、18.8%、60.6%,改良材料的发病株率比京 24 降低了 40%以上。根据丝黑穗病病情分级标准及茎腐病病情分级标准,所有改良材料均表现抗病水平,受体京 24 对两种病均表现为高感。可见,改良材料的抗性得到了显著提高。

1.3 单抗改良材料与京 24 的穗部性状和田间表型性状比较分析

对单抗丝黑穗病改良材料、单抗茎腐病改良材料以及分别与两种改良材料相同条件下种植的对照京 24 进行穗部性状（包括穗长、穗行数、行粒数、穗粗、出籽率和百粒重）和田间表型性状（包括株高和穗位高）比较分析，结果见表 2 和表 3。由两表可知，改良材料的上述穗部性状和田间表型性状表现均与京 24 无显著性差异。

表 2 单抗丝黑穗病改良材料与京 24 表型性状比较分析
Table 2 The comparative analysis of phenotypic characters between improved single-resistance head smut materials and Jing24

材料 Materials	穗长（cm）Ear length（cm）	穗行数（行）Rows per ear（rows）	行粒数 Kernels per row	穗粗（cm）Ear diameter（cm）	出籽率（%）Produced grain percentage（%）	百粒重（g）100-kernels weight（g）	株高（cm）Plant height（cm）	穗位高（cm）Ear height（cm）
京 24 Jing24	12.96a	10.40a	23.00a	4.40a	77.44 a	30.40 a	184.80 a	56.80 a
A75	13.04 a	10.80 a	23.80 a	4.06 a	79.20 a	31.38 a	187.20 a	57.80 a
A77	13.52 a	11.20 a	26.00 a	4.20 a	78.52 a	31.94 a	185.20 a	57.20 a
A78	13.72 a	11.20 a	24.20 a	4.30 a	78.60 a	31.54 a	184.00 a	57.20 a
A79	13.26 a	11.20 a	25.00 a	4.30 a	78.74 a	31.62 a	186.20 a	57.00 a
A80	13.30 a	11.60a	24.00 a	4.22 a	79.42 a	31.64 a	184.60 a	56.40 a

注：数字后的小写字母表示在 0.05 水平上具有显著性差异
Note: The small letters at the back of the figures represent significant difference at 0.05 probability level

表 3 单抗茎腐病改良材料与京 24 表型性状比较分析
Table 3 The comparative analysis of phenotypic characters between improved single-resistance stalk rot materials and Jing24

材料 Materials	穗长（cm）Ear length（cm）	穗行数（行）Rows per ear（rows）	行粒数 Kernels per row	穗粗（cm）Ear diameter（cm）	出籽率（%）Produced grain percentage（%）	百粒重（g）100-kernels weight（g）	株高（cm）Plant height（cm）	穗位高（cm）Ear height（cm）
京 24 Jing24	11.88 a	11.60 a	22.20 a	3.98 a	76.50 a	24.86 a	168.40 a	56.80 a
A124	11.78 a	11.20 a	20.80 a	3.82 a	77.02 a	25.36 a	167.40 a	58.60 a
A130	12.12 a	12.00 a	22.20 a	3.86 a	75.08 a	23.90 a	171.40 a	52.00 a

2 讨论

京 24 单抗改良材料保持了京 24 原有的优良性状和较高的遗传相似度，并且在 40 个核心位点上达到了较高的纯合度。在此前提下，单抗改良材料经过分子标记辅助回交转育，分别在丝黑穗病和茎腐病抗性上有了明显提高，具有较高的应用价值。一方面，可以作为抗病基因进一步聚合的基础材料。目前抗丝黑穗病和抗茎腐病改良材料的抗病基因/主效 QTL 已经纯合，将这两种改良材料杂交，再经一步自交，将会实现京 24 的抗病基因聚合。另一方面，这些单抗改良材料可以作为组配抗病高产品种的优良亲本。受体京 24 由于具有抗倒、早熟、高配合力等优点，作为亲本已经组配了京单 28

等系列玉米品种。本研究通过遗传背景、抗病性、表型性状分析证实，利用分子标记辅助京24抗病性状改良获得的两种单抗改良材料在基本保留京24遗传背景的基础上具有较高的抗丝黑穗病或抗茎腐病能力，有望组配出抗性较强的优良杂交种，为培育抗病品种带来了契机。

根据《中华人民共和国农业部，NY/T1432—2007，玉米品种鉴定DNA指纹方法》规定，累计检测40个核心引物位点，如果品种间差异位点数≥2，判定两品种为不同品种；如果品种间差异位点数=0，判定两品种为相同品种或极近似品种。本研究选用的7份单抗改良材料经40对核心引物扩增检测，有两份材料与京24的差异位点数≥2。跟据上述标准判断，这两份材料为京24抗病改良新种质，可作为新系在育种中使用。7份材料中与京24间差异位点数=0的材料也有两份，在育种中可代替原来的感病京24使用。

周洪昌等（2011）、李少博等（2012）分别对吉1037/京24和1145/京24的回交BC_2代群体抗病性进行了前景标记检测和田间抗性接种鉴定，结果表明杂合抗病植株（携带杂合的主效抗性QTL）对丝黑穗病和茎腐病的抗病率分别比纯合感病植株（不含主效QTL）高22%和36%。本研究结果显示，纯合单抗丝黑穗病改良材料（携带纯合的主效抗性QTL）的发病株率比京24降低了37.3%~51.2%，纯合单抗茎腐病改良材料的发病株率比京24降低了40.0%以上。由此可见，无论是在主效抗性QTL位点上纯合的抗病株还是杂合的抗病株与纯合感病株相比，在抗病性上均有明显提高，利用分子标记辅助前景选择进行自交系抗病性状改良是行之有效的。

本研究的供试材料7份京24单抗改良材料均是经3代回交、2代自交获得的，其中在BC_1~BC_3代均利用分子标记辅助背景选择，以加快京24遗传背景的回复速度（周洪昌，2011；李少博，2012）。基于40对核心SSR引物的遗传背景检测结果表明，单抗改良材料与受体京24间的遗传相似度均在95%以上。可见，利用分子标记辅助背景选择结合回交转育，仅需回交3代即可以获得遗传背景与轮回亲本京24高度相似的改良材料，与单纯利用回交转育常规育种方法相比，可以节约2~4代的选育时间。由于目前分子标记辅助目标性状前景选择的有效应用受到可用标记少等诸多因素的限制，利用分子标记辅助背景选择加快遗传背景回复进程可能成为分子标记辅助选择育种的一项重要内容。

3 材料与方法

3.1 供试材料

自交系京24，5份单抗丝黑穗病京24改良材料A75、A77、A78、A79和A80，2份单抗茎腐病京24改良材料A124、A130：种植于北京市农林科学院玉米研究中心小汤山育种基地，并于大喇叭口期叶片取样提取DNA。

3.2 40对核心引物信息

用于检测单抗改良材料和京24遗传背景的40对SSR核心引物信息详见（Wang et al, 2011）。

3.3 DNA提取和引物扩增

DNA提取：CTAB法（Rogers and Bendich, 1985）。

引物扩增反应体系：反应液总体积20μl，其中包括：DNA模版4μl、ddH_2O 12.35μl、

10×Buffer（2.5mmol/L MgCl$_2$）2μl、10mmol/L dNTP 1.2μl、5U/μl *Taq* DNA Enzyme 0.2μl、Primer（F+R）0.25μl。SSR扩增反应程序为：94℃预变性5min，1个循环；94℃变性40s，60℃退火35s，72℃延伸45s，共35个循环；72℃延伸10min；4℃保存。PCR扩增在Veriti 384 Well Thermal Cycler上完成。

3.4 扩增产物检测

利用AB 3730XL DNA分析仪对PCR产物进行毛细管荧光10重电泳，即按照扩增片段大小和荧光颜色差异将10对引物的扩增产物混于同一单孔中进行电泳。96孔电泳板每孔分别加入4μl扩增产物的混合物、7.92μl甲酰胺和0.08μl GS3730-500分子量内标（Applied Biosystems, USA）。95℃变性5min，4℃保存10min，2 500rpm离心1min后，于AB 3730XL DNA分析仪上进行电泳。15kV预电泳2min；15kV电泳20min。用Date Collection v1.0软件收集原始数据。

利用GeneMarker-BMSTC v1.0软件（由北京市农林科学院玉米研究中心与北京华生恒业科技有限公司共同研发）对收集的原始数据进行分析。分子内标选择GS500LIZ_3730，分析类型选择Fragment（Plant），得出扩增片段大小的数据，excel文档保存。

3.5 丝黑穗病和茎腐病田间抗性接种鉴定及抗性评价标准

用丝黑穗病病菌接种单抗丝黑穗病改良材料和对照京24，抗病对照Mo17、感病对照黄早四。接种方法：在材料播种前一周将通风干燥保存的玉米丝黑穗病病穗破碎，捻碎冬孢子团，过筛得到纯的病原菌菌粉。将菌粉与过筛的细土充分拌匀，配制成0.1%菌土用于接种。接种时，每个穴先播下种子，覆盖菌土100g，再覆盖田土。丝黑穗病病穗采集于上一季节典型病株上。

用茎腐病病菌接种单抗茎腐病改良材料和对照京24，抗病对照齐319、感病对照掖478。接种方法：播种前将病原菌菌剂在高温灭活的玉米籽粒上培养繁殖。播种时，将带有病原菌的玉米籽粒若干均匀盖在种子上面，覆土。接种后灌溉1次，此后保持充足的土壤水分。茎腐病病原菌由河北省农林科学院植物保护研究所提供。

丝黑穗病和茎腐病的调查方法：发病期对单抗改良材料、京24逐株调查记录抗感情况。单抗改良材料及京24的发病株率（%）计算公式如下：（感病株数/总株数）×100%。玉米丝黑穗病和茎腐病田间抗性评价标准见表4。

表4 玉米丝黑穗病和茎腐病田间抗性评价标准
Table 4 The standards of resistance evaluation of maize head smut and stalk rot in field

丝黑穗病 Head smut		茎腐病 Stalk rot	
发病株率（%）Incidence rate（%）	抗性 Resistance	发病株率（%）Incidence rate（%）	抗性 Resistance
0~5.0	高抗 Highly resistant	0~5.0	高抗 Highly resistant
5.1~20.0	抗 Resistant	5.1~10.0	抗 Resistant
20.1~50.0	感 Susceptible	10.1~30.0	中抗 Moderately resistant
50.1~100.0	高感 Highly susceptible	30.1~40.0	感 Susceptible
		40.1~100.0	高感 Highly susceptible

3.6 表型性状调查及数据分析、遗传相似度和纯合度计算方法

散粉后，分别测量单抗改良材料和京 24 的株高和穗位高，每份材料随机测量 5 株。收获后，将果穗置于通风处晾干，水分在 14% 以下时对单抗改良材料和京 24 考种，每份材料随机选取 5 个果穗，测量穗长、穗行数、行粒数、穗粗、出籽率和百粒重。利用 IBM SPSS Statistics 19 软件进行表型性状数据处理和统计分析。

单抗改良材料与京 24 的遗传相似度（%）计算公式如下：$(1-b/a) \times 100\%$。其中，a 为总等位基因数，即总核心位点数的 2 倍；b 为有差异的等位基因数。

单抗改良材料的纯合度（%）= 40 个核心位点中纯合位点数/总位点数×100。

参考文献（略）

本文原载：分子植物育种，2014，12：(1)：56-61

京724玉米自交系S型细胞质雄性不育系分子标记辅助选育研究

宋 伟 苏爱国 邢锦丰 吴金凤 赵久然

(北京市农林科学院玉米研究中心/玉米DNA指纹及分子育种北京市重点实验室,北京 100097)

摘 要:以自育的S型细胞质雄性不育系MD32为供体,京科968的母本京724为受体,利用回交转育结合分子标记辅助背景选择,仅回交三代即获得了京724遗传背景的S型细胞质雄性不育系。经花粉I_2-KI染色鉴定,京724不育系花粉败育完全。

关键词:玉米;分子标记辅助选择;细胞质雄性不育

Study on Breeding of New S-type Cytoplasmic Male Sterile Material from Maize Inbred Line Jing724 with Molecular Marker Assisted Selection

Song Wei Su Aiguo Xing Jinfeng Wu Jinfeng Zhao Jiuran

(*Maize Research Center, Beijing Academy of Agriculture and Forestry Sciences, Beijing Key Laboratory of maize DNA Fingerprinting and Molecular Breeding, Beijing 100097, China*)

Abstract: MD32 wocs used as donor, male sterile trait was transferred to the inbred line Jing724 using backcross breeding with molecular marker assisted background selection, which is female parent of Jingke968. Backcross only three generations, we obtained cms-Jing724 that genetic background close to Jing724. The male sterility characteristics of cms-Jing724 were stable, no anthers exposed on tassel and pollen aborted completely through I_2-KI staining.

Key words: Maize; Molecular marker assisted selection; Cytoplasmic male sterility

强优势杂交种的选育和推广是提高玉米产量的重要途径。利用常规的制种方法配制杂交种需要对母本进行人工去雄,不仅耗费大量的劳动力,增加种子生产成本,且存在由于去雄不及时、不彻底影响种子质量的潜在风险。利用雄性不育系制种是保证种子纯度提高

基金项目:国家科技支撑项目(2014BAD01B09)、北京市农林科学院科技创新能力建设专项(KJCX20140202、QNJJ201302)、北京市科技计划课题(Z141100002314016)、北京市博士后基金(2014ZZ-68)

作者简介:宋伟(1979—),女,山东人,副研究员,博士,从事玉米分子育种工作。Tel:010-51503983 E-mail:songwei1007@126.com 苏爱国为本文共同第一作者
赵久然为本文通讯作者

玉米产量的一种有效方法。早在20世纪五六十年代，国内外就开始了对玉米不育化制种的研究。细胞质雄性不育由于容易实现不育系、保持系和恢复系的配套，是玉米育种中利用的主要类型。

实现玉米细胞质雄性不育化制种的关键是不育系、保持系和恢复系的选育。选育不育系最常用的方法是回交转育法，即用稳定的不育系作非轮回亲本，选择优良自交系作轮回亲本，进行多代回交，育成不育性稳定的优良不育自交系，而原轮回亲本便是该不育系的保持系。但仅仅利用传统的回交转育方法，一般需回交5~6代，转育周期较长，延缓了不育化制种技术在玉米杂交种上的应用。利用回交转育方法进行恢复系的选育比不育系更为复杂，在回交的同时要进行测交，以测交结果作为恢复系的选择标准；或在回交转育中利用不育细胞质提供育性的指标性状。由于选育过程相对繁琐，在恢复系尚未育成或只有部分恢复性的情况下，须将不育化的杂交种子与正常杂交种种子按适当比例掺合使用，这就给种子生产在技术和管理上提出了更高的要求。因此，如何加快不育系的选育速度、简化恢复系的选育过程，是目前雄性不育化制种技术亟需解决的问题。

分子标记辅助背景选择能够有效提高回交转育过程中轮回亲本的背景回复速度。MD32是北京市农林科学院自育的综合性状优良、育性稳定彻底的S型细胞质雄性不育系。本研究以MD32为不育基因的供体，利用回交转育结合分子标记辅助背景选择，将不育性状转移到自交系京724遗传背景中。京724是农业部2012—2015年主导品种京科968的母本。因此，选育成功育性稳定、败育彻底的京724 S型细胞质雄性不育系，将为实现国审品种京科968（杂交组合：京724/京92）的雄性不育化制种奠定基础。

1 材料与方法

1.1 供试材料

以雄性不育系MD32、京科968母本京724、杂交F_1代，及以京724为轮回亲本得到的回交群体为研究材料。京724与MD32室内沙培发苗，3~5叶期提取DNA。2012年夏，在北京市农林科学院玉米研究中心小汤山试验基地利用MD32与京724组配F_1代。同年冬，海南田间种植F_1代，植株花药不外露，不育性表现彻底。以F_1代植株做母本，与轮回亲本京724回交得到BC_1代。海南田间种植BC_1代群体，通过表型选择和分子标记辅助背景选择确定入选BC_1代单株，与京724回交得到BC_2代。2013年夏，在小汤山试验基地播种BC_2代群体，通过表型选择和分子标记辅助背景选择确定入选BC_2代单株，与京724继续回交获得BC_3代群体，即为京724遗传背景的S型胞质雄性不育系。

1.2 分子标记辅助背景选择

1.2.1 背景选择引物

利用40对多态性高、在染色体上均匀分布的SSR核心引物进行分子标记辅助背景选择，引物信息详见（Wang F et al, 2011）。

1.2.2 PCR扩增反应体系

反应液总体积20μl，其中包括：DNA模版4μl、ddH_2O 12.35μl、10×Buffer（2.5mmol/L $MgCl_2$）2μl、10mmol/L dNTP 1.2μl、5U/μl Taq DNA Enzyme 0.2μl、Primer（F+R）0.25μl。SSR扩增反应程序为：94℃预变性5min，1个循环；94℃变性40s，60℃退火35s，72℃延伸45s，共35个循环；72℃延伸10min；4℃保存。PCR扩增在Veriti 384

Well Thermal Cycler 上完成。

1.2.3 扩增产物检测

根据引物荧光颜色差异和扩增产物片段大小，利用 AB 3730XL DNA 分析仪对 PCR 产物进行毛细管荧光多重电泳。96 孔电泳板每孔分别加入 4μl 扩增产物的混合物、7.92μl 甲酰胺和 0.08μl GS3730-500 分子量内标（Applied Biosystems，USA）。95℃变性 5min，4℃保存 10min，2 500rpm 离心 1min 后，于 AB 3730XL DNA 分析仪上进行电泳。15kV 预电泳 2min；15kV 电泳 20min。用 Date Collection v1.0 软件收集原始数据。

利用 Genemarker 软件对收集的原始数据进行分析，统计每个单株的基因型。

1.2.4 遗传相似度

计算公式：$\{1-[$差异等位基因数$/(2\times$比较总位点数$)]\}\times 100\%$

其中，与京 724 PCR 扩增图谱谱带不一致的等位基因数目为差异等位基因数；比较总位点数为 40。

1.3 花粉育性鉴定

每份供试材料散粉期随机选取 5 株，取雄穗上、中、下 3 个部位的小穗用卡诺固定液固定。用镊子剥取花药置于干净的载玻片上，挤出花粉粒后，用 1%的 I_2-KI 染色，于 OLYMPUS 倒置荧光显微镜 4×10 倍下观察。每个小穗观察 3 个视野，观察花粉染色情况。

2 结果与分析

2.1 分子标记辅助京 724 雄性不育回交转育

2.1.1 分子标记辅助背景选择引物筛选

利用 40 对 SSR 核心引物对 MD32 和京 724 进行扩增，经荧光毛细管电泳检测从中筛选到了 24 对在二者之间表现多态性的 SSR 引物（表 1）。

表 1 MD32 和京 724 间存在多态性的 SSR 引物

Table 1 Polymorphic SSR markers between MD32 and Jing724

编号 Serial No.	引物名称 Primer name	染色体位置 Chr. position	编号 Serial No.	引物名称 Primer name	染色体位置 Chr. position
1	bnlg439w1	1.03	13	bnlg249k2	6.01
2	bnlg1671y17	1.10	14	bnlg1702k1	6.05
3	phi96100y1	2.00	15	umc1545y2	7.00
4	umc2007y4	2.04	16	umc2160k3	7.01
5	bnlg1940k7	2.08	17	umc1936k4	7.03
6	bnlg1520k1	2.09	18	umc1125y3	7.04
7	umc2105k3	3.00	19	bnlg2235y5	8.02
8	phi053k2	3.05	20	umc2084w2	9.01
9	phi072k4	4.01	21	phi065k9	9.03
10	bnlg2291k4	4.06	22	umc1492y13	9.04
11	umc2115k3	5.02	23	umc1231k4	9.05
12	bnlg161k8	6.00	24	umc1506k12	10.05

2.1.2 BC_1 代分子标记辅助背景选择

以 MD32（非轮回亲本）做供体、京 724（轮回亲本）做受体进行雄性不育性状回交转育。田间观察发现，F_1 代植株花药均不外露，不育性表现彻底。从种植的 2000 株 BC_1 代群体中选择雄花完全不育、植株性状（如株高、穗位高、株型、雄穗分枝数和果穗性状等）与京 724 接近的 616 个 BC_1 代单株挂牌标记、提取叶片 DNA。利用在回交亲本间具有多态性的 24 对背景选择引物对 BC_1 代入选单株进行毛细管电泳检测，统计扩增谱带与轮回亲本京 724 存在差异的等位基因数目，计算 BC_1 代各单株与京 724 间的遗传相似度。结果表明，616 个单株与京 724 的遗传相似度在 75%~95%；选取遗传相似度 ≥92.5% 的单株与京 724 继续回交。遗传相似度排在前 10 位的单株检测结果详见表 2。

表 2 与京 724 遗传相似度最高的前 10 份 BC_1 代单株
Table 2 The top 10 BC_1 plants with highest genetic similarity to Jing724

单株编号 Plant No.	引物总数 Total number of primers	差异等位基因数 number of differential alleles	遗传相似度（%） Genetic similarity
B97	40	4	95.00
B120	40	4	95.00
B202	40	4	95.00
B166	40	5	93.75
B257	40	5	93.75
B300	40	5	93.75
B331	40	5	93.75
B393	40	5	93.75
B411	40	5	93.75
B490	40	5	93.75

2.1.3 BC_2 代分子标记辅助背景选择

从种植的 1 500 株 BC_2 代群体中选择雄花完全不育，植株性状与京 724 接近的 300 个 BC_2 代单株挂牌标记，提取叶片 DNA。利用 BC_2 代单株对应的母本 BC_1 代与京 724 扩增带型不同的 SSR 引物对其进行 PCR 扩增；以京 724 为对照比较扩增谱带，计算遗传相似度。结果表明，300 个单株与京 724 的遗传相似度在 95%~100%。

2.1.4 S 京 724 的获得

选取遗传相似度为 100% 的单株与京 724 继续回交得到 BC_3 代群体，田间表现不育性彻底、植株性状与京 724 基本一致，即为京 724 遗传背景的 S 型胞质雄性不育系，命名为 S 京 724。

2.2 S 京 724 花粉活力检测

利用 I_2-KI 染色法检测京 724 和不育系 S 京 724 的花粉活力（图 1）。正常胞质京 724 的花粉粒饱满呈圆形，由于积累了较多的淀粉，经 I_2-KI 染色后呈蓝色；S 京 724 所有待测单株的镜检小穗花粉粒均表现败育，不含淀粉或淀粉积累较少，呈畸型，经 I_2-KI 染色后呈黄褐色，表明 S 京 724 的花粉败育彻底，属于无花粉型的雄性不育系。

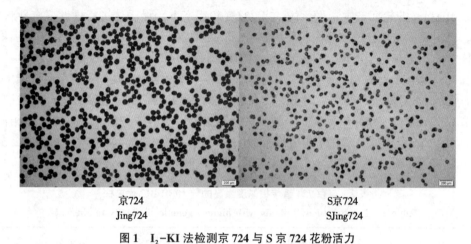

京724　　　　　　　　　　　S 京724
Jing724　　　　　　　　　　SJing724

图 1　I$_2$-KI 法检测京 724 与 S 京 724 花粉活力

Figure 1　Pollens of Jing724 and SJing724 tested by I$_2$-KI

3　结论与讨论

细胞质雄性不育系可划分为 T、S 和 C 型。S 型细胞质不育系是 3 种类型中最大的一个组，已有研究证明，昌 7-2 等黄改系材料是其天然强恢复系。国内优良玉米杂交种的父本多属黄改种质，因此充分利用 S 型胞质不育系可以节省恢复系的转育工作，是快速实现不育化制种研究和应用的有效途径。MD32 是北京市农林科学院创制的综合性状优良的 S 型细胞质雄性不育材料。经多年多点观察，其败育彻底、育性稳定，具有较大的利用潜力。本研究利用该系作为雄性不育性状的供体将京科 968 母本京 724 转育成了 S 型细胞质雄性不育系；后续研究结果证实，父本京 92（黄改系材料）是 S 京 724 不育系的强恢复系，节省了恢复性的转育工作。

刘纪麟、马春红等研究表明，雄性不育系转育过程中仅通过表型选择，一般需回交至 BC$_6$ 或 BC$_7$ 代时，植株性状才能与轮回亲本基本一致。本研究利用 40 对 SSR 核心引物进行分子标记辅助背景选择，仅回交 3 代即获得了不育性彻底、植株性状与京 724 基本一致的雄性不育材料，大幅提高了转育效率。同时也证明，利用多态性高、染色体上均匀分布的 40 对核心引物进行分子标记辅助背景选择是行之有效的。通过背景选择，从 616 个 BC$_1$ 代单株中即可筛选到与京 724 遗传相似度达 95% 的个体。可见，在回交早代进行分子标记辅助选择，只要能够保证一定的群体数量，即可获得比较理想的选择效果。

目前，国内科研育种单位开展分子标记辅助前景选择面临的瓶颈主要是可用于目标性状鉴定的标记比较少，且集中在少数抗病和品质性状上。因此，分子标记辅助背景选择是现阶段分子标记在育种实践中发挥作用的重要方面，可以显著加快新种质材料的创制速度和效率，具有广泛的推广应用价值。

参考文献（略）

本文原载：玉米科学，2016，24（1）：33-36，42

SSR 和 SNP 标记在玉米分子标记辅助背景选择中的应用比较

宋 伟　赵久然　王凤格　田红丽　葛建镕　王元东　赵衍鑫

（北京市农林科学院玉米研究中心/玉米DNA指纹及分子育种北京市重点实验室，北京　100097）

摘　要：利用SSR和SNP两种标记对H2321/J076/J076的BC_1F_1代群体进行分子标记辅助背景选择，比较两种标记在分子标记辅助背景选择上的应用效果。经筛选表明，40对SSR引物中在回交亲本间具有多态性的引物有6对。利用多态性引物对BC_1代群体进行毛细管电泳检测结果表明，157个单株轮回亲本背景回复率在50%~100%，背景回复率为100%的单株有1株，背景回复率91.67%的单株有14株。利用MaizeSNP3072芯片分析后发现，104个多态性SNP位点可以将157株个体进行精细排序，背景回复率在51.92%~93.75%。比较SSR和SNP标记的背景选择结果，两种标记呈中等程度相关。在回交早代将SSR和SNP标记结合使用进行背景选择，可以实现大群体严选择，提高选择效率。

关键词：玉米；分子标记辅助背景选择；SSR；SNP

Comparison of SSR and SNP Markers in Maize Molecular Marker Assisted Background Selection

Song Wei　Zhao Jiuran　Wang Fengge
Tian Hongli　Ge Jianrong　Wang Yuandong　Zhao Yanxin

(*Maize Research Center, Beijing Academy of Agriculture and Forestry Sciences, Beijing Key Laboratory of Maize DNA Fingerprinting and Molecular Breeding, Beijing 100097, China*)

Abstract: SSR and SNP were used for the molecular marker assisted background selection of H2321/J076/J076 BC_1F_1 population. Six pairs of polymorphic primers screened between donor and receptor parents from 40 pairs of SSR primers were analyzed in BC_1 plants by capillary electrophoresis. The result showed that, proportion of recurrent parent genome (PRPG) of 157 BC_1 plants was from 50% to 100%, only one individual with PRPG 100% and 14 plants with 91.67%. Using MaizeSNP3072 microarray analysis, 157 plants could be sorted finely with 104 polymorphic SNP loci,

---

基金项目：国家科技支撑项目（2014BAD01B09）、北京市农林科学院科技创新能力建设专项（KJCX20140202）、北京市科技计划课题（Z141100002314016）

作者简介：宋伟（1979—　），女，山东人，副研究员，博士，从事玉米分子育种工作。Tel：010-51503983　E-mail：songwei1007@126.com

赵久然为本文通讯作者

and PRPG of BC$_1$ plants was from 51.92% to 93.75%. Comparison of SSR and SNP in the background selection showed that the two types of markers were correlative moderately. So a combination of SSR and SNP markers in the backcross early generation was suggested to improve the background selection efficiency, so as to realize large population and strict selection.

Key words: Maize; Molecular marker assisted background selection; SSR; SNP

玉米常规育种是通过表型观察选择目标个体，不但依赖育种家的经验，而且经常受到基因型和环境互作的影响，很难选择到理想的基因型。分子标记辅助选择是将分子标记应用于作物品种改良过程中进行选择的一种辅助手段，主要包括对目标性状的前景选择和对遗传材料的背景选择。在回交转育过程中利用分子标记辅助背景选择，可以显著加快背景回复的速度，缩短育种时间。

适用于背景选择的分子标记需要满足简单、快捷和低成本3个基本条件。SSR和SNP标记是目前玉米遗传育种中应用最为广泛的两种分子标记，二者各有优缺点。以PCR为基础的SSR标记具有多态性高、操作简单、重复性好等优点，但是难以实现位点的高通量；以序列为基础的SNP标记分布密度高、遗传稳定性好、易于实现高通量、自动化分析，但是检测成本相对较高。本研究分别利用SSR和SNP标记对玉米回交转育群体材料进行遗传背景分析，比较两种标记在分子标记辅助背景选择上的应用效果，为今后开展相关工作提供参考。

1 材料与方法

1.1 供试材料

玉米自交系J076配合力高、综合性状优良，但存在穗位偏高、抗倒性稍差的缺点。以低穗位且抗倒性好、与J076亲缘关系较近的自交系H2321作为供体亲本，以J076为轮回亲本，通过回交转育改良J076的抗倒性。选用经田间表型选择初步入选的BC$_1$F$_1$代、BC$_2$F$_1$代回交群体为研究材料。

1.2 DNA提取

H2321与J076室内沙培发苗，3~5叶期提取DNA。BC$_1$F$_1$代田间单株挂牌标记取样。采用CTAB法提取基因组DNA，去除RNA。琼脂糖电泳检测DNA的完整性，利用Nano-Drop2000微量紫外分光光度计检测DNA的质量和浓度。

1.3 SSR标记

1.3.1 引物

利用40对多态性高、在染色体上均匀分布的SSR核心引物进行分子标记辅助背景选择，引物信息参照文献（Wang F et al, 2011）。

1.3.2 PCR扩增

PCR总反应体系为20μl，DNA模版4μl，ddH$_2$O 12.35μl，10×Buffer（2.5mmol/L MgCl$_2$）2μl，10mmol/L dNTP 1.2μl，5U/μl Taq DNA Enzyme 0.2μl，Primer（F+R）0.25μl。PCR反应程序为，94℃预变性5min，94℃变性40s，60℃退火35s，72℃延伸45s，共进行35个循环；72℃延伸10min，4℃保存。PCR扩增在Veriti 384 Well Thermal Cycler上完成。

1.3.3 扩增产物检测

根据引物荧光颜色差异和扩增产物片段大小，利用 AB 3730XL DNA 分析仪对 PCR 产物进行毛细管荧光多重电泳。96 孔电泳板每孔分别加入 4μl 扩增产物的混合物、7.92μl 甲酰胺和 0.08μl GS3730-500 分子量内标（Applied Biosystems，USA）。95℃变性 5min，4℃冷却 10min，2 500r/min 离心 1min 后，于 DNA 分析仪上进行电泳。用 Date Collection v1.0 软件收集原始数据。利用 Genemarker 软件对收集的原始数据进行分析，统计每个单株的基因型。

1.4 SNP 标记

1.4.1 位点信息

利用 MaizeSNP3072 芯片对供试材料 DNA 进行 SNP 基因分型，该芯片包含 3 072 个均匀分布于玉米全基因组的 SNP 位点，位点信息详见 http：//www.illumina.com.cn 和 http：//www.panzea.org 网站。本研究中选用的 1 202 个 SNP 位点是根据分型效果、数据缺失率等从上述 3 072 个位点中优选出来的。

1.4.2 芯片检测

基于 Illumina 公司的 GoldenGate 技术平台对 MaizeSNP3072 芯片进行检测。待提取的基因组 DNA 与激活的生物素充分结合后，通过离心沉淀使 DNA 纯化；将探针与纯化的目的 DNA 链杂交，漂洗杂交产物除去非特异结合试剂或过量试剂后进行延伸连接反应；延伸后产物经漂洗变性后作为模板进行 PCR 扩增，PCR 产物与磁珠结合后过滤纯化；将纯化的 PCR 产物与芯片杂交，杂交结束后进行芯片清洗、真空抽干；风干后的芯片立刻在 iScan 仪上进行扫描，最后利用 GenomeStudio 软件进行数据分析。试验流程参照 GoldenGate© Genetyping Assay 试验操作指南。

1.5 轮回亲本背景回复率（proportion of recurrent parent genome，PRPG）

PRPG=［1-差异等位基因数/（2×有效位点数）］×100%；式中，差异等位基因数是指 BC_1 单株与 J076 表现不一致的等位基因数目；有效位点数是指在双亲中存在多态性的位点总数。

1.6 田间种植及性状调查

于北京市农林科学院玉米研究中心小汤山试验基地种植自交系 J076、H2321 和 BC_1F_1 代，BC_2F_1 代种植在海南三亚南滨育种基地，行长 5m、行距 60cm、株距 25cm。其中入选单株及果穗测量株高、穗位高、穗长、穗粗、穗行数、行粒数、单穗粒重、百粒重等农艺及产量性状。以 J076 作为对照，随机选取均匀一致的 10 个单株进行上述性状的调查。

2 结果与分析

2.1 SSR 标记辅助背景选择

2.1.1 多态性引物筛选

利用 40 对 SSR 核心引物对 H2321 和 J076 的 DNA 样品进行扩增，经荧光毛细管电泳检测从中筛选到 6 对在二者之间表现多态性的 SSR 引物（表1）。

表 1 H2321 和 J076 间存在多态性的 SSR 引物

Table 1 Polymorphic SSR markers between H2321 and J076

编号 Serial No.	引物名称 Primer name	染色体位置 Chr. position
1	umc1147y4v	1.07
2	bnlg1940k7	2.08
3	bnlg1520k1	2.09
4	phi053k2	3.05
5	umc1999y3	4.09
6	umc1545y2	7.00

2.1.2 BC₁ 代分子标记辅助背景选择

从种植的 H2321/J076//J076 BC₁ 代群体中选择穗位高较低、植株性状与 J076 接近的 157 个 BC₁ 代单株挂牌标记，提取叶片 DNA。利用在回交亲本间具有多态性的 6 对背景选择引物对 BC₁ 代单株进行毛细管电泳检测，统计扩增谱带与轮回亲本 J076 存在差异的等位基因数目，计算 BC₁ 代各单株轮回亲本背景回复率。结果表明，157 个单株背景回复率集中在 50%、58.33%、66.67%、75%、83.33%、91.67%、100%这 7 个数值上，背景回复率为 66.67%的单株数量最多，为 45 株；背景回复率为 100%的单株数量最少，仅有 1 株（图 1）。背景回复率≥91.67%的 15 个单株见表 2。

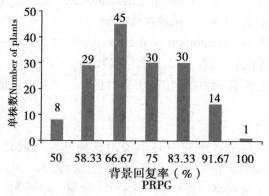

图 1 基于 SSR 标记的 BC₁F₁ 代单株背景回复率分布

Figure 1 PRPG of BC₁F₁ plants based on SSR markers

表 2 背景回复率≥91.67%的 BC₁ 代单株（基于 SSR 标记）

Table 2 The BC₁ plants with PRPG above 91.67% based on SSR markers

单株编号 Plant No.	有效标记数（个） Number of effective Primers	差异等位基因数（个） Number of differential alleles	背景回复率（%） PRPG
83	6	0	100.00
7	6	1	91.67
24	6	1	91.67
34	6	1	91.67

(续表)

单株编号 Plant No.	有效标记数(个) Number of effective Primers	差异等位基因数(个) Number of differential alleles	背景回复率(%) PRPG
47	6	1	91.67
56	6	1	91.67
74	6	1	91.67
81	6	1	91.67
113	6	1	91.67
115	6	1	91.67
119	6	1	91.67
123	6	1	91.67
128	6	1	91.67
143	6	1	91.67
144	6	1	91.67

2.2 SNP标记辅助背景选择

利用MaizeSNP3072芯片对157株BC_1代单株及双亲进行SNP基因分型，在1 202个分析位点中有104个SNP位点在双亲间存在多态性，统计分型结果与轮回亲本J076存在差异的等位基因数目，计算BC_1代各单株的轮回亲本背景回复率。结果表明，157个单株的背景回复率在51.92%~93.75%，其中，具有相同回复率的单株数最多不超过5个，即利用1 202个SNP位点可以将157个单株轮回亲本背景回复率进行更加精细的排序。图2显示，BC_1F_1代单株背景回复率从高到低的分布情况，回复率≥86.54%的15个单株见表3。

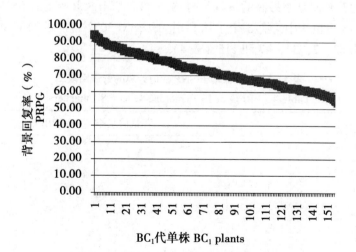

图2 基于SNP标记的BC_1F_1代单株背景回复率分布
Figure 2 PRPG of BC_1F_1 plants based on SNP markers

表3 背景回复率≥86.54%的 BC₁ 代单株（SNP 标记）
Table 3 The BC₁ plants with PRPG above 86.54% based on SNP markers

单株编号 Plant No.	有效标记数（个）Number of effective Primers	差异等位基因数（个）Number of differential alleles	背景回复率（%）PRPG
83	104	13	93.75
11	104	16	92.31
115	104	17	91.83
116	104	18	91.35
7	104	21	89.90
23	104	22	89.42
93	104	22	89.42
47	104	23	88.94
34	104	25	87.98
49	104	26	87.50
26	104	26	87.50
96	104	26	87.50
140	104	27	87.02
25	104	28	86.54
157	104	28	86.54

2.3 SSR 和 SNP 标记辅助背景选择结果比较

将 157 个 BC₁ 代单株分别基于 SSR 标记和 SNP 标记计算的背景回复率进行相关性分析，相关系数为 0.475，说明二者存在中等程度相关。由表2、表3可以看出，利用两种标记方法检测的背景回复率最高的单株编号均为 83。基于 SSR 标记检测结果，背景回复率≥91.67%的 15 个单株中，有 5 个单株利用 MaizeSNP3072 检测的回复率排在前 10 位，单株编号分别为 83、7、34、47 和 115（表4）。

表4 15 个单株基于 SNP 标记计算的背景回复率
Table 4 PRPG of 15 Plants based on SNP markers

| 单株编号 Plant No. | 背景回复率（%）PRPG | |
	SSR	SNP
83	100.00	93.75
7	91.67	89.90
24	91.67	65.87
34	91.67	87.98
47	91.67	88.94
56	91.67	78.85
74	91.67	83.17

(续表)

单株编号 Plant No.	背景回复率（%）PRPG	
	SSR	SNP
81	91.67	78.37
113	91.67	81.25
115	91.67	91.83
119	91.67	77.88
123	91.67	82.69
128	91.67	83.65
143	91.67	81.73
144	91.67	74.52

2.4 分子标记辅助背景选择的效果

选取 SSR 与 SNP 标记检测背景回复率均表现最高的第 83 号单株，与 J076 继续回交。BC_2F_1 代种子田间种植后，选取穗位偏低、其他表型性状与 J076 基本一致的 10 个植株进行株高、穗位高、穗长、穗粗、穗行数、行粒数、单穗粒重、百粒重等性状的调查。结果表明，入选单株的平均表现与 J076 相比，除穗位高明显降低外，二者在株高等其他性状上的表现在均无显著性差异。

3 结论与讨论

传统的回交育种方法，往往需要至少 6 个世代才能使轮回亲本基因组比率达到 99% 以上。利用分布在全基因组的分子标记在后代群体中对遗传背景进行选择能够快速回复轮回亲本的遗传组成。夏军红等研究表明，采用同样数量的分子标记所得出的遗传背景回复率的变异系数，在 BC_1F_1 代总是比 BC_2F_1 代的要大，出现极端个体的可能性更高，说明在 BC_1F_1 代进行分子标记辅助选择效率比 BC_2F_1 代更高。随着世代数增加，背景选择效率会逐渐降低。

RAPD 标记是显性标记，且重复性差，AFLP 标记操作复杂，费时耗材。本研究选用 SSR 和 SNP 两种标记方法，SSR 标记采用荧光毛细管多重电泳检测，与前人利用 PAGE 电泳进行 SSR 标记辅助背景选择相比，检测效率大大提高；SNP 标记利用芯片平台可以实现位点的高通量检测，由于位点不能自由组合，灵活性稍差。

本研究供体亲本 H2321 和轮回亲本 J076 由于亲缘关系较近，在 40 对 SSR 引物中仅筛选到 6 对多态性引物。利用这 6 对引物对 BC_1F_1 代 157 个单株进行基因型分析，获得 1 株轮回亲本背景回复率达到 100% 的个体，14 个单株背景回复率为 91.67%。因此，对于双亲亲缘关系较近的回交群体，需要适当增加多态性引物的数量，才能实现对回交个体的精细选择。MaizeSNP3072 芯片分析结果表明，利用 104 个双亲间存在多态性的 SNP 位点可以对 157 个单株进行更加精细的排序，其中，背景回复率最高的为 93.75%。但是由于芯片上的位点是固定的，无法针对特定的群体筛选多态性位点进行遗传背景检测，而且对于单个样品来说，芯片的检测成本相对较高，直接利用芯片对整个群体的各个单株进行背景选择的费用高，很难实现广泛普及。比较 SSR 和 SNP 标记的背景选择结果，两种标记呈中等程度相关，未达到强相关，可能是因为本研究使用的多态性 SSR 引物数目较少，对遗传背景分析的精确性不够。

  为达到更为理想的分子标记辅助背景选择效果，可以将 SSR 和 SNP 两种标记方法结合起来使用。对于特定的回交 BC_1 代群体，首先利用适量 SSR 标记对所有表型选择入选个体进行初步筛选，然后利用 SNP 芯片对少数个体进行精细选择，这样既可以实现选系过程中的大群体，又能够做到严选择。计算机模拟研究证实，当群体很大时，利用分子标记辅助选择能减少误差，提高选择效率，选择效率随着群体增加而增大。因此，如果在 BC_1 代大群体中可以选择到遗传背景基本回复到轮回亲本的个体，在以后的回交世代中可以不进行分子标记辅助，从而简化育种程序，提高分子育种的可操作性。

参考文献（略）

本文原载：玉米科学，2016，24（3）：57-61

基于SNP标记的玉米容重QTL分析

许理文[1,2]　段民孝[2]　田红丽[2]　宋　伟[2]
王凤格[2]　赵久然[2]　刘保林[2]　王守才[1]

(1. 中国农业大学农学与生物技术学院，北京　100193；2. 北京市农林科学院玉米研究中心/玉米DNA指纹及分子育种北京市重点实验室，北京　100097)

摘　要：以240个DH系为作图群体，利用SNP芯片对DH群体进行基因型分析，构建连锁图谱，对DH群体进行容重性状鉴定。结果表明，采用复合区间作图法进行QTL定位分析，在4个环境下共检测到5个控制玉米容重的QTL，联合贡献率为23.61%；位于第1染色体的qTw1-1介于SNP标记PZE-101032246与SYN13192之间，位于第9染色体的qTw9-1介于SNP标记PZE-109011840和SYN6085之间，分别解释11.9%和7.8%的表型变异，表明这些区域可能包含调控玉米容重性状关键基因。

关键词：玉米；容重；数量性状位点；单核苷酸多态性；加倍单倍体

QTL Identification for Test Weight Based on SNP Mapping in Maize

Xu Liwen[1,2]　Duan Minxiao[2]　Tian Hongli[2]　Song Wei[2]
Wang Fengge[2]　Zhao Jiuran[2]　Liu Baolin[2]　Wang Shoucai[1]

(1. College of agriculture and biotechnology, China Agricultural University, Beijing, 100193;
2. Maize Research Center Beijing Academy of Agriculture & Forestry Sciences, Beijing Key Laboratory of Maize DNA Fingerprinting and Molecular Breeding, Beijing, 100097)

Abstract: A mapping population consisting of 240 DH lines was constructed, and significant genotypic variation for maize test weight was observed in 4 environments. Based on the genetic map containing, QTL for maize test weight were detected by composite interval mapping. Five QTLs were identified on chromosomes 1, 4, 8 and 9, and together explained 23.61% of the phenotypic variation. qTw1-1 was detected on chromosome 1, locating in SNP of interval PZE-101032246-SYN13192, qTw9-1 was detected in SNP of interval PZE-109011840-SYN6085 on chromosome 9,

---

基金项目：国家科技支撑项目（2011BAD35B09，2014BAD01B09）、北京市农林科学院科技创新能力建设专项（KJCX20140202）、"863"计划"玉米产量功能基因研究与克隆"（2012AA10A305）

作者简介：许理文（1981—　），男，河北邯郸人，博士，从事玉米遗传与育种研究。Tel：010-51503987，E-mail：xulw0408@126.com

赵久然和王守才为本文通讯作者，E-mail：maizezhao@126.com；E-mail：wangshoucai678@sina.com with explained 11.9% and 7.8% of the phenotypic variation respectively. The results showed that some key genes for test weight are possibly contained in these regions.

Key words：Maize；Test Weight；QTL；SNP；Doubled haploid（DH）

容重是玉米籽粒在单位容积内的重量，容重能够真实地反映玉米的成熟度、完整度和使用价值。2001年，我国将纯粮率定等改为容重定等。容重定等适应了粮食由数量扩张型向质量效益型转变和加入WTO与国际接轨的需要，作为国际贸易中质量定级的重要因素，关系到等级和种植效益等方面，引起了国内外育种和栽培工作者的重视。前人对容重的大量研究主要集中在容重与农艺性状、物理性状、营养成分及外界环境之间的关系，关于容重遗传机制的研究较少。Pixley研究了燕麦优良种质容重的遗传变异性，结果表明，容重有较高的遗传力，容重有75%源自一般配合力（GCA），容重基本不具有杂种优势。

自Paterson等利用分子标记定位QTL的技术，已定位了大量玉米农艺性状相关的QTL。Beavis等利用B73和Mo17构建的$F_{2:4}$群体和96个RFLP分子标记，定位了4个与玉米容重相关的QTL，分布在玉米的第3、5、7、10号染色体上。Ajmone-Marsan等利用B73和A7构建的$F_{2:3}$群体和72个RFLP标记定位到6个和玉米容重相关的QTL，分布在玉米的第1、2、3、5、9号染色体上。丁俊强等利用郑58和昌7-2构建的$F_{2:3}$群体和188个SSR分子标记，在玉米的1、2、3、4、5号染色体上各定位1个QTL，能够解释玉米容重25.2%的遗传变异。

目前，发掘QTL主要是基于SSR标记或RFLP标记等传统分子标记进行定位，其所构建的连锁图谱平均图距10~20cM，标记密度很低，大量的大间隔区段无多态性标记覆盖，定位的QTL置信区间较大，而且这些标记的基因型分析过程复杂、自动化程度低，已远远不能满足高密度遗传连锁图谱构建和QTL精确定位的需求。单核苷酸多态性（SNPs）是广泛存在于基因组中的一类DNA序列变异，是目前为止分布最为广泛、存在数量最多，且标记密度最高的一种遗传多态性标记。研究发现，在玉米基因组上平均每48~104bp就有1个SNP。随着高通量SNP芯片和测序技术进步，自动化程度更高的SNP标记将迅速取代SSR、STS、RFLP等传统标记，广泛应用于植物遗传连锁图谱构建、生物物种起源和亲缘关系分析以及生物多态性等方面的研究。

玉米容重是复杂性状，受多基因控制和环境影响，多年多点的表型鉴定对容重QTL的定位至关重要。本研究利用高密度SNP连锁图谱和基于优良玉米单交种先玉335构建的双单倍体（DH）群体，通过两年两点的容重表型鉴定，对玉米容重QTL进行复合区间作图分析，以发掘出新的容重QTL，为进一步认识玉米容重的遗传机制提供参考。

1 材料和方法

1.1 试验材料

DH群体由北京市农林科学院玉米研究中心采用京科诱006诱导优良玉米单交种先玉335产生单倍体，经秋水仙素处理加倍，获得DH系群体。

1.2 试验方法

1.2.1 田间试验

2010年和2011年春，DH系群体及其双亲PH6WC和PH4CV分别在北京市农林科学院小汤山试验基地（E1 和 E3 环境）和农安试验基地种植（E2 和 E4 环境），试验采用随机区组设计，3 次重复，双行区，行长 5m，行距 60cm，株距 25cm，田间管理同大田。开放授粉，成熟后每小区收获 30 个果穗，在自然条件下晾晒至水分含量小于 14%，小区果穗混合脱粒，采用玉米容重国家标准 GB 1353—1999 推荐的 GHCS-1000 型容重器按照国标要求进行玉米容重的测定，每个小区的玉米容重随机测定 3 次，试验允许误差不超过 3g/L。

1.2.2 数据分析

利用 SPSS16.0 软件的 GLM 程序进行单环境、全部环境联合的方差分析，并按照 Hallauer 和 Miranda（1988）的公式 $h^2 = \sigma_g^2 / (\sigma_g^2 + \sigma_{ge}^2/e + \sigma_\varepsilon^2/re)$ 计算广义遗传力，其中 σ_g^2、σ_{ge}^2 和 σ_ε^2 分别为遗传方差、基因型与环境互作的方差和误差方差，r 为重复数；e 为环境数。

1.2.3 连锁图谱构建

DH 群体基因型分析采用 MaizeSNP3072 芯片按照 Illumina 公司提供的操作指南，在北京市农林科学院玉米研究中心完成。对所有 SNP 标记的孟德尔预期分离比（1∶1）进行卡方检测，采用 Joinmap4.0 软件进行遗传连锁图谱构建，LOD＝3.0 作为判断两个标记是否连锁的阈值。采用 Haldane's 函数换算成图距，用缩写 Chr. 加染色体序号命名染色体。

1.2.4 QTL 定位

采用 WinQTL Cartographer v2.5 软件及复合区间作图法（Composite Interval Mapping，CIM），对玉米容重在 2010—2011 两个年度的表型观测数据分别进行 QTL 检测，QTL 定位过程中，每个环境下的每个性状均单独在 $\alpha = 0.05$ 显著水平下进行 500 次排列检验，确定性状 QTL 的 LOD 阈值，窗口大小默认为 10cM，背景标记 5 个，采用正向-反向逐步回归法控制背景，步长设为 1cM。然后根据各性状的阈值进行 QTL 定位，并依据 Stuber 等提出的判别标准确定 QTL 作用方式。QTL 命名基本遵照 McCouch 等方法并稍作改动：q 加性状英文缩写名称（首字母大写）加连锁群的序号，同一连锁群上多个 QTL 用 1、2、3 来表示，且本文中 QTL 名称均用斜体表示。例如，"$qTw1-2$" 表示在连锁群 1 上检测到的第 2 个关于容重（Test weight）的数量性状基因位点。

2 结果与分析

2.1 容重性状的田间表现

两个亲本（PH6WC 和 PH4CV）及其 DH 群体的容重性状在 4 个环境的表现见表 1。两个亲本的容重有较大差异，亲本 PH6WC 的容重较高，籽粒为硬粒型，胚乳含角质多；亲本 PH4CV 籽粒为半马齿型，胚乳含角质相对较少，说明籽粒胚乳的角质含量可能对籽粒的容重有影响。Lee 等研究发现，粒重与胚乳角质化程度相关，籽粒体积相同时，角质胚乳占的比例越大，粒重越大。DH 群体容重 4 个环境的方差分析（ANOVA）结果见表 1，DH 群体的容重在 4 个环境下的遗传力都高于 0.7；4 个环境的联合方差分析增加了基因型与环境的互作引起遗传力下降，但仍然高于 0.6，表明玉米容重性状主要受遗传因素控制，与前人的报道一致，适合进行 QTL 分析。

表1 DH群体及亲本表型值的统计分析
Table 1 Statistical analysis of DH population and parental phenotypic value

年份 Year	环境 Env	PH6WC Mean±SD	PH4CV Mean±SD	DH population Mean±SD	σ_g^2	σ_{ge}^2	σ_e	H^2
2010	E1（北京）	775.3±7.5	741.5±5.4	753.6±36.4	479.1		177.2	0.730
	E2（农安）	768.9±6.8	738.2±6.1	749.2±31.7	411.7		163.5	0.716
2011	E3（北京）	771.4±8.1	734.7±7.4	751.4±38.7	394.4		161.1	0.712
	E4（农安）	774.8±7.6	733.5±6.3	749.5±33.6	417.6		175.6	0.704
联合		773.8±7.2	736.7±6.8	751.8±34.6	617.2	233.8	127.4	0.631

表2 运用复合区间法4个环境条件下检测到的影响玉米容重的QTL
Table 2 QTL for test weight determined by CIM using phenotypic data obtained in four environmental conditions

QTL	染色体 Chrom	位置 Pos	左标记 Left marker	右标记 Right marker	北京2010 LOD	Add	R^2%	农安2010 LOD	Add	R^2%	北京2011 LOD	Add	R^2%	农安2011 LOD	Add	R^2%
qTw1-1	1	56	PZE-101032246	SYN13192	6.33	4.17	9.56	4.89	3.51	7.73	5.59	4.67	9.86	5.27	3.05	11.94
qTw9-1	9	44	PZE-109011840	SYN6085	2.42	-2.16	7.68	4.17	-2.98	5.84	3.02	-2.62	6.77	4.10	-3.03	7.83
qTw1-2	1	78	PZE-101069879	SYN35005										3.03	2.78	3.81
qTw4-1	4	79	PZE-104114286	PZE-104114501				3.89	2.10	3.55						
qTw8-1	8	94	SYN13209	SYN4439							3.07	-1.08	4.89			

2.2 连锁图谱构建

利用MaizeSNP3072芯片对DH群体进行基因型分析，获得1 337个在先玉335的双亲间有多态的SNP数据，对SNP进行卡方检测（$P<0.05$），发现516个SNP表现偏分离，占38.6%。这些偏分离SNP中有285个偏向母本PH6WC，占55.2%；231个偏向父本PH4CV，占44.8%。偏分离SNP可以分为两类。一类为少量、独立、随机分布于染色体的偏分离SNP，发现17个这样的SNP；另一类为广泛分布于不同染色体上的偏分离SNP呈连续聚集，而且偏分离方向一致，形成染色体上的偏分离热点区（Segregation Distortion Region，SDR）。利用Joinmap4.0软件，"DH1"模型构建玉米遗传图谱，将1 337个SNP的数据加载到Joinmap4.0，对SNP进行"Similarity of loci"分析，将相似度为大于0.99的成对标记仅保留一个（Joinmap随机去除），由于作图群体不够大，相似度小于0.99的SNP之间没有交换，对连锁图谱构建没有贡献，去除了325个SNP；运用"groupings"命令，当LOD值为8时SNP被连锁到玉米的10条染色体上，运行"map"完成连锁图谱构建，发现第一类偏分离SNP，如PZE-107136612和SYN6674等12个SNP对连锁图谱的干扰比较大（"Similarity of loci"分析去除了5个），使连锁图谱上SNP的顺序与其在B73基因组序列上的位置顺序差异较大，说明这类偏分离SNP对遗传连锁图谱构建影响较大，去除该类SNP后重复以上操作，构建成包含1 000个SNP的连锁图谱，SNP的顺序与其在B73基因组上的排列有较高的一致性；连锁图谱覆盖玉米基因组全长的1 489.3cM，平均图距为1.5cM；其中，第1染色体最长，为208.9cM，拟合SNP标记最多，为154个；第

10染色体最小，为111.2cM，包含的SNP标记最少，为65个。

2.3 QTL分析

采用复合区间作图法对4个环境下容重性状分别进行QTL分析，全基因组扫描北京点检测到3个增效基因QTL，农安点检测到4个增效基因QTL，共检测到5个增效基因QTL，其中3个来自母本PH6WC，2个来自父本PH4CV，QTL标记区间、LOD值、效应值以及贡献率列于表2。QTL分布在第1、4、8、9染色体上（图1），5个QTL共能解释容重表型变异的23.61%，单个贡献率为3.6%~11.9%。位于第1染色体的 qTw1-1 介于SNP标记PZE-101032246与SYN13192之间，贡献率最大，为11.9%，其基因效应来自母本PH6WC；位于第9染色体的 qTw9-1 介于SNP标记PZE-109011840和SYN6085之间，贡献率次之，为7.8%，其基因效应来自父本PH4CV；qTw1-1 和 qTw9-1 是4个环境都能检测到的QTL，是环境钝感QTL，推断这两个QTL可能为控制容重的主效QTL。qTw8-1 在2011年北京点检测到，位于SNP标记YN13209和SYN4439之间，贡献率为4.9%；qTw4-1 和 qTw1-2 分别在2010年和2011年农安点检测到，贡献率分别为3.6%和3.8%，这3个QTL可能受环境的影响较大。

3 结论与讨论

3.1 标记密度对QTL定位的影响

不同类型标记在玉米基因组上的分布频率决定了标记在遗传连锁图谱上分布密度，从而影响QTL定位的精度。从第1代分子标记到第3代分子标记——SNP，标记在玉米基因组上的分布越来越丰富，从而大大提高了构建遗传图谱的标记密度。本研究构建了包含1 000个SNP的连锁图，标记间的平均图距为1.5cM，同传统的RFLP、SSR标记构建的连锁图谱相比，所构建的SNP连锁图谱全长与丁俊强和张伟强的相当，SNP标记数目远高于其他研究，大大提高了连锁图谱的标记密度，标记间的平均图距显著小于同类研究的SSR图谱，而且图谱中标记间的大间隔区段数目远低于上述研究结果，增加了QTL检测精度。

本研究虽然没有发现新的QTL，但验证并精确了QTL的置信区间，qTw1-1 位于Ajimone-marsan发现的 qtest10 的置信区间；qTw9-1 落在Beavis所定位 qtest7 的置信区间，相比 qtest10 和 qtest7 的18cM和21cM的定位区间。本研究将 qTw1-1 和 qTw9-1 定位在2.7cM和2.4cM的置信区间，将容重QTL定位在较小的范围，获得紧密连锁的标记，可用于分子标记辅助育种。qTw1-1 和 qTw9-1 能够解释的表型变异较大，推测是控制玉米容重的主效QTL。qTw1-2、qTw4-1 和 qTw8-1 也都是前人的研究中曾经检测到的QTL，本研究进一步缩小了定位区间，发现这3个QTL受环境的影响较大，推测与环境有相互作用。

3.2 玉米容重与其他性状间的关系

玉米籽粒容重是复杂数量遗传性状，与粒型、粒重、籽粒比重、胚乳成分、硬度和水分含量等性状有密切的关系。籽粒形状和比重是决定容重的重要因素。籽粒形状受遗传控制，不同玉米品种之间粒型有差异，扁平的玉米籽粒容重普遍低于圆形玉米籽粒，亲本PH4CV籽粒扁平，容重低于粒型介于扁平与圆形之间的亲本PH6WC。比重是籽粒单位体积的重量，而容重是容积密度，不仅反映籽粒比重大小，也与籽粒盛装效率有关；硬度与玉米籽粒的容重和比重都有相关性（r>0.6），硬度反映了角质胚乳与粉质胚乳与的比率，

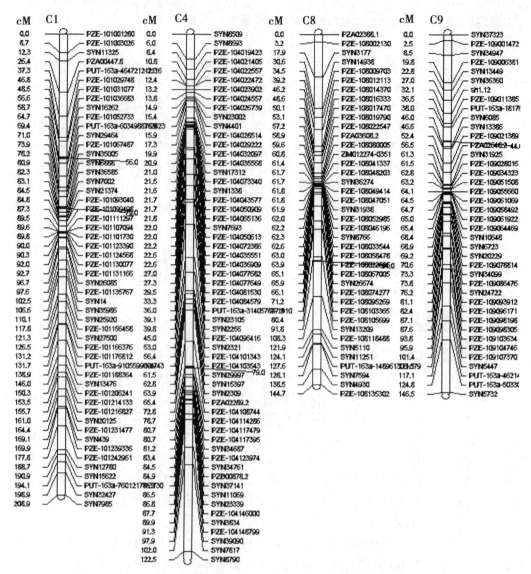

图 1 基于 SNP 标记的玉米 QTL 分析图谱

Figure 1 Analytical profile of maize QTL that based on SNP markers

角质胚乳含量高，籽粒硬度高，粒重高。容重是一定体积内籽粒的重量，与粒重的关系显而易见。水分含量也影响玉米的容重，水分越高，容重越低，但是容重与水分并不具有线性相关性，对容重的影响比较复杂。因此，本研究 DH 系的容重测量都是在籽粒干燥到水分含量低 14% 时进行，将水分对容重测量的影响降到最低。此外，环境对玉米容重也有影响，籽粒在田间自然成熟并干燥的过程中，有下雨、潮湿或雾等重新返潮引起收获前萌发，消化吸收籽粒中的营养成分，降低容重。

参考文献（略）

本文原载：玉米科学，2015，23（5）：21-25

玉米生物诱导孤雌生殖 DH 群体的 SNP 偏分离分析

许理文[1,2]　段民孝[2]　宋　伟[2]　田红丽[2]　王凤格[2]　赵久然[2]　王守才[1]

(1. 中国农业大学农学与生物技术学院，北京　100193；
2. 北京市农林科学院玉米研究中心，北京　100097)

摘　要：以单交种先玉 335 作母本，与京科诱 006 单倍体诱导系杂交，产生单倍体及秋水仙素加倍形成双单倍体（DH）群体，用 SNP 芯片对 DH 群体进行基因型分析，获得 1 337 个在先玉 335 的双亲间有多态的 SNP 数据，对 SNP 进行卡方检测（$P<0.05$），发现 516 个 SNP 表现偏分离，占 38.6%，这些偏分离 SNP 中有 285 个偏向母本 PH6WC，占 55.2%；231 个偏向父本 PH4CV，占 44.8%。在玉米的 8 条染色体上，发现 10 个偏分离热点区域（SDR），分析 SNP 的 χ^2 值与其在染色体上的位置之间的关系发现，χ^2 值最大的 SNP 往往在 SDR 的中心位置。

关键词：玉米；SNP；偏分离；加倍单倍体

Segregation Distortion study on Bio-induced Parthenogenesis-derived Doubled Haploid Lines of Maize by SNP

Xu Liwen[1,2]　Duan Minxiao[2]　Song Wei[2]　Tian Hongli[2]
Wang Fengge[2]　Zhao Jiuran[2]　Wang Shoucai[1]

(1. College of Agriculture and Biotechnology, China Agricultural University, Beijing 100193;
2. Maize Research Center, Beijing Academy of Agricultural and Forestry Sciences, Beijing 100097, China)

Abstract: Haploids originating from the hybrid, Xianyu335 (PH6WC×PH4CV) were crossed with the haploid inducer, Jingkeyou006 and the resulting doubled haploid lines (DHs) were obtained by chromosome doubling with colchicine. A total of 1 337 SNP markers detected polymorphism between PH6WC and PH4CV. Among 1 337 SNP, covering whole maize 10 chromosomes, five hundred and

基金项目：北京市科技计划课题（D131100000213002）、国家科技支撑项目（2011BAD35B09）、北京市农林科学院科技创新能力建设专项（KJCX20140202）、"863" 计划 "玉米产量功能基因研究与克隆"（2012AA10A305）

作者简介：许理文（1981—　），男，河北邯郸人，博士，从事玉米遗传育种研究。Tel：010-51503987　E-mail：xulw0408@126.com
赵久然、王守才为本文通讯作者。E-mail：maizezhao@126.com，wangshoucai678@sina.com

sixteen SNP (38.6%) showed the genetic distortion ($P<0.05$). Of the total segregation distortion SNP, two hundred and eighty-five SNP (55.2%) deviated toward female parent, PH6WC, while 231 SNP (44.8%) deviated toward male parent, PH4CV. Totally, tensegregation distortion regions (SDRs) were detected among 8 different chromosome, the relationship between X2 valueand position in chromosome of SNP was analyzed.

Key words: Maize; SNP; Segregation distortion; Doubled haploid (DH)

偏分离是指在一个分离群体中观察到的基因型比例偏离孟德尔分离规律的现象。Mangelsdof 和 Jones 在研究玉米配子体因子 Gal 与胚乳淀粉 Su 基因间的遗传连锁时最早报道了偏分离。随着 RFLP、RAPD、AFLP、SSR 等各种分子标记在遗传图谱的构建中的应用,在玉米 F_2、BC、永久 F_2、RIL 和花粉培养 DH 群体的遗传研究中发现存在偏分离现象。XU 等认为,几乎所有类型的杂交分离群体都能发现偏分离现象。遗传偏分离是自然界非常普遍的现象,并被认为是生物进化的动力之一。前人研究发现,偏分离标记在连锁群上成簇分布,而且多数偏分离方向一致,称为偏分离的热点区域(SDR,Segregation Distortion Regions)。严建兵等研究发现,4 个 SDR 与定位的配子体基因的位置相近,由此表明,配子体基因是导致偏分离的部分原因。偏分离标记会影响标记间遗传距离的估计和标记的相对顺序,即影响连锁图的精确度与准确度,进而影响偏分离标记的遗传分析。因此有必要基于标记在玉米染色体的物理位置对偏分离标记进行分析。单核苷酸多态性(SNPs)是广泛存在于基因组中的一类 DNA 序列变异,是目前为止分布最广、存在数量最多、标记密度最高的一种遗传多态性标记。研究发现,在玉米基因组上平均每 48~104bp 有一个 SNP。随着高通量 SNP 芯片检测技术的发展,SNP 应用于高分辨率遗传图谱的构建和 QTL 定位中。玉米测序已经完成,能够获得每个 SNP 在玉米染色体上的位置信息,为基于标记在玉米染色体的物理距离进行偏分离分析奠定基础。本研究利用 SNP 芯片分析,采用孤雌生殖诱导方式获得的 DH 群体的基因型,分析 DH 群体 SNP 标记的分离比例,借助 SNP 在玉米染色体上的物理距离数据,构建 SNP 的物理图谱,分析偏分离 SNP 标记在玉米染色体分布规律,检测 SNP 标记在染色体上的偏分离热点区域(SDR),寻找可能引起偏分离的偏分离位点(SDL),为利用分子标记构建连锁图谱提供有益的信息。

1 材料与方法

1.1 试验材料

单倍体诱导试验于 2007 年春在北京市农林科学院小汤山试验基地进行。单倍体孤雌生殖诱导系为北京市农林科学院玉米研究中心选育的京科诱 006,是基于 R-nj 遗传标记系统,籽粒的紫色冠顶和紫色胚为显性标记。种植玉米单交种先玉 335(PH6WC×PH4CV,F_1),在开花期以孤雌生殖诱导系京科诱 006 为父本进行杂交,杂交果穗晾晒后,利用 R-nj 遗传标记逐粒鉴定挑选,带紫色冠顶白色胚的籽粒为单倍体,带紫色冠顶紫色胚的籽粒为杂交二倍体,不显任何紫色标记的籽粒为受花粉污染的二倍体种子,另外还有少量无胚或死胚种子。获得单倍体籽粒于 2007 年冬,在北京市农林科学院南繁试验基地利用秋水仙素进行加倍,获得 DH 系。DH 系于 2008 年在北京市农林科学院小汤山试验基地进行人工自交授粉、扩繁，获得240份DH系。

1.2 DNA准备

在玉米的幼苗期田间采集叶片，每个DH系采集10个单株的叶片，混合磨样，采用CTAB法提取基因组DNA。DNA质量和浓度用NanoDrop 2000（Thermo Scientific）微量紫外分光光度计进行测定；利用吸光值λ260/λ280的比值检测DNA的质量，λ260的值反映DNA的含量，根据测量值调节工作液浓度约50ng/μl。

1.3 SNP基因型鉴定

利用本实验室开发基于IlluminaGoldenGate平台的MaizeSNP3072芯片，对DH群体进行基因型分析。MaizeSNP3072芯片是用于玉米品种真实性检测和玉米分子育种而开发的芯片，包含307个均匀分布于全基因组的SNP位点，SNP的详细信息可以从http://www.illumina.com.cn和http://www.panzea.org网站查询。

试验参照Illumina公司提供的操作指南，在北京市农林科学院玉米研究中心完成。原理与具体过程为：每个SNP标记包括3条引物，一条引物为通用的分析条形码，另外两条为标记有不同荧光信号的等位基因特异引物；每个检测样品包括5μl浓度为50ng/μl的基因组DNA；在96孔的SentrixArray Matrices中放置95个DNA样品（其中含检验样品2份）和1份5ml的ddH$_2$O样品作为空白对照，然后与芯片上的3 072个SNP杂交。杂交反应第一步为基因组DNA与3条引物、DNA聚合酶、连接酶共同温育，进行引物的延伸反应，得到PCR反应的模板；第二步以用荧光标记的等位基因特异引物进行PCR扩增，与固定相载体上的磁珠进行杂交，并读出结果，每个样品能同时检测3 072个SNP位点；最后利用Illumina软件GenomeStudio进行基因型数据分析，剔除杂交质量较差、基因型鉴定不可靠的SNP位点，同时对一些基因型鉴定不确切的标记进行手工调整。

1.4 SNP标记的统计分析

利用SNP标记分析各DH系的基因型，与母本基因型相同的标记为"AA"，与父本基因型相同标记为"BB"，缺失记为"-"。将实际观察值逐一按孟德尔分离的理论比率（1:1）进行卡方检测（$P<0.05$），推断被检测标记位点是否存在偏分离，并对照亲本基因型确定偏分离的方向。

1.5 SNP物理图谱构建

从http://www.panzea.org网站获得SNP位点在玉米染色体上的物理位置信息，利用Mapchart把双亲间表现多态的SNP映射在玉米染色体上，构建SNP标记的物理图谱。

2 结果与分析

2.1 SNP基因型分析

利用IlluminaGenomeStudio基因分析软件对所有基因型数据进行分析。所有检测的DH系被清晰地分成了两类，分别代表AA、BB基因型，DH系没有AB基因型。每个SNP标记根据杂交的质量进行评分，可设计性分级评分值（Designability rank score）介于0.1~1.0，在3 072个SNP标记中有2 707个SNP标记评分值大于0.6，占88.1%；评分值介于0.4~0.6的SNP标记194个，占6.3%；低于0.4的SNP标记171个。如果SNP标记的评分值低于0.4，软件一般不能很好地自动识别材料的基因型，即使手工调节也难以保证数据的准确性，因此，此类SNP标记将被去除。本试验中选用评分值大于0.6的2 707个

SNP 的数据用于后续分析。

2.2 DH 群体基因型组成分析

2 707 个 SNP 中有 1 337 个 SNP 在 DH 群体的双亲间有多态，多态率 49.4%。分析在双亲间表现多态性的 1 337 个 SNP 的基因型分布，结果表明，SNP 的基因型来源于母本 PH6WC 的在 14.2%~85.8%，平均为 50%；来源于父本 PH4CV 的在 14.2%~85.8%，平均为 50%。对群体的所有株系在全部标记位点上双亲基因型的分析结果为，来源于母本 PH6WC 的是 164 390，占 51.2%；来源于父本 PH4CV 的是 154 690，占 48.2%；缺失数据 1 800，占 0.6%。χ^2 测验结果表明（$\chi^2 = 147.89$，$>\chi^2_{0.05} = 3.84$），在整个 DH 系群体中，基于 SNP 标记的分析，双亲染色体同源片段分离符合 1:1 的理论分离比例。

2.3 SNP 的统计分析

对 1 337 个多态 SNP 进行 χ^2 测验显著性分析，当 $P<0.05$ 时，516 个 SNP 偏分离达显著水平，占 38.6%，其中，285 个 SNP 偏向母本 PH6WC，占 55.2%；231 个 SNP 偏向父本 PH4CV，占 44.8%。偏分离 SNP 在玉米的 10 条染色体都有分布，但是分布很不均匀，在 1 号、3 号、8 号、4 号、2 号染色体上发生偏分离的 SNP 标记较多，分别是 116 个、97 个、95 个、79 个、55 个，总计达到 442 个，占偏分离 SNP 总数的 82.2%；10 号染色体发生偏分离的 SNP 最少，只有 3 个（表1）。

表1 玉米 10 条染色体上的总 SNP 数目和发生偏分离的 SNP 数目
Table1 Summary of SNPs and segregation distortion SNP of 10 chromosomes of maize

染色体 Chr.	偏分离 SNPSegregation distortionSNP	偏向母本 Deviated toward female	偏向父本 Deviated toward female	染色体的 SNP 数 All SNP on Chr.
1	106	96	10	198
2	55	49	6	133
3	97	5	92	152
4	74	72	2	173
5	27	25	2	131
6	14	14	0	135
7	21	21	0	118
8	95	0	95	116
9	24	0	24	95
10	3	3	0	86
总和	516	285	231	1 337

2.4 偏分离 SNP 在染色体上分布和方向

基于 SNP 在染色体上的位置信息，利用 Mapchart 把在双亲间表现多态的 1 337个 SNP 映射在玉米染色体上，建立 SNP 的物理图谱，把偏分离达显著水平的 516 个 SNP 标记出来。根据偏分离 SNP 在染色体上的分布是否聚集分为两类：一类为少量独立、随机分布于染色体的偏分离 SNP，本研究中发现 17 个这样的 SNP（图 2 箭头所指），这些偏分离

SNP 的 χ² 值往往比较大，其中，SYN6674 和 SYN21851 的 χ² 值最大，都是 122.3，但是偏分离方向不同，SNP 位点 SYN6674 有 34 个父本基因型，206 个母本基因型，偏向于母本 PH6WC；SYN21851 则正好相反，偏向于父本 PH4CV。SYN6674 的严重偏分离并没有引起上下相邻 SNP 位点的跟随，更有如 7 号染色体上的 PZE-107136612 和 PZE-107136972 上下相邻的 SNP 位点偏分离方向是父本，而这两个 SNP 位点偏分离方向是母本；相似的还有 2 号染色体的 SYN11378 和 SYN30559，3 号染色体上的 PZE-103095468 和 PZE103095475。另一类为广泛分布于不同染色体上的偏分离 SNP 呈连续聚集，而且偏分离方向一致，形成染色体上的偏分离热点区（SDR）。本研究以超过连续 10 个偏分离 SNP 聚集区域称为偏分离热点区域，共检测到 10 个偏分离热点区域，分别命名为 SDR1-1、SDR1-2、SDR2、SDR3-1、SDR3-2、SDR4、SDR5、SDR7、SDR8 和 SDR9（图1）。其中，5 个偏分离热点区 SDR1-1、SDR3-1、SDR3-2、SDR8 和 SDR9 偏向父本 PH4CV；另外 5 个偏分离热点区偏向母本 PH6WC。研究发现，SDR1-2、SDR2、SDR3-1、SDR4 和 SDR8 这 5 个 SDR 在染色体上的跨度很大，达半条染色体甚至几乎整条染色体，如 SDR8，8 号染色体上共有多态性的 SNP116 个，其中，95 个 SNP 的偏分离达显著程度，这些偏分离的 SNP 在染色体是连续，而且都是向父本 PH4CV 偏分离（表2）。3 号染色体上的如果不是由于 PZE-103095468 和 PZE-103095475 向母本 PH6WC 的偏分离，SDR3-1 和 SDR3-2 形成一个大的 SDR。SNP 的 χ² 值反映了 SNP 位点的偏分离程度，χ² 值越高，SNP 位点的偏分离越厉害，5 号染色体上的 SYN6674 和 SYN21851 的 χ² 值都是 122.3，是偏分离程度最大的 SNP，但是 SYN6674 和 SYN21851 的偏分离方向相反，SYN6674 偏向于母本 PH6WC，而 SYN21851 偏向于父本 PH4CV，SNP 的 χ² 值的不能反映出 SNP 的偏分离方向。

表2 偏分离热点区的染色体位置、SNP 数和偏分离方向
Table2 Summary of chromosomal position, segregation distortion, and SNP No. insegregation distortion regions

偏分离热点区 SDR	染色体 Chr.	SDR 内 SNP 数（个） SNP No. in SDR	偏向 Direction	SDR 跨度（Mb） Span o fSDR
SDR1-1	1	10	父本	12.0~35.0
SDR1-2	1	88	母本	121.0~288.0
SDR2	2	47	母本	32.0~173.2
SDR3-1	3	77	父本	7.9~151.8
SDR3-2	3	14	父本	156.0~178.3
SDR4	4	70	母本	152.5~241.8
SDR5	5	22	母本	2.1~17.8
SDR7	7	18	母本	7.3~30.2
SDR8	8	95	父本	8.7~164.3
SDR9	9	24	父本	1.0~23.7

为了更加形象地反映 SNP 的偏分离信息，以 SNP 的 χ² 值×（母本等位基因数目-父

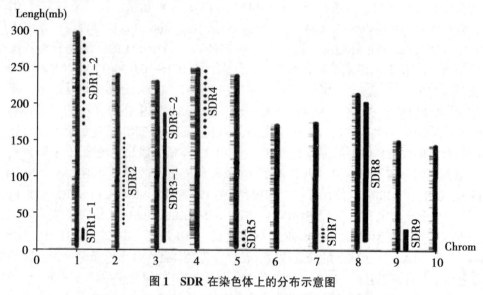

图 1　SDR 在染色体上的分布示意图
Figure 1　Schematic of distribution of SDR in chromosome

注：灰色圆柱代表染色体及其上的 SNP 分布；黑色点线表示 SDR 偏向于母本 PH6WC；黑色实线表示 SDR 偏向父本 PH4CV。下图同

Note：Gray column is chromosome, black dotted line is SDR deviated toward female, black solid line is SDR deviated toward male. The same below

本等位基因数目），当 SNP 位点的母本等位基因数目大于父本等位基因数目时，是正值，表示 SNP 偏分离方向是母本；当父本等位基因数目大于母本时，是负值，表示该 SNP 的偏分离方向是父本。处理后的 χ^2 值与 SNP 在染色体上的位置信息进行散点图分析（图2），处理后 SNP 的 χ^2 值连线类似抛物线，顶点位置的 SNP 的 χ^2 值最大，即偏分离最严重的 SNP，而且抛物线轨迹与染色体上的 SDR 对应，抛物线的顶点往往是 SDR 的中心，SNP 的偏分离程度随着远离 SDR 中心偏分离程度降低，χ^2 值减小，这在 SDR1-1、SDR1-2、SDR2、SDR3-2、SDR4、SDR5、SDR7 和 SDR9 表现明显。

3　结论与讨论

3.1　DH 群体偏分离的原因

目前，在多种作物的不同类型分离群体上，均有大量关于偏分离热点区域的报道，认为偏分离热点区域是由偏分离位点（SDL）引起。本研究中，SDR 区域 χ^2 值最大的 SNP 位点可能是偏分离位点或偏分离位点的作用点，其两侧 SNP 偏分离则是基因在染色体上的线性排列和遗传连锁的关系。Vogl 等认为，如果一个 SDL 在一个群体中发生分离，与其连锁的相邻标记就会表现一定的偏分离。标记离 SDL 越近（连锁紧密），偏分离越严重，表现为 χ^2 值增大。Cheng 等认为，可以利用 χ^2 值与遗传距离的分布图初步推测 SDL 的位置。根据偏分离位点基因起作用时间可分为配子选择和合子选择，分别在受精前和受精后受偏分离位点的控制。配子体选择分为花粉致死、花粉管竞争和选择性受精。孤雌生殖诱导单倍体通过自然加倍产生 DH 系，DH 系的基因型与减数分裂后形成的雌配子的基因型是一样的，即一个雌配子发展成为一个 DH 系，因此，诱导系的花粉的选择性受精

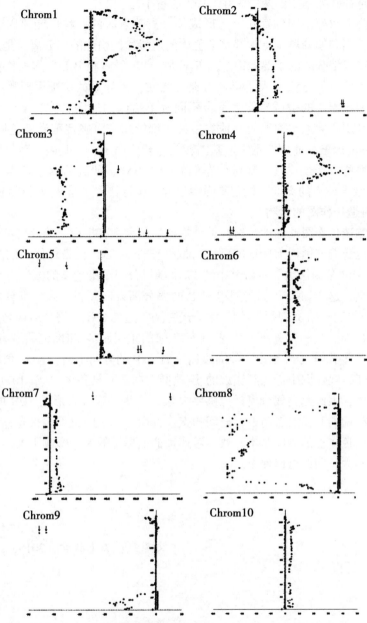

图 2 玉米 10 条染色体上处理后 SNP 的 χ^2 值与物理距离散点图分析
Figure 2　Scatter diagram of the processed χ^2 value against location of SNP on chromosome

(诱导产生单倍体)可能与 DH 系群体偏分离相关,而花粉致死和花粉管生长速度的竞争在孤雌生殖诱导产生单倍体过程中是不起作用的。合子选择是授粉后起作用的选择,往往是合子生存能力竞争的选择,在孤雌生殖诱导生产 DH 系的过程中,这一选择更多地存在于单倍体时期。单倍体时期玉米基因组由 2n 变成 n,导致隐性致死基因大量暴露,由于遗传搭车效应(genetic hitch-hiking effect),致死因子两侧的染色体都随着发生偏分离,致死因子的位置是偏分离最为严重的地方,因此,SNP 的 χ^2 值在其物理距离上呈类似抛物线的曲线，越往致死基因的两侧 SNP 的 χ^2 值越小。

染色体上隐形致死基因的存在，是形成偏分离热点区的一种可能原因。前人研究认为，隐形致死基因和偏分离位点等都是染色体的遗传片段引起的偏分离，是造成偏分离的遗传因素，是造成偏分离的主要原因。与 F_2 群体、BC 群体和 RIL 群体相比，通过诱导系孤雌生殖产生单倍体的方法引入了第三方遗传物质，诱导系的雄配子对雌配子是否有选择，不同基因型雌配子的被诱导能力是否不同及单倍体的加倍是否对偏分离有影响。因此，需要通过利用相同亲本构建的其他类型群体继续研究，确定玉米染色体上的 SDR 和 SDL。此外，环境因素、非同源重组、基因转换、转座因子、转基因沉默等都有可能导致偏分离。对于那些缺乏规律性、随机分布的偏分离 SNP，研究认为，亲本存在剩余杂合性、DNA 引物结合位点的突变和标记基因型划分、统计错误等是造成此类偏分离的原因。

3.2 偏分离对遗传作图的影响

偏分离影响遗传作图准确性，这在多个物种的研究中有报道。偏分离可以影响标记间的重组距离，也影响连锁群上标记的顺序。依据偏分离 SNP 在染色体上的分布是否有规律，把偏分离 SNP 分为两类。本研究中的 17 个随机分布于染色体的偏分离 SNP 对遗传作图有较大的影响，这类 SNP 依据与周边 SNP 的偏分离方向是否一致，细分为两类：与周边偏分离 SNP 的方向相同，只是 χ^2 值远大于左右两边的 SNP，偏分离很严重，如 5 号染色体上的 SYN6674，这类偏分离 SNP 会对 SNP 间的重组距离和标记顺序有较大的影响，作图中应该去除此类偏分离 SNP；与周边偏分离 SNP 的方向不同，且 χ^2 值较大，偏分离比较严重的 SNP，在作图中进入到错误的连锁群，如 7 号染色体 PZE-107136612 作图中进入 2 号染色体，因此此类偏分离标记也要去除。那些成簇分布形成 SDR 的 SNP 在作图中往往是相邻标记的互换或者附近标记间的交换，而且发生的也不是很多，标记在连锁群上的顺序与在染色体上的顺序基本一致，对连锁群上标记的顺序影响不大，对标记间的重组距离影响也不大，因此可以保留。

参考文献（略）

本文原载：玉米科学，2015，23（3）：8-14

同一来源玉米 DH 系的性状分离分析

邢锦丰　张如养　赵久然　段民孝　刘新香　王元东　王文广

(北京市农林科学院玉米研究中心，北京　100097)

摘　要：利用单倍体诱导系对遗传背景相近的玉米自交系 MC_1 和 MC_2 所组配的 F_1 进行杂交诱导获得单倍体，经过自然加倍得到181份纯系（DH 系），分析 DH 系农艺性状的分离情况。结果表明：181个 DH 系株高、穗位高、穗长、秃尖长、穗行数、百粒重和单穗粒重等数量性状的次数分布符合正态分布；所得 DH 系各数量性状的变异类型丰富，且在产量性状上存在优于两亲本的分离；两亲本的21对 SSR 核心引物分子标记带谱在 DH 系中的分布都近似于1∶1比率，未出现偏分离。因此，通过遗传诱导的方法所得 DH 系类型多样，可以满足育种上的需要。单倍体育种是一种选育优良玉米纯系的有效方法。

关键词：玉米；DH 系；性状；分离；育种

The Separation of Characters of Doubled Haploid Lines in Maize

Xing Jinfeng　Zhang Ruyang　Zhao Jiuran　Duan Minxiao
Liu Xinxiang　Wang Yuandong　Wang Wenguang

(*Maize Research Center, Beijing Academy of Agriculture & Forestry Sciences, Beijing* 100097)

Abstract: Maize doubled haploid (DH) breeding was able to generate pure inbred lines rapidly. Two inbred lines with high combining ability, MC_1 and MC_2, were used for the study. Crosses were made between those two lines. The F_1 plants were crossed by pollen from haploid induction line and 181 doubled haploid lines were obtained. The plant-height, ear-height, ear-length, bald-length, rows per ear, 100-kernel weight and ear-weight of DH lines were analyzed. The results indicated that the distribution of Quantitative traits was in normal. The significant variations of characters of DH lines were observed. Compared with parents, 181 DH lines were likely to produce a beneficial separation in ear-length, rows per ear 100-kernel weight and ear-weight. The results also showed that, there was no significant segregation distortion in DH lines by SSR markers test. Therefore, haploid breeding method is an effective way for breeding elite inbred lines in maize.

Key words: Maize (*Zea Mays* L.); Doubled haploid line; Characters; Separation; Breeding

---

基金项目：首都现代农业育种服务平台建设：玉米 DH 工程化育种研究与利用
　　　　（D111100001311001）
作者简介：邢锦丰，副研究员，研究方向为玉米育种
　　　　赵久然为通讯作者，研究员，主要从事玉米研究

利用单倍体育种技术可以快速、大规模的创制玉米纯系（Double Haploid Line，DH系），并可直接应用于育种实践。DH系的利用价值研究受到育种工作者的广泛关注，如果所创制的DH系变异类型丰富，且存在优于两个亲本的分离，说明单倍体育种可以有效的选育稳定纯系。如果所创制的DH系在性状上不存在优于亲本的表现，说明单倍体育种可能不是选育优良纯系的有效方法。国内很多单位都已开展玉米单倍体育种研究，但目前有关DH系的分离和利用价值研究报道较少。本研究通过田间性状调查和室内考种，分析利用单倍体育种方法所得的181份DH系株高、穗位高、穗长、秃尖长、穗行数、百粒重和单穗粒重等数量性状的分离情况，比较DH系与两个亲本在同一种植环境下的性状表现，同时利用SSR（Simple Sequence Repeats）核心引物分子标记的方法分析所得DH系的分离特点，以期为玉米DH工程化育种的深入开展提供参考。

1 材料与方法

1.1 试验材料

所用的181份DH均由北京市农林科学院玉米研究中心以遗传背景相近、高配合力的自交系MC_1和MC_2为亲本组配的F_1经单倍体诱导和自然加倍所得。

1.2 田间试验设计与调查项目

181份DH系和2个亲本自交系于2009年6月在北京市农林科学院玉米研究中心小汤山育种基地种植，采用随机区组设计，3次重复，双行区，每行20株，行距0.65m，株距0.265m，密度为66 000株/hm^2。由于微效多基因所控制的数量遗传性状，受土壤肥水以及气候变化、地理差异影响较大，变化幅度较大，当土壤肥力不均及田间管理不一致时，会出现植株高低不齐、长势长相不同、果穗大小不均等现象。本试验所选择的试验地肥力水平较高，地力较均匀，按高产栽培管理，减少环境条件的差异，创造良好的生长条件。同时所有材料均开放授粉，10月收获。每个小区取中间12株进行田间调查和室内考种，田间调查株高、穗位高，室内考查穗长、秃尖长、穗行数、百粒重、单穗粒重等数量性状。

1.3 SSR分子标记

在181个DH系和2个亲本自交系的苗期，每个材料随机剪取6株幼苗的嫩叶，储存于超低温冰箱，用于提取DNA和SSR分析。采用CTAB方法提取DH系和亲本的DNA。引物合成参照IBM2 Neighbors 2005 SSR标记的遗传图谱，根据扩增效果和引物多态性水平，选择分布在10条染色体上的40对核心引物用于DH和亲本的多态性检测，并从中筛选出21对在2个亲本间都具有不同的带型的引物用于分析，引物均由美国应用生物系统公司合成。

1.4 统计分析方法

利用小区平均值对DH系和亲本的农艺性状进行方差分析，利用最小显著极差法比较各农艺性状在DH系间的显著性差异性，利用Microsoft Excel软件进行数据处理和分析。SSR结果分析参照Senior等方法进行。

2 结果与分析

2.1 DH系群体各性状的次数分布

由各性状的次数分布图（图1）和正态分布峰度与偏度及P值表（表1）可知，181个DH系各性状次数分布的峰度和偏度绝对值均小于1，P值都大于0.05，符合正态分布。就偏度而言，秃尖长和穗行数的偏度值较大，分别为0.97和0.67，说明这两个性状的个体分布向大于平均数的方向偏斜，其他性状如株高、穗位、单穗粒重、穗长和百粒重都向小于平均数的方向偏斜。针对峰度来说，穗长、秃尖长、穗行数的峰度较大分别为0.65、0.98和0.98，表明它们的分布比较陡峭，峰态明显，总体变数分布比较集中；其他性状的峰度都较小，分布比较平缓，总体变数分布比较散。

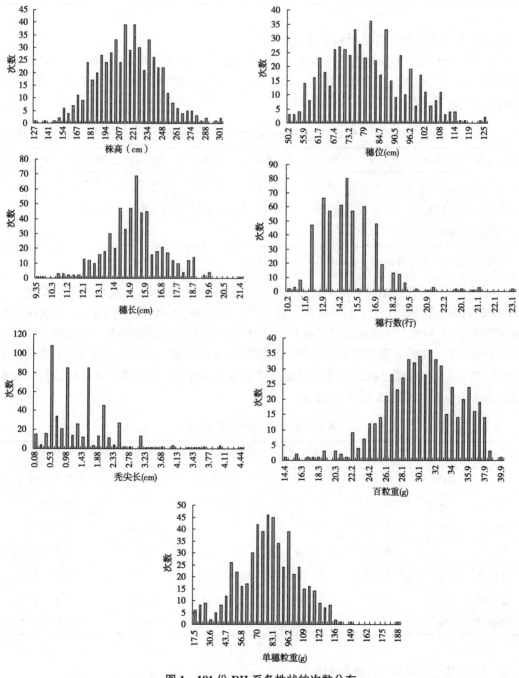

图1 181份DH系各性状的次数分布

表1 DH系各性状次数分布的峰度与偏度及P值

性状	株高	穗位高	穗长	秃尖长	穗行数	百粒重	单穗粒重
偏度	0.08	0.37	0.07	0.97	0.67	-0.87	-0.08
峰度	-0.04	-0.34	0.65	0.98	0.98	0.39	0.43
P值	0.2937	0.1524	0.2134	0.0832	0.0986	0.1572	0.2418

2.2 DH系各性状的方差分析和变异系数

表2中DH系各性状的差异显著性检验结果表明，181个DH系在株高、穗位高、单穗粒重、穗长、秃尖长、穗行数和百粒重等7个主要数量性状上的差异均达极显著水平，说明各性状在DH系间存在真实的差异性。

表2 DH系各性状的差异显著性检验

变异来源	自由度	株高	穗位高	穗长	秃尖长	穗行数	百粒重	单穗粒重
区组	2	2.12	0.19	0.11	5.24*	0.35	1.29	1.1
处理	180	5.232**	5.628**	8.428**	4.032**	8.359**	10.937**	6.959**
误差	360							

注：* 和 ** 分别表示在0.05和0.01水平上差异显著

从表3可知，DH系间变异系数最大的是秃尖长，达61.91%，其次为单穗粒重，达32.99%，性状变异系数较小的有株高、穗行数，分别为13.40%、13.65%，各性状平均变异系数为23.81%。说明通过两亲本自交系经单倍体诱导和加倍所得DH系间各性状的变异类型丰富。同时，各性状的变异范围都涵盖了两亲本值，且存在产量性状上表现优于亲本的DH系。在穗部性状上，穗长能超高亲2.8cm，穗行数能超高亲4.2行，百粒重能超高亲3.3g，单穗粒重能超高亲30.8g。

表3 DH系各性状的变异系数

性状	均值	标准差	变异系数（%）	亲本MC_1	亲本MC_2	变异范围
株高	213.74	28.64	13.40	247.9	208.0	150.3~276.9
穗位高	79.93	14.77	18.47	103.6	79.1	56.1~108.6
穗长	15.13	1.70	11.25	16.7	14.6	9.9~19.5
秃尖长	1.23	0.76	61.91	1.1	1.5	0.2~3.0
穗行数	14.70	2.01	13.65	15.4	15.8	10.7~20
百粒重	30.33	4.55	15.01	34.9	24.6	15.1~37.6
单穗粒重	79.03	26.08	32.99	105.9	76.4	16.1~136.7

2.3 SSR分子标记在DH系间的分布

本研究所筛选的21个核心引物分子标记在2个亲本间都具有较好的多态性，所有标记在 2 个亲本间都具有不同的带型，DH 系具有亲本之一的带型。由表 4 可知，2 个亲本在 21 对核心引物 phi96100、umc2105、bnlg2305、phi102228、umc1125、bnlg1940、phi032、umc1492、phi065、phi053、umc1545、bnlg490、bnlg249、bnlg2291、umc1429、bnlg1671、umc1061、umc1524、umc1999、umc1936、umc2163 所呈现的带型在 181 个 DH 系中都没有表现出明显偏离 1∶1 比率。卡方检验结果表明所用的 21 对核心引物分子标记的 χ^2 值在 0.01~3.45 之间，未达到显著水平，所以在 DH 群体中两亲本的 SSR 标记在 DH 中不存在明显偏分离的现象。

表 4 SSR 分子标记在 DH 系间的分布

亲本 MC_1 带型数	亲本 MC_2 带型数	SSR 引物分子标记结果	∣O-E∣	χ^2 值
80	101	phi96100	9.5	2.44
82	99	umc2105	2.5	1.60
83	98	bnlg2305、phi102228	3.5	1.24
87	94	umc1125	8.5	0.27
90	91	bnlg1940	12.5	0.01
92	89	phi032、umc1492	8.5	0.05
93	88	phi065、phi053	10.5	0.14
94	87	umc1545	6.5	0.27
97	84	bnlg490	10.5	0.93
99	82	bnlg249	11.5	1.60
100	81	bnlg2291	11.5	1.99
101	80	umc1429、bnlg1671	1.5	2.44
102	79	umc1061、umc1524	2.5	2.92
103	78	umc1999、umc1936、umc2163	7.5	3.45

3 结论与讨论

一个基础材料通过单倍体育种方法可以快速得到稳定的 DH 系，但 DH 系的培育过程仅经过一次减数分裂，不利于基因连锁的打破，而常规方法选育自交系经多次减数分裂重组，有利于打破不利基因的连锁，所以通过单倍体育种方法所得 DH 系是否具有育种价值将直接关系到单倍体技术在玉米育种中的利用。本研究结果表明 181 个 DH 系株高、穗位高、穗长、秃尖长、穗行数、百粒重、单穗粒重等数量性状都存在较大变异，且存在性状表现优于亲本的 DH 系群体，在穗部性状上能产生穗长超高亲 2.8cm，穗行数超高亲 4.2 行，百粒重超高亲 3.3g，单穗粒重超高亲 30.8g 的分离，说明通过单倍体育种的方法所得的纯系具有丰富的变异类型，且一些重要农艺性状能够得到显著的改善，这与 Chase 等研究通过单倍体方法培育纯系与常规选系途径获得的自交系在农艺性状和遗传多样性等方面具有同样效果的结果具有一定的相似性。另外，从性状的次数分布上看，所得具有优良农艺性状的DH系个体占总数的比例较小，且前人研究表明大部分材料的自然加倍率都较低，所以在单倍体育种过程中用于诱导的基础材料的群体应具有足够大，这样才能得到丰富基因型的单倍体和DH系。比如要从一个基础材料中获得200个DH系，按该材料的加倍率为5%，诱导率为5%，每个果穗为200个籽粒计算，则大约需要杂交果穗400穗。因此，将两个具有优良互补基因的自交系进行组配，再通过单倍体育种的方法产生大量单倍体和加倍获得纯系，这样就有可能互相改善对方不良的农艺性状，得到符合育种目标的个体。

  DH系的获得是经单倍体诱导和加倍两个环节，不同基础材料因遗传背景的差异，单倍体诱导和加倍的频率也存在一定的差异。如果研究所用的两个亲本材料之间的诱导率或自然加倍率存在差异，都可能使得获得在性状上偏向于较高诱导率和加倍率亲本的DH系数量大于诱导率和加倍率较低的亲本。本研究结果表明的亲本MC_1和MC_2的SSR核心引物分子标记带型比例在181份DH系中没有表现出明显偏离1:1比率，可以说明本研究所用两个自交系在单倍体诱导和加倍过程不存在偏分离，可能是由于研究所用两个基础材料的遗传背景相近，单倍体诱导率和自然加倍率差异较小。综上所述，利用单倍体育种技术所得的DH系各性状间存在较大差异，且可得性状优于亲本的分离，不存在人为淘汰有利用潜力材料的问题，能够为大规模创制有利用价值的纯系提供有效的途径。

参考文献（略）

本文原载：作物杂志，2012（6）：25-28

秋水仙素浸芽法处理玉米单倍体的加倍效果研究

段民孝[1] 刘新香[1] 赵久然[1] 张如养[1] 王元东[1] 邢锦丰[1] 陈绍江[2]

(1. 北京市农林科学院玉米研究中心，北京 100097；
2. 中国农业大学，国家玉米改良中心，北京 100193)

摘 要：试验采用28个育种材料的玉米单倍体籽粒，使用0.06%秋水仙素溶液浸泡幼芽上部和幼芽全部，研究秋水仙素对玉米单倍体的加倍效果的影响。结果表明，3年中不同秋水仙素浸芽方法处理玉米单倍体的成活率、授粉率和结实率之间的差异达到显著或极显著水平。浸泡幼芽上部处理能够明显减轻药害，缩短缓苗时间，提高单倍体植株的加倍效果，成活率、授粉率和结实株率均是浸泡幼芽全部处理的1.5倍左右，而且浸泡幼芽上部处理的授粉率和结实率是分别是对照的2倍和4倍以上。

关键词：玉米；单倍体；秋水仙素；加倍效果

Study on the Doubling Effect Using Colchicine in Maize Haploid

Duan Minxiao[1] Liu Xinxiang[1] Zhao Jiuran[1] Zhang Ruyang[1]
Wang Yuandong[1] Xing Jinfeng[1] Chen Shaojiang[2]

(1. *Maize Research Center, Beijing Academy of Agriculture and Forestry Sciences, Beijing 100097, China*; 2. *China Agricultural University, China National Maize Improvement Center, Beijing 100193, China*)

Abstract: Haploids, which were induced from 28 different genetic backgrounds, were used to study on the double effect by using 0.06% Colchicine Treatment. Two methods, including the entire immersion and the buds soaking, were performed to investigate the double effect. The results indicated that The difference of the survival rate, pollination rate and the seed setting rate of haploid plants treated by different colchicine dip bud methods in 3 years, reached significant or extremely significant level. The buds soaking can significantly reduce phytotoxicity and shorten the resting time, which survival rate, pollen appearance rate and the setting rate were 1.5 times or so as the buds soaking method, and which pollen appearance rate and the setting rate were more than 2 times and 4 times respectively.

Key words: Maize; Haploid; Colchicines; Double effect

基金项目：北京市科委项目（D111100001311001）、国家863项目（2012AA101203）、国家863项目（2011AA10A103）、北京市农林科学院青年基金（QN201106）
作者简介：段民孝（1972— ），男，山西平顺人，博士，副研究员，从事玉米遗传育种研究。Tel：010-51503772。E-mail：duanminxiao2006@yahoo.com.cn
通讯作者：赵久然（1962— ），男，北京人，博士，研究员，主要从事玉米研究。Tel：010-51505328。E-mail：maizezhao@126.com

玉米中采用双单倍体育种（Doubled haploid，DH）可以缩短自交系选育时间，加快育种进程，已经成为玉米育种的主要方法。获得足够单倍体和加倍获得DH系是DH育种的基本环节。据研究，玉米单倍体自然加倍比较困难，许多材料的自然加倍率低于5%，有些材料甚至不发生自然加倍，因此采取人工处理提高加倍率来满足育种与相关研究的需要显得尤为重要。

人工处理加倍的方法较为多样，其中秋水仙素是处理常用的加倍试剂，加倍效果由于所采用的药剂浓度、处理方式、处理时间等的不同而存在较大差异。多数研究认为，秋水仙素浸种法和浸根法加倍效果有限，而注射法和浸芽法的加倍效果相对较好。但是研究发现，采用注射法在田间处理单倍体幼苗时，由于单倍体幼苗大小不一，叶面张力与伤流的存在以及人为因素等，致使很难保证处理时期和注射点的一致，稳定性不高，无法保证加倍效果。另外，玉米单倍体幼苗植株矮小，田间注射十分困难，操作人员很难持久完成大量材料。采用浸芽法处理，操作方法简单易行，而且具有较高的加倍效率。Gayen等、Eder等和Bordes等分别在不同的年份通过不同的玉米单倍体加倍试验研究，发现秋水仙素浸芽法效果最好，显著高于对照自然加倍的结实率。据报道，中国农业大学陈绍江实验室成功建立了以浸芽法为基础的高效加倍方法，并在此基础上申请了秋水仙素加倍管理办法专利，雄穗散粉率一般为30%以上，远远高于作为对照的雄穗散粉率（一般5%以下）。笔者在采用浸芽法处理玉米单倍体中发现，经处理后的单倍体幼苗容易受药害，缓苗期较长，甚至大量幼苗死亡。分析原因，可能整个幼苗都浸泡在秋水仙素溶液中，不仅幼芽吸收秋水仙素，而且幼苗根部也吸收了大量的秋水仙素溶液，产生药害，影响了根系的生长，从而造成幼苗死亡。因此，探索一种既能减少幼苗死亡又能提高玉米单倍体加倍率的加倍方法显得尤为重要。

基于以上分析，本研究采用秋水仙素溶液对玉米单倍体幼芽进行两种处理，比较不同处理的加倍效果，旨在探索一种更加高效低毒的加倍技术，为玉米单倍体育种技术规模化、工程化提供依据。

1 材料和方法

1.1 试验材料

采用引进的诱导率在8%~10%的玉米单倍体诱导系作为父本，自配的杂交组合和收集的杂交种作为母本，进行授粉。收获后根据母本籽粒的颜色标记挑选出单倍体籽粒（胚乳顶端紫色，胚无色），作为试验材料。2009年试验材料5个，2010年试验材料15个，2012试验材料8个（表1）。

表1 秋水仙素处理单倍体材料

年份	单倍体来源
2009	IPT、ST1、ST2、ST3、ST4
2010	1324/ph4cv、2163/吉853、PH6WC/5003、WZF08、XY802、XY804、XY806、XY808、XY955、XY957、XY959、XY960、XY962、XY963、XY965
2012	10N40558-1-1、10N6405-2-1、4055724/4055、5401/D08-23-1、C92H、M9/M54-11-1、WC92、京35/335F

1.2 试验方法

1.2.1 秋水仙素溶液配制

配制 0.6mg/ml 的秋水仙素溶液备用。配制方法：0.6g 的秋水仙素溶解于 2.0% 的二甲基亚砜（DMSO）中，不断摇动以便充分溶解，然后加入蒸馏水并定容至 1 000ml，摇匀之后，倒进棕色的试剂瓶中，并将试剂瓶放置在阴暗避风处保存。

1.2.2 单倍体幼芽处理

处理流程如下：①将单倍体种子放置在 28℃ 左右的培养箱中暗光培养。②待玉米幼芽长至 2cm 左右时，切除胚芽鞘顶端备用。③以清水浸泡幼苗全部为对照（CK），秋水仙素溶液浸泡处理单倍体芽苗，采用 2 种不同处理方式：浸泡幼芽上部（即仅将幼芽浸泡在秋水仙素溶液中）；浸泡幼芽全部（即将整个幼苗全部浸泡在秋水仙素溶液中），时间为 8h。④8h 后，将药液倒掉，用清水冲洗幼芽 30min。⑤将幼芽栽在育苗钵中，放置在阴凉通风处进行缓苗。⑥待 2~3 叶期，及时将幼苗移栽在试验田中，并加强管理。试验在北京市农林科学院玉米研究中心位于昌平区的小汤山试验基地进行。

1.3 调查项目

室内进行秋水仙素处理玉米单倍体幼芽的时候，统计各处理的幼苗数量，即单倍体处理数。

玉米植株抽雄后，观察并记载单倍体植株的雄花育性恢复情况。凡是能显露花药的植株，标记为显药株数。将显露花药的玉米单倍体采取正常授粉或人工辅助多次授粉，标记授粉株树。成熟时，统计单倍体成活数，并收获自交授粉后结实单倍体植株，标记为收获 DH 数。把各项调查结果整理汇总之后，计算成活率、显药率、授粉率和结实率。

各项指标计算方法如下：

成活率（%）= 单倍体成活数/单倍体处理数×100

显药率（%）= 显药株数/单倍体处理数×100

授粉率（%）= 授粉株数/单倍体处理数×100

结实率（%）= 收获 DH 数/单倍体处理数×100

数据按照朱军等的 QGAStation 软件进行整理，采用线性混合模型，用 MINQUE（1）法估算各项随机效应的方差分量，并进一步估算各性状平均数间差异及其标准误，从而进行相应的统计检验。

2 结果与分析

表 2 列出了不同秋水仙素浸芽方法处理玉米单倍体的加倍效率情况。从表中可以看出，3 年中不同秋水仙素浸芽方法处理玉米单倍体的成活率、授粉率和结实率之间的差异达到显著或极显著水平。对于成活率来说，秋水仙素浸泡幼芽上部处理和对照处理差异不显著，但是秋水仙素浸泡幼芽全处理的成活率却远远低于对照和浸泡幼芽上部处理，且达到极显著水平。这说明浸泡幼芽上部处理能够减少秋水仙素对单倍体幼苗的药害作用，从而提高单倍体植株的成活率，其成活率是浸泡幼芽全部处理的 1.5 倍以上。2009 年和 2010 年对照处理的单倍体成活率低于秋水仙素浸泡幼芽上部处理，这可能是用于对照处理的幼芽个体发育较差，在营养钵育苗后逐渐死亡，造成成活率偏低。

3 年中，秋水仙素浸泡幼芽上部的平均显药率在 26.44%~32.63%，浸泡幼芽全部处理的平均显药率在 19.24%~20.82%，低于浸泡幼芽上部，而对照的平均显药率在年际间变化较大，变异范围在 11.61%~18.45%。从3年的平均显药率可以看出，对照处理的单倍体显药率最低，浸泡幼芽全部处理居中，显药率略高于 20%，而浸泡幼芽上部处理的显药率最高，为 28.09%，但是相互之间都没有达到显著水平。这说明秋水仙素处理能够提高单倍体植株的显药率，浸泡幼芽上部处理的效果更加明显一些。但由于年份之间气候差异，可能影响提高显药率的稳定性。

表2 不同秋水仙素浸芽方法处理玉米单倍体的加倍效果
Table 2 The double effects of haploid plants treated by different colchicine dip bud methods

年份 Year	材料名称 Test material	成活率			显药率 PNAAHP			授粉率 PNPHP			结实率 PDH		
		浸泡幼芽上部 SBCS	浸泡幼芽全部 EICS	对照 CK	浸泡幼芽上部 SBCS	浸泡幼芽全部 EICS	对照 CK	浸泡幼芽上部 SBCS	浸泡幼芽全部 EICS	对照 CK	浸泡幼芽上部 SBCS	浸泡幼芽全部 EICS	对照 CK
2009	IPT#	58.17	19.05	46.15	31.99	21.07	20.55	24.74	7.97	14.55	11.48	4.58	3.85
	ST1	65.67	15.96	46.46	30.35	17.87	16.54	19.40	6.38	5.51	7.96	2.34	1.57
	ST2	70.00	41.67	46.84	35.00	22.73	20.25	28.57	17.42	12.66	13.57	8.59	2.53
	ST3	73.49	35.33	44.59	39.76	24.44	22.97	32.53	21.33	21.62	14.46	11.11	1.35
	ST4	64.71	44.60	33.50	26.05	18.01	11.93	15.97	10.46	7.87	5.04	7.20	2.54
	均值	66.41	31.32	43.51	32.63	20.82	18.45	24.24	12.71	12.44	10.50	6.76	2.37
2010	1324/PH4CV	81.09	42.54	78.57	47.64	30.94	32.14	29.09	22.10	20.24	12.36	11.05	9.52
	2163/吉853	82.18	47.37	78.52	32.36	20.47	22.82	19.27	12.48	10.74	7.27	6.43	2.01
	PH6WC/5003	78.84	41.90	78.57	24.60	18.10	16.07	15.34	8.57	14.29	3.17	2.86	1.79
	WZF08	84.58	48.10	83.21	9.74	7.34	6.57	6.09	4.62	4.38	2.64	1.63	0.73
	XY802	88.42	65.63	72.92	24.21	28.13	14.58	14.74	14.06	10.42	5.26	4.69	0.00
	XY804	89.09	54.92	77.06	22.08	9.84	15.60	10.39	7.25	8.26	4.68	3.11	2.75
	XY806	86.79	64.04	70.27	32.08	15.73	16.22	13.21	3.93	8.11	5.66	1.12	0.00
	XY808	84.98	63.64	78.95	31.92	30.68	27.63	15.49	15.34	14.47	10.33	6.82	2.63
	XY955	81.01	39.22	75.00	18.40	16.99	11.25	10.68	8.50	8.75	3.26	1.96	1.25
	XY957	90.94	66.91	88.46	31.01	26.91	17.31	19.51	11.27	10.90	7.32	3.27	2.56
	XY959	83.42	68.97	79.69	32.64	24.90	17.19	17.10	16.86	9.38	3.63	3.83	0.00
	XY960	80.23	37.78	80.82	18.02	13.33	15.07	9.88	8.89	8.22	6.98	5.93	1.37
	XY962	78.49	49.44	78.05	21.51	20.22	7.32	13.95	11.24	2.44	5.81	5.62	0.00
	XY963	84.97	39.13	77.14	33.68	30.43	14.29	18.13	17.39	2.86	9.33	8.70	0.00
	XY965	69.08	37.68	66.13	16.78	14.49	8.06	12.83	11.59	3.23	5.26	4.35	0.00
	均值	82.94	51.15	77.56	26.44	20.57	16.14	15.05	11.61	9.11	6.20	4.76	1.64

（续表）

年份 Year	材料名称 Test material	成活率			显药率 PNAAHP			授粉率 PNPHP			结实率 PDH		
		浸泡幼芽上部 SBCS	浸泡幼芽全部 EICS	对照 CK	浸泡幼芽上部 SBCS	浸泡幼芽全部 EICS	对照 CK	浸泡幼芽上部 SBCS	浸泡幼芽全部 EICS	对照 CK	浸泡幼芽上部 SBCS	浸泡幼芽全部 EICS	对照 CK
2012	10N40558-1-1	48.32	32.53	62.79	14.09	10.84	5.81	8.72	7.23	3.49	4.70	3.61	1.16
	10N6405-2-1	70.99	42.47	75.21	28.40	17.12	14.10	16.05	9.59	8.97	6.48	4.79	0.85
	4055724/4055	63.88	32.47	66.67	22.81	17.21	16.92	16.73	11.36	11.28	4.94	2.60	2.05
	5401/D08-23-1	88.00	56.58	94.54	36.00	29.82	10.38	27.27	25.88	5.46	14.91	14.47	0.00
	C92H	68.55	41.25	73.79	36.29	25.00	8.74	31.45	24.38	7.77	25.81	13.75	5.83
	M9/M54-11-1	86.08	50.72	90.00	45.36	26.81	20.00	28.35	21.74	13.64	10.31	7.97	4.55
	WC92	74.23	32.51	75.98	32.65	16.61	6.37	31.27	15.19	4.90	20.27	9.54	0.98
	京35/335F	50.74	23.79	64.04	11.03	10.48	10.53	10.29	8.87	1.75	2.21	2.02	0.00
	均值	68.85	39.04	75.38	28.33	19.24	11.61	21.27	15.53	7.16	11.20	7.34	1.93
	总平均	75.96bB	44.14aA	70.85bB	28.09aA	20.23aA	15.26aA	18.47bB	12.93aAB	9.15aA	8.40bB	5.85bB	1.86aA

注：总平均中小写字母表示5%水平下的差异显著性；大写字母表示为1%水平下的差异显著性；相同字母表示差异不显著

3年的平均授粉率从高到低依次是 B2>B1>CK，即浸泡幼芽上部处理最高，达到18.47%，显著高于浸泡幼芽全部处理和对照，而且与对照处理的差异达到极显著水平。其次是浸泡幼芽全部处理，为12.93%，对照平均授粉率最低，仅有9.15%，后两者之间的差异不显著。说明秋水仙素浸泡幼芽上部处理不仅能够极显著地提高单倍体植株的授粉率，而且与浸泡幼芽全部处理、与对照处理的差异也达到了显著水平。2009年浸泡幼芽全部处理的授粉率略高于对照处理，分别为12.71%和12.44%，而浸泡幼芽上部处理的授粉率则达到24.24%，是其他两种处理方法的近2倍。2010年浸泡幼芽上部的平均授粉率仍然高于浸泡幼芽全部和对照，分别是15.05%、11.57%和9.11%。2012年对照处理的平均授粉率为7.16%，浸泡幼芽全部的平均授粉率为15.53%，是对照的2倍多，而浸泡幼芽上部的平均授粉率最高，达21.27%，是对照的近3倍。这充分说明了秋水仙素浸泡幼芽上部处理在提高单倍体植株的授粉率方面具有极大的优势。

从结实率情况来看，秋水仙素浸泡幼芽上部处理和浸泡全部处理的结果分别是8.40%和5.85%，均高于对照处理（为1.86%），且差异达到极显著水平，但是两者之间的差异不显著。浸泡幼芽上部处理的结实率是浸泡幼芽全部处理的近1.5倍，是对照的4倍以上。3年中，秋水仙素浸泡幼芽上部处理的结实率均为最高，分别是10.50%、6.20%和11.20%，秋水仙素浸泡幼芽处理居中，分别为6.76%、4.76%和7.34%，而对照处理的结实率则均在2.5%以下，远远低于秋水仙素处理的结果。说明与对照相比，秋水仙素浸泡幼芽上部处理和秋水仙素浸泡幼芽全部处理均能够极显著地提高单倍体植株的结实率，而且秋水仙素浸泡幼芽上部处理的效果更好一些。同时，从另一个侧面也说明，单倍体的自然加倍率很低，在单倍体育种工作进程中，进行单倍体加倍处理的必要性以及探索更高更好加倍处理方法的紧迫性。

3 讨论与结论

3.1 秋水仙素浸泡幼芽上部处理方法的加倍效果

很多研究表明，利用秋水仙素溶液均可以提高玉米单倍体的加倍率，但是采用药剂浓度、处理方式、处理时间不尽相同，结果也不一致。秋水仙素浸芽法使用较多且加倍效果相对较好。Gayen等研究发现以胚芽切口、0.06%的秋水仙素溶液浓度下浸泡12h效果最好，有18%的种子加倍成功。Eder等0.06%秋水仙素配以0.5%的DMSO浸芽处理的结实率能达到27.3%，显著高于对照自然加倍的结实率。而Bordes等用0.15%的秋水仙素溶液浸泡3叶期的单倍体幼苗3h，使3组不同来源单倍体材料的雄花可育率达到30%~60%，而3组未处理的雄花全部不育。陈绍江等使用浸芽法处理玉米单倍体的雄穗散粉率一般达到30%以上。

本研究中利用0.06%秋水仙素溶液配以2.0%的DMSO浸泡幼芽上部和幼芽全部，研究对玉米单倍体的加倍效果。从3年的平均结果可以看出，浸泡幼芽上部处理的成活率、显药率、授粉率和结实率分别为75.96%、28.09%、18.47%和8.40%，而浸泡幼芽全部处理仅为44.15%、20.23%、12.92%和5.85%，各项指标均明显低于前者。说明采用秋水仙素浸泡幼芽上部处理玉米单倍体，不仅可以提高单倍体植株的成活率、显药率和授粉率，而且具有较高的结实率，得到更多的DH系。

3.2 浸泡幼芽上部处理方法具有更大的实用性

秋水仙素是白色或黄色粉末，无味的剧毒物质，3~13mg即会引起人体窒息而死亡，而且价格昂贵。因此，在试验过程中，应该尽量减少其使用剂量。按照浸泡幼芽全部的方法，如果处理100株单倍体幼苗，使用浸泡幼芽全部处理的时候，大约需要300ml秋水仙素溶液，而采用浸泡幼芽上部方法进行处理，50ml秋水仙素溶液即可，减少了秋水仙素的使用量5~6倍。从这个角度来说，采用浸泡幼芽上部处理方法，不仅降低了秋水仙素对环境的污染程度，而且大大节省了试验成本，比浸泡幼芽全部处理具有更大的实用性和更广阔的应用前景。但是在具体操作中，大批量处理芽苗、添加药剂以及育苗移栽等环节还有待完善。目前，这方面的研究工作正在进行中。

参考文献（略）

本文原载：种子，2013，32（2）：19-23

单倍体技术在玉米种质改良和育种中的应用方向

张如养　段民孝　赵久然　刘新香　邢锦丰　王元东

（北京市农林科学院玉米研究中心，北京　100097）

摘　要：单倍体技术为玉米种质改良和新品种选育提供了重要的技术支撑。介绍了利用生物杂交诱导方法获得玉米单倍体的基本流程，详细阐述了单倍体技术在玉米种质改良和育种中的主要应用方向，同时就单倍体技术体系中值得注意的问题进行了总结，并展望了单倍体技术在玉米领域中的应用前景。

关键词：玉米；单倍体；种质改良；育种

The Application of Haploid Technology on Germplasm Improving and Breeding in Maize

Zhang Ruyang　Duan Minxiao　Zhao Jiuran
Liu Xinxiang　Xing Jinfeng　Wang Yuandong

(Maize Research Center, Beijing Academy of Agriculture & Forestry Sciences, Beijing 100097)

Abstract: Haploid technology provided technology support for germplasm improving and breeding in maize. The basic technological process of haploid produced by bio-induced was introduced and the application of haploid on germplasm improving and breeding in maize was expounded in this paper. Meanwhile the problem which deserved to be considered in haploid breeding was summed up. Perspectives on haploid breeding were also mentioned.

Key words: Maize (*Zea mays* L.); Haploid; Germplasm improving; Breeding

单倍体技术因其快速、高效的特点，在我国玉米育种领域中逐步得到普及。"十二五"期间，国家将"玉米 DH（Double Haploid）工程化育种技术与新品质选育研究"项目列入国家高技术研究发展计划（即 863 计划）。首先利用遗传诱导方法获得单倍体——依据诱导系的标记性状对诱导系与普通玉米材料的杂交籽粒进行单倍体筛选，再经自然或人工加倍即可获得稳定纯系应用于新品种选育或利用其进行种质改良和遗传研究等。这样

基金项目：国家 863 项目"玉米 DH 工程化育种技术与新品质选育研究"（2012AA101203）、北京农业育种基础创新平台Ⅱ（D08070500690802）、北京市科委"玉米 DH 工程化育种研究与利用"（D111100001311001）、北京农林科学院青年基金（QN201106）

作者简介：张如养，在读硕士，研究方向为玉米种质资源创新
　　　　　段民孝为通讯作者，副研究员，主要从事玉米遗传育种研究

将单倍体育种技术与种质的扩增有机的结合起来，先对目标诱导材料进行有效的重组改良，累积足够的有利基因位点，再利用单倍体技术获得纯系或改良群体，能够为玉米种质的改良和强优势品种的选育提供重要的技术支撑。针对我国玉米单倍体技术的现状，介绍了遗传诱导获得单倍体的流程，并重点阐述了单倍体群体的主要应用方向，以利于充分发挥单倍体技术在玉米种质改良和新品种选育中的作用，为促进单倍体技术在玉米 DH 工程化育种中深入、有效的应用提供参考。

1 单倍体的获得及其鉴定

1.1 利用单倍体诱导系获得单倍体籽粒

目前，玉米是唯一利用遗传诱导培育单倍体的作物，而遗传诱导所利用的大部分单倍体诱导系都具控制籽粒糊粉层和形成胚芽色素的 R-nj 基因及控制不定根、叶鞘和茎秆色素形成的 $ABPI$ 基因，使之诱导产生的单倍体具有籽粒和植株双显性遗传标记。单倍体的获得主要是通过以具有广泛遗传基础和优良基因的群体作为母本，以单倍体诱导系作为父本进行杂交，获得母本群体与诱导系的杂交果穗，并依据诱导系的遗传标记系统对杂交籽粒进行单倍体筛选。杂交所产生的籽粒可以分为以下几种类型：①胚芽和籽粒顶部均为紫色是杂合二倍体籽粒。②籽粒顶部紫色，胚芽白色，为拟单倍体籽粒。③胚芽和籽粒顶部均无紫色的，属其他花粉污染所致。④胚芽紫色，顶部无紫色的，可能由异受精引起。

1.2 单倍体的植株鉴定

将籽粒顶部紫色和白色胚芽的拟单倍体籽粒进行单粒播种，通过单倍体与正常二倍体在苗期、成株期形态上的区别对其做进一步的鉴定。①苗期鉴定：单倍体籽粒发芽后的生长势较弱，植株矮小，而二倍体籽粒发芽后生长势较强，植株较高。因单倍体诱导系具有紫色叶鞘标记的 $ABPI$ 基因，凡幼苗叶鞘绿色者为单倍体，紫色者为杂交二倍体。②成株期鉴定：单倍体植株叶片狭窄、上冲，茎秆细小；正常二倍体植株叶片宽大、披散，茎秆粗壮。其中值得注意的是部分单倍体个体可能早期就完成加倍，田间种植表现为株型较高、紧凑且叶片上冲，与正常自交系相似，对于这些植株不应作为杂株去除，需严格自交和进一步鉴定。

2 单倍体技术在种质改良和育种中的应用方向

2.1 利用单倍体技术选育 DH 系

单倍体是具有配子体染色体遗传组成的个体、组织或细胞，经自然或者人工处理加倍，即可产生纯合二倍体。通过单倍体加倍得到遗传稳定的纯系只需两个世代，能有效缩短培育纯系的年限，加速育种进程。对单倍体进行加倍研究选育 DH 系是目前研究的重点方向，然而单倍体间雌雄的育性差异较大，雄穗育性是单倍体自交结实的关键因素，而雄穗育性加倍的方法主要有自然加倍和人工加倍。

第一，自然加倍的研究方向主要是选择不同的加倍地点和同一地点不同播期对加倍率的影响。刘志增等研究发现不同环境（播期、地点）的加倍率有明显差异；刘治先等研究报道济南春播的单倍体苗自然加倍率最低，夏播种植的单倍体苗自然加倍率相对较高，海南秋播玉米单倍体苗自然加倍率最高。段民孝等研究表明在甘肃春播的加倍率最高，其次是海南冬季种植，北京春播的加倍率最低。

第二，目前，人工加倍处理主要以单一药剂、低浓度、短时间（8~12h）、低温度（11~20℃）处理单倍体芽期或6叶期较为适宜。通过不同的化学药剂如秋水仙素（Colchicine）、AMP（Amiprophosmethyl）、拿草特（Pronamide）、安磺灵（Oryzalin）、氟乐灵（Trifluralin）和 N_2O 等，在不同的浓度梯度下（处理时或以二甲基亚砜为助渗剂），通过不同的药剂介入方法如浸泡、注射、喷施等，以不同的处理时间（浸泡法）或次数（注射法、喷施法）对单倍体的不同组织（如种子、根尖、芽尖）和生产阶段（6叶期、拔节期）等进行处理，并以单一药剂处理的效果最佳。

2.2 利用单倍体技术创造和筛选新型突变体

由于玉米单倍体只含有一套染色体，对外界环境的变化较为敏感，容易发生基因突变。只要当染色体上的一个基因位点发生突变，就有可能产生形态或生理上的变异。目前有关利用单倍体的易突变性进行突变研究的报道较少，但在田间较大群体的情况下（大于 50 000 株），就不难发现玉米单倍体有许多突变类型。Chang 报道在一块有着 5 万~10 万个玉米单倍体在苗期有许多突变类型，如无叶舌、有光泽的、棕色叶脉等。如果能从突变体中筛选有利的变异（抗病、抗虫和抗除草剂）并保持，那么对育种工作将是非常有利的；同时可利用单倍体突变的广泛性进行相关性状的遗传研究。

2.3 利用单倍体技术进行种质改良

轮回选择是玉米群体改良的主要方法，包括混合选择、家系选择、半同胞轮回选择、全同胞轮回选择等多种方法。但上述方法均需较多世代，群体改良周期较长，效率较低。如果在选择阶段结合利用单倍体技术，可以消除显性基因对隐性基因的掩盖作用，增加选择的准确性，提高轮回选择的效率。据刘志增等报道玉米单倍体的雌穗的育性接近100%，且母本材料经过诱导后所得单倍体基因存在分离，穗部性状存在较大的差异；Chang 报道利用单倍体诱导系所得单倍体的基因型是随机的；刘玉强等研究报道单倍体群体中 SSR 分子标记呈随机分离，没有发现明显的偏分离现象等，上述研究结果都为通过单倍体雌穗的广泛分离进行选择和种质改良提供了可能。利用单倍体群体进行轮回选择流程如图1所示。

Chalyk 对两个人工合成群体 SA 和 SP 进行诱导获得单倍体，然后用正常二倍体花粉对单倍体进行授粉，成熟后根据目标性状（穗部性状）在单倍体时期进行选择，结果发现通过两轮的选择所得群体在产量性状如单株穗重、穗长、穗粗、千粒重等方面与原始群体相比得到了显著提高。可能原因是：第一，外界环境对单倍体的选择，淘汰含有较差基因型的个体。第二，依据目标性状的人为选择，较高的世代拥有丰富的有利于植株活力的基因。同样 Rotarenco 报道利用单倍体群体通过三轮的轮回选择所得群体，在籽粒产量、穗长、穗粗、株高、穗位高等性状上与原始群体相比得到了显著的提高，且在第三轮所得群体比第二轮所得群体的性状改良更为显著，可以有效的提高育种效率。另外，对在单倍体时期田间农艺性状较好但无法加倍的植株，可以利用回交的方法将其有利基因保留，并进一步利用，以提高单倍体的利用率，增加获得更多有利基因型的可能性。

2.4 利用单倍体技术在玉米育种早期进行选择

单倍体加倍所得后代的 DH 系数量较多，那么对其后期的测配工作量将是巨大的，给种质改良和育种工作带来较多的困难。然而单倍体与加倍后代所含的基因型是一致的，且单倍体时期无等位基因的互作效应，更有利于提高选择的准确性。Chalyk 研究报道单倍体

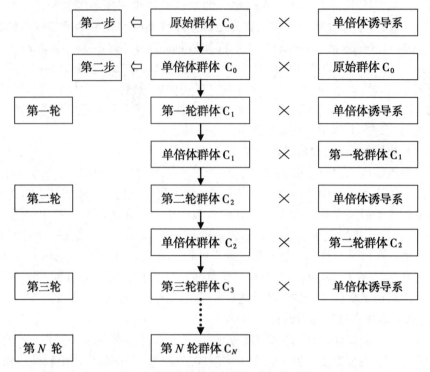

图1 利用单倍体进行轮回选择的流程

注：第一步，产生原始群体（C_0）的单倍体；第二步，将原始群体（C_0）的花粉给单倍体授粉，并根据目标性状在单倍体时期进行选择

时期雌穗的大小与加倍后的二倍体植株雌穗大小存在着正向的相关性。Rotarenco 对单倍体时期农艺性状的相关性与加倍后的二倍体农艺性状的相关性进行比较研究，发现在单倍体时期株高与穗位高相关性及叶长与叶宽的相关性比二倍体时期高，因此在单倍体时期依据目标性状进行选择是具有可行的。如果能够在单倍体时期对进行适当的选择和淘汰，不仅可以减少工作量，而且更有利于 DH 系鉴定等后续工作的开展，以更好地应用于新品种选育。

2.5 利用单倍体加倍所得 DH 系进行育种和遗传研究

加倍后所得的 DH 系所有基因位点都是纯合的，在主要农艺性状上均表现为高度的一致性，是选配高产、优质、强优势组合的宝贵材料。与传统系谱法所选育的自交系相比，单倍体育种方法劣基因淘汰明显，所得 DH 系具有相同的利用潜力。第一，需对 DH 系进行广泛测配，研究种质的利用潜力。配合力研究是玉米 DH 系工作的重点也是难点，也是衡量该种质在育种中的应用价值的标准之一。由于获得的 DH 系数量较多，因此应根据基础材料的来源，选拔出符合目标性状且农艺性状优良的纯系并淘汰一部分性状较差的，最后选择恰当的测验种进行广泛测配，以期选育出具有强优势的组合。第二，将 DH 系应用于群体改良，聚合更多有利基因。依据基础材料来源，将测配后所得表现较优良 DH 系在组合累积更多有利基因，再通过诱导获得纯系，可有效的加快育种进程。同时可利用 DH 系进行轮回选择，以进一步改良 DH 系群体和创新玉米种质。

DH 群体是永久性群体，每一品系都是纯合的，可稳定繁殖，且无显性和隐性互作的干扰，能准确地估计所研究性状的遗传本质。DH 群体的遗传结构直接反映了 F_1 配子中的基因分离和重组，可直接进行遗传作图，提高作图效率。另外，DH 群体可以反复使用，特别适合于一些性状的 QTL 定位研究。

3 问题探讨

利用单倍体技术进行玉米种质改良和新品种选育，每一个环节均影响到种质改良和育种效率，因此需建立起一套完善的技术体系。下面总结了在单倍体技术体系中值得注意的问题，供同行探讨。第一，关于高频诱导系的选育。选育标记明显、诱导率高的玉米新诱导材料是研究的重点方向。第二，基础材料的选择。对基础材料的选择将是单倍体育种工作成败的基础环节，它决定了后续育种工作的意义。限于目前诱导率和加倍率较低的问题，用于诱导的基础材料群体应足够大，提高获得更多有利基因型的单倍体和 DH 系的可能性；同时单倍体诱导率和加倍率与基础材料也紧密相关。第三，单倍体种植地点和密度的问题。选择不同环境地点种植单倍体直接影响其自然加倍频率，同时通过在高密度的条件下筛选，单倍体植株间存在选择和竞争压力，有利优良基因的选择和不良基因的淘汰。第四，加倍方法的研究。选择甘肃和海南为种植地点，加倍效果较好。第五，单倍体精细授粉的问题。由于单倍体一般表现雄性不育特性，只有较少部分雄穗的花药饱满外露，能形成正常的花粉，因此仔细观察少数恢复育性有正常花药或能正常散粉的植株进行人工辅助授粉是加倍成功的关键；另外，对于株型较高、紧凑、叶片上冲且与正常自交系相似的植株应妥善自交和进一步鉴定，避免无意识的淘汰早期加倍的单倍体个体。第六，拟 DH 系扩繁。在单倍体时期可能存在自交不严格的现象，加倍后所得籽粒存在混杂的问题，对拟 DH 系进行扩繁和鉴定，可以有效的淘汰混杂材料，避免测配的盲目性，减少测配成本。第七，DH 系的测配，重点是选择适当的测验种。由于 DH 系杂交组合比较试验的工作量巨大，需依据育种目标对 DH 系进行选择和选择适当的测验种有针对性的测配。

4 展望

与分子标记相结合的单倍体育种技术是现代生物技术育种的两大支柱之一。与转基因育种相比，DH 育种不存在生物安全问题，并且效率高、成本低、见效快；与传统"小而散，各立山头，单兵作战"的育种模式相比，DH 育种是链条式流水线作业的工程化育种模式，能有效地加快育种进程，特别适合我国目前育种实际的需要。然而，在单倍体技术中最基础的应用环节就是单倍体的利用，对其合理有效的利用显得尤为重要。因此，以单倍体技术为核心，重点研究单倍体群体和 DH 系的应用方向，并结合杂种优势利用技术、形态改良技术、分子生物学技术，能为大规模、快速创制玉米的优良 DH 系种质和大量选育出高产、优质、多抗、广适、易制种的强优势品种提供技术保障，同时也为壮大民族种业，发展良种工程，保障国家粮食安全和农产品有效供给做出更大的贡献。

参考文献（略）

本文原载：作物杂志，2012（5）：4-8

不同基因型玉米材料的单倍体诱导和加倍研究

张如养[1] 段民孝[2] 赵久然[2] 刘新香[2] 邢锦丰[2]

(1. 北京农学院植物科学技术学院,北京 102206;
2. 北京市农林科学院玉米研究中心,北京 100097)

摘 要:利用12份不同基因型 F_1 进行单倍体杂交诱导与自然加倍研究,分析不同材料单倍体诱导率和加倍结实率的差异性。结果表明:12份 F_1 的单倍体诱导率和加倍结实率都存在显著的差异,诱导率最高达11.03%,最低为3.34%;加倍结实率最高达13.20%,最低为2.61%。两组正反交组合的单倍体诱导率差异较小,而其中一组正反交组合的加倍结实率存在显著差异,最高相差3.34%。单倍体的雄穗加倍率与加倍结实率存在极显著的相关性,且雄穗正常散粉植株加倍获得稳定纯系的比例较高,达89.52%。

关键词:玉米;单倍体;诱导率;加倍结实率

Research on Haploid Induction and Doubling Efficiency in Maize

Zhang Ruyang[1] Duan Minxiao[2]
Zhao Jiuran[2] Liu Xinxiang[2] Xing Jinfeng[2]

(1. *Plant Science and Technology*, *Beijing University of Agriculture*, *Beijing* 102206;
2. *Maize Research Center*, *Beijing Academy of Agriculture & Forestry Sciences*, *Beijing* 100097)

Abstract: Twelve different genotype F_1 maize materials were crossed by pollen from haploid induction line to test their ability to produce haploid and to spontaneous chromosome doubling. The difference of haploid induction rates and self-seed rates was analyzed in this paper. Results indicated that the haploid induction rates and self-seed rates among different genotype differ greatly. The highest haploid induction rate is 11.03% while the lowest is 3.34%. The highest self-seed rate is 13.20% while the lowest is 2.61%. The results also showed that the difference of haploid induction rates was insignificant in the two reciprocal cross, but the self-seed rate was significant in one reciprocal cross, up to 3.34%. At the same time, there was significant correlation between the haploid tassel doubling rates and the self-seed rates. The anther scattered pollen normally was in the majority in self-seed plant, up to 89.52%.

基金项目:国家863项目"玉米DH工程化育种技术与新品质选育研究"(2012AA101203)、北京市科委项目(D111100001311001)、北京农林科学院青年基金(QN201106)

作者简介:张如养(1987—),男,福建人,在读硕士,研究方向为玉米种质资源创新与利用。E-mail:ruyangzhang2009@sina.com

Key words: Maize (*Zea mays* L.); Haploid; Haploid induction rates; Self-seed rates

利用单倍体技术选育新品种是玉米育种发展的趋势，该技术能有效的将种质扩增和育种材料的改良相结合，累积丰富的有利基因位点，为选育出高产、优质、多抗、广适、易制种的强优势品种提供重要的技术支撑。据有关报道，国内已有研究单位利用单倍体技术选育出优良的玉米 DH（Double Haploid）系，并成功的应用于育种实践。单倍体的诱发获得和加倍是玉米单倍体育种技术中的两个中心环节，直接影响到 DH 系的获得数量及其应用价值。本研究利用玉米单倍体诱导系对 12 份不同基因型 F_1 进行杂交诱导和自然加倍研究，比较不同材料的单倍体诱导率和加倍结实率的差异性，分析单倍体诱导率、雄穗加倍率和加倍结实率三者之间的相关性及其加倍获得稳定纯系个体的雄穗育性恢复情况，并总结有利于开展单倍体育种的基础材料，为优化单倍体技术流程提供参考。

1 材料和方法

1.1 试验材料

试验所用的玉米单倍体诱导系京科诱 044 为北京市农林科学院玉米研究中心引进国外基础材料经多代自交选育而成。该诱导系具有控制籽粒顶层和形成胚芽色素的 $R-nj$ 基因及控制不定根、叶鞘和茎秆色素形成的 *ABPI* 基因，使诱导产生的后代具有籽粒和植株双显性遗传标记。用于诱导单倍体的基础材料为塘四平头杂种优势群的 7 个优良玉米自交系京 92、HC7、C632-1、C632-2、京 179、京 2416 和京 24 相互杂交组配的 12 个 F_1（包括 2 个反交组合），具体材料见表 1。

表 1 基础材料

编号	组合名称	类型	编号	组合名称	类型
01	京 92/C632-1	单交种	07	HC7/C632-1	单交种
02	京 92/C632-2	单交种	08	HC7/C632-2	单交种
03	京 92/京 2416	单交种	09	HC7/京 2416	单交种
04	京 92/京 24	单交种	10	HC7/京 24	单交种
05	京 92/京 179	单交种	11	HC7/京 179	单交种
06	京 179/京 92	单交种	12	京 179/HC7	单交种

注：京 92/京 179 和京 179/京 92，HC7/京 179 和京 179/HC7 为两组正反交组合

1.2 试验方法

单倍体诱导系与 12 份 F_1 材料的杂交于 2011 年夏季在北京市农林科学院玉米研究中心小汤山育种基地进行。为了保证诱导系与 F_1 的花期能够相遇，将 F_1 一次播种，单倍体诱导系京科诱 044 分别在 3d 后和 5d 后分两期播种。在花期以诱导系为父本分别对 12 个 F_1 进行杂交诱导，每个组合杂交 20 穗以上。成熟后收获杂交果穗，并依据 Navajo 标记挑选杂交的籽粒，将顶部为紫色、胚部无色的籽粒视为拟单倍体籽粒。当选的拟单倍体籽粒于 2011 年冬季在北京市农林科学院海南玉米南繁育种基地单粒点播种植，精细管理，在苗期依据 ABPI 紫色植株标记进行进一步鉴定。淘汰幼苗叶鞘为紫色的植株，选择幼苗叶鞘为绿色、生长发育慢、植株较矮、叶片上冲和叶色较浅的单倍体植株进行统计。

抽雄后，每天调查雄穗的散粉情况，将花药散露的植株（即雄穗加倍株）分为 A、B、C 三种类型进行标记（A 型：花药大、散粉正常；B 型：花药小、散粉正常；C 型：花药小、散粉不正常）。同时，通过人工辅助的手段将花药散露的单倍体植株自交，其中花药小、散粉不正常的植株用小刀将花药切开再进行自交。成熟后，将结实植株的果穗收获标记为结实株，并登记结实株的花药散露情况。2012 年夏季在北京市农林科学院玉米研究中心小汤山育种基地将结实的籽粒按照穗行单粒点播进行纯合度鉴定，淘汰植株高大、生长势旺盛的穗行，保留个体间植株性状典型一致的穗行，并标记为 DH 系。

1.3 统计项目

单倍体诱导率 = 单倍体数 / 杂交籽粒数（带标记）× 100%

雄穗加倍率 = 雄穗加倍株数 / 单倍体数 × 100%

加倍结实率 = DH 系数 / 单倍体数 × 100%

1.4 数据处理

利用 DPS7.05 数据处理软件和 Microsoft Excel 2007 进行数据处理和统计分析。

2 结果与分析

2.1 单倍体诱导率的差异性分析

由表 2 可知，本研究杂交诱导的基础材料数量较大，杂交总粒数最多的为 19 998 粒，平均为 12 881.4 粒，统计所得的诱导率具有一定的代表性。12 份不同基因型 F_1 通过玉米单倍体诱导系杂交所得的单倍体诱导率存在显著的差异，最高为 11.03%，最低为 3.34%。5 个自交系与京 92 和 HC7 组配组合的单倍体诱导率均值由高到低依次为京 179（9.09%）＞京 24（7.37%）＞京 2416（7.05%）＞632-1（5.50%）＞632-2（5.36%）。正反交组合京 92/京 179 和京 179/京 92 的诱导率相差 0.15%，HC7/京 179 和京 179/HC7 的诱导率相差 0.14%，都不存在显著的差异。

表 2　不同基因型材料诱导率的基本统计

基础材料	杂交籽粒数	拟单倍体籽粒数	杂株数	单倍体数	单倍体诱导率（%）
京 92/632-1	4 230	333	8	324	7.66 abcd
京 92/632-2	5 960	316	20	294	5.70 cde
京 92/京 2416	18 400	1 203	90	1 109	6.04 cde
京 92/京 24	14 136	996	57	934	6.60 cd
京 92/京 179	17 565	1 413	79	1 323	7.60 abcd
京 179/京 92	15 557	1 278	70	1 201	7.75 abc
HC7/632-1	8 214	328	48	274	3.34 e
HC7/632-2	5 629	305	18	283	5.01 de
HC7/京 2416	17 930	1 555	98	1 450	8.05 abc
HC7/京 24	19 998	1 880	247	1 624	8.13 abc
HC7/京 179	19 727	2 276	233	2 038	10.43 ab
京 179/HC7	7 231	905	110	762	10.57 a

注：同一列数字后小写字母标记分别表示处理间差异达 0.05 水平差异显著。下同

2.2 单倍体加倍结实率的差异性分析

单倍体在大田栽培条件下任其自然恢复加倍，有利于自然选择和人工选择有机结合，淘汰不利基因型，选出优良的基因型个体。然而在自然条件下单倍体植株一般无法正常减数分裂，不能形成正常的花粉，主要表现雄性不育，只有少部分单倍体植株能发生染色体自然加倍，花药外露，产生少量或极少量的花粉。因此需每天仔细观察雄穗的加倍情况，发现有外露的花药时，及时进行人工辅助自交。

由表 3 可知，12 个 F_1 的单倍体植株的雄穗加倍率和加倍结实率存在较大的差异，雄穗加倍率最高为 29.31%，最低为 10.37%；加倍结实率最高为 13.20%，最低为 2.61%，均达显著水平的差异。5 个自交系分别与京 92 和 HC7 组配组合的平均雄穗加倍率和平均加倍结实率由高到低依次为京 179（11.10%）>C632-1（6.16%）>C632-2（5.74%）>京 24（3.71%）>京 2416（3.55%）。正反交组合京 92/京 179 和京 179/京 92 的雄穗加倍率相差较小，为 0.57%，差异不显著；加倍结实率相差 1.03%，差异也不显著。正反交组合 HC7/京 179 和京 179/HC7 的雄穗加倍率和加倍结实率相差较大，分别为 5.66% 和 3.34%，都表现出显著水平的差异。

单倍体田间授粉时需每天进行观察雄穗的育性恢复情况，同时个体间的雄穗雄穗育性恢复的情况存在一定的差异性，因此有必要进一步分析加倍结实株的花药散露情况。将加倍雄穗的花药分为 A（花药大、散粉正常）、B（花药小、散粉正常）、C（花药小、散粉不正常）三种类型（图1）进行数据统计分析。结果表明：A 型花药和 B 型花药占加倍结实株的比例较高，平均占 67.20% 和 22.32%，而 C 型花药的比例最低，为 10.48%。

表 3 不同基因型材料单倍体加倍率的基本统计

基础材料	单倍体数	雄穗加倍株数	雄穗加倍率（%）	加倍结实株数	DH系数	加倍结实率（%）	A	比率（%）	B	比率（%）	C	比率（%）
京92/C632-1	324	93	28.70a	21	21	6.48c	18	85.71	3	14.29	0	0.00
京92/C632-2	294	65	22.11bc	15	14	4.76cde	11	78.57	2	14.29	1	7.14
京92/京2416	1 109	115	10.37g	31	29	2.61e	21	72.42	5	17.24	3	10.34
京92/京24	934	136	14.56ef	34	29	3.10de	12	41.38	12	41.38	5	17.24
京92/京179	1 323	356	26.91a	156	148	11.19ab	80	54.05	52	35.14	16	10.81
京179/京92	1 201	330	27.48a	130	122	10.16b	66	53.95	43	34.85	13	10.30
HC7/C632-1	274	44	16.06ef	27	16	5.84cd	12	75.00	2	12.50	2	12.50
HC7/C632-2	283	55	19.43cd	21	19	6.71c	10	52.63	5	26.32	4	21.05
HC7/京2416	1 450	199	13.72f	67	65	4.48cde	48	73.85	11	16.92	6	9.23
HC7/京24	1 624	277	17.06de	72	70	4.31cde	46	65.72	18	25.71	6	8.57
HC7/京179	2 038	482	23.65b	212	201	9.86b	146	72.64	39	19.40	16	7.96
京179/HC7	795	233	29.31a	114	105	13.20a	76	71.95	21	20.00	8	8.05

 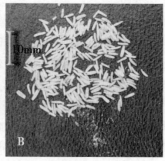 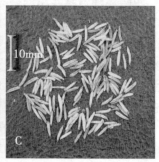

图 1 雄穗花药散露类型

2.3 相关性分析

如表 4 所示,12 个 F_1 的诱导率与雄穗加倍率相关系数为 0.34,未达到显著水平;诱导率与结实率相关系数为 0.37,未达到显著水平;而雄穗加倍率与结实率的相关系数为 0.77,达到极显著水平。说明在玉米单倍体育种中,基础材料的诱导和加倍是两个相对独立的环节,诱导率与雄穗加倍率和加倍结实率之间相互影响较小;而在加倍环节中,雄穗加倍率与加倍结实率紧密相关。

表 4 相关系数

变量 1	变量 2	相关系数
诱导率	雄穗加倍率	0.34
诱导率	加倍结实率	0.37
雄穗加倍率	加倍结实率	0.77**

注:* 和 ** 分别表示在 0.05 和 0.01 水平上差异显著

3 讨论

单倍体诱导率、雄穗加倍率和加倍结实率受基础材料遗传组成影响。本研究 12 个 F_1 的亲本来源均为塘四平头杂种优势群,但由于不同组合基因型的差异,单倍体诱导率和雄穗加倍率及加倍结实率表现出显著的差异,这与前人研究的结果相一致。另外,两组正反交组合的单倍体诱导率都不存在显著的差异,其中一组正反交组合的雄穗加倍率和加倍结实率差异较大,达显著水平,而另一组正反交组合的雄穗加倍率和加倍结实率不存在显著的差异。因此在前期基础材料的组配时应充分考虑这一影响因子,同时组配正反交材料用于单倍体诱导和加倍研究。但由于本研究所用正反交组合数量较少,对于其他材料的正反交组合的雄穗加倍率和加倍结实率是否也存在较大差异,还需进一步增加试验材料来证明。

玉米单倍体育种是快速选育纯系的有效方法。但有些基础材料通过大量的单倍体诱导,由于其单倍体诱导率和加倍结实率较低等原因,所得 DH 系的数量十分有限,无法满足育种目标的需求。因此,分析不同材料的单倍体诱导率和加倍结实率的差异性,并总结有利于开展单倍体育种的材料,对单倍体诱导规模的控制及提高单倍体育种效率具有重要的意义。本研究所用的自交系京179与两个自交系京92和HC7所组配F_1的单倍体诱导率、雄穗加倍率和加倍结实率都相对较高,可以在一定程度上说明该自交系更适合于单倍体育种。因此,可利用其高单倍体诱导率和高加倍结实率的特性,进一步与其他材料组配用于单倍体诱导和加倍,以为选育出更加优良的纯系奠定基础。而自交系京2416和京24与两个自交系京92和HC7所组配F_1的单倍体雄穗加倍率和加倍结实率都相对较低,所以在早期需进一步加大基础材料的单倍体杂交诱导数量,以期更好的获得符合育种目标的优良纯系。

另外,本研究结果还表明单倍体的诱导率与雄穗加倍率和加倍结实率之间的相关性不显著,说明两个环节之间的影响较小。加倍结实率与雄穗加倍率之间存在极显著的相关性,说明单倍体加倍结实与雄穗的育性紧密相关,这与徐国良等研究表明单倍体雄穗的育性是单倍体加倍结实的限制因素结果具有相似性。由于单倍体个体能够产生外露雄穗的比率较高,花期较短,人工辅助授粉工作量大,然而加倍结实率又较低,且个体间雄穗育性的恢复程度存在较大差异,如果能总结加倍结实株的雄穗育性恢复的主要类型,将可以在很大程度上提高育种效率。本研究分析雄穗花药散露的不同类型对加倍结实株的贡献,结果表明在加倍获得纯系的单倍体中雄穗育性恢复程度为正常散粉的大花药比率最高,雄穗育性恢复程度为正常散粉的小花药次之,而不能正常散粉的小花药比率最小,可能原因是不能正常散粉小花药中的花粉量极少或花粉的活力较低。因此在进行单倍体田间授粉时,应着重观察雄穗具有正常散粉花药的单倍体植株,并及时人工辅助授粉,适当的放弃不能正常散粉的小花药的植株,以在授粉同等数量单倍体的情况下增加获得更多纯系的可能性。

参考文献(略)

本文原载:种子,2013,32(1):1-4

6个玉米单倍体诱导系诱导率的差异性研究

张如养　段民孝　赵久然　刘新香　邢锦丰
王元东　张华生　Chang Mingtang

（北京市农林科学院玉米研究中心，北京　100097）

摘　要：利用6个玉米单倍体诱导系对6个玉米自交系通过杂交进行单倍体诱导，分析不同单倍体诱导系诱导率的差异性。结果表明，6个不同诱导系的诱导率均值存在显著差异，依次为京科诱043（6.06%）＞京科诱044（5.45%）＞京科诱045（5.24%）＞京科诱006（4.11%）＞京科诱041（2.45%）＝京科诱005（2.45%）。6个自交系与同一诱导系杂交得到的单倍体诱导率均值也存在显著差异，依次为自交系京724（5.53%）＞农系531（5.42%）＞京007（5.28%）＞京24（3.78%）＞京92（3.37%）＞郑58（2.38%）。

关键词：玉米；单倍体；诱导系；诱导率

Difference in the Haploid Induction Rates of Six Maize Inducers

Zhang Ruyang　Duan Minxiao　Zhao Jiuran　Liu Xinxiang　Xing Jinfeng
Wang Yuandong　Zhang Huasheng　Chang Mingtang

(*Maize Research Center*, *Beijing Academy of Agriculture & Forestry Sciences*, *Beijing* 100097, *China*)

Abstract: In order to analyze the differences in haploid induction rates of the different inducers, six maize inducers with purple crown and purple plumula color as the male parents to pollinate the six inbred lines. The results showed that the mean haploid induction rates were significant differences among inducers, from high to low was Jingkeyou043（6.06%）＞Jingkeyou044（5.45%）＞Jingkeyou045（5.24%）＞Jingkeyou006（4.11）＞Jingkeyou041（2.45）＝Jingkeyou005（2.45）; and also indicated that the haploid induction rates of the different inbred lines were significant difference, from high to low was Jing724（5.53%）＞Nongxi531（5.42%）＞Jing007（5.28%）＞Jing24（3.78%）＞Jing92（3.37%）＞Zheng58（2.38%）. Therefore, the haploid inducer line JingKeYou044, which had high haploid induction rates, low coefficient variation and markedness, was an elite haploid inducer.

Key words: Maize (Zea mays L.); Haploid; Haploid inducer; Haploid induction rate

---

基金项目：国家863项目"玉米DH工程化育种技术与新品质选育研究"（2012AA101203）、北京市科委项目（D111100001311001）、北京农林科学院青年基金（QN201106）

作者简介：张如养（1987—　），男，硕士，研究方向为玉米种质资源创新与利用。Tel：15201196859，E-mail：ruyangzhang2009@sina.com

赵久然和段民孝为本文通讯作者

单倍体的获得是玉米 DH（Double Haploid）工程化育种的一个重要环节，其获得数量的多少直接影响到工程化育种的效率。目前，利用单倍体诱导系进行遗传诱导是获得玉米单倍体最为有效的方法，即利用玉米单倍体诱导系对基础材料通过杂交进行诱导获得单倍体的方法。从 1959 年 Coe 发现单倍体诱导系 Stock6 以来，经国内外专家学者的研究和选育，玉米单倍体诱导系的研究取得了重大进展，先后选育出了一批具有应用价值的诱导系。其中应用较广的主要有 ZMS（Zarodyshevy Marker Saratovsky）、KMS（Korichnevv Marker Saratovsky）、MHI（Moldovian Haploid Inducer）、农大高诱号系列、吉高诱系 3 号和单倍体诱导系 H01~H05 等。国内外研究表明同一诱导系对不同基础材料的单倍体诱导率存在较大差异，因此选择诱导率高且稳定的诱导系进行单倍体诱导显得尤为重要。本研究利用 6 个玉米单倍体诱导系对 6 个自交系通过杂交进行单倍体诱导，分析不同诱导系对同一自交系诱导率的差异性及不同自交系与同组诱导系杂交得到的单倍体诱导率的差异性，探讨诱导系的选用策略，并总结标记明显、诱导率较高且稳定的诱导系，以期为单倍体的大批量获得及其育种效率的提高提供参考。

1 材料与方法

1.1 供试材料

试验所用的 6 个玉米单倍体诱导系京科诱 005、京科诱 006、京科诱 041、京科诱 043、京科诱 044 和京科诱 045 均由北京市农林科学院玉米研究中心引进国外基础材料经多代自交选育而成，具有花粉量大、结实性较好、抗病性强等优良性状。其中京科诱 005、京科诱 041、京科诱 043、京科诱 044 具有控制籽粒顶层和形成胚芽色素的 $R\text{-}nj$ 基因（籽粒具有 Navajo 标记），还含有控制不定根、叶鞘和茎秆色素形成的 $ABPI$ 基因，使诱导产生的后代具有籽粒和植株双显性遗传标记。而京科诱 006、京科诱 045 具有控制籽粒顶层和形成胚芽色素的 $R\text{-}nj$ 基因，但没有植株色泽标记。受体材料均为玉米骨干自交系，京 24、京 724、京 92、京 007、郑 58、农系 531。具体材料见表 1、表 2。

表 1 玉米单倍体诱导系
Table 1 The maize hapliod inducers

序号 No.	材料类型 Type	材料名称 Material name	种质来源 Germplasm origin	性状标记特征 Marker trait
A01	诱导系	京科诱 005	Stock6 衍生系	籽粒和植株双显性
A02	诱导系	京科诱 006	Stock6 衍生系	紫色籽粒显性
A03	诱导系	京科诱 041	Stock6 衍生系	籽粒和植株双显性
A04	诱导系	京科诱 043	Stock6 衍生系	籽粒和植株双显性
A05	诱导系	京科诱 044	Stock6 衍生系	籽粒和植株双显性
A06	诱导系	京科诱 045	Stock6 衍生系	紫色籽粒显性

表 2 基础材料

Table 2 Based materials

序号 No.	材料类型 Type	材料名称 Material name	基础种质类型 Based groud germplasm
B01	自交系	京 24	塘四平头群
B02	自交系	京 724	兰卡斯特群
B03	自交系	京 92	塘四平头群
B04	自交系	京 007	瑞得群
B05	自交系	郑 58	瑞得群
B06	自交系	农系 531	BSSS

1.2 试验设计

本试验 6 个自交系的杂交诱导于 2011 年冬季在北京市农林科学院海南玉米育种育种基地进行。为了保证诱导系与自交系的花期能够相遇，将 6 个自交系一次播种，将 6 个诱导系分别在 4d 后和 6d 后分两期播种。在花期取诱导系京科诱 005、京科诱 006、京科诱 041、京科诱 043、京科诱 044、京科诱 045 的花粉分别授予 6 个自交系（完全吐丝），每一杂交组合授粉 18 穗左右，成熟后收获。依据诱导系的 Navajo 标记挑选收获的杂交籽粒，将顶部紫色、胚部无色的籽粒视为拟单倍体籽粒，并统计不同杂交组合籽粒的总粒数。

2012 年夏季将挑选的拟单倍体籽粒全部种植在北京市农林科学院玉米研究中心小汤山玉米育种基地，采用高密度单粒点播，密度为 150 000 粒/hm^2。田间精细管理，拔节后依据植株色泽标记和生长势进一步鉴定，植株出现紫色或生长势强的植株均为杂合二倍，而生长发育慢、植株较矮、叶片短且上冲植株为单倍体，并统计相关数据。

1.3 统计方法

杂株率（%）= 杂株数/拟单倍体粒数×100

单倍体诱导率（%）= 田间鉴定单倍体数/总粒数×100

2 结果与分析

2.1 杂交诱导所得籽粒的主要类型

通过具有 Navajo 标记诱导系杂交诱导所得籽粒主要有以下两种类型：①籽粒顶部紫色，胚芽白色，为拟单倍体籽粒；②籽粒胚芽和顶部均为紫色，为杂合二倍体籽粒。具体籽粒类型如图 1 所示：A 图表示标记明显的拟单倍体籽粒；B 图表示标记不明显的拟单倍体籽粒，籽粒顶部标记有些退化，这种类型的籽粒在挑选时需仔细观察；C 图表示标记明显的杂交二倍体籽粒。

2.2 同一自交系由不同诱导系杂交诱导的诱导率

本研究杂交诱导所得的籽粒数量较大，杂交总籽粒数最多的为 6 217 粒，最少的为 1 514，平均为 3 293 粒，统计所得的诱导率具有一定的代表性。由于单粒点播的拟单倍体籽粒中可能存在发育不正常的籽粒，导致无法正常出苗和成株，因此所得单倍体数与杂株

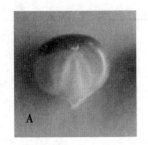 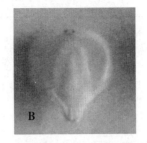 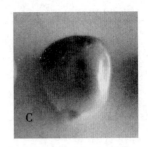

图1 杂交籽粒 Navajo 标记的表现型
Figure 1 The phenotype of Navajo marker in kernels

数的总和与播种的拟单倍体籽粒数不相等。另外，由于诱导系的标记退化等原因，可引起挑选的拟单倍体籽粒中存在一些杂合二倍体籽粒，而杂合二倍体的发芽势和生长势较强，几乎都能正常生长，依据田间植株标记和生长势可进一步鉴别。由表3可知，同一自交系由6个不同诱导系杂交诱导所得拟单倍体田间鉴定的平均杂株率，从低到高依次为京科诱044（30.76%）＜京科诱043（33.19%）＜京科诱005（37.39%）＜京科诱041（38.00%）＜京科诱045（39.45%）＜京科诱006（39.93%）。说明诱导系京科诱044和京科诱043的籽粒标记较为显著，而其他4个诱导系的籽粒标记都相对较差。

同时6个自交系京724、农系531、京007、京24、京92、郑58通过相同单倍体诱导系杂交诱导，所得单倍体诱导率均值存在显著差异，分别5.53%、5.42%、5.28%、3.78%、3.37%、2.38%。

表3 同一自交系由不同诱导系杂交诱导的诱导率
Table 3 The haploid induction rates in the same inbred lines

自交系 Inbred lines	诱导系 Inducers	总粒数（粒）Total kernel number	拟单倍体籽粒数（粒）Pre-hapliod number	田间鉴定杂株数 Hybrid number	杂株率（%）The rate of hybird	单倍体数 Hapliod number	单倍体诱导率（%）Hapliod induction rate	诱导率均值（%）Mean value
京24	京科诱005	3 941	204	96	0.47	101	2.56cd	
京24	京科诱006	3 969	260	130	0.50	117	2.95c	
京24	京科诱041	1 817	127	80	0.63	37	2.04d	
京24	京科诱043	2 274	245	111	0.45	124	5.45a	
京24	京科诱044	3 460	223	79	0.35	133	3.84b	
京24	京科诱045	3 120	484	312	0.64	182	5.83a	3.78ab
京724	京科诱005	2 119	116	18	0.16	90	4.25c	
京724	京科诱006	1 514	183	78	0.43	93	6.14b	
京724	京科诱041	3 081	128	35	0.27	85	2.76d	
京724	京科诱043	2 051	255	89	0.35	153	7.46a	

（续表）

自交系 Inbred lines	诱导系 Inducers	总粒数 （粒） Total kernel number	拟单倍体籽粒数 （粒） Pre-hapliod number	田间鉴定杂株数 Hybird number	杂株率 （%） The rate of hybird	单倍体数 Hapliod number	单倍体诱导率 （%） Hapliod induction rate	诱导率均值 （%） Mean value
京724	京科诱044	2 140	181	42	0.23	132	6.17b	
京724	京科诱045	3 325	331	110	0.33	213	6.41b	5.53a
京92	京科诱005	2 992	102	57	0.56	39	1.30d	
京92	京科诱006	2 979	242	112	0.46	122	4.10b	
京92	京科诱041	2 989	164	68	0.41	90	3.01c	
京92	京科诱043	4 382	301	156	0.52	129	2.94c	
京92	京科诱044	2 569	260	113	0.43	136	5.29c	
京92	京科诱045	2 457	180	78	0.43	88	3.58bc	3.37ab
京007	京科诱005	3 908	144	48	0.33	88	2.25e	
京007	京科诱006	4 605	408	187	0.46	216	4.69d	
京007	京科诱041	4 150	121	21	0.17	96	2.31e	
京007	京科诱043	5 744	573	61	0.11	500	8.70a	
京007	京科诱044	4 862	369	69	0.19	289	5.94c	
京007	京科诱045	6 217	647	151	0.23	485	7.80b	5.28a
郑58	京科诱005	2 986	74	18	0.24	50	1.67c	
郑58	京科诱006	4 740	172	38	0.22	98	2.07bc	
郑58	京科诱041	3 188	84	36	0.43	24	0.75d	
郑58	京科诱043	2 941	138	36	0.26	84	2.86b	
郑58	京科诱044	4 220	324	104	0.32	188	4.46a	
郑58	京科诱045	2 579	120	44	0.37	64	2.48b	2.38b
农系531	京科诱005	2 434	139	67	0.48	65	2.67d	
农系531	京科诱006	2 895	220	72	0.33	137	4.73c	
农系531	京科诱041	4 510	278	100	0.36	171	3.80e	
农系531	京科诱043	1 773	247	75	0.30	159	8.97a	
农系531	京科诱044	1 974	218	69	0.32	138	6.99b	
农系531	京科诱045	3 673	328	117	0.36	197	5.36c	5.42a

注：小写字母标记表示处理间差异达0.05水平差异显著。下同

Note：The lowercase letters indicate significant difference at 5% level. The same below

2.3 同一诱导系对不同自交系的诱导率及其变异系数

由表4可知，诱导系京科诱043对6个自交系的诱导率均值最高，为6.06%，但变异系数较大，为45.29%，说明该诱导系在不同自交系间诱导率的稳定性较差；京科诱044的诱导率均值次之，为5.45%，但变异系数最低，为21.32%，说明该诱导系在不同自交系间诱导率的稳定性较好；诱导系京科诱006、京科诱045的诱导率均值和变异系数都居中；而诱导系京科诱005、京科诱041的诱导率均值最低，且变异系数较大，说明两个诱导系在诱导率和稳定性方面都存在一定的不足。6个诱导系对6个自交系的单倍体诱导率的变化趋势呈一定的相似性。6个诱导系对自交系京724、京007和农系531的诱导率均值都相对偏高，对自交系京24、京92和郑58的诱导率均值都相对偏低。

表4 同一诱导系对不同自交系的诱导率
Table 4 The haploid induction rates in the difference inbred lines

自交系 Inbred line	诱导系 Inducer					
	京科诱005	京科诱006	京科诱041	京科诱043	京科诱044	京科诱045
京24	2.56b	2.95b	2.04c	5.45c	3.84e	5.83c
京724	4.25a	6.14a	2.76b	7.46b	6.17b	6.41b
京92	1.30d	4.10c	3.01b	2.94d	5.29c	3.58d
京007	2.25b	4.69b	2.31bc	8.70a	5.94b	7.80a
郑58	1.67c	2.07e	0.75d	2.86d	4.46d	2.48d
农系531	2.67b	4.73b	3.80a	8.97a	6.99a	5.36c
均值	2.45	4.11	2.45	6.06	5.45	5.24
变异系数（%）	41.79	35.01	42.11	45.29	21.32	36.86

3 结论与讨论

目前，利用单倍体诱导系通过杂交进行诱导所得籽粒挑选的依据主要有Navajo标记和高油分花粉直感效应。本研究中单倍体籽粒挑选是依据Navajo标记进行，标记的明显程度代表该诱导系Navajo标记鉴别单倍体的效果。由于诱导系标记基因不稳定或诱导系与基础材料之间互作等原因，可引起杂交诱导的籽粒Navajo标记退化或不明显，从而在田间鉴定时会存在部分杂合二倍体植株。因此，应选用Navajo标记显著的诱导系进行杂交诱导。本研究表明，利用诱导系京科诱044、京科诱043进行诱导，在所挑选的拟单倍体籽粒中杂合二倍体的比例较低，说明两个诱导系的籽粒标记较强，而其他4个诱导系籽粒标记较差。

同时，利用遗传诱导方法获得单倍体的频率受亲本基因型影响。本研究结果表明，不同诱导系对同一自交系进行杂交诱导所得单倍体诱导率存在显著的差异，这与刘志增等研究父本基因型对同一材料的单倍体诱导率有影响的结果相一致。另外，不同自交系由相同诱导系杂交诱导所得的单倍体诱导率也存在差异性，且不同诱导系对不同自交系诱导率的变化趋势呈一定的相似性，但变化程度存在差异。诱导系京科诱006、京科诱044、京科诱045对不同自交系诱导率的变异系数较小，说明这3个诱导系诱导率的稳定性较高，而其他 3 个诱导系京科诱 005、京科诱 041、京科诱 043 诱导率的稳定性较差。因此,诱导系的诱导能力不仅受自身遗传背景的影响,也受其诱导的基础材料遗传背景的影响,但由于本研究所用的基础材料有限,还需进一步增加研究材料来说明。

  在诱导系选用时,可根据基础材料的来源选择两个及以上诱导率较高、相对稳定且标记明显的诱导系对其进行杂交诱导,以在杂交诱导同等数量基础材料的情况下获得更多的单倍体,提高单倍体育种效率,减少诱导成本。本研究中诱导系京科诱 044 的籽粒标记显著、诱导率较高且稳定,是一个优良的玉米单倍体诱导系,可以作为重点材料进行深入的研究和利用。而诱导率均值最高、标记显著但变异系数较大的诱导系京科诱 043,可作为辅助诱导系应用于基础材料的杂交诱导。而其他 4 个诱导系在籽粒标记明显程度、诱导率或稳定性等方面存在一定的缺陷。

参考文献(略)

本文原载:玉米科学,2013,21(2):6-10

8个玉米单倍体诱导系诱导率的配合力研究

张如养　段民孝　赵久然　刘新香　邢锦丰　宋　伟
王元东　张华生　张雪原　张春原

(北京市农林科学院玉米研究中心/玉米DNA指纹及分子育种北京市重点实验室，北京　100097)

摘　要： 利用8个玉米单倍体诱导系，按照NCⅡ方法进行杂交组配得28个诱导系 F_1，与6个骨干自交系京724、京92、京X005、昌7-2、郑58和农系531杂交，进行诱导率测验，分析诱导系诱导率的配合力及其遗传特性。结果表明诱导系JKY044、JKY043、JKY042的一般配合力效应较高，是优良的育种材料。特殊配合力分析表明3个诱导系组合JKY042/JKY044、JKY005/JKY041、JKY042/JKY046诱导率的特殊配合力效应较高。遗传参数分析表明诱导率性状主要以加性效应为主，狭义遗传力较高，宜在早代进行选择。

关键词： 玉米；单倍体；诱导系；诱导率

Combining Ability of Haploid Induction Rate of Eight Inducers in Maize

Zhang Ruyang　Duan Minxiao　Zhao Jiuran　Liu Xinxiang
Xing Jinfeng　Song Wei　Wang Yuandong　Zhang Huasheng
Zhang Xueyuan　Zhang Chunyuan

(*Maize Research Center, Beijing Academy of Agriculture & Forestry Sciences / Beijing Key Laboratory of Maize DNA Fingerprinting and Molecular, Beijing* 100097)

Abstract： Twenty-eight inducer F_1 were obtained by he eight haploid inducers crossed by NCⅡ method. The eight haploid inducers and the twenty-eight inducer F_1 were used as the male parent. The six inbred lines, Jing 724, Jing 92, Jing 464, Chang 7-2, Zheng 58, Nongxi531 were used as the

---

基金项目：国家863项目"玉米DH工程化育种技术与新品种选育研究"（编号：2012AA101203）、北京市农林科学院青年科研基金"玉米单倍体自然加倍适宜区域的研究"（编号：QNJJ201418)、国家科技计划课题"华北西北春夏播区强优势玉米杂交种的创制与应用"（编号：2011AA10A103-3）、2014年现代农业产业技术体系"北京市创新团队建设-岗位专家"、北京市农林科学院科技创新能力建设专项"玉米种质资源发掘、收集整理与创新"（编号：KJCX20140107）

作者简介：张如养（1987—　），男，硕士，研究方向为玉米种质资源创新与利用。Tel：15201196859，E-mail：ruyangzhang2009@126.com
　　　　　赵久然、段民孝为本文通讯作者

female parent. The cross between male and female was made manually both in Hainan and Beijing. The Combining ability of haploid induction rate and genetic parameters were analyzed in inducers. The results indicated that the general combining ability of JKY044, JKY043, JKY042 were higher. The special combining ability analysis indicated that the inducer JKY042/JKY044、JKY005/JKY041、JKY042/JKY046 were outstangding. At the same time, the genetic parameters analysis showed that the haploid induction rate was mainly effected by additive effects, which have high narrow heritability, should be selected in early generations.

Key words: Maize; Haploid; Inducer; Haploid induction rate

玉米单倍体育种技术可以有效缩短育种年限、加快育种进程，为种质的改良和新品种的选育提供重要的技术支撑。利用单倍体诱导系通过遗传诱导方法获得单倍体是一个重要的技术环节，因此研究单倍体诱导系的首要性状就是其诱导率的高低。从最早发现的单倍体诱导系 Stock6 到目前国内应用较为广泛的中国农业大学选育的高油系列诱导系、吉高诱 3 号，JKY 系列玉米单倍体诱导系，单倍体诱导系的诱导率有了较大的提高，通过遗传改良的方法提高单倍体诱导率是行之有效的。

诱导率的遗传改良需要采用在诱导率性状上具有较高利用潜力的诱导系来构建基础群体，并采取优化的选育方法和策略来进行。玉米单倍体诱导系诱导率的配合力高低，是衡量该诱导系是否具有进一步利用价值的重要标准。国内外已有单位研究利用不同诱系导间的杂种优势进一步提高诱导率。因此，研究不同诱导系诱导率的配合力及其遗传特性，对较高诱导率的诱导系选育和杂种优势利用具有重要的意义。本研究利用 8 个不同来源的优良单倍体诱导系，按照 NC Ⅱ 方法进行杂交组配 28 个诱导系 F_1，以 6 个自交系京 724、京 92、京 X005、昌 7-2、郑 58 和农系 531 杂交，进行单倍体诱导率测验，通过诱导率的配合力测定和遗传参数分析，了解诱导率遗传特性以及诱导系的利用潜力，探讨单倍体诱导系的选育方法和策略，为较高诱导率的诱导系选育和杂种优势利用提供参考。

1 材料与方法

1.1 试验材料

试验材料为 28 个玉米诱导系 F_1 和 6 个测验种。28 个诱导系 F_1 是由 8 个玉米单倍体诱导系经不完全双列杂交而来，8 个玉米单倍体诱导系均为北京市农林科学玉米研究中心引进国外不同来源的优良单倍体诱导材料经多代自交和测验选育而成。诱导系 JKY005、JKY041、JKY042、JKY043、JKY044 和 JKY046 都具有控制籽粒顶层和形成胚芽色素的 R-nj 基因（籽粒具有 Navajo 标记），还含有控制不定根、叶鞘和茎秆色素形成的 ABPI 基因，使诱导产生的后代具有籽粒和植株双显性遗传标记；JKY006、JKY045 具有控制籽粒顶层和形成胚芽色素的 R-nj 基因，没有植株色泽标记。测验种为 6 个玉米骨干自交系京 724、京 92、京 x005、郑 58、昌 72 和农系 531，其中，京 724、京 92 和京 x005，为北京市农林科学院玉米研究中心自主选育；郑 58、昌 72 和农系 531 为其他单位引进。

1.2 试验设计

1.2.1 诱导系与测验种杂交

28个单倍体诱导系F_1与6个骨干自交系测验种于2012年冬季在北京市农林科学院海南玉米南繁育种基地进行杂交。为了保证诱导系材料与测验种的花期能够相遇，将测验种一次播种，28个单倍体诱导系F_1分别在3d后和5d分两期播种，在花期以28诱导材料为父本分别对6个测验种进行杂交诱导，每个组合杂交15穗左右。

1.2.2 单倍体统计

成熟后收获杂交果穗，脱粒后统计顶部紫色籽粒总数，即为杂交籽粒数。依据Navajo标记将顶部为紫色、胚部无色的籽粒视为候选单倍体籽粒。将候选单倍体籽粒于2013年夏季在小汤山育种基地进行单粒点播种植，精细管理，在苗期依据ABPI紫色植株标记和植株生长势进行进一步的鉴定。淘汰幼苗叶鞘为紫色和生长势强的植株，将幼苗叶鞘为绿色、生长发育慢、植株较矮、叶片上冲和叶色较浅的植株确定为单倍体。诱导系的诱导率以对6个测验种的均值进行整理分析。2013年夏季在北京市农林科学院玉米研究中心小汤山育种基地进行重复诱导试验，并在海南进行单倍体鉴定。

1.3 诱导率统计

$$单倍体诱导率（\%）=单倍体数/杂交籽粒数\times100\%$$

1.4 数据分析

采用NCⅡ方法，进行配合力方差分析，并进一步估算诱导率一般配合力（GCA）和特殊配合力（SCA）的相对效应值及其遗传参数估计。用SAS 9.0和Microsoft Excel 2007等数据软件进行数据处理和分析。

$$GCAg_i = \overline{X}_i - \overline{X}_{总}$$

$$SCAg_{ij} = \overline{X}_{ij} - \overline{X}_{总} - GCAg_i - GCAg_j$$

2 结果与分析

2.1 诱导率的配合力方差分析

研究现有诱导材料配合力，可以初步探索其利用价值和应用潜力。诱导率配合力方差分析如表1所示。结果表明，诱导率的GCA在不同单倍体诱导系间的差异均达到极显著水平；同时，诱导率SCA在不同的组合间差异也都达极显著水平。

表1 诱导率配合力效应值方差分析

Table 1　The variance analysis of combining ability value

变异来源 Variation	自由度 Df	平方和 SS	均方 MS	F值 F-value
一般配合力	7	39.53	5.65	131.63**
特殊配合力	27	22.26	0.80	18.53**

2.2 一般配合力效应值分析

不同单倍体诱导系诱导率性状GCA效应值存在极显著的差异，进一步进行多重比较

（表2）。多重比较分析表明，GCA效应值最高的是JKY044（0.8897），最低的是JKY005（-0.0339）。

表2 诱导率一般配合力效应值多重比较
Table 2 The multiple comparisons of GCA values in induction rate

诱导系 Inducers	GCA效应值 GCA-value	5%显著水平 5% Significant level	1%极显著水平 1% Significant level
JKY044	0.8897	a	A
JKY043	0.6193	b	B
JKY042	0.5201	b	B
JKY045	0.4924	b	B
JKY046	-0.0272	c	C
JKY006	-0.4763	d	D
JKY041	-0.9841	e	E
JKY005	-1.0339	e	E

2.3 特殊配合力的效应值分析

由特殊配合力的效应值分析可知，杂交组合JKY042/JKY044、JKY005/JKY041、JKY042/JKY046、JKY005/JKY045、JKY041/JKY042诱导率的SCA效应值相对较高，均在1以上，分别为1.6429、1.1792、1.1667、1.1559、1.1454。由特殊配合力正向效应的不同组合的亲本比较可以看出，诱导率GCA效应值较低的诱导系（如JKY005）组配组合中，存在SCA正向效应值较高的组合，如JKY005/JKY041（1.1792）和JKY005/JKY045（1.1559）。说明诱导率GCA值较低的诱导系，通过正确选择组配亲本，也可出现特殊配合力为正向效应值较高的组合。既具有高GCA，同时与多个材料组配组合的SCA也较高的材料是JKY042和JKY004（表3）。

表3 特殊配合力效应值分析
Table 3 The analysis of SCA value

杂交组合 Hybrid	诱导率 Induction rate	杂交组合 Hybrid	诱导率 Induction rate
JKY005/JKY006	0.4413	JKY041/JKY043	0.6074
JKY005/JKY041	1.1792	JKY041/JKY044	-0.1118
JKY005/JKY042	0.1785	JKY041/JKY045	-1.0884
JKY005/JKY043	-0.2883	JKY041/JKY046	-0.3421
JKY005/JKY044	-0.081	JKY042/JKY043	0.4073
JKY005/JKY045	1.1559	JKY042/JKY044	1.6429
JKY005/JKY046	-1.0912	JKY042/JKY045	-0.3429

(续表)

杂交组合 Hybrid	诱导率 Induction rate	杂交组合 Hybrid	诱导率 Induction rate
JKY006/JKY041	-0.643 7	JKY042/JKY046	1.166 7
JKY006/JKY042	-0.800 9	JKY043/JKY044	0.933 7
JKY006/JKY043	-0.176 3	JKY043/JKY045	-0.150 8
JKY006/JKY044	-1.209 9	JKY043/JKY046	-0.479 7
JKY006/JKY045	0.729 4	JKY044/JKY045	0.551 3
JKY006/JKY046	0.827 4	JKY044/JKY046	0.319 7
JKY041/JKY042	1.145 4	JKY045/JKY046	0.193 9

2.4 遗传参数分析

为进一步了解各性状的遗传表现，根据方差分析结果估算性状的遗传参数。从表4可以看出诱导率性状的加性方差大于显性方差，说明该性状的遗传主要以加性效应为主，显性基因的作用较小。因此通过亲本的选择，可以有效地改良该性状。遗传方差高于环境方差，也高于表型方差，说明诱导率性状主要由遗传决定，受环境的作用较小。狭义遗传力较高，为73.36%，说明在早代诱导率性状可以进行选择。

表4 遗传参数估计
Table 4 The genetic parameter estimation

遗传参数 Genetic parameter	诱导率 Induction rate	遗传参数 Genetic parameters	诱导率 Induction rate
加性方差/AV	0.970 5	环境方差/EV	0.085 8
显性方差/DV	0.751 9	表型方差/PV	1.322 9
遗传方差/GV	1.722 0	狭义遗传力/NH	73.36%

3 结论与讨论

诱导率是玉米单倍体诱导系研究最重要的性状，其高低直接影响到杂交诱导效果的好坏，而对于构建选育优良诱导系的基础材料，最重要指标是其原始亲本诱导率一般配合力（GCA）的高低。前人研究表明，一般配合力表现是基因的加性效应，是能够稳定遗传的部分。因此诱导率GCA效应值高的材料比低的材料在后期育种研究中具有更高的价值。本研究筛选出诱导系JKY044、JKY043、JKY042、JKY045诱导率的一般配合力较高，具有较高的利用潜力，可以用这些诱导系作供体改良其他诱导系，为选育诱导率进一步提高的诱导系提供可能。也可将这些诱导系同其他一些诱导率性状GCA优良的材料杂交组配所得的优良诱导系F_1直接应用与单倍体诱导实践。

特殊配合力（SCA）是杂交组合与其双亲平均表现基础上的预期结果的偏差，不能在上下代间稳定遗传，但它可以指导杂种优势的利用和杂交种的选育。特殊配合力分析表明，3个诱导系组合JKY042/JKY044、JKY005/JKY041、JKY042/JKY046诱导率的特殊配合力效应较高。通过正确的选择组配亲本，诱导率一般配合力效应值较低的诱导系能够组配出特殊配合力为正向效应值较高的组合。可能原因是特殊配合力为正向效应值较高组合的双亲组配后聚合了较多的微效基因位点。

对玉米单倍体诱导系诱导率性状遗传力的分析，有助于选用不同的育种策略，这对诱导率性状的改良和提高育种效率具有重要意义。狭义遗传力的大小反映了变异可以真实遗传的部分，具有可传递性，在育种中早代选择有效。本研究结果表明，该诱导率遗传主要以加性效应为主，通过有利基因的不断累加，可以有效地提高诱导率。同时诱导率主要由遗传决定，狭义遗传力较高，可在早代进行选择。

在新型玉米单倍体诱导系的选育中，首先要选择具有较高利用潜力的材料构建基础群体，再通过利用多个遗传背景差异较大的材料为测验种，在早代对基础群体进行诱导率测验，筛选诱导率性状优良的材料进一步研究，并及时淘汰诱导能力较差的材料，以提高诱导系的诱导率。

参考文献（略）

本文原载：玉米科学，2015，23（5）：12-15

玉米单倍体与其加倍后代性状的相关性研究

张如养　段民孝　赵久然*　刘新香　邢锦丰

(北京市农林科学院玉米研究中心，北京　100097)

摘　要：利用 8 个不同基因型 F_1 材料，经诱导和加倍获得大量单倍体和二倍体植株。研究单倍体与其二倍体之间 12 个形态性状的相关性。结果表明：植株个体从单倍体水平到二倍体，穗轴长、株高、穗位高和雄穗主轴长呈增大的趋势，穗行数保持相对稳定，雄穗一级分支数呈减小的趋势；除穗轴粗与穗行数的相关性外，其余多数性状间的相关性在单倍体和二倍体时期都具有相似性；穗轴长、穗行数、株高、穗位高、雄穗一级分支数和雄穗主轴长的相关性达显著或极显著水平；因此，在单倍体时期对性状进行适当的选择具有可行性。

关键词：玉米；单倍体；加倍单倍体；性状；相关性

　　单倍体是指含有亲本单拷贝染色体的个体，不论是显性还是隐性基因都能发挥其对性状发育的调控作用。单倍体的田间表现主要为植株紧凑、矮小、叶片较细薄、花器较小、高度不育等，而其加倍后代（即 DH 系，Double Haploid line），可恢复正常的形态和生理机能并能稳定遗传，可直接应用于育种实践。据有关报道，在一些作物中单倍体的一些性状在二倍体时期并没有表现，植物从单倍体到二倍体性状的表达涉及基因计量效应等影响。目前，有关玉米单倍体与其加倍后代的表型是否具有一定的相关性以及不同性状之间的相关性 2 个水平上是否存在一定差异的研究报道较少。针对这一问题，本研究利用不同遗传背景的单倍体为基础材料，调查了单倍体及其加倍后代 12 个数量性状的表现，分析单倍体与其加倍后代性状的相关性以及不同性状之间在单倍体和二倍体时期相关性的差异性，以期为在单倍体时期进行适当的选择提供理论依据。

1　材料与方法

1.1　试验材料

　　试验材料为 B19X/郑 58、京 724/京 464、昌 72/W13、X35/X96、京 92/京 179、

基金项目：国家 863 项目"玉米 DH 工程化育种技术与新品种选育研究"（编号：2012 AA 101203）、北京市农林科学院院青年基金（编号：QN 201106）、国家 863 项目"强优势玉米杂交种的创制与应用"（编号：2011 AA 10 A 103）、北京市农林科学院院青年基金（编号：QN 201405）

作者简介：张如养（1987—　），男，福建人；研究实习员，研究方向：玉米种质资源创新与利用；E-mail：ruyangzhang2009@126.com

* 通讯作者：赵久然（1962—　），男，北京人；研究员，博士，主要从事玉米研究；E-mail：maizezhao@126.com

NK72/X35、HK84/X35、D507/S902 共 8 个不同基因型的 F_1 利用玉米单倍体诱导系京科诱 044 授粉，诱导获得单倍体，并经加倍获得二倍体后代。

1.2 试验设计和调查项目

将所获得的单倍体籽粒单粒点播，良好的田间管理，依靠自然加倍，并对有育性恢复的植株进行人工授粉自交。以雄穗花药能正常散粉的植株为主要调查对象，2012 年调查单倍体数为 297 株，最后得到具有结实籽粒的单倍体为 230 株，平均结实粒数为 8 粒。将所得籽粒按穗行种植，其中有 31 个系不发芽，12 个系为杂合个体，最后所调查性状典型一致的 DH 系 187 个。2013 年调查结实的单倍体数为 149 株，平均结实粒数为 14 粒，播种后有 23 个系不发芽，18 个系为杂合个体，最后调查的性状典型一致的 DH 系为 108 个。

在生长发育中后期调查单倍体及其加倍后代植株的株高、穗位高、穗上功能叶数、穗数、穗位叶长、穗位叶宽、雄穗一级分支数、雄穗主轴长、气生根层数等 9 个数量性状；在果穗成熟后进行室内考种，调查穗轴长、穗轴粗、穗数等 3 个数量性状。统计分析植株各性状在单倍体和二倍体之间的相关性及不同性状之间的相关性在单倍体和二倍体水平的差异性。

1.3 数据处理

利用 SAS 9.0 数据处理软件和 Microsoft Excel 2003 进行数据处理和统计分析。

2 结果与分析

2.1 单倍体与二倍体时期数量性状的比较

由表 1 可知，单倍体水平与二倍体水平各数量性状值存在一定的差异。从单倍体到二倍体，穗轴长、穗轴粗、株高、穗位高、穗位叶长、穗位叶宽和雄穗主轴长 7 数量性状的均值都表现出增大的趋势，平均增大倍数分别为 1.46、1.36、1.57、1.85、1.38、1.61 和 1.40 倍；穗行数和穗上功能叶数在 2 个水平之间的变化较少；而雄穗一级分支数和穗数都呈现出减少的趋势；气生根层数的变幅不稳定，2012 年变化趋势较小，2013 年表现出减少趋势。

表 1 单倍体与其加倍后代性状比较

Table 1 The comparison of traits in two levels of ploidy

性状 Traits	年份 year					
	2012 年			2013 年		
	单倍体 Haploid	二倍体 Double Haploid	比值 ratio	单倍体 Haploid	二倍体 Double Haploid	比值 ratio
穗轴长	8.25	13.88	1∶1.68	9.93	12.23	1∶1.23
穗轴粗	1.66	2.68	1∶1.62	1.84	2.03	1∶1.10
穗行数	13.44	14.33	1∶1.07	13.25	14.50	1∶1.09
株高	136.23	175.51	1∶1.29	112.43	207.92	1∶1.85
穗位高	42.55	55.51	1∶1.30	28.88	68.92	1∶2.39
雄穗一级分支数	11.53	8.00	1∶0.69	6.75	6.11	1∶0.90

(续表)

性状 Traits	年份 year					
	2012年			2013年		
	单倍体 Haploid	二倍体 Double Haploid	比值 ratio	单倍体 Haploid	二倍体 Double Haploid	比值 ratio
气生根层数	1.41	1.45	1:1.03	1.66	1.16	1:0.70
穗上功能叶数	6.16	6.13	1:1.00	6.32	6.61	1:1.05
穗数	2.56	1.69	1:0.66	1.71	1.38	1:0.80
穗位叶长	55.92	72.63	1:1.30	52.29	76.40	1:1.46
穗位叶宽	5.60	9.96	1:1.78	5.83	8.46	1:1.45
雄穗主轴长	20.91	26.46	1:1.27	19.72	30.35	1:1.54

2.2 单倍体与二倍体水平性状间相关性的分析

由表2可知，单倍体与二倍体时期各性状之间的相关性存在差异。穗轴长与株高的相关性、穗轴长与雄穗主轴长的相关性、株高与穗位高的相关性、株高与穗上叶数的相关性、株高与雄穗主轴长的相关性、气生根数与穗数的相关性、穗位叶长与穗位叶宽的相关性以及穗位叶长于雄穗主轴长的相关性在单倍体水平和二倍体水平都表现型出显著或极显著的水平。穗轴粗与穗行数的相关性在二倍体水平达极显著的水平，而在单倍体水平相关性不显著。穗位高与雄分支数的相关性在单倍体水平达极显著的水平，而在二倍体水平的相关性不显著。穗轴长与穗轴粗的相关性在二倍体水平达显著的水平，而在单倍体水平相关性不稳定。

表2 各性状之间相关性在单倍体和二倍体间的差异
Table 2 The difference of the relations between traits in two levels of ploidy

相关变量1 Variance one	相关变量2 Variance two	相关系数 Correlation Coefficient			
		2012年		2013年	
		单倍体 Haploid	二倍体 Double Haploid	单倍体 Haploid	二倍体 Double Haploid
穗轴长	穗轴粗	0.19*	0.33*	0.18	0.28*
穗轴长	株高	0.17*	0.42**	3.34**	0.33*
穗轴长	雄穗主轴长	0.41**	0.50**	0.43**	0.52**
穗轴粗	穗行数	0.11	0.25**	0.11	0.55**
株高	穗位高	0.50**	0.68**	0.42**	0.44**
株高	穗上功能叶数	0.45**	0.55**	0.49**	0.43**
株高	雄穗主轴长	0.35**	0.49**	0.46**	0.39**
气生根层数	穗数	0.31**	0.25**	0.29*	0.38**
穗位叶长	穗位叶宽	0.46**	0.32**	0.39**	0.43**
穗位叶长	雄穗主轴长	0.20*	0.49**	0.47**	0.49**

注：* 和 ** 分别表示在0.05和0.01水平上差异显著。下同

Note: * and **. Significant difference at 0.05 and 0.01 probability levels, Respectively. The same below

2.3 单倍体与二倍体性状的相关性

配子体选择是一种有力的手段，它的潜力体现在双单倍体技术在育种上的应用。由于数量性状的表达受微效多基因控制，且不同性状的遗传存在差异，对数量性状进行配子体选择应更加慎重。由表3可知，从单倍体到二倍体性状的相关性存在差异。单倍体与其加倍后代的穗轴长、穗行数、株高、穗位高、雄穗一级分支数、穗上功能叶数和雄穗主轴长等8个数量性状存在显著或极显著的相关性；穗轴粗、穗数和穗位叶长等性状的相关性不显著，而气生根层数和穗位叶宽的相关性在2年试验中不稳定。

表3 单倍体与二倍体性状的相关性
Table 3 The relations between the traits of haploids and double haploids lever

性状 Traits	相关系数 Correlation Coefficient	
	2012 年	2013 年
穗轴长	0.43**	0.22*
穗轴粗	-0.04	-0.05
穗行数	0.23*	0.35**
株高	0.25**	0.35**
穗位高	0.22*	0.55**
雄穗一级分支数	0.53**	0.43**
气生根层数	0.34**	0.1
穗上功能叶数	0.17	0.28*
穗数	0.07	0.06
穗位叶长	0.09	0.06
穗位叶宽	0.04	0.32*
雄穗主轴长	0.23*	0.24*

3 讨论

植株个体从单倍体水平到二倍体水平，体细胞的染色体数目增加了一倍，而植株的农艺和穗部等数量性状的变化趋势研究报道较少。本研究结果表明，玉米从单倍体到二倍体水平，穗轴长、穗轴粗、株高、穗位高、穗位叶长、穗位叶宽和雄穗主轴长等7个数量性状的均值都变大，原因可能是由于这些性状受细胞染色体从单倍体到二倍体水平加倍的变化影响较为明显。穗行数和穗上功能叶数在2个水平之间的变化较少，可能原因是由于2个性状受细胞染色体水平变化的影响较小。雄穗一级分支数和穗数都呈现出减少的趋势，穗数减少的可能原因是单倍体授粉结实性较差，从源-库理论分析，后期源的供应大于库需求，从而增加了后期腋芽分化成雌穗。而二倍体水平穗部结实正常，源-库可以达到一个较好的平衡；气生根层数的变幅不稳定，可能原因为气生根的生长发育受种植的土壤透气性的影响，需进一步调查研究。

性状间相关是由于基因的多效性或控制不同性状的基因紧密连锁。本研究结果表明，多数性状间的相关性在单倍体时期和二倍体时期具有相似性，而穗轴粗与穗行数之间的相关性在二倍体时期达极显著水平，单倍体时期则不存在。可能原因为控制穗轴粗和穗行数微效多基因在单倍体时期与二倍体时期的表达存在差异。因此在单倍体时期利用穗轴粗和穗行数性状间的相关性进行辅助选择时应谨慎。

单倍体与加倍后代基因型是一致的，且单倍体水平无等位基因的互作效应，如果能够在单倍体时期进行适当的选择和淘汰，将有利于玉米单倍体育种效率的提高。研究结果表明，对于受微效多基因所控制的数量遗传性状，穗轴长、穗行数、株高、穗位高、雄穗一级分支数、穗上功能叶数和雄穗主轴长等在单倍体水平与二倍体体水平上存在显著或极显著的相关性，说明这些性状在单倍体时期进行适当的选择能起到一定的效果，这与 Chalyk 研究报道单倍体与其加倍后代穗部性状存在相关性的结果具有一致性；同时，穗轴长、穗位高和雄穗主轴长等性状具有增大的趋势，穗行数保持相对稳定，而雄穗分支数有减小的趋势，在选择时应给予关注。而穗轴粗、穗数和穗位叶长等性状的相关性不显著，不适宜在单倍体时期进行选择。由于本研究只统计到 2 年的数据，对于相关性不稳定的性状还需进一步增加试验材料来证明。综述所述，在单倍体时期对在单倍体与二倍体时期存在显著或极显著的相关的性状适当的选择具有可行性。

参考文献（略）

本文原载：种子，2014，33（11）

玉米单倍体诱导系诱导性状的杂种优势分析

张如养　段民孝　赵久然　刘新香　邢锦丰

(北京市农林科学院玉米研究中心/玉米 DNA 指纹及分子
育种北京市重点实验室，北京　100097)

摘　要：以 8 个玉米单倍体诱导系和按照 NC Ⅱ 方法进行杂交组配得 28 个诱导系 F_1，分析株高、雄穗小花数的杂种优势；并利用 6 个骨干自交系京 724、京 92、京 X005、昌 7-2、郑 58 和农系 531 为测验种进行单倍体诱导测验，分析诱导率的杂种优势。结果表明，25 个诱导系 F_1 的株高表现出高亲优势，最高超高亲 36.7cm；28 个诱导系 F_1 的雄穗小花数都表现出高亲优势，最高超高亲 265.4 个；15 个诱导系 F_1 诱导率表现出高亲优势，最高超高亲 2.30%；综合诱导率、株高、雄穗小花数三个性状分析，表现优良的组合有 JKY042/JKY044、JKY043/JKY044、JKY044/JKY045、JKY042/JKY046、JKY042/JKY043。因此通过正确选用不同来源诱导系进行杂交组配，可综合利用诱导率、株高、花粉量上的杂种优势用于单倍体的规模化人工隔离诱导。

关键词：玉米；单倍体；诱导系；杂种优势

杂种优势是指 2 个遗传性不同的亲本杂交产生的杂种一代在生长势、抗逆性、产量等方面优于其亲本的现象。在玉米常规育种中杂种优势得到广泛地应用，通过将 2 个遗传背景存在一定差异的自交系进行杂交获得具有强优势的组合应用于生产实践。国内外研究结果表明，不同玉米单倍体诱导系诱导率、株高等性状之间也存在杂种优势，因此如何将诱导性状杂种优势有效的应用于玉米单倍体育种实践中，不断的提高育种效率是玉米育种者需要深入研究的内容。提高单倍体诱导材料的诱导率可在同等诱导规模下获得更多的单倍体；增加单倍体诱导材料的株高和增大诱导材料的雄穗小花数可以满足 DH 工程化大规模隔离诱导材料的需求。

玉米单倍体诱导系间杂种优势的利用可为诱导率的进一步提高及其农艺性状的改良提供可能，其主要目的：①筛选诱导率提高，株高和花粉量等性状表现都优良的诱导系 F_1；

基金项目：国家 863 项目"玉米 DH 工程化育种技术与新品种选育研究"（编号：2012AA101203）、北京市农林科学院青年科研基金"玉米单倍体自然加倍适宜区域的研究"（编号：QNJJ201418）、国家科技计划课题"华北西北春夏播区强优势玉米杂交种的创制与应用"（编号：2011AA10A103-3）、2014 年现代农业产业技术体系"北京市创新团队建设-岗位专家"、北京市农林科学院科技创新能力建设专项"玉米种质资源发掘、收集整理与创新"（编号：KJCX20140107）

作者简介：张如养（1987—　），男，硕士，研究方向为玉米种质资源创新与利用。Tel：15201196859，E-mail：ruyangzhang2009@126.com

通讯作者：赵久然（1962—　），男，北京人；研究员，博士，主要从事玉米研究

②获得诱导率与高亲接近，而株高、花粉量得到明显改善的诱导系 F_1；③利用诱导率性状提高的 F_1 进行构建基础群体，为后期新型高频诱导系的选育奠定良好基础。本研究利用 8 个玉米单倍体诱导系和按照 NC Ⅱ 方法进行杂交组配的 28 个诱导系 F_1，分析株高、雄穗小花数的杂种优势，并利用 6 个骨干自交系京 724、京 92、京 X005、昌 7-2、郑 58 和农系 531 为测验种进行诱导率测验，分析不同单倍体诱导系的杂交 F_1 诱导率的杂种优势，以期获得诱导率性状和农艺性状显著改善的诱导系 F_1，满足大规模隔离种植诱导玉米单倍体的需要，同时也为后期新型高频诱导系的选育提供参考，以期提高玉米单倍体育种效率。

1 材料与方法

1.1 试验材料

试验材料为 36 个玉米诱导系和 6 个测验种。其中 36 个诱导系分别是 8 个玉米单倍体诱导系，以及 28 个单倍体诱导系 F_1。试验中用到的 8 个玉米单倍体诱导系均为北京市农林科学玉米研究中心从国外引进的不同来源的优良单倍体诱导材料，并经多代自交和测验选育而成，分别是：诱导系 JKY005、JKY041、JKY042、JKY043、JKY044、JKY046，都具有控制籽粒顶层和形成胚芽色素的 R-nj 基因，以及控制不定根、叶鞘和茎秆色素形成的 $ABPI$ 基因，2 个基因同时存在，可使诱导产生的后代具有籽粒和植株双显性标记；而 JKY006、JKY045 只具有控制籽粒顶层和形成胚芽色素的 R-nj 基因。另外 28 个诱导系 F_1 是由上述 8 个玉米单倍体诱导系，经过不完全双列杂交而来。6 个测验种为玉米自交系，分别是京 724、京 92、京 x005、郑 58、昌 72 和农系 531。

1.2 试验设计

1.2.1 调查诱导材料的株高和雄穗小花数

2012 年冬季在海南和 2013 年夏季在北京分别调查 8 个诱导系和 28 个诱导系 F_1 的株高和雄穗小花数性状。每个材料调查 5 株，3 次重复，取均值进行整理分析。

1.2.2 诱导材料与测验种杂交

36 个单倍体诱导材料与 6 个自交系测验种于 2012 年冬季在北京市农林科学院海南玉米南繁育种基地进行杂交。为了保证诱导系材料与测验种的花期能够相遇，将测验种一次播种，36 个单倍体诱导系材料分别在 3d 后和 5d 分 2 期播种。同时，在花期，以 36 诱导材料为父本分别对 6 个测验种进行杂交诱导，平均每个组合杂交 15 穗。

1.2.3 单倍体数的统计

成熟后收获杂交果穗，脱粒后统计顶部紫色籽粒总数，即为杂交籽粒数。依据 Navajo 标记将顶部为紫色、胚部无色的籽粒视为拟单倍体籽粒。当选的拟单倍体籽粒于 2013 年夏季在小汤山育种基地进行单粒点播种植，精细管理，在苗期依据 APBI 紫色植株标记和植株生长势进行进一步的鉴定。淘汰幼苗叶鞘为紫色和生长势强的植株，选择幼苗叶鞘为绿色、生长发育慢、植株较矮、叶片上冲和叶色较浅的单倍体植株，即为单倍体数。2013 年夏季在北京市农林科学院玉米研究中心小汤山育种基地进行重复诱导试验，并于 2013 年冬季在海南进行单倍体鉴定。诱导系的诱导率以对 6 个测验种的 2 年试验数据的均值进行整理分析。

1.3 统计项目

单倍体诱导率（%）= 单倍体数/杂交籽粒数 × 100%

中亲优势（%）=（F_1-亲平均值）/双亲平均值 × 100%

超亲优势（%）=（F_1-高亲值）/高亲值 × 100%

低亲优势（%）=（F_1-低亲值）/低亲值 × 100%

2 结果与分析

2.1 8个诱导系性状的方差分析

由于单倍体诱导率的高低与基础材料遗传背景紧密相关，在测定诱导率时利用了遗传背景存在较大差异的6个自交系为测验种，所得诱导率具有一定的代表性。通过方差分析发现：①不同诱导系间的诱导率均值、株高和雄穗小花数性状存在极显著的差异（表1）。②不同地点间诱导率均值和雄穗小花数差异不显著。③不同地点间株高性状存在显著差异，可能原因是海南基地日均气温较高，在一定程度上缩短了个体的生长发育进程，使株高降低。

由表2可知，对于诱导率性状，诱导系JKY043、JKY044、JKY045较高，JKY006、JKY042、JKY046居中，而JKY005、JKY041的诱导率较低；对于株高性状，由高到低依次为JKY005、JKY046、JKY042、JKY044、JKY045、JKY041、JKY043、JKY006，最高相差30cm；对于雄穗小花数性状，由多至少分别为JKY045（358.3）>JKY043（294.0）>JKY006（271.3）>JKY046（268.0）>JKY041（237.3）>JKY044（225.3）>JKY042（209.0）>JKY005（143.7）。8个单倍体诱导系的诱导率、株高和雄穗小花数性状都存在较大差异，这为杂种优势的利用奠定了良好的基础。

表1 8个诱导系诱导性状均值的方差分析（F值）

Table 1 The variance analysis of mean value of Eight Inducer's traits (F value)

变异来源 Variation	诱导率（%） Mean induction rate	株高 Phant height	雄穗小花数数 Tassel pollen
诱导系	10.907**	582.332**	324.215**
地点	0.275	81.344*	11.218
诱导系×地点	0.727	4.203	4.011

注：* 和 ** 分别表示在0.05和0.01水平上差异显著

Note: * and **. Significant difference at 0.05 and 0.01 probability levels, Respectively

表2 8个诱导系诱导性状均值的多重比较

Table 2 The multiple comparisons of mean value of Eight Inducer's traits

诱导系 Inducers	诱导率（%） Mean induction rate	株高 Phant height	雄穗小花数数 Tassel pollen
JKY005	2.789b	182.0a	143.7d
JKY006	4.618ab	152.0c	271.3bc
JKY041	2.869b	166.3b	237.3c

(续表)

诱导系 Inducers	诱导率（%） Mean induction rate	株高 Phant height	雄穗小花数数 Tassel pollen
JKY042	4.764a	172.3b	209.0c
JKY043	5.498a	163.0bc	294.0b
JKY044	6.064a	169.0b	225.3c
JKY045	5.606a	169.0b	358.3a
JKY046	5.503a	174.7ab	268.0bc

注：小写字母标记表示处理间差异达 0.05 水平差异显著

Note：The lowercase letters indicate significant difference at 5% level

2.2 株高和雄穗小花数的杂种优势分析

鉴于目前大多数玉米单倍体诱导系都存在株高较低和花粉量偏少的问题，本研究利用不同诱导系相互杂交方法，通过利用株高和雄穗小花数的杂种优势来克服其不足，从而满足单倍体大规模隔离诱导的要求。对 8 个不同单倍体诱导系相互杂交后的 28 个杂交种及其父母本的株高和雄穗小花数进行进行杂种优势分析。由图 1 可知，不同诱导系间杂交 F_1 在株高（左）和雄穗小花数（右）性状上存在着明显的杂种优势。从表 3 可以看出，对于株高性状，大部分 F1 在株高性状上都存在一定的杂种优势株，高位于前 5 位的组合为 JKY005/JKY042、JKY044/JKY046、JKY005/JKY044、JKY044/JKY045、JKY005/JKY043。只有 3 个组合（JKY006/JKY045、JKY006/JKY046、JKY042/JKY044）没有表现出超高亲优势。雄穗小花数的多少可以在一定程度上代表雄穗个体花粉量的大小。对于雄穗小花数性状，所有组合的雄穗小花数与其高值亲本相比都有不同程度的提高，位于前 5 位的组合为 JKY045/JKY046、JKY006/JKY042、JKY042/JKY046、JKY043/JKY046、JKY041/JKY046。因此，利用不同来源的诱导系杂交就能克服单倍体诱导系株高较低和花粉量较少的不足。

表 3 株高和雄穗小花数的杂种优势分析

Table 3 The analysis of heterosis in plant height and tassel pollen number

杂交组合 Hybrid	株高 Plant height			雄穗小花数 Tassel pollen		
	F_1均值 Mean	高亲优势 H-parental heterosis	中亲优势 M-parental heterosis	F_1均值 Mean	高亲优势 H-parental heterosis	中亲优势 M-parental heterosis
JKY005/JKY006	199.3	9.5	19.4	410.3	51.2	97.7
JKY005/JKY041	201.7	10.8	15.8	393.0	65.6	106.3
JKY005/JKY042	217.0	19.2	22.5	404.7	93.6	129.5
JKY005/JKY043	204.3	12.3	18.5	340.0	15.6	55.4
JKY005/JKY044	205.7	13.0	17.2	468.7	108.0	154.0
JKY005/JKY045	199.3	9.5	13.6	392.0	9.4	56.2
JKY005/JKY046	201.3	10.6	12.9	458.0	70.9	122.5

（续表）

杂交组合 Hybrid	株高 Plant height			雄穗小花数 Tassel pollen		
	F_1均值 Mean	高亲优势 H-parental heterosis	中亲优势 M-parental heterosis	F_1均值 Mean	高亲优势 H-parental heterosis	中亲优势 M-parental heterosis
JKY006/JKY041	198.3	19.2	24.6	500.3	84.4	96.7
JKY006/JKY042	183.3	6.4	13.1	580.0	113.8	141.5
JKY006/JKY043	192.3	18.0	22.1	474.0	61.2	67.7
JKY006/JKY044	203.7	20.5	26.9	492.0	81.3	98.1
JKY006/JKY045	164.3	-2.8	2.4	494.7	38.1	57.1
JKY006/JKY046	174.0	-0.4	6.5	514.0	89.4	90.6
JKY041/JKY042	177.0	2.7	4.5	388.0	63.5	73.9
JKY041/JKY043	187.0	12.4	13.6	506.0	72.1	90.5
JKY041/JKY044	202.0	19.5	20.5	502.7	111.8	117.3
JKY041/JKY045	187.3	10.8	11.7	440.0	22.8	47.7
JKY041/JKY046	185.0	5.9	8.5	528.0	97.0	109.0
JKY042/JKY043	177.0	8.6	5.6	468.0	59.2	86.1
JKY042/JKY044	168.0	-0.6	-1.6	416.3	84.7	91.7
JKY042/JKY045	188.7	11.6	10.5	390.0	8.8	37.5
JKY042/JKY046	182.3	4.4	5.1	560.0	109.0	134.8
JKY043/JKY044	171.7	1.6	3.4	382.0	29.9	47.1
JKY043/JKY045	193.0	14.2	16.3	509.0	42.0	56.1
JKY043/JKY046	186.7	6.9	10.6	544.0	85.0	93.6
JKY044/JKY045	205.7	21.7	21.7	469.3	31.0	60.8
JKY044/JKY046	212.0	21.4	23.4	488.0	82.1	97.8
JKY045/JKY046	193.3	10.7	12.5	608.0	69.7	94.1

图1 株高和雄穗小花数的杂种优势

Figure 1 The heterosis in plant height and tassel pollen number

2.3 诱导率的杂种优势分析

对玉米单倍体诱导系杂交后所得 28 个 F_1 诱导率的超高亲优势、中亲优势、低亲优势进行分析（表4），结果表明，15 个 F_1 的诱导率与其高亲值相比都有不同程度的提高，最高超高亲值 2.296，平均超高亲值 0.8741。另外有 8 个组合的 F_1 诱导率没有表现出中亲优势，2 个 F_1 组合的诱导率没有表现出低亲优势，整体来看，诱导率主要表现高亲和中亲优势。同时，表现出超高亲优势组合的 2 个亲本的诱导率值都普遍较高，如 JKY042/JKY046（6.690），两亲本值分别为 4.372 和 4.678；JKY042/JKY044（8.083），两亲本值分别为 4.372 和 5.787；JKY043/JKY044（7.473），两亲本值分别为 5.842 和 5.787；因此，通过正确选择双亲，利用其杂种优势来提高杂交组合的诱导率是可能的。诱导率均值前 5 位的有 JKY042/JKY044、JKY043/JKY044、JKY044/JKY045、JKY042/JKY046、JKY042/JKY043，可在利用杂种优势的基础上，结合自身表现进一步重点研究利用。

表4 诱导率的杂种优势分析
Table 4 The heterosis of hapliod induction rate

杂交组合 Hybrid	F_1均值 Mean	高亲优势 H-parental heterosis	中亲优势 M-parental heterosis	低亲优势 L-parental heterosis	杂交组合 Hybrid	F_1均值 Mean	高亲优势 H-parental heterosis	中亲优势 M-parental heterosis	低亲优势 L-parental heterosis
JKY005/JKY006	3.961	−11.854	18.087	78.83	JKY041/JKY043	5.273	−9.746	23.613	96.10
JKY005/JKY041	4.191	55.875	70.936	89.21	JKY041/JKY044	4.824	−16.643	13.825	79.40
JKY005/JKY042	4.695	7.390	42.551	111.96	JKY041/JKY045	3.450	−37.165	−15.642	28.30
JKY005/JKY043	4.327	−25.930	7.413	95.35	JKY041/JKY046	3.677	−21.408	−0.186	36.74
JKY005/JKY044	4.805	−16.972	20.090	116.93	JKY042/JKY043	6.577	12.576	28.782	50.43
JKY005/JKY045	5.645	2.803	46.502	154.85	JKY042/JKY044	8.083	39.669	59.127	84.88
JKY005/JKY046	2.878	−38.487	−16.506	29.93	JKY042/JKY045	5.700	3.808	15.584	30.38
JKY006/JKY041	2.926	−34.889	−18.527	8.81	JKY042/JKY046	6.690	42.991	47.835	53.02
JKY006/JKY042	4.273	−4.915	−3.605	−2.26	JKY043/JKY044	7.473	27.913	28.519	29.13
JKY006/JKY043	4.997	−14.467	−3.311	11.19	JKY043/JKY045	5.991	2.550	5.730	9.11
JKY006/JKY044	4.234	−26.843	−17.641	−5.79	JKY043/JKY046	5.143	−11.974	−2.237	9.94
JKY006/JKY045	5.776	5.192	15.692	28.53	JKY044/JKY045	6.964	20.329	23.492	26.83
JKY006/JKY046	5.354	14.443	16.743	19.14	JKY044/JKY046	6.212	7.348	18.721	32.79
JKY041/JKY042	5.712	30.646	61.783	112.42	JKY045/JKY046	5.689	3.618	11.894	21.61

3 结论与讨论

随着目前玉米 DH 工程化育种进程的不断推进，对单倍体获得的数量和效率有了更高的要求，而提高玉米单倍体诱导率，是实现大规模获取单倍体的有效途径。因此，对于新型玉米单倍体诱导系的选育应以提高诱导率为重点，同时兼备优良的农艺性状。诱导率提高可以大大增加获得玉米单倍体的数量，而农艺性状的改良可以满足规模化隔离诱导单倍体的需求。由本研究结果可知，大部分诱导系 F_1 在株高和雄穗小花数性状上都存在超亲优势，通过利用诱导系间的杂种优势就能较好的克服株高较低和花粉量较少的不足，因此需重点研究诱导率性状的变化。对诱导率的杂种优势分析发现，整体表现以中亲优势为主，同时存在一定的超亲优势。有些表现出超高亲优势组合的两个亲本诱导率均值都较高，如 JKY042/JKY046、JKY042/JKY044、JKY043/JKY044。因此，在杂种优势利用时应选择高诱导率的材料进行应用，以期组配和筛选出在诱导率和农艺性状上都存在较大优势的 F_1 应用于 DH 工程化育种进程。通过杂种优势利用，获得了在诱导率、株高和花粉量性状表现都较为优良的组合 JKY043/JKY044、JKY044/JKY045、JKY042/JKY046、JKY042/JKY043；而组合 JKY042/JKY044 株高没有超亲优势，但诱导率最高、花粉量较大，具有较高的利用价值，可在后期的育种实践中进一步研究利用。

参考文献（略）

本文原载：种子，2016，35（1）：77-80

高产、高淀粉玉米品种 NK718 的选育与栽培技术

王卫红　王元东　张春原　杨国航

（北京市农林科学院玉米研究中心，北京　100097）

摘　要：玉米品种 NK718 系北京市农林科学院玉米研究中心选育并于 2011 年通过内蒙古自治区审定的玉米品种，审定编号：蒙审玉 2011003 号。该品种产量较高、耐密植，抗性较好，适应性较强，淀粉含量高，粗淀粉含量达到 75.32%。适宜在内蒙古自治区巴彦淖尔市、赤峰市、通辽市 ≥10℃ 活动积温 2 900℃ 以上适宜区和类似生态区春播种植；也可在北京、天津、河北、山东、山西、河南等省夏播种植。

关键词：玉米；NK718；选育；栽培技术

1　品种选育

选种目标：以高产、耐密植，抗病虫、抗倒伏为育种目标。

亲本来源：以京 464 为母本，京 2416 为父本选育而成。母本来源于国外新种质 X 群体选系；父本来源于京 24 与 5237 杂交选系。母本京 464 具有一般配合力高和综合抗性好的特点。株形紧凑，株高 267cm，穗位 100cm，穗长 17cm，穗行数 16 行，穗粗 4.5cm，果穗长筒形，白轴。半硬粒型黄粒。穗粒数 540 粒。抗大、小斑病、茎腐病和矮花叶病毒病。父本京 2416 株型紧凑，株高 162cm，穗位 83cm，雄穗分支数 3~5 个，穗长 15cm，穗行数 12~14 行，穗粗 4.5cm，果穗圆锥形，白轴，硬粒型，黄粒，千粒重 377g，出籽率 86%，抗大、小斑病、茎腐病和矮花叶病毒病。

2　品种特征特性

幼苗叶片绿紫色，叶鞘紫色。植株半紧凑型，株高 295cm，穗位 124cm，20 片叶。雄穗护颖绿色，花药紫色。花丝紫色。果穗筒形，白轴，穗长 17.3cm，穗粗 5.2cm，秃尖 0.9cm，穗行数 16，行粒数 37，穗粒数 587，单穗粒重 190.2g，出籽率 83.6%。籽粒偏马齿型，橙黄色，百粒重 33.1g。

3　产量表现

2008 年参加内蒙古自治区玉米预备试验，6 点平均产量 12 402kg/hm²，比对照四单 19 增产 13.16%，6 点全增，平均生育期 130d。

基金项目：北京市农林科学院科技创新能力建设专项
作者简介：王卫红（1966—　），推广副研究员，主要从事农作物品种推广示范工作
通讯作者：杨国航

2009年参加内蒙古自治区玉米区域试验，6点平均产量13 918.5kg/hm²，比对照郑单958增产8.2%，6点4增2减，平均生育期131.5d，比对照晚0.6d（表1）。

2010年参加内蒙古自治区玉米晚熟组生产试验，6点平均产量12 465kg/hm²，比对照郑单958增产3.9%，6点4增2减，平均生育期131.3d，比对照晚0.8d（表2）。

表1 2009年在内蒙古自治区中熟组生产试验中产量相关性状表现

试点	产量(kg/hm²)	比CK(±%)	位次	穗长(cm)	穗粗(cm)	秃尖(cm)	穗行数(行)	穗粒数(粒)	单穗粒重(g)	百粒重(g)	出籽率(%)	空秆率
金山	12 712.5	-0.5	12	17.9	5.3	1.5	16.8	631	203.1	36.8	80.5	0.0
赤院	12 165.0	-0.6	13	17.8	5.2	2.2	16.6	641	190.7	33.5	81.8	1.4
巴院	15 163.5	15.7	6	17.6	5.1	0.2	15.7	634	225.3	35.5	88.6	1.7
开鲁	15 897.0	6.1	8	18.6	5.1	0.7	15.7	564	231.0	40.7	82.8	0.0
赤站	12 756.0	13.6	2	19.0	5.2	1.6	16.5	667	197.5	31.4	83.7	0.7
巴站	14 820.0	15.2	4	17.8	4.9	0.6	15.8	654	228.7	36.5	87.1	0.0
平均	13 918.5	8.2	6	18.1	5.1	1.1	16.2	632	212.7	35.7	84.1	0.6

表2 2010年在内蒙古自治区中熟组生产试验中产量相关性状表现

试点	产量(kg/hm²)	比CK(±%)	位次	穗长(cm)	秃尖(cm)	穗行数(行)	穗粒数(粒)	单穗粒重(g)	百粒重(g)	出籽率(%)	空秆率
金山	11 941.5	5.6	8	15.6	1.4	14.8	468	184.9	37.2	81.9	0.0
开鲁	11 197.5	-1.3	9	15.6	0.8	16.2	432	171.6	39.2	83.0	0.0
丰田	10 087.5	-9.9	10	16.2	0.6	16.4	613	156.6	26.5	80.1	0.0
敖汉	12 166.5	4.0	6	17.2	1.3	17.0	612	188.7	30.8	86.5	0.0
科河	14 452.5	8.3	6	19.7	1.0	16.7	740	211.1	27.4	85.1	0.0
巴站	14 941.5	14.2	2	19.4	0.2	15.2	660	227.8	37.5	84.9	0.1
平均	12 465.0	3.9	9	17.3	0.9	16.1	587	190.2	33.1	83.6	0.0

4 抗性

4.1 抗病虫性

经吉林省农业科学院植保所人工接种、接虫抗性鉴定表明，该品种中抗大斑病（MR），感弯孢病（S），中抗丝黑穗病（5.7%，MR），高抗茎腐病（0，HR），抗玉米螟（4.0，R）。2009—2010年两年参加试验，各点的病虫害田间调查显示，田间抗病虫良好，对当地的主要病虫害多达到高抗水平。

4.2 抗倒性

2008—2010年参加内蒙古自治区预备试验、区域试验、生产试验，3年平均倒伏率均为0，平均倒折率分别为0.2%、0.62%和1.65%，抗倒性好（表3、表4）。

表3 2009年内蒙古自治区区试各点抗性情况

试点	倒伏率（%）	倒折率（%）	大斑病（级）	小斑病（级）	灰斑病（级）	弯斑菌（级）	锈病（级）	纹枯病（级）	玉米螟（级）	红蜘蛛（级）	丝黑穗病（%）	黑粉病（%）
金山	0.0	0.0	1	1	1	1	1	1	1	1	0.0	0.0
赤院	0.0	0.0	1	1	1	1	1	1	1	1	0.0	0.0
巴院	0.0	3.7	1	1	1	1	1	1	1	1	0.0	0.5
开鲁	0.0	0.0	3	1	3	1	3	5	1	1	0.0	0.0
赤站	0.0	0.0	1	1	1	1	1	1	1	1	0.0	0.0
巴站	0.0	0.0	1	1	1	1	1	1	2	3~5	0.0	0.7

表4 2010年内蒙古自治区中熟组生产试验田间抗性表现

试点	大斑病（级）	丝黑穗病（%）	黑粉病（%）	茎腐病（%）	倒伏率（%）	倒折率（%）	≥7级其他病虫害	整齐
金山	1	0.0	0.0	0.0	0.0	0.0	—	整齐
开鲁	1	0.0	0.0	0.0	0.0	0.0	0	整齐
松山	1	0.0	0.0	9.6	0.0	9.6	无	整齐
敖汉	3	0.0	0.0	0.0	0.0	0.0	—	整齐
科河	1	0.0	0.0	0.0	0.0	0.0	—	整齐
巴站	1	0.0	0.0	0.0	0.0	0.3	—	整齐

5 品质

经农业部谷物及制品质量监督检验测试中心（哈尔滨）测定，属于高淀粉品种，粗淀粉含量达到75.32%，容重比较高，达到770g/L，粗蛋白含量8.46%，粗脂肪含量3.97%，赖氨酸含量0.29%。

6 适宜种植区域

适宜在内蒙古自治区巴彦淖尔市、赤峰市、通辽市≥10℃活动积温2 900℃以上适宜区和类似生态区春播种植；也可在北京、天津、河北、山东、山西、河南等省夏播种植。

7 制种技术

隔离条件：250~300m，种植密度：67 500株/hm²，最高密度不超过75 000株/hm²。播种日期：4月中下旬。错期：第1期父本晚播8d，第2期父本晚播12d。母本由于株高较高，节间拉开较长，抽雄时带3~4片叶。这样既方便抽雄，又降低株高，便于父本授粉。施肥纯氮150~225kg/hm²，可全部播前深翻底施或1/2底施，1/2在小喇叭口期追施，注意磷钾肥配合；防治玉米螟。脱粒加工时注意清选机器，严防机械混杂。

8 栽培技术要点

8.1 确保播种质量

春播适宜播期一般为 4 月下旬到 5 月上旬前后。夏播一般在 6 月中旬前后。足墒播种。使用包衣种子，可有效防治地下害虫，促进萌发。

8.2 合理施肥

施肥纯氮 150~225kg/hm²，可全部播前深翻底施或 1/2 底施，1/2 在小喇叭口期追施，注意磷钾肥配合。施足底肥可保证前期早生快发，后期不缺肥，利于形成大穗，籽粒饱满，提高结实率。底肥要根据土壤肥力、产量水平和品种需肥特点综合考虑底肥数量、肥料元素，进行平衡施肥。有条件的地块可适当施入有机肥，选用腐熟农家肥或精制有机肥，腐熟农家肥一般用量 15 000kg/hm² 以上，精制有机肥一般用量 6 000kg/hm² 以上，有机肥于播前均匀撒入田间，随农事操作翻入耕层。如果选用缓释长效肥料，采用全生育期肥料一次底深施技术，要选好播种机型，避免肥料距离种子过近，造成烧种。

8.3 合理密植

合理的留苗密度是实现高产的关键。适宜密度 60 000株/hm²，高肥水地块种植密度可增加到 67 500株/hm²。把握好间苗、定苗的时间，一般最好在晴天下午，因为病苗、虫咬苗及发育不良的幼苗在下午较易萎蔫，便于识别。对苗矮叶密、下粗上细、弯曲、叶色黑绿的丝黑穗浸染苗，应彻底剔除。

8.4 注意防治病虫害

应以预防为主，综合防治。苗期注意防止飞虱、蓟马，喷施阿维菌素药剂；玉米螟为害严重地区，可大喇叭口拔节期利用辛硫磷拌沙灌心进行防治。注意控制杂草，在播后苗前使用 38% 阿特拉津 1 875ml+50% 乙草胺 1 125~1 500ml/hm² 进行土壤封闭处理，一般使用大型喷药车常规喷雾，对水不得少于 35kg/667m²（注意阿特拉津合理用量，以免对下茬冬小麦造成药害）。玉米出苗后，注意防治黏虫，具体措施可用 BT 乳油 50g 加上高效氯氢菊酯 200g 对水 30kg 喷洒。

8.5 适期收获

不能简单以苞叶变黄发白作为成熟的标志，应当在籽粒基部与穗轴连接处出现"黑层"后，玉米才真正进入完熟期，可确保籽粒品质，减少霉变几率，同时也可减少晾晒环节的工作量，提高效率。适时早种、晚收，籽粒乳腺消失后收获，充分发挥该品种的增产潜力，一般可增产 5% 以上。

本文原载：种子，2012，31（11）：105-107

适于全程机械化生产的玉米新品种选育探讨

王元东 张华生 段民孝 张雪原 张春原 陈传永 赵久然

(北京市农林科学院玉米研究中心，北京 10007)

摘 要：通过分析国外玉米品种的主要特征特性，提出了我国适于全程机械化生产的育种目标；分析了我国原有种质和引进的国外新种质在适应全程机械化生产的玉米新品种育种潜力，提出了相应的利用策略。

关键词：玉米；全程机械化；生产；育种；目标；新种质

玉米生产全程机械化作业技术是推进现代农业发展的重要举措，该技术不仅可以大幅度减轻农民的劳动强度、降低生产成本、解放劳动力，而且可以显著的提高玉米单产的功效，同时还能实现玉米种植的标准化、规模化，进而大幅度提高我国玉米市场的竞争力。美国是世界上最早实现机械化生产的国家，德国、法国等欧洲国家的玉米全程机械化生产也已经实现，同时在育种上紧紧围绕适应机械化生产的育种目标持续不断的改良创新种质、选育新杂交种。近几年，我国玉米机械化生产得到快速发展，品种选育也开始注重适宜机械化生产的一些农艺性状。而要达到2020年基本实现玉米生产机械化的目标，则需要创新型玉米新品种。因此，玉米育种者应在坚持高产、优质、多抗、适应性广等传统育种目标的基础上，注重选育出适宜全程机械化生产要求的品种。

1 适于全程机械化生产的玉米品种的主要特征特性

分析我国原有推广面积较大的品种如掖单13、丹玉2号、丹玉13、掖单2号、农大108以及现在主推品种郑单958、浚单20、中科4号和吉单27发现：这些品种普遍存在成熟期偏长，抗倒伏能力差，籽粒脱水速度慢，收获时籽粒水分含量高等问题，从而阻碍了玉米实现全程机械化生产。先玉335和德美亚1号分别是是美国先锋种子公司和德国KWS公司在中国大面积推广的适宜机械化生产的优良品种，这些品种突出特点是熟期早、种子适宜单粒点播、成熟时籽粒脱水速度快、收获时茎秆站立性好以及籽粒含水量低。调查数据显示，正常情况下先玉335机械收获的果穗播净率达90%，而郑单958是50%~60%。实际考察发现，美国玉米品种株高在240~250cm，穗位在110cm，播种至成熟110~120d，苞叶蓬松，高抗倒伏，有利于机械收获。

2 适于全程机械化生产的玉米品种主要育种目标

根据我国玉米生产的实际情况，高产目标是永恒的主题。考虑到适应全程机械化生产

基金项目：国家"863"项目（2011AA10A103）、粮食作物产业技术体系北京市创新团队专项经费资助（BITG7）、北京市农林科学院科技创新能力建设专项（KJCX20140107）

的要求：一是适合机械化单粒播种，二是适合机械化田间管理，三是适合机械化收获，因此选育的创新型品种同时还应注重适宜单粒精量播种、成熟早、耐密植、强抗倒（折）、苞叶松、轴细硬、半硬粒、籽粒脱水速度快等育种目标，尤其更加注重抗倒伏农艺性状的目标。

2.1 种子商品率高，适宜机械化单粒精量点播

机械化单粒精量播种生产效率高，节约用种，节省人工，增产效果显著。要求种子商品率高，其中发芽率不低于98%，发芽势强，纯度不低于98%，种子饱满一致。具体性状目标是种子半硬粒型、角质多、大小适宜、耐低温深播。

2.2 株型理想、耐密植、高抗倒（折），适宜田间机械化管理

品种的抗倒（折）性是仅次于产量的最重要目标之一，玉米使用各种机械进行田间管理的便利与否与玉米抗倒（折）密切相关。此外，植株对除草剂不敏感、株型理想、耐密植也便于田间机械化管理作业，提高生产效率，降低生产成本。具体性状目标是植株株型通透、茎秆细而坚韧、根系发达。

2.3 早熟、籽粒脱水速度快、含水量低，茎秆站立性好，适宜田间机械化收获

生育期偏长、成熟时籽粒脱水速度慢是导致收获时籽粒含水量高的重要因素，造成收获时籽粒破损严重以致减产。而收获时茎秆倒伏或折断则严重影响机械化生产效率，同时容易漏收果穗。为了提高机械化收获的质量，除了要求品种本身抗倒（折）外，还应注意品种的抗茎腐、抗玉米螟和抗穗腐能力。具体性状目标是中小果穗、苞叶薄、苞叶层数少而短、成熟后松散、穗轴细而坚硬、抗玉米螟、茎腐病和穗腐病。

3 适于全程机械化生产的新品种种质基础与杂优模式

分析研究认为，20世纪末期我国玉米种质主要为改良瑞德、改良兰卡斯特、唐四平头、旅大红骨和P群（代表自交系分别为掖478、Mo17、黄早四、丹340和齐319）。其中改良瑞德、改良兰卡斯特和P群为国外种质，为母本自交系选育来源；唐四平头和旅大红骨为国内种质，为父本自交系选育来源。进入21世纪，随着美国先锋种子公司、孟山都公司和法国利马格兰以及德国KWS等公司品种在中国大面积的推广应用，国外种质衣阿华（Iowa）坚秆综合种（BSSS）、Maiz Amargo、兰卡斯特（Lancaster）、Oh43、LH82、衣阿华瑞德黄马牙（Iodent）、Oh07-Midland、明尼苏达13（Minn13）、欧洲早熟硬粒在中国有着越来越重要的影响，其中，衣阿华（Iowa）坚秆综合种（BSSS）和Maiz Amargo属于BSSS杂优群，为母本群；其余为non-BSSS杂优群，为父本群。多年育种实践和品种种植经验表明，我国原有种质以及根据杂优模式选育的品种普遍存在成熟期偏长，抗倒伏能力差，籽粒脱水速度慢、收获时籽粒含水量高等问题，而新种质的引进以及从这些新种质选育出来的优良自交系对选育适宜全程机械化生产的玉米新品种奠定了良好基础。

3.1 衣阿华（Iowa）坚秆综合种（BSSS）和Maiz Amargo利用潜力

新引进的外源种质衣阿华（Iowa）坚秆综合种（BSSS）和Maiz Amargo具有配合力高、株型理想、高抗倒伏和制种产量高、种子商品率高、适宜单粒精量播种等优良特点。多年育种实践表明，直接利用衣阿华（Iowa）坚秆综合种（BSSS）和Maiz Amargo等新种质与我国原有5大种质组配创制新组合均不理想。其中与改良瑞德组配由于血缘上相近，

杂交种配合力不高，优势不大；由于我国在 Mo17 改良上总体进展不大，因此组配出的组合在产量和抗性上与国外优秀杂交种相比差距较大；与国内唐四平头种质、旅大红骨种质以及 P 群种质组配出的杂交种具有较高的杂种优势，但熟期偏晚、植株繁茂、籽粒脱水速度较慢。因此这批种质经过改良后应该主要与从国外引进的改良 non-BSSS 种质进行组配，该模式组配组配的杂交种具有较强的杂种优势，同时具备适宜全程机械化生产的目标性状。

3.2 新引进的 non-BSSS 种质利用潜力

新引进的兰卡斯特（Lancaster）、Oh43、LH82、衣阿华瑞德黄马牙（Iodent）、Oh07-Midland、明尼苏达 13（Minn13）等 non-BSSS 种质具有耐密、早熟、脱水速度快、出籽率高等优点。该类种质与我国改良瑞德和改良兰卡斯特种质杂种优势不大，与国内唐四平头种质、旅大红骨种质以及 P 群种质具有较强的杂种优势，丰产性好、籽粒脱水速度相对较快，但是株型不理想、植株繁茂、易倒伏、抗病虫害能力较差。因此育种实践认为，新引进的 non-BSSS 种质应重点与国外的改良 BSSS 种质进行组配。

3.3 来源于国外商业杂交种 X1132X 选系的利用潜力

利用国外商业杂交种杂交种 X1132X 选育出的自交系含有 BSSS 和 non-BSSS 血缘。在近几年成功选育出京 724、京 MC01、京 464、京 725、DH382、M54、NH60 等几个具有影响力的商业自交系。这些自交系与国内唐四平头骨干系京 2416、京 92 等组配出京科 968（京 724/京 92）、京科 665（京 725/京 92）、NK718（京 464/京 2416）、京农科 728（京 MC01/京 2416）等强优势杂交组合；与国内改良瑞德系 DH351 组配出登海 605（DH351/DH382），与旅大红骨自交系 S122、S121 组配出良玉 88（M54/S122）、农华 101（NH60/S121）等组合。分析近几年选出的这些新品种可以看出，国外商业杂交种杂交种 X1132X 选系与国内唐四平头和旅大红骨种质具有较强的杂种优势，组配出的杂交种不仅具有本土化品种的优势（抗病性强、适应性好、苗势壮），而且还具备某些适宜机械化生产的指标，如满足适宜机械化精量单粒播种，籽粒成熟时脱水速度相对较快，登海 605 高抗倒伏（折），京农科 728 则早熟、耐密、高抗倒、脱水速度快，适宜单粒播种，基本符合京津唐玉米区全程机械化生产的各项指标。综上分析说明，尽管目前选育出的组合还达不到全程机械化生产的全部指标，但是已经取得了一定进步。适宜京津唐全程机械化生产的玉米新品种京农科 728 选育的成功则表明 X1132X 种质的选系潜力和组配潜力的有待于进一步发掘。

4 结语

综上所述，实现玉米生产全程机械化是现代农业发展要求，随着我国玉米生产全程机械化的快速发展，急需调整育种目标，发掘现有种质和新引进种质的育种潜力，选育出适合全程机械化生产的玉米新品种。长久以来，我国玉米育种目标没有或很少考虑到适应全程机械化生产的需求，因此导致缺乏符合机械化育种目标的关键材料、自交系，面对国外大型种子公司如美国先锋公司、孟山都公司、德国 KWS 公司、法国利马格兰公司在种质创新、育种技术和研发经费上的巨大优势，国内种业巨头和育种者应该既要有紧迫感，同时更应沉着应对。一方面，应加强研究创新近几年新引进的如衣阿华（Iowa）坚秆综合种（BSSS）、Maiz Amargo、兰卡斯特（Lancaster）、Oh43、LH82、爱华马齿（Iodent）、

Oh07-Midland、明尼苏达13（Minn13）和欧洲早熟硬粒等种质，这些种质除了具有优良的产量基因外，还累积了适应全程机械化生产的众多优良基因，同时杂种优势模式明确，国内大企业可以在BSSS、Maiz Amargo/non-BSSS模式下快速进行商业化育种，同时注意扬长避短，注重利用国内抗叶斑病优良P群种质提高其抗病性。另一方面，国内农业公益院所育种者加强国外商业杂交种X1132X的利用潜力，深入发掘创制出新的种质材料和自交系，使之与国内唐四平头和旅大红骨种质组配，从而选育出适应机械化育种目标的新品种。

参考文献（略）

本文原载：中国种业，2014（11）：23-25

利用外来新种质 X1132x 选育优良玉米自交系的研究

王元东　张华生　段民孝　张雪原　张春原　陈传永　赵久然

(北京市农林科学院玉米研究中心，北京　10007)

摘　要：利用外来新种质 X1132x，通过高密度、大群体、变换地、强胁迫、严选择，选育出若干优良自交系。研究了部分优良自交系的生物学特征特性以及抗逆性和适应性，并进行了配合力的分析。还探讨了这些优良自交系的杂种优势模式，并对其利用潜力进行了分析。结果表明：(1) 京724、京725、京464 等优良自交系生物学特性优良，自身产量高，易制种，抗病性好，适应性广，配合力高；(2) 该选系与京92、京2416 等黄改系组配构成主要杂种优势模式，其中组合京 724×京92 (审定名称京科968)、京725×京92 (审定名称京科665)、京464×京2416 (审定名称 NK718) 在推广示范中均有较好表现，表现出良好的适应性和丰产性，说明这些自交系的选育是成功的。

关键词：新种质 X1132x；优良自交系；杂种优势

育种要有大的突破，首先要有新的优异种质和以新种质为基础的核心优良自交系的选育。其中，外来新种质在我国玉米育种实践中一直占有重要地位。据统计，外来种质对我国玉米生产的贡献率已接近 60%，其中，美国外来种质的利用率每一个百分点，全国玉米平均产量增加约 $10kg/hm^2$。

20 世纪 70—80 年代，我国从美国直接引进 Mo17 自交系、利用美国杂交种选育出的铁 7922、沈 5003、U8112 以及它们之间相互杂交选育出的掖 478、C8605-2、丹 9046 等优良玉米自交系对我国玉米生产起了重要作用。其中，Mo17 及其衍生系构成我国兰卡斯特杂种优势群核心种质，组配出中单 2 号、丹玉 13 等优良品种；掖 478 及其改良系郑 58 则是我国改良瑞德杂种优势群核心种质，选育出掖单 13 和郑单 958 等优良品种。20 世纪 90 年代，利用美国杂交种 P78599 及其同类杂交种选育出 X178、齐 319、87-1 等优良自交系，构成了我国另一杂种优势群 P 群核心种质，组配出农大 108、鲁单 981、豫玉 22 等优良品种。

由于现代玉米生产的需求，我国玉米育种需要有进一步的跨越和突破。利用近年来从国外引入我国的优异外来种质作为选育优良自交系的基本材料，坚持在高密度、大群体、变换地、强胁迫等条件下进行严格选择，有望选育出新的优良骨干自交系。本研究以外来新种质 X1132x 杂交种为基础选系材料，基于上述选择方法，选育出若干优良自交系，对其中 8 个优良自交系的生物学特征特性、抗逆性和配合力进行了观察研究，并进行了强优

基金项目：国家 "863" 项目 (2011AA10A103)、粮食作物产业技术体系北京市创新团队专项经费资助 (BITG7)、北京市农林科学院科技创新能力建设专项 (KJCX20140107)

通讯作者：赵久然

势杂交组合的组配，探讨了其主要杂种优势模式，同时展望了这些优良自交系的利用潜力。

1 材料与方法

1.1 材料

试验利用外来新种质 X1132x 优良单交种，通过市场上销售商品种子得到。经鉴定具有株型清秀、出子率高、后期籽粒脱水速度快、容重高、千粒重高、综合抗性好（尤其是对一些主要病害的水平抗性较为突出）等特性。在株型上除紧凑外，其雄穗小、雄穗分支数少、开叶距等特点非常鲜明，累积了丰富的株型优良基因和产量性状基因。

1.2 方法

2003 年春在其 S_0 植株经混合授粉获得约 30 000 粒 S_1 种子，2004 年将获得的 S_1 种子平均分成 3 份，每份约 10 000 粒，分别在北京昌平、吉林四平和河南郑州种植（北京昌平、吉林四平代表温度较低、较为干旱、病虫害尤其是丝黑穗和大斑发病严重的环境，河南郑州代表阴雨寡照、高温高湿病虫害严重的环境），3 点试验种植小区均为 10m 行长，55 行区，行距 0.6m，株距 0.18m，密度为 9 万株/hm^2。整个试验约 3 000 穴，每穴播种 3 粒，留苗 1 株，共有约 3 000 个基本株。定苗前，淘汰弱病株，留健康长势强植株；自交授粉前，通过严格选择，淘汰大量的不符合要求的基因型单株，入选单株挂牌；授粉期间，选择雌雄协调、吐丝快而集中的植株进行一次性自交授粉；收获前，在各生态区淘汰病虫害植株、空秆株、倒伏（倒折）株、籽粒败育和果穗畸形株，对入选果穗进行室内考种，淘汰畸形、结实差的果穗，留下单穗粒重大于 100g、出子率大于 88%、穗行数大于 14 行的优良果穗。

2004 年春将三地得到的 S_2 果穗均混合脱粒。并相互变换试验点进行种植，种植和筛选的方式同 2003 年。2005 年春将三地得到的 S_3 果穗进行穗行种植，继续进行变换试验。各穗行种植密度为 9 万株/hm^2，授粉前，淘汰不良穗行，选取优良穗行进行自交授粉；收获前，淘汰感病虫害穗行；收获后，决选出单穗粒重大于 100g、出子率大于 88%、穗行数大于 14 行的穗行 80 个。2005 年冬，将得到的 S_4 优良果穗在海南进行测配并加代。

2006 年春在北京昌平试验地，将得到的 S_5 优良果穗进行穗行种植，各穗行种植密度为 67 500 株/hm^2，授粉前，淘汰不良穗行，选取优良穗行进行自交授粉；收获前，淘汰感病虫害穗行；收获后，决选出但穗粒重大于 100g、出子率大于 88%、穗行数大于 14 行的穗行。结合配合力测配结果，留取穗行 15 个。2006 年冬，将入选穗行继续在海南加代为 S_6，同时继续测配。

2007 年春，将穗行继续进代为 S_7，各穗行基本稳定。同时，对上年海南测配组合进行比较试验，测出小区产量超 10 000kg/hm^2，超过对照郑单 958 10%以上的组合 8 个。这些选系被命名京 724、京 725、京 464、MC0304、京 719、90110-2、京 465 和 HNBXJL22。

2 结果与分析

2.1 生物学特征特性

根据田间观察，8 个 X1132x 杂交种选育优良自交系在苗期叶鞘均为深紫色，其中，京 464、京 724 和京 725 苗期长势强，叶色深绿，叶片波曲，心叶上冲。从表 1 中看出，8

个自交系生育期适中，株高相对较高，变幅在220.7~278.3cm，株型紧凑，雄穗分支数除了90110-2和京464外均少于3个，ASI（抽雄至吐丝间隔期）值表明所有8个自交系吐丝快而集中，抗倒伏能力均较强。

在产量性状上，8个优良自交系的单穗粒重均较高，单穗粒数也很多，穗行数在14~18行，均为硬粒或半硬粒型，适合用作母本自交系。其中，京724单穗粒重最重，达到140.6g，由于其百粒重相对小些，单穗总粒数达到453粒，籽粒商品性好，适于单粒精量播种。

表1 利用X1132x杂交种新选育的8个优良自交系主要特征特性

材料	株高（cm）	穗位高（cm）	雄穗分枝数	ASI（d）	倒伏率（%）	单穗粒重（g）	单穗粒数	穗行数（个）	穗长（cm）	穗粗（cm）
京719	245.6	60.5	2.2	2.6	0.0	130.8	396.4	17.8	17.1	4.5
京464	243.8	61.2	2.7	2.6	0.0	128.6	428.7	16.7	13.8	4.2
京724	278.3	83.6	1.2	3.7	1.9	140.6	453.5	17.1	16.8	4.6
京725	267.8	68.9	1.9	1.9	1.1	135.8	424.4	17.2	15.5	4.5
90110-2	223.4	55.7	5.1	3.3	0.0	128.2	377.1	15.2	17.8	4.1
MC0304	276.7	76.5	1.1	2.8	0.0	118.6	370.6	16.8	16.4	4.6
京465	220.7	56.3	5.2	2.6	0.0	127.6	375.3	15.2	18.1	4.1
HNBXJL22	235.6	58.3	1.8	1.7	0.0	138.5	420.0	17.2	16.5	4.6
PH6WC	270.6	80.4	1.4	3.7	0.0	119.5	362.1	15.7	15.9	3.9
PH4CV	182.1	51.8	3.4	3.1	11.7	89.6	289.0	14.2	11.8	3.8
郑58	160.3	45.7	4.7	3.7	0.0	109.4	312.6	13.6	15.1	3.6

2.2 配合力

配合力是一个自交系的主要特性，也是选育优良自交系的主要目标之一。在2007—2008年间，8个优良选育自交系以及PH6WC、PH4CV和郑58共13个系为母本，分别与京24、昌72、京92、吉853、Lx9801、京2416等6个测验种杂交，比较8个优良自交系与PH6WC、PH4CV和郑58在产量配合力效应的优势，同时进一步比较8个优良自交系的内部产量配合力效应的大小。

从表2可以看出，8个优良自交系的一般配合力除了京719外，其余7个均大于PH6WC，均强于PH4CV和骨干系郑58，其中，京724和京725两个自交系具有很高的一般配合力。在6个测验种中，京24、京2416、京92和Lx9801的一般配合力较高，其中京92的配合力最高。在特殊配合力方面，京724与京92配合力最高，其次是京725与京92，表现出较高的产量潜力。

2.3 抗病虫害

8个优良自交系自育成以来，在北京、吉林、河南、海南等种植鉴定，均表现出良好的抗性。其中，京724、京725和HNBXJL22均中抗多种病害，京464、京724和京725

还表现出良好的抗玉米螟能力（表3）。

表2 8个优良自交系产量配合力效应

母本	父本（测验种）						GCA-m
	京24	京92	京2416	吉853	昌72	Lx9801	
京719	-3.5	1.3	-1.8	-30.7	-91.8	8.1	-19.6
京464	2.7	21.7	29.7	17.5	2.6	28.4	17.2
京724	68.2	118.5	42.1	35.6	14.5	95.2	62.5
京725	30.0	114.5	38.5	28.7	7.3	77.6	49.5
90110-2	6.2	-21.3	28.2	-24.8	-8.9	-2.8	-3.8
MC0304	-42.8	93.0	-18.4	4.4	-47.2	51.7	6.9
京465	8.0	-27.2	38.1	-21.7	2.7	2.5	0.5
HNBXJL22	25.2	48.3	-71.3	30.5	-39.5	42.1	6.0
PH6WC	-5.6	8.4	21.5	-6.4	-53.7	1.5	-5.6
PH4CV	-73.8	-80.3	-81.5	-128.4	-93.8	-98.7	-92.7
郑58	-44.8	-41.9	-34.6	-51.7	36.2	15.0	-20.2
GCA-f	-2.6	21.5	-0.8	-13.3	-24.6	20.2	

表3 8个优良自交系在多年多点的抗病虫方面的抗性评价

材料	小斑病	大斑病	弯孢霉叶斑病	茎腐病	丝黑穗病	玉米螟
京719	MR	MR	R	R	MR	-
京464	R	R	MR	R	MR	R
京724	R	R	R	R	R	R
京725	R	R	R	R	R	R
90110-2	MR	MR	MR	S	MR	-
MC0304	MR	S	MR	MR	MR	-
京465	MR	MR	MR	S	R	-
HNBXJL22	R	R	R	R	R	-

3 讨论及其利用潜力分析

3.1 讨论

经分析，外来玉米新种质X1132x杂交种具有丰富的优良植株农艺性状和产量性状基因，其优良的株型、后期灌浆速度快以及丰产性状是国内推广的许多优良品种所缺少的，同时也具有熟期合适、籽粒脱水速度快等适宜机械化收获的优良性状，受到企业青睐。但该种质也具有感大斑病、前期抗倒伏能力差、高感玉米螟等抗性缺陷。如何利用该种质成功选育出优良玉米自交系是育种者必须考虑的问题。

高密度可以人为创造出多种非生物胁迫高压逆境，是一种提高种质耐多种非生物胁迫（包括抗旱和耐低 N）的育种策略。在高密度选择压力下，选择群体内植株间竞争加剧，大量植株表现出空秆、秃尖、结实性差、倒伏（倒折）、发病严重等现象，因此，在早代可以淘汰大量不良基因型，选择出适应性强和农艺性状优良的基因型。同时，增大基础选择群体的样本容量，可以使选择优异基本单株机会增加，给严格淘汰选择留够空间。与传统育种方法比较，高大严（高密度、大群体、严选择）选择的准确率大大提高。

在大群体、高密度选择过程中，坚持以单穗粒重和出子率为核心选择指标，同时注意加强一些耐高密的次级性状的选择，如 ASI、雄穗大小、雄穗分支数、吐丝快慢和是否集中、穗下部叶片的衰老程度等。ASI 在高密度条件下与果穗粒重高度相关，可以作为次级性状来选择果穗的产量。

通过变换选育地点和生态环境，可以提高材料的广适性，具有更丰富的抗性。在玉米主产区黄淮海区，其生态环境主要特征是阴雨寡照、高温、高湿、多风、病虫害严重；在东华北主产区为温度较低，较为干旱，丝黑穗发病严重的生态环境。变换地选择，可以将适合几大主产区的优良抗性基因不断累积起来，后代选系适应性更广。

根据上述原则，自交系京 724、京 725、京 464、MC0304、HNBXJL22 等系选育比较成功，这些优良自交系自身产量均在 7 500kg/hm² 左右，适应性好，配合力高，而且在选择过程中通过基因重组具有抗大斑、抗倒伏、抗玉米螟等优良特性。表 4 是比对照郑单 958 增产的 12 个杂交组合，分析组合杂优模式发现 X1132x 的选系分别与京 92、Lx9801、京 24、京 2416 等黄改系具有很强的杂种优势。

表 4 比对照郑单 958 增产的 12 个优良组合

位次	组合模式	产量（kg/hm²）	SCA	对照优势（%）
1	京 724×京 92	13 359.2	118.5	10.2
2	京 725×京 92	13 299.1	114.5	9.7
3	京 724×Lx9801	13 009.5	95.2	7.3
4	MC0304×JC73	12 976.5	93.0	7.0
5	京 725×Lx9801	12 745.5	77.6	5.1
6	京 724×京 24	12 604.5	68.2	4.0
7	MC0304×Lx9801	12 357.3	51.7	1.9
8	HNBXJL22×京 92	12 306.1	48.3	1.5
9	京 724×京 2416	12 213.5	42.1	0.7
10	HNBXJL22×Lx9801	12 213.0	42.1	0.7
11	京 725×京 2416	12 159.5	38.5	0.3
12	京 464×京 2416	12 153.0	38.1	0.2

3.2 利用潜力分析

京 724×京 92 的组合（审定后为京科 968）在 2009—2010 年参加国家东华北春播区

试,2009 年参加国家东华北玉米品种区域试验,21 点次增产,2 点次平产,平均亩产 807.7kg,比对照郑单 958 增产 8.07%,综合性状表现列该区组第 1 位;2010 年参加国家东华北玉米品种区域试验,平均单产 734.4kg,比对照(总平均值)增产 6.08%。2010 年生产试验,20 点次全部增产,平均单产 716.3kg,比对照郑单 958 增产 10.45%。经辽宁省丹东农业科学院与吉林省农业科学院植物保护研究所 2 年接种鉴定,中抗-抗大斑病、中抗-抗灰斑病、中抗-抗弯孢叶斑病、中抗-高抗丝黑穗病、中抗-高抗茎腐病、高抗玉米螟。经农业部谷物及制品质量监督检验测试中心(哈尔滨)测定,籽粒容重 767g/L,粗蛋白含量 10.54%,粗脂肪含量 3.41%,粗淀粉含量 75.42%,赖氨酸含量 0.3%。从结果上看,京科 968 的丰产性、抗病性和品质均较为优良,离不开优良自交系京 724 的优良基因效应。组合京 725×京 92(审定名称京科 665),京 464×京 2416(审定名称 NK718)在 2013—2014 年示范推广中均有较好表现,表现出良好的适应性和丰产性,表明这些自交系的选育是成功的。

参考文献(略)

本文原载:中国种业,2015(2):41-44

高产早熟耐密抗倒伏宜机收玉米新品种"京农科 728"的选育与配套技术研究

段民孝 赵久然 李云伏 王元东 邢锦丰
张华生 刘新香 刘春阁 张雪原 张春原

（北京市农林科学院玉米研究中心，北京 100097）

摘　要：为实现中国粮食增产 500 亿 kg，玉米作为中国粮食增产的主力军，选育并应用优良玉米新品种是提高玉米单产保证总产持续增加的主要途径。通过利用玉米优异资源为基础，集成"高大严"、IPT、DH 等多种育种方法，进行选育优良自交系和杂交种研究。以自育优良自交系"京 MC01"为母本，"京 2416"为父本杂交选育成玉米新品种"京农科 728"。试验结果表明，该品种耐密植，抗倒伏，综合抗性好，产量高，增产潜力大，稳产性好，脱水快，早熟，适宜机械化收获。该品种于 2012 年通过国家审定（国审玉 2012003），2014 年通过北京市审定（京审玉 2014006），适宜在北京、天津两市，河北省的唐山、廊坊、保定北部、沧州中北部夏玉米种植区种植。该品种的种子生产技术简便，产量高，降低生产成本。在生产中结合科学管理，提高单产，带动总产增加，为保障粮食安全奠定基础。

关键词：玉米；高产；早熟；密植；抗倒伏；机械化收获

Study on the Breeding and Supporting Technology of New Maize Variety "Jingnongke 728"

Duan Minxiao　Zhao Jiuran　Li Yunfu　Wang Yuandong
Xing Jinfeng　Zhang Huasheng　Liu Xinxiang　Liu Chunge
Zhang Xueyuan　Zhang Chunyuan

(*Maize Research Center, Beijing Academy of Agriculture & Forestry Sciences, Beijing 100097, China*)

Abstract: For the aim of 100 billion pounds grain output in China, the application of elite maize

---

基金项目：国家"863"计划"玉米 DH 工程化育种技术与新品种选育研究"（2012AA101203）、国家"863"计划"强优势玉米杂交种的创制与应用"（2011AA10A103）、北京市农林科学院院青年基金"玉米单倍体自然加倍适宜区域的研究"（QNJJ201418）

第一作者：段民孝，男，1972 年出生，山西人，副研究员，博士，主要从事玉米育种研究。通讯地址：100097 北京市海淀区曙光花园中路 9 号北京市农林科学院玉米研究中心，Tel：010-51503772，E-mail：duanminxiao@126.com

通讯作者：赵久然，男，1962 年出生，北京人，研究员，博士，主要从事玉米研究。通讯地址：100097 北京市海淀区曙光花园中路 9 号北京市农林科学院玉米研究中心，Tel：010-51503936，E-mail：maizezhao@126.com varieties, which was the first major grain crops in China, was a key way to improve the per unit area yield and the total production in maize. Based on the rich genetic resources, the elite hybrid was selected through the integration of "high standard, large population and severe selection", IPT, DH and other breeding techniques. The elite variety "Jingnongke728", "Jing MC01" and "Jing 2416" as the female and male parent respectively, which had been authorized by National Crop Variety Committee (No. 2012003) and Beijing Crop Variety Committee (No. 2014006). The elite hybrid showed better lodging resistance, high density tolerance, high yield better yield stability, high dehydrate rate, early maturity. The performance of suitable for mechanical harvesting had been verified in many places such as Beijing, Tianjin, Tangshan, Langfang, Baoding and Cangzhou in China. The hybrid seeds production technology was easily and higher yield to get, so reducing the cost of production. It would be also convinced that the combination of elite hybrid and scientific managements will increase the per unit area yield and the total production in maize, and which will lay the foundation for the food security in China.

Key words: Maize; High Yield; Earliness; High Population Planting; Lodging Tolerance; Mechanical Harvesting

玉米已成为中国第一大作物，2013年总产达到2.1775亿t，实现了2004年以来的十连增，是中国粮食增产的主力军，贡献率超过50%。根据国务院《国家粮食安全中长期发展规划2008—2020年》所制定的《全国新增1 000亿斤粮食生产能力规划（2009—2020年）》，玉米承担53%的增产份额。鉴于土地资源有限，通过提高单产来保持玉米持续增产是唯一途径。自20世纪70年代以来，中国玉米已经经历了4～5轮的品种更替，选用良种对单产提高的贡献率近50%。中国玉米种植区域广阔，生态类型复杂，需求优良品种多样化。全国农业技术推广服务中心优化试验区域布局、完善试验管理，筛选优良品种和示范新品种。新世纪以来，"郑单958"、"浚单20"、"先玉335"等优良品种的成功选育与大面积推广，对促进全国玉米生产整体水平提高发挥了重要作用。

国家京津唐夏播早熟玉米组试验主要为东华北春玉米种植区界限南缘和黄淮海夏玉米种植区界限北缘的广大区域选拔合适的夏播玉米品种。此区域夏玉米种植面积超过66.67万hm^2，夏播玉米生育期约95天（有效积温2 400～2 500℃）。20世纪90年代，"唐抗5号"作为该区的主要品种得到大面积种植。新世纪以来，"京单28"、"京科25"等新品种在该区域得到大面积种植。随着"一增四改"技术的推广，生产上急切需求耐密性好、抗倒伏性强、脱水快、早熟性好、适宜机械化的玉米新品种。选育株型理想、节间拉开、耐密植、不空秆、抗倒伏、抗、耐多种主要病虫害、中早熟、产量高、稳产性好、综合抗性好的玉米新品种成为主要目标。为此，笔者利用杂交种X1132新种质（具有高产稳产、农艺性状特点鲜明、综合抗性优良、籽粒商品性好、出籽率高以及脱水速度快等优点），采用"高大严"选系方法（即高密度、大群体、严选择并变换不同地点）、DH（双单倍体）技术等方法，开展优良自交系选育，组配杂交组合，结合多点鉴定技术，以期选育到具有本土杂优模式的优良杂交种，并对杂交种制种技术和栽培技术要点进行研究，更好的为生产服务提供理论指导。

1 品种选育

"京农科728"是北京农科院种业科技有限公司以自育自交系"京MC01"为母本，

"京 2416"为父本杂交选育而成的高产早熟耐密植且适宜机械化收获的玉米新品种,先后参加国家京津唐夏播早熟玉米组试验(区试代号 ZY728)和北京市夏播早熟玉米组试验(区试代号"京科早 728")。试验结果表明,该品种综合表现优良,适宜在北京、天津两市,河北省的唐山、廊坊、保定北部、沧州中北部夏玉米种植区种植。2012 年通过国家农作物品种审定(编号:国审玉 2012003),2014 年通过北京市农作物品种审定(编号:京审玉 2014006),2013 年申请植物新品种权保护(申请号为 CNA010431E)。

1.1 亲本选育

母本"京 MC01"以杂交种 X1132 为选系基础材料,按照"高大严"选系方法(即高密度、大群体、严选择并变换多个不同地点),同时经配合力测定等程序,在 S4 代优良穗行中选择优良植株利用 DH 育种技术进行诱导,获得单倍体籽粒,种植后将自然加倍植株自交获得结实果穗(DH 系),再经扩繁和多点鉴定及严格筛选,获得了优良的 DH 纯系"京 MC01"。"京 MC01"具有一般配合力高和综合抗性好的特点。株型紧凑,花药深紫色,花丝淡红色。北京地区夏播生育期 100d,春播 110d。株高 198cm,穗位 65cm,穗长 18cm,穗行数 14 行,穗粗 4.5cm,长筒形果穗,轴红色。籽粒半硬粒型,黄粒。抗大、小斑病、茎腐病、矮花叶病毒病和穗粒腐病。

父本"京 2416"来源于"京 24"与 5237 杂交组成的基础材料的选系,按照"高大严"选系技术并变换不同生态区经多代自交选育而成。"京 2416"具有较高的一般配合力。株型紧凑,幼苗叶鞘紫红色,淡紫色花药,绿色花丝。北京地区春播生育期 100d,夏播 95d。株高 160cm,雄穗分支 3~5 个,花粉量大,穗位 75cm,穗长 16cm,穗粗 4.8cm,穗行数 12 行,锥形穗,白轴,硬粒型,黄粒。该自交系抗玉米大、小斑病、花叶病毒病等。

1.2 杂交组合配制与筛选

依据杂种优势群和杂种优势模式理论,将亲缘关系较远的自交系进行组配易选育到强优势杂交组合。2007 年冬在海南初配京 MC01×京 2416 杂交组合,2008 年在本单位测比试验中,平均小区产量折 12 829.5kg/hm^2,比对照"京单 28 号"增产 12.9%,在 2009 年本单位品比试验中,该杂交种的平均小区产量折 12 061.5kg/hm^2,比对照"京单 28 号"增产 14.2%。2 年试验均表现突出。该组合产量潜力高,抗病抗倒,早熟,籽粒品质好,后期脱水速度快。2009 年参加国家京津唐夏播早熟玉米组预备试验,2010—2011 年参加国家玉米区域试验。2012—2013 参加北京市夏播早熟玉米组试验。

2 品种特征特性

在京津唐夏播区出苗—成熟 98d,比对照"京单 28 号"早熟 1d;北京地区夏播出苗至成熟 98d,比对照"京玉 7 号"晚 1d;需有效积温 2 500℃左右;幼苗叶鞘紫色,叶片绿色,叶缘淡紫色,花药淡紫色,颖壳淡紫色;株型紧凑,株高 276cm,穗位高 95cm,成株叶片数 19 片,花丝淡红色,果穗筒形,穗长 17.7cm,穗行数 14~16 行,穗轴红色,籽粒黄色,半马齿型,百粒重 38.3g。平均倒伏(倒折)率 0.45%。

3 产量表现

"京农科 728"于 2009 年参加国家京津唐夏播早熟玉米组预备试验(区试代号

ZY728），在 7 个试验点全部比相临对照"京玉 7 号"增产，按临标比增产列参试品种第 3 位，进入 2010—2011 参加国家京津唐夏播早熟玉米区域试验，2011 年参加国家京津唐玉米生产试验。试验表明，该品种丰产性、稳产性、抗性和生育期均达标。2012 年通过国家审定（表 1）。

2012—2013 年参加北京市夏播早熟玉米组试验（区试代号"京科早 728"），2013 年参加同组的生产试验。试验表明，该品种比对照显著增产，符合审定，2014 年通过北京市审定。

表 1　"京科早 728"参加玉米各级试验的产量表现

年份	试验区域	试验类型	增产试验点数/试点数	平均产量（kg/hm²）	比对照增减（%）	排名/试验品种数
2009	国家京津唐夏玉米组	预备试验	7/7	9904.5	15.87	3/64
2010	国家京津唐夏玉米组	区域试验	8/8	10144.5	5.55	1/13
2011	国家京津唐夏玉米组	区域试验	10/10	11310.0	12.06	1/15
2011	国家京津唐夏玉米组	生产试验	11/11	10356.0	7.38	1/3
2012	北京市夏播早熟玉米组	区域	3/4	11266.2	11.54	1/9
2013	北京市夏播早熟玉米组	区域	4/4	10021.35	21.55	2/15
2013	北京市玉米早熟组	生产	5/5	9623.55	17.20	1/2

"京农科 728"列为北京市 2013 年玉米更新换代品种，在北京市密云县河南寨镇两河村种植晚春播（5 月 29 日播种）2hm² 进行示范。生产表明，"京农科 728"株型紧凑，穗位低，早熟，综合性状优良；田间病害极轻，遇大风强降雨（7 月 31 日）未发生倒伏，综合抗性好。田间实地测产穗数 66 690 穗/hm²、穗粒数 561 粒、千粒重 408g，产量达 12 990kg/hm²，建议在北京晚春播和夏播地区作为主栽品种推广应用。

4　抗性

4.1　抗病虫性

4.1.1　接种鉴定结果

经中国农业科学院作物研究所与河北省农林科学院植物保护研究所 2010 年、2011 年接种鉴定，中抗—高抗大斑病、小斑病（变幅 1~5 级），中抗—抗茎腐病（变幅 5.4%~15.4%），感—中抗弯孢叶斑病（变幅 5~7 级），感—高感玉米螟（变幅 8.0%~9.0%）。2012 年、2013 年中国农业科学院作物研究接种结果见表 2。

表 2　2012 年、2013 年中国农业科学院作物科学研究所病害接种鉴定结果

年份	大斑病	小斑病	弯孢菌叶斑病	矮花叶病	腐霉茎腐病
2012	HR	R	S	HS	HR
2013	MR	MR	HS	HS	HR

4.1.2 田间表现

抗病性表现为：2012年北京区试田间，1个试点出现丝黑穗病为1%。2013年区试田间1个试点出现大斑病为3级；1个试点出现弯孢叶斑病为3级；1个试点出现青枯病为3%；2个试点出现粗缩病分别为1.1%、0.4%。

4.2 抗倒性

2010—2011国家京津唐玉米区域试验，平均倒伏（倒折）率0.45%。

2012年北京市玉米区域试验，田间各试点均无倒伏、倒折。2013年北京市玉米区试田间1个试点发生倒伏，比例为0.4%；1个点发生倒折，比例为0.8%。区试两年平均倒伏率0%、平均倒折率0.1%。

5 品质

经农业部谷物及制品质量监督检验测试中心（哈尔滨）测定，"京农科728"的籽粒粗蛋白含量9.03%，粗脂肪含量4.12%，粗淀粉含量73.33%，赖氨酸含量0.31%，容重757g/L。

6 种子生产制种技术

"京农科728"的制种区域隔离条件为250~300m，种植密度67 500株/hm^2，最高密度不超过75 000株/hm^2。在甘肃适于4月上旬播种，父本和母本同期播种，行比为5∶1。母本自交系节间拉开较长，植株较高，抽雄时带3~4片叶，可以降低株高，便于父本授粉。

施用肥料要求纯氮150~225kg/hm^2，可全部播前深翻底施，或1/2底施和1/2在小喇叭口期追施，磷钾肥配合。苗期喷施阿维菌素药剂防治飞虱、蓟马，大喇叭口期利用辛硫磷拌沙灌心，防治玉米螟。

成熟果穗脱粒加工时注意清选机器，严防机械混杂。

7 适宜种植区域与栽培技术要点

"京农科728"丰产性好，适应性广，稳产性高，抗病性强，生育期较短，品质优良，适宜北京、天津两市，河北省的唐山、廊坊、保定北部、沧州中北部等有效积温2500℃左右地区种植，生产上进行逐步试验示范，以得到大面积推广。

为保证品种的潜力得到发挥，要制定配套的栽培措施，包括种子、田间肥水管理和病虫害防治等。

7.1 适时播种，确保播种质量

在京津唐等区域夏播，适宜播种期为6月中旬至下旬。精选种子，使用包衣种子，进行播前晾晒，可较好防治苗期地下害虫，促进萌发，起到促根壮苗作用。在中等肥力以上地块，足墒播种，保证播种质量，达到苗齐、苗全、苗壮。

7.2 适当密植，平衡施肥

"京农科728"株型紧凑，适宜密度60 000株/hm^2，在高肥水地块种植密度可增加到67 500株/hm^2。在其他地区种植参照当地种植方式进行。

综合考虑土壤肥力、产量水平和品种需肥特点来计划肥料数量、肥料元素，进行平衡施肥。施足底肥可保证玉米前期早生快发，后期不缺肥，利于形成大穗，籽粒饱满。可使用腐熟农家肥 15 000kg/hm² 以上，精制有机肥 6 000kg/hm² 以上，于播前均匀撒入田间，随着耕作翻入耕层。纯氮 210~240kg/hm²，可全部播前深翻底施，或 1/2 底施和 1/2 在小喇叭口期追施。氮磷钾复合肥作种肥 150kg/hm²。拔节后追尿素 225kg/hm² 和硫酸钾 600kg/hm²，促茎秆健壮；大喇叭口期浇水以攻穗增粒；开花期结合浇水追尿素 75~120kg/hm²，以养根保叶，提高光合生产力，增加粒重。

7.3 加强田间管理，适时收获

加强田间管理，保证苗全苗壮。3 叶期间苗，5 叶期定苗。病苗、虫咬苗及发育不良的幼苗在晴天下午较易萎蔫，便于识别。彻底剔除苗矮叶密、下粗上细、弯曲、叶色黑绿的丝黑穗浸染苗。

肥水管理应该以促为主，重施穗肥，酌施粒肥。拔节期追施氮肥，大喇叭口期追施尿素 150kg/hm²。

防治病虫应以预防为主，综合防治。苗期喷施阿维菌素药剂防治飞虱、蓟马；防治二代黏虫，可用 2.5% 功夫乳油或 20% 速灭杀丁乳油 300ml/hm² 对水 450kg/hm² 进行喷雾防治。大喇叭口期利用辛硫磷拌沙灌心或 BT 乳剂（200~300 倍液）防治玉米螟。

控制杂草。使用 38% 莠去津悬浮剂 2 250~3 000ml 或 50% 乙草胺乳油 1 500~2 250ml/hm²，对水不少于 600kg/hm²，使用常规喷雾，进行土壤封闭处理。

玉米成熟后适时收获。玉米籽粒乳腺消失出现"黑层"后，才真正进入完熟期。尽量晚收，可以充分发挥该品种的增产潜力，增加籽粒产量，减少霉变机率，提高品质。"京农科 728"脱水快，早熟性好，抗倒性强，适宜机械化收获。

8 结论与讨论

针对玉米生产上亟需优良新品种并加以配套技术来获得高产的现实要求，利用优异新种质 X1132 为基础材料，通过集成"高大严"选系方法、DH 技术等方法选育优良自交系，以"京 MC01"为母本，"京 2416"为父本杂交育成玉米杂交种"京农科 728"，具有高产、稳产、早熟、耐密、抗倒、籽粒脱水速度快等特点，适合玉米全程机械化生产。父母本分属于不同种质类群，遗传差异较大，杂种优势较强，符合杂交种选育原则和杂种优势模式。杂交种种子生产技术简便，母本"京 MC01"具有良好农艺性状，综合抗性好，父本"京 2416"散粉性好，制种产量高，风险和生产成本低。该品种由北京龙耘种业独家代理开，2013 年列为北京市玉米更新换代品种，列入 2014 年北京市主推品种。在生产上，结合具体耕作条件，采取配套栽培技术措施，通过科学管理，合理使用肥水，发挥品种潜力，容易获得高产。"京农科 728"将对京津冀玉米区实现全程机械化起到很大推动作用，对落实京津冀协同发展国家战略，共同支撑与引领京津冀现代农业发展，实现粮食持续增产和保障粮食安全具有重要意义。

参考文献（略）

本文原载：农学学报，2015，5（2）：10-14

玉米新品种 NK971 的选育及栽培制种技术要点

王元东　张华生　段民孝　张春原　张雪原　刘新香　陈传永　赵久然

(北京市农林科学院玉米研究中心，北京　100097)

黄淮海夏玉米区是我国玉米主产区之一，近几年种植面积在 1 000万 hm² 左右，被农业部列为我国玉米生产优势产业带。该地区气候复杂多变，温度高、湿度大、阴雨寡照，病虫害严重，为促进该地区夏玉米生产的可持续发展，保证国家粮食安全，生产中需要不断选育出高产、稳产、优质、抗病、耐密植等的新品种。玉米新品种 NK971（国审玉2014016）是以国内核心种质为基础，导入合理比例的国外优良种质，利用"高大严"技术手段，结合 DH 育种新技术，选育而成的适于黄淮海夏播区种植的新品种。在缩短玉米品种的生育期，提高籽粒灌浆与脱水速度，提高耐密植、抗倒伏和适应机械收获等方面有所创新。

1　亲本选育及其特征特性

1.1　母本

京 388 为北京市农林科学院玉米研究中心自选系，来源于自交系郑 58 与 4CV 杂交，又回交一次郑 58 所构成的基础群体材料。在"高大严"（高密度、大群体和严选择）条件下经过 1 代自交和选择，再利用 DH 育种技术选优株进行诱导和加倍，并对优良的 DH 系进行多点鉴定及严格筛选所获得，命名为京 388。幼苗叶鞘浅紫色，叶片绿色，叶缘绿色，雄穗分支多且主枝长，花药黄色，颖壳淡紫色；株型紧凑，株高 160cm，穗位高 92cm，花丝浅紫色，果穗穗柄长度中等，果穗长筒形，苞叶长度中等，穗长 16.8cm，穗行数 14 行，秃尖 0.1cm，单穗粒重 120g 左右，穗轴白色，籽粒黄色，半马齿型，百粒重 341g。适于制种用作母本，制种产量 500kg/666.7m² 左右。

1.2　父本

京 372 为北京市农林科学院玉米研究中心自选系，来源于自交系昌 7-2 与自选系京 24 杂交，又回交一次昌 7-2 所构成的基础群体材料，在"高大严"条件下经过 1 代自交选择，再利用 DH 育种技术选优株进行诱导和加倍，并对优良的 DH 系进行多点鉴定及严格筛选所获得，命名为京 372。幼苗叶鞘紫色，叶片绿色，叶缘绿色，雄穗分支多且主枝长中等，花药浅紫色，颖壳淡紫色；株型紧凑，株高 150cm，穗位高 62cm，花丝红色，果穗穗柄长度中等，果穗锥形，苞叶长度中等，穗长 12.8cm，穗行数 16 行，秃尖 0.2cm，单穗粒重 100g 左右，穗轴白色，籽粒黄色，半硬粒型，百粒重 311g。适于制种用作父本，花粉量大，散粉持续时间长。

基金项目：粮食作物产业技术体系北京市创新团队专项（BITG7）、北京市农林科学院科技创新能力建设专项（KJCX20140107）

2 杂交种选育及其特征特性

2.1 选育过程

2009年冬季南繁时组配该组合，2010年进行多点鉴定和品比试验。结果表明该组合田间表现突出。2011年参加国家主产区预试，2012—2013年参加国家黄淮海夏玉米区试，2013年参加国家黄淮海夏玉米生产试验，2014年通过农业部国家农作物审定委员会审定，定名为NK971。

2.2 特征特性

幼苗叶鞘浅紫色，叶片绿色，叶缘绿色，雄穗分支多且枝长中等，花药浅紫色，颖壳淡紫色；株型紧凑，株高262cm，穗位高112cm。花丝浅紫色，果穗穗柄短，果穗长筒形，苞叶长度中等，穗长17.9cm，穗行数14行，秃尖0.4cm，单穗粒重180g，穗轴白色，籽粒黄色，半马齿型，百粒重34.1g。籽粒商品性好。在黄淮海夏播区出苗至成熟102天，与对照郑单958熟期相同，全生育期叶片数20片。

2.3 抗性级品质

经辽河北省农林科学院植物保护研究所、中国农业科学院作物科学研究所2012年、2013年2年接种鉴定，中抗小斑病（5级），感大斑病（5~7级），高感腐霉茎腐病（31.3%~48.5%），感廉孢茎腐病（39.5%）。2013年经农业部谷物及制品质量检验检测中心（北京）检测，籽粒容重786g/L，粗蛋白含量10.18%，粗脂肪含量3.23%，粗淀粉含量73.07%，赖氨酸含量0.33%；籽粒容重较高，品质优良。

3 产量表现

2011年在黄淮海夏玉米主产区预备试验每666.7m^2平均产量为578.4kg，比对照郑单958增产5.81%；2012年参加国家黄淮海夏播区域试验，平均产量为704.9kg，比对照增产3.3%，增产极显著；2013年参加国家黄淮海夏播区域试验，平均产量为666.4kg，比对照增产5.6%，增产极显著；2013年参加国家黄淮海夏播生产试验，平均产量为627.0kg/666.7m^2，比对照郑单958增产7.2%，增产极显著。

2014年NK971在山东、河北大面积连片示范种植中测产验收，每666.7m^2平均产645.2kg，较当地主推品种郑单958增产6.5%；在2015年河北沧州、衡水晚春播节水新品种大面积示范，平均产689.3kg，比郑单958增产10.1%。近2年的示范推广说明，该品种不仅产量高，而且稳定性好，在河北、山东等晚春播地区推广潜力很大。

4 栽培技术要点

在黄淮海夏玉米区以及河北山东晚春播玉米等区域种植。根据各地种植技术要求合理设置株距和行距，4 000~4 500株/666.7m^2，合理密植栽培。

结合地力条件，测土配方施肥。底肥每666.7m^2施用三元复合肥15kg，促苗期生长；小喇叭口—大喇叭口期机械深施尿素20kg，攻穗增粒；开花期结合浇水追尿素5~8kg，以养根保叶，提高光合生产力，增加粒重。玉米籽粒乳腺消失后收获。应注意防治腐霉茎腐病，避开腐霉茎腐病重发区种植。种子包衣，预防土传性病害发生。

5 制种技术要点

5.1 保持品种种性技术要点

亲本的种性保持应严格按照育种家种子（原原种）—原种的种子生产程序进行生产。繁殖隔离要求在500m以上，并严格去除杂株。杂交种的种性保持应是使用高纯度的亲本种子，确保制种区的安全隔离，并严格去除亲本杂株，及时干净、彻底去除母本雄穗，授粉结束后及时砍除父本，确保种子纯度。

5.2 种子生产技术要点

选择土质肥沃、地势平坦、排灌方便的制种基地，隔离距离不小于300m。在甘肃适于4月上旬播种，先播种母本，第1期父本晚播3d，第2期父本晚播6d，父本独立成行，行比1∶5，母本适宜种植密度6 000株/666.7m^2。加强田间肥水管理，适时收获。脱粒加工时注意清选机器，严防机械混杂。

本文原载：中国种业，2016（2）：54-55

玉米新品种 MC220 的选育、特征特性及栽培技术要点

赵久然　王元东　王荣焕

（北京市农林科学院玉米研究中心，北京　100097）

摘　要：MC220 是北京市农林科学院玉米研究中心采用"高大严"选系法（即高密度、大群体、严选择）等育种新方法，经过多年研究育成的玉米新品种，具有熟期适宜、株型紧凑、耐密抗倒、粒大粒重、高产优质等特点。2013 年通过国家农作物品种审定委员会审定，适宜在京津唐及类似生态区夏播种植。

关键词：玉米；新品种；MC220

1　亲本及杂交组合的选育

1.1　亲本自交系的选育

MC220 单交种的 2 个亲本自交系均是北京市农林科学院玉米研究中心采用"高大严"（即高密度、大群体、严选择）等选系方法选育的自选系。其中，母本京 X220 来源于 X1132 杂交种混粉构建的基础群体选系；父本京 C632 来源于（昌 7-2×京 2416）×京 2416 构建的基础群体选系。

母本自交系的选育过程：2005 年，在经过之前多点鉴定筛选确定的基础上，在北京育种基地完成新种质 X1132x 混粉群体的构建；2006 年在北京育种基地采取 90 000 株/hm² 的高密度种植 S1 代，并进行严格选择；2007 年，在吉林四平种植选择 S2 代；2008 年，在河南郑州种植选择 S3 代，并于冬季在海南进行 S4 代的自交纯合和配合力测配；2009 年，在北京对 S5 代进行进一步自交纯合，同时进行组合产量、综合抗性测试；2009 年，在海南对 S6 代进行继续自交纯化，育成稳定自交系京 X220。

父本自交系的选育过程：父本京 C632 属于自选优良骨干自交系京 2416 的相近自交系或派生品种自交系。京 2416 自交系是由北京市农林科学院玉米研究中心利用京 24 与 5237 杂交构建基础选系群体，按照"高大严"选系法选育出的优良骨干自交系。利用京 2416 优良骨干自交系已经组配出京单 58、京单 68、京科 389、京农科 728 等多个国审和省审优良品种正在生产上大面积推广种植。利用（昌 7-2×京 2416）×京 2416 组配和构建的基础群体作为选系材料，并采用优系聚合法和高大严选系法等育种新方法，经过多代自交和严格选择，所选育出的优良自交系，正式定名为京 C632。

基金项目：国家高技术研究发展计划（863 计划）（2012AA101203）、北京市科技计划（D121100003512001）

作者简介：赵久然，研究员，主要从事玉米遗传育种及高产栽培研究

1.2 杂交组合的选育

2007年开始组配（京X220×京C632）组合。经过多年多点试验筛选，鉴定出该组合（京X220×京C632）产量潜力高、抗病抗倒、早熟、籽粒品质好，并且生育期适宜机械化收获。2008年和2009年在北京顺义、昌平，天津蓟县、西青、武清，河北玉田、廊坊、保定、沧州、定兴、容城和易县等试点进行广泛测试，平均产量11 445kg/hm^2，比对照品种京单28增产12%。

2 品种特征特性

MC220熟期适宜，在京津唐夏播区生育期101d，需有效积温2 500℃左右，在京津唐地区夏播种植可正常成熟。株型紧凑，株高275cm，穗位高111cm，耐密性好，适宜密植。果穗筒形，穗长17cm，穗行数14~16行，穗轴红色。籽粒大、黄色，半马齿型，百粒重42.4g。

经中国农业科学院作物科学研究所与河北省农林科学院植物保护研究所2011年和2012年连续两年接种鉴定，MC220对黄淮海区域夏播玉米的主要病害茎腐病和小斑病都在中抗以上。品质优良，商品性好，属于一级高淀粉品种。经农业部谷物及制品质量监督检验测试中心（哈尔滨）测定，籽粒容重732g/L，淀粉含量高达76.15%。

3 区试产量表现

2011—2012年，MC220参加国家京津唐夏播早熟玉米组品种区域试验，2年平均产量11 040.0kg/hm^2，比对照品种京单28增产8.4%。其中，2011年区试平均产量10 645.4kg/hm^2，比对照品种京单28增产5.48%，10个试点9增1减；2012年区试平均产量11 433.0kg/hm^2，比对照品种京单28增产11.4%，10个试点全部增产。2012年参加国家京津唐夏播早熟玉米组生产试验，平均产量11 095.5kg/hm^2，比对照品种京单28增产10.8%。

4 适宜种植区域

经过多年多点的国家玉米新品种区域试验、生产试验以及大面积生产示范种植，结果表明，MC220适宜在北京、天津以及河北省的唐山、廊坊、保定、沧州等类似生态区域夏播种植。

5 栽培技术要点

为充分挖掘和发挥MC220玉米新品种的高产潜力，在生产实践中应根据该品种的特征特性，重点把握好适期早播和合理密植两项栽培要点。此外，该品种耐密抗倒性好，后期适宜机械化收割作业。

5.1 适期早播

前茬作物收获后，尽量减少农耗，抢早播种，以争取更多积温，实现高产优质。一般适宜播种期为6月上、中旬，应在6月20日前完成播种。若遇土壤墒情不足，可先播种，播后再及时补浇"蒙头水"，保证种子尽早萌发和出苗。

5.2 合理密植

MC220株高和穗位较低，耐密性强，抗倒性好，适宜密植栽培。在大田生产中，各地可根据具体的土壤肥力、管理水平、生产条件等进行合理密植。一般大田生产条件下，种植密度以 67 500株/hm^2 左右为宜。高肥力地块可适当增加种植密度。

5.3 防治病虫害

注意防治大斑病、弯孢叶斑病和玉米螟，避免或尽量减少因病虫为害造成的产量损失。

5.4 机械适时晚收

建议在不耽误下茬小麦播种的情况下尽量晚收，以最大限度利用光热资源，确保籽粒充分灌浆和成熟，充分发挥品种产量潜力，增加产量并提高籽粒品质。该品种适宜机械化收割作业，可在9月底至10月上旬期间，玉米籽粒达到生理成熟后，采用机械进行收获。当籽粒含水率降至30%以下时，也可以采用适宜的机械直接收获玉米籽粒。

本文原载：作物杂志，2014（4）：156-157

机收玉米新品种 MC812 的选育

张华生　段民孝　陈传永　张春原　张雪原
刘新香　毛振武　张　亮　王元东　赵久然

(北京市农林科学院玉米研究中心，北京　100097)

随着我国城镇化发展和人口结构老龄化的变化，农村劳动力越来越短缺，实行玉米生产全程机械化收获是当前玉米生产发展的迫切要求。选育早熟、耐密、抗倒和适宜机收的玉米新品种是当前的育种方向。玉米新品种 MC812（京审玉 2015003）是以优良自交系杂交聚合作为选系材料，利用"高大严"选系方法（即高密度、大群体、严选择）和 DH（双单倍体）育种技术，选育而成的适于机收的新品种。

1 亲本选育及其特征特性

1.1 母本

京 B547 为北京市农林科学院玉米研究中心自选系，是京 MC01 和京 4055 两个优良自交系的杂交聚合选系，其中优良自交系京 MC01 是适于机收玉米品种京农科 728（国审玉 2012003、京审玉 2014006）的母本，具有耐密、早熟、株型理想、籽粒脱水速度快等优点，缺点是抗倒伏能力略差、抗病性略差。另一个优良自交系京 4055 是玉米品种 MC220 的母本，具有高抗倒伏、抗性强、综合农艺性状优良等优点，缺点是晚熟、耐密性差。2006 年在北京组配两个优良自交系杂交组合获得 F_1 种子，同年在海南自交获得 F_2 籽粒，2007—2008 年利用"高大严"选系技术在北京经过 F_2 和 F_3 等 2 代自交和选择，实现优良遗传基因快速、有效地累积聚合，筛选出综合农艺性状优良的 F_4 株系；2008 年在海南在 F4 株系自交纯合的同时，利用京 2416、京 92 等测验种对其进行配合力测定，进一步筛选出配合力较高的 F_5 株系；2009 年在北京利用 DH 育种技术将 F_5 株系进行诱导，获得单倍体；2009 年在海南将单倍体进行加倍获得的 DH 系；2010 年在北京对 DH 系扩繁并进行严格筛选评价，选中后命名为京 B547。

京 B547 幼苗叶鞘紫色，叶片、叶缘绿色，雄穗分支少（2~3 个）且主枝长，花药深紫色，颖壳淡紫色；株型紧凑，株高 164cm，穗位高 82cm，花丝红色，果穗穗柄长度中等，果穗长筒形，苞叶长度长，穗长 16.0cm，穗行数 14 行，秃尖 0.9cm，单穗粒重 110g 左右，穗轴红色，籽粒黄色、半马齿型，百粒重 351g。制种生产适于用作母本，每 666.7m^2 制种产量 450kg 左右。

基金项目：粮经作物产业技术体系北京市创新团队专项经费资助（BITG7）、北京市农林科学院科技创新能力建设专项（KJCX20151406）、北京市农林科学院科技创新能力建设专项（KJCX20140107）

1.2 父本

京 2416 为北京市农林科学院玉米研究中心自选系,为自交系 5237 与自选系京 24 杂交聚合选系,其中优良自交系 5237 是玉米品种掖单 22 的母本,具有耐密、早熟、株型理想等优点,缺点是抗倒伏能力差,组配的杂交种籽粒具有黄白顶。另一个优良自交系京 24 是玉米品种京单 28 的母本,具有高抗倒伏、抗性强、综合农艺性状优良等优点,缺点是籽粒品质差、穗行数少、籽粒大。2002 年在北京组配两个优良自交系杂交组合获得 F1 种子,同年在海南自交获得 F_2 籽粒,2003—2004 年利用"高大严"(高密度、大群体和严选择)选系技术在北京经过 F_2 和 F_3 等 2 代自交和选择,实现优良遗传基因快速、有效地累积聚合,筛选出综合农艺性状优良的 F_4 株系;2004 年在海南在 F_4 株系自交纯合的同时,利用 CH3、CH8 等测验种对其进行配合力测定,进一步筛选出配合力较高的 F5 株系;2005 年在北京和海南继续自交纯合,稳定后命名为京 2416。

京 2416 幼苗叶鞘淡紫色,叶片、叶缘绿色,雄穗分支 3~5 个,花药浅紫色,颖壳淡紫色;株型紧凑,株高 160cm,穗位高 75cm,花丝绿色,果穗穗柄长度中等,果穗锥形,苞叶长度中等,穗长 16.0cm,穗行数 12 行,秃尖 0.2cm,单穗粒重 102g 左右,穗轴白色,籽粒黄色、半硬粒型,百粒重 371g。制种生产中常用作父本,花粉量大,散粉持续时间中等。

2 杂交种选育及其特征特性

2.1 选育过程

2010 年冬季南繁时组配该组合,2011 年进行多点鉴定和品比试验。结果表明该组合田间表现突出。2012 年参加北京市夏播预备试验,2013—2014 年参加北京市夏播玉米区试试验,2014 年参加北京市夏播玉米生产试验,2015 年通过北京市农作物审定委员会审定,定名为 MC812。

2.2 特征特性

幼苗叶鞘浅紫色,叶片、叶缘绿色,雄穗分支 4~6 个,花药淡紫色,颖壳淡紫色;株型紧凑,株高 268cm,穗位高 109cm。花丝淡红色,果穗穗柄长度中等,果穗筒形偏锥,苞叶长度中等,穗长 17.2cm,穗行数 14.7 行,秃尖 0.9cm,单穗粒重 168g,穗轴红色,籽粒黄色、半马齿型,百粒重 39.4g。籽粒商品性好。北京地区夏播出苗至成熟 103d,与对照京单 28 号相同。全生育期叶片数 20 片。

2.3 抗性级品质

经中国农业科学院作物科学研究所 2013 年、2014 年 2 年接种鉴定,中抗-抗大、小斑病,中抗弯孢菌叶斑病,抗-高抗腐霉菌茎腐病,高感矮花叶病毒病。2014 年经农业部谷物品质监督检验测试中心检验(干基),该品种籽粒中含粗淀粉 74.35%,粗脂肪 4.29%,粗蛋白 8.61%,赖氨酸 0.31%,容重 754g/L。

3 产量表现

2013 年在北京市夏播区域试验,每 666.7m² 平均产 561.2kg,比对照京单 28 增产 19.6%;2014 年北京市夏播区域试验,平均产 635.7kg,比对照京单 28 增产 14.7%,2014 年北京市夏播生产试验平均产 682.5kg,比对照京单 28 增产 21.6%。

2015年MC812在山东、河北、天津以及内蒙古等地进行大面积示范种植，收获时籽粒机械化直收，每666.7m^2平均产649.8kg，较当地主推品种郑单958增产5.2%，籽粒含水量28.7%，比对照郑单958降低3%，籽粒杂质率3.1%，籽粒破损率3.7%。通过大面积的示范推广说明，该品种不仅产量高，而且抗病性（高抗茎腐病和大斑病）和抗倒性好，耐密植，籽粒脱水速率快，适于机械化收获籽粒，在黄淮海夏播区和东华北春播区机收潜力很大。

4 栽培技术要点

在黄淮海夏玉米区以及东华北春播玉米等区域种植。根据各地种植技术要求合理设置株距和行距，4 500~5 000株/667m^2，合理密植栽培，单粒精量播种。结合地力条件，测土配方施肥。底肥每666.7m^2施用三元复合肥15kg，促苗期生长；小喇叭口—大喇叭口期机械深施尿素20kg，攻穗增粒；开花期结合浇水追尿素5~8kg，以养根保叶，提高光合生产力，增加粒重。玉米籽粒乳腺消失后收获。应注意防治腐霉茎腐病，避开腐霉茎腐病重发区种植。种子包衣，预防土传性病害发生。

在生理成熟后（玉米籽粒黑层出现），籽粒水分含量降至30%以内进行机械化籽粒直收。黄淮海夏播区机械化籽粒直收在10月12—18日，东华北春播区在10月20日左右。

5 制种技术要点

5.1 保持品种种性技术要点

亲本的种性保持应严格按照育种家种子（原原种）—原种的种子生产程序进行生产。繁殖隔离要求在500m以上，并严格去除杂株。杂交种的种性保持应是使用高纯度的亲本种子，确保制种区的安全隔离，并严格去除亲本杂株，母本带2~3片叶去雄，及时花检，授粉结束后及时砍除父本，确保种子纯度。

5.2 种子生产技术要点

选择土质肥沃、地势平坦、排灌方便的制种基地，隔离距离不小于300m。在甘肃适于4月上旬播种，先播种母本，第1期父本晚播5d，第2期父本晚播9d，父本独立成行，行比1:5，母本适宜种植密度6 000株/666.7m^2。加强田间肥水管理，适时收获。脱粒加工时注意清选机器，严防机械混杂。

本文原载：中国种业，2016（2）：62-64

玉米新品种 MC703 选育与配套技术

张华生　范会民　张雪原　陈传永　段民孝
张春原　刘新香　王元东　赵久然

（北京市农林科学院玉米研究中心，北京　100097）

优良品种在玉米增产中的贡献率约占 40%，对我国粮食安全和增产发挥着重要的作用。玉米新品种 MC703 的选育是针对我国现代玉米生产和市场中的突出问题，通过对亲本材料在大群体、高密度、变换不同生态区严格选择条件下进行遗传改良，增加品种的适应性、抗逆性，具有高产、稳产、优质、脱水快、适合机收、综合抗性强等特点，适合目前及未来几年玉米生产对新品种的要求。目前 MC703 已通过河南省（豫审玉 2015004）、北京市（京审玉 2015001）、内蒙古自治区（蒙审玉 2015007）3 个不同生态区的审定，并在国家东北春播中熟区、国家西北春播区和河北夏播区区域试验中进入续试试验。

1　亲本选育及其特征特性

1.1　母本

母本京 X005，来源于国外杂交种 X1132x，按照"高大严"选系方法（即高密度、大群体、严选择并变换多个不同地点）于 2007 年选育而成。2003 年春天在北京按照 6 000 株/667m^2 的密度种植 X1132x 的 F$_2$ 群体 3 000 株左右，在苗期淘汰弱病株，留健康长势强植株；自交授粉前，淘汰杂合性强、株型松散、叶片宽大等大量不符合要求的单株；在授粉期间，选择雌雄协调、吐丝快而集中的植株进行一次性自交授粉；收获前，淘汰病虫害植株、倒伏（倒折）株、籽粒败育和果穗畸形株，对入选果穗进行室内考种，淘汰结实性差的果穗，留下单穗粒重大于 100g，出籽率大于 88%，穗行数大于 14 行的约 50 个优良果穗。2003 年冬季在海南对这 50 个果穗再进行高密度穗行种植（8 000 株/667m^2），每个果穗种植 4 行（5m 行长），按照对 F$_2$ 群体选择时的标准，每个穗行选择 4~6 个果穗，又根据果穗性状（结实性、穗行数、籽粒品质）淘汰 10 个穗行左右，共收获 30 个穗行，约 150 个果穗，室内考种选出约 120 个果穗。

2004 年春天在河南和吉林 2 个地点同时按照穗行种植，2 行区（5m 行长），种植密度为 6 000 株/667m^2，选择河南、吉林两地均表现优良的穗行，每个穗行收获 3~5 穗，共收获 300 个果穗左右，室内考种选出约 200 个果穗。2004 年冬季在海南进行自交加代，种植密度 4 500 株/667m^2，同时用改良兰卡斯特血缘自交系进行配合力测定。2005 年春季在北京队优良果穗继续进行自交纯合，种植密度 4 500 株/667m^2，根据配合力测定结果，选择配合力高的穗行。2005 年冬季在汉南继续自交纯合，同时复配组合。2006 年春季在北

基金项目：粮经作物产业技术体系北京市创新团队专项经费资助（BITG7）、北京市农林科学院科技创新能力建设专项（KJCX20140107）

京扩繁该自交系，同时定名为京 X005。

"京 X005"具有一般配合力高和综合抗性好的特点。苗期叶鞘紫色，株型半紧凑，叶片绿色，雄穗分支 2~4 个，花药紫色，颖壳淡紫色，花丝淡红色，株高 191cm，穗位高 105cm，穗长 18cm，穗行数 14~16 行，穗粗 4.9cm，果穗长筒形，白轴，硬粒型，黄粒。抗大斑病、小斑病、茎腐病、矮花叶病毒病和穗粒腐病。适宜用作母本，制种产量在 500kg/667m² 左右，种子商品性好，适宜单粒精量播种。

1.2 父本

父本京 17 来源于国外多个兰卡斯特血缘系混合授粉组成的群体与 MO17 杂交，选优株与郑 58 杂交后获得的基础选系材料。按照"高大严"选系方法于 2010 年选育而成。2005 年冬天在海南种植国外 6 个兰卡斯特血缘系的混粉群体，选择优良单株与郑 58 杂交；2006 年春天在北京高密度（6 000株/667m²）种植 3 000 个单株的大群体，严格选择雌雄协调、株型好、抗倒、抗病、花粉量大的单株进行自交授粉，后期选择单株抗病性好，果穗的穗行数在 14 行以上，结实性好、脱水速度快和籽粒深的果穗。2006 年冬季在海南继续进行高密度种植（8 000株/667m²）自交选择。2007 年春季同时在河南、吉林 2 个地点分别按照穗行种植，2 行区（5m 行长），种植密度为 6 000株/667m²，选择河南、吉林两地均表现优良的穗行。2007 年冬季在海南进行自交加代，种植密度 4 500株/667m²，同时用改良瑞德血缘自交系京 X005 进行配合力测定。2008 年春季在北京对优良果穗继续进行自交纯合，种植密度 4 500株/667m²，根据配合力测定结果，选择配合力高的穗行。2009 年冬季在海南继续自交纯合，同时复配组合。2010 年春季在北京扩繁该自交系，同时定名为京 17。

"京 17"具有较高的一般配合力和综合抗性好的特点。苗期叶鞘淡紫色，株型紧凑，叶片绿色，雄穗分支 2~4 个，花药黄色，颖壳黄色，花丝淡红色，株高 181cm，穗位高 99cm，穗长 16cm，穗粗 5.9cm，穗行数 14~16 行，果穗长筒形，红轴，马齿型，黄粒，出籽率高。中抗玉米大斑病、小斑病、茎腐病和矮花叶病毒病，抗穗粒腐病。适宜用作父本，花粉量大。

2 杂交种选育及其特征特性

2.1 选育过程

2009 年冬在海南初配京 X005×京 17 杂交组合，2010 年在本单位测比试验中，平均小区产量折合每 667m² 产 859.5kg，比对照郑单 958 增产 12.9%；在 2012 年本单位品比试验中，该杂交种的平均小区产量折合每 667m² 产 804.1kg，比对照郑单 958 增产 14.2%。2 年试验均表现突出，综合农艺性状优良。

2.2 植物学性状

苗期叶鞘淡紫色，叶片绿色，叶缘绿色。成株株型紧凑，护颖淡紫色，花药淡紫色，花丝淡红色；株高 305.9cm，穗位 115.1cm，总叶片数 21 片，雄穗一级分枝 4 个。果穗长筒形，穗轴红色，穗长 19cm，穗粗 4.9cm，穗行数 16~18 行，行粒数 38.1 粒，秃尖 1.1cm，出籽率 88.9%。籽粒半马齿型，黄色，千粒重 346.6g。

2.3 抗性

在内蒙古等华北春玉米地区感大斑病，中抗弯孢叶斑病、茎腐病、玉米螟，高抗丝黑穗病。在河南等黄淮海夏玉米地区高抗瘤黑粉病，抗锈病和穗腐病，中抗小斑病和弯孢菌叶斑病和茎腐病。

2.4 品质

据2014年农业部农产品质量监督检验测试中心（郑州）对该品种多点套袋果穗的籽粒混合样品品质分析检验报告：容重741g/L，粗蛋白质10.81%，粗脂肪4.35%，粗淀粉74.48%，赖氨酸0.35%。

3 产量表现

2013年在河南省区域试验，每667m^{2}平均产670.8kg，比对照郑单958增产10.71%，与对照相比差异达极显著水平，丰产稳产性好；2014年续试，平均产660.5kg，比对照郑单958增产6.21%，与对照相比差异达极显著水平，丰产稳产性较好。2014年河南省生产试验平均产660.9kg，比对照增产5.0%。

2013年在北京市区域试验，每667m^{2}平均产771.9kg，比对照郑单958增产6.0%；2014年续试，平均产799.3kg，比对照郑单958增产9.3%。2014年北京市生产试验，每667m^{2}平均产718.7kg，比对照郑单958增产1.7%。

2012年内蒙古晚熟组预备试验，7点每667m^{2}平均产877.7kg，比对照郑单958增13.3%。2013年内蒙中晚熟组区试试验，每667m^{2}平均产876.9kg，比对照组均值增产5.85%。2013年中晚熟组生产试验，每667m^{2}平均产957.6kg，比对照增产9.03%。

4 栽培技术要点

MC703适合内蒙古等东华北春玉米区和黄淮海夏玉米区种植，种植密度3 800~4 500株/667m^2。注意旱薄低水肥地宜稀，高肥水地宜密。春播玉米田抓住良好墒情，当土壤5cm地温通过10~12℃时即可播种；夏玉米田可在小麦收获前7~10d播种，灰飞虱发生较重的区域可进行麦后直播。在玉米抽雄至吐丝期出现干旱应及时浇水保持正常授粉受精，提高结实率和千粒重。黄变白松散、籽粒变硬发亮、乳腺消失、籽粒尖端出现黑白层时，为收获的最佳时期。

5 保持品种种性及制种技术要点

5.1 保持品种种性技术要点

亲本的种性保持应严格按照育种家种子（原原种）—原种的种子生产程序进行生产。繁殖隔离要求在500m以上，并严格去除杂株。杂交种的种性保持应是使用高纯度的亲本种子，确保制种区的安全隔离，并严格去除亲本杂株，母本带2~3片叶去雄，及时花检，授粉结束后及时砍除父本，确保种子纯度。

5.2 种子生产技术要点

种子生产密度6 000株/667m^2，在甘肃适于4月上旬播种，父本分期播种，第1期晚播5d，第2期晚播4d，行比为5∶1；注意父本不抗盐碱，母本不耐高温。

本文原载：中国种业，2016（6）：62-64

玉米新品种 MC1002 的选育

张雪原[1]　赵久然[1]　王元东[1]　段民孝[1]　张华生[1]
邢锦丰[1]　张春原[1]　刘新香[1]　解　强[2]

（1. 北京市农林科学院玉米研究中心，北京　100097；
2. 河南省现代种业有限公司，郑州　450000）

玉米新品种 MC1002 是北京市农林科学院玉米研究中心和河南省现代种业有限公司联合选育的极早熟玉米杂交种，具有高产、稳产、抗病、耐密、抗倒、宜机收等特性，2016 年通过内蒙古自治区审定，审定编号：蒙审玉 2016005。适宜在内蒙古自治区兴安盟、呼伦贝尔盟、赤峰市、通辽市、乌兰查布市等 ≥10℃ 有效积温 2 100℃ 以上的极早熟区域种植。

1　亲本来源及选育经过

1.1　母本

京 D102M 是来自美国的，具有兰卡斯特血缘的早熟杂交种选系，于 2012 年选育而成。2007 年在北京按照 8 000 株/667m² 的密度种植 F_2 分离群体，授粉时严格选择茎秆坚韧抗倒、吐丝早、吐丝畅快的基本株；收获时，选择结实良好、苞叶疏松及穗行数 14 行以上的 S_1 果穗。2007 年在海南种植 S_1 穗行，开花时选择吐丝早且集中的穗行进行行内混合花粉授粉，获得 S_2 果穗；2008 年在北京选择优良穗行混合花粉授粉获得 S_3，同年在海南种植并以同样的方式授粉获得 S_4。2009 年在北京选择优良 S_4 穗行利用 DH 育种技术诱导，同年在海南进行加倍，并获得 12 个 DH 系。2010 年在北京扩繁并取花粉与测验种杂交，经 2011 年和 2012 年在极早熟区布点鉴定，筛选出 DH 系 H14008，定名为京 D102M。

京 D102M 具有一般配合力高和综合抗性好的特点。苗期叶鞘、叶缘深紫色；成株株型半紧凑，叶片绿色，雄穗分支 5~7 个，花药、颖壳、花丝淡紫色，株高 150cm，穗位高 70cm，穗长 12cm，穗行数 14~16 行，穗粗 4.6cm，果穗筒形，白轴，偏硬粒型、黄粒。中感大、小斑病，高抗茎腐病和矮花叶病毒病，抗穗粒腐病。适宜用作母本，制种产量在 370kg/667m² 左右，种子商品性好，适宜单粒精量播种。

1.2　父本

京 D1F 来源于欧洲极早熟杂交种混粉群体选系，于 2010 年育成。2005 年在北京按 8 000 株/667m² 的密度种植混粉群体，早代 S_1~S_3 在高密度、大群体、强胁迫条件严格选择吐丝早、吐丝快而集中、雄穗发达、花粉量大、籽粒硬粒型的穗行，穗行内混合花粉授粉，连续自交 3 代，在 S_4 代选择优良穗行利用 DH 育种技术诱导和加倍，共获得 8 个 DH

基金项目：粮经作物产业技术体系北京市创新团队专项经费资助（BITG7）、北京市农林科学院科技创新能力建设专项（KJCX20140107）

系，再经扩繁和配合力多点鉴定，获得了该 DH 纯系 D1001，定名为京 D1F。

京 D1F 具有一般配合力高和综合抗性好的特点。苗期叶鞘淡紫色，幼叶绿色；成株株型半紧凑，叶片绿色，雄穗分支 5~8 个，花药淡紫色，颖壳紫色，花丝绿色，株高 70cm 左右，穗位 30cm 左右，穗长 10cm，穗行数 10 行，穗粗 3.5cm，果穗锥形，白轴，硬粒型、黄粒。中感大、小斑病，抗茎腐病和丝黑穗病，抗穗粒腐病。适宜用作父本，花粉量大。

1.3 品种选育过程

根据自交系京 D102M 和京 D1F 血缘关系以及特征特性，2011 年在海南组配京 D102M×京 D1F 杂交组合。2011 年在内蒙古乌兰浩特玉米极早熟区进行鉴定，经观察测量，每 667m² 平均产 731.2kg，比对照九玉五增产 14.6%，熟期合适，耐密，抗倒伏；同年在海南继续复配该组合。2012 年在内蒙古、黑龙江玉米早熟区进行多点品比试验，经鉴定每 667m² 平均产 720.5kg，比对照九玉五增产 13.9%，表现突出，综合农艺性状优良；同年大量组配该组合参加内蒙古极早熟区预备试验。2013 年进入区试试验；2014 年升级进入生产试验；2015 年该组合达到极早熟玉米品种审定标准，通过审定。

2 品种特性

2.1 植物学特性

MC1002 幼苗叶片绿色，叶鞘深紫色。成株株型半紧凑，株高 261cm，穗位高 99cm，17 片叶。雄穗一级分枝 8~12 个，护颖绿色，花药、花丝浅紫色；果穗长筒形，白轴，穗长 18.9cm，穗粗 4.6cm，秃尖 0.5cm，穗行数 14~16，行粒数 37，百粒重 33.4g，单穗粒重 192.8g，出籽率 79.6%。籽粒橙黄色、硬粒型。

2.2 生物学特性

在内蒙古自治区极早熟区，春播生育期 115d 左右，需≥10℃活动积温 2 100℃。

2.3 品质

2015 年农业部谷物及制品质量监督检验测试中心（哈尔滨）测定，容重 794g/L，粗蛋白 8.95%，粗脂肪 5.57%，粗淀粉 73.20%，赖氨酸 0.28%。

2.4 抗性

2015 年生产试验中，田间表现玉米大斑病 1~5 级，灰斑病 1~3 级，平均黑穗病株率 0.3%，黑粉病株率、茎腐病株率、倒伏率均为 0，倒折率 0.3%。2015 年吉林省农业科学院植保所人工接种、接虫抗性鉴定，感大斑病（7S），抗弯孢菌叶斑病（3R），中抗丝黑穗病（5.6%MR），中抗玉米螟（5.4MR），高抗茎腐病（0%HR）。

3 产量表现

2013 年参加内蒙古极早熟组预备试验，每 667m² 平均产 712.5kg，比对照九玉五增产 12.1%；2014 年参加内蒙古极早熟组区域试验，平均产 759.6kg，比组均值增产 1.31%；2015 年参加极早熟组生产试验，每 667m² 平均产 737.6kg，比对照九玉五增产 3.05%。2015 年在内蒙古乌兰浩特进行小面积示范，每 667m² 平均产 729.8kg，比生产上推广的品种德美亚 1 号增产 4.12%。

4 栽培技术要点

种植密度为 5 000~5 500 株/667m^2,注意干旱瘠薄土地宜稀,高水肥地块宜密,春播玉米抓住良好墒情。当 5~10cm 耕层温度稳定通过 10℃ 时即可播种,播前种子需要包衣处理,预防土传病害的发生并促进苗齐、苗全、苗壮。最好结合测土配方施肥,一般肥力地块底肥每 667m^2 施用三元复合肥 20kg,促进苗期生长;小喇叭口至大喇叭口期每 667m^2 追施尿素 25kg,攻穗增粒;开花期结合浇水每 667m^2 追施尿素 10~15kg,以养根保叶,提高光合生产力,增加粒重。当玉米籽粒乳线消失后收获。

本文原载:中国种业,2016(8):73-74

玉米新品种 MC4592 选育及栽培制种技术要点

王元东　张华生　段民孝　张春原　张雪原　刘新香　陈传永　赵久然

(北京市农林科学院玉米研究中心，北京　100097)

MC4592 是以国外优良种质与国内核心种质相结合，利用"高大严"技术手段，结合 DH 育种新技术，选育而成的适于东华北春播区种植的玉米新品种。在增强玉米品种抗倒伏能力、抗病能力（中抗—高抗多种主要病害）、提高籽粒灌浆与脱水速率、提高种植密度和适应机械收获等方面有所创新。2014 年通过北京市农作物审定委员会审定，审定编号：京审玉 2014001；2015 年通过辽宁省农作物审定委员会审定，审定编号：辽审玉 2015035。

1 亲本选育及特征特性

1.1 母本

京 4055 为北京市农林科学院玉米研究中心自选系，该系以国外杂交种 X1132x 为选系基础材料，按照"高大严"（高密度、大群体和严选择）选系方法并变换多个不同地点，同时经配合力测定等程序，在 S_4 优穗行中选优株，再利用 DH 育种技术进行诱导和加倍，并对该 DH 系进行多点鉴定及严格筛选所获得，命名为京 4055。该自交系具有一般配合力高、籽粒灌浆快、脱水速率快和综合抗性好的特点。幼苗叶鞘浅紫色，叶片、叶缘绿色，雄穗分支少（1~2 个）且主枝长，花药紫色，颖壳淡紫色；株型紧凑，株高 193cm，穗位高 85cm，花丝绿色，果穗穗柄长度中等，果穗长筒形，苞叶长度中等，穗长 18cm，穗行数 14 行，穗粗 4.5cm，穗轴红色，籽粒黄红色、半硬粒型。该自交系高抗倒伏，中抗大斑病、小斑病、茎腐病和矮花叶病毒病，抗穗粒腐病，易制种，制种产量 500kg/667m^2 左右。

1.2 父本

京 92 为北京市农林科学院玉米研究中心自选系，来源于自交系昌 7-2 与自选系京 24 杂交，再与 Lx9801 杂交所构成的基础群体材料，在"高大严"条件下经过多代自交选择所获得。该自交系一般配合力较高，幼苗叶鞘紫红色，叶片、叶缘绿色，花药紫色，花丝红色；株型半紧凑，株高 160cm，穗位高 75cm，雄穗分枝 5~7 个，花粉量大，果穗穗柄长度中等，果穗锥形，穗长 16cm，穗行数 12 行，穗粗 4.8cm，穗轴白色，籽粒黄色、半硬粒型。该自交系高抗玉米大斑病、小斑病、花叶病毒病，抗倒，适于制种用作父本，花粉量大，散粉持续时间长。

基金项目：粮食作物产业技术体系北京市创新团队专项经费资助（BITG7）、北京市农林科学院科技创新能力建设专项（KJCX20140107）

2 杂交种选育及特征特性

2.1 选育过程

2009年冬季南繁时组配该组合，2010年进行多点鉴定和品比试验。结果表明该组合田间表现突出。2011年参加国家东华北区域试验，2012—2013年参加北京市春播玉米区试，2013年参加北京市春播玉米生产试验，2013年参加辽宁省区试初试试验，2014年参加辽宁区试复试试验和生产试验。2014年通过北京市农作物审定委员会审定；2015年通过辽宁省农作物审定委员会审定。

2.2 特征特性

幼苗叶片绿紫色，叶缘绿色，雄穗分支多且枝长中等，花药、颖壳淡红色；株型紧凑，株高289cm，穗位高107cm，叶片数20。花丝淡红色，果穗穗柄短，果穗锥形，苞叶长度中等，穗长18.3cm，穗粗5.1cm，穗行数16行，秃尖0.9cm，单穗粒重199.2g，出子率87.6%，穗轴白色，籽粒黄色、半马齿型，百粒重38.1g，籽粒商品性好。

2.3 抗性及品质

经中国农业科学院作物科学研究2012年、2013年2年接种鉴定，中抗大斑病、弯孢菌叶斑病、丝黑穗病、玉米螟，高抗茎腐病。2013年经农业部谷物及制品质量检验检测中心（北京）检测，籽粒容重786g/L，粗蛋白含量8.41%，粗脂肪含量3.97%，粗淀粉含量74.32%，赖氨酸含量0.29%。

3 产量表现

2012年在北京市春播试验，每667m^2平均产740.26kg，比对照郑单958增产5.64%；2013年北京市春播区域试验，平均产769.51kg，比对照郑单958增产5.65%，2013年北京市春播生产试验平均产710.02kg，比对照郑单958增产7.38%；2013—2014年参加辽宁省春播区试初试和复试试验，2年平均产803.7kg，比对照郑单958增产5.6%，2014年辽宁省生产试验，平均产863.1kg，比对照郑单958增产9.6%。

2015年MC4592在北京、辽宁大面积连片示范种植中测产验收，每667m^2平均产946.8kg，较当地主推品种郑单958增产7.2%。通过大面积的示范推广说明，该品种不仅产量高，而且抗病性（高抗茎腐病和大斑病）和抗倒性好，耐密植，籽粒脱水速率快，在该地区推广潜力很大。

4 栽培技术要点

该品种适宜在北京地区及辽宁境内≥10℃活动积温在2 857℃以上的中晚熟玉米区域种植。根据各地种植技术要求合理设置株距和行距，每667m^2种植4 500株，合理密植栽培。结合地力条件，测土配方施肥。底肥每667m^2施用三元复合肥18kg，促苗期生长；小喇叭口至大喇叭口期机械深施尿素25kg，攻穗增粒；开花期结合浇水追尿素10~15kg，以养根保叶，提高光合生产力，增加粒重。玉米籽粒乳腺消失后收获。播种前进行种子包衣，预防土传性病害发生。

5 制种技术要点

5.1 保持品种种性技术要点

亲本的种性保持应严格按照育种家种子（原原种）—原种的种子生产程序进行生产。繁殖隔离要求在500m以上，并严格去除杂株。杂交种的种性保持应是使用高纯度的亲本种子，确保制种区的安全隔离，并严格去除亲本杂株，及时干净、彻底去除母本雄穗，授粉结束后及时砍除父本，确保种子纯度。

5.2 种子生产技术要点

选择土质肥沃、地势平坦、排灌方便的制种基地，隔离距离不小于300m（由于高温影响慎重在新疆地区制种生产）。在甘肃适于4月上旬播种，先播种母本，第1期父本晚播5d，第2期父本晚播8d，父本独立成行，行比1∶6，母本适宜种植密度6 200株/$667m^2$，父本因有苗期苗枯而死亡现象，因此适当加大播种量。加强田间肥水管理，适时收获。脱粒加工时注意清选机器，严防机械混杂。

本文原载：中国种业，2016（2）：69-70

玉米新品种京科 193 的选育及配套技术

邢锦丰 段民孝 王元东 宋 伟 赵久然

(北京市农林科学院玉米研究中心，北京 100097)

摘 要：京科193是北京市农林科学院玉米研究中心采用单倍体育种手段与分子标记辅助育种技术相结合选育的高产、优质、适应性广的杂交玉米新品种，2013年通过北京市农作物品种审定委员会审定，适宜北京及周边地区夏播种植。

关键词：玉米；杂交种；京科193；选育；栽培技术；制种技术

1 亲本的选育

母本 DH07019 以杂交种 X1132 为选系基础材料，按照"高大严"选系方法，即高密度、大群体、严选择并变换多个不同地点，同时经配合力测定等程序，在 S4 优穗行中选优株利用 DH 育种技术诱导和加倍，再经扩繁和多点鉴定及严格筛选，于 2007 年获得了该 DH 纯系。

DH07019 具有一般配合力高和综合抗性好的特点。株型半紧凑，花药、花丝紫色。北京地区夏播生育期 100d，春播 110d。株高 190cm，穗位高 70cm，穗长 18cm，穗行数 14 行，穗粗 4.5cm，果穗长筒形，红轴。硬粒型，黄粒。抗大斑病、小斑病、茎腐病和矮花叶病毒病，较抗穗粒腐病。

父本京 24Ht 是以北京市农林科学院自育自交系京 24 为基础，利用吉 1037 为丝黑穗抗原，与京 24 杂交导入抗原，并通过多代回交聚合及分子标记辅助选择选育而成。京 24Ht 遗传背景与京 24 一致，只是在丝黑穗病抗性等方面有较大提高。

京 24Ht 与京 24 自交系一样，具有较高的一般配合力。株型半紧凑，幼苗叶鞘紫红色，花药淡紫色，花丝绿色。北京地区夏播生育期 95d，春播 100d。株高 160cm，穗位 75cm，穗长 16cm，穗粗 4.8cm，穗行数 12 行，锥形穗，红轴，硬粒型黄粒。该自交系抗玉米大、小斑病、花叶病毒病、丝黑穗病等。

2 杂交组合的配制与筛选

2007 年冬在海南初配 DH07019×京 24Ht 杂交组合。2008 年在本单位测比试验中，平均小区折合产量 12 142.5kg/hm^2，比对照京单 28 号增产 10.9%；在 2009 年本单位品比试

基金项目：北京市科委项目（D111100001311001）、国家 863 项目（2012AA101203）、国家 973 项目（2009CB118405）

作者简介：邢锦丰，副研究员，从事玉米育种工作

赵久然为通讯作者，研究员，主要从事玉米研究

验中，该杂交种的平均小区折合产量 11 761.5kg/hm²，比对照京单 28 增产 10.2%。两年试验均表现突出。

3 京科 193（MC193）选育系谱图

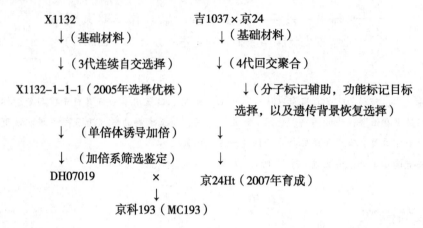

4 区试产量表现

2011 年参加北京市夏播区试，在 15 个参试品种中居第 1 位，产量 10 960.50kg/hm²，比对照京单 28 增产 9.5%，增产极显著，5 个试点中 4 个点增产、1 个点减产。2012 年北京市夏播区试，产量 9 981.45kg/hm²，比对照京单 28 增产 6.87%，增产显著，4 个试点全部增产。两年区试平均产量 10 471.05kg/hm²，比对照京单 28 增产 8.26%。2012 年生产试验平均产量 9 300.30kg/hm²，比对照京单 28 增产 11.31%，居 6 个参试品种的第 3 位，4 个试点中 3 个点增产、1 个点减产。

5 品种特征特性

京科 193 属早熟玉米单交种，在北京地区夏播生育期平均 104.0d，比对照京单 28 早 0.6d。该品种幼苗拱土能力强，株型半紧凑，群体内果穗均匀一致，边际效应小，耐密性好，活秆成熟，籽粒品质好。株高 273.6cm，穗位 109.1cm，果穗筒形，穗轴红色，穗长 17.9cm，穗粗 5.2cm，秃尖长 0.4cm，穗行数 12~16 行，行粒数 36.9 粒。穗粒重 175.4g，出籽率 82.6%。籽粒黄色，半硬粒型，粒深 1.2cm，千粒重 402.3g。

品质分析结果：经农业部谷物品质监督检验测试中心检验（干基），该品种籽粒中含粗淀粉 74.85%，粗脂肪 3.76%，粗蛋白 8.99%，赖氨酸 0.30%，容重 740g/L。

抗病性调查结果：2011 年区试田间 1 个试点出现大斑病为 3 级；2 个试点出现虫害分别为 5%、15%。2012 年区试田间 1 个试点出现大斑病为 3 级；1 个试点出现小斑病为 3 级；1 个试点出现弯孢菌叶斑病为 3 级；1 个试点出现粗缩病为 1%；1 个试点出现丝黑穗病为 1.2%。

6 栽培技术要点

一般种植密度 67 500 株/hm² 为宜，播期在 6 月中旬。采取直播栽培方式，玉米播种器精良点播或机器播种，播种后 3~4d 进行药剂封闭除草。田间管理以促为主，施好基肥、种肥，确保苗全、苗匀、苗壮；重施穗肥，酌施粒肥，确保籽粒灌浆充分；在生理成熟后适时收获。

7 适宜种植区域

经多年区试、生产试验以及多点示范结果表明，京科193适宜北京及周边地区夏播种植。

8 制种技术要点

在甘肃张掖选择海拔相对较低、热量充足、无霜期大于150d、土壤条件为沙壤土或壤土的区域生产；选择地力及肥水条件上等、灌溉条件便利的地块种植。依照国家玉米杂交种制种规程，结合实际，将时间、空间，以及人为设置障碍物等自然条件合理结合运用，杜绝外源花粉对制种区的影响。播种时间：5cm地温稳定通过10℃（甘肃张掖一般在4月15日之前）播种，母本一次播种完成。种植形式：在甘肃张掖标准地膜覆盖种植；父母本按行比1：5加父本满天星种植。错期指标：先播母本；当50%母本扎根（在张掖7~9d）单行等株距均匀种植第一期父本，父本播量应适当加大；当50%母本出苗（母本播种后10~12d）在膜床中间点播种植第二期父本，株距100cm（满天星种植）。种植密度：父母本总保苗密度82 500株/hm^2；其中父本15 000株/hm^2，母本67 500株/hm^2。去杂：出苗后至抽雄前分3次去杂，根据父本、母本自交系的特征特性及生长状况，严格组织去杂。去雄：带4片叶摸苞超前去雄。在整个去雄期间，做到及时、彻底，不留断枝，风雨无阻。父本散粉结束后，及时割除父本并清出田间。在9月25日前采收，并及时烘干、脱水加工。无烘干条件的及时收获晾晒，严防冻害，及时脱水降至安全水分，直至达到国家标准，进行加工、包装、成品。

本文原载：作物杂志，2013（3）：155-156

青贮玉米品种利用现状与发展

杨国航[1]　吴金锁[2]　张春原[1]　刘春阁[1]

（1. 北京市农林科学院玉米研究中心，100097　北京；
2. 北京市延庆县种子管理站，102100　北京）

摘　要：通过对比我国与畜牧业发达国家青贮玉米的应用情况，对我国青贮玉米品种的选育利用和发展进行分析，指出了青贮玉米品种选育相对滞后的现状。2004—2012年9年间通过国家审定的青贮玉米品种仅27个，而且品种数量、水平均有待提高，粮饲兼用型玉米品种偏多，品种更新速度慢。借鉴国内外青贮玉米应用实际，为在畜牧业发达地区大力推广种植青贮玉米，提出了在重点发展粮饲通用型青贮玉米的同时，需要兼顾专用青贮玉米品种的选育和发展。

关键词：青贮玉米；品种利用；现状与发展

Silage Maize Variety Utilization and Development in China

Yang Guohang[1]　Wu Jinsuo[2]　Zhang Chunyuan[1]　Liu Chunge[1]

(1. Maize Research Centre, Beijing Academy of Agricultural & Forestry Sciences, Beijing 100097; 2. Beijing Yanqing County Seed Management Station, Beijing 102100, China)

Abstract: By comparing the application of silage maize between developed countries in Animal Husbandry and China, the breeding and progress of silage maize variety in our country was analyzed, and the current situation of silage maize breeding was described in this paper. Only 27 silage maize varieties were authorized by the National Certification in 9 years spanning from 2004 to 2012. Furthermore, the quantity and quality of the varieties are all remained to be raised. The ratio of grain raise general-purpose silage maize is relatively large. The speed of variety upgrade in silage maize needs to be raised. After learning lessons from domestic and international silage maize in practical application, this paper suggested that to promote the cultivation of silage corn, the breeding of silage maize should be concentrated on. Accompanied with the focused development of grain raise general-purpose silage maize, the breeding and progress of silage maize varieties also need to be considered.

Key words: Silage maize; Variety; Utilization; Present situation and development

我国虽是农业大国，但与发达国家相比，畜牧业发展相对落后，制约畜牧业发展的主

基金项目：北京市农林科学院科技创新能力建设专项（KJCX201101008）
作者简介：杨国航，研究员，研究方向为玉米品种评价鉴定、高产配套栽培技术、节水农业等

要原因之一是优质饲草料供应不足,尤其是北方冷季饲草料短缺,要解决我国北方冷季家畜饲草严重缺乏的局面,抵御自然灾害的威胁,必须大力发展人工种植饲草饲料,并进行大量青贮,以备冬季饲料不足时使用。青贮玉米饲料营养价值完善,适口性好,易消化,满足反刍动物冬春季节的营养需要,使家畜终年保持高水平的营养状态和生产水平。发展青贮玉米可有效地解决秸秆过腹还田和农牧交错带的粮饲争地问题,解决实行牛羊圈养,减轻过度放牧,草场退化等生态问题。青贮玉米易于机械化栽培,能充分发挥机械化水平高的优势。发展青贮玉米可促进农业种植结构调整,大幅度提高农牧民收入,实现粮、饲、经三元结构的有机结合,为发展"两高一优"饲料生产闯出了一条新路子,对畜牧业生产起到积极的促进作用,对实现农业由数量型增长向优质高效转变有很大意义。

1 青贮玉米品种的类型

青贮玉米品种主要有专用青贮型、粮饲兼用型和粮饲通用型3种类型。

专用青贮型指收获玉米的鲜绿植株全株,经过切碎发酵,用于牛羊等草食牲畜饲料的玉米。一般籽粒产量较低,在北方果穗多发育不良。分单秆或多秆,多秆如墨西哥玉米、科多4号、科多8号等。近期通过审定的品种,尤其是国审品种,多数都属于单秆品种。

粮饲兼用型指收获玉米籽粒用作粮食或配合饲料,然后再收获青绿的茎叶用作青贮,是国内以前应用较多的一种类型。

粮饲通用型是指既可作为普通玉米品种收获籽粒,用于食物或配合饲料,也可作为青贮玉米品种,收获果穗和茎叶在内的全株,用于青饲料或青贮饲料。

2 青贮玉米品种利用现状

欧美等畜牧业发达国家非常重视青贮玉米的加工和利用,广泛使用青贮饲料。欧盟国家每年种植青贮玉米约占玉米种植面积的80%,达到400万hm^2以上。其中,法国158万hm^2,德国133万hm^2。英国、丹麦、卢森堡、荷兰等国种植的玉米基本是青贮玉米。美国青贮玉米年播种面积最高曾达到460万hm^2,现约300万hm^2,收获的青贮玉米多是粮饲通用型品种,耐密植,有较高生物产量潜力,同时也有较高籽粒产量潜力,既可以在乳熟末期作青贮玉米收获,也可到完熟后收获籽粒。

目前,国内对于3种类型青贮玉米品种均有利用,但其中粮饲通用型青贮玉米应用偏少。1985年通过北京市审定的京多1号是我国首次审定的青贮玉米品种,该品种是多秆多穗类型。"七五"期间我国将青贮玉米育种列入国家科技攻关计划,以多秆多穗、青枝绿叶、茎叶多汁、富含糖分、适口性好和生物产量高为主要育种目标。辽宁省农业科学院于1988年育成了辽原1号。1989年,中国科学院遗传所育成的科多4号通过审定。随后各地先后育成了太多1号、太穗枝1号、科多8号、辽原白、龙牧1号、辽青85、沪青1号、黑饲1号、科青1号、中原单32等专用或兼用青贮玉米品种。这些品种多为分蘖型的,青贮产量比较高,但抗性不足,尤其是抗倒伏倒折能力差。

总体看,我国青贮玉米发展滞后,品种的培育重视不足。在一些地区,管理粗放,密度甚至可以达到15万株/hm^2以上,品种的选择无从谈起,有些地方甚至采用商品粮直接播撒、不间苗的方式种植,产量、品质均无法得到保证。

根据我国青贮玉米生产需求和科研育种发展的状况,国家有关部门于2002年正式启动了国家青贮玉米品种区域试验。截至2012年，国家青贮玉米区试共对各地育成的250多个不同类型青贮玉米组合进行鉴定，其中已有27个品种通过审定并陆续在生产上示范推广。新审定的品种大多是单秆或分蘖少的品种，抗性较好，保绿性强。2002—2011年通过国家或省级审定的青贮玉米品种约有160多个，表1为2004—2012年国家审定的青贮玉米品种。

表1 2004—2012年国家审定的青贮玉米品种

序号	品种名称	序号	品种名称	序号	品种名称	序号	品种名称
1	中北青贮410	8	中农大青贮4515	15	登海青贮3930	22	桂青贮1
2	奥玉青贮5102	9	三北青贮17	16	强盛青贮30	23	雅玉青贮04889
3	辽单青贮625	10	辽单青贮529	17	登海青贮3571	24	铁研青贮458
4	中农大青贮67	11	京科青贮301	18	金刚青贮50	25	津青贮0603
5	晋单青贮42	12	雅玉青贮27	19	京科青贮516	26	豫青贮23
6	屯玉青贮50	13	郑青贮1	20	辽单青贮178	27	雅玉青贮79491
7	雅玉青贮8	14	雅玉青贮26	21	锦玉青贮28		

专用青贮玉米品种的利用在一些畜牧业发达地区比较普遍。如东陵白，2008年在内蒙古自治区仍有9.8万hm^2种植，列当年全国推广面积第48位，2010年有6万hm^2，2011年有5.33万hm^2；同时2010年英红在内蒙古有1.2万hm^2的种植面积；在河北省，2010年青贮白马牙的种植面积达到3万hm^2，以上品种籽粒产量低，但青贮产量较高。郑青贮1号、青贮巡青518、雅玉青贮8号、京科青贮516等新的专用青贮玉米品种面积同样较大，类似的品种生产还有很多，呈现上升势头。青贮品种的种植利用多以大型畜牧、奶业公司订单生产，尤其在畜牧业发达地区多为乳熟期收获青贮；但天津等大城市周边，普通玉米种植较多的区域，收获玉米果穗后，青贮余下秸秆的现象也很普遍。

鉴于目前专用青贮品种种植面积最大的还是东陵白、英红等农家种或综合种，产量品质均一般，且品种退化比较严重，专用青贮玉米杂交种仍有很大的发展空间，更新换代工作任务和意义巨大。

3 青贮玉米品种选育存在的问题

从2002—2012年国家青贮玉米区试方案中，可大致体现目前我国青贮玉米品种选育情况。

3.1 青贮玉米育种水平不高

3.1.1 生物产量年际间变化很小

从2003—2012年，参试品种平均产量没有明显变化，均维持在北方区22 500 kg/hm^2左右，黄淮海区18 000 kg/hm^2左右，南方区13 500~15 000 kg/hm^2水平，年际间差距很小。

在农大108作为对照时，参试品种平均产量均超过对照，自2007年更换专用青贮品种作为对照后，对照品种生物产量明显增加，参试品种平均产量均低于对照。

3.1.2 参试品种数量减少

国家青贮玉米审定品种的分布基本体现了目前青贮玉米育种水平的情况。截至2012年，通过国家审定的专用青贮玉米品种27个，涉及17个单位，9个省、市、自治区，集中于北方。

截至目前，国家青贮玉米平均审定通过率接近10%，较国家玉米品种审定通过率（3%~5%）高很多，但从审定时间可以看出，通过率高主要是因为前期对照农大108为非专用青贮品种，青贮产量较低造成的，更换对照后，审定品种数量即急剧减少，近几年间没有品种通过审定。

3.2 育种单位积极性不高

从表2可以看出，参试品种数量在2004年达到顶峰，之后逐渐下降，尤其是2007年以后，下降幅度极大，目前仅仅保持在10个左右。究其原因，有如下几个方面。

首先，国家青贮玉米区试组开设初期，因品种直接参加区域试验，感觉选择压力小，相对容易通过。

表2 国家青贮玉米区试中不同年份参试品种数量

年份（年）	2002	2003	2004	2005	2006	2007	2008	2009	2010	2011	2012
品种数（个）	17	36	45	34	34	22	20	12	13	9	10

其次，育种单位选拔目标不清晰。试验之初，育种单位多不清楚其选择的标准，盲目参试，部分品种生物量达不到要求，随着试验的进行，参试单位目标逐渐清晰，品种选择逐渐慎重。

再次，对照标准逐渐变高。试验之初对照农大108属于青贮产量较高的普通玉米品种，因此只能设定专用青贮玉米只有比其超过8%，产量才能过关，即使这样，达标也相对比较容易。2007年，分区域（北方、南方）采用了雅玉青贮26（北方）、雅玉青贮8号（南方）作为对照品种，参试品种达标难度加大。

最后，育种单位的重视不够，国家政策支持少，限制了专用青贮玉米发展。青贮玉米虽是玉米，但用途与普通玉米截然不同，专门为青贮玉米设立的鼓励政策少。同时，青贮玉米品种本身的销售价值也不如普通玉米品种，种植多采取订单农业或自种自用形式，研究人员和单位对此重视不够，投入少。

4 青贮玉米品种类型发展方向探讨

1954年，我国研究出利用玉米籽粒收获后的秸秆进行青（黄）贮，在全国"三北"地区大面积推广，为我国草食家畜的发展起到了重要的推动作用。但那时是基于当时玉米生产十分落后的产物，目前来看单纯考虑收获籽粒后进行植株青贮的粮饲兼用型品种是很原始的低层次的青贮形式，不合乎现代农业高产、高效、优质的发展方向，仅可作权宜之计考虑，因为果穗收获时，植株的纤维化程度已经过高，很多营养成分已经损失，适口性差，并不适合作为饲料喂牲畜。借鉴国内外青贮玉米品种的实际应用情况，出于品质和降低种植风险角度，一般考虑用粮饲通用型玉米来替代专用青贮玉米。粮饲通用型青贮玉米品种利用灵活，弹性大，可根据当年玉米生产情况，及时调节收获形式和内容，在籽粒效益高时，收获籽粒，当青贮效益高时，收获青贮，可最大程度减少利用风险。青贮玉米的发展方向应有利于刺激育种人员的积极性。

粮饲通用型玉米品种是未来的发展方向，但中国玉米种植区域情况比较复杂，受生态条件、畜牧业发展、农民种植习惯等因素制约，生物产量高、分蘖型、多穗型玉米也有其利用的空间。如前所述，目前专用青贮玉米品种东陵白、英红、白马牙等的利用也很普遍，因此不能单纯强调粮饲通用，也要兼顾专用青贮玉米。

参考文献（略）

本文原载：作物杂志，2013（2）：13-16

青贮玉米新品种京科 932 选育及配套技术

邢锦丰 段民孝 王元东 刘新香 宋 伟 赵久然

(北京市农林科学院玉米研究中心/玉米 DNA 指纹及分子育种北京重点实验室,北京 100097)

摘 要:青贮玉米京科 932 是北京市农林科学院玉米研究中心采用单倍体育种手段与分子标记辅助育种技术相结合,以 MX1321 为父本、京 X005 为母本,选育的高产、优质、适应性广的青贮玉米新品种。该品种生物产量高、抗倒性好、抗病性强、纤维品质优,2015 年通过北京市农作物品种审定委员会审定,适宜北京及周边地区夏播种植。介绍了京科青贮 932 的选育及配套技术,供参考。

关键词:青贮玉米;杂交种;京科 932;选育;栽培技术

近年来,随着我国国民经济的发展和人民生活水平的提高,人们的饮食结构正在发生着巨大变化,对肉、蛋、奶的需求不断增长。为满足人们对畜产品日益增长的需要,我国农业现代化建设应大力发展畜牧业。畜牧业的发展必然导致对饲草料需求的大大增加,目前我国饲草料的发展速度滞后,制约了畜牧业的快速发展。

玉米是饲料之王,在饲料中占有不可替代的重要地位。它不仅是优质的高能量精饲料,而且是最佳的青贮饲料,是发展畜牧业的重要饲料来源。发展优质青贮玉米和粮饲兼用是畜牧业发展的重要保障,优质、高产饲用玉米品种的选育势在必行,已成当务之急。

京科 932 是北京市农林科学院玉米研究中心针对我国当前对优质、高产饲用玉米的需求选育的青贮玉米新品种。该品种以自选系京 X005 为母本、自选系 MX1321 为父本,于 2013、2014 年参加北京市夏播青贮玉米区域试验,2015 年通过北京市农作物审定委员会审定(审定编号:京审玉 2015008)。该品种品质好、生物产量高、抗倒性好、综合抗病性强,适宜在京津冀地区夏播种植。

1 亲本选育及特征特性

京科青贮 932 杂交种的组合为京 X005/MX1321。父母本均为自选系,其中母本京 X005 来源于国外杂交种 X1132x,按照"高大严"选系方法,即高密度、大群体、严选择并变换多个不同地点,同时经配合力测定等程序,在 S4 在优穗行中利用分子标记鉴定并选择血缘偏向 Reid 群的优株,利用 DH 育种技术诱导和加倍,再经扩繁和多点鉴定及严格筛选,获得了自交系京 X005。该自交系株型半紧凑。北京地区夏播生育期 105d,春播

基金项目:国家科技支撑项目(2014BAD01B09)、北京市农林科学院科技创新能力建设专项(KJCX20140202)、北京市科技计划课题(Z141100002314016)

作者简介:邢锦丰(1974—)男,汉族,北京大兴人,本科,副研究员,从事玉米育种工作

通讯作者:赵久然(1962—),男,汉族,北京平谷人,博士,研究员,主要从事玉米研究

123d，株高210cm，穗位80cm，穗长18cm，穗行数14~16行，穗粗4.5cm，果穗长筒形，白轴，籽粒黄色偏硬粒。抗大斑病、小斑病、茎腐病和矮花叶病毒病，一般配合力高。

父本MX1321以国内P群自交系MC30、1145和JN15组成混合群体，按照"高大严"选系育种技术，并变换多个不同生态区，同时经IPT配合力测定等程序，多代自交选育，直至纯合稳定。该系幼苗叶色深绿，株型半平展。株高210cm左右，穗位100cm左右。果穗长筒形，穗长20cm，穗粗5cm，穗行数14~16行，行粒数40粒左右，籽粒深黄色、硬粒，穗轴红色。雄穗发达花粉量大，分枝8~10个，花药紫色，花丝红色，穗柄长3~4cm。该系高抗大小斑病、矮花叶病毒病、茎腐病、南方锈病、黑粉病等，保绿性好，活秆成熟。

2010年冬在海南组配京X005/MX1321杂交组合，2011年和2012年在北京市农林科学院玉米研究中心试验点鉴定，该组合连续两年表现综合抗性好，生物产量高，命名为京科青贮932。

2 品种特征特性

京科青贮932在京津冀地区夏播播种至青贮收获期105d，需有效积温2 900℃左右。幼苗第一叶鞘花青甙紫色、第一叶尖端勺状、叶片边缘紫色、穗上叶与茎秆角度30°左右、穗位95cm、穗位与株高比约1:3、株型半紧凑、茎粗2.5cm左右。雄穗抽出期55d左右，最上面一个节间的长15cm左右，雄穗抽出剑叶的长度30cm、雄穗分枝长8~12cm、雄穗侧枝上冲、雄穗小穗颖片紫色、花药（新鲜花药）红色、雄穗主轴与分枝夹角30°左右、雄穗小穗密度中等、雄穗最高位侧枝以上的主轴长15cm左右，花丝红色、果穗向上、穗柄长5cm左右、苞叶长度25cm左右、穗形长筒形、穗长23cm左右、穗粗5cm、穗轴红色、穗轴中部直径2.5cm左右、果穗籽粒14~16行。硬粒型、粒色黄，粒顶部黄色、籽粒近楔形、籽粒中等、穗轴颖片红色。

经中国农业科学院作物科学研究所病害接种鉴定，该品种抗大小斑病，高抗腐霉茎腐病。2013年平均倒伏率为0.5%，平均倒折率为1.0%；2014年平均倒伏率为0.3%，无倒折。

经农业部谷物品质监督检验测试中心检验，该品种植株中中性洗涤纤维含量40.57%，酸性洗涤纤维含量18.11%，粗蛋白含量7.83%，等级为一级。

3 区试产量表现

2013年参加北京市夏播青贮玉米区域试验，在6个参试品种中居第3位，产量16 950kg/hm^2，比对照农大108增产7%，增产显著。2014年北京市夏播青贮玉米区域试验，在9个参试品种中居第5位，产量13 365kg/hm^2，比对照农大108增产6.6%。两年区域试验平均产量15 157.5kg/hm^2，比对照农大108增产6.81%。2014年生产试验平均产量17 655kg/hm^2，比对照农大108增产6.7%。

4 栽培技术要点

京科青贮932种植密度4 500株/667m^{2}为宜，在北京及周边地区播期在6月上中旬。

可使用玉米精良播种机直播，出苗前进行药剂封闭除草。注意肥水管理，施好基肥、种肥，确保苗全、苗匀、苗壮；追肥以尿素为主，350kg/hm² 左右，确保籽粒灌浆充分；在生理成熟后适时收获。

5 适宜种植区域

经多年区试、生产试验以及多点示范结果表明，京科青贮 932 适宜北京及周边地区夏播种植。

6 制种技术要点

6.1 制种地选择

京科青贮 932 的父母本生育期较长，在甘肃等地应选择海拔相对较低、热量充足、无霜期大于 150d、土壤条件为沙壤土或壤土的区域、地力及肥水条件上等、灌溉条件便利的地块种植。

6.2 播种时间及花期调节

甘肃地区播种时间一般在 4 月 15 日之前，当 5cm 地温达到 10℃时播种，采取地膜覆盖种植。母本一次播种完成，父本分两期播种，第一期父本与母本同期，单行等株距均匀种植，适当加大父本播种量。当 50%母本扎根（在张掖 7~9d），在膜床中间点播（满天星种植）第二期父本，株距 100cm。

6.3 种植密度

父母本总保苗密度 5 500 株/667m²；其中母本 4 500 株/667m²，父本 1 000 株/667m²。出苗后至抽雄前分 3 次去杂，根据父本、母本自交系的特征特性及生长状况，严格组织去杂。带 4 片叶摸苞超前去雄。在整个去雄期间，做到及时、彻底，不留断枝，风雨无阻。父本散粉结束后，及时割除父本并清出田间。

6.4 种子收获机加工

在 9 月 25 日前采收，并及时烘干、脱水加工。无烘干条件的及时收获晾晒，严防冻害，及时脱水降至安全水分，直至达到国家标准，之后进行加工、包装。

本文原载：种子科技，2016，34（7）：59

杂交玉米新品种京科968种子生产技术

冯培煜

(北京市农林科学院玉米研究中心，北京 100097)

摘 要：京科968是由北京市农林科学院玉米研究中心组配选育的高产、优质杂交种。为使该品种更快速推向市场，介绍杂交种子生产技术。

关键词：京科968；制种技术；抽雄

京科968是由北京市农林科学院玉米研究中心利用自选自交系京724为母本，京92为父本组配育成的玉米新品种，经国家农作物品种审定委员会审定（编号：国审玉2011007）。该品种2009—2010年参加东华北春玉米品种区域试验，2年平均单产771.1kg/667m^2，比对照增产7.1%。2010年生产试验，平均单产716.3kg/667m^2，比对照郑单958增产10.5%。在品种区试及生产试验中产量均位列第一，且高抗玉米螟，为2012年农业部推介的玉米种植主导品种。在甘肃规模化种子生产中取得了良好的效益与经验，种子质量全部达到国标一级以上标准，制种产量最高单产超过了600kg/667m^2，种子生产省工省时，技术易被群众掌握和接受。其亲本主要性状及关键制种技术介绍如下。

1 组合及双亲性状

1.1 母本

春播生育期146d左右，比先玉335母本PH6WC熟期稍长；幼苗苗势强壮，叶片浓绿，叶鞘紫色；抗多种病害。株高260~300cm，穗位高80~120cm，穗上叶片数5~7片，穗上叶上举，穗及穗下叶宽大平展；全生育期18~20片叶，平均19片叶；雄穗分支1~3个，雄穗主枝粗长，分支平展、短小或没有；花药紫色，花粉量较大；花丝颜色开始时为黄绿色，完全吐出后变为红色，花丝吐丝快、集中，雌雄花期协调；果穗穗长16~18cm，穗粗约5cm，穗行数16~18行，行粒数26~35粒；籽粒偏硬粒型，橙黄色，千粒重290~320g；白穗轴。

1.2 父本

春播生育期140d左右；发苗快，叶鞘淡紫色；株高230~260cm，穗位高80~100cm；全生育期17~19片叶，平均18片叶；雄穗分支5~8个，雄穗发达，花药淡紫色，花粉量大，散粉持续时间长，花丝颜色为红色；果穗长16cm左右，穗粗约4.5cm，穗行数12~14行，籽粒偏硬粒型，黄色，千粒重360g左右；白穗轴。

作者简介：冯培煜（1963— ），男，河北省l临城县人；高级农艺师，主要从事玉米种子生产及加工技术研究及管理工作

2 杂交种子生产技术

2.1 选地与隔离区设置

选地：选择海拔相对较低、热量充足、保证生长期在150d以上、土壤条件为沙壤土或壤土的区域生产；选择地力及肥水条件上等、灌溉条件便利的地块种植。

隔离：依照国家玉米杂交种制种规程，依据实际，将时间、空间，以及人为设置障碍物等自然条件合理结合运用，杜绝外源花粉对制种区的影响。

2.2 调节播期与密度

播种时间：5cm地温稳定通过10℃播种母本，一次播种完成。

种植形式：地膜覆盖种植，即用幅宽80cm超薄膜覆盖，形成膜床面宽70cm、床面之间30cm的标准膜床；父母本按行比1∶6，并在膜床中间加父本（株距100cm）满天星种植。

错期指标：先播母本；当母本50%扎根，单行等株距均匀种植第1期父本；当母本50%出苗（母本播种后10d左右）在膜床中间点播种植第2期父本，播种后注意用湿土封坡。

种植密度：父母本总保苗密度6 000株/667m^2；其中父本850株，母本5 150株。

2.3 去杂及去雄

去杂：出苗后至抽雄前分3次去杂，根据父本、母本自交系的特征特性及生长状况，严格组织去杂。

去雄：当母本倒数第5片展开时即开始（摸苞）带叶超前去雄，带4片叶（未展开）一次完成抽雄为宜，以利降低母本植株高度，同时也有利于提高母本授粉结实率。在整个去雄期间，做到及时、彻底，不留断枝。

2.4 田间管理

整地覆膜时用玉米专用肥25kg/667m^2作底肥；播种前种子包衣处理；3叶期间苗，5叶期定苗，母本定苗时去除病苗、弱苗和畸形苗，留生长均匀一致的健壮苗。拔节期追肥，尿素40kg/667m^2；及时灌溉浇水；中后期特别注意防治红蜘蛛。

2.5 割除父本

父本散粉结束后，及时割除父本并清出田间。

3 收获加工

在9月25日前采收，并及时烘干、脱水加工。无烘干条件的及时收获晾晒，严防冻害，及时脱水降至安全水分，直至达到国家标准，进行加工、包装、成品。

本文原载：种子，2013，32（3）：116-117

京科 968 等系列玉米品种"易制种"性状选育与高产高效制种关键技术研究

王元东　赵久然　冯培煜　段民孝　张华生　王荣焕　陈传永

（北京市农林科学院玉米研究中心，北京　100097）

摘　要：分析京科968等系列玉米品种与易制种相关的农艺性状、典型制种高产田的产量及产量构成要素。结果表明，京科968等系列玉米品种具有良好的易制种特性，单位面积制种产量大幅高于先玉335，与郑单958相当或略高。按单位面积有效商品籽粒数计，京科968玉米品种则较郑单958和先玉335具有明显优势。先进的育种指导思想和技术路线是易制种京科968等系列玉米品种选育成功的关键，集成和优化制种关键技术措施是实现高产高效制种和产业化推广应用的基础。

关键词：玉米；京科968；"易制种"性状

Characteristics Related to "Easy Seed Production" and the Key Technology of High Yield and High Efficiency Seed Production of Commercial Hybrids Jingke 968 et al

Wang Yuandong　Zhao Jiuran　Feng Peiyu　Duan Minxiao
Zhang Huasheng　Wang Ronghuan　Chen Chuanyong

（*Maize Research Center*, *Beijing Academy of Agricultural & Forestry Sciences*, *Beijing* 100097）

Abstract: Characteristics related to "easy seed production" of Jingke 968 et al, and their yield and yield components in typical seed high-yield plots were analyzed. It was showed that Jingke 968 et al possessed good traits for easy seed production, and their seed yield per unite area was as high as Zhengdan 958 and higher than Xianyu 335. Compared to Zhengdan 958 and Xianyu 335 there were obvious advantages in the total effective commodity grain number per unit area for Jingke 968 et al, especially for Jingke 968. Advancing breeding guidelines and technical routes were the key to the success for Jingke 968 et al in easy seed production, and integrated and optimized the key technical measures

---

基金项目：国家863项目"强优势玉米杂交种的创制与应用（2011AA10A103）"、粮食作物产业技术体系北京市创新团队专项经费资助（BITG7）、北京市农林科学院科技创新能力建设专项"玉米种质资源发掘、收集整理与创新"（KJCX20140107）

作者简介：王元东，男，副研究员，研究方向为玉米遗传育种，E-mail：wyuandong@126.com
　　　　　赵久然为本文通讯作者。E-mail：maizezhao@126.com

were the basis for high yield and high efficiency in seed production.

Key words：Maize；Jingke 968；Characteristics of easy seed production

玉米品种的"易制种"性状是提高品种市场竞争力的核心因素之一。易制种性状包括母本耐密性、穗粒重、穗粒数以及父本散粉持续期、父母本花期相遇性等。近年来北京市农林科学院玉米研究中心选育并审定了京科968、京科665、NK718和京农科728等系列玉米品种，并已经在生产上实现大面积推广与应用。在甘肃等主要玉米制种基地，显示出了良好的易制种特性，大面积制种平均7 500kg/hm²以上，高产地块可达9 000kg/hm²以上，高于先玉335，与郑单958相当或略高。京科968的易制种特性主要源自其母本生长势强、父本散粉期长、穗粒数多、单穗粒重高等因素。通过各项配套技术，实现了种子发芽率、纯度、净度和含水量等各项指标全部达到国标一级以上标准。通过分析京科968等系列品种易制种性状的选育方法、亲本特征特性、易制种的产量潜力及构成因素，可进一步集成和优化制种关键技术措施，指导高产高效制种生产，并对玉米新品种的选育提供参考。

1 选育指导思想和技术路线

根据我国玉米主产区生产发展和西北玉米制种的实际情况，选育母本注重利用优新种质，在高密度（90 000株/hm²）、大群体（S_1代群体3 000个基本株以上）、强胁迫（早期采取低温早播、深播、干旱、高密度等胁迫，后期进行耐贫瘠、耐干旱胁迫）、变换地（S_1选择在吉林、S_2在河南、S_3在甘肃）条件下进行严选择（加大基本株的淘汰力度，选择吐丝快而集中的单株），在果穗选择过程中坚持果穗籽粒大小适中、单穗籽粒数大于400粒、穗粒重大于100g、籽粒硬粒-半硬粒、出籽率大于88%的标准。父本以黄改群骨干系构建基础选系材料，注重选择雄穗发达、花粉量大、散粉持续时间长的单株。杂种优势模式为国外选系×国内黄改系。

京科968等系列玉米品种的母本均来自于X1132x国外优新种质所构建的基础群体选系；父本来源于昌7-2、Lx9801、京24和5237等骨干黄改系所构建的基础材料选系（表1）。X1132x新种质株型较紧凑，出籽率高，后期籽粒脱水速度较快，容重高，商品性好，对一些主要病害的水平抗性较好，是构建选育母本系的优新种质材料。

表1 京科968等系列玉米品种亲本、选系材料基础及杂种优势模式

Table 1 The parents and source of commercial hybrids Jingke 968 et al

品种 hybrid	审定编号 Admitted number	亲本名称 Parents name		亲本来源 Sources of parents		杂种优势模式 Heterotic pattern	适应区域 Growing environment
		母本	父本	母本	父本		
京科968	国审玉2011007	京724	京92	X1132x	（昌7-2×京24）×Lx9801	国外X种质群体选系×国内黄改种质群体选系	北方春玉米区及黄淮海夏玉米区等
京科665	国审玉2013003	京725	京92	X1132x	（昌7-2×京24）×Lx9801	国外X种质群体选系×国内黄改种质群体选系	北方春玉米区及黄淮海夏玉米区等

(续表)

品种 hybrid	审定编号 Admitted number	亲本名称 Parents name		亲本来源 Sources of parents		杂种优势模式 Heterotic pattern	适应区域 Growing environment
		母本	父本	母本	父本		
京农科728	国审玉2012003	京MC01	京2416	X1132x	京24×5237	国外X种质群体选系×国内黄改种质群体选系	东北中熟春玉米区及京津唐夏玉米区
NK718	蒙审玉2011003	京464	京2416	X1132x	京24×5237	国外X种质群体选系×国内黄改种质群体选系	北方春玉米区及黄淮海夏玉米区等

2 母本、父本与易制种相关的特征特性

由表2可以看出，京科968等系列玉米品种母本熟期为中晚熟，植株生长势强，株高平均269.8cm；穗位中等，雄穗分支较少，平均1.7个；花药紫色，花丝红色，果穗较大，但籽粒大小适中，千粒重相对较低，均小于300g，平均290.8g，均低于郑58和PH6WC；单穗粒数较多，均接近500粒，平均491.7粒，均高于郑58和PH6WC；粒型为硬粒-半硬粒型，单穗粒重均在140g以上，平均144.0g，均高于郑58和PH6WC。由此说明，针对易制种性状进行有针对性严选择是有效的。

从理论和实践两方面来看，京724等母本熟期偏晚、植株相对高大，是保证自身产量较高的生物学基础；雄穗分支少、雄穗主枝较长有利于去雄彻底和不留断枝；穗位高/株高系数小，穗上部叶片较多，可以带3~4片叶超前去雄。京724等花药花丝颜色较深为田间去杂提供便捷的辨别标记。果穗较长，穗行数较多，同时千粒重较小，粒型偏硬粒，籽粒粒色商品性好，穗型为筒形，籽粒大小一致，可以大大提高有效穗粒数，制种效益较高。此外京科系列玉米品种母本京724等在抗性上突出，耐低温播种、耐密植、空秆率低，抗旱耐涝耐盐碱，对地块要求较低，从而降低生产成本。

京科968等系列玉米品种父本熟期为早熟-中熟，株型为半紧凑-紧凑，较耐密植，便于提供充足的花粉量；京92株高较高，雄穗分支数最多，散粉时间持续时间最长，为5.1d，比国外系PH4CV有较大优势，和昌7-2相比也有明显改善，充分发挥出作为父本花粉量大、散粉时间长的优势。

表2 2011—2013年北京市田间京科968等系列玉米品种亲本主要特征特性
Table 2 The parents characteristics of commercial hybrids Jingke 968 et al

自交系 Inbred line	熟期 Growth days	株型 Plant type	株高（cm）Plant height	穗位（cm）Ear height	雄穗分支数 Anthesis branch number	雄穗散粉持续期（day）Tassel pollen duration	穗长（cm）Ear length
京724	中晚熟	紧凑	278.3	90.9	1.2	-	18.1
京725	中晚熟	紧凑	267.8	81.5	1.9	-	16.5
京464	中晚熟	紧凑	263.4	78.8	2.1	-	16.6

（续表）

自交系 Inbred line	熟期 Growth days	株型 Plant type	株高（cm） Plant height	穗位（cm） Ear height	雄穗分支数 Anthesis branch number	雄穗散粉持续期（day） Tassel pollen duration	穗长（cm） Ear length
京科系列平均	中晚熟	紧凑	269.8	83.7	1.7	-	17.1
郑58	中晚熟	紧凑	161.2	57.2	4.6	-	16.2
PH6WC	中晚熟	紧凑	275.7	87.3	2.1	-	17.1
京92	中熟	半紧凑	172.7	70.4	5.3	5.5	15.4
京2416	早熟	紧凑	163.5	65.2	3.8	4.5	14.8
昌7-2	中熟	半紧凑	167.4	85.9	5.1	5.0	14.8
PH4CV	中熟	半紧凑	170.1	81.5	4.1	4.0	16.1

自交系 Inbred line	穗行数 Ear row	轴色 Cob color	千粒重（g） 1000-kernel weight	穗粒数 Grain number per era	穗粒重 Kernel (g) weight per ear	穗型 Ear type	粒型 Kernel type
京724	17.1	白	290.6	497.6	144.6	筒	硬粒
京725	16.8	红	287.8	490.2	141.1	筒	半硬
京464	16.4	白	299.9	487.3	146.2	筒	半硬
京科系列平均	16.8	白或红	292.8	491.7	144.0	筒	半硬-硬粒
郑58	12.4	白	404.1	309.8	125.2	筒	马齿
PH6WC	15.4	白	313.5	396.9	124.4	筒	硬粒
京92	13.5	白	318.4	-	-	锥	半硬
京2416	12.8	白	337.3	-	-	锥	硬粒
昌7-2	15.1	白	288.6	-	-	-	硬粒
PH4CV	14.8	红	317.8	-	-	筒	马齿

3 京科系列玉米品种制种生产高产田产量结构分析

由表3可以看出，京科968品种制种在母本78 900株/hm²密度情况下，有效穗为78 075穗/hm²，穗粒数501.4粒，千粒重284.3g，单穗粒重142.7g，理论产量达到9 490.1kg/hm²，理论种籽粒数3 327.5万粒/hm²，实际产量达到9 135.0kg/hm²。

郑单958和先玉335是我国近几年推广面积和制种面积较大的两个品种，制种优势明显。京科系列品种的制种产量与郑单958相当或略高，明显高于先玉335；单位面积的有效粒数均大大高于郑单958和先玉335。在目前单粒精量播种时代，粒数比穗粒重或制种产量更有意义。因此，京科968等系列玉米品种的制种效益非常突出。

表3 2012—2014年甘肃张掖高产田京科968等系列玉米品种典型高产田制种产量及产量结构
Table 3 Yield and yield components of typical seed high-yield plots of Jingke 968 et al

品种 Hybrid	密度 （株/hm²） Plant density	母本有效穗数 （穗/hm²） Ear number	穗粒数 （粒） Grain number per era	千粒重（g） 1 000-kernel weight	穗粒重（g） Kernel weight per ear	理论产量 （kg/hm²） Theoretical yield	理论粒数产量 （×10⁴/hm²） Theoritial kernel number	实际产量 （kg/hm²） Real yield
京科968	78 900	78 075	501.4	284.3	142.7	9 470.1	3 327.5	9 135.0
京科665	79 950	78 795	492.7	278.4	137.2	9 189.1	3 299.9	8 347.5
NK718	74 700	72 255	477.8	304.7	139.7	8 579.9	2 934.5	8 275.5
京农科728	87 000	85 650	442.3	280.6	124.1	9 034.8	3 220.1	8 233.5
京科系列品种平均	80 137.5	78 693.8	478.6	287.0	135.9	9 068.5	3 195.5	8 497.9
先玉335	84 885	81 585	394.8	310.4	122.5	84 95.0	2 737.8	7 735.5
郑单958	90 750	87 045	311.6	401.3	125.1	9 255.9	2 305.5	8 488.5

4 京科系列玉米品种高产制种关键技术

4.1 选择生态条件良好的地区，适时早播，后期控水

优越的生态条件是高产制种的重要条件。甘肃张掖制种区属于温带大陆性干旱气候，≥10℃年有效积温2 941~3 088℃·d；甘州区≥10℃年有效积温3 076.1℃·d。年均降雨量113~200mm，无霜期138~179d，海拔1 474m。该区日照时数长，昼夜温差大，灌溉条件优良，玉米制种产量高，是我国最大最集中的玉米种子生产基地。京科系列玉米品种生育期偏长，应发挥母本耐低温萌发优势适时早播，后期快要成熟时适当控水，有利于籽粒脱水，提早收获，保障种子发芽率和活力。

4.2 合理密植

京科系列玉米品种母本果穗大、粒重高、穗型筒形、单穗商品种子数高，合理密度保持在70 000~90 000株/hm²，增加有效穗数，实现穗大粒多。

4.3 准确把握花期调节，适时超前去雄

根据父母本总叶片数来确定和调节播种期。先播母本，当5cm地温稳定通过10℃播种母本，一次播种完成；当母本50%次生根长出扎根时，开始播种父本一期（一、二期各占50%）；当母本50%出苗（母本播后10d左右）再播种父本二期。

当50%母本倒数第4片展开时即开始带叶超前去雄，带3~4叶一次完成抽雄为宜，或者过2~3d再次去雄，一般在3~5d内进行两次去雄率可达到99.9%以上，真正做到去雄不见雄，去雄去（长）势的作用。带3~4片叶超前去雄不仅能够防止母本自交，保证种子质量，也降低了母本植株高度，一般可降低母本株高100~130cm，由于母本去雄后的株高与父本株高基本持平，更有利于父本花粉的传播，可有效地提高母本授粉结实率。同时，采取此技术去雄在时间上为组织劳力提供了充裕空间，也大大降低了劳动强度，降低了劳动成本。在整个去雄期间，做到及时、彻底，不留断枝。

5 结语

一个品种能否开发成功,一方面其自身的良好生产表现是重要基础,另一方面其易制种特性及制种产量潜力大也是关键因素。京科系列玉米品种实现在生产上大面积推广与应用与其易制种特性密切相关。京科968等系列玉米品种在单位面积制种产量上超过先玉335,与郑单958相当或略高,但目前商品玉米杂交种子正在朝按粒数包装出售的市场策略,按单位面积有效商品籽粒数计,京科968等系列玉米品种较郑单958和先玉335具有明显优势,尤其是京科968。

易制种京科968等系列玉米品种选育成功与先进的育种指导思想密切相关,其中,在品种选育过程中明确母本、父本自交系,并针对所需性状要求选择理想种质,制定相应的育种方案,在实施过程中严格按照既定方案工程化进行,确保入选材料符合方案要求。在变换地选择过程中,将S_3世代放在甘肃进行实施就是为了选择出适宜将来易制种的优良单株;其次在选育方案中一些性状更加量化,而不是凭感觉进行筛选,如单穗粒数、单穗粒重、出籽率等性状。创新杂优模式,母本来源于X1132x国外种质,选育出的母本自交系京724等具有高配合力、高抗、高产等优点,适于用作母本;父本来源于国内黄改系,选育出的京92等自交系配合力高、雄穗发达、花粉量大、散粉持续时间长,适于用作父本。

参考文献(略)

本文原载:玉米科学,2016,24(2):11-14

雄性不育制种京科 968 的遗传背景及其表型分析

宋 伟 赵久然 邢锦丰 张如养 苏爱国 易红梅 田红丽
王继东 王向鹏 石 子

（北京市农林科学院玉米研究中心/玉米 DNA 指纹及分子育种北京市
重点实验室，北京 100097）

摘 要：玉米品种京科 968（京 724/京 92）是由北京市农林科学院玉米研究中心选育并通过国家审定的优良品种。为进一步提高京科 968 的制种效率和降低成本，基于 S 型细胞质雄性不育系研发建立了该品种的三系配套杂交种制种方法。本研究利用 40 对 SSR 核心引物和 MaizeSNP3072 芯片对京科 968 雄性不育及常规制种的母本、杂交种进行遗传背景分析后发现，雄性不育制种母本不育系 S 京 724 与常规制种母本京 724 之间、雄性不育制种京科 968（简称 S 京科 968）与常规制种京科 968（简称 N 京科 968）之间均未检测到差异位点。表型比较结果表明，S 京 724 与京 724 在株高等农艺性状、单穗粒重等产量性状上的表现均无显著性差异。经花粉 I_2-KI 染色鉴定，S 京科 968 花粉表现约 50% 可染，50% 败育，与 S 型细胞质的恢复型 F_1 代植株花粉育性吻合。河南、河北、北京、辽宁、吉林多点鉴定结果表明，S 京科 968 与 N 京科 968 在株高、穗位高、雄穗分支数、雄穗主轴长等农艺性状和穗长、穗粗、穗行数、行粒数、单穗粒重、容重等产量性状上的表现均无显著性差异。上述结果证明，利用 S 型细胞质雄性不育系进行京科 968 三系配套杂交种制种是切实可行的。

关键词：玉米；京科 968；S 型细胞质雄性不育；遗传背景；表型

Analysis on the Genetic Background and Phenotype of Maize Hybrid Jingke968 Produced with Male Sterility Lines

Song Wei Zhao Jiuran Xing Jinfeng Zhang Ruyang Su Aiguo
Yi Hongmei Tian Hongli Wang Jidong Wang Xiangpeng Shi Zi

(*Maize Research Center*, *Beijing Academy of Agriculture and Forestry Sciences*, *Beijing Key Laboratory of maize DNA Fingerprinting and Molecular Breeding*, *Beijing*, 100097, *China*)

Abstract: The maize hybrid Jingke968 (Jing 724/Jing 92) is a superior variety, developed by

---

基金项目：国家科技支撑项目（2014BAD01B09）、北京市农林科学院科技创新能力建设专项（KJCX20140202）、北京市科技计划课题（Z141100002314016）、北京市博士后基金（2014ZZ-68）

作者简介：宋伟（1979— ），女，山东人，副研究员，博士，从事玉米分子育种工作。Tel：010-51503983；E-mail：songwei1007@126.com

赵久然为本文通讯作者

the Maize Research Center in Beijing Academy of Agriculture and Forestry Sciences. To improve the production efficiency ofJingke968, the three-line seed production method was developed and established based on the S-type cytoplasmic male sterility (CMS) line. In the present study, the genetic background of the normal and sterile female parent of Jingke 968 and their resulted hybrids were evaluated with 40 core SSR loci and MaizeSNP3072 chip. No loci difference was observed between the sterile female parent S-Jing724 and normal Jing724 nor between the hybrid obtained by S-Jing724 (S-Jingke968) and normal Jingke968 (N-Jingke 968).In addition, the phenotypic analysis demonstrated that no significant difference was observed in the agronomic traits such as plant height nor in the yield traits such as grain weight between S-Jing 724 and Jing724. The pollen staining with I_2-KI demonstrated that half of the pollens from S-jingke968 was viable and half was sterile, which is consistent with the pollen of restored S-CMS F_1 plants. The multiple location evaluation results from Henan, Hebei, Beijing, Liaoning and Jilin indicated that no substantial difference was present in the agronomic traits, including plant height, ear height, tassel branch number andprincipal axis length, and yield traits, including ear length, ear diameter, number of rows, kernels per row, bare tip length, grain weight per ear andtest weight, between S-Jingke968 and N-Jingke968. Our data testify the feasibility of three-line seed production system of Jingke968 using S-type CMS line of Jing724.

Keywords:Maize;Jingke968;S-type cytoplasmic male sterility;Genetic background;Phenotype

玉米是杂交良种应用最早的作物。利用常规的制种方法配制杂交种需要对母本进行人工去雄,费工费时,增加种子生产成本。目前在甘肃等主要玉米制种基地,每亩仅人工去雄环节至少需要两个人工日,同时存在由于去雄不及时、不彻底影响种子质量的潜在风险。改变人工去雄方式的主要途径包括机械去雄和雄性不育化制种技术。其中机械去雄存在一定局限性:一是易对植株造成机械损伤,影响制种产量;二是难以实现超前"摸包"去雄,影响制种质量;三是部分制种田地块不适合机械化作业等。利用雄性不育系制种是降低成本、提高质量、增加效益的有效技术手段,是种子产业发展的必然趋势。但是我国目前实现雄性不育制种的杂交种寥寥无几。

玉米杂交种京科968(京724/京92)是由北京市农林科学院玉米研究中心选育的优良品种,2011年通过国家审定(审定编号:国审玉2011007),具有高产、优质、多抗、广适、易制种等优点。为提高京科968的制种效率,北京市农林科学院自2012年起开展了该品种的雄性不育制种研究,结合利用分子标记辅助选择,快速获得了母本京724的S型细胞质雄性不育系S京724,同时以京724做保持系,京92做恢复系(由于京92是S京724的天然强恢复系,节省了恢复系的转育工作),研发建立了京科968三系配套杂交种制种方法。本研究通过对雄性不育制种京科968进行遗传背景和表型分析,比较S京724与京724、雄性不育制种京科968(简称S京科968)与常规制种京科968(简称N京科968)在DNA水平和农艺性状上的差异,验证京科968 S型细胞质雄性不育制种的可行性,为该品种进行大面积三系配套杂交种制种应用提供数据支撑。

1 材料与方法

1.1 供试材料

京724、S京724、N京科968和S京科968。其中,本研究中使用的S京724种子是以京724为受体,以S型细胞质雄性不育系MD32为供体,结合分子标记辅助背景选择,回交4代获得的雄性不育系;S京科968种子是以上述S京724做母本、京92做父本组配的杂交种。

1.2 DNA提取

将京724、S京724、N京科968和S京科968室内发苗,分别混株提取DNA,提取方法参考CTAB法。

1.3 SSR标记

1.3.1 引物

利用农业行业标准《玉米品种鉴定技术规程SSR标记法NY/T 1432—2014》中推荐使用的40对多态性高、在染色体上均匀分布的SSR核心引物进行遗传背景分析,引物信息详见Wang F et al（2011）。

1.3.2 PCR扩增和扩增产物检测

参照李少博等的方法。

1.4 SNP标记

1.4.1 位点信息

利用北京市农林科学院玉米研究中心研发并获得国家发明专利的MaizeSNP3072芯片对供试材料DNA进行SNP基因分型。该芯片包含3 072个均匀分布于玉米全基因组的SNP位点,位点信息详见http：//www.illumina.com.cn和http：//www.panzea.org。

1.4.2 芯片检测

基于Illumina公司的GoldenGate技术平台对MaizeSNP3072芯片进行检测。具体试验操作流程参照吴金凤等的方法。

1.5 田间种植及性状调查

2015年在北京市农林科学院院内试验田种植自交系京724和S京724,行长5m、行距60cm、株距25cm,2行区,3次重复。每个小区随机选取均匀一致的10个单株及果穗测量株高、穗位高、穗长、穗粗、穗行数、行粒数、秃尖长、单穗粒重、容重等农艺及产量性状。

2015年于河南、河北、北京、辽宁、吉林5个试验点分别种植杂交种N京科968和S京科968。行长5m、行距60cm、株距32cm,6行区,3次重复。每个小区随机选取均匀一致的30个单株及果穗测量株高、穗位高、雄穗分支数、雄穗主轴长、穗长、穗粗、穗行数、行粒数、秃尖长、单穗粒重、容重等农艺及产量性状。

数据统计分析使用SPSS软件进行。

1.6 花粉育性鉴定

散粉期,每份供试材料随机选取5株,取雄穗上、中、下3个部位的小穗用卡诺固定液固定。用镊子剥取花药置于干净的载玻片上,挤出花粉粒后,用1%的I_2-KI染色,于OLYMPUS倒置荧光显微镜4×10倍下观察。每个小穗观察3个视野,观察花粉染色情况。

2 结果与分析

2.1 遗传背景分析

利用40对SSR核心引物对S京724和京724的DNA进行扩增,经荧光毛细管电泳检测后发现，二者在40对引物上的指纹完全一致，差异位点数为0。利用MaizeSNP3072芯片对S京724、京724进行SNP基因分型。3 072个位点中，前期根据分型效果、数据缺失率、染色体均匀分布等优选出1 202个SNP位点用于数据分析。结果显示，S京724和京724在有效判定的1 202个SNP位点上未检测到差异。利用上述SSR和SNP标记方法对S京科968的遗传背景进行分析，结果表明S京科968与N京科968间在40对SSR核心引物和MaizeSNP3072芯片上的基因型一致，未检测到差异位点。

2.2 S京724与京724间表型比较分析

通过比较S京724与京724的部分农艺性状和产量性状后发现，二者在株高、穗位高、穗长、穗粗、穗行数、行粒数、秃尖长、单穗粒重、容重等性状上的表现虽存在细微差别，但是T检验结果表明，在上述性状上S京724与京724间均无显著性差异（详见表1）。

2.3 S京科968花粉I_2-KI染色

分别在河南、河北、北京、辽宁、吉林等多地试验点田间种植S京科968和N京科968，观察鉴定雄性不育制种杂交种S京科968植株雄穗的育性恢复情况，S京科968表现为育性恢复，花药外露，正常散粉，花粉量较N京科968略有减少。利用I_2-KI染色法检测S京科968和N京科968的花粉活力（图1）。其中常规制种获得的杂交种N京科968的绝大多数花粉粒均饱满呈圆形，由于积累了较多的淀粉，经I_2-KI染色后呈蓝色，仅有极个别花粉粒呈畸形，由于花药组织发育不同步，不含淀粉或淀粉积累较少，经I_2-KI染色后呈黄褐色。而S京科968所有待测单株的镜检小穗花粉粒均表现约50%可染，50%败育。

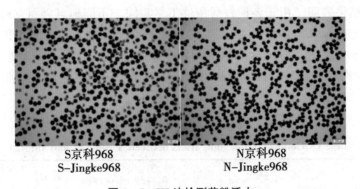

S京科968　　　　　　N京科968
S-Jingke968　　　　　N-Jingke968

图1　I_2-KI法检测花粉活力

Figure 1　Pollens of S-Jingke968 and N-Jingke968 tested by I_2-KI

2.4 S京科968与N京科968间表型比较分析

在河南、河北、北京、辽宁、吉林5个鉴定点上，分别以N京科968作对照，对S京科968进行表型调查分析。通过比较株高、穗位高、雄穗分支数、雄穗主轴长等农艺性状及穗长、穗粗、穗行数、行粒数、单穗粒重、容重等产量性状后发现，S京科968与N京科968之间在上述性状上均无显著性差异（详见表2）。

3 结论与讨论

本研究利用40对SSR核心引物和MaizeSNP3072芯片对S京724和京724进行遗传背

景比较分析，结果表明，二者在40对SSR引物和1 202个SNP位点上均无差异。其中40对SSR引物为农业行业标准《玉米品种鉴定技术规程 SSR标记法 NY/T 1432—2014》规定推荐的引物。说明S京724通过分子标记辅助背景选择和四代回交转育后，其遗传背景与京724极度近似，因此从基因型上来说，S京724可以代替京724进行京科968三系配套杂交种制种。通过比较株高等农艺性状和穗长等产量性状的数据可以看出，S京724与京724在上述性状上均不存在显著性差异，因此从表型上来说，S京724同样可以代替京724在京科968杂交种制种过程中使用。对雄性不育制种京科968进行遗传背景分析后发现，在检测的40对SSR引物和MaizeSNP3072芯片上，S京科968与N京科968间差异位点数为0，说明雄性不育制种京科968在DNA水平上已与常规制种京科968高度近似。

由于S型雄性不育属于配子体雄性不育，育性反应由配子体的基因型控制，其恢复型F_1代植株的花粉50%表现可育，50%表现败育。而C型、T型雄性不育性属于孢子体雄性不育，其恢复型F_1代植株的花粉全部可育。本研究中，S京科968的花粉经过I_2-KI染色后表现为约一半可染，一半败育，与S型细胞质的恢复型F_1代植株花粉育性吻合。这一结果也验证了我们使用的雄性不育系属于S型不育系。S型不育系由于败育时期较晚，育性的稳定性相对较差且易受核背景基因型的影响，在某些核背景中不育性高度稳定。我单位通过多年多点育性筛选鉴定，选育获得的京科968母本不育系S京724育性稳定，败育彻底，目前已在杂交种生产中得到了很好的应用。

为了验证三系配套杂交种制种京科968与常规方法制种京科968的性状表现是否一致，本研究在河南、河北、北京、辽宁、吉林5个试验点安排多点鉴定对S京科968进行表型分析。结果表明，由于各地气候条件不尽相同，不同地点之间京科968的性状表现势必存在一定差异。但是在同一地点内，S京科968在重要农艺性状和产量性状上的表现与N京科968无显著性差异。说明利用S型细胞质雄性不育系S京724进行京科968的三系配套杂交种制种，技术方法可行，同时又可以降低种子生产成本，具有广阔的应用前景。

表1 S京724与京724部分性状比较
Table 1 Comparison of some traits between S-Jing724 and Jing724

品种 Variety	株高（cm） Plant height	穗位高（cm） Ear height	穗长（cm） Ear length	穗粗（cm） Ear diameter	穗行数 Ear row number	行粒数 Kernels per row	秃尖长（cm） Bald tip length	单穗粒重（g） Single ear grain weight	容重（g/L） Test weight
京724（CK）	242.87	61.00	16.20	4.10	17.07	25.87	1.34	140.63	739.77
S京724	240.75	63.50	16.28	4.19	18.00	27.07	0.74	144.23	750.42
P-Value	0.264	0.504	0.872	0.396	0.079	0.209	0.055	0.704	0.089

表2 S 京科968与 N 京科968部分性状比较
Table 2 Comparison of some traits between S-Jingke968 and N-Jingke968

地点 Site	品种 Variety	株高（cm）Plant height	穗位高（cm）Ear height	雄穗分支数 Tassel branch number	雄穗主轴长（cm）Principal axis length	穗长（cm）Ear length	穗粗（cm）Ear diameter	穗行数 Ear row number	行粒数 Kernels per row	单穗粒重（g）Grain weight per ear	容重（g/L）Test weight
河南	N 京科968	295.41	124.71	11.59	44.07	19.35	5.62	17.02	39.38	276.06	759.76
	S 京科968	291.92	122.23	11.59	43.84	19.77	5.54	16.21	40.41	276.79	766.08
	P-Value	0.584	0.421	0.987	0.363	0.629	0.084	0.051	0.516	0.971	0.111
河北	N 京科968	291.36	107.38	11.83	44.55	21.30	5.38	17.01	43.50	293.73	742.17
	S 京科968	282.49	105.77	12.07	45.81	21.17	5.51	17.20	42.74	287.82	738.22
	P-Value	0.125	0.735	0.563	0.29	0.783	0.158	0.315	0.269	0.509	0.718
北京	N 京科968	287.90	108.42	10.06	39.20	19.15	5.26	16.46	40.99	253.97	749.76
	S 京科968	284.90	108.58	9.82	39.32	18.93	5.24	16.27	40.54	241.10	756.08
	P-Value	0.532	0.964	0.648	0.889	0.696	0.708	0.593	0.571	0.348	0.269
辽宁	N 京科968	302.96	152.31	12.57	39.58	17.37	5.16	17.02	36.51	216.16	772.41
	S 京科968	293.20	151.29	12.53	40.25	17.64	5.22	16.93	37.07	216.72	764.21
	P-Value	0.345	0.82	0.967	0.403	0.687	0.388	0.395	0.688	0.967	0.509
吉林	N 京科968	303.42	127.29	14.02	39.93	21.27	5.55	16.68	43.36	275.70	656.78
	S 京科968	302.61	127.62	13.82	39.79	21.00	5.47	16.73	42.63	285.18	669.32
	P-Value	0.814	0.868	0.623	0.866	0.487	0.104	0.829	0.252	0.225	0.168

参考文献（略）

本文原载：玉米科学，2017（4）：7-11

玉米品种京农科 728 北京密云地区制种技术

王荣焕[1] 王元东[1] 赵久然[1] 徐田军[1] 陈传永[1] 刘新香[1] 崔铁英[2]

(1. 北京市农林科学院玉米研究中心/玉米 DNA 指纹及分子育种北京市重点实验室，北京 100097；2. 北京龙耘种业有限公司，101500)

摘　要：京农科 728 具有早熟优质、抗旱节水、耐密抗倒、高产稳产、适宜机收籽粒和全程机械化等突出优势，2012 年以来相继通过国家、北京市、内蒙古和黑龙江省审（认）定，近年种植面积呈强劲上升势头。为加快京农科 728 在生产中的大面积推广应用，为生产提供高质量种子，并充分利用北京密云地区传统规模化玉米制种基地，解决农村剩余劳动力就业，带动农民增收致富，根据京农科 728 亲本特征特性并经多年试验研究和实践探索，提出了以"反交制种、地膜覆盖、父本分期播种、节水灌溉"为核心的适宜北京密云地区的京农科 728 高产高效制种技术。

关键词：玉米；京农科 728；制种技术

京农科 728 由北京市农林科学院玉米研究中心选育，具有早熟优质、抗旱节水、耐密抗倒、高产稳产、适宜机收籽粒和全程机械化等突出优势，2012 年通过国家审定（国审玉 2012003）、2014 年通过北京市审定（京审 2014007）、2016 年通过内蒙古认定（蒙认玉 2016001）和黑龙江省审定（黑审玉 2016017）。被农业部推荐为玉米主导品种，被北京市列为更新换代新品种和高产创建主导品种，并被列为京津冀一体化主推玉米新品种。京农科 728 由北京龙耘种业有限公司进行产业开发，通过实施科企合作实现了产学研结合、育繁推一体化，加快了京农科 728 的大面积推广应用，2014 年推广面积 40 多万亩、2015 年达到近 200 万亩、2016 年迅速增至近 500 万亩，且上升势头强劲，推广应用前景广阔。

为生产提供高质量种子是实现玉米品种大面积推广的关键。北京密云冯家峪镇番字牌地区地处深山，自然隔离条件好，降雨和积温条件适宜，是理想的玉米制种基地。当地发展玉米制种产业有 30 多年历史，是目前北京仅存的规模化玉米制种基地，约 5 000 亩。位于北京密云的北京龙耘种业有限公司充分利用当地区位优势，组织番字牌地区农村剩余劳动力就近开展玉米品种京农科 728 制种和加工。但当地玉米制种面临干旱缺水、水源不能保证、露地种植等突出问题，甘肃地区的制种技术在密云不适用。我们根据北京市节水农业发展需求、番字牌地区的自然气候特点及京农科 728 父母本特征特性，并经多年试验研究和生产实践探索，提出了以"反交制种、地膜覆盖、父本分期播种、节水灌溉"为核心的适宜北京密云地区的京农科 728 高产高效制种技术。

基金项目：北京市科技计划课题（Z151100001015014）
通讯作者：赵久然

1 亲本选育及特点

1.1 母本自交系 MC01

以国外杂交种 X1132x 为选系基础材料,按照"高大严"选系方法(即高密度、大群体、严选择),变换多个不同地点,经配合力测定等在 S_4 优良穗行中选择优良植株利用 DH 育种技术诱导和加倍,再经扩繁和多点鉴定及严格筛选,获得了 DH 纯系京 MC01,具有一般配合力高,抗大、小斑病、茎腐病、矮花叶病毒病和穗粒腐病,综合抗性好特点。株型紧凑,株高 198cm,穗位 65cm;果穗长筒形,穗轴红色;穗长 18cm,穗粗 4.5cm,穗行数 14 行;籽粒半硬粒型,黄色。

1.2 父本自交系京 2416

由京 24 与 5237 杂交组成的基础材料选系,按照"高大严"选系技术,变换不同生态区,经多代自交选育而成,具有较高的一般配合力,抗玉米大、小斑病、矮花叶病毒病等。株型紧凑,株高 160cm,穗位 75cm;雄穗分支 3~5 个,花粉量大;果穗锥形,穗轴白色;穗长 16cm,穗粗 4.8cm,穗行数 12 行;籽粒硬粒型,黄色。

2 适宜北京密云地区的制种技术

2.1 父、母本

在正常情况下,和甘肃大面积制种基地均采取以 MC01 为母本、京 2416 为父本进行杂交种生产。但根据密云冯家峪镇番字牌地区制种基地的气候、土壤特点、生产习惯和节水需求,建议采取反交制种,即以京 2614 为母本(抗旱性更强)、MC01 为父本。

2.2 选地与隔离

在密云选择热量充足、无霜期大于 150d、土壤条件为沙壤土或壤土的区域进行生产;选择地力(肥水条件)较好地的块种植。利用密云得天独厚的自然隔离条件(山、树木、村庄等),并依照国家玉米杂交种制种规程,将时间、空间以及人为设置障碍物等自然条件合理结合运用,杜绝外源花粉对制种区的影响。

2.3 播期与密度

5cm 地温稳定通过 10℃ 时播种。母本一次播种完成,一般在 4 月底至 5 月初开始播种。根据自然降雨情况,也可推迟到 5 月中旬进行适墒播种。地膜覆盖种植可根据墒情提早播期,或等雨晚播种需要加快生育进程时,应采用地膜覆盖栽培进行制种。母本行距 50cm,株距 23cm,种植密度 5 800 株/亩,保苗成株 4 600 株/亩。父本株距 23cm,种植密度(行比兼满天星)1 700 株/亩,保苗成株 1 200 株/亩。

父母本按行比 1∶5 兼父本满天星种植。播 5 行母本,紧接着播 1 行父本,依次播种。在母本的第 2 或第 3 行中间适当加大行距(留 60cm),用于第二期满天星方式种植父本。母本一次性播种、父本分两期播种。第一期父本与母本同期播,第二期父本当 50% 母本在土壤中膨胀刚要顶芽(母本播种后 5~6d)时,在留下的第 2 行或第 3 行中间播(满天星按 50cm 点播或穴播)。

2.4 去杂及去雄

出苗后至抽雄前,分三次去杂。第 1 次:结合定苗,根据幼苗的长相、长势、叶色、叶型等,先去掉肥、大、弱和可疑苗,父本定苗要掌握"留大、中、小苗"或

50%双苗，以便延长花期。第2次：在拔节后，根据叶色、叶型、株高、株型及叶片宽窄等，去掉优势株、异形株等杂株，保留性状一致的典型株。第3次：在抽雄前，根据株高、株型、叶缘、叶型、叶片宽窄进行。因父本杂株对种子纯度影响很大，父本一旦散粉，损失无法挽回。因此，在散粉前必须根据父本的雄穗形状、分枝多少、护颖颜色等严格去杂。

另外，种子脱粒前要进行穗选，根据穗型、轴色、粒型、粒色等进行最后一次去杂。田间杂株率母本不超2‰、父本不超1‰。注意京农科728在大喇叭口期一般株高1.1~1.3m，对小株判断比较准确，结合去杂去除小苗对提高种子质量有较大好处。

去雄分3个时期进行，初花期、盛花期、末花期。密云地区的玉米制种田大多为梯田，虽然播种集中，但母本长势不整齐一致，抽雄持续时期较长，3个时期要30d左右全面结束。

另外，抽雄结束后去除3类苗（弱、病、小未结雌穗苗）很重要，3类苗雄穗抽出较晚，不易管理，但花粉生命力强很易自交，也是影响产量的关键时期。因此，及时去除3类苗并带出田间掩埋也是保证制种质量的重要环节。

2.5 田间管理

整地时，每亩施复合肥或玉米专用肥25kg作底肥；播种前，进行种子包衣处理；3叶期间苗、5叶期定苗，母本定苗时去除病苗、弱苗和畸形苗，留生长均匀一致的健壮苗。拔节期追肥，每亩追施尿素40kg。采用滴灌等节水灌溉技术，确保制种田关键生育时期不受旱。

2.6 割除父本

父本散粉结束后，割掉父本是保证质量、增产和提高种子商品性的有效途径。要及时割除父本，并清出田间。

2.7 收获加工

10月5日前采收，及时收获晾晒，严防冻害，及时脱水降至安全水分。水分未达标准时进行烘干，直至达到国家标准。然后进行加工、包装、成品。

3 在北京密云地区的制种表现

近年来，北京龙耘种业有限公司以北京市农林科学院玉米研究中心为技术支撑，采取"科研+公司+农户"方式在密云冯家峪镇番字牌地区组织开展了京农科728规模化制种，并从整地、施肥、播种、去杂、管理、收获等环节进行全程技术指导服务。通过采取配套高产高效高质制种技术，番字牌地区京农科728制种基地普通示范区（4 000余亩）平均制种产量可达350kg/亩、核心制种示范区（500亩）则高达400kg/亩，带动了整个基地实现高质量制种。且制种质量稳步提高，精选后种子纯度不低于99.7%、净度不低于99.0%、发芽率不低于95%、水分不高于13%，达到了单粒精播质量标准。

2015年和2016年，该基地5 000亩京农科728制种田产种量为175万kg（按亩产350kg保守计算），籽种7.4元/kg，总产值1 295万元；比种普通玉米（亩产600kg，0.8元/kg）增收815万元，产值增加一倍以上。所生产的京农科728种子在满足当地用种的前提下，还辐射到河北、内蒙古、黑龙江、吉林等地。为加快京农科728在京津冀及东华北区大面积推广应用、充分利用农村剩余劳动力、带动农民增收致富以及推动北京节水农业发展和农业种植结构调整发挥了重要作用。

参考文献（略）

本文原载：中国种业，2017（2）：60-63

京科青贮 516 玉米高产制种技术

冯培煜 宋瑞连

(北京市农林科学院玉米研究中心 100097)

玉米品种京科青贮 516 审定编号：国审玉 2007029，是北京市农林科学院玉米研究中心于 2005 年利用 MC0303 为母本，MC30 为父本选育而成的粮饲兼用型优良玉米品种。该品种自审定推广以来已在北京、内蒙古、河北北部、辽宁东部、吉林中南部、黑龙江第一积温带、山西北部春播区作专用青贮或粮饲兼用玉米品种大面积种植。目前在西北春播区甘肃、新疆以及华北夏播区亦有引进种植，生产均表现生长抗性强，粮饲产量高。由于该品种双亲生育期长，植株长势强，在一般地区制种往产量低，发芽率也难于保障，经过几年在不同地区的种子生产实践，形成了在我国玉米种子主产地甘肃河西张掖地区制种产量达到 7 500 kg/hm² 以上的制种技术。

1 自交系特征特性

母本 MC0303，穗长、不秃尖、穗行整齐、籽粒深、抗性强，幼苗叶鞘紫色，叶边缘紫红色，叶色浅黄色。株型半紧凑。株高 220m 左右，穗位 100cm 左右，穗柄中等。果穗长筒形，穗长 16~18cm，穗粗 4cm 左右，无秃尖。穗行数 14~16 行，行粒数 30~35 粒。籽粒浅黄色。马齿型。穗轴红色。雄穗分枝 10 个左右并向下弯曲。花药浅紫色。该自交系一般配合力高，具有茎秆坚硬，抗倒伏能力强，耐旱性强，抗黑粉病、粗缩病的优点。

父本 MC30，幼苗叶片深绿色，叶鞘紫色。株型半平展。株高 240cm 左右，穗位 110cm 左右。果穗长筒形，穗长 24cm 左右，穗粗 5cm，穗行 14~16 行，行粒数 45 粒左右。籽粒深黄色，半硬粒，轴色为红色。雄穗发达多分枝，有 12~14 个分枝。花药紫色。雌穗花丝红色，穗柄长 5~7cm。该自交系具有植株高大，叶片浓绿、花粉量大、散粉时间长、高抗玉米大、小斑花叶病毒病、茎腐病、黑粉病等，保绿性好，活秆成熟等优点。

2 种子生产技术

2.1 选地与隔离区设置

选地：在甘肃河西地区选择海拔相对较低、热量充足、无霜期大于 150d、土壤条件为沙壤土或壤土的区域。

生产：选择地力及肥水条件上等、灌溉条件便利的地块种植。

隔离：依照国家玉米杂交种制种规程，结合实际，将时间、空间隔离以及人为设置障碍物等自然条件合理结合运用，杜绝外源花粉对制种区的影响。

2.2 调节播期与密度

播种时间：5cm 地温稳定通过 10℃播种，母本一次播种完成。

种植形式：在甘肃河西地区标准地膜覆盖种植；膜床母本双行种植，父本在母本双行中间（地膜中间）满天星种植。

错期指标：当母本播种后种子破胸后播第1期父本（50%），当母本扎根（7~9d）播第2期父本（50%）。

种植密度：父母本总保苗密度82 500株/hm^2，其中父本7 500株/hm^2，母本75 000株/hm^2。

2.3 去杂及去雄

去杂：出苗后至抽雄前分3次去杂，根据父本、母本自交系的特征特性及生长状况，严格组织去杂。去雄：当70%母本倒数第3片叶展开时，带3~4片叶超前去雄，真正做到去雄不见雄。在整个去雄期间，做到及时、彻底，不留断枝，风雨无阻。一般地块在3~5d内抽两遍雄就能达到99.9%以上的去雄率。

2.4 田间管理

整地覆膜时用玉米专用肥375kg/hm^2作底肥；播种前种子包衣处理；三叶期间苗，五叶期定苗，母本定苗时去除病苗、弱苗和畸形苗，留生长均匀一致的健壮苗。拔节期追肥，尿素600 kg/hm^2；及时灌溉浇水；中后期特别注意防治红蜘蛛。

2.5 割除父本

父本散粉结束后，及时割除父本并清出田间。

3 收获加工

在9月25日前采收，并及时烘干、脱水加工。无烘干条件的及时收获晾晒，严防冻害，及时脱水降至安全水分，直至达到国家标准，进行加工、包装、制成品。

本文原载：种子世界，2015（1）

玉米制种单粒播种技术及相关配套措施

冯培煜 宋瑞连

(北京市农林科学院玉米研究中心，北京 100097)

摘 要：玉米制种应用单粒播种技术比大田种植应用此技术更为省种省工，增产增收增效。玉米制种应用单利点播技术主要要求是更高质量的亲本，纯度要保证在99.9以上，发芽率保证在95以上，亲本种子活力高，特别是做母本的自交系不仅要求活力高还要均匀一致，做到同源来、同年产、同地块出的、同等质量的一批种子；亲本种子加工质量及处理要求更加精细，对不同形状及大小的自交系采用不同的风选、筛选、比重选、粒型选等手段进行精加工；严格实行针对不同自交系、不同地块的特性特点使用专制种衣剂进行种子包衣制度，做到有的放矢；制种地要求严格，地块平整，土质结构一致，茬口一致，地力均匀，整地细致、无坷垃、无暗埇，排灌条件一致；栽培管理制度及播种工具的统一与一致。总之，在玉米制种生产中推广运用此项技术的效果，除与以上几项注意的技术措施有关外，还与种子生产公司的负责人或老总认知度与推广组织的推动力有密切关系。

关键词：玉米；制种；技术；单粒直播

单粒直播技术在玉米大田生产中已不是新生事物，但在玉米种子生产中由于受传统"有钱买籽，无钱买苗"的思想影响，以及在自交系繁殖中"一年繁多年用"、贮藏、播种等因素的影响和制约，此项技术仍难以在玉米种子生产中广泛应用推广。公认，玉米大田生产单粒直播不仅省籽，节约生产成本，还有利于苗齐、苗壮，有利于降耗，增产增收。经过几年在京单28、京单38、京科968、京科665等京科系列玉米种子生产中的实践，应用推广单粒播种技术制种，取得了比大田更为"省种省工，苗齐苗壮，降低生产成本，增产增收增效"的结果，特别是对种子质量的提高、减少抽雄用工，以及花期调整更具有益的效果。将几年来在应用推广玉米制种单粒播种技术及相关配套措施的经验予总结，供参考。

1 亲本种子质量要求严格

1.1 纯度

纯度要保证在99.9%以上。如玉米制种要求保苗5 500~6 000株/667m^2，即出现杂株不超过10株/667m^2。因此，要求在自交系繁殖中要更加严格，只有繁殖出高纯度的自交系，玉米制种单粒播种技术才有可能实现。

1.2 发芽率

发芽率是和纯度同等重要的要求指标，保证在95%以上。采用单粒直播技术，只要按常规制种技术要求的保苗密度增加10%的播种密度就可以了。玉米种子生产的产量所

作者简介：冯培煜（1963— ），男，高级农艺师，主要从事玉米 种子繁育与生产加工工作
通讯作者：宋瑞连（1965— ），女，高级农艺师，主要从事玉米种子繁育与检验工作

得是由单株个体和群体发育生长相互协调作用产生的。

1.3 亲本种子活力

亲本种子活力高，特别是做母本的自交系不仅要求活力高还要均匀一致，做到同源来、同年产、同地块出的、同等质量的一批种子。运用这样的亲本制种，才有可能保证播种后出苗齐、出苗壮。为此在选择繁殖自交系的地块时一定要选择隔离条件好、地力均匀、平整、灌溉有保障、肥力高的地块；如果地力不均匀，可选择分收、分加工的办法，分开批次进行生产用种。

1.4 亲本种子加工

亲本种子加工质量要求更精细。特别是从粒重、粒型上做到更精细精准加工，做到亲本种子群体中的个体更加趋于一致。特别要求对种子的加工器械应进行深入研究，使不同粒型的种子籽粒分离或区分，确实做到不仅能按种子大小（重力、比重、风筛选）进行分级分批播种，还能够按种子的形状（长或圆）进行分级分别，实现分批播种。

1.5 加工

严格实行针对不同亲本、不同地块的特性特点使用专制种衣剂进行种子包衣制度，确实做到专防、专制、专治、专包、专用，有的放矢。

2 制种地要求严格

制种地要求地块平整、土质结构一致；茬口一致，地力均匀；整地细致、无坷垃、无暗墒，排灌条件一致。

3 栽培管理制度及播种工具的统一与一致

3.1 栽培管理制度要统一

如地膜玉米制种自1987年首先在掖单二号制种试验成功以来，现今已在广大玉米制种区推广运用，甘肃张掖、新疆等地以用80~110cm宽幅地膜覆盖，膜上2个边缘种植母本或父本行，这样的种植耕作方式使得母本个体间发育均匀一致，有利于父本与母本花期调节，有利于增产；而甘肃武威等地区由于受种植蔬菜覆地膜的习惯影响，在180~2 400cm的宽幅地膜上种植4行母本，结果往往是靠膜中间的2行亲本比两边行的亲本发育早且长势强，从而会影响双亲花期调节的完全相遇性，也对制种产量易于造成不利影响，应当改之。

3.2 播种器械要适应单粒直播要求

主要是要使用能够适应不同亲本的专用或可调节的播种器械，如长形与圆形、大型与小型籽粒的区别。

总之，在玉米制种生产中推广运用此项技术，除以上几项注意的技术措施外，关键是种子生产公司的负责人或老总认知度与推广组织的推动力。一般而言，在种子生产型公司老总在生产管理中对种子质量要求考核是偏重的，一票否决，这无疑是正确的；在生产技术上往往认为只要父本与母本花期相遇，制种就成功了，而对其他对种子质量、产量有利的技术往往重视不够。其实，制种专业公司间的效益差异就在这一点一点边际效应的累加效果的不同。

本文原载：种子，2015，34（10）：131-132

春玉米区玉米制种预防高温危害的方法与措施

冯培煜 宋瑞连 王晓光

(北京市农林科学院玉米研究中心，北京 100097)

摘 要：2015年，春玉米制种区遭遇了高温危害，致使玉米制种授粉结实不良，造成减产，对国内玉米种子市场供给也造成巨大影响。高温危害，一是发生在7月20日即"入伏"前后10d左右，气温超过39℃将会影响玉米花粉的活力，会使玉米制种母本结实率降低，造成减产，甚至绝收；二是发生在入秋后的10d左右时段的"秋老虎"，这时的高温不会影响玉米制种授粉，一般也不会造成玉米制种田绝收。预防高温危害措施和方法，一是采取合理的栽培措施躲避或降低高温危害，如适时早播躲避高温，培育壮苗抗高温，调节好花期提高母本结实几率等管理手段；二是通过育种手段提高自交系的耐高温能力；三是运用综合措施，防患未然，事半功倍。

关键词：玉米制种；高温；预防；技术

2015年我国春玉米制种区特别是新疆遭遇了前所未有的高温，致使玉米制种授粉结实不良，造成减产，有的地块甚至绝收，制种农户和企业损失惨重，对国内玉米种子市场供给也造成巨大影响。玉米种子生产是自然（天、地）、物（玉米品种组合双亲）以及与人（栽培管理）活动的相互作用的农业生产，其成功与否主要是看人与物结合作用的结果能否降低或躲避自然危害影响度。春玉米制种区发生高温危害，一般，一是发生在7月20日即"入伏"前后10d左右，气温超过39℃将会影响玉米花粉的活力，会使玉米制种母本结实率降低，造成减产，甚至绝收；二是发生在入秋后的10d左右时段的"秋老虎"，这时的高温不会影响玉米制种授粉，一般也不会造成玉米制种田绝收。通过调查分析总结，归结以下几点供同行在玉米制种生产实践中参考。

1 合理的栽培措施躲避或降低高温危害

（1）适时早播躲避高温危害期。适时早播是躲避夏季高温危害最有效，也是最简便的措施。我们知道，在8℃以上就能满足玉米萌发的温度条件，通过实施玉米制种地膜覆盖技术，一般能够比裸地种植提早15d播种，提早出苗10d左右，夏季抽雄吐丝时间也能提早7d以上；实施地膜覆盖制种技术，即使玉米制种出苗后遇到低温冷害，由于地膜的增温保温作用，以及玉米苗期其生长点没有裸露在地表以上的特点，一般春季冷害对玉米制种产量不会造成影响。实施早整地、早覆膜、早播种，即使没有遇到高温灾害性天气，对制种产量也是百利而无一害。

（2）调节好父母本花期，提高结实率，降低高温危害程度。这是遭遇灾害性天气与

作者简介：冯培煜（1963— ），男，高级农艺师，主要从事玉米种子生产加工技术及管理工作
通讯作者：宋瑞连（1965— ），女，高级农艺师，主要从事玉米种子繁育与检验工作

否都要强调的必要技术,这里主要强调的是在保证父本盛花期与母本吐丝高峰期相遇的前提下,想方设法拉长父本群体的散粉时间,以减少遭遇不良自然气候条件变化对母本吐丝期变化的影响。一般在实践中,即使是多年生产的制种技术成熟的组合也应采取适当增加父本密度,以及播 3 期父本的措施,以防止不良气候的发生。

(3)推广玉米制种单粒播种,培育壮苗抗高温。玉米单粒播种制种是省籽、省工、省时,有利于增产的技术,特别是在地膜制种田实施效果更加显著。在我国北方春播区,早春地膜覆盖后田间表现是气温明显低于地温,此时播种出苗的玉米表现是根比苗(地上)长得快、根扎得深、且根粗,如果采取传统多粒穴播方法,须必在出苗 3 叶后采取间苗定苗措施,此过程的结果会严重影响所留苗(受伤根、降温的影响)的生长,延缓其生长发育,留下的壮苗反而会变成弱苗,使其抗逆性减弱。实施单粒播种必须要有更高质量的亲本种子,以及高质量精准的播种器械等措施相配套,从而才能达到理想效果。

(4)选择排灌便利的地块,在高温到来之时及时灌溉,使田间小气候增湿降温。

(5)其他常规栽培管理措施,如精细整地、增施底肥培育壮苗抗高温,种植父本采粉备用(繁育)区,人工辅助授粉,人工降雨,结合喷施农药增大田间空气湿度,起到降温作用等人工措施。

2 选育和组配耐高温、抗逆性强的自交系和组合

(1)做母本用的自交系要求具备果穗产量高、有效粒数比例(种子商品率)高、吐丝畅、花丝活力强、雄穗分枝少、花粉量少,植株根系发达、穗位低、抗多种病虫害等特点;特别是在选育自交系时应当在高温干旱的我国西北地区做 1~2 生育周期的耐高温耐干旱强胁迫选择。京科 968 的母本京 724 就是按照这一标准选育成功的一个优良自交系。

(2)做父本用的自交系要求具备花粉量大、散粉时间长、花粉活力高、根系发达、株高适宜、自交结实率高、与母本配合力高等特点。目前以国内血缘黄早 4 为基础材料改良的玉米自交系,如昌 7-2、京 92 等都是适宜做父本、耐高温、适应性强的自交系。

(3)适宜的父母本组合。一般育种家在选配组合时都注重双亲的遗传配合力,对 F_1 代的测定鉴定也比较多,在选配双亲其他性状时往往局限于母本产量高、父本花粉量大、植株高大等几个性状要求,而忽视了品种或组合在制种阶段在制种基地的适应适宜程度,因此,一个组合或品种在审定或推广之前应当在制种区进行试制、试繁,以便摸透弄通种子生产技术。对父母本性状除上述要求外,对两者株高比例要求,一般以母本去完雄后的株高与父本株高相当,较为适宜,这样既有利于母本授粉,也有利于父本个体发育;相对父本植株高度过于超过母本植株高度的组合,也可避免当遇到高温干旱时父本雄穗受到阳光的直接烤灼,遇到干热风时也会对其危害小一些;在选配双亲组配时,对 2 个自交系生育期要求,以相同或相近生育期为佳。

3 运用综合措施,防患未然,事半功倍

玉米种子生产环节多,时间长,受环境影响大,制种成功无疑是各种因素完美集成统一的结果;失败或有瑕疵,没有达到理想结果,肯定是某环节或那几个环节有问题,而这

些问题一旦出现，对其采取的人工干预措施往往效果有限。因此，在生产实施前就应当制定切实可行的、细致的、严密的、综合的技术及组织实施方案，以防为主，防患未然，达到预期目标。

本文原载：种子，2016，5（3）：127-128

以玉米籽粒粒形进行分离精选种子的意义及其新方法

冯培煜 宋瑞连

(北京市农林科学院玉米研究中心，北京 100097)

摘　要：玉米果穗中部的籽粒遗传势比两端的高，且中部籽粒以长方形或方形为主，两端粒形以接近圆形为主。差速循环式玉米种子籽粒分离精选机（专利号：ZL 2015 20032748, X）的设计理论是，利用玉米不同粒形的种子在重力作用下在运动的皮带上滚动速度不同（差），实现差速分离长、圆形种子，进一步精选种子。此项新技术不仅可提高玉米种子精选质量，降低加工成本，且对延长品种遗传势及品种使用生命周期都具有重大意义；本技术具有结构简单、使用方便、市场通用件多、制造成本低等优点，其应用范围可推广至其他籽粒形似长圆形与圆形的作物种子。

关键词：玉米种子；粒形；差速循环；精选机

玉米种子籽粒形状主要分为长方形、方形和近圆形。通常在玉米果穗上不同形状的籽粒分布为：生长于基部或顶部的为近圆形；生长于中下部及中上部的为长方形或方形；生长于中部的籽粒更深，更近似于长方形。一般玉米果穗籽粒越深出籽率越高，产量也越高。近些年，随着育种技术对玉米出籽率的要求不断提高，对生产种子用的自交系籽粒性状要求也越来越高，即要求中部籽粒的粒型更长。由于玉米果穗顶部籽粒的成熟度相对较差，质量小，比重轻，一般在经过风选、比重选后都能够被清选出去，但果穗基部的种子不容易分离。运用"差速循环式玉米种子籽粒分离精选机"（专利号：ZL 2015 2 0032748, X）可实现其分离精选，现将其实现意义及原理介绍如下。

1　玉米种子精选加工历史现状

较早实现玉米种子深加工、精加工的是美国和欧洲国家，距今已有100余年的历史，其技术设备先进，并且形成了垄断。国内的玉米种子生产加工清选分级设备大多是引进消化吸收国外的基础上发展而来，仅有50年时间，且主要是以种子籽粒的容重、质量、大小或颜色，利用风力、筛片的筛孔大小或色泽识别，进行风筛选、比重选或色泽选；这些加工技术与设备还不能从玉米种子籽粒的形态、形状上进行区分并分离达到精选。国外的窝眼筛筛选技术虽然能够实现玉米种子按粒形（长、圆）分离精选，因机械制造技术工艺高而复杂，设备造价高，推广困难；"差速循环式玉米种子籽粒分离精选机"从另一个新的技术理论角度，更经济地实现了按玉米种子粒形分离分级精选。

作者简介：冯培煜（1963—　），男，高级农艺师，主要从事玉米种子生产加工技术及管理工作
通讯作者：宋瑞连（1965—　），女，高级农艺师，主要从事玉米种子繁育与检验工作

2 主要工作原理及设计

差速循环式玉米种子籽粒分离精选的设计技术理论是：利用玉米不同粒形的种子在重力作用下在运动的皮带上的滚动速度不同（差），实现差速分离。设计实施，它包括入料斗、分离带和支架，所述入料斗安装在分离带上方的支架上，入料斗下部设有入料斗出口，料斗调节装置分别安装在入料斗的两侧，支架的两端分别安装有动力滚筒和从动滚筒，支架一端的侧面安装有变速电机，变速电机与动力滚筒驱动连接，支架底部还安装有支腿，支腿底部安装有调位装置，分离带为环状，并分别装在支架两端的动力滚筒和从动滚筒上，分离带上设有交错有序的八字型凸起，在分离带边缘设置有皱褶挡边。本技术的应用范围还可以推广至籽粒形似相近的长圆形与圆形的作物，如花生、蓖麻或豆类等，进行进一步分离精选。

3 技术实现的意义与作用

玉米果穗不同部位的籽粒对该品种特征特性的遗传传递能力是有差异的，中部的籽粒具有较强的遗传势，最能代表该品种的遗传特性，如不加选择地把自交系果穗的上、中、下部不同遗传势的种子混合作为扩繁用种，就会造成自交系种性退化，影响品种的一致性、丰产性，并导致其抗逆性减弱。孙庆海对鲁玉2号杂交种上、中、下部籽粒进行田间种植试验，结果表明：中部籽粒表现发芽快、幼苗健壮、生长势强、植株偏高、籽粒产量也高。傅兆麟选用玉米果穗中部和中下部籽粒做种，比全穗做种混用增产10%以上。北京农学院生物系孙祎振等研究试验也证明，玉米果穗上不同部位的籽粒具有不同的遗传势，对自交和杂交后代的产量性状影响很大，以中部和中部杂交后代产量最高，较群体平均值提高34.5%，不同遗传势的玉米籽粒混合扩繁，可以造成自交系退化；并建议采用玉米果穗中、下部的籽粒作为自交和杂交用种，以大幅度提高自交系和杂交种产量。李忠南等以先玉335的亲本及2个美系材料和自选系为试验材料，组配16个杂交组合，在保苗61 530株/hm²的条件下进行产量比较，同时分析籽粒水分、穗轴重、出籽率等相关数据，进行简单相关和遗传分析。结果表明，杂交种穗轴重与出籽率呈极显著负相关；出籽率与双亲平均穗轴重、父本穗轴重呈极显著负相关，与双亲平均出籽率、父本出籽率呈极显著正相关；百粒重与产量呈显著正相关，与籽粒水分呈显著负相关，与父本百粒重呈显著正相关；产量与父本百粒重呈极显著正相关，与双亲平均百粒重呈显著正相关。穗轴重、百粒重遗传累加效应极显著，超高亲效应明显；出籽率以遗传累加效应为主，存在超亲效应；建议组配新组合途径可以直接利用 PH 6WC 及 Reid 系统的美系材料与自选系组配，从而获得高出籽率的品种。因此研制以玉米种子籽粒的长、方、圆形状进行进一步分离分及精选的生产加工设备是科技与市场的迫切需求。

差速循环式玉米种子籽粒分离精选机是种子加工由清选、精选向更加精选的粒型选的技术升级，利用此项新型技术分离精选的玉米种子籽粒在粒重、形状上更均匀一致，有利于机械精量、单粒播种，降低播种空穴率，提高播种质量，从而有利于苗齐、苗匀和苗壮，有利于玉米种植增产增收。利用该技术精选出的果穗中部的长形种子繁殖的亲本或生产的杂交种，遗传能力强，有利于延长品种寿命期和增加产量。当杂交玉米种子生产制种中因自然等因素造成的父、母本花期相遇不良、母本接受父本花粉率降低、母本果穗结实性差、圆粒种子增多时，非本品种外的花粉或本品种母本的自身花粉授粉几率增大、结实率低的情况下，运用此实用新型加工精选技术，可提高种子的纯度。此外，本技术具有结构简单、使用方便、市场通用件多、制造成本低等优点。

参考文献（略）

玉米杂交制种测产方法及应用

冯培煜　宋瑞连

(北京市农林科学院玉米研究中心，北京　100097)

摘　要：测产是玉米杂交制种生产的重要环节，是基地技术员深入熟悉亲本特征特性的必要手段，其数据的精准与可靠，无论对改进生产技术，还是提高制种产量以及提高企业效益均具有现实意义。测产与预产、估产是有区别的；测产是利用仪器或工具来度量，即测绘、测量、测控得出的产量数据，其主要方法步骤有测产前准备、产量层级划分及测产取样点数（位）确定、田间测产取样、室内考种及汇总等。

关键词：玉米杂交制种；测产；样点；平均数

测产是玉米杂交制种生产中不可缺少的一个环节，是技术员深入熟悉亲本特征特性的必要手段，其数值的精准与可靠，无论对改进生产技术，提高制种产量，提高企业效益，还是对国家宏观决策，对高产、超高产制种产量验收的技术评价都有着现实意义和作用。在工作实践中它常与预产、估产相混淆。

1　基本含义及其与预产、估产、实产的关系

1.1　基本含义

测产：利用仪器或工具来度量，即测绘、测量、测控，得出的产量数值。预产：是在玉米种子落实生产之前的计划或对正在生产阶段中，凭经验、历史数据或田间感观对产量（实产）有预先、事先的预计、预见。估产：对玉米制种田产量（实产）的揣测，大致地推算，估计。实产：符合客观真实的收获产量。

1.2　测产预产、估产及实产的相互关系

从生产的起始过程看，预产、估产、测产、实产是计划生产开始到生产结束各环节对产量信息的及时把控，是产量逐渐趋真、趋实、趋于精准的产量表达的技术术语。相对与实产，预产的误差值比估产、测产都要大，测产的数值结果应最接近实产的数值；相对测产，预产、估产两者的含义更接近；在生产实践中，预产、估产、测产常常被群众混用。

2　测产前工作准备

2.1　测产时间

一般在玉米制种母本蜡熟期末进行。

2.2　测产前的基本资料掌握

准确掌握制种田的种植区域、规模、面积、品种、父本与母本种植（行比）形式以

作者简介：冯培煜（1963—　），男，本科，高级农艺师，主要从事种子生产及加工技术研究与推广

及栽培管理等条件。

2.3 人员组织

一般由生产基地负责人具体负责本基地（村）的测产工作，组织 2~3 人，将记录、丈量等环节分工协作。

2.4 工具准备

卷尺（>15m）、秤、笔、网袋、记录表（本）、标签、水分测试仪等。

2.5 登记表样式

登记表样式参照表 1。

召开测产准备、培训工作会议。

表 1 品种制种测产田间取样登记表（单位：667m^2、cm、kg）

产量层级	代表面积	点次	父母本行比	行距	株距	样点母本密度	测产区样本密度	20株鲜穗重	标准水分籽粒重量	测产区单产	百粒重	备注
高级产田		01										
		02										
		03										
	平均											
	本级合计											
中级产田		01										
		02										
		03										
	平均											
	本级合计											
低级产田												
测产区平均												
测产区总计产量												

3 产量层级划分及测产取样点数（位）确定

3.1 划分产量层级

在同一个村或同品种"百亩""千亩"的规模化制种区内，不同地块的地力和生产水平是不同的，一般在正式田间取样测产前，应全面探查所测产区域内的不同地力、地块的分布情况基础上，依据不同地力、不同种子田间长相、不同生产水平的地块进行层次分级。如，按种子田间的长相、长势以及产量性状分高产、中产、低产3级或更多层分级。划分层级原则是，测产区域内的地块种子长相、长势、产量性状均匀一致的划为同一层级，长相、性状越均匀层级数越少，面积越小层级数应越少。

3.2 确定取样点位及点数

当划分出不同生产水平层级并确定面积及地块分布后，再根据不同地块的形状确定取样点位置及样点数，样点位置可采用梅花、对角线、十字形、S弯形等方式确定；取样点数，一般同一层级的不低于3个点次，面积大可适当增加点次。注意，取样点应避开道路、水渠、地埂边缘，以及树阴、电线杆、水井等对产量有影响的设施或物体，以防样点间测产数据出现大的偏差。

4 田间测产取样方法及步骤

4.1 确定样点及密度测量

4.1.1 确定取样起始点

在进入预定所取样点区域的地块时，一般远离道路、地埂、水渠地边边行10行或行株数20株，并实际数到10行（20株）止，以下一行（株）为本样点开始测量的第1（行）株，即确定为本样点的实测及取样起始株（点）。注意，数到父本行时，应以下一行母本为起始行，同样连续顺行数20株，以第21株为取样点起始株。

4.1.2 株距测量

以样点起始株顺行丈量到51株，测得长度（cm），计算株距。

株距（cm）= 1~51株之间长度（cm）/（51-1）。

4.1.3 行距测量

以起始行丈量到第21行，测量宽度（cm），计算株距。

株距（cm）= 21行间宽度（cm）/（21-1）。

4.1.4 样点母本密度计算

$$母本密度（株/667m^2）= \frac{667m^2 \times 10\,000}{株距（cm）\times 行距（cm）}$$

4.1.5 制种测产区样点密度的计算

单设父本行的计算公式：

$$密度（株/667m^2）= \frac{667m^2 \times 10\,000}{株距（cm）\times 行距（cm）} \times \frac{母本行数}{母本行数+父本行数}$$

没有单设父本行的计算公式：

$$母本密度（株/667m^2）= \frac{667m^2 \times 10\,000}{株距（cm）\times 行距（cm）}$$

4.2 取样

以样点起始株为第1株顺行连续取20株的果穗（单穗、双穗或空秆都应计入样品株数）并称重，而后将记录有和登记表信息（样点层级、点次、地块、鲜重等）一致的标签同样本一同放入网袋并扎好，带回晾干、考种。当同一产量层级的样点之间，产量水平差异大时，一般相差 5 以上时，应加大取样密度或复测取样点；如复测结果和第 1 次测产结果相近，也可将此地块单设为一个重点产量层级，并进行重点测产和考种。

4.3 其他需田间调查的内容

空株率、倒伏株率、双穗株率、单株果穗率、折断株率、发病率、株高、穗位等，均以样点起始株为起点，连续数取 50 株或 100 株，以实际调查折算为百分率并记录。

5 室内考种及汇总

5.1 基本穗部性状考种

样品晒干后，从样品（点）内随机连续选取 10～20 个果穗或全部果穗，调查穗长、秃顶长度、籽粒行数及每行籽粒数等性状。

5.2 实测样点重量、出籽率及百粒重

各样点果穗晒干后，称其果穗重（g）；脱粒并筛簸干净，称其籽粒重量（g）；随机数 500 粒，称重并折算成百粒重。脱粒后所测得数据，一定要用水分仪测得种子含水率，并将数值折算成标准水分的种子重量后才可计算产量。

5.3 产量结果计算

5.3.1 测产区样点单产（$kg/667m^2$）

$$样点单产（kg/667m^2）= \frac{样点（20 株）籽粒产量（g）}{20 \times 1\,000} \times 测产区样点密度（株/667m^2）$$

5.3.2 同一层级的产量计算

本层级平均单产（$kg/667m^2$）= 本层级内的各样点单产（$kg/667m^2$）之和/本层级内的样点数。

注意：本层级平均单产是用算数平均数计算方法。

本层级总产（kg）= 本层级平均单产（$kg/667m^2$）× 本层级代表面积（$667m^2$）

5.3.3 测产区的总产（各层级总产之和 kg）计算

测产是实测的产量，不是用百粒重×穗粒数×穗数×面积，计算出来的，有时所得 2 个结果极其相近；不能以往年（历史）的百粒重用于当年测产计算数据。

5.3.4 测产区的平均单产计算

测产区平均单产（$kg/667m^2$）= 测产区的总产（kg）/测产区总面积（$667m^2$）。

注意：测产区的平均单产是用加权算数平均数计算。

5.3.5 测产区产量的矫正

一般因田间受畦渠、地埂、井台、树阴等影响，测得的产量一般比实际产量高出 5% 以上，为此，测产数据要减掉 5%（校正系数），即测产区校正后总产（kg）= 测产区的总产（kg）×（1-5%）。

6 撰写测产报告

本文原载：种子，2017，36（3）：132-134

适贮条件下不同基因型玉米种子活力及生理特性研究

张海娇[1]　成广雷[2]　赵久然[2]　刘春阁[2]　王晓光[2]　高春双[3]　赵静峰[4]

(1. 北京农学院植物科学技术学院，102206；2. 北京市农林科学院玉米研究中心，100097；3. 内蒙古林西县农业局，025250；4. 吉林省梨树县梨树镇农业站，136500)

摘　要：以5个不同基因型玉米种子为试验材料，设置种子含水量11%、贮藏温度25℃的适贮条件，贮藏一年时间，每月测定比较供试材料的生活力及生理变化。结果表明：各基因型玉米种子在适贮条件下随着贮藏时间的延长，发芽势、发芽率、发芽指数、活力指数、可溶性蛋白及可溶性糖含量、脱氢酶活性均出现了不同程度的下降，种子电导率及丙二醛含量均出现不同程度的升高。贮藏一年后各基因型玉米种子的发芽率仍保持在较高水平，但活力指数下降幅度较大，活力指数可作为反映种子劣变及老化程度的重要指标。同一贮藏条件下，基因型是影响生理特性的决定因素，参试材料京科968种子生活力及生理指标均变化较小，表现出较强的稳定性和耐贮性。

关键词：玉米；种子贮藏；种子活力；种子生理

中国是世界第一用种大国和第二大种子消费国，种子在农业生产中占有举足轻重的地位。但种子从成熟收获到生产应用都或长或短的经历一段贮藏时间，种子含水量、贮藏温度、贮藏时间等因素均会对种子质量造成一定影响，直接关系到生产的成效与收益。玉米种子作为产业化程度最高的作物，市场大，生产贮藏量多，每年都面临巨大的贮藏损失。玉米种子贮藏是种子产业化的重要一环，好的贮藏条件可以极大程度地降低对种子的伤害，减少对种子活力的不利影响。种子贮藏的根本目的是使种子保持旺盛的生命力，具有良好的播种质量。而玉米种子的储藏环境直接影响种子的生命力。对适贮条件下不同基因型玉米种子的活力及生理特性进行系统研究，为筛选优良的耐贮性玉米品种，了解玉米种子耐贮性生理生化机制，改善玉米种子贮藏技术提供理论依据，不仅具有较高的学术价值，而且具有广阔的市场前景，必将产生巨大的经济和社会效益。

1 材料与方法

1.1 试验材料

采用国内玉米生产中种植面积较大、有一定代表性的5个品种郑单958、农大108、先玉335、正大619及京科968等作为试验材料。

1.2 试验设计

试验在北京市农林科学院玉米研究中心实验室内进行，测定5种基因型玉米种子的初

基金项目：北京市博士后科研经费资助项目、北京市农林科学院博士后基金

通讯作者：成广雷，赵久然

始含水量,将每种基因型玉米种子回湿至11%,再将其装入塑封袋放入25℃的恒温培养箱中贮藏一年。每隔1个月测定其发芽势、发芽率、发芽指数、活力指数、电导率以及丙二醛含量(MDA浓度)、可溶性蛋白含量、可溶性糖含量、脱氢酶活性等。

1.3 测定方法

采用烘干减重法测定参试基因型玉米种子初始含水量;将种子在高温、高湿环境下进行回湿处理,将处理后的种子进行生活力及生理指标测定。按照农作物检验规程(GB/T 3543—1995),采用砂培法进行发芽试验;丙二醛含量测定采用硫代巴比妥酸方法;利用考马斯亮蓝法测定可溶性蛋白含量;蒽酮乙酸乙酯法测定可溶性糖含量;氮蓝四唑染色法测定脱氢酶含量。采用 Sigmaplot 11.0,DPS 3.01,Microsoft Excel 2003 进行数据统计与分析。

2 结果与分析

2.1 不同基因型玉米种子初始状况

对不同基因型玉米种子初始含水量、生活力及相关生理指标测定的结果如表1所示,结果表明:不同基因型玉米种子的初始含水量均处于较低水平,在7.394 4%~8.705 7%。在初始状态下,各基因型玉米种子的发芽势及发芽率均在90%以上,具有较强的萌发能力。先玉335的电导率最高,农大108及正大619具有较高的MDA浓度,郑单958具有较高的可溶性蛋白及可溶性糖含量,脱氢酶活性也较高,各基因型间生理指标差异较大,且没有明显的规律性。

表1 不同基因型玉米种子初始指标鉴定

	郑单958	农大108	先玉335	京科968	正大619
初始含水量(%)	7.394 4	7.687 0	8.705 7	8.237 9	7.821 6
发芽势(%)	96.666 7	97.333 3	93.333 3	97.333 3	90.000 0
发芽率(%)	99.333 3	98.000 0	94.000 0	97.333 3	90.666 7
发芽指数	53.111 9	52.855 6	50.892 1	52.630 2	64.904 0
活力指数	5.850 8	4.955 0	4.167 4	5.236 7	3.759 9
电导率(μs/cm·g)	2.368 2	3.186 0	3.446 6	3.010 6	3.236 7
MDA浓度(μmol/L)	1.228 1	1.660 3	1.306 3	1.198 1	1.688 9
可溶性蛋白含量(mg/g)	7.577 2	6.324 0	5.407 4	4.479 2	5.733 4
可溶性糖含量(%)	11.370 7	7.134 4	10.458 2	8.272 8	8.241 2
脱氢酶含量	0.547 3	0.304 7	0.459 3	0.682 5	0.465 0

2.2 不同基因型玉米种子贮藏过程中生活力的变化

通过比较在适贮条件下玉米种子的生活力变化(图1),结果表明:随着贮藏时间的延长,各基因型的玉米种子生活力均出现不同程度下降,但不同基因型玉米种子发芽势、发芽率的高低与发芽指数的高低并不一定呈正比,初始发芽势、发芽率较低的正大

619 发芽指数明显高于初始发芽势、发芽率较高的郑单 958。各基因型玉米种子在贮藏过程中发芽指数、活力指数与发芽势、发芽率变化规律及趋势基本一致，但变幅差异较大，各基因型玉米种子在含水量 11%、贮藏温度 25℃条件下贮藏一年发芽势、发芽率变化较小，但活力指数下降幅度很大。

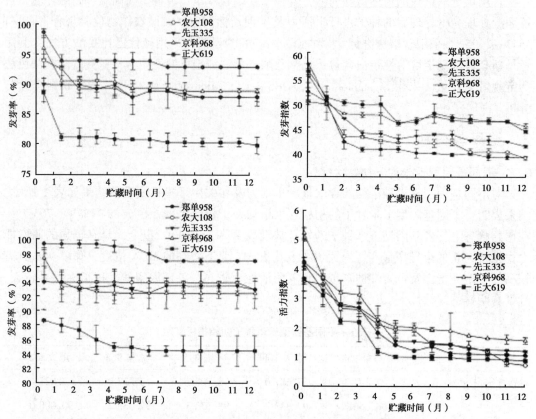

图1 适贮条件下不同基因型玉米种子各阶段生活力比较

2.3 适贮条件下不同基因型玉米种子膜透性及脂质过氧化变化

通过比较在适贮条件下玉米种子的电导率及丙二醛变化（图2），结果表明：随着贮藏时间的延长，不同基因型的玉米种子电导率及丙二醛含量均出现了不同程度的上升，但升高幅度均不大。其中农大 108 的电导率上升幅度最小，郑单 958 前 4 个月电导率升高较快，电导率上升幅度最大。农大 108 及正大 619 丙二醛含量明显高于其他基因型玉米种子，郑单 958、先玉 335 及京科 968 丙二醛含量差异不明显，在适贮条件下各基因型玉米种子的丙二醛含量上升趋势较为一致。

2.4 适贮条件下不同基因型玉米种子有关贮藏物质的变化

比较适贮条件下不同基因型玉米种子的可溶性蛋白及可溶性糖含量变化（图3），结果表明：随着贮藏时间的延长，不同基因型的玉米种子可溶性蛋白及可溶性糖含量均出现了不同程度的下降。各基因型玉米种子的可溶性蛋白含量在第一个月下降较快，之后趋于平缓。郑单 958、先玉 335 在贮藏前 2 个月可溶性糖含量下降明显，其他玉米种子在适贮条件下的可溶性糖含量下降趋势较缓慢。参试材料中郑单 958 可溶性蛋白含量明显高于其

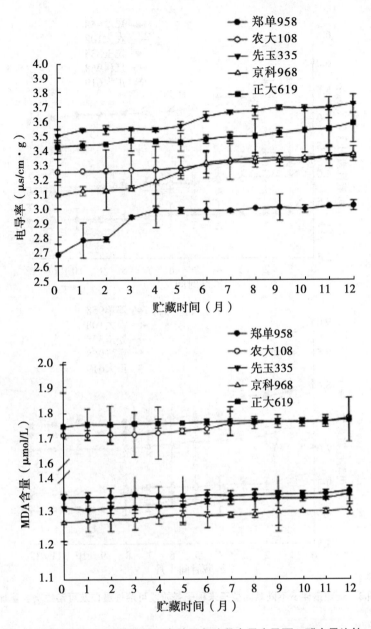

图2 适贮条件下不同基因型玉米种子各阶段电导率及丙二醛含量比较

他基因型玉米种子,京科968在贮藏过程中可溶性蛋白及可溶性糖含量下降幅度均最小。

2.5 适贮条件下不同基因型玉米种子脱氢酶活性变化

通过比较在适贮条件下玉米种子的脱氢酶活性变化(图4),结果表明:随着贮藏时间的延长,不同基因型的玉米种子脱氢酶活性均出现了不同程度的下降,脱氢酶含量由大至小为:郑单958>京科968>先玉335>正大619>农大108。农大108及先玉335的脱氢酶含量在贮藏前5个月下降较快,之后趋于平缓,其他基因型玉米种子下降较慢。

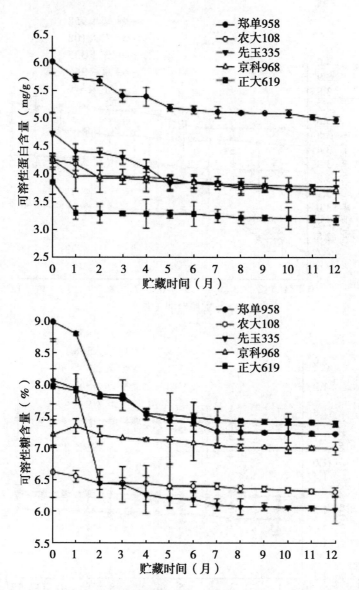

图3 适贮条件下不同基因型玉米种子各阶段可溶性蛋白及可溶性糖含量比较

3 结论与讨论

研究表明随着贮藏时间的延长，不同基因型玉米种子发芽势、发芽率、发芽指数及活力指数均出现不同程度的下降。不同基因型玉米种子各阶段发芽势和发芽率升降规律及趋势一致，呈显著正相关关系。各基因型玉米种子在贮藏过程中发芽指数、活力指数与发芽势、发芽率变化规律及趋势基本一致，但变幅差异较大，各基因型玉米种子在含水量11%、25℃条件下贮藏一年发芽势、发芽率变化较小，但活力指数下降幅度很大。本研究中，各基因型玉米种子活力变化总体规律与可溶性糖、可溶性蛋白、脱氢酶活性均呈正相关关系，与电导率、丙二醛均呈负相关关系，这与前人研究基本一致。

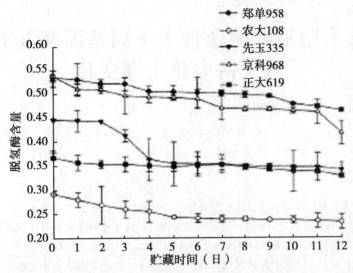

图 4 不同贮藏条件下不同基因型玉米种子各阶段脱氢酶含量比较

　　电导率、可溶性糖含量、可溶性蛋白含量、丙二醛含量、脱氢酶活性可以作为种子活性和劣变程度等生理指标的研究内容，但种子含水量、贮藏温度、贮藏时间对种子生活力及生理指标均产生不同程度的影响，玉米种子耐贮性生理机制较为复杂，且各基因型间差异较大，在贮藏生理指标上很难制定统一的评价标准。种子在失去发芽力之前，已发生了劣变，活力指数和发芽指数将种子生活力趋势进行了放大，故二者比发芽率能更加灵敏地反映种子的劣变程度及老化程度，但是均不能直接反映田间种子的出苗情况，种子发芽率、发芽势仍然是表现种子生活力最直接和稳定的指标。同一贮藏条件下，基因型是影响生理特性的决定因素，综合试验指标情况，参试材料中京科968种子生活力及生理指标均变化较小，表现出较强的稳定性和耐贮性。

参考文献（略）

本文原载：中国种业，2014（5）：42-45

临界胁迫贮藏条件下不同基因型玉米种子活力及生理变化

成广雷　张海娇　赵久然　刘春阁　王元东　王晓光
王荣焕　陈传永　徐田军

（北京市农林科学院玉米研究中心，北京　100097）

摘　要：【目的】了解不同基因型玉米种子在临界胁迫贮藏条件下的生活力及生理变化规律，对不同基因型玉米种子的耐贮性做出客观评价，为玉米种子耐贮性生理生化机制提供理论依据。【方法】选用目前主栽玉米杂交种郑单958、农大108、先玉335、正大619、京科968种子为试验材料，分别测定供试材料的初始含水量、活力水平及相关生理指标。将供试材料的种子含水量回湿至14%，在种子含水量14%和贮藏温度35℃（14% & 35℃）的临界胁迫贮藏条件下，贮藏一年时间，每月测定种子发芽势、发芽率、发芽指数、活力指数、电导率、MDA浓度、可溶性蛋白含量、可溶性糖含量及脱氢酶活性。采用砂培法进行发芽试验，用DDS-ⅡA型电导率仪测定种子电导率，利用硫代巴比妥酸（TBA）溶液法测定丙二醛含量，可溶性蛋白含量的测定利用考马斯亮蓝 G-250 法，可溶性糖含量的测定利用蒽酮比色法，脱氢酶活性的测定采用氯蓝四唑（TTC）染色法。比较不同基因型玉米种子在临界胁迫贮藏条件下的活力及生理指标变化。【结果】在同一贮藏条件下，基因型是影响种子活力及生理特性的决定性因素。参试玉米种子的初始含水量均处于较低水平，为 7.39%~8.71%。在初始状态下，各基因型玉米种子的发芽势及发芽率均在 90% 以上，具有较强的萌发能力。贮藏一年后，京科 968 发芽率、发芽势为 50%~60%，发芽指数为 25%，种子活力指数为 0.3；农大 108、先玉 335、正大 619 发芽率、发芽势为 15%~25%，发芽指数 8%~25%，活力指数 0~0.08；郑单 958 发芽率、发芽势、发芽指数活力指数在贮藏 1 年后均下降为 0。临界胁迫贮藏条件下不同基因型玉米种子随着贮藏时间的延长，发芽势、发芽率、发芽指数、活力指数与可溶性蛋白含量、可溶性糖含量、脱氢酶活性均出现不同程度的下降，种子可溶性蛋白含量和可溶性糖含量的变化与种子活力的变化关系密切。发芽势、发芽率、发芽指数、活力指数的变化趋势及规律基本一致，但变幅差异较大，不同基因型玉米种子因发芽速率不同发芽指数差异较大，而种子活力指数的降低总是先于种子发芽率的下降，可以真实体现种子的老化及劣变程度。不同基因型玉米种子随着贮藏时间的延长，膜透性变差，种子电导率及丙二醛含量均出现了不同程度的升高，其变化趋势及种子生活力变化呈负相关关系（R^2 = 0.752），各基因型间细胞膜系统功能存在差异。在本试验中，不同基因型玉米种子活力与丙二醛含量无显著相关性（R^2 =-0.171~-0.094），与可溶性蛋白含量、可溶性糖含量以及脱氢酶活性呈显著正相关（R^2=0.284~0.517），但各基因型间玉米种子生理变化机制复杂，差距较大。【结论】不同基因型玉米种子对临界胁迫贮藏条件表现均较敏感，但基因型间存在较

基金项目："十二五"农村领域国家科技计划（2011AA10A103）、北京市科技计划（D12110500350000）、北京市博士后科研经费资助项目（2012ZZ-83）、北京市农林科学院博士后基金、国家星火计划（2013GA600001）
作者简介：成广雷，E-mail：cglseed@163.com。通讯作者赵久然，E-mail：maizezhao@126.com

大差异,在临界胁迫贮藏条件下,郑单958的活力下降最快,京科968活力及各项生理指标变化相对稳定,表现出较强的耐贮性。

关键词: 玉米;临界胁迫;种子活力;种子生理;耐贮性

Vigor and Physiology Changes of Different Genotypes Maize Seed (*Zea mays* L.) Under Critical Stress Storage Condition

Cheng Guanglei Zhang Haijiao Zhao Jiuran Liu Chunge Wang Yuandong
Wang Xiaoguang Wang Ronghuan Chen Chuanyong Xu Tianjun

(*Beijing Academy of Agriculture and Forestry Sciences*, Beijing, 100097)

Abstract:【Objective】This experiment was carried out to understand the viability and physiological variation rule of different genotypes maize seeds under critical stress storage conditions, evaluate its of storability objectivity, provide a theoretical basis for clearing the physiological and biochemical mechanisms of maize seed storability. 【Method】Major maize hybrids (ZD958, ND108, XY335, ZD619, JK968) were used as experimental materials. Seed initial moisture content, vigor and physiological indicators was measured. The moisture content of experimental materials wet back to 14% and stored at temperature of 35℃, for one year, then the germination energy, germination rate, germination index, vigor index, electrical conductivity, MDA concentration, soluble protein and soluble sugar content and dehydrogenase activity of materials were measured every month. The germination experiment was conducted by using sand culture method, DDS-ⅡA conductivity meter was used to measure the seed electrical conductivity, the TBA was used to measure the MDA content, coomassie brilliant blue G-250 method was used to analyze the soluble sugar content, and dehydrogenase activity was determined by TTC method. The vigor and physiological indicators change of different genotypes maize seeds were compared under the critical stress storage conditions. 【Result】Seed genotype was the determinative factor under the same storage condition. The initial moisture content of experimental materials was at a low level, between 7.39% and 8.71%. The germination potential and germination rate of different genotypes maize seed were over 90%, showing a stronger germination ability. After storage for one year, the germination potential and germination rate of JK968 was 50%~60%, germination index was 25%, and vigor index was 0.3. The germination potential and germination rate of ND108, XY335, ZD619 was 15%~25%, germination index was 8%~25%, and vigor index was 0~0.08. The germination potential, germination rate, germination index and vigor index of ZD958 decreased to 0 after storage for one year. Different degrees in decline of germination potential, seed soluble protein content and soluble sugar content changes showed a close relationship with seed vigor. Germination rate, germination index, vigor index were almost the same, however, there was a number of differences in amplitude of variation. Different germination speeds lead to different germination indexes of each genotype of maize seed, seed vigor index declined was always ahead of seed germination ability, which reflected the degree of real seed aging and deterioration. With the storage time extension, the membrane permeability variation of different genotypes of maize seeds, electrical conductivity and MDA content were increased and the change trend was negatively related with seed vigor

($R^2 = 0.752$). A membrane permeability difference existed among different genotypes of maize seeds. The vigor of different genotypes of maize seed had no significant correlation with the MDA ($R^2 = -0.171 \sim -0.094$), it significantly positively correlated with soluble protein, soluble sugar and dehydrogenase ($R^2 = 0.284 \sim 0.517$), but the mechanism of physiological changes of different genotypes of maize seed was complex, the gap was bigger. 【Conclusion】 Different genotypes of maize seed were more sensitive to critical stress storage condition, ZD958 was sensitive to critical stress, the vigor and physiological indicator changes of JK968 were stable, showed a higher storability.

Key words: Maize; Critical stress; Vigor; Seed physiology; Storability

引言

【研究意义】玉米种子是产业化程度较高的作物,市场大,生产贮藏量多,每年都面临较大的贮藏损失。另外,随着农业机械化程度的提高,单粒播种面积迅速扩大,但作为单粒播种的种子,对种子活力和发芽率要求很高,这对种子贮藏提出了更高的要求。对临界胁迫贮藏条件下不同基因型玉米种子的活力及生理特性进行系统研究,进一步了解玉米种子耐贮性生理生化机制,不仅具有较高的学术价值,而且具有广阔的市场前景。【前人研究进展】种子贮藏特性(seed storage behavior)是指由物种遗传基因所决定的种子贮藏生理特性,综合表现在不同物种的种子对贮藏前及贮藏期间环境的需求及适应能力的不同。贮藏温度、贮藏时间和种子含水量一直被认为是影响种子寿命的主要因素。Ellis 和 Hong 认为种子含水量与种子密封贮藏条件下的温度存在对数曲线关系,降低含水量可以适当的增加贮藏温度。Barton 等指出,水分对种子劣变极其重要,种子的劣变过程随着水分的增加而愈趋严重。种子含水量少,生理活性微弱,呼吸缓慢,能量消耗维持在最低水平。大多数情况下,温度和含水量越低,种子的生活力保存的越长。陈晓玲等认为种子的耐贮性与种子的活力有关,近年来,许多学者用高温高湿人工老化的方法对许多种子劣变的生理生化规律进行了研究,其中包括洋葱、小麦、油菜、水稻、苜蓿、花生等种子。高温高湿处理能在较短时间内使种子的膜透性增加、脂质过氧化产物含量升高、酶的活性降低、蛋白质和核酸的合成能力下降、有毒代谢物积累、贮藏养料消耗等。Elena 研究认为种子在贮藏期间会积累有毒物质,膜结构受到损伤,脂质过氧化是导致种子劣变的主要原因之一。王煜等研究认为,随着贮藏时间的延长,油菜种子可溶性蛋白含量均有不同程度的下降,且品种间存在差异。伴随着贮藏时间的增加,呼吸作用不断分解可溶性糖,导致总糖含量随着贮藏时间的增加而不断下降。Sheikh 等对花生种子的可溶性糖含量研究中发现,贮藏过程中,温度对可溶性糖含量的影响不大,但是随着贮藏种子含水量增高,可溶性糖含量逐渐下降。张永娟等研究认为脱氢酶活性对老化处理反应比较敏感。由于种子的活力高低直接决定了贮藏过程中的生理变化,因此不同材料经过相同的处理后,种子活力越高,则耐贮性越好。【本研究切入点】过去的研究大都通过人工老化使种子快速劣变测定种子的活力和生理变化指标,本研究立足实际应用,模拟现实商品种子的贮藏状况设置临界胁迫贮藏条件,通过对影响玉米种子萌发及贮藏的主要因素——贮藏温度、种子水分的控制,分阶段对不同基因型玉米种子活力及生理生化指标进行测定比较。【拟解决的关键问题】旨在了解不同基因型玉米种子在临界胁迫贮藏条件下的活力差异,以期为明确保持玉米种子活力的贮藏方法,改善玉米种子贮藏技术提供理论依据。

1 材料与方法

1.1 供试材料

试验于2012—2013年在北京市农林科学院玉米研究中心实验室内进行。本试验采用目前国内主栽的5个玉米杂交种，郑单958、农大108、先玉335、正大619及京科968等作为试验材料，试验材料来源于甘肃制种基地收获的当年种子。

1.2 试验设计

基于GB 4404.1—2008规定和实践经验，长城以南（高寒地区除外）玉米种子水分不得高于13%，且在0~30℃贮藏范围内，温度每增高5℃，种子安全贮藏水分就相应降低1%。因此本试验设定种子含水量14%，贮藏温度35℃为玉米种子的临界胁迫贮藏条件。在试验中将各基因型玉米种子的含水量回湿至14%，再将含水量为14%的玉米种子塑封后放入35℃的恒温培养箱中贮藏一年。每隔1个月分别测定其发芽势、发芽率、发芽指数、活力指数、电导率以及丙二醛含量、可溶性蛋白及可溶性糖含量、脱氢酶活性等。

1.3 测定项目与方法

种子含水量的测定采用烘干减重法，取试样15g放铝盒内，盖盖称重后放入130℃烘箱中，直至温度恒定不变，称重。对供试材料进行回湿处理，将种子置于尼龙网袋中，放入由饱和$CaCl_2$溶液、饱和NH_4Cl溶液以及水所造成的相对种子含水量环境的干燥器中，密封，将回湿至14%的种子装入密封袋至35℃下恒温培养箱中保存，待用。

按照农作物检验规程（GB/T 3543—1995），采用沙培法进行发芽试验。第4d进行发芽势测定，第7d测定发芽率。7d后将根取出，烘干，称取干重。计算发芽势、发芽率、发芽指数、活力指数。具体计算公式：发芽势（%）=第4d正常发芽种子数/供试种子总数×100，发芽率（%）=第7d正常发芽种子数/供试种子总数×100，发芽指数=Σ第n d正常发芽种子数/相应发芽天数，活力指数=发芽指数×根重。用DDS-ⅡA型电导仪测定种子电导率，以纯无离子水作为对照。利用硫代巴比妥酸（TBA）溶液法在450nm、532nm、600nm处的吸光值，计算MDA含量。利用蒽酮比色法测定可溶性糖含量。脱氢酶活性测定采用氮蓝四唑（TTC）法，测定485nm OD值。

1.4 统计分析

采用DPS3.01，Microsoft Excel 2003进行数据统计与分析。

2 结果与分析

2.1 不同基因型玉米种子初始状况

不同基因型玉米种子初始含水量、生活力及相关生理指标测定结果的统计分析表明（表1）：不同基因型玉米种子的初始含水量均处于较低水平，在7.39%~8.71%。在初始状态下，各基因型玉米种子的发芽势及发芽率均在90%以上，具有较强的萌发能力。各基因型玉米种子的电导率差异不显著，先玉335的电导率最高；农大108及正大619的MDA浓度显著高于其他基因型玉米种子；郑单958具有较高的可溶性蛋白及可溶性糖含量，脱氢酶活性也较高，总体来说，在供试材料中郑单958的初始活力处于较高水平，正大619的活力水平相对较低。各基因型间生理指标差距较大，没有明显的规律性。

2.2 不同基因型玉米种子贮藏过程中生活力变化

2.2.1 发芽势

发芽势是计算播种量的因子之一，也是检测种子质量的重要指标之一。比较临界胁迫贮藏条件下玉米种子的发芽势变化表明（图1），随着贮藏时间的延长，参试品种的发芽势均出现不同程度的下降，郑单958发芽势下降幅度最大，贮藏第11个月后发芽势降为0。农大108、先玉335、正大619的发芽势下降趋势相对一致。参试品种中京科968的发芽势较为稳定，下降幅度较小，贮藏6个月后，发芽势下降趋势趋于平缓，表现出了突出的耐贮性。

表1 不同基因型玉米种子初始含水量、生活力及生理指标鉴定
Table 1 Identification of the initial moisture content, vigor and physiological indicators of different genotypes of maize seeds

品种 Varieties	初始含水量 Initial moisture content (%)	发芽势 Germination potential (%)	发芽率 Germination rate (%)	发芽指数 Germination index	活力指数 Vigor index	电导率 Electrical conductivity [μs/(cm·g)]	丙二醛 MDA (μmol/g)	可溶性蛋白 Soluble protein (mg/g)	可溶性糖 Soluble sugar (%)	脱氢酶 Dehydrogenase [μgTTF/(g·h)]
郑单958 ZD958	7.39dD	98.67aA	99.33aA	53.11bB	5.85aA	2.37aA	1.23bB	7.58aA	11.37aA	0.55bAB
农大108 ND108	7.69cCD	97.33abA	98.00aA	52.86bB	4.96abA	3.19aA	1.66aA	6.32abAB	7.13bA	0.30cC
先玉335 XY335	8.71aA	93.33bcAB	94.00aAB	50.89bB	4.17abA	3.45aA	1.31bAB	5.41bcAB	10.46aA	0.46bB
京科968 JK968	8.24bB	97.33abA	97.33aA	52.63bB	5.24abA	3.01aA	1.20bB	4.48cB	8.27abA	0.68aA
正大619 ZD619	7.82cC	90.00cB	90.67bB	64.90aA	3.76bA	3.24aA	1.69aA	5.73bcAB	8.24abA	0.47bB

注：同列数据后对应的大、小写字母分别表示在α=1%、5%水平（SSR法比较）差异显著

Note: Different capital and lowercase letters in the same column indicates the α=1% and 5% level differences significant by SSR, respectively

2.2.2 发芽率

种子发芽率是国家标准中规定的种子检验的4项指标（净度、发芽率、纯度、含水量）之一，是影响田间出苗率的内在因素。由图2可知，随着贮藏时间的延长，临界胁迫贮藏条件下的玉米种子发芽率出现了不同程度的下降，尤以郑单958下降最为显著。各基因型间发芽率下降规律及趋势基本与发芽势一致，呈显著正相关。农大108、先玉335、正大619在14%&35℃条件下发芽率下降明显。京科968在贮藏5个月后发芽率下降趋势趋于稳定，在临界胁迫贮藏条件下表现出较高且稳定的萌发能力。

2.2.3 发芽指数

发芽指数是发芽率指标的细化和深化，它放大了种子活力的特征。临界胁迫贮藏条件下玉米种子的发芽指数变化表明（图3），不同基因型玉米种子发芽势、发芽率的高低与

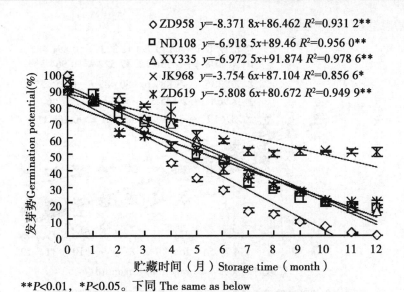

**P<0.01，*P<0.05。下同 The same as below

图1 临界胁迫贮藏条件下不同基因型玉米种子的发芽势变化

Figure 1 Germination potential changes of different genotypes maize seed under critical stress storage conditions

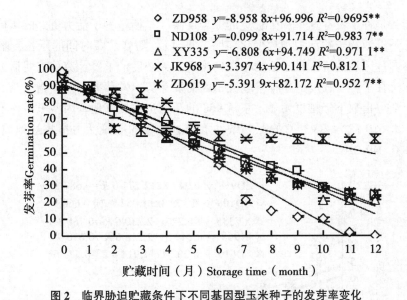

图2 临界胁迫贮藏条件下不同基因型玉米种子的发芽率变化

Figure 2 Germination rate changes of different genotypes maize seed under critical stress storage conditions

发芽指数的高低并不一定呈正比，初始发芽势、发芽率低的正大619发芽指数明显高于初始发芽势、发芽率较高的郑单958。随着贮藏时间的延长，临界胁迫贮藏条件下的不同基因型的玉米种子发芽指数均出现了不同程度的下降，与发芽势、发芽率贮藏变化规律及趋势基本一致。农大108在贮藏初期，发芽指数下降缓慢，从第3个月开始，下降幅度明显。京科968发芽指数在临界胁迫贮藏条件下，贮藏前5个月发芽指数下降较快，之后趋于平缓，表现出较高的稳定性。

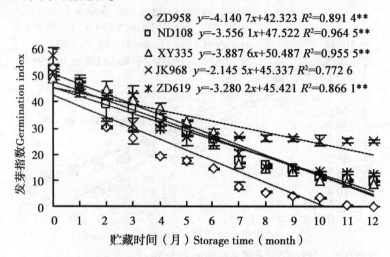

图3 临界胁迫贮藏条件下不同基因型玉米种子的发芽指数变化

Figure 3 Germination index changes of different genotypes maize seed under critical stress storage conditions

2.2.4 活力指数

活力指数是种子发芽速率和生长量的综合反映，是衡量种子活力更好的指标。临界胁迫贮藏条件下玉米种子的活力指数变化表明（图4），随着贮藏时间的延长，临界胁迫贮藏条件下不同基因型的玉米种子活力指数均出现了大幅度的下降，充分表明活力指数是一个很好的评价种子劣变程度的指标。农大108的发芽势、发芽率、发芽指数及活力指数互为正相关，活力指数下降幅度明显。京科968的活力指数较其他基因型玉米种子更高。郑单958及正大619在14%&35℃的贮藏条件下经过一年的贮藏活力指数达到或接近于0，分别为0和0.049 226。

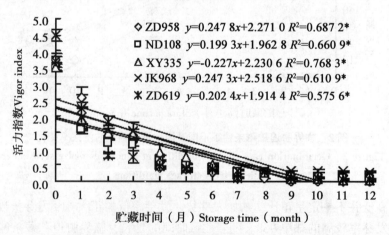

图4 临界胁迫贮藏条件下不同基因型玉米种子的活力指数变化

Figure 4 Vigor index changes of different genotypes maize seed under critical stress storage conditions

2.3 不同基因型玉米种子贮藏过程中电导率变化

测定种子浸出液的电导率,能够间接地评价种子批质量、评价不同基因型玉米种子的贮藏耐性。如图5不同基因型玉米种子电导率差异较大,随着贮藏时间的延长,临界胁迫贮藏条件下的不同基因型的玉米种子电导率均出现了不同程度的上升。不同基因型玉米种子的膜系统稳定性有明显差异,临界胁迫贮藏条件对农大108的膜系统伤害最大,贮藏一年后电导率上升了29%,而先玉335及正大619在临界胁迫贮藏条件下贮藏一年后电导率上升幅度较小。

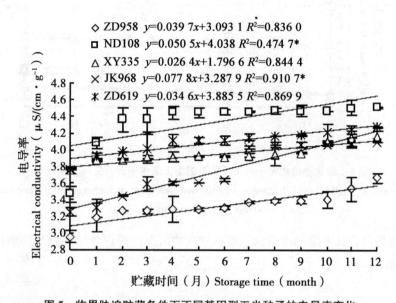

图5 临界胁迫贮藏条件下不同基因型玉米种子的电导率变化

Figure 5 Electrical conductivity changes of different genotypes maize seed under critical stress storage conditions

2.4 不同基因型玉米种子贮藏过程中丙二醛含量变化

丙二醛是膜脂过氧化的最终分解产物,它在细胞内浓度的大小表示脂质过氧化强度和膜系统伤害的程度,是逆境生理中一个重要指标。临界胁迫贮藏条件下玉米种子的丙二醛含量变化表明(图6),随着贮藏时间的延长,不同基因型的玉米种子丙二醛含量均出现了不同程度的上升,但升高幅度均不大。农大108在临界胁迫贮藏条件下的丙二醛含量要明显高于其他基因型玉米种子,郑单958和京科968在临界胁迫贮藏条件下的丙二醛含量差异不明显。农大108在贮藏1~3个月,先玉335在贮藏第4~6个月丙二醛含量明显上升,正大619、郑单958和京科968的丙二醛含量在整个贮藏周期内无明显变化。

2.5 不同基因型玉米种子贮藏过程中可溶性蛋白含量变化

在贮藏过程中,种子将大分子的贮藏物质降解为可溶性的小分子物质以提供给代谢和生长所需,例如蛋白质和脂肪水解提供生理活性物质和能量。临界胁迫贮藏条件下玉米种子的可溶性蛋白含量变化表明(图7),随着贮藏时间的延长,不同基因型的玉米种子可溶性蛋白含量均出现了不同程度的下降,贮藏6个月后各基因型玉米种子的可溶性蛋白含量趋于稳定。在临界胁迫贮藏条件下郑单958的可溶性蛋白含量最高。农大108及先玉

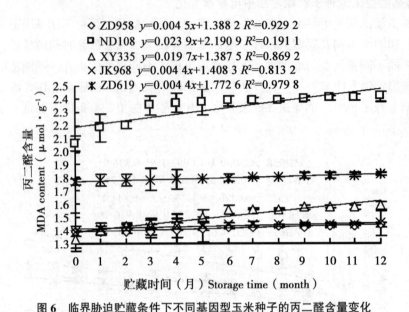

图 6 临界胁迫贮藏条件下不同基因型玉米种子的丙二醛含量变化

Figure 6 MDA content changes of different genotypes maize seed under critical stress storage conditions

335 的可溶性蛋白含量下降幅度最大。京科 968 的可溶性蛋白含量在临界胁迫贮藏条件下的下降幅度最小。

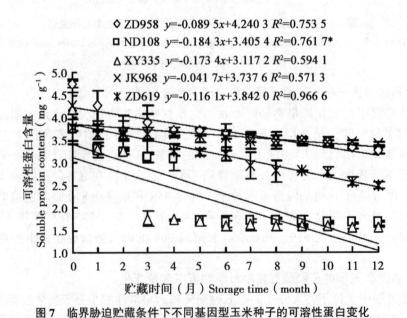

图 7 临界胁迫贮藏条件下不同基因型玉米种子的可溶性蛋白变化

Figure 7 Soluble protein content changes of different genotypes maize seed under critical stress storage conditions

2.6 不同基因型玉米种子贮藏过程中可溶性糖含量变化

可溶性糖是种子的主要呼吸底物,对种子萌发与胚的生长有着极重要的作用。临界胁迫贮藏条件下玉米种子的可溶性糖含量变化表明(图8),随着贮藏时间的延长,不同基因型的玉米种子可溶性糖含量均出现了不同程度的下降。郑单958、京科968、正大619的可溶性糖含量较高,农大108、京科968及正大619在临界胁迫贮藏条件下的可溶性糖含量变化不明显。其中,京科968在临界胁迫贮藏条件下可溶性糖含量变化较小,郑单958、先玉335、正大619在贮藏前5个月可溶性糖含量下降较明显。在临界胁迫贮藏条件下,郑单958的可溶性糖含量下降幅度最大,贮藏一年后达到了6.47%。

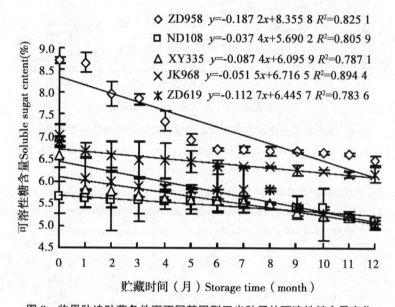

图8 临界胁迫贮藏条件下不同基因型玉米种子的可溶性糖含量变化
Figure 8 Soluble sugar content changes of different genotypes maize seed under critical stress storage conditions

2.7 不同基因型玉米种子贮藏过程中脱氢酶活性变化

脱氢酶可以反映种子萌发后的代谢强度,而代谢强弱直接影响种子的活力。临界胁迫贮藏条件下玉米种子的脱氢酶活性变化表明(图9),随着贮藏时间的延长,不同基因型的玉米种子脱氢酶活性均出现了不同程度的下降。临界胁迫贮藏条件下不同基因型玉米种子的脱氢酶含量下降趋势表现较一致,表现为郑单958>京科968>先玉335>正大619>农大108。除郑单958第7个月下降较显著外,其他基因型玉米种子在贮藏2~4个月后下降趋势趋于平缓。

2.8 临界胁迫贮藏条件下种子活力与生理指标的相关性分析

由表2可知,在临界胁迫贮藏条件下,不同基因型玉米种子活力与丙二醛含量无显著相关性,与可溶性蛋白、可溶性糖含量及脱氢酶活性呈显著正相关;电导率与发芽势、发芽率呈显著负相关,与发芽指数无显著相关性,与可溶性糖含量及脱氢酶活性呈显著正相关,与其他指标表现为极显著正相关。由此可以看出,不同基因型玉米种子在临界胁迫贮藏条件下,种子活力与生理指标变化关系密切,呈极显著的相关性。

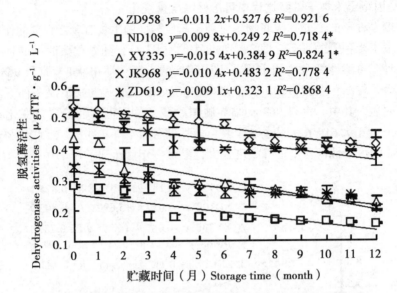

图9 临界胁迫贮藏条件下不同基因型玉米种子的脱氢酶活性变化

Figure 9 Dehydrogenase activity changes of different genotypes maize seed under critical stress storage conditions

表2 临界胁迫贮藏条件下不同基因型玉米种子活力与生理指标的相关性分析

Table 2 Correlation analysis between vigor and physiological indexes of different genotypes maize seeds under critical stress storage conditions

	发芽势 Germination potential	发芽率 Germination rate	发芽指数 Germination index	活力指数 Vigor index	电导率 Electrical Conductivity	MDA含量 MDA content	可溶性蛋白 Soluble protein	可溶性糖 Soluble Sugar	脱氢酶 Dehydrogenase
发芽势 Germination potential	1								
发芽率 Germination rate	0.987**	1							
发芽指数 Germination index	0.982**	0.965**	1						
活力指数 Vigor index	0.821**	0.789**	0.855**	1					
电导率 Electrical conductivity	-0.290*	-0.285*	-0.229	-0.448**	1				
MDA含量 MDA content	-0.153	-0.131	-0.094	-0.171	0.752**	1			
可溶性蛋白 Soluble protein	0.504**	0.484**	0.446**	0.517**	-0.673**	-0.482**	1		

(续表)

	发芽势 Germination potential	发芽率 Germination rate	发芽指数 Germination index	活力指数 Vigor index	电导率 Electrical Conductivity	MDA含量 MDA content	可溶性蛋白 Soluble protein	可溶性糖 Soluble Sugar	脱氢酶 Dehydrogenase
可溶性糖 Soluble sugar	0.372**	0.385**	0.307*	0.466**	-0.837**	-0.566**	0.760**	1	
脱氢酶 Dehydrogenase	0.357**	0.363**	0.284*	0.416**	-0.906**	-0.822**	0.782**	0.869**	1

注：* 表示在0.05水平上显著相关，** 表示在0.01水平上极显著相关

Note: * Correlation is significant at the 0.05 level. ** Correlation is significant at the 0.01 level

3 讨论

不同基因型玉米种子在临界胁迫贮藏条件下，随着贮藏时间延长，种子的萌发能力及活力出现了显著下降。这与前人在大豆、小麦、及玉米和水稻种子老化研究中的结果相同。研究证明玉米种子的萌发特性及耐贮性可以通过发芽势、发芽率、发芽指数和活力指数真实可靠地反映出来。本试验中，不同基因型玉米种子的发芽势、发芽率在各贮藏阶段变化趋势及变幅均呈极显著正相关关系，但与发芽指数、活力指数的变幅不呈正比，差距较大，且不同基因型间存在差异。在本研究参试玉米种子中，正大619在临界胁迫贮藏条件下的发芽势及发芽率最低，但发芽指数较高，这说明正大619具有出苗较快速的萌发特性。京科968在种子含水量较高且贮藏温度也较高的双因素复合胁迫条件下种子活力相对稳定，表现出了较好的耐贮性。

殷换弟等认为丙二醛（MDA）可以作为逆境胁迫程度的指示指标。刘明久、乔燕祥等通过对人工老化的玉米种子的研究，结果表明相对电导率与丙二醛含量随种子老化时间而升高，和发芽指标、活力指标呈显著的负相关。但杨剑平、陈晓玲等研究认为MDA含量在衰老与正常种子中的差异均不显著，并不随老化程度的加剧而增加。本研究得到了和刘明久、乔燕祥等基本一致的结果，说明临界胁迫贮藏条件对不同基因型玉米种子的正常膜结构生理功能已经造成了一定的影响，但基因型间差距较大，且分析结果显示不同基因型玉米种子活力与电导率呈显著负相关，但与丙二醛含量无显著相关性。

贮藏物质的变化对玉米种子萌发与胚的生长有着至关重要的作用。在本研究中随着贮藏时间的不断延长，玉米种子的可溶性蛋白含量呈现不断下降的趋势，这与刘霞、孙青春、马跃青研究结果一致。但与范国强在对花生种子老化过程发芽率和蛋白质的变化关系研究结论不一致。在本试验中，各基因型玉米种子随着贮藏时间的延长，可溶性糖含量均出现了不同程度的下降，这与前人的研究结果基本一致彭建等认为温度高或含水量高使种子在贮藏过程中呼吸作用加强，糖代谢加快。因此在贮藏温度较高时，种子的可溶性糖含量下降程度较快，低温处理条件下可以显著延缓总糖下降。在本试验中各基因型间的可溶性糖含量下降幅度差异较大，京科968下降幅度最小，郑单958的可溶性糖含量下降幅度最大，这与种子的活力下降趋势一致，呈显著正相关关系。

脱氢酶在种子萌发阶段的作用至关重要。杨建肖试验也表明，种子老化处理后，脱氢

酶活性下降，并且不同品种脱氢酶活性在老化过程中的变化幅度存在差异。Chis 等人的研究结果表明种子脱氢酶活性的强弱与种子本身的生活力强弱呈正相关。本研究中，玉米种子经过贮藏后，脱氢酶活性下降，且不同基因型玉米种子的脱氢酶含量的下降程度不同。这可能是因为种子贮藏过程中萌发种子呼吸速率下降，种子萌发及幼苗生长所需要的能量供应减少，导致种子萌发速率变缓，从而使得脱氢酶活性降低。另外玉米种子的脱氢酶含量与种子本身的特性相关，可能受种籽粒型及种胚大小影响较大。

本研究中，各基因型玉米种子在不同贮藏条件下的活力变化总体规律与可溶性糖、可溶性蛋白、脱氢酶活性均呈正相关关系，与电导率、丙二醛含量均呈负相关，这与前人研究基本一致。电导率、可溶性糖、可溶性蛋白、脱氢酶活性等可以作为种子活性和劣变程度等生理指标的研究内容，但这些指标均易受到外部环境影响，且各基因型间差异较大，玉米种子耐贮性生理机制较为复杂，在生理指标上很难制定统一的评价标准。从本试验中可以看到，种子活力指数的降低总是先于种子发芽力的下降。说明种子在失去发芽力之前已发生了劣变，活力指数和发芽指数将种子生活力趋势进行了放大，故二者比发芽率能更能灵敏地反映种子的劣变程度及老化程度，但是均不能直接反映田间种子的出苗情况，种子发芽率、发芽势仍然是表现种子生活力最直接和稳定的指标。

4 结论

不同基因型玉米种子均对临界胁迫贮藏条件表现敏感。随着贮藏时间的延长，各基因型玉米种子在临界胁迫贮藏条件下的发芽势、发芽率与发芽指数、活力指数均出现不同程度的降低，变化规律及趋势基本一致。电导率及丙二醛含量随贮藏时间延长逐渐上升，其变化趋势与生活力变化趋势呈负相关。可溶性蛋白及可溶性糖含量、脱氢酶活性均出现了不同程度的下降，其变化趋势与生活力变化趋势呈正相关。综合试验分析结果可以看出，参试品种京科 968 在临界胁迫贮藏条件下的种子活力保持在较高水平，各项生理指标相对稳定，表现出了较好的耐贮性。

参考文献（略）

本文原载：中国农业科学，2015，48（1）：33-42

北京优质杂交玉米种子内生细菌种类多样性

李南南[1,2]　刘洋[2,3]*　赵燃[2]　王荣焕[3]　肖明[1]*　赵久然[3]　程池[2]

(1. 上海师范大学生命与环境科学学院，上海　200234；2. 中国食品发酵工业研究院中国工业微生物菌种保藏管理中心，北京　100015；3. 北京市农林科学院玉米研究中心，北京　100097)

摘　要：通过传统微生物培养方法研究北京市自主培育并具有遗传相关性的优质杂交玉米种子内生细菌种类的多样性。试验表明来自共同父本京2416的杂交玉米京华8号、京农科728、京单68和NK718种子的内生细菌分别有10个、14个、13个和18个OTUs；Microbacterium maritypicum DSM 12512T（AJ853910）是京农科728种子的第一优势种，是京单68种子的第二优势种，也是NK718种子的并列第三优势种；Microbacterium saperdae IFO 15038T（AB004719）是京农科728和京单68的并列第三优势种。来自共同父本京92的杂交玉米京科968和京科665种子的内生细菌分别有14个和7个OTUs；京科968种子的第二优势种Bacillus horneckiae DSM 23495T（FR749913）也是京科665的第一优势种。结果表明玉米种子基因型对其内生细菌的种类多样性具有一定的影响，来自共同父本的杂交玉米内生菌在OTU数目上相近，并且具有相同的优势菌种；不同父本的玉米种子内生菌OTU数目不同，且内生菌群落结构也有所差异。这是首次以具有遗传相关性的北京优质玉米杂交种子为试验材料研究玉米种子内生细菌种类多样性与其基因型的关系。

关键词：杂交玉米；种子；内生细菌；种类多样性

Diversity of Endophytic Bacteria in Beijing High-quality Hybrid Maize (*Zea mays* L.) Seed

Li Nannan[1,2]　Liu Yang[2,3]*　Zhao Ran[2]
Wang Ronghuan[3]　Xiao Ming[1]*　Zhao Jiuran[3]　Cheng Chi[2]

(1. *College of Life and Environmental Sciences*, *Shanghai Normal University*, *Shanghai* 200234, *China*; 2. *China Center of Industrial Culture Collection*, *China National Research Institute of Food and Fermentation Industries*, *Beijing* 100015, *China*; 3. *Maize Research Center*, *Beijing Academy of Agriculture and Forestry Sciences*, *Beijing* 100097, *China*)

基金项目：杂交水稻国家重点实验室开放课题（2014KF06）、公益性行业（农业）科研专项（201403002）、北京市科技新星计划项目（Z141105001814095）

作者简介：李南南，女，硕士研究生，研究方向为微生物区系分析与分类鉴定

* 刘洋，男，高级工程师，博士，主要从事微生物分类鉴定与植物内生菌方面的研究。通讯作者

* 肖明，男，教授，博士，主要从事微生物与植物的相互作用及微生物分子生物学方面的研究。通讯作者

Abstract: The objective of this study was to investigate the endophytic bacterial communities in seeds of Beijing high-quality hybrid maize varietie, which had genetic relationships used microbial culture-dependent methods. The results of this study indicated that the hybrid maize, which had same male parent Jing 2416, Jinghua 8, Jingnongke 728, Jingdan 68 and NK718 had 10, 14, 13 and 18 OTUs, respectively. The first dominant species (Microbacterium maritypicum DSM 12512^T) (AJ853910) of the hybrid Jingnongke 728 was consistent with the second dominant species of the hybrid Jingdan 68 and consistent with the tied for third dominant species of the hybrid NK718. The tied for second dominant species (Microbacterium saperdae IFO 15038^T) (AB004719) of the hybrid Jingnongke 728 was consistent with the tied for second dominant species of the hybrid Jingdan 68. Jingke 968 and Jingke 665 with the same male parent Jing 92 had 14 and 7 OTUs, The second dominant species [Bacillus horneckiae DSM 23495^T (FR749913)] of the hybrid Jingke 968 was consistent with the first dominant species of the hybrid Jingke665. The result indicated that the endophytic bacterial diversity had certain relevance in the seeds of hybrid maize which were genetically related. Hybrid maize with the same male parent had same dominant species and had similar OTU number while hybrid maize with different male parents had different OTU numbers and community structures. Accrording to the available literature, this was the first research reported in world on the diversity of endophytic bacterial communities in seeds of Beijing high-quality hybrid maize varieties.

Key words: Hybrid maize; Seed; Endophytic bacteria; Diversity

玉米不仅是世界第一大粮食作物，也是我国第一大粮食作物，是中国近年来粮食作物中产量和面积增长最快的。玉米具有适应性广、产量潜力高等优点，并具有粮、经、果、饲、能等很多用途，发展玉米生产对于国家粮食安全的保障、畜牧业的加速发展、城乡人民膳食结构的改善及人民生活水平的提高具有重要意义。杂交玉米具有果穗大、籽粒饱、抗逆性强、抗病高产等优良特性。到目前为止，国家和省级审定的玉米新品种就已达到4 000多个，但其中已真正转化为现实生产力的品种却很少。内生细菌是一类内共生微生物，能够定殖在健康植物组织内并且对植物组织无害，可通过直接或者间接作用影响植物生长和发育。诸多研究证实在种子表面和种子内部有丰富的微生物群落，包括各种病菌和有益的细菌。有益内生细菌能够对植物的健康生长起到直接或者间接促进作用，如生物防治、植物促生、内生固氮等。目前关于有益内生细菌生态功能的报道很多，其中对植物的直接促生作用主要包括生物固氮、分泌和诱导产生植物生长调节物质、改善植物对矿物质的利用率等；对植物的间接促生作用主要是植物内生细菌的生防作用，主要包括分泌抗菌物质、竞争生态位与营养物质、诱导系统抗性等。近年来，随着植物微生态学研究的快速发展，人们逐步认识到与植物相联合的细菌，特别是其中的有益细菌对植物的健康生长发挥着重要的生物学作用。虽然与植物相联合细菌的相关研究报道越来越多，但相对于水稻而言，与玉米相联合细菌的研究报道相对较少，且主要集中于与玉米根系相联合细菌的研究。与玉米种子相联合的细菌的研究鲜有报道，有关不同基因型玉米种子内生细菌的研究则更少。在本课题组对植物种子相联合的微生物研究基础上，用传统微生物纯培养方法研究具有遗传相关性的北京优质玉米杂交品种间的内生细菌种类的多样性，可为植物基因型与定殖在其内的内生细菌群落间的相关性的研究提供理论依据，为指导玉米种子生产与贮藏过程中的微生物控制与强化及保障种子的健康与活力提出科学依据，这将对推进种子科学研究及种业发展具有重要理论依据和实践意义。

1 材料和方法

1.1 试验材料

1.1.1 玉米种子的采集

试验样品为北京农林科学院玉米研究中心自主培育并提供的6种杂交玉米品种（京华8号、京农科728、京单68、NK718、京科968、京科665），样品具有遗传相关性，见表1，于2013年采自位于甘肃张掖的北京农林科学院试验基地（北纬39°04′57.94″，东经100°12′49.40″），4℃保存。

表1 样品间遗传相关性
Table 1 Genetic relationships among samples

试验组	杂交品种	父本	母本
第一组（父本为京2416）	京华8号	京2416	京X005
	京农科728	京2416	京MC01
	京单68	京2416	CH8
	NK718	京2416	京464
第二组（父本为京92）	京科968	京92	京724
	京科665	京92	京725

1.1.2 培养基及试剂

LB、TSA和R2A成品培养基，购于北京陆桥公司；细菌基因组DNA提取试剂盒和PCR相关试剂购于TIANGEN公司；试验所用PCR引物由北京诺赛生物公司合成。

1.2 试验方法

1.2.1 样品表面灭菌

先用无菌水清洗杂交玉米种子，之后依次用体积分数为70%的乙醇溶液浸泡3min，次氯酸钠溶液（氯离子浓度为2.5%）浸泡5min，70%乙醇浸泡30s，最后用无菌水淋洗5~7次，在无菌滤纸无菌条件下晾干；同时取最后一次淋洗水120μl涂于LB固体平板上，28℃恒温培养72h，检测玉米种子表面灭菌效果。表面灭菌彻底的种子用于试验。

1.2.2 内生细菌的分离与纯化

玉米内生细菌群落多样性研究采用传统分离培养技术。取5.0g表面灭菌的玉米种子用无菌研钵研磨成粉末，采用梯度稀释法制备$1×10^{-1}$g/ml到$1×10^{-3}$g/ml的系列稀释液。每一梯度稀释液分别取200μl涂布于R2A、LB、TSA平板上，每个处理组设置3个平行，30℃培养3d后根据平板上菌落的形态（大小、颜色、形状、表面光泽度、透明度、边缘整齐度等）随机挑取具有代表性单菌落，纯化后4℃保存备用，同时保存于中国工业微生物菌种保藏中心（CICC）。

1.2.3 内生细菌的16S rDNA序列分析

用细菌基因组DNA提取试剂盒（TIANGEN）提取细菌基因组DNA。采用正向引物27F（5′-AGAGTTTCATCTGGCTCAG-3′）和反向引物1492R（5′-GGTTACCTTGTT AC-

GACTT-3')扩增细菌16SrDNA。PCR反应体系（50μl）为DNA模板3μl、10×buffer 5μl、引物27F（10mmol/L）1μl、引物1492R（10mmol/L）1μl、dNTP（2.5mmol/L）4μl、Taq酶（5 U/L）0.25μl，最后用ddH₂O补足至50μl。反应程序为94℃预变性5min，94℃变性1min，55℃复性1min，72℃延伸1min，共30个循环后72℃延伸10min。用1%琼脂糖凝胶电泳检测PCR扩增产物。扩增出的PCR产物用ABI 3730型DNA测序仪测序（ABI, USA），将所有序列信息提交到NCBI，测序得到的结果在EzTaxon server 2.1进行比对，确定与已知序列同源关系，比对后序列相似性达到98.65%以上的归为同一个运算的分类单位（Operational Taxonomic Units，OTU）。

2 试验结果

通过传统分离培养方法从杂交玉米品种京华8号、京农科728、京单68、NK718、京科968和京科665种子中分别挑取20株、29株、26株、39株、30株、16株细菌，分别有10个、14个、13个、18个、14个、7个代表菌株（表2~表7），将代表菌株序列信息提交到GenBank，并获得登录号。

表2 玉米京华8号种子内生细菌分布
Table 2 Distribution of endophytic bacteria in seed of maize Jinghua 8

类群	OTU数目	属	代表菌株	克隆数目	克隆百分比（%）	NCBI数据库中相似度最高的菌种	相似度（%）	登录号
	1	Citrobacter	1L1	2	10	Citrobacter youngae CECT 5335^T（AJ564736）	98.84	KP843007
	2	Pantoea	1R3	1	5	Pantoea brenneri LMG 5343^T（EU216735）	100	KP843012
			1L2	2	10	Pantoea agglomerans DSM 3493^T（AJ233423）	100	KP843008
Proteobacteria	1	Pseudomonas	1L3	9	45	Pseudomonas taiwanensis BCRC 17751^T（EU103629）	98.89	KP843009
	1	Leclercia	1R2	1	5	Leclercia adecarboxylata GTC 1267^T（AB273740）	99.89	KP843011
	1	Prolinoborus	1T2	1	5	Prolinoborus fasciculus CIP 103579^T（JN175353）	99.71	KP843014
	1	Moraxella	1T5	1	5	Moraxella osloensis NCTC 10465^T（X74897）	98.92	KP843017
Firmicutes	1	Bacillus	1T1	1	5	Bacillus horneckiae DSM 23495^T（FR749913）	97.46	KP843013
Actinobacteria	1	Zhihengliuella	1T3	1	5	Zhihengliuella aestuarii DY66^T（EU939716）	99.86	KP843015
	1	Microbacterium	1T4	1	5	Microbacterium aoyamense KV-492^T（AB234028）	98.88	KP843016

表3 玉米京农科728种子内生细菌分布

Table 3 Distribution of endophytic bacteria in seed of maize Jingnongke 728

类群	OTU数目	属	代表菌株	克隆数目	克隆百分比（%）	NCBI数据库中相似度最高的菌种	相似度（%）	登录号
Proteobacteria	1	Brevundimonas	2L7	2	6.90	Brevundimonas vesicularis LMG 2350^T (AJ227780)	98.97	KP843023
	1	Enterococcus	2L3	2	6.90	Enterococcus lactis BT159^T (GU983697)	99.79	KP843020
	2	Leuconostoc	2M2	2	6.90	Leuconostoc mesenteroides subsp. suionicum LMG 8159^T (HM443957)	99.44	KP843025
			2M3	4	13.79	Leuconostoc mesenteroides subsp. dextranicum NRIC 1539^T (AB023246)	99.79	KP843026
	1	Exiguobacterium	2R6	2	6.90	Exiguobacterium sibiricum 255-15^T (CP001022)	98.98	KP843027
	1	Enterococcus	2T7	1	3.45	Enterococcus mundtii CECT972^T (AJ420806)	99.32	KP843031
Bacteroidetes	1	Chryseobacterium	2T2	1	3.45	Chryseobacterium indoltheticum LMG 4025^T (AY468448)	99.51	KP843029
Actinobacteria	1	Curtobacterium	2L1	1	3.45	Curtobacterium flaccumfaciens LMG 3645^T (AJ312209)	99.02	KP843018
	5	Microbacterium	2L2	5	17.24	Microbacterium maritypicum DSM 12512^T (AJ853910)	99.50	KP843019
			2L4	1	3.45	Microbacterium oleivorans DSM 16091^T (AJ698725)	98.74	KP843021
			2L5	3	10.34	Microbacterium saperdae IFO 15038^T (AB004719)	98.57	KP843022
			2L8	3	10.34	Microbacterium paraoxydans CF36^T (AJ491806)	98.53	KP843024
			2R7	1	3.45	Microbacterium oxydans DSM 20578^T (Y17227)	99.37	KP843028
	1	Curtobacterium	2T5	1	3.45	Curtobacterium ammoniigenes NBRC 101786^T (AB266597)	98.79	KP843030

表4 玉米京单68种子内生细菌分布
Table 4 Distribution of endophytic bacteria in seed of maize Jingdan 68

类群	OTU数目	属	代表菌株	克隆数目	克隆百分比（%）	NCBI 数据库中相似度最高的菌种	相似度（%）	登录号
Proteobacteria	1	Acinetobacter	3L4	3	10.71	Acinetobacter lwoffii NCTC 5866^T (AIEL01000120)	100	KP843035
	1	Stenotrophomonas	3L7	1	3.57	Stenotrophomonas rhizophila DSM 14405^T (CP007597)	98.76	KP843037
	1	Prolinoborus	3R2	1	3.57	Prolinoborus fasciculus CIP 103579^T (JN175353)	99.57	KP843039
	1	Brevundimonas	3T8	1	3.57	Brevundimonas nasdae GTC 1043^T (AB071954)	99.54	KP843044
Firmicutes	1	Bacillus	3R4	1	3.57	Bacillus anthracis ATCC 14578^T (AB190217)	99.39	KP843041
	1	Paenibacillus	3T2	1	3.57	Paenibacillus lautus NRRL NRS-666^T (D78473)	96.78	KP843043
	1	Enterococcus	3L6	9	32.14	Enterococcus lactis BT159^T (GU983697)	99.79	KP843036
Actinobacteria	5	Microbacterium	3L1	4	14.29	Microbacterium maritypicum DSM 12512^T (AJ853910)	99.50	KP843032
			3L2	3	10.71	Microbacterium saperdae IFO 15038^T (AB004719)	98.57	KP843033
			3L3	1	3.57	Microbacterium paraoxydans CF36^T (AJ491806)	98.60	KP843034
			3R1	1	3.57	Microbacterium oxydans DSM 20578^T (Y17227)	99.51	KP843038
			3R8	1	3.57	Microbacterium natoriense TNJL143-2^T (AY566291)	98.59	KP843042
	1	Okibacterium	3R3	1	3.57	Okibacterium fritillariae VKM Ac-2059^T (AB04209)	98.32	KP843040

表5 玉米 NK718 种子内生细菌分布
Table 5 Distribution of endophytic bacteria in seed of maize NK718

类群	OTU数目	属	代表菌株	克隆数目	克隆百分比（%）	NCBI 数据库中相似度最高的菌种	相似度（%）	登录号
Proteobacteria	2	Pantoea	4L14	4	10.26	Pantoea vagans LMG 24199^T（EF688012）	99.77	KP843055
			4T2	1	2.56	Pantoea agglomerans DSM 3493^T（AJ233423）	100	KP843061
	2	Brevundimonas	4L7	1	2.56	Brevundimonas nasdae GTC 1043^T（AB071954）	100	KP843051
			4R4	1	2.56	Brevundimonas nasdae GTC 1043^T（AB071954）	98.90	KP843057
Firmicutes	1	Enterococcus	4L1	3	7.69	Enterococcus lactis BT159^T（GU983697）	99.79	KP843045
	2	Exiguobacterium	4L4	3	7.69	Exiguobacterium sibiricum 255-15^T（CP001022）	98.78	KP843048
			4T3	1	2.56	Exiguobacterium mexicanum 8N^T（AM072764）	99.28	KP843062
	1	Paenibacillus	4L13	2	5.13	Paenibacillus hordei RH-N24^T（HQ833590）	97.19	KP843054
	2	Leuconostoc	4R9	5	12.82	Leuconostoc mesenteroides subsp. dextranicum NRIC 1539^T（AB023246）	99.10	KP843059
			4R15	2	5.13	Leuconostoc mesenteroides subsp. suionicum LMG 8159^T（HM443957）	98.75	KP843060
Actinobacteria	3	Curtobacterium	4L2	2	5.13	Curtobacterium flaccumfaciens LMG 3645^T（AJ312209）	99.09	KP843046
			4L6	1	2.56	Curtobacterium ammoniigenes NBRC 101786^T（AB266597）	98.93	KP843050
	3	Microbacterium	4L3	4	10.26	Microbacterium maritypicum DSM 12512^T（AJ853910）	99.29	KP843047
			4L5	1	2.56	Microbacterium paraoxydans CF36^T（AJ491806）	98.61	KP843049
			4L11	1	2.56	Microbacterium saperdae IFO 15038^T（AB004719）	98.57	KP843053
	1	Leucobacter	4L15	1	2.56	Leucobacter chromiiresistens JG 31^T（AGCW01000231）	98.61	KP843056
	1	Rathayibacter	4R5	1	2.56	Rathayibacter iranicus DSM 7484^T（AM410684）	99.37	KP843058
Bacteroidetes	1	Chryseobacterium	4L8	5	12.82	Chryseobacterium indoltheticum LMG 4025^T（AY468448）	98.36	KP843052

表6 玉米京科968种子内生细菌分布
Table 6 Distribution of endophytic bacteria in seed of maize Jingke 968

类群	OTU数目	属	代表菌株	克隆数目	克隆百分比(%)	NCBI数据库中相似度最高的菌种	相似度(%)	登录号
Proteobacteria	1	Pantoea	5L5	2	6.67	Pantoea vagans LMG 24199^T (EF688012)	99.77	KP843067
	1	Moraxella	5R2	1	3.33	Moraxella osloensis NCTC 10465^T (X74897)	99.17	KP843072
	1	Brevundimonas	5R5	1	3.33	Brevundimonas vesicularis LMG 2350^T (AJ227780)	98.98	KP843074
Firmicutes	2	Enterococcus	5L1	2	6.67	Enterococcus lactis BT159^T (GU983697)	99.58	KP843063
			5L2	2	6.67	Enterococcus mundtii CECT972^T (AJ420806)	99.32	KP843064
	1	Paenibacillus	5L8	1	3.33	Paenibacillus jamilae CECT 5266^T (AJ271157)	98.49	KP843070
	1	Leuconostoc	5M1	9	30	Leuconostoc mesenteroides subsp. dextranicum NRIC 1539^T (AB023246)	99.38	KP843071
	1	Bacillus	5T3	3	10	Bacillus horneckiae DSM 23495^T (FR749913)	97.06	KP843077
Actinobacteria	1	Arthrobacter	5T4	2	6.67	Arthrobacter bambusae GM18^T (KF150696)	99.01	KP843078
	3	Microbacterium	5L7	1	3.33	Microbacterium maritypicum DSM 12512^T (AJ853910)	99.57	KP843069
			5R4	1	3.33	Microbacterium oxydans DSM 20578^T (Y17227)	99.51	KP843073
			5R8	2	6.67	Microbacterium saperdae IFO 15038^T (AB004719)	98.79	KP843075
	1	Plantibacter	5T2	1	3.33	Plantibacter auratus IAM 18417^T (AB177868)	99.01	KP843076
Bacteroidetes	1	Chryseobacterium	5L4	2	6.67	Chryseobacterium indoltheticum LMG 4025^T (AY468448)	99.36	KP843066

表7 玉米京科665种子内生细菌分布
Table 7 Distribution of endophytic bacteria in seed of maize Jingke 665

类群	OTU数目	属	代表菌株	克隆数目	克隆百分比(%)	NCBI数据库中相似度最高的菌种	相似度(%)	登录号
Proteobacteria	1	Moraxella	6R1	3	18.75	Moraxella osloensis NCTC 10465^T (X74897)	99.18	KP843083

(续表)

类群	OTU 数目	属	代表菌株	克隆数目	克隆百分比（%）	NCBI 数据库中相似度最高的菌种	相似度（%）	登录号
Firmicutes	2	Bacillus	6L5	6	37.50	Bacillus horneckiae DSM 23495^T（FR749913）	97.12	KP843081
			6R3	1	6.25	Bacillus methylotrophicus-CBMB205^T（EU194897）	99.58	KP843085
	1	Paenibacillus	6R2	1	6.25	Paenibacillus lautus NRRL NRS-666^T（D78473）	96.79	KP843084
Actinobacteria	3	Paenibacillus	6L1	3	18.75	Microbacterium maritypicum DSM 12512^T（AJ853910）	99.72	KP843079
			6L3	1	6.25	Microbacterium saperdae IFO 15038^T（AB004719）	98.86	KP843080
			6R6	1	6.25	Microbacterium marinum H101^T（HQ622524）	99.16	KP843086

父本为京 2416 的杂交玉米品种在 OTU 数目相近，京华 8 号，京农科 728，京单 68 和 NK718 种子分别有 10 个，14 个，13 个和 18 个 OTUs，京农科 728、京单 68 的 OTU 数目最相近；父本为京 92 的杂交玉米品种京科 968 和京科 665 种子分别有 14 个和 7 个 OTUs。父本为京 2416 的杂交玉米品种种子内生细菌的 OTU 数目多于父本为京 92 的杂交玉米品种。研究玉米京华 8 号种子内生细菌分布，见表 2，由试验可知，本试验中的 6 种杂交玉米品种优势菌种不同。京华 8 号第一优势种是 Pseudomonas taiwanensis BCRC 17751T（EU103629）；Pantoea agglomerans DSM 3493T（AJ233423）和 Citrobacter youngae CECT 5335T（AJ564736）是并列第 2 优势种。

研究玉米京农科 728 种子内生细菌分布，见表 3，由试验可知，京农科 728 第一优势种和第二优势种分别是 *Microbacterium maritypicum* DSM 12512^T（AJ853910）和 *Leuconostoc mesenteroides* subsp. *dextranicum* NRIC 1539^T（AB023246）；*Microbacterium saperdae* IFO 15038^T（AB004719）和 *Microbacterium paraoxydans* CF36^T（AJ491806）是并列第三优势种。

研究玉米京单 68 种子内生细菌分布，见表 4，由试验可知京单 68 第一和第二优势种分别是 *Enterococcus lactis* BT159^T（GU983697）和 *Microbacterium maritypicum* DSM 12512^T（AJ853910）；*Acinetobacter lwoffii* NCTC 5866^T（AIEL01000120）和 *Microbacterium saperdae* IFO 15038^T（AB004719）是并列第三优势种。

研究玉米 NK718 种子内生细菌分布，见表 5，由试验可知 *Leuconostoc mesenteroides* subsp. *dextranicum* NRIC 1539^T（AB023246）和 *Chryseobacterium indoltheticum* LMG 4025^T（AY468448）是 NK718 并列第一优势种；Microbacterium marityicum DSM 12512^T（AJ853910）和 Pantoea vagans LMG 24199^T（EF688012）是并列第二优势种。

研究玉米京科 968 种子内生细菌分布，见表 6，由试验可知京科 968 第一和第二优势种分别是 *Leuconostoc mesenteroides* subsp. *dextranicum* NRIC 1539^T（AB023246）和 *Bacillus horneckiae* DSM 23495^T（FR749913）；*Pantoea vagans* LMG 24199^T（EF688012），*Enterococcus lactis* BT159^T（GU983697），*Enterococcus mundtii* CECT972^T（AJ420806），*Ar-*

throbacter bambusae GM18^T（KF150696），*Microbacterium saperdae* IFO 15038^T（AB004719）和 *Chryseobacterium indoltheticum* LMG 4025^T（AY468448）是并列第三优势种。

研究玉米京科 665 种子内生细菌分布，见表 7，由试验可知京科 665 第一优势种是 *Bacillus horneckiae* DSM 23495^T（FR749913）；*Moraxella osloensis* NCTC 10465^T（X74897）和 *Microbacterium maritypicum* DSM 12512^T（AJ853910）是并列第二优势种。

虽然各品种的优势菌种存在差异，但是来自共同父本有遗传相关性的杂交玉米品种存在共同的优势菌种。父本为京 2416 的杂交玉米品种种子内生菌中，京农科 728 的第一优势种 *Microbacterium maritypicum* DSM 12512^T（AJ853910）是京单 68 的第二优势种，也是 NK718 的并列第三优势菌种；京农科 728 和京单 68 的并列第三优势菌种均为 *Microbacterium saperdae* IFO 15038^T（AB004719）。父本为京 92 的杂交玉米品种种子内生菌中京科 968 的第二优势菌种 *Bacillus horneckiae* DSM 23495^T（FR749913）也是京科 665 的第一优势种。

3 结果与讨论

种子是种子植物遗传信息的传递者与保存者，是植物的延续器官。同时种子是农业生产的基础，其开发及利用水平是衡量一个国家种植业发展水平的重要标志。种子表面和内部存在多种微生物，是多种有益微生物和病原菌的传递载体。目前对根际微生物的研究已较深入，但与种子相联合微生物的研究相对较少，其中对杂交玉米种子内生细菌群落结构多样性的研究较水稻少。本研究以北京市自主培育，并具有遗传相关性的优质玉米杂交品种为研究对象，通过传统微生物培养方法，对玉米种子内生细菌种类多样性进行研究，旨在揭示具有遗传相关性的不同玉米品种对其种子内生细菌种类多样性的影响。

杂交玉米品种在遗传上具有相关性。由表 1 可知，京华 8 号、京农科 728、京单 68 和 NK718 杂交玉米品种来共同父本，内生菌在 OTU 数目上基本相同且存在相同的优势菌种，如京华 8 号、京农科 728、京单 68 和 NK718 种子内生菌分别有 10 个，14 个，13 个和 18 个 OTUs，京农科 728、京单 68 和 NK718 种子内生菌中都有 *Microbacterium maritypicum* DSM 12512^T（AJ853910），京农科 728 和京单 68 种子内生菌中都存 *Microbacterium saperdae* IFO 15038^T（AB004719）。来自共同父本的京科 968 和京科 665 种子内生菌也表现出相似的结果，它们分别有 14 个和 7 个 OTUs 并且都存在 *Bacillus horneckiae* DSM 23495^T（FR749913）。虽然种植土壤、空气和水分等环境因素对杂交后代内生菌群结构具有一定影响，本试验为规避环境因素差异对内生菌群落结构的影响，玉米材料均从相同环境条件下采集得到，因此杂交玉米品种内生菌在 OTU 数目、优势菌种的相似性与其父本基因型相同有关联。植物的基因型对与其相联合的细菌群落结构具有重要的影响。Weinert 等研究马铃薯的根际细菌群落时发现不同品种之间及遗传转化的马铃薯与其亲本植株的根际细菌群落结构有明显差异。不同水稻品种根面定殖的细菌种类和数量有明显差异，抗病性强的品种定殖的具有拮抗性的细菌数量比抗病性弱的品种多。种子是携带亲本基因的雏形植物体，种子的基因型对相联合的细菌有一定的影响。Adams 等对不同棉花品种的内生菌研究中发现，棉花品种不同，遗传关系、生理学和外部形态学特征方面的差异会使定殖在种子中的内生菌群落结构有明显差异。不同基因型番茄种子内生细菌群落结构不同。邹媛媛等对具有遗传相关性杂交水稻组合及其各自亲本的种子固有细菌群落结构多样性进行研究发现杂交子代种子的优势菌属及其丰度与亲本种子的基因型具有一定的相关性。Liu 等在对具有遗传相关性的玉米杂交组合及其亲本种子内生细菌群落结构进行研究时发现子代与亲本种子的内生细菌类群数量及种类存在着明显差异，但是子代中优势菌大都能在亲本种子中检测到，同时发现具有遗传相关性的子代内生优势菌种类与其亲本具有一定关联性。据报道，本研究中的很多菌属（*Citrobacter*、*Pantoea*、*Pseudomonas*、*Bacillus*、*Acinetobacter*）属于植物促生菌。研究发现 *Citrobacter* 能够超表达植酸酶并且引发植物生长，这是由于 *Citrobacter braakii phytasegene*（appA）基因具有促进植物生长特性，appA 的超表达能够增加它们作为生物菌剂的潜能。非致病菌 *Pantoea* spp. 在很多植物扮演着生物肥料和生物菌剂角色，能够增加作物产量，除此之外还有多种作用如固氮，拮抗病菌等。Chickpea（*Cicer arietinum* L.）中的 *Pseudomonas putida* 和 *Bacillus* 能够改善干旱胁迫。水稻根中分离到的 *Bacillus* 能够参与抗菌、促进植物生长和系统性抗性活动。植物促生菌 *Acinetobacter* 能增加单子叶植物 Lemna minor（duckweed）和双子叶植物 Lactuca sativa（lettuce）的叶绿素含量。*Bacillus subtilis* 可以快速定殖于玉米苗中，对玉米纹枯病菌具有明显抑制作用并能够促进玉米生长，并且对玉米小斑病菌有明显拮抗作用。*Pseudomonas geniculata* 和 *Arthrobacter* sp. 是新的玉米内生固氮菌资源，能够促进玉米生长。这些植物促生菌能够增加植物产量，提高质量，增强耐受性，固氮和改善环境条件等。

4 结论

这是首次用传统分离培养方法研究北京市自主培育并具有遗传相关性的优质玉米杂交品种种子内生细菌群落多样性，对推进首都种子科学研究及种业发展具有积极的理论与实践价值。为进一步挖掘植物与共生内生细菌之间的互作关系提供参考依据，可为深入研究种子内生微生物区系结构与植物基因型相关性提供理论依据。本试验首次分离到的北京杂交玉米品种种子内生细菌资源，能为今后合理开发新的植物益微菌剂奠定资源基础。

参考文献（略）

本文原载：食品科学技术学报，2016，34（5）：55-63

常压室温等离子体对玉米种子及花粉萌发的影响

骆美洁 赵衍鑫 宋 伟 赵久然

(北京市农林科学院玉米研究中心/玉米DNA指纹及分子育种北京市重点实验室,北京 100097)

摘 要：利用常压室温等离子体（ARTP）处理玉米种子和花粉,研究其对玉米种子萌发、幼苗生长和花粉萌发的影响。结果表明：ARTP处理5min、15min、30min对玉米种子出芽率没有作用,但是影响了其幼苗的生长：随着处理时间的延长,幼苗干重逐渐增加；同对照相比,ARTP处理30min幼苗显著增长,而幼苗根长/苗高比值降低。随着ARTP对玉米花粉处理时间的增加（0s、18s、54s、90s、144s、180s、270s、360s),花粉萌发率大幅度降低,处理360s时,花粉萌发率接近为零。本研究将有利于ARTP在玉米诱变育种中的应用。

关键词：玉米；ARTP；种子萌发；花粉萌发

Effects on Maize Seed and Pollen Germination by Atmospheric and Room Temperature Plasma

Luo Meijie Zhao Yanxin Song Wei Zhao Jiuran

(Maize research center of Beijing academy of agriculture & forestry sciences, Beijing Key Laboratory of Maize DNA Fingerprinting and Molecular Breeding, Beijing, 100097)

Abstract: Atmospheric and Room Temperature Plasma (ARTP) was used to treat maize seed and pollen, and its effect on maize seed germination, seedling growth and pollen germination was studied. Results showed that after treating maize seed for 5min, 15min and 30min, ARTP had no effect on maize seed germination rate, but it influenced the growth of their seedlings the dry weight of maize seedlings increased as the treatment time extended; compared with the control, treated by ARTP for 30min, maize seedling length significantly increased while the ratio of root length to seedling height declined. As the increase of ARTP treatment time (0s, 18s, 54s, 90s, 144s, 180s, 270s, 360s) on maize pollen, the germination rate of pollen obviously decreased, and under 360s ARTP treatment, the pollen germination rate nearly was zero. This study will contribute to the application of ARTP in mutation breeding of maize.

Key words: Maize; ARTP; Seed germination; Pollen germination

玉米是我国重要的饲料和粮食作物（孙卫永等,2015),其种植面积和总产量位居我

基金项目：本研究由国家科技支撑项目（2014BAD01B09）和北京市农林科学院科技创新能力建设专项共同资助

通讯作者：maizezhao@126.com

国第一，为我国粮食安全和农产品供给提供保障。除常规育种技术外，大量新的生物技术（官春云，2002）、物理（马海林等，2009）和化学诱变（彭波等，2007）技术等已用于作物新种质的创制和培育来不断提高玉米的产量和品质，如转基因技术、细胞工程育种技术、航天辐射育种、甲基磺酸乙酯诱变育种（EMS）。

常压室温等离子体（ARTP）是一种新型的诱变育种手段，因其具有射流温度低（可室温操作）、所产生的等离子体均匀、对操作人员安全、操作简便、成本低廉等优点，已应用广泛于微生物如细菌（Bao et al, 2014）、真菌（自振滔等，2013）、酵母（康富帅等，2014）等诱变育种中。同时，一些研究表明，利用 ARTP 处理大豆（Li et al, 2014）、小麦（Jiang et al, 2014）、番茄（周筑文等，2010）、黄瓜（李怀智和庞金安，2003）等作物的种子可以促进其萌发、增加幼苗干重、抗逆性。本研究利用 ARTP 育种仪对玉米优良自交系京 724（王元东等，2015）的种子和花粉进行等离子体处理，探讨 ARTP 对京 724 玉米种子萌发、幼苗生长和花粉管萌发的影响，为玉米材料的诱变育种提供筛选材料和理论依据。

1 结果与分析

1.1 ARTP 处理对玉米种子萌发的影响

胚是种子萌发的核心器官，所以我们将玉米种子有胚面朝上进行 ARTP 处理。处理完后，立即进行种子萌发试验。室温萌发 5d 后，大部分种子生长成为二叶期的幼苗。ARTP 处理 0min、5min、15min 和 30min 的出芽率分别为 95%±4%、96%±1%、92%±5% 和 95%±3%（图1），无显著性区别。说明短时间内 ARTP 处理未对玉米种子出芽率产生显著影响。

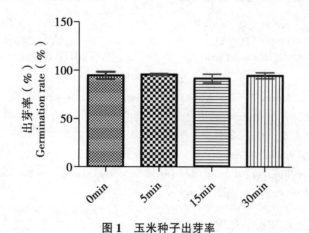

图 1 玉米种子出芽率
Figure 1 Maize seed germination rate

1.2 ARTP 处理对玉米幼苗生长的影响

ARTP 处理的种子，室温萌发 5d 后，对其幼苗的根长和苗高进行测定和统计分析，发现 min、5min、15min 和 30min ARTP 处理的种子其幼苗根长分别为（14.7±0.4）cm、（15.1±0.3）cm、（14.5±0.4）cm 和（14.0±0.4）cm（图2A），苗高分别为（10.7±0.21）cm、（11.3±0.20）cm、（11.1±0.27）cm、（12.9±0.25）cm（图2B）。30min 处

理组的苗高显著性高于未处理组（0min）（$P<0.001$）。计算其根长/苗高比，结果表明 ARTP 处理 30min 组的根长/苗高比值为 1.1 ± 0.03，其显著性低于 0min 处理组（比值为 1.4 ± 0.04）（$P<0.001$）（图 2C）。同时我们对幼苗的干重进行了测定，发现同未处理组相比，随着 ARTP 处理时间的延长，幼苗的平均干重显著增加，从 0min 的（230 ± 3.2）mg 增加到 5min 的（244 ± 2.4）mg（$P<0.05$）、15min 的（246 ± 2.4）mg（$P<0.05$）以及 30min 的（250 ± 5.8）mg（$P<0.01$）（图 2D）。

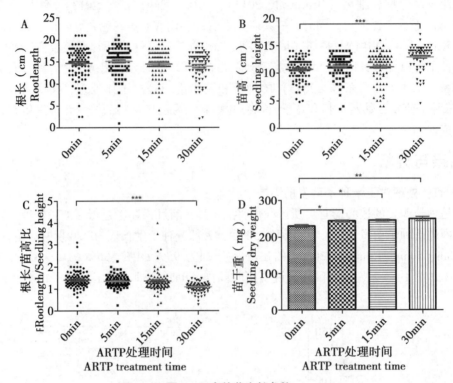

图 2　玉米幼苗生长参数

Figure 2　The growth parameters of maize seedlings

注：A. 根长；B. 苗高；C. 根长/苗高比；D. 苗干重；A-C. 每个点代表一颗幼苗，0min、5min 和 15min 的数据来自于 5 次重复，30min 的数据来自于 3 次重复

Note：A. root length；B. seedling height；C. root length to seedling height；D. seedlings dry weight；A-C. Each dot represents a seedling. Data of 0min, 5min and 15min were pooled from five times repeats; data of 30min were pooled from three times repeats

1.3　ARTP 处理对玉米花粉的影响

田间收集的新鲜花粉立即进行 ARTP 处理，处理后的花粉放入液体浅层花粉萌发培养基中萌发，30min 后进行成像观察和统计分析。正常未处理玉米花粉直径为（90.0 ± 2.2）μm（n=15），萌发后，其内含物吐出，颜色变浅。花粉管直径为（14.5 ± 0.8）μm（n=12）。萌发 30min 后，花粉管长为（187.1 ± 20.6）μm（n=18），约为花粉直径的 2 倍（图 3B）。

统计分析表明（表 1），随着 ARTP 处理时间从 0s 逐渐延长到 360s，花粉萌发率从 45.1%大幅度降低到 0%（图 3A）。处理 9s 的花粉萌发率仅为未处理组（0min）的 1/3，处理270s后，花粉仅有少数萌发，在ARTP处理360s时，花粉萌发率接近零（图3C）。可见，相较于种子，花粉对ARTP的处理更为敏感。

表1 不同处理时间下统计的花粉总数及萌发数
Table 1 Total number and the germination number of maize pollen under different treatment time

ARTP 处理时间 ARTP treatment time	0s	18s	54s	90s	144s	180s	270s	360s
萌发数 Germination number	1 662	1 010	790	389	215	47	38	3
花粉总数 Total number	3 682	2 668	4 504	2 178	2 258	2 779	1 119	1 965
萌发率（%） Germination rate（%）	45.1	37.9	17.5	17.9	9.5	1.7	3.4	0

注：数据来自3次独立性重复
Note：Data were pooled from three times independent repeats

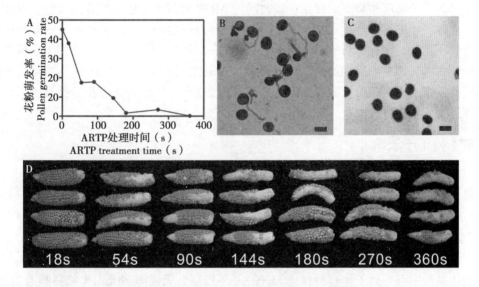

图3 ARTP处理对玉米花粉萌发率的影响
Figure 3 Effect on maize pollen germination rate by ARTP treatment

注：A. 不同ARTP处理时间下玉米花粉萌发率；
B-C. 玉米花粉体外萌发；D. ARTP处理的花粉授粉后结实情况；
B. ARTP 处理 0s；C. ARTP 处理 360s Bar＝100μm

Note：A. Maize pollen germination rate under different ARTP treatment time；B-C. Maize pollen germination in vitro；D. Maize cobs obtained from ARTP treated pollen；
B. ARTP treated for 0s；C. ARTP treated for 360s Bar＝100μm

此外，我们将ARTP处理后的花粉进行田间授粉，观察授粉果穗的结实率。结果显示，与花粉萌发率变化趋势一致，随着ARTP处理时间的延长，玉米果穗的结实率逐渐降低（图3D）。尤其是ARTP处理360s条件下，玉米结实率最低且果穗瘦小。可见，长时间的ARTP处理（270s和360s），可能导致花粉遗传物质发生变异，此时萌发的花粉可用于玉米新种质的创制和筛选。

2 讨论

目前，大量的物理和化学诱变技术已用于微生物和植物的诱变育种，如温度处理、磁处理和辐射处理等（尹美强等，2010）。对植物种子或花粉进行诱变处理，可以增加植物的抗虫性（Selcuk et al, 2008）、提高产量（王敏等，2007；边少锋等，2005）（如增加植株的干重、增加株高）、改善品质（Chen et al, 2012）等。

本研究对种子幼苗的多个生长参数进行分析，结果发现，ARTP 处理种子 30min，增加了其幼苗生长高度，缩小了根长/苗高比值，同时促进了幼苗干物质积累，增加了干重。说明 ARTP 处理可以改善玉米幼苗的生长。类似的结果已在小麦（Jiang et al, 2014；Dobrin et al, 2015）和大豆（Li et al, 2014）等作物中被研究报道。

ARTP 中的高浓度的化学活性粒子可直接直接作用于生物大分子，从而造成遗传物质损伤，细胞膜通透性改变、酶结构改变和酶活性升高（张雪等，2014）。当 ARTP 处理植物种子时，由于果皮和种皮的存在，短时间的 ARTP 处理不能够直接作用于胚，可能仅对果皮种皮和胚外层细胞产生影响，故用 ARTP 短时间处理种子很难得到可遗传的稳定变异。而对于花粉，仅有花粉壁和细胞膜，短时间的 ARTP 处理可以直接作用于生殖细胞内的大分子物质，如 DNA 和蛋白质等。短时间的 ARTP 处理（270s 和 360s）即可显著降低花粉的萌发率。说明花粉更适合 ARTP 处理的诱变育种。ARTP 处理花粉授粉所产生的籽粒的筛选工作正在进行中。

3 材料和方法

3.1 试验材料

供试材料为玉米（*Zea mays* L.）自交系 724，由北京市农林科学院玉米研究中心提供。

3.2 试验方法

3.2.1 ARTP 诱变方法

利用常压室温等离子体育种仪（ARTP-Ⅱ平板型，无锡源清天木生物科技有限公司）对玉米种子及花粉进行处理。本研究所用 ARTP 诱变育种仪以 99.99%高纯氦气作为工作气体，工作气流量为 15SLM，额定功率为 120W，等离子体发射源距待处理样品 3mm。处理玉米种子时，种子胚面朝上，处理时间分别为 0min、5min、15min 和 30min（Dobrin et al, 2015）。除 30min 处理为三次重复外，其余处理均为 5 次重复，每次处理的种子数为 20~25 粒。玉米花粉处理时间为 0s、18s、54s、90s、144s、180s、270s 和 360s，试验进行 3 次重复。

3.2.2 玉米种子体外萌发

玉米种子播于 2cm 深的沙床上，盖沙 2cm，室温光照培养 5d（张同祯等，2014）。

3.2.3 玉米花粉体外萌发

玉米花粉体外萌发试验参照 Suen 等（Der and Huang, 2007）。田间收集的新鲜花粉进行 ARTP 处理后立即放入液体浅层花粉萌发培养基中，室温培养 30min 后观察并统计花粉萌发率。

3.2.4 测定指标及其方法

本研究主要考查玉米种子的出芽率、根长、苗高、根长/苗高比、干重和玉米花粉的萌发率共6个指标。其测定方法为：

出芽率：培养5d正常发芽的种子数/供试种子数×100%。

根长：用直尺分别测量幼苗的主根长度，单位为（cm）。

苗高：用直尺测量幼苗地上部分的植株长度，单位为（cm）。

根长/苗高比：根长/苗高。

干重：将整株幼苗放入纸袋，置于烘箱中60℃烘干至重量恒定（约7d），用天平称其重量，单位为（mg）（石海春等，2005）。

花粉萌发率：显微镜下观察花粉管萌发，每个样品随机挑取10个视野进行统计，花粉管伸长大于花粉直径计为花粉萌发（靳芳等，2011；叶利民，2012）。花粉萌发率=（花粉萌发数/花粉总数）×100%。

3.2.5 数据统计及分析方法

数据采用均值±标准差来表示，组间比较使用单因素方差分析（one way ANOVA）方法。* 代表 $P<0.05$，** 代表 $P<0.01$，*** 代表 $P<0.001$。

参考文献（略）

本文原载：分子植物育种，2016（5）：1 262-1 267

转基因玉米种子萌发期抗旱性鉴定

冷益丰[1]　张　彪[1]　赵久然[2]　杨俊品[1]　刘　亚[2]　康继伟[1]
陈　洁[1]　唐海涛[1]　谭　君[1]　何文铸[1]

(1. 四川省农业科学院作物研究所，成都　610066；
2. 北京市农林科学院，北京　100097)

摘　要：采用聚乙二醇（PEG）高渗溶液，对13个转基因玉米自交系进行了种子萌发期模拟水分胁迫发芽试验，研究了渗透胁迫模拟干旱对抗旱转基因玉米种子萌发状况的影响。结果表明：干旱胁迫对玉米发芽势、发芽率、根数、胚根长、胚芽长、贮藏物质转化率等均有不同程度的影响，不同转基因玉米自交系在抗旱性上存在明显差异。同时，使用隶属函数法对参试转基因玉米自交系的种子萌发抗旱性进行了分析评价，其中SD10为萌发期抗旱性较强的自交系；CBF-3-4、SD13、SD05、CBF-1-4、CBF-6-5、SD-P3、SD-P5为萌发期抗旱性中等的自交系；SD06、CBF-3-1、SD07、SD-P1、SD-P4为萌发期抗旱性较弱的自交系。种子萌发抗旱指数与隶属函数对玉米自交系抗旱性分析结果比较一致。

关键词：转基因玉米；种子萌发；渗透胁迫；隶属函数

Identification of Drought Resistance of Transgenc Maize During Seed Germination Stage

Leng Yifeng[1]　Zhang Biao[1]　Zhao Jiuran[2]　Yang Junpin[1]　Liu Ya[2]
Kang Jiwei[1]　Chen Jie[1]　Tang Haitao[1]　Tan Jun[1]　He Wenzhu[1]

(1. *The Crop Research Institute*, *Sichuan Academy of AgricalturalSciences*, *Chengdu* 610066, *China*; 2. *Being Academy of Agricultural&Forestry Science*, *Being* 100097, *China*)

Abstract: The experiment was carried out with 13 transgenic inbred lines of maize under the polyethylene glycol (PEG) simulated water stress to study the impact of osmotic stress on seed germination. The results showed that the germination potential, germination rate, root number, radicle length, plumule length and transforming rate of storage substance in maize were affected by drought stress to different extend, and there existed significant differences in drought resistance among different inbred lines. The quantitative evaluation by using subordinate function method indicated that SD10 was

---

基金项目：转基因生物新品种培育重大专项（2009ZX08003-007B）、国家玉米支撑计划（2011BAD35B01）、国家玉米863计划（2011AA10A103-2）、四川省农科院青年基金（201lQNJJ-008）

作者简介：冷益丰（1986—　），男，重庆垫江人，硕士，主要从事玉米分子育种。Email：yifeng_71@163.com

通讯作者：何文铸，副研究员，主要从事玉米育种工作。E-mail：wenzu-he@163.com

the inbred line with high drought-resistance; CBF-3-4, SD13, SD05, CBF-1-4, CBF-6-5, SD-P3 and SD-P5 were the inbred lines with mid drought-resistance; and SD06, CBF-3-1, SD07, SD-P1 and SD-P4 were the inbred lines with low drought resistance. The analysis results obtained by grain drought resistant index (GDRI) were similar to those by subordinate function method.

Key words: Transgenic maize; Seed germination; Osmotic stress; Subordinate function

干旱已成为影响世界粮食减产的重要因素之一，近3年统计数据表明，因干旱造成的损失在15.6%~48.5%，严重影响农业生产的可持续发展。近年来我国西南地区伏旱天气出现频繁，玉米从播种到收获整个生育期均受到不同程度干旱的影响，从而导致玉米减产甚至绝收。培育抗旱玉米品种是减少干旱制约玉米生产的有效途径。不同玉米自交系或品种的抗旱性差异较大，目前，国内外学者从不同方面开展了玉米抗旱性鉴定研究，已形成规范的利用PEG模拟干旱胁迫试验方法，如《国际种子检验规程》（ISTA）。本试验通过高渗溶液法，利用-0.6MPa PEG-6000水溶液模拟干旱，对13个抗旱转基因玉米自交系进行了种子萌发期抗旱性筛选试验，采用隶属函数法对玉米自交系的耐旱性进行综合评价，以期筛选出抗旱性较强的玉米材料。

1 材料与方法

1.1 试验材料

供试材料为13个转基因玉米自交系SD05、SD06、SD07、SD10 SD13、SD-P1、SD-P3、SD-P4SD-P5、CBF-3-4、CBF-3-1、CBF-1-4、CBF-6-5和3个非转基因玉米自交系（对照）478、S17、81565，其中478为SD系列野生型对照，S17为CBF系列野生型对照，81565为周树峰等鉴定的耐旱玉米自交系。试验于2011年5—8月在四川省农科院作物所实验室进行。CBF系列、SD和SD-P系列分别是以拟南芥CBF4基因、盐生植物盐芥 *TsDREB2A* 基因和玉米PIS基因为目标基因，构建高效表达载体，采用基因枪法或农杆菌介导法转入玉米受体先早17和478中，批量获得转基因植株，并经多代连续筛选分子检测呈阳性且遗传稳定纯合的转基因玉米株系。

1.2 试验方法

选取整齐一致、无破损的玉米种子，用7%漂白粉溶液消毒2~3min，灭菌蒸馏水冲洗3遍后再用滤纸吸干种子表面附着水。将种子置于直径为9cm的灭菌培养皿中，每个培养皿50粒，双层滤纸做发芽床，加入20ml -0.6MPa PEG-6000水溶液做干旱胁迫处理，同时进行蒸馏水正常培养，干旱胁迫与正常均设3次重复，种子萌发于人工气候箱中25℃黑暗条件下进行。种子萌发期间培养皿加盖，每天补充适量蒸馏水，避免水分蒸发以保持渗透势不变。

1.3 指标测定

种子萌发期间每隔2d调查1次种子发芽情况（以胚根或胚芽突破种皮2mm为发芽标准计算发芽种子数），持续调查至第8d。在种子萌发试验进行的第4d调查种子发芽势，第7d调查种子发芽率；第8d统计发芽种子根数，测量胚芽和胚根长度，将胚芽、胚根以及籽粒剩余部分分别用纸包好置于105℃烘箱中杀青5min后，于80℃恒温烘至恒重，分别称量干重。

分别采用以下式（1）、式（2）、式（3）、式（4）、式（5）计算种子萌发率、萌发指数、萌发抗旱指数、相对发芽率以及贮藏物质转化率：

$$nd = \frac{X_{Ger}}{X_{TS}} \times 100\% \quad (1)$$

$$PI = 1.00nd_2 + 0.75nd_4 + 0.50nd_6 + 0.25nd_8 \quad (2)$$

$$GDRI = \frac{PI_S}{PI_C} \quad (3)$$

式中，nd 为种子萌发率，nd_2、nd_4、nd_6 和 nd_8 分别为第 2d、4d、6d、8d 的种子萌发率；X_{Ger} 为在特定时间的种子萌发数；X_{TS} 为种子总数；PI 为种子萌发指数，PI_s 和 PI_c 分别为胁迫和正常条件下种子萌发指数；GDRI 为种子萌发抗旱指数。

$$相对发芽率（\%）= \frac{胁迫发芽率}{正常发芽率} \times 100 \quad (4)$$

$$贮藏物质转化率（\%）= \frac{（芽+根）干重}{（芽+根+籽粒剩余部分）干重} \times 100 \quad (5)$$

隶属值计算参照侯建华等的方法进行，隶属值（U_{ij}）=（$X_{ij}-X_{imin}$）/（$X_{imax}-X_{imin}$），其中 U_{ij} 为某玉米自交系对于第 i 项指标的隶属值，X_{ij} 为某玉米自交系第 i 项指标测定值，X_{imin} 为全部玉米自交系 i 项指标的最小值，X_{imax} 为全部玉米自交系 i 项的最大值，i 为某项指标，j 为某个玉米自交系。并求出各玉米自交系隶属值平均值。

1.4 数据处理

采用 Excel 进行数据录入与整理，使用 SPSS16.0 软件进行数据的相关统计分析。

2 结果与分析

2.1 种子萌发抗旱指数与玉米抗旱性

种子萌发抗旱指数是种子萌发期进行抗旱性鉴定的重要指标，抗旱性强的材料种子萌发抗旱指数较大。早在 1984 年 Bouslama 等在研究大豆的抗逆性时提出种子萌发胁迫指数的概念，孙彩霞等将其称为种子萌发抗旱指数（GDRI）。本试验结果表明（表 1），不同的转基因玉米种子在萌发时期表现出不同的抗旱性，种子萌发抗旱指数从 0.031 9~0.272 7。CBF 系列的 CBF-3-4、CBF-6-5 两个自交系的种子萌发抗旱指数大于其野生型对照 S17；SD 系列的 SD-P5、SDlO、SD13、SD05、SD-P3 五个自交系的种子萌发抗旱指数大于其野生型对照 478。CBF-3-4、CBF-6-5 的种子萌发抗旱指数大于抗旱材料 81565，属于可利用的适应四川生态区抗旱性较强的玉米材料。

表 1 不同玉米自交系在干旱胁迫下的种子萌发抗旱指数

Table 1 Grain drought resistant index (GDRI) of different transgenic inbred lines of maize under drought stress

材料 Material	PIs	PIc	GDRI
SD05	0.105 0	1.246 7	0.084 2
SD06	0.071 7	1.508 3	0.047 5
SD07	0.056 7	1.318 3	0.043 0

（续表）

材料 Material	*PIs*	*PIc*	*GDRI*
SD10	0.128 3	0.995 0	0.129 0
SD13	0.171 7	1.370 0	0.125 3
CBF-3-4	0.241 7	0.991 7	0.243 7
CBF-3-1	0.050 0	0.741 7	0.067 4
CBF-1-4	0.086 7	0.995 0	0.087 1
CBF-6-5	0.181 7	0.863 3	0.210 4
SD-P1	0.045 0	0.921 7	0.048 8
SD-P3	0.090 0	1.251 7	0.071 9
SD-P4	0.050 0	1.566 7	0.031 9
SD-P5	0.110 0	0.801 7	0.137 2
S17	0.081 7	0.816 7	0.100 0
478	0.061 7	1.011 7	0.061 0
CK	0.106 7	0.548 3	0.194 5

注：PIS 和 PIC 分别为干旱胁迫和正常萌发条件下的种子萌发指数；GDRI 为种子萌发抗旱指数；CK（对照）为四川生态区耐旱玉米自交系材料 81565。下同

Note：PIS and PIC mean seed germination index under drought stress and normal germination, respectively; GDRI means grain drought resistant index; and CK means 81565, a drought-tolerant inbred line of maize in Sichuan province. The same below

2.2 干旱胁迫对不同玉米自交系发芽率的影响

干旱胁迫处理下各玉米自交系的种子萌发明显受到抑制（图1A），种子发芽势弱、发芽率低，萌发的种子胚根、胚芽短小，有的种子甚至仅有胚根，而胚芽很难突破种皮，同时萌发的种子大多仅有胚根长成的主根而无侧根（图1B）。

 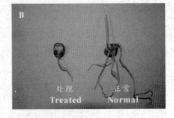

图1 干旱胁迫处理下的种子萌发表现

Figure 1　Seed germination performance under drought stress

干旱胁迫下的种子发芽率可以说明玉米自交系的抗旱发芽能力。干旱胁迫下的玉米种子萌发率较正常萌发明显下降，且不同转基因玉米自交系间差异显著（$F=70.83>F_{0.05}=2.00$，图2）。CBF 系列 CBF-3-4、CBF-6-5、CBF-1-4 的种子发芽率超过其野生型对照 S17；SD 系列 SD13、SD-P5、SD10、SD05、SD-P3 的种子发芽率超过其野生型对照 478，

同时种子发芽率均比对照材料81565高，说明抗旱基因的导入增强了玉米的耐旱性。CBF-3-4在干旱胁迫处理下的种子发芽率最高，且与正常萌发时的差异最小，说明干旱对其萌发影响最小，其抗旱能力较强，在后续育种中可对其加以利用。

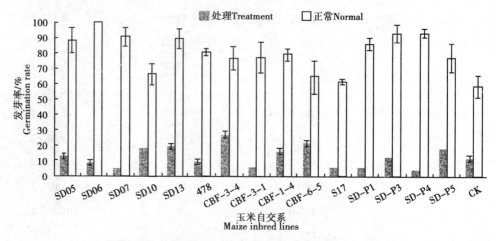

图2　不同转基因玉米自交系在干旱胁迫下的种子萌发率

Figure 2　Seed germination of different transgenic inbred lines of maize under drought stress

2.3　干旱胁迫对不同玉米自交系胚根、胚芽生长的影响

PEG-6000渗透胁迫对各玉米自交系的胚根、胚芽生长均有影响，自交系间敏感程度存在明显差异。种子根系是响应干旱环境较明显的应激部位，胁迫条件下不同转基因玉米自交系的种子根系生长差异显著（$F=108.350>F_{0.05}=2.00$，表2），SD07、SD-P4等5个自交系种子萌发后仅有胚根长成的主根，其余11个自交系除主根外还有1~4条侧根。CBF系列转基因玉米的根系均比其野生型对照发达，而SD系列只有SD10的根系生长超过其野生型对照478。6个转基因玉米自交系根数超过对照81565，其中SD10种子根系最发达，平均根数达到3.9，且其在干旱胁迫处理下的种子根数与正常萌发最为近，CBF-1-4也占到正常萌发时的50%以上。

抗旱能力强的材料其种子萌发时胚根或胚芽生长较快，与正常条件下的发芽表现差异愈小。本试验中干旱胁迫下萌发的种子胚根都能够突破种皮进行根的营养生长，但自交系间种子萌发后的胚根长差异显著（$F=130.351>F_{0.05}=2.00$，表2）。SD系列SD10、SD-P3、SD-P5、SD05的胚根长超过其野生型对照，同时也超过对照81565，而CBF转基因系列只有CBF-3-4、CBF-1-4的胚根长超过其野生型对照，同时也超过对照81565。CBF-3-4种子萌发后的胚根最长，平均达到3.87cm；材料SD-P5的胚根长度与正常萌发差异最小，二者相差36%，材料CBF-3-4和CBF-1-4次之，约为正常萌发时的63%和59%。

玉米种子在干旱胁迫下胚芽的生长受到明显抑制，自交系间种子萌发后的胚芽长差异显著（$F=29.648>F_{0.05}=2.00$，表2），这以自交系SD06、SD07、SD-P1、SD-P4的萌发表现最为明显，其种子胚芽不能突破种皮而仅有胚根出现，其余自交系的胚芽能够突破种皮生长，但胚芽长度均不超过1cm，其中SD10种子萌发后的胚芽最长，平均为0.95cm，其在干旱处理下的种子胚芽长度与正常萌发差异最小，为正常萌发胚芽长度的20%。

表2 不同转基因玉米自交系在干旱胁迫下的种子胚根、胚芽生长
Table 2 Seed radicle and plumule growth of different transgenic inbred lines of maize under drought stress

材料 Material	根数 Number of roots		胚根长（cm）Radicle length		胚芽长（cm）Plumule length	
	处理 Treated	正常 Normal	处理 Treated	正常 Normal	处理 Treated	正常 Normal
SD05	2.7±0.2d	5.7±0.6de	2.89±0.19g	7.74±0.80de	0.37±0.05e	3.64±0.95abc
SD06	1.0±0.0a	6.2±0.3ef	1.93±0.22d	7.54±0.94cde	0.02±0.00a	4.31±0.86abc
SD07	1.0±0.0a	5.9±0.2de	0.82±0.11b	6.40±0.23bcde	0.00±0.00a	2.98±0.25ab
SD10	3.9±0.1f	6.6±0.4ef	3.68±0.03i	7.94±0.73e	0.94±0.27g	4.67±1.09abc
SD13	2.3±0.0c	5.0±0.3bcde	2.81±0.09g	5.52±0.85abcd	0.34±0.12e	3.77±1.07abc
CBF-3-4	1.8±0.2b	6.1±0.8ef	3.87±0.10i	6.11±1.02abcde	0.09±0.01ab	4.58±0.32abc
CBF-3-1	1.3±0.5a	3.8±0.2abc	1.73±0.03d	4.65±0.30ab	0.10±0.07a	2.44±0.80a
CBF-1-4	2.9±0.2d	5.0±1.4bcde	2.47±0.10f	4.14±0.91ab	0.71±0.07f	4.30±0.19abc
CBF-6-5	1.1±0.1a	4.3±0.8abcd	1.86±0.08d	5.66±0.32abcde	0.06±0.01ab	4.54±0.91abc
SD-P1	1.0±0.0a	3.4±1.0ab	1.20±0.27c	4.79±0.86ab	0.00±0.00a	2.72±0.65ab
SD-P3	3.5±0.2e	7.6±0.6f	3.29±0.33h	7.49±0.96cde	0.41±0.08e	4.77±0.56abc
SD-P4	1.0±0.0a	6.2±0.4ef	0.52±0.08a	5.33±0.33abc	0.00±0.00a	5.37±0.21bc
SD-P5	2.1±0.1c	3.7±0.9abc	3.34±0.10h	5.24±0.14abc	0.26±0.05cde	2.55±0.49ab
S17	1.0±0.0a	4.9±0.5bcde	1.01±0.02bc	5.61±1.00abcde	0.00±0.00a	6.34±0.80c
478	3.5±0.2e	5.1±0.4cde	2.90±0.22g	6.20±0.24abcde	0.33±0.01de	4.15±0.36abc
CK	2.1±0.2c	2.7±0.6a	2.20±0.18e	3.93±0.34a	0.21±0.01bcd	4.38±0.24abc

注：同列不同小写字母表示SSR0.05水平差异显著。下同

Note：Different small letters in the same column show significant difference at SSR 0.05. the same below

2.4 干旱胁迫对不同玉米自交系种子萌发期贮藏物质转化率的影响

不同转基因玉米种子萌发后的胚根干重差异显著（$F=194.012>F_{0.05}=2.00$，表3），SD系列SD06、SD10的胚根干重超过其野生型对照478，CBF转基因系均比野生型强，13份转基因玉米材料只有SD-P4种子萌发后的胚根干重小于对照81565。SD06的胚根干物质量最大，平均每粒种子胚根重0.0148g，且其与正常萌发时差异最小，达到正常萌发时的61.6%。不同转基因玉米种子萌发后的胚芽干重差异显著（$F=110.641>F_{0.05}=2.00$，表3），SD06种子萌发后的胚芽干重最重，平均每粒种子胚芽重0.0050g，SD10和CBF-1-4次之，其中超过对照81565的自交系有5个；供试13个转基因玉米自交系中以SD06种子萌发后的胚芽干重与正常萌发时差异最小，为正常萌发时的25.6%。种子发芽是一个籽粒贮藏物质从胚乳转运到胚根、胚芽进行营养生长的消耗过程。在干旱胁迫处理下，不同转基因玉米种子萌发时籽粒贮藏物质转化率差异显著（$F=267.362>F_{0.05}=2.00$，表3），SD10的贮藏物质转化率最高，胚根和胚芽干重占到整个籽粒干重的7.33%，SD06和SD05次之；贮藏物质转化率超过对照81565的转基因自交系有7个，包括3个SD系列

(SD10、SD06、SD05)和全部 CBF 系列。自交系 SD06 与正常萌发时的差异最小,为正常萌发的 39.5%,SD-P5 次之。

2.5 隶属法评价玉米种子萌发期抗旱性

隶属函数均值是玉米抗旱性鉴定的重要指标,抗旱性越强的玉米材料其均值越高,目前隶属函数法在评价玉米抗旱性中得到了广泛应用。本研究从测定指标中选出发芽势、发芽率、根数、胚根长、胚芽长、贮运物质转化率、种子萌发抗旱指数 7 个指标进行隶属值分析,利用隶属函数均值对 13 个转基因玉米自交系种子萌发期的抗旱性进行综合评价。

根据侯建华等隶属值计算方法和付凤玲等的评价标准,隶属值≥0.7 的为高抗自交系,0.7>隶属值≥0.4 为中抗自交系,隶属值<0.4 为弱抗自交系。本试验中转基因玉米材料 SD10 各指标的隶属值均值最高,为萌发期抗旱性较强的自交系;CBF-3-4、SD13、SD05、CBF-1-4、CBF-6-5、SD-P3、SD-P5 为萌发期抗旱性中等的自交系;SD06、CBF-3-1、SD07、SD-P1、SD-P4 均值在 0.4 以下,为萌发期抗旱性较弱的自交系(表 4)。

表 3 不同转基因玉米自交系在干旱胁迫下的种子胚根干重、胚芽干重
Table 3 Dry weight of radicle and plumule of different transgenic inbred lines of maize under drought stress

材料 Material	胚根干重(g)Dry weight of radicles		胚芽干重(g)Dry weight of plumule		贮藏物质转化率(%)Transformation ratio of storage substance	
	处理 Treated	正常 Normal	处理 Treated	正常 Normal	处理 Treated	正常 Normal
SD05	0.008 5g	0.024 5abc	0.003 3e	0.018 0bc	3.98g	16.95de
SD06	0.014 8j	0.024 1abc	0.005 0g	0.019 5bc	6.38h	16.18cde
SD07	0.002 3bc	0.023 6abc	0.000 0a	0.017 1bc	0.87b	17.17de
SD10	0.013 5i	0.028 8bc	0.004 6f	0.023 9c	7.33i	25.91f
SD13	0.0059e	0.041 8c	0.002 2d	0.018 4bc	3.16ef	24.98ef
CBF-3-4	0.007 4f	0.018 8ab	0.000 2ab	0.015 4abc	2.97e	15.88cde
CBF-3-1	0.004 2d	0.011 8ab	0.000 4ab	0.006 0a	1.31c	5.28a
CBF-1-4	0.005 1e	0.012 8ab	0.003 4e	0.016 6bc	2.85e	11.07abcd
CBF-6-5	0.004 1d	0.013 0ab	0.000 2ab	0.017 7bc	1.57c	12.80abcd
SD-P1	0.002 7c	0.007 8a	0.000 0a	0.013 7abc	0.89b	7.81ab
SD-P3	0.007 2f	0.024 2abc	0.001 6c	0.021 2c	2.45d	12.76abcd
SD-P4	0.000 6a	0.011 8ab	0.000 0a	0.019 0bc	0.34a	12.62abcd
SD-P5	0.005 5e	0.009 1ab	0.000 6b	0.009 4ab	2.25d	7.33ab
S17	0.001 6b	0.011 9ab	0.000 0a	0.015 8abc	0.55ab	10.59abcd
478	0.009 5h	0.020 0ab	0.001 5c	0.023 1c	3.43f	15.00bcd
CK	0.001 4b	0.007 3a	0.001 9cd	0.015 7abc	1.26c	9.47abcd

表4 不同转基因玉米自交系的隶属函数值

Table 4 Subordinate function values of different transgenic inbred lines of maize

材料 Material	发芽势 Germination potential	发芽率 Germination rate	根数 Number of roots	胚根长 Radicle length	胚芽长 Plumule length	转化率 Transformation efficiency	GDRI	平均值 Mean value	抗旱性 Drought resistance
SD05	0.23	0.39	0.6	0.7	0.4	0.52	0.217 2	0.44	中抗 Mid resistance
SD06	0.08	0.21	0.0	0.4	0.0	0.86	0.064 8	0.23	弱抗 Low resistance
SD07	0.31	0.03	0.0	0.1	0.0	0.08	0.046 1	0.08	弱抗 Low resistance
SD10	0.46	0.61	1.0	0.9	1.0	1.00	0.403 2	0.77	高抗 High resistance
SD13	0.69	0.70	0.4	0.7	0.3	0.40	0.387 9	0.52	中抗 Mid resistance
CBF-3-4	1.00	1.00	0.3	1.0	0.1	0.38	0.879 6	0.66	中抗 Mid resistance
CBF-3-1	0.08	0.06	0.2	0.4	0.1	0.14	0.147 4	0.15	弱抗 Low resistance
CBF-1-4	0.00	0.52	0.7	0.6	0.7	0.36	0.229 2	0.44	中抗 Mid resistance
CBF-6-5	0.77	0.79	0.0	0.4	0.1	0.18	0.741 3	0.43	中抗 Mid resistance
SD-P1	0.15	0.06	0.0	0.2	0.0	0.08	0.070 2	0.08	弱抗 Low resistance
SD-P3	0.08	0.36	0.9	0.8	0.4	0.30	0.166 1	0.43	中抗 Mid resistance
SD-P4	0.23	0.00	0.0	0.0	0.0	0.00	0.000 0	0.03	弱抗 Low resistancew
SD-P5	0.15	0.64	0.4	0.8	0.3	0.27	0.437 3	0.43	中抗 Mid resistance
S17	0.54	0.09	0.0	0.1	0.0	0.03	0.282 8	0.16	弱抗 Low resistancew
478	0.00	0.24	0.9	0.7	0.3	0.44	0.120 8	0.38	弱抗 Low resistance
CK	0.46	0.33	0.4	0.5	0.2	0.13	0.675 2	0.38	弱抗 Low resistancew

3 讨论

作物的耐旱性与遗传和环境密切相关，表现为多基因控制的数量性状。在选择抗旱材料或品种上目前还没有发现具体某一性状可以作为唯一的、可靠的耐旱性鉴定指标。玉米抗旱性的鉴定方法主要是田间鉴定法，但田间鉴定法周期较长，需要的人力物力较多，而且水分难以控制,而室内模拟鉴定法有利于玉米材料的早期鉴定,环境因素容易控制,可以同时鉴定较多的材料,是一种比较理想的鉴定方法。在评价方法上,因为玉米抗旱性是一个综合性状,用单一性状评价可靠性不高。本研究采用隶属函数法对13个转基因玉米自交系进行了种子萌发期抗旱性综合评价,消除了单一指标评价的片面性,隶属值平均值越大表明抗旱性越强。本试验结果表明,SD10为萌发期抗旱性较强的自交系;CBF-3-4、SD13、SD05、CBF-1-4、CBF-6-5、SD-P3、SD-P5为萌发期抗旱性中等的自交系;SD06、CBF-3-1、SD07、SD-P1、SD-P4为萌发期抗旱性较弱的自交系。在各抗旱筛选指标中,种子萌发抗旱指数与隶属函数对玉米抗旱性分析结果较为一致,生产实践中需要将两者结合起来对参试材料进行抗旱性综合评价。

目前,玉米育种中对耐旱玉米材料的筛选利用主要表现在对常规自交系和不同杂交种的鉴定和评价上,而对新型抗旱转基因材料的利用则较少。本研究对转入抗旱基因CBF4、TsDREB2A和PIS的玉米自交系进行筛选评定,研究结果表明不同转基因玉米自交系在萌发期的抗旱性上具有显著差异。种子萌发期抗旱性鉴定是玉米整体抗旱性评价的重要内容,接下来我们将利用鉴选的抗旱性较强的自交系与四川生态区优良玉米材料组配,同时也将进一步对材料在苗期、开花期及灌浆期等不同生育时期的抗旱性表现进行鉴定,全方面评价真正适合于生产上育种利用的抗旱玉米新材料,进而选育出抗旱性优良的玉米新品种。

参考文献(略)

本文原载:干旱地区农业研究,2013,31(1):177-182

玉米萌发幼苗期的抗旱性鉴定评价

徐田军[1]　吕天放[1]　赵久然[1]　王荣焕[1]　刘月娥[1]
张连平[2]　叶翠玉[2]　刘秀芝[1]

(1. 北京市农林科学院玉米研究中心/玉米 DNA 指纹及分子育种北京市重点实验室,
北京　100097；2. 北京市种子管理站, 北京　100088)

摘　要：以不同浓度 PEG-6000 水溶液模拟 4 个玉米种子萌发环境条件,对 18 个玉米生产主推品种进行萌发幼苗期抗旱性鉴定与评价。结果表明：①在 4 个萌发环境条件下,京科 968、京科 665、农华 101、联科 96、京农科 728、京单 38、京单 68 和旺禾 8 号相对发芽率(培养后 8 d 萌发率)相对较高；郑单 958、先玉 335 和辽单 565 等品种中等；农大 108、中单 28、浚单 20 和纪元 1 号较低。②相对幼苗高度、相对幼苗鲜重、相对幼苗干重、相对根系鲜重、相对根系干重、相对根系长度和相对根条数在 4 个干旱环境条件下均小于对照；不同品种间降幅不同,其中春播品种京科 968、京科 665、农华 101 等和夏播品种京农科 728、京单 58、京单 68 降幅较小。春播品种农大 108、中单 28 等以及夏播品种浚单 20 和纪元 1 号降幅较大。③综合分析得出,春播品种中京科 968、京科 665、农华 101、先玉 335 和联科 96 萌发幼苗期抗旱性强,郑单 958、农大 108 和辽单 565 萌发幼苗期抗旱性中等,中单 28 萌发幼苗期抗旱性最差；夏播品种中京农科 728、京单 38、京单 68、京单 58 和京科 528 萌发幼苗期抗旱性强,京单 28、纪元 1 号和旺禾 8 号萌发幼苗期抗旱性中等,浚单 20 萌发幼苗期抗旱性最差。

关键词：玉米；萌发幼苗期；抗旱性；鉴定评价

玉米是目前我国种植面积最大、总产量最高的第一大作物,在保障国家粮食安全中发挥着主力军作用。干旱是影响我国玉米生产的最主要因素,我国约 60% 的玉米种植面积均受到不同程度的干旱胁迫,由此造成的玉米减产幅度高达 20%~30%。种植抗旱高产玉米品种是我国现代农业绿色可持续发展的必然趋势和方向。

玉米不同生育时期的干旱胁迫对其生长发育和产量的影响不同,萌发幼苗期的干旱胁迫主要影响种子的萌发过程及出苗质量；拔节期之后干旱胁迫则导致叶片生长受到抑制、干物质积累转运受阻、根条数变少等,从而影响玉米的生长发育,在一定程度上造成玉米减产；开花期是玉米的需水临界期,干旱胁迫常导致散粉吐丝间隔增大,花粉和花丝活力降低,减产最严重；灌浆期干旱胁迫导致叶片早衰,光合能力下降,进而减产。大田生产

基金项目：北京市科技计划课题（Z151100001015014）；北京市优秀人才培养资助项目（2014000020060G177）；国家玉米产业技术体系（CARS-02-18）；北京市农林科学院科技创新能力建设专项（KJCX20170708）；国家重点研发计划项目（2016YFD0300106）

徐田军、吕天放为共同第一作者

通讯作者：赵久然, 王荣焕

中，玉米萌发幼苗期的抗旱性与田间出苗率及幼苗质量密切相关，进而直接影响留苗密度及群体整齐度。本研究以18个玉米生产主推品种为试验材料，采用PEG-6000溶液模拟不同干旱胁迫环境，在砂培模式下研究不同浓度PEG-6000胁迫处理对玉米种子萌发及幼苗质量相关指标的影响，旨在筛选萌发幼苗期抗旱性强的玉米品种。

1 材料与方法

1.1 试验材料

以18个玉米生产主推品种为试验材料。其中，春播和夏播品种各9个（表1）。

表1 试验品种信息

品种播期类型	品种名称	审定编号
春播	郑单958	国审玉2009009
	先玉335	国审玉2006026
	农大108	国审玉2001002
	农华101	国审玉2010008
	京科968	国审玉2011007
	京科665	国审玉2013003
	辽单565	国审玉2004003
	联科96	京审玉2008001
	中单28	京审玉2005001
夏播	浚单20	国审玉2003054
	京单28	国审玉2007001
	京单68	国审玉2010003
	京单58	国审玉2010004
	京农科728	国审玉2012003
	旺禾8号	京审玉2011003
	纪元1号	京审玉2005008
	京单38	京审玉2009005
	京科528	京审玉2008008

1.2 试验方法

每品种选取大小均匀一致、饱满的种子1 000粒，用7%次氯酸钠溶液消毒10min，再用蒸馏水冲洗5次，用滤纸吸干后摆放在长19cm、宽9cm、盛有一定质量细沙的发芽盒中，种子间距离为粒长1~2倍，各处理250粒，3次重复，每个品种各处理共750粒。PEG-6000水溶液质量浓度分别为40g/L、60g/L、75g/L和87.5g/L，与之对应的溶液水势分别为-0.05MPa、-0.10MPa、-0.15MPa和-0.20MPa，以蒸馏水为对照。将发芽盒置于 RXZ-1000B 型人工气候箱，其昼/夜温度为 28℃/25℃，湿度为 60%，光照/黑暗为 12h/12 h，光照强度为 134μmol /m²·s，连续培养 10d。在第 2d、4d、6d、8d 调查发芽数，并于第 8d 测定根系长度、根系条数、根系鲜重、根系干重、幼苗鲜重和幼苗干重，并计算相对发芽率、相对幼苗高度、相对幼苗鲜重、相对幼苗干重、相对根系鲜重、相对根系干重、相对根系长度、相对根系条数和相对种子萌发抗旱指数。为了减少各玉米品种间固有的差异，对各测定指标均采用干旱胁迫处理和对照测定的相对值，相对值比绝对值能更好地反映不同玉米品种的抗旱性。

1.3 数据分析

采用 Excel 2007 软件进行数据整理与作图。

2 结果与分析

2.1 相对发芽率

由表 2 可知，相对发芽率在干旱胁迫下总体呈下降趋势。随胁迫程度加大，相对发芽率降幅增大。4 个水势下，参试品种的相对发芽率介于 0.70~0.92。其中，春播品种京科 968、京科 665、农华 101 和联科 96，夏播品种京农科 728、京单 38、京单 68 和旺禾 8 号等品种相对发芽率（培养后 8d 萌发率）相对较高；郑单 958、先玉 335 和辽单 565 等品种中等；农大 108、中单 28、浚单 20 和纪元 1 号较低，平均值为 0.79。

表 2 不同 PEG-6000 水溶液浓度胁迫下参试玉米种子相对发芽率的变化

品种播期类型	品种	水势（MPa）			
		-0.05	-0.10	-0.15	-0.20
春播	郑单 958	0.88	0.83	0.83	0.79
	先玉 335	0.88	0.83	0.83	0.79
	农大 108	0.83	0.79	0.75	0.75
	农华 101	0.92	0.88	0.83	0.83
	京科 968	0.92	0.88	0.84	0.84
	京科 665	0.92	0.88	0.88	0.88
	辽单 565	0.87	0.78	0.78	0.78
	联科 96	0.92	0.88	0.83	0.83
	中单 28	0.83	0.74	0.74	0.70
夏播	浚单 20	0.87	0.83	0.78	0.74
	京单 28	0.91	0.87	0.83	0.78
	京单 68	0.92	0.92	0.88	0.88
	京单 58	0.92	0.92	0.88	0.86
	京农科 728	0.92	0.92	0.89	0.89
	旺禾 8 号	0.91	0.91	0.88	0.87

(续表)

品种播期类型	品种	水势（MPa）			
		-0.05	-0.10	-0.15	-0.20
夏播	纪元1号	0.88	0.83	0.79	0.75
	京单38	0.92	0.92	0.92	0.88
	京科528	0.9	0.88	0.88	0.84

2.2 种子萌发抗旱指数

种子萌发抗旱指数是玉米萌发幼苗期抗旱性鉴定的重要指标。由表3可知，相对种子萌发抗旱指数随干旱胁迫程度增加总体呈下降趋势。4个水势下，参试品种的相对种子萌发抗旱指数介于0.39~0.98；春播品种相对种子萌发抗旱指数（0.68）低于夏播品种（0.76），其中春播玉米品种京科665、京科968、联科96和农华101相对种子萌发抗旱指数相对较高，郑单958、先玉335、辽单565和农大108中等，中单28最低，为0.57；夏播品种中京农科728、京科528、旺禾8号和京单68相对种子萌发抗旱指数相对较高，京单38、京单58、京单28和纪元1号中等，浚单20最低。

表3 不同PEG-6000水溶液浓度胁迫下参试相对玉米种子萌发抗旱指数的变化

品种播期类型	品种	水势（MPa）			
		-0.05	-0.10	-0.15	-0.20
春播	郑单958	0.84	0.71	0.64	0.53
	先玉335	0.84	0.71	0.64	0.53
	农大108	0.79	0.66	0.54	0.48
	农华101	0.86	0.73	0.66	0.60
	京科968	0.90	0.77	0.66	0.60
	京科665	0.90	0.77	0.71	0.64
	辽单565	0.83	0.64	0.57	0.50
	联科96	0.89	0.76	0.64	0.58
	中单28	0.78	0.59	0.52	0.39
夏播	浚单20	0.83	0.69	0.57	0.45
	京单28	0.95	0.76	0.69	0.59
	京单68	0.92	0.82	0.73	0.71
	京单58	0.89	0.81	0.69	0.63
	京农科728	0.98	0.90	0.79	0.71
	旺禾8号	0.93	0.84	0.78	0.70

（续表）

品种播期类型	品种	水势（MPa）			
		-0.05	-0.10	-0.15	-0.20
夏播	纪元1号	0.94	0.75	0.62	0.50
	京单38	0.90	0.82	0.75	0.64
	京科528	0.93	0.85	0.79	0.67

2.3 萌发性状的相对值

由表4可知，春播玉米品种京科665、京科968、农华101和联科96的根长、幼苗高度、幼苗鲜重、根系鲜重、幼苗干重和根系干重相对值较大；郑单958、农大108、辽单565和中单28各项指标的相对值比较小，对干旱胁迫比较敏感。夏播玉米品种中京农科728、旺禾8号、京单68和京科528的根长、幼苗高度、幼苗鲜重、根系鲜重、幼苗干重和根系干重相对值较大；而纪元1号和浚单20的各项指标的相对值比较小。

表4 干旱胁迫条件下参试玉米品种萌发性状的相对值

品种	水势（MPa）	性状相对值						
		X1	X2	X3	X4	X5	X6	X7
郑单958	-0.05	0.87	0.86	0.96	0.90	0.92	0.93	0.95
	-0.10	0.84	0.77	0.93	0.80	0.76	0.77	0.73
	-0.15	0.84	0.71	0.74	0.76	0.64	0.70	0.62
	-0.20	0.84	0.64	0.67	0.66	0.52	0.60	0.51
先玉335	-0.05	0.96	0.98	0.93	0.97	0.99	0.94	0.96
	-0.10	0.96	0.87	0.86	0.87	0.87	0.86	0.87
	-0.15	0.91	0.82	0.83	0.87	0.83	0.82	0.82
	-0.20	0.91	0.76	0.79	0.76	0.76	0.76	0.76
农大108	-0.05	0.85	0.87	0.95	0.90	0.81	0.88	0.76
	-0.10	0.78	0.81	0.83	0.80	0.76	0.81	0.68
	-0.15	0.74	0.73	0.83	0.80	0.66	0.74	0.59
	-0.20	0.74	0.68	0.67	0.65	0.63	0.63	0.54
农华101	-0.05	0.98	0.97	0.95	0.97	0.96	0.97	0.96
	-0.10	0.95	0.92	0.85	0.92	0.93	0.93	0.94
	-0.15	0.93	0.87	0.85	0.84	0.92	0.85	0.92
	-0.20	0.86	0.83	0.81	0.81	0.89	0.82	0.84

(续表)

品种	水势（MPa）	性状相对值						
		X1	X2	X3	X4	X5	X6	X7
京科968	-0.05	0.95	0.96	0.96	0.90	0.94	0.89	0.96
	-0.10	0.95	0.91	0.93	0.86	0.93	0.86	0.90
	-0.15	0.92	0.86	0.92	0.81	0.93	0.81	0.80
	-0.20	0.9	0.83	0.91	0.78	0.89	0.79	0.78
京科665	-0.05	1.00	0.90	0.95	0.97	0.99	0.97	0.98
	-0.10	0.97	0.90	0.95	0.93	0.96	0.94	0.96
	-0.15	0.95	0.86	0.90	0.89	0.93	0.90	0.92
	-0.20	0.92	0.84	0.87	0.85	0.89	0.84	0.86
辽单565	-0.05	0.97	0.64	0.79	0.77	0.74	0.82	0.71
	-0.10	0.83	0.63	0.79	0.69	0.71	0.76	0.65
	-0.15	0.79	0.57	0.78	0.63	0.65	0.73	0.61
	-0.20	0.59	0.51	0.73	0.58	0.59	0.64	0.52
联科96	-0.05	1.00	0.95	0.99	0.97	0.98	0.98	0.97
	-0.10	1.00	0.92	0.99	0.93	0.77	0.93	0.90
	-0.15	0.97	0.78	0.90	0.88	0.67	0.86	0.74
	-0.20	0.77	0.73	0.70	0.73	0.60	0.70	0.68
中单28	-0.05	1.00	0.98	0.95	0.87	0.94	0.91	0.93
	-0.10	1.00	0.92	0.91	0.74	0.76	0.76	0.75
	-0.15	0.97	0.89	0.88	0.62	0.73	0.64	0.75
	-0.20	0.93	0.77	0.83	0.56	0.57	0.58	0.61
浚单20	-0.05	0.81	0.93	0.94	0.90	0.87	0.82	0.78
	-0.10	0.81	0.65	0.88	0.71	0.68	0.64	0.61
	-0.15	0.78	0.6	0.82	0.56	0.51	0.48	0.43
	-0.20	0.69	0.57	0.70	0.46	0.40	0.42	0.37
京单28	-0.05	0.97	0.85	0.94	0.95	0.92	0.95	0.93
	-0.10	0.90	0.82	0.88	0.93	0.89	0.91	0.88
	-0.15	0.87	0.77	0.82	0.89	0.80	0.81	0.79
	-0.20	0.83	0.75	0.70	0.82	0.66	0.77	0.71

（续表）

品种	水势（MPa）	性状相对值						
		X1	X2	X3	X4	X5	X6	X7
京单68	-0.05	0.97	0.96	0.98	0.96	0.95	0.98	0.97
	-0.10	0.91	0.91	0.92	0.92	0.93	0.92	0.93
	-0.15	0.88	0.80	0.91	0.82	0.86	0.81	0.89
	-0.20	0.88	0.76	0.83	0.79	0.80	0.78	0.81
京单58	-0.05	0.93	0.93	0.95	0.93	0.91	0.93	0.89
	-0.10	0.86	0.91	0.90	0.87	0.88	0.91	0.81
	-0.15	0.86	0.89	0.85	0.85	0.75	0.80	0.75
	-0.20	0.83	0.74	0.79	0.72	0.66	0.76	0.68
京农科728	-0.05	0.98	0.91	0.93	0.99	0.93	0.97	0.94
	-0.10	0.94	0.86	0.92	0.97	0.80	0.95	0.93
	-0.15	0.93	0.83	0.89	0.92	0.79	0.92	0.91
	-0.20	0.92	0.81	0.81	0.84	0.65	0.9	0.89
旺禾8号	-0.05	0.97	0.94	0.94	0.93	0.97	0.95	0.95
	-0.10	0.90	0.86	0.90	0.91	0.96	0.89	0.93
	-0.15	0.90	0.84	0.88	0.89	0.73	0.85	0.84
	-0.20	0.89	0.80	0.82	0.82	0.64	0.82	0.82
纪元1号	-0.05	0.91	0.98	0.86	0.93	0.81	0.95	0.91
	-0.10	0.86	0.91	0.81	0.85	0.81	0.91	0.87
	-0.15	0.82	0.87	0.77	0.83	0.69	0.71	0.69
	-0.20	0.82	0.72	0.51	0.69	0.59	0.66	0.63
京单38	-0.05	1.00	0.97	0.98	0.92	0.93	0.92	0.95
	-0.10	0.97	0.90	0.94	0.90	0.88	0.90	0.93
	-0.15	0.93	0.90	0.81	0.83	0.83	0.83	0.89
	-0.20	0.83	0.77	0.77	0.80	0.81	0.78	0.88
京科528	-0.05	1.00	0.94	0.92	0.98	0.91	0.97	0.95
	-0.10	0.94	0.79	0.87	0.88	0.90	0.86	0.93
	-0.15	0.87	0.75	0.83	0.78	0.85	0.76	0.91
	-0.20	0.84	0.75	0.73	0.77	0.74	0.76	0.86

注：X1. 相对根条数；X2. 相对根系长度；X3. 相对幼苗高度；X4. 相对幼苗鲜重；X5. 相对根系鲜重；X6. 相对幼苗干重；X7. 相对根系干重

3 讨论与结论

玉米的抗旱性在品种间存在较大的遗传差异。用高渗溶液进行干旱模拟可代替土壤水分胁迫处理并获得比较可靠的结果。玉米生产中，萌发出苗期是实现苗全、苗齐、苗壮的关键时期。玉米种子萌发期抗旱性评价是节水农业研究中的热点问题之一。通过简便、快速而准确的萌发期抗旱性鉴定分析方法，筛选出萌发期抗旱性强的品种，对指导生产品种布局具有重要参考和指导意义。本研究表明，玉米种子的发芽率、种子萌发胁迫指数、初生根条数、根系长度、苗高、幼苗和根系鲜重、干重适用于评价鉴定玉米萌发期抗旱性，基于该性状的抗旱性排序与试验中的观察结果基本一致，能够有效地对试验品种的抗旱性进行评定。

4个萌发环境条件下，相对发芽率、相对种子萌发抗旱指数、相对根系长度、相对根系条数、相对根系鲜重、相对根系干重、相对幼苗鲜重和相对幼苗干重都有不同程度的降低，这说明干旱胁迫抑制了玉米种子的萌发过程。春播玉米品种京科665、京科968、农华101和联科96的根长、幼苗高度、幼苗鲜重、根系鲜重、幼苗干重和根系干重相对值较大，郑单958、农大108、辽单565和中单28各项指标的相对值比较小；夏播玉米品种中京农科728、旺禾8号、京单68和京科528的根长、幼苗高度、幼苗鲜重、根系鲜重、幼苗干重和根系干重相对值较大，而纪元1号和浚单20的各项指标的相对值比较小。

综合分析玉米萌发幼苗期的抗旱性表明，春播品种中京科968、京科665、农华101、先玉335和联科96萌发幼苗期抗旱性强，郑单958、农大108和辽单565萌发幼苗期抗旱性中等，中单28萌发幼苗期抗旱性最差；夏播品种中京农科728、京单38、京单68、京单58和京科528萌发幼苗期抗旱性强，京单28、纪元1号和旺禾8号萌发幼苗期抗旱性中等，浚单20萌发幼苗期抗旱性最差。

参考文献（略）

本文原载：中国种业，2017（4）：42-46

转 TsVP 基因玉米抗旱性鉴定研究

张志方[1] 刘 亚[2*] 任 雯[2] 戴陆园[3]

(1. 云南农业大学农学与生物技术学院,昆明 650201;2. 北京市农林科学院玉米研究中心,北京 100097;3. 云南省农业科学院生物技术与种质资源研究所,昆明 650223)

摘 要:以转 TsVP 基因玉米材料为研究对象,采用反复干旱法测定玉米苗期存活率和反复干旱存活率,并对株高、根冠比、叶片相对含水量、叶绿素含量、丙二醛含量、相对电导率、籽粒产量等指标进行抗旱性考察。另外,对试验结果进行方差分析、灰色关联度分析,并计算抗旱隶属度。结果显示,P 值<0.05,关联系数大小为株高>REC>根冠比>MDA 含量>RWC>Chl 含量,隶属度大小为 TsVP5>TsVP6>Y478;由此得出结论:苗期存活率受品种和干旱处理次数等因素的影响,差异显著;Chl 含量为辅助指标,其他为主要指标;品种抗旱性大小为 TsVP5>TsVP6>Y478。

关键词:转基因玉米;抗旱性鉴定;灰色关联度分析;隶属度

Identifying the Drought Resistance of TsVP Transgenic Maize

Zhang Zhifang[1] Liu Ya[2*] Ren Wen[2] Dai Luyuan[3]

(1. College of Agronomy and Biotechnology, Yunnan Agriculture University, Kunming 650201, China; 2. Maize Research Center, Beijing Academy of Agricultural & Forestry Sciences, Beijing 100097, China; 3. Institute of Biotechnology and Germplasm Resources, Yunnan Academy of Agricultural Sciences, Kunming 650223, China)

Abstract: Taking TsVP transgenic maize as material, we identified the survival rate as well as repeated drought survival rate after repeated drought stress. Some indexes has been picked for drought resistance identification, and the indexes included plant height, root shoot ratio, relative water content, chlorophyll concentration, MDA content, relative electric conductivity and the output. After that the job that Variance Analysis, Gray correlation degree Analysis and calculating the drought degree of membership would be done. In the end, results displayed that P value>0.05, and the result of correlation coefficient was that plant height>REC>root shoot ratio>MDA content>RWC>Chl content, then, the results of degree of membership was TsVP5>TsVP6>Y478. Therefore, the experiment came to conclusions that the difference of survival rate was obvious for the influence of variety and times of treatment, furthermore, Chl content merely be selected for the assistant index while the other items can be selected for the primary index, moreover, the drought resistance of TsVP5 was the best, then, the drought resistance of TsVP6 was better than Y478.

Key words: Genetically modified maize; Drought resistance; Gray correlation degree; Degree of membership

玉米是我国的第三大粮食作物,在国民经济中占有举重轻重的地位,因此,不断提高玉米产量一直以来都是育种家们努力的目标。然而,受到管理,品种,土质,病害,虫害,洪涝,干旱,气候等诸多因素的影响,提高玉米产量困难重重。近年来,由于人口增长、水资源污染、水源利用率低、温室效应等原因,再加上水资源本身就有限,使得水资源日益紧缺,干旱引起地玉米减产现象也越发地突出,干旱已经成为阻碍玉米产量提高的重要因素之一。赵美玲等认为,玉米的抗旱性主要体现在玉米对干旱的适应性和抵抗能力方面,因此,进行玉米转基因抗旱育种,培育出对干旱有更强适应能力的品种势在必行。抗旱鉴定是抗旱育种中很关键的环节,国内外学者在玉米抗旱性育种和鉴定方面也已做了大量研究。但受试验材料自身因素的影响,进行实际试验时,选择哪些指标能更有效地检测作物的抗旱性,仍然值得研究的问题。

本试验采用转 $TsVP$ 基因玉米材料进行研究。$TsVP$ 基因即氢离子焦磷酸酶(H^+-PPase)基因,是从盐芥(Thellungiella halophila)中克隆获得的,初步研究表明,该基因具有抗旱耐盐的功能。其抗旱机理在于,H^+-PPase 的表达,能促进生长素的极性运输,进而增强植物根系的发育,利于植物品种在干旱条件下对水分的吸收利用。利用 $TsVP$ 基因构建高效双元表达载体 pCAMBIA1300-Ubi-TsVP-als,采用农杆菌介导法,将重组载体转入玉米自交系掖 478(Y478)的芽尖中,获得了大量的转 $TsVP$ 基因植株,再经过多代连续筛选、分子检测等手段,得到了呈阳性且稳定遗传的、纯合的转 $TsVP$ 基因玉米株系。

苗期是玉米生长环节中很关键的一个时期,玉米在苗期生命力相对比较弱,容易受到外界因素干扰和毁坏。干旱影响种子的萌发、出苗,使玉米造成缺苗断垄,严重年份缺苗可达 40%~50%;同时干旱胁迫使幼苗生长受阻,芽势弱,发根量少且根短,苗弱,成活率低,对植株后期的的生长发育及产量造成严重影响。因此,本研究将主要在苗期这一关键的生育期进行,并借鉴前人在番茄,甜菜,向日葵,饲用玉米,小麦等作物的苗期展开的工作,对试验材料进行抗旱性鉴定,快速高效地筛选出抗旱性强的玉米材料,为育种工作的进一步展开提供更好的种子资源,以及科学的试验依据;对所用检测指标进行相关性分析,确定指标在抗旱鉴定中的作用和地位,为以后抗旱鉴定工作的设计和开展提供可靠的试验依据。

1 材料和方法

1.1 试验材料

转 $TsVP$ 基因玉米 TsVP5、TsVP6,非转基因受体材料玉米自交系品种掖 478(Y478),均由山东大学提供。

1.2 试验方法

试验在北京市农林科学院温室展开,各材料于 330mm×260mm 花盆中种植,每盆播 6 株,每份材料有 10 盆,对照和处理各 5 盆,平行做 3 个重复;每天下午 4:30,对幼苗适量补水至田间持水量,于两叶期间苗,每盆留 3 株;4 叶 1 心期,采用反复干旱法,对处理组进行水分胁迫,测定幼苗成活率和反复干旱存活率;反复干旱试验完成后,取样测定指标,然后恢复正常供水,于成熟收获期测定产量;对照组和处理组除了水分胁迫外,其他管理措施一致。

1.3 反复干旱法

安排3次干旱处理，每次处理调查完数据后，即进行下次处理。植株以中午萎蔫，早、晚舒展为度作为处理标准，经过处理，所有试验材料的叶片在中午前后萎蔫，50%出现不同程度枯萎，少数材料出现整株"枯死"。达到处理结果后，恢复正常供水，3d后调查成活茎数（以幼苗或分蘖叶片转为鲜绿色为成活标准）。根据统计结果，计算存活率和反复干旱存活率。

1.4 指标测定与统计计算

存活率与反复干旱存活率按下述公式计算：

$$存活率 = \frac{每个重复存活幼苗数}{每个重复幼苗总数} \times 100\% ; 反复干旱存活率(\%) = \frac{L1 + L2 + L3}{3}$$

其中，$L1$为第一次干旱后的存活率，$L2$为第二次干旱后存活率，$L3$为第二次干旱后存活率。

其他指标包括：株高、根冠比、叶片相对含水量（RWC）、叶绿素含量、丙二醛（MDA）含量、相对电导率及产量，采用常规方法测定。

1.5 数据分析与处理

数据的统计、方差分析、灰色关联度分析以及隶属度计算等工作，通过Excel软件和DPS软件完成。

隶属度指标参照王黄英等的方法计算，隶属函数公式如下：

$$U_{ij} = \frac{X_{ij} - X_{ij\min}}{X_{ij\max} - X_{ij\min}}$$

其中，U_{ij}为品种j对于指标i的隶属度；X_{ij}为品种j第i项指标测定值；$X_{ij\min}$为品种j第i项指标的最小值；$X_{ij\max}$为品种j对第i项指标的最大值；i为某项指标；j为某个品种。

抗旱胁迫指数参照齐华等的方法进行计算，公式为：

$$D_{ij} = \frac{X_{ij} - Y_{ij}}{X_{ij}} \times 100$$

其中，D_{ij}为品种j第i项指标的抗旱胁迫指数，X_{ij}为品种j第i项指标在对照组的测定值，Y_{ij}为品种j第i项指标在处理组的相应测定值。

2 结果与分析

2.1 反复干旱条件下的品种苗期存活率

利用Excel软件对幼苗存活率进行分析，未经干旱处理的对照组的存活率接近100%，干旱处理组存活率结果如图1所示，呈现以下特点：随着干旱处理次数地增加，处理组三个品种的幼苗存活率均成下降趋势，即第一次>第二次>第三次；第二次处理和第三次处理间存活率下降幅度，较第一次和第二次之间的降幅显著，此特点在材料Y478和TsVP6上表现的更为明显；品种不同，处理组每次处理的幼苗存活率均呈现TsVP5>TsVP6>Y478；第三次处理后材料TsVP5的存活率仍保持在较高的水平，在60%以上，而Y478的存活率相对最小，接近20%，TsVP6的最终存活率在40%以上。

经历三次干旱处理，幼苗反复干旱存活率仍保持着较高的水平，结果如表1所示。三

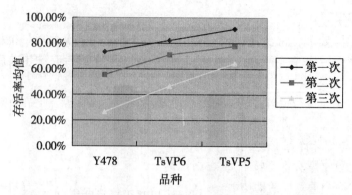

图 1 不同玉米试验材料的幼苗存活率
Figure 1 Seeding survival rate among different maize materials

个品种的反复干旱存活率大小都在 50% 以上，说明三个品种都有一定的抗旱性；但是材料 TsVP5 和 TsVP6 的幼苗反复干旱存活率均在 66% 以上，较非转基因玉米材料 Y478 的反复干旱存活率有显著提高，说明转 *TsVP* 基因玉米材料的抗旱性较非转基因材料的抗旱性优势明显；三个材料的反复干旱存活率大小依次为 TsVP5>TsVP6>Y478，说明 TsVP5 的抗旱性相对最强，TsVP6 次之，Y478 最弱。

表 1 幼苗反复干旱存活率
Table 1 Seedling survival rate after repeated drought stress

品种	第一次	第二次	第三次	均值
Y478	73.33%	55.56%	26.67%	51.85%
TsVP6	82.22%	71.11%	46.67%	66.67%
TsVP5	91.11%	77.78%	64.44%	77.78%

对三个品种的存活率进行可重复双因素方差分析，取 $\alpha=0.05$，结果显示，在品种间，处理间以及品种与处理间，P 值大小分别为 1.49E-09、5.16E-12 和 0.011，都远小于 0.05，说明受品种和处理次数因素的影响，幼苗存活率出现了显著的差异，且这种差异是可靠的。所以，可将幼苗存活率作为抗旱性鉴定的参考指标。

2.2 指标抗旱隶属度分析

本试验共选用 7 个指标进行考察，结果见表 2、表 3。对于每个品种而言，不同指标的隶属度均值的大小排名不尽相同，说明单个指标的隶属函数值不能准确地反应材料的抗旱性，必须对各项指标进行综合评价；对各项指标进行综合分析显示，各品种检测指标隶属度综合平均值大小排序为：TsVP5>TsVP6>Y478，由此认为，材料 TsVP5 抗旱性最强，更适合被选作后续试验材料，TsVP6 次之，Y478 相对最弱。

表 2 不同品种苗期形态指标及产量

Table 2 The average values of seedling morphological indexes and yield of different varieties

材料	组别	根冠比	株高（cm）	RWC	MDA 含量（μmol/g）	Chl 含量（mg/g）	REC	产量（g）
TsVP5	处理	0.630	46.180	0.778	0.015	0.003	0.704	150.463
	对照	0.529	68.260	0.969	0.009	0.004	0.281	267.883
TsVP6	处理	0.535	44.993	0.841	0.013	0.003	0.885	139.527
	对照	0.423	63.487	0.967	0.012	0.004	0.682	254.720
Y478	处理	0.549	41.700	0.787	0.015	0.003	0.644	115.030
	对照	0.450	60.140	0.972	0.009	0.004	0.305	216.304

表 3 各品种抗旱性指标隶属度

Table 3 Degree of membership of drought resistance indexes of different varieties

材料	根冠比	株高（cm）	RWC	MDA 含量（μmol/g）	Chl 含量（mg/g）	REC	产量（g）	均值
TsVP5	0.450	0.513	0.619	0.515	0.603	0.565	0.565	0.547
TsVP6	0.346	0.420	0.548	0.419	0.589	0.676	0.509	0.501
Y478	0.483	0.380	0.376	0.414	0.568	0.459	0.521	0.457

2.3 灰色关联度分析

以灰色关联度分析品种抗旱性，计算抗旱胁迫指数，利用软件 DPSv7.05 进行数据标准化，结果见表 4、表 5。

表 4 不同品种苗期形态指标及产量胁迫指数

Table 4 The average values of stress indexes of seedling morphological indexes and yield of maize varieties

材料	REC	株高（cm）	RWC	MDA 含量（μmol/g）	Chl 含量（mg/g）	根冠比	产量（g）
TsVP5	−165.49	31.93	19.73	−68.31	22.01	−21.98	42.30
Y478	−139.90	30.71	19.07	−59.63	26.96	−23.45	45.61
TsVP6	−32.12	29.02	13.01	−12.34	26.20	−28.33	44.42

表 5 不同品种苗期形态指标及产量胁迫指数均值标准化

Table 5 Standardization of the average values of stress indexes of seedling morphological indexes and yield of maize varieties

材料	REC	株高（cm）	RWC	MDA 含量（μmol/g）	Chl 含量（mg/g）	根冠比	产量（g）
TsVP5	−0.7486	0.9402	0.6645	−0.7154	−1.1429	0.7846	−1.0790
Y478	−0.3871	0.1104	0.4855	−0.4273	0.7141	0.3414	0.8957
TsVP6	1.1357	−1.0506	−1.1501	1.1427	0.4288	−1.1260	0.1833

本研究以产量为参照标准 X_0，和其他指标测定值初值构成关联矩阵，利用软件

DPSv7.05版对各项指标进行灰色关联度分析。以关联度大小排序，结果见表6。关联度从大到小顺序为：株高>REC>根冠比>MDA含量>RWC>Chl含量。由关联分析可知，关联系数的大小体现了其他数列跟参考数列的关系，二者呈正相关。Chl含量与产量的关联系数较小，说明其在抗旱鉴定时的权重位置较小，只能作为辅助指标；其他指标的关联系数较大，说明与材料的抗旱性关系相对密切，可以作为抗旱鉴定的主要测定指标。

表6 产量与其他各项指标关联系数

Table 6 Correlation coefficient between output and other indexes

关联矩阵	REC	株高(cm)	RWC	MDA含量(μmol/g)	Chl含量(mg/g)	根冠比
产量	0.6862	0.6959	0.656	0.6579	0.5985	0.6779
关联序	2	1	5	4	6	3

3 讨论

抗旱鉴定就是对作物的抗旱能力进行筛选、评价的过程，通过指标的抗旱鉴定可以体现作物的抗旱性，本试验选用存活率、反复干旱存活率、一些形态指标和生理生化指标进行研究，充分证明了转 $TsVP$ 基因玉米材料明显比非转基因材料Y478具有更强的抗旱性，在干旱情况下表现出了很强的抗逆性，利于在水分胁迫条件下存活。也进一步验证了，在苗期这一生长阶段，$TsVP$ 基因的抗旱性能显著，并且在转入玉米体内后，$TsVP$ 基因得到了充分的表达，在干旱情况下增强了根系的生长，使根冠比得到了显著提高。

不同作物在不同生长时期通常会选择不同的指标。油菜种子萌发期可以选择萌发抗旱指数和相对发芽率作为抗旱性鉴定指标，而在苗期则用叶片萎蔫指数作为抗旱鉴定的关键指标。和叶片养分相关的生理性状可以作为冬小麦抗旱鉴定的主要指标；株高可以作为青贮玉米抗旱研究的主要指标；株型和抗旱性状的关联关系较为复杂，没有ASI的关联度大；百粒重与一些玉米地方品种产量关联密切。在玉米抗旱性鉴定中，常选苗期存活率作为苗期的抗旱性鉴定指标。另外，株高，根冠比、叶片相对含水量、叶绿素含量、丙二醛含量、相对电导率、产量也可以很好地体现材料的抗旱性，比较适合用来进行转 $TsVP$ 基因材料的抗旱性鉴定。但是，各个指标在鉴定中所起的作用略有不同，叶绿素含量可被选作辅助指标，其他指标可以作为主要指标。

本研究只证明了转 $TsVP$ 基因玉米在苗期的抗旱性得到了一定程度的提高，但是在玉米生长的过程中，干旱情况时有发生，在任一个生长阶段都会出现。因此，对转 $TsVP$ 基因玉米在其他生长时期的抗旱性鉴定工作也很有必要展开，只有这样，所得到的转基因玉米品种才能真正地适应干旱的环境条件，并用于推广种植。并且，只有在各个生长阶段得到了充分的验证，才能说明 $TsVP$ 基因确实是在玉米基因组中发挥出了应有的功能，并稳定地表达，使玉米植株的抗逆性得到了增强、抗旱性能得到了提高。

从结果来看，用单个指标评判和用多个指标综合评判所得出的结果是不完全一致的，结果与毛培春等在无芒雀麦中的结果一致。因此，进行抗旱鉴定时不能只看个别指标，而应综合评价。这是因为，作物的抗旱机理十分复杂，进行植物抗旱性研究时需要综合多个性状，全面评价分析，才能提高抗旱鉴定的准确性。本文通过抗旱隶属度分析，对各项指标综合评价，鉴定出了试验材料的抗旱性强弱，也体现了全面分析和评价的重要性。

参考文献（略）

本文原载：生物技术进展，2013（3）：206-210

远红外成像技术在植物干旱响应机制研究中的应用

刘 亚

(北京市农林科学院玉米研究中心,北京 100097)

摘 要:植物干旱响应机制非常复杂,气孔开闭是植物适应干旱逆境的机制之一。一般气孔行为的改变会直接反映在一些生理指标(如气孔导度、蒸腾强度等)的改变上,继而影响叶片的温度。一旦遇到外界胁迫(如干旱)的影响,叶温发生变化将被用来监测诊断植株的受胁迫情况。远红外热成像是一种可将目标物体红外热辐射转化成热像彩图的技术,具有高分辨率、非接触、高通量的特征。为此,对植物干旱响应机制与远红外成像技术在植物上的应用研究进行综述,阐明远红外成像技术应用于植物干旱响应研究的机制,提出了远红外热成像技术应用于植物抗旱育种的可能性。

关键词:干旱;响应机制;叶温;远红外成像

Application of Infrared Thermography in the Research of Plant Mechanism Response to Drought

Liu Ya

(*Maize Research Center, Beijing Academy of Agricultural and Forestry Sciences, Beijing 100097*)

Abstract: Plant drought response mechanism is very complicated, stomatal movement is one of the mechanisms of plant adaptation to drought stress. General stomatal behavior change will directly reflect on some physiological indexes such as stomatal conductance, transpiration intensity change, thus resulting in the increase of leaf temperature. Once encounter stresses such as drought, leaf temperature changes will be used to monitor and diagnosis of plant stress condition. Infrared thermography technology can convert thermal radiation of a target object into color pictures we can see, which have characteristic with high resolution, non-contact, high-throughput. In this study, the mechanism of plant response to drought, as well as the infrared technology application in plant drought resistance research was reviewed to clarify the infrared thermography to study the mechanisms of plant response to drought, proposed the possibility of infrared thermography technology applying in the plant breeding for drought-resistance.

Key words: Drought; Response mechanism; Leaf temperature; Infrared thermography

---

基金项目:北京农业育种基础研究创新平台建设(Ⅱ)(Z08070500690802)

作者简介:刘亚,男,博士,研究方向为玉米分子育种。通讯地址:北京市海淀区曙光花园中路9号(100097),E-mail:srlyyd@gmail.com

引言

干旱是一种世界性的重大农业灾害，被认为是非常重要的非生物胁迫因素，仅次于土壤贫瘠。植物的抗旱机理十分复杂，抗旱性是植物受许多形态解剖和生理生化特性控制的复合遗传性状，不同形态解剖和生理特性之间既相互联系又相互制约。植物的气孔是一个与植物抗旱性密切相关的性状，在控制光合作用中二氧化碳的吸收以及植物蒸腾作用过程中的水分散失方面起着至关重要的作用。

在对植物的耐旱性研究中，气孔对干旱胁迫的响应行为得到了大量的研究，如引起气孔在干旱胁迫时关闭的激素 ABA 的作用机制以及它在植物体内的一系列的信号传导过程目前已成为了研究热点之一。在正常情况下植物的叶片温度通过蒸腾失水来维持相对的稳定性，一旦遇到外界胁迫（如干旱）的影响，叶温发生变化将被用来监测诊断植株的受胁迫情况。一般气孔行为的改变会直接反映在一些生理指标（如气孔导度、蒸腾强度等）的改变上，而蒸腾强度的改变通常会改变叶片表面热量损失的程度大小，继而反应在叶温的改变上。

日益灵敏的远红外成像系统为广大科研人员研究叶片表面的气孔变化及其动态的高分辨率研究提供了可行性。远红外热成像已被认为是一种非接触、高通量的大规模筛选与野生型拟南芥表现出不同叶温的突变体植株的工具。为此，笔者就植物适应干旱的响应机制、远红外技术的发展以及远红外技术应用于植物抗旱研究等方面进行了简要的阐述与探讨，以期为远红外技术应用于植物抗旱育种提供参考。

1 植物适应干旱的响应机制

1.1 植物适应干旱的机理分类

植物的抗旱性是指在干旱条件下植物生存的能力，而作物的抗旱性不仅指在干旱条件下作物的存活能力，而且强调能使作物产量稳定在一定水平的能力，不同种类的作物或同一作物的不同品种之间对干旱往往表现出不同的抗旱性。作物可以通过不同的途径来抵御或适应干旱，Levitt 认为，作物适应干旱的机理可以分为 3 类：避旱、御旱和耐旱，其中又把御旱性和耐旱性统称为抗旱性。Turner 在对作物适应干旱的机理进行了进一步的分析之后指出，避旱、高水势下耐旱和低水势下耐旱是作物适应干旱的 3 种方式。避旱是通过调节生长发育进程避免干旱的影响，如气孔关闭来节水保水，加快根系吸收水分等；高水势下耐旱是通过减少失水或维持吸水达到其维持高水势的目的，低水势下耐旱的途径则是维持膨压或者是耐脱水或干化。Hall 将作物适应干旱的机理也划分为 3 种，即御旱、耐旱和高水分利用效率，其中御旱主要通过扩展根系和调节气孔来维持体内的高水势，耐旱的机制主要是通过渗透调节，高水分利用效率的品种则能够在缺水的条件下形成较高的产量。

1.2 有关植物抗旱的形态特征

为了适应干旱的逆境条件，植物在形态上发生一些变化来保证植株正常生长，如株型紧凑、叶直立；根系发达、根冠比增大；叶片被蜡质、角质层增厚、气孔开度变小等。从形态学的角度来讲，抗旱性强的植物品种表现为根系较为发达、密集茸毛型、蜡质多、叶片厚、株型小、叶直立。对叶片的解剖结构发现，其维管束排列紧密、束内系列导管较多，导管直径较大。对禾本科作物而言，一般叶片较小、窄而长、叶片薄、叶色淡绿、叶片与茎秆夹角小、干旱时卷叶等被认为是抗旱的形态结构指标。Weerathaworn 等认为，根系与植株干重的比率可以作为玉米受水分胁迫强度的标志，在干旱条件下，抗旱品种的根与植株干重比率更高。

1.3 有关植物抗旱的生理生化特性

1.3.1 叶片水势

叶片水势是反映植物水分状况的一个可靠指标。沈维良等用 SPAC 理论分析大豆、花生、玉米、甘薯 4 种作物水势分布特点及耐旱性结果表明，植物叶片水势是植物水分状况的较好生理指标，不同作物叶片水势差异显著，作物叶片水势越高，其耐旱性越强。Kramer 认为，当植物水势和膨压减少至足以干扰植物的正常功能时即发生水分胁迫，叶片水势是植物水分状况的最好度量。大量研究表明，在干旱条件下，植物叶片水势高低是反映抗旱性强弱的重要标志。

1.3.2 光合作用

水分胁迫条件下，植物的光合作用也受到一定程度的抑制，这种抑制可能在很大程度上与 CO_2 扩散受到限制有关。植株光合作用能力的降低，与遭受干旱胁迫时叶面积的伸展、叶绿体的光化学和生物化学活性均受到很大影响之间存在紧密的联系，干旱条件下光合速率受到抑制程度越低的品种更有利于抗旱。

1.3.3 气孔导度

在遭受干旱胁迫条件下，植物的气孔趋向关闭，而当水分胁迫得到缓解时，气孔则又趋向开放。不同植物种类的气孔调节有一定的差异，一些植物则变化范围较大，可由完全开放到完全关闭。气孔开度小的品种可以减少通过气孔蒸发的水量，控制水分丢失。从短期看，不利于光合作用和一定的代谢水平；从长期看，有利于叶片水分的保持能力，从而有利于植物的光合和代谢。气孔开度大的品种，有利于光合作用和维持一定的代谢水平，但气孔水分蒸发量加大。植株失水快，不利于植株自身的生理代谢，最终反应在生物量的减少上。因此，可以以气孔导度评价品种的耐旱性。干旱条件下，气孔导度小，气孔阻力大的作物品种更有利于抗旱。

1.3.4 内源激素 ABA

植物在响应外界各种逆境包括干旱、低温、组织缺氧等，根和茎叶中合成 ABA。ABA 作为植物的内源激素，是一种小亲脂性植物激素，它在植物发育、种子休眠萌发、细胞分裂和由于干旱、寒冷、盐害、病原菌侵染以及紫外辐射等环境胁迫所诱发的细胞反应中起调节作用。水分胁迫时 ABA 的累积可能由生物合成增强与生物损害降低引起。ABA 合成被认为是首先从玉米黄质的环氧化作用开始的。土壤干旱时，植物根系可以通过合成 ABA 来感知土壤的干旱程度并作为植物根系与地上部通讯的化学信号。因此，根系 ABA 产生速度的大小直接反映根系对土壤干旱的反映敏感性的大小，而叶中 ABA 浓度的大小又反映了根系产生及向地上部运输的水平。Ivanevic 等对实验室内快速干旱胁迫下的玉米幼苗叶片和大田旱作条件下的花期叶片中 ABA 的合成进行分析，发现干旱胁迫处理使 ABA 含量成倍增加。ABA 含量的增加可诱导气孔关闭，阻止水分蒸腾，使植物本身能够保持自身代谢所必需的水分，这是植物对干旱胁迫的反应。

1.3.5 渗透调节

渗透调节是植物忍耐和抵御干旱的一种重要生理机制，渗透调节物质种类很多，主要是无机离子、游离氨基酸、可溶性糖、有机酸碱、醇类物质、可溶性蛋白分子、激素类物质等。有研究发现，渗透调节能力因水分胁迫速度而异，在缓慢干旱条件下，渗透调节能力随干旱强度加剧而增强，且耐旱品种较敏感品种增强更甚；但在快速干旱条件下，耐旱品种的渗透调节能力降低甚至丧失。

1.4 植物的气孔与抗旱性

植物叶片表皮上密布的气孔是植物与环境间进行水分和气体交换的门户，组成气孔的保卫细胞对环境条件非常敏感，可以通过水分的出入来调节气孔孔径大小，进而调控植物的蒸腾作用和光合作用等重要生理过程，气孔的启闭运动是植物适应环境变化的一种主动调节。Raschke提出哑铃形保卫细胞比肾形保卫细胞在张开时有更高的效率，即在单位孔径宽度变化时需要的体积变化更小。

气孔对土壤水分胁迫反应的传统观点认为，气孔开度受植物水分状况调节，是一种反馈式反应。当土壤变干时，植物的水分供应减少，叶水势下降，膨压随之降低而引起气孔关闭。但在另外一些禾谷类作物上，当植株的水分状况并未受到土壤干旱的影响时，气孔即已开始关闭，如 Jones 等在苹果幼苗上发现受旱植株在中午的叶水势甚至高于对照，而且这种高水势与较低的气孔导度有关，因此他认为在这些植物上与其说是水分状况控制了叶片的气孔运动，还不如说是气孔运动控制了植物的水分状。有一种理论认为，植物的气孔开度受起源于受旱根并且通过植物体内的水流传递到气孔复合体的化学信号控制，已有许多证据支持这一理论。

2 远红外成像技术的发展及其在植物抗旱研究方面的应用

2.1 远红外成像技术的发展

红外线，又称红外辐射，是指波长为 $0.78 \sim 1\,000\mu m$ 的电磁波。其中波长为 $0.78 \sim 2.0\mu m$ 的部分称为近红外，波长为 $2.0 \sim 1\,000\mu m$ 的部分称为热红外线。自然界温度处于绝对零度（-273℃）以上的物体都能发射红外电磁波谱，这些能量一般从物体表面大约 2.5×10^{-5} mm 处发射出来，不为人眼所见。环境温度里所有物体都能发射大约 $10\mu m$ 波长的远红外线。对 $8 \sim 14\mu m$ 波长敏感的探测器可以将这种射线转换成可读的温度值，这种探测器是非成像红外温度计的基础，它们可以测定物体表面的平均温度值大小。通过增加一个扫描系统远红外探测器则能产生红外热图，图中每个点的测量通过被探测物体辐射的假彩值来进行分配。红外探测器将物体红外热辐射的功率信号转换成电信号后，经成像装置的放大处理输出信号就可以完全一一对应地模拟扫描物体表面温度的空间分布，形成视频信号传至显示屏或监视器上，得到与物体表面热分布相应的热像图。物体的热辐射能量的大小，直接和物体表面的温度相关。红外热成像技术是一种非接触、无损的、热能人眼所能见的技术，人们可以利用这项技术来对目标物体进行无接触、无损的温度测量和热状态分析，从而为科学研究、工农业生产、环境保护等方面提供了一个重要的检测手段和诊断工具。

根据物体能够发射红外线的特点，各国竞相开发出各种红外热成像仪器。美国德克萨斯仪器公司（TI）在 1964 年首次研制成功第一代的热红外成像装置，叫红外前视系统，至今仍是军用飞机、舰船和坦克上的重要装置。20世纪60年代中期瑞典AGA公司和瑞典国家电力局开发了具有温度测量功能的热红外成像装置。这种第二代红外成像装置，通常称为热像仪。20世纪70年代法国汤姆逊公司研制出，不需致冷的红外热电视产品。20世纪90年代出现致冷型和非致冷型的焦平面红外热成像产品，这是一种最新一代的红外电视产品，可以进行大规模的工业化生产，把红外热成像的应用提高到一个新的阶段。近年来，新一代的热成像系统已被商业化。阵列探测器的应用排除了进一步对那些昂贵的高速扫描系统的应用。并且能在环境温度下操作的传感器也得到了发展，而先前的探测器需要冷却到低温下才能取得0.1℃的温度分辨率。这些技术的发展使得热成像的价位更能为大众接受，使用起来也更加方便。目前民用领域FLIR公司生产的ThermoCAM SC3000采用尖端的制冷型量子阱红外光电探测器（Quantum Well Infrared Photodetector）技术，温度分辨率已达到0.03℃，能够提供优异图像分辨率和无比精确温度测量能力。

2.2 远红外成像技术应用于植物响应抗旱研究机制

红外热成像技术最早应用于军事领域，近年来民用需求的不断增长也成为红外成像技术的发展的新的助推剂，其在民用方面的重要作用日益凸显。在工业、农业、环境控制、森林管理、电力、消防、石化以及医疗等方面应用也逐步完善和成熟。随着红外探测器在技术上的不断突破，热成像仪换代时间也大为缩短，应用的领域不断拓宽，测量的精度和准确度日益提高，红外热成像技术的应用发展更是日新月异。植物的能量收支及叶片的温度决定于环境因子，充足的阳光和水分是能量收支的主要影响因素及获得产量的前提。

研究植物遗传机制是当前的研究热点之一。植物耐旱方式之一是通过增强根系的吸水能力而达到抗旱的目的，另外一种方式是利用减少叶片水分蒸腾的方式来达到抗旱的目的，这种方式既使作物实现了抗旱，又节约了水资源，从而达到了抗旱与节水的一个有效的平衡，使作物能更加有效的抵御干旱灾害，真正从农业可持续发展的方向上实现作物抗旱节水的突破。

植物的气孔是植物进行气体交换的主要门户，也是水分蒸腾的主要通道，气孔开度大小对气体和水分进出气孔速度有着直接的影响，气孔的开闭在调节光合作用与水分代谢方面意义重大。环境缺水导致玉米吸水不足或不能弥补蒸腾失水而使组织脱水，植株可借自身的生理和生长调节去增加根量或根系下扎来增加吸收水分或通过气孔行为减少蒸腾而抵御干旱。通常气孔蒸腾可占总蒸腾量的80%~90%，因此减少气孔蒸腾是植物控制失水和抗旱的一个关键。在干旱胁迫条件下，气孔开度大小受到植物体的内源信号或外界环境的影响而变小，同一作物不同品种之间气孔对干旱的反应存在着一定的差异。气孔调节往往可以反应一个作物或品种对干旱的抵御能力。

植物叶片温度与蒸腾之间经多个研究证实具有较好的相关性。Kümmerlen等开发出的一种同时监测同一片叶子热成像及其气体交换的装置，证实了叶温与蒸腾之间的线性关系。Jones等也报导用热成像测量校正得到的气孔导度与用常规的气孔计得到的气孔导度值表现出了非常好的相关性。在缺水条件下，由于水分吸收无法补足水分蒸腾而使植物丢失的水分，因此气孔为了维持植物体内的水分平衡将会趋于关闭，继而导致蒸腾的减少以及叶温的升高。基于这一原理，红外成像技术在植物的生理领域得到了很大应用。

2.3 远红外成像技术在植物抗旱研究方面的应用

2.3.1 植物蒸腾的预测

植物蒸腾作用是植物以气态散失水分的生理过程，受植物蒸腾器官的形态结构与生理机能所调节。蒸腾作用不但能够促进植物体对矿质元素的吸收和传导，使溶于水的盐类迅速地输送到植物体的各个部位，同时也能够降低叶片的温度，对植物的正常生长具有非常重要的意义。植物叶片吸收的光能大部分转变为热能，通过蒸腾作用能将大量的热散发出去，否则叶片的温度将会远远高于叶片所能承受的正常温度而被灼伤枯死。植物蒸腾作用的测量一般用气孔计夹住叶片持续一段时间来进行测量，若操作不当将会导致植株叶片受到损伤。利用远红外热成像无接触、无损的特点，对植物的蒸腾以及水分状态的监测将成为可能。因为植物的叶片温度取决于天气条件，Jones 等提出应用热成像来评价田间的水分胁迫状态并结合精准矫正方法，来补偿由于环境条件变化引起的偏差。利用热成像测量校正得到的气孔导度与用常规的气孔计得到的气孔导度值表现出了非常好的相关性。目前已开发出一种同时监测同一片叶子热成像及其气体交换的装置，证实了叶温与蒸腾之间的线性关系。

2.3.2 突变体的筛选

在外界逆境（如干旱等）胁迫条件下，气孔通过由 ABA 控制的一系列信号传导过程而关闭。通过气孔开闭程度大小的调节，植物叶片的气孔导度以及蒸腾强度大小发生改变，利用热成像技术能够较易检测到蒸腾冷却效应的变化，因此，热成像技术成为一种筛选植株群体中与正常蒸腾植株呈显著差异突变体植株的有效工具。应用热成像筛选对 ABA 非敏感的突变体最初应用在大麦上。近年来这项技术已被用于模式植物拟南芥突变体的筛选。Merlot 等将正常生长 10d 的拟南芥植株持续干旱胁迫 3d 后，气孔蒸腾功能发生突变的植株与正常植株通过远红外热成像仪可以清楚的显示出来。利用可见光拍摄完全"正常"的突变体植株（图 1A）在远红外热图中叶片温度与周围的正常植株有着显著的差异（图 1B）利用这一方法筛选到了 abi1-1、OST1 和 OST2 拟南芥突变体植株，OST1 并已得到了克隆，被证明是一个由 ABA 激活的非 Ca^{2+} 依赖型的蛋白激酶。对于热成像筛选来说，激素调节的植物蒸腾是一个非常合理的目标。Song 等利用远红外成像仪，建立起气孔反应的筛选体系，在不伤害植物的前提下，对经化学诱变的拟南芥幼苗进行多种胁迫信号（干旱、H_2O_2 及 CO_2 等）单独或复合处理，以幼苗叶片温度高于或低于正常植株 0.5℃以上为筛选指标，通过对大约 6 万株诱变后的拟南芥 M_2 代幼苗进行筛选，得到拟南芥气孔突变体超过 40 株。经过对这些突变体后代进行生理和杂交遗传分析，发现所得突变体为隐性单基因突变所致，并且突变体对气孔关闭的调节上都与野生型有明显的差异。

2.3.3 植物抗旱性监测与筛选

高通量的表型鉴定技术远红外热成像对于分析与抗旱想密切相关的性状能够提供长时间的高分辨率的监测，将植物响应干旱复杂综合性状剖析成一系列的细分性状进行分析它将是一个非常有价值的工具。Giuseppe 等利用远红外热成像技术在国际玉米小麦改良中心（CIMMYT）对 92 个亚热带玉米冠层进行了分析，发现灌水条件下冠层温度明显低于干旱胁迫条件下的冠层温度，在干旱条件下不同玉米基因型的冠层温度表现出了极显著的差异，认为远红外热成像技术是一个非常有潜力加速筛选对干旱适应增强的玉米表型的工具。刘亚等以 83 个优良玉米自交系为材料，利用远红外热成像仪检测干旱胁迫条件下苗

 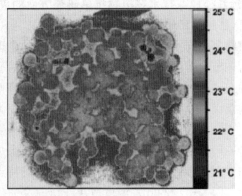

图1 远红外热图区分野生型植株中的突变体植株
A 为可见光拍摄图，B 为与可见光拍摄图一一对应的红外热成像图

期叶片温度的变化。结果发现遭受干旱胁迫时，玉米苗期叶片温度变化可以显著反映玉米苗期的耐旱性，叶温差可以作为玉米苗期耐旱性初步筛选的一个指标，将远红外热成像技术运用于玉米耐旱育种存在可行性。Inagaki 等通过对 11 个小麦在中度水分胁迫条件下的冠层温度利用远红外热成像仪进行了比较分析，发现不同基因型见表现出了轻微的差异，认为冠层温度可以作为在水分亏缺情况下小麦叶片热变化的一个可靠的指标，远红外热成像仪可以迅速的将它们之间的差异区分开来。Marcelo 等将叶温作为筛选耐旱品种的指标之一对 8 个甘蔗品种进行了研究，发现在干旱胁迫条件下，耐旱品种的叶温平均较不耐旱品种低 2.2℃，这个结果与应用其他 3 个生理抗旱指标叶绿素荧光 Fv/Fm、叶绿素相对含量 SPAD 指数、叶片相对含水量 RWC 对耐旱与非耐旱品种的划分是一致的，表明远红外热成像技术在耐旱性筛选方面是较为可靠地，而且可以做到无损害、快速的进行评价。

3 结语

植物抗旱性是一个由多基因控制的复杂性状，其中气孔则在植物抗旱过程中扮演者一个非常重要的角色。通过气孔开闭调节使植物更加适应干旱环境，可以使植物有效的抵御干旱逆境。植物叶片温度与蒸腾呈显著相关性的特点使红外成像技术在研究植物响应干旱机制方面成为可能。远红外热成像仪能够输出高分别率的热图，是整个叶片温度的一个整体精确的反映，它提供了一个非常有用的研究叶温响应水分胁迫的工具。通过远红外热成像技术的应用，植物抗旱机制有了一种新的研究手段和方法，其过程也将会更加快速高效。

参考文献（略）

本文原载：中国农学通报，2012，28（3）：17-22

2002—2009年东北早熟春玉米生育期及产量变化

王玉莹[1,2]　张正斌[2]　杨引福[1]　王 敏[1,2]　赵久然[3]　杨国航[3]

（1. 西北农林科技大学农学院，陕西杨凌　712100；2. 中国科学院遗传与发育生物学研究所农业资源研究中心，石家庄　050021；3. 北京市农林科学院玉米研究中心，北京　100097）

摘　要：【目的】分析气候变暖背景下中国东北春玉米生育期及产量的变化趋势，为品种栽培和育种目标方向提供科学依据。【方法】利用2002—2009年国家东北早熟春玉米品种区域试验中对照品种生育期、产量和气候资料等相关数据，分析气候变暖对东北早熟春玉米生产及育种的影响。【结果】对照品种的生育期、营养生长期、花粒期与相应期间的活动积温呈较强的正相关，随着活动积温的增加，生育期和花粒期均有延长的趋势。对照品种产量和年平均气温呈显著正相关（$r=0.647^*$，Sig.$=0.041$，$\alpha=0.05$）。随着年平均气温的增加，对照品种产量有进一步提高的趋势。同时，随着生产水平、栽培方式等因素的变化，东北玉米种植密度呈增加趋势。【结论】气候变暖显著影响中国东北早熟春玉米的生育期、产量、生产栽培和育种。选育总生育期长、营养生长期短但生殖生长期长、耐密植的高产品种是东北春玉米适应气候变暖的育种方向。

关键词：气候变暖；东北地区；春玉米；生育期；产量

Growth Period and Yield of Early-Maturing Spring Maize in Northeast China from 2002-2009

Wang Yuying[1,2]　Zhang Zhengbin[2]　Yang Yinfu[1]
Wang Min[1,2]　Zhao Jiuran[3]　Yang Guohang[3]

（1. College of Agronomy, Northwest Agricultural and Forestry University, Yangling 712100, Shaaxi; 2. Center of Agricultural Resources Research, Institute of Genetics and Developmental Biology, Chinese Academy of Sciences, Shijiazhuang 050021; 3. Maize Research Center, Beijing Academy of Agricultural & Forestry Sciences, Beijing 100097）

Abstract：【Objective】In the context of global climate change, the objective of this research is to analyze the phenophases and yield changes of early-maturing spring maize in Northeast China and provide scientific evidence for the variety cultivation and breeding direction of maize. 【Method】The phenological phase data, yield traits and meteorological data from 2002 to 2009 at 10 stations in 4

基金项目：国家"973"项目（2010CB951500）
联系方式：王玉莹，E-mail：wyy0624@126.com
通讯作者：张正斌，E-mail：zzb@sjziam.ac.cn；杨引福，E-mail：yinfuyang@163.com

provinces were analyzed. The possible effects of climate warming on spring maize production and breeding in Northeast China were analyzed. 【Result】Growth period, vegetative period and anthesis kernel stage of check varieties showed a strong positive correlation with the active accumulated temperature. Along with the increase of active accumulated temperature, the growing period and anthesis kernel stage tended to be longer. There is a significant positive correlation between check varieties yield and annual average temperature ($r = 0.647^*$, Sig. = 0.041, α = 0.05). With the increase of the average temperature, the check varieties production presented an increasing trend. Meanwhile, spring maize cultivation methods have been changed, the optimum density was changed from low to high. 【Conclusion】These results mean that climate warming has a significant positive impact on spring maize growth period, production, cultivation and breeding direction in the Northeast China. Therefore, breeding direction needs to be adjusted to fit the climate warming, intensive planting, longer growth period, pest and disease resistance, drought resistance, high quality and high yield should be the breeding objective of spring maize in the future in Northeast China.

Key words: Climate warming; Northeast China; Spring maize; Growth period; Yield

【研究意义】目前，气候变暖已成为国内外研究热点，尤其是气候变暖对农业生产影响的研究最为突出。农业是最容易受气候变化影响的系统之一，因为气象变量控制着作物生长和发育的基本过程。气候变暖直接导致中国粮食生产的气候资源条件变化，直接影响作物布局和农业生产结构的调整。在气候变化的背景下，植物育种必须重新定向。联合国粮食与农业组织（FAO）最近指出，作物品种的改良应该适应气候变化以保证粮食产量的可持续增长。【前人研究进展】玉米起源于南美洲地区，是 C_4 高光效喜温作物，由于气候变暖以及玉米作为食品、饲料等化工原料的广泛应用，世界玉米种植面积和产量都大幅度增加。玉米成为全球第一大谷物，这种作物严重依赖气候条件。一般情况下，玉米在墨西哥会受极端气候（如干旱、洪水）和潜在气候变化的影响减产，但在墨西哥的高海拔地区，气候变化对玉米不一定有负面影响。Meza 等的动态模拟试验结果表明，在智利玉米受气候变化的影响产量减少 5%~10%。Tubiello 等利用哈德利中心模型、加拿大中心气候模型和农业技术转移决策支持系统（DSSAT）对美国 45 个具有代表性的试点进行分析，发现气候变化有利于美国北部地区的粮食生产而对美国的南部地区的粮食生产不利。近年来，中国气候变暖，玉米种植面积和总产急剧上升，已成为中国第一大粮食作物。2011 年玉米种植面积已达 0.34 亿 hm^2，成为中国种植面积最大的粮食作物，其产量达 1 928 亿 kg，占全国粮食总产量的 33.7%，而其产量的增加主要来自中国东北地区。东北早熟春玉米区是中国玉米主产区之一，包括黑龙江第一、二积温带、吉林东部和北部、辽宁东北部、内蒙古兴安盟、通辽、赤峰、呼和浩特等地区。2002 年辽宁、吉林、黑龙江、内蒙古四省（区）玉米播种面积分别为 159.88 万 hm^2、290.15 万 hm^2、217.95 万 hm^2、167.56 万 hm^2。截至 2012 年，辽宁种植面积增至 215.14 万 hm^2，增幅为 34.56%；吉林玉米播种面积增至 316.62 万 hm^2，增幅为 9.12%；黑龙江种植面积增至 536.04 万 hm^2，增幅为 145.95%；内蒙古种植面积增至 325.09 万 hm^2，增幅为 94.01%。辽宁、吉林两省播种面积小幅增长，内蒙古增加幅度较大，黑龙江是东北玉米扩种的主要区域。近百年东北地区气温变化的趋势与全球及中国的气温变化的总趋势一致，是 20 世纪中国变暖趋势最为明显的地区之一。气候变暖使中国东北地区热量资源增加，农作物生长季延长，农业种植界线向北移动，农作物种植面积扩大，有利于玉米产量增加。贾建英等对东北地区近46年（1961—2006年）玉米气候资源变化进行研究指出，与20世纪60年代相比，东北地区玉米气候资源（≥10℃初日、初霜日、生育期天数、≥10℃有效积温）在2001—2006年增加较显著，不同熟型的玉米分布界线在2001—2006年北移东扩很显著。高永刚等的模拟研究发现，气温变化趋势的增高是1961—2003年黑龙江省玉米模拟产量变化趋势增加的主要气候因素。Tao等的研究结果表明，在1981—2000年间，气温变化对农作物的物候期和产量产生了一定的影响，气温升高使得哈尔滨地区玉米的播种期、开花期提前，产量显著提高。随着气候明显增温，产量较高的玉米有迅速扩种之势，这也是近20年来东北地区玉米显著增产的重要原因。在未来气候变暖背景下，在不考虑CO_2浓度升高对作物生长发育影响的前提下，东北三省春玉米不同熟型品种种植北界不同程度向北移动，玉米生育期延长，干物质积累增加，可以提高东北三省春玉米产量。Wang等指出从长远角度来看，玉米品种的引种要符合未来气候变暖。【本研究切入点】从东北气候变暖对玉米种植范围扩大和玉米总产提高的影响来看，气候变暖确实对东北玉米生产影响很大，但目前国内外还未见用玉米品种区域试验多年结果来研究品种农艺性状变化适应气候变化的报道。农作物品种区域试验是品种审定和推广的基础和依据，对于促进种植业结构调整、实施农产品优势区域布局具有重要意义。在区域试验中，对照品种起到了示范和标杆的作用，对照品种的选择均以当时大面积推广的品种为依据，且随着品种试验和育种水平的不断发展，对照品种将适时地进行更迭。品种区域试验中对照品种可能在中长期保持不变，同时连续多年在多个区域试验点种植，研究对照品种多年多点适应气候变化的趋势，就能够说明品种适应气候变化的结果。【拟解决的关键问题】本研究通过对2002—2009年中国东北早熟春玉米区域试验对照品种的生育期和产量及相应期间的气温性状变化趋势进行相关分析研究，以期为气候变暖条件下中国东北春玉米品种栽培和育种目标方向提供科学依据。

1 材料与方法

1.1 材料来源及预处理

生育期和产量资料来源于2002—2009年国家东北早熟春玉米组品种区域试验，研究区域包括辽宁、吉林、黑龙江和内蒙古4省（区），共10个连续参与区域试验试点，这些区域试验点按照地理纬度由南到北依次排列：辽宁本溪、辽宁新宾、内蒙古赤峰、吉林磐石、内蒙古通辽、黑龙江五常、吉林洮南、黑龙江双城、黑龙江阿城、黑龙江肇东。2002—2005年以四单19（CK1）和本玉9（CK2）为统一对照，2006—2008年以吉单261（CK）为统一对照；2009年的对照品种为先玉335，要求参试品种生育期比对照先玉335长2 d左右。所使用的气温观测资料数据为国家气象局提供的中国地面气候资料日值数据集，资料根据195年以来各省、市、自治区气候资料处理部门逐月上报的《地面气象记录月报表》整理而得。本研究使用了东北春玉米区4个省（区）7个气象观测台站2002—2009年期间的气温观测记录，涉及的数据项包括营养生长期、花粒期、生育期（定义为出苗期与成熟期之间的天数）的活动积温和年平均气温。7个观测台站分别是辽宁本溪、辽宁清原、吉林磐石、吉林白城、内蒙古通辽、内蒙古巴林左旗、黑龙江哈尔滨。

1.2 分析方法

出苗至抽雄期间为玉米主要营养生长时期。花粒期主要是指从雄穗开花到籽粒成熟,包括开花、吐丝和成熟3个时期。试验中所用到的生育期、产量性状等数据均是按照国家玉米区域试验统一要求进行田间测定的。其中2002—2005年对照品种四单19和本玉9的营养生长期和花粒期期间的活动积温是按照区域试验种提供的具体日期将气象资料中日平均气温累加得到的。采用Excel软件进行数据处理,相关性分析采用SPSS数据处理软件。

2 结果

2.1 对照品种所需活动积温和全生育期的变化

根据2002—2009年国家东北玉米区域试验10个试点的资料,东北春玉米区域试验对照品种由四单19→吉单261→先玉335的依次更替,对照品种所需活动积温的变化也由2 500 ℃→2 600℃→2 700℃逐步增加。东北春玉米区对照品种的生育期呈现出极显著的增加趋势,由2002年的120.9d延长到2009年的133.9d,年平均增加速率为1.625d。对照品种的玉米熟型向着晚熟品种方向发展(图1)。对照品种生育期的活动积温和生育期的相关性分析表明,生育期与相应期间的活动积温呈显著正相关($r = 0.740^*$, Sig. = 0.018, $\alpha = 0.05$)。由此可见,东北气候变暖对玉米育种和生产的影响是明显的,3次更换的对照品种生育期显著延长。随着东北地区气候变暖,无霜期增加,有利于增加玉米生长所需的积温,有利于延长玉米的生长季,选育生育期长的玉米品种。

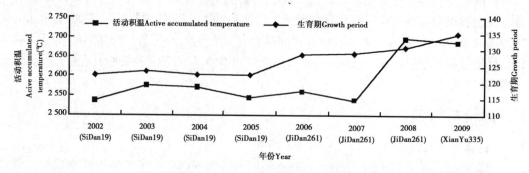

图1 对照品种所需活动积温和全生育期的变化

Figure 1 Changes of the growth period and active accumulated temperature of check varieties

2.2 对照品种营养生长期和花粒期的变化

由于东北春玉米品种比较区域试验记载数据资料有限,只能对2002—2005年连续4年10个区域试验点的2个对照品种四单19和本玉9的营养生长期和花粒期天数平均值以及其占当年全生育期的百分比、营养生长期和花粒期的活动积温进行比较和相关分析(图2、图3)。四单19营养生长期、花粒期与其期间的活动积温的相关系数分别为0.756、0.864;本玉9花粒期与其期间的活动积温的相关系数分别为0.632、0.866。虽然没有达到显著水平,但是从其相关系数的大小可以看出四单19和本玉9营养生长期、花粒期与其期间的活动积温呈较强正相关。四单19和本玉9的营养生长期分别由2002年的66.95d和67.25d缩短到2005年的61.8d和62.7d,分别缩短了5.15d和4.55d;其营养生长期占当年全生育期的百分比有减少的趋势:55.11%、53.97%(200年)>54.86%、

53.57%（2003年）>53.54%、52.08%（2004年）>51.35%、48.65%（2005年）。四单19和本玉9的花粒期分别由2002年的54.5d和57.35d延长到2005年的58.65d和60.65d，分别延长了4.15d和3.3d；其花粒期占当年全生育期的百分比有增加的趋势：44.89%、46.03%（2002年）<45.14%、46.43%（2003年）<46.46%、47.92%（2004年）<48.65%、49.12%（2005年）。

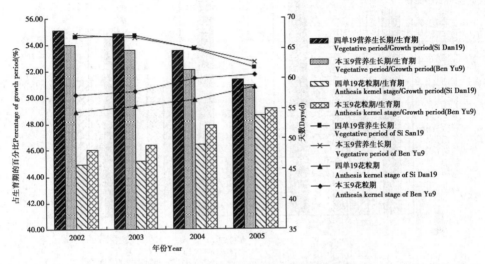

图2　对照品种营养生长期和花粒期的变化

Figure 2　Changes of vegetative period and anthesis kernel stage of check varieties

由此可见，随着气候变暖，2个对照品种的平均营养生长期均缩短，由占原来全生育期的54.54%，下降到51.12%；而平均花粒期则是逐年延长，由占原来全生育期的45.46%，增加到48.89%。特别是花粒期的延长有利于玉米延长灌浆时间，获得高产，同时可以降低籽粒水分含量。随着先玉335等种子脱水快品种的推广，一方面提高了东北玉米品种成熟产量和种子品质，同时还减少了东北玉米储藏和北粮南运期间玉米种子因为水分含量多，容易发霉变质的风险，提高了玉米的生产效益。

2.3　对照品种产量和年平均气温的变化

从图4可以看出2002—2009年中国东北春玉米区的年平均气温大体上呈上升趋势，同时东北早熟玉米区对照品种的单位面积产量也有缓慢上升的趋势。单位面积产量和年平均气温相关性分析表明，气候变暖对中国东北早熟春玉米的产量有显著地正效应影响（$r=0.647^*$，Sig.$=0.041$，$\alpha=0.05$）。2005年大幅度减产主要是由气温骤降和供种质量问题导致。随着生产水平的提高，对照品种要进行更替，其品种农艺特性将要发生变化，同时栽培方式也相应要配套，由原来大株大穗稀植改变为矮秆密植。在区域试验中种植密度也相应跟随生产种植方式也相应改变，对照品种四单19、吉单261、先玉335的适宜种植密度分别为3 000株/667m²、3 300~3 600株/667m²、4 000~4 500株/667m²。中国东北早熟春玉米的选育正向着耐密型品种方向发展，由此可见，东北玉米总产增加主要是靠提高玉米的种植密度，扩大玉米种植面积。各省（区）的平均单产变化不尽相同（图5），辽宁、吉林、内蒙古自治区、黑龙江4省（区）的单位面积产量变化依次为10.12~9.95t/hm²、9.13~11.26t/hm²、10.42~10.72t/hm²、10.76~9.39t/hm²，增减幅度分别为

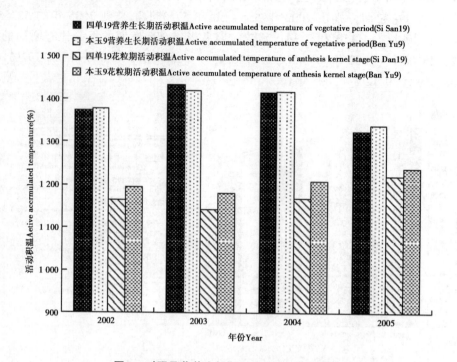

图3 对照品营养生长期和花粒期的活动积温

Figure 3　Changes of active accumulated temperature in vegetative period and anthesis kernel stage of check varieties

-1.72%、18.94%、2.79%、-14.48%。可见,吉林、内蒙古自治区2省(区)的单位面积产量呈增加的趋势,辽宁省对照品种的单位面积产量增加趋势不是很明显,而黑龙江省的单位面积产量却呈减产的趋势。黑龙江其单位面积产量的变化可能是黑龙江省气候变化异常不稳定等原因造成的。

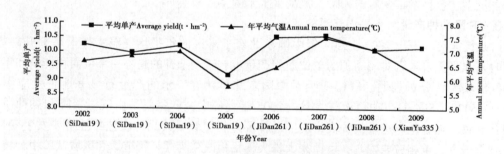

图4 对照品种的年平均温度和平均单产的变化趋势

Figure 4　Changes of annual mean temperature and average yield of check varieties

3 讨论

利用8年的国家级玉米品种区域试验报告,通过对10个固定区域试验点的生育期和

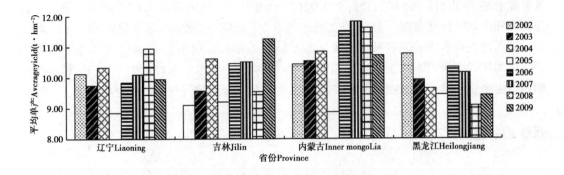

图5 对照品种平均单产在辽宁、吉林、内蒙古、黑龙江的变化趋势
Figure 5 Changes of average yield of check varieties in Liaoning, Inner Mongolia, Jilin, and Heilongjiang during 2002-2009

产量等相关数据分析，发现东北早熟春玉米区域试验对照品种的生育期天数呈现出极显著的增加趋势，说明东北春玉米生育期的延长是由于气候变暖导致的结果。这与李正国等得出的分析结果基本一致。因为气候变暖延长了东北春玉米的生长季，在今后育种过程中应选择生育期适当偏长的春玉米品种。

东北地区纬度较高，温度是粮食作物生长季内主要气候限制因子，气候变暖最有利于该地区粮食产量的提高。随着热量资源的不断增加，东北地区玉米播种范围不断扩大，种植面积稳步增加。2002—2009年东北早熟玉米区对照品种的单位面积产量大体上呈较小上升趋势，可见，近些年来东北玉米总产量的增加主要是依靠种植面积的扩大，而单位面积产量对总产的贡献相对较小。这与国家发改委农村经济司副司长方言的观点是一致的。就辽宁、吉林、内蒙古、黑龙江这4个省（区）来说，吉林、内蒙古两省（区）的单位面积产量呈增加的趋势，而黑龙江省的单位面积产量却呈减产的趋势。说明气候变化在不同的国家和地区有不同的影响，在中国范围内，气候变化的影响也不是全国统一的。目前中国的玉米单产比欧美国家的平均水平低了100多千克，这表明中国的玉米单产还有巨大的潜力，因此，应加强玉米高产品种的选育，发掘玉米的增产潜力。

2002—2009年东北早熟春玉米区区域试验要求玉米品种的适宜种植密度呈上升趋势，由2002年的3 000株/667m²向2009年的4 000~4 500株/667m²发展，这说明东北春玉米区正向着耐密型品种、依靠密度增加产量的方向发展。

气温变化首先是对作物物候期的产生较大的影响。通过对2002—2005年连续4年10个固定试点的对照品种四单19和本玉9营养生长期和花粒期的研究分析，结果表明，随着东北地区气候的变暖，玉米的营养生长期的生长速率加快，营养生长期变短，这与王琪等对气候变化下气温对玉米生长速率研究的试验结果相似。花粒期则是在东北地区暖干气候条件下呈现逐年延长的趋势，秋季气温升高，晚霜延迟可能是造成早熟春玉米花粒期延长的主要原因。

4 结论

对照品种的生育期、营养生长期、花粒期与相应期间的活动积温呈较强正相关，随着活动积温的增加，生育期和花粒期均有延长的趋势。气候变暖对中国东北早熟春玉米的产量有显著地影响（$r=0.647^*$，Sig. $=0.041$，$\alpha=0.05$），使得单位面积产量有一定的上升趋势。在东北气候向着暖干化趋势发展的背景下，2002—2009年东北早熟春玉米区域试验对照品种更换3次，适宜种植密度呈增加趋势，玉米品种的选育向着耐密型品种方向发展。说明气候变暖对东北玉米生产和育种方向影响较为明显。选育总生育期长、营养生长期短但生殖生长期长，耐密植的高产品种是东北春玉米适应气候变暖的育种方向。

参考文献（略）

本文原载：中国农业科学，2012，45（24）：4959-4966

不同种植方式对玉米农艺性状和产量的影响

赵 杨 钱春荣[1] 王俊河[1] 于 洋[1] 宫秀杰[1] 姜宇博[1] 杨国航[2]

(1. 黑龙江省农业科学院耕作栽培研究所,哈尔滨 150086;
2. 北京市农林科学院玉米研究中心,北京 100097)

摘 要:为选取有效种植方式指导玉米生产,以先玉335为试验材料,研究不同种植方式对玉米的产量和农艺性状的影响。结果表明:平作的玉米产量高于平作垄管和垄作,前者与后两者产量差异达极显著水平。平作种植方式较垄作种植方式增产7.2%。采取平作种植的玉米在棒三叶叶面积、株高、穗粗和穗行数等性状上表现突出。

关键词:玉米;耕作方式;产量;农艺性状

我国是世界第二大玉米生产国,黑龙江省是我国玉米主产区,常年播种面积在270万hm^2以上。其高产与稳产在保障国家粮食安全和实现黑龙江省政府提出的千亿斤粮食产能工程中具有举足轻重的作用。在玉米生产中,种植方式是协同高密度条件下,个体通风透光条件、营养状况并最终影响产量的因素之一,目前黑龙江省的玉米种植主要采取垄作的方式。但是随着玉米施肥水平、种植密度的提高以及品种的不断变化,单一的垄作种植方式能否为玉米的生长发育提供良好条件也是玉米栽培面对的一个严峻问题。现通过田间试验探讨垄作、平作垄管、平作3种种植方式对玉米各农艺性状和产量的影响,为玉米种植管理提供理论依据和技术指导。

1 材料与方法

1.1 材料
供试材料为当地主栽玉米品种先玉335。

1.2 方法
1.2.1 试验设计
试验于2010年在黑龙江省双城市幸福乡永支村进行,该地区年平均温度在6.5℃左右,年降雨量在430mm左右,该试验地前茬作物为玉米。试验采取随机区组设计,3种耕作方式:垄作T3、平作垄管T2、平作T1,设3次重复,随机排列,3行区,行长3m,行距0.68m,小区面积为17m^2。

供试肥料为尿素(含N46%)、磷酸二铵(含P48%,N18%)、氯化钾(K_2O60%),

基金项目:"973"计划资助项目(2009CB118601)
第一作者:赵杨(1985—),女,黑龙江省哈尔滨市人,学士,研究实习员,从事作物耕作栽培研究。E-mail:mmzymy@163.com
通讯作者:钱春荣(1973—),女,黑龙江省延寿县人,硕士,副研究员,从事玉米育种与栽培研究。E-mail:qianjianyi318@163.com

磷肥、钾肥和1/3氮肥作基肥,2/3氮肥在拔节期作追肥,施肥量为：尿素375kg·hm^{-2}、氯化钾125kg·hm^{-2}、磷酸二铵150kg·hm^{-2}。于5月18日施肥,5月20日播种,6月24日追肥,10月4日收获。田间管理为垄作处理结合追肥扶垄,平作垄管处理结合追肥起垄,平作整个生育期内均保持平作状态。

1.2.2 调查项目

生育期内测定植株最大叶面积、棒三叶叶面积、株高、穗位高和茎粗。收获时,选取小区内均匀一致不缺苗断垄的2行全收,按14%含水量折算小区产量。室内考种项目测定穗长、穗粗、穗行数、穗粒数和百粒重。所有数据采用DPS7.01软件进行统计分析。

2 结果与分析

2.1 不同种植方式对玉米农艺形状的影响

由表1可知,不同种植方式对玉米最大叶面积指数无显著差异,但平作与垄作的棒三叶叶面积存在显著差异,平作的棒三叶叶面积较垄作棒三叶叶面积增大3.3%,在株高方面平作与其余两者差异均达到显著水平,其中平作株高较垄作株高增高6.0%,不同种植方式的植株穗位高表现为平作>平作垄管>垄作,平作与平作垄管差异显著,平作与垄作间差异极显著。

2.2 不同种植方式对玉米产量及产量构成因子的影响

从表2可看出,平作种植方式较平作垄管与垄作种植方式产量差异达极显著水平,平作较平作垄管和垄作分别增产6.7%和7.2%,表明在试验条件下平作耕作方式优于其余两种耕作方式。在产量构成因素方面来看,平作种植的玉米在穗粒数、百粒重上表现突出,但未达显著水平,平作与垄作的穗粗与穗行数差异均达到显著水平。

表1 种植方式对玉米农艺性状的影响
Table 1 Effect of planting patterns onagronomic traits

处理 Treatments	最大叶面积指数 Maximum LAI	棒三叶叶面积 （cm^2）Leaf area Of three ear leaves	株高（cm） Plant height	穗位（cm） Ear position	茎粗（cm） Stem diameter
T1	4.52aA	2 113.84aA	326.19aA	125.77aA	2.30aA
T2	4.51aA	2 063.74abA	307.38bA	118.70bAB	2.26aA
T3	4.50aA	2 046.85bA	307.86bA	114.84bB	2.26aA

表2 不同处理对产量及产量构成因素的影响
Table 2 Effect of different treatments on yield and its component

处理 Treatments	最大叶面积指数 Maximum LAI	棒三叶叶面积 （cm^2）Leaf area Of three ear leaves	株高（cm） Plant height	穗位（cm） Ear position	茎粗（cm） Stem diameter
T1	4.52aA	2 113.84aA	326.19aA	125.77aA	2.30aA
T2	4.51aA	2 063.74abA	307.38bA	118.70bAB	2.26aA
T3	4.50aA	2 046.85bA	307.86bA	114.84bB	2.26aA

2.3 不同种植方式对收获指数的影响

收获指数（Harvest Index，HI）是作物收获时经济产量（籽粒、果实等）与生物产量之比，能够反映作物同化产物在籽粒和营养器官上的分配比例，是评价栽培成效的重要指标。由图1看出不同种植方式的收获指数顺序为平作>平作垄管>垄作，可见平作种植有利于作物的源器官将光合产物转运到库器官。

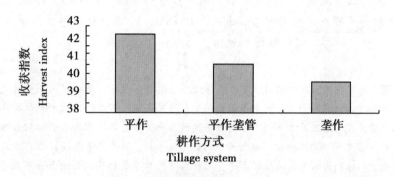

图 1 不同处理对收获指数的影响
Figure 1 Effect of different treatments on harvest index

3 结论

研究结果表明，玉米的种植方式对其产量和农艺性状均产生影响。选用合理的种植方式、是增加产量的有效措施。在该试验中平作的棒三叶、收获指数与产量均明显优于平作垄管与垄作，平作的玉米产量分别比平作垄管和垄作高6.7%和7.2%。另外平作种植玉米不仅是提高产量，其不需要起垄与封垄从而节约成本，增加经济效益，符合当下农业发展的低碳原则，虽然现在广大农户的种植模式中垄作是较为普遍的，但有针对性地推广玉米平作种植，将为农户带来更高的产量与收益。

参考文献（略）

本文原载：黑龙江农业科学，2012（1）：32-33

不同种植模式对玉米产量与农艺性状影响分析

赵 杨[1]　钱春荣[1]　王俊河[1]　于 洋[1]　宫秀杰[1]
姜宇博[1]　杨国航[2]　左 辛[3]

(1. 黑龙江省农业科学院耕作栽培研究所，哈尔滨　150086；2. 北京市农林科学院玉米研究中心，北京　100089；3. 黑龙江省农业科学院农村能源研究所，哈尔滨　150086)

摘　要：为了探明玉米种植的最优方式，试验采用裂区设计，研究不同耕作方式、种植密度及施 N 肥量对玉米的产量和农艺性状的影响。结果表明：不同的耕作方式和不同施 N 肥量玉米的产量差异均达显著水平，不同种植密度的玉米产量差异不显著。在各组合中，以平作耕作方式、密度 67 500 株/hm^2、施 N 量为 112.5 kg/hm^2，玉米产量最高；耕作方式对玉米棒三叶叶面积和株高影响差异达显著水平，对最大叶面积指数和茎粗影响差异不显著；密度对玉米最大叶面积指数、棒三叶叶面积和茎粗影响差异显著，对株高影响差异不显著；施氮与不施氮之间最大叶面积指数、棒三叶叶面积、茎粗和株高差异显著。

关键词：玉米；耕作方式；密度；N 肥；产量；农艺性状

Influence of Plant Arrangement, Density and the Rate of Fertilizer Applied on Maize Yield and Agronomic Traits Efficiency

Zhao Yang[1]　Qing Chunrong[1]　Wang Junhe[1]　Yu Yang[1]
Gong Xiujie[1]　Jiang Yubo[1]　Yang Guohang[2]　Zuo Xin[3]

(1. *Crop Cultivation Institute of Heilongjiang Academy of Agricultural Science*, *Harbin* 150086; 2. *Maize Research Center*, *Beijing Academy of Agricultural and Forestry Sciences*, *Beijing* 100089; 3. *Rural Energy Sources Institute*, *Heilongjiang Academy of Agricultural Sciences*, *Harbin* 150086)

Abstract: To investigate the optimal way of corn planting, a split-plot design was adopted in

---

基金项目："973" 计划 "东北北部春玉米密植高产与水热高效协调模式构建与验证"（2009CB118601）

第一作者：赵杨，女，1985 年出生，黑龙江哈尔滨人，本科，研究方向为玉米育种与栽培。地址：黑龙江省哈尔滨市南岗区学府路 368 号黑龙江省农业科学院耕作栽培研究所（150086），Tel：0451-86678615，E-mail：mmzymy@163.com。

通讯作者：钱春荣，女，1973 年出生，黑龙江延寿人，副研究员，研究方向为玉米育种与栽培。地址：黑龙江省哈尔滨市南岗区学府路 368 号黑龙江省农业科学院耕作栽培研究所（150086），Tel：0451-86678615，E-mail：qianjianyi318@163.com。

this study. The effects of plant arrangement, plant density and N fertilizer levels on the yield of maize and agronomic traits were studied. The results showed that: among three factors, the plant arrangement and the N fertilizer was significantly impact on the yield. But there was no significant difference for yield in three kinds of densities.

The most optimum treatment was flat planting, 67 500 plants per hm^2, 112.5kg/hm^2 N fertilizer. The plant arrangement reached significant level to leaf area of three ear-leaves and the plant height, and not reached significant level to the largest of leafarea index and stem diameter. The plant density was reached significant level to the largest of leaf area index, leaf area of three ear-leaves and stem diameter, and not reached significant level to the plant height. With and without application of N fertilizer, the largest of leaf area index, three ear-leaves, stem diameter and the plant height was reached significant level.

Key words: Maize; Plant arrangement; N fertilizer; Plant density; Yield; Agronomic traits

黑龙江省作为中国玉米主产区，常年播种面积在 270 万 hm^2 以上，占全省粮食播种面积的 30%，年产玉米 120 亿 kg 以上，占全省粮食总产量的 40%，约占全国总产的 11%，是中国重要的玉米商品粮生产基地。黑龙江省春玉米种植多以垄作为主，密度大多在 45 000 株/hm^2 左右，东北地区平均施 N 肥量在 200kg/hm^2。已有研究结果显示，种植方式、种植密度及施肥措施均会影响玉米的生产效益：适当增加种植密度可以增产增收；提高种植密度可以获得高产，平作种植不会破坏土壤结构，有利于蓄水保墒。但前人的研究往往是单项技术研究，综合技术研究较少。将栽培措施、品种特性与肥料施用有机结合，充分发挥作物生产潜能，达到高产、优质、低耗和环境友好的目的一直是玉米发展战略面临的重大课题。为此，笔者通过田间试验探讨 3 种种植方式（垄作、平作垄管、平作）、3 个种植密度（45 000 株/hm^2、67 500 株/hm^2、90 000 株/hm^2）、4 个 N 肥施用水平（0、112.5kg/hm^2、225kg/hm^2、337.5kg/hm^2）对玉米产量和各农艺性状的影响，为玉米大田生产获得高产和低碳种植提供借鉴和参考。

1 材料与方法

1.1 试验材料

供试品种为"先玉 335"。

1.2 试验设计

试验于黑龙江省双城市幸福乡永支村进行，该地区年平均温度在 6.5℃ 左右，年降水量在 430mm 左右，该试验地前茬作物为玉米。试验采取裂区设计，主处理为 3 种耕作方式（垄作、平作垄管、平作），副处理为 3 个种植密度（45 000 株/hm^2、67 500 株/hm^2、90 000 株/hm^2）。每种种植密度设 4 个氮肥水平：分别为 N0（不施 N 肥），N1（施 N 量为 112.5kg/hm^2），N2（施 N 量为 225kg/hm^2），N3（施 N 量为 337.5kg/hm^2），设 3 次重复，随机排列，试验小区面积为 5m × 3.25m = 17.25m^2，每小区种植 5 行，平均行距为 0.65m。供试肥料为尿素（含 N 46%）、二铵（含 P_2O_5 46%、N 18%）、氯化钾（含 K_2O 60%）和过磷酸钙（含 P_2O_5 44%），磷肥、钾肥、和 1/3 氮肥作基肥，2/3 氮肥在拔节期作追肥。基肥施肥日期为 2010 年 5 月 18 日，播种时间为 5 月 20 日，追肥日期为 6 月 24 日，收获时间为 10 月 5 日。田间管理为垄作处理结合追肥扶垄，平作垄管处理结合追肥起垄，平作整个生育期内均保持平作状态。

1.3 调查项目

在生育期内测定植株的最大叶面积、棒三叶叶面积、株高、茎粗；收获为小区全收，按14%含水量折算小区产量。室内考种项目：穗长、穗粗、穗行数、穗粒数、百粒重。

1.4 数据分析方法

所得数据采用DPS7.01软件进行数据的统计分析。

2 结果与分析

2.1 不同种植模式对玉米农艺形状的影响

2.1.1 耕作方式对玉米农艺性状的影响

在玉米叶片中棒三叶的叶面积最大，光合作用最强，在相同的时间内形成光合作用产物最多，对玉米干物质的积累非常重要。由表1可见耕种方式不同，玉米的最大叶面积指数无显著差异，但平作与垄作的棒三叶叶面积存在显著差异，平作的棒三叶叶面积较垄作棒三叶叶面积增大3.3%。平作的耕作方式与其他2种耕作方式的株高差异均达到显著水平，平作株高较垄作株高增高6.0%。

表1 耕作方式对玉米农艺性状的影响

种植方式	最大叶面积指数	棒三叶叶面积（cm^2）	株高（cm）	茎粗（cm）
平作	4.52 aA	2 113.84 aA	326.19 aA	2.30 aA
平作垄管	4.51 aA	2 063.74 abA	307.38 bA	2.26 aA
垄作	4.50 aA	2 046.85 bA	307.86 bA	2.26 aA

注：小写的不同字母表示0.05水平差异显著，大写的不同字母表示0.01水平差异极显著。下同

2.1.2 种植密度对玉米农艺性状的影响

由表2可见，3种种植密度玉米的最大叶面积指数存在极显著差异，玉米的最大叶面积指数随着密度的增加而增大，密度为45 000株/hm^2与67 500株/hm^2的棒三叶叶面积差异显著；密度为45 000株/hm^2与密度为90 000株/hm^2的棒三叶叶面积差异极显著；随着密度的增加植株茎粗逐渐减小，密度为45 000株/hm^2与2个较高的密度的茎粗差异达显著水平。

表2 种植密度对玉米农艺性状的影响

密度（株/hm^2）	最大叶面积指数	棒三叶叶面积（cm^2）	株高（cm）	茎粗（cm）
45 000	3.09 cC	2 153.86 aA	319.36 aA	2.44 aA
67 500	4.51 bB	2049.39 bAB	311.77 aA	2.21 bB
90 000	5.93 aA	2021.17 bB	310.30 aA	2.17 bB

2.1.3 施 N 肥对玉米农艺性状的影响

由表 3 可知,不施 N 肥与 3 种施 N 肥处理在最大叶面积指数上差异达极显著水平,施 N 水平为 N2、N3 与不施 N 的棒三叶叶面积差异显著,施 N 水平的棒三叶叶面积显著高于不施 N 处理。不施 N 与施 N 的株高差异极显著。可见 N 肥促进了植株叶片和茎秆的生长。

表 3　不同施 N 量对玉米农艺性状的影响

施肥	最大叶面积指数	棒三叶叶面积（cm^2）	株高（cm）	茎粗（cm）
N0	4.21 bB	2 020.97 bA	293.03 cB	2.18 aA
N1	4.61 aA	2 081.92 abA	316.33 aA	2.27 aA
N2	4.55 aA	2 099.74 aA	326.11 aA	2.30 aA
N3	4.67 aA	2 096.60 aA	324.77 aA	2.34 aA

2.2 不同种植模式对玉米产量及产量构成因子的影响

2.2.1 不同耕作方式对玉米产量及产量构成因子的影响

表 4 显示,平作种植的玉米在穗粒数、百粒重方面表现突出,平作与垄作在穗粗和穗行数差异均达到显著水平,其中平作较垄作的穗粗增加 1.7%,穗行数增加 4.3%。平作与平作垄管和垄作的产量差异达极显著水平,平作比平作垄管和垄作分别增产 6.7% 和 7.2%。

2.2.2 不同种植密度对玉米产量及产量构成因子的影响

表 5 显示,种植密度对产量构成因子——穗长、穗粗、穗行数、百粒重均有显著影响。随着密度的增加,穗长、穗粗、穗行数、穗粒数、百粒重在逐渐降低;密度为 67 500 株/hm^2 的玉米产量最高,比密度为 45 000 株/hm^2 和 90 000 株/hm^2 分别高出 0.86% 和 1.1%。

2.2.3 N 肥施用量对玉米产量及产量构成因子的影响

合理施肥是玉米高产的关键技术,20 世纪 50 年代后,氮肥在农业生产上开始发挥重要作用。表 6 显示,施 N 肥处理在穗长、穗粗、穗粒数、百粒重上表现突出;施 N 肥处理产量明显高于不施 N 肥处理,且产量差异极显著。施 N 量为 N1 时产量最高,较不施 N 肥处理高出 17.1%,但施 N 的 3 种水平产量差异不显著,就施 N 肥处理来看,并不是施 N 量越多产量越高。

表 4　耕作方式对玉米产量构成因子的影响

耕作方式	籽粒产量（kg/hm^2）	穗长（cm）	穗粗（cm）	穗行数（行）	穗粒数（粒）	百粒重（g）
平作	10 051.16 aA	18.07 aA	4.90 aA	15.97 aA	607.91 aA	36.67 aA
平作垄管	9 417.55 bB	18.15 aA	4.85 abA	15.52 bAB	599.75 aA	36.56 aA
垄作	9 378.19 bB	17.50 aA	4.81 bA	15.30 bB	587.69 aA	36.45 aA

表5 种植密度对玉米产量构成因子的影响

密度/ （株/hm²）	籽粒产量 （kg/hm²）	穗长 （cm）	穗粗 （cm）	穗行数 （行）	穗粒数 （粒）	百粒重 （g）
45 000	9 595.19 aA	18.86 aA	4.98 aA	15.80 aA	634.75 aA	39.31 aA
67 500	9 677.94 aA	17.87 bAB	4.82 bB	15.63 abA	610.47 aA	36.02 bB
90 000	9 573.77 aA	16.99 cB	4.76 bB	15.36 bA	550.13 bB	34.35 cC

表6 氮肥施用量对玉米产量构成因子的影响

施N肥水平	籽粒产量/ （kg/hm²）	穗长/cm	穗粗/cm	穗行数/行	穗粒数/粒	百粒重/g
N0	8 613.59 bB	17.95 aA	4.78 bB	15.48 bB	573.85 bA	36.46 aA
N1	10 084.96 aA	18.19 aA	4.86 aA	15.55 bAB	617.18 aA	36.61 aA
N2	9 888 aA	17.85 aA	4.88 aA	15.40 bB	590.55 abA	37.03 aA
N3	9 876 aA	17.64 aA	4.90 aA	15.96 aA	612.22 abA	36.14 aA

2.2.4 不同耕作方式对收获指数的影响

收获指数是作物收获时经济产量（籽粒、果实等）与生物产量之比，是评价栽培成效的重要指标。由图1可见，平作的收获指数在不施N和3种施N水平下均高于平作垄管与垄作；平作垄管的收获指数在不施N和3种施N水平下均高于垄作。

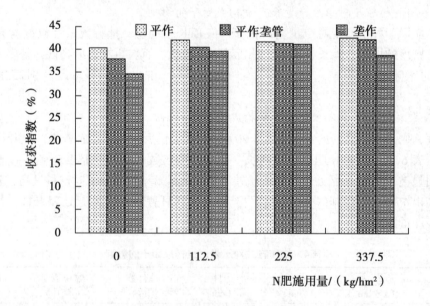

图1 不同种植方式收获指数

3 结论

在耕作方式方面，本试验中平作种植方式提高了玉米的棒三叶叶面积、株高、穗粗、穗粒数、收获指数，且平作种植的产量高于平作垄管和垄作。目前黑龙江省在农业生产中主要以垄作耕作方式为主，垄作可以提高土壤温热，有利于抗旱防涝。但传统垄作耕法存在土壤频繁耕翻，机械作业次数多，压实土壤，破坏土壤结构，使耕地退化，平作技术体系以免耕播种为主，在春季不搅动和破坏土壤结构，同样有利于蓄水保墒。试验区全年无灾害性气候，且平作种植无倒伏现象发生，这应该也是平作获得高产的主要原因，所以今后可以有针对性的在气候适宜地区推广平作种植。

在种植密度方面，以密度为 67 500 株/hm^2 产量最高，试验所设的 3 种密度对于产量的影响不显著。随着种植密度的提高，棒三叶叶面积、茎粗、穗长、穗粗、穗粒数、百粒重等均明显下降，虽然高密度种植提高了单位面积内果穗的数目，但同时降低了植株的综合表现，表明适宜的种植密度即可增产。

在 N 肥施用量方面，不施 N 肥与施 N 肥的玉米产量差异达极显著，但 3 种不同的施 N 水平间不存在显著差异，这说明：过量的 N 肥施用并不能带来产量的提高，而且造成资源的浪费，合理施用 N 肥既可获得高产提高经济效益。

该试验研究结果表明，改进耕作方式、合理密植和有效的 N 肥施用量都是增加产量的有效措施。本试验结果显示，采取平作耕作方式、密度为 67 500 株/hm^2 和施 N 量为 112.5kg/hm^2 为最佳种植方式。

参考文献（略）

本文原载：中国农学通报，2012，28（9）：106-109

应对及预防京郊玉米倒伏的技术措施

王荣焕　赵久然　徐田军　陈传永　刘春阁　张爱武

(北京市农林科学院玉米研究中心，北京　100097)

摘　要：近年来，倒伏已成为影响京郊玉米生产的重要不利因素。本文结合京郊玉米生产实际，提出了应对及预防玉米倒伏的抗灾减灾技术措施，旨在为指导京郊玉米生产和降低倒伏灾害损失提供参考和指导。

关键词：玉米；倒伏；应对及预防；技术措施

倒伏已成为近年影响京郊玉米生产的重要不利因素，玉米生长季降雨量大并伴有强风是导致玉米倒伏的最直接原因之一。北京市为典型的暖温带半湿润大陆性季风气候，全年降水分配很不均匀，其中80%集中在夏季的6—8月。近年来，特别是7月中下旬至8月上旬，京郊玉米生产常遭遇暴风雨袭击，造成玉米不同程度倒伏倒折。如2011年7月11—25日，京郊玉米主产区平均降雨量190.2mm，较常年同期偏多44%；24日晚至25日上午，全市遭遇大到暴雨（平均降雨量93.1mm，多地超过200mm）并伴有6级以上强阵风，许多玉米田发生倒伏。2012年7月21—22日，北京市出现61年来最强连续降水过程（平均降雨量190.3mm，春、夏玉米主产区平均降水量分别为140.9mm、200.8mm，为大暴雨量级）且个别地区伴有龙卷风，局部玉米倒伏严重。2013年，全市6—7月降雨量288.7mm，较常年同期偏多18.6%，持续降雨不利于"蹲苗"，导致茎秆质脆易折。7月31日至8月1日，全市平均降雨量19.6mm并伴短时大风（最强风力达9级），导致1.33万公顷（20万亩）的玉米倒伏倒折。这段时间，正值春玉米吐丝灌浆期、夏玉米大喇叭口至抽雄吐丝期，倒伏倒折对玉米生产不利。此外，栽培农艺技术措施不当等也是造成京郊玉米倒伏倒折的重要不利因素。

本文结合近年京郊玉米生产实际，提出了玉米遭受暴风雨发生倒伏后的抗灾减灾措施及预防玉米倒伏的技术措施，旨在为指导京郊玉米生产和降低倒伏灾害损失提供参考和指导。

1　玉米倒伏后的抗灾减灾技术措施

不同播期和地块的玉米因所处的生育时期不同，倒伏灾害的类型及程度也存在较大差异。因此，生产中玉米倒伏后，应视玉米植株的生育进程和倒伏类型进而采取相应的有效抗灾减灾技术措施，力争将灾害损失降至最低。

基金项目：国家玉米产业技术体系（nycytx-02）；北京市农林科学院青年科研基金（QN201115）、北京市农林科学院科技创新基金（CXJJ201309）

作者简介：王荣焕，副研究员，主要从事玉米高产生理及配套栽培技术研究

　　　　　赵久然为通讯作者，研究员，主要从事玉米研究

1.1 倒伏地块

对处于大喇叭口期之前的根倒和茎倒地块，因大部分植株可在较大程度上自行恢复，因此不必人工扶正，避免毁坏根系，对恢复生长不利。待玉米植株站立后，要抓紧进行中耕培土和补施适量尿素。对已进入抽雄吐丝期倒伏的春玉米地块，因植株高大，自行恢复能力减弱，需根据情况及时人工扶正、培土，重新将植株固定，并补施粒肥。玉米倒伏后，因植株间相互倒压将严重影响光合作用，如不及时采取有效技术措施，将会造成不同程度减产。

1.2 倒折地块

对于大面积连片严重发生茎折的玉米地块，可酌情收获作青饲用，也可根据具体情况复种时令蔬菜，尽量减少损失。

此外，近年京郊玉米生产中遭遇的玉米大面积严重倒伏多是由暴风雨袭击所致，对于因强降雨而导致发生涝害的地块，则应尽快排水降渍。排水后待能下地时，要抓紧时间进行中耕培土，并追施速效氮肥。改善土壤水、肥、气、热严重失调的状况，避免土壤板结，提高土壤通透性，促进玉米根系恢复生长和尽快缓苗。

2 预防京郊玉米倒伏的主要技术措施

2.1 选择抗倒性强的品种

因地制宜，选择植株抗倒伏能力强的玉米品种是预防倒伏的最有效途径。抗倒伏能力强的品种具有根系发达、茎秆粗壮且韧性好、矮秆、株高穗位适中、抗病虫能力强等特点。近年京郊玉米生产实践证明，玉米主导品种郑单958、京单28，以及京科528、京单68和京单38等玉米新品种具有较强的抗倒伏能力。

2.2 土壤深松（耕）

针对因小型农机具大面积和长期使用而导致土壤耕层变浅、犁底层不断加厚的春玉米地块，可通过进行土壤深松（耕）作业（深度30cm以上）打破犁底层，改善深层土壤结构，促进玉米根系发育和下扎，进而提高植株的抗倒能力。京郊春播玉米主要分布在延庆、密云、怀柔等区县，在实际生产中实施土壤深松作业地块的春玉米因耕层深、培土多、根系入土深等，较旋耕春玉米抗倒性强。

2.3 夏玉米免耕贴茬直播

夏播玉米采用免耕贴茬直播技术可提高玉米植株的抗倒性。夏玉米免耕贴茬直播不仅可减少作业环节争取农时、提高播种质量和幼苗整齐度，并且因播前不动土、播时动土较少，种子与实土接触，玉米根系扎得实、扎得牢，主根发达且表层根量多，将玉米植株牢牢地固定，大大增强了玉米植株的抗倒伏能力。

2.4 合理密植

为构建合理群体结构，促进玉米植株群体和个体协调发展，避免因密度过大而增加倒伏风险，应根据品种特性、土壤肥力、栽培方式和管理水平等确定适宜的种植密度，进行合理密植。一般生产条件下，郑单958、京单28、京科528、京单68等高产耐密型玉米品种的适宜密度为每亩60 000~67 500株/hm^2，中单28、纪元1号、先玉335等适宜稀植的品种适宜密度为每亩52 500株/hm^{2}左右。高肥力地块和高产创建地块可较一般大田适当增加种植密度。

2.5 科学肥水管理

根据玉米的需肥规律和不同品种的需肥特性，进行科学肥水管理，避免因肥水不当增加玉米倒伏风险。李波等研究表明，施钾肥可显著提高玉米茎秆的穿刺强度和抗倒伏能力。而目前京郊玉米生产在施肥方面普遍存在P肥用量偏多、K肥用量偏少的问题，一般亩施底肥600kg/hm^{2}左右（15-15-15的复合肥或高氮复合肥），拔节后亩追施氮肥（尿素）375kg/hm^{2}左右。因此，建议适当增施钾肥，以提高玉米植株的抗倒性能。此外，对肥水供应充足、苗期长势过旺地块，可在拔节前进行控水肥蹲苗，以促进根系下扎和茎秆健壮。建议适当延迟追施氮肥时期，可由拔节期追肥改为小喇叭口期。

2.6 适时化控防倒

采取化学调控技术措施可有效降低玉米株高和穗位高，促进根系生长和植株健壮，提高植株抗倒能力和产量。建议对品种抗倒性差、密度过大、生长过旺地块，以及风灾倒伏频发地区，可在拔节至小喇叭口期喷施玉米健壮素或金得乐等植物生长调节剂，以有效降低株高和穗位高，增强茎秆强度，提高植株抗倒伏能力。

参考文献（略）

本文原载：作物杂志，2013（6）：132-133

国审玉米品种京单 68 的主要特点及高产栽培技术

刘春阁　路明远　刘海伍　袁二虎　成广雷

（北京市农林科学院玉米研究中心/北京农科院种业科技有限公司　100097）

京单 68 是由北京市农林科学院玉米研究中心自主选育的中早熟玉米新品种。该品种具有熟期适宜、高产稳产、综合性状优、抗倒性强、适应性广等特点。于 2010 年通过国家审定（审定编号：国审玉 2010003），适宜京津唐区夏播种植。

1　主要特点

1.1　熟期适宜

经国家玉米区域试验及北京、河北等区域试验多年多点观测，在京津唐区夏播出苗至成熟平均 98d；在实际大田生产中，比目前我国夏播玉米主栽品种郑单 958 早熟 3~5d。在北京和京津唐地区 6 月中下旬播种，9 月底至 10 月初收获，籽粒能够充分成熟，对确保玉米高产优质及协调玉米－小麦上下两茬合理搭配有明显优势。

1.2　高产稳产

京单 68 具有良好的丰产性和稳产性。2008—2009 年参加国家京津唐组玉米品种区域试验平均亩产 643.8kg，比对照品种京玉 7 号增产 8.7%。2009 年参加国家京津唐组玉米生产试验，平均亩产 674.7kg，比对照品种京玉 7 号增产 15.2%。在一般大田种植条件下，夏播亩产 650kg 左右。

1.3　综合性状优

播种后出苗快且发棵早，群体整齐度高。株型紧凑、穗位整齐。果穗大小均匀，封顶性好，无秃尖，籽粒灌浆饱满。穗轴白色。籽粒黄色、半马齿型。经农业部谷物及制品质量监督检验测试中心（哈尔滨）测定，籽粒容重 730g/L，粗蛋白含量 8.78%，粗脂肪含量 3.90%，粗淀粉含量 73.65%，赖氨酸含量 0.26%。

1.4　抗倒性强

根系发达，茎秆健壮，株高和穗位较低，在国家京津唐组玉米区试中株高平均 247cm，穗位高平均 99cm。在国家玉米区域试验及大田生产中表现出了良好的抗倒性。2012 年 7—8 月，北京市以及河北省的保定北部、沧州北部等地发生多次强降雨及大风天气，导致郑单 958、先玉 335 等品种严重倒伏，而京单 68 因抗倒性强，直立未倒，经受住了生产实践的考验。

1.5　适应性广

在京津唐区具有良好的适应性，适宜在北京、天津和河北的唐山、廊坊、保定北部及沧州北部夏播种植。并且，对土壤条件要求不高，在不同地块和年份均表现出了良好的适应性。

2 高产栽培技术

实行良种良法配套,有利于充分挖掘和发挥高产品种的产量潜力。京单68的高产栽培技术主要包括以下几个方面:

2.1 麦茬及秸秆处理

京津唐区夏玉米一般是在前茬小麦收获后进行免耕贴茬直播,前茬小麦的秸秆量和留茬高度对夏玉米播种和出苗影响较大。为提高京单68的播种和出苗质量,应在前茬小麦收获时将麦秸直接粉碎还田,麦秸粉碎长度不宜超过10cm,粉碎效果差的地块应进行2次粉碎。玉米播种前,将麦秸铺散均匀,不成堆、不成垄。如前茬小麦秸秆量大或割茬太高,播前可用秸秆粉碎机械进行秸秆处理。

2.2 抢时早播,提高播种质量

京津唐夏玉米区热量资源紧张,抢时早播对夏玉米高产优质具有重要意义。收获前茬小麦后要抢早播种,一般适宜播种期为6月中旬,争取6月20日前完成播种。直接选用包衣种子,采用单粒精量点播机进行贴茬精量播种,60cm等行距种植,播深5cm左右,做到播深一致、行距一致、覆土一致、镇压一致。播种同时施底肥,将氮肥总量的1/3和全部磷、钾肥作底肥,确保种、肥分开,以免烧种或烧苗。若土壤墒情不足,为提早播种,可先播种播后及时补浇"蒙头水",以保证种子尽早萌发和出苗。

2.3 合理密植

京单68虽是耐密型品种,但在实际大田生产中并不是种植密度越大越好。应根据该品种的具体特性、地力条件、管理水平和各地的生产条件等确定适宜的种植密度。根据多年生产实践,京单68的适宜种植密度为每亩4 000~4 500株。高产田适宜种植密度为每亩4 500株左右。

2.4 化学除草

播后苗前,土壤墒情适宜时进行土壤封闭除草。也可在玉米出苗后,用玉米苗后除草剂进行苗后除草。喷药时,确保不重喷、不漏喷,并注意用药安全。

2.5 适时追肥

小喇叭口期(9叶展)前后,将剩余的2/3氮肥进行机械开沟侧深施(深度10cm左右),避免地表撒施。玉米生育后期如出现脱肥现象,则应根据植株的具体长势,补施适量速效氮肥。

2.6 旱灌涝排

如播种时土壤墒情不足,播后要及时补浇"蒙头水"。苗期耐涝性差,如遇暴雨导致田间出现积水,要及时排水。孕穗至灌浆期是玉米需水的关键时期,此期如遇旱应及时灌溉,避免因干旱严重减产。

2.7 防治病虫

黏虫是夏玉米苗期的主要害虫之一,免耕覆盖夏玉米播后苗前,可结合玉米化学除草加入杀黏虫药剂进行喷药防治。大喇叭口期及时防治玉米螟,可用BT乳剂灌心,也可在成虫产卵始盛期释放赤眼蜂进行防治。及时防治锈病、叶斑病等病害。

2.8 适时晚收

根据京单 68 的籽粒灌浆进程及籽粒乳线情况，在不严重影响下茬小麦播种的情况下要尽量晚收，以保证籽粒充分灌浆和成熟，增加籽粒产量、提高品质。一般于 9 月底至 10 月初收获。

本文原载：中国种业，2013（11）：69-70

不同播期条件下"京单68"和"郑单958"的籽粒灌浆特性研究

徐田军　王荣焕　赵久然　陈传永　刘秀芝
王元东　刘春阁　成广雷　王晓光

(北京市农林科学院玉米研究中心，北京　100097)

摘　要：研究不同播期条件下"京单68"和"郑单958"的籽粒灌浆特性。以"京单68"和"郑单958"为试材，设置春播（5月14日）和夏播（6月13日）2个播期处理，用Richard方程模拟籽粒灌浆过程，并得出相关参数。结果表明：春播条件下玉米灌浆期间的平均温度、最高温度、最低温度和积温均比夏播高。春播条件下"京单68"和"郑单958"的最终百粒干物重分别为43.03g和39.02g，夏播分别为41.17g和34.80g。不同播期间，参试品种灌浆速率最大时的生长量（W_{max}）、活跃灌浆期积温（P）及到达最大灌浆速率时的积温（T_{GDD}）均表现为春播高于夏播；而最大灌浆速率（G_{max}）、灌浆起始势（R_0）呈相反趋势。同一播期不同品种间，W_{max}、G_{max}、R_0表现为"京单68"高于"郑单958"，而T_{GDD}和P则呈相反趋势。该研究表明，"京单68"较"郑单958"灌浆启动早，积累起始势大，最大灌浆速率高，达到最大灌浆速率、活跃灌浆期积温少，"京单68"最终百粒干物重高于"郑单958"。

关键词：播期；"京单68"；'郑单958'；籽粒灌浆特性

The Study on the Grain Filling Characteristics of "Jingdan68" and "Zhengdan958" Under Different Sowing Dates

Xu Tianjun　Wang Ronghuan　Zhao Jiuran　Chen Chuanyong　Liu Xiuzhi
Wang Yuandong　Liu Chunge　Cheng Guanglei　Wang Xiaoguang

(*Maize research center*, *Beijing Academy of Agriculture and Forestry Sciences*, *Beijing* 100097)

Abstract: In order to study the change of grain filling rate of "ingdan68" and "Zhengdan958"

基金项目：国家玉米产业技术体系（nycytx-02）；北京市农林科学院青年科研基金"京郊玉米品种生态适应及耐密性研究与应用"（QN201115）；北京市农林科学院科技创新基金项目"玉米新品种配套栽培技术研究及应用"（CXJJ201309）

第一作者：徐田军，男，1982年出生，山东临沂人，研究实习员，硕士，主要从事玉米高产栽培与生理生态研究。通讯地址：北京市海淀区曙光花园中路9号北京市农林科学院玉米研究中心（100097），Tel：010-51503703，E-mail：xtjxtjbb@163.com

通讯作者：赵久然，男，1962年出生，北京平谷人，研究员，博士，主要从事玉米遗传育种与高产栽培研究。通讯地址：北京市海淀区曙光花园中路9号 北京市农林科学院玉米研究中心（100097），E-mail：maizezhao@126.com under different sowing dates,"Jingdan68" and "Zhengdan958" were used in field experiments with two sowing dates (14th May, and 13th June). The average temperature, maximum and minimum temperature, accumulated temperature during the grain filling stage under spring sowing were higher than the summer sowing condition. Under spring sowing condition, the 100 - grain dry weight of "Jingdan68" and "Zhengdan958" were 43.03 g and 39.02 g, and 41.17 g and 34.80 g under summer sowing condition; The dry weight of 100 - grain at maximum grain filling rate (W_{max}), the thermal time needed during the active grain filling period (P) and the thermal time reaching the maximum filling rate (T_{GDD}) under spring sowing were higher than that of summer sowing condition. Thus, the maximum grain filling rate (G_{max}), initial grain filling potential (R_0) showed opposite trend. The T_{GDD} and P of "Zhengdan958" under the same sowing date were higher than "Jingdan68", but the W_{max}, G_{max}, R_0 presented opposite trend. The study showed that grain filling of "Jingdan68" was started as early as "Zhengdan958", initial grain filling potential and grain filling rate were higher than "Zhengdan958" and the accumulated temperature required for its active grain-filling period was less than "Zhengdan958", so final grain dry matter weight was significantly higher than "Zhengdan958".

Key words: Sowing Date; "Jingdan68"; "Zhengdan958"; Grain Filling Characteristics

引言

玉米籽粒灌浆期长短和灌浆强度高低决定了玉米籽粒干物重大小，而玉米籽粒灌浆过程受基因型和环境的互作影响，研究环境条件对不同基因型玉米籽粒灌浆进程的影响对提高玉米产量具有重要意义。适宜播期是玉米在生育期内充分利用生态环境中的有利光热资源的保障。郑洪建等指出积温是影响玉米产量的重要因子。解文孝研究表明播期影响水稻灌浆进程的起始灌浆量、灌浆速率和有效灌浆时间等灌浆参数。李丰等研究发现早播处理的小麦灌浆持续时间最长，快增期较短且平均灌浆速率、最大灌浆速率均较小，所以导致千粒重不高。而晚播处理的小麦灌浆持续时间短，渐增期持续时间、缓增期持续时间较短，所以不能获得较高粒重。肖荷霞等研究表明，夏播玉米穗粒数和粒重受日平均气温的影响较大，灌浆期平均气温与粒重呈负线性关系，气温日较差与粒重呈正线性关系。前人针对水稻、小麦、玉米籽粒灌浆影响方面做了相关研究，且多数研究籽粒干物重随时间的变化规律，而在授粉后积温对玉米籽粒建成的影响方面研究较少，因此本试验以2个不同基因型玉米品种"京单68"和"郑单958"为试材，设置春播和夏播共2个播期，用Richard方程模拟籽粒灌浆进程得出籽粒干物重随授粉后积温变化情况及相应的籽粒灌浆参数，分析其籽粒灌浆特性，为适时收获和挖掘不同品种的籽粒产量潜力提供理论依据和指导。

1 试验条件

1.1 试验地点

试验在北京昌平小汤山试验基地进行，试验田耕层土壤（0~20cm）含有机质1.07%、全氮0.06%、全磷0.05%、全钾0.9%、碱解氮93.28mg/kg、速效磷24.20mg/kg、速效钾141.00mg/kg。春播玉米生育期积温为3 400℃左右，降水量565.60mm，总日照时数1 000h；夏播玉米生育期间积温2 600℃左右，降水量483.00mm左右，总日照时数715h左右。

1.2 材料与方法

供试品种为"京单68"（北京市农林科学院玉米研究中心选育的玉米新品种，适宜在华北地区晚春播或夏播种植）和"郑单958"。设2个播期：春播（5月14日播种）和夏播（6月13日播种）。试验采用裂区设计，播期为主区，品种为副区，3次重复，6行区，行距0.6m，行长7.5m，小区面积27m²，种植密度为52 500株/hm²。田间管理同当地生产。

1.3 测定项目与方法

在玉米大喇叭口期选择有代表性的植株标记，抽雄吐丝期套袋统一授粉，春播处理分别在授粉之日起第15d、22d、29d、36d、43d、50d、57d、64d、71d、78d；夏播处理分别在第15d、22d、29d、36d、43d、50d、57d、64d进行取样。每次各处理分别取3个果穗，每果穗取中部籽粒100粒，称鲜重后，在105℃烘箱中杀青30min，80℃烘干至恒量，称百粒干物重（天平精度为0.001g）。以授粉后积温（℃）为自变量，百粒干物重为因变量（W），参照朱庆森等和顾世梁的方法，用Richards方程$W=A(1+Be^{-CT})^{-1/D}$模拟籽粒灌浆过程，籽粒的灌浆速率：$F=ACBe^{-CT}/(1+Be^{-CT})^{(D+1)/D}$，式中：$W$为粒重（g）；$A$为最终粒重（g）；$T$为授粉后积温（℃）；$B$、$C$、$D$为回归方程所确定的参数，其中，$B$为初值参数，$C$为生长速率参数，$D$为形状参数，当$D=1$时，即为Logistic方程。计算下列灌浆特征参数：达最大灌浆速率时的天数$T_{max}=(\ln B-\ln D)/C$，灌浆速率最大时的生长量$W_{max}=A(D+1)^{-1/D}$，最大灌浆速率$G_{max}=(CW_{max}/D)[1-(W_{max}/A)D]$，灌浆活跃期（约完成总积累量的90%）$P=2(D+2)/C$。

1.4 数据处理

通过CurveExpert 3.0软件进行灌浆动态拟合，用Excel进行数据计算和作图。

2 结果与分析

2.1 玉米灌浆期的温度和积温变化

由表1可知，与春播处理相比，夏播处理玉米籽粒灌浆期平均温度、最高温度、最低温度及积温呈降低趋势。其中，春播条件下灌浆期平均温度、最高温度、最低温度和积温分别比夏播高1.22℃、0.54℃、2.37℃和480.54℃。

表1 不同播期条件下籽粒灌浆期温度和积温的情况

播期（月/日）	平均温度（℃）	最高温度（℃）	最低温度（℃）	积温（℃）
5/14	23.01	29.39	18.41	1 872.41
6/13	21.79	28.86	16.04	1 391.87

由表2可知，授粉后15~78d内平均气温、最高气温、最低气温呈下降趋势且春播高于夏播。其中，春、夏播条件下，平均气温分别在花后57d、36d后均低于18℃。春播条件下灌浆期平均气温、最高气温、最低气温分别较夏播高3.76℃、2.36℃、4.70℃。

表 2　不同播期玉米籽粒灌浆期取样间隔内的温度变化

播期（月/日）	温度	取样间隔（d）							
		15~22	22~29	29~36	36~43	43~50	50~57	57~64	64~71	71~78
5/14	AT	24.47	24.04	23.06	22.28	22.56	20.21	17.78	15.98	14.30
	HT	28.24	29.30	27.40	26.97	28.51	24.73	23.13	24.61	22.39
	LT	19.74	18.03	17.97	16.86	17.10	14.03	12.11	6.47	6.21
6/13	AT	22.32	20.47	18.12	16.25	15.04	12.97	12.12		
	HT	28.25	25.17	23.38	24.46	22.65	21.92	20.63		
	LT	16.65	14.36	12.39	7.21	7.46	4.92	4.05		

注：AT. 平均气温；HT. 最高气温；LT. 最低气温。

2.2　播期对"京单68"和"郑单958"籽粒干物重和灌浆速率的影响

由图1可知，春夏播条件下"京单68"和"郑单958"的籽粒干物重随授粉后积温均呈"慢-快-慢"的S型变化趋势，且春播条件下"京单68"和"郑单958"的最终百粒干物重分别为43.03g和39.02g，夏播条件下分别为41.17g和34.80g。春播条件下"京单68"和"郑单958"的百粒干物重分别较夏播高4.96%和12.13%。

由图2可知，春夏播条件下，"京单68"和"郑单958"的籽粒灌浆速率呈"先升后降"的单峰变化。其中春播条件下"京单68"和"郑单958"到达峰值所需授粉后积温均为800℃左右，而夏播条件下分别为800℃和730℃左右。春、夏播条件下分别在1 200℃、1 080℃左右，"京单68"的灌浆速率高于"郑单958"，之后则呈相反趋势。

2.3　播期对"京单68"和"郑单958"灌浆参数的影响

由表3可知，随播期推迟，2个品种到达最大灌浆速率的所需积温（T_{GDD}）呈减少趋势，其中春播条件下"京单68"和"郑单958"的 T_{GDD} 较夏播高74.62℃和11.75℃；籽粒灌浆速率最大时的生长量（W_{max}）、籽粒活跃灌浆积温（P）呈降低趋势，而籽粒最大灌浆速率（G_{max}）和灌浆起始势（R_o）呈升高趋势。春播条件下"京单68"和"郑单958"的 W_{max} 较夏播分别增加0.85和1.09g/100grain，增幅为4.52%和5.08%；P 值增加130.47和52.15℃，增幅为9.08%和4.33%。春播条件下"京单68"和"郑单958"的 G_{max}、R_o 较夏播下降3.70%和5.00%、13.04%和20.69%。同一播期，不同品种间 T_{GDD} 和 P 表现为"郑单958"高于"京单68"，而 W_{max}、G_{max}、R_o 表现为"京单68"高于"郑单958"。

表 3　播期对"京单68"和"郑单958"的灌浆参数的影响

播期（月/日）	品种	T_{GDD}（℃）	W_{max}（g/100grain）	G_{max}[g/(100grain·℃)]	R_o（g/100grain）	P（℃）
5/14	ZD958	814.83	19.65	0.038	0.0040	1 566.69
	JD68	803.81	21.91	0.052	0.0046	1 256.84
6/13	ZD958	803.08	18.80	0.040	0.0046	1 436.22
	JD68	729.19	20.82	0.054	0.0058	1 204.69

注：T_{GDD} 为到达最大灌浆速率所需积温；W_{max} 为灌浆速率最大时的生长量；G_{max} 为最大灌浆速率；R_o 为起始势；P 为活跃灌浆期积温。

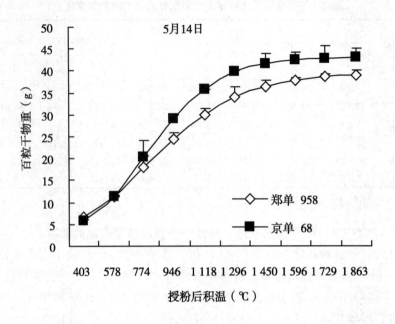

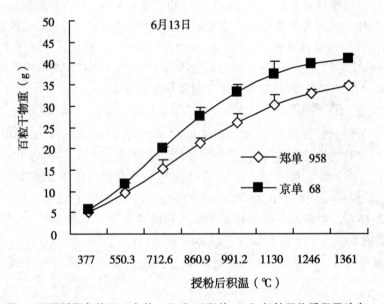

图1 不同播期条件下"京单68"和"郑单958"籽粒干物质积累动态

2.4 玉米籽粒灌浆参数与籽粒粒重的通径分析

对影响玉米籽粒百粒重（Y）的因子：品种、播期、G_{max}、P、授粉后积温做通径分析，以明确各因子对玉米籽粒百粒重的影响。由表4、表5可知，品种、P、G_{max}直接影响籽粒干物重大小，其直接通径系数分别为2.186 8、1.400 0、0.001 3。播期主要通过积温和活跃灌浆积温间接影响籽粒干物重大小，播期越早，花后积温越大（$r_{25}=-0.996\ 2^{**}$），活跃灌浆所需积温越少（$r_{24}=-0.316\ 7$），灌浆速度越快，粒重越大。

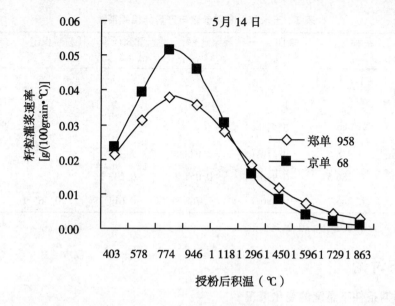

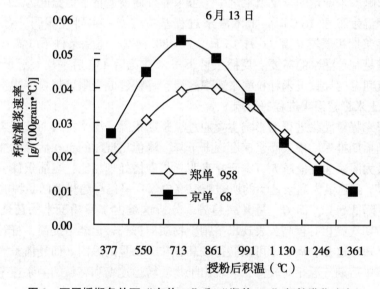

图2 不同播期条件下"京单68"和"郑单958"籽粒灌浆速率

表4 玉米籽粒灌浆参数与籽粒粒重的通径分析

	品种（X_1）	播期（X_2）	灌浆速率（X_3）	活跃灌浆所需积温（X_4）	授粉后积温（X_5）	百粒重（Y）
X_1		2.186 8	0	0.001 3	-1.314 3	-0.027
X_2	0		-0.364	0.000 2	-0.443 3	0.311 2
X_3	2.164 8	-0.051 5		0.001 3	-1.363 8	0.017 3
X_4	-2.052 9	0.115 3	-0.001 2		1.4	-0.072 9
X_5	0.189 2	0.362 6	-0.000 1	0.326 5		-0.312 4
Y						

表5 玉米籽粒灌浆参数与产量的相关系数

	品种 (X_1)	播期 (X_2)	灌浆速率 (X_3)	活跃灌浆所需积温 (X_4)	授粉后积温 (X_5)	百粒重 (Y)
X_1	1					
X_2	0	1				
X_3	0.989 9**	0.141 4	1			
X_4	-0.938 8**	-0.316 7	-0.974 1**	1		
X_5	0.086 5	-0.996 2**	-0.055 2	0.233 2	1	
Y	0.846 7**	-0.496	0.768 1**	-0.611 7*	0.565 9*	1

注：*、** 分别表示0.05、0.01显著水平。

3 结论与讨论

3.1 不同播期条件下温度的变化情况

播期是影响玉米产量的重要因素之一。玉米灌浆期最适日平均温度为22~24℃。当玉米籽粒灌浆期温度低于16℃时，玉米灌浆过程基本停止。本研究表明，春播（5月14日）条件下灌浆期积温较夏播（6月13日）高480.54℃。春播条件下气温在灌浆后期仍较高，可以满足籽粒灌浆的需要。夏播条件下玉米在花后43d后平均气温低于18℃，特别是过低的夜间温度导致玉米籽粒灌浆速率缓慢，籽粒建成受阻，籽粒干物重下降。

3.2 播期对玉米籽粒灌浆进程的影响

播期对玉米籽粒灌浆进程的影响主要通过灌浆期积温、灌浆持续期及灌浆速率。前人研究表明，随播期推迟，最大灌浆速度逐渐下降，峰值出现时间也在后移。播期越早灌浆时间越长，最大灌浆速度值越大。本研究表明，与春播处理相比，夏播条件下"京单68"和"郑单958"到达最大灌浆速率的所需积温（T_{GDD}）呈偏早趋势而两品种的籽粒灌浆速率W_{max}、P呈降低趋势，而G_{max}呈升高趋势。分析这是由于春播玉米的苗期气温相对较低，灌浆中后期气温相对较高，表现出茎鞘干物质向籽粒转运能力提高，籽粒灌浆速率升高，灌浆期缩短，这是春播玉米可以获得较高籽粒干物重的原因。而夏播玉米苗期温度相对较高，灌浆中后期气温相对较低，表现出干物质转运速率和籽粒的灌浆速率降低，灌浆持续期延长。这是夏播玉米籽粒干物重较低的原因。

同一播期，"京单68"到达最大灌浆速率时所需积温少于"郑单958"，且"京单68"灌浆启动时间早、灌浆速率高，灌浆进程短。这说明，"京单68"比"郑单958"灌浆快，更易获得较高的粒重，并且该优势在积温偏少年份表现更为明显。

4 结论

播期主要通过调节玉米生育期内光热资源来影响生长发育。本研究通过设置春夏播2个播期处理，研究不同播期条件下"京单68"籽粒灌浆特性，得出以下结论：在6月13日（夏播）条件下，玉米灌浆后期气温偏低，有效灌浆时间短，在有限的积温条件下未能达到完全成熟。春播条件下"京单68"和"郑单958"的百粒干物重分别较夏播高4.96%和12.13%。与夏播处理相比，春播处理下两品种达到最大灌浆速率时的所需积温多，灌浆速度慢，灌浆起始势低，而活跃灌浆期的积温高，即灌浆持续期长。同一播期不同品种间，"京单68"较"郑单958"灌浆启动早，灌浆速度快，百粒干物重高。

参考文献（略）

本文原载：农学学报，2013，3（10）：1-5

密度和播期对京单 68 冠层结构和产量的影响

徐田军　刘秀芝　赵久然　陈传永　王荣焕
王元东　刘春阁　成广雷　王晓光

（北京市农林科学院玉米研究中心，北京　100097）

摘　要：以玉米品种京单68为研究材料，通过设置密度（52 500株/hm²、60 000株/hm²、67 500株/hm²、75 000株/hm²、82 500株/hm²）和播期（5/15、6/13）两个处理因素，研究种植密度和播期对京单68冠层结构和产量的影响。结果表明，京单68的群体叶面积指数（LAI）随生育进程呈先升高后降低的单峰曲线变化，峰值出现在吐丝期。随种植密度增加，京单68的群体LAI和群体光合势呈增加趋势，但增幅逐渐减小。不同播期间，群体LAI表现为春播高于夏播，但LAI到达峰值的时间则表现为春播较夏播晚5d。就产量而言，随密度增加，京单68的群体产量呈升高趋势，但增幅逐渐减小。不同播期间，春播条件下京单68的穗长、穗粗、穗粒数及产量均高于夏播，而秃尖长则呈相反趋势。密度和播期处理对京单68产量的影响主要体现在穗粒数和千粒重方面。

关键词：京单68；密度；播期；叶面积指数；产量

Effect of Sowing Date and Plant Density on Canopy Structure and Yield of Jingdan68

Xu Tianjun　Liu Xiuzhi　Zhao Jiuran　Chen Chuanyong
Wang Ronghuan　Wang Yuandong　Liu Chunge
Cheng Guanglei　Wang Xiaoguang

（*Maize research center*, Beijing Academy of Agriculture
and Forestry Sciences, Beijing 100097）

Abstract: In order to study the effect of sowing date and plant density on canopy structure and yield of Jingdan68 (JD68), the experiment was carried out with five plant densities (52 500plant/hm², 60 000plant/hm², 67 500plant/hm², 75 000plant/hm², 82 500plant/hm²) and two sowing date (5/15, 6/13) treatments. The results showed that the leaf area index (LAI) of JD68 was showed the trend of a single peak curve with "first increase and then decrease" and reached the peak in silking stage. The LAI and photosynthetic potential of JD68 showed an increase trend, but

---

基金项目：国家玉米产业技术体系（nycytx-02）、北京市科技计划（D121100003512001）、北京市农林科学院科技创新基金（CXJJ201309）

作者简介：徐田军，硕士，主要从事玉米高产栽培与生理生态研究
　　　　　赵久然为通讯作者。E-mail：maizezhao@126.com

the trend reduced gradually with plant density increased. The difference of LAI between sowing date was that the LAI in spring sowing date was higher than summer sowing date, while the time of reaching peak lated 5d than the summer sowing date. With the increase of plant density, the yield of JD68 was increased, but the trend reduced gradually. The ear length, ear diameter, grain number per spike and grain yield of JD68 were higher than that of summer sowing date, while bald tip length showed the opposite trend. The effect of plant density and sowing date on the yield of JD68 was mainly in the grain number per spike and 1 000-grain weight.

Key words: Jingdan68; Plant density; Sowing date; Leaf area index; Yield

玉米是目前我国第一大粮食作物，对确保我国粮食安全具有重要战略意义。随着畜牧业和生物能源制造业的发展，玉米需求量急剧增加。我国耕地面积日益减少，提高玉米产量必须依靠单产。玉米干物质绝大部分来源于光合作用，叶片是进行光合作用的主要场所，合理的冠层结构和群体结构有利于玉米叶片光合作用的提高，进而利于光合产物的转化和产量提高。叶面积指数是反映玉米群体结构的重要指标之一，是衡量群体生长状况和光能利用率的重要指标。前人研究表明，种植密度可改变作物群体结构进而改变作物光能利用率。播期主要通过日照时数和光合辐射量等生态因子影响玉米的生育进程。肖荷霞等研究表明，适时早播可增加有效积温，延长玉米的有效生长期，使籽粒灌浆期相应延长，从而增加玉米植株干物质积累量，进而利于高产。陈素英等研究表明，随播期延迟小麦叶面积指数和冠层截获的光合有效辐射量呈降低趋势。综上所述，密度和播期是影响作物生长发育、形态构建和产量构成的重要因素。合理的种植密度和适宜的播期是玉米获得较高产量的必要条件。前人研究多针对密度或播期一个因子，而对密度和播期耦合对玉米叶面积指数和产量影响的研究较少。

京单 68 是适宜京津唐地区夏播种植的玉米新品种，2010 年通过国家审定，2013 年被北京市列为更新换代夏播玉米品种。本试验通过设置不同密度及播期处理，研究和探索播期和密度互作对京单 68 冠层结构和产量的影响，以期为京单 68 的生长调控和产量提高提供理论依据和技术支持。

1 试验条件

1.1 试验地点

在前期预备试验的基础上，2013 年试验在北京市昌平区小汤山试验基地进行，试验田耕层土壤（0~20cm）含有机质 1.07%、全氮 0.06%、全磷 0.05%、全钾 0.9%、碱解氮 93.28mg/kg、速效磷 24.20mg/kg、速效钾 141.00mg/kg。春播玉米生育期内积温 3 400℃左右，降水量 565.60mm，总日照时数 1 000h；夏播玉米生育期内积温 2 600℃左右，降水量 483.00mm 左右，总日照时数 715h 左右。

1.2 材料与方法

供试玉米品种为京单 68（JD68），由北京市农林科学院玉米研究中心选育并提供。试验设 5 个密度水平：52 500株/hm^2（A）、60 000株/hm^2（B）、67 500株/hm^2（C）、75 000株/hm^2（D）、82 500株/hm^2（E）；2 个播期水平：春播（5/15，5 月 15 日播种）、夏播（6/13，6 月 13 日播种）。试验采用裂区设计，播期为主区，密度为副区，3 次重复。小区行距 0.6m，行长 7.5m，小区面积 27m^2。田间管理同当地大田生产。

1.2.1 叶面积及光合势测定

观察记载京单68在各处理条件下的生育进程。三叶期，在各小区分别选取生长健壮整齐一致的3株进行挂牌标记，分别在三叶期、拔节期、大喇叭口期、开花期和成熟期测量叶长和叶宽，计算叶面积指数和群体光合势。采用Ration模型$y=(a+bx)/(1+cx+dx^2)$进行 LAI 和相对出苗后天数的动态模拟。

$$单叶面积 = 长 \times 宽 \times 0.75$$

$$叶面积指数（LAI）= 单株叶面积 \times 单位土地面积株数/单位土地面积$$

群体光合势 $LAD=(LA2+LA1)/2\times(t2-t1)$，LA1、LA2 分别为测定时间 t_1、t_2 时单位土地面积上的叶面积。

1.2.2 产量测定

在每个处理小区中间3行取6m共10.8m²收获全部玉米果穗。计数玉米果穗数目后，称鲜果穗重 Y1（kg），按平均穗质量法取30个果穗，其中15个果穗用于测定出籽率和含水率，15个果穗用于考种及测产。

$$鲜果穗重 Y（kg/hm^2）=（Y1/20）\times 10\,000$$

$$实测产量（kg/hm^2）= 鲜穗重（kg/hm^2）\times 出籽率（\%）\times$$
$$[1-籽粒含水率（\%）]/（1-14\%）$$

1.3 数据处理和分析

数据采用DPS6.5软件进行方差分析，其中处理平均数间差异显著性采用LSD进行检验（$P<0.05$）。用CurveExpert 1.4 软件进行叶面积指数动态变化模拟。用Excel 2003 进行数据计算和作图。

2 结果与分析

2.1 气象因子

由表1可知，京单68自三叶期到吐丝期，春播条件下的平均气温、最高气温和最低气温均低于夏播，而在灌浆期则呈相反趋势。苗期到大喇叭口期的有效积温在春夏播条件下差别不大，而开花期到灌浆期春播条件下的有效积温大幅高于夏播。京单68各生育时期的日照时数均表现为春播高于夏播，春播条件下降雨量在拔节期、开花期、灌浆期高于夏播，而三叶期、大喇叭口期呈相反趋势。

2.2 不同处理京单68 LAI 变化

京单68在不同种植密度和播期条件下，群体 LAI 随生育进程呈先升高后降低的单峰曲线变化（图1），吐丝期到达峰值，之后开始下降且下降幅度表现为A>B>C>D>E。随种植密度增加，京单68的群体叶面积指数呈增加趋势，但增幅逐渐减小。不同密度条件下，群体 LAI 在吐丝期差异最大，而在三叶期、拔节期及成熟期差异较小。不同播期条件下，群体 LAI 的差异主要表现为春播高于夏播，而到达峰值的时间表现为夏播早于春播。

表 1 不同播期条件下京单 68 生育阶段气象因子

Table 1 Meteorological factors in the growing period of JD68 under different sowing date

播期 Sowing date	生育阶段 Growing period	平均气温 Average temperature (℃)	最高气温 Maximum temperature (℃)	最低气温 Minimum temperature (℃)	有效积温 Accumulated temperature (℃)	日照时数 Sunshine duration (h)	降水量 Rainfall (mm)
5/15	三叶期	22.2	28.8	15.4	195.7	125.5	3.3
	拔节期	22.4	27.3	18	235.6	115.6	77.3
	大喇叭口期	26.1	31.5	21.3	321.6	128.1	64.8
	吐丝期	26.2	31	22.4	210.3	73.5	98
	灌浆期	24.8	30.8	19.9	900.3	436.7	184.3
6/13	三叶期	25	30.7	20.3	195	87.9	12.9
	拔节期	26.5	31.8	21.5	231.2	92.6	58.5
	大喇叭口期	26.8	32.1	22.3	318.7	125.3	96.1
	吐丝期	26.3	32.6	21.7	162.5	80.7	81.8
	灌浆期	22.7	28.7	17.8	685.1	357.6	104.6

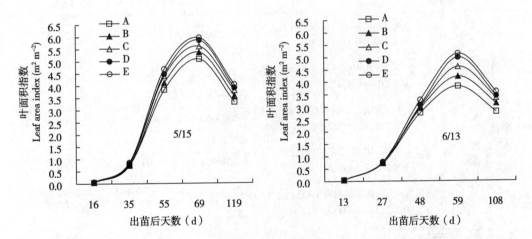

图 1 不同播期和密度下京单 68 的 LAI

Figure 1 Leaf area index of JD68 under different sowing date and plant density conditions

2.3 不同处理京单 68 LAI 的特征参数

玉米的 LAI_{max} 能够反映群体的最大同化能力强弱。表 2 表明,Ration 模型 $y=(a+bx)/(1+cx+dx^2)$ 可以很好地模拟 LAI 的变化情况,决定系数 R^2 为 0.9859~0.9960。由表 2 可知,随种植密度增加,LAI_{max} 呈增加趋势,其中春、夏播条件下京单 68 的 LAI_{max} 变化范围分别为 5.10~6.07、3.96~5.33,而种植密度对达 LAI_{max} 所需时间的影响不显著。随播期延迟,LAI_{max} 呈降低趋势,其中春播条件下京单 68 的 LAI 较夏播高 0.95,增幅为 20.17%;达 LAI_{max} 所需时间为夏播早于春播,早 5d。

表2 不同种植密度和播期条件下京单68的 LAI 特征参数
Table 2 LAI characteristic parameters of JD68 under different plant density and sowing date conditions

播期 Sowing date	密度 Planting density	Ration 模拟方程 Ration model	R^2	LAI_{max} Maximum of LAI	达 LAI_{max} 所需时间（d） Time for the maximum of leaf area index（d）
5/15	A	$y=(-0.9323+0.0386x)/(1-0.0250x+0.0002x^2)$	0.9875	5.10	72.70
	B	$y=(-0.9125+0.0392x)/(1-0.0252x+0.0002x^2)$	0.9863	5.47	71.70
	C	$y=(-1.0541+0.0443x)/(1-0.0248x+0.0002x^2)$	0.9865	5.71	72.70
	D	$y=(-1.1027+0.0458x)/(1-0.0251x+0.0002x^2)$	0.9859	5.97	72.70
	E	$y=(-1.1033+0.0466x)/(1-0.0248x+0.0002x^2)$	0.9862	6.07	72.70
	Average			5.66	72.50
6/13	A	$y=(-0.4920+0.0324x)/(1-0.0249x+0.0002x^2)$	0.9960	3.96	67.50
	B	$y=(-0.4928+0.0318x)/(1-0.0258x+0.0002x^2)$	0.9898	4.50	67.50
	C	$y=(-0.5398+0.0341x)/(1-0.0258x+0.0002x^2)$	0.9902	4.75	68.50
	D	$y=(-0.5322+0.0341x)/(1-0.0264x+0.0002x^2)$	0.9882	5.01	66.50
	E	$y=(-0.7383+0.0425x)/(1-0.0257x+0.0002x^2)$	0.9941	5.33	67.50
	Average			4.71	67.50

2.4 不同处理京单68光合势变化

光合势是衡量作物群体绿叶面积大小及光合持续期长短的重要生理指标。由表3可知，随密度增加，京单68各生育时期的群体光合势呈升高趋势，在吐丝期之后达到最大值。其中，春、夏播条件下82 500株/hm²的总光合势较52 500株/hm²密度下高123.96（m²·d）/m²、167.42（m²·d）/m²，增幅分别为16.90%、29.20%。不同播期间，京单68各生育期的光合势表现为春播条件下高于夏播。其中，春播条件下京单68总光合势较夏播高137.38（m²·d）/m²，增幅为17.09%。

表3 不同种植密度和播期条件下的京单68光合势 [(m²·d)/m²]
Table 3 Photosynthetic potential of JD68 under different plant densities and sowing date

播期 Sowing date	密度 Plant density	出苗-三叶期 Sowing-Three-leaf	三叶期-拔节期 Three-leaf-shooting	拔节期-大喇叭口期 Shooting-big trumpet	大喇叭口期-吐丝期 Big trumpet-silking	吐丝期-成熟期 Silking-maturity	总LAD Total LAD
5/15	A	7.87	45.03	57.53	226.16	396.95	733.54
	B	8.01	48.46	61.53	238.43	416.83	773.26
	C	8.71	51.94	64.97	252.34	438.42	816.38
	D	8.89	52.87	67.13	260.30	449.68	838.87
	E	9.40	55.37	69.16	264.55	459.02	857.5
6/13	A	7.66	34.68	42.69	171.50	316.88	573.41
	B	7.82	36.61	46.48	194.15	345.44	630.5
	C	8.15	37.60	49.47	212.77	368.37	676.36
	D	8.24	38.17	52.30	226.86	385.99	711.56
	E	8.41	40.48	54.70	236.60	400.64	740.83

2.5 不同处理京单68产量及产量构成要素

由表4可知,随密度增加,京单68的穗长、穗粗、穗粒数、千粒重呈降低趋势。其中,春播条件下,82 500株/hm²的密度下较52 500株/hm²分别低14.58%、8.04%、15.14%、10.08%;夏播条件下,分别低9.36%、2.73%、13.93%、12.62%。而产量随密度提升呈增加趋势,但增幅逐渐降低。春、夏播条件下,82 500株/hm²密度下京单68的产量较52 500株/hm²分别高4 765.95 kg/hm²、3 633.9 kg/hm²,增幅分别为31.77%、38.63%。不同播期间,春播条件下京单68的穗长、穗粗、穗粒数及产量均高于夏播,而秃尖长则呈相反趋势。方差分析表明,不同密度间及播期间穗长、秃尖长、穗粒数、千粒重及产量差异均达到显著水平,而穗粗的差异不显著。由此可见,高密度下产量的提高主要得益于单位面积穗数的增加,进而弥补了因穗粒数与千粒重随密度增加而降低造成的单株产量下降。

表4 不同种植密度和播期条件下的京单68产量及产量构成要素
Table 4 Yield and yield component factors of JD68 under different plant densities and sowing date

播期 (月/日) Sowing date	密度 Plant density	穗长 Ear length (cm)	穗粗 Ear diameter (cm)	秃尖长 Bald tip length (cm)	穗粒数 Grain numbers per spike	千粒重 1 000-grain weight (g)	产量 Yield (kg/hm²)
5/15	A	20.23a	5.35a	0.07c	542.82a	407.16a	9 648.30e
	B	20.13b	5.18a	0.08c	533.32b	396.36b	11 610.95d
	C	18.90c	5.10a	0.10b	521.31c	387.32c	13 168.35c
	D	17.87d	4.95b	0.15b	487.39d	377.77d	13 394.95b
	E	17.28e	4.92c	0.23a	460.61e	366.12e	14 414.20a
	Average	18.88	5.10	0.13	509.09	386.95	12 447.35

(续表)

播期 （月/日） Sowing date	密度 Plant density	穗长 Ear length （cm）	穗粗 Ear diameter （cm）	秃尖长 Bald tip length （cm）	穗粒数 Grain numbers per spike	千粒重 1 000-grain weight（g）	产量 Yield （kg/hm²）
6/13	A	18.37a	5.12a	0.10c	528.87a	397.72a	9 407.85e
	B	17.58b	5.12a	0.18b	517.87b	388.07b	10 034.05d
	C	17.40c	5.08a	0.25b	492.26c	375.62c	11 111.40c
	D	17.13d	5.07a	0.37a	471.33d	360.02d	12 094.80b
	E	16.65e	4.98b	0.38a	455.20e	347.51e	13 041.70a
	Average	17.43	5.07	0.26	493.11	373.79	11 137.96
均方 Mean squares (ANOVA)	播期 Sowing date	383.13**	0.24	53.27**	55.82**	127.35**	1048.50**
	密度 Plant density	132.99**	4.08*	21.30**	182.56**	193.38**	1326.10**
	播期*密度 Sowing date × Plant density	23.15**	1.52	3.00*	3.13*	3.32*	54.34**

注：*，** 分别表示在5%和1%水平上显著

Note：*，** mean significant at the 0.05 and 0.01 probability levels, respectively

2.6 播期、密度、LAI_{max}、光合势与产量的相关性分析

由表5可知，有效积温与 LAI_{max} 和群体光合势呈极显著正相关（相关系数分别为0.76、0.79），与产量之间存在正相关关系但未达到显著性水平。密度、LAI_{max} 及光合势与产量存在极显著的正相关关系（相关系数分别为0.89、0.87和0.85）。此外，LAI_{max} 与密度亦呈极显著正相关（相关系数为0.64）。由此说明，在一定的范围内，适当早播以增加玉米生育期内有效积温和增加种植密度有利于提高京单68的群体产量。

表5 有效积温、密度、LAI_{max}、光合势与产量的相关性分析

Table 5 Correlation analysis between yield and the effective accumulated temperature, plant density, LAI_{max} and photosynthetic potential

相关系数 Correlation coefficient	有效积温 The effective accumulated temperature	密度 Plant density	LAI_{max}	光合势 Photosynthetic potential	产量 Yield
有效积温 The effective accumulated temperature	1				
密度 Plant density	0	1			
LAI_{max}	0.76**	0.64*	1		

(续表)

相关系数 Correlation coefficient	有效积温 The effective accumulated temperature	密度 Plant density	LAI_{max}	光合势 Photosynthetic potential	产量 Yield
光合势 Photosynthetic potential	0.79**	0.6	1.00**	1	
产量 Yield	0.4	0.89**	0.87**	0.85**	1

3 讨论

3.1 播期和密度对京单68 LAI和光合势的影响

播期主要通过温度、光照时数、降雨量影响玉米植株的生长发育。合理的种植密度利于玉米充分利用光热资源进而构建良好群体结构。前人研究表明，合理的 LAI 可提高玉米产量。LAI 可通过播期和种植密度进行调控。本研究表明，随种植密度增加，京单68的 LAI 和光合势呈增加趋势，但增幅逐渐降低；并且，LAI_{max} 与密度呈极显著正相关关系。随密度增加，LAI 和光合势的增加是由单株叶面积减少和群体 LAI 增加共同作用的结果。在不同播期条件下，群体 LAI 的差异主要表现为春播高于夏播。随播期推迟，LAI_{max} 呈降低趋势，到达 LAI_{max} 的时间表现为夏播较春播提早5d。本结论与于吉琳等的研究结果不一致，这与玉米灌浆期春播条件下的光照和降雨量高于夏播有关。

3.2 播期和密度对京单68产量及产量构成要素的影响

适宜的播期和合理的种植密度是玉米获得高产的必要条件。前人研究表明，在一定范围内产量随着密度增加而提高，到达一定密度后产量达到最高，继续增加密度产量反而下降。播期对玉米产量的影响主要是生育期间生态因子共同作用的结果。随播期推迟，玉米穗粒数和千粒重减少，产量降低。适时早播可充分利用生育期光热资源，延长玉米生育期，有利于玉米高产。本研究表明，在种植密度 52 500~82 500株/hm^2 范围内，京单68产量随着密度提升而增加，但增幅越来越小。在本研究的5个密度条件下，京单68均未发生倒伏，这说明该品种具有较高的抗倒伏能力和密植增产潜力。不同播期间，春播条件下京单68的穗长、穗粗、穗粒数及产量均高于夏播，而秃尖长则呈相反趋势。播期对玉米产量的影响显著大于密度。这与春播条件下灌浆中后期光热资源条件相对较好，延缓了叶片衰老和 LAI 降低，进而能够维持较高的光合势有关。

参考文献（略）

本文原载：作物杂志，2014（6）：76-80

玉米新品种京科968配套高产栽培技术

赵久然

(北京市农林科学院玉米研究中心，北京 100097)

京科968是由北京市农林科学院玉米研究中心选育的玉米新品种，2011年通过了国家审定。适宜在北京、天津、山西中晚熟区、内蒙古赤峰和通辽、辽宁中晚熟区（丹东除外）、吉林中晚熟区、陕西延安以及河北承德、张家口、唐山等地区春播种植。2012—2013年，连续被农业部推荐为玉米主导品种。

该品种在东华北地区从出苗至成熟共计128d，与郑单958相当。株型下部平展、上部紧凑，株高3m左右，穗位高约1.2m。穗行数16~18行，穗轴白色。据多年多点试验和示范证明，京科968在种子活力、幼苗顶土能力、植株长势、抗逆性、抗病虫能力和产量潜力等方面具有明显优势。京科968种子活力强，田间成苗率高，且苗期发棵快，叶色浓绿，茎秆粗壮，可为中后期的生长发育和高产打下坚实基础。从抗性来看，京科968抗旱性好、耐涝性强，且抗多种玉米生产主要病害，特别是高抗玉米螟，丰产稳产性好。2012年，我国东北地区暴发玉米大斑病和黏虫为害，且玉米螟发生较重，相比其他品种，京科968依然能够保持活秆成熟、青枝绿叶，在大灾之年表现出了很好的抗病性和抗虫性，经受住了生产的考验。京科968籽粒商品性好，容重高（767g/L），并且是高淀粉品种（淀粉含量75.42%），籽粒容重和淀粉含量两项指标都高于国家一级玉米种子质量标准。

选用优良品种和优质种子，并坚持良种良法配套，才能保证品种的高产潜力充分发挥出来。玉米新品种京科968的配套高产栽培技术主要包括以下方面。

1 选地

根系是玉米吸收水分和养分的重要器官，而土壤是玉米扎根生长的场所，可为玉米植株的生长发育提供水分、空气及养分。土层深厚，耕层疏松绵软，土壤大小空隙比例适当，上虚下实，速效养分含量高且比例适宜，水、肥、气、热协调的土壤环境才利于玉米根系的生长发育和水肥吸收。要实现京科968高产，最好选择地势平坦、土层深厚、土壤疏松且通透性好，耕层有机质和速效养分高，保水、保肥能力好，且灌排方便的地块。

2 整地

为实现玉米生长健壮和高产稳产，生产中应注意通过合理耕作和培肥地力等措施为玉米的生长和发育创造良好的土壤条件。整地可为玉米创造合理的耕层结构，使水、肥、气、热状况适宜根系和植株的生长发育。而土壤深耕则可有效改善土壤的水、热、气状况，利于微生物活动和养分的分解、转化，从而提高地力，促进根系发育。若深耕配合增施有机肥料则效果更加明显。如果是当年秋季进行翻地整地，土壤熟化时间较长，有利于接纳早秋和晚秋的雨水；如果是秋季翻地，待第二年春季玉米播种前再整地，则要特别注意春耕后应立即采取镇压提墒等保墒措施。

特别需要说明的是，对于生产中因中小型农机具的大面积和长期使用而导致土壤耕层变浅、犁底层不断加厚的地块，可在秋季进行全面深松，或在玉米生育期间进行条带深松作业，以打破犁底层，改善深层土壤结构，提高土壤通透性和蓄水保墒能力，为促进玉米根系发育和获得高产创造良好的土壤条件。

3　一播全苗，苗齐苗匀苗壮

俗话说"豆打长秸麦打齐，玉米缺苗不用提"。玉米是单秆作物，不像小麦、水稻那样可以分蘖，一般每棵玉米只结1个果穗，缺苗断垄易造成穗数不足，进而减产。即使是采取点籽补种、拔苗移栽等补救措施，也会产生小株，甚至是空秆。因此，为获得高产，要确保一次播种拿全苗，实现苗全、苗齐、苗匀、苗壮。其中，苗全是高产的基础，苗齐、苗匀、苗壮是高产的关键。

播种是保障玉米苗全苗齐苗壮、提高群体整齐度和奠定良好丰产稳产基础的最重要环节。为实现玉米生产的"四苗"要求，在选用优良品种和优质种子的同时，还要严把播种质量关，做到"七分种、三分管"，努力提高播种质量和群体整齐度。

从播种时间来看，各地应根据当地的自然气候条件和品种特性等把握好播种时期，一般以4月中旬至5月上旬为宜。如果播种过早，易造成低温烂种和病虫为害，特别是地下害虫和丝黑穗病等土传病害的侵害，导致出苗不齐；如果播种过晚，则影响植株的正常生长发育及籽粒后期的灌浆和脱水，导致籽粒不能正常成熟，进而减产。

在播种技术方面，应大力推广机械播种，以加快播种进度，将播种工作集中在最佳播期，提高播种质量和出苗整齐度。京科968种子活力高，适宜单粒播种，可采用机械单粒精量播种技术进行播种。对没有灌溉条件或者灌溉条件较差地区，可采取坐水种、地膜覆盖、膜下滴灌、育苗移栽等抗旱保墒播种技术措施。但无论采取哪种播种方式，都要力争一次播种保全苗和提高群体整齐度。

4　合理密植

玉米亩产量=亩株数×平均单株产量（或亩穗数×平均单穗重）。因此，要提高产量，一方面是增加株数，另一方面是提高和稳定单穗重。种植密度过稀或过密都不利于获得高产。种植密度过稀，不能充分利用土地、空间、养分和阳光。虽然单株生长发育好，穗大、籽粒饱满，但因减少了总穗数，因此产量降低；如果种植密度过大，虽然总穗数增加了，但全田荫蔽，通风透光不良，不利于单株的生长发育，易造成秃尖、小穗甚至空秆，最大的风险就是容易造成倒伏，进而减产。只有合理密植，才能充分有效地利用光、气、水、热和养分，使玉米植株群体和个体协调发展，进而提高单位面积的产量。

合理密植要根据品种特性、当地的自然生态条件、土壤肥力、施肥水平和管理水平等综合确定。京科968穗大、粒多，单株产量潜力高，适宜留苗密度以每亩3 800株为宜。

在实际生产中，为确保合理留苗密度，应注意抓好以下环节：一是提高播种质量，力争一次播种苗全、苗齐、苗壮；二是对于非单粒精量播种地块，在间苗、定苗留苗环节，要严格按照京科968的适宜密度要求进行间苗和定苗；三是及时防治苗期病虫草鼠害，减少因病虫草鼠害造成的株数损失。

5 科学水肥运筹

俗话说"有收无收在于水，收多收少在于肥"，这充分说明了水肥对玉米生产，特别是对玉米高产的重要性。

5.1 合理灌溉

玉米一生中，不同生育时期对水分的需求及其对干旱的反应存在较大差异。播种期至出苗期，虽然需水量很小，但非常关键和重要。充足适宜的土壤墒情有利于种子萌发和出全苗，因此要求播前底墒充足；如墒情不足，播后要及时浇水补墒。大喇叭口期至灌浆高峰期是需水量最多的时期，特别是吐丝前后为水分敏感期。大喇叭口期如遇旱，则会造成"卡脖旱"，导致雄穗或雌穗抽不出来，雌雄不协调，形成大量缺粒与秃尖，影响粒重和产量；抽雄吐丝期如遇旱，则授粉结实不良，造成穗粒数减少，甚至空秆；籽粒灌浆期如遇旱，即遭遇"秋吊"，则导致粒重降低，进而减产。玉米生产中就有"开花不灌、减产一半"、"前旱不算旱、后旱减一半"等说法。因此，在大田生产中，玉米生长后期要根据天气和土壤墒情等遇旱及时浇水。实现京科968高产，要保证关键生育期不受旱。

5.2 科学施肥

从需肥特性来看，玉米不同生育时期对肥料的需求数量和种类也存在较大差异。其中，玉米苗期需肥较少；拔节至开花期吸收养分速度快、数量多，是需肥关键时期；而生育后期对养分的吸收速度减慢且吸收量也少。从需肥种类来看，玉米对氮的需求量最大，其次是钾，而对磷的需求量相对较小。

目前，玉米生产中普遍存在化肥地表撒施和"一炮轰"等不合理的施肥方法。化肥地表撒施不仅引起肥料大量挥发和流失，造成肥料浪费、利用率低，并且还污染环境。而"一炮轰"施肥则容易造成烧种、烧苗和后期植株早衰脱肥。

为提高肥料利用率，生产中应做到"有机肥与无机肥配合、氮磷钾与微肥配合、基肥与追肥配合"，建议有条件的地方采用测土配方平衡施肥。并且，要重施基肥、氮肥分期追施，注意化肥深施，种、肥隔离，以实现培肥地力、肥水耦合、以肥调水、以肥保水，充分发挥肥水的增产效应。

6 加强田间管理

6.1 苗期

苗期指的是从播种期至拔节期的这段时间，包括种子萌发、出苗根系建成和幼苗生长等。苗期是玉米营养生长阶段，以根系生长为主，田间管理应以促进根系发育、控制地上部徒长、培育壮苗为中心。

这期间的玉米田间管理，一是要及早进行中耕铲趟，并要做到多铲多趟，以提高土壤通透性，增温保墒，促进玉米根系发育和植株生长。而对低洼湿渍地块，则要注意排水，并在适宜条件下中耕。二是若基肥、种肥及底墒不足，严重影响幼苗生长，则应早施重施苗肥并浇水。

6.2 穗期

穗期指的是从拔节期至雄穗开花期的这段时间。穗期是玉米营养生长和生殖生长同时并进阶段，也是玉米一生中生长发育最旺盛的阶段，是决定穗数、粒数多少的关键时期，也是吸收肥水量最多、田间管理最关键的时期。

这期间的玉米田间管理措施主要是协调水肥供应，调节植株生育状况，促叶、壮秆，并保证雌雄穗发育良好，建成壮株，为穗大、粒多、粒重打好基础。可在拔节期至小喇叭口期，侧开沟、深追氮肥（深10cm左右，每亩追施尿素15~20kg或等氮量其他氮肥），并可根据测土结果补施适量的钾肥和微肥等。为避免脱肥，还可在大喇叭口期根据植株长势，酌情补施适量氮肥，以促进棒三叶，攻大穗。此外，还可结合追肥进行土壤深松作业。做到化肥深施，提高肥效，清除田间杂草，打破土壤犁底层，增加土壤通透性，提高土壤蓄水能力，促进玉米植株生长。同时，可结合追施穗肥进行浇水。

6.3 花粒期

花粒期指的是从雄穗开花期至籽粒成熟期这一阶段，这期间主要进行开花、授粉、受精、籽粒形成及灌浆成熟等。该阶段的生长中心是籽粒。花粒期是玉米生殖生长阶段，是玉米开花、授粉及籽粒形成的关键时期。

这期间的玉米田间管理应保证水肥供应，防止叶片早衰，促早熟，争取穗大、粒多、粒重。一是看长相，巧追攻粒肥。为维持和延长中上部叶片功能期，争取制造更多的光合产物，促进籽粒形成和籽粒饱满，如后期发生脱肥应立即补施攻粒肥。可采用0.3%的磷酸二氢钾+2%的尿素混合液（每亩用1.5kg尿素+250g磷酸二氢钾，对水50kg）进行叶面喷肥。二是若遇"秋吊"，及时灌攻粒水。三是防早霜，促早熟。根据具体生产情况，采取积极应对措施，如放秋垄、拿大草、割除空秆及病株、打底叶、喷施磷酸二氢钾、站秆剥皮晾晒等，改善田间通风透光条件，加速籽粒成熟，避免遭遇早霜冻害影响产量和品质。

7 适时收获

玉米绝大部分的籽粒产量是在灌浆期间形成的，玉米只有在完全成熟的情况下粒重最重、产量最高，才能实现高产优质。在实际生产中，有些地方有玉米早收的习惯，也就是在果穗苞叶刚开始变白时就进行收获，而实际上此时籽粒灌浆仍在继续进行，玉米籽粒并没有真正成熟。收获过早导致玉米籽粒的灌浆时间缩短，籽粒成熟度差、粒重轻、籽粒含水量高，不仅降低产量而且还不利于脱粒和贮藏。因此，适期收获对玉米高产优质具有重要意义。

为实现京科968高产优质，建议在适期早播的情况下，坚持待籽粒乳线消失、黑层出现，即生理成熟时再进行收获，以充分利用当地的光热资源，确保玉米籽粒充分成熟，降低籽粒含水率，增加粒重，提高籽粒产量和品质。

本文原载：农民科技培训，2013（8）：32-34

京科 968 的灌浆特征与产量性能分析

陈传永　赵久然　王荣焕　陈国平　王元东　成广雷　王晓光

（北京市农林科学院玉米研究中心，北京　100097）

摘　要：以京科 968 为材料，采用春播、夏播 2 种播期，设置 45 000 株/hm^2、60 000 株/hm^2、75 000 株/hm^2 共 3 种密度处理，对其灌浆特征与产量性能进行分析。结果表明，粒重与灌浆速率最大时的生长量（W_{max}）、最大灌浆速率（G_{max}）、活跃灌浆天数（P）呈极显著正相关，与达到最大灌浆速率的天数（T_{max}）、灌浆起始势（R_0）呈显著或极显著负相关。平均叶面积指数、穗数、产量与密度呈正相关；平均净同化率、穗粒数、千粒重与密度呈负相关；平均叶面积指数与穗数与产量都呈极显著正相关。

关键词：玉米；灌浆特征；产量性能

Grain Filling Characteristics and Yield Capability Analysis of Jingke968

Chen Chuanyong　Zhao Jiuran　Wang Ronghuan et al

(*Maize research center, Beijing Academy of Agriculture and Forestry Sciences, Beijing 100097, China*)

Abstract: The Jingke968 was planted at the density of 45 000 plants/hm^2, 60 000 plants/hm^2, 75 000plants/hm^2 on May 11st and June 19th, respectively. The main objective is tostudy the grain filling characteristics and yield analysis capability of Jingke968. The results showed that the grain weight was positively related with weight reaching maxium grain filling rate (W_{max}), maximum grain filling rate (G_{max}), active grain filling period (P), significantly; and negatively related with time reaching maxium grain filling rate (T_{max}) and Initial grain filling potential (R_0), significantly. Mean leaf area index, ear number and yield were positively related with density, significantly; and mean net assimilation rate (NAR), grains per ear and grain weight were negatively related with density, significantly. Mean leaf area index and ear number were positively related with yield, significantly, despite of spring sowing or summer sowing.

Key words: Maize; Grain filling characteristics; Yield capability analysis

京科 968（国审玉 2011007）是由北京农林科学院玉米研究中心选育的强优势玉米新

基金项目：国家玉米产业体系（CARS-02）、北京市农林科学院青年科研基金（QN201115）
作者简介：陈传永（1978—　），男，助理研究员，博士，主要从事玉米高产栽培与生理生态研究。
　　　　　Tel：010-51503703　E-mail：youngsirchen@sohu.com
　　　　　赵久然为本文通讯作者。E-mail：maizezhao@126.com

品种。适宜在北京、天津、山西中晚熟区、内蒙古赤峰和通辽，辽宁中晚熟区（丹东除外）、吉林中晚熟区、陕西延安和河北承德、张家口、唐山地区春播种植，2010年在国家东华北区试和生产试验中产量排名均在第一位，具有良好的增产潜力。

粒重受灌浆速率，持续灌浆时间等多种因素的影响，产量性能分析将光合性能参数与产量构成参数有机结合，揭示了产量形成的源库关系变化。本文通过分析京科968在不同播期、密度条件下的籽粒灌浆特征，明确不同产量水平下产量性能参数，以期为京科968高产栽培技术方案的制定以及品种推广提供参考。

1 材料与方法

1.1 试验地点

2011年在北京市昌平区小汤山国家精准农业示范基地进行试验，试验田耕层土壤（0~20cm）含有机质1.07%、全氮0.06%、全磷0.05%、全钾0.9%、碱解氮93.28mg/kg、速效磷24.20mg/kg、速效钾141.00mg/kg。春播玉米生育期间积温3 400℃左右，降水量565.60mm，总日照时数1 000h；夏播玉米生育期间积温2 600℃左右，降水量483.00mm，总日照时数715h。

1.2 材料与设计

供试品种为京科968，随机区组设计，3次重复。设3个密度水平，分别为4 500株/hm²、6 000株/hm²和7 500株/hm²。6行区，行距0.6m，行长7.5m，小区面积27m²。春播5月11日播种，夏播6月19日播种，春、夏播均10月8日收获，每小区收中间3行测产、考种。整地时底肥用量：烘干鸡粪7 500kg/hm²、尿素375kg/hm²，春、夏玉米拔节期均追施尿素375kg/hm²。其他管理同当地大田生产。

1.3 测定内容及方法

1.3.1 籽粒灌浆过程的模拟

以开花后天数（t）为自变量，以开花后每隔7d测得的百粒重为因变量（w），用Logistic方程$y=A/(1+Be^{-Cx})$对籽粒灌浆过程进行模拟，通过CurveExpert 1.3软件进行拟合，得到Logistic方程参数A、B、C（其中，A为终极生长量，B为初值参数，C为生长速率参数。计算下列灌浆特征参数：达最大灌浆速率时的天数$T_{max}=(\ln B)/C$，灌浆速率最大时的生长量$W_{max}=A/2$，最大灌浆速率$G_{max}=(C\times W_{max})/2$，积累起始势$R_0=C$，灌浆活跃期$P$（大约完成总积累量的90%）$=2(3/C)$。

1.3.2 产量性能分析

利用产量性能方程，产量=穗数×穗粒数×千粒重=平均叶面积指数×天数×平均净同化率×收获指数，分析产量形成过程中光合性能参数与产量性能参数的变化。其中平均叶面积指数（MLAI）根据张宾的方法计算，平均净同化率根据候玉虹的方法计算。

1.3.3 叶面积测定

长宽系数法，即叶片最大长度×最大宽度×0.75。

1.3.4 考种内容

选取20穗平均穗进行考种，测定穗长、穗粗、秃尖长度、穗行数、行粒数和千粒重。

1.4 统计分析

所有数据均采用 DPS3.01 和 Microsoft Excel 2003 进行统计分析。

2 结果与分析

2.1 灌浆速率特征分析

用 Logistic 模型可较好的模拟京科 968 的灌浆过程，决定系数 R^2 在 0.9886~0.9978，分析粒灌浆特征参数发现（表1），京科 968 的 T_{max}、R_0 与密度呈正相关；W_{max}、G_{max}、P 与密度呈负相关；在相同密度时，T_{max}、W_{max}、P 春播条均高于夏播，G_{max}、R_0 春播低于夏播。

密度间比较，取春、夏播相同密度均值，在 45 000株/hm²、60 000株/hm²、75 000株/hm² 条件下，T_{max} 分别为 27.71d、27.83d、29.44d，W_{max} 分别为 176.26、166.25、155.31g/（1 000粒），G_{max} 分别为 9.71、9.38、8.98g/（1 000粒·d），R_0 分别为 0.1103d、0.1132d、0.1160d，P 为 54.49d、53.16d、52.02d；75 000株/hm² 比 60 000株/hm² 与 45 000株/hm² 条件下，T_{max} 分别高 0.42%、6.21%，R_0 分别高 2.63%、5.08%，45 000株/hm² 比 60 000株/hm² 与 75 000株/hm² 条件下，W_{max} 分别高 6.02%、13.49%，G_{max} 分别高 3.49%、8.14%，P 分别高 2.49%、4.74%。

春、夏播比较，取各密度均值，春播条件下 T_{max}、W_{max}、P 较夏播分别高 20.89%、8.58%、11.32%，G_{max}、R_0 较夏播低 2.64%、11.46%。

表 1 不同密度春、夏播玉米籽粒灌浆特征参数

Table 1 Characteristic parameters of spring and summer maize hybrids with different densities during grain-filling stage

播期 Sowing date	密度（万株/hm²）Density	达最大灌浆速率时的天数（d）T_{max}	灌浆速率最大时的生长量 [g/（1 000粒）] W_{max}	最大灌浆速率 [g/（1 000粒·d）] G_{max}	积累起始势 R_0	灌浆活跃期（d）P
春播	4.5	30.03 ab	180.88 a	9.62 a	0.1064 a	56.39 a
	6.0	30.51 ab	174.98 ab	9.36 a	0.1072 a	55.97 a
	7.5	32.49 a	162.44 b	8.72 ab	0.1074 a	55.87 a
夏播	4.5	25.15 a	171.64 a	9.79 a	0.1141 a	52.59 a
	6.0	25.40 a	157.52 b	9.38 a	0.1189 a	50.46 ab
	7.5	26.40 a	148.18 b	9.25 a	0.1248 a	48.08 b

注：数值后不同字母表示密度内处理达 0.05 差异水平

Note: Values followed by different letters are significantly ($P<0.05$) different among density treatments

2.2 产量性能分析

对京科 968 进行产量性能方程分析表明（表2），平均叶面积指数（MLAI）、单位面积穗数（EN）、产量与密度呈正相关；平均净同化率（MNAR）、穗粒数（GN）、千粒重（GW）与密度呈负相关；各密度间收获指数无显著差异；相同密度条件下，光合性能参

数中除 MNAR 春播低于夏播外，MLAI、生育天数（D）、HI 春播均高于夏播；产量构成参数中 EN 春播高于夏播，GN 在 45 000株/hm² 春播高于夏播，60 000株/hm² 与 75 000株/hm² 时春播低于夏播，但 GW 春播显著高于夏播。结合产量与产量性能参数相关分析（表3），无论春播还是夏播，MLAI 与 EN 与产量都呈极显著正相关。

密度间比较，取春、夏播相同密度均值，在 45 000株/hm²、60 000株/hm²、75 000株/hm² 条件下，MLAI 分别为 2.15、2.64、3.30，；MNAR 分别为 8.63 万 m²·d/hm²、7.97 万 m²·d/hm²、6.93 万 m²·d/hm²；HI 分别为 0.50、0.50、0.49；EN 分别为 4.50 万穗/hm²、5.99 万穗/hm²、7.47 万穗/hm²；GN 分别为 668.39 粒、597.56 粒、545.89 粒，GW 分别为 668.39g、597.56g、545.89g，产量分别为 9 889.25kg/hm²、11 326.67kg/hm²、12 161.25kg/hm²。75 000株/hm² 比 60 000株/hm² 与 45 000株/hm² 条件下 MLAI 分别高 22.91%、43.77%，EN 分别高 33.15%、49.71%，产量分别高 14.54%、20.06%。45 000株/hm² 比 60 000株/hm² 与 75 000株/hm² 条件下 MNAR 分别高 8.35%、24.62%，GN 分别高 11.85%、22.44%，GW 分别高 7.63%、13.66%。

春、夏播比较，取各密度均值，MLAI、D、HI、GN、GW、产量春播较夏播分别高 16.36%、28.87%、0.68%、0.91%、15.46%、18.59%，MNAR、EN 春播较夏播低 27.14%、0.22%。

就产量而言，春、夏播条件下均以 75 000株/hm² 产量最高，春播条件下分别比 60 000株/hm² 与 45 000株/hm² 条件下高 23.77%、11.57%；夏播条件下分别高出 22.01%、2.63%。在夏播条件下，由 60 000株/hm² 与 75 000株/hm² 产量增益较小。

表2 不同密度下春、夏玉米产量性能方程参数
Table 2 The parameters of yield capability equation of spring and summer maize with different densities

播种时期 Sowing date	密度 (万株/hm²) Density	光合性能参数 Photosynthetic parameter				产量构成参数 Yield component			产量 (kg/hm²) Yield
		平均叶面积指数 MLAI	生育天数 (d)	平均净同化率 (万 m²·d/hm²) MNAR	收获指数 HI	单位面积穗数 (万穗/hm²) EN	穗粒数 (粒) GN	千粒重 (g) GW	
春播	4.5	2.35 c	125	7.65 a	0.50 a	4.49 c	689.78 a	369.67 a	10 820.50 b
	6	2.72 b	125	7.10 a	0.50 a	5.98 b	592.44 b	340.33 b	12 003.33 ab
	7.5	3.62 a	125	5.96 b	0.49 a	7.46 a	537.78 b	323.33 b	13 392.50 a
夏播	4.5	1.94 b	97	9.61 a	0.50 a	4.50 c	647.00 a	316.83 a	8 283.75 b
	6	2.56 a	97	8.83 a	0.49 b	5.99 b	602.67 a	297.50 ab	10 650.48 ab
	7.5	2.98 a	97	7.89 b	0.49 b	7.48 a	554.00 b	280.67 b	10 930.00 a

注：数值后不同字母表示密度内处理达 0.05 差异水平
Note：Values followed by different letters are significantly （$P < 0.05$）different among density treatments

表 3 京科 968 产量与产量性能方程参数的相关性分析

Table 3 Correlation coefficient between yield and the parameters of yield capability equation of Jingke968

项目 Effect	春播 spring seeding	夏播 Summer Seeding
MLAI	0.982 0**	0.960 7**
MNAR	-0.988 4**	-0.902 3**
HI	-0.888 2**	-0.991 4**
EN	0.998 8**	0.924 2**
GN	-0.978 7**	-0.913 5**
GW	-0.980 3**	-0.938 7**

注:** 表示在 0.01 水平差异显著

Note:** indicated significant difference at 1% probabilitylevel

2.3 灌浆速率特征参数与产量构成参数相关分析

分析产量性能方程参数与灌浆速率特征参数的相关性(表 4)表明,京科 968 在春、夏播条件下,在密度增加过程中,产量性能参数与灌浆速率特征参数的相关性基本一致。GW 是灌浆速率各参数综合作用的结果,并且与 GN、EN 决定产量,GW、GN 与 W_{max}、G_{max}、P 呈极显著正相关,与 T_{max}、R_0 呈显著或极显著负相关,而与产量呈极显著正相关的产量构成参数 EN 与光合性能参数 MLAI 则与 W_{max}、G_{max}、P 呈极显著或显著负相关,与 T_{max}、R_0 呈显著或极显著正相关。

表 4 产量性能方程参数与灌浆速率特征参数的相关性分析

Table 4 Correlation coefficient between characteristic parameters of grain-filling stage and the parameters of yield capability equation of spring and summer maize

播期 Sowing date	参数 parameter	T_{max}	W_{max}	G_{max}	R_0	P
春播	EN	0.942 6**	-0.978 7**	-0.957 7**	0.757 2*	-0.762 3*
	GN	-0.878 0**	0.933 9**	0.900 2**	-0.851 0*	0.855 1*
	GW	-0.881 8**	0.936 8*	0.903 8*	-0.846 6*	0.850 8*
	MLAI	0.994 8**	-0.999 5**	-0.998 6**	0.581 6*	-0.588 0*
	MNAR	-0.990 3**	0.999 9**	0.995 9**	-0.861 6*	0.617 9*
	HI	-0.982 9**	0.949 7**	0.972 8**	-0.654 7*	0.334 8
	Y	0.957 6**	-0.987 4**	-0.970 5**	0.724 8*	-0.730 3*

(续表)

播期 Sowing date	参数 parameter	T_{max}	W_{max}	G_{max}	R_0	P
夏播	EN	0.755 9*	-0.993 2**	-0.946 2**	0.998 2**	-0.999 5**
	GN	-0.773 3*	0.989 7**	0.937 1**	-0.999 5**	0.999 9**
	GW	-0.729 2*	0.997 0**	0.958 3**	-0.995 1**	0.997 4**
	MLAI	0.679 1*	-0.999 9**	-0.976 1**	0.985 6**	-0.989 9**
	MNAR	-0.789 9*	0.985 5**	0.927 4**	-0.999 9**	0.999 8**
	HI	-0.327 3	0.918 5**	0.981 3**	-0.834 9**	0.849 6*
	Y	0.448 5	-0.962 5**	-0.998 1**	0.899 9**	-0.911 5**

注：* 在 0.05 水平差异显著，** 在 0.01 水平差异显著
Note：* Significant at the 5% probability level. ** Significant at the 1% probability level

3 结论与讨论

玉米粒重受玉米籽粒灌浆与有效灌浆时间等多方面因素的影响。李绍长等研究发现，同一品种粒重的差异是由灌浆速度决定的，不同品种粒重的差异是灌浆持续期的长短造成的，且在同一生态区，播期主要是通过灌浆期温度和灌浆持续期来影响粒重。李玉玲等则认为不同品种间粒重差异受籽粒灌浆速率的高低和有效灌浆期的长短影响较小。王小春研究表明，最大百粒重与平均灌浆速率、最大灌浆速率出现时间、最大灌浆速率呈显著正相关。黄振喜等发现，籽粒灌浆启动快且高灌浆速率持续时间和生长活跃期（50d 以上）长的杂交种更容易实现高产。本试验通过对京科 968 籽粒产量形成的 Logistic 方程解析发现，在同一生态区，京科 968 在密度增加过程中千粒重与 W_{max}、G_{max}、P 呈极显著正相关，与 T_{max}、R_0 呈显著或极显著负相关。春播产量高于夏播主要得益于春条件下 T_{max}、G_{max}、P 显著高于夏播。不同产量水平在产量性能参数配置上不同，要进一步挖掘玉米高产潜力，还须优化产量性能参数构成。本试验通过产量性能分析表明，增加种植密度是提高京科 968 产量的有效途径，但夏播条件下密度对产量的增益效应要低于春播条件，春播产量高于夏播主要得益于平均叶面积指数、生育天数增加导致千粒重高于夏播。分析春、夏播玉米在密度增加过程中产量的增益效应发现，春播条件下适宜提高种植密度以增加单位面积穗数与平均叶面积指数的结构性增产途径，夏播条件下适宜在适当增加密度的条件下改善农田生态条件以增加平均叶面积指数、增强平均净同化率以提高千粒重的结构性增产与功能性增产并重的途径。综合分析表明，在以增加群体密度提高京科 968 产量的过程中，须协调产量构成参数中单位面积穗数增加与千粒重、穗粒数减少，光合性能参数中平均叶面积指数提高与平均净同化率、收获指数降低之间的矛盾，提高 W_{max}、G_{max}、P。由于不同生态区光温资源配置、地力条件以及种植方式不同，灌浆速率与产量性能参数有所差异，因此，在其他生态区京科 968 的灌浆特征与产量性能表现还需要进一步研究。

参考文献（略）

京单 38 不同播期籽粒灌浆特性研究

王荣焕　赵久然　陈传永　徐田军　刘秀芝　王元东　刘春阁

(北京市农林科学院玉米研究中心，北京　100097)

摘　要：以郑单958为对照，设置春播（5月14日播种）和夏播（6月13日）2个播期处理，研究玉米品种京单38在不同生态条件下的籽粒灌浆特性。结果表明，京单38的百粒干物重（42.75g）高于对照品种郑单958（36.41g），春、夏播条件下增幅分别为14.74%和20.26%；且春播百粒干物重较夏播高1.87%。京单38的籽粒灌浆速率高于郑单958，春、夏播最大灌浆速率分别较对照高42.27%和49.50%；且夏播最大灌浆速率较春播高9.42%。相关和通径分析表明，最大灌浆速率与粒重呈极显著正相关。与郑单958相比，京单38籽粒灌浆速率高且灌浆速度快，春、夏播条件下均可获得较高的粒重。

关键词：玉米；京单38；籽粒灌浆；播期

Grain Filling Characters of Maize Hybrid Jingdan38

Wang Ronghuan　Zhao Jiuran　Chen Chuanyong　Xu Tianjun
Liu Xiuzhi　Wang Yuandong　Liu Chunge

(*Maize research center*, Beijing Academy of Agriculture and Forestry Sciences, Beijing 100097, China)

Abstract: The grain filling characters under spring and summer sowing conditions (5/14, T_1 and 6/13, T_2) of maize hybrid Jingdan38 (JD38) were studied by using Zhengdan958 (ZD958) as the control. The results showed that the 100-grain dry weight of JD38 under T_1 and T_2 treatment was 14.74% and 20.26% higher than ZD958 respectively, and the T_1 treatment was 1.87% higher than T_2. The grain filling rate of JD38 was higher than ZD958, the G_{max} of JD38 was 42.27% and 49.50% higher than ZD958, the G_{max} of JD38 under T_2 was 9.42% higher than T_1. Correlation and path analysis showed that the kernel dry weight was positively correlated with the G_{max}. Compared to ZD958, JD38 was characterized with higher grain filling rate and faster grain development, and could gain higher grain weight under spring and summer sowing conditions.

Key words: Maize; Jingdan38; Grain filling; Sowing date

基金项目：国家玉米产业技术体系（nycytx-02）、北京市农林科学院青年科研基金（QN201115）、北京市农林科学院科技创新基金（CXJJ201309）

作者简介：王荣焕（1980— ），女，博士，副研究员，主要从事玉米高产栽培与生理生态研究。
　　Tel：010-51503703，E-mail：ronghuanwang@126.com
　　赵久然为本文通讯作者。E-mail：maizezhao@126.com

大量研究表明，玉米产量可分解为单位面积收获的籽粒数和平均单粒重，粒重是玉米产量的主要构成因素之一。粒重的高低取决于籽粒库容的大小、灌浆持续期的长短和灌浆速率的高低，不同品种间粒重的差异主要取决于籽粒的灌浆特性。籽粒灌浆特性不仅是玉米生长后期影响产量的主要因素，而且还关系到玉米生长后期籽粒能否正常成熟，与籽粒的商品品质密切相关，特别是在热量资源紧张、积温不足的地区和年份更为突出。

玉米籽粒的灌浆过程不仅与品种的基因型有关，而且还受环境条件的影响。前人曾对不同熟期、基因型、粒重类型以及不同生态地区、耕作方式、氮肥运筹等条件下的玉米籽粒灌浆特性开展了大量研究，但对同一品种在同一生态区不同播期条件下的籽粒灌浆特性研究较少。京单38是由北京市农林科学院玉米研究中心选育的玉米新品种，适宜在华北地区晚春播或夏播种植。研究和明确京单38在不同播期条件下的籽粒灌浆特性，可为挖掘籽粒产量潜力进而实现玉米高产优质提供理论依据和指导。

1 材料与方法

1.1 试验地概况

试验于北京昌平小汤山试验基地（N 40°12′，E 116°26′，海拔40m）进行。试验地0~20cm耕层土壤含有机质1.07%、全氮0.06%、全磷0.05%、全钾0.9%、碱解氮93.28mg/kg、速效磷24.20mg/kg、速效钾141.0mg/kg。春播玉米生育期内积温3 400℃左右，降水量565.60mm，总日照时数1 000h；夏播玉米生育期内积温2 600℃左右，降水量483.00mm左右，总日照时数715h左右。玉米籽粒灌浆至成熟期的气温和日照时数见表1（气象数据由北京市气象局提供）。

表1 玉米籽粒灌浆至成熟期间的气温和日照时数
Table 1 Temperature and sunshine conditions during maize grain filling to maturity period

月份 month	平均气温（℃） Average Temperature	最高气温（℃） Highest Temperature	最低气温（℃） Lowest Temperature	日照时数（h） Sunshine Time
7月	26.9	31.0	22.8	178.8
8月	25.8	30.9	20.6	221.8
9月	20.4	26.3	15.1	215.0
10月	17.3	24.3	11.5	91.1

1.2 试验设计

试验品种为北京市农林科学院玉米研究中心选育的玉米品种京单38，对照品种为郑单958。设置春播（T_1）和夏播（T_2）共2个播期，春播处理于5月14日（5/14）播种，夏播处理于6月13日（6/13）播种。种植密度为3 500株/亩。试验采用裂区设计，其中播期为主区、品种为副区，3次重复，6行区，行距0.6m，行长7.5m，小区面积27m^2。田间管理同当地大田生产。

1.3 测定项目与方法

吐丝前分别选择生长健壮一致的果穗进行挂牌标记。各播期处理均在授粉后15d起，每隔一周取样1次直至收获。其中，春播处理共取样10次，分别于授粉后第15d、22d、29d、36d、43d、50d、57d、64d、71d和78d取样；夏播处理共取样8次，分别于授粉后第15d、22d、29d、36d、43d、50d、57d和64d取样。每次取样各小区分别取3穗，取果穗中部籽粒100粒，在105℃烘箱中杀青30min后80℃烘干至恒量，测定各品种不同播期条件下的最终百粒干物重。

1.4 数据分析

采用Microsoft Excel 2007和SAS 9.0处理数据。以授粉后天数（x）为自变量、粒重（Y）为依变量，运用CurveExpertPro 1.5.0软件的Richard模型，模拟参试品种在不同播期条件下的籽粒灌浆过程，并推算该方程系列相关参数，分析籽粒灌浆过程。

2 结果与分析

2.1 不同播期条件下的籽粒干物重变化

不同播期条件下，京单38和郑单958的籽粒干物重均随灌浆时间的延长而不断增加，粒重增长动态符合"S"形生长曲线（图1）。

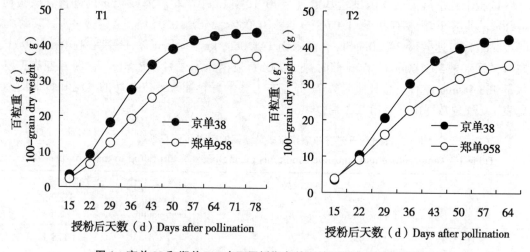

图1 京单38和郑单958在不同播期条件下的籽粒干物重增长情况

Figure 1 Grain growth dynamics of JD38 and ZD958 under differentsowing conditions

由表2可见，春播条件下京单38和郑单958的最终百粒干物重分别为43.14g和37.60g；夏播条件下的最终百粒干物重分别为42.35g和35.22g。方差分析表明，不同品种间及播期间玉米最终百粒干物重的差异均达到极显著水平。不同播期间的最终百粒干物重表现为春播（40.37g）高于夏播（38.79g），其中京单38春播较夏播最终百粒干物重高0.79g，增幅1.87%；而郑单958春播较夏播最终百粒干物重高2.38g，增幅6.76%。不同品种间的最终百粒干物重则表现为京单38（42.75g）高于郑单958（36.41g），其中春播条件下京单38较郑单958最终百粒干物重高5.54g，增幅14.74%；夏播条件下则高7.14g，增幅高达20.26%。

表2 京单38和郑单958在不同播期条件下的最终百粒干物重及方差分析

Table 2 100-grain dry weight of JD38 and ZD958 under differentsowing conditions

试验因素 Experiment factor		百粒干物重（g） 100-grain dry weight		
		T_1	T_2	平均 Average
品种	京单38	43.14±0.12	42.35±0.10	42.75 a
	郑单958	37.60±0.14	35.22±0.09	36.41 b
	平均	40.37a	38.79b	
均方	播期			10.07**
	品种			160.72**
	播期×品种			2.54**

2.2 不同播期条件下的籽粒灌浆速率

参试品种的籽粒灌浆特征表明，京单38和郑单958在不同播期条件下的籽粒灌浆速率变化均呈现出以籽粒灌浆速率最大值为峰值的单峰曲线变化趋势（图2）。春、夏播条件下，京单38在籽粒发育前期的灌浆速率均大幅高于郑单958，但自授粉后43d左右起郑单958籽粒灌浆速率高于京单38。就平均灌浆速率而言，京单38在春、夏播条件下分别为0.63g/（百粒·d）和0.80g/（百粒·d），郑单958在春夏播条件下分别为0.56g/（百粒·d）和0.64g/（百粒·d）。籽粒平均灌浆速率的总体趋势，相同品种间春播高于夏播，不同品种间为京单38高于郑单958。

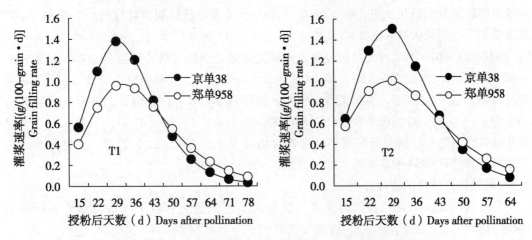

图2 京单38和郑单958在不同播期条件下的籽粒灌浆速率

Figure 2 Grain filling rate of JD38 and ZD958 under differentsowing conditions

2.3 籽粒灌浆参数分析

以授粉后的天数（x）为自变量、粒重（Y）为依变量，用Richard生长方程分别对参试品种在不同播期条件下的籽粒灌浆过程进行模拟，得到Richard模拟方程（表3）。由表3可看出，Richard方程可以很好地拟合参试品种在不同播期条件下的籽粒灌浆过程（系

数 R^2 介于 0.995 6~0.999 9），故可用该方程的次级参数模拟和分析籽粒灌浆过程。

表 3 京单 38 和郑单 958 在不同播期条件下的籽粒灌浆参数
Table 3 Grain filling parameters of JD38 and ZD958 under different conditions

播期 Sowing date	品种 Hybrid	Richard 方程 Richard Equation $W=A(1+Be^{-Ct})^{-1/D}$	R^2	T_{max} (d)	W_{max} (g/百粒)	G_{max} [g/ (百粒· d)]	P (d)
T_1	京单 38	$W=44.09(1+7.66e^{-0.1011t})^{-1/0.39}$	0.9956	29.57	18.93	1.38	47.22
	郑单 958	$W=38.25(1+1.45e^{-0.074t})^{-1/0.14}$	0.9983	31.60	15.01	0.97	57.90
T_2	京单 38	$W=43.03(1+8.38e^{-0.1133t})^{-1/0.37}$	0.9998	27.56	18.37	1.51	41.83
	郑单 958	$W=37.21(1+1.64e^{-0.080t})^{-1/0.18}$	0.9999	27.87	14.81	1.01	54.25

注：T_{max} 为达到最大灌浆速率的时间；W_{max} 为灌浆速率最大时的生长量；G_{max} 为最大灌浆速率；P 为活跃灌浆期

Note：T_{max}. Time reaching the maximum filling rate；W_{max}. The growing mass at maximum filling rate；G_{max}. The maximum filling rate；P. The active grain filling period

由 Richard 方程的次级参数（表 3）可看出，京单 38 在春、夏播条件下的灌浆活跃期（P）分别为 47.22d 和 41.83d，郑单 958 分别为 57.90d 和 54.25d，两品种春播较夏播籽粒灌浆活跃期分别延长 5.39d 和 3.65d，且郑单 958 灌浆活跃期相对较长。京单 38 在春、夏播条件下达到最大灌浆速率的天数（T_{max}）分别为 29.57d 和 27.56d，郑单 958 分别为 31.60d 和 27.87d，两品种春播较夏播达到最大灌浆速率的天数分别延迟 2.01d 和 3.73d，春播条件下京单 38 较郑单 958 提早 2d 达到最大灌浆速率，而夏播时达到最大灌浆速率的时间相当。京单 38 在春、夏播条件下的最大灌浆速率（G_{max}）分别为 1.38g/（百粒·d）和 1.51g/（百粒·d），郑单 958 分别为 0.97g/（百粒·d）和 1.01 g/（百粒·d），两品种夏播条件下的籽粒最大灌浆速率分别较春播高 9.42% 和 4.12%，且春、夏播条件下京单 38 分别较郑单 958 的最大灌浆速率高 42.27% 和 49.50%。

2.4 籽粒灌浆特性的通径分析

为明确相关因子与粒重间的关系，对参试品种的最终百粒干物重与灌浆特征参数（灌浆持续期、最大灌浆速率）、播期和气象因子（日照时数、平均气温）进行了相关分析和通径分析（表 4、表 5）。

结果表明，最大灌浆速率与粒重呈极显著正相关关系，相关系数为 0.906 6；而灌浆活跃期与粒重呈极显著负相关关系，相关系数为 -0.815 4。最大灌浆速率对粒重的直接通径系数为 1.483 5，所起的直接效应最大；其次是灌浆活跃期，对粒重的直接通径系数为 0.521 8，灌浆活跃期主要是通过最大灌浆速率的较大负效应（-1.458 8）来间接影响粒重，灌浆活跃期与最大灌浆速率呈极显著负相关，相关系数为 -0.983 4。

表4 各因子与粒重的相关分析
Table 4 Relative efficient of the relative factors on grain weight

性状 Trait	X_1	X_2	X_3	X_4	X_5	Y
X_1	1					
X_2	0.363 5	1				
X_3	−0.190 9	−0.983 4**	1			
X_4	1**	0.363 5	−0.190 9	1		
X_5	1**	0.363 5	−0.190 9	1**		
Y	0.240 9	−0.815 4**	0.906 6**	0.240 9	0.240 9	1

注:** 在0.01水平下极显著相关。X_1. 播期;X_2. P;X_3. G_{max};X_4. 日照时数;X_5. 平均气温;Y. 粒重

Note:** means significant difference at 0.01 level. X_1. Sowing date; X_2. P; X_3. G_{max}; X_4. Sunshine time; X_5. Average temperature; Y. Grain weight

表5 各因子与粒重的通径分析
Table 5 Path efficient of the relative factors on grain weight

性状 Trait	直接作用 Direct effect	总和 Total	间接作用 Indirect effect				
			→X_1	→X_2	→X_3	→X_4	→X_5
X_1	−0.428 1	0.037 3		−0.189 7	0.283 2	0.188 4	−0.244 6
X_2	0.521 8	−1.282 8	0.155 6		−1.458 8	−0.068 5	−0.088 9
X_3	1.483 5	−0.605 5	−0.081 7	−0.513 1		0.036 0	−0.046 7
X_4	−0.188 4	0.579 2	0.428 1	0.189 7	−0.283 2		0.244 6
X_5	0.244 6	0.146 2	0.428 1	0.189 7	−0.283 2	−0.188 4	

注:X_1. 播期;X_2. P;X_3. G_{max};X_4. 日照时数;X_5. 平均气温;Y. 粒重

Note: X_1. Sowing date; X_2. P; X_3. G_{max}; X_4. Sunshine time; X_5. Average temperature; Y. Grain weight

3 结论与讨论

玉米籽粒灌浆至成熟期是决定玉米产量和品质的重要阶段。播期主要通过调节光热资源进而影响籽粒灌浆过程。从自然生态条件来看,灌浆期间的气温变化对玉米籽粒灌浆影响较大。玉米灌浆期最适宜的日均气温为22~24℃,灌浆后期气温低于16℃即不再灌浆。本研究表明,随播期提早,玉米籽粒灌浆时间延长,运输到籽粒内的干物质逐渐增多,进而百粒重逐渐增加,与刘文成、马国胜等的研究结果一致。本试验中,春播处理于5月14日播种,夏播处理于6月13日播种。春播条件下,籽粒灌浆期间光温条件较好,灌浆持续期较长;而夏播籽粒灌浆期间气温较春播降低,籽粒灌浆活跃期缩短且灌浆高峰期提早。其中,8月的平均气温、最高和最低气温分别较7月低1.1℃、0.1℃和2.2℃,9月的平均气温、最高和最低气温分别较7月低6.5℃、4.7℃和7.7℃,且夜间气温下降较快

（平均最低气温仅 15.1℃），对籽粒后期灌浆不利。京单 38 春播较夏播籽粒活跃灌浆期延长了 5.39d，夏播达到最大灌浆速率的时间较春播缩短了 2d，春播百粒干物重较夏播高 0.79g。而对照品种郑单 958 春播较夏播籽粒活跃灌浆期延长了 3.65d，夏播达到最大灌浆速率的时间较春播缩短了 3.73d，春播百粒干物重较夏播高 2.38g。

不同基因型玉米籽粒最终粒重的高低主要取决于灌浆速率的大小，而与灌浆持续期长短关系不大。灌浆持续期易受基因型和环境条件的影响，而灌浆速率则相对稳定。本试验结果表明，与郑单 958 相比，京单 38 在籽粒灌浆中前期一直保持较高的灌浆速率。授粉后 43d 时，京单 38 春播和夏播处理的百粒干物重已分别达到最终百粒干物重的 73.1% 和 85.5%，而郑单 958 的百粒干物重分别为最终百粒干物重的 64.2% 和 79.1%。这说明，京单 38 的籽粒产量形成阶段主要集中在灌浆中前期。就最大灌浆速率而言，春、夏播条件下京单 38 的最大灌浆速率分别较郑单 958 高 42.27% 和 49.50%。并且，相关分析和通径分析表明，最大灌浆速率与粒重呈极显著正相关。虽然京单 38 的籽粒灌浆活跃期较郑单 958 短 10d 左右，但该品种灌浆速度快且灌浆强度高，有利于积累更多的籽粒干物质。京单 38 灌浆强度大的优势弥补了其灌浆时间较郑单 958 短的缺陷，进而在不同播期条件下均可获得较高的粒重。因此，在生产实践中，积极采取有效技术措施，特别是加强灌浆中前期的肥水供应，维持京单 38 的籽粒灌浆强度，获得较高粒重。

参考文献（略）

本文原载：玉米科学，2013（6）：59-63

播期对玉米干物质积累转运和籽粒灌浆特性的影响

徐田军* 吕天放* 陈传永 赵久然** 王荣焕**
刘月娥 刘秀芝 王元东 刘春阁 成广雷

(北京市农林科学院玉米研究中心/玉米DNA指纹及分子育种北京市重点实验室,北京 100097)

摘 要:以郑单958和京单58为试验材料,设置春播(5月14日播种)和夏播(6月13日播种)2个播期处理,研究播期对玉米干物质转运和籽粒灌浆特性的影响。结果表明:春播条件下,两品种的粒重、茎鞘、叶片干物质转运率及产量均高于夏播。京单58茎鞘干物质向籽粒的转运能力、粒重及产量均高于郑单958;其中,2013年郑单958和京单58的百粒重分别较夏播高17.2%和15.1%;京单58(42.4g)比郑单958(35.0g)高20.9%;夏播条件下郑单958和京单58达到最大灌浆速率的时间(T_{max})较春播分别缩短了5.4d和3.2d。京单58的灌浆持续期(P)略短,但灌浆起始势(R_0)和最大灌浆速率(G_{max})高于郑单958;粒重与灌浆期平均气温呈显著正相关(相关系数分别为0.78*),与G_{max}和R_0均呈极显著正相关(相关系数分别为0.91**和0.93**)。由此说明,播期主要是通过温度条件影响玉米干物质积累、转运及籽粒灌浆特性;茎秆干物质向籽粒中转运速率快、灌浆速率高是京单58在春夏播条件下粒重和产量较高的原因。

关键词:播期;玉米;干物质积累转运;籽粒灌浆特性

Effect of Sowing Date on Maizedry Matter Accumulation, Transformation and Grain Filling Characters

Xu Tianjun* Lv Tianfang* Chen Chuanyong Zhao Jiuran**
Wang Ronghuan** Liu Yuee Liu Xiuzhi Wang Yuandong
Liu Chunge Cheng Guanglei

(*Beijing Key Laboratory of maize DNA Fingerprinting and Molecular Breeding, Maize research center, Beijing Academy of Agriculture and Forestry Sciences, Beijing, 100097*)

Abstract: To study the effect ofdifferent sowingdates ondry matter transportation and the process

---

基金项目:国家玉米产业技术体系项目(NYCYTX-02)、北京市农林科学院科技创新基金项目(CXJJ201309)资助

* 作者简介:徐田军与吕天放为本文共同第一作者。徐田军,助理研究员,主要从事玉米高产栽培与生理生态研究。E-mail:xtjxtjbb@163.com。吕天放,研究实习员,主要从事玉米高产栽培与生理生态研究。E-mail:314565358@qq.com

** 通讯作者:赵久然,研究员,主要从事玉米高产栽培和育种。E-mail:maizezhao@126.com
王荣焕,副研究员,主要从事玉米高产栽培与生理生态研究。E-mail:ronghuanwang@126.com

of grain filling, Jingdan 958 (ZD958) and Jingdan 58 (JD58) were used in the field experiments with two sowingdate (14th May and 13th June). The result showed that grain weight, yield, the export percentage and transformation percentage of the matter in stem, sheath and leaves of ZD958 and JD58 were higher than that of summer sowingdate, and the transport capacity, grain weight and yield of JD58 was higher than that of ZD958; the 100-graindry weight of ZD958 and JD58 under spring sowingdate were17.2% and 15.1% higher than that of summer sowingdate. Among thedifferent varieties, 100-graindry weight of JD58 (42.4g) was 20.9% higher than ZD958 (35.0 g). The time reached the maximum grain filling rate (T_{max}) of ZD958 and JD58 under summer sowingdate were 5.4d and 3.2d ahead of spring sowingdate. The grain fillingduration (P) of the JD58 was relatively shorter than ZD958. The G_{max} and R_0 of JD58 were higher than that of ZD958, but the P showed the opposite trend. The graindry weight significantly and positively correlated with average temperature at grain filling Stage (correlation coefficients were 0.78*), significantly positively correlated with Gmax and R0 (correlation coefficients were 0.91** and 0.93**). Compared with the ZD958, the JD58 had higher 100-graindry weight and grain filling rate, higher export percentage and transformation percentage of the matter in stem, so JD58 can obtain higher graindry weight underdifferent sowingdate.

Key words: Sowingdate; Maize; Dry matter accumulation and transformation; Grain filling characters

籽粒灌浆是玉米产量形成的主要过程，适期播种是玉米在生育期内高效利用光热资源的保障和获得高产的重要栽培措施。关于播期对籽粒灌浆特性和产量的影响，前人做了相关研究。郑洪建等研究了积温与玉米产量的关系，指出积温是影响玉米产量的重要环境因子。Cirilo 等研究发现在温带地区，播期影响了玉米籽粒灌浆期的生长环境，随播期推迟，粒重降低，从而导致产量降低。李建奇等研究发现随着播期的推迟，籽粒灌浆期缩短，籽粒未能达到正常生理成熟，从而使籽粒质量下降，进而造成减产。宋羽等研究表明，渐增期灌浆速率与千粒重呈极显著负相关；快增期持续天数、缓增期灌浆速率和灌浆活跃期与千粒重均呈极显著正相关。肖荷霞等研究表明，夏播玉米穗粒数和粒重受日平均气温影响较大，灌浆期平均气温与粒重呈负线性关系，气温日较差与粒重呈正线性关系。

适宜播期是玉米实现高产的必要条件之一，玉米产量受灌浆期干物质积累、分配及转运速率、和灌浆特性的影响。灌浆期干物质积累、分配及转运对玉米籽粒的贡献方面前人研究发现，叶片和苞叶的干物质转运量对籽粒的贡献率较大，茎鞘和穗轴的贡献率相对较小。因此，研究播期对玉米籽粒灌浆、干物质转运和分配过程的影响对于实现玉米稳产具有重要意义。本试验以郑单958和京单58为试验材料，设置春播和夏播共2个播期处理，分析玉米干物质积累、转运和籽粒灌浆特性的影响，为适期播种和高产栽培提供理论依据。

1 试验条件

1.1 试验地点

试验在北京昌平小汤山试验基地进行。试验田耕层土壤（0~20cm）含有机质 10.70mg/kg、全氮6.78mg/kg、速效磷24.20mg/kg、速效钾141.00mg/kg。

1.2 材料与方法

供试品种为郑单958（ZD958）和京单58（JD58），分别由北京德农种业科技有限公司和北京农科院种业有限公司。设2个播期处理：春播（5月14日播种，S_1）和夏播（6月

13日播种,S_2)。2012年和2013年春播试验均于9月25日收获,夏播试验均于10月11日收获。试验采用裂区设计,播期为主区,品种为副区,3次重复,种植密度为52 500株/hm^2。小区行距0.6m,8行区,行长7.5m,小区面积36m^2。田间管理同当地大田生产。

1.2.1 籽粒灌浆速率测定

吐丝前分别选择生长健壮一致的果穗进行挂牌标记,统一授粉。各播期处理均在授粉后14d起,每隔一周取样1次直至收获。其中,春播分别于授粉后第15d、22d、29d、36d、43d、50d、57d、64d、71d和78d取样;夏播处理分别于授粉后第15d、22d、29d、36d、43d、50d、57d和64d取样,每次取样各小区分别取3穗,取果穗中部籽粒100粒,在105℃烘箱中杀青30min后,80℃烘干至恒量,测定各品种不同播期条件下的最终百粒重。以授粉后天数(d)为自变量、授粉后每隔7d测得的百粒重为因变量(W),参照朱庆森等的方法,利用Richards方程$W=A(1+Be^{-Ct})^{-1/D}$模拟籽粒灌浆过程。籽粒灌浆速率:$F=ACBe^{-Ct}/(1+Be^{-Ct})^{(D+1)/D}$,式中:$W$为粒重(g);$A$为最终粒重(g);$t$为授粉后天数(d);$B$、$C$、$D$为回归方程所确定的参数,其中,$A$为终极生长量,$B$为初值参数,$C$为生长速率参数,$D$为形状参数,当$D=1$时,即为Logistic方程。计算下列灌浆特征参数:

灌浆起始势 $R_0=C/D$

达最大灌浆速率时的天数 $T_{max}=(\ln B-\ln D)/C$

灌浆速率最大时的生长量 $W_{max}=A(D+1)^{-1/D}$

最大灌浆速率 $G_{max}=(CW_{max}/D)[1-(W_{max}/A)^D]$

灌浆活跃期(约完成总积累量的90%) $P=2(D+2)/C$

1.2.2 茎鞘干物质积累及转运测定

在吐丝期和生理成熟期,每小区随机取3株生长一致的植株用于测定干重,吐丝期将样品分为茎、叶和鞘3部分;成熟期将样品分为茎、叶、籽粒、穗轴和苞叶5部分,所有样品均在80℃下烘干至恒重后称量,获得干重并计算茎鞘和叶片物质输出率和转运率。

茎鞘物质输出率=(开花期单株茎鞘重-成熟期单株茎鞘重)/开花期单茎鞘重×100%

茎鞘物质转运率=(开花期单株茎鞘重-成熟期单株茎鞘重)/成熟期单穗粒重×100%

叶片物质输出率=(开花期单株叶片重-成熟期单株叶片重)/开花期单茎鞘重×100%

叶片物质转运率=(开花期单株叶片重-成熟期单株叶片重)/成熟期单穗粒重×100%

1.3 数据处理

采用DPS 6.5软件对数据进行方差分析,其中处理间差异显著性采用LSD法进行检验($\alpha=0.05$)。通过CurveExpert 3.0软件进行灌浆动态拟合,用Excel进行数据计算和图表的制作。

2 结果与分析

2.1 不同播期条件下京单58和郑单958产量变化

由图1、表1可知,京单58和郑单958的产量在品种和播期间差异显著。2012年春播和夏播条件下京单58产量高于郑单958,分别较郑单958高156.04kg/hm^2和251.20kg/hm^2。2013年春播和夏播条件下京单58产量高于郑单958,分别较郑单958高110.95kg/hm^2和146.05kg/hm^2。春播条件下京单58和郑单958的平均产量2012年和2013年显著高于夏播,分别高952.80kg/hm^2和882.15kg/hm^2。

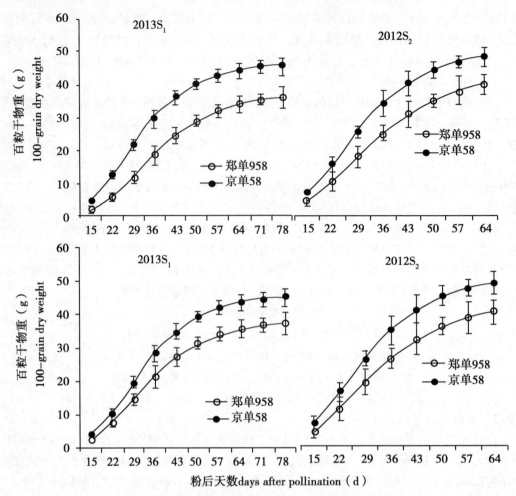

图1 不同播期条件下郑单958和京单58的百粒干物重增长动态（2012年和2013年）
Figure 1 Increasingdynamics of 100-graindry weight for ZD958 and JD58 atdifferent sowingdate conditions (2012 and 2013)

表1 郑单958和京单58在不同播期条件下的产量
Table 1 The yield of ZD958 and JD58 under different sowing conditions

年际 Year	品种 Cultivar	S1	S2
2012	JD58	10 001.54±110.2	9 096.32±99.3
	ZD958	9 845.5±92.1	8 845.12±86.5
2013	JD58	10 060.75±109.2	9 196.15±114.3
	ZD958	9 949.8±97.5	9 050.1±92.1
变异来源 Sources of variation	播期 Sowingdate		**
	品种 Cultivar		**
	播期×品种 Sowingdate×Cultivar		NS

注：** 表示在 $P<0.01$ 水平差异显著
Note: ** indicates significant difference at $P<0.01$ level

2.2 干物质积累及茎鞘物质转运

干物质积累是生物学产量形成的基础，干物质的运转决定着营养物质的流向和产量的高低。随着生育进程的推进，玉米生长中心发生转移，干物质在各器官中的分配也随之变化。由表2可知，2012年和2013年开花期茎鞘重和叶片干重、收获期茎鞘重、叶片干重和单穗粒重在播期间表现为春播大于夏播，品种间表现为京单58大于郑单958，年际间表现为2013年大于2012年，鞘物质输出率和转运率亦表现为相同的变化趋势，其中2013年春夏播条件下京单58茎鞘物质转运率较郑单958分别高14.3%和10.5%。叶片转运率较郑单958分别高这说明0.55%和0.73%，这说明在不同年际和播期间，京单58茎鞘和叶片干物质向籽粒的转运能力均高于郑单958。

表2 不同播期条件下郑单958和京单58的茎鞘和叶片干物质积累和转运情况

Table 2 Dry matter accumulation and transformation of ZD958 and JD58 in stem, leaves sheath under different sowing date conditions

年际 Year	播期 Sowing Date	品种 Cultivar	叶片干重 (g) Dry leaves weight (g)		茎鞘干重 (g) Dry Single stem and sheath weight (g)		成熟期单穗粒重 (g) Kernel weight of single ear at mature stage	输出率 (%) Export percentage of the matter (%)		转运率 (%) Transformation percentage of the matter	
			开花期 Flowering stage	成熟期 Maturing	开花期 Flowering stage	成熟期 Maturing		叶片 leaves	茎鞘 stem and sheath	叶片 leaves	茎鞘 stem and sheath
2012	S1	ZD958	41.50b	33.19ab	110.67bc	67.09bc	236.76bc	20.02c	39.38c	3.51d	18.41c
		JD58	44.56a	34.56a	129.98a	72.34a	270.43a	22.44a	44.35a	3.70c	21.31a
	S2	ZD958	39.93b	31.47b	99.98b	62.95b	216.98d	21.19b	37.04d	3.90b	17.07d
		JD58	42.15ab	32.54b	118.54b	69.87b	240.32b	22.8a	41.06b	4.00a	20.25b
2013	S1	ZD958	43.98b	34.06b	113.52c	63.17cd	240.60c	22.56b	44.35b	4.12c	20.93c
		JD58	48.88a	35.95a	143.62a	78.21a	276.65a	26.45a	45.54a	4.67a	23.64a
	S2	ZD958	40.89c	32.98b	102.42d	60.30d	222.50d	19.34c	41.12c	3.56d	18.93d
		JD58	45.35b	34.79b	123.78b	72.03b	246.43b	23.29b	41.81c	4.29b	21.00b

注：同一列内相同因素处理下不同字母表示差异达 $P<0.05$ 显著水平

Note: Date of same treatment in the same columns, sharing with the lifferent letter, are significantly different at $P<0.05$

2.3 百粒重

由图2可知，2012年和2013年京单58和郑单958的百粒干重在春夏播条件下均随籽粒发育呈慢-快-慢的S型变化趋势。其中，2012年不同播期间表现为春播（40.8g）>夏播（35.9g），平均增幅为13.6%；不同品种间表现为京单58（42.3g）比郑单958（34.5g）高22.6%。2013年不同播期间表现为春播（41.3g）>夏播（36.1g），平均增幅为14.3%；不同品种间表现为京单58（42.4g）比郑单958（35.0g）高20.9%。

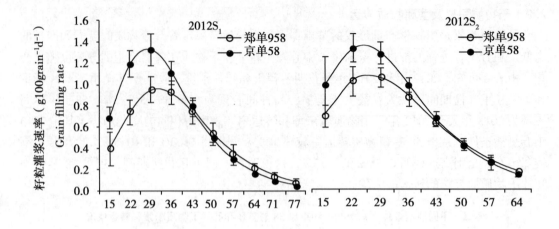

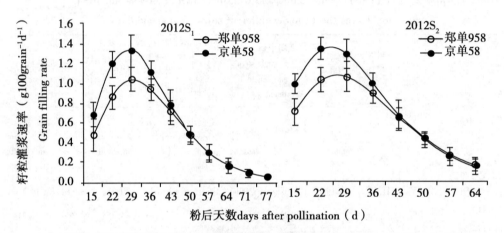

图2 不同播期条件下郑单958和京单58的籽粒灌浆速率（2012年和2013年）
Figure 2 Grain filling rate of ZD958 and JD58 under different sowing date conditions (2012 and 2013)

2.4 籽粒灌浆速率

由图2可知，2012和2013年，不同播期条件下，郑单958和京单58籽粒的灌浆速率均随籽粒的发育呈现先增加后降低的单峰曲线变化，且表现为春播大于夏播。春、夏播条件下，灌浆中前期京单58的灌浆速率大幅高于郑单958，在授粉后43d、50d左右则表现为相反趋势。

2.5 籽粒灌浆特征参数

用Richard方程可较好地模拟不同播期下郑单958和京单58籽粒的灌浆过程，决定系数R^2介于0.994 0~0.996 2。分析籽粒灌浆特征参数发现（表3），2012年和2013年夏播条件下郑单958、京单58达到最大灌浆速率的时间（T_{max}）较春播分别缩短了4.16d和5.41d、0.6d和3.19d，灌浆速率最大时的生长量（W_{max}）和最大灌浆速率（G_{max}）表现为春播高于夏播，而活跃灌浆期（P）表现为相反趋势，其中，京单58的R_0、W_{max}和G_{max}高于郑单958，而P呈相反趋势；这说明京单58较郑单958灌浆启动早、灌浆速率大，粒重的建成主要表现在灌浆的中前期。

表3 不同播期条件下郑单958和京单58的籽粒灌浆特征参数

Table 3 Grain filling parameters of ZD958 and JD58 under different sowing date conditions

年际 Year	播期 Sowingdate	品种 Cultivar	R^2	Richard方程 Richard equation	T_{max} (d)	W_{max} (g/100 kernels)	R_0	P (d)	G_{max} (g/100 kernel·d)
2012	S_1	ZD958	0.9988	$W=37.69(1+0.88e^{-0.073t})^{-11.20}$	31.43	14.46	1.65	58.48	0.97
		JD 58	0.9981	$W=46.82(1+0.32e^{-0.079t})^{-23.58}$	25.53	17.58	1.86	51.70	1.33
	S_2	ZD958	0.9962	$W=35.25(1+0.28e^{-0.069t})^{-23.65}$	27.27	13.24	0.48	56.89	0.88
		JD 58	0.9981	$W=40.79(1+1.39e^{-0.084t})^{-5.77}$	24.93	16.22	1.64	52.97	1.16
2013	S_1	ZD958	0.9982	$W=38.30(1+1.45e^{-0.074t})^{-7.12}$	31.62	15.02	0.53	58.01	0.97
		JD 58	0.9940	$W=45.89(1+0.94e^{-0.083t})^{-10.48}$	27.55	17.66	0.87	50.57	1.34
	S_2	ZD958	0.9974	$W=34.66(1+0.23e^{-0.074t})^{-30.69}$	26.21	12.95	0.48	54.77	0.93
		JD 58	0.9983	$W=41.16(1+0.90e^{-0.082t})^{-8.25}$	24.36	16.02	0.68	51.58	1.17

2.6 粒重与相关参数的相关性分析

由表4可知,粒重与茎鞘物质转运率(Z)和灌浆期间平均温度(T)呈显著正相关;与灌浆起始势(R_0)和最大灌浆速率(G_{max})呈极显著正相关,相关系数分别为0.93和0.91。因此,适当提早播期,通过播期调节温度等环境效应,提高籽粒灌浆速率,最终可获得较高的粒重。

表4 粒重与相关参数的相关性

Table 4 Correlation analysis between graindry weight and related factors

	T	Z	T_{max}	R_0	P	G_{max}	G_{wt}
Z	1						
T	0.70*	1					
T_{max}	−0.07	0.64	1				
R_0	0.11	0.05	−0.09	1			
P	−0.03	0.02	0.07	−0.99**			
G_{max}	−0.04	0	0.02	0.97**			
G_{wt}	0.78*	0.45	−0.35	0.93**	0.15	0.91**	1

注:Z. 茎鞘物质转运率;T. 灌浆期平均温度;G_{wt}. 粒重。*和**分别表示在5%和1%水平上显著

Note:Z. transformation percentage of the matter in stem and sheath;T. mean temperature of grain filling stage;G_{wt}. grain weight. * and ** mean significant at the 0.05 and 0.01 probability levels, respectively

3 讨论

播期主要是通过温度、光照时数、降雨量来影响玉米生长发育，适期播种能使玉米充分利用光热资源来构建良好群体结构。前人研究表明，玉米灌浆期最适宜的日均气温为22~24℃，灌浆后期气温低于16℃灌浆基本停止。本研究中，春播条件下籽粒灌浆期间的平均温度（23.0℃）、最高温度（29.4℃）、最低温度（18.4℃，均优于夏播，分别比夏播高1.2℃、0.5℃、2.4℃。相关分析表明，籽粒百粒重与灌浆期平均温度呈显著正相关（相关系数为0.78*），与王荣焕等研究结果一致。由此可见，在春播条件下在灌浆后期日平均气温仍然能够满足两品种的灌浆需求，而夏播条件下玉米灌浆后期日平均气温低于最适灌浆温度，过低的温度造成玉米灌浆速率缓慢，甚至停止。分析粒重与气象因素的相关性发现，春夏播条件下，京单58和郑单958籽粒干物重与灌浆期平均气温呈显著相关（表4），这也是夏播条件两品种粒重较春播低的主要原因。

植株的干物质积累、转运及籽粒灌浆状况是决定粒重进而影响产量的重要因素。不同播期下玉米灌浆期温度存在较大差异，籽粒灌浆起始势（R_0）、最大灌浆速率（G_{max}）、灌浆速率最大时的生长量（W_{max}）、灌浆速率达峰值的时间（T_{max}）及灌浆持续期（P）等是反映玉米籽粒建成及灌浆特征的重要参数。前人研究表明随播期推迟玉米籽粒灌浆期推迟，灌浆速率下降，且T_{max}推迟，粒重减轻。玉米粒重的高低主要取决于灌浆速率的大小，与灌浆持续时间的长短无关。本试验表明，粒重与灌浆期平均气温呈显著正相关（相关系数分别为0.78*），与G_{max}和R_0均呈极显著正相关（相关系数分别为0.91**和0.93**）。由此说明，与夏播相比，春播条件下较高的温度条件更有利于植株干物质积累和籽粒灌浆，进而获得较高粒重。京单58较强的茎鞘干物质向籽粒物质转运能力强、灌浆起始势和灌浆速率高，从而粒重和产量显著高于郑单958。

参考文献（略）

本文原载：中国农业科技导报，2016，18（6）：112-118

玉米籽粒灌浆特性对播期的响应

徐田军　吕天放　赵久然*　王荣焕　陈传永　刘月娥　刘秀芝　王元东

(北京市农林科学院玉米研究中心/玉米DNA指纹及分子育种北京市重点实验室，北京　100097)

摘　要：以"京科528"和"郑单958"为试验材料，设置早春播（4月10日播种）和春播（5月14日播种）2个播期处理，研究不同播期条件下玉米的籽粒灌浆特性。结果表明：JK528、ZD958在早春播条件下的最终百粒重和产量显著高于春播，增幅分别为6.8%和10.1%、17.8%和9.2%；籽粒最大灌浆速率（G_{max}）、平均灌浆速率（G_{ave}）表现为早春播高于春播，而籽粒活跃灌浆期（P）呈相反趋势；不同品种间，京科528百粒重和产量显著高于郑单958，其中京科528百粒重和产量较郑单958高7.4g和1 189.6kg/hm^2，增幅分别为21.6%和10.8%；T_{max}和P表现为郑单958大于京科528，而W_{max}、G_{max}、G_{ave}表现为京科528大于郑单958；同一播期，京科528在灌浆前、中期的平均灌浆速率高于后期，且高于郑单958。相关分析表明，籽粒干物质积累量与平均气温和积温间呈极显著正相关。可见，充分利用光热资源，提高灌浆速率，有利于获得较高粒重，从而提高玉米产量。京科528在灌浆前中期灌浆速率快的优势弥补了其灌浆活跃期略短的情况，进而在不同播期条件下均可获得较高的粒重。

关键词：播期；玉米；籽粒灌浆特性

Response of Grain Filling Characteristics of Maize to Sowing Date

Xu Tianjun　Lv Tianfang　Zhao Jiuran*　Wang Ronghuan
Chen Chuanyong　Liu Yuee　Liu Xiuzhi　Wang Yuandong

(Maize research center, Beijing Academy of Agriculture and Forestry Sciences, Beijing Key Laboratory of Maize DNA Fingerprinting and Molecular Breeding, Beijing 100097)

Abstract: To study the effect of different sowing dates on the process of grain filling characteristics, Jingke 528 (JK528) and Zhengdan 958 (ZD958) were used in the field experiment with two sowing dates (10th April and 14th May). The results showed that the grain dry weight and yield of JK528 and ZD958 with the early spring sowing were significantly higher than that with the spring sowing date, increased by 6.8% and 10.1%, 17.8% and 9.2%, respectively; The maximum grain

---

基金项目：由国家玉米产业技术体系（nycytx-02）和北京市农林科学院科技创新基金（CXJJ201309）资助

* 通讯作者：E-mail：maizezhao@126.com

filling rate (G_{max}) and the average grain filling rate (G_{ave}) were higher with the early spring sowing than that with the spring sowing date, while the P showed an opposite trend. The 100-grain weight and yield of JK528 were significantly improved by 7.4g and 1 189.6kg/hm² compared to ZD958, respectively, with the increase of 21.6% and 10.8%; The P and T_{max} of ZD958 were higher than that of JK528, while W_{max}, G_{max} and G_{ave} of JK528 were higher than that of ZD958. The average grain filling rate of JK528 during early and mid grain filling stages was higher than that of late grain filling stages, which was also higher than that of ZD958 at the same sowing date. Furthermore, correlation analysis revealed that the mean temperature and the accumulate temperature were significantly positively correlated with the grain dry matter accumulation. Therefore, the full use of hot-ray resources and the increase of the mean grain-filling rate can improve maize grain yield. The higher grain filling rate during early and mid grain filling stages in JK528 can compensate for the short active grain filling period, leading to a higher yield under different sowing dates.

Key words: Sowing date; Maize; Grain filling characteristics

玉米的产量和品质受遗传因素和环境条件的双重影响。播期是影响玉米产量形成和品质最主要的环境因素之一，适宜播期种植是实现玉米高产的必要条件。玉米籽粒灌浆期是决定玉米籽粒产量和品质的重要阶段，籽粒灌浆特性是决定粒重和产量的重要生理性状，播期对玉米粒重和灌浆特性的影响主要通过光温状况来实现。李绍长等认为，同一生态区播期主要是通过灌浆期温度和灌浆持续期来影响粒重。Wiloson等认为，低温有利于延长生育期，尤其是延长籽粒的灌浆期，有利于玉米的增产。然而很多研究报道认为，玉米灌浆期低温严重影响籽粒建成，连续5d的10℃低温处理即可对籽粒生物膜系统产生直接伤害，阻碍籽粒灌浆，降低粒重。王若男等研究表明，灌浆期低温会加速植株衰老，叶面积指数下降，叶片光合生理功能受到限制，造成"源"的不足，同时减缓光合物质由叶片向籽粒的转运过程，从而影响籽粒的灌浆，致使千粒重降低，最终造成产量降低。本试验立足北京地区，比较了播期对京科528和郑单958籽粒灌浆特性的影响，为京科528的推广和高产栽培提供理论指导和技术支持。

1 材料与方法

1.1 试验地概况

试验在北京昌平小汤山试验基地进行，试验田耕层土壤（0~20cm）含有机质10.7g/kg、碱解氮93.3mg/kg、速效磷24.2mg/kg、速效钾141.0mg/kg。2013年早春播玉米生育期积温为3 386℃，降水量539.6mm，总日照时数1 021h；春播玉米生育期积温2 700℃，降水量491.0mm，总日照时数722h。2014年早春播玉米生育期积温为3 400℃，降水量565.6mm，总日照时数1 100h；春播玉米生育期积温2 720℃，降水量483.0mm，总日照时数735h。

1.2 供试材料与试验设计

供试品种为"郑单958"（ZD958）和"京科528"（JK528）。试验设置2个播期处理：早春播（4月10日播种，S_1）和春播（5月14日播种，S_2）。试验采用随机区组设计，3次重复，6行区，行距0.6m，行长7.5m，小区面积27m²，种植密度为52 500株/hm²。田间管理同当地生产。

1.3 测定项目与方法

吐丝前分别选择生长健壮一致的果穗进行挂牌标记。各播期处理均在授粉后 14d 起,每隔一周取样 1 次直至收获。其中,早春播、春播处理分别取样 10 次,分别于授粉后第 14d、21d、28d、35d、42d、49d、56d、63d、70d 取样。每次取样各小区分别取 3 穗,取果穗中部籽粒 100 粒,在 105℃烘箱中杀青 30min 后,80℃烘干至恒量,测定各品种不同播期条件下的最终百粒重。以授粉后天数(d)为自变量、授粉后每隔 7d 测得的百粒重为因变量(W),参照顾世梁等的方法,利用 Richards 方程 $W=A(1+Be^{-Ct})^{-1/D}$ 模拟籽粒灌浆过程。籽粒灌浆速率:$F=ACBe^{-Ct}/(1+Be^{-Ct})^{(D+1)/D}$,式中:$W$ 为粒重(g);A 为最终粒重(g);t 为授粉后天数(d);B、C、D 为回归方程所确定的参数,其中,A 为终极生长量,B 为初值参数,C 为生长速率参数,D 为形状参数,当 $D=1$ 时,即为 Logistic 方程。计算下列灌浆特征参数:达最大灌浆速率时的天数 $T_{max}=(\ln B-\ln D)/C$,灌浆速率最大时的生长量 $W_{max}=A(D+1)^{-1/D}$,最大灌浆速率 $G_{max}=(CW_{max}/D)[1-(W_{max}/A)D]$,灌浆活跃期(约完成总积累量的 90%)$P=2(D+2)/C$。

1.4 数据处理

采用 DPS 6.5 软件对数据进行方差分析,其中处理间差异显著性采用 LSD 法进行检验($\alpha=0.05$)。通过 CurveExpert 3.0 软件进行灌浆动态拟合,用 Excel 进行数据计算和图表的制作。2013 年和 2014 年的试验结果趋势基本一致,本文均采用 2014 年数据进行分析。

2 结果与分析

2.1 玉米生育期内日气象因素的变化

小汤山试验基地日均气温随着生育期的推进呈先增加后降低的趋势(图1),平均气温为 22.4℃,峰值出现在 7 月中旬(31.3℃),8 月中旬之后气温逐渐下降。总降雨量为 329.8mm,其中 52.0%集中在 7 月,8 月中旬之后雨量偏少。平均相对湿度为 66.1%。平均辐射强度为 175.3MJ/m²。

2.2 早春、春播条件下两品种灌浆期取样间隔内气象因素情况

由表 1 可知,在早春、春播条件下,JK528 和 ZD958 在灌浆期内降雨量差别不大,均在 110mm 左右。早春播条件下,JK528 和 ZD958 灌浆期平均气温均为 26.20℃,在灌浆前中期平均气温均为 27.2℃,灌浆后期平均气温均为 23.8℃;而春播条件下 JK528 和 ZD958 灌浆期平均气温均为 24.9℃,在灌浆前中期平均气温分别为 26.5℃和 26.3℃,灌浆后期平均气温分别为 21.1℃和 20.3℃。早春播条件下,JK528 和 ZD958 在灌浆期内辐射强度均为 1 280MJ/m²,春播条件下分别为 12 292MJ/m² 和 11 788MJ/m²。

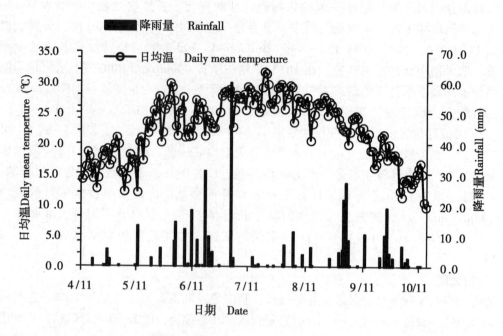

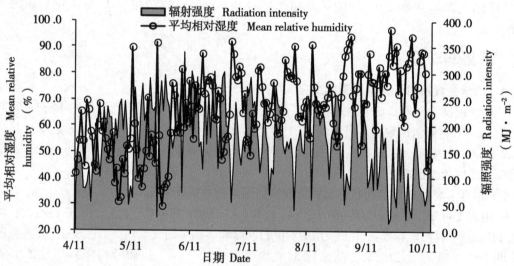

图1 玉米生育期内日均温、降雨量、平均相对湿度及辐射强度的变化

Figure 1 Change of daily mean temperature, rainfall, mean relative humidity and radiation intensity during maize growing period

表1 不同播期条件下两品种灌浆期降雨量、平均气温和辐射强度的变化

Table 1 Change of rainfall, average temperature and radiation intensity during maize grain filling period under the different sowing condition

播期 Sowing date	品种 Cultivar	气象因子 Meteorological factor	取样间隔 Sampling interval (d)								
			0~7	7~14	14~21	21~28	28~35	35~42	42~49	49~56	56~63	63~70
S$_1$	JK528	RA	0.0	4.0	0.2	7.8	11.4	10.0	0.4	2.4	67.8	0.2
		AT	27.0	27.7	28.7	28.0	27.2	25.8	26.0	25.8	22.3	23.3
		RI	1 173	1 690	1 039	1 410	1 016	1 286	1 569	1 385	854	1 385
	ZD958	RA	0.0	4.0	0.2	7.8	11.4	10.0	0.4	2.4	67.8	0.2
		AT	27.0	27.7	28.7	28.0	27.2	25.8	26.0	25.8	22.3	23.3
		RI	1 173	1 690	1 039	1 410	1 016	1 286	1 569	1 385	854	1 385
S$_2$	JK528	RA	4.0	0.4	7.6	11.4	10.4	0.0	7.2	63.0	10.4	2.8
		AT	27.2	28.5	28.4	27.0	25.0	26.5	25.2	22.4	20.4	18.4
		RI	1 413	1 162	1 354	1 034	1 444	1 327	1 396	1 021	1 131	1 010
	ZD958	RA	0.2	7.8	11.4	10.0	0.4	2.4	67.8	0.2	10.2	2.8
		AT	28.7	28.0	27.2	25.8	26.0	25.8	22.3	23.3	19.2	18.4
		RI	1 039	1 410	1 016	1 286	1 569	1 385	854	1 385	833	1 010

注：RA. 降雨量；AT. 平均温度；RI. 辐射强度；S$_1$. 早春播；S$_2$. 春播；JK528. 京科；ZD958. 郑单。下同

Note：RA. rainfall；AT. mean temperature；RI. radiation intensity；S$_1$. the early spring sowing date；S$_2$. spring sowing date；JK528. jingke 528；ZD958. zhengdan 958. The same below

2.3 早春、春播条件下JK528和ZD958籽粒百粒重变化

在不同播期条件下，JK528和ZD958的籽粒重随授粉后天数呈"慢-快-慢"的"S"形动态曲线变化趋势（图2）。在早春播和春播条件下，JK528的最终百粒重分别为42.9和40.2g，显著高于ZD958（37.5g和31.8g），增幅分别为14.4%和26.2%（平均增幅为21.6%）。分析不同播期之间玉米品种百粒重的变化发现：JK528和ZD958在早春播条件下的最终百粒重显著高于春播，增幅分别为6.8%和17.8%（平均增幅为10.3%）。

2.4 早春、春播条件下JK528和ZD958籽粒灌浆速率变化

由图3可知，早春、春播条件下，JK528和ZD958的籽粒灌浆速率呈"先升后降"的单峰变化。在灌浆前期，JK528的灌浆速率显著高于ZD958，其中早春播条件下，JK528和ZD958到达峰值所需时间相当，均在授粉后24d左右，峰值分别为1.5g/d和1.4g/d。JK528的最大灌浆速率较ZD958高8.2%。早春播条件下授粉后49d前表现为JK528的灌浆速率高于ZD958，之后表现相反趋势；而春播条件下JK528的灌浆速率一直高于ZD958。分析不同播期条件下玉米籽粒灌浆速率发现：JK528和ZD958在早春播条件下的最大灌浆速率高于春播，增幅分别为19.2%和25.9%。

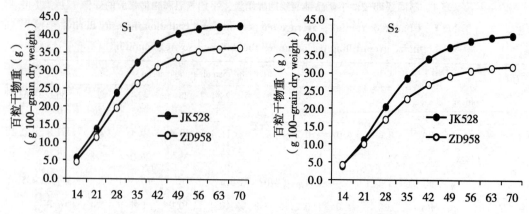

图 2 JK528 和 ZD958 在不同播期条件下的籽粒干物重变化

Figure 2 Dynamics of grain dry matter of Jingke 528 and Zhengdan 958 under different sowing dates.

注：JK528. 京科；ZD958. 郑单 958。下同

Note：JK528. jingke 528；ZD958. zhengdan 958. The same below

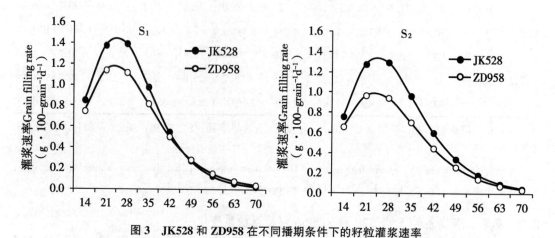

图 3 JK528 和 ZD958 在不同播期条件下的籽粒灌浆速率

Figure 3 Grain filling rate of Jingke 528 and Zhengdan 958 under different sowing dates

2.5 早春、春播条件下 JK528 和 ZD958 籽粒干物质积累量与温度的相关性

由表 2 可知，早春播条件下，JK528 和 ZD958 的籽粒干物质积累量与平均气温和积温间存在极显著正相关（相关系数分别为 0.75** 和 0.75**、0.74** 和 0.74**）。春播条件下 JK528、ZD958 籽粒干物质积累量与平均气温和积温间呈显著正相关关系（相关系数分别为 0.64* 和 0.64*、0.61* 和 0.64*）。早春、春播条件下两品种的籽粒干物质积累量与降雨量和辐照强度呈负相关，达到显著水平。由此可见，适当提早播期，充分利用积温条件才能获得较高的粒重。

表2 灌浆期降雨量、平均温度、积温和辐射强度与籽粒干物质积累的相关系数
Table 2 The correlation coefficient among rainfall, average temperature, accumulated temperature and radiation intensity during grain filling period and grain dry matter accumulation

播期 Sowing date	品种 Cultivar	相关性 Correlation	X1	X2	X3	X4
S_1	JK528	X_2	-0.62*			
		X_3	-0.63*	1.00**		
		X_4	-0.6	0.24	0.24	
		X_5	-0.24	0.75**	0.75**	-0.08
	ZD958	X_2	-0.62*			
		X_3	-0.63*	1.00**		
		X_4	-0.6	0.24	0.24	1
		X_5	-0.25	0.74**	0.74**	-0.08
S_2	JK528	X_2	-0.3			
		X_3	-0.3	1.00**		
		X_4	-0.41	0.52	0.52	
		X_5	-0.19	0.64*	0.64*	-0.29
	ZD958	X_2	-0.23			
		X_3	-0.23	1.00**		
		X_4	-0.44	0.54	0.53	
		X_5	-0.21	0.61*	0.64*	-0.31

注：X_1. 降雨量；X_2. 平均温度；X_3. 积温；X_4. 辐射强度；X_5. 籽粒干物质积累量

Note: X_1. rainfall; X_2. mean temperature; X_3. accumulated temperature; X_4. radiation intensity; X_5. grain dry matter accumulation

2.6 早春、春播条件下JK528和ZD958籽粒灌浆参数变化

由表3可知，春播条件下JK528和ZD958到达最大灌浆速率时所需的时间（T_{max}）较早春播分别提前0.61d和0.73d；籽粒灌浆速率最大时的生长量（W_{max}）、籽粒最大灌浆速率（G_{max}）、平均灌浆速率（G_{ave}）表现为早春播大于春播，而籽粒活跃灌浆期（P）呈相反趋势。早春播条件下JK528和ZD958的W_{max}较春播分别增加3.96g/100kernel和3.07g/100kernel，增幅分别为26.4%和23.6%；早春播条件下JK528、ZD958的G_{max}和G_{ave}较春播分别高19.2%和25.9%、19.2%和26.1%。同一播期不同品种间，T_{max}和P表现为ZD958高于JK528，而W_{max}、G_{max}和G_{ave}表现为JK528高于ZD958。

表3 JK528 和 ZD958 在不同播期条件下的籽粒灌浆参数
Table 3 Grain filling parameters of Jingke 528 and Zhengdan 958 under different sowing dates

播期 Sowing date	品种 Cultivar	Richard 方程 Richard equation $W=A(1+Be^{-Ct})^{-1/D}$	R^2	灌浆参数 Grain filling Parameter				
				T_{max} (d)	W_{max} g·100 kernel^{-1}	P (d)	G_{max} (g·100 kernel^{-1}·d^{-1})	$G_{ave.}$ (g·100 kernel^{-1}·d^{-1})
S_1	JK528	$W=42.89(1+8.16e^{-0.115t})^{-1/0.49}$	0.9956	24.61	18.99	43.42	1.46	0.99
	ZD958	$W=36.96(1+2.36e^{-0.0970t})^{-1/0.23}$	0.9983	24.60	16.10	44.80	1.35	0.92
S_2	JK528	$W=41.13(1+1.42e^{-0.095t})^{-1/0.14}$	0.9998	24.00	15.03	46.01	1.18	0.80
	ZD958	$W=32.41(1+1.80e^{-0.0920t})^{-1/0.20}$	0.9999	23.87	13.03	47.84	1.00	0.68

2.7 播期对玉米籽粒灌浆阶段及其灌浆强度的影响

根据灌浆曲线上灌浆速率变化的拐点时间，将玉米籽粒灌浆过程划分为3个阶段。由表4可知，早春播条件下两品种灌浆前期时间较春播明显延长，而中后期持续时间均明显缩短。早春播条件下，JK528和ZD958前期分别比春播延长11.2%和11.4%，中期分别比春播缩短7.8%和6.0%，后期分别比春播缩短39.4%和26.5%。同一播期条件下，JK528在灌浆前、中期的平均灌浆速率高于后期，且高于ZD958；籽粒干物质积累量对总灌浆进程的贡献率（RGC）表现为中期大于后期大于前期。平均灌浆速率和灌浆天数的乘积可以看作籽粒在一定阶段的干物质积累量，进一步比较播期对玉米灌浆各时期籽粒干物质积累的影响可知，灌浆前期早春播条件下干物质积累值明显高于春播，而中后期差异不显著。同一播期下，JK528干物质积累值在灌浆前期和后期高于ZD958，而中期ZD958高于JK528。

表4 不同播期条件下玉米籽粒灌浆前、中、后期持续天数、平均速率及贡献率（RGC）
Table 4 Grain filling period, average grain filling rate and ratio of the grain filling contributed to the final grain weight in the early, middle and late stage of grain filling stage under different sowing dates

播期 Sowing date	品种 Cultivar	前期 Early stage			中期 Middle stage			后期 Late stage		
		天数 Day	平均速率 G_{ave} (g·100grain^{-1}·d^{-1})	贡献率 RGC	天数 Day	平均速率 G_{ave} (g·100grain^{-1}·d^{-1})	贡献率 RGC	天数 Day	平均速率 G_{ave} (g·100grain^{-1}·d^{-1})	贡献率 RGC
S_1	JK528	14.53	0.40	0.14	20.15	1.09	0.51	30.07	0.47	0.33
	ZD958	13.90	0.28	0.10	21.41	1.17	0.62	37.54	0.31	0.29
S_2	JK528	13.07	0.02	0.85	21.85	0.80	47.78	49.59	0.21	28.79
	ZD958	12.48	0.02	0.91	22.77	0.60	42.76	51.10	0.17	27.93

2.8 早春、春播条件下 JK528 和 ZD958 的产量

由表 5 可知，JK528 和 ZD958 的产量在播期和品种间存在极显著差，而在年际间差异不显著。播期间，早春播条件下 JK528 的产量较春播高 1 179.1 kg/hm²，ZD958 较春播高 973.5 kg/hm²。品种间，JK528 的产量较 ZD958 高 2 379.3 kg/hm²。年际间，2014 年两品种的产量较 2013 年高 265.2 kg/hm²。

表 5　JK528 和 ZD958 在不同播期条件下的产量
Table 5　Yield of Jingke 528 and Zhengdan 958 under different sowing dates

年份 Year	品种 Cultivar	产量 Yield (kg/hm²)		
		S_1	S_2	平均 Average
2013	JK528	12 701.8±424.0	11 473.9±623.0	12 087.9
	ZD958	11 462.4±734.9	10 466.2±410.7	10 964.3
2014	JK528	12 984.0±354.0	11 854.5±324.0	12 419.3
	ZD958	11 638.5±648.0	10 687.5±397.5	11 163.0
	平均 Average	12 196.7	11 120.6	
均方 Mean square	年份 Year			421 986.2
	播期 Sowing date			6 950 530.1**
	品种 Cultivar			8 491 364.8**
	年份×播期 Year× Sowing date			7 646.9
	年份×品种 Year× Cultivar			26 188.8
	播期×品种 Sowing date×Cultivar			63 365.9
	年份×播期×品种 Year× Sowing date× Cultivar			1024.4

3　讨论

籽粒灌浆状况对产量的形成至关重要。前人研究表明，籽粒灌浆不仅决定于品种的遗传特性，也受土壤、氮素营养、水分、播期等因素的影响。播期主要通过调节光热资源进而影响籽粒灌浆过程。生态因子中吐丝后日均温度和有效积温对玉米产量的影响较大。因此，灌浆期温度的高低对玉米的产量影响显著。王忠孝和王春乙等指出，玉米灌浆期的最适日平均温度为 22~24℃；当玉米籽粒灌浆期温度低于 16℃ 时，玉米灌浆过程基本停止。本研究表明，早春播条件下两品种的籽粒灌浆速率、百粒干物重及产量均高于春播。分析不同播期条件下粒重和灌浆速率变化的原因发现：早春播条件下，JK528 和 ZD958 在灌浆前中期平均气温均为 27.2℃，灌浆后期平均气温均为 23.8℃，而春播条件下两品种在灌浆前中期平均气温分别为 26.5℃ 和 26.3℃，灌浆后期平均气温分别为 21.1℃ 和 20.3℃。由此可见，在早春播条件下在灌浆后期日平均气温仍然能够满足两品种的灌浆需求，而春

播条件下玉米灌浆后期日平均气温略低于最适灌浆温度。分析粒重与气象因素的相关性发现，早春、春播条件下，JK528 和 ZD958 籽粒干物重与平均气温、积温间存在极显著和显著相关，这也是春播条件两品种粒重和产量较早春播低的主要原因。

玉米籽粒灌浆至成熟期是决定玉米产量和品质的重要阶段。粒重的高低取决于籽粒库容的大小、灌浆持续期的长短和灌浆速率的高低，不同品种间粒重的差异主要取决于籽粒的灌浆特性。籽粒的灌浆特性是影响粒重的主要原因，而播期对玉米籽粒灌浆进程的影响主要通过灌浆期积温、灌浆持续期及灌浆速率。前人研究表明，随播期推迟，最大灌浆速率逐渐下降，峰值出现时间也在后移。播期越早灌浆时间越长，最大灌浆速度值越大。本研究表明，随播期推迟，两品种到达最大灌浆速率的所需时间（T_{max}）提前，其中早春播条件下 JK528 和 ZD958 较春播提前 0.61d 和 0.73d；籽粒灌浆速率最大时的生长量（W_{max}）、籽粒最大灌浆速率（G_{max}）、平均灌浆速率（G_{ave}）呈降低趋势，而籽粒活跃灌浆期（P）呈升高趋势，与刘文成、马国胜等的研究结果一致。

根据灌浆曲线灌浆速率变化的拐点时间，将玉米籽粒灌浆过程划分为灌浆前期、中期和后期 3 个阶段。JK528 在灌浆前、中期的平均灌浆速率高于后期且高于 ZD958，干物质积累值在灌浆前期和后期高于 ZD958，而在灌浆中期 ZD958 高于 JK528。同一播期下，不同品种间 P 表现为 ZD958 略高于 JK528，早春、春播条件下分别高 1.38d、1.8d，而 G_{max}、G_{ave} 表现为 JK528 高于 ZD958。JK528 灌浆速率快的优势弥补了其灌浆活跃期略较 ZD958 短的缺陷，进而在不同播期条件下均可获得较高的粒重和产量。

因此，在生产实践中，根据品种特性，适期早播，以充分利用生育后期积温资源，同时减弱灌浆期低温对玉米生育后期的不利影响，延长灌浆天数，加强灌浆中期的田间管理，从而提高粒重，最终实现丰产稳产的目标。

参考文献（略）

本文原载：应用生态学报，2016，27（8）：2 513-2 519

不同熟期玉米品种的籽粒灌浆特性及其与温度关系研究

钱春荣[1,2] 王荣焕[1] 赵久然[1] 于洋[2] 郝玉波[2]
徐田军[1] 姜宇博[2] 宫秀杰[2] 李梁[2] 葛选良[2]

(1. 北京市农林科学院玉米研究中心/玉米DNA指纹及分子育种北京市重点实验室,北京 100097;2. 黑龙江省农业科学院耕作栽培研究所,农业部东北地区作物栽培科学观测实验站,哈尔滨 150086)

摘 要:玉米籽粒成熟度与吐丝至成熟阶段热量条件显著相关,明确不同熟期玉米品种籽粒灌浆特性及其与灌浆阶段温度的关系,可为玉米籽粒发育调控和熟期选择提供理论依据。研究以极早熟、中早熟和中晚熟3种熟期类型的9个玉米品种为试验材料,于2014年和2015年在哈尔滨实施田间试验,研究不同熟期玉米品种的籽粒灌浆特性及其与温度的关系。结果表明,3种熟期类型品种相比,中早熟品种的最大灌浆速率(G_{max})、达到最大灌浆速率时的生长量(W_{max})和平均灌浆速率(G_{mean})最大,极早熟品种次之,中晚熟品种最小;灌浆活跃期(P)和有效灌浆时间(t_3)随品种熟期延长而增加,增加部分主要在灌浆缓增期。活跃灌浆期、有效灌浆期与快增期≥10℃有效积温显著正相关,与灌浆期日平均气温显著负相关;最大灌浆速率和平均灌浆速率与渐增期≥10℃有效积温、灌浆期日平均气温显著正相关,与快增期≥10℃有效积温和渐增期日平均气温显著负相关。在热量资源有限的生态区,中晚熟品种进入灌浆缓增期后日平均气温较低,不利于籽粒灌浆,导致中晚熟品种粒重低于中早熟和极早熟品种。因此,在热量资源有限生态区,一味地延长熟期并不利于粒重增加,中早熟品种具有明显的灌浆优势,从而获得较高的产量。

关键词:玉米;熟期;灌浆;温度

Study on the Grain Filling Characteristics and Their Relationship with Temperature of Maize Hybrids Differing in Maturities

Qian Chunrong[1,2] Wang Ronghuan[1] Zhao Jiuran[1]
Yu Yang[2] Hao Yubo[2] Xu Tianjun[1] Jiang Yubo[2]
Gong Xiujie[2] Li Liang[2] Ge Xuanliang[2]

(1. *Beijing Key Laboratory of Maize DNA Fingerprinting and Molecular Breeding, Maize Research Center, Beijing Academy of Agriculture & Forestry Science, Beijing 100097; 2. Scientific Observing and Experimental Station of Crop Cultivation in Northeast China, Ministry of Agriculture; Institute of Crop Cultivation and Farming, Heilongjiang Academy of Agricultural Sciences, Harbin 150086, China*)

Abstract:Maize grain maturity was significantly correlated with the heat condition during silking to mature stage. Clarifying the grain filling characteristics of different maize hybrids and their relationship with the temperature of grain filling stage could provide theoretical basis for maize grain development regulation and maturity selection. Field experiments were conducted in Harbin in 2014 and 2015 with 9 maize hybrids to study the grain filling characteristics of different maturity maize hybrids, and the relationship with temperature. The results showed that the maximum grain filling rate (G_{max}), grain weight at the time of maximum grain filling rate (W_{max}) and average grain filling rate (G_{mean}) of middle-early maturity hybrids were the highest among the 3 maturity hybrids, while those of super-early maturing hybrids were the second. The active filling stage (P) and effective filling time (t_3) increased with the extension of ripening period, and the increase was mainly in the period of late grain filling stage. The active filling stage (P) and effective filling time (t_3) was significantly positive correlated with the effective accumulated temperature during middle stage of grain filling period, and significantly negative correlated with the daily average temperature during grain filling stage. The maximum and average grain filling rate were positively correlated with the effective accumulated temperature ≥ 10℃ during the early stage of grain filling period and daily average temperature during the filling period, which was significantly negative correlated with the effective accumulated temperature ≥ 10℃ during the middle stage of grain filling period and daily mean temperature during the early stage of grain filling period. In the ecological area with limited caloric resources, the average daily temperature during the late stage of grain filling period of the middle-late maturing hybrids was lower, which was not conducive to grain filling, leading to the grain weight of middle-late maturing hybrids was less than that of the middle-early and super-early maturing hybrids. The prolonged maturity was not conducive to grain weight increase under the condition with limited caloric resources. The middle-early maturing hybrids had obvious advantage in grain filling, and then could obtain higher yield.

Key words: Maize; Maturity; Grain filling; Temperature

玉米产量的高低决定于库容的大小和灌浆充实的程度。灌浆特性是玉米生长后期影响产量的关键因素之一，灌浆期和灌浆活跃期的长短、灌浆强度的高低决定了玉米籽粒干物质积累的多少。前人从生态条件、施肥水平以及种植密度等方面对玉米籽粒灌浆特性开展了大量研究，比较一致的观点是粒重由灌浆时间和灌浆速率共同决定。

温度是影响玉米生长发育的重要生态因子，通过影响生育期，进而影响光有效辐射截获率和生长发育，最终影响产量。玉米籽粒成熟度与种植区热量条件显著相关，以吐丝至成熟阶段地区间差异对成熟度影响最大。董桂芳等研究指出，即使在同一生态区内，玉米品种因其自身基因型的差异，决定了各自对气象条件反应有所不同。王晓慧等研究表明，中熟品种灌浆启动快，灌浆活跃期和有效灌浆时间短，中晚熟、晚熟和超晚熟品种灌浆启动慢，灌浆活跃期和有效灌浆时间长。黑龙江省是我国最大的玉米主产区，国家统计局数据显示，2015年黑龙江省玉米播种面积582万hm^2，玉米产量3 544万t，分别占全国玉米面积与总产的15.27%和15.78%。该区域无霜期短，热量资源有限，随着玉米全程机械化种植方式的转变，生育期长、熟期晚的品种越来越不适应生产要求，需要选择相对早熟的品种加以替代。但针对该区域不同熟期玉米品种灌浆规律研究较少，不同熟期玉米品种籽粒灌浆过程有何差异，灌浆参数与灌浆阶段温度关系如何，上述科学问题的研究尚未见相关报道。

本研究选用极早熟、中早熟、中晚熟3个熟期段9个玉米品种，于2014年和2015年两个生长季在哈尔滨生态区种植。运用Richard生长模型对其籽粒灌浆特性进行解析,探明黑龙江省不同熟期玉米品种籽粒灌浆特性及其与灌浆阶段温度的关系,以期为黑龙江省玉米生产合理选用相应熟期品种提供理论指导。

1 材料与方法

1.1 试验地概况

试验于2014年和2015年在黑龙江省哈尔滨市民主乡黑龙江省农业科学院农业科技园区进行(N 45°50′,E 126°50′)。黑龙江省哈尔滨市属于黑龙江省第一积温带,年均气温3.5~4.5℃,年降雨量400~600mm,年无霜期135~145d。试验地土壤为碳酸盐黑钙土,试验区碱解氮160.40mg/kg,速效磷19.22mg/kg,速效钾675.90mg/kg,有机质19.6g/kg,pH5.95。

1.2 试验设计

试验选用黑龙江省目前大面积推广、新审定或新育成的表现突出的9个玉米品种为试验材料,供试品种见表1。其中德美亚1号和3号种子由黑龙江垦丰种业提供,克单14由黑龙江省龙科种业提供,其余6个试验品种由北京市农林科学院玉米研究中心提供。试验采取随机区组设计3次重复,每个小区6行,行长5m,行距0.65m,小区面积19.5m^2。

表1 供试品种及生育期

Table1 Maize hybrids and growth stages

熟期类型 Maturity	品种 Maize hybrids	生育期(d) Growth stage(d)
极早熟 Supper-early maturity	德美亚1号 Demeiya 1	108
	克单14 KD14 Kedan 14	108
	德美亚3号 Demeiya 3	117
中早熟 Middle-early maturity	吉单27 Jidan 27	126
	京农科728 Jingnongke 728	126
	京单28 Jingdan 28	130
中晚熟 Middle-late maturity	先玉335 Xianyu 335	>130
	郑单958 Zhengdan 958	>130
	京科968 Jingke 968	>130

1.3 田间管理

试验于2013年和2014年秋季耕翻整地,秋施肥,每公顷施入控释掺混肥料(N:P_2O_5:K_2O=26:10:12)600kg。2014年和2015年分别于5月8日和5月10日人工播种,每穴3粒,种植密度每公顷52 500株,出苗后于三叶期按设计密度定苗,2014年和2015年分别于10月11日和10月13日人工收获,其他管理措施与当地大田高产培技术措施一致。

1.4 测定项目与方法

产量和产量构成:每小区收获中间2行,按每小区平均鲜穗重从中随机选取10个样本果穗,用于考察产量构成及穗部性状。实际籽粒产量:各小区所收获的果穗全部脱粒后称重,测定籽粒含水量,按14%的含水量进行折算。

于玉米吐丝期选择生长一致且同一天吐丝的植株挂牌标记，自吐丝14d起，每7d取果穗一次，每次每个小区取3个挂牌标记的果穗，取其中下部（自穗底部第8~20环）籽粒100粒，分别称其鲜重，于80℃烘干至恒重后称籽粒干重。

以授粉后天数（t）为自变量，授粉后每隔7d测得的籽粒干重为因变量（W），参照顾世梁等的方法，利用Richards方程 $W=A(1+Be^{Ct})^{-1/D}$ 模拟籽粒灌浆过程。

籽粒灌浆速率 $G=ACBe^{-Ct}/D(1+Be^{-Ct})^{(D+1)/D}$

式中，W 为百粒重（g），A 为最终百粒重（g），t 为授粉后天数（d），B、C、D 为回归方程所确定的参数。

由方程一阶导数和二阶导数推导出灌浆参数。

积累起始势 $R_0=C/D$

达最大灌浆速率时的天数：$T_{max}=(\ln B-\ln D)/C$

灌浆速率最大时的生长量 $W_{max}=A(D+1)^{-1/D}$

最大灌浆速率 $G_{max}=(CW_{max}/D)[1-(W_{max}/A)^D]$

灌浆活跃期（约完成总积累量的90%）$P=2(D+2)/C$

平均灌浆速率 $G_{mean}=AC/2(D+2)$

灌浆高峰开始日期 $t_1=-\ln\{[D^2+3D+D(D^2+6D+5)^{1/2}]/2B\}/C$，对应此时粒重为 $W_1=A(1+Be^{-ct1})^{-1/D}$

灌浆高峰结束日期 $t_2=-\ln\{[D^2+3D-D(D^2+6D+5)^{1/2}]/2B\}/C$，对应此时粒重为 $W_2=AA(1+Be^{-ct1})^{-1/D}$

花后粒重 W 达99%时为有效灌浆期，$t_3=-\ln\{[(100A/99)^D-1]/B\}/C$，对应此时粒重为 W_3。

1.5 数据分析

采用Microsoft Excel 2013进行数据计算，用CurveExpert1.4软件进行方程拟合，SAS 9.0进行数据分析。

2 结果与分析

2.1 不同熟期玉米品种的产量及其产量构成

不同熟期类型品种间的产量差异，以及年际间的产量差异均达到极显著水平（表2）。中晚熟品种较中早熟和极早熟品种分别增产13.98%和59.46%，中早熟品种较极早熟品种增产39.91%，2015年中晚熟和中早熟品种产量差异不显著；不同熟期品种的百粒重差异达极显著水平，年际间百粒重差异不显著，中早熟品种百粒重最高，极早熟品种次之，中晚熟品种百粒重最低。不同熟期类型品种穗粒数差异，以及年际间穗粒数的差异均达到极显著水平，中晚熟品种穗粒数最多，其次是中早熟品种，极早熟品种穗粒数最少。同一熟期内的不同品种百粒重与穗粒数也存在差异，其中极早熟品种"德美亚3"和中早熟品种"京单28"属于大粒型品种，其百粒重明显高于同熟期内的其他品种，而穗粒数却显著低于其它品种。极早熟品种虽然百粒重显著高于中晚熟品种，但其穗粒数却显著低于中晚熟品种，从而导致极早熟品种产量显著低于中晚熟品种，由此可见，不同熟期类型品种产量差异主要来源于穗粒数（表3）。

表2 不同熟期类型玉米品种产量性状统计分析 F 值

Table 2 *F* values of ANOVA for yield traits in maize hybrids differing in maturities

差异来源 Source	自由度 *f*	穗粒数 Kernel number per ear	百粒重 100-kernel weight	产量 Yield
年际 Year	1	28.51**	0.56	33.65**
熟期 Maturity	2	412.22**	28.78**	200.62**
年际×熟期 Year × maturity	2	54.98**	1.5	26.59**

注：** 表示 $P<0.01$ 水平差异显著

Note: ** represents significant difference at $P<0.01$ level

表3 不同熟期类型玉米产量性状

Table 3 The yield traits of maize hybrids differing in maturities

熟期类型 Maturity	品种 Maize hybrids	穗粒数 Kernel number per ear		百粒重 (g) 100-kernel weight (g)		产量 (kg/hm²) Yield (kg/hm²)	
		2014	2015	2014	2015	2014	2015
极早熟 Super-early maturity	德美亚1号 Demeiya 1	527	493	34.70	33.23	8 415.04	7 890.32
	克单14 Kedan 14	447	491	35.82	35.50	8 355.37	7 500.58
	德美亚3号 Demeiya 3	389	537	43.86	41.61	7 910.49	10 498.92
	均值 Mean	454 Cc	507 Cc	38.13 Aa	36.78 Bb	8 226.97 Cc	8 629.94 Bb
中早熟 Middle-early maturity	吉单27 Jidan 27	675	595	37.12	36.85	12 794.01	11 158.55
	京农科728 Jingnongke 728	657	569	35.97	38.92	12 878.93	12 096.42
	京单28 Jingdan 28	572	510	44.70	45.07	11 527.58	10 296.48
	均值 Mean	635 Bb	558 Bb	39.29 Aa	40.28 Aa	12 400.18 Bb	11 183.82 Aa
中晚熟 Middle-late maturity	先玉335 Xianyu335	772	719	35.86	35.60	15 126.05	13 571.60
	郑单958 Zhengdan 958	726	618	31.37	35.12	14 085.41	11 092.24
	京科968 Jingke 968	768	677	33.78	34.46	15 480.07	11 284.86
	均值 Mean	755 Aa	671 Aa	33.67 Bb	35.06 Bb	14 897.17 Aa	11 982.90 Aa

注：数据后不同大小写字母分别表示同一指标同一年度间差异达 $P<0.05$ 和 $P<0.01$ 显著水平

Note: For the same item, capital and small letters within the same column indicate significant difference between hybrids in the same year at $P<0.05$ and $P<0.01$ level, respectively

2.2 不同熟期玉米品种灌浆过程粒重的变化

粒重变化动态呈 S 形增长曲线（图 1），在籽粒灌浆粒重渐增期，年际间不同熟期品种表现不同，2014 年极早熟和中早熟品种籽粒增重快，2015 年中晚熟品种籽粒增重快，可见粒重渐增期容易受环境条件影响，年际间变化较大；粒重快增期和缓增期，2 个年度都表现为中晚熟品种粒重增长落后于中早熟和极早熟品种；由此说明，不同熟期品种灌浆动态的差异主要取决于灌浆快增期与缓增期。

2.3 不同熟期玉米品种灌浆过程中灌浆速率的变化

3 类熟期品种籽粒灌浆速率均呈单峰曲线变化（图 2），随着灌浆日数增加，3 种熟期类型品种灌浆速率的差异逐渐增大，3 种熟期类型品种均在授粉后 25d 左右达到灌浆高峰，至灌浆高峰时品种间的差异达到最大，中早熟品种灌浆速率峰值最大，其次是极早熟品种，中晚熟品种峰值最小。灌浆高峰后，各熟期品种灌浆速率逐渐下降，极早熟品种下降幅度较快，中晚熟品种下降最慢。

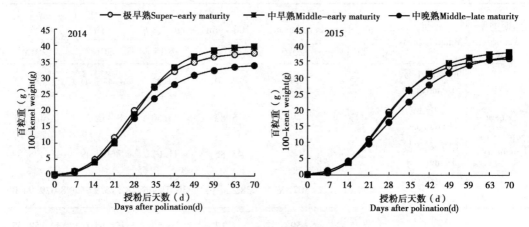

图 1　不同熟期玉米粒重的变化

Figure 1　Grain weight of maize hybrids differing in maturities

2.4 不同熟期玉米品种的籽粒灌浆参数

以授粉后天数（t）为自变量，粒重（W）为因变量，运用 Richard 方程对籽粒灌浆过程进行模拟，获得灌浆参数（表 4）。极早熟和中早熟品种的积累起始势（R_0）年度间变化较大，中晚熟品种相对稳定，两年平均积累起始势（R_0）中晚熟品种>极早熟>中早熟；最大灌浆速率（G_{max}）、达到最大灌浆速率时的生长量（W_{max}）和平均灌浆速率（G_{mean}）3 个灌浆参数表现为中早熟品种最大，极早熟品种次之，中晚熟品种最小；灌浆活跃期（P）和有效灌浆时间（t_3）则以中晚熟品种最长，中早熟品种次之，极早熟品种最短。

2.5 不同熟期玉米品种的籽粒灌浆阶段特征

依据 Richard 曲线将玉米灌浆时期划分为 3 个阶段，分别为灌浆渐增期、快增期和缓增期。灌浆各阶段的持续时间为缓增期最长，其次是快增期，渐增期最短（表 5）；各阶段籽粒累积量高低依次为快增期、缓增期和渐增期；各阶段平均灌浆速率大小依次为快增期、渐增期和缓增期。3 类熟期品种中，中晚熟品种的渐增期持续时间最短，但其快增期

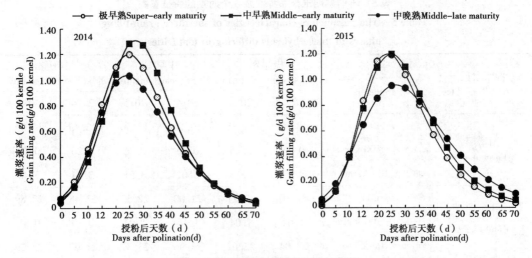

图 2 不同熟期玉米灌浆速率的变化
Figure 2 Grain filling rate of maize hybrids differing in maturities

和缓增期持续时间却最长；中早熟品种的渐增期和缓增期持续时间均比极早熟品种长，快增期持续时间2类品种相近。各个灌浆阶段的平均灌浆速率都表现为中早熟品种>极早熟品种>中晚熟品种。渐增期和快增期籽粒累积量为中早熟品种>极早熟品种>中晚熟品种，缓增期籽粒累积量为中晚熟品种>中早熟品种>极早熟品种。

表 4 不同熟期玉米品种的灌浆特征参数
Table 4 The parameters of grain-filling characteristics in maize hybrids differing in maturities

参数 Parameters	极早熟 Super-early maturity		中早熟 Middle-early maturity		中晚熟 Middle-late maturity	
	2014	2015	2014	2015	2014	2015
A	38.11	36.15	40.06	38.29	34.48	37.87
B	1.29	-0.16	2.35	0.06	-0.13	-0.26
C	0.10	0.09	0.11	0.09	0.09	0.07
D	0.32	0.09	0.51	0.12	0.11	0.11
R_0	0.31	1.02	0.22	0.78	0.77	0.64
G_{max}	1.20	1.20	1.31	1.20	1.04	0.96
T_{max}	24.66	23.46	27.26	24.77	23.89	26.50
W_{max}	15.99	13.90	17.86	14.86	13.36	14.68
G_{mean}	0.81	0.82	0.88	0.81	0.70	0.65
P (d)	46.88	44.26	45.35	47.12	49.05	58.27
t_3 (d)	71.17	72.08	68.78	76.01	77.30	89.93

注：R_0. 积累起始势；G_{max}. 最大灌浆速率；T_{max}. 灌浆速率最大时时间；W_{max}. 灌浆速率最大时生长量；G_{mean}. 平均灌浆速率；P. 灌浆活跃期；t_3. 有效灌浆时间

R_0. initial grain-filling potential； G_{max}. maximum grain-filling rate； T_{max}. the time reaching the maximum grain-filling rate； W_{max}. grain weight at the time of maximum grain-filling rate； G_{mean}. mean grain-filling rate； P. active grain-filling period； t_3. effective grain-filling time

表5 不同熟期玉米品种灌浆3个时期的特征参数
Table 5 The parameters characteristics of the three grain-filling phases in maize hybrids differing in maturities

灌浆阶段 Grain filling stage	参数 Parameters	极早熟 Super-early maturity		中早熟 Middle-early maturity		中晚熟 Middle-late maturity	
		2014	2015	2014	2015	2014	2015
渐增期 Early stage	T_1 (d)	13.61	12.85	16.76	13.50	12.15	12.55
	v_1 (g/d·100-kernel)	0.34	0.25	0.36	0.26	0.26	0.28
	w_1 (g/100-kernel)	4.67	3.18	6.02	3.51	3.14	3.45
快增期 Middle stage	T_2 (d)	22.10	21.21	21.00	22.55	23.48	27.89
	v_2 (g/d·100-kernel)	1.04	1.04	1.14	1.03	0.89	0.83
	w_2 (g/100-kernel)	23.04	22.02	23.96	23.31	21.00	23.06
缓增期 Late stage	T_3 (d)	35.46	38.03	31.02	39.96	41.67	49.48
	v_3 (g/d·100-kernel)	0.28	0.28	0.31	0.28	0.24	0.22
	w_3 (g/100-kernel)	10.02	10.58	9.69	11.09	10.00	10.98

注：T_1. 灌浆渐增期持续时间；v_1. 渐增期平均灌浆速率；w_1. 渐增期粒重增量；T_2. 灌浆快增期持续时间；v_2. 快增期平均灌浆速率；w_2. 快增期粒重增量；T_3. 灌浆缓增期持续时间；v_3. 缓增期平均灌浆速率；w_3: 缓增期粒重增量

Note: T_1. grain-filling duration of early stage; v_1. mean grain-filling rate of early stage; w_1. increased grain weight of early stage; T_2. grain-filling duration of middle stage; v_2. mean grain-filling rate of middle stage; w_2. increased grain weight of middle stage; T_3. grain-filling duration of late stage; v_3. mean grain-filling rate of late stage; w_3. increased grain weight of late stage

2.6 不同熟期玉米品种籽粒灌浆阶段的温度变化

不同熟期品种各自灌浆阶段所经历的日平均气温不同，其完成灌浆所需求的≥10℃有效积温也不同（表6和表7），极早熟品种灌浆期日平均气温和≥10℃有效积温需求最高，中早熟品种次之，中晚熟品种最低；不同熟期品种灌浆期有效积温需求在灌浆各阶段的分配比例不同，极早熟品种在渐增期、快增期和缓增期的积温需求比例分别为24%、38%和39%，中早熟品种在各阶段的分配比例分别为30%、39%和31%，中晚熟品种在各阶段的分配比例分别为26%、45%和28%。

表6 籽粒灌浆阶段日平均气温
Table 6 Average daily temperature in each filling stage

品种 Maize hybrids	渐增期 Early stage		快增期 Middle stage		缓增期 Late stage		灌浆期 Grain filling stage	
	2014	2015	2014	2015	2014	2015	2014	2015
德美亚1号 Demeiya 1	22.35	22.59	22.52	24.09	18.92	18.85	20.56	21.09

（续表）

品种 Maize hybrids	渐增期 Early stage		快增期 Middle stage		缓增期 Late stage		灌浆期 Grain filling stage	
	2014	2015	2014	2015	2014	2015	2014	2015
克单 14 Kedan 14	22.54	22.63	22.05	24.12	20.33	18.61	21.48	21.23
德美亚 3 号 Demeiya 3	23.16	23.47	22.16	23.68	15.19	16.86	18.71	19.88
均值 Mean	22.68	22.90	22.24	23.96	18.15	18.11	20.25	20.73
吉单 27 Jidan 27	22.19	23.97	22.26	23.14	17.39	17.58	20.27	20.44
京农科 728 Jingnongke 728	22.74	24.61	21.87	22.27	17.36	15.57	20.03	23.75
京单 28 Jingdan 28	22.3	24.66	22.12	21.16	13.89	13.66	18.37	18.07
均值 Mean	22.41	24.41	22.08	22.19	16.21	15.60	19.56	20.75
先玉 335 Xianyu335	23.04	24.55	21.96	21.65	14.73	13.68	18.01	18.14
郑单 958 Zhengdan 958	22.38	24.66	22.24	20.53	14.12	13.57	18.24	18.07
京科 968 Jingke 968	23.34	24.68	21.78	21.45	13.41	14.38	17.49	17.96
均值 Mean	22.92	24.63	21.99	21.21	14.09	13.88	17.91	18.06

表 7 籽粒灌浆阶段≥10℃有效积温（℃）
Table 7　Effective accumulated temperature（℃）above 10℃ in each filling stage

品种 Maize hybrids	渐增期 Early stage		快增期 Middle stage		缓增期 Late stage		灌浆期 Grain filling stage	
	2014	2015	2014	2015	2014	2015	2014	2015
德美亚 1 号 Demeiya 1	148.19	163.65	262.84	295.98	338.89	327.5	749.92	787.13
克单 14 Kedan 14	200.69	189.48	241.04	268.2	258.34	249.7	700.07	707.38
德美亚 3 号 Demeiya 3	185.88	161.64	301.43	301.04	244.25	291.25	731.56	753.93
均值 Mean	178.25	171.59	268.44	288.41	280.49	289.48	727.18	749.48
吉单 27 Jidan 27	219.38	181.65	232.89	262.73	184.7	265.35	636.97	709.73

（续表）

品种 Maize hybrids	渐增期 Early stage		快增期 Middle stage		缓增期 Late stage		灌浆期 Grain filling stage	
	2014	2015	2014	2015	2014	2015	2014	2015
京农科728 Jingnongke 728	178.38	189.88	237.34	245.3	206.09	225.75	621.81	660.93
京单28 Jingdan 28	233.73	190.58	266.69	312.45	162.25	168.45	662.67	671.48
均值 Mean	210.50	187.37	245.64	273.49	184.35	219.85	640.48	680.71
先玉335 Xianyu335	130.38	174.58	274.99	337.75	231.7	172.9	637.07	685.23
郑单958 Zhengdan 958	184.14	190.58	268.94	326.35	174.35	154.55	627.43	671.48
京科968 Jingke 968	146.79	190.58	294.39	263.25	174.35	192.6	615.53	646.43
均值 Mean	153.77	185.25	279.44	309.12	193.47	173.35	626.68	667.71

2.7 玉米籽粒灌浆参数与灌浆期温度的相关分析

对不同熟期玉米品种籽粒灌浆参数和灌浆期、灌浆不同阶段≥10℃有效积温和日平均气温进行相关分析。活跃灌浆期、有效灌浆期、快增期和缓增期持续时间与快增期≥10℃有效积温、渐增期日平均气温极显著正相关，与快增期、缓增期、灌浆期日平均气温显著或极显著负相关；最大灌浆速率、平均灌浆速率、快增期和缓增期平均灌浆速率与渐增期≥10℃有效积温、缓增期和灌浆期日平均气温显著正相关，与快增期≥10℃有效积温和渐增期日平均气温显著负相关（表8）。

表8 灌浆参数与灌浆期≥10℃有效积温和日平均气温的相关系数
Table 8 Correlation of filling parameters and effective accumulated temperature ≥10℃ and average daily temperature

因子 Factors	x_1	x_2	x_3	x_4	x_5	x_6	x_7	x_8
P	-0.13	0.01	0.82**	-0.56*	-0.67**	0.58**	-0.68**	-0.72**
t_3	-0.04	-0.14	0.89**	-0.45	-0.63**	0.67**	-0.57*	-0.77**
T_1	-0.19	0.88**	-0.49*	-0.29	0.16	-0.50*	0.02	0.19
T_2	-0.08	-0.09	0.87**	-0.51*	-0.67**	0.63**	-0.64**	-0.75**
T_3	0.04	-0.36	0.89**	-0.29	-0.55*	0.71**	-0.46*	-0.70**
G_{max}	0.14	0.50*	-0.60**	0.22	0.54*	-0.48*	0.41	0.48*
G_{mean}	0.15	0.48*	-0.59**	0.24	0.53*	-0.47*	0.43	0.47*

(续表)

因子 Factors	x_1	x_2	x_3	x_4	x_5	x_6	x_7	x_8
v_1	−0.18	0.59**	−0.4	−0.21	0.07	−0.43	−0.15	0.3
v_2	0.11	0.52*	−0.63**	0.2	0.53*	−0.49*	0.39	0.48*
v_3	0.07	0.59**	−0.65**	0.16	0.50*	−0.53*	0.35	0.49*

注：x_1. 灌浆期≥10℃有效积温；x_2. 渐增期≥10℃有效积温；x_3. 快增期≥10℃有效积温；x_4. 缓增期≥10℃有效积温；x_5. 灌浆期日平均气温；x_6. 渐增期日平均气温；x_7. 快增期日平均气温；x_8. 缓增期日平均气温。P. 灌浆活跃期；t_3. 有效灌浆时间；T_1. 灌浆渐增期持续时间；T_2. 灌浆快增期持续时间；T_3. 灌浆缓增期持续时间；v_1. 渐增期平均灌浆速率；v_2. 快增期平均灌浆速率；v_3. 缓增期平均灌浆速率。* 和 ** 分别表示 $P<0.05$ 和 $P<0.01$ 水平显著相关

Note: x_1. effective accumulated temperature ≥10℃ in grain filling stage; x_2. effective accumulated temperature ≥10℃ in grain filling early stage; x_3. effective accumulated temperature ≥10℃ in grain filling middle stage; x_4. effective accumulated temperature ≥10℃ in grain filling late stage; x_5. average daily temperature in grain filling stage; x_6. average daily temperature in grain filling early stage; x_7. average daily temperature in grain filling middle stage; x_8. average daily temperature in grain filling late stage; P. active grain-filling period; t_3: effective grain-filling time; T_1. grain-filling duration of early stage; T_2. grain-filling duration of middle stage; T_3. grain-filling duration of late stage; v_1. mean grain-filling rate of early stage; v_2. mean grain-filling rate of middle stage; v_3. mean grain-filling rate of late stage. * and ** indicate significant correlation at $P<0.05$ and $P<0.01$ levels, respectively

3 讨论

籽粒灌浆特性是玉米品种自身的固有特征，与产量形成密切相关。李绍长等研究认为不同品种的粒重主要受灌浆持续期影响，而同一个品种的粒重则主要受灌浆速度的影响。张丽等研究表明，灌浆后期保持相对较高的灌浆速率是粒重较大的主要原因，同时较长的灌浆持续期有利于粒重提高。本研究表明，中晚熟品种灌浆持续期比中早熟和极早熟品种长，主要体现在灌浆缓增期比较长；而缓增期平均灌浆速率与缓增期日平均气温显著正相关。在哈尔滨生态区，中晚熟品种灌浆缓增期内日平均气温不足14℃，致使中晚熟品种灌浆速率比较低，因此，中晚熟品种尽管灌浆持续时间比较长，但其间籽粒干物质积累量却不高，最终导致中晚熟品种的粒重较中早熟和极早熟品种低。由此可见，在热量资源有限的生态区，一味地延长品种熟期并不利于粒重增长。王晓慧等研究认为，提高渐增期的灌浆速率，可以缩短渐增期库容建成时间，相对延长快增期和缓增期持续日数，从而增加玉米产量。本研究表明，中早熟品种在灌浆各个阶段的灌浆速率均高于中晚熟和极早熟品种，渐增期和快增期籽粒累积量也高于中晚熟和极早熟品种，缓增期籽粒累积量高于极早熟品种。由此说明，在哈尔滨这种热量资源有限的地区，中早熟品种比中晚熟和极早熟品种表现更多的灌浆优势，从而获得较高的产量。

理论上，作物产量和生育期在一定范围内成正相关关系，适当延长生育期可增加光能利用时间，从而获得较高产量。所以，长期以来，在生产上为了获得高产，越区种植中晚熟和晚熟品种在黑龙江省大部分地区很普遍。本研究表明，在黑龙江省哈尔滨地区，3种

熟期类型品种。在同等种植密度下，中晚熟和中早熟品种产量显著高于极早熟品种，但中晚熟和中早熟品种产量差异并不显著。已有研究也表明，东北地区玉米品种在130~140d时间段对产量贡献不大，通过延长品种生育期来提高产量的潜力有限。本研究表明，中早熟品种的最大灌浆速率、平均灌浆速率和达到最大灌浆速率时的粒重生长量都高于中晚熟品种。黑龙江省是全国农业机械化水平最高的省份，随着玉米全程机械化技术的推广，生产上需要熟期相对早，脱水快的品种。因此，在新的技术推动生产需求下，晚熟品种不宜再占主导地位，应以中早熟品种为主，既可充分利用当地热量资源，又可加速籽粒脱水利于机械化收获。本试验中的中早熟品种京农科728年际间产量变异系数只有3%，稳产性很好，且生理成熟后脱水速率快，是一个适于哈尔滨区域种植，且有利于实施机械化收获的高产品种。

参考文献（略）

本文原载：中国农业科技导报，2017,，19（8）：：105-114

玉米果穗不同位势籽粒灌浆特性分析

钱春荣[1,2]　王荣焕[1]　赵久然[1]　于　洋[2]　郝玉波[2]
姜宇博[2]　宫秀杰[2]　李　梁[2]　葛选良[2]

(1. 北京市农林科学院玉米研究中心，北京 100097；2. 黑龙江省农业科学院耕作栽培研究所/农业部东北地区作物栽培科学观测实验站，哈尔滨，150086)

摘　要：以 8 个玉米品种为试验材料，研究了玉米灌浆期果穗不同部位籽粒灌浆特性。结果表明：供试 8 个玉米品种，中下部籽粒的百粒重变幅范围为 34.02~51.38g，极差为 17.36g，品种之间的变异系数为 13.76%；上部籽粒的百粒重变幅范围为 28.61~35.13g，极差为 6.52g，变异系数为 7.29%；从灌浆开始至灌浆快增期结束上部籽粒百粒重变异系数为 66.13%~13.18%，中下部籽粒百粒重变异系数为 54.38%~12.10%，随灌浆进程籽粒百粒重变异系数呈减小趋势；中下部籽粒平均灌浆速率始终高于上部籽粒，上部籽粒灌浆速率的变异系数大于中下部籽粒，并且灌浆速率的变异系数随灌浆进程呈先降后增的趋势，吐丝后 56d 灌浆速率的变异系数快速增长。品种间各灌浆参数的变异系数，均为上部籽粒大于中下部籽粒。因此，选用籽粒灌浆粒位差异小的品种，采取有效措施促进上部籽粒灌浆，提高粒重是实现玉米高产栽培的技术途径。

关键词：玉米；籽粒灌浆；灌浆特性

Analysis on Characteristics of Grain-Filling at Different Positions on a Maize Ear

Qian Chunrong[1,2]　Wang Ronghuan[1]　Zhao Jiuran[1]　Yu Yang[2]
Hao Yubo[2]　Jiang Yubo[2]　Gong Xiujie[2]　Li Liang[2]　Ge Xuanliang[2]

(1. *Maize Research Center, Beijing Academy of Agriculture and Forestry Sciences, Beijing* 100097, *China*; 2. *Institute of Crop Cultivation and Farming, Heilongjiang Academy of Agricultural Sciences/Scientific Observing and Experimental Station of Crop Cultivation in Northeast China, Ministry of Agriculture, Harbin* 150086, *China*)

Abstract: 8 maize hybrids were planted to observe and analyze their characteristics of grain-filling at different positions on a maize ear. The results showed that the 100-kernel weight of middle and basal kernels changed from 34.02g to 51.38g, range was 17.36g and variable coefficient was

---

基金项目：北京市农林科学院博士后科研基金、北京市博士后科研活动经费资助项目
作者简介：钱春荣（1973—　），女，博士，副研究员，主要从事玉米高产栽培与生理生态研究。
　　　　　E-mail：qcr3906@163.com
　　　王荣焕为本文共同第一作者。赵久然为本文通讯作者，E-mail：maizezhao@126.com

13.76%; The 100-kernel weight of upper changed from 28.61 to 35.13g, range was 6.2g and variable coefficient was 7.29%. During the early and middle grain filling stage the variation coefficient of 100-grain weight at the upper was in the range from 13.18% to 66.13% and at the middle and basal was in the range from 12.10% to 54.38%, coefficient of variation of grain weight per 100grains was between 54.38% and 12.10%, andthe variation coefficient of 100-grain weight decreased with grain filling going. Variable coefficient of grain-filling rate at the upper kernel was greater than that at middle and basal. Variable coefficient of grain-filling rate decreased at earlier grain filling stage ant then gradually increased with grain filling going, and when 56 days after silking variable coefficient of grain-filling rate rapidly grown. Variable coefficient of characteristics parameters was greater at the upper kernels than that at the middle and basal. Selecting the hybrids that difference between kernels at different position on an ear was tiny and promoting the upper kernel grain filling were a technical approach to increase kernel weight for high yield.

Key words: Maize; Grain filling; Filling characteristic

玉米产量的高低决定于库容的大小和灌浆充实的程度。灌浆特性是玉米生长后期影响产量的关键因素之一，灌浆期和灌浆活跃期的长短、灌浆强度的高低决定了玉米籽粒干物质积累的多少。玉米果穗上的粒重因着生位置不同而有很大的差异，玉米籽粒的最大灌浆速度和平均灌浆速率也因着生位置存在显著差异。相对于果穗中下部籽粒，上部籽粒是弱势粒，灌浆充实度差，粒重偏低。对于玉米、小麦和水稻等禾本科作物弱势粒灌浆差的原因，国内外作了大量的研究。品种的多样性导致粒重的差异，不同基因型之间粒重差异很大。同一品种籽粒重的差异是由灌浆速度决定的，而不同品种粒重差异则是由灌浆持续期的长短造成的。玉米籽粒灌浆特性和粒重变化的研究目前多集中于生态条件和生理机制差异等方面，有关玉米不同品种、不同部位籽粒灌浆特性及粒重的变异特征的研究报道较少。

本研究以黑龙江省大面积推广和北京市农林科学院玉米研究中心新审定或新育成的16个高产玉米品种为研究对象，运用生长模型对其不同部位籽粒灌浆特性进行分析，探明不同春玉米品种、不同部位籽粒灌浆共性及差异，为玉米籽粒发育调控及品种选择提供理论依据。

1 材料与方法

1.1 试验地概况

试验于2015年在黑龙江省哈尔滨市民主乡黑龙江省农业科学院农业科技园区进行（N 45°50′，E 126°50′）。本试验生长季节（5月10日至10月4日）活动积温2 915℃·d，平均气温19.7℃，降雨量390.75mm。试验地土壤为碳酸盐黑钙土，试验区碱解氮160.40mg/kg，速效磷19.22mg/kg，速效钾675.90mg/kg，有机质19.6g/kg，pH值5.95。

1.2 试验设计

试验选用黑龙江省目前大面积推广和新审定或新育成表现突出的8个玉米品种为试验材料，供试品种见表1。其中产量为2015年试验结果。试验采取随机区组设计，3次重复，每个小区6行，行长5m，行距0.65m，小区面积19.5m^2。

表 1　供试玉米品种、产量及生育期
Table 1　Maize hybrids, experimental yield and growth stages

品种 Hybrids	产量（kg/hm²） Trial yield	生育期（d） Growth stage
吉单 27	10 905±462	126
京农科 728	11 822±268	126
京单 28	10 063±204	130
NK718	11 779±630	130
京华 8 号	11 418±639	130
利民 33	10 412±423	130
农华 101	10 021±949	130
先玉 335	13 263±766	132

1.3　田间管理

试验于 2014 年秋季耕翻整地，秋施肥，施入控释掺混肥料（N-P_2O_5-K_2O 比例为 26-10-12）600kg/hm²。2015 年 5 月 10 日人工播种，每穴 3 粒，种植密度为 52 500 株/hm²，出苗后于三叶期按设计密度定苗。其他管理措施与当地大田高产栽培技术措施一致。

1.4　测定项目与方法

吐丝期选择生长一致且同一天吐丝的植株挂牌标记，自吐丝 14d 起，每 7d 取果穗一次，每次每个小区取 3 个挂牌标记的果穗，取其上部（自顶部第 3~12 环）及中下部（自穗底部第 8~20 环）籽粒各 100 粒，分别称其鲜重，于 80℃ 烘干至恒重后称籽粒干重。

以授粉后天数（d）为自变量，授粉后每隔 7d 测籽粒干重为因变量（W），参照顾世梁等的方法，利用 Richards 方程 $W=A(1+Be^{-ct})^{-1/D}$ 模拟籽粒灌浆过程。籽粒灌浆速率 $G=ACBe^{-ct}/D(1+Be^{-ct})^{(D+1)/D}$，式中：$W$ 为粒重（mg），A 为最终粒重（mg），t 为授粉后天数（d），B、C、D 为回归方程所确定的参数。由方程一阶导数和二阶导数推导出灌浆参数。

积累起始势 $R_0 = C/D$

达最大灌浆速率时的天数：$T_{max} = (\ln B - \ln D)/C$

灌浆速率最大时的生长量 $W_{max} = A(D+1)^{-1/D}$

最大灌浆速率 $G_{max} = (CW_{max}/D)[1-(W_{max}/A)D]$

灌浆活跃期（约完成总积累量的 90%）$P = 2(D+2)/C$

平均灌浆速率 $G_{mean} = AC/2(D+2)$

灌浆高峰开始日期 $t_1 = -\ln\{[D^2+3D+D(D^2+6D+5)^{1/2}]/2B\}/C$，对应此时粒重为 $W_1 = A(1+Be^{-ct_1})^{-1/D}$

灌浆高峰结束日期 $t_2 = -\ln\{[D^2+3D-D(D^2+6D+5)^{1/2}]/2B\}/C$，对应此时粒重为 $W_2 = A(1+Be^{-ct_2})^{-1/D}$

花后粒重 W 达 99% A 时为有效灌浆期，$t_3 = -\ln\{[(100A/99)^D-1]/B\}/C$，对应此时粒重为 W_3。

1.5 数据分析

采用 Microsoft Excel 2013 进行数据计算，用 CurveExpert1.4 软件进行方程拟合，SAS 9.0 进行数据分析。

2 结果与分析

2.1 果穗上不同部位的籽粒重

用 Richards 方程拟合了供试玉米品种果穗不同部位最终百粒重，从表2可见，中下部粒重较上部粒重平均增重 32.51%，供试 8 个玉米品种中下部籽粒百粒重的变异范围为 34.02~51.38g，极差为 17.36g，品种之间的变异系数为 13.76%；上部籽粒百粒重的变异范围为 28.61~35.13g，极差为 6.52g，变异系数为 7.29%，说明不同品种间中下部籽粒的变异幅度更大，品种间粒重的差异主要来自于中下部籽粒。不同品种间果穗不同部位粒重的差异程度存在多态性，不同品种中下部粒重较上部粒重增加幅度为 6.75%~62.13%，其变异系数高达 63.51%，说明不同品种籽粒灌浆的粒位效应差异显著。

表2 玉米果穗上不同部位的百粒重
Table 2 100-kernel weight of kernels at different positions on a maize ear

品种 Hybrids	中下部（g）Middle and basal	上部（g）Upper	中下部粒重增幅（%）Increment percentage of middle and basal to upper
吉单 27	34.02	28.61	18.91
京农科 728	36.89	30.21	22.11
京单 28	45.31	30.06	50.73
NK718	42.23	31.92	32.30
京华 8 号	44.90	29.20	53.77
利民 33	51.38	31.69	62.13
农华 101	37.50	35.13	6.75
先玉 335	38.48	33.94	13.38
平均值 Average	41.34	31.35	32.51
标准差 Std.	5.69	2.28	20.65
变异系数 CV（%）	13.76	7.29	63.51

2.2 玉米灌浆过程中果穗不同部位粒重变化

由图1和表3可知，在灌浆进程中，中下部籽粒重始终高于上部籽粒，随着灌浆进程推进，不同位置的粒重差异逐渐加大。比较灌浆进程中各个阶段籽粒重量的变异系数，发现无论是中下部籽粒还是上部籽粒，均表现为随灌浆进程品种间的粒重变异系数呈减小趋势，说明不同品种在灌浆前期差异更明显；同时发现在灌浆中前期（7~42d）上部籽粒重的变异系数大于中下部籽粒,而在灌浆后期(49d 以后),中下部籽粒重的变异系数大于上部籽粒,说明,不同品种最终粒重的差异还是决定于灌浆后期的中下部籽粒。

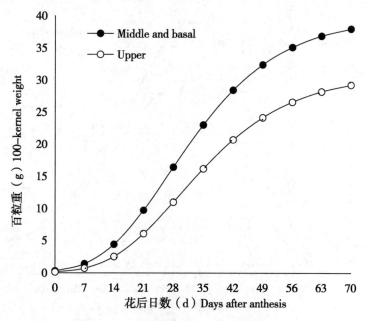

图 1 玉米果穗不同部位籽粒重量的变化

Figure 1 Changes in kernel weight at different positions on a maize ear

表 3 吐丝后玉米果穗不同部位粒重的变化

Table 3 Changes in kernel weight and CV of kernel weight at different positions on a maize ear

| 花后日数（d） | 百粒干重（g） | | 变异系数（%） | |
| Days after anthesis | 100-kernel weight | | CV | |
	中下部 Middle and basal	上部 Upper	中下部 Middle and basal	上部 Upper
7	1.46	1.17	54.38	66.13
14	4.38	2.89	20.72	31.36
21	9.78	6.19	14.35	23.23
28	16.53	10.86	16.05	21.11
35	23.00	16.06	14.54	16.72
42	28.31	20.74	12.1	13.18
49	32.28	24.25	10.23	10.14
56	35.12	26.61	9.31	6.94
63	37.1	28.16	9.21	4.01
70	38.46	29.2	9.59	2.39

2.3 玉米灌浆过程中不同部位籽粒灌浆速率的变化

从图 2 可见,果穗不同部位籽粒的灌浆速率均呈单峰曲线变化,中下部籽粒的灌浆速率始终高于上部籽粒,花后 21d 中下部与上部籽粒灌浆速率的差异幅度达到最大,二者灌浆速率相差 0.27g100-kernel^{-1} d^{-1},至灌浆末期,不同部位籽粒的灌浆速率逐渐趋于一致。

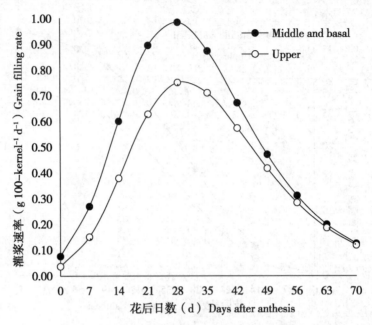

图 2 玉米果穗不同部位籽粒灌浆速率的变化
Figure 2 Changes in kernel filling rate at different positions on a maize ear

对比灌浆过程中各阶段玉米品种灌浆速率的变异系数,整体表现为上部籽粒灌浆速率的变异系数大于中下部籽粒,这与上部籽粒重的变异系数较高相一致;并且灌浆速率的变异系数随灌浆进程呈先降后增的趋势,花后 56d 灌浆速率的变异系数快速增长,由此说明,玉米品种间灌浆速率的差异在灌浆后期表现的更明显,而且上部籽粒灌浆速率的变异幅度更大(表4)。

表 4 玉米果穗不同部位籽粒灌浆速率的变化
Table 4 Changes in kernel filling rate and CV of kernel filling rate at different positions on a maize ear

授粉后天数(天) Days after anthesis (d)	灌浆速率(g 100-kernel^{-1} d^{-1}) Kernel filling rate		灌浆速率的变异系数(%) CV	
	中下部 Middle and basal	上部 Upper	中下部 Middle and basal	上部 Upper
0	0.08	0.04	69.78	70.11
7	0.27	0.15	24.24	26.91
14	0.60	0.38	21.03	32.36
21	0.89	0.63	25.28	31.58
28	0.98	0.75	17.54	20.54
35	0.87	0.71	12.83	20.52
42	0.67	0.57	20.94	20.82

（续表）

授粉后天数（天）Days after anthesis (d)	灌浆速率（g 100-kernel^{-1} d^{-1}）Kernel filling rate		灌浆速率的变异系数（%）CV	
	中下部 Middle and basal	上部 Upper	中下部 Middle and basal	上部 Upper
49	0.47	0.42	32.57	23.66
56	0.31	0.28	45.26	42.76
63	0.20	0.19	59.55	67.91
70	0.12	0.12	75.63	92.21

2.4 玉米灌浆过程中不同部位籽粒灌浆参数

由表5可知，中下部籽粒生长势（R_0）略低于上部籽粒，中下部籽粒最大灌浆速率出现的时间早于上部籽粒，中下部籽粒的最大灌浆速率和平均灌浆速率均高于上部籽粒，中下部籽粒灌浆活跃灌浆期（P）和有效灌浆期（T）均小于上部籽粒，中下部籽粒灌浆速率最大时的籽粒生长量（W_{max}）高于上部籽粒。分析不同部位籽粒的阶段灌浆特征可见（表6），中下部籽粒的渐增期、快增期和缓增期分别比上部籽粒少1~2d，中下部籽粒各时期的平均灌浆速率均高于上部籽粒，而且中下部籽粒每个时期的籽粒积累量也均高于上部籽粒。分析不同部位籽粒灌浆参数的变异系数，发现除灌浆速率最大时的籽粒生长量（W_{max}）和快增期的籽粒生长量（W_2）以外，品种间各灌浆参数的变异系数，均为上部籽粒大于中下部籽粒，由此说明，品种间灌浆特性的差异更多表现在上部籽粒。

表5 不同部位玉米籽粒的灌浆特征参数
Table 5 The parameters of grain-filling characteristics in kernel at different positions on a maize ear

参数 Parameters	平均值 Average value		变异系数（%）CV	
	中下部 Middle and basal	上部 Upper	中下部 Middle and basal	上部 Upper
R_0	0.51	0.53	62.57	79.78
T_{max} (d)	27	29	16.57	16.20
G_{max} (g 100-kernel^{-1} d^{-1})	0.99	0.75	18.56	21.14
G_{mean} (g 100-kernel^{-1} d^{-1})	0.67	0.51	18.59	20.78
W_{max} (g 100-kernel^{-1})	15.61	12.12	14.46	14.44
P (d)	59	61	31.25	36.48
T (d)	91	95	27.82	33.60

注：R_0. 积累起始势；T_{max}. 灌浆速率最大时时间；G_{max}. 最大灌浆速率；W_{max}. 灌浆速率最大时生长量；G_{mean}. 平均灌浆速率；P. 灌浆活跃期；T. 有效灌浆期

Note：R_0. initial grain-filling potential；T_{max}. the time reaching the maximum grain-filling rate；G_{max}. maximum grain-filling rate；W_{max}. grain weight at the time of maximum grain-filling rate；G_{mean}. mean grain-filling rate；P. active grain-filling period；T. effective grain-filling time

表6 不同部位玉米籽粒的阶段灌浆特征
Table 6 The characteristics of different stage of grain-filling in kernel at different positions on a maize

灌浆阶段 Grain filling stage	灌浆参数 Grain filling parameter	灌浆参数值 Value of grain filling parameter		灌浆参数的变异系数（%） CV	
		中下部籽粒	上部籽粒	中下部籽粒	上部籽粒
渐增期	T_1 (d)	13	15	13.34	26.34
	v_1 (g·100-kernel^{-1}·d^{-1})	0.29	0.19	20.88	25.74
	w_1 (g·100-kernel^{-1})	3.81	2.93	23.68	45.41
快增期	T_2 (d)	28	29	31.45	39.03
	v_2 (g·100-kernel^{-1}·d^{-1})	0.85	0.65	18.51	21.41
	w_2 (g·100-kernel^{-1})	24.2	18.86	14.27	12.84
缓增期	T_3 (d)	50	51	33.74	48.84
	v_3 (g·100-kernel^{-1}·d^{-1})	0.23	0.18	18.42	22.5
	w_3 (g·100-kernel^{-1})	11.36	8.89	16.13	24.6

注：T_1. 灌浆渐增期持续时间；v_1. 渐增期平均灌浆速率；w_1. 渐增期粒重增量；T_2. 灌浆快增期持续时间；v_2. 快增期平均灌浆速率；w_2. 快增期粒重增量；T_3. 灌浆缓增期持续时间；v_3. 缓增期平均灌浆速率；w3. 缓增期粒重增量

Note：T_1. grain-filling duration of early stage；v_1. mean grain-filling rate of early stage；w_1. increased grain weight of early stage；T_2. grain-filling duration of middlestage；v_2. mean grain-filling rate of middle stage；w_2. increased grain weight of middle stage；T_3. grain-filling duration of late stage；v_3. mean grain-filling rateof late stage；w_3. increased grain weight of late stage

3 结论与讨论

玉米产量是由穗数、穗粒数和粒重3要素构成，在穗数、穗粒数一定的条件下，粒重对玉米产量起着至关重要的作用。不同玉米品种间粒重差异很大，而粒重的差异主要是由于灌浆过程不同引起的。关于灌浆过程对粒重的影响，大多数观点认为粒重由灌浆时间和灌浆速率共同决定。籽粒灌浆特性是玉米品种自身所固有特征，与玉米产量形成密切相关，因此，明确品种的灌浆特性，才能充分发挥品种的增产潜力。

有关小麦、水稻和玉米穗上不同部位籽粒灌浆和粒重的差异，国内外已有大量研究。但对于玉米灌浆期果穗不同部位籽粒重及灌浆参数品种间变异情况的研究鲜有报道。对于水稻和小麦穗上强势粒与弱势粒的划分已有明确的标准。但玉米穗上强、弱势粒的划分还没有明确的方法，徐云姬等建议将玉米果穗顶部籽粒（约占玉米果穗总长度的1/3）作为弱势粒，将玉米果穗中、下部籽粒（约占玉米果穗总长度的2/3）作为强势粒。本研究中籽粒取样部位与徐云姬等基本一致，也观察到果穗不同部位籽粒的生长势、最终粒重和灌浆速率，均表现为果穗下部籽粒高于上部籽粒，与徐云姬等研究结果一致，因此，笔者赞同将果穗中下部籽粒和上部籽粒分别定性为强势粒和弱势粒。本研究中试验品种8个，不同品种强、弱势粒的差异程度存在显著差异，农华101强、弱势粒差异最小，强势粒较弱势粒粒重仅增加6.75%，利民33强、弱势粒差异最大，强势粒较弱势粒粒重增加

62.13%，可见玉米果穗不同粒位粒重的差异程度是品种的自身特性，因此，在玉米高产栽培中，宜选择强、弱势粒差异较小的品种。

本研究观察到，玉米果穗中强势粒粒重的变异系数大于弱势粒，灌浆前期粒重的变异系数大于后期，籽粒灌浆速率的变异系数表现为后期大于前期、弱势粒大于强势粒；籽粒灌浆参数的变异系数均表现为弱势粒大于强势粒，因此，在玉米高产栽培中采取有效措施，比如增施氮肥、喷施化学调控剂等有效措施，促进弱势粒灌浆，从而增加玉米粒重。可见，深入探讨促进玉米弱势粒灌浆的调控途径和关键栽培技术，对于促进弱势粒灌浆和提高玉米产量具有重要意义。

参考文献（略）

本文原载：玉米科学，2017，25（3）：80-86

不同生育时期遮光对玉米籽粒灌浆特性及产量的影响

陈传永[1]**　王荣焕[1]**　赵久然[1]*　徐田军[1]　王元东[1]
刘秀芝[1]　刘春阁[1]　裴志超[2]　成广雷[1]　陈国平[1]

(1. 北京市农林科学院玉米研究中心，北京　100097；2. 北京市农技推广站，北京　100029)

摘　要：为探索不同时期遮光对玉米籽粒灌浆特性和产量的影响，2012—2013年以玉米品种京科968与郑单958为试验材料，在大田条件下用透光率50%的遮阳网分别于13叶全展期（T1）、吐丝期（T2）、吐丝后15d（T3）遮光处理，每期遮光7d，以自然光照为对照，并利用Logistic方程$y=A/(1+Be^{-Cx})$比较不同遮光处理玉米籽粒灌浆过程。结果表明，不同时期遮光均导致玉米穗粒数、千粒重和产量不同程度降低，且遮光时期越晚降幅越大，其中产量差异显著；遮光导致灌浆高峰持续期与活跃灌浆天数（P）缩短，最大灌浆速率（G_{max}）与平均灌浆速率均下降，籽粒终极生长量（A）降低；品种特性决定粒重、穗粒数与灌浆参数的关系，同一品种的粒重与灌浆速率最大时的生长量（W_{max}）呈显著正相关，品种间的粒重差异由活跃灌浆天数（P）决定，同一品种的穗粒数与最大灌浆速率（G_{max}）呈显著正相关，品种间的穗粒数差异由灌浆速率最大时的生长量（W_{max}）决定。选择适宜品种，提高籽粒灌浆速率最大时的生长量（W_{max}）与最大灌浆速率（G_{max}），并延长活跃灌浆天数（P）是玉米在光胁迫环境下获得高产、稳产的重要途径。

关键词：玉米；遮光；灌浆特性；产量

Effects of Shading on Grain-Filling Properties and Yield of Maize at Different Growth Stages

Chen Chuanyong[1]**　Wang Ronghuan[1]**　Zhao Jiuran[1]*
Xu Tianjun[1]　Wang Yuandong[1]　Liu Xiuzhi[1]　Liu Chunge[1]
Pei Zhichao[2]　Cheng Guanglei[1]　Chen Guoping[1]

(1. Maize research center, Beijing Academy of Agriculture and Forestry Sciences, Beijing 100097, China; 2. Beijing Agricultural technology Extension, Beijing 100029, China)

Abstract: The objective of this study was to explore the effect of shading at different growth

---

基金项目：本研究由北京市农林科学院科技创新基金项目（CXJJ201309）、国家玉米产业技术体系项目（nycytx-02）、北京市科技计划（D121100003512001）和北京市农林科学院青年科研基金项目（QN201115）资助

* 通讯作者：赵久然，E-mail：maizezhao@126.com，Tel：010-51503876

第一作者联系方式：E-mail：chuanyongchen@126.com

** 同等贡献

stages on grain filling properties and yield of maize (Zea mays L.). The Field experiments were carried out using maize cultivars (Jingke968, Zhengdan958) with shading of 50% at the 13th leaf fully expanded (T1), the silking (T2) and the 15th day after silking (T3) for seven days in 2012—2013. The natural light condition without shading was used as the control. The Logistic equation was used to analyze the grain filling process. The results showed that grains per ear, 1 000-kernel weight and yield were all decreased to a different degree under shading conditions at different growth stages, and the decrease rate of grain yield was increased with lasting shading duration. The yield was significantly different under different shading treatments. The duration of the maximum grain filling and active grain-filling period (P) were shortened, the maximum and mean grain-filling rate (G_{max}) decreased and the final grain weight (A) reduced under shading conditions. The relation among grain weight, grain number per ear and grain yield was determined by characteristics of maize cultivar. The grain weight of the same cultivar was significantly positively correlated with grain weight with the maximum grain-filling rate (W_{max}), and the grain weight of different maize cultivars was determined by active grain-filling period (P). The grain number per ear of the same maize cultivar was significantly positively correlated with G_{max}, and the grain number per ear of different maize cultivars was determined by W_{max}. It is an important way to keep a high or stable yield by selecting the suitable cultivar, improving W_{max} and G_{max} and prolonging active grain-filling period (P) under weak light stress.

Key words: Maize; Shading; Grain-filling properties; Grain yield

玉米是C_4作物，具有较高的光饱和点，充足的光照和适宜的温度是玉米产量形成和品质改善的前提。近年来，由于气候条件与种植区域变化，玉米生产中因晚播、阴雨、越区种植等因素造成的玉米生育期内光照不足的现象时有发生。光照不足引起的弱光胁迫影响玉米的生长发育、光合特性及产量构成。前人研究表明：遮光限制玉米营养器官和生殖器官发育，降低光合生产能力，干物质积累减少，败育粒数显著增多，籽粒体积和粒重降低，产量下降，且产量降幅取决于遮光时期和遮光强度。其中，花粒期是籽粒灌浆形成期，也是影响玉米产量及其构成因素的关键时期。利用Logistic方程可准确地拟合玉米籽粒灌浆过程，对玉米籽粒灌浆进行生长分析。目前，对籽粒灌浆的研究多集中在自然光照条件下，而对遮光条件下玉米籽粒灌浆特性的研究鲜有报道。本文选取适应性广、具有良好推广前景的玉米新品种京科968与近年来我国推广面积最大的玉米品种郑单958为试验材料，在不同生育时期短期遮光处理，参考朱庆森等、黄振喜等和王育红等的方法，利用Logistic拟合方程推导出一系列次级参数，分析玉米籽粒灌浆特征参数，系统比较遮光条件下玉米籽粒灌浆过程，探讨不同时期遮光对玉米灌浆特性的影响，以期为玉米高产、稳产栽培，制定适时调控栽培管理措施提供理论依据。

1 材料与方法

1.1 试验地点

试验于2012年、2013年5—9月，在北京市昌平区小汤山国家精准农业研究示范基地进行。试验田0~20cm耕层土壤养分见表1。2012年、2013年玉米生育期积温分别为3 237℃、3 536℃，降水量分别为539.30mm、447.10mm，总日照时数分别为976h、1 001h；吐丝后积温分别为1 527℃、1 745℃，降水量分别为389.90mm、330.45mm，总日照时数分别为445h、506h。

表 1 土壤养分含量状况

Table 1 Soil nutrient in the experiment

年份 (Year)	有机质 Organic matter (g kg^{-1})	全氮 Total N (g kg^{-1})	全磷 Total P (g kg^{-1})	全钾 Total K (g kg^{-1})	碱解氮 Alkali-hydrolysable N (mg kg^{-1})	速效磷 Available P (mg kg^{-1})	速效钾 Available K (mg kg^{-1})
2012	10.90	0.70	0.52	9.56	94.36	25.37	145.15
2013	10.92	0.78	0.50	9.67	95.00	25.28	152.39

1.2 材料与设计

供试品种为京科968和郑单958。以全生育期自然光照为对照（CK），采用透光率为50%的遮光网分别于13叶全展（T1）、吐丝期（T2）和吐丝后15d（T3）遮光，每时期连续遮光7d，随机区组设计，3次重复，6行区，行长7.5m，每小区面积27m^2，密度52 500株/hm^2。整地时施用烘干鸡粪6 000kg/hm^{2}作底肥，拔节期追施尿素270kg/hm^2，其他管理同当地大田生产。

1.3 测定项目及方法

1.3.1 粒重测定

在玉米吐丝期，选取长势、穗型基本一致的植株挂牌标记。自吐丝后15d起，每隔7d在每小区标记的植株上取3个果穗，每穗取中部籽粒100粒，于105℃烘箱中杀青30min，80℃烘至恒重后称重。

1.3.2 灌浆速率模拟

以开花后天数（x）为自变量，开花后每隔7d测得的百粒重为因变量（y），用Logistic方程 $y=A/(1+Be^{(-Cx)})$ 模拟籽粒灌浆过程，得到Logistic方程参数 A、B、C（其中，A 为终极生长量，B 为初值参数，C 为生长速率参数）。达到最大灌浆速率的天数 $T_{max}=(\ln B)/C$，灌浆速率最大时的生长量 $W_{max}=A/2$，最大灌浆速率 $G_{max}=(C \times W_{max})/2$，灌浆活跃期 P（大约完成总积累量的90%）$=6/C$。

1.3.3 测产及考种内容

以每小区收获中间2行测产，选取20个平均穗考种。测定穗长、穗粗、秃尖长度、穗行数、行粒数、千粒重和籽粒含水率。

1.4 统计分析

采用CurveExpert 1.3软件拟合灌浆方程，使用DPS软件进行相关统计分析，Microsoft Excel 2003软件计算数据、绘制图表。

2 结果与分析

2.1 不同时期遮光对玉米产量及产量构成因素的影响

遮光条件下2个品种的产量、穗粒数和千粒重均低于同时期对照（表2），其中各处理间产量差异显著，穗粒数与千粒重在部分处理间差异显著。产量、穗粒数和千粒重在T1、T2、T3条件下较CK分别下降8.38%、14.67%、29.02%、4.62%、8.65%、19.55%、3.32%、6.27%和10.75%。品种间比较，京科968的产量、穗粒数和千粒重在

T1、T2、T3 条件下较 CK 分别下降 6.20%、13.97%、24.03%、2.77%、8.35%、15.76%、2.96%、5.89%和 10.02%；郑单 958 的产量、穗粒数和千粒重在 T1、T2、T3 条件下较 CK 分别下降 10.55%、15.37%、34.02%、6.47%、8.95%、23.33%、3.68%、6.65%、11.47%。说明不同时期遮光均导致玉米产量及其构成因素下降，且遮光时期越晚降幅越大，产量及构成因素的降幅为京科 968<郑单 958。

表 2　不同时期遮光条件下的玉米产量与产量构成因素
Table 2　Yield and its components of maize under different shading treatments

品种 Cultivar	遮光处理 Shading treatment	2012			2013		
		穗粒数 Grain No. per ear	千粒重 1 000-kernel weight (g)	产量 Grain yield (kg hm^{-2})	穗粒数 Grain No. per ear	千粒重 1 000-kernel weight (g)	产量 Grain yield (kg hm^{-2})
京科 968 Jingke968	CK	563.88 Aa	327.74 Aa	9 473.10 Aa	568.80 Aa	320.00 Aa	9 550.35 Aa
	T1	546.90 Aab	320.35 Aab	8 906.01 Ab	554.40 ABa	308.30 B	8 937.00 Bb
	T2	511.62 ABbc	312.76 ABb	8 167.09 Bc	526.50 Bb	296.94 Cc	8 198.10 Cc
	T3	466.24 Bc	296.95 Bc	7 097.58 Cd	488.00 Cc	285.92 Dd	7 355.70 Dd
郑单 958 Zhengdan958	CK	518.07 A a	318.87 A a	8 453.21 A a	520.50 A a	315.79 A a	8 564.64 A a
	T1	464.48 B b	315.97 A ab	7 430.93 ABb	507.00 B b	295.40 B b	7 792.43 B b
	T2	447.72 B b	310.41 A b	7 033.69 B c	498.00 C c	282.17 C c	7 370.40 C c
	T3	361.39 C c	285.66 B c	5 055.81 C d	435.00 D d	276.22 C c	6 179.34 D d

注：大、小写字母分别表示在 1%和 5%水平上差异显著。CK. 自然光照；T1. 13 叶期遮光；T2. 吐丝期遮光；T3. 吐丝后 15d 遮光

Note: Values followed by different letters are significantly different at the 1% (capital) and the 5% (lower-case) probability levels, respectively. CK. natural light condition without shading; T1. shading treatment at the 13th leaf fully expanded; T2. shading treatment at the silking; T3. shading treatment at the 15th day after silking

2.2　不同时期遮光对玉米籽粒灌浆特性的影响

2.2.1　籽粒灌浆动态

不同时期遮光处理条件下，玉米籽粒增重仍符合慢—快—慢的"S"形生长曲线（图 1），随着遮光时期推迟，粒重增加趋势变缓，籽粒灌浆速率呈单峰曲线变化（图 2）。不同时期遮光处理下籽粒灌浆速率达到最大的时间基本一致，且最终粒重、平均灌浆速率在 T1、T2、T3 条件下较 CK 分别降低 3.51%、6.54%、11.05%；3.15%、6.40% 和 10.96%。在 CK、T1、T2、T3 条件下，京科 968 最终粒重、平均灌浆速率较郑单 958 分别高 2.41%、2.98%、1.89%、3.55%；2.64%、2.77%、1.64%和 3.37%。

2.2.2　籽粒灌浆特征参数

利用 Logistic 方程拟合不同时期遮光条件下的玉米籽粒灌浆过程，计算得出拟合方程参数值和决定系数（表 3），其决定系数均在 0.99 以上，说明 Logistic 方程可以较好地描述玉米籽粒灌浆过程。由表 3 可知，处理间比较，不同时期遮光均导致籽粒灌浆速率最大

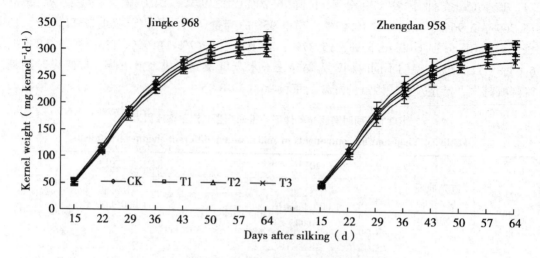

图 1 不同处理条件下的玉米籽粒增重动态曲线

Figure 1 Dynamics curve of grain weight-increasing of maize under different treatments at different stages

注：CK. 自然光照；T1. 13 叶期遮光；T2. 吐丝期遮光；T3. 吐丝后 15d 遮光。

Note: CK. natural light condition without shading; T1. shading treatment at the 13th leaf fully expanded; T2. shading treatment at the silking; T3. shading treatment at the 15th day after silking.

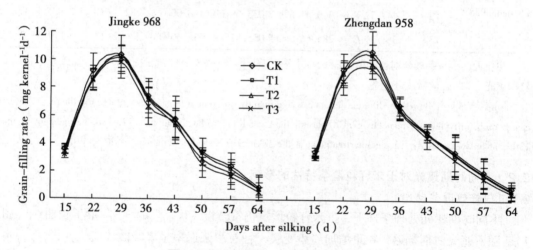

图 2 不同处理条件下的玉米灌浆速率

Figure 2 The grain-filling rate of maize under different treatments

注：缩写同图 1

Note: Abbreviations are the same as given in Figure 1.

时的生长量（W_{max}）、最大灌浆速率（G_{max}）和活跃灌浆天数（P）不同程度降低，在 T1、T2、T3 条件下较 CK 分别降低 3.42%、6.71%、9.40%；2.10%、3.54%、5.12%；1.34%、3.29%、4.51%；达到最大灌浆速率的天数（T_{max}）无明显变化。与郑单 958 比较，京科 968 达到最大灌浆速率的天数（T_{max}）推迟 0.5d 左右，在 CK、T1、T2、T3 条

件下，活跃灌浆天数（P）分别延长 1.61d、2.02d、1.00d 和 0.38d；籽粒灌浆速率最大时的生长量（W_{max}）分别高 2.64%、3.14%、2.05% 和 0.19%；最大灌浆速率（G_{max}）分别低 0.85%、1.29%、0.19% 和 0.67%。

对 Logistic 拟合方程求导得到 2 个拐点，将籽粒灌浆过程分为籽粒灌浆缓增期，高峰持续期和后期（表3和图2）3个阶段。不同处理条件下，玉米籽粒灌浆高峰起始期差别不大，均为授粉后 16.50d 左右；灌浆高峰结束期随遮光时期推迟而提前，CK、T1、T2 和 T3 分别在授粉后 37.16d、37.13d、36.55d 和 36.37d。品种间比较，在 CK、T1、T2、T3 条件下，京科 968 和郑单 958 灌浆高峰起始时间分别在授粉后 16.70d、16.82d、16.77d、16.71d；16.49d、16.84d、16.54d、16.74d，结束期则分别在授粉后 37.62d、37.56d、36.88d、36.44d；36.70d、36.69d、36.21d、36.30d；灌浆高峰持续期分别为 20.93d、20.74d、20.11d、19.73d；20.22d、19.85d、19.68d 和 19.56d，京科 968 籽粒灌浆高峰期持续期较郑单 958 分别延长 0.71d、0.89d、0.44d 和 0.17d。

表3和图2表明，在不同时期遮光条件下，同一品种灌浆高峰起始时间无明显变化，但结束时间提前，导致灌浆高峰持续期缩短，且灌浆强度降低，最大灌浆速率与平均灌浆速率均下降，籽粒终极生长量降低。相同处理下，尽管郑单 958 最大灌浆速率（G_{max}）较高，但因籽粒灌浆高峰持续期与活跃灌浆天数（P）相对较短，导致粒重和产量低于京科 968。

表3 不同玉米品种各处理条件下的籽粒灌浆特征参数
Table 3 Characteristic parameters of maize at grain-filling stage under different treatments

品种 Cultivar	遮光处理 Shading treatment	决定系数 Coefficient of determination	方程参数 Parameter of equation			籽粒灌浆参数 Parameter of grain filling			
			A	B	C	T_{max} (d)	W_{max} (g 100-kernel^{-1})	G_{max} (g 100-kernel^{-1} d^{-1})	P (d)
京科 968 Jingke 968	CK	0.9966	32.72	30.54	0.1259	27.16	16.36	1.0296	47.67
	T1	0.9968	31.68	31.62	0.1270	27.19	15.84	1.0057	47.24
	T2	0.9972	30.44	33.54	0.1309	26.83	15.22	0.9964	45.82
	T3	0.9978	29.29	34.75	0.1335	26.57	14.65	0.9778	44.93
郑单 958 Zhengdan 958	CK	0.9982	31.88	31.98	0.1303	26.60	15.94	1.0383	46.05
	T1	0.9984	30.71	34.88	0.1327	26.77	15.36	1.0187	45.22
	T2	0.9946	29.83	34.16	0.1339	26.38	14.91	0.9983	44.82
	T3	0.9986	29.23	35.57	0.1347	26.52	14.62	0.9843	44.55

注：A. 终极生长量；B. 初值参数；C. 生长速率参数；T_{max}. 达到最大灌浆速率的天数；W_{max}. 籽粒灌浆速率最大时的生长量；G_{max}. 最大灌浆速率；P. 活跃灌浆天数；CK. 自然光照；T1. 13 叶期遮光；T2. 吐丝期遮光；T3. 吐丝后 15d 遮光

Note：A. the final grain weight；B. initial parameter；C. growth rate parameter；T_{max}. the time reaching the maximum grain-filling rate；Wmax. weight of a kernel at the time of maximum grain-filling rate；G_{max}. maximum grain-filling rate；P. active grain-filling period；CK. natural light condition without shading；T1. shading treatment at the 13th leaf fully expanded；T2. shading treatment at the silking；T3. shading treatment at the 15th day after silking

2.3 籽粒灌浆参数、穗粒数、千粒重与产量的通径分析

相关分析表明（表4），灌浆速率最大时的生长量（x_3）、穗粒数（x_5）、千粒重（x_6）与产量均呈显著或极显著正相关。通径分析表明，不同品种灌浆参数与产量构成参数对产量的直接通径系数除活跃灌浆天数（x_2）和最大灌浆速率（x_4）为负向效应外，均为正向效应，各参数对产量的直接作用因品种而异。京科968表现为灌浆速率最大时的生长量（x_3）>活跃灌浆天数（x_2）>穗粒数（x_5）>千粒重（x_6）>最大灌浆速率（x_4）>达到最大灌浆速率的天数（x_1），且灌浆速率最大时的生长量（x_3）通过穗粒数（x_5）、千粒重（x_6）对产量的间接作用正相关程度较高，其他各参数通过另外参数对产量的间接作用总和也均为正相关；郑单958各参数对产量的直接作用表现为，穗粒数（x_5）>活跃灌浆天数（x_2）>灌浆速率最大时的生长量（x_3）>最大灌浆速率（x_4）>千粒重（x_6）>达到最大灌浆速率的天数（x_1），穗粒数（x_5）通过活跃灌浆天数（x_2）、灌浆速率最大时的生长量（x_3）、千粒重（x_6）对产量的间接作用为正相关，除达到最大灌浆速率的天数（x_1）外，其他各参数通过另外参数对产量的间接作用总和也均为正相关。分析结果表明，活跃灌浆天数（x_2）、灌浆速率最大时的生长量（x_3）、穗粒数（x_5）对产量直接贡献较大，活跃灌浆天数（x_2）、最大灌浆速率（x_4）、千粒重（x_6）的间接作用较大，且间接通过灌浆速率最大时的生长量（x_3）、穗粒数（x_5）的正效应较大。进一步分析发现，籽粒灌浆参数与千粒重、穗粒数的相关性存在品种间差异（表5），京科968千粒重与籽粒灌浆速率最大时的生长量（W_{max}）呈极显著相关，与活跃灌浆天数（P）呈显著相关，穗粒数与籽粒灌浆速率最大时的生长量（W_{max}）呈极显著相关，与最大灌浆速率（G_{max}）呈显著相关；郑单958千粒重与籽粒灌浆速率最大时的生长量（W_{max}）呈极显著相关，穗粒数与最大灌浆速率（G_{max}）呈显著相关。

表4 籽粒灌浆参数、穗粒数、千粒重与产量的通径分析
Table 4 Path and regression analysis of effect of grain-filling parameters, grains per ear, 1 000-kernel weight with grain yield

作用因子 Effect factors	相关系数 Correlation coefficient	直接作用 Direct effec	间接作用 Indirect effect					
			总和 Total	→x1	→x2	→x3	→x4	→x5	→x6

作用因子	相关系数	直接作用	总和 Total	→x1	→x2	→x3	→x4	→x5	→x6
京科968 Jingke 968									
x1	0.180 4	0.007 1	0.173 3		-0.368 2	0.284 5	0.059 8	0.023 6	0.173 6
x2	0.341 2	-0.505 9	0.847 0	0.005 2		0.409 4	0.075 7	0.109 1	0.247 7
x3	0.873 6**	0.546 2	0.327 5	0.003 7	-0.379 2		-0.009 9	0.386 2	0.326 7
x4	0.541 4	-0.126 0	0.667 4	0.003 4	0.303 8	0.042 7		0.303 6	0.020 7
x5	0.986 7**	0.481 9	0.504 8	0.000 4	-0.114 6	0.437 6	-0.079 4		0.260 8
x6	0.867 1**	0.330 8	0.536 3	0.003 7	-0.378 7	0.539 3	-0.007 9	0.379 9	

(续表)

作用因子 Effect factors	相关系数 Correlation coefficient	直接作用 Direct effec	间接作用 Indirect effect						
			总和 Total	→x1	→x2	→x3	→x4	→x5	→x6
郑单958 Zhengdan958									
x1	-0.268 4	0.043 6	-0.311 9		-0.641 5	0.236 0	0.371 0	-0.330 3	0.052 8
x2	-0.138 6	-0.689 2	0.550 6	0.040 6		0.318 4	0.368 8	-0.251 8	0.074 7
x3	0.660 0*	0.519 2	0.140 8	0.019 8	-0.422 6		0.083 0	0.326 9	0.133 6
x4	0.534 4	-0.411 2	0.945 6	-0.039 3	0.618 2	-0.104 8		0.489 9	-0.018 3
x5	0.965 5**	0.725 9	0.239 6	-0.019 8	0.239 1	0.233 8	-0.277 5		0.063 9
x6	0.685 0*	0.134 6	0.550 4	0.017 1	-0.382 7	0.515 2	0.056 0	0.344 8	

注：x1. 达到最大灌浆速率的天数；x2. 活跃灌浆天数；x3. 灌浆速率最大时的生长量；x4. 最大灌浆速率；x5. 穗粒数；x6. 千粒重。*，** 分别表示在0.05和0.01水平上显著

Note: x1. the time reaching to maximum grain filling rate (T_{max}); x2. active grain filling period (P); x3. weight of a kernel at the time of maximum grain-filling rate (W_{max}); x4. maximum grain filling rate (G_{max}); x5. grain number per ear; x6. 1 000-kernel weight. *, ** mean significance at 0.05 and 0.01 probability levels, respectively

表5 不同玉米品种灌浆特征参数与千粒重和穗粒数的相关系数

Table 5 Correlations coefficient of grain-filling parameters with 1 000-grain weight and grains per ear of maize

品种 Cultivar	籽粒灌浆参数 Parameter of grain filling	千粒重 1 000-kernel weight	穗粒数 Grain number per ear
京科968 Jingke 968	T_{max}	0.524 8	0.048 9
	W_{max}	0.987 4**	0.801 3**
	G_{max}	0.056 0	0.621 9*
	P	0.748 6*	0.226 5
郑单958 Zhengdan 958	T_{max}	0.390 8	-0.455 0
	W_{max}	0.992 2**	0.450 4
	G_{max}	-0.146 9	0.66 62*
	P	0.555 3	-0.346 9

注：*，** 分别表示在5%和1%水平上显著

Note: *, ** mean significant at the 0.05 and 0.01 probability levels, respectively. Abbreviations are the same as given in Table 3

3 讨论

3.1 不同生育时期遮光对玉米产量及其构成因素的影响

不同时期遮光导致玉米籽粒产量不同程度降低，但遮光对产量构成因素影响的研究结

果不尽一致。有研究认为不同时期遮光均降低穗粒数与千粒重，也有研究认为营养生长阶段遮光导致穗粒数和产量下降，对千粒重无显著影响；开花期遮光导致穗粒数下降，千粒重上升，产量显著下降；灌浆期遮光导致千粒重和穗粒数均下降，产量显著下降。关于不同时期遮光对产量降幅的影响，贾士芳等研究发现，授粉前0~14d遮光处理的产量降幅最大；张吉旺等发现不同时期遮光对玉米产量降幅的影响为花粒期>穗期>苗期，其中花粒期遮光主要影响玉米籽粒发育，导致败育粒增加，籽粒体积和干重降低。本研究表明，不同时期遮光均导致产量及其构成因素下降，其中穗粒数降幅高于千粒重，产量下降主要是穗粒数与千粒重共同作用的结果，这与张吉旺等研究的结果不同；遮光时期对产量及构成因素下降幅度的影响均为吐丝后15d>吐丝期>13叶全展，这与张吉旺等的研究结果类似。品种间比较，在相同遮光条件下，产量及构成因素均表现为京科968>郑单958，且产量及构成因素降幅表现为京科968<郑单958，说明玉米对遮光引起的弱光胁迫响应存在品种间差异。本研究结果表明，京科968对弱光胁迫的耐受性优于郑单958。

3.2 不同生育期遮光对玉米籽粒灌浆及灌浆参数的影响

玉米粒重与产量取决于籽粒灌浆过程，对籽粒灌浆过程进行方程拟合，推导出具有生物学意义的特征参数，能更好地解释籽粒灌浆过程。关于粒重与籽粒灌浆特征参数的关系有多种观点，刘霞等研究认为，玉米粒重与缓增持续期、快增期灌浆速率、缓增期灌浆速率和灌浆持续期呈显著正相关，受平均灌浆速率影响较小；张海艳等认为，玉米粒重形成取决于灌浆速率，而不是灌浆持续时间；杨青华等指出，粒重由平均灌浆速率和灌浆持续时间共同决定；刘明等认为，粒重与平均灌浆速率和最大灌浆速率显著正相关，与灌浆活跃期相关较小；陈晨等认为，玉米粒重与活跃灌浆期、平均灌浆速率显著相关。本试验研究表明，品种特性决定粒重与灌浆参数的关系，京科968粒重与籽粒灌浆速率最大时的生长量和活跃灌浆天数呈极显著或显著相关，郑单958粒重与籽粒灌浆速率最大时的生长量呈极显著相关，品种间粒重差异由活跃灌浆天数决定，这与李绍长等的研究结果一致。综合分析表明，不同时期遮光对籽粒达到最大灌浆速率的天数无明显影响，产量降低主要是籽粒灌浆速率最大时的生长量降低、最大灌浆速率减小、灌浆高峰持续期与活跃灌浆天数缩短，灌浆强度降低导致穗粒数减少，千粒重降低所致。

4 结论

不同时期遮光均导致玉米籽粒灌浆高峰持续期与活跃灌浆天数缩短、灌浆强度降低，最大灌浆速率、籽粒灌浆速率最大时的生长量与平均灌浆速率均下降，籽粒终极生长量减少，致使玉米穗粒数与千粒重下降，产量降低，且遮光时期越晚降幅越大。品种特性决定粒重、穗粒数与灌浆参数的关系，粒重与灌浆速率最大时的生长量呈显著正相关，品种间的粒重差异由活跃灌浆天数决定，穗粒数与最大灌浆速率呈显著正相关，品种间的穗粒数差异由灌浆速率最大时的生长量决定。在遮光条件下，京科968较郑单958适应性强。因此，选择适宜品种，通过栽培调控措施提高籽粒灌浆速率最大时的生长量与最大灌浆速率，延长活跃灌浆天数是玉米在光胁迫环境下获得高产、稳产的重要途径。

参考文献（略）

本文原载：作物学报，2014，40（9）：1 650-1 657

遮光对玉米干物质积累及产量性能的影响

陈传永　王荣焕　赵久然　徐田军　刘秀芝　王元东　成广雷　王晓光

(北京市农林科学院玉米研究中心，北京　100097)

摘　要：选用玉米品种郑单958与先玉335，在大田条件分别于7叶全展期（T1）、13叶全展期（T2）、吐丝期（T3）、吐丝后15d（T4）进行遮光处理（50%），研究不同时期遮光对玉米干物质积累与产量性能的影响。结果表明，不同时期遮光均导致玉米终极生长量（a）、干物质积累速率最大时的生长量（W_{max}）、最大干物质积累速率（G_{max}）、平均叶面积指数（MLAI）、平均净同化率（MNAR）和收获指数（HI）降低，致使干物质积累与产量不同程度降低，其中吐丝后15d遮光（T4）对干物质与产量影响最大；在产量构成因素中，遮光处理对穗粒数与千粒重影响较大，其中穗粒数减少占主导，其次是千粒重的降低。

关键词：玉米；遮光；干物质；产量性能

Effects of shading on Dry Matter Accumulation and Yield Capability of maize

Chen Chuanyong　Wang Ronghuan　Zhao Jiuran　Xu Tianjun
Liu Xiuzhi　Wang Yuandong　Cheng Guanglei　Wang Xiaoguang

(*Maize research center*, *Beijing Academy of Agriculture and Forestry Sciences*, *Beijing 100097*, *China*)

Abstract: The objective of the experiment was to explore the effect of low light stress on dry matter accumulation characteristics and yield capability of maize (*Zea mays* L.). Field experiments were conducted in Changping Beijing using maize cultivars (zhengdan958, xianyu335) with shading of 50% at different stages. Shading treatments were conducted during the 7th leaf fully open (treatment T1), and the 13th leaf fully open (treatment T2), and the silking (treatment T3), and the15[th] day after silking (treatment T4) and natural light condition without shading was used as the control. Each shading treatment would lasted 7 days. The results showed that, the maximum growth biomass (a), weight reaching maximum dry matter accumulation rate (W_{max}), the maximum dry matter accumulation rate (G_{max}), mean leaf area index (MLAI), mean net assimilation rate (MNAR) and harvest index (HI) decreased under shading conditions, the yield and dry matter accumulation

---

基金项目：国家玉米产业技术体系（nycytx-02）、北京市农林科学院青年科研基金（QN201115）、北京市农林科学院科技创新基金（CXJJ201309）

作者简介：陈传永（1978—　），男，博士，主要从事玉米高产栽培与生理生态研究。Tel：010-51503703，E-mail：chuanyongchen@126.com

赵久然为本文通讯作者

then reduced in different degree correspondingly. Generally, the treatment T4 had the most important influence on yield and dry matter accumulation than other treatments. The results suggested that Grain number per ear than 1 000-kernel weight were greatly influence by shading than other yield components.

Key words: Maize; shading; Dry matter; Yield capability

生育期遮光可以显著改变作物生长环境,遮光时期及持续时间长短、遮光程度均在不同程度上影响作物的生长发育。玉米是 C_4 作物,具有较高的光饱和点和补偿点,生长发育与产量形成需要充足的光照。近年来,在玉米生长中经常遭受持续阴雨天气,减少了玉米光照时间,对玉米生长发育与产量形成造成了较大影响。

研究表明,不同时期遮光均会降低玉米籽粒产量,即使是短期遮光也可以降低植株生产能力与籽粒产量。其中,穗期遮光显著影响玉米的穗分化,导致花丝数和雄穗分枝数显著降低。花粒期遮光则影响生殖器官的发育与干物质分配方式,且遮光后干物质积累减少,导致千粒重下降、收获指数减小、产量降低。前人在盆栽或大田条件下对不同时期遮光或长期遮光对玉米光合特性、产量形成的影响做了诸多研究,但对不同时期遮光处理下干物质积累动态及产量性能分析的研究鲜有报道。本试验以郑单958与先玉335为试验材料,通过在不同生育时期进行短期遮光处理,利用Logistic方程模拟各处理条件下的干物质积累动态、利用产量性能方程分析光合性能与产量构成参数的关系,探讨不同时期遮光对玉米干物质积累及产量性能的影响,以期为玉米高产、稳产栽培,调控栽培措施提供理论依据与技术支持。

1 材料与方法

1.1 试验地点

2012年,在北京市昌平区小汤山国家精准农业研究示范基地进行试验。试验田耕层土壤(0~20cm)含有机质1.09%、全氮0.07%、全磷0.05%、全钾1.0%、碱解氮94.36mg/kg、速效磷25.37mg/kg、速效钾145.15mg/kg。玉米生育期间积温3 237℃左右,降水量539.30mm,总日照时数976h。

1.2 材料与设计

供试品种为郑单958、先玉335。共设4个遮光处理:采用透光率为50%的遮光网搭建可拆卸式遮光棚,分别于7叶全展(6月7日、T1)、13叶全展(6月20日、T2)、吐丝期(7月4日、T3)和吐丝后15d(7月19日、T4)时开始进行遮光,每时期遮光时间为7d,以自然光照作为对照(CK)。随机区组设计,3次重复,密度5 250株/hm^2。6行区,行距0.6m,行长10m,小区面积36m^2。整地时施用烘干鸡粪6 000kg/hm^2做底肥、拔节期追施尿素270kg/hm^2,其他管理同当地大田生产。

1.3 测定内容及方法

1.3.1 干物质测定

每个重复取代表性植株3株,105℃杀青0.5h,80℃烘干至恒重后测定干物质。

1.3.2 叶面积指数的计算

$$叶面积 = 叶长 \times 叶宽 \times 0.75$$

$$LAI = 单位土地上面积上的叶片总面积/单位土地面积$$

1.3.3 干物质积累动态模拟

以出苗后天数（t）为自变量，出苗13d后每隔7d测得的单株干物质重为因变量（w），用Logistic方程 $y = A/(1 + Be^{(-CX)})$ 对干物质积累动态进行模拟，通过CurveExpert 1.3 软件进行拟合，得到Logistic方程参数A、B、C（其中，A为终极生长量，B为初值参数，C为生长速率参数）。计算下列干物质积累特征参数：干物质积累高峰（干物质积累量达到50%时所用的时间）$T_{max} = (\ln B)/C$，干物质积累速率最大时的生长量 $W_{max} = A/2$，最大干物质积累速率 $G_{max} = (C \times W_{max})/2$，达到活跃生长期时的天数 P（大约完成总积累量的90%）$= 2 \times (3/C)$。

先玉335在吐丝后15d（T4）遮光处理结束后，陆续出现青枯，干物质测定数据取至出苗后104d。

1.3.4 产量性能分析

利用产量性能方程"产量（Yield）= 穗数（EN）×穗粒数（GN）×千粒重（GW）=（平均叶面积指数）MLAI×（天数）D×平均净同化率（MNAR）×收获指数（HI）"，分析产量形成过程中光合性能参数与产量构成参数的变化。其中平均叶面积指数（MLAI）根据张宾的方法计算。

1.3.5 考种内容

选取20穗平均穗进行考种。测定穗长、穗粗、秃尖长度、穗行数、行粒数和千粒重。

1.4 统计分析

所有数据均采用DPS3.01和Microsoft Excel 2003进行统计分析。

2 结果与分析

2.1 遮光处理对玉米干物质积累特征分析

2.1.1 遮光处理对玉米干物质积累的影响

不同时期遮光均导致玉米吐丝前、吐丝后及总干物质积累量不同程度降低，且生育后期遮光较前期遮光对干物质积累影响更显著（表1）。处理1、处理2主要影响玉米吐丝前干物质积累，郑单958较对照差异显著，分别下降5.71%、8.49%；先玉335较对照在处理1条件下差异不显著，在处理2条件下差异显著，分别下降3.90%、8.45%；而处理3、处理4则显著降低花后干物质积累量与总干物质积累量，且较对照差异显著，其中花后干物质积累量表现为，郑单958较对照下降6.21%、13.87%；先玉335较对照下降10.78%、35.22%；总干物质积累量表现为，郑单958较对照下降5.04%、10.72%；先玉335较对照下降7.84%、23.49%。不同品种间比较，则表现为先玉335较郑单958受遮光影响较大。

表 1 遮光处理对玉米干物质积累量的影响

Table 1 The effect of shading on dry matter accumulation (DMA) of maize

g/株

品种 Cultivar	遮光处理 Shading treatment	干物质积累量 DMA			吐丝后干物质/总干物质（%）Post silking proportion（%）
		吐丝前 DMA before silking	吐丝后 DMA after silking	总干物质 Total DMA	
郑单 958 Zhengdan958	CK	102.03 a	225.52 a	327.55 a	68.85 ab
	T1	96.20 b	225.50 a	321.70 b	70.09 a
	T2	93.37 b	223.81 a	317.18 b	70.56 a
	T3	99.52 ab	211.52 b	311.03 c	68.01 bc
	T4	98.20 ab	194.23 c	292.43 d	66.42 c
先玉 335 Xianyu335	CK	114.18 a	218.24 a	332.42 a	65.63 a
	T1	109.73 a	215.74 a	325.47 a	66.29 a
	T2	104.53 b	208.36 a	312.90 b	66.59 a
	T3	111.63 a	194.72 b	306.35 b	63.57 a
	T4	112.97 a	141.38 c	254.34 c	55.58 b

2.1.2 遮光处理下玉米干物质积累动态模拟

用 Logistic 方程对其进行拟合，得出不同遮光处理下不同品种的干物质积累参数（表2），决定系数均大于 0.99，拟合效果良好。从拟合情况分析，不同品种，处理 1、处理 2、处理 3、处理 4 的终极生长量、干物质积累速率最大时的生长量、最大干物质积累速率均低于对照，其中不同品种的终极生长量与干物质积累速率最大时的生长量变化一致，郑单 958 与先玉 335 较对照均分别低 2.57%、2.91%、6.11%、11.72%；1.82%、5.61%、6.98%、24.04%；最大干物质积累速率则表现为郑单 958 与先玉 335 较对照分别低 2.82%、4.51%、9.21%、14.85%；3.26%、7.25%、8.88%、12.86%。不同品种间比较干物质积累参数，在相同处理下，干物质积累高峰期、干物质积累速率最大时的生长量、达到活跃生长期时的天数表现为郑单958>先玉335，最大干物质积累速率表现为先玉335>郑单958。

比较不同遮光处理下的干物质积累高峰期，两品种处理 1、处理 2 处理条件下的干物质积累高峰期较对照推迟 1d 左右，其中郑单 958 与先玉 335 在自然光照条件下达到干物质积累高峰期的时间分别为苗后 70d、67d 左右（先玉 335 由于处理 4 处理后期发生青枯，故处理 4 处理中各参数均为无效数据）。进一步分析，不同玉米品种对照处理干物质积累高峰起始时间、高峰终止时间均表现为处理 1、处理 2 处理较对照多一天，其中对照达到干物质积累高峰起始时间、高峰终止时间分别为，郑单 958 为 48d、92d，先玉 335 为 46d、88d，结合生育期调查，干物质积累高峰起始时间出现在大喇叭口期，高峰期出现在吐丝后 15d 左右，终止时间出现在吐丝后 37d 左右，其中干物质积累高峰期，玉米发育即进入籽粒线性增长时期，恰是粒重形成关键期，其粒重变化还需结合籽粒灌浆动态分析。干物质积累高峰期结束后 8 d 左右到达即结束玉米生长活跃期，其后干物质积累进入缓增期。

表2 不同玉米品种和遮光处理间干物质积累特征参数
Table 2 Characteristic parameters of maize dry matter accumulation under different shading treatments

品种 Cultivar	遮光处理 Shading treatment	决定系数 coefficient of determination R^2	方程参数 Parameter of equation			干物质积累参数 Parameter of dry matter accumulation			
			A	B	C	T_{max} (d)	W_{max} (g plant^{-1})	G_{max} (g plant^{-1} d^{-1})	P (d)
郑单958 Zhengdan 958	CK	0.995 8	353.87	66.53	0.060 1	70	176.94	5.32	99.83
	T1	0.996 6	344.77	72.19	0.060 0	71	172.39	5.17	100.00
	T2	0.995 8	343.56	67.22	0.059 2	71	171.78	5.08	101.35
	T3	0.995 8	332.26	57.31	0.058 2	70	166.13	4.83	103.09
	T4	0.995 2	312.41	55.21	0.058 0	69	156.21	4.53	103.45
先玉335 Xianyu 335	CK	0.996 0	349.07	69.43	0.063 2	67	174.54	5.52	94.94
	T1	0.996 0	342.72	67.53	0.062 3	68	171.36	5.34	96.31
	T2	0.996 6	329.50	73.07	0.062 1	69	164.75	5.12	96.62
	T3	0.995 6	324.71	63.56	0.062 0	67	162.36	5.03	96.77
	T4	0.993 4	265.17	74.02	0.072 6	59	132.59	4.81	82.64

注：R^2为决定系数；A. 终极生长量；B. 初值参数；C. 生长速率参数；T_{max}. 干物质积累高峰期；W_{max}. 干物质积累速率最大时的生长量；G_{max}. 最大干物质积累速率；P. 达到活跃生长期时的天数

Note: R^2. the coefficient of determination; A. the maximum growth biomass; B. the initial parameter; C. the growth rate parameter; T_{max}. the time reaching maxium dry matter accumulation rate; W_{max}. weight reaching maxium dry matter accumulation rate; G_{max}. maximum dry matter accumulation rate; P. the days of development to the active growing stage.

2.2 遮光处理对玉米产量性能的影响

2.2.1 不同遮光处理下玉米产量性能方程参数分析

遮光对玉米产量的影响因遮光时期不同而异（表3），两个品种产量均表现为对照与处理1，处理2与处理3差异不显著。对照、处理1与处理2、处理3及处理4之间差异显著，产量高低顺序均表现为对照>处理1>处理2>处理3>处理4，其中处理1、处理2、处理3、处理4条件下郑单958与先玉335的产量与对照相比，降幅分别为1.52%、14.05%、18.07%、41.52%、1.54%、25.93%、32.49%、50.35%。

从产量性能参数分析，两品种的平均叶面积指数均表现为对照与处理4差异显著，与其他处理差异不显著；单位面积穗数表现为对照与处理1、处理2间差异显著，与处理3、处理4差异不显著；其他参数显著性差异因品种与遮光时期不同而异，郑单958的平均净同化率、收获指数、穗粒数与均表现为对照、处理1与其他处理间呈差异显著；先玉335的平均净同化率、千粒重均表现为对照、处理1与其他处理间差异显著，且不同遮光处理间差异显著，收获指数表现为对照与处理4差异显著，与其他处理差异不显著，穗粒数表现为对照、处理1与其他处理之间差异显著。综合分析,除生育天数外,不同时期遮光均导致玉米产量性能参数不同程度降低,且遮光时期越晚,平均叶面积指数、平均净同化率、收获指数、穗粒数、千粒重等参数降低程度越显著,郑单958与先玉335的平均叶面积指数、平均净同化率、收获指数、单位面积穗数、穗粒数、千粒重与对照相比,最大降幅分别达到7.06%、20.80%、18.65%、4.17%、31.63%、11.76%;5.58%、32.26%、15.70%、2.81%、24.70%、35.62%,说明遮光影响玉米雌穗分化与籽粒建成,导致穗粒数减少与千粒重降低进而影响产量,且遮光时期越晚,对产量影响越大。

表3 不同遮光处理下玉米产量性能方程参数
Table 3 The parameters of yield capability equation of maize with different shading treatments

品种 Cultivar	遮光处理 Shading treatment	光合性能参数 Photosynthetic parameter				产量构成参数 Yield component parameter			产量 Yield (kg hm^{-2})
		MLAI	D (Day)	MNAR (m^{2}d·m^{-2})	HI	EN (Ear No.·m^{-2})	GN (Grain No. per ear)	GW (1 000-kernel weight)	
郑单958 Zhengdan 958	CK	2.55 a	124	3.89 a	0.50 a	5.51 a	528.33 a	323.86 a	8 645.70 a
	T1	2.53 a	124	3.85 a	0.49 a	5.32 bc	522.32 a	319.55 a	8 514.30 a
	T2	2.47 ab	124	3.75 b	0.45 b	5.28 c	464.32 b	316.60 a	7 431.00 b
	T3	2.45 ab	124	3.74 b	0.44 b	5.38 ab	447.72 b	311.74 b	7 083.75 b
	T4	2.37 b	124	3.08 c	0.41 c	5.38 ab	361.23 c	285.78 b	5 055.75 c
先玉335 Xianyu 335	CK	2.51 a	123	4.24 a	0.47 a	5.33 a	537.93 a	318.67 a	8 520.45 a
	T1	2.49 a	123	4.23 a	0.45 a	5.21 b	526.46 a	316.38 a	8 389.35 a
	T2	2.46 ab	123	3.40 b	0.43 ab	5.18 b	415.61 b	306.97 b	6 311.40 b
	T3	2.42 ab	123	3.28 c	0.42 b	5.28 ab	411.38 b	282.12 c	5 752.35 b
	T4	2.37 b	110	2.87 d	0.39 b	5.30 a	405.09 b	205.16 d	4 230.75 c

注:MLAI. 平叶面积指数;D. 出苗至成熟期天数;MNAR. 平均净同化率;HI. 收获指数;EN. 单位面积穗数;GN. 穗粒数;GW. 千粒重

Note:MLAI. mean leaf area index; D. growth period; MNAR. mean net assimilation rate; HI. harvest index; EN. ear No. m^{-2}; GN. grains per ear; GW. 1 000-kernels weight

2.2.2 遮光处理下不同玉米品种产量及产量性能参数的通径及相关分析

对遮光条件下玉米产量性能参数与产量进行通径与相关分析发现(表4),各参数对产量的影响因品种而异。其中郑单958表现为:穗粒数(x_5)对产量直接作用的正效应最大,虽然间接效应总和较小,但也表现为正效应,以致直接相关系数最高,决定系数高达0.909 1。平均净同化率(x_2)对产量的直接作用较小,仅为0.060 2,由于其间接作用的正效应总和较大,达到0.884 5,正效应累加以致决定系数为0.892 3。平均叶面积指数(x_1)和收获指数(x_3)对产量的直接作用均有微弱负效应,但其间接作用的正效应较大,分别达到0.704 7与0.977 0,正负效应抵消后,直接相关系数仍显著,决定系数为分别为

0.398 2、0.660 4。千粒重（x_6）对产量的直接作用和间接作用总和均表现为较小的正效应，最终决定系数为 0.541 8。先玉 335 表现为：平均叶面积指数（x_1）、平均净同化率（x_2）、穗粒数（x_5）、千粒重（x_6）对产量的直接和间接作用总和均表现为正效应，且间接作用总和对产量的正效应高于直接作用对产量的正效应，直接相关系数均显著，决定系数分别为 0.467 9、0.959 1、0.767 7、0.754 3。收获指数（x_3）对产量的直接作用有负效应，但间接作用总和的正效应较大，正负效应抵消后，直接相关系数仍显著，决定系数为 0.393 6。以上结果表明，对于郑单 958，穗粒数对产量的直接贡献最大，千粒重、平均叶面积指数、平均净同化率、收获指数的间接作用较大；对于先玉 335，穗粒数与千粒重对产量的直接贡献最大，平均叶面积指数、平均净同化率、收获指数的间接作用较大。综上分析，在寡照条件下，应选择穗粒数较多，保绿性好的玉米品种，以提高平均叶面积指数与平均净同化率，以此增加千粒重与收获指数，达到稳产的目的。

表 4 产量性能参数素（x）对其产量（y）的通径与回归分析

Table 4 Path and regression analysis of the yield effect of parameters of yield capability (x) on grain yield (y) of maize

品种 Cultivar	性状 Trait	相关系数 Correlation coefficient	直接作用 Direct effect	总和 Total	$x_1 \to y$	$x_2 \to y$	$x_3 \to y$	$x_4 \to y$	$x_5 \to y$	$x_6 \to y$
郑单 958 Zhengdan 958	x_1	0.632 0**	-0.072 8	0.704 7**		0.041 3	-0.110 1	-0.002 4	0.582 9	0.193 0
	x_2	0.944 6**	0.060 2	0.884 5**	-0.050 0		-0.129 7	-0.017 0	0.815 4	0.265 6
	x_3	0.812 7**	-0.164 4	0.977 0**	-0.048 8	0.047 5		0.013 9	0.714 1	0.250 3
	x_4	-0.130 3	0.085 1	-0.215 4	0.002 0	-0.012 0	-0.026 8		-0.100 7	-0.077 9
	x_5	0.953 5**	0.883 9**	0.069 7	-0.048 0	0.055 5	-0.132 5	-0.009 7		0.204 6
	x_6	0.736 1**	0.347 8	0.388 3	-0.040 4	0.046 0	-0.118 3	-0.019 1	0.520 0	
先玉 335 Xianyu 335	x_1	0.684 0**	0.118 9	0.565 2*		0.230 5	-0.115 7	-0.004 7	0.181 3	0.273 7
	x_2	0.977 3**	0.333 4	0.643 9**	0.082 2		-0.141 5	-0.001 1	0.368 8	0.335 5
	x_3	0.632 1*	-0.199 6	0.831 7**	0.068 9	0.236 4		-0.001 8	0.263 5	0.265 1
	x_4	-0.114 1	0.016 4	-0.130 5	-0.033 6	-0.022 3	0.022 4		0.046 4	-0.143 4
	x_5	0.876 2**	0.413 7*	0.462 5*	0.052 1	0.297 3	-0.127 0	0.001 8		0.238 3
	x_6	0.868 5**	0.405 1*	0.463 4*	0.080 3	0.276 1	-0.130 6	-0.005 8	0.243 4	

注：x_1. 平均叶面积指数；x_2. 平均净同化率；x_3. 收获指数；x_4. 穗数；x_5. 穗粒数；x_6. 千粒重
Note：x_1. mean leaf area index (MLAI)；x_2. mean net assimilation rate (MNAR)；x_3. harvest index (HI)；x_4. ear No. m^{-2} (EN)；x_5. grains per ear (GN)；x_6. 1000-kernels weight (GW)

3 讨论与结论

3.1 遮光对干物质积累的影响

遮光显著降低叶面积指数和净光合速率，减少同化物积累量，导致干物质积累与籽粒产量不同程度降低，收获指数减小。本研究表明，不同时期遮光均影响玉米干物质积累，随着遮光时期推迟，物质积累量降低、收获指数下降，产量影响降低，且遮光时期越晚对产量影响越大。Logistic 方程模拟分析表明，不同时期遮光对玉米单株生长潜力即终极生长量、干物质积累速率最大时的生长量、最大干物质积累速率均有较大影响，且遮光时期越晚影响越大；不同遮光处理均使达到活跃生长期时的天数推迟 1~4d；最大干物质速率出现时间仅受处理 1、处理 2 处理影响，延迟 1d 左右。

在处理 1、处理 2、处理 3 中，先玉 335 与郑单 958 干物质积累特征参数对不同遮光处理的反应基本一致，在处理 4 中，先玉 335 在遮光后相继出现青枯，其原因是否与其遗传背景有关，还需进一步研究。

3.2 遮光对产量性能的影响

关于不同时期遮光对产量构成因素的影响，研究结果不尽一致。有研究认为，不同时期遮光均降低穗粒数与千粒重，其中穗期遮光穗粒数降低占主导，花粒期遮光千粒重降低占主导。也有研究认为，营养生长阶段遮光导致粒数和产量下降，而对粒重无显著影响；开花期遮光，穗粒数下降，而粒重则稍有上升，但产量仍显著下降；灌浆期遮光，粒重和粒数均不同程度降低，产量也显著下降。本试验通过产量性能分析表明，遮光对光合性能参数平均叶面积指数、平均净同化率、收获指数与产量性能参数穗粒数、千粒重均有影响，且生育后期遮光处理较前期遮光影响更大，遮光主要通过影响平均净同化率与收获指数导致穗粒数、千粒重降低，进而降低产量。通径分析表明，遮光显著降低平均叶面积指数、平均净同化率、收获指数等光合性能参数所形成的光合同化物转化效率，引起单位面积穗数、千粒重下降，导致产量降低。比较郑单 958 与先玉 335 的产量性能参数发现，在遮光条件下，郑单 958 的综合性状优于先玉 335。综合分析，遮光处理导致产量降低的原因以穗粒数减少占主导，其次是千粒重的降低。因此，通过选择适宜品种，调整栽培措施有效提高、平均叶面积指数与平均净同化率以减少籽粒败育，增加穗粒数，提高粒重是光胁迫环境下减少玉米产量损失的有效途径。

参考文献（略）

本文原载：玉米科学，2014，22（2）：70-75

玉米新品种京农科 728 全程机械化生产技术

王元东[1]　张华生[1]　段民孝[1]　赵久然[1]　李云伏[1]
刘春阁[1]　陈传永[1]　成光雷[1]　崔铁英[2]

(1. 北京市农林科学院玉米研究中心　100097；2. 北京龙耘种业有限公司　101500)

　　京农科 728 是北京市农林科学院选育，并于 2012 年通过国家审定的玉米新品种（国审玉 2012003），2013 年申请了植物新品种保护，2014 年通过北京市审定（京审玉 2014007）。适宜在北京、天津两市，河北省的唐山、廊坊、保定北部、沧州中北部夏玉米种植区种植。该品种早熟、耐密、脱水快、产量高、抗性好、后期站立性好、制种产量高，非常适合机械化精量播种和收获，是京津冀目前农业机械化推广过程中迫切需要的品种类型。

1　选育背景

　　全国农业技术推广中心在华北春玉米种植区界限南缘和黄淮海夏玉米种植区界限北缘的广大区域（包括北京、天津，河北省廊坊、唐山、沧州、保定等）设置了国家京津唐夏播早熟玉米组试验，本种植区夏播玉米生育期约 95d（有效积温 2 400~2 500℃），夏玉米种植面积超过 66.67 万 hm^2。该区域是"麦玉两熟"种植区，玉米生长期短，同时又有黄淮海气候特征，气候复杂多变，温度高、湿度大、阴雨寡照、病害严重等。20 世纪 90 年代，唐抗 5 号曾经为该区的主要推广种植品种，但由于易倒伏、籽粒商品性差而被淘汰。进入 21 世纪，该区经过品种更新换代，京单 28 因其良好的丰产性和抗倒性，得到大面积种植，但是由于生育期偏长、籽粒脱水速度慢等问题满足不了全程机械化生产的要求。因此，迫切需要选育创新型玉米新品种。北京市农林科学院玉米研究中心育种团队致力于选育适于全程机械化生产的早熟、高产、稳产的优良玉米新品种，将选育目标分解为：中早熟，产量高，株型理想，节间拉开，耐密植，不空秆，高抗倒伏，抗、耐多种主要病虫害，适于单粒精量点播和机械化收获，并在此选育目标的基础上进行了大量育种研究。

2　品种遗传基础

　　京农科 728 以京 MC01 为母本、京 2416 为父本杂交选育而成，具有 50% 的国外血缘和 50% 的国内血缘，遗传基础丰富。母本京 MC01 具有国外优良血缘，以国外杂交种 X1132X 为选系基础材料，按照"高大严"选系方法（即高密度、大群体、严选择），并变换多个不同地点，同时经配合力测定等程序，在 S4 优良穗行中选择优良植株利用 DH

基金项目：国家"863"项目（2012AA101203）、粮食作物产业技术体系北京市创新团队专项经费资助（BITG7）、北京市农林科学院科技创新能力建设专项（KJCX20140107）

育种技术诱导和加倍，再经扩繁和多点鉴定及严格筛选，获得了 DH 纯系京 MC01，具有一般配合力高和综合抗性好的特点。株型紧凑，适于作母本。在西北制种产量可达到 14 835kg/hm²，种子商品率 98%，全部达到单粒精量机播的要求。父本京 2416 具有国内核心种质黄早四的血缘，来源于京 24 与 5237 杂交组成的基础材料的选系，按照"高大严"育种技术并变换不同生态区经多代白交选育而成，具有较高的一般配合力。株型紧凑，花粉量大，适于作父本。

3 产量表现

2009 年国家京津唐夏播早熟玉米组预备试验，7 个试点中全部增产，每公顷平均产量 9 904.5kg，较对照增产 15.87%，列参试品种第 3 位。2010 年国家京津唐夏播早熟玉米区域试验，8 个试点中全部增产，每公顷平均产量 10 144.5kg，较对照增产 5.55%，达极显著水平，列本区试组第 1 位；2011 年续试，10 个试点全部增产，平均产量 11 310kg，较对照增产 12.06%，达极显著水平，列本区试组第 1 位。2011 年国家京津唐玉米生产试验，11 个试点全部增产，每公顷平均产量 10 356kg，较对照增产 7.38%。2012 年北京市玉米早熟组区域试验，每公顷平均产量 11 266.2kg，比对照增产 11.54%，列于参试 9 个品种中第 1 位；2013 年续试，平均产量 10 021.35kg，比对照增产 21.55%，列于参试 15 个品种中第 2 位。2012 年北京市玉米早熟组生产试验，5 个试点全部增产，每公顷平均产量 9 623.55kg，比对照增产 17.20%。

4 适于全程机械化生产的主要农艺性状

4.1 制种产量高

种子半硬粒型，半圆，大小适宜，千粒重在 330g 左右。在西北制种每公顷平均产量 10 835kg，最高产量 12 012kg，种子商品率 98%，全部达到单粒精量机播的要求。

4.2 早熟、籽粒后期脱水速度快

在京津唐夏播区出苗至成熟 97d，比对照京单 28 早熟 1d；比郑单 958 早熟 4d；需有效积温 2 500℃左右。吐丝集中，授粉结实性好，成熟后苞叶变黄并松散，籽粒半硬粒型，脱水速度快，可以直接机械化收获籽粒。

4.3 耐密、高抗倒

一般地块最佳种植密度在 67 500株/hm²，肥水高地块最佳种植密度 75 000株/hm²。抗倒伏能力强，后期植株站立性好。2010 年、2011 年国家京津唐玉米区域试验，平均倒伏（倒折）率 0.45%。2012 年北京市玉米区域试验，田间各试点均无倒伏、倒折。2013 年、2014 年北京市玉米区试 2 年平均倒伏率 0、平均倒折率 0.1%。

4.4 适应性好、中抗玉米主要病害

由于该品种具 50% 的国外血缘和 50% 的国内血缘，遗传基础丰富，因此适应性好，综合抗性好。经中国农业科学院作物科学研究所与河北省农林科学院植物保护研究所 2010 年、2011 年接种鉴定，中抗至高抗大斑病（变幅 1~5 级），中抗至高抗小斑病（变幅 1~5 级），中抗至抗茎腐病（变幅 5.4%~15.4%），感至中抗弯孢菌叶斑病（变幅 5~7 级）。

5 全程机械化生产栽培技术要点

5.1 机械化单粒精量播种技术

(1) 精量、适时播种在京津唐等区域夏播,适宜播种期为6月中旬至下旬,土壤含水量在20%左右时即可播种。选用包衣种子,播前晾晒。精密播种理论上要求每穴下种量为1粒,但由于播种机性能、质量及种子形状的差异,无法绝对达到这一要求,为此国家规定合格的精密播种作业标准是:单粒率≥85%,空穴率<5%,碎种率≤1.5%。

(2) 播深准确。种子播在耕层土壤中的位置保证在镇压后种子至地表的距离为5cm,误差不能大于±1cm。

(3) 株距、行距均匀。种植密度应控制在 67 500~75 000株/hm^2,精密播种要求株距均匀一致,误差量应≤20%,种子落在种床后的左右偏差要小,以种床中心线为基准,左右偏差不大于4cm,出苗后一条线,以利于田间管理。

(4) 种肥深施。在种下5~8cm、种侧5cm左右,并尽可能分层施肥。分层施肥能提高化肥的利用率,上层肥在种子下方3~5cm,占种肥量的1/3;下层肥在种子下方12~15cm,占种肥量的2/3。生产中多采用二铵或氮磷钾复合肥作种肥,用量50~100kg/hm^2。

5.2 机械化收获技术

摘收玉米果穗时应尽量减少损失和损伤,落穗率不大于3%、籽粒破碎率小于1%、落粒损失不大于2%;机器带有剥皮装置时,玉米苞叶的剥净率应大于70%;留茬要整齐,留茬高度10cm以下,秸秆粉碎长度小于10cm,漏切率小于3%;机器的使用可靠性大于90%。

本文原载:中国种业,2014(10):68-69

不同播期对玉米品种京农科 728 产量及机收籽粒相关性状的影响

王元东[1] 王荣焕[1] 张华生[1] 张春原[1] 段民孝[1]
赵久然[1] 陈传永[1] 刘新香[1] 叶翠玉[2]

(1. 北京市农林科学院玉米研究中心，北京 100097；
2. 北京市种子管理站，北京 100088)

摘 要：以京农科 728 为试验材料，采用随机区组设计，设 5 个播期处理，研究不同播期对京农科 728 产量及机收籽粒相关性状的影响。结果表明：(1) 不同播期处理条件下，单粒精量播种田间出苗率均达到 95% 以上；生育期随播期推迟而缩短，但均可正常成熟；株高、穗位高随播期推迟呈增高趋势；5 月 25 日播期条件下穗长最长、穗粒数最多。(2) 过早或过晚播种，病虫害发生较重且结实性降低。(3) 产量以 5 月 25 日晚春播处理为最高。随播期推迟，收获时籽粒含水量和机收籽粒破碎率增加，进而机收难度增加。

关键词：播期；密度；京农科 728；产量

京津冀地区是我国水资源最为短缺的地区之一。为降低农业用水，京津冀区粮田作物种植结构有所调整，小麦等高耗水作物将逐步减少，北京、河北多地种植方式已由小麦-玉米 1 年 2 茬转变为 1 年种植 1 茬春玉米，通过采用高产抗旱玉米品种、调整播期、抢墒等雨播种等，实施雨养旱作，达到节水生产的目的。机收是玉米高效生产的基础和必然方向，不仅可大幅减轻农民劳动强度、降低生产成本、解放劳动力，同时还有利于实现玉米种植标准化和生产规模化，进而大幅提高我国玉米的竞争力。播期是影响玉米产量和机收效果的重要栽培因素，研究播期对玉米产量以及与机收相关农艺性状的影响具有重要的理论和现实意义。

京农科 728 是北京市农林科学院玉米研究中心选育的玉米新品种，2012 年通过国家审定（国审玉 2012003），2014 年通过北京市审定（京审玉 2014007）。该品种除高产外（2 年区试和生产试验产量均超对照 15% 以上），还具备早熟、耐密、籽粒脱水快、高抗倒伏和制种产量高的优点，适合机械化单粒精量播种和收获，是京津冀目前农业机械化推广过程中迫切需要的品种类型，2014 年被列为京津冀一体化重点推广品种。为加速京农科 728 的推广应用，充分发挥其增产潜力，本研究针对不同播期处理对其产量以及与机收相关农艺性状的影响进行了探讨，以确定该品种在京津冀地区适宜机收的最佳播期，为生产推广应用和高产高效栽培提供科学依据。

基金项目：国家 863 项目（2012AA101203）、粮食作物产业技术体系北京市创新团队专项（BITG7）、北京市农林科学院科技创新能力建设专项（KJCX20140107）

通讯作者：赵久然

1 材料与方法

1.1 试验材料

供试玉米品种为京农科728，由北京龙耘种业有限公司提供。

1.2 试验设计

2014年，在北京市通州区国际种业科技园区北京市农林科学院玉米研究中心试验基地开展试验。试验地为黏壤土，地势平坦，肥力中等偏上。设5个播期处理。分别为：5月5日、5月15日、5月25日、6月5日、6月15日。随机区组设计，3次重复，小区行长5m，行距0.6m，6行区等行距种植。单粒精量播种。基肥为氮磷钾复合肥500kg/hm^2。5叶期定苗，大喇叭口期追施尿素300kg/hm^2。管理同当地大田生产。

1.3 测定项目与方法

生育时期：准确记载播种期、出苗期、拔节期、大喇叭口期、抽雄期、吐丝期和成熟期。农艺性状：苗期调查出苗率，灌浆期测量玉米株高、穗位高。田间抗性：准确记载不同生育期玉米螟、大小斑病、茎腐病、弯孢菌叶斑病等病虫害发生等级。在成熟期统计每个小区的倒伏（折）株数、空秆株数，计算空秆率与倒伏（折）率。产量及机收籽粒性状：成熟期，每小区选取10株长势一致植株，测量果穗穗长、穗粗、穗行数、行粒数、秃尖长。成熟后，收获中间4行，利用奥地利产的Wintersteiger小区籽粒收获机收获，实收测产，测定收获籽粒含水量、籽粒破碎度与千粒重。

1.4 数据分析

应用Excel2010软件进行数据处理与分析。

2 结果与分析

2.1 不同处理条件下的农艺及产量相关性状

由表1可知，不同播期处理条件下京农科728的田间出苗率均达到95%以上，说明京农科728种子发芽势强，适宜单粒精播。其中，5月5日早春播出苗率最低，6月5日和6月15日夏播出苗率最高，这主要是早春播（5月5日）播种前后降水少，土壤墒情不足所致，而夏播（6月5日和6月15日）播种前后降水较多，土壤墒情理想。随播期推迟，京农科728的生育期缩短，夏播（6月15日）播期处理生育期仅98.1d，但仍能正常成熟，说明该品种播期灵活，具有良好的播期适应性。果穗性状以5月25日播期处理的果穗最长、穗粒数最多、秃尖最小且粒重较高，说明该播期对该品种穗部性状有利；因此在实际生产过程中，选择适宜播期是发挥京农科728产量潜力的有效途径。

表1 不同播期处理对京农科728农艺及产量相关性状的影响

播期	出苗率（%）	生育期（d）	株高（cm）	穗位（cm）	穗长（cm）	穗粗（cm）	穗粒数（粒）	千粒重（g）	秃尖长（cm）
5月5日	95.2	112.4	273.4	93.5	17.8	4.9	511.5	337.2	1.1
5月15日	96.3	108.2	275.7	96.4	18.3	4.6	513.2	336.9	0.9
5月25日	96.8	102.3	278.5	98.3	19.5	4.9	523.5	336.7	0.4

(续表)

播期	出苗率（%）	生育期（d）	株高（cm）	穗位（cm）	穗长（cm）	穗粗（cm）	穗粒数（粒）	千粒重（g）	秃尖长（cm）
6月5日	98.6	102.4	279.4	95.1	17.8	4.5	509.4	360.7	0.8
6月15日	98.8	98.1	276.5	96.6	16.5	4.5	503.7	317.4	1.3

2.2 不同处理条件下的田间抗性表现

由表2可知，京农科728在5月25日播期处理条件下的综合抗性最好，过早或过晚播种其病虫害发生较重且结实性降低。

表2 不同播期处理对京农科728田间抗性的影响

播期	抗病（虫）性（级）					空秆率（%）	倒伏（折）率（%）
	玉米螟	大斑病	小斑病	茎腐病（%）	弯孢斑叶斑病		
5月5日	3	3	1	17.1	1	0.5	0.1
5月15日	3	1	1	10.6	1	0.5	0.5
5月25日	1	1	1	6.7	1	0.1	0.3
6月5日	1	1	3	10.4	3	0.7	1.8
6月15日	3	1	3	11.7	5	0.7	5.5

2.3 不同处理对产量及机收性状的影响

表3表明，京农科728在5月25日晚春播条件下产量最高，达到11 283.9 kg/hm²，极显著高于其他播期处理；其次是6月5日处理。过早（5月5日）或过晚播种（6月15日）产量相对较低，播期在5月15日至6月5日产量较高。

对影响机械化收获的两个关键农艺性状籽粒含水量和籽粒破碎率调查发现，随着播期推迟，籽粒的含水量和籽粒破碎率增加。除6月15日处理籽粒含水量超过25%外，其余播期处理均在25%以内，适于机械化收获。说明播期是影响京农科728机收效果的重要因素。

表3 不同播期、密度处理对京农科728产量及机收性状表现

播期	籽粒含水量（%）	籽粒破碎率（%）	产量（kg/hm²）
5月5日	19.7	1.8	9 654.7 dD
5月15日	20.4	2.0	10 408.7 cdCD
5月25日	21.3	2.1	11 283.9 aA
6月5日	23.7	3.4	11 178.6 bB
6月15日	26.9	7.5	10 654.8 cC

注：不同小写字母表示在0.05水平上差异显著，大写字母表示在0.01水平上差异显著

3 结论与讨论

京农科728在玉米生产中产量高,综合性状优良,适宜单粒精量播种,尤其是早熟、耐密植、高抗倒、籽粒后期脱水速度快等性状非常适宜机械化收获。本研究发现,采用单粒精量播种技术,京农科728在京郊不同播期处理条件下的田间出苗率均达到95%以上。生育期随播期推迟而缩短。株高、穗位高随播期推迟呈增高趋势。5月25日播期处理条件下的穗长最长、穗粒数最多且粒重较高。播期过早或过晚病虫害发生较重且结实性降低,综合抗性随密度增加呈降低趋势。产量以5月25日晚春播处理条件下最高,过早或过晚播种产量相对较低;机收籽粒含水量和籽粒破碎率随播期推迟呈增加趋势。

京农科728是适宜京津冀区种植的早熟宜机收玉米新品种,产量高,综合抗性好,播期跨度时间长,在该区早春播、晚春播、早夏播和晚夏播均可,并且早播不早衰、更易机收,晚播也能正常成熟且可获得较高产量。本研究表明,该品种在晚春播5月25日播种最佳,其次是早夏播6月5日。早春播注意适当增加密度,以密度增大抵消品种早熟对产量的影响,进而利用早熟性换取机收籽粒含水量和籽粒破碎率的降低;晚夏播要注意最大限度利用该品种籽粒灌浆速度快、后期脱水速度快的特点,注意合理控制密度。

参考文献(略)

本文原载:中国种业,2015(9):43-45

播期和密度对玉米籽粒机收主要性状的影响

王荣焕[1]　徐田军[1]　赵久然[1]　王元东[1]　吕天放[1]
邢锦丰[1]　刘月娥[1]　刘秀芝[1]　刘春阁[1]　张一弛[1]　叶翠玉[2]

(1. 北京市农林科学院玉米研究中心/玉米 DNA 指纹及分子育种北京市重点实验室,北京　100097;2. 北京市种子管理站,北京　100088)

摘　要:以京农科 728 为试验材料,设置 5 个播期、5 个密度处理,研究播期和密度对玉米籽粒机收主要性状影响。结果表明,不同播期条件下,京农科 728 产量 6 723.0~7972.5kg/hm^2,6 月 10 日、6 月 15 日、6 月 20 日 3 个播期产量达 7 500kg/hm^2 以上,籽粒机收主要质量指标(籽粒含水率、杂质率)达到国家玉米机收籽粒标准。不同密度条件下,以 75 000 株/hm2 密度处理籽粒产量最高,达到国家玉米机收籽粒标准。京农科 728 在 5 个播期处理条件下均能正常成熟,生育期随播期推迟呈缩短趋势;随种植密度增加,其生育期呈延长趋势。株高和穗位高随播期推迟及种植密度增加呈升高趋势;倒伏率随播期推迟呈降低趋势,随种植密度增加呈升高趋势。播期和密度与玉米籽粒机收性状密切相关,京农科 728 在北京夏播以 6 月 10—20 日为最佳播期,75 000 株/hm^2 为最佳密度,可实现籽粒直接机械收获。

关键词:玉米;播期;密度;籽粒机收;京农科 728

Effects of Sowing Date and Planting Density on Maize Grain Mechanical Harvesting Related Traits

Wang Ronghuan[1]　Xu Tianjun[1]　Zhao Jiuran[1]
Wang Yuandong[1]　Lv Tianfang[1]　Xing Jinfeng[1]
Liu Yuee[1]　Liu Xiuzhi[1]　Liu Chunge[1]　Zhang Yichi[1]　Ye Cuiyu[2]

(1. Maize Research Center, Beijing Academy of Agriculture & Forestry Sciences, Beijing Key Laboratory of Maize DNA Fingerprinting and Molecular Breeding, Beijing 100097, China; 2. Beijing Seed Management Station, Beijing 100088, China)

Abstract: In order to illustrate the effects of sowing date and planting density on maize grain me-

---

基金项目:国家玉米产业技术体系(CARS - 02 - 18)、北京市优秀人才培养资助项目(2014000020060G177)、中央财政支农项目、北京市农林科学院科技创新能力建设专项(KJCX20170708)

作者简介:王荣焕(1980—　),女,博士,副研究员,主要从事玉米栽培生理及配套技术研究。
　　　　　E-mail:ronghuanwang@126.com, Tel:010-51503703
　　　　徐田军为共同第一作者。E-mail:xtjxtjbb@163.com

通讯作者:赵久然。E-mail:maizezhao@126.com

chanical harvesting related traits, a field experiment was conducted with five sowing dates and five planting densities by using Jingnongke728 at Tongzhou research station. The results showed that the yield of Jingnongke728 was 6 723. 0~7 972. 5kg/hm² under different sowing date conditions, and exceeded 7 500kg/ha during June 10~20th sowing treatments with the grain mechanical harvesting quality reaching the national standard; the yield under different planting densities was the highest under 75 000 plant/hm², which reaching the national maize grain mechanical harvesting standard. Jingnongke728 could mature normally under different sowing conditions, and the growth period shortened with the delay of sowing date, but delayed with the increase of planting density. The plant height and ear height of Jingnongke728 were increased with the delay of sowing date and the increasing of planting density; the lodging rate was decreased with the delay of sowing date, but increased with the increasing of planting density. Thus, sowing date and planting density were closely related with maize grain mechanical harvesting traits. Jingnongke728 can realize grain mechanical harvesting production with sowing during June 10~20th and planting with 75 000 plant/hm² in Beijing summer-cultivated conditions.

Key words: Maize; Sowing date; Density; Grain mechanical harvesting; Jingnongke728

近年来,随着我国农村劳动力进一步转移、土地流转进程加快,种粮大户、家庭农场、专业合作社等新型农业生产主体和经营方式大量涌现,规模化、机械化和轻简化成为现代农业发展的必然趋势。我国玉米生产中整地、播种环节已基本实现机械化,机械收获成为影响玉米生产全程机械化的主要技术瓶颈。目前我国玉米生产中收获环节以机收果穗为主,果穗收获后还需晾晒、脱粒等程序才能入库,需花费更多时间和精力,劳动强度大且生产成本高。玉米机械直收籽粒技术是实现玉米生产节本增效、增产优质的一项技术,有利于推广适时晚收技术并进一步增加粒重、提高品质,且籽粒收获后直接烘干、入库,可减少晾晒、脱粒等环节,节省劳动力节约成本,机械直收籽粒是我国玉米机械收获的发展趋势。

玉米机收籽粒作业质量与收获机械性能、品种特性(生育期、株高穗位、抗倒性、脱水速率、籽粒含水率等)以及播期、密度等种植条件密切相关。对于同一品种而言,机收时的籽粒含水率直接影响籽粒的破碎率、杂质率和商品品质。本文以符合京津冀调结构转方式发展高效节水农业和全程机械化特别是机收籽粒需求的玉米品种京农科728为试验材料,研究并明确播期和密度对玉米籽粒机收性状的影响,旨在为玉米籽粒机收高产高效生产提供理论指导和参考。

1 材料与方法

1.1 试验设计

2016年,在北京市通州区国际种业科技园区北京市农林科学院科研基地以品种京农科728(国审玉2012003)为试验材料,播期处理设置5个,分别为6月5日(S1)、6月10日(S2)、6月15日(S3)、6月20日(S4)、6月25日(S5);密度试验设置5个,分别为60 000株/hm²(D1)、67 500株/hm²(D2)、75 000株/hm²(D3)、82 500株/hm²(D4)和90 000株/hm²(D5)。小区面积均为667m²,3次重复。足墒播种后全生育期雨养旱作。密度试验于6月18日适墒播种。播期试验密度为75 000株/hm²。其他管理同当地大田生产。

1.2 性状调查

生育期性状：调查记载各处理的播种期、出苗期、拔节期、吐丝期、生理成熟期、收获期。

株高、穗位高：吐丝期，在各小区分别取代表性植株共10株，测定株高和穗位高，并求均值。

倒伏率：收获前，调查植株倾斜度大于45°但未折断总株数，计算倒伏植株占小区总株数的百分率。

收获及测产：于10月15日（各处理小区籽粒均达生理成熟）采用福田雷沃玉米籽粒收获机（4YZ-1J），进行全区机械直收籽粒，并立即称小区籽粒重量。

籽粒机收性状：在收割机舱内，随机取2kg籽粒样品，采用PM-8188A水分测定仪测定各小区的籽粒含水率，重复3次；然后，手工分拣完整粒、破损粒和杂质，分别称重（KW_1为完整粒重量、KW_2为破碎粒重量、NKW为杂质重量）。

1.3 数据统计与分析

采用Microsoft Excel 2007和DPS6.5软件进行数据整理和统计分析。

籽粒杂质率、破碎率及籽粒产量的计算公式：

$$籽粒杂质率（\%）=[NKW/(KW_1+KW_2+NKW)]×100$$

$$籽粒破碎率（\%）=[KW_2/(KW_1+KW_2)]×100$$

$$籽粒产量（kg/hm^2）=[小区鲜籽粒重（kg）×15]×[1-杂质率（\%）]×[1-收获时籽粒含水率（\%）]/[1-籽粒标准含水率（14\%）]$$

2 结果与分析

2.1 不同播期和密度条件下的籽粒机收性状及产量

表1结果表明，不同播期条件下京农科728籽粒机收时的含水率介于23.4%~35.8%，籽粒含水率随播期推迟呈增加趋势，其中6月5—15日播期条件下机收籽粒含水率均处于25%以下，6月20日播期处理条件下籽粒含水率增至28%，待播期推迟至6月25日时籽粒含水率高达35.8%。不同密度条件下，籽粒机收时的籽粒含水率随种植密度增加呈升高趋势，变幅为25.4%~26.5%。

不同播期条件下，京农科728机收籽粒破碎率介于6.6%~17.6%，均高于5%的国家标准（GBT-21961-2008）；籽粒杂质率介于0.0%~1.6%，均大幅低于3%的国家标准。不同密度条件下，机收籽粒破碎率变幅为7.3%~9.5%，均高于5%的国家标准（GBT-21961-2008）；籽粒杂质率以82 500株/hm^2和90 000株/hm^2密度条件下最高（0.5%、0.4%），所有密度条件下的机收籽粒杂质率均低于3%的国家标准。

从产量性状来看，不同播期条件下京农科728产量介于6 723.0~7 972.5kg/hm^2，其中6月10—20日播期条件下产量均可达到7 500kg/hm^2以上，并以6月20日播期产量为最高。不同密度条件下，以90 000株/hm^2密度产量最低（7 653.0kg/hm^2）、75 000株/hm^2密度产量最高（8 416.5kg/hm^2），该密度条件下籽粒含水率为25.7%且籽粒破碎率和杂质率相对较低。

表1 不同播期和密度条件下京农科728籽粒含水率、破碎率、杂质率和产量

Table1 Grain moisture content, broken rate, impure rate and yield of Jingnongke728 under different sowing and density conditions

处理 Treatment		籽粒含水率 Grain moisture content (%)	籽粒破碎率 Grain broken rate (%)	籽粒杂质率 Grain impure rate (%)	产量 Yield (kg/hm^2)
播期 Sowing date	S1	23.4d	9.2b	1.6a	6 723.0c
	S2	24.7c	6.6d	0.0b	7 660.5ab
	S3	24.9c	6.7d	0.1b	7 818.0ab
	S4	28.0b	8.1c	0.2b	7 972.5a
	S5	35.8a	17.6a	0.3b	7 432.5b
密度 Planting density	D1	25.4c	9.5a	0.2bc	7 761.0bc
	D2	25.6bc	8.3b	0.1c	7 924.5b
	D3	25.7abc	7.6c	0.0c	8 416.5a
	D4	26.3ab	7.3c	0.5b	7 929.0b
	D5	26.5a	7.8c	0.4b	7 653.0c

2.2 不同播期和密度条件下的生育进程

由表2可见，京农科728在6月5—25日播期处理条件下籽粒均能够正常成熟，说明该品种在北京夏播具有良好的播期适应性。随播期推迟，其生育期呈缩短趋势，其中以6月5日播期处理生育期最长（96d）、6月25日播期处理生育期最短（91d），二者相差5d。

与播期处理相比，京农科728在不同密度条件下的生育期变幅不大，总体表现为随种植密度增加生育期呈延长趋势。其中，以60 000株/hm^{2}密度条件下生育期最短（95d）、90 000株/hm^{2}密度条件下生育期最长（97d），二者相差2d。

表2 不同播期和密度条件下京农科728的生育进程

Table2 The growth and development process of Jingnongke728 under different sowing and density conditions

处理 Treatment		出苗期 （月·日） Emergence stage	拔节期 （月·日） Jointing stage	吐丝期 （月·日） Silking stage	生理成熟期 （月·日） Physiological maturity	收获期 （月·日） Harvesting date	生育期 (d) Growth period
播期 Sowing date	S1	6·11	7·5	7·30	9·15	10·15	96
	S2	6·16	7·8	8·3	9·19	10·15	95
	S3	6·21	7·11	8·5	9·24	10·15	95
	S4	6·25	7·15	8·11	9·25	10·15	92
	S5	6·29	7·20	8·15	9·28	10·15	91

（续表）

处理 Treatment		出苗期（月·日）Emergence stage	拔节期（月·日）Jointing stage	吐丝期（月·日）Silking stage	生理成熟期（月·日）Physiological maturity	收获期（月·日）Harvesting date	生育期（d）Growth period
密度 Planting density	D1	6·24	7·12	8·6	9·27	10·15	95
	D2	6·24	7·12	8·6	9·27	10·15	95
	D3	6·24	7·13	8·7	9·28	10·15	96
	D4	6·24	7·13	8·7	9·29	10·15	97
	D5	6·24	7·13	8·7	9·29	10·15	97

2.3 不同播期和密度条件下的农艺性状

从农艺性状来看，株高和穗位高随播期推迟呈升高趋势，其中6月25日播期处理条件下的株高和穗位高分别比6月5日播期处理高42.7cm和22.2cm，增幅为17.0%和25.6%（表3）。不同密度条件下，株高和穗位高随种植密度增大呈升高趋势。其中，90 000株/hm^2密度条件下的株高和穗位高均较60 000株/hm^2密度处理高18.2cm，增幅分别为6.7%和20.0%。

7月19—20日，因遭遇连续下雨及大风不利天气，部分小区出现倒伏。不同播期条件下，京农科728的倒伏率随播期推迟呈降低趋势，其中以6月5日播期处理倒伏率最高（正值大喇叭口期，倒伏率8.1%）、6月25日播期处理倒伏率最低（正值拔节期，倒伏率0.0%）。在不同密度条件下，倒伏率则随种植密度增大呈升高趋势，其中以90 000株/hm^2密度条件下倒伏率最高（12.5%）、60 000株/hm^2密度条件下倒伏率最低（1.7%）。由此说明，同一玉米品种因所处的生育时期不同，其植株的抗倒伏能力存在较大差异；并且，同一品种在相同播期条件下随种植密度增加，其植株的倒伏风险亦随之增加。

表3 不同播期和密度条件下京农科728的株高、穗位高和倒伏率性状
Table 3 Plant height, ear height and lodging rate of Jingnongke728 under different sowing and density conditions

处理 Treatment		株高（cm）Plant height	穗位高（cm）Ear height	倒伏率（%）Lodging rate
播期 Sowing date	S1	250.8b	86.7c	8.1a
	S2	261.4b	90.9b	3.6b
	S3	284.3a	106.6a	2.5c
	S4	290.6a	107.3a	1.4d
	S5	293.5a	108.9a	0.0e

(续表)

处理 Treatment		株高（cm） Plant height	穗位高（cm） Ear height	倒伏率（%） Lodging rate
密度 Planting density	D1	271.2b	91.0b	1.7e
	D2	272.3b	93.0b	2.4d
	D3	286.0a	106.0a	3.5c
	D4	287.9a	107.7a	9.7b
	D5	289.4a	109.2a	12.5a

3 讨论和结论

玉米品种的籽粒机收作业质量受收获机械、品种特性及栽培措施等因素影响。本试验结果表明，在相同收获机械和品种条件下，播期和密度与玉米籽粒机收相关性状（籽粒的含水率、破碎率、杂质率，植株性状及生育期等）密切相关。

玉米品种的籽粒机收作业质量受收获机械、品种特性及栽培措施等多种因素的影响。本试验结果表明，在相同收获机械和品种条件下，播期和密度条件与玉米籽粒机收相关性状（籽粒含水率、破碎率、杂质率，植株性状及生育期等）密切相关。试验品种京农科728在京郊及京津冀玉米生产中以冬小麦-夏玉米一年两茬种植方式为主，夏播集中在6月上中旬，部分晚播地块于6月下旬播种。该品种在北京通州6月5—25日夏播范围内，生育期随播期推迟呈缩短趋势，在5个不同播期处理条件下籽粒均能正常成熟，播期适应性良好。随播期推迟，其株高和穗位高呈升高趋势。本试验过程中，7月19、20日出现连续下雨及大风不利天气，造成部分小区植株倒伏。此时，京农科728在最早播种的6月5日播期处理条件下正处于大喇叭口期，倒伏率高达8.1%，其他播期处理倒伏率均为5%以下。从籽粒机收质量来看，京农科728在北京通州夏播6月5—25日播期范围内，10月15日籽粒机收时含水率随播期推迟呈增加趋势，6月10—25日播期范围内籽粒破碎率和杂质率随播期推迟呈增加趋势。播种过早（6月5日播种），虽然籽粒含水率最低（23.4%），但籽粒破碎率（9.2%）和杂质率（1.6%）较高，产量最低；播期过晚（6月25日播种），籽粒含水率（高达35.8%）和破碎率（17.6%）均最高，籽粒产量较低。

近年来，随着耐密植玉米品种的选育及推广应用，密植增产技术成为挖掘单位面积玉米产量潜力的重要技术途径，但不同密植群体的机收性能指标存在较大差异。本研究结果表明，随种植密度增加，试验品种京农科728籽粒机收时的含水率变化不大，但株高和穗位高均呈上升趋势，植株抗倒伏能力随之下降，倒伏率呈增加趋势。密度过高或过低均不利于实现京农科728籽粒机收质量和高产目标的协调统一。在本试验条件下，综合产量、倒伏率和籽粒机收指标，京农科728以75 000株/hm^{2}为籽粒机收最佳种植密度。

参考文献（略）

本文原载：玉米科学，2017，25（3）：94-98

品种和氮素供应对玉米根系特征及氮素吸收利用的影响

程乙[1]　王洪章[1]　刘鹏[1]　董树亭[1]　赵久然[2]
王荣焕[2]　张吉旺[1]　赵斌[1]　李耕[1]　刘月娥[2]

(1. 山东农业大学农学院/作物生物学国家重点实验室，山东泰安，271018；
2. 北京市农林科学院玉米研究中心/玉米DNA指纹及分子育种北京市
重点实验室，北京，100097)

摘要：【目的】研究玉米根系特性与氮素吸收利用及其与地上部生物量和产量形成的关系，探明根系形态特征与氮素吸收能力对玉米高产性能的影响，为玉米高产高效生产提供理论依据。【方法】试验于2014—2015年在山东农业大学黄淮海区域玉米技术创新中心（36°18′N，117°12′E）和作物生物学国家重点实验室进行，以京科968（JK968）、郑单958（ZD958）和先玉335（XY335）为试验材料，采用土柱栽培，设置两个氮素水平，施氮量分别为1.5g/plant（LN）和4.5g/plant（HN），在抽雄期（VT）和完熟期（R6）进行根系及植株取样，测定根系相关指标（根系干重、根系长度、根系表面积、根系体积），干物质及氮素积累与分配规律，探究品种和氮素供应对玉米根系特征及氮素吸收利用的影响。【结果】两个氮素水平下JK968单株籽粒产量、生物量、根系各指标和植株氮素积累量、氮转运率、氮素收获指数、氮素利用效率均显著高于XY335和ZD958（$P<0.05$）。JK968单株生物量、籽粒产量、植株氮素积累量较XY335和ZD958在低氮水平下分别增加15.2%、17.7%、9.0%和31.6%、44.1%、31.4%，在高氮水平下分别增加5.4%、12.9%、8.9%和13.5%、26.8%、23.5%；高氮水平下JK968、XY335、ZD958的单株生物量、单株籽粒产量和植株氮素积累量较低氮水平下分别增加15.7%、10.2%、33.9%，26.5%、14.8%、34.0%和34.3%、25.1%、42.5%。抽雄期JK968根系干重、根系长度、根系表面积、根系体积较XY335和ZD958在低氮水平下分别增加41.8%、9.0%、47.1%、24.0%和63.2%、41.6%、60.4%、105.1%，在高氮水平下分别增加24.3%、6.0%、35.2%、19.7%和40.3%、30.0%、49.3%、78.7%；高氮水平下JK968、XY335、ZD958的根系干重、根系长度、根系表面积、根系体积较低氮水平下分别增加48.3%、37.3%、36.4%、12.7%，69.1%、41.3%、48.4%、16.7%和72.5%、49.7%、46.5%、29.3%。相关分析表明，吸氮量与根系干重、根系长度、根系表面积、根系体积呈显著线性正相关，但品种的响应程度不同。在抽雄前，JK968植株吸氮量对根系干重、根系长度、根系表面积、根系体积增长的响应度要高于XY335和ZD958；而抽雄后的响应度则低于XY335和ZD958。【结论】JK968整个生育期的根系各项指标均显著高于XY335和

基金项目：国家自然科学基金（31371576，31401339）、国家重点研发计划（2016YFD0300106）、国家"十二五"科技支撑计划（2013BAD07B06-2）、国家公益性行业（农业）科研专项经费项目（201203100，201203096）、山东省现代农业产业技术体系项目（SDAIT-02-08）、国家现代农业产业技术体系建设项目（CARS-02-20）、山东省农业重大应用技术创新课题
联系方式：程乙，E-mail：chengyi722@126.com。通讯作者刘鹏，E-mail：liupengsdau@126.com。通讯作者王荣焕，E-mail：ronghuanwang@126.com

ZD958,且氮素吸收能力强,生物量大,低氮条件下优势更加明显。JK968 较发达的根系,保证了植株对氮素的吸收,具有较高的氮素转运效率、贡献率和氮素利用效率,有利于进行物质生产,因而获得更高的籽粒产量。

关键词:玉米;品种;施氮量;根系;氮素吸收

Effect of Different Maize Varieties and Nitrogen Supply on Root Characteristics and Nitrogen Uptake and Utilization Efficiency

Cheng Yi[1]　Wang Hongzhang[1]　Liu Peng[1]　Dong Shuting[1]
Zhao Jiuran[2]　Wang Ronghuan[2]　Zhang Jiwang[1]
Zhao Bin[1]　Li Geng[1]　Liu Yuee[2]

(1. College of Agronomy, Shandong Agricultural University, State Key Laboratory of Crop Biology, Taian 271018, Shandong; 2. Maize Research Center, Beijing Academy of Agriculture & Forestry Sciences, Beijing Key Laboratory of Maize DNA Fingerprinting and Molecular Breeding, Beijing 100097, China)

Abstract:【Objective】Through the study of the relationship of maize root characteristics and nitrogen uptake and utilization efficiency, shoot biomass and yield formation to prove the effect of root morphological characteristics and nitrogen uptake capacity of maize yield. And then provide theoretical basis for the high yield and efficiency of maize production. 【Method】The experiments were conducted in 2014-2015 at the Technological Innovation Center of Maize in Huang-Huai-Hai Region (36°18′N, 117°12′E) and the State Key Laboratory of Crop Biology, located at Shandong Agricultural University in Taian. To explore the effect of different maize varieties and nitrogen supply on root characteristics and nitrogen uptake and utilization efficiency, using Jingke 968 (JK968), Zhengdan 958 (ZD958) and Xianyu 335 (XY335) as the experimental materials, setting two nitrogen levels, 1.5g/plant (low nitrogen, LN) and 4.5g/plant (high nitrogen, HN), sampling root and shoot of plant at tasseling stage and maturity stage for determination of the related indexes of root system (the root dry weight, root length, root surface area, root volume), dry matter and nitrogen accumulation and distribution in soil column tests. 【Result】The results showed that grain yield and biomass per plant, each index of root, nitrogen accumulation amount per plant, nitrogen translocation rate, nitrogen harvest index and use efficiency of JK968 were all significantly higher ($P<0.05$) than those of XY335 and ZD958. The biomass, grain yield, N accumulation amount per plant of JK968 were higher than those of XY335 and ZD958 by 15.2%, 17.7%, 9.0% and 31.6%, 44.1%, 31.4%, respectively, under LN level, 5.4%, 12.9%, 8.9% and 13.5%, 26.8%, 23.5%, respectively, under HN level. Compared with LN level, the biomass, grain yield, N accumulation amount per plant at the HN level of JK968, XY335 and ZD958 increased by 15.7%, 10.2%, 33.9% and 26.5%, 14.8%, 34.0% and 4.3%, 25.1%, 42.5%, respectively. The root dry weight, root length, root surface area, root volume of JK968 at tasseling stage are higher than those of XY335 and ZD958 by 41.8%, 9.0%, 47.1%, 24.0% and 63.2%, 41.6%, 60.4%, 105.1%, respectively, under LN level, 24.3%, 6.0%, 35.2%, 19.7% and 40.3%, 30.0%, 49.3%, 78.7%, respectively, under HN level. Compared with LN level, the root dry weight, root length, root surface area, root volume at the HN level of JK968, XY335 and ZD958 increased by 48.3%, 37.3%, 36.4%, 12.7% and 69.1%, 41.3%, 48.4%, 16.7% and 72.5%, 49.7%, 46.5%, 29.3%, respectively. The correlation analysis indicated that the amount of N-uptake showed a significant positive linear correlation with the root dry weight, root length, root surface area and root volume. The root index of different cultivars showed different responses to nitrogen, which the responsivity of each root index of JK968 to nitrogen was higher before tasseling but lower after tasseling than XY335 and ZD958. 【Conclusion】As for JK968, the root indexes of the whole growth period were significantly higher than those of XY335 and ZD958, it had a stronger nitrogen uptake ability and larger biomass, which were more distinct at the low nitrogen level. All these indicate that the larger root system of JK968 can ensure the nitrogen uptake to have higher nitrogen transportation efficiency, nitrogen contribution rate and nitrogen utilization efficiency, making for its material production, and finally obtaining a higher grain yield.

Key words: Maize; Varieties; Nitrogen application rate; Root; Nitrogen uptake

引言

【研究意义】氮素是限制植物生长发育和产量形成的首要因素，增施氮肥已成为作物获得高产的重要措施。但随着施氮量的持续增加，出现报酬递减现象，其增产效果和利用效率逐渐下降，且土壤中大量盈余的氮素也会因氨挥发、淋洗而损失，造成资源浪费和环境污染。玉米氮吸收效率、氮利用效率和氮效率具有明显的基因型差异。根系作为玉米吸收氮素的主要器官，其形态结构以及在土壤中的时空分布是造成氮素吸收效率差异的重要因素。因此，比较玉米对氮吸收利用能力的基因型差异，探明玉米根系特性与氮素吸收利用的关系，对玉米高产与资源高效利用具有重要意义。【前人研究进展】根系是作物生长发育的基础，其数量、活性与光合产物合成与运转分配、籽粒结实、叶片衰老等密切相关。根系发达和长时间保持高活力不仅是地上部生长发育、氮素吸收利用和产量形成的重要保证，也是高产玉米品种的一个显著特点。根系生长发育不仅取决于品种，还受土壤类型、质地、肥力等土壤环境条件，以及耕作、种植、施肥等农业管理措施的制约。研究表明，氮素供应水平对根系的生长发育有明显的调控效应，对根系形态建成和氮素吸收利用产生间接或直接的影响。增施氮肥能增加玉米根系干重、根长密度和表面积，低氮有利于根系纵向伸长，而高氮有利于根系横向伸展。【本研究切入点】前人针对玉米利用水培、盆栽和田间种植等多种方式对玉米根系与氮素吸收利用间的关系进行了较多的研究，但多数集中在生长发育前期根系的重量与形态指标上，而不同品种抽雄前、后根系特性与氮素吸收利用的关系及对不同氮素水平的响应差异的研究相对较少。【拟解决的关键问题】本研究选用不同玉米品种，设置不同施氮量，研究高产玉米根系特征及其对氮素用量的响应，从根系发育和功能方面阐明玉米氮素高效利用的生理机制，为提高玉米产量和氮素利用效率提供理论依据。

1 材料与方法

1.1 试验设计

试验于 2014—2015 年在山东农业大学黄淮海区域玉米技术创新中心（36°18′N，

117°12′E）和作物生物学国家重点实验室进行。供试材料为京科968（JK968）、郑单958（ZD958）和先玉335（XY335）。将高100cm，直径35cm的PVC管，填埋于90cm深的土坑中进行土柱试验，设置两个氮素水平（LN：1.5g/plant；HN：4.5g/plant），4次重复。按照表层土（50%）、沙子（40%）、蛭石（5%）、珍珠岩（5%）的比例混匀基质后装入柱中，沉实后在距土表10cm处施肥。基质中有机质含量10.50g·kg^{-1}、全氮0.70g·kg^{-1}、碱解氮50.35mg·kg^{-1}、速效磷24.17mg·kg^{-1}、速效钾108.20mg·kg^{-1}。2014—2015年玉米生长期内降雨量分别为293.7mm和279.4mm。从出苗到小喇叭口期每隔一天浇200ml水，小喇叭口期以后每天浇300ml水。2014年6月7日播种，9月28日收获；2015年6月10日播种，9月30日收获。

1.2 测定项目及方法

1.2.1 地上部生物量

分别于抽雄期（VT）、完熟期（R6）取样，4次重复。按照茎秆（含雄穗和穗轴）、叶片（含苞叶）、籽粒分开，于105℃杀青30min后80℃烘干后称重。

1.2.2 根系形态测定

将完整的根系在低水压下冲洗干净后，低温下保存，利用Epson PerfectionTM V700 Photo彩色图像扫描仪扫描根系，采用Win RHIZO2016根系分析系统分析得到根系长度（RL）、根系表面积（RSA）和根系体积（RV）。各指标测定完成后将根系放入烘箱80℃下烘至恒重，测定根系干重（RDW）。

1.2.3 植株样品含氮量

植株样品烘干磨碎后用Rapid N III氮素分析仪（Elementar Analysensysteme, Germany）测定全氮含量。

1.3 数据处理与分析

植株总氮素积累量（TNAA, g/plant）= 完熟期单株干重×完熟期单株含氮量（%）；

营养器官氮素转运量（NTA, g/plant）= 抽雄期营养器官氮素积累量—完熟期营养器官氮素积累量；

抽雄后氮素同化量（AANAT, g/plant）= 完熟期籽粒氮素积累量—营养器官氮素转运量；

氮素转运效率（NTE,%）= 营养器官氮素转运量/抽雄期营养器官氮素积累量×100；

氮素转运对籽粒的贡献率（NTCP,%）= 营养器官氮素转运量/完熟期籽粒氮素积累量×100；

氮素同化对籽粒的贡献率（NACP,%）= 抽雄后氮素同化量/完熟期籽粒氮素积累量×100；

氮素收获指数（NHI, kg·kg^{-1}）= 籽粒氮素积累总量/植株氮素积累总量；

氮素利用效率（NUE, kg·kg^{-1}）= 籽粒产量/植株氮素积累总量。

采用DPS 11.0统计软件LSD法进行统计分析（$a=0.05$），Sigmaplot 10.0作图。

2 结果

2.1 单株产量、生物量及收获指数

由表1可见，在两氮素水平下，JK968的单株生物量和籽粒产量均显著高于XY335和

ZD958（$P<0.05$），收获指数也较高。在低氮水平下，JK968 的单株生物量和籽粒产量较 XY335、ZD958 分别提高了 15.2% 和 17.7%、31.6% 和 44.1%，而在高氮水平分别增加 5.4% 和 12.9%、13.5% 和 26.8%。施氮显著提高了各品种的单株生物量和籽粒产量（$P<0.05$）；高氮水平下 JK968、XY335、ZD958 的单株生物量和单株籽粒产量较低氮水平下分别增加 15.7% 和 10.2%、26.5% 和 14.8%、34.3% 和 25.1%。

除氮肥水平与基因型互作对玉米生物量及收获指数的影响分别为显著水平（$P<0.01$）和无显著影响（$P>0.05$）之外，氮肥水平、基因型、氮肥水平与基因型互作对玉米单株生物量、产量及收获指数的影响均达到极显著水平（$P<0.001$）。

2.2 根系特性
2.2.1 根系干重

由图 1 可知，在两个氮素水平下，VT 和 R6 期各品种的根系干重（RDW）间差异均达到显著水平，表现为 JK968>XY335>ZD958（$P<0.05$）。与 VT 期相比，各品种在 R6 期的 RDW 均有所降低，其中低氮水平下 JK968、XY335、ZD958 的降幅分别为 16.9%、17.4% 和 20.3%，在高氮水平下分别为 17.4%、21.5% 和 25.6%。在低氮水平下，VT 期 JK968 的 RDW 较 XY335 和 ZD958 分别提高了 41.8% 和 63.2%，而在高氮水平分别增加 24.3% 和 40.3%。施氮显著提高了各品种的 RDW（$P<0.05$），与低氮水平相比，高氮水平下 VT 期 JK968、XY335 和 ZD958 的 RDW 分别增加 48.3%、69.1% 和 72.5%。

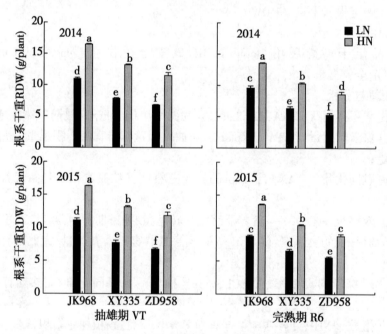

图 1　施氮水平对不同品种玉米根系干重的影响

Figure 1　Effect of N application rates on root dry weight of different maize varieties

表1 施氮水平对不同品种玉米单株生物量、产量及收获指数的影响
Table 1 Effect of N application rates on grain yield and harvest index of different maize varieties

年份 Year	处理 Treatment	品种 Variety	单株生物量 Biomass (g/plant)	单株籽粒产量 Grain yield (g/plant)	收获指数 Harvest index
2014	LN	JK968	367.64 c	194.43 b	0.529 a
		XY335	313.97 d	162.45 d	0.517 ab
		ZD958	276.13 e	135.43 e	0.490 c
	HN	JK968	424.46 a	214.15 a	0.505 b
		XY335	405.38 b	191.15 c	0.472 d
		ZD958	378.80 c	171.56 d	0.453 e
2015	LN	JK968	367.43 c	194.79 b	0.530 a
		XY335	324.35 d	168.27 d	0.527 a
		ZD958	282.32 d	134.70 e	0.478 bc
	HN	JK968	426.32 a	214.64 a	0.496 b
		XY335	401.90 b	188.59 c	0.488 c
		ZD958	371.02 c	166.49 d	0.462 c
变异来源 Source of variation					
氮水平 Nitrogen level (N)			***	***	***
基因型 Genotype (G)			***	***	***
氮水平×基因型 N×G			**	***	NS

同列数据后不同小写字母表示品种间差异达0.05显著水平;*** 代表0.001显著水平,** 代表0.01显著水平,* 代表0.05显著水平,NS 代表0.05水平不显著。下同

Values followed by different lowercase letters in a column are significantly different between varieties at the 0.05 level; *** represents significant at the 0.001 probability level, ** represents significant at the 0.01 probability level, * represents significant at the 0.05 probability level, NS represents no significant at the 0.05 probability level. The same below

2.2.2 根系形态

由表2可知,在两个氮素水平下,VT和R6期各品种的根系长度(RL)、根系表面积(RSA)和根系体积(RV)间差异均达到显著水平,表现为JK968>XY335>ZD958($P<0.05$)。低氮水平下,VT期JK968的RL、RSA和RV较XY335和ZD958分别增加9.0%、47.1%、24.0%和41.6%、60.4%、105.1%,而在高氮水平分别增加6.0%、35.2%、19.7%和30.0%、49.3%、78.7%。施氮显著提高了各品种的RL、RSA和RV($P<0.05$),与低氮水平相比,高氮水平下VT期JK968、XY335、ZD958的RL和RSA分别增加37.3%和36.4%、41.3%和48.4%、49.7%和46.5%,而RV分别增加12.7%、16.7%和29.3%。

与VT期相比,各品种在R6期的RL、RSA和RV均有所降低,其中JK968、XY335、ZD958的RL降幅在低氮水平下分别为14.7%、26.9%和11.2%,在高氮水平下分别为

20.5%、33.0%和22.7%；RSA降幅在低氮水平下分别为7.4%、3.2%和5.6%，在高氮水平下分别为16.5%、23.6%和19.9%；而RV降幅在低氮水平下分别为40.2%、42.0%和28.3%，在高氮水平下分别为23.6%、42.4%和24.3%。

氮肥水平、基因型对RL、RSA和RV的影响均达到极显著水平（$P<0.001$），而氮肥水平与基因型互作对RL和RSA无显著影响（$P>0.05$），对RV的影响达到显著水平（$P<0.01$）。

表2 不同品种玉米根系形态特征
Table 2 Root morphological characteristics of different maize varieties

年份 Year	处理 Treatment	品种 Variety	根系长度 RL (m/plant)		根系表面积 RSA (m^2/plant)		根系体积 RV (cm^3/plant)	
			VT	R6	VT	R6	VT	R6
2014	LN	JK968	732.13 d	618.52 b	1.17 b	1.07 b	337.36 b	201.25 b
		XY335	662.47 e	485.09 d	0.77 d	0.71 d	269.11 d	157.88 d
		ZD958	515.95 f	446.61 e	0.70 e	0.67 d	164.05 f	119.26 e
	HN	JK968	989.25 a	797.82 a	1.61 a	1.32 a	379.56 a	288.47 a
		XY335	938.85 b	629.84 b	1.15 b	0.88 c	317.10 c	180.16 c
		ZD958	770.18 c	592.65 c	1.04 c	0.83 c	211.03 e	157.83 c
2015	LN	JK968	723.52 d	623.50 bc	1.14 bc	1.07 b	331.48 b	198.83 b
		XY335	672.69 e	490.85 d	0.80 d	0.81 c	270.41 d	155.11 d
		ZD958	511.94 f	466.58 d	0.74 d	0.69 d	162.12 f	114.73 e
	HN	JK968	1010.00 a	792.05 a	1.54 a	1.31 a	374.05 a	286.92 a
		XY335	947.72 b	633.84 b	1.18 b	0.90 c	312.52 c	182.19 c
		ZD958	768.72 c	597.50 c	1.07 c	0.86 c	210.76 e	161.37 d
变异来源 Source of variation								
氮水平 Nitrogen level (N)			***		***		***	
基因型 Genotype (G)			***		***		***	
氮水平×基因型 N×G			NS		NS		*	

RL：根系长度 Root length；RSA：根系表面积 Root surface area；RV：根系体积 Root volume

2.3 植株氮素吸收与利用

2.3.1 氮素积累量

由表3可知，在两个氮素水平下JK968氮素积累量均显著高于XY335和ZD958（$P<0.05$）。在低氮水平下，JK968的氮素积累量较XY335和ZD958分别提高了9.0%和31.4%，而在高氮水平分别增加8.9%和23.5%。施氮显著提高了各品种的氮素积累量（$P<0.05$），高氮水平下JK968、XY335和ZD958的氮素积累量较低氮水平下分别增加33.9%、34.0%和42.5%。从抽雄前、后氮素积累所占比例可知，氮素大部分是在抽雄期前吸收的，其所占比例在（61.76±3.77）%；JK968抽雄前氮素吸收所占比例高于XY335和ZD958。

表3 施氮水平对不同品种玉米植株氮积累量的影响

Table 3 Effect of N application rates on nitrogen accumulation amount of different maize varieties

年份 Year	处理 Treatment	品种 Variety	氮素积累量 NAA（g/plant）			抽雄前比例 Rate of pre-VT （%）	抽雄后比例 Rate of post-VT（%）
			完熟期 R6	抽雄前 Pre-VT	抽雄后 Post-VT		
2014	LN	JK968	3.80 d	2.25 d	1.55 bc	59.25	40.75
		XY335	3.59 e	2.07 e	1.53 bc	57.46	42.54
		ZD958	2.85 f	1.72 f	1.12 d	60.47	39.53
	HN	JK968	5.09 a	3.42 a	1.67 a	67.15	32.85
		XY335	4.70 b	3.12 b	1.58 b	66.38	33.62
		ZD958	4.13 c	2.65 c	1.48 c	64.23	35.77
2015	LN	JK968	3.86 d	2.23 d	1.63 c	57.84	42.16
		XY335	3.44 e	1.99 e	1.45 e	57.79	42.21
		ZD958	2.98 f	1.71 f	1.27 f	57.34	42.66
	HN	JK968	5.17 a	3.39 a	1.78 a	65.54	34.46
		XY335	4.72 b	3.01 b	1.71 b	63.77	36.23
		ZD958	4.18 c	2.67 c	1.51 d	63.85	36.15

NAA：氮素积累量 Nitrogen accumulation amount

2.3.2 氮素转运

由表4可以看出，在两个氮素水平下JK968籽粒氮素总积累量均显著高于XY335和ZD958（$P<0.05$）。在低氮水平下，JK968的籽粒氮素总积累量较XY335和ZD958分别提高了16.6%和33.3%，而在高氮水平分别增加13.7%和28.3%。各处理间氮素转运量和同化量差异均达到显著水平（$P<0.05$）。JK968的氮素转运率最高，其氮素转运量均显著高于XY335和ZD958（$P<0.05$），而氮素同化量在高、低氮素水平下较XY335和ZD958分别增加了4.9%和15.4%、6.7%和33.1%。从氮素转运和同化对籽粒的贡献率可知，JK968的氮素转运对籽粒的贡献率显著高于XY335和ZD958（$P<0.05$），而氮素同化贡献率则显著低于XY335和ZD958（$P<0.05$）。

表4 施氮水平对不同品种玉米植株抽雄后氮素向籽粒中的转移和氮素同化对籽粒贡献的影响

Table 4 Effect of N application rates on nitrogen translocation and assimilation amount to grains after tasseling of different maize varieties

年份 Year	处理 Treatment	品种 Variety	籽粒氮素总积累量 TNAAG （g/plant）	氮转运量 NTA （g/plant）	抽雄后氮同化量 AANAT （g/plant）	氮转运率 NTR（%）	氮素转运对籽粒的贡献率 NTCP（%）	氮素同化对籽粒的贡献率 NACP（%）
2014	LN	JK968	2.18 c	0.63 c	1.55 bc	27.85 a	28.85 bc	71.15 bc
		XY335	1.93 e	0.40 e	1.53 bc	19.56 c	20.92 d	79.08 a
		ZD958	1.61 f	0.48 d	1.12 d	27.93 a	29.96 bc	70.04 bc
	HN	JK968	2.59 a	0.92 a	1.67 a	26.80 a	35.40 a	64.60 d
		XY335	2.29 b	0.70 b	1.58 bc	22.56 b	30.82 b	69.18 c
		ZD958	2.02 d	0.54 d	1.48 c	20.52 bc	26.93 c	73.07 b

(续表)

年份 Year	处理 Treatment	品种 Variety	籽粒氮素 总积累量 TNAAG (g/plant)	氮转运量 NTA (g/plant)	抽雄后氮 同化量 AANAT (g/plant)	氮转运率 NTR (%)	氮素转运 对籽粒的 贡献率 NTCP (%)	氮素同化 对籽粒的 贡献率 NACP (%)
2015	LN	JK968	2.18 c	0.56 b	1.63 c	25.01 a	25.56 bc	74.44 bc
		XY335	1.81 e	0.36 c	1.45 e	18.13 c	19.92 d	80.08 a
		ZD958	1.66 f	0.39 c	1.27 f	22.76 ab	23.41 c	76.59 b
	HN	JK968	2.63 a	0.85 a	1.78 a	25.03 a	32.21 a	67.79 d
		XY335	2.30 b	0.59 b	1.71 b	19.44 c	25.48 bc	74.52 bc
		ZD958	2.05 d	0.54 b	1.51 d	20.40 bc	26.48 b	73.52 c

TNAAG：籽粒氮素总积累量 Total nitrogen accumulation amount of grains；NTA：氮转运量 Nitrogen translocation amount；AANAT：抽雄后氮同化量 Assimilating amount of nitrogen after tasseling；NTR：氮转运率 Nitrogen translation rate；NTCP：氮素转运对籽粒的贡献率 Nitrogen translocation contribution proportion；NACP：氮素同化对籽粒的贡献率 Nitrogen assimilation contribution proportion

2.3.3 氮素利用效率

由图2可知，在两个氮素水平下，JK968的氮素收获指数（NHI）和利用效率（NUE）均显著高于XY335和ZD958（$P>0.05$）。与低氮水平相比，高氮水平下JK968、ZD958和XY335的NHI和NUE均显著下降（$P<0.05$）。

2.4 根系指标与植株吸氮量之间的关系

由图3可知，无论在抽雄前还是抽雄后，JK968、XY335和ZD958植株吸氮量与RDW、RL、RSA、RV均呈显著线性正相关，且相关性系数较大，但品种的响应程度不同。在抽雄前回归系数JK968>XY335>ZD958，抽雄后回归系数JK968<XY335<ZD958。这说明在抽雄前JK968植株吸氮量对RDW、RL、RSA、RV增长的响应度均高于XY335，而XY335植株吸氮量对RDW、RL、RSA、RV增长的响应度均高于ZD958；在抽雄后，JK968植株吸氮量对RDW、RL、RSA、RV降低的响应度均低于XY335，XY335植株吸氮量对RDW、RL、RSA、RV增长的响应度均低于ZD958。与XY335和ZD958相比，JK968植株吸氮量在抽雄前能保持对RDW、RL、RSA、RV增长高的响应度，而在抽雄后却能对RDW、RL、RSA、RV的降低保持较低响应度。

3 讨论

根系作为玉米水分和养分吸收的主要器官，影响整株的生长发育，氮素吸收依赖于根系的大小和吸收性能。本研究中，两氮素水平下JK968在生育期中维持了较大的根系生物量，且在低氮水平下优势更为明显；而在根系形态上表现为根系长度、根系表面积、根系体积较大，整个生育期内均显著高于XY335和ZD958。前人研究指出，植株的生长发育是地上和地下部分协调发展的结果，地上部分可为根系提供充足的光合产物，有利于根系良好的形态结构的建成和生理功能的维持，本文研究发现JK968在抽雄至完熟期根系干重、长度、表面积的降幅较小，说明其花后有较强的光合生产能力，维持了较强的根系功能。而强大的根系可以吸收更多的水分养分供给地上部生长，从而获得较高的生物产量和

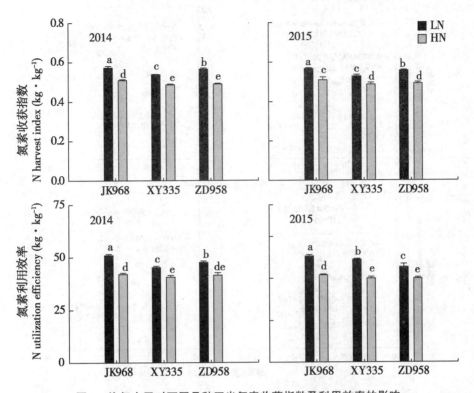

图 2 施氮水平对不同品种玉米氮素收获指数及利用效率的影响

Figure 2　Effect of N application rates on N harvest index and N utilization efficiency of different maize varieties

籽粒产量，达到根冠协调。本研究结果表明，两个氮水平下 JK968 的根系生物量、单株生物产量和籽粒产量均显著高于 XY335 和 ZD958，其产量与根系发育程度密切相关，发达的根系总是伴随着地上部的旺盛生长并获得高产。由此可见，JK968 优良的根系特性，较长的根系活性高值持续期，延缓了生育后期的根系衰老，使根系吸收更多的水分和养分以满足籽粒形成期对氮素的需求。同时，地上部健康生长也可为根系提供充足的光合产物，从而更有利于维持较高的根系活性，保证后期籽粒充实过程的养分供给。

抽雄至成熟期是玉米氮素吸收运转分配的重要时期。籽粒中的养分，一部分来自根系直接吸收，另一部分来自营养器官的养分再转移。供氮不足可能导致营养体氮素向外调运过多而引起叶片早衰，供氮充足时植株的吸氮量主要受生长"库"的需求所调节。本研究表明，JK968 具有较高的籽粒产量和收获指数，这说明其库容量和库活性比其他两个品种高，从而有利于促进植株吸收更多的氮素，拥有较高的氮素利用效率。与 XY335 和 ZD958 相比，JK968 在两氮素水平下营养体向籽粒中转运的氮素绝对量较多，具有较高的氮素转运效率和贡献率，对充分利用营养体中的营养物质，促进籽粒灌浆充实和产量提高起到了重要作用。JK968 在低氮水平下氮素积累量增加幅度和花后氮素转运对籽粒贡献率大于高氮水平，说明低氮水平有利于 JK968 植株氮素吸收和促进花后营养体氮素转运。

前人对玉米根系性状与地上部之间的相关性研究发现，根系与株高之间存在明显的同伸关系，且在水肥条件较差时表现更明显；根系干重与地上部干重、绿叶面积、氮素积累

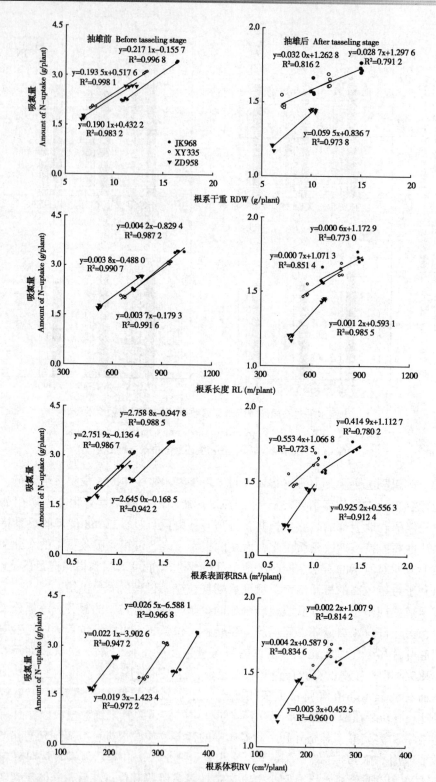

图 3 根系干重、根系长度、根系表面积、根系体积与植株吸氮量的关系

Figure 3 The relationship between root dry weight, RL, RSA, RV and amount of N-uptake

量之间均呈极显著正相关；根系总长度与氮素积累量在低氮水平下表现为极显著线性相关，而高氮下无显著相关性。本研究表明，植株吸氮量与根系干重、根系长度、根系表面积、根系体积均呈显著线性相关，而不同品种对氮素的响应度不同。在抽雄前，JK968植株吸氮量对RDW、RL、RSA、RV增长的响应度均高于XY335，而XY335植株吸氮量对RDW、RL、RSA、RV增长的响应度均高于ZD958；在抽雄后，JK968植株吸氮量对RDW、RL、RSA、RV降低的响应度均低于XY335，XY335植株吸氮量对RDW、RL、RSA、RV增长的响应度均低于ZD958。这说明JK968植株吸氮量在抽雄前能保持对RDW、RL、RSA、RV增长高的响应度，而在抽雄后却能对RDW、RL的降低保持较低响应度，这有利于其在整个生育时期保持对氮素的稳定和高效吸收。与XY335和ZD958相比，JK968可通过发育良好的根系形态来提高根系对氮素的吸收和利用能力，从而提高全株氮素积累量和生物量，获得较高籽粒产量。

4 结论

JK968整个生育期的根系各指标均显著高于XY335和ZD958，且氮素吸收能力强，生物量大，低氮条件下优势更加明显。相关分析表明植株吸氮量与根系指标呈显著线性正相关，品种间根系指标对氮素的响应度不同，抽雄前JK968根系指标对氮素的响应度高于XY335和ZD958，而抽雄后JK968根系指标对氮素的响应度低于XY335和ZD958。JK968较发达的根系，保证了植株对氮素的吸收，具有较高的氮素转运效率、贡献率和氮素利用效率，有利于进行物质生产、获得更高的籽粒产量。

参考文献（略）

本文原载：中国农业科学，2017，50（12）：2 259-2 269

我国糯玉米育种及产业发展动态

赵久然　卢柏山　史亚兴　徐　丽

（北京市农林科学院玉米研究中心/玉米 DNA 指纹及分子育种北京市重点实验室，北京　100097）

摘　要：糯玉米也称蜡质玉米、粘玉米，是玉米传入我国之后变异产生的一种新类型。国外糯玉米以美国研究和利用最多，但主要用于淀粉加工和饲料，基本不作为鲜食玉米。我国糯玉米育种起步较晚，但发展迅速，在糯玉米种质创新、组配模式优化、新品种培育、品质改良等方面均具有我国特色，在某些方面具有全球领先优势。京科糯 2000 等系列品种的选育与推广极大地促进了我国糯玉米产业的整体发展。本文综述了国内外糯玉米育种及产业发展现状、我国糯玉米育种创新点和新亮点，并为我国糯玉米产业在玉米调结构、转方式形势下如何进一步发展提出建议：以鲜果穗销售为主，加工为辅；加强甜加糯新类型品种的选育和推广；加强糯玉米秸秆有效利用；高端特色，打造品牌。

关键词：糯玉米；育种；京科糯；产业；动态

Development Trends of Waxy Corn Breeding and Industry in China

Zhao Jiuran　Lu Baishan　Shi Yaxing　Xu Li

（*Maize Research Center, Beijing Academy of Agriculture and Forestry Sciences, Beijing Key Laboratory of Maize DNA Fingerprinting and Molecular Breeding, Beijing, 100097, China*）

Abstract: Waxy corn, which was also being called Chinese glutinous corn, was a new variant type generated in common maize. In the countries abroad, America has been studying and using the most of waxy corn, but mainly in starch processing industry and fodder, barely nofresh corn. Waxy corn breeding started late in China, but developed quite quickly, and now China has formed its own features and hasbeen the global leader in waxy germplasminnovation, crossing mode optimizing, variety developingand quality improvement. Developing and application of series of Jingke waxy corn such as Jingkenuo2000 has greatly driven the development of waxy corn in China. Development status of waxy corn breeding and industry in and abroad, innovation and new highlights in China were summarized and advice on how waxy corn industry moving forward were offered in this paper: give priority to fresh ears and supplement processing products; enhancebreeding and apply sweet-waxy corn; strengtheneffective use of waxy corn straw; develop high-end, distinctive varieties and create waxy

---

作者简介：赵久然（1962—　），男，北京人，研究员，博士，主要从事玉米遗传育种等研究。
Tel：010-51503936；E-mail：maizezhao@126.com

corn brand.

Key words：Waxy corn；Breeding；Jingkenuo；Industry；Trend

糯玉米（*Zeamays* L. *certainaKulesh*）也称蜡质玉米、粘玉米，是玉米传入我国之后，在栽培种植过程中发生变异而产生的一种新类型。与普通玉米相比，糯玉米籽粒所含淀粉几乎全为支链淀粉，食用消化率高；营养物质丰富，其所含蛋白质、氨基酸等也高于普通玉米，尤其赖氨酸含量比普通玉米高30%~60%。糯玉米具有绵软适口、甜粘清香的独特口感和风味，不仅可蒸煮鲜果穗用于青食，还可用于速冻加工、制作糯玉米罐头等，深受消费者喜爱和青睐，在我国及东南亚地区消费较为集中。此外，糯玉米中的支链淀粉粘滞性强，透明度高，可作为食品加工业、纺织、造纸等工业的重要原料。糯玉米产业在我国发展潜力巨大。

1 糯玉米生产基本情况

1.1 玉米家族中的重要成员

从生物学角度（主要依据籽粒形态和成分），玉米可分为马齿型、硬粒型、半马齿型、粉质型、甜质型、甜粉型、爆裂型、蜡质型、有稃型9种类型。糯玉米（蜡质型）则是9种类型之一，是玉米家族的重要成员。从收获物和用途上划分，玉米可分为籽粒用玉米、青贮玉米、鲜食玉米三大类。其中鲜食玉米是指可象水果蔬菜一样收获和食用鲜嫩果穗的玉米类型，主要有甜玉米、糯玉米和笋玉米。目前我国鲜食玉米年种植面积近130万 hm^2，其中糯玉米种植面积约80万 hm^2，占总面积的60%以上，是鲜食玉米的主力军。

1.2 糯玉米的起源及 *wx* 基因遗传研究

1492—1496 年，哥伦布发现美洲大陆，将玉米从美洲带回欧洲，陆续传遍世界各地。糯玉米是玉米在16世纪传入中国种植后，发生变异而形成的一种新玉米类型，中国西南地区（云南、广西壮族自治区一带）是公认的糯玉米起源中心。我国史籍中在1760年之前就有黏玉米的记载，在我国有很长的种植历史。糯玉米的糯质特性由位于第9染色体短臂上的单隐性基因 *wx* 控制。*wx* 基因是一个包括至少31个异点等位基因的复合基因座，由编码颗粒凝结型淀粉合成酶的基因突变而来，该基因活性被抑制，不能合成直链淀粉，即表现为糯质。20世纪80年代，国外学者对糯玉米种质的遗传基础进行了研究，表明 *wx* 基因包含14个外显子和13个内含子，长约3 718bp。田孟良等对我国西南地区糯玉米种质 *wx* 基因进行了部分核苷酸序列测定，推测我国糯玉米是由硬粒型玉米单基因突变产生；*wx* 基因第10外显子中，糖基转移酶结构域起始位置核苷酸的缺失是中国糯玉米独特的突变机制。陈亭亭利用糯玉米和普通玉米材料对 *wx* 基因进行了更深入研究，表明不同糯玉米材料 *wx* 基因的启动子序列均发生了变异，该变异减弱了束缚态淀粉合成酶（合成直链淀粉）的活性，导致其 *wx* 基因转录水平显著低于普通玉米；同时发现 *wx* 基因在第2、4、7外显子处均有不同程度的碱基缺失，且均与淀粉合成有关。

2 国外糯玉米的发展与现状

国外糯玉米以美国研究和利用最多。1908年，美国传教士（J. M. W. Farnharm）通过上海领事从云南征集了几个玉米地方品种，寄交美国农业部国外引种处（the U. S. office of

Foreign Seed and Plant Introduction）。并附言说"这是一些特殊的玉米,有几种颜色。中国人说它们都是同一品种,比其他玉米要黏得多,可能会发现它有新的用途"。1908年5月,植物学家柯林斯（G. N. Collins）把从中国寄去的糯玉米种子种在华盛顿附近,其中有52株成熟,并将结果在1909年12月的《美国农业新闻简报》上发表。通过种植观察,他把起源于中国的糯玉米隐性突变基因定名为"wx",并给它定名为"中国蜡质玉米"。但相当一段时间内,只是出于好奇在遗传试验中作为标记基因而种植,对糯玉米育种研究较少。1936年,依阿华州立大学发现糯玉米支链淀粉的性质与当时进口日益困难的木薯块根淀粉相似,遂开始大规模杂交育种计划,并于1942年育成第一个粘玉米杂交种投入生产。随着研究深入,美国玉米育种家将wx性状转育到高产杂交种或自交系,使蜡质基因不再是高产的限制因子,糯玉米和普通玉米有相近的产量,且糯玉米的抗病性也与普通玉米相似。70年代初,美国玉米带遭受小斑病重创后,由于糯玉米的优良抗性和独特品质,使其更成为一个研究重点。目前,美国有多家种子公司从事糯玉米种子销售推广工作,种植面积约30万hm^2,主要集中在Illinois和Indiana中部,Iowa北部,Minnesota南部和Nebraska等地,种植面积稳中有升。美国糯玉米品种几乎全部是由大田玉米转育而来,即在选育出一个有商业化推广价值的杂交种和骨干自交系后,将其亲本自交系转育为同型糯质自交系（其他同型系还有甜、雄不育、转基因等）。美国糯玉米基本不作为鲜穗和速冻加工类产品消费,其中70%以上用于淀粉加工,将近30%的糯玉米用作饲料。美国糯玉米淀粉年生产量在160万~203万t,占整个湿磨淀粉产量的8%~10%,利用糯玉米支链淀粉制作的食品已达400多种。随糯玉米淀粉用途的不断扩大,加拿大、欧洲等地目前也有较稳定的糯玉米种植面积。

3 我国糯玉米发展概况

糯玉米虽起源于我国,在我国西南也已长期传统种植,但都是以零星种植农家种为主,我国糯玉米育种及产业起步较晚,直到20世纪70年代,我国才开始糯质玉米的杂交育种工作。烟台市农业科学研究所从非糯品种衡白多穗中得到糯质突变体衡白522,并于1975年育成我国第一个糯玉米单交种烟单5号（白粒）。1989年山东省农业科学院育成鲁糯玉1号,但受当时生产条件和市场需求所限,大都未能系统而深入地进行,导致我国糯玉米育种长期处于低迷状态,生产上零星种植,品种数量少,种类单一。至20世纪90年代,我国糯玉米育种才逐渐加快,陆续选育出一些品种,如中糯1号、苏玉糯1号、垦粘等。但此阶段我国糯玉米育种总体上仍是品种类型少,产量和品质矛盾突出,影响了糯玉米产业的发展。

进入20世纪以来,糯玉米品种选育和产业得到快速发展。糯玉米种植面积由20世纪初的7万多hm^2增长至目前约80万hm^2。这得益于此阶段大批优良种质资源的创新和一批优良糯玉米品种的选育及推广。特别是优良糯玉米品种京科糯120及京科糯2000的出现,实现了我国糯玉米高产与优质的结合,极大提升了我国糯玉米品种水平,促进了糯玉米产业发展。京科糯2000由北京市农林科学院玉米研究中心于2000年组配而成。其母本自1996年开始,利用硬粒型基础材料经北京、海南两地4年8代选育而成,品质与产量均表现突出;父本BN2利用马齿粉质型种质为基础材料,同样经过北京、海南两地4年8代选育而成,与京糯6表现出极强的特殊配合力。京科糯2000品种的育成创新了我国糯玉米杂优模式，该模式目前已成为我国糯玉米育种主导模式。京科糯2000于2006年通过国家审定，之后陆续通过北京、上海、福建、吉林等近20个省市区审定，并成为我国第一个通过国外审定的玉米品种。京科糯2 000亩产鲜穗可达1 000kg以上，口感糯中带甜，绵软适口，实现了产量与品质的结合。适采期可达10d，比之前的对照品种大大延长；适合企业大规模种植和加工。目前京科糯2000年播种面积保持在40万hm^2左右，约占全国糯玉米种植总面积的一半，对我国糯玉米产业链延伸和产业整体发展起到了极大的推动作用。

与普通玉米相比，糯玉米籽粒淀粉为分枝型，在籽粒特性、用途以及育种材料、目标、方法等方面明显不同于普通大田玉米。鲜食糯玉米不但要像普通玉米一样具有高产稳产、多抗广适的优良特性，更重要的是还要满足营养品质高、口感风味好、消费者喜欢吃，以及在成熟期、籽粒颜色、果穗加工等方面的多样化要求。因此，选育出优良糯玉米新品种比普通玉米育种难度要更大。在长期糯玉米选育过程中，我国育种家将不同种质的优良性状聚合，重点选择具有产量高、品质优、抗性强、口感好等特点的优良材料，同时根据市场需求，选育专用化品种，用于加工、鲜食、速冻等不同用途。

经过长期选择和创新，我国糯玉米育种形成了鲜明的中国特色，在全球范围具有明显的领先优势：①作为糯玉米起源地，我国糯玉米种质资源丰富，且具有不同地方特色。②持续创新和选育出优良糯玉米品种。2000—2014年我国审定糯玉米品种共计432个，育成通过国家审定的糯玉米新品种共计69个，包含不同类型、颜色、用途、熟期等。③我国已成为世界糯玉米种植面积最大的国家，并仍保持增长势头。④品种和产品输出。不但玉米生产上使用的全部是我国自育的品种，而且品种和加工产品均实现了向国外输出和出口，从品种到产品都具有国际市场竞争力。

4 目前市场主推品种类型

4.1 白色，糯玉米的主流色

白色糯玉米是目前我国所选育糯玉米中的主流色。如中国农业科学院作物育种栽培研究所选育的中糯1号，江苏沿江地区农业科学研究所选育的苏玉糯1号，北京市农林科学院玉米研究中心选育的京科糯2000等，重庆农科院选育的渝糯7号等。白色糯玉米品种有早、中熟类型，适宜种植区域广，品质优良，产量高，目前白色糯玉米在我国年播种面积约53万hm^2，占糯玉米总种植面积的70%左右。

4.2 金黄色，糯玉米的经典色

金黄色糯玉米是糯玉米中的经典色。与白色糯玉米不同之处在于其籽粒含有胡萝卜素。我国黄色糯玉米品种数量较少，主要代表品种是垦粘1号，由黑龙江省农科院作物所玉米育种室育成，以该类型品种为代表的黄糯玉米主要集中在我国东北及西北的早熟区域种植，由于更新品种较少，黄糯玉米在我国的种植面积也逐渐减少，目前推广面积约占总种植面积的10%。国外的玉米籽粒加工及作饲料的糯玉米基本上都是黄色糯玉米，黄色糯玉米有很大潜力。

4.3 花彩色，糯玉米的多样化

糯玉米中除籽粒颜色为白色、黄色的品种外，为满足市场多样化需求，我国还选育出好吃好看的五彩糯玉米品种，包括紫色、黑色、多种颜色混合的彩色等，如京紫糯218、天紫23、沪紫黑糯1号、京花糯2008。花色糯玉米在市场中占有10%左右的份额，由于市场需求的改变，种植面积呈现逐年增加的态势。

4.4 甜加糯，糯玉米的新类型

甜加糯品种是我国自主创新的一种鲜食玉米新类型，是目前鲜食玉米育种的新方向。其组配模式中的亲本之一为甜加糯种质，同时含有甜、糯双隐性基因（$shshwxwx$），在与另一亲本糯性种质（$wxwx$）杂交后，F_1籽粒表现为糯质（$shshwxwx$），F_2果穗中则同时含有甜粒和糯粒，且甜质籽粒与糯质籽粒之比为1∶3。我国在2000年左右就有双隐性甜玉米纯合体和甜糯纯合体材料的研究，但当时多是利用普甜基因（$susu$），超甜基因（$shsh$）较少，由于种质资源的限制，品种产量难以突破，未能进行大面积推广。近年来，随育种技术的提高和种质改良的深入，我国甜加糯（$shshwxwx$）双隐性种质逐渐丰富，甜加糯型品种也逐渐增多。目前甜加糯类型代表品种有北京市农林科学院玉米研究中心选育的京科糯928、京科糯2010、农科玉368，海南绿川种苗有限公司选育的美玉3号、美玉13号、美玉16号，北京金农科种子科技有限公司金糯628等。甜加糯型玉米聚合了甜玉米"甜、脆、鲜"和糯玉米"糯、绵、香"两方面优点，口感香味浓郁、甜糯相宜、柔软度好。并具有高产稳产、多抗广适、易制种等综合优点。深受种植户和消费者喜欢。目前该甜加糯玉米在我国属于起步阶段，种植面积还较少，约10万hm^2。

4.5 高叶酸等，糯玉米的新亮点

叶酸是一类水溶性B族维生素，又称维生素M、维生素Bc，对维持人体正常生理功能必不可少。研究证明，叶酸摄入不足会引发巨幼红细胞贫血和胎儿神经管发育缺陷等病症，因此叶酸对孕妇和胎儿的健康尤其重要。人体自身不能合成叶酸，只能从饮食中摄取，成人每日叶酸摄入量为400μg，孕妇则需要每日摄入600μg左右。为满足正常需要，目前孕妇主要靠服用人工合成叶酸。经中国农业科学院生物技术研究所范云六院士团队检测研究发现，京科糯928（甜加糯型）等糯玉米中叶酸含量要大大高于普通玉米和一般糯玉米、甜玉米。其中京科糯928中叶酸含量可高达294.88μg/100g鲜籽粒，高于菠菜中叶酸含量（196μg/100g）。而一般的籽粒玉米（百克干重含量），以及糯玉米、甜玉米（百克鲜重含量）的叶酸含量都在100μg以下，有些甚至低至10μg左右。高叶酸含量是鲜食糯玉米的新亮点，将会有很大的开发和利用价值。目前，糯玉米中叶酸含量的合成及调控机制仍在研究中。

4.6 玉米籽粒糯质化

美国Custom公司用5个糯玉米与其同型普通大田杂交种进行产量对比，两者间产量没有明显差异；另一个试验用14个糯玉米与其同型大田普通杂交种在两个地点进行产量对比，同样证明产量差异不明显；这至少说明转换成糯质同型系后籽粒产量可以保持普通大田玉米的籽粒产量。另据资料，糯玉米的消化率高于普通玉米，以糯玉米为饲料喂养家畜，有更好的营养。因此，选育高产的籽粒用糯玉米，在淀粉加工、食品加工、饲料以及青贮玉米等多方面都大有潜力。

5 我国糯玉米产业发展概况

我国糯玉米产业发展大致可分4个阶段：2000年以前，我国品种审定体系不完善，各科研单位对糯玉米育种多以公益性研究为主，商业化育种进程缓慢。1994年，吉林天

景食品有限公司成立,成为我国糯玉米速冻加工企业的开端。之后在我国东北、山西等地又成立了一些糯玉米加工企业,但总体数量较少。此阶段我国糯玉米种植面积约 10 万 hm^2。2001—2005 年,随着我国种子法的出台和国家糯玉米审定制度的建立,我国糯玉米商业化育种快速发展,各地不断有省级审定和国审品种出现,品种品质明显提高。糯玉米种植面积快速上升至 27 万 hm^2。糯玉米速冻加工企业达到上百家,企业年生产规模由原来年加工几百万穗上升为数千万穗,糯玉米速冻产品出口量也呈快速增长趋势,主要出口至韩国、东南亚等国家。2006—2010 年,糯玉米国审和省审品种逐渐增多,京科糯 2000 等高产优质型品种引领了这一时期的糯玉米加工和鲜售市场。彩色糯玉米品种有了一定的推广面积,市场出现了明显的细化。此时,我国糯玉米速冻加工企业数量迅速增加,导致我国糯玉米种植总面积也达到 50 多万 hm^2。但其间 2008 年左右,由于金融危机对我国整体经济形势的影响,出现国内市场需求下滑,出口量滞销的现象,导致我国加工企业数量及加工量明显减少,糯玉米鲜售量也大幅下滑。但随后二三线城市市场的开拓,激发了我国内需市场的巨大需求量,鲜食糯玉米与加工糯玉米市场快速回升。2011—2015 年,随着我国经济的发展,糯玉米市场需求量持续增大,出口市场回暖,糯玉米加工企业数量有所恢复,数量达到上千家,年加工量达到 100 亿穗。此阶段我国糯玉米总种植面积由原来的 50 多万 hm^2 增加至目前的约 80 万 hm^2,其中用于速冻加工的糯玉米约 25 万 hm^2,用于鲜果穗销售的糯玉米面积约 55 万 hm^2。品种类型细化,突出品质与品牌是我国糯玉米此时发展的主导思路。

6 对我国糯玉米产业发展的几点建议

6.1 调结构,转方式,打造绿色有机都市型糯玉米产业

目前我国居民消费观念正在从生存型消费向享受型消费过度。糯玉米营养物质丰富,口感独特,以观光采摘模式将糯玉米与自然、生态相结合,可极大丰富和满足消费者需求,又可促进一二三产业的融合发展,带动农业产业链的延伸,增加农民收入。

6.2 以鲜售为主,加工为辅

从目前消费形式和消费数量来看,我国糯玉米在一定时期内的发展趋势仍将以消费鲜穗为主,种植户和生产商应根据市场供需形式,合理安排种植时间和方式,就近种植,避免在果穗保鲜、储运等方面对品质造成影响,保证为消费者直接提供高档优质鲜果穗。

6.3 将甜加糯类型品种作为发展重点

近年来,为提升鲜食玉米消费档次,实现品种类型的多样化,我国自主创新了甜加糯型鲜食玉米品种,经推广种植,深受消费者青睐。继续选育和推广该类型品种将是鲜食玉米产业的发展趋势。

6.4 糯玉米秸秆有效利用

糯玉米除主要利用果穗以外,其茎叶、穗轴等都有较高的利用价值。糯玉米秸秆多汁鲜嫩、营养丰富,通过颗粒或发酵可做成青贮饲料。利用糯玉米秸秆饲喂奶牛可增加产奶量,改善乳品质同时可减少精料的饲喂量,降低成本。

6.5 细分市场,突出高端,发展特色,打造品牌

糯玉米作为鲜食玉米,具有较高的营养价值和独特的口感风味,正逐渐成为人们日常消费的蔬菜或食品。继续糯玉米品种类型、花色、品质和口感风味的创新,形成品种高

端、独具特色的糯玉米产业将会有利带动我国特色农业的发展。另一方面，应对我国高端糯玉米产品加大宣传力度，提高市场认知度，在糯玉米规模化和产业化的基础上，实现品牌化，提高我国糯玉米产品的竞争力。

参考文献（略）

本文原载：玉米科学，2016，24（4）：67-71

京科甜系列水果型优质玉米品种选育及应用

卢柏山　史亚兴　徐　丽　樊艳丽　陈　哲　霍庆增　张爱武

（北京市农林科学院玉米研究中心/玉米DNA指纹及分子育种北京市重点实验室，北京　100097）

摘　要：我国甜玉米育种起步较晚，与国外水平相比，我国甜玉米产业发展还存在品种类型单一、品质差、加工专用型品种缺乏等问题。针对甜玉米尤其是水果型甜玉米生产及市场发展中的问题，近十年通过引进200多份国外和国内优良种质资源，并进行种质创新和优化杂优模式。按照"高、大、严"选系原则，选育出20余个骨干自交系；组配选育出京科甜115、京科甜116、京科甜183、京科甜158、京科甜2000、京科甜168、京科甜189等京科甜系列水果型玉米品种，口感甜脆、商品性明显提高，同时具有抗性强、产量高的特点。京科甜系列品种已分别通过国家和省级农作物品种审定委员会审定，在我国华北和东北地区得到大面积推广应用。

关键词：水果型甜玉米；选育；杂优模式；京科甜系列品种；应用

Breeding and Application of Jingke series of Fruit Sweet Corn

Lu Baishan, Shi Yaxing, Xu Li, Fan Yanli, Chen Zhe, Huo Qingzeng, Zhang Aiwu

（Maize Research Center, Beijing Academy of Agriculture and Forestry Sciences, Beijing Key Laboratory of maize DNA Fingerprinting and Molecular Breeding, Beijing, 100097, China）

Abstract: Compared with other countries, the development of national sweet corn industry, which started late, still had some deficiencies, such as lacking of variety type, poor quality, and processing-type deficiency. Based on the production and market situation of sweet corn, especially the fruit type, we introduced more than 200 germplasm home and abroad, which were innovated and applied about heterotic pattern. Seven sweet corn varieties, Jingketian126, Jingketian115, Jingketian116, Jingketian183, Jingketian158, Jingketian2000, Jingketian189, which had a higher quality, high yield, and strong resistant, were developed. So far varieties of Jingke series innovated heterotic pattern of national sweet corn breeding, had been authorized by national or province Crop variety approval committee, and they were widely planted in the northern part of our country.

Key words: Fruit sweet corn; Breeding; Heterotic pattern; Jingke series varieties; Application

---

作者简介：卢柏山，推广研究员，硕士，主要从事鲜食玉米遗传育种研究

甜玉米籽粒乳熟期含糖量高，含有人体必需的氨基酸、蛋白质及多种维生素，因此也称作果蔬型玉米，而那些皮薄无渣、口感甜脆可以生吃的优质甜玉米也称作水果玉米，并深受消费者青睐。随着我国种植结构调整及北京市建设现代化都市型农业的定位方向，在我国尤其是北京地区大力发展甜玉米产业将有重要意义及广阔前景。

近年来，在市场需求的推动下，全球甜玉米产业化进展迅速。美国是最早进行甜玉米遗传研究和育种的国家，也是目前世界上最大的甜玉米生产国和出口国。品种多具有品质优良、适宜加工的特点。我国甜玉米育种起步较晚。20世纪60年代我国第一个甜玉米品种"北京白砂糖"的育成，开启了我国甜玉米育种事业的发展。随研究的不断深入，市场需求的不断增加，全国越来越多的科研单位和企业投入到甜玉米发展中，甜玉米品种不断涌现，种植面积也逐渐增大，目前已经成为亚洲第一大甜玉米生产国和出口国。与国外甜玉米产业相比，我国甜玉米产业在发展的同时，还并存着一些问题。通过调查研究发现，生产上大面积种植的甜玉米品种仍是以产量为主要目标，缺少对品质、口感、颜色、用途等的研究利用，导致品种类型单一、品质差、果皮厚、皮渣多、不适宜生产加工，能被称作水果型甜玉米的品种则更少，因此远不能满足甜玉米市场发展的需要，甜玉米品种的选育还急需改进和提高。

近几年，北京市农林科学院玉米研究中心一直致力于甜玉米的选育研究与推广应用工作。针对目前市场上甜玉米品种类型单一尤其是水果型甜玉米，不能满足生产发展需要的现状，提出了"以市场需求为导向，综合围绕甜玉米产量性状、品质性状、商品性状"的育种方向，选育出京科甜系列优质水果型甜玉米品种。本文在甜玉米种质改良创新、自交系选育利用等方面分析了京科甜系列水果型优质玉米品种的选育特点，同时介绍了其推广应用情况，以期为甜玉米育种及产业发展提供参考。

1 种质扩增、改良与创新

首先在种质资源创新上，先后收集了国内外200多份种质资源，按照"高品质、大群体、严选择"的选系原则，选育出20多个甜玉米骨干自交系，组配出10余个优良杂交种，部分已通过国家或省级农作物品种审定委员会审定。

具体选育过程为：在$S_0 \sim S_1$代一般不进行育种性状的选择，$S_2 \sim S_3$分离世代快速使种质群体扩大，到S_3世代种质扩增至3 000~5 000株，在必要的情况下可扩增至10 000株，以提高优良品质与产量基因出现并聚合的频率。扩增至3 000~5 000株群体时，严格筛选抗性好、雌雄协调、果穗颜色品质较好的自交材料，并与测验种组配成测交种。测交种在田间进行抗性、品质、一般配合力的筛选与测定，筛选出10%的优良组合，对应的约30~50个穗行的测配果穗进行自交保种。到S_4代，继续进行穗行选择，最终获得2~3个穗行长势较好，整齐一致的高代自交材料。S_6代后一般仅存一到二个具有优良目标基因的高代自交穗行，稳定纯合成系。

S_3代测配选择的过程是最为重要的阶段，尤其是对品质的选择。在选育京科甜系列水果型甜玉米品种时，总结出以下几点原则。

（1）高品质。定位于高品质育种目标，选择产量、品质优势明显、优点突出的种质资源，锁定优质品质基因。

（2）大群体。通过放大群体数量，增加优良基因重组频率。

（3）严选择。对测配的 F_1 进行产量、品质、抗性的严格筛选，大量淘汰，明确产量、品质等优势基因重组频率高的穗行。

目前骨干甜玉米自交系 T68 是通过上述原则与方法，以美国甜玉米杂交种麦哥娜姆为基础种质选育而成的。麦哥娜姆是美国高品质甜玉米杂交种的代表品种，自身属于温带高品质杂交甜玉米品种，具有高产，品质优良的性状。将该材料在 $S_2 \sim S_3$ 世代扩增至 3 000 株左右。利用国内优良甜玉米自交系 SH-251 对扩增优势穗行组配测验，组配组合量达到 500 多个，测配同时保留组配穗行果穗种质资源，其余穗行严格淘汰。通过对 F_1 的田间测试，将优质基因锁定在第 T68、T179、T813 穗行上。通过进一步对测配杂交种品尝、口感，以及果穗性状的筛选，选择了第 T68 穗行，取名 T68。再通过连续两代的纯化过程，成为高品质骨干自交系。目前该自交系已组配出 4 个已审品种，以及数个优势组合。

2 主要杂种优势模式

甜玉米是杂种优势利用成功的作物之一。北京市农林科学院玉米研究中心在选育优良自交系的同时，也创新优化了甜玉米育种的杂种优势模式。目前，主要创新优化的模式有国外种质改良系与国内种质改良系杂交；国外种质改良系与国外种质改良系杂交，这两种模式的主要创新点在于聚合了国外种质优良的品质性状，增加了甜玉米品质与产量优势基因重组的频率，提高了品质和口感，果皮薄脆、糖分增加、籽粒更加柔嫩。

国外种质改良系与国内种质改良系杂交的模式是目前应用最多、也是应用最成功的模式。以京科甜 158 的选育为例，母本 SH-251 是以杂交种超甜一号为选系材料，经过 6 代自交于 2002 年选育而成，为纯白色籽粒超甜玉米自交系。该自交系抗病性强、早熟、穗型好、口感好。父本 T68 是以从美国引进的早熟甜玉米品种麦哥娜姆经过 6 代自交于 2004 年选育而成的，为纯黄色籽粒超甜玉米自交系。T68 自交系具有很高的一般配合力和特殊配合力。甜度 18 度左右，鲜籽粒亮黄色，皮薄，口感好。高抗玉米大、小叶斑病、活秆成熟。

2006 年冬在海南初配 SH-251×T68 杂交组合，经过多年多点试验，鉴定出该组合产量高、高抗病性、品质好等特点。2007—2008 年在北京多点品比试验，两年平均亩产 775kg，平均比对照京科甜 183 增产 10%。该品种于 2011 年通过北京市农作物品种审定委员会审定，并被定为北京市玉米更新换代品种。其他水果型甜玉米品种组配模式详见表 1。

表1 其他水果型甜玉米品种组配模式

品种名称	母本种质基础	父本种质基础	审定编号	适应区域	推广面积
京科甜 115	母本"T9"是以国内加强甜玉米群体为基础材料选育而成	父本"T2"是利用国外甜玉米杂交种"3243"为最初选系材料，自交两代后与国内甜玉米自交系 YT83 组配成基础材料	京审玉 2002010	北京，河北，等我国北方地区	50 万亩

(续表)

品种名称	母本种质基础	父本种质基础	审定编号	适应区域	推广面积
京科甜116	母本"T9"是以国内加强甜玉米群体为基础材料选育而成	父本"F101"是利用国外杂交种SW1连续自交二代后，再与国内地方甜玉米品种杂交组配成基础材料	京审玉2003004	北京，河北，等我国北方地区	20万亩
京科甜183	母本"双金11"以国外双色超甜玉米品种为基础材料选育而成	父本"SH-251"是以国内超甜1号杂交种为基础材料连续自交6代而得。	京审玉2005014 鄂审玉2010003	北京，河北，等我国北方地区，湖北省春播地区	40万亩
京科甜2000	母本"金杂11"来源于日本杂交种金银穗自交2代后的材料×杂甜1号获得种质基础选育而成	父本"SH-251"是以国内超甜1号杂交种为基础材料选育而成白色自交系	京审玉2007007	北京，河北，等我国北方地区	20万亩
京科甜168	母本"U-SC-1"以澳大利亚超甜玉米材料U-SC为基础材料选育而成	父本"水玉"以国内超甜玉米品种甜102为基础材料选育而成	京审玉2008010	北京，河北，等我国北方地区	10万亩
京科甜189	母本"Sh-241"是以超甜1号为基础材料选育而成纯黄色自交系	父本"T68"以美国超甜玉米麦哥娜姆为基础材料选育而成		北京，河北，等我国北方地区	50万亩
京科甜158	母本"SH-251"是以国内超甜1号杂交种为基础材料选育而成白色自交系	父本"T68"以美国超甜玉米麦哥娜姆为基础材料选育而成	京审玉2011008	北京，河北，等我国北方地区	10万亩

3 京科甜系列水果型优质玉米品种的生产应用

3.1 京科甜系列种子生产技术要点

甜玉米是控制胚乳碳水化合物代谢的纯合隐性突变体。成熟的甜玉米种子胚乳皱缩凹陷，碳水化合物积累少，导致发芽率低。除了正常的播种质量，错期，抽雄等常规制种要求外，如何提高甜玉米种子发芽率以及如何提高制种产量是甜玉米应用中的难题。针对这一问题，根据京科甜系列品种自身的特点，提出"前促、中供、后控"的田间水肥管理原则，同时配合采收期的控制、种子挑选分级等，提高了京科甜系列水果型优质甜玉米种子质量和产量，明显提高了种子发芽率，每亩制种产量达到150kg。

"前促、中供、后控"的田间水肥管理原则主要针对京科甜系列高品质品种。具体为：前期施足底肥，省去蹲苗期，促使其快速生长。大喇叭口至灌浆初期，不间断保证肥水供应。灌浆至完熟期，逐步减少肥水供应，防止贪青徒长，避免产生穗发芽现象。

根据京科甜系列品种果皮薄，含糖量高的特点，应在授粉后30d左右将果穗苞叶撕开，加快果穗脱水速度，防止种子霉变。在35~40d根据天气和降水情况及时采收、烘干至安全水分含量。

3.2 品种推广

京科甜系列品种自审定以来主要在我国北方地区推广种植，一直是北京京郊地区的主栽品种，并成为我国华北、东北地区的主推品种。其中京科甜189为高品质，高产类型品种，适宜果穗脱粒速冻加工，在河北秦皇岛地区、黑龙江、吉林等加工厂聚集的地区被广泛种植。京科甜158为高品质高产类型品种，适宜鲜果穗上市，在北京京郊、河北等地被广泛种植。同时，京科甜系列品种中京科甜183被定为北京市甜糯玉米区域试验对照品种，京科甜158被定为北京市玉米更新换代品种，均被大面积推广应用。目前我们主要以联系鲜果商、速冻玉米加工厂，通过与农户签订订单的形式进行推广，既可以使京科甜系列品种快速占领高端市场，又可保证种植户长期稳定的效益。据统计，京科甜系列玉米品种在全国地区累计推广种植200万亩；在北京累计推广种植几十万亩，京科甜系列品种每年可为北京京郊种植户创造上千万元的直接经济效益，同时满足了市场对高端甜玉米的需求。

通过京科甜系列水果型优质玉米品种的选育，创制出具有我国自主知识产权的优良自交系，拓宽了我国甜玉米种质资源。同时，建立了适合我国甜玉米发展的杂种优势模式，提高了我国甜玉米育种的整体水平。

参考文献（略）

本文原载：作物杂志，2015（1）：46-48

鲜食玉米的发展与前景——探索我国甜玉米的北方市场

史亚兴[1]　张保民[2]

(1. 北京市农林科学院玉米研究中心，北京　100097；
2. 北京保民种业有限公司，北京　100024)

1　我国鲜食玉米种植现状

我国鲜食玉米在过去已经形成"南甜北糯"的种植格局，但随着广大育种工作者对高端温带品种的选育以及新品种类型的不断出现，我国鲜食玉米的种植格局逐步发生了变化。在种植区域方面，目前，我国鲜食玉米的种植区域由北到南主要分布在东北三省、京津冀、甘肃、南宁、江浙沪、四川、重庆、贵州、云南、广西壮族自治区和广东等地，从各省种植面积分析来看，我国鲜食玉米总种植面积约 1 267 300hm^2（图1）。其中，中部区域已经形成了以浙江省为中心的甜加糯玉米种植区，在我国北方，高端甜玉米的种植面积也呈现了明显的上升趋势。在种植类型方面，随着冷链物流及交通的不断发展，鲜果穗流通越来越快捷，鲜食玉米反季节种植技术也越来越成熟；在以往甜玉米种植的基础上，我国南方糯玉米的种植面积也有所增加，经济效益逐渐提高。

在全国范围内，鲜食玉米种植面积最大的区域为两广地区，总面积约 300 150hm^2，其中90%为甜玉米。该地区一般早春1月中旬至2月中旬即可种植，可进行春、夏、秋一年三季种植，种植面积超过 200 100hm^2，占全国鲜食玉米种植面积的50%以上，广东省从发展初期的珠江三角洲不断向东南两翼扩展渗透，形成了从梅州、潮州、惠州、广州到肇庆、阳江、茂名、湛江的鲜食玉米优势产业带，目前已经形成以广东为中心的甜玉米优势产区，且产业化发展不断完善。两广地区主要种植的鲜食玉米品种包括华珍、先甜5号、仲鲜甜3、新美夏珍、金中玉、夏王、瑞珍；甜糯玉米包括美玉系列、京科糯928、彩甜糯6号等。

种植面积居全国第二位的是东北地区，面积在 266 800hm^2，其中以吉林省的种植面积最大，约占东北地区总种植面积的一半。该地区主要选择在5、6月初进行播种，于8月底至9月中旬采收，年种植一茬，60%的果穗被用于速冻加工。目前，东北地区从事鲜食玉米加工的注册厂家有近300家，生产的产品包括甜玉米罐头、速冻果穗、速冻玉米粒、玉米饮料等，一些知名品牌出口日本、美国、韩国及东南亚等地，东北产区鲜食玉米整体产量、产值和市场占有率等主要经济指标在全国均名列前茅。东北地区主要种植的品种包括糯玉米垦粘1号、吉农糯7、黄糯262、京科糯2000、天紫23；籽粒加工型甜玉米米哥、脆玉、金菲以及鲜食型甜玉米库普拉和双色先蜜等。

京津冀地区鲜食玉米的种植面积在 133 400hm^2，90%为鲜食型品种，且由于该地区聚人口优势、交通优势、技术优势于一体，对品种的质量和多样化要求较高。速冻加工厂多集中于万全、秦皇岛、昌黎和徐水等地,年果穗加工量可达4亿穗,其中一半产品销往国内大中城市,一半出口国外。该地区一般于每年3月底至6月初进行春播、7月上旬进行夏播,主要种植品种包括鲜食糯玉米京科糯2000、加工型白糯玉米万糯2000、加工型花糯玉米天紫23、京花诺2008、籽粒加工型甜玉米米哥、金菲;鲜食甜玉米金冠218、万甜2000;甜糯玉米京科糯928、彩甜糯6等。

2008—2015年,我国鲜食玉米国家区试参试品种共749个,甜玉米262个,糯玉米(含甜加糯类型)共487个;南方甜玉米参试品种数为174个,北方为88个,南方糯玉米参试品种数为278个,北方为209个(图2);通过国家审定的鲜食玉米品种共42个,甜玉米11个,糯玉米(含甜加糯类型)31个(图3)。可以看出我国近年来参加区试的甜玉米新品种产量呈增高趋势(图4,表1),甜玉米新品种的口感和品质显著提高(图5),皮渣率显著下降(图6,表2),尽管参加国家区试和通过审定的甜玉米品种远没有糯玉米多,但是甜玉米的研究和发展已经受到了很大重视。

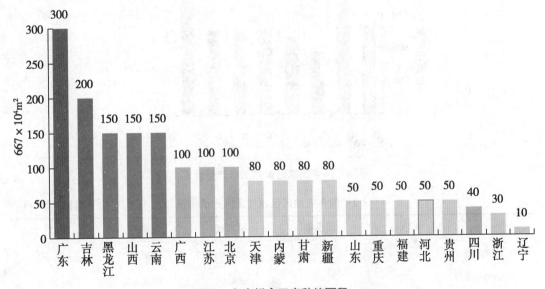

图1 各省鲜食玉米种植面积

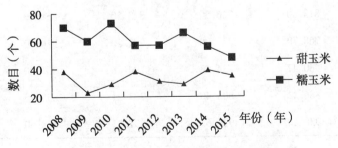

图2 2008—2015年国家区试鲜食玉米品种数目

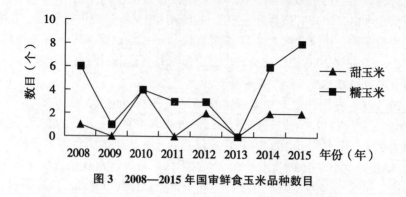

图3　2008—2015年国审鲜食玉米品种数目

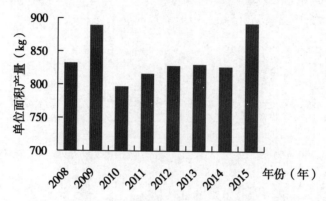

图4　历年国家区试甜玉米产量情况

表1　历年各地区区试甜玉米产量情况　　　　　　　　　　　单位面积产量（kg）

年份	东华北	黄淮海	西南区	东南区	平均
2008	—	—	875.60	789.00	832.30
2009	—	—	868.50	910.10	889.30
2010	841.80	713.90	811.40	821.80	797.23
2011	813.60	726.50	873.50	851.40	816.25
2012	768.70	726.00	901.50	916.70	828.23
2013	930.47	734.98	794.18	861.20	830.21
2014	886.42	803.40	783.21	835.16	827.05
2015	938.88	926.63	837.21	868.60	892.83

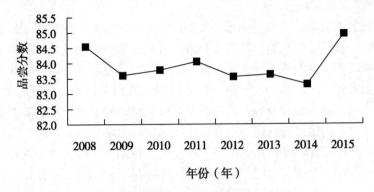

图 5 历年国家区试甜玉米口感品质评价

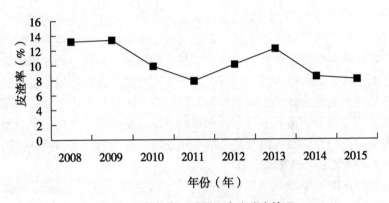

图 6 历年国家区试甜玉米皮渣率情况

表 2 历年各地区区试甜玉米皮渣率情况   %

年份	东华北	黄淮海	西南区	东南区	平均
2008	—	—	13.20	13.20	13.20
2009	—	—	12.90	13.90	13.40
2010	5.60	10.00	12.10	11.74	9.86
2011	4.80	12.40	11.50	2.80	7.88
2012	5.21	13.00	11.90	10.20	10.08
2013	5.78	16.25	12.17	14.53	12.18
2014	3.72	7.51	10.47	12.09	8.45
2015	4.91	8.84	11.13	7.37	8.06

2 国家政策调整后鲜食玉米的发展机遇

2016 年,农业部《全国种植业结构调整规划》的通知指出:适当发展鲜食玉米。适应居民消费升级的需要,扩大鲜食玉米种植,为居民提供营养健康的膳食纤维和果蔬。

2.1 我国玉米产业的现状

玉米主要分为 3 大类，籽粒用玉米：主流定位是高淀粉能量型作物，收获玉米的成熟籽粒，用于口粮、食品、精饲料、加工原料等；青贮玉米：收获玉米的鲜绿全株，经切碎发酵，用于牛羊等食草牲畜饲料的玉米；鲜食玉米：即像水果蔬菜一样收获和食用的玉米鲜穗。籽粒型玉米作为全球第一大作物，其种业市值最大，是种业竞争的焦点和制高点，是杂交优势技术应用最早最普及的作物，也是我国粮食增产的主力军。随着目前玉米高产量、高库存、高成本的生产现状形成以及我国玉米国际市场竞争力减弱、进口量增加的情况出现。国家临储政策和补贴政策也进行了调整，籽粒用玉米的种植面积和发展方式都亟待调整和转型，2016 年，农业部在《全国种植业结构调整规划》中指出，要适当发展鲜食玉米，扩大鲜食玉米种植，这意味着鲜食玉米的种植和发展从国家政策层面得到了支持和重视。

我国目前已经成为全球最大的鲜食玉米种植、生产、加工和消费的国家，不同于欧美各国以及全球甜玉米消费量最大的美国，我国鲜食玉米的发展独具优势和特色。糯玉米起源于我国，目前我国是世界上糯玉米种植面积最大的国家，我国生产使用的品种均为自主选育的品种，而且这些品种已经走出国门，成为了东南亚、韩国等国的主导品种，速冻果穗等糯玉米产品更是远销全球多个国家。与此同时，广大育种工作者依托我国糯玉米产业发展这一优势，结合对全球市场的考察，选育出了甜加糯等新的鲜食玉米类型，并进行大规模产业化应用，进一步加强我国鲜食玉米产业发展，并向周边各国及欧美国家传播。

2.2 改革背景下鲜食玉米的发展机遇

随着互联网行业的不断发展，"互联网+"的形式同各行各业紧密联系在了一起，随着"互联网+农业"的不断发展，各种各样的农产品有了新的产业发展模式。鲜食玉米的消费形式要随着互联网等行业的发展实现多元化，随着城乡居民生活水平的不断提高适时发展鲜食玉米观光采摘项目，从单一的农贸市场销售到农贸市场与电商销售相结合等多种形式，如北京房山某镇种植的鲜食玉米品种京科糯928，即采用电商销售的形式，电商以 1.5 元/根的价格收购果穗，再以 2.99 元/根进行 24h 内配送销售，以确保产品的品质，符合都市消费者对质量的需求，也迎合了消费者对便捷消费形式的青睐，同时实现了农户、经纪人、电商利润的上升。

在传统营销模式方面，我国鲜食玉米鲜果商已有上百家，他们引导了广大种植户，进行鲜穗收购，在全国范围内配送，占有全国鲜食玉米消费 40% 的市场，对全国鲜食玉米产业发展的影响举足轻重，因此，鲜穗经纪人在鲜食玉米产业下一步大发展中仍将起到十分重要的作用。而随着冷链物流和加工工艺的发展，可加工的鲜食玉米品种更加多样化，品质更加高端，鲜食玉米的加工产品类型将更加丰富，质量也会不断提高，发展高端鲜食玉米加工业成为行业发展的必然趋势。目前，自秦皇岛以北至黑龙江地区，主要生产的速冻加工产品由中低端向高端温带甜玉米品种转变，已经初步实现了甜玉米品种高端化的发展，在这一基础上，未来鲜食玉米高端产品加工产业化的道路将更加顺遂。

2.3 鲜食玉米产业面临的问题

目前我国鲜食玉米的发展仍存在着季节性产能过剩的问题，造成上市集中，库存压力大，经济效益低等问题。在生产方面，地区范围内品种单一，同质化问题严重，不能满足广大消费者多元化的消费需求。同时，随着生产成本的逐年增高，尤其是人工成本的不断提高，鲜食玉米产业利润空间逐年降低，需要寻求转型发展途径。在加工工艺上，加工水平亟待提高，产品质量问题难以确保，需要制定统一的标准来进行监测。在消费层面，仍有部分民众受到鲜食玉米转基因谣言的影响，对鲜食玉米产业的推动和发展存在着不利的影响，这就需要更多的宣传和政府以及权威机构的发声，同时确保育种工作严格遵循非转基因的要求，让广大消费者更加放心的进行消费。

2.4 在供给侧方面对鲜食玉米产业进行改革

在鲜食玉米种植总面积上，应避免盲目扩张，需要从过去盲目种植向计划种植方向转变，并实现与不同作物轮作倒茬种植。在土地成本降低的基础上，播种、收获和加工环节上需加大先进设备的投入使用，减少人工用量，使用高品质节约型化肥农药，从而降低生产成本。对于广大鲜食玉米种植户以及加工企业来说，生产过程中，可通过选择新优品种或生产上较稳定的品种进行搭配种植，以增加市场竞争力，对产品加工工艺进行持续不断的创新，可以利用不同品种的特点开发新的产品，改善加工工艺，提高产品品质。广大种子企业包括产品加工企业应该着重提高产品质量，降低次品率，生产过程中做到不盲目，不盲从，稳扎稳打，提高核心竞争率，淘汰落后产能，引进新设备、新技术，由粗放管理模式向精细化管理转变，不断引进新型人才，将发展目标放长远。

3 培育、引导甜玉米消费方式，打造以北京为中心的甜玉米北方市场

3.1 甜玉米的定位

甜玉米是世界范围内第三大蔬菜作物，在国际间农产品贸易中的定位为大宗蔬菜作物，地位排在马铃薯、番茄之后。甜玉米因其甜、嫩、香、脆的口感以及富含糖、膳食纤维、多种矿物元素、多种维生素、蛋白质、脂质和18种氨基酸等对人体有益成分的特点而深受消费者的喜爱。研究表明，长期食用甜玉米具有保健作用，可降低人体内胆固醇、饱和脂肪酸和纳的含量，从而降低高血压、冠心病以及某些癌症的发病率，其富含的核黄素和玉米黄质可预防、延缓中老年视神经黄斑恶化和白内障的发病率。甜玉米是欧美、韩国和日本等发达国家的主要蔬菜之一，美国是全世界最大的甜玉米生产国，也是最大的消费国。世界范围内的甜玉米产业发展总的趋势是稳步上升，各类甜玉米的产品分别有较大数量的出口，同时也有较大规模的进口，所有发达国家都是甜玉米生产和消费的主要国家，发展中国家的甜玉米生产基本处于出口大于内销的情况。

3.2 世界范围内的甜玉米产业方兴未艾

对于传统的甜玉米生产国——美国以及多数发达国家，甜玉米的生产和消费已基本稳定，基于降低生产成本的目标，越来越多的发达国家会把甜玉米的产品加工放在发展国家进行，这一过程就会引导许多原本不消费甜玉米的国家逐渐成为甜玉米的消费国，正如我国的南方，早年作为发达国家甜玉米的生产和加工基地，目前已经成为我国甜玉米消费量最大、产业最发达的地区。同样的，毗邻我国的越南、缅甸、泰国等国，现在成为了我国甜玉米鲜穗生产的反季节基地，这些国家也会同我国的南方一样，在不久的将来成为甜玉米的消费国。在将甜玉米种植和生产加工引进某些发展中国家时，由于地域和环境的影响，培育更适合该地区生产的优势品种就成为了必须要面临的一个问题，在先正达等一批知名的美国甜玉米种子公司多年的努力下，东南亚、非洲、南美等发展中国家和地区的甜

玉米产业也有了较快的发展。

3.3 中国甜玉米发展现状

据不完全统计，截止到2015年我国甜玉米的种植面积在366 850~400 200hm², 全国各地甜玉米加工厂有数百家，年加工速冻籽粒在400 000~500 000t, 甜玉米罐头40 000~50 000t, 真空包装穗2亿~3亿穗。以广东省为中心的南方市场在播种面积、鲜食加工比例、甜玉米消费方式以及人均消费量和消费频率等方面已经相当或高于美国水平，全国各地的甜玉米消费和加工产品的生产消费已经由四通八达的物流网络全面紧密地连接为一体，鲜食果穗消费的迅速发展，也极大的带动了甜玉米鲜穗种植，尤其是高中端品质品种的种植，给我国甜玉米产业未来的发展奠定了基础。

3.4 中国甜玉米未来的发展趋势

3.4.1 以北京为中心的北方市场呼之欲出

中国甜玉米市场正在进入一个快速发展的高速路，最重要的标志是以鲜穗鲜食消费成为主流的消费方式，甜玉米鲜穗开始进入千家万户。在南方市场中，城市居民甜玉米的消费水平已经接近甚至超过发达国家，未来5~7年时间里，南方市场的增长，一要靠发展品质优良的高端品种和符合南方居民消费放式的中高端品种，继续扩大城市居民甜玉米的消费频率和消费量；二是深入开发小城镇和广大农村近4个亿人口的日常消费。可以预见在甜玉米消费有着广泛基础的南方市场，未来5~7年消费面积仍有十分可观的增长量。目前中国甜玉米发展现状还是南方市场独大，这一方面是由于传统的饮食习惯促成，更是因为早期甜玉米反季节订单生产给当地居民较早接触甜玉米的机会，以及在之后足够长的时间里消费市场培育形成的结果，现在很多原来为广东省生产甜玉米鲜穗的省份也已经成为了甜玉米消费大省，据此，我们可以预言，未来5~7年甜玉米的发展将会以北方市场为主，一个以北京为中心、毗连京津冀、囊括十几个省市的北方市场呼之欲出。

3.4.2 中国北方甜玉米中心轮廓

中国北方市场所涉及的省市大致有：北京、上海、天津、河北、河南、山东、辽宁、江苏、安徽、陕西、山西、吉林、内蒙古自治区、黑龙江等十几个省市，这其中人口排序前50个城市居民人口数目在7 000万左右，比南方市场多出2 000多万，这是甜玉米鲜穗及产品的主要消费群体；北方市场十几个省市城乡总人口6.5亿左右，比南方市场多1.5个亿，这也是未来北方市场发展、扩大的潜力所在。北方市场有很多独特优：温带气候，昼夜温差较大，适合生产高品质的甜玉米；土肥地广，可大规模集约化进行生产；加工便捷，拥有多家大规模现代化甜玉米加工厂；地理位置优越，以北京为辐射点带动消费等。目前，这十几个省市甜玉米的播种面积仅66 700hm²左右，同南方市场形成巨大反差，且大多数为加工速冻籽粒和真空包装等加工产品而种植，有些还以鲜穗的形式直接供应南方市场。从未来8~15年的长期发展来看，考虑到北方人口消费、产品加工需求、部分南方市场反击供应需求以及出口需求，北方市场甜玉米种植面积预计达到与目前南方市场相当水平。

3.4.3 打造北方中心市场的条件、方式和切入点

培养消费习惯是拉动消费增长最重要的手段，通过各种形式的宣传活动，让更多的消费者接触到高品质的甜玉米，促进消费欲望产生，针对不同消费人群，采取不同的形式，提供不同层次和种类的产品，促进消费朝着多元化的方向发展，可大力发展城市周边的家

庭农场、采摘园、农家院、示范农场、生态农场等，这些都是可以对甜玉米进行示范、采摘、品尝和宣传的场所。鲜穗经纪人在产业的发展中扮演着十分重要的角色，他们将市场与生产紧密结合在一起，甜玉米鲜穗生产联盟是未来发展的方向，鲜穗经纪人凭着他们对市场的敏锐判断力，通过统一计划、统一安排、统一调运，达到甜玉米鲜穗的生产、采收、运输和销售均按照计划有条不紊进行的目的，从而规避各类生产和销售过程中的风险，这一联盟的实现将对甜玉米市场的发展起到不可估量的作用。

随着国家对种植业的调整，甜玉米的种植也受到国家的重视，这对甜玉米产业的发展十分有利。在这一背景下，广大甜玉米产业工作者应该着眼于甜玉米鲜穗市场的打开，在品种选择上建议从高端甜玉米品种入手，通过培养一部分甜玉米消费者拉动消费市场，进而建设成以北京为中心的北方甜玉米市场。

本文原载：蔬菜，2016（12）：1-6

不同温度胁迫对双隐性甜糯玉米出苗的影响

卢柏山[1]　史亚兴[1]　樊艳丽[1]　律宝春[2]

(1. 北京市农林科学院玉米研究中心，北京　100097；2. 北京市种子管理站，北京　100088)

摘　要：以含双隐性基因 sh_2sh_2wxwx 的甜糯玉米自交系和含单隐性基因 sh_2sh_2 的甜玉米自交系为材料，研究在不同温度处理下各材料的萌发及出苗情况。结果表明：单隐性甜玉米自交系 SH-251 在 10℃低温处理下发芽率最高为 91%，在 35℃高温处理下发芽率最高为 93%，其耐低温和高温的能力要高于双隐性甜糯玉米自交系。双隐性甜糯玉米自交系中，D6644 和京甜糯 6 的耐低温和高温能力较强，D6644 在 35℃的发芽率最高为 92.3%，活力指数为 7.76，京甜糯 6 在 35℃的发芽率最高为 79.3%，活力指数为 8.83。

关键词：甜玉米；双隐性甜糯玉米；发芽率；活力指数

鲜食玉米是指在乳熟后期至蜡熟初期收获的玉米，也称为水果玉米或蔬菜玉米，主要是指甜玉米、糯玉米。近年来，随着居民生活水平提高，营养结构的改善，人们对鲜食玉米需求量急剧增加的同时，对其品质和口感也有了更高的要求。甜加糯型玉米是近年来通过杂交育种将甜、糯玉米的优良特性组合到一起而培育出来的一类新品种，是一种含有双隐性基因的新型育种材料，其商品果穗上同时包含甜粒和糯粒，极大提高了品质和口感，在现有市场上正发挥越来越重要的作用。适宜的温度是种子发芽与出苗的重要条件，种子萌发时的冷害和高温均会延迟发芽，降低发芽率，造成出苗困难、缺苗等现象，对玉米生长发育和产量影响很大。基于此，以甜加糯型及甜质型玉米自交系为材料，研究不同温度胁迫下的发芽率及生长活力，以期指导育种、生产。

1　材料与方法

1.1　试验材料

本试验共选用 5 个不同自交系，分别为 D6624、D6644、甜糯 2、京甜糯 6、SH-251。其中 D6624、D6644、甜糯 2、京甜糯 6 为甜糯玉米自交系，含有双隐性基因 sh_2sh_2wxwx，SH-251 为甜玉米自交系，含单隐性基因 sh_2sh_2。5 份自交系均为北京市农林科学院玉米研究中心选育。

1.2　试验设计

本试验共设置 5 个处理，记为 T1~T5。T1~T3 分别为在 7℃、10℃、12℃低温、黑暗条件下，将供试材料发芽 10d 后，转移至 25℃光照下进行发芽。T4（对照处理，CK）为 25℃光照条件下进行发芽。T5 为 35℃高温光照条件下进行发芽。所有处理均采用砂培法，

作者简介：卢柏山（1972—　），研究员，从事鲜食玉米遗传育种工作

选用无任何化学物质污染的细沙，在160℃高温下灭菌3h后，装入不同苗盆中，制成3cm厚的砂床，并使播前细砂相对含水量为75%。精选大小一致、无病害、正常成熟的不同材料种子各100粒，分别播种至细砂表面，然后在种子表面覆3cm湿砂，于种子发芽恒温箱中进行发芽。每个处理均设3次重复。待发芽至7d时，测定各项指标。

1.3 测定指标

处理T1~T3在25℃下发芽至第7d，T4、T5在各自条件下发芽至第7d时，分别统计发芽率，每处理随机选取代表性的20株幼苗测定其株高，并计算简化活力指数。发芽不到20株的全部测量。

$$发芽率（\%）= 正常发芽种子数/供试种子总数 \times 100\%$$

$$简化活力指数 = 发芽率 \times 株高$$

1.4 试验数据分析

本试验中数据采用EXCEL 2003和SAS8.0软件进行处理和方差分析。

2 结果与分析

2.1 不同温度胁迫下各材料的发芽率统计结果

由表1可以看出，在不同温度处理下，双隐性甜糯玉米自交系的发芽率（35%~92.3%）均低于对照甜玉米自交系SH-251（83%~93%），且在低温处理T1~T3中，有显著差异。在双隐性甜糯玉米自交系中，D6644的发芽率（69%~92.3%）要显著高于其他材料（35%~79.3%），且在10℃低温下处理后，发芽率达到80.7%，在35℃高温下，发芽率为92.3%，与25℃处理（90%）相比，分别降低9.3%和增高2.3%。在不同温度处理下，D6624的发芽率均最低，在7℃低温时发芽率为35%，在25℃时发芽率为59%，35℃时为61%。甜糯2、京甜糯6的发芽率介于D6644与D6624之间，不同温度处理下的变幅为39.3%~79.3%。所有材料在25~35℃时的发芽率高于低温处理的发芽率，说明此为甜玉米自交系及双隐性甜糯玉米自交系的适宜发芽温度。

表1 不同温度下不同材料的发芽率（%）

	T1	T2	T3	T4	T5
D6624	35.0d	41.3d	32.7d	59.0b	61.0d
D6644	69.0b	80.7b	68.7b	90.0a	92.3a
甜糯2	46.7c	60.0c	39.3cd	66.0b	72.0c
京甜糯6	46.7c	62.0c	45.0c	66.0b	79.3b
SH-251	89.0a	91.0a	83.0a	90.3a	93.0a

2.2 不同温度胁迫下各材料的株高变化情况

由株高调查结果（表2）可以看出，在7℃、10℃、12℃低温处理、35℃高温处理中，对照甜玉米自交系SH-251的株高均要高于双隐性甜糯玉米自交系。双隐性甜糯玉米自交系料中，D6644和京甜糯6在所有处理中的株高均高于D6624和甜糯2。此结果说明D6644和京甜糯6在苗期的长势要好于D6624和甜糯2。

表2 不同温度下不同材料的株高（cm）表现

	T1	T2	T3	T4	T5
D6624	9.7b	8.1d	8.7c	7.6ab	9.2b
D6644	11.8a	9.6bc	10.3b	8.4a	8.4b
甜糯2	9.9b	9.2c	9.1c	7.4b	10.4a
京甜糯6	9.9b	10.2ab	10.7b	7.5b	11.1a
SH-251	12.6a	10.8a	12.0a	6.4c	11.1a

2.3 不同温度胁迫下各材料的活力指数变化情况

活力指数是种子发芽速率和生长量的综合反映，是体现种子活力的重要指标。本文用简易活力指数即种子发芽率与幼苗株高的乘积来体现本试验材料的种子活力。由表3可以看出，在7℃、10℃、12℃低温处理、35℃高温处理中，对照甜玉米自交系SH-251的简易活力指数高于双隐性甜糯玉米自交系，且有显著差异，说明在低温和高温胁迫之下，SH-251的种子活力要显著高于双隐性甜糯玉米自交系，SH-251有较强的耐低温和高温能力。双隐性甜糯玉米自交系中，D6644和京甜糯6在所有处理中的简易活力指数要高于D6624和甜糯2，说明种子活力要高于后两者。在7℃、10℃、12℃低温处理，25℃正常条件下，D6644的简易活力指数显著高于京甜糯6，在35℃高温处理下，D6644的简易活力指数显著低于京甜糯6，说明D6644耐低温的能力较京甜糯6高，耐高温的能力较京甜糯6低。

表3 不同温度下不同材料的活力指数

	T1	T2	T3	T4	T5
D6624	3.4d	3.5d	2.8d	4.5c	5.6d
D6644	8.1b	7.7b	7.1b	7.6a	7.8c
甜糯2	4.6c	5.5c	3.6c	4.9bc	7.5c
京甜糯6	4.6c	6.4c	4.8c	4.9bc	8.8b
SH-251	11.2a	9.9a	10.0a	5.8b	10.3a

3 结论与讨论

鲜食玉米种子在播种出苗期遭遇春寒冷冻，或在夏播时遭遇高温天气，是鲜食玉米种子发芽率低，出苗弱的重要原因之一。有研究表明，玉米种子发芽率随着温度的提高而增大，甜玉米种子在22~35℃的发芽率在86.3%~87%，为适宜种子发芽的温度。本研究结果表明：在25~35℃时，各材料的发芽率在59%~93%，要高于低温处理的发芽率。在低温胁迫和高温胁迫处理中，甜玉米自交系SH-251耐低温能力均显著高于双隐性甜糯玉米自交系。双隐性甜糯玉米自交系中，D6644耐低温能力最强，京甜糯6耐高温能力最强，D6624耐低温和高温能力均最弱。因此，含双隐性基因的甜糯玉米自交系应适当延迟播种，提高种子发芽温度，可保证种子发芽率，同时应选用耐低温的品种，促进农业生产。

参考文献（略）

本文原载：北京农业，2014（27）：25

不同播种深度对双隐性甜糯玉米出苗的影响

史亚兴[1]　卢柏山[1]　樊艳丽[1]　律宝春[2]

(1. 北京市农林科学院玉米研究中心，北京　100097；2. 北京市种子管理站，北京　100088)

摘　要：以4个双隐性（sh_2sh_2wxwx）甜糯玉米自交系和1个甜玉米自交系（sh_2sh_2）为材料，研究在不同播种深度下各材料的出苗情况，并比较不同类型材料的耐深播能力。结果表明：随播种深度增加玉米发芽率和活力指数逐渐降低，双隐性甜糯玉米自交系D6644的适宜播深为3~6cm，甜玉米自交系SH-251，双隐性甜糯玉米自交系（D6624、京甜糯6、甜糯2）的适宜播深为3cm。发芽率、活力指数与播种深度呈极显著负相关，中胚轴长度与播种深度呈显著正相关。

关键词：双隐性；甜糯玉米；发芽率；活力指数

鲜食玉米是指在乳熟后期至蜡熟初期收获的玉米，也称为水果玉米或蔬菜玉米，主要是指甜玉米、糯玉米。与普通玉米相比，鲜食玉米不仅在口感上鲜香甜嫩，其所含的蛋白质、氨基酸、维生素等营养物质均高于普通玉米。近年来，随着居民生活水平提高，营养结构的改善，人们对鲜食玉米需求量急剧增加的同时，对其品质和口感也有了更高的要求。甜加糯型玉米是近年来通过杂交育种将甜、糯玉米的优良特性组合到一起而培育出来的一类新品种，是一种新型的鲜食玉米育种材料，其商品果穗上同时包含甜粒和糯粒，极大提高了品质和口感，在现有市场上正发挥越来越重要的作用。

播种深度是影响种子出苗的重要因素。深播能保证种子萌发所需要的水分，且深播的品种根系生长较深，因此抗旱和抗倒伏能力也较强。但研究表明，盲目深播会延迟种子发芽时间，降低种子发芽率，而甜玉米籽粒干燥成熟后，胚乳皱缩，种子淀粉含量显著低于普通玉米，顶土能力弱，活力相对较低。因此，对于含有隐性突变基因的甜玉米种子或甜糯玉米种子，适宜的播种深度对其出苗尤为重要。本研究以4个骨干双隐性甜糯玉米自交系和1个甜玉米自交系为材料，研究不同播种深度下其发芽率及幼苗生长情况，比较不同类型双隐性甜糯玉米自交系的耐深播能力，明确适宜播种深度，为农业生产提供科学依据。

1　材料与方法

1.1　试验材料

本试验共选用5个不同自交系，分别为D6624、D6644、甜糯2、京甜糯6、SH-251。其中D6624、D6644、甜糯2、京甜糯6为甜糯玉米自交系，含有双隐性基因 sh_2sh_2wxwx，SH-251为甜玉米自交系，含单隐性基因 sh_2sh_2。5份自交系均为北京市农林科学院玉米研究中心选育。

1.2 试验设计

试验采用盆栽、沙培的方法进行。选用无任何化学物质污染的细沙,在160℃高温下灭菌3h后,装入不同苗盆中,制成3cm厚的沙床,并使播前细沙相对含水量为75%。精选大小一致、无病害、正常成熟的不同材料种子各100粒,分别播种至细沙表面,然后按照3cm、6cm、9cm的播种深度覆盖同样的沙土,并浇水,保持沙土湿润,于25℃室温内发芽,并控制光照12h/d,每处理设3次重复。待发芽至7d时,测定各项指标。

1.3 测定指标

每处理发芽至7d时,统计各处理中的发芽种子数,并计算发芽率;每处理随机选取20株去砂土的幼苗,测定其高度,及中胚轴长度,并计算简易活力指数。发芽不到20株的全部测量。

$$发芽率(\%) = 正常发芽种子数/供试种子总数 \times 100\%$$
$$简化活力指数 = 发芽率 \times 株高$$

1.4 试验数据分析

本试验中数据采用 EXCEL 2007 和 SAS8.0 软件进行处理和方差分析。

2 结果与讨论

2.1 不同播种深度对甜糯玉米发芽率的影响

由图1可以看出,在3~9cm播种深度下,所有材料均能发芽,并且随着播种深度的增加发芽率呈现降低的趋势。所有材料在3cm播深的情况下,甜玉米自交系SH-251的发芽率最高,为94.3%。双隐性甜糯自交系中,D6644发芽率最高,为90%,D6624发芽率最低,为63.3%,京甜糯6与甜糯2的发芽率分别为66%、70%。在6cm播深处理下,双隐性甜糯玉米自交系D6644的发芽率最高,为80.3%,较3cm播深时降低了9.7%;D6624的发芽率最低,为33%;SH-251,京甜糯6,甜糯2的发芽率分别为73.3%、42.7%、52.3%。播深在9cm时,SH-251,D6624,甜糯2,京甜糯6的发芽率均降至20%以下,D6644的发芽率降至49.7%。此结果说明双隐性甜糯玉米自交系D6644的发芽率要高于其他材料,其适宜播种深度为3~6cm,而其他自交系的适宜播种深度为3cm。

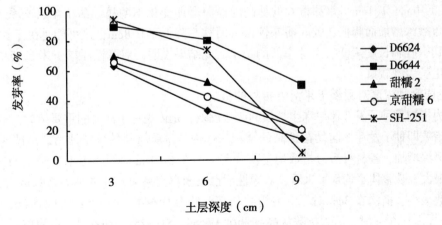

图1 不同播种深度对甜糯玉米发芽率的影响

2.2 不同播种深度对甜糯玉米中胚轴长度的影响

中胚轴长度是指玉米根部节间至芽鞘节之间的长度。玉米种子发芽时的顶土能力与中胚轴长度密切相关，播种深时，中胚轴较长的种子，顶土能力较强。由图2可以看出，在所有播种深度处理下，双隐性甜糯玉米自交系D6644的中胚轴长度均要大于其他自交系，播深为3cm时，5个自交系的中胚轴长度范围为1.8~2.7cm。随播种深度增加，双隐性甜糯自交系D6644，京甜糯6，D6624的中胚轴长度随之变长。甜糯2中胚轴长度表现为先增加后降低的趋势，在9cm播深时的中胚轴长度（3.1cm）较6cm时（3.2cm）降低了0.1cm。甜玉米自交系SH-251在6cm播深时最长为2.2cm，而在9cm播深时降低了0.6cm。此结果说明双隐性甜糯自交系D6644在发芽时顶土能力最强，甜玉米自交系SH-251顶土能力最弱，与发芽率结果一致。

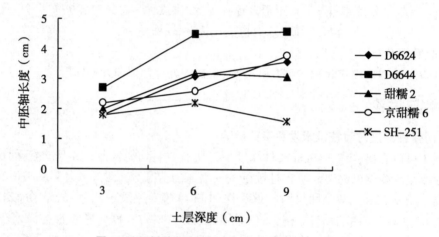

图2 不同播种深度对甜糯玉米中胚轴长度的影响

2.3 不同播种深度对甜糯玉米株高的影响

由图3可以看出，在所有播种深度处理下，双隐性甜糯玉米自交系D6644的株高均要高于其他自交系，其在9cm播深时的株高为12.8cm。D6644、D6624、甜糯2的株高随着播种深度的增加有增高的趋势，与3cm播深相比，其在9cm播深时的株高分别增高了2.1cm、1.4cm、1.1cm。京甜糯6的株高随播深增加变化不明显。甜玉米自交系SH-251的株高随播深增加而降低，9cm播深较3cm时株高降低了0.8cm，这可能是由于在9cm播深下，其中胚轴长度降低、发芽率低的结果。此结果说明，双隐性甜糯玉米自交系D6644的幼苗生长表现较好。

2.4 不同播种深度对甜糯玉米活力指数的影响

活力指数是种子发芽速率和生长量的综合反映，是体现种子活力的重要指标。本文用简易活力指数即种子发芽率与幼苗株高的乘积来体现本试验材料的种子活力。由图4可以看出，随播深增加，各材料活力指数均呈下降趋势，在6~9cm时，甜玉米自交系SH-251下降幅度最大。从整体平均水平来看，双隐性甜糯玉米自交系D6644的活力指数最大，D6624活力指数最小。播深在3cm时，活力指数从高到低为D6644，SH-251，京甜糯6，甜糯2，D6624；播深在6cm时，活力指数从高到低为D6644，SH-251，甜糯2，京甜糯6，D6624；播深在9cm时，活力指数从高到低为D6644，京甜糯6，甜糯2，D6624，SH-251。

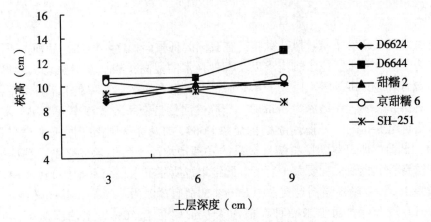

图3 不同播种深度对甜糯玉米株高的影响

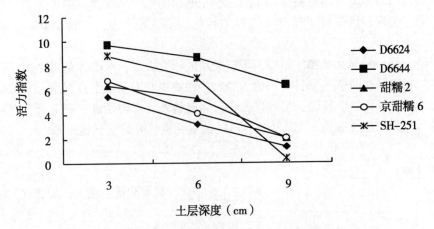

图4 不同播种深度对甜糯玉米活力指数的影响

2.5 不同播种深度对甜糯玉米出苗指标的影响

对供试材料的各项指标与播种深度之间的相关性分析表明（表1）：发芽率、活力指数与播种深度呈极显著负相关，中胚轴长度与播种深度呈显著正相关，即随播种深度增加，发芽率、活力指数越低，中胚轴长度越长。株高与播种深度呈正相关性。发芽率与活力指数，中胚轴长度与株高均呈极显著正相关，发芽率越高，活力指数越大；中胚轴长度越长，株高值越大。其他各指标直接则无明显相关性。

表1 出苗指标与播种深度的相关性

	播种深度	发芽率	株高	中胚轴长度	活力指数
播种深度	1				
发芽率	-0.818**	1			
株高	0.327	0.063	1		
中胚轴长度	0.546*	-0.176	0.831**	1	
活力指数	-0.733**	0.981**	0.240	-0.026	1

3 讨论

鲜食玉米因其营养丰富、风味独特,在我国的种植面积逐年增加,品种主要以甜玉米和糯玉米为主。近年来,应市场需求以及育种技术的提高,甜加糯型鲜食玉米迅速发展,并深受消费者欢迎,市场潜力巨大。本文通过对双隐性甜糯玉米和甜玉米自交系在不同播种深度下的研究表明:在播深为3cm时,双隐性甜糯自交系的发芽率低于甜玉米自交系,但随播深增加至6cm时,双隐性甜糯自交系D6644的发芽率则高于甜玉米自交系,播深为9cm时,试验中所有双隐性甜糯自交系的发芽率均高于甜玉米自交系,且从平均水平看,前者株高、中胚轴长度要大于后者。此结果说明甜玉米自交系SH-251不耐深播,其最适播深为3cm,双隐性甜糯自交系中D6644的适宜播深为3~6cm,其他材料为3cm。岳丽杰等的研究表明:普通玉米品种在播深为9cm时仍有50%以上的出苗率,最高可达81.7%,本文研究播深在9cm时,D6644的发芽率最高位49.7%,其余材料则在20%以下,这可能是因为本文所用材料为含单隐性突变基因的甜玉米或含双隐性突变基因的甜糯玉米自交系,突变基因阻碍了糖分向淀粉的转化,使胚乳皱缩,淀粉积累减少,故发芽率低。

吴子恺,郝小琴等对双隐性甜糯玉米自交系的研究表明,双隐性基因之间存在基因互作效应,使得双隐性甜糯玉米自交系的一些农艺性状要好于甜玉米自交系。本试验中双隐性甜糯自交系的株高、中胚轴长度,以及在深播时的发芽率均要好于甜玉米自交系,验证了前人的研究。双隐性甜糯自交系中,D6644的发芽特性要好于其他材料。

参考文献(略)

本文原载:蔬菜,2015(3):15-18

基于SNP标记技术的糯玉米种质遗传多样性分析

史亚兴 卢柏山 宋伟 徐丽 赵久然

(北京市农林科学院玉米研究中心,北京 100097)

摘 要:为揭示糯玉米种质的遗传多样性,利用新一代分子标记技术-SNPs(Single Nucleotide Polymorphisms)对39份不同基因型的糯玉米自交系进行了基因型分析。结果表明,1 059个SNP标记在39份自交系中的多态性信息含量(PIC)为0.05~0.38,平均含量为0.31;最小等位基因频率(MAF)为0.03~0.50,平均值为0.29;期望杂合度变化范围为0.05~0.50,平均期望杂合度为0.39。39份自交系之间的亲缘关系(kinship)系数在0.00~0.91,京6与京糯6之间亲缘关系最近。通过Neighbor-joining(NJ)聚类分析,将39份自交系划分为5大类群,其中系谱来源不清晰的糯玉米自交系均被划分至不同类群中,明确了其亲缘关系。因此,SNP标记适用于糯玉米自交系的遗传多样性分析及亲缘关系研究,可为糯玉米种质资源利用及品种选育提供参考。

关键词:糯玉米;SNP标记;遗传多样性;京糯6;白糯6

Genetic Diversity Analysis of Waxy Corn Inbred Lines by Single Nucleotide Polymorphism (SNP) Markers

Shi Yaxing Lu Baishan Song Wei Xu Li Zhao Jiuran

(*Maize Research Center, Beijing Academy of Agriculture and Forestry Sciences, Beijing, 100097, China*)

Abstract: In order to explore genetic diversity of waxy corn varieties, a novel molecular marker technology-Single Nucleotide Polymorphism (SNP) Markers were adopted in genotypic analysis of 39 elite waxy corn inbred lines. Polymorphic Information Content (PIC) of 1059 SNP markers on these inbred lines were 0.05~0.38, with average value being 0.31; Minor Allele Frequency (MAF) were 0.03~0.50, with average value being 0.29; Expected Heterozygosity were 0.05~0.50, with an average of 0.39. Kinship coefficient between 39 inbred lines ranged from 0.00 to 0.91. Jing6 and Jingnuo6 had the strongest similarity. Five groups were divided from the 39 lines by Neighbor-joining (NJ) clustering method. Inbred lines with unclear pedigree were divided into different groups, so their relatedness with other lines were defined. Therefore, SNP markers were applicable to the analysis of genetic diversity and genetic relationships of waxy corn inbred lines, could provide data for utilization and de-

---

作者简介:史亚兴(1977—),北京人,副研究员,硕士,主要从事鲜食玉米遗传育种研究
赵久然为本文通讯作者

veloping new varieties in breeding program.

Key words：Waxy corn；SNP marker；Genetic diversity；Jingnuo6，Bainuo6

糯玉米（*Zea mays* L. *certaina* Kulesh）是普通玉米发生基因隐性突变（Wx 突变为 wx）而形成的一种特殊玉米类型。糯玉米胚乳中淀粉主要为支链淀粉，分子量小，食用消化率高，其所含蛋白质比普通玉米高 3%~6%，赖氨酸、色氨酸含量均较高，是一种高营养价值和经济价值的特用玉米。糯玉米起源于中国，也是玉米各类型中唯一起源于我国的类型，因此在我国遗传资源十分丰富。近几十年来，育种家们通过多种育种方法和技术，创制培育了大量优良糯玉米种质，拓宽了我国糯玉米种质遗传基础。因此，分析现有糯玉米种质资源的遗传多样性，研究种质之间的亲缘关系，对选育优良糯玉米品种有重要作用。

近年来，分子标记技术在玉米种质遗传多样性分析中已得到广泛应用。RFLP 标记是最早发展并被应用的一种标记技术，随后 RAPD 和 AFLP 标记也相继发展和应用。20 世纪 80 年代末，SSR 标记迅速发展起来，是目前玉米育种研究中应用较多的分子标记。但 SSR 标记在操作时需要利用凝胶电泳对基因位点分型，较费时费力，难以实现大规模、高通量的分析。

1996 年 Lander 首次提出 SNP（Single Nucleotide Polymorphism，单核苷酸多态性）作为新一代分子标记。SNP 是广泛存在于基因组中的一类 DNA 序列变异，其变异频率为 1% 或更高。它是由单个碱基的转换或颠换引起的点突变，遗传稳定性更高、可靠性更强，且摆脱了凝胶电泳的瓶颈，通过序列比对来分析遗传物质差异，易于实现高技术含量、大规模的基因型检测。

随着现代分子生物学及基因芯片的发展，SNP 分子标记因其优点受到国内外育种家越来越多的重视，并在黄瓜、苹果、大豆、小麦、水稻、玉米等作物的品种鉴定、遗传作图、基因挖掘与 QTL 定位等方面均有应用。在玉米种质遗传多样性研究方面，Yang 等利用 884 个 SNP 标记对 154 份普通玉米自交系进行了遗传多样性分析；Wu 等利用 1 015 个在基因组中平均分布的 SNP 标记，对 367 份普通玉米自交系进行了群体遗传多样性的分析并进行聚类；吴金凤等利用 1 041 个 SNP 标记对 51 份玉米自交系进行了遗传多样性分析及类群划分，划分结果与系谱来源一致。但目前利用 SNP 分子标记技术分析特用玉米种质遗传多样性的研究还未见报道。因此，本研究选用 39 份我国常用及自选糯玉米自交系，利用覆盖全基因组的 SNP 标记进行遗传多样性分析和亲缘关系研究，旨在为糯玉米优良品种选育、杂优模式开发与利用等提供理论参考。

1　材料和方法

1.1　试验材料

供试材料分为 2 部分：37 份常用及自选糯玉米自交系，编号为 N1~N37；2 份普通玉米自交系 B73、Mo17，作为标准测验种，分别代表 Reid 群和 Lancaster 群，编号为 N38-N39。各自交系名称及系谱来源见表 1。

表1 39份自交系名称及系谱来源
Table 1 List of 39 inbred lines and their pedigrees

编号 No.	名称 Name	系谱来源 Pedigree	编号 No.	名称 Name	系谱来源 Pedigree
N1	京糯6	选自中糯1号	N21	紫糯419	不详
N2	白糯6	选自紫糯3号	N22	紫糯680	不详
N3	京6	选自中糯1号	N23	黄糯705	不详
N4	BN2	选自紫糯3号	N24	黑糯732	不详
N5	N601	京糯968母本	N25	紫糯961	不详
N6	N39	选自中糯309	N26	黄糯850	不详
N7	紫糯264	不详	N27	紫糯736	不详
N8	N8131	不详	N28	京糯5	选自黄糯种质
N9	京糯72	昌7-2转糯选系	N29	黄糯6	选自紫糯3号
N10	紫糯5	不详	N30	京糯2	选自白早糯种质
N11	紫糯3	选自紫糯3号	N31	中玉04	中糯1号母本
N12	紫玉2	选自紫香糯1号	N32	糯1	垦粘1号母本
N13	Z09	引自美国DH系	N33	糯2	垦粘1号父本
N14	糯70	昌7-2转糯选系	N34	京甜糯2	选自SH-251×白糯6
N15	京糯255	不详	N35	D6644	选自SH-251×白糯6
N16	CB1	选自美国双隐性种质	N36	京甜糯69	选自SH-251×白糯6
N17	紫糯258	不详	N37	D6644-6	选自SH-251×白糯6
N18	紫糯263	不详	N38	B73	BSSSC5
N19	紫糯877	不详	N39	Mo17	187-2×103
N20	紫糯196	不详			

1.2 试验方法

1.2.1 DNA提取

将供试种子在恒温培养箱内发芽,玉米幼苗长至4~5片叶时,取鲜嫩玉米叶片,参照Rogers和Bendich的方法提取叶片基因组DNA。所提DNA要求条带单一,完整无降解;紫外分光光度计检测A260/280介于1.8~2.0(DNA样品没有蛋白、RNA污染),A260/230介于1.8~2.0(DNA样品盐离子浓度低),稀释样品DNA浓度至100ng/μl。

1.2.2 SNP位点选取

利用北京市农林科学院玉米研究中心开发的MaizeSNP3072芯片对供试自交系进行SNP基因分型。根据以下指标选用1 059个SNP位点:数据质量评估值≥0.7;AA、AB、BB这3种基因型具备明显界限;位点特异性探针设计评估值≥0.7;在全基因组上均匀分布,用于本研究。

1.2.3 基因芯片（Golden gate 技术）试验操作流程

将提取的玉米基因组 DNA 与激活的生物素充分结合，通过离心沉淀使 DNA 纯化；将探针与纯化的目的 DNA 链杂交，漂洗杂交产物后进行延伸连接反应；延伸的产物经漂洗变性处理后作为模板进行 PCR 扩增，PCR 产物与磁珠结合后过滤纯化；将纯化的 PCR 产物与芯片杂交，杂交结束后进行芯片清洗、真空抽干；风干后的芯片立刻在 iScan 仪上进行扫描，最后利用 Genome Studio 软件进行数据分析（详细试验流程参照 Illumina 公司提供的试验操作指南 Golden Gate® Genetyping Assay，Manual Part#15004070 Rev. B）。

1.2.4 数据统计分析

本研究数据利用 PowerMarker V3.25 软件计算其最小等位基因频率（MAF）、多态性信息含量（PIC）、期望杂合度（Expected Heterozygosity, He）、观测杂合度（Observed Heterozygosity, Ho）等。利用 SPAGeDi（Spatial Pattern Analysis of Genetic Diversity）进行自交系间亲缘关系（kinship）分析；采用 Nei's 1972 算法计算自交系间遗传距离，并按照邻接法（Neighbor-joining）构建 NJ 聚类图，进行杂种优势群的划分。

2 结果与分析

2.1 SNP 多态性分析

利用在全基因组中平均分布、数据质量评估值高的 1 059 个 SNP 标记，并通过基因芯片的方法检测了 39 份糯玉米自交系的基因型，并用 PowerMarker V3.25 对基因分型数据进行分析。结果表明，1 059 个 SNP 标记在 39 份自交系中共检测到 2 118 个等位基因，缺失率在 0~21% 之间，平均为 4%。平均期望杂合度为 0.39，范围为 0.05~0.50；平均观测杂合度为 0.04，范围为 0.00~0.29；平均多态性信息含量为 0.31，变化范围在 0.05~0.38（表 2）。

表 2　1 059 个 SNP 标记在 39 份自交系中的多态性
Table 2　Polymorphisms of 1 059 SNPs in 39 inbred lines

多态性 Polymorphisms	位点数 No. of loci	等位基因数 Allels	期望杂合度 He	观测杂合度 Ho	缺失率 Missing Rate	PIC 值 PIC Value
范围 Range	1 059	2 118	0.05~0.50	0.00~0.29	0.00~0.21	0.05~0.38
平均值 Average	—	2	0.39	0.04	0.04	0.31

PIC 值是衡量 DNA 座位变异程度高低的重要指标。某 DNA 座位的 PIC 值>0.50 时，表明其为高度多态性座位；0.25<PIC 值<0.50 时，表明其为中度多态性座位；PIC 值<0.25 时，表明为有低度多态性座位。本研究中，1 059 个 SNP 标记在 39 份糯玉米自交系上的平均 PIC 值为 0.31，说明所用 SNP 标记为中度偏高多态性。期望杂合度常被用来衡量群体的遗传多样性，平均期望杂合度越高，表明群体遗传一致性越低，遗传多样性越高。由表 2 可知，本研究中平均期望杂合度为 0.39，表明 39 份自交系有较好的遗传多态性。

最小等位基因频率（MAF）也叫做次要等位基因频率。本研究中，1 059 个 SNP 标记的 MAF 值在 0.03~0.50，平均值为 0.29，其中≥0.20 的位点有 835 个，占 79%。由图 1

可以看出，MAF 在 0.10~0.50 均匀分布。

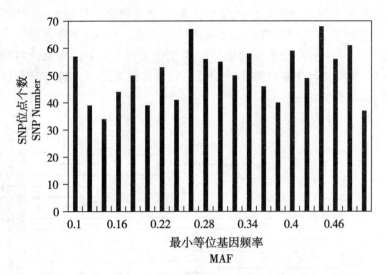

图 1 1 059 个 SNP 标记在 39 份自交系中的最小等位基因频率
Figure 1 MAF of 1 059 SNPs in 39 inbred lines

2.2 不同供试基因型间的亲缘关系（kinship）分析

39 份自交系的 kinship 分析表明（图 2），98.8%的自交系间亲缘关系系数（kinship 系数）分布在 0.00~0.50。其中 60%的 kinship 系数为 0，表明 60%的成对自交系间无亲缘关系。23.8%的 kinship 系数分布在 0.00~0.10，说明这些成对自交系之间的亲缘关系较弱。15%的 kinship 系数分布在 0.10~0.50，表明这些自交系之间有不同程度的亲缘关系。另有 1.2%的 kinship 系数分布在 0.50~0.91，说明这些成对自交系间有很大的亲缘相关性。39 份自交系之间，京 6 与京糯 6 之间 kinship 系数最大为 0.91，白糯 6 与 BN2 之间 kinship 系数次之，为 0.82，说明这两对自交系之间亲缘关系最近。此结果与系谱来源一致，也表明我国糯玉米种质资源有较广泛的遗传基础。

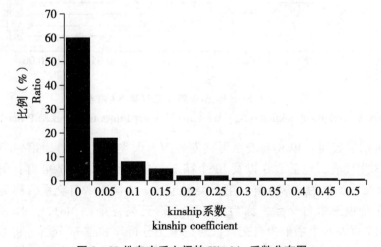

图 2 39 份自交系之间的 Kinship 系数分布图
Figure 2 Distribution of kinship coefficient of 39 lines

2.3 不同供试基因型的聚类分析

根据39份自交系在1 059个SNP位点的基因分型数据,通过PowerMarker V3.25软件,并应用Nei's 1972算法计算自交系之间的遗传距离,构建NJ聚类图,进行聚类分析(图3)。结果表明,39份自交系间的遗传距离变化范围为0.00~0.72,自交系间平均遗传距离为0.47,其中,自交系白糯6与BN2之间遗传距离最小,B73与Mo17之间遗传距离最大,与kinship分析结果一致。

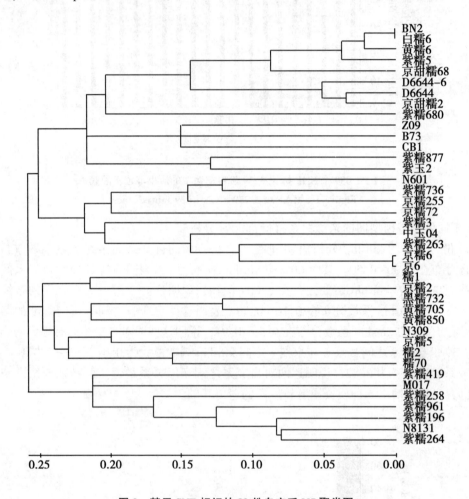

图3 基于SNP标记的39份自交系NJ聚类图
Figure 3 Neighbor-joining (NJ) trees for 39 inbred lines based on SNP markers

聚类分析结果表明,39份自交系可被划分为5大类群(表3)。第一类群以普通玉米自交系B73为代表,包括B73以及13个糯玉米自交系,称为Reid群;第二类群包括9个糯玉米自交系,称为京糯6群;第三类群包括2个糯玉米自交系,称为糯1群;第四类群包括7个糯玉米自交系,称为糯2群;第五类群是以Mo17为代表的Lancaster群,包括Mo17以及6个糯玉米自交系。通过聚类分析和kinship分析,将系谱不清晰的糯玉米自交系划分至不同的类群中,明确了与其余自交系间的亲缘关系,可提高其利用效率。

表3 39份自交系类群划分明细
Table 3 List of 39 lines of different heterotic groups

编号 No.	类群名称 Groups	自交系数目 Number	自交系名称 Inbreds
I	Reid 群	14	B73、Z09、CB1、BN2、白糯6、黄糯6、紫糯5、京甜糯68、D6644-6、D6644、京甜糯2、紫糯680、紫糯877、紫玉2
II	京糯6群	9	京糯6、京6、N601、紫糯736、京糯255、京糯72、紫糯3、中玉04、紫糯263
III	糯1群	2	糯1、京糯2
IV	糯2群	7	糯2、糯70、黑糯732、黄糯705、黄糯850、N309、京糯5
V	Lancaster 群	7	Mo17、紫糯419、紫糯258、紫糯961、紫糯196、N8131、紫糯264

3 结论与讨论

3.1 SNP 遗传多态性

SNPs 是指基因组 DNA 中由于单个核苷酸的替换而引起的多态性。以这样的 DNA 序列多态性特征做标记即可区分两个个体之间遗传物质的差异。SNP 标记广泛存在于基因组中，多态性数量丰富，与 SSR 标记相比，SNP 标记稳定性更好，与基因芯片技术结合，可快速、高通量的检测个体遗传多样性。

Yang 等利用在全基因组分布的 926 个 SNP 标记分析了 527 份普通玉米自交系的遗传多样性以及亲缘关系，结果表明 926 个 SNP 标记的平均 PIC 值为 0.31，平均 MAF 为 0.3；Lu 等利用 1 034 个 SNP 标记对 770 份国内外玉米自交系进行遗传多样性分析的研究中，多态性信息含量平均值为 0.26。本研究中，1 059 个 SNP 标记在 39 份糯玉米自交系中的 PIC 值变化范围为 0.05~0.38，平均值为 0.31，平均 MAF 为 0.29。与前人研究结果相比较，本研究所用 SNP 标记有较好多态性，属于中度偏高多态性座位，同时表明糯玉米自交系有较好的遗传多样性。Yang 等利用 884 个 SNP 标记在对 154 份普通玉米自交系的研究中，平均期望杂合度为 0.44。本研究中期望杂合度变化范围为 0.05~0.50，平均值为 0.39，低于 Yang 等的结果。这可能是由于本研究所选糯玉米自交系数量较少的原因。

3.2 糯玉米种质的聚类分析

新品系的鉴定和筛选是作物育种中的重要环节，利用聚类分析将不同种质资源进行分类，可为品系（种）间的遗传差异比较、杂交育种亲本组配提供参考依据。Wu 等利用 1 015 个 SNP 标记将 367 份普通玉米自交系分为 2 个大群和 5 个亚群，结果与系谱一致；在糯玉米种质类群划分研究上，前人利用 SSR 标记做了较多研究。王慧等利用 65 对 SSR 引物将 165 份糯玉米材料划分为 3 大类，划分结果基本符合材料来源；程宇坤等[31]利用 49 对 SSR 引物将 58 份糯玉米自交系划分为 5 个类群。本研究中，利用覆盖玉米全基因组的 1 059 个 SNP 标记，通过 Neighbor-joining（NJ）聚类分析，将 39 份自交系划分为 5 个类群，划分结果与系谱来源基本一致，说明利用 SNP 分子标记可用于糯玉米种质遗传多样性分析及类群划分，并能取得良好效果。

  本研究类群划分结果表明，我国优良糯玉米品种垦粘 1 号的父母本糯 2 和糯 1 被划分至不同类群中，因此，糯 1 群×糯 2 群的杂种优势模式可供借鉴；国审糯玉米品种京科糯 2000（京糯 6×BN2）、京紫糯 218（紫糯 5×紫糯 3）的双亲均分别来自京糯 6 群和 Reid 群，说明京糯 6 群×Reid 群也是我国糯玉米品种选育的主要杂优模式，在亲本选配中可进行利用。本研究中系谱来源不清晰的糯玉米自交系被划分至不同类群中，依据此结果可指导杂交种组配，避免盲目测配，提高了其利用效率。

参考文献（略）

本文原载：华北农学报，2015，30（3）：77-82

利用SNP标记划分甜玉米自交系的杂种优势类群

卢柏山　史亚兴　宋　伟　徐　丽　赵久然

(北京市农林科学院玉米研究中心/玉米DNA指纹及分子育种北京市重点实验室,北京　100097)

摘　要:利用1 031个SNP标记对39份甜玉米自交系进行基因型分析,结果表明1 031个SNP标记在供试材料中的平均多态性信息含量(PIC)为0.290,最小等位基因频率(MAF)平均值为0.275。39份自交系间遗传距离变化范围为0.032~0.678,平均值为0.430。通过Neighbor-joining(NJ)聚类分析,将供试材料划分为5个类群,分别为华珍母本群,京甜糯2群,彩甜糯群,温带种质群,华珍父本群。5个类群间遗传距离变化范围较小,在0.394~0.445。其中,华珍父本群与温带种质群之间的遗传距离为0.394;彩甜糯群与华珍父本群、温带种质群之间的遗传距离为0.445。

关键词:甜玉米;自交系;SNP标记;类群划分

Heterotic Grouping of Sweet Corn Inbred Lines by SNP Markers

Lu Baishan　Shi Yaxing　Song Wei　Xu Li　Zhao Jiuran

(*Maize Research Center, Beijing Academy of Agriculture and Forestry Sciences, Beijing, 100097, China*)

Abstract: 1 031 Single Nucleotide Polymorphism (SNP) Markers were adopted in heterotic grouping of 39 elite sweet corn inbred lines. Polymorphic Information Content (PIC) and Minor Allele Frequncy (MAF) of these SNP markers was 0.290 and 0.275 on average, respectively. Genetic distance between 39 inbred lines ranged from 0.032 to 0.678, with an average value being 0.430. Five groups were divided from the 39 lines by Neighbor-joining (NJ) clustering method which are Huazhen maternal group, Jingtiannuo2 group, Colorful sweet-waxy group, Temperate group and Huazhen paternal group. Genetic distance between the groups ranged from 0.394 to 0.445. Huazhen paternal group and Temperate group have the minimum genetic distance being 0.394, while Huazhen paternal and Colorful sweet-waxy group, Temperate group and Colorful sweet-waxy group have the

---

作者简介:卢柏山(1972—　),男,北京人,推广研究员,硕士,主要从事鲜食玉米遗传育种研究。Tel: 010-51503562. E-mail: maizelu@126.com

史亚兴(1977—　),北京人,副研究员,硕士,主要从事鲜食玉米遗传育种研究。Tel: 010-51503400. E-mail: syx209@163.com

卢柏山和史亚兴为本文共同第一作者

赵久然为本文通讯作者

maximum genetic distance being 0.445.

Key words: Sweet corn; Inbred lines; SNP marker; Genetic grouping

甜玉米（*Zea mays* L. *saccharata* Sturt）是玉米属（*Zea mays* L.）的一个玉米亚种，即甜质型玉米亚种，是籽粒胚乳中控制糖分转化的基因发生隐性突变而产生的突变体。甜玉米籽粒乳熟期含糖量高，并富含多种氨基酸、维生素，是一种具有高营养价值和经济价值的特用型玉米。甜玉米原产于南美洲，至今已有100多年的栽培历史。我国甜玉米育种始于20世纪50年代，最初的育种材料引自美国。据研究，目前我国育种资源仍主要来自美国、泰国和我国台湾等地，育种技术上主要以选育二环系为主，基础较薄弱，整体水平与国外先进水平相比还有一定差距。因此，搜集与培育遗传基础复杂、适应性强的甜玉米育种资源，并挖掘新的杂种优势群和杂种优势模式，才能从整体上提高甜玉米育种的工作效率。

优异的种质是新品种选育的基础，而成功选择种质的一个重要依据是杂种优势群。近几十年来，研究者利用亲本形态差异、地理来源、系谱、配合力表现以及同工酶标记等划分杂种优势群；之后随着分子标记技术的飞速发展，在分子水平上研究玉米的杂种优势群及其模式成为新的手段。Smith等利用131对SSR引物对58份玉米自交系进行聚类分析，并与RFLP标记进行比较，表明二者高度相关，分类结果与系谱分析基本吻合；袁立行等利用RFLP、SSR、AFLP及RAPD 4种分子标记对15个玉米自交系进行聚类分析，并指出SSR更适合玉米种质遗传多样性分析。钟昌松等利用15对SSR引物对20份特用玉米自交系的亲缘关系进行了研究；胡俏强等利用64对SSR引物对92份甜玉米自交系的遗传变异进行研究，划分了杂种优势群，分类结果与系谱基本吻合。

随着人类SNP研究的快速发展，以及基因组测序和DNA芯片技术的发展，SNP分子标记目前已迅速取代SSR、RFLP等传统标记，开启了新的分子标记时代。SNP标记具有遗传稳定性好、位点丰富且分布广泛、代表性强、和易于实现自动化分析检测等优点，在高密度遗传图谱构建、基因精确定位、群体遗传结构分析以及系统发育等方面均已被应用。其中，Wu等利用1 015个在基因组中平均分布的SNP标记，对367份普通玉米自交系进行了群体遗传多样性的分析并进行聚类，将367份自交系分为2个大群和5个亚群，结果与系谱一致，表明SNP标记在玉米种质资源遗传多样性分析和类群划分上的可行性。本研究利用1 031个SNP标记，对39份甜玉米自交系进行杂种优势群划分的研究，旨在为甜玉米种质改良与创新、杂种优势模式的开发与利用提供理论参考。

1 材料与方法

1.1 试验材料

本研究选取39份甜玉米自交系为试验材料，自交系编号为T1-T39，其名称及系谱来源见表1。

表1 139份甜玉米自交系名称及系谱来源
Table1 List of 39 sweet corn lines and their pedigrees

编号 No.	名称 Name	系谱来源 Pedigree	编号 No.	名称 Name	系谱来源 Pedigree
T1	T68W	选自美国甜玉米杂交种麦哥娜姆	T21	BKL	选自杂交种巴卡拉
T2	T8367	选自美国甜玉米杂交种520	T22	T268	选自华珍
T3	SH-251	选自中农大超甜1号	T23	T784-3	选自ZBM二环系
T4	双金11	选自美国杂交种库普拉	T24	YC08	外引375种质
T5	矮大华	锦甜10母本	T25	YC09	选自美国杂交种脆王
T6	美白3	锦甜10父本	T26	YC10	外引101种质
T7	1118	美国温带种质选系	T27	YC27-1	选自美国超白种质
T8	26	美国温带种质选系	T28	T2497-4	选自瑞士杂交种
T9	TF60	选自温带种质	T29	T764-6	选自阿根廷杂交种
T10	T16	选自华珍母本群体	T30	T2501	选自美国杂交种奥弗兰
T11	T8345	选自美国甜玉米杂交种520	T31	T1093-1	选自品试27
T12	A203	华宝甜8号父本	T32	T1104-2	选自中农大甜419
T13	KY99	台湾种质华珍父本	T33	T1133-2	选自斯达206
T14	KY188	华珍母本	T34	TN98	选自彩甜糯6号
T15	CB1	选自美国超白种质	T35	ZN6	选自紫糯3号
T16	SH-241	选自中农大超甜1号	T36	京甜糯2	选自SH-251×白糯6
T17	金杂11	选自美国杂交种库普拉	T37	D6644	选自SH-251×白糯6
T18	TN2965	选自彩甜糯4号	T38	京甜糯68	选自SH-251×白糯62
T19	TN68	选自彩甜糯6号	T39	D6644-6	选自SH-251×白糯6
T20	T9	选自suse材料			

1.2 试验方法

1.2.1 DNA提取

玉米基因组DNA提取参照Rogers和Bendich提出的CTAB法进行。所提DNA质量需满足以下要求：①条带单一，完整无降解；②紫外分光光度计检测A260/280介于1.8~2.0（DNA样品没有蛋白、RNA污染）；③A260/230介于1.8~2.0（DNA样品盐离子浓度低）；④样品DNA浓度至100ng/μl。

1.2.2 SNP位点选取

利用北京市农林科学院玉米研究中心开发的MaizeSNP3072芯片对供试自交系进行SNP基因分型。根据以下指标选用1 031个SNP位点：①数据质量评估值≥0.7；② AA、

AB、BB 这 3 种基因型具备明显界限；③位点特异性探针设计评估值≥0.7；④在全基因组上均匀分布，用于本研究。

1.2.3 GoldenGate 技术芯片试验操作流程

待提取的基因组 DNA 与激活的生物素充分结合，通过离心沉淀使 DNA 纯化；将探针与纯化的目的 DNA 链杂交，漂洗杂交产物，除去非特异结合试剂或过量试剂后进行延伸连接反应；延伸后产物经漂洗变性后作为模板进行 PCR 扩增，PCR 产物与磁珠结合后过滤纯化；将纯化的 PCR 产物与芯片杂交，杂交结束后进行芯片清洗、真空抽干；风干后的芯片立刻在 iScan 仪上进行扫描，最后利用 GenomeStudio 软件进行数据分析。

1.2.4 数据统计分析

本研究数据利用 PowerMarker V3.25 软件进行分析，计算其最小等位基因频率（MAF）、多态性信息含量（PIC）、杂合率等；采用 Nei's 1972 算法计算自交系间遗传距离，并利用 Mega4.0 软件构建 NJ 聚类图，进行杂种优势群的划分。

2 结果与分析

2.1 SNP 位点总体评价

利用 1 031 个 SNP 标记检测了 39 份甜玉米自交系的基因型，并用 PowerMarker V3.25 分析其基因分型数据，对 1 031 个 SNP 位点进行评价。结果表明，1 031 个 SNP 位点在 39 份甜玉米自交系中的杂合率在 0~30.8%，平均值为 3%；在 39 份自交系中的缺失率在 0~20.5%，平均值为 4.6%，说明 1 031 个 SNP 位点有较高的品种区分能力。多态性信息含量（PIC）的变化范围为 0.000~0.375，平均值为 0.290。最小等位基因频率（MAF）的变化范围为 0.000~0.500，平均值为 0.275，其中≥0.1 的位点有 917 个，占 88.9%（图 1）。

PIC 值是衡量 DNA 座位变异程度高低的重要指标。某 DNA 座位的 PIC 值>0.50 时，表明其为高度多态性座位；0.25<PIC 值<0.50 时，表明其为中度多态性座位；PIC 值<0.25 时，表明为低度多态性座位。本研究中，1031 个 SNP 标记在 39 份糯玉米自交系上的平均 PIC 值为 0.29，说明为中度多态性，39 份自交系遗传多样性较好。

2.2 自交系聚类分析结果

根据 39 份自交系在 1 031 个 SNP 位点的基因分型数据，通过 PowerMarker V3.25 软件，并应用 Nei's1972 算法计算两两自交系之间的遗传距离，构建 NJ 聚类图，进行聚类分析（图 2）。结果表明，自交系间的遗传距离变化范围为 0.032~0.678，平均值为 0.430。其中，自交系 TN68 与 T1104-2 之间遗传距离最大，自交系 TN68 与 TN98 之间遗传距离最小。

类群划分结果表明，供试 39 份自交系可被划分为 5 大杂种优势类群。矮大华，T16，KY188，T1133-2 等 4 个自交系被划分为第一类群，因均含有华珍母本血缘，又称为华珍母本群。D6644，京甜糯 2，D6644-6，京甜糯 68，ZN6 等 5 个自交系被划分为第二类群，又称为京甜糯 2 群。TN98，TN68，TN2965 等 3 个自交系被划分为第三类群，因均选自于彩甜糯杂交种，又称为彩甜糯群。第四类群由 T2497-4，YC09，SH-241，SH-251，T1093-1，1118，26，T9，双金 11，金杂 11，TF60，BKL，T68W，YC09，T1104-2，YC10，美白 3，A203，CB1，YC27-1，T8367，T8345，T784-3，T2501，T764-6 等共 25

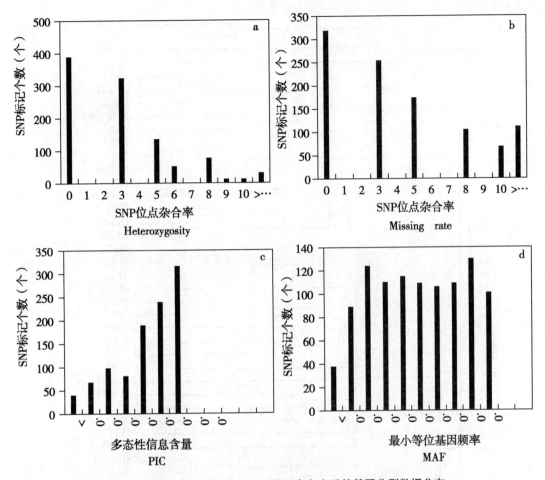

图 1　1 031 个 SNP 位点对 39 份甜玉米自交系的基因分型数据分布
Figure 1　Genotypic statistics of 39sweet corn inbred lines based on 1031 SNPs

注：a. SNP 位点杂合率分布；b. SNP 位点缺失率分布；c. 39 份自交系基于 1 031 个 SNP 位点的多态性信息含量；d. 39 份自交系基于 1 031 个 SNP 位点的最小等位基因频率分布

Note：a. Heterozygosity of SNPs; b. Missing rate of SNPs; c. PIC of 39 lines based on 1 031 SNPs; d. MAF of 39 lines based on 1 031 SNPs

个自交系组成，因其中大部分自交系为温带种质，又称为温带种质群（图 2）。自交系 KY99，T268 被划分为第五类群，又称为华珍父本群。

根据类群划分结果，计算出 5 个杂种优势群的群体间遗传距离（表 2）。由表中可以看出，5 个类群之间的遗传距离变化范围较小，在 0.394~0.445，其中华珍父本群与温带种质群之间的遗传距离为 0.394，彩甜糯群与华珍父本群、彩甜糯群与温带种质群之间的遗传距离最远，均为 0.445。

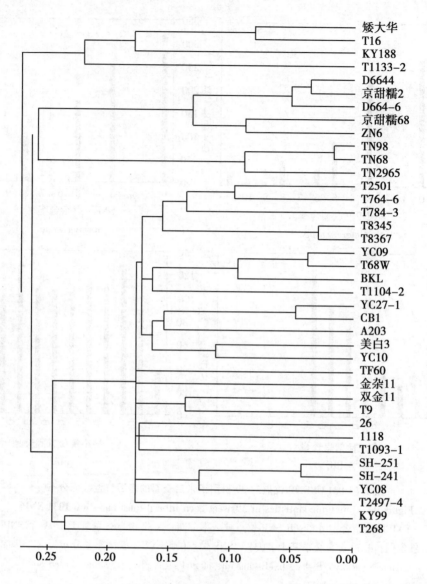

图 2 基于 SNP 标记的 39 份自交系 NJ 聚类图
Figure 2　Neighbor-joining (NJ) trees for 39 inbred lines based on SNP markers

表 2　杂种优势群群体间的遗传距离
Table 2　Genetic distance between different heterotic groups

	华珍母本群 Huazhen maternal	京甜糯 2 群 Jingtiannuo2	彩甜糯群 Colorful sweet-waxy	温带种质群 Temperate
京甜糯 2 群 Jingtiannuo2	0.423			
彩甜糯群 Colorful sweet-waxy	0.437	0.410		
温带种质群 Temperate	0.419	0.395	0.445	
华珍父本群 Huazhen paternal	0.410	0.422	0.445	0.394

3 讨论

分子标记的发展使得在分子水平上研究玉米遗传多样性成为可能。国内外诸多学者利用 SSR 分子标记研究了不同来源甜玉米种质的遗传多样性并划分了类群。近年来，SNP 标记因其位点丰富、自动化程度高等优点，逐渐取代了传统标记，并在普通玉米种质遗传组成分析上有了广泛的应用。

本研究选取 1 031 个 SNP 位点对 39 份甜玉米自交系进行基因型分析，数据结果表明，1 031 个 SNP 位点的杂合率平均值为 3%，缺失率平均值为 4.6%，多态性信息含量的平均值为 0.290。Lu 等利用 SNP 标记对 770 份国内外玉米自交系进行遗传多样性分析的结果表明，1 034 个 SNP 位点的多态性信息含量平均值为 0.259，这说明本研究选用的 1 031 个 SNP 位点具有较高的多态性和品种区分能力。杂种优势类群划分结果表明：供试 39 份自交系被划分为 5 个杂种优势群，划分结果与系谱来源有较好的一致性，如 D6644，京甜糯 2，D6644-6，京甜糯 68 均为二环系（SH-251×白糯 6）选育而来，被聚为同一类群；自交系 T8345 与 T8367 均选自美国杂交种 520，二者遗传距离为 0.057，被聚到一起，说明利用 SNP 标记对甜玉米种质资源进行杂种优势群划分能够获得比较理想的结果。

本研究所划分的 5 个杂种优势群分别为华珍母本群，京甜糯 2 群，彩甜糯群，温带种质群，华珍父本群。通过聚类分析发现，我国优良超甜玉米品种华珍的父母本被划分到不同类群，甜玉米品种锦甜 10 号（矮大华×美白 3）的母本来自华珍母本群，父本来自温带种质群，因此华珍母本群×华珍父本群、华珍母本群×温带种质群是甜玉米品种选育的优良杂优模式，而华珍母本群均为亚热带种质，因此亚热带种质×温带种质也是值得探索的主要模式。另外，本研究中彩甜糯群与华珍父本群、温带种质群间的遗传距离最远，因此可探讨它们之间新的杂优模式，以提高育种效率。

参考文献（略）

本文原载：玉米科学，2015，23（1）：58-68

不同收获期糯玉米杂交种的种子萌发和幼苗生长

卢柏山[1]　史亚兴[1]　徐　丽[1]　樊艳丽[1]　律宝春[2]　云晓敏[2]　牛　茜[2]

(1. 北京市农林科学院玉米研究中心，北京　100097；
2. 北京市种子管理站，北京　100088)

摘　要：选用糯玉米杂交种京科糯2000、京紫糯218，甜加糯类型玉米杂交种京科糯928为试验材料，研究新疆制种基地糯玉米杂交种种子成熟度对种子萌发及幼苗生长的影响，确定高活力种子的适宜收获期。结果表明：不同品种、授粉后不同收获期对种子发芽势、发芽率、活力指数、生长势、干物质积累、抗冷性等的影响均达到显著水平。在授粉后20～55d的不同收获期，三个品种种子各指标总体均呈现先升高后降低的趋势，京科糯2000在授粉后40～50d收获，京科糯928在授粉后45d收获、京紫糯218在授粉后50d收获时，种子活力指标和生长指标值均达最大、抗冷性最强，为最适收获期；品种之间，最适收获期的发芽率为京科糯928（98.67%）>京科糯2000（94%）>京紫糯218（92.33%），三个品种的抗冷性强弱表现为京科糯928>京科糯2000>京紫糯218。

关键词：糯玉米；种子成熟度；收获期；种子活力；幼苗生长

Seed Germination and Seeding Geowth of Hybrid Waxy Corn in Different Harvesting Time

Lu Baishan[1]　Shi Yaxing[1]　Xu Li[1]
Fan Yanli[1]　Lv Baochun[2]　Yun Xiaomin[2]　Niu Qian[2]

(1. Maize Research Center, Beijing Academy of Agriculture and Forestry Sciences, Beijing 100097; 2. The Seed Station of Beijing, Beijing 100088, China)

Abstract: Effects of waxy corn Jingkenuo2000, Jingzinuo218 and Jingkenuo928 (waxy and sweet) with different seed maturation degree on seed germination and seedling growth were studied, in order to determine the suitable harvesting time. The results showed that variety, harvesting time both have significant effects on seed germination potential, germination rate, vigour index, growth potential, amount of dry matter, and cold resistance. Generally all the index above showed increase firstly and then decreased in 20～55days after pollination (DAP). Seed vigour index and growth index both reached a maximun level when harvested 40～50DAP for Jingkenuo2000, 45DAP for Jingkenuo928 and 50DAP for Jingzinuo218, which were the suitable harvesting time. Relatively, germination rate of seeds on suitable harvesing time was Jingkenuo928（98.67%）>Jingkenuo2000（94%）>Jingzinuo218（92.33%）.Cold resistance among three varieties showed Jingkenuo928>Jingkenuo2000>Jing-

---

作者简介：卢柏山（1972—　），男，推广研究员，硕士，主要从事鲜食玉米遗传育种研究

zinuo218.

Key words: Waxy corn; Seed maturity; Harvesting time; Seed vigour; Seedling growth

糯玉米（*Zea mays* L. *certaina* Kulesh）起源于我国，是普通玉米发生基因隐性突变而形成的一种特殊玉米类型。糯玉米籽粒胚乳中所含淀粉100%为支链淀粉，分子量小，食用消化率高；籽粒中富含蛋白质、氨基酸、维生素等营养物质，尤其是人体必需氨基酸之一赖氨酸含量比普通玉米高16%~74%，是一种高营养价值和经济价值的特用玉米。近年来随着糯玉米品种的改良和推广，其种植面积逐年增大，据统计2014年我国糯玉米种植总面积已近1 000万亩，并仍将保持增长势头。

种子是农业生产的基本生产资料，种子质量的高低直接影响农作物的产量和品质，而种子活力是一个能够全面衡量种子质量的指标，种子活力与种子成熟度密切相关。樊廷录等研究认为授粉后不同采收期对籽粒发芽率、种子活力等的影响达到极显著水平；李爱玲等研究认为，授粉后55~61d的玉米种子活力指数最大，61d以后活力指数降低；张建成等的研究表明成熟度高的花生种子发芽整齐迅速，活力高；刘旭欢等研究表明，成熟度高的小麦种子可有效抵抗种子老化。目前，关于玉米种子成熟度对其活力影响的研究多针对于普通玉米，对糯玉米的研究较缺乏。而实际生产中，糯玉米以其丰富营养和独特口感已成为一种新型果蔬食品，被大面积推广种植。因此，分析研究糯玉米制种时高活力种子的适宜收获期具有重要意义。本文以纯糯玉米和甜加糯玉米类型为试验材料，以制种过程中种子收获期代表种子的成熟度，研究了不同收获期对糯玉米种子萌发、幼苗生长的影响，同时对不同收获期种子进行抗冷性测定，确定品种的最适收获期，避免早秋霜冻和低温等不良气候影响，提高糯玉米制种子质量，为糯玉米高活力种子生产技术提供理论指导。

1 材料与方法

1.1 试验材料

供试品种为京科糯2000，京科糯928，京紫糯218。其中京科糯928为甜加糯鲜食玉米类型，即同一果穗上既有甜粒又有糯粒；京科糯2000和京紫糯218为纯糯玉米类型，前者籽粒表现白色，后者籽粒表现紫色。3个品种均由北京市农林科学院选育并通过国家或省级农作物品种审定委员会审定。

1.2 种植与取样

试验于2014年在新疆省昌吉市军户农场制种基地进行。该地位于北纬43°~45°，东经86°~87°，年均气温6.6℃，年均降水183~200mm，有效积温3 400~3 584℃，属半干旱大陆性气候。4月15日进行播种，种植密度为5 000株/亩。父母本种植比例为1∶4，即1行父本与4行母本交替种植。母本吐丝前，及时抽掉雄穗以防止自交，同时选取500株母本植株在吐丝前套住雌穗，吐丝3~4cm时收集父本花粉进行人工授粉，并记录授粉日期。自授粉后20d起，每5d收获标记果穗50穗，至授粉后55d。收获后的果穗在烘干塔中以30℃温度进行烘干至安全水分（12%）。果穗脱粒精选后，随机选取100粒称重，3次重复。其余种子保存备用。

1.3 种子发芽试验

发芽试验采用沙培法。选用无污染的细沙过2cm孔径筛，用清水冲洗后在160℃高温下灭菌3h,并在标准发芽盒内制成3cm厚的沙床。取不同处理的种子各100粒,播种至沙床,表面覆2~3cm湿沙。标准发芽率测定中,将发芽盒置于25℃恒温发芽室12h光照+12h黑暗条件下培养;抗冷性发芽率测定中,将发芽盒先置于7℃恒温发芽箱内5d,再转移至25℃恒温发芽室12h光照+12h黑暗条件下培养。以上每处理均3次重复。

1.4 指标测定

标准试验中,发芽势和发芽率分别指发芽开始第4d和第7d发芽的种子数占总种子数的百分比。发芽第7d,在各处理中随机选取10株有代表性的幼苗测定苗高、根长,然后在70℃烘箱中烘干至恒重,分别称取苗干重和根干重,计算简化活力指数。简化活力指数=发芽率×苗高。抗冷性发芽试验中,指标测定时间以经低温处理后转移至标准发芽环境中时为第1d,测定方法同上。

1.5 数据统计分析

本试验数据采用SAS8.0软件进行分析,采用EXCEL2007进行数据整理和图表制作。

2 结果与分析

2.1 不同收获期对糯玉米杂交种的籽粒百粒重

不同时期收获处理条件下,杂交种籽粒百粒重随收获期的推后呈上升趋势,籽粒增重基本符合慢-快-慢的"S"形曲线(图1)。授粉后20~30d,籽粒百粒重快速增大,随灌浆天数增加,百粒重增加趋势变缓。京科糯2000杂交种在授粉后50d时百粒重达最大,为20.29g,较授粉后20d增加78.56%,达到生理成熟;京科糯928和京紫糯218在授粉后50~55d时百粒重仍在增加,55d时的百粒重分别为21.03g和21.68g,较授粉后20d分别增加88.64%、70.93%。表明京科糯928和京紫糯218的灌浆持续期较京科糯2000要长。

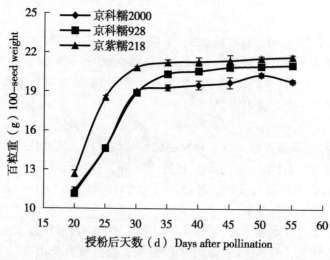

图1 不同收获期(授粉后天数)糯玉米杂交种籽粒的百粒重

Figure 1 100-seed weight of waxy corn at different harvest time (days after pollination)

2.2 不同收获期对糯玉米杂交种的种子萌发

授粉后不同时期收获处理条件下,3个品种的发芽势、发芽率和活力指数在授粉后

20~55d 均呈现先升高后降低的趋势，尤其在 20~30d 增加显著（表1）。京科糯 928 和京紫糯 218 在授粉后 30d 的种子发芽率已达到 90% 以上（92.33% 和 90%），京科糯 2000 在授粉后 35d 的发芽率达 90% 以上（92%）。这表明，试验品种种子在授粉 30d 以后已有较高的发芽率。随着灌浆天数的增加，种子发育逐渐成熟，3 个品种的发芽率均继续增大，京科糯 928、京科糯 2000、京紫糯 218 的发芽率分别在授粉后第 45d、40d、50d 达到最高，为 98.67%、94%、92.33%，与授粉后 20d 相比分别增加 20.82%、20.51%、19.91%；之后，3 个品种的发芽率均有所降低，京科糯 928 和京科糯 2000 降低差异不显著，变化范围分别在 94.33%~98.67% 和 93%~94%。京紫糯 218 在授粉后 50~55d 的发芽率降低显著，55d 时降至 83.33%。

京科糯 928、京紫糯 218 的活力指数变化情况与发芽率趋势一致，分别在授粉后 45d 和 50d 时最大；京科糯 2000 活力指数在授粉后 50d 最大，与发芽率达最大日期相比延迟 10d。品种间，3 个品种的发芽率和活力指数均有显著差异，发芽率为京科糯 928>京科糯 2000>京紫糯 218，活力指数为京科糯 2000>京科糯 928>京紫糯 218。因此，不同基因型品种、同一品种不同收获期对种子的活力特征参数均有显著影响。从发芽情况看，京科糯 928 杂交制种最适宜收获期为授粉后 45d，京科糯 2000 为授粉后 40~50d，京紫糯 218 为授粉后 50d，收获期过晚，种子发芽率和活力指数显著降低。

表1 不同收获期糯玉米杂交种的种子萌发
Table 1　Seed germination of waxy corn at different harvest time

授粉后天数（d） Days after pollination	芽势（%） Germination potential			芽率（%） Germination rate			活力指数 Vigour index		
	京科糯 928	京科糯 2000	京紫糯 218	京科糯 928	京科糯 2000	京紫糯 218	京科糯 928	京科糯 2000	京紫糯 218
20	74.67c	60.00c	68.33a	81.67c	78c	77.00d	9.06e	10.16d	8.96f
25	78.00b	70.00b	74.67ab	87.67b	85.33b	84.33bc	11.08d	12.36c	10.42e
30	83.33abc	81.00a	79.67a	92.33ab	88.00ab	90.00ab	12.80c	12.94c	12.05c
35	86.00ab	83.67a	81.67a	96.00a	92.00a	91.33a	12.78c	13.95b	12.83b
40	88.33ab	88.33a	81.33a	95.67a	94.00a	92.00a	13.74b	13.74b	13.55ab
45	89.00a	87.00a	80.67a	98.67a	93.67a	91.33a	14.47a	13.68b	13.31ab
50	87.33ab	85.67a	82.00a	94.33a	93.00a	91.33a	13.54bc	14.88a	13.94a
55	88.00ab	84.33a	79.67a	96.67a	93.00a	83.33c	13.34bc	12.94c	11.19d
平均 Average	84.33a	80.00b	78.50b	92.88a	89.63b	87.71c	12.60b	13.08a	12.03c

注：表中字母代表 0.05 水平差异显著性。最后一行同一指标下的不同字母表示不同品种之间的差异（横向），其余数据后面的不同字母表示同一指标下同一品种不同收获期间的差异，下同

Note: Letters in the table represent difference significance on 0.05 level. Letters in the last row under the same index mean difference among three varieties (horizontal), the other letters mean difference of one index of the same variety. The same below

2.3 不同收获期糯玉米杂交种的幼苗生长

选用苗高、根长、苗干重、根干重等生长指标,对参试3个糯玉米品种在不同收获期条件下的种子幼苗进行测定,结果表明,杂交种种子成熟度不同,幼苗生长各指标均有显著差异。

由表2可以看出,京科糯928、京紫糯218、京科糯2000的苗高和根长分别在授粉后45d、50d、50d达到最大,分别为14.67cm和17.65cm、15.09cm和17.85cm、16.00cm和19.33cm。与授粉后20d相比,三品种苗高和根长分别增加32.16%和29.68%、29.75%和36.36%、22.89%和20.74%。授粉后45~55d,三品种苗高和根长均有所降低。品种间来看,3个品种的苗高为京科糯2000〉京紫糯218>京科糯928,前者与后两者差异显著;主根长为京科糯2000>京科糯928>京紫糯218,且三者之间差异显著,表明在不同收获期条件处理下,京科糯2000杂交种幼苗长势最强;京科糯928、京紫糯218幼苗长势与其活力指标变化趋势一致,均在同一收获期达到最大;京科糯2000种子在授粉40d以后虽发芽率有所降低,但幼苗长势继续增强,至授粉后50d苗高和主根长达最大值。

表2 不同收获期糯玉米杂交种的幼苗高和根长度
Table 2 Root length and seeding height of waxy corn seedlings at different harvest time

授粉后天数 (d) Days after pollination	苗高 (cm) Seedling height			根长 (cm) Root length		
	京科糯928	京科糯2000	京紫糯218	京科糯928	京科糯2000	京紫糯218
20	11.10c	13.02c	11.63c	13.61b	16.01d	13.09c
25	12.64bc	14.48abc	12.35bc	14.82b	16.07d	14.92b
30	13.87ab	14.70abc	13.39ab	16.87a	17.62bc	14.96b
35	13.32ab	15.17ab	14.05ab	16.88a	17.30bcd	15.61b
40	14.37ab	14.62abc	14.73a	14.99b	17.78bc	15.43b
45	14.67a	14.60abc	14.57a	17.65a	18.63ab	16.11b
50	14.35ab	16.00a	15.09a	16.82a	19.33a	17.85a
55	13.80ab	13.91bc	13.42ab	16.72a	16.82cd	16.21b
平均 Average	13.51b	14.56a	13.65b	16.05b	17.44a	15.52c

在不同收获期处理条件下,3个品种苗、根干物质积累变化趋势与其苗高和根长一致(表3),京科糯928种子的苗干重和根干重在授粉后20~45d显著增加,45d时达最大,分别为0.33g和0.37g。与授粉后20d相比分别增加135.71%和68.18%。45~55d时有所降低,差异不显著。京科糯2000、京紫糯218种子的苗干重和根干重在授粉后20~50d显著增加,50d时达到最大,分别为0.35g和0.40g、0.46g和0.40g,与授粉后20d相比分别增加133.33%和233.33%、119.05%和42.86%。授粉后50~55d时降低,差异不显著。品种间来看,3个品种的苗干重为京紫糯218>京科糯2000>京科糯928,前者与后两者之间差异显著;3品种之间根干重无显著差异。综合本试验参试品种生长情况来看,京科糯928在授粉后45d左右收获幼苗长势好,干物质积累量最大;京科糯2000和京紫糯218在

授粉后50d收获时幼苗长势好，干物质积累量最大。

表3 不同收获期糯玉米杂交种幼苗的干物质积累
Table 3 Dry matter accumulation of waxy corn seedings at different harvest time

授粉后天数（d）Days after pollination	苗干重（g/10株）Seedling dry weight			根干重（g/10株）Root dry weight		
	京科糯928	京科糯2000	京紫糯218	京科糯928	京科糯2000	京紫糯218
20	0.14d	0.15c	0.21c	0.22c	0.12e	0.28b
25	0.19cd	0.20bc	0.32b	0.28bc	0.26d	0.29b
30	0.22bc	0.27ab	0.36b	0.29abc	0.31bcd	0.37ab
35	0.29ab	0.34a	0.44a	0.34ab	0.29cd	0.34ab
40	0.33a	0.29a	0.35b	0.32ab	0.31bcd	0.31b
45	0.33a	0.32a	0.36b	0.37a	0.39ab	0.35ab
50	0.33a	0.35a	0.46a	0.32ab	0.40a	0.40a
55	0.31a	0.33a	0.35b	0.34ab	0.36abc	0.33ab
平均 Average	0.27b	0.28b	0.36a	0.31a	0.31a	0.33a

2.4 低温处理糯玉米杂交种的种子发芽率

经低温处理后，不同收获期处理条件下种子的发芽率均降低（图2）。与标准发芽率相比，授粉后20~30d的种子发芽率降低显著，京科糯928、京科糯2000、京紫糯218的降低幅度分别为2.33%~16.67%，5%~34.33%和18.67%~40.67%。授粉后30~55d的种子发芽率降低幅度减小，差异不显著，京科糯928、京科糯2000、京紫糯218分别降低1.33%~4.34%，1.67%~4.67%，7.67%~14.33%。低温处理条件下京科糯928、京科糯2000、京紫糯218在最适合收获期的发芽率分别为94.33%、92%、84%，与标准发芽率相比分别降低了4.34%、2%、8.33%。这表明，成熟度低的种子抗冷性差；标准发芽率达最大时，种子的抗冷性也最强；低温对糯玉米种子萌发的影响存在品种间差异，京紫糯218的抗冷性要低于京科糯928和京科糯2000。

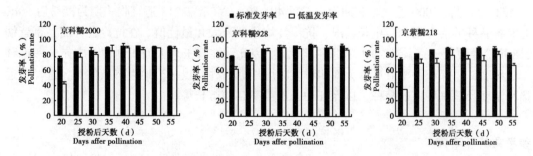

图2 低温处理不同收获期糯玉米杂交种的种子发芽率
Figure 2 Seed germination rate of waxy corn with different harvest time at low temperature

3 讨论与结论

种子活力形成于种子发育过程，随着种子体内蛋白质、淀粉等贮藏物质的积累，种子不断成熟，种子的发芽率和活力等也逐渐提高。樊廷录等利用普通玉米先玉335、郑单958等研究了种子成熟期与种子活力的关系，表明先玉335、郑单958杂交种分别在授粉后第67d、70d达到生理成熟，而二者杂交种种子的发芽率和活力较高值均出现在达到生理成熟之前15~20d。本研究中，京科糯2000种子在授粉后50d的百粒重最大，达到生理成熟，而其发芽率和活力指数则分别在授粉后40d和50d达最大值，这是因为在授粉后50d收获的京科糯2000种子发芽率虽有所降低，但其幼苗长势并没有减弱，苗高反而继续增大，使得活力指数最高值对应的收获期推迟；京科糯928、京紫糯218种子在授粉后50~55d的百粒重仍在增加，而两者发芽率和活力指数则分别在授粉后45d和50d达最大。这与前人研究结果类似，表明成熟度高的种子发芽率和活力相应提高，但种子发芽率和活力等指标达到最大时并不一定是种子完全达到生理成熟时，而是在生理成熟之前；京科糯928和京紫糯218灌浆持续期比京科糯2000要长，普通玉米籽粒灌浆持续期比糯玉米长。

种子发芽时遭遇的低温冷害和早霜冻等不良天气可显著降低种子发芽率和活力。本研究选用目前推广应用面积较大的不同类型糯玉米品种进行制种试验，在授粉后不同时期收获，进行萌发和生长情况测定，同时进行抗冷性试验，结果表明，3个品种的标准发芽率在任一收获期均要高于低温发芽率。授粉后20d，品种平均低温发芽率为48.33%，较标准发芽率降低了38.74%；随种子成熟度的增加，种子的抗冷性也逐渐提高，授粉后35d时，平均低温发芽率为83%，较同期收获种子的标准发芽率降低了10.86%；平均标准发芽率达最大时，相应收获期种子的平均低温发芽率也达到最大，为90.11%，较标准发芽率降低了5.15%。3个品种之间抗冷性不同，京科糯928、京科糯2000、京紫糯218的平均低温发芽率分别为87.13%、82.29%、71.21%，三者之间差异显著。京科糯928的抗冷性最强、京紫糯218的抗冷性较弱。这与郝楠等、卢柏山等研究结果一致，表明低温冷害可显著降低种子发芽率。这可能是因为低温胁迫降低了种子体内一些消化酶如α-淀粉酶的活性，减缓了种子中贮藏物质的降解，影响了种子萌发所需要的物质与能量的供给，导致发芽率降低。

综合以上研究结果，新疆昌吉地区糯玉米制种时，京科糯928种子适宜收获期为授粉后45d，京科糯2000适宜收获期为授粉后40~50d，京紫糯218适宜收获期为授粉后50d，此时各品种的活力指标、生长情况、抗冷性均达到最大和最优值。在抗冷性方面，京紫糯218抗冷性要低于京科糯2000和京科糯928。因此，在糯玉米大面积制种时应充分考虑种子成熟度，以保证种子质量；大面积种植时应根据品种特点，当地生态条件等，选择合适播种温度及适宜种植的品种，以保证种子出苗率和幼苗长势。

参考文献（略）

甜玉米杂交种种子成熟度对种子萌发和幼苗生长的影响

卢柏山[1]　史亚兴[1]　徐 丽[1]　樊艳丽[1]　律宝春[2]　云晓敏[2]　牛 茜[2]

(1. 北京市农林科学院玉米研究中心, 北京　100097；
2. 北京市种子管理站, 北京　100088)

摘　要：为确定新疆制种基地甜玉米种子适宜收获期，为高活力甜玉米种子生产提供依据，本文选用4个甜玉米杂交种为试验材料，研究新疆制种基地甜玉米杂交种不同成熟度种子的萌发及幼苗生长情况。结果表明：在授粉后20~45d的不同收获期，参试品种种子发芽势、发芽率、活力指数、生长势、干物质积累、抗冷性等均呈现先升高后降低的趋势。京科甜179和京科甜533在授粉后35d收获，京科甜183和京科甜158在授粉后40d收获时，各品种种子活力指标和生长指标值均达最大，抗冷性最强，为最适收获期；品种之间，最适收获期的发芽率为京科甜183（92.00%）>京科甜158（89.00%）>京科甜533（87.00%）>京科甜179（85.33%）。在抗冷性方面，京科甜183抗冷性最强，京科甜179抗冷性最弱。

关键词：甜玉米；种子成熟度；收获期；种子活力；幼苗生长

Effects of Seed Maturation of Sweet Corn on Seed Germination and Seedling Growth

Lu Baishan[1]　Shi Yaxing[1]　Xu Li[1]　Fan Yanli[1]
Lv Baochun[2]　Yun Xiaomin[2]　Niu Qian[2]

(1. Maize Research Center, Beijing Academy of Agriculture and Forestry Sciences, Beijing, 100097; 2. The Seed Station of Beijing, Beijing, 100088, China)

Abstract: To determine the suitable harvesting timeof sweet corn seed in Xinjiang test base, seed germination and seedling growthof four varieties with different maturation degree were studied. The results showed thatseed germination potential, germination rate, vigour index, growth potential, amount of dry matter, and cold resistance of the four varieties displayed increasing firstly and then decreasing in 20~45days after pollination (DAP). Seed vigour index and growth index both reached a maximun level when harvested 35DAP for Jingketian179 and Jingketian533, 40DAP for Jingketian183 and Jingketian158, which were the suitable harvesing time. Relatively, germination rate of seeds on suitable harvesing time was Jingketian183（92.00%）>Jingketian158（89.00%）>Jingketian533（87.00%）>Jingketian179（85.33%）. Jingketian183 showed the strongest resistanceto cold, while

---

作者简介：卢柏山（1972—　），男，汉族，北京。硕士，推广研究员，主要从事鲜食玉米遗传育种研究

Jingketian179 showed the weakest.

Key words: Sweet corn; Seed maturity; Harvesting time; Seed vigour; Seedling growth

甜玉米（Zea mays L. saccharata Sturt）是玉米属（Zea mays L.）的一个玉米亚种，即甜质型玉米亚种，其籽粒在乳熟期富含果糖、蔗糖、氨基酸等，可生食或作蔬菜食用，因此又被称为水果玉米或蔬菜玉米。近年来，甜玉米在我国种植规模逐渐扩大，新品种选育、产品加工贮藏等水平不断提升，甜玉米已成为我国发展潜力巨大的果蔬兼用型农作物。

种子是农业生产的基本生产资料，种子质量的好坏可直接影响农作物的产量和品质，而种子活力是一个能够全面衡量种子质量的指标。大量研究表明，种子成熟度与种子活力密切相关，成熟度高的种子发芽整齐迅速，并可有效抵抗低温、种子老化等对生产的影响。在玉米作物上，周朋等的研究表明，种子成熟度越高，种子发芽率、发芽势等活力指标也相应提高。樊廷录等研究表明，种子发芽率和田间出苗率随种子成熟度的提高而增大，并在授粉后57~72d达到最大。沈雪芳等研究发现，普通甜玉米授粉后30d收获的种子发芽率显著高于20d收获种子的发芽率，并且随授粉后天数的增加，发芽率逐渐升高。目前，关于玉米种子成熟度与种子活力关系的研究多针对于普通玉米，缺乏对甜玉米的研究。而甜玉米作为一种蔬果型食品已被越来越多的消费者认可和喜爱，推广种植面积逐年增大，因此，研究甜玉米杂交种种子的适宜收获期具有重要意义。本文以超甜玉米为试验材料，以制种过程中种子收获期代表种子的成熟度，研究了不同收获期对甜玉米种子萌发、幼苗生长的影响，明确不同品种的最适收获期，同时对不同收获期甜玉米种子进行抗冷性测定，避免早秋霜冻和低温等不良气候影响，为甜玉米高活力种子生产技术提供理论指导。

1 材料与方法

1.1 试验材料

本研究供试品种为京科甜179、京科甜183、京科甜158、京科甜533。四个品种均为超甜类型，并均由北京市农林科学院选育并通过省级农作物品种审定委员会审定。

1.2 试验方法

1.2.1 种植与取样

试验于2014年在新疆省昌吉市制种基地进行。该地年均气温6.6℃，年均降水183~200mm，有效积温3 400~3 584℃，属半干旱大陆性气候。试验材料于4月中旬播种，种植密度为6 000株/亩，1行父本与6行母本交替种植。母本吐丝前及时抽掉雄穗，同时选取600株母本植株在吐丝前套住雌穗，吐丝3~4cm时收集父本花粉进行人工授粉，并记录授粉日期。自授粉后20d起，每5d收获标记果穗70穗，至授粉后45d。收获后的果穗在烘干塔中以30℃温度烘干至安全水分（12%）后进行脱粒，去除发霉和破损籽粒，随机选取100粒称重，3次重复。其余种子保存备用。

1.2.2 种子发芽试验

选用无污染的细沙过2cm孔径筛，用清水冲洗后在160℃高温下灭菌3h，并在标准发芽盒内制成3cm厚的沙床。取不同处理的种子各100粒，播种至沙床表面后再覆2~3cm

湿沙。标准发芽率测定中，将发芽盒置于25℃恒温发芽室12h光照+12h黑暗条件下培养；抗冷性发芽率测定中，将发芽盒先置于7℃恒温发芽箱内5d，再转移至25℃恒温发芽室12h光照+12h黑暗条件下培养。以上每处理均3次重复。

1.2.3 指标测定

标准试验中，发芽势和发芽率分别指发芽开始第4d和第7d发芽的种子数占供试种子总数的百分比。发芽第7d，在各处理中随机选取10株有代表性的幼苗测量苗高、根长，然后在70℃烘箱中烘干至恒重，分别称取苗干重和根干重，计算活力指数。活力指数=发芽率×第7d苗高。抗冷性发芽试验中，指标测定时间以经低温处理后转移至标准发芽环境中时为第1d，测定方法同上。

1.3 数据统计分析

本试验数据采用SAS8.0软件进行分析，采用EXCEL2007进行数据整理和图表制作。

2 结果与分析

2.1 不同收获期对甜玉米杂交种籽粒百粒重的影响

不同时期收获处理条件下，杂交种籽粒百粒重随收获期的推后呈上升趋势（图1）。在授粉后20d之前，京科甜179和京科甜158的百粒重高于京科甜183和京科甜533。京科甜533杂交种在授粉后35d时百粒重达最大，为15.85g，较授粉后20d增加85.12%，京科甜179和京科甜158杂交种在授粉后40d时百粒重达最大，分别为15.50g和14.15g，较授粉后20d分别增加33.29%和32.94%，达到生理成熟。京科甜183在授粉后40~45d时百粒重仍在增加，45d时百粒重为13.87g，较授粉后20d增加62.57%。表明，京科甜179和京科甜158籽粒前期（授粉后20d之前）灌浆速度快；从籽粒整个发育过程来看，京科甜533灌浆持续期较短，京科甜183灌浆期持续期较长。

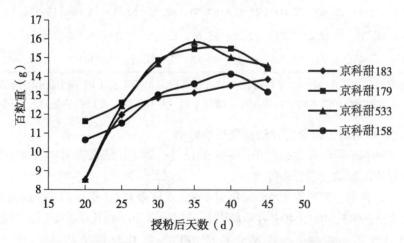

图1 不同收获期（授粉后天数）下甜玉米杂交种籽粒的百粒重

2.2 不同收获期对甜玉米杂交种种子萌发的影响

对参试品种在授粉后不同时期收获的种子进行发芽试验，结果表明，4个甜玉米品种种子的发芽势、发芽率和活力指数在授粉后20~45d均呈现先升高后降低的趋势，尤其在20~35d升高显著（表1）。各品种种子在灌浆前期（授粉后20d之前）发芽率均较低，在30%以下，随着灌浆天数的增加，种子发育逐渐成熟，参试品种的发芽率均继续增大。京科甜179和京科甜533在授粉后第35 d时发芽率达到最大，分别为85.33%和87%，与授粉后20 d相比，分别增加312.9%和203.49%；京科甜183和京科甜158在授粉后第40 d时发芽率达最大，分别为92%和89%，较授粉后20 d分别增加688.57%和260.81%。之后各参试品种发芽率有所降低，但降低不显著。品种间，四个甜玉米品种发芽率表现为京科甜183>京科甜533>京科甜158>京科甜179，前者与后三者之间差异显著。

各参试品种的活力指数变化情况与发芽势、发芽率趋势一致。京科甜179、京科甜183、京科甜158和京科甜533分别在授粉后35 d、40 d、35 d和40 d时最大，之后随发芽率的降低，活力指数也降低。品种间，4个甜玉米品种活力指数表现为京科甜158>京科甜183>京科甜533>京科甜179，前两者与后两者之间差异显著。因此，品种类型、收获期对种子活力特征参数均有显著影响。从发芽情况看，京科甜179和京科甜533杂交制种时最适收获期为授粉后35 d，京科甜183和京科甜158为授粉后40 d。过早或过晚收获，种子发芽率和活力指数均显著降低。

表1 不同收获期对甜玉米杂交种种子萌发的影响

授粉后天数(d)	发芽势（%）				发芽率（%）				活力指数			
	京科甜179	京科甜183	京科甜158	京科甜533	京科甜179	京科甜183	京科甜158	京科甜533	京科甜179	京科甜183	京科甜158	京科甜533
20	5.00e	18.75b	2.00c	4.75b	20.67d	11.67c	24.67e	28.67d	1.30e	0.93d	2.78e	2.62e
25	15.25d	23.00b	4.50c	9.50b	74.33bc	70.33b	66.67d	65.67c	8.85d	6.99c	8.93d	7.86d
30	16.25cd	41.00a	7.00bc	14.25b	80.33ab	86.00a	73.00cd	72.67b	10.06c	11.91b	10.69c	9.40c
35	57.00a	43.25a	16.25ab	29.50a	85.33a	88.00a	78.00bc	87.00a	12.61a	12.26b	11.69b	12.37a
40	41.75b	47.00a	25.00a	28.00a	78.00bc	92.00a	89.00a	83.00b	11.09b	13.53a	13.45a	12.30a
45	26.00c	45.75a	16.00ab	7.00b	72.00c	85.67a	83.33ab	81.33a	9.98b	11.99b	11.50b	10.84b
平均	26.88b	36.46a	11.79c	15.50c	68.44b	72.88a	69.11b	69.72ab	8.98b	9.60a	9.84a	9.23b

注：表中字母代表0.05水平差异显著性。最后一行同一指标下的不同字母表示不同品种之间的差异（横向），其余数据后面的不同字母表示同一指标下同一品种不同收获期间的差异。下同

2.3 不同收获期对甜玉米杂交种幼苗生长的影响

对参试品种在不同收获期条件下的种子幼苗进行测定，结果表明，幼苗生长情况随杂交种种子成熟度不同均表现显著差异。

由表2可以看出，京科甜179、京科甜183、京科甜158、京科甜533的苗高和根长分别在授粉后35 d、40 d、40 d、40 d达到最大。与授粉后20 d相比，4个甜玉米品种苗高分别增加135.00%和240.44%、84.89%和75.90%；根长分别增加34.00%和52.00%、62.17%和115.37%。授粉后35~45 d，苗高和根长均有所降低。品种间来看，参试品种的苗高为京科甜158>京科甜533>京科甜183>京科甜179，前者与后三者差异显著；品种主根长表现与苗高一致。表明不同收获期处理下，京科甜158杂交种的幼苗长势最强，京科甜179长势较弱；京科甜179、京科甜183和京科甜158的幼苗长势与其活力变化趋势一致，均在同一收获期达到最大；京科甜533种子在授粉35 d以后发芽率有所降低，但幼苗

长势继续增强，在授粉后第 40d 时苗高和主根长达到最大值。

表 2　不同收获期对甜玉米杂交种幼苗苗根长度的影响

授粉后天数(d)	苗高（cm）				根长（cm）			
	京科甜179	京科甜183	京科甜158	京科甜533	京科甜179	京科甜183	京科甜158	京科甜533
20	6.28b	7.96b	11.28b	9.14b	4.30c	9.33b	12.46b	9.44d
25	11.90a	9.94b	13.40ab	11.97a	11.13b	10.85b	13.86b	12.20c
30	12.52a	13.85a	14.64a	12.93a	14.27a	14.68a	17.49a	16.31b
35	14.78a	13.93a	14.98a	14.22a	14.64a	15.50a	17.58a	18.78ab
40	14.22a	14.71a	15.12a	14.82a	14.22a	16.41a	19.00a	20.32a
45	13.87a	14.00a	13.80ab	13.33a	14.18a	14.97a	17.45a	19.05a
平均	12.26b	12.40b	13.87a	12.73b	12.12c	13.62b	16.31a	16.02a

　　4 个甜玉米品种苗、根干物质积累均随授粉后天数的增加呈现先升高后降低的变化趋势（表 3）。京科甜 179、京科甜 183、京科甜 158、京科甜 533 的苗和根干物质积累分别在授粉后第 35d、40d、40d、40d 达到最大值，与其苗高和根长生长规律一致。与授粉后 20d 相比，参试品种的幼苗总干物质积累量均有显著增加，并分别增加 388.89%、213.33%、73.08% 和 280.00%。在授粉后 35~45d，京科甜 183、京科甜 158 的苗、根干物质积累均有所降低，京科甜 179 和京科甜 533 两个品种则变化不明显。品种间，4 个甜玉米品种的总干物质积累量比较为京科甜 158>京科甜 183>京科甜 179>京科甜 533。综合来看，京科甜 179 在授粉后 35d 左右收获幼苗长势好，干物质积累量最大；京科甜 158、京科甜 183、京科甜 533 在授粉后 40d 收获时幼苗长势好，干物质积累量最大。

表 3　不同收获期对甜玉米杂交种幼苗干物质积累的影响

授粉后天数(d)	苗干重（g/10 株）				根干重（g/10 株）			
	京科甜179	京科甜183	京科甜158	京科甜533	京科甜179	京科甜183	京科甜158	京科甜533
20	0.06c	0.08c	0.18c	0.07c	0.03b	0.07b	0.09b	0.03b
25	0.12b	0.14b	0.20bc	0.15b	0.07b	0.09b	0.08b	0.06ab
30	0.16b	0.27a	0.28a	0.16b	0.13a	0.15a	0.12ab	0.08ab
35	0.28a	0.27a	0.27a	0.25a	0.16a	0.15a	0.12ab	0.10a
40	0.26a	0.27a	0.29a	0.26a	0.15a	0.20a	0.16a	0.12a
45	0.26a	0.25a	0.26ab	0.26a	0.15a	0.17a	0.13ab	0.12a
平均	0.19b	0.21b	0.25a	0.19b	0.11b	0.14b	0.12a	0.08b

2.4　低温对糯玉米杂交种种子发芽率的影响

　　经低温处理后，任一成熟期收获的甜玉米杂交种种子发芽率均降低（图 2）。与标准

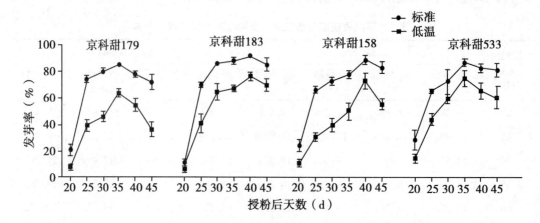

图2 低温对甜玉米杂交种种子发芽率的影响

发芽率相比,授粉后20d的种子低温发芽率降低不明显;随种子成熟度的提高,两种发芽条件下种子发芽率均在增大,但低温发芽率显著低于标准发芽率。京科甜179、京科甜533杂交种种子在授粉后35d时低温发芽率达最大,分别为63.67%、75.33%,比同期标准发芽率分别降低42.74%、13.41%;京科甜183、京科甜158杂交种种子在授粉后40d时低温发芽率达最大,分别为76.33%、73.33%,比同期标准发芽率分别降低17.03%、17.60%。这表明,成熟度低的种子发芽率低,抗冷性差;标准发芽率达最大时,种子的抗冷性也最强;低温对甜玉米种子萌发的影响存在品种间差异,京科甜183的低温发芽率要高于其他三个品种,抗冷性最强,京科甜179抗冷性最弱。

3 讨论与结论

种子是生物遗传的主要载体,高质量和高活力种子具有明显的生长优势和生产潜力,对农业生产具有十分重要的意义。本研究选用生产中推广应用面积较大的不同籽粒颜色甜玉米品种进行不同成熟度种子活力的测定,结果表明,京科甜179杂交种在授粉后40d时百粒重达最大,达到生理成熟,而其发芽率和活力指数则均在授粉后35d达最大值;京科甜183杂交种种子在授粉后40~45d时仍未达到生理成熟,百粒重仍在增加,而其发芽率和活力指数则在授粉后40d时已达最大值。这与时小红、毛笈华等、沈雪芳等的研究结果相似,表明成熟度高的种子发芽率和活力相应提高,但种子发芽率和活力等指标达到最大时并不一定是种子完全达到生理成熟时,而是在生理成熟之前;甜玉米杂交制种时种子适宜收获期要比普通玉米早。

在所有收获期处理条件下,京科甜183的最高发芽率为92%,京科甜179、京科甜158、京科甜533三个品种的最高发芽率分别为85.33%、89%、87%,均在90%以下;参试品种种子的最高发芽势在28%~57%,幼苗高度在14.71~15.12cm,种子百粒重在13.87~15.85g之间。这与张海艳等研究结果相似,而与樊廷录等研究结果不同。后者利用普通玉米进行种子成熟期与种子活力关系的研究,表明种子达到生理成熟时的百粒重均在25g以上,最适收获期的种子发芽率均在95%以上。此结果说明甜玉米在种子活力、生长势等方面均低于普通玉米,因此在种植管理中更要精细,以保证出苗良好。另外,本研究抗冷性试验结果表明，参试品种的低温发芽率在任一收获期均低于其标准发芽率；标准发芽率达最大时，同期种子的低温发芽率也达到最大，说明温度过低会影响甜玉米种子发芽和幼苗生长，成熟度高的种子抗冷性强。这可能是因为种子萌发过程中所需要的养分和能量均来自于种子自身贮藏物质的转化，而低温会降低贮藏物质降解所需要的酶的活性，因此种子淀粉含量少、萌发时能量供给不足就会导致发芽率降低。

综合来看，新疆昌吉地区甜玉米制种时，京科甜 179 和京科甜 533 适宜收获期为授粉后 35 d，京科甜 183 和京科甜 158 适宜收获期为授粉后 40 d，此时各品种的活力指标、生长情况、抗冷性均达到最大和最优值。在抗冷性方面，京科甜 183 抗冷性最强，京科甜 179 抗冷性较弱。因此，甜玉米大面积制种时应充分考虑种子成熟度，避免过早或过晚收获，防止种子成熟度不够或后期种子霉变和穗发芽等现象。

参考文献（略）

本文原载：种子，2016，35（10）：95-99

优质高产超甜玉米品种京科甜158的选育及栽培技术

史亚兴　卢柏山　樊艳丽　徐　丽　席胜利

（北京市农林科学院玉米研究中心，北京　100097）

摘　要：京科甜158是北京市农林科学院玉米研究中心以SH-251为母本，以T68为父本组配而成的超甜玉米杂交种，具有优质、高产、高抗、早熟的特点。该品种于2011年通过北京市农作物品种审定委员会审定，2013年、2014年被北京市种子管理站列为北京市玉米更新换代品种，适宜在北京、河北及类似生态区推广种植。

关键词：超甜玉米；京科甜158；选育；栽培技术

京科甜158是北京市农林科学院玉米研究中心按照"高品质、大群体、严选择"的选系原则，组配选育出的高品质超甜玉米杂交种。2011年通过北京市农作物品种审定委员会审定（审定编号：京审玉2011008）。

1　选育经过

京科甜158是由北京市农林科学院玉米研究中心以SH-251为母本，以T68为父本杂交选育而成的超甜玉米品种。母本SH-251是以国内超甜玉米杂交种"超甜一号"为选系材料，经过6代连续自交和选择于2002年选育而成。SH-251为超甜类型，果穗口感好、穗型漂亮，籽粒纯白色，具有优质、抗病性强、早熟的特点。父本"T68"是以从美国引进的早熟甜玉米杂交种为基础选系材料，经过6代连续自交和选择于2004年选育而成。T68为超甜类型，鲜籽粒亮黄色，皮薄，有风味，甜度18度左右；具有很高的一般配合力和特殊配合力，高抗玉米大、小叶斑病、活秆成熟。

2006年冬在海南初配SH-251×T68杂交组合，经过2007年、2008年在海南、北京小汤山等多地多年多点品比试验，鉴定出该组合具有品质好、产量高、高抗病性等特点。两年平均产量为11 625kg/hm^2，比对照京科甜183增产10%。2009年、2010年参加北京市甜糯玉米区域试验，2011年通过北京市农作物品种审定委员会审定，定名为京科甜158。

2　品种特征特性

2.1　植株性状

北京地区种植播种至鲜穗采收期平均83d，株高178cm，穗位高58cm，单株有效穗数1.01个，空秆率2.2%，穗长18.8cm，穗粗4.6cm，穗行数14~16行，行粒数39，秃尖长0.3cm。粒色黄白，粒深1.1cm，鲜籽粒千粒重386g，出籽率58.6%。

2.2　品质性状

北京市2009—2010年甜糯玉米区试品尝结果表明，京科甜158品尝得分两年均高于对照；经农业部谷物品质监督检验测试中心检验，该品种鲜样籽粒中含水分：77.2%，粗蛋白：3.17%，粗脂肪：1.43%，粗淀粉：3.72%，总糖：10.42%，蔗糖：7.5%，还原糖：2.52%。干基中含粗蛋白：13.89%，粗脂肪：6.29%，粗淀粉：16.31%，总糖：45.69%，蔗糖：32.88%，还原糖：11.07%，赖氨酸：0.50%。因此该品种具有含糖量高，脆嫩香醇，皮薄无渣的特点，尤其是赖氨酸含量高达0.5%，达到国家规定高赖氨酸玉米水平。

2.3 产量表现

2009—2010年，京科甜158参加北京市甜糯玉米区试，两年平均鲜穗产量为11 257.5kg/hm^2，比对照京科甜183增产0.8%；两年平均鲜粒产量为7 399.9 kg/hm^2，比对照京科甜183增产6.4%。2010年生产试验结果平均鲜穗产量为9 346.5kg/hm^2，比对照京科甜183增产1.0%；平均鲜粒产量为6 162kg/hm^2，比对照京科甜183增产6.4%。

3 栽培技术

3.1 隔离种植

甜玉米与其他玉米要有空间或时间隔离，以免接受其他玉米花粉造成花粉直感，影响品质。在空间隔离上，如没有障碍物的平原地区需300m的隔离带；在时间隔离上，需与其他玉米种植时期错开25d以上。如大面积成片种植京科甜158，可适当降低隔离标准。

3.2 择优选地

选择地力中等、平整、有排灌条件的地块，以菜地最好，应适当集中连片，以利于隔离和管理。也可以套种，间种相结合种植京科甜158，提高土地使用率。

3.3 精细整地

首先要把地整平整细。大面积平作可将底肥均匀播撒于地表，然后旋耕翻下与土壤充分混匀。底肥数量：重施底肥，有条件的可以使用缓释肥，每亩施磷二胺25kg；有条件的地方最好施一定数量的有机肥，每亩鸡粪500kg。潮湿易涝地区应起垄种植以利排水。

3.4 播种时期

在北京平原地区3月下旬初至7月上旬均可播种。早春播种（3月下旬）应盖膜栽培。夏播最晚应在7月15日前播完。具体播种日期应根据生产加工安排，大面积种植应错期播种，错期采收鲜穗。

北京地区冬季大棚种植应在9月底前下种，10—12月播种需要提供加温装置的大棚，否则无法正常生长。早春大棚可以安排在1月下旬以后播种，可以在5月上市。个别地区大棚套种草莓，葡萄效果很好，可以实现立体栽培，有效利用空间。

3.5 把好播种关

播前一定要保证有充足适宜的底墒，田间最大持水量在60%~70%。播深4cm左右，每亩用种量1kg左右。

3.6 适时定苗，密度合理

4~5叶期定苗，行距65cm，株距30cm，平均每亩留苗3 000~3 300株。

3.7 田间管理

提高播种质量。有条件可以育苗移栽，保证苗全，苗齐，保证采收期一致，提高商品率；机械播种尽量选取精量播种机提高播种质量。播种后及时封地表，减少田间操作强度。拔节至小喇叭口（7~10叶展）期间适时追肥，每亩施尿素15kg，追肥后及时浇水。小喇叭口至灌浆期是需水关键期绝对不能受旱，遇旱一定要及时浇水。病虫害防治尽量采取生物方法，比如赤眼蜂、白僵菌控制玉米螟等虫害，减少药物残留。

3.8 适时收获

甜玉米适时采收特别重要，一般为吐丝授粉后21~23d为宜。在此特别指出春播糯玉米收获时正处在夏季高温时期，其灌浆速度快，种植户与加工厂家要注意观察适时收获，不然会影响籽粒品质；夏播糯玉米收获时为秋季低温时期，其灌浆速度慢，收获时间可适当拉长。

参考文献（略）

本文原载：农业科技通讯，2015（5）：271-272

果蔬型甜玉米新品种京科甜 179 的选育

史亚兴　卢柏山　徐　丽　赵久然

（北京市农林科学院玉米研究中心，玉米 DNA 指纹及分子
育种北京市重点实验室，北京　100097）

摘　要：京科甜 179 是以超甜玉米自交系 T68 为母本，以 T8867 为父本组配而成的果蔬型甜玉米杂交种，春播生育期 85d 左右，株型平展，植株长势强。鲜果穗筒形，果穗长 18~20cm，直径 4.5~4.9cm，大小均匀，籽粒黄白相间。平均每 667m² 鲜果穗产量 800~1 000kg，鲜籽粒含可溶性糖 23%~35%，还原糖 7%~10%，皮渣率 5.6%，口感甜脆。适宜在北京、河北、山西、内蒙古、辽宁、吉林、黑龙江、新疆作鲜食甜玉米春播种植，适宜北京、天津、河北、山东、河南、江苏淮北、安徽淮北、陕西关中灌区作鲜食甜玉米品种夏播种植。

关键词：甜玉米；京科甜 179；一代杂种；果蔬型

A New Fruit-vegetable Sweet Corn Cultivar— "Jingketian No. 179"

Shi Yaxing　Lu Baishan　Xu Li　Zhao Jiuran

(*Maize Research Center*, *Beijing Academy of Agriculture and Forestry Sciences*, *Beijing Key Laboratory of Maize DNA Fingerprinting and Molecular Breeding*, *Beijing* 100097, *China*)

Abstract: "Jingketian No. 179" is a new fruit-vegetable type sweet corn variety developed by crossing 'T68' as female parent and "T8867" as male parent. Its growth period is 85~88 days, plant shape is flat, and plant growth vigor is strong. It has cylindric ears with equal size, which is 18-20cm in length and 4.5~4.8cm in diameter, and yellow-white kernels on the same ear. Its yield is 800~1 000kg · ($667m^{2})^{-1}$, soluble sugar, reducing sugar, and testa dregs rate in fresh kernels is 23%~35%, 7%~10%, and 5.6%, respectively, which forms sweet and succulent taste. It is suitable for cultivation in Beijing, Hebei, Shanxi, Neimengu, Liaoning, Jilin, Heilongjiang, Xinjiang Province in spring, and suitable for cultivation in summer in Beijing Tianjn, Hebei, Shandong, Henan, north Jiangsu, north Anhui province, and Shani Guanzhongguan District.

作者简介：史亚兴，男，副研究员，专业方向：鲜食玉米遗传育种，电话：010-51503400，E-mail：syx209@163.com

通讯作者：卢柏山，男，硕士，研究员，专业方向：鲜食玉米遗传育种
　　　　　E-mail：maizelu@126.com

Key words: Sweet corn; "Jingketian 179"; F_1 hybrid; Fruit-vegetable type

甜玉米是在乳熟期食用鲜果穗的一类特用玉米，其籽粒营养丰富，口感甜、脆、嫩、香，可作为水果和蔬菜食用，是一种兼具休闲型与保健型的现代食品，近年来各地陆续选育出一些新品种（韩云等，2012；王颢等，2015）。根据甜玉米生产及市场需求，北京市农林科学院创新甜玉米育种杂优模式，以高品质为重点筛选指标，以T68为母本、T8867为父本杂交选育出果蔬型超甜玉米品种京科甜179，具有品质优、产量高、商品率高、适应性广等突出特点。目前该品种已在我国华北、黄淮海地区如北京、天津、吉林、河北、山西、山东等地推广种植，累计种植面积20 000hm²（30万亩）。

1 选育过程

母本T68以引自国外早熟超甜玉米杂交种为选系材料，利用二环系方法经8代自交并严格选择选育而成。植株生长势强，叶片浅绿色，叶中脉白色。株高120cm，穗位高50cm，雌雄花期协调，雄花小穗着生密度中等，花粉量较大，果穗结实性较好。果穗圆筒形，穗长10cm，穗粗4cm，籽粒黄色，生育期中等，配合力强，抗大斑病、小斑病、青枯病等病害。父本T8867以国外中熟超甜玉米杂交种为选系材料，经连续8代自交并严格选择选育出的自交系。植株长势旺，叶色深绿色，叶中脉白色。株高140cm，穗位高60cm，雄穗发达，花粉量大，雌雄花期协调，散粉持续时间7d左右，适合做父本，果穗结实性好。果穗圆筒形，穗长9cm，穗粗4.1cm，籽粒白色，生育期中等，配合力强，对大斑病、矮花叶病、黑粉病有很好的抗性。

2009年在海南三亚试验基地配置杂交组合，2010年于北京进行杂交种鉴定，组合T68×T8867表现出极强的配合力，果穗产量高、商品性好、品质优良，口感甜脆清香，适合作为水果和蔬菜食用；果穗黄白色相间，抗丝黑穗病、小斑病、黑粉病等病害。2011—2012年进行品种比较试验，2012—2014年分别年参加北京市、国家东华北区、国家黄淮海区区域试验，2013—2014年参加北京市、国家东华北区、国家黄淮海区生产试验，在产量、品质、抗性等方面均表现突出，于2014年通过北京市农作物品种审定委员会审定（审定编号：京审玉2014007），2015年通过国家农作物品种审定委员会审定（审定编号：国审玉2015040），定名为京科甜179。

2 选育结果

2.1 丰产性

2.1.1 品系比较试验

2011—2012年在北京小汤山国家精准农业试验站进行品种比较试验。采用随机区组排列，3次重复，小区面积为15m²，5行区，行距0.6m，行长5m，其他管理同北京市甜糯玉米区试管理办法。两年试验结果表明（表1），京科甜179平均每666.7m²鲜穗产量854.7kg，比对照京科甜183高7.9%；鲜籽粒产量566.1kg，比对照高12.5%。京科甜179生育期与对照相当，播种至鲜果穗采收约86d。

表1 京科甜179品种比较试验结果

年份	品种名称	生育期(d)	鲜穗产量[kg·(667m²)⁻¹]	比对照±%	鲜籽粒产量[kg·(667m²)⁻¹]	比对照(±%)
2011	京科甜179	87	823.0	5.5	541.0	12.0
	京科甜183（CK）	84	780.3	—	482.9	—
2012	京科甜179	85	886.4	10.4	591.2	13.0
	京科甜183（CK）	85	802.6	—	523.3	—
平均	京科甜179	86	854.7	8.0	566.1	12.5
	京科甜183（CK）	85	791.5	—	503.1	—

2.1.2 区域试验

2013—2014年同时参加东华北和黄淮海两大区区域试验，试验无重复排列，小区面积24m²，6行区，密度为3 500株·(667m²)⁻¹。为防止花粉直感造成相互影响，每个品种在2边行套袋自交或互交20穗，套袋隔离直至采收，按国标甜玉米（NY/T 523—2002）进行品尝鉴定和评价。授粉后20~23d为最佳采收期，收中间4行计产。试验结果表明，京科甜179在东华北区10个试验点中，两年平均每667m²鲜穗产量933.24kg，与对照中农大甜413产量相当；在黄淮海区13个试验点中，两年平均每667m²鲜穗产量786.7kg，较对照增产4.5%。其中在河北、内蒙古自治区、新疆石河子等地表现突出，两年平均产量均在1 000kg·(667m²)⁻¹以上。京科甜179是甜玉米组参试品种中唯一进入第二年区试和生产试验的品种。

2.1.3 生产试验

2014年参加国家北方区（东华北和黄淮海）生产试验。试验采用间比法排列，不设重复，种植行数为10~15行，面积≥100m²，全区收获计产。品种密度为3 500株·(667m²)⁻¹。试验结果表明，京科甜179在东华北区平均每667m²鲜穗产量889.9kg，比对照中农大甜413增产1.6%；在黄淮海区平均每667m²鲜穗产量820.9kg，比对照增产5.5%。其中在河北省万全县试验点，京科甜179综合表现突出，鲜穗产量达1 280.2kg·(667m²)⁻¹，较对照增产11.4%，鲜果穗大小均匀，籽粒排列整齐，商品性好（表2）。

表2 京科甜179生产试验产量结果

试验点（东华北区）	鲜穗产量[kg·(667m²)⁻¹]		比对照（±%）	试验点（黄淮海区）	鲜穗产量[kg·(667m²)⁻¹]		比对照（±%）
	京科甜179	中农大甜413（CK）			京科甜179	中农大甜413（CK）	
北京中农	653.0	880.3	-25.8	安徽界首	741.7	643.4	15.3
河北承德	911.0	1 030.7	-11.6	石家庄	882.4	827.9	6.6
内蒙赤峰	930.5	779.7	19.3	江苏盐城	790.3	770.5	2.6
黑龙江肇东	645.4	584.4	10.4	山东登海	837.6	763.4	9.7

(续表)

试验点 (东华北区)	鲜穗产量 [kg·(667m²)⁻¹]		比对照 (±%)	试验点 (黄淮海区)	鲜穗产量 [kg·(667m²)⁻¹]		比对照 (±%)
	京科甜179	中农大甜413（CK）			京科甜179	中农大甜413（CK）	
河北万全	1 280.2	1 149.1	11.4	陕西杨凌	1 096.7	884.5	23.9
山西屯留	918.4	868.7	5.7	天津武清	576.6	770.7	-25.1
平均	889.8	882.2	1.6	平均	820.9	776.7	5.5

2.2 品质

经吉林农业大学农学院品质鉴定分析，京科甜179鲜果穗中含可溶性糖类35%，还原糖11.14%，均高于对照中农大甜413，果穗皮渣率为3.97%，与对照相当。经中国农科院作物科学研究所、河北省承德市农业科学研究所、吉林农业大学农学院按照国标甜玉米（NY/T 523-2002）品尝鉴定，京科甜179品尝品质优于对照（表3），总评分高于中农大甜413。京科甜179颜色黄白相间，大小均匀，适于甜玉米加工厂进行速冻加工。

表3 京科甜179品质性状鉴定结果

品种	还原糖含量(%)	可溶性总糖含量(%)	皮渣率(%)	品尝品质							总评分
				感官	气味	色泽	风味	糯性	柔嫩性	皮薄厚	
京科甜179	11.1	35.0	3.9	25.3	6.1	6.0	7.9	16.1	8.2	16.5	86.1
中农大甜413	8.9	32.8	3.8	25.0	6.0	6.0	8.0	16.0	8.0	16.0	85.0

注：还原糖、可溶性总糖、皮渣率均为鲜基含量；品尝以85分为对照，分值越高口感越好

2.3 抗病性

根据东华北区抗病虫接种鉴定单位吉林省农业科学院和黄淮海区接种鉴定单位河北省农林科学院植物保护研究所2013年、2014年田间鉴定的调查结果，京科甜179对弯孢菌叶斑病、大小斑病、丝黑穗病、瘤黑粉病和茎腐病均表现出较强抗性（表4），达到国家玉米品种审定中的抗性标准。

表4 京科甜179田间抗病性调查结果

年份	品种	大斑病（级）	小斑病（级）	灰斑病（级）	弯孢菌叶斑病（级）	丝黑穗病（%）	瘤黑粉病（%）	茎腐病（%）	玉米螟（%）
2013	京科甜179	1.6	0.5	0.7	0.7	1.0	0.5	0	0.4
	中农大甜413（CK）	1.3	0.5	0.7	0.7	0.8	0.4	0	0.1
2014	京科甜179	2.5	0.8	—	0.7	0.6	0.9	0	2.8
	中农大甜413（CK）	1.4	0.8	—	0.7	0.1	0.2	0	5.2

3 品种特征特性

京科甜 179 在东华北春玉米区出苗至鲜穗采摘 82d。株高 224cm，穗位高 82.6cm，花丝绿色，果穗筒形，穗长 19.9cm，穗粗 4.9cm，穗行数 14~16 行，穗轴白色，籽粒黄白色，鲜籽粒百粒重 38.0g。品尝鉴定 86.6 分，皮渣率 4.5%，还原糖含量 9.9%，水溶性糖含量 33.6%。在黄淮海夏玉米区出苗至鲜穗采摘 72d。株高 207.8cm，穗位高 66.9cm，穗长 18.7cm，穗粗 4.8cm，鲜籽粒百粒重 39.2g。品尝鉴定 86.8 分，皮渣率 11.2%，还原糖含量 7.8%，水溶性糖含量 23.5%。田间抗丝黑穗病。在我国北方区平均每 667m² 产量 900kg 左右。适宜北京、河北、山西、内蒙古、辽宁、吉林、黑龙江、新疆作鲜食甜玉米春播种植；适宜北京、天津、河北、山东、河南、江苏淮北、安徽淮北、陕西关中灌区作鲜食甜玉米品种夏播种植。

4 栽培技术要点

北方平原地区 3 月下旬初至 7 月上旬均可播种。早春播种（3 月下旬）应盖膜栽培，中等肥力以上地块隔离种植。夏播最晚在 7 月中旬之前。播深约 4cm，每 667m² 用种量 1kg 左右，种植密度 3 500 株·$(667m^2)^{-1}$。拔节至小喇叭口（7~10 展叶）期适时追肥施尿素 15kg·$(667m^2)^{-1}$，追肥后及时浇水。春播采收期一般为吐丝授粉后 21~23d。夏播甜玉米收获时为秋季低温时期，其灌浆速度慢，采收期一般在授粉后 22~25d。

本文原载：中国蔬菜，2016，1（12）：64-67

甜糯玉米新品种"京科糯928"

卢柏山* 史亚兴 赵久然 徐 丽 樊艳丽 席胜利

(北京市农林科学院玉米研究中心，北京 100097)

摘 要："京科糯928"是利用半双隐杂交法选育出的新一代优质高产甜糯玉米新品种。同一果穗上包含甜、糯两种籽粒，甜糯籽粒比为1∶3，香味浓郁，甜糯相宜，柔软度好；鲜籽粒含叶酸 $1.72\mu g \cdot g^{-1}$，鲜果穗产量约 $13.7t \cdot hm^{-2}$，综合抗性好，适宜在中国北方及重庆等地栽培种植。

关键词：甜糯玉米；优质；品种

A New Excellent Sweet-waxy Corn Cultivar "Jingkenuo 928"

Lu Baishan* Shi Yaxing Zhao Jiuran Xu Li Fan Yanli Xi Shengli

(*Maize Research Center*, *Beijing Academy of Agriculture and Forestry Sciences*, *Beijing* 100097, *China*)

Abstract:"Jingkenuo 928" is a new sweet-waxy corn with good quality and high yield, which could be used as fruit and vegetable, developed by the method of crossing between single recessive and double recessive lines. Sweet and waxy kernels which is 1∶3 quantitatively are contained on the same ear, which tastes sweet-scented and soft; The content of folic acid in fresh kernels is $1.72\mu g \cdot g^{-1}$; The yield of fresh ears is $13.7t \cdot hm^{-2}$; The new cultivar is with good combined resistance, extensive adaptability, and is suitable for cultivation in Northern China and Chongqing.

Key words: Sweet-waxy corn; Excellent quality; Cultivar

培育甜中带糯、糯中有甜的优质新型甜糯玉米是中国鲜食玉米市场发展的需要，也是育种者的育种目标（史亚兴，等，2015；刘亚利，等，2011；张静、彭海，2010）。

甜糯玉米"京科糯928"（图1）是利用半双隐杂交法（郝小琴，等，2011）育成的新型甜加糯玉米品种。母本"京糯6"是以杂交种"中糯1号"为选系材料，于1996—1999年经过连续6代分离筛选、自交定向选育而成，为单隐性纯糯（$wxwx$）自交系。父本"甜糯6"为双隐性（sh_2sh_2wxwx）自交系，2005年由纯糯自交系"白糯6"（SH_2SH_2wxwx）与超甜自交系"SH-251"（sh_2sh_2WXWX）组配二环系，获得 S_0 代（SH_2sh_2WXwx），自交后获得 S_1 代种子，共有16种基因型，3种表现型；从3种表现型中淘汰普通粒（$SH_2\_WX\_$）和糯粒（SH_2_wxwx），选择甜粒（$sh_2sh_2WX\_$、sh_2sh_2wxwx），对 $S_1 \sim S_6$

---

* E-mail: maizelu@126.com

图 1　甜糯玉米新品种"京科糯 928"
Figure 1　A new sweet-waxy corn cultivar "Jingkenuo 928"

代进行优株（系）的单株测配同时进行自交；根据测配组合的表型，鉴定筛选出配合力高、优质高产的纯合双隐株系（sh_2sh_2wxwx）；根据该对应纯合双隐株系自交后代的表现，从中筛选优良植株，最终从 6 个世代约 170 000 株材料中选出 1 个最优株系，命名"甜糯 6"。2008 年于海南省进行组配并鉴定，同一果穗甜糯籽粒比为 1∶3，既有糯香，又不失甜味，口感甜糯相宜，鲜香脆嫩。2011—2013 年参加北京、重庆的甜、糯玉米区域试验，平均比对照增产 7.1%，表现抗大斑病、小斑病、矮花叶病等。2013 年 5 月和 2014 年 5 月分别通过北京市、重庆市农作物品种审定委员会审定。

1　品种特征特性

株型半紧凑，株高 251.7cm，穗位高 106.7cm。叶色绿色，花药黄色，花丝粉红色。果穗锥形，穗长 21.7cm，穗粗 4.9 cm，穗行数 12~14 行，行粒数 41 粒，单鲜果穗质量 450~500g，鲜果穗产量 13.7t·hm^{-2}。籽粒白色，鲜籽百粒质量 37.8g；含粗淀粉 15.5%；支链淀粉/粗淀粉为 100%；赖氨酸 0.1%；粗蛋白 3.5%；粗脂肪 1.7%；叶酸 1.72μg·g^{-1}，最高可达 2.95μg·g^{-1}，高于普通玉米数倍。早熟，北京地区播种至鲜穗采收期平均 89d。经两年人工接种鉴定，抗大斑病、小斑病及矮花叶病等主要生产病害。采收期长，蒸煮后冷食不回生。

2　栽培技术要点

适宜在北京及类似生态区，以及重庆地区海拔 800m 以下作为鲜食或菜用玉米种植，同时也适用类似气候地区设施栽培。露地栽培：北京地区可在 4 月初至 7 月中旬播种；重庆地区以 3 月上中旬播种育苗为宜。栽植密度控制在 48 000~52 500株·hm^{-2}。前期底肥以有机肥为主，15 000~22 500kg·hm^{-2}。大喇叭口至灌浆初期适时追肥，施尿素 225kg·hm^{-2}，并及时浇水，保证土壤墒情，以促进孕穗灌浆。春播收获时正处在夏季高温期，灌浆速度快，注意观察，适时采收；夏播收获时为秋季低温时期，灌浆速度慢，采收时间可适当延长，以提高籽粒品质、口感及加工产品质量。

参考文献（略）

本文原载：园艺学报，2015，42（S2）：2 933-2 934

优质甜糯玉米新品种京科糯2010的选育及栽培技术要点

史亚兴　卢柏山　赵久然　徐　丽　席胜利　樊艳丽

(北京市农林科学院玉米研究中心，北京　100097)

摘　要：京科糯2010是利用半双隐杂交法选育出的新一代甜加糯类型杂交种。介绍了京科糯2010的选育过程、特征特性、配套栽培技术等。该品种表现早熟，品质优良，同一果穗上甜糯籽粒比例为1∶3，口感绵软，甜糯相宜。鲜果穗产量平均12.3t/hm²，抗倒性强。适宜在我国北方进行露地种植，也可进行设施温室栽培。

关键词：优质；甜糯；玉米；选育；栽培

近年来，鲜食玉米在我国的种植面积逐年增加，已达93.3万hm²左右。目前生产上大面积种植的鲜食玉米品种主要是超甜和纯糯两类，而随着鲜食玉米市场的发展和消费者需求的不断变化，对鲜食玉米育种工作提出了更高的要求。

京科糯2010是利用半双隐杂交法育成的甜加糯类型玉米品种，属于一种我国自主创新的鲜食玉米类型。半双隐杂交法是指利用单隐性自交系与双隐性自交系进行杂交组配，使双隐性基因在同一果穗上得以同时表达的方法。京科糯2010果穗上同时包含甜、糯两种类型籽粒，甜糯籽粒比例为1∶3，口感绵软，甜糯相宜。该品种于2014年通过北京市农作物品种审定委员会审定（京审玉2014010）。

1　选育过程

京科糯2010是以优良单隐性纯糯自交系"N39"（wxwx）为母本，以双隐性纯糯自交系"CB1"（sh_2sh_2wxwx）为父本组配而成的新一代甜糯玉米杂交种。

母本"N39"是2004年从自有糯玉米群体中选育的优良单株，经过6代连续自交和选择于2006年育成，为纯糯（wxwx）自交系。该自交系植株紧凑，抗倒抗病性强。父本"CB1"是从美国引进的白色双隐杂交种中经连续多代自交、筛选、测配定向选育获得的，为双隐性甜糯（sh_2sh_2wxwx）自交系。

2010年配制杂交组合N39×CB1，表现早熟、籽粒口感好、品质优良、产量潜力高、抗倒性强。2011年在北京多个试验点进行品种比较试验，2012—2013年参加北京市甜糯玉米区域试验，产量与对照相当，均表现出较高的增产潜力。经品尝试验，品质和口感均高于对照。

2　品种特征特性

2.1　植株性状

京科糯2010属新一代甜加糯类型鲜食玉米品种。熟期表现为早熟，在北京地区春播播种至鲜穗采收平均86d，比对照京科糯2000平均早7d。植株株型半紧凑，株高246.5cm，穗位高91.4cm。穗型筒形，穗长19.5cm，穗粗4.7cm，穗行数16~18行，行粒数36.6粒，秃尖长度1.8cm，粒行整齐。粒色白色，鲜籽粒千粒重370.0g，出籽率61.7%。籽粒（干基）含粗蛋白11.1%，粗脂肪4.8%，粗淀粉47.6%，支链淀粉/粗淀粉为98.9%，赖氨酸0.33%。

2.2 品质、产量及抗性表现

京科糯2010品质优良，口感柔嫩，可以同时品尝甜、糯两种类型籽粒，蒸煮后冷食不回生。区域试验中经双盲法品尝试验，该品种果穗性状和品质性状两年平均得分为84分，均要高于对照（80分）。2011年在北京怀柔、顺义、昌平、海淀、海南等5个点次进行品种比较试验，按点次计平均鲜果穗产量为12.3t/hm²；2012—2013年北京市甜糯玉米区域试验，按年份计平均鲜果穗产量为12.4t/hm²；两年抗倒性均为0级，抗倒性极强。

3 栽培技术要点

3.1 隔离种植

京科糯2010种植时需与其他类型玉米有空间或时间隔离。平原地区种植需300m以上隔离带；或播种时间与其他类型玉米错开25d以上，以保证籽粒品质和口感不受影响。

3.2 精细播种

京科糯2010适宜在我国北方地区种植。播前精细整地，并根据气候及土壤条件、市场和加工需要确定播期。京科糯2010露地栽培种植可在4月初至7月中旬，采用地膜覆盖时可适当早播；有加温设施的温室大棚可在每年9月至第二年3月分期播种，以实现鲜果穗周年不间断供应。足墒精量下种，适宜播深2~3cm。密度以每666.67m²种植3 500株为宜。

3.3 田间管理

播后苗前，及时进行土壤封闭除草。前期施足底肥，结合播前整地，每666.67m²施有机肥1 000~1 500kg。大喇叭口至灌浆初期，适时追肥，施尿素225kg·hm^{-2}，并及时浇水，保证土壤墒情，以促进孕穗灌浆。

3.4 适时采收

京科糯2010在授粉后25d左右即进入采收期。因此在开花授粉后，应注意观察籽粒灌浆进度，做到适时采收。春播收获时正处在夏季高温期，果穗灌浆速度快，应适时及早采收；夏播收获时为秋季低温时期，灌浆速度慢，采收时间可适当延长，以提高籽粒品质、口感及加工产品质量。

参考文献（略）

文本原载：蔬菜，2015（11）：71-72

新型甜加糯鲜食玉米品种农科玉 368 的选育

卢柏山　史亚兴　徐　丽　赵久然

(北京市农林科学院玉米研究中心/玉米 DNA 指纹及分子育种北京市重点实验室，
北京　100097)

摘　要：农科玉 368 是以糯质（wxwx）自交系京糯 6 为母本，以纯合双隐性（同时含有 shrunken-2 和 waxy 隐性纯合基因）自交系 D6644 为父本组配而成的新型甜加糯鲜食玉米杂交种，果穗上同时包含甜、糯两种籽粒类型，且甜粒：糯粒≈1：3，口感甜糯相宜、绵软适口。鲜果穗平均每 667m² 产量 800~1 100kg，鲜籽粒含支链淀粉 97.1%，皮渣率 4.5%。综合抗性好，适宜在我国北方及西南地区种植。

关键词：优质；甜加糯；农科玉 368

鲜食玉米主要包括甜玉米和糯玉米两种类型。甜玉米口感清甜爽脆、嫩滑无渣，但缺少糯性；糯玉米中淀粉几乎全为支链淀粉，黏软清香，但甜味不足。随着经济发展，市场需求不断变化，糯中有甜、甜中带糯的新型鲜食玉米成为目前我国鲜食玉米市场发展的新方向。

北京市农林科学院玉米研究中心根据市场需求，以优良糯玉米骨干自交系京糯 6 为母本，以甜糯双隐性自交系 D6644 为父本，组配出果蔬型优质甜加糯鲜食玉米新品种农科玉 368，通过全国范围内试验和示范，表现出适应区域广，品质优，产量高，抗性强等优点，同一果穗上包含甜和糯两种籽粒。目前该品种已通过国家黄淮海区审定（国审玉 2015034），通过江苏（苏审玉 201504）、福建（闽审玉 2015003）、北京（京审玉 2015011）、宁夏（宁审玉 2015029）4 个省市的审定。

1　选育经过

1.1　亲本来源

母本京糯 6 是以糯玉米杂交种中糯 1 号为选系基础材料，经连续 6 代自交和严格选择选育而成的糯玉米自交系。株型半紧凑，株高 130cm，穗位高 50cm，籽粒白色，抗性强，经测试，表现出极强的一般配合力。

父本 D6644 为同时含有纯合隐性甜质基因（shrunken-2，sh）和纯合隐性糯质基因（waxy，wx）的双隐性自交系。它是以纯糯自交系白 6（wxwx）和超甜自交系 SH251（shsh）为基础材料，利用二环系方法，并配合半姊妹单株测交技术选育而成的。植株株型平展，株高 135cm，穗位高 55cm，籽粒白色，有较强的一般配合力和特殊配合力。

基金项目：半双隐型甜加糯玉米种质创新及新品种示范（编号：QNJJ2016）

作者简介：卢柏山（1978—　），男，研究员，主要从事鲜食玉米育种工作；E-mail：maizelu@126.com

1.2 选育经过

2009年在海南组配杂交组合京糯6×D6644，并于2010—2011年在北京进行鉴定和品种比较试验，均表现出极高的配合力与抗性。同一果穗上甜粒与糯粒约为1∶3。

2013年开始，参加国家黄淮海区甜糯玉米区试，以及北京、宁夏回族自治区、福建、江苏等省市的甜糯玉米区试。2014年进行续试，同时进行生产试验，并推荐审定。

2015年，通过国家黄淮海区审定，通过江苏、福建、北京、宁夏4个省市的审定。

2 特征特性

2.1 产量表现

2013—2014年参加国家黄淮海区鲜食糯玉米区试，平均鲜穗产量848.7kg·$(667m^2)^{-1}$，比对照苏玉糯2号平均增产9.26%。13个试点中，平均11试点表现增产，增产率84.6%。鲜籽粒平均产量为591.3kg·$(667m^2)^{-1}$。2014年生产试验中，平均鲜穗产量927.2kg·$(667m^2)^{-1}$，比对照苏玉糯2号增产8.0%，6个试验点中5增1减，增产率83.3%（表1）。

表1 农科玉368参加国家黄淮海区试产量表

年份	品种	鲜穗产量 [kg·$(667m^2)^{-1}$]	比对照±%	增产点数	减产点数	鲜籽粒产量 [kg·$(667m^2)^{-1}$]
2013	农科玉301	811.8	9.1	11	2	539.9
	苏玉糯2号	747.6	0	—	—	496.4
2014	农科玉301	885.6	9.4	10	3	642.6
	苏玉糯2号	809.3	0	—	—	543.0

2.2 品质性状

农科玉368为甜加糯鲜食玉米，果穗上同时含有甜粒和糯粒，品质优良，口感绵软适口，甜糯相宜。经中国农业科学院、河北省种子站、河南农业大学的检测，农科玉368平均支链淀粉含量为97.1%，总淀粉含量为62.5%，皮渣率为4.5%。品尝结果表明，农科玉368在口感、风味上均优于对照，总评分达87.5分（表2）。

表2 农科玉368品质性状

品种	粗淀粉（%）	支链淀粉（%）	皮渣率（%）	品尝品质							总评分
				感官品质	气味	色泽	风味	糯性	柔嫩性	皮薄厚	
农科玉368	62.5	97.1	4.5	25.3	6.4	6.1	11.1	13.9	8.3	16.4	87.5
苏玉糯2号	63.0	98.3	4.6	25.0	6.0	6.0	10.7	13.3	8.0	16.0	85.0

注：粗淀粉、支链淀粉为干基含量；皮渣率为鲜基含量；品尝以85分为对照，分值越高口感越好

2.3 生物学特性

农科玉368属中早熟甜加糯鲜食玉米品种，黄淮海夏玉米区出苗至鲜穗采收期76d。株型半紧凑，株高233.2cm，穗位高97.5cm，成株叶片数19片。花丝淡紫色，果穗锥形，穗长18.6cm，穗行数12~14行，穗轴白色，籽粒白色，硬粒质型，鲜籽粒百粒重38.7g。品尝鉴定87.5分；粗淀粉含量62.5%，直链淀粉占粗淀粉的4.6%，皮渣率

4.5%。鲜果穗采收后，秸秆持绿性好，可作为青贮饲料，实现鲜食玉米综合利用。

2.4 抗性表现

根据国家黄淮海糯玉米区域试验2013年、2014年调查结果，农科玉368综合抗性表现与对照苏玉糯2号相当，高抗茎腐病，瘤黑粉病，抗矮花叶病和粗缩病（表3）。

表3 农科玉368抗病性鉴定结果

年份	大斑（级）	小斑（级）	黑粉（%）	茎腐（%）	矮花（级）	粗缩（%）	螟虫（%）
2013	1.4	1.9	0.2	0.1	0.5	1	2.3
2014	1.4	2	0.8	0.3	0.6	0.4	2.3

3 栽培技术要点

3.1 隔离种植，精细播种

农科玉368适宜北京、天津、河北、山东、河南、江苏淮北、安徽淮北、陕西关中灌区作鲜食糯玉米夏播种植；适宜江苏各地春播种植；适宜福建省春播种植。春播时间一般在4月中旬至5月底，采用地膜覆盖可适当早播；夏播时间在6—7月中旬；温室大棚设施栽培可在每年8月至第二年3月分期播种育苗，可实现鲜果穗周年不间断供应。足墒精量下种，每667m^2用种量1kg左右。为防止串粉，应设置500m以上的空间隔离，或分期播种。

3.2 合理密植

为保证果穗商品性，应适当稀植，密度以4.5万~5.25万株/hm^2为宜。

3.3 适时采收

甜加糯玉米果穗上同时含有甜粒和糯粒，因甜粒灌浆持续期短，后期易出现籽粒回缩，因此甜加糯玉米应适当早收，授粉后21~23d为宜，以保证果穗品质和商品性。夏播种植时，灌浆期温度变低，应注意观察籽粒灌浆进度，避免过早或过晚收获。

3.4 秸秆青贮

甜加糯玉米果穗采收后，其秸秆仍具有良好的持绿性，且较普通玉米的秸秆营养丰富、适口性好，可作为优质的青贮饲料用于畜牧业，进一步提高全株利用率和生产附加值。

参考文献（略）

本文原载：种子，2016，35（12）：106-107

中国种子企业研发模式与发展策略研究

杨海涛[1,2]　赵久然[2]　吕波[1]　邹奎[1]　杨洋[1]　陈红[1]　赵军[1]

(1. 农业部种子管理局，北京 100125；2. 北京市农林科学院玉米研究中心，北京 100097)

摘　要：中国种子企业以中小微类型居多，这些企业受资金、研发设备、人才等因素制约，技术创新能力不强，研发模式的选择与策略的制定显得尤其重要。为此，以中国种子企业根据战略发展需要确定研发模式和制定策略为研究对象，分析了中国种子企业主要研发模式与问题。研究认为，中国种子企业主要有自主研发、合作研发和技术购买3种研发模式，主要问题是缺乏适合企业的研发策略，缺少保证研发的机制与制度，研发人才相对匮乏，信息化管理水平低等问题。提出了中国种子企业提高研发能力的策略性建议。

关键词：种业；种子企业；研发模式；发展；策略

Researches on R & D Mode and Development Tactics of Seed Enterprises in China

Yang Haitao[1,2]　Zhao Jiuran[2]　Lv Bo[1]　Zou Kui[1]
Yang Yang[1]　Chen Hong[1]　Zhao Jun[1]

(1. Seed Management Bureau of Agricultural Ministry, Beijing 100125；
2. Maize Research Center of Beijing Agriculture and Forestry Sciences, Beijing 100097)

Abstract: The purpose of this paper was to select R&D mode and institute tactics based on the need of development of strategy, and analyzed the R&D problems of seed enterprises. It considered that independent R&D, cooperative R&D, and purchased R&D were the three main modes of enterprises of China, for enterprises, the major problems were lack of proper R&D tactics, mechanism and institutions to ensure R&D, few talents for R&D, lowing level of information management. Some tactic suggestions were proposed to improve R&D abilities of enterprises of China.

Key words: Seed industry; Seed enterprise; Cooperative R & D; Mode; Strategy

农作物种业作为国家战略性和基础性的核心产业，处于农业产业链的前端，对农业的稳定与发展具有至关重要的作用，而随着我国种子产业的初步形成与发展，种子市场价值的逐年增加，市场竞争也日趋激烈，促使种子企业不断加大研发投入，以实现企业的可持续发展。目前有关种子企业根据战略发展要求确定企业研发模式和制定相应策略等方面的研究较少，而我国种子企业又以中小微类型居多，这些企业存在受资金、研发设备、人才等因素制约，技术创新能力不强，研发模式的选择与策略的制定显得尤其重要，笔者开展了此方面的研究，以期得到一些有助于种子企业有效开展研发的策略性建议。

1 种子企业研发模式概述

Normann R. Ramirezr 最早将企业的研发活动归类为自主研发、研发外包和合作研发三种模式。于成永从企业边界角度将研发模式分为内部研发、企业并购、技术购买、合作研发、研发联盟等形式,以技术获取方式将研发模式分作内部研发、合作研发、委托研发、技术购买。刘延秋认为企业研发模式有自主研发、委托研发和合作研发3种典型研发模式。按照组织边界,从产品或技术方式的获得进行研发模式划分,企业研发模式主要包括自主研发、合作研发、研发外包和技术购买这4种研发模式。研发外包存在研发结果的不确定性与外包费用问题,种子企业很少采用研发外包模式。我国种子企业主要采用自主研发模式、合作研发模式、技术购买模式3种类型。

1.1 自主研发模式

自主研发又称为内部研发,指企业通过自身努力,在完全依靠自身力量的条件下取得技术进展或突破,并成功实现商业化的研发模式。自主研发可以使研发成果快速、顺利实现转化、建立核心竞争优势、提高企业人员研发、管理与学习能力、获得超额垄断利润等好处,但是自主研发投资成本大,面临的不确定性大,投资回收期长,沉淀成本大,失败的风险非常大,研究成果具有外部性,对企业人员的素质、管理水平等要求高。跨国种子企业采用自主研发模式,或将研发的核心活动自主完成,这主要是因为跨国种子企业拥有较为完备的开展研发的各项资源,如资金、技术、人才、设备等,因此能够独立进行研发,一般集新品种研发、生产、加工、销售、技术服务于一体。我国仅有少数在资金、人才、技术、服务等方面具备一定优势的大型种子企业采用自主研发模式,如上市公司登海种业股份有限公司前身-登海种子公司发展初期是比较典型的自主研发种子企业,企业主要依靠自身资源开展研发,选育出一系列紧凑型杂交玉米,实现了育繁推一体化,企业得到稳步发展。

1.2 合作研发模式

合作研发是两个以上的企业或组织之间通过投入资金、技术、人员、研发设备等资源,建立项目研发团队、研发机构等组织形式,实行共同面对研发风险与实行利益共享的合作原则,合作开展研发活动的一种研发方式。技术创新的难度日益增加,企业完全依靠自身的资源实现研发目标变得日益困难,自主研发企业在研发过程中往往要投入数量巨大的资金、人员、仪器设备等资源,面临研发成本与失败风险,而合作研发可以有效降低投资风险,充分发挥各合作方的比较优势。我国种子产业发展相对较晚,种子企业数量众多,大部分企业规模小,受资金、种质资源、研发人才等因素的制约,这些企业往往采取合作的方式进行研发,一般与科研院所或其他企业开展研发合作。合作研发做得比较好的企业是北京金色农华种业公司,该公司成立以来,与科研院所开展深入合作,引进中国农业大学、北京市农林科院、浚县农科所等多家单位的育种材料与重点组合,开展多点试验鉴定,向品种所有权单位提供科研经费、品种权提成等方式,吸引合作者进行合作,缩短了研发周期,实现了企业跨越式发展。

1.3 技术购买模式

技术购买一般包括专利技术、成套设备和关键设备等,技术购买具有简单、一次性交易的特点,但是购买的技术多是成熟的技术。技术购买作为一种研发模式,对于种子企业，不只是简单的技术购买行为，也包含大量的研究与分析工作，这些研究与分析工作一般要在企业内部完成，由研发人员完成。采用技术购买的种子企业没有完整的研发体系及必要的研发条件，但拥有相当数量的资金，组织生产和市场营销能力较强，通过外购获得品种所有权或经营权。奥瑞金种业公司、德农种业公司在发展初期，主要采用技术购买模式，尽管企业为获得植物新品种所有权或品种经营权支出了大笔费用，但也减少了自主研发所要求的专业技术人员、育种材料、科研设备等各项投入，研发周期大为缩短，也降低了研发失败的风险。

2 种子企业研发主要问题

2.1 缺乏适合企业的研发策略

研发模式的选择要与企业发展战略相适应。我国种子企业数量众多，规模相对较小，相当数量的种子企业重视短期目标的实现和经济效益的提高，但对于研发模式的选择与发展策略的制定重视不足，主要表现为：一是企业内部缺少研发部门或研发力量薄弱，研发人员在企业中的地位不高，待遇一般，此类现象在科研院所办种子企业较为普遍，主要原因是企业在研发方面过于依赖科研单位，自身研发力量或能力不足。二是研发模式与企业发展战略不适应，一些种子企业尽管经营状况良好，管理规范，但企业研发模式的缺失或不适合企业的发展战略，如一些种子企业营运资金管理水平不高，大量资金处于闲置状态，使企业失去了实现跨越式发展的机遇。三是研发模式受企业决策水平的制约，企业缺少必要的决策支持系统，使企业作出各项决策时，还主要依赖于经验进行判断，如对于采用技术购买的种子企业，一些种子企业仅以 1~2 个指标就决定购买某项技术或者品种，例如只关注产量，而忽略了其他一些重要的指标，如用户评价、审定区域、抗病性、商品性等，因而在经营过程中，容易出现产品滞销，使企业面临巨大的经营风险。

2.2 缺少保证研发的机制与制度

对于以合作研发为主要模式的种子企业，有效保证合作研发的机制与制度的建立非常关键，因为不论是企业间的合作研发，还是企业与科研院所开展的合作，如果缺少相应的机制与制度的保证，合作很难有效深入开展和长期保持，实际上，多数企业只是进行简单的育种材料的互换，或付出一定量的资金，购买企业所需要的育种材料，没有真正实现优势互补，相互促进的合作效果。一些科研院所办种子企业习惯于将研发单位培育的品种直接拿来开发，将企业一小部分利润返给育种单位，由于缺少相应的合作机制与制度保证，育种单位和育种人利益经常难以得到保证，造成育种单位不愿将好品种让企业来开发，企业面临和育种单位合作难以长期保持的问题。

2.3 研发人才相对匮乏

种子企业研发过程中包括基础研究、育种、制种、试验、示范、市场调查等多个环节，涉及大量的数据收集、分析与整理工作，对种子企业研发能力有很高的要求，我国相当数量的种子企业研发能力不强，不仅是缺少必要的研发条件，研发人才的缺乏也是阻碍企业发展的重要因素，而一些企业尽管比较重视人才引进，但缺少人才培养机制和相关制度，对企业职业技能培训重视不足，阻碍了企业的进一步发展。

2.4 信息化管理水平低

信息技术贯穿于企业研发的各个环节和各个层面，不同企业对信息技术应用能力与需求存在一定差异，但信息技术对促进企业研发，提升企业的核心竞争力起着非常关键的作用。我国种子企业以中小微企业为主要构成，信息技术应用能力普遍较低，企业对信息技术的重视不足，技术装备水平较差，人才相对匮乏。多数种子企业没有建立涵盖生产、加工、销售、试验与示范的企业数据库，缺少应用信息技术对公司业务数据的科学分析，因而不能及时发现客户的需求变化，没有建立共同的软件平台来实现部门间的信息共享，造成企业的创造力、学习能力与竞争力都较差，致使企业的研发和经营风险逐年累积。

3 种子企业主要研发策略

3.1 自主研发种子企业发展策略

目前，我国自主研发种子企业数量相当少，以大型种子企业为主，这些企业开展研发时间较早，有较为完善的研发部门，资金实力也相对较强，但随着我国种子市场的逐渐放开，跨国种子企业加大了在我国的研发投入和市场开发力度，自主研发种子企业面临着越来越大的市场竞争压力，这类企业在坚持自主研发的同时，也可以针对企业自身存在的劣势与不足，通过合作研发、技术购买等方式，进一步提升企业的研发能力。这些种子企业可加强与科研院所的研发合作，在研究设备、材料、人员等多方面开展合作，也可以寻求与跨国种子企业开展研发合作，学习最先进的研发技术与方法。如登海种业股份有限公司与杜邦-先锋共同出资，成立登海先锋种业公司，实际上是更为紧密的的一种合作方式，有助于企业学习跨国种子企业在研发、生产、经营、服务等方面的先进理念，丰富企业研发品种的类型，增加种子企业产品数量，扩大了市场份额。对于有大量闲置资金的种子企业，在具备一定的技术购买条件下，可以考虑技术购买的研发模式，评价与筛选适于企业需要的品种与技术，尤其是外购一些市场潜力大的品种，填补企业市场空白，缩短研发周期，发挥企业的资金效用，使企业获得进一步的发展。

3.2 合作研发种子企业发展策略

合作研发是不具备自主研发能力的中小微种子较多采用的研发模式，是这些企业长期发展的基础，企业加强合作研究策略包括签订长期合作协议、建立合作激励制度和实现合作广度与深度。

3.2.1 签订长期合作协议

作为合作研发模式，由于合作关系中包括二个以上的合作主体，风险共担与利益共享是基本的合作原则，种子企业与合作对象签订长期研发合作协议，协议应约定双方的具体合作内容，重点界定品种经营权的获得保证方式与有偿使用办法，通过签定协议的方式来保证双方的长期稳定合作。

3.2.2 建立合作激励制度

创新资源的获取、整合和创新动力的提供、支持，还是创新机制的展开、运转和创新功能的发挥，都离不开激励制度系统的支撑。合作激励制度的建立是长期合作的基础，因此，建立公平合理的合作激励制度非常必要，作为一个有机整体，合作激励制度具体内容可以包括奖励与分配制度、岗位职责制度、绩效考核与评价制度，其中，奖励制度可以根据每个品种的效益情况，定期对品种的选育者给予一定比例的物质奖励，提高其开展创新的动力。

3.2.3 实现合作广度与深度

一些种子企业存在合作研发对象少，合作范围窄，不够深入等问题，使研发存在很大的局限性，应有针对性地增加合作对象，扩大合作范围，实现深入合作，以实现企业提高研发能力，增加品种和技术来源，丰富品种类型等目标。

3.3 技术购买种子企业发展策略

具备良好现实型吸收能力的企业，善于购买外部技术并能及时地把它们转换到企业的生产系统之中加以利用，进而迅速为企业创造利润。对于技术购买模式种子企业，作物新品种的鉴定与识别能力对于种子企业非常重要，是研发能力的重要体现，多数技术购买型种子企业逐渐重视品种鉴定与评价，但种子企业在此还需要加强，建立企业的品种鉴定与评价体系非常关键。决策者综合考虑品种的产量、抗病性、商品性、用户评价等多种因素，可以利用解决复杂的多准则决策问题的层次分析法（AHP），确定这些因素的权重，制定品种评价矩阵，对备选品种进行评价与筛选，提高企业决策水平，减少经营风险。

3.4 提高研发人员素质与能力

人才是知识与技术的载体，新世纪企业之间竞争实际上是人才资源的竞争，种子企业对于人才的培养与使用关系到企业的长远发展。企业要引进和培养人才并重，建立企业人才引进和培养的机制和管理制度，做好人才培养规划与具体方案。研发人员素质与能力的提高的方法与形式多样，如联创种业科技公司对于营销人才的培养非常重视，每年7月销售季节结束，公司投入一笔费用，开展各种技能培训，请专业的培训专家对销售经理进行演讲、沟通能力的培训，这些经理们在产品推介会上表现出了很高的业务素质，由于针对性强，取得了很好的效果，促进了企业的经营。金色农华种业公司则努力构建一种人才培养的机制与文化，每年都定期让员工参加技术与能力拓展培训，培养员工的吃苦精神，沟通技能，整合资源意识，努力营造一种学习文化氛围。因此，种子企业可以学习与借鉴一些先进公司的方法，加大人才培养方面的投入，形成一支业务素质出众的人才队伍，提高企业的研发水平，增强提高企业活力与核心竞争力。

3.5 信息技术辅助研发

信息时代，信息技术的应用对企业发挥正起着越来越大的作用，具有对于种子企业，可在三个方面加强信息管理水平，从而促进企业研发，改善企业经营。

3.5.1 品种试验与示范信息化管理

品种试验与示范是企业研发中的重要活动，涉及大量数据的收集、分析与整理工作，因此，品种试验与示范的数据调查、数据库建立、数据分析的信息技术环节还需要加强，统计分析示范必要，适用的软件包括 SAS、SPSS、BMDP、Excel 等。

3.5.2 种子生产信息化管理

作为种子企业，实现以销定产是每个种子企业的理想经营目标，如何进行市场定位，分析顾客需求与偏好，科学预测每一个品种的销量十分关键，通过提高销售预测水平，有利于企业降低产品库存，减少资金占用，降低企业经营风险。因此，通过适用的信息技术，做好市场调查，分析每个经营品种的市场信息，确定企业最佳生产量，减少库存积压，改善经营状况。

3.5.3 产品加工与配送信息化管理

建立企业生产档案数据库，调查与分析生产各关键时期的数据，及时发现种子生产中存在问题，降低制种风险，提高种子质量。优化产品条码技术，便于企业统计产品销量，跟踪产品动向，使企业能及时调整销售布局，减少经营费用。

参考文献（略）

本文原载：北京国际种子产业发展高峰论坛，2012

种子企业技术购买研发模式的决策研究

杨海涛[1]　赵久然[1]　鲁利平[2]　刘春阁[1]　聂晓红[3]　李瑞媛[1]　王卫红[1]

(1. 北京市农林科学院玉米研究中心，北京　100097；2. 北京市延庆县农业技术推广站，北京　102100；3. 北京市顺义区种植业服务中心，北京　101300)

摘　要：为了提高种子企业进行技术或品种购买时，对购买对象的筛选、判断与决策水平，筛选出符合市场发展要求的品种或技术，通过采用AHP法，选取品种比较排名、用户评价、审定区域、抗病性、商品性等作为评价指标，得到这些指标的权重，开展了对拟购买的技术或品种进行比较和决策研究。结果表明，AHP法可较好地解决种子企业的多准则决策问题，得出技术购买模式是种子企业获取技术的重要方式，采用技术购买研发模式的种子企业要具备一定数量的资金、对技术的识别与获得能力和很强的市场开发能力，使企业能选择适宜的方法来提高决策水平，从而降低经营风险。

关键词：种业；种子企业；技术购买研发；模式；决策

Researches on Decision of Technology Purchased R&D Mode of Seed Enterprise

Yang Haitao[1]　Zhao Jiuran[1]　Lu Liping[2]　Liu Chunge[1]
Nie Xiaohong[3]　Li Ruiyuan[1]　Wang Weihong[1]

(1. Maize Research Centre, Beijing Academy of Agriculture and Forestry Sciences, Beijing 100097; 2. Agricultural Technology Extension Station of Yanqing County, Beijing 102100; 3. Planting Industry Service Centre of Shunyi District of Beijing City, Beijing 101300)

Abstract: The purpose of this paper was to improve the technology purchased problem of multi-criteria decision of seed enterprise that was a process of selecting, judging and decision making. The author selected ranking of comparison trial of breeds, user evaluation, area examined and approved, disease resistance, and product properties as evaluation index, utilized AHP method and obtained the weight of these index for comparison of technology or breed purchasing intended and decision mak-

---

基金项目：北京市科技新星计划（2006A38）

第一作者：杨海涛，男，1972年出生，北京人，副研究员，注册咨询工程师（投资），硕士，主要从事作物耕作与栽培研究、农业产业规划与咨询。通讯地址：100097 北京海淀区曙光中路9号 北京市农林科学院玉米研究中心，Tel：010-51503986，E-mail：yht3190@163.com

通讯作者：赵久然，男，1962年出生，北京人，研究员，博士，主要从事玉米育种与栽培研究。通讯地址：100097 北京海淀区曙光中路9号 北京市农林科学院玉米研究中心，Tel：010-51503936，E-mail：maizezhao@126.com ing. The results of this research suggested that problem of multi-criteria decision of seed enterprise could be resolved by applying AHP method, the author drew the conclusions that the technology purchased R&D mode was a important way to obtain technology, the enterprise generally should have much money, the abilities of high level of distinguishing and obtaining technology, strong ability of market development, and mean while, the enterprise should chose proper method to promote the level of decision making, and decreased the management risk.

Key words：Seed Industry；Seed Enterprise；Technology Purchased R & D；Mode；Strategy

研发具有竞争优势的作物品种的种子是种子企业生存与发展的基础，种子企业获取品种或技术的途径包括自主研发、合作研发和技术购买，对于采用技术购买的种子企业，由于需要投入相当数量的资金用于获取种子的所用权或经营权，组织生产与市场开发，因而企业面临一定的经营风险，实际上，确实有相当数量的种子企业由于仅以单一指标确定购买技术或品种，一般以购买品种的产量作为指标，而忽略了其他一些也很重要的指标，如用户评价、审定区域、抗病性、商品性等，因而在经营过程中，常常出现品种滞销，给企业带来很大的经济损失，甚至有的企业因此而破产倒闭。企业作为设计、生产、营销、交货以及对产品起辅助作用的各种活动的集合，研发是其开展创新活动的重要内容，资金实力和市场开发能力较强的种子企业，通过技术购买方式来获得品种的所有权或经营权，对购买对象有一个筛选、判断与最终决策的过程，而这方面的研究相对较少，而技术购买决策需要考虑多项指标，实质上是一个多准则决策问题，AHP 法在评价与决策中有广泛应用，有较好的应用效果，但在种子企业技术购买决策方面的研究很少。因此，笔者主要使用 AHP 法开展了此方面的研究，以期为种子企业在进行技术购买决策提供一定的借鉴作用。

1 技术购买模式

企业是创新的主体，企业的研发由依靠自身资源自主研发逐渐向合作研发、研发外包、技术购买等方向发展。Normann R Ramirezr 认为企业的研发活动主要包括自主研发、研发外包和合作研发。于成永从企业边界角度将研发模式分为内部研发、企业并购、技术购买、合作研发、研发联盟等形式，以技术获取方式将研发模式分为内部研发、合作研发、委托研发、技术购买。笔者认为从组织边界和获得产品或技术方式划分，企业研发模式主要有自主研发、合作研发、研发外包和技术购买模式。而技术购买通常指一个组织或企业，向技术所有权人支付一定数量的资金，获得技术的所有权或经营权，对技术进行消化、吸收与利用的行为。农业种子企业技术购买主要指作物新品种的所有权、经营权的取得，包括相关配套材料与技术，如亲本、制种技术等。

1.1 技术购买模式的动因分析

随着竞争的激化，事业和研究开发的规模增大，价值链的全部都在一个企业内完成逐渐变得困难。从 20 世纪 90 年代开始，从风险企业购买新技术，已经成为获取新技术的主流。在经营者视野中，技术并购至少和技术研发具有同样的战略意义和地位。由于植物新品种的选育不仅需要投入大量资金，更为关键的是需要有育种、评价、筛选等专业技术人才、种质资源、科研条件与设备，一个成功的品种，还需要经历相当漫长的研发周期，面临研发失败的巨大风险。中国自主培育的玉米品种"农大108"于1998—2002年累计推广800万 hm²,育种者许启凤教授从1973年开始育种,到1993年筛选鉴定出"农大108",共用了20年时间,比"农大108"推广面积更大的玉米品种"郑单958"也用了12年时间选育而成,因此,许多研发单位尝试采用南方繁育、引进育种材料、生物技术的应用等措施缩短新品种选育过程,但是一个玉米品种从选育到推广一般需要的时间至少在8年以上。自主研发与合作研发模式存在研发结果的不确定性,这些不确定性主要包括技术自身因素和技术环境因素。如从事农业技术基础理论研究和应用理论研究,以及对技术成果进行实用性或工艺性再开发活动中出现的各种不利的可能性和不确定性;如研发过程中的失误风险、中断风险、时间风险、流失风险等。长期以来,由于体制、机制、观念等方面的束缚,育种大部分集中在科研机构。一方面农业科研人员花费大量精力和经费研究开发出来的新品种少有人问津;另一方面,由于企业自身缺乏研发能力,又必须依靠购买科研机构的研发成果。

技术购买尽管付出了相当数量的资金用于获得技术的所有权或经营权,但可以规避自主研发所要求的专业技术人才、种质资源、科研条件与设备等硬性条件,缩短了研发周期,降低了研发失败的风险,对于企业快速发展具有重要作用。在市场机制的推动和国家产业政策的指引下,大量技术通过技术市场流向企业,以企业为主体的技术转移是提高企业自主创新能力和国际竞争力的重要保证。随着技术交易规模不断扩大,技术市场在发挥资源配置、促进经济发展方面具有重要的作用,科技成果转化、产业化需要技术市场有效地配置资源。利用市场机制,组织和协调其他生产要素,实现科技成果向生产力的转化。北京技术市场经过二十几年的培育和发展,技术合同成交额每年以10%以上的速度增长,从1991年的22.4亿元增加到2010年的1 579.5亿元,增长近70倍。在农业领域,玉米新品种所有权或使用权的技术转让交易日益活跃,越来越多的企业通过技术市场来进行品种所有权或经营权的交易。

1.2 技术购买成功案例

万向德农股份有限公司为农业高科技上市企业,农业产业化国家重点龙头企业。种子业务由北京德农种业有限公司和黑龙江德农种业有限公司2个子公司承担,主要经营玉米、油葵、马铃薯、棉花、牧草等农作物种子。"郑单958"玉米杂交种的品种权人河南省农业科学院。北京德农种业科技发展公司2001年支付许可费200万元,许可期限为2001年7月至2010年7月,"郑单958"市场表现良好,为企业带来了丰厚的利润,北京德农种业有限公司与河南省农业科学院续签"郑单958"经营许可合同,许可费高达2 000万元,期限延至2016年12月。北京德农种业有限公司还通过技术购买这种方式,还独家经销或合作经销的浚单、辽单、沈单、兴单等系列玉米品种,这些品种的所有权或经营权的获得与商业化运作,使公司只经过短短几年,便获得了丰厚的回报,取得了快速的发展。而北京金色农华种业科技有限公司、北京奥瑞金种子科技开发有限公司等规模较大、资金实力较强的企业,也采用技术购买模式获得品种的所有权或经营权,结合公司自身很强的开发能力,使企业在短时期内收回了购买技术的投入,企业在短时期内获得了快速发展。

2 技术购买决策

2.1 提高技术购买决策水平的必要性

开发一个品种需要大量投入，也意味着要承担很大的风险，需要企业提高技术购买决策水平。具备良好现实型吸收能力的企业，善于购买外部技术并能及时地把它们转换到企业的生产系统之中加以利用，进而迅速为企业创造利润。一个成功的商用作物品种一般应具有高产、抗病性好、适应性强、品质优异等特性，需要很多技术分析工作，对于一个品种认识与评价非常关键，对于技术购买企业，向技术转让方出一笔品种所有权或经营权的转让费，进而生产、加工、销售，如果对于品种评价的偏差过大，则会出现产品滞销，使企业生产成本上升，进而影响企业的经营，因此，技术购买的量化评价非常必要。对于种子企业，购买品种是其主要的技术购买行为之一，是一个多目标决策问题，决策者综合考虑品种的产量、抗病性、商品性、用户评价等因素。因此，在进行技术购买时，要综合考虑这些因素，提高决策的水平，减少经营风险。

2.2 技术购买条件分析

作为技术购买者，具有对所购技术拥有其专业的分析与决策能力非常关键，一般来说，技术购买主要为一些资本运作能力强，对所购技术或品种分析评价能力强的企业所采用。采用技术购买研发模式一般应具备的一些基本条件有：拥有一定数量的资金，应包括购买技术的资金、种子生产与加工资金；有市场竞争优势的品种识别与获得能力；很强的市场开发能力。

种子企业在进行技术或品种购买时，应具备一定的资金数量，用于购买技术或品种的所有权与经营权，对企业的财务可持续性的评价非常必要，因此，通过计算公司的流动比率、速动比率、资产负债率等财物指标，评价企业财务可持续性，一般情况下，要求公司的流动比率大于2，速动比率均大于1，资产负债率低于60%作为基本的标准。

2.3 技术购买决策主要步骤

首先，通过分析与调查的方式，通过与具有市场经验的技术人员、销售人员进行交流的方式，确定玉米品种市场前景的关键因素，即选取区试排名、抗病性、用户评价、商品性、审定区域5个指标，利用Thomas L Satty开发的用于解决复杂的多准则决策问题的层次分析法（AHP），确定这些因素的权重，制定了品种评价矩阵，对被选品种评价筛选，做出决策。虽然品种评价表认为人为因素影响较大，不能精确评价一个品种，但是通过数量化一个品种的表现，有助于提高决策水平，降低决策偏差给企业带来的风险。

2.3.1 构建层次

整体目标：选出最优品种获组合；确定标准：品种比较排名、抗病性、用户评价、商品性、审定区域；决策方案：A1品种、A2品种……An品种中确定所要选择的品种。

2.3.2 AHP法确定各项标准的权重

用AHP法对各项标准的重要性进行量化，采用1~9的尺度来衡量（表1）。

表1 标准重要性的比较尺度

语言描述	数值等级
极重要	9
非常重要至极重要	8
非常重要	7
很重要至非常重要	5
很重要	5
比较重要至很重要	4
比较重要	3
同等重要至比较重要	2
同等重要	1

评价标准的进行两两比较,结果见表2。

表2 评价标准两两比较汇总

比较对象	更重要的标准	重要程度	数值等级
品种比较排名 vs. 抗病性	区试排名	较重要	3
品种比较排名 vs. 用户评价	区试排名	同等重要至较重要	2
品种比较排名 vs. 商品性	区试排名	较重要	3
品种比较排名 vs. 审定区域	审定区域	同等重要至较重要	2
抗病性 vs. 用户评价	用户评价	较重要	3
抗病性 vs. 商品性	商品性	同等重要	1
抗病性 vs. 审定区域	审定区域	较重要	3
用户评价 vs. 商品性	用户评价	较重要	3
用户评价 vs. 审定区域	审定区域	同等重要	1
商品性 vs. 审定区域	审定区域	较重要	3

建立两两比较矩阵,将结果输入表3,计算每一列总数的和。

表3 两两比较矩阵

	品种比较排名	抗病性	用户评价	商品性	审定区域
品种比较排名	1	3	2	3	2
抗病性	1/3	1	1/3	1	1/3
用户评价	1/2	3	1	3	1
商品性	1/3	1	1/3	1	1/3
审定区域	1/2	3	1	3	1
总和	2.67	11	4.67	11	4.67

将矩阵的每一项除以它所在列的总和,计算每行平均数,得到各项指标的权重(优

先级），结果如表 4 所示。

表 4　各项指标权重

	品种比较排名	抗病性	用户评价	商品性	审定区域	权重（优先级）
品种比较排名	0.37	0.27	0.43	0.27	0.43	0.35
抗病性	0.12	0.09	0.07	0.09	0.07	0.09
用户评价	0.19	0.27	0.21	0.27	0.21	0.23
商品性	0.12	0.09	0.07	0.09	0.07	0.09
审定区域	0.19	0.27	0.21	0.27	0.21	0.23

2.3.3　建立品种评价表

对备选品种的各项评价指标进行打分，分值为范围为 1~5，品种某一方面表现越好，获得的分值越高，5 代表最优，4 代表次优，3 代表良好，2 代表一般，1 代表较差。例如 A 品种在黄淮海玉米区域排名第一，得分为 5，抗病性中等，得分为 3，审定区域在玉米主产区，得分为 5 等（表 5）。

表 5　品种评价表

评价指标	权重	A 品种		B 品种		C 品种	
		得分	加权值	得分	加权值	得分	加权值
品种比较排名	0.35	5	1.75	4	1.4	4	1.4
用户评价	0.23	5	1.15	5	1.15	5	1.15
审定区域	0.23	4	0.92	3	0.69	3	0.69
抗病性	0.09	4	0.36	4	0.36	4	0.36
商品性	0.09	4	0.36	4	0.36	3	0.27
总分	1		4.54		3.96		3.87

2.3.4　进行决策

根据企业发展战略，选择一批拟购买品种进行评价，对每个品种每项指标进行打分，得分乘以权重获得的加权值，计算总得分，对备选品种但总得分由高到低进行最终排名，选择排名靠前的品种，并针对中选品种进行资金筹措、谈判、技术获得、技术开发等后续工作。

3　结论

3.1　技术购买模式是种子企业获取技术的重要方式

农业种子企业进行自主研发或合作研发，由于研发周期长、研发失败的风险等问题，而很多企业还由于缺少研发的所需的技术人员、育种材料、研发设备等条件，因此，采用技术购买是其获取技术的重要方式，种子企业必须要增强对技术购买的重视与利用。

3.2　实行技术购买的种子企业应具备一定的条件

采用技术购买研发模式的种子企业一般要具备一定数量的资金，竞争优势强的技术识别与获得能力和很强的市场开发能力，具备了这些基本条件，有利于企业降低技术购买失败的风险。

3.3 种子企业应不断提高技术购买决策水平

企业根据自身资源状况，选择适宜的决策方式或方法，不断地提高其决策水平。种子企业在进行技术或品种购买是一个多准则决策问题，可选取品种比较排名、用户评价、审定区域、抗病性、商品性等作为评价指标，用 AHP 法得到这些指标的权重，对拟购买的技术或品种进行比较并最终决策，使种子企业提高技术购买决策水平，降低其经营风险。

4 讨论

本研究结果表明，AHP 法应用于种子企业技术购买的决策，有助于提高决策水平，从而降低经营风险。这主要是在企业进行技术购买时，仅凭经验或单一指标都会对决策质量造成影响，增加企业经营风险，因而，企业应对拟购买的技术或品种进行综合评价与选择，提高决策水平。AHP 法在项目评价与决策中有较多应用，但在种子企业技术购买决策方面的应用很少，这是本研究的主要创新点。

综合的主要是针对种子企业技术购买决策中存在的问题，即重视程度的应用意义在于技术购买不仅是简单的购买技术，还涉及对技术的消化与吸收，有时还需要在引进技术的基础上进行再创新，这些都需要得到企业的重视，使企业保持一定的市场竞争力。另外，资金筹措、技术获得、市场开发等后续工作也会对技术或品种的开发效果产生重要影响，这些方面都有待于进一步研究。

参考文献（略）

本文原载：农学学报，2012，2（1）：74-78

我国玉米品种权保护现状、问题与建议

杨海涛[1]　赵久然[1]　杨凤玲[2]　王凤格[1]　宋伟[1]　陈红[3]

(1. 北京市农林科学院玉米研究中心/玉米DNA指纹及分子育种北京市重点实验室，北京　100097；2. 内蒙古通辽市农科院，内蒙古　028000；3. 农业部科技发展中心，北京　100122)

摘　要：简要分析了我国玉米品种权保护现状，对1999—2013年玉米品种权申请量变化、玉米品种权申请主体和不同类型单位申请量变化等进行了统计分析。认为品种权保护中存在审查批准周期长、规划保护滞后、维权难度大、信息收集利用少和激励制度不健全等问题。提出改进DUS测试方法、开展技术协作与合作、加强市场秩序维护、挖掘利用新品种保护信息、完善品种权激励制度和加强知识产权人才培养和技术培训等建议。

关键词：玉米；品种权；植物新品种权；知识产权

Present Condition, Problems and Advices for Protection of Maize Variety Rights in China

Yang Haitao[1]　Zhao Jiuran[1]　Yang Fengling[2]
Wang Fengge[1]　Song Wei[1]　Chen Hong[3]

(1. Maize Research Center of Beijing Academy of Agriculture and Forestry Sciences, Beijing Key Laboratory of Maize DNA Fingerprinting and Molecular Breeding, Beijing 100097; 2. Tongliao Academy of Agriculture Sciencies of Neimenggu Autonomous Region, Neimenggu 028000; 3. Science and Technology Development Center of Agriculture Ministry, Beijing 100122)

Abstract: This thesis briefly analyze the present condition of protection of maize variety rights in China, analyze and summarize the quantity variation of applying for maize variety rights, the main bodies of applying for maize variety rights and the quantity variation of applying for maize variety rights of different unit types from1999 to 2013. Conclude that the main problems are long period of examination and approval, lagging programme of protection, difficulty of protecting variety rights, poor utilizing information and lacking incentive institution. Advise to innovate the DUS method, develop techno-

---

基金项目：国家科技支撑项目（2014BAD01B09）

作者简介：杨海涛（1972—　），男，北京人；硕士，副研究员，主要从事植物新品种保护、农业产业规划与咨询，E-mail：yht3190@sina.com

通讯作者：赵久然，E-mail：maizezhao@126.com

logical collaboration and corporation, strength the market order, dig and utilize information of variety protection, improve variety rights incentive institution and enhance talent cultivation of intellectual property and technology training.

Key Words: Maize; Variety rights; Rights in new varieties of plants; Property rights

玉米用途广泛，市场需求量大，是我国种植面积最大、总产量最高的粮食作物。随着我国进入玉米消费快速增长阶段，受耕地、水等资源条件制约，满足玉米需求增长将主要依靠科技创新，提高玉米单产。鼓励玉米育种创新，提高新品种保护水平，对于促进玉米生产和实现粮食增产具有重要作用。

1 我国玉米品种权保护现状

我国于1997年10月1日起施行《中华人民共和国植物新品种保护条例》、1999年4月23日加入《国际植物新品种保护联盟公约》以来，植物新品种保护制度和体系逐步完善，建立起包括审查测试机构、执法结构、中介服务结构和自律性维权组织的植物新品种保护体系，植物新品种保护意识显著增强，事业发展迅速，2012年，植物新品种权年申请量已跃居国际植物新品种保护联盟成员第二位。玉米种子需求量大，市场价值高，随着玉米种子向市场化和产业化发展，玉米品种权保护呈较快发展趋势。对年度间、不同类型单位玉米品种权申请量进行分析（数据来源于农业部植物新品种保护办公室网站），玉米品种权保护有以下特点。

1.1 品种权申请数量显著增加

对1999—2013年玉米品种权申请量统计，如图1所示，1999年玉米品种权年申请量仅94件，2005年已增长至334件，2012品种权申请量达到393件峰值，是历年申请量中最多的一年，2013年申请量较2012年略有减少。1999—2005年玉米品种权申请量增速较快，2006年以来出现一些的波动，但总体趋势是稳步增加。

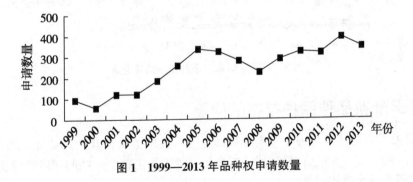

图1　1999—2013年品种权申请数量

1.2 品种权申请主体是企业和科研机构

1999—2013年，共有736家单位申请玉米新品种权3 669件，如果将申请单位按内资种子企业、科研机构、外资种子企业、个人、高校等分类，如图2所示，内资种子企业申请数量所最多，占总申请量的52%，科研机构申请数量居第2位，占总申请量的31%，外资种子企业申请数量居第3位，占总申请量的7%。由此可见，内资种子企业和科研机构是申请玉米品种权的主体，占申请量的80%以上。

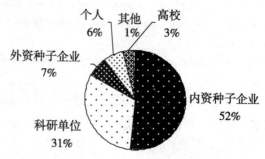

图 2 各类型单位玉米品种权申请数量占比

1.3 不同类型单位品种权申请数量变化显著

比较 1999 年以来不同类型单位年玉米品种权申请量变化，如图 3 所示，1999—2002 年，科研机构申请量要远多于其他类型单位，到 2003 年，种子企业申请量已与科研机构持平，2004—2013 年，种子企业申请量一直处于领先位置。外资种子企业 2008 年以来申请数量呈显著增加趋势，2013 年申请量为 60 件，科研机构为 69 件，外资种子企业年申请数量已接近科研机构。不同类型单位申请量的变化反映了技术创新主体的变化与发展，随着我国种业市场化和产业化进一步发展，企业作为技术创新主体地位与作用将进一步显现。

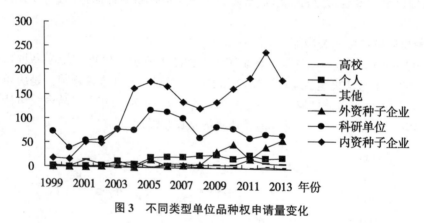

图 3 不同类型单位品种权申请量变化

2 我国玉米新品种保护存在的问题

规范良好的市场环境是种业发展的基础，随着我国玉米种业向产业化、市场化发展，市场规模逐步扩大，竞争也日趋激烈，玉米新品种保护出现的一些问题亟需解决。

2.1 品种权审查批准周期长

植物新品种权的授予要经过申请、受理、初步审查、实质审查、批准等程序，实质审查包括提交繁殖材料、测试机构作 DUS 测试等步骤，完成玉米品种权授予程序一般需经过 3 年左右的时间。随着种业市场环境的变化，玉米品种的经济寿命则呈缩短趋势，一般只有 5 年左右，如果一个品种经过审定后申请植物新品种权，获得授权的玉米品种常常已到生命周期的成熟期或衰退期，一些育种单位也因此而放弃品种权申请，不利于植物新品种权的及时有效保护和品种权人的经济利益维护。

2.2 品种权规划保护滞后

在激烈的种业市场竞争环境中，拥有自主知识产权是育种单位、种子企业获取优势的关键。一些育种单位、种子企业关注短期的育种目标与经济效益，重视品种审定和推广，对品种权保护重视不足，缺少知识产权规划与发展战略，没有相应的知识产权部门和人才，忽视植物新品种权的申请和保护，在遭受品种侵权后，往往缺少有效的法律、技术等手段来维权。多数种子企业的目标市场只是限于国内市场，对欧美等国家地区的知识产权制度、市场准入规则等不熟悉，不具备国际市场开拓与发展的能力。

2.3 品种权维权难度大

植物品种权的有效保护是培育良好的市场环境，促进育种创新和种业发展的基础。目前种业市场存在的侵权、套牌、仿冒等问题仍然严重，2010 年，8 700 多家种子企业中，有 30% 的种子企业存有或无证经营，或仿冒种子，或套牌种子的侵权违法行为。套牌侵权对于供种安全、种业科技创新和市场经营秩序都具有很大的危害。如何进一步完善植物新品种保护体系，从根本上解决这些问题非常必要，尤其是从根源上杜绝套牌侵权种子的生产和销售，创造良好的种业环境，保护品种权人利益。

2.4 品种权信息收集利用少

申请保护的玉米新品种与骨干自交系是育种家经过多年辛勤培育的智力成果，包括育种材料的来源、选育过程和育种方法、杂优模式利用等有价值的信息，这些信息对于种质资源利用、育种创新具有很好的借鉴价值，但目前还没有得到种子企业、研发机构相应的重视，对于品种权保护相关信息的收集、整理、分析不足，缺少国外植物品种权和植物专利等相关信息的收集与利用。

2.5 品种权激励制度不健全

激励制度是技术创新、科技进步的制度性保障，激励制度的不完善和缺失在科研院所和种子企业普遍存在，是影响玉米育种创新和新品种保护主要问题。科研院所申请审定的玉米品种和品种权作为职务研究成果，品种权归属于育种者所在单位或者上级主管部门，育种者的利益被忽视或难以保证，育种者的积极性会受到极大影响。种子企业由于缺少完善的激励制度，尤其是中小种子企业，带来人才吸引力差，人员流动性大，育种创新能力弱等一系列问题。

3 我国玉米新品种保护发展建议

3.1 改进 DUS 测试方法

植物新品种保护联盟（UPOV）的《测试准则的发展》文件中，对于测试周期数的表述是：在所观察品种间差异特别明显的情况下可以不必多于一个生长周期，一个测试试验中，不少于两个生长周期是确保某个性状差异充分一致的方法。在《DUS 测试的试验设计与技术》文件中，一个年度内完成两个独立的生长周期测试的方法有：两个试验在同一地点和年份，试验间有一段合适的时间，则可以提供两个独立的生长周期；在两个种植周期实施在同一年份和时间，两个试验地点有一个合适的距离或者差异。因此，在满足 DUS 测试要求的条件下，选择适当的方法，如在一个年度将两个试验安排在同一试验点，如安排在观热资源丰富的海南省，或者在不同生态区实现同期播种，都可以在一年内完成 DUS 测试，减少测试时间，从而加快品种权审查进程，实现品种权的及时授予和保护。

3.2 开展技术协作与合作

植物新品种权是对玉米育种成果的一种权益保护方式，也是创新成果的重要体现。创新实践往往需要汇集多方面的力量，创新过程是在多主体互动与整合的有效合作中不断发展。加强合作与协作是促进育种创新，加快成果转化的重要途径。科研单位聚集大部分育种资源和专业人才，科技创新能力强，种子企业对于市场需求更了解，在种子生产和营销等方面更具有优势，二者有很好的合作基础，科企合作有利于发挥协同效应，实现优势互补。逐步完善品种权交易机制，促进授权品种从科研单位向企业的转移应用。加强国际技术交流与合作，通过合作研发、技术购买等方式，开展商业化育种，提升品种权质量，促进新品种保护。

3.3 加强市场秩序维护

良好的市场秩序是种业发展的基础，为营造良好的市场环境，农业行政主管部门已开展多项整治行动，如农业部启动农作物种子打假护权专项行动，2010—2012年连续三年的种子执法年活动，地方政府的农业、工商等部门开展联合执法，加大市场检查力度，取得了一定的整治效果，市场秩序得到有效改善。由于我国种子市场分散，企业数量多，在加大执法检察和惩罚力度的同时，还需要进一步完善知识产权保护体系，改进种业市场管理体系和方式，包括建立种子生产信息登记管理系统，实现种子生产的可追溯制度，加强对市场占有率高、需种量大的玉米品种抽查和检查等，扶持发展专业性强的中介服务组织，进一步协调行政保护和司法保护，促进人民法院与农业行政主管部门的沟通和配合，建设良好的种业市场发展环境。

3.4 挖掘利用新品种保护相关信息

信息管理是新品种保护的基础性工作，新品种保护信息包括公知公用品种信息、DUS测试技术、育种材料的来源和选育过方法，国外植物品种权和植物专利等，需要做好相应的信息收集和整理。玉米品种权的保护期限是15年，第一批获得授权的玉米品种至2015年将到期，实际上，由于商业价值不大、没有续缴年费等原因，多数品种权在远未达到保护期前已经失效，其中也包括一些优良骨干自交系。对于科研机构和种子企业，尤其是科技创新能力相对弱、种质资源匮乏的中小种子企业，挖掘利用新品种保护相关信息，有助于提高品种权申请质量，了解国内外玉米育种发展动态，拓宽种质资源利用渠道。

3.5 完善品种权激励制度

种子企业和科研机构需要结合发展规划，完善品种权激励制度，制定相应的产权激励、收益分配、组织激励等制度。随着《关于深化中央级事业单位科技成果使用、处置和收益管理改革试点的通知》《关于进一步创新体制机制加快全国科技创新中心建设的一件》科技成果等国家、地方政策性文件的出台，对科技成果的使用、处置和收益进行改革，制约科技成果转化的制度性障碍逐步破除，研究机构、高校等事业单位需要完善相应的激励制度，协调品种权人、培育人和公共利益的关系，完善品种权使用与交易制度，促进品种权的应用。种子企业可以采用技术持股、技术入股或技术成果分成的激励方式，吸引科技人才，增强员工归属感，推动企业技术创新。

3.6 加强知识产权人才培养和技术培训

《国家知识产权战略纲要》提出加强知识产权人才队伍建设，广泛开展知识产权培训。我国植物新品种保护制度建立实施较晚，农业知识产权人才相对不足，需要制定知识产权人才培养计划，创造良好的成长环境，培养具有法律、新品种保护、遗传育种等复合型知识结构的人才，竞争力强的大型种子企业还需要培养熟悉国外知识产权制度、相关法律法规、品种权保护和专利申请程序的高级人才，增强开拓国际市场的能力。政府各级知识产权主管部门组织开展多层次的技术培训，扶持农业知识产权中介服务机构，提高专业化服务水平，加强知识产权国际交流合作，积极参与知识产权国际组织相关事务，提升植物新品种保护水平。

参考文献（略）

本文原载：种子，2015，34（9）：71-73

基于案例分析的玉米品种权维权问题、启示与建议

杨海涛[1] 赵久然[1] 陈 红[2] 宋 伟[1] 王凤格[1] 邢锦丰[1]
张如养[1] 王元东[1] 段民孝[1] 杨凤玲[3]

(1. 北京市农林科学院玉米研究中心/玉米DNA指纹及分子育种北京市重点实验室，北京 100097；2. 农业部科技发展中心，北京 100122；3. 内蒙古通辽市农科院，内蒙古 028000)

摘 要：近几年，植物新品种权许可使用中的侵权案件、权属纠纷案件逐渐增多，相当多的维权人由于诉讼败诉而蒙受巨大经济损失。本文以三个较为典型的玉米品种权纠纷案件为对象，分析案例所反映的品种权申请、保护、风险防控、价值评估和维权等问题，提出相应的解决办法和建议，为权利人维护自身权益和规避法律风险提供参考。

关键词：案例；玉米；品种权；保护；问题；启示；建议

Problems, Enlightment and Suggestions of right-defending of Maize Variety Rights Based on the Analysis of Cases

Yang Haitao[1]   Zhao Jiuran[1]   Chen Hong[2]   Song Wei[1]
Wang Fengge[1]   Xing Jinfeng[1]   Zhang Ruyang[1]   Wang Yuandong[1]
Duan Minxiao[1]   Yang Fengling[3]

(1. *Maize Research Center of Beijing Academy of Agriculture & Forestry Sciences Beijing* 100097；2. *Science and Technology Development Center of Agriculture Ministry*，*Beijing* 100122；3. *Tongliao Academy of Agriculture Sciencies of Neimenggu Autonomous Region*，*Neimenggu* 028000)

Abstract：In recent years, along with the increasing of the cases of infringement and ownership dispute of the rights of varieties, many prosecutors suffered great economic losses for the failure of the lawsuits. Three typical cases were analyzed in this paper, and the cases reflected the problems of applications for rights, protection, risks prevention, value appreciation and rights protection, proposed the solutions and suggestions accordingly, offered the references for the prosecutors to protect their

---

基金项目：国家科技支撑项目（2014BAD01B09）
作者简介：杨海涛（1972— ）男，北京人，副研究员，硕士，主要从事植物新品种保护、农业产业规划与咨询。E-mail：yht3190@sina.com
通讯作者：赵久然，E-mail：maizezhao@126.com

rights and benefits and evade the legal risks.

Key Words：Case；Maize；Variety right；Protection；Problem；Enlightment；Suggestion

近几年，随着我国玉米种子产业化、规模化与市场化发展，植物新品种权转让和许可的数量显著增加，品种权的侵权案件、权属纠纷案件也逐渐增多，相当多的权利人由于维权失败而蒙受巨大经济损失，权利人亟需加强品种权维护和法律风险规避工作。本文选取了三个较为典型的玉米品种权属纠纷案件，分析案例所反映的品种权管理和维权问题，提出了相应的解决办法和建议，为种子企业和科研单位的品种权维权工作提供参考。

1 品种权纠纷案件回顾

1.1 吉祥1号玉米品种权维权案

2012年12月22日，甘肃省敦煌种业股份有限公司（以下简称"敦煌种业"）发布公告称已与武威市农业科学研究院签署《玉米杂交品种吉祥1号生产经营权转让合同》，敦煌种业以2 680万元受让吉祥1号独家生产经营权。然而，由于市场上多家公司生产经营未经授权的吉祥1号玉米品种，敦煌种业的权益受到极大损害。2012年1月起，敦煌种业将河南弘展农业科技有限公司、武威市武科种业科技有限责任公司、河南省大京九种业有限公司、郑州赤天种业有限公司等涉嫌侵犯吉祥1号品种权的种子企业提起三起诉讼，一审、二审均败诉，2014年5月，最高人民法院驳回敦煌种业三起诉讼的再审申请，诉讼请求没有得到法院支持，维权失败使敦煌种业蒙受了巨大经济损失。

1.2 大丰30玉米品种权侵权案

2015年3月，山东登海先锋种业有限公司（以下简称"登海种业"）控告陕西农丰种业有限责任公司（以下简称"农丰种业"）和山西大丰种业有限公司（以下简称"大丰种业"）侵犯"先玉335"品种权案二审败诉。大丰30是利用先玉335的母本PH6WC和Mo17配制组合"PH6WC/Mo17//PH6WC"后，经多代选育而成，父本与先玉335的父本相同，都是PH4CV，因此，大丰30与先玉335的遗传背景非常接近。最终，陕西省高级人民法院以农业部植物新品种保护办公室出具的DUS测试报告为依据，即大丰30与先玉335比较，大丰30具备特异性，大丰30和先玉335不是一个品种，判定大丰公司生产、农丰公司销售的涉案"大丰30"不存在侵权行为。

1.3 郑单958玉米品种权纠纷

2015年10月26日，万象德农股份有限公司发布公告：河南金博士种业股份有限公司（下文简称"金博士种业"）于2014年8月19日以侵害植物新品种权为由对该公司控股子公司北京德农种业有限公司（下文简称"德农种业"）及河南省农业科学院（以下简称"河南农科院"）向河南省郑州市中级人民法院（下文简称"郑州中院"）提起民事诉讼。郑州中院对金博士诉北京德农及河南农科院侵害植物新品种纠纷案作出一审判决，认定北京德农公司侵犯了金博士公司对"郑58"玉米品种享有的植物新品种权，判令其赔偿金博士公司4 950万余元，被告河南农科院对上述赔偿在300万元范围内承担连带责任。

2 品种权纠纷案件反映的主要问题

2.1 品种权申请保护不及时

随着种子市场上玉米品种数量增加，市场竞争加剧，多数玉米品种从推广到退市的品种生命周期呈缩短趋势，一个品种从申请到授权要需要经过3~4年时间，如果没有及时申请和获得品种权，如果品种生产经营过程中遭受侵权，会不利于品种维权工作开展。申请保护不及时还可能会使品种丧失新颖性，而不能获得授权。大丰30玉米品种权侵权案中，先玉335的父本PH6CV、母本PH6WC都没有申请保护，如果这些亲本申请了保护并获得授权，那么大丰种业直接利用PH4CV生产大丰30的行为就违反了《中华人民共和国种子法》、《中华人民共和国植物新品种保护条例》的相关规定，即"任何单位或者个人未经植物新品种权人许可，不得生产、繁殖或者销售该授权品种的繁殖材料，不得为商业目的将该授权品种的繁殖材料重复使用于生产"，那么就可以通过法律手段维护品种权益。

2.2 品种保护工作基础薄弱

在郑单958玉米品种权纠纷案例中，河南农科院是郑单958的品种权人，但不是郑单958母本——郑58的品种权人，郑58最初的品种权人是河南省荥阳市飞龙种子有限公司（以下简称飞龙公司），该公司于2002年1月1日获得授权，2007年7月1日，郑58品种权由飞龙公司转至张发林，2009年5月1日，又由张发林转至河南金博士种业股份有限公司。由于郑58的品种权不属于河南农科院，因此，受让人不仅要获得郑单958的生产经营权，还要经过郑58品种权人的授权许可，才能合法生产销售郑单958。案例暴露了受让方知识产权意识不强，对郑单958相关的品种权信息收集掌握不足，没有意识到到郑单958繁制过程中重复使用郑58所涉及的侵权问题。

2.3 缺乏品种权价值评估与决策能力

对于品种权的转让方和受让方，品种权的价值评估都是一项基础性工作，相对客观、准确的价值评估是进行科学决策的重要依据，也有助于受让方降低财务风险和生产经营风险，促进品种权市场的持续稳定发展。而在实际的品种权转让过程中，品种权价值评估工作容易被忽视，主要依赖经验来确定品种权价值，缺少结合市场环境、企业发展战略、品种特性等因素的品种权价值分析与评估。本文前述的"吉祥1号玉米品种权维权案"、"郑单958玉米品种权纠纷"二个案例中，受让方都支付了数额巨大的品种权转让、许可费用，蒙受了巨大的经济损失，客观反映出品种权价值评估和决策方面所存在的问题。

2.4 缺少品种权许可、转让风险防控措施

具有高产、优质、多抗、广适等优良特性的品种是市场的稀缺资源，这些品种的品种权转让、许可费用数额巨大，如敦煌种业获得吉祥1号的独家生产经营权，德农公司和中种集团两家单位得到郑单958生产经营权都支付了巨额费用，因此，作好品种权转让和许可的风险分析与防控非常必要。三个案例均暴露出受让方在风险管理方面存在的问题，如敦煌种业只是获得了武威市农业科学研究所的吉祥1号生产经营许可，而吉祥1号的品种权人还包括自然人黄文龙，最高人民法院驳回敦煌种业的主要依据是"敦煌种业公司仅取得部分品种权人许可、不具备合法的诉讼主体资格"。因不具备诉讼主体资格，致使权利主张得不到法院支持，从而蒙受了巨大的经济损失。

2.5 品种权法律维权业务能力弱

利用司法手段维护权益过程中,一般要经过调查取证、提起诉讼、法院审理、法庭辩论、法院判决等法律程序,品种权人一般委托法律服务机构代理维权,要求法律服务机构具需要具有很强的法律咨询和服务能力,维权人则需要在调查取证、证据保护等方面提供专业技术支持。调查取证工作是维护权益的基础性工作,证据的获取和保全对于权益有效维权非常关键。大丰30玉米品种侵权案法律纠纷中,法院以农业部品种保护办公室出具的 DUS 检测报告作为主要的判定依据,登海种业的诉讼请求遭到法院驳回。案件焦点之一是登海种业所获取证物需要进一步进行田间测试时,涉案被扣押的种子已丢失,即证物没有得到妥善地保全,由此使登海种业不能进一步申请司法鉴定。

3 启示与建议

3.1 建立规范的品种权管理制度

品种权管理工作内容多,既有申请前的田间调查、申请材料撰写等技术性工作,也有费用及时缴纳、著录项目申请变更、品种权转让、品种权终止等事务性工作,品种权人要注意新品种选育与保护的有机协调,建立起规范的品种权管理制度,包括品种权申请、品种权事务、工作流程、协调机制、激励制度等内容,提升品种权管理水平。根据育种目标,选择配合力高、多抗、广适的骨干自交系,适时申请保护。玉米作为主要农作物,需要经过审定才能应用于生产,审定与保护的同步完成是维护品种权的重要基础,对于参加区域试验的重点组合,及时收集、分析试验结果,提早申请植物新品种权。通过田间观察、实验室检测等方式,判断品种是否具备特异性、一致性和稳定性。根据形态特征、遗传背景、分子检测结果等确定近似品种,及时扩繁种子,尽早提交申请保护文件材料和繁殖材料。

3.2 提高品种权价值评估与决策能力

品种权价值评估与决策是品种权管理的重要工作,对于品种权人和品种权许可受让人,相对科学、准确的价值评估工作都十分必要,是双方达成实施许可,实现互利双赢的基础。由于品种权是一种知识产权,也是一类无形财产权,具有创造性、风险性、先进性、垄断性和期权性等特征,知识产权的评估方需要从多角度、多方面、多层次分析知识产权的价值,确定合理的分成率,选择适合的评估方法。《资产评估准则——无形资产》规定的评估方法包括成本法、收益法、市场法等。实际应用中,以收益法和以收益法为基础的改进、补充方法已有较多的研究与应用,可以作为价值评估的主要方法,也可以利用基于这些方法的价值评估应用软件、品种权评估平台供利用和参考。品种权价值评估是对于未来收益的量化,受国家产业政策、国际农产品价格、市场需求、市场竞争、利益分成等多种因素影响,进行价值评估时,需要综合分析这些因素,建立相应的评价指标体系,作好市场预测和分析。随着品种权交易数量的增加,信息的公开化与透明化,市场法也可以作为品种权价值评估方法,提高品种权价值评估能力与决策水平。

3.3 做好品种权许可、转让和维权的风险评估与防控工作

权利人需要针对品种权转让、许可和维权过程中存在的不确定因素与潜在风险,识别主要风险因素,建立相应的风险评估机制和制度,制定必要和完善的风险防控措施。要注意履约过程中主要风险节点,规范维权过程的取证、检测、证据保全、法律诉讼程序等法

律行为，制订技术性工作规范。品种权许可、转让过程中容易出现信息不对称问题，需要作好包括亲本、杂交种、品种权人、品种权法律状态、品种权存续期等信息收集、分析和管理工作，对于非独家所有的品种权，签订的品种权转让协议要取得所有品种权人同意的意思表示。注意完善品种权转让、许可协议内容，规避信息不对称引起的合同法律纠纷和品种权维权障碍，在转让协议签订后，及时向审批机关登记，由审批机关予以公告。

3.4 增强品种权法律维权的业务能力

品种权维权工作专业性强，维权过程中涉及取证、证据保全、提起诉讼、法庭辩论等环节，维权人需要熟悉、了解维权相关法律和诉讼程序，作好工作计划，制定详细的工作方案。品种权维权过程中的取证和证据保全是重点工作，重视证据收集，可以定期在专业性报刊、杂志上刊载维权公告，征集维权线索。作好调查取证，注意证据保全程序的合法和有效。《最高人民法院关于审理侵犯植物新品种权纠纷案件具体应用法律问题的若干规定》第五条规定："品种权人或者利害关系人向人民法院提起侵犯植物新品种权诉讼时，同时提出先行停止侵犯植物新品种权行为或者保全证据请求的，人民法院经审查可以先行作出裁定。"因此，品种权人或利害关系人需要作好侵权行为、证物的调查和取证，妥善保存品种权证书、生产经营权转让、许可合同、维权授权书、公安侦查记录、扣押清单等相关材料，作好法律诉讼的必要准备工作。

3.5 专业化的法律咨询与服务

随着品种权纠纷诉讼案件的增多，越来越多的育种单位、种子企业意识到品种权转让和维权过程中需要注意的法律问题和风险，客观上增加了品种权法律咨询、维权的专业服务需求。科研单位和种子企业可以通过引进、培养具有农业、知识产权和法律等复合型知识结构的专业人才，增强品种权管理和维权的业务能力和技术水平，实力强、规模大的种子企业还可以通过建立专门的法律事务部门，聘请有经验的法律顾问。为了能更好地满足种子企业和育种机构的品种权维权需要，法律服务机构需要进一步提高服务水平和业务能力，可以根据品种权的特殊属性，与权利人明确分工、密切协作，作好品种权的信息收集、调查取证、诉讼前各项材料准备工作，提出具有建设性的法律建议和指导，提供更加有效和专业的法律咨询服务。

参考文献（略）

本文原载：种子，2017，36（5）：131-133